THE OXFORD HANDBOOK OF

THE HISTORY OF PHYSICS

THE OXFORD HANDBOOK OF

THE HISTORY OF PHYSICS

Edited by

JED Z. BUCHWALD

AND

ROBERT FOX

UNIVERSITY PRESS

Great Clarendon Street, Oxford, OX2 6DP,
United Kingdom

Oxford University Press is a department of the University of Oxford.
It furthers the University's objective of excellence in research, scholarship,
and education by publishing worldwide. Oxford is a registered trade mark of
Oxford University Press in the UK and in certain other countries

© Oxford University Press 2013

The moral rights of the authors have been asserted

First Edition published in 2013

Impression: 1

All rights reserved. No part of this publication may be reproduced, stored in
a retrieval system, or transmitted, in any form or by any means, without the
prior permission in writing of Oxford University Press, or as expressly permitted
by law, by licence or under terms agreed with the appropriate reprographics
rights organization. Enquiries concerning reproduction outside the scope of the
above should be sent to the Rights Department, Oxford University Press, at the
address above

You must not circulate this work in any other form
and you must impose this same condition on any acquirer

Published in the United States of America by Oxford University Press
198 Madison Avenue, New York, NY 10016, United States of America

British Library Cataloguing in Publication Data
Data available

Library of Congress Control Number: 2013940065

ISBN 978–0–19–969625–3

Printed and bound in Great Britain by
CPI Group (UK) Ltd, Croydon, CR0 4YY

Contents

List of Contributors	viii
Introduction *Jed Z. Buchwald and Robert Fox*	1

PART I. PHYSICS AND THE NEW SCIENCE

1.	Was There a Scientific Revolution? *John L. Heilbron*	7
2.	Galileo's Mechanics of Natural Motion and Projectiles *N. M. Swerdlow*	25
3.	Cartesian Physics *John A. Schuster*	56
4.	Physics and the Instrument-Makers, 1550–1700 *Anthony Turner*	96
5.	Newton's *Principia* *Chris Smeenk and Eric Schliesser*	109
6.	Newton's Optics *Alan E. Shapiro*	166
7.	Experimentation in the Physical Sciences of the Seventeenth Century *Nico Bertoloni Meli*	199
8.	Mathematics and the New Sciences *Niccolò Guicciardini*	226

PART II. THE LONG EIGHTEENTH CENTURY

9. The Physics of Imponderable Fluids 267
   Giuliano Pancaldi

10. Physics on Show: Entertainment, Demonstration, and Research in the Long Eighteenth Century 299
    Larry Stewart

11. Instruments and Instrument-Makers, 1700–1850 326
    Anita McConnell

12. Mechanics in the Eighteenth Century 358
    Sandro Caparrini and Craig Fraser

13. Laplace and the Physics of Short-Range Forces 406
    Robert Fox

14. Electricity and Magnetism to Volta 432
    Jed Z. Buchwald

PART III. FASHIONING THE DISCIPLINE: FROM NATURAL PHILOSOPHY TO PHYSICS

15. Optics in the Nineteenth Century 445
    Jed Z. Buchwald

16. Thermal Physics and Thermodynamics 473
    Hasok Chang

17. Engineering Energy: Constructing a New Physics for Victorian Britain 508
    Crosbie Smith

18. Electromagnetism and Field Physics 533
    Friedrich Steinle

19. Electrodynamics from Thomson and Maxwell to Hertz 571
    Jed Z. Buchwald

20.	From Workshop to Factory: The Evolution of the Instrument-Making Industry, 1850–1930 *Paolo Brenni*	584
21.	Physics Textbooks and Textbook Physics in the Nineteenth and Twentieth Centuries *Josep Simon*	651
22.	Physics and Medicine *Iwan Rhys Morus*	679
23.	Physics and Metrology *Kathryn M. Olesko*	698

PART IV. MODERN PHYSICS

24.	Rethinking 'Classical Physics' *Graeme Gooday and Daniel Jon Mitchell*	721
25.	The Emergence of Statistical Mechanics *Olivier Darrigol and Jürgen Renn*	765
26.	Three and a Half Principles: The Origins of Modern Relativity Theory *Daniel Kennefick*	789
27.	Quantum Physics *Suman Seth*	814
28.	The Silicon Tide: Relations between Things Epistemic and Things of Function in the Semiconductor World *Terry Shinn*	860
29.	Physics and Cosmology *Helge Kragh*	892

Name Index   923
Subject Index   936

List of Contributors

Nico Bertoloni Meli Professor of History and Philosophy of Science, Indiana University

Paolo Brenni President of the Scientific Instrument Commission of the International Union of History and Philosophy of Science, and President of the Scientific Instrument Society

Jed Z. Buchwald Doris and Henry Dreyfuss Professor of History, California Institute of Technology

Sandro Caparrini Department of Mathematics, University of Turin

Hasok Chang Hans Rausing Professor of History and Philosophy of Science, University of Cambridge

Olivier Darrigol Research Director, Centre National de la Recherche Scientifique

Robert Fox Emeritus Professor of the History of Science, University of Oxford

Craig Fraser Professor at the Institute for the History and Philosophy of Science and Technology, Victoria College, University of Toronto

Graeme Gooday Professor of History of Science and Technology, University of Leeds

Niccolò Guicciardini Associate Professor of History of Science, University of Bergamo

John L. Heilbron Professor of History and Vice Chancellor Emeritus, University of California at Berkeley, and Honorary Fellow of Worcester College, Oxford

Daniel Kennefick Assistant Professor, Department of Physics at the University of Arkansas

Helge Kragh Professor of the History of Science, Department of Science Studies, University of Aarhus

Anita McConnell Affiliated Research Scholar, University of Cambridge

Daniel Jon Mitchell Department of History and Philosophy of Science, University of Cambridge

LIST OF CONTRIBUTORS   ix

Iwan Rhys Morus Professor in the Department of History and Welsh History, Aberystwyth University

Kathryn M. Olesko Professor in the Department of History, Georgetown University

Giuliano Pancaldi Professor of the History of Science and Head of the International Centre for the History of Universities and Science, University of Bologna

Jürgen Renn Honorary Professor of the History of Science, Humboldt University, Berlin

Eric Schliesser BOF Research Professor of Philosophy, Ghent University

John A. Schuster Honorary Research Fellow, Unit for the History and Philosophy of Science, and the Sydney Centre for the Foundations of Science, University of Sydney

Suman Seth Associate Professor, Department of Science and Technology Studies, Cornell University

Alan E. Shapiro Professor Emeritus, School of Physics and Astronomy, University of Minnesota

Terry Shinn Historian and sociologist of science, Centre national de la recherche scientifique, Paris, France

Josep Simon Assistant Professor, GESCTP, Universidad del Rosario, Bogota

Chris Smeenk Associate Professor of Philosophy, University of Western Ontario, and interim Director of the Rotman Institute of Philosophy

Crosbie Smith Professor of the History of Science, University of Kent

Friedrich Steinle Professor at the Institut für Philosophie, Literatur-, Wissenschafts- und Technikgeschichte, TU Berlin

Larry Stewart Professor of History, University of Saskatchewan

N. M. Swerdlow Professor Emeritus, Department of Astronomy and Astrophysics, University of Chicago

Anthony Turner Independent historian based in Le Mesnil-le-Roi, France

INTRODUCTION

JED Z. BUCHWALD AND ROBERT FOX

Modern trends in the historiography of science have led to a rethinking of many of the classic areas of the history of physics. A history that might have been cast, in the days of Florian Cajori's *History of Physics* (1899) or W. F. Magie's *Source Book in the History of Physics* (1935), as a progressive uncovering of the truths of the physical world by a succession of great thinkers has now to be interpreted as a far more complex process. Writing a satisfactory history has come to require a sensitivity both to the content of a physicist's individual contribution and to the cultural, institutional, and economic context in which the contribution was made. *The Oxford Handbook of the History of Physics* seeks to explore this complexity through examples of cutting-edge writing in what has come to be recognized as a particularly vibrant area of historical research. By presenting a wide diversity of studies in a single volume, it gives a flavour of scholarly contributions that have tended to be dispersed in journals and books not easily accessible to the general reader.

The *Handbook* brings together chapters on key aspects of physics from the seventeenth century to the present day. The chapters are grouped in four sections, following a broadly chronological structure. While there is always something unsatisfactory about temporal boundaries, the arrangement in sections allows the character of each period to be identified and major trends to be highlighted. Chapters in the first section explore the place of reason, mathematics, and experiment in the age of what we still conventionally, if uneasily, refer to as the scientific revolution of the seventeenth century. The contributions of Galileo, Descartes, and Newton are central to this section, as is the multiplicity of paths to the common goal of understanding. Some of these paths represented a continuation of discussions of method and logic with roots in medieval and renaissance philosophy, others a reinforcement of attempts to mathematize long-known natural phenomena, and others again reflected the turn to Thomas Kuhn's category of 'Baconian' sciences—newer, more empirical investigations focused on heat, electricity, magnetism, optics, and chemistry. The second section treats the 'long' eighteenth century—a period (extending into the early nineteenth century) that is often regarded as synonymous with the 'age of Newton'. The influence of Newton was indeed powerful, and it bore fruit in mathematical and experimental approaches to scientific enquiry, following the models set in, respectively, the *Principia* and the *Opticks*. But the Newtonian character of the age was tempered in important ways by traditions whose Newtonian pedigree was either absent or was coloured by other approaches, including some with clear Cartesian roots. Eighteenth-century speculations about a universal ether and the widely shared beliefs in the existence of property-bearing imponderable fluids underline the point, reminding us of the snare of simple characterizations that risk masking the tensions and controversy inherent to physics in most periods. The contributions to the third section are broadly concerned with the nineteenth century, during which the definition of what constituted physics and demarcated it from chemistry and older traditions of natural philosophy came to assume a form that we should recognize today. It was now that the familiar subcategories of heat, light, electricity, sound, electricity, and magnetism established themselves, along with mechanics, as the canonical areas constituting any university-level curriculum in the discipline. Through the nineteenth century, all these areas became the focus for the deployment of varying degrees of mathematical techniques and of experimental procedures made possible by the emergence of the new wave of laboratories that institutions of higher education and advanced training everywhere had come to regard as essential to their calling by the end of the century. The chapters in this third section convey the unprecedented intensity and theoretical and experimental sophistication of work at the recognized frontiers of physics, especially from mid-century, when the concept of energy lent the discipline a unity that it had not had before and both optics and electromagnetism raised problems only capable of investigation in the academic setting. Finally, a fourth section of chapters takes us into the age of 'modern physics'. Here, the canonical landmarks are well known. Among them were the discovery of the photoelectric effect in 1887, Planck's work on the quanta of radiation, Einstein's special theory of relativity of 1905, and the elaboration of the various facets of what we now know as quantum physics between 1900 and 1930. But the profound conceptual changes that took shape in the first three decades of the twentieth century and the new departures in experimental physics that went hand in hand with them, largely focused on the microworld of atoms, radioactivity, and X-rays, sit uneasily with the neat designation on which any sharp distinction between the 'classical' and 'modern' phases of physics rests. One reason for this, which emerges clearly from chapters in the fourth section, was the simultaneous, ever closer integration of physics in the worlds of industry and the military. The result was a disciplinary porosity that opened physics to demands and practices distinct from the abstract, theoretically driven priorities of, say, the first Solvay conference in physics in 1911. By the early twentieth century, physics was firmly set on its way to a central position in the 'big science' of our own day and the source of economic profit in settings typified in Silicon Valley and other high-tech locations throughout the world.

Throughout the volume our aim as editors has been to give authors as much freedom as possible in the handling of their chosen topics. While the profile of the *Handbook* emerged from discussion with and between authors, the volume was never intended to stand as an exhaustive history. Within the broad general plan, authors were asked to select leading themes and to address them at a level accessible not only to specialists but also to practising scientists, students, and the wider reading public for which the *Oxford Handbook* series is intended. No particular approach or style of writing was favoured, and we were glad to see work drawing on diverse historiographical traditions and with an appropriately wide vision of what constitutes the history of physics. The content of physics, of course, remains paramount and provides an essential thread to the volume. But the *Handbook* also contains chapters on other dimensions that have their place in any rounded history. These include the work of instrument-makers in the seventeenth and eighteenth centuries and instrument-making companies in the nineteenth and twentieth centuries—work that helped to bring rigorous experimental technique to the centre of physicists' concerns. Communication too is a recurring theme. Correspondence, journals, and conferences are self-evidently essential elements in the passage of knowledge between experts, and they are referred to frequently in the volume. But communication had other guises too, including popular manuals and demonstrations of physical phenomena in public lectures in the eighteenth century and the growth, in the nineteenth century, of formal courses and the concomitant rise of examinations and the emergence of the essentially new genre of the physics textbook.

It would be extravagant to attempt to draw general conclusions from a collection of essays conceived and executed in this way. But certain trends are clear enough. Changes in the national setting for the performance of the most creative work are one of these. From Renaissance Italy and then England in the early and mid-seventeenth century, the leading centre for experimental physics and the mathematical tools of physics moved to France in the eighteenth century, then gradually during the nineteenth century to Germany, and by the beginning of the twentieth century to the USA as well. Another clear trend is the move to ever more rigorous standards of measurement. What began in the later eighteenth century as a new 'quantifying spirit', first manifested in mechanics, optics, and astronomy, developed as a quest for precision that increasingly drew impetus from areas of practical activity as disparate as geodesy, steam-power technology, telegraphy, medicine, and more generally a quest for dependable information that would facilitate the need for regulation in modern society. Such applications for precision all played their part in raising the status of physics and its consolidation as a global discipline in which metrology and the standardization of units and terminology were essential to the transfer of information relevant to the modern world. The implications for economic or national supremacy were evident, and they resulted in a new readiness of public bodies, from governments to universities, to provide as generously as possible for escalating science budgets in which physics was a conspicuously voracious element. It is often said that the cutting edge of science has begun to shift away from physics towards the life sciences. And recent patterns in the focus for investment in research, as well as in the career choices of the young, suggest that this may be so. But of the centrality of physics in fashioning the industrial–military complex that has done so much to shape today's world there can be no doubt. For that reason, as much as for the intellectual interest of physics as an academic pursuit, the history whose course is reflected in the volume deserves our attention.

PART I
PHYSICS AND THE NEW SCIENCE

CHAPTER 1

WAS THERE A SCIENTIFIC REVOLUTION?

JOHN L. HEILBRON

1.1 AN ANALYTICAL TOOL

Was there a Scientific Revolution (SR) at any time between 1550 and 1800? Yes, indeed, if SR signifies a period of time; perhaps, if it is taken as a metaphor. According to the findings presented here, the metaphor can be useful, that is, productive of insights, if it is taken in analogy to a major political revolution. The analogy suggests a later onset, and a swifter career, for the Scientific Revolution than is usually prescribed, and reveals Newton not as its culmination but as its counter. Whether as a period or as a metaphor, SR is an historian's category, an analytical tool, and must be judged by its utility.

The label 'Scientific Revolution' to indicate a period in the development of natural knowledge in early modern Europe has secured a place in historiography as secure as 'Renaissance' and 'Enlightenment'. Like them, its nature and duration vary in accordance with the concerns of historians who write about it. An early approach, developed under the powerful influence of Alexandre Koyré, took cosmology and the physics of motion as constitutive of science and accordingly began the revolution with Copernicus and ended it with Newton—a period of 150 years. Koyré's approach, though still followed here and there, receded before the more generous programme of A. Rupert Hall, who admitted all respectable sciences and so lengthened the 'revolution' to three centuries.[1] The most recent and most ambitious account of the SR yet attempted, by H. Floris Cohen, limits the subject matter to 'modes of nature-knowledge', that is, to transformations in method. With this specification

Cohen managed to confine the SR to the seventeenth century and to little more than 800 pages.[2] Recent surveys tend to prefer this tighter schedule.[3]

Like 'Renaissance' and 'Enlightenment', 'SR' is a metaphor as well as a shorthand for particular developments within an historical period. The metaphor has given rise to much debate between those who regard these developments primarily as an abrupt transformation and those who see mainly continuity of pace and content. The debaters also deliberate whether all or most advances in natural knowledge, or only those subject to mathematics, underwent revolutionary transformation. Much information of value came to light through efforts to demonstrate one or the other of these alternatives. One reason that the debate has been fruitful is that its participants left fuzzy the concept of revolution against which they measured the SR.

All the low-lying fruit may now have been gathered, however, since most historians have lost interest in deciding the appropriateness of the metaphor. And so a recent survey opens, apparently wittily, 'There was no such thing as the Scientific Revolution and this is a book about it'. The book does not give reasons for thinking there was none, as might be expected, but goes on to describe the same sort of material, in much the same way, as other surveys do.[4] In effect, the author employs SR in two different senses: as the name of certain aspects of a historical period, which is the subject of his book, and as a metaphor, which he regards as empty.

The rejection of the metaphor does not imply that the period labelled the SR lacks the unity that would qualify it for a special name to set it off from periods on either side. A gradualist image, say of a bend in the road, may suit the case better, especially if the period called 'revolutionary' is too long to administer a short sharp shock. Approaching a bend, the driver sees nothing of the vista soon to appear; on entering it, he sees the new vista gradually unfold; and, looking back on emerging from it, he no longer can see where he was. The image has the advantage that different drivers can negotiate the bend at different speeds.

Floris Cohen's masterwork, *How Modern Science Came into the World*, is of this type. Its subtitle, *Four Civilizations, One Seventeenth-Century Breakthrough*, telegraphs the story line: four times in history, civilizations have stood poised to break through to modern science but have succeeded only once. And then the process took a century, in six overlapping 'transformations'. These concerned 'realist–mathematical science' (Kepler, Galileo); 'kinetic–corpuscular philosophy' (Beeckman, Descartes); experimental philosophy (Bacon, Gilbert, Harvey); geometrized corpuscular philosophy (Huygens, young Newton); 'the Baconian brew' (Boyle, Hooke); 'the Newtonian synthesis'. Although revolution seems an inappropriate descriptor for the series of loosely linked standard episodes that drove seventeenth-century natural philosophers and their fellow travellers around the bend, Cohen perforce names his distinctive period 'SR'. The wine is new, the bottle old, the label misleading.

Metaphors ordinarily are not branded as true or false but as striking, compelling, insightful, beautiful, useful, or their opposites. As analytical tools, their value lies in their utility. How should we judge the SR, regarded as a metaphor, on this criterion? An obvious way would be to seek strict analogies between the SR and an episode accepted as a *bona fide* revolution, say the upheaval in France beginning in 1789. If the analogies produce new knowledge or a new way to view acquired knowledge, we might declare the metaphor alive. If nothing new comes out, we should regard it merely as a label for certain developments within an historical period of uncertain extent.

None of the sixty-three authors whose opinions Floris Cohen canvassed in his chrestomathy on the SR seems to have considered the SR from this point of view.[5] No more did Bernard Cohen, though now and again he glanced that way when filling his cornucopia of references to Scientific Revolution over the last four centuries. He observed that political revolutions are violent and suggested the conquest by belligerent Cartesians and, following them, Newtonians, of the institutes of science as comparable acts of violence in the SR. This observation, together with Cohen's hint at a connection between political disturbances and natural knowledge in the mid-seventeenth century, pointed in a promising direction; but his primary interest did not allow him to handle the concept of Scientific Revolution as an analytical (or historian's) tool.[6] Built up from occurrences of the term in the writings of historical actors, historians, and modern working scientists, Cohen's inventory does not support or describe an exploitable metaphoric interpretation of the SR.

The following sketch, adopted from an earlier essay, illustrates how the deployment of the metaphor to seventeenth-century natural knowledge might be accomplished.[7] The exercise brings some unexpected results.

1.2 The Metaphor

'A sudden, radical, or complete change', 'the overthrow or renunciation of one government or ruler and the substitution of another by the governed'. Thus the dictionary sets the bounds for our metaphor. Did the SR witness a sudden, radical, or complete change analogous to the overthrow of an established government? Here it is important to distinguish revolutionary ideas and revolutionary situations from revolutions. As for revolutionary ideas, no more need be said than that where they are encouraged and rewarded there is no end to them. Revolutionary situations also are commonplace. According to the great authority on political revolutions, R. R. Palmer, a revolutionary situation can develop when people lose confidence in existing law and authority, when they reject obligations as impositions, regard respect for superiors as humiliation, and condemn privilege as unfair and government as irrelevant. A people or nation thus afflicted has lost its sense of community and may be ripe for revolution. A revolution need not ensue, but the social and institutional base must undergo some significant change to avoid one.[8]

We must not take a revolutionary situation for a revolution. Thus when Galileo cried out while lecturing on Aristotle's theories of motion, 'Good Lord! I am so tired and ashamed of having to use so many words to refute such childish objections as

Aristotle['s]',[9] he expressed the rejection of authority, and the feeling of imposition and humiliation in having to respect it, typical of a revolutionary situation. And when he described himself as a singular eagle and the established philosophers of his time as flocks of starlings who 'fill the skies with shrieks and cries wherever they settle, and befoul the earth beneath them', he expressed the truculent bitterness of an oligarch forced to suffer equality in a democracy, not a call, let alone a programme, for revolution.[10] Galileo was a Voltaire rather than a Robespierre—witty, cutting, brilliant, a master of language, an ambitious courtier, an incisive critic, but not a fierce destroyer of the old with a grand vision of the new.

A popular question with those who search for the origins of modern science is: 'Why was there no Scientific Revolution in China, or in ancient Greece, or in the Abbassid Empire'? A more pertinent question is: 'Why was there no Scientific Revolution in Europe in the late sixteenth century'? Many of the primary ingredients in the standard histories of science were by then available: Copernicus' *Revolutions* and Vesalius' *Fabrica*, which set the bases for a new astronomy and a new anatomy; the warlike bombast of Paracelsus and the new logic of Petrus Ramus; the mathematical ways of Archimedes and Plato; the secrets of Hermes; the natural magic of della Porta; and the technologies of Agricola and Biringuccio. These and other novelties—notably reports from the new world of exotic floras, faunas, and people—were disseminated quickly in standard editions by a printing industry then a hundred years old. And yet no revolution in natural knowledge resulted. The schools took cognizance of the new facts, patched their teachings where necessary, and became more dogmatic as experience, information, and expectations at odds with their principles accumulated.

One reason for the failure of the revolutionary situation of 1550 to become a revolution was the engagement of many of the best minds in Europe in doctrinal disputes and the wars of the Reformation. Thomas Sprat, the historian–apologist of the Royal Society of London, observed that for many years before he wrote in 1667, an 'infinite number of Wits . . . [had] been chiefly taken up about . . . the *Writings* of the *Antients*: or the *Controversies* of *Religion*: or *Affairs* of *State*'. The engagement of all those wits in antiquities, controversies, and politics had the consequence, as Sprat put it, that 'knowledge of nature has been very much retarded'.[11] Whatever their preoccupations, the wits were not numerous; perhaps no more than a thousand mobilizable ones existed when the putative SR was in full swing. A main subject of our inquiry, on which the value of our metaphor may ride, must be the origin and recruitment of sufficient cadres to support and sustain a revolution.

The Peace of Augsburg of 1555 did not open a period of leisure for the engaged learned. It was but a pause. Where Rome could exercise power, the new machinery of the Counter Reformation, especially the decrees of the Council of Trent and the operations of the Society of Jesus, enforced a doctrinal conformity little conducive to innovation in natural philosophy. The retrospective appointment of Thomas Aquinas as doctor of the Roman Catholic Church created an unprecedented and unfortunate unity of philosophy and theology. As Descartes put the point in 1629, 'theology . . . has been so subjected to Aristotle that it is almost impossible to set out another philosophy without its appearing at first contrary to faith'.[12] The Tridentine interdisciplinary synthesis made it more difficult than ever to topple the teachings of the schools in Catholic countries—especially since the Jesuits ran the best schools. The situation came to a head with the trial and condemnation of Galileo in 1633. That made him a martyr of the ensuing revolution, but not its leader.

Again, a look at the political situation during the first half of the seventeenth century suggests why the pursuit of natural knowledge did not have high priority in Europe then. The Thirty Years War had devastated the German States and absorbed the energies of what would be the Dutch Republic. England went from Civil War to Puritan crackdown to, in 1660, the Stuart restoration. France ruined itself in the Thirty Years War, the subsequent Frondes, and its private war with Spain, which dragged on until 1659, the year when Louis XIV reached his majority and began to reign on his own. Not until the second half of the seventeenth century was an exhausted Europe able to devote what energy it had left to improving and disseminating natural knowledge. After a century of war and strife, an investment in the arts and sciences, especially in sciences removed from theology, and a measure of tolerance in their pursuit, seemed a promising way simultaneously to soothe and advance society.

Bishop Sprat stressed the importance of a place where people who might not agree on politics or religion could meet civilly and productively over a common interest in what the Royal Society's charter called 'natural knowledge'. The membership agreed not to take up contentious topics—religion specifically—likely to produce altercation and disintegration.[13] In their book, *Leviathan and the Air Pump*, Steven Shapin and Simon Schaffer developed this theme from Sprat into a model of what they call the 'experimental life'. In their interpretation, the circle around Robert Boyle, the Society's most prolific aristocratic experimentalist, aimed to create matters of fact beyond dispute, and unassertive, eclectic theories about them that did not provoke unpleasant confrontations.[14]

Another connection between the coming of peace and the promotion of natural philosophy was the air pump, which became the most important scientific instrument of the later seventeenth century and the most powerful single piece of hardware in the revolutionary arsenal. Its inventor, Otto von Guericke—an engineer who served as mayor of Magdeburg—brought this fruit of his scarce leisure to the Imperial Diet convened in Regensburg in 1654 to plan the rebuilding of the Empire after the catastrophic Thirty Years War. He employed the machine to entertain his fellow delegates while they rested from their labours.

Guericke was a Lutheran. Among the delegates at Regensburg was the Catholic Bishop of Würzburg. He took a great interest in Guericke's spectacular demonstrations of the power of nothing, bought the pump, and presented it to the Jesuit College in Würzburg. There the professor of mathematics, Gaspar Schott, set about analyzing and improving the pump. His account of what he called the Magdeburg experiments, published with Guericke's permission, carried news of the machine to England, where the Anglicans Boyle and Hooke made their version. This vignette, like Sprat's apology for the Royal Society, directs our attention to the 1650s and 1660s as a period of dramatic change in the material circumstances in which natural knowledge was cultivated, and in the level of religious toleration in the Republic of Letters. Many wits turned from controversy and politics to safer subjects. As Henry Oldenburg, a comrade-in-arms of Sprat, summed up the new situation in his capacity of Secretary of the Royal Society of London, 'mathematics and natural science [*physica*] are now developing and flourishing everywhere'.[15]

Four years after the restored Stuart King Charles II chartered the Royal Society of London, the recently mature and secure Louis XIV set up, as a sort of government bureau, the Académie Royale des Sciences. The establishments of the French and English kings followed by only a few years the Accademia del Cimento, created in 1657 by the Grand Duke of Tuscany and his brother to carry out experiments with a minimum of theorizing. The Florentine academicians were to proceed by trial and error, *provando e riprovando*, testing and retesting matters of fact.[16] Unlike the royal academies of experiment in France and England, the Cimento did not survive very long, but its example nonetheless inspired many imitators in Italy and the German States. Before 1650 there were at most two or three short-lived academies of experiments, and these quite insignificant despite Galileo's pride in belonging to the Lincei; in the second half of the seventeenth century, many were founded, some with royal or high ecclesiastical patronage.

Relative peace and effective organization do not make a revolution, however. Indeed, they seem rather opposed to a metaphor suggesting violent change. But then, the institutions that support the body of knowledge do not define it. We can have a violent alteration in ideas and practices that make use of organizations that were not set up to change anything violently. That is what happened in natural knowledge in the second half of the seventeenth century, when ideas opposed to established learning took root in experimental academies. A possible comparison is the revolution caused by the rapid invasion of the newly created universities of the thirteenth century by the then newly recovered *libri naturales* of Aristotle. Let us accept this comparison as the first fruit of our inquiry. It suggests that the same coincidence of novel philosophy and fresh institutions that put Aristotle's system at the centre of European philosophy in the thirteenth century removed it to the periphery in the seventeenth century.

1.3 The Ingredients

The preceding analysis also suggests that our revolution in knowledge awaited the advent of political and social peace and reconstruction in Western Europe. By good luck, the ingredients that came together at this unusual juncture were just those normally required to make a good political revolution: a powerful programme to supplant established ways and teachings, the existence of vigorous well-educated cadres devoted to the programme, and the creation of new institutions and instrumentalities with which to preserve the gains of the cadres.

1.3.1 The Programme

The revolutionary programme appeared in its definitive form in 1644, as the *Principia philosophiae* of René Descartes, which was soon translated into French, reduced to manuals for teachers and, insofar as it pertained to natural knowledge, reworked into physics texts and demonstrations that attracted wide audiences outside the schools. The distinctive Cartesian universe of stars and tourbillons, of comets, moons, and planets swimming in vortices, of particulate streams that cause magnetism, of levers and springs and bellows that make up animal bodies, had the security of the existence of Descartes' doubting self and of a God unable to deceive him. All this is as intelligible now as it was then, though no doubt it is not as compelling to us as it was to the expectant disciples Descartes had begun to collect on the strength of his *Discours de la méthode* (1637), his directions for the correct manner of reasoning.

Three mutually reinforcing factors gave Cartesian natural philosophy its revolutionary power. For one, it did away in one stroke with the scholastic bric-a-brac of forms, essences, potentialities, acts, sympathies, antipathies, and *qualitates occultae*. Where Aristotle populated the world with irreducible natures, Descartes insisted on explaining all the phenomena that fall under our senses as consequences of matter and motion. To be sure, the transfer of motion in collisions may be no more intelligible than an occult quality.[17] But even if we do not understand the metaphysics of collisions, we have intuitions about mobility, rest, size, and shape that enable us to reason further about the constitution and activity of the world—if we assume that it consists only of matter and motion. An Aristotelian, in contrast, could not deduce from his concepts of 'dogginess' and magnetism whether or not a dog is a magnet.

One of Descartes' ambitions was to live long enough to put medicine on as firm a foundation as the physics of the solar system.[18] He conceded that a great amount of experiment was necessary to determine which of the many ways he could conceive for the mechanical operation of the body God in fact had chosen. Once these ways were known, reason, exploiting the intelligibility of mechanical interactions, would swiftly find the means to promote health and longevity. This intelligibility came not only from everyone's intuition of extension and motion, but also from Descartes' mastery of mathematics. The geometrical ingredients of his vortical universe all lent themselves to mathematics.

Although few of Descartes' mechanisms ever in fact felt the yoke of mathematics, the possibility he revealed of a mathematical physics was as radical and exciting as the premiss that everything is matter and motion. The school philosophy held in general that mathematics could not attain to truth, and hence had no significant use in natural philosophy. As exact description it had its utility, for example, in computing the positions of the sun, moon, and planets for navigational or calendrical or astrological purposes; but since the machinery for astronomical calculation was arbitrary and opportunistic, neither its successes nor its failures could confirm or refute physical principles. Nor was mathematics considered a useful tool of exploration in physics. 'After a great deal of calculation ... I do not see any fruit other than a chimerical proportion, which reveals nothing about the things that are real in nature'. Thus André Graindorge, co-founder of the Académie Royale de Physique of Caen, expressed the widespread view that mathematics could at best describe accurately the least interesting features of the physical world.[19] In Descartes' system, mathematics was neither opportunistic nor irrelevant. He derived the existence and nature of the world-moving tourbillons, and also their laws of motion, from first principles. Philosophers who were also mathematicians could develop Cartesian physics and bring to bear the great advances in mathematics achieved during the seventeenth century, not least by Descartes himself, confident that they reasoned from a true world picture.

The third thread in Cartesian natural philosophy (besides its rejection of scholastic forms and its privileging of quantifiable concepts) was its comprehensiveness. Here a comparison with the writings of Francis Bacon is useful. As a critic of scholastic philosophy, Bacon was more effective than Descartes; and, as a methodologist he offered a fuller and, as it turned out, a better long-range plan for achieving a new philosophy. But he did not get there himself, and made no use of Galileo's physics, Gilbert's magnetism, or Kepler's optics and astronomy. And even if Bacon had known how to appreciate these fragments, he would not have been able to unseat the school philosophy. Quantified *disjecta membra* do not make a physics nor even, in the eyes of the seventeenth century, a start on one.

We know that there is no truth in the application of mathematics in the absence of physical principles known to be sound prior to quantification. But that is not the main reason that Bacon could not have made a sound physics from the pieces that Galileo, Gilbert, and Kepler supplied. The main reason is that philosophers abhor a vacuum even more than nature does. 'I have no reason to believe that vacuum does not exist', says Guarino Guarini, 'except that I have not seen completely convincing proofs; and without them, I have seldom left my Aristotle'.[20] That was to follow Aristotle's advice. 'It is wrong to remove the foundations of a science unless you can replace them with others more convincing'. Galileo made Simplicio, the hack peripatetic in the *Dialogue on the Two Chief World Systems* (1632), run to the same asylum. 'Who would there be to settle our controversies if Aristotle were to be deposed? What other author should we follow in the schools, the academies, the universities?' The Galilean characters in the *Dialogue* continued to scare Simplicio with the *horror vacui philosophici* until he cried out to his tormentors to name an author—any author—capable of replacing Aristotle. They could not do it.[21]

It was just this failure that Descartes seized on in his evaluation of Galileo's *Discourses on Two New Sciences* (1638)—one of the key texts of the usual account of the Scientific Revolution:

I find that he philosophizes much better than is usual because he rejects the errors of the schools as far as he can and tries to examine physical matters with mathematical reasoning. I agree entirely with him in that, and believe that there is no other way of finding the truth. But it seems to me that he lacks a lot since he continually digresses and does not stop to explain anything thoroughly, which shows that he has not examined things in order; without investigating the first causes of nature he has only tried to give reasons for some effects and so he has built without foundations ... Every thing he says about the velocity of bodies falling freely in void has no basis: for he first should have determined what weight is, and then he would have known that velocity is zero in a vacuum.[22]

This is of course nonsense, as is much else in Descartes' physics, but you do not have to be right to make a revolution. You have to have a plausible and comprehensive programme.

1.3.2 The Cadres

Cartesianism quickly gained influential recruits throughout the Republic of Letters. At the top was Queen Christina of Sweden, whose interest proved as fatal as it was flattering. A lesser and safer princess was Elizabeth of Bohemia, a good mathematician and metaphysician in her own right, who helped disseminate Descartes' ideas. In an instructive letter she described to him the conversion process of a doctor Weis, whose confidence in Aristotle and Galen had been shaken by reading Bacon. That had not made him abandon the ancients. But, Elizabeth reported, 'your Méthode made him reject it altogether and [moreover] convinced him of the circulation of the blood, which destroys all the ancient principles of medicine'.[23]

There are at least three significant points about this short story. Firstly, unwilling to face the *horror vacui philosophici*, Weis could not abandon Aristotle without an adequate replacement. Secondly, Weis' *itinerarium ad veritatem*, beginning with Bacon's condemnations and compilations and ending in Cartesian truth, became a standard route, and was the path recommended by Descartes himself.[24] Thirdly, Descartes' concept of a living body as a machine perfectly accommodated the discovery of the circulation of the blood, although Descartes rejected Harvey's identification of the heart with a pump.[25] Up-to-date physicians, especially younger ones, found Cartesian philosophy a welcome weapon in their fight against Galenic physiology and professional conservatism.

These modernizing doctors made a strong and effective cadre. They pushed Descartes' ideas into the medical schools beginning with Utrecht and Leiden in the 1640s and 1650s, whence Cartesianism spread throughout Europe.[26] An early and persistent example is Naples, where physicians marched under Descartes' banner against the educational and political establishment on and off throughout the seventeenth century.[27] Despite its deserved reputation for conservatism, the medical profession played an important part in fomenting Scientific Revolution. So did lawyers. In France, Cartesian doctors made common cause with lawyers, with the *noblesse de robe*, from whom Descartes sprang. A group supported by Henri de Montmor, a lawyer in Paris, pushed the Cartesian philosophy to the doorstep of the Académie Royale des Sciences, albeit with a litigious spirit perhaps too lively for the Republic of Letters.[28] Popular lecturers on the physics of tourbillons made a good living catering to doctors, lawyers, and ladies.[29]

The universities also furnished cadres for the revolution. The English—who tend to be oversusceptible to French philosophy until, on more mature judgement, they reject it altogether—provide an extreme example. Smitten with Descartes in the 1650s, Oxbridge dons extolled him as 'incomparable', 'precious and incomparable', 'most excellent', and 'the miracle of men'.[30] According to the Cambridge philosopher Henry More, Descartes was the 'very secretary of nature'—far more penetrating than Galileo, indeed, than everyone and anyone: 'all those who have attempted anything in natural philosophy hitherto are mere shrimps and fumblers in comparison'.[31] Cartesian physics had made its way to Sweden by the time that More was rhapsodizing over it in England. Petrus Hoffwenius, having drunk directly from the Cartesian spring in Leiden, sprinkled the mechanical philosophy around the medical school at the University of Uppsala, beginning in the early 1660s.[32]

The professional people and professors who made up the Cartesian cadres went to work, as revolutionaries do, on colleagues without fixed allegiances who were dissatisfied by all else on offer. These potential fellow travellers adopted various forms of mechanical philosophy, paid lip-service at least to the rhetoric of mathematics, recommended and sometimes even practiced experiment, and brought forward useful things from the ancients. Their hero was Robert Boyle, and their motto 'Amicus Plato, amicus Aristoteles, amicus Cartesius, sed magis amica veritas'.

Among the large and diverse group of fellow travellers even Jesuits could be found—for example, Gaston Pardies, whose colourful characterization of his position deserves to be preserved: 'Just as formerly God allowed the Hebrews to marry their captives after many purifications, to cleanse them of the traces of infidelity, so after having washed and purified the philosophy of M. Descartes, I could very well embrace his opinions.' Even on the Cartesian side, purifications might be needed. As the perpetual secretary of the Paris Academy, Bernard le Bovier de Fontenelle, put it, Descartes' 'new way of philosophizing [is] much more valuable than the philosophy itself, a good part of which is false, or very uncertain.' 'Il faut admirer toujours Descartes, et le suivre quelquefois'.[33]

The cadres and fellow travellers met strong resistance. The curators of the Dutch universities under attack by Cartesian doctors prohibited the teaching of any philosophy but Aristotle's.[34] The Paris Parlement reaffirmed an edict it had passed in 1624, prohibiting the teaching of any philosophy but Aristotle's.[35] The Archbishop of Paris shut down public lectures on Cartesian physics; the Holy Office in Rome condemned Descartes' *Principia philosophiae*; the Jesuits proscribed Descartes' writings again and again within their own order and initiated the proceedings that placed his works on the Index of Prohibited Books.[36] The Papal and Spanish authorities occasionally jailed the Cartesian doctors and lawyers of Naples.[37] Under pressure from theologians, the Swedish Rigsdag condemned Cartesian philosophy in 1664, just after the Roman Catholic Church had indexed it and largely for the same reason: its bearing on the doctrine of the Eucharist.[38]

Our revolutionary model requires that we follow these battles and report the changing fortunes of the two sides, expecting, as was the case, different sorts and levels of conflict in different theatres of war. The most dramatic and instructive confrontations occurred where the Roman Catholic Church was strongest, the least exciting and instructive where representative government had a grip. In England, the Scientific

Revolution was as bloodless as the Glorious Revolution, which deposed the Catholic King James II almost without firing a shot. Our model directs our attention away from England. It is a mistake to found general lessons about the Scientific Revolution on the doings of Robert Boyle.

The smoke cleared around 1700. By then, or shortly after, even the Jesuits allowed the teaching of Cartesian physics, albeit as an hypothesis, and by 1720, according to an old expert on the intellectual development of Italy, the moderns could anticipate a full and imminent victory.[39] On the other side of the aisle, the rector of the Calvinist Academy of Geneva urged Descartes' approach on his students and faculty. It was 'incomparable', he said, 'the best human reason has found', no less than 'royal', a great compliment in a republic.[40]

About the same time, in 1699, the Paris Academy of Sciences reorganized under its Cartesian secretary Fontenelle. In his éloges of deceased members, Fontenelle developed a standard account of enlightenment beginning with the discovery of Descartes and ending with admission to the pantheon of science. The Oratorian priest Nicolas Malebranche was exemplary. At first, owing to his scholastic education, he could read Descartes' natural philosophy only in homeopathic doses, as it gave him palpitations of the heart, 'which obliged him sometimes to interrupt his reading'. With perseverance he learned to handle stronger doses, and so advanced from scholastic darkness to the light of mathematics and true physics and, in good time, to the comfort of a seat in the Paris Academy.[41]

1.3.3 Institutions and Instrumentalities

We come to the academies—the institutions in which the gains of the revolution were preserved and multiplied. The Royal Society of London and the Académie Royale des Sciences of Paris are to the movement what Queen Christina and Princess Elizabeth were to the cadres—the peaks of the groups that formed during the second half of the seventeenth century to perform and witness experiments, learn and impart natural knowledge, and exchange the news and rumours of the Republic of Letters. There were cadres around princes and prelates, librarians and lawyers, and, perhaps most significant of all, professors.

Representative examples from the main theatre of war, Italy, are the Accademia Fisico-Matematica founded in Rome by Giovanni Giusti Ciampini,[42] the groups that gathered around Geminiano Montanari in Bologna and Padua,[43] and the Accademia degli Inquieti set up in Bologna by Eustachio Manfredi.[44] Ciampini, the impresario of the Rome academy, was a monsignore high up in the papal bureaucracy; his academy lasted twenty years, until his death in 1698. Montanari and Manfredi were professors of mathematics. Montanari's academy at Bologna was a forerunner of Manfredi's, which became the nucleus of the Bologna Academy of Sciences, one of the most important of the old academies. It still exists.

For a long time the leading member of Ciampini's Accademia Fisico-Matematica Romana was Francesco Eschinardi, professor of mathematics at the Jesuit College in

Rome. He developed Galileo's mathematics, experimented usefully on optics, and abhorred Galileo's and Descartes' cosmologies.[45] It may not be far-fetched to compare his conservative moderating role in the academy to the effective collaboration between representatives of the lower clergy and the third estate during the early days of the French Revolution.

Montanari was a second-generation Galilean and sometime collaborator of the Accademia del Cimento. He began teaching mathematics and astronomy at Bologna in the early 1660s, and was soon offering instruction, in his own academy, in the sort of experiments that he had seen the Florentine academicians perform. As the capital of the Papal States, Bologna faced a censorship as strict as Rome's; yet astronomy, mathematics, and physical experiments flourished there provided they were not exploited noisily in favour of novelty. At Padua, Montanari had a freer hand. He started a regular programme of observations, which resulted in the discovery of several comets and the variability of stars, and he gave a course of what he called 'physical–mathematical experiments', which he interpreted in the manner of Descartes. The participants in this course became the nucleus of an academy.[46]

Montanari's successor at Bologna, Manfredi, taught the same sort of things and in much the same way, but his situation in the Papal States bred in him a caution, even timidity, so great that his aptly named Accademia degli Inquieti had no need of external censorship.[47] In contrast, the short-lived Accademia degli Aletofili of Verona, established by Montanari's prize student Francesco Bianchini and some doctors dissatisfied with the usual teaching in the medical schools, asserted a vigorous corpuscularian research programme. It languished not because of its advocacy of the mechanical and experimental philosophy, but because Bianchini moved to Rome, where he became a leading light in Ciampini's academy.[48]

These indications lack the whiff of gun smoke. A few examples of the guerrilla warfare around the academies may therefore be in order. A priest was recommended to the Accademia del Cimento as sufficiently open-minded that 'he sometimes rejects Aristotle, in favour of some modern opinion ... although he will hear nothing about the mobility of the Earth; [still] I would guess that he could be persuaded quickly not to rail at that ingenious system so much, since he seems a man of honour [*Galantuomo*]'. This suggests recruitment on the basis of a proven record of eclecticism and gentlemanly behaviour, rather than on the basis of subscription to a particular world system. On the other hand, a 'rotten and mouldy [*marcio e muffo*] peripatetic', who wanted to appear 'a free and modern philosopher',[49] is black-balled; a revolutionary cadre must guard against infiltration. Or, to take a subtler means of co-optation, the censor who passed Montanari's major work in physico-mathematics, which describes experiments on capillary rise from a corpuscular point of view, judged that it disposed of many 'fallacies and discrepancies of the various philosophical schools ... That obliges me to confess [he was a priest!] that sometimes experiment satisfies minds searching for truth better than speculation'. Decoded, this signifies that Aristotle is not to be preferred to modern explanations founded on or illustrated by experiments, and suggests that the censor had enlightened himself by reading books as a consultant to the Index of Prohibited Books.[50]

Against this process of recruitment and cooptation, sterner censors and other guardians of the past could try to mobilize the church's formidable machinery of repression, which was not limited to the Index and the Inquisition. At Bologna in 1661 the medical faculty decided to require an oath of their students not to stray from Aristotle, Galen, and Hippocrates. In a later crackdown, a professor of medicine led an attack on the home of the modernizing doctor and naturalist Marcello Malpighi, a member and correspondent of the Royal Society of London, broke his instruments, and vandalized his library. It was not an isolated incident. But as princes like the Grand Duke of Tuscany appointed men more open to modern ideas to chairs of medicine, the counter-revolution of the Galenists was turned; in 1714 thirteen cardinals and fifty prelates celebrated the gift to a teaching hospital in Rome of the large up-to-date library of the pope's modernizing physician, Giovanni Maria Lancisi.[51]

Around 1700, leading Italian academicians, supposing that the wider the academic movement the faster it would move, designed pan-peninsular institutions. Ciampini contemplated setting up affiliates beyond Rome in the manner developed by the literary Accademia degli Arcadi founded in 1690. Plans were floated to form a league of academies to realize Bacon's programme. Ludovico Antonio Muratori, an historian with an interest in experimental physics, promoted a pan-peninsular academy to advance science and to ensure that Italians obtained due credit for their inventions.[52] Pursuing these tactics, the academic movement had almost completed the process of the Scientific Revolution in Italy by the time of the death of Clement XI in 1721.

Like Italy, the German States had groups that cultivated experimental philosophy with one foot inside and the other outside a university. On the inside the doctoral thesis could serve as the vehicle for the student and professor to develop ideas or arguments around a set of experiments also viewed and perhaps performed by people from outside. An example was the group around Johann Andreas Schmidt, who taught physics at the University of Helmstedt and accumulated a set of experimental apparatus that Leibniz, who viewed it in action, thought exemplary. Schmidt taught an eclectic natural philosophy that built on mathematics and mechanics, and above all, on experiment. 'Follow the watchword of the Royal Society [he told his students], and swear in no one's words'. Schmidt praised four or five people for their exemplary practice of this method. They included Descartes and Johann Christoph Sturm, author of an eclectic *Physica electiva* (1697–1722). Read and reread Sturm's eclectic dissertations, Schmidt bid his audience. *Neminem sequere, et omnes*—'follow no one and everyone'.[53]

Sturm had imbibed Cartesianism from the master's first disciples in Leiden in the 1650s. He developed a course of lectures at the University of Altdorf and set up a Collegium experimentale there, in emulation of the Accademia del Cimento. It was perhaps the first such course and academy in Germany.[54] Through Sturm and Schmidt the methods and instruments of the Dutch universities and Italian academies made their way to Germany, and also through Leibniz, who attended meetings of the Accademia Fisico-Matematica Romana during his visit to Italy in 1690. Leibniz's proposals for an academy of sciences in Berlin mix together the courses of experiments offered by Sturm and Schmidt, the wider programmes of the Italian academies, and the royal sponsorship of the scientific institutions of London and Paris. His German academy was to incorporate the best practices of all the others, and of the Society of Jesus too, and so perfect a prime instrument for the advancement of science and society.[55]

1.4 A Fruitful Concept?

The attempt to fit the metaphor of revolution to the development of natural knowledge during the seventeenth century produced the following results:

- From the 1660s we can speak of a continuous reinforcing institutionalized pursuit of experimental philosophy within a progressive programme. The timing had to do with relative peace, which abetted the transformation of a revolutionary situation into a revolution.
- The characteristic marks of this revolution are corpuscular philosophies and academies of experiment; its favourite weapon was the air pump; its slogans, 'libertas philosophandi', 'provando e riprovando', and 'nullius in verba'.
- The academies and groups that met to do and witness experiments tended to emphasize the features of the new science deemed useful to the newly stabilized states. These features—applied mathematics and practical experiments—were also the cardinal characteristics of the new science.
- In England, the revolution was quiet—the intellectual parallel to the bloodless Glorious Revolution. The main battlefields were Italy and the Northern States, including Sweden, which have been neglected in our historiography owing to a confusion of product and process. England led the way in discovery largely because it did not suffer the guerrilla fighting of the politically divided peninsula and the religiously divided continent. In Germany, as in Italy and for some of the same reasons, the spotlight falls on small experimental academies associated with universities.
- The Scientific Revolution is not just, or primarily, a story of Galileo, Descartes, and Newton. Eclectics, érudits, doctors, waverers, trimmers, compromisers, and fellow travellers played an important part in the revolutionary process.
- The model directs attention to the histories of medicine and education, printing and communication, the recruitment and activities of cadres, and to general history. It suggests that comparisons of the pursuits of natural knowledge under various religious, political, and social regimes can still yield something of interest, and it integrates internal and external factors and approaches.

Our model has a place even for Newton. He is the Napoleon of the piece, the Prince of Physics, the Emperor of Science. Like Napoleon, he consolidated the gains of a revolution fought by others and extended it beyond their wildest dreams. But to extend it

Newton had to deny the strict mechanical programme on which it was founded, just as Napoleon rejected the republic that had given him his start. Newton exhibited additional imperial characteristics, such as denying Leibniz a share in the invention of the calculus and, in his capacity of Master of the Mint, prosecuting to execution clippers of the coin of England.

There is one final revolutionary conclusion from our play with the metaphor of the SR: it was less significant to its participants than we usually reckon. That is because as it progressed its opponents came to face a much graver danger than the substitution of a world of pushes and pulls for the rich diversity of the scholastic universe. In the later seventeenth century, serious savants began to question the inspired authorship of the Bible, made the Old Testament the work of several anonymous hands, reduced its stories to imaginative literature, and contemplated the existence of men before Adam. Here were the makings of a revolution! It caused confrontation everywhere, even in England. Its outcome was glorious, but not its process.

These conclusions, hints, and suggestions are not unique to the revolutionary model. Nor, of course, is this model the only productive one for exploring the development of early modern science. An evolutionary type would be better for bringing out connections on the level of ideas. A renaissance type would link natural knowledge more closely to other rebirths, notably in literature and art (*the* Renaissance) and religion (the Reformations). Further afield there is the model of pathocenosis. Yes, pathocenosis—the collection of diseased states that characterizes a biotype.[56] A pathocenosis strives for equilibrium, which may be prevented or ruptured by changes in the environment or in the disease agents. The dynamic equilibrium between diseases and their hosts nicely represents the intellectual state of learned people in Europe around 1550. In the schools, most of them caught an Aristotelian rash, which one or other Christian doctrine exacerbated. At first this infection protected them from competing afflictions, such as atomism, hermeticism, and Pythagoreanism. But then the equilibrium broke, and many new intellectual diseases, singly and in combination (the eclectics!), ravaged the body of knowledge. The model directs attention to the ways in which the learned lost their immunity, and to when and how equilibrium was regained.

No doubt these few suggestions do not exhaust available useful models. The essential point is to take a model seriously if at all, follow it through, and discard it when, as now, it is squeezed dry.

Notes

1. Alexandre Koyré, *La révolution astronomique. Copernic, Kepler, Borelli* (Paris: Hermann, 1961), and 'A Documentary History of the Problem of Fall from Kepler to Newton', in American Philosophical Society, *Transactions*, 45 (1955), 329–95; A. R. Hall, *The Scientific Revolution 1500–1800* (London: Longmans Green, 1954).
2. H. Floris Cohen, *How Modern Science Came into the World* (Amsterdam: Amsterdam University Press, 2010).

3. E.g., Peter Dear, *Discipline and Experience: The Mathematical Way in the Scientific Revolution* (Chicago: Chicago University Press, 1995).
4. Steven Shapin, *The Scientific Revolution* (Chicago: University of Chicago Press, 1996), which runs from Galileo to Newton.
5. H. F. Cohen, *The Scientific Revolution: A Historiographical Inquiry* (Chicago: University of Chicago Press, 1994).
6. I. Bernard Cohen, *Revolution in Science* (Cambridge: Harvard University Press, 1985), 12–13, 41–4, 77.
7. J. L. Heilbron, *Coming to Terms with the Scientific Revolution* (Uppsala University: Office for History of Science, 2006), reprinted with small changes and deletions in *European Review*, 15 (2007), 473–89.
8. R. R. Palmer, *The Age of the Democratic Revolution: A Political History of Europe and America, 1760–1800* (2 vols., Princeton: Princeton University Press, 1959–64), I, 21. An alternative set of criteria, in J. A. Goldstone, *Revolution and Rebellion in the Early Modern World* (Berkeley: University of California Press, 1991), xxiii, requires a financial crisis, divisions among the elite, and potential to mobilize a sizeable population already aggrieved. A Scientific Revolution modelled on these criteria would emphasize resources and instruments, fights over school curricula, and the belief that traditional learning was worthless.
9. Adapted from Galileo Galilei, *On Motion and Mechanics* (c. 1590), ed. J. E. Drabkin and Stillman Drake (Madison: University of Wisconsin Press, 1960), 58.
10. Aristotle, *Politics*, 1303b5–8, in Richard McKeon, ed., *The Basic Works of Aristotle* (New York: Random House, 1941), 1237; Galileo Galilei, *The Assayer*, in Stillman Drake and C.D. O'Malley, *The Controversy on the Comets of 1618* (Philadelphia: University of Pennsylvania Press, 1960), 151–336, on 189.
11. Thomas Sprat, *History of the Royal Society* (1667), ed. J. I. Cope and H. W. Jones (St Louis: Washington University Press, 1958), 23, 25, resp.
12. Descartes to Marin Mersenne, 18 December 1629, in René Descartes, *Tutte le lettere 1619–1650. Testo francese, latino e olondese*, ed. Giulia Belgioioso (Bologna: Bonpiani, 2005), 100.
13. Sprat (ref. 11), 32–3, 55–6, 82.
14. Steven Shapin and Simon Schaffer, *Leviathan and the Air Pump: Hobbes, Boyle, and the Experimental Life* (Princeton: Princeton University Press, 1985), chap. 7.
15. Oldenburg to Marcello Malpighi, 15 March 1670/1, in Henry Oldenburg, *Correspondence*, VIII, ed. A. R. Hall and M. B. Hall (Madison: University of Wisconsin Press, 1970), 517.
16. W. E. K. Middleton, *The Experimenters: A Study of the Accademia del Cimento* (Baltimore: Johns Hopkins University Press, 1971), 52–3.
17. René Dugas, *La mécanique au XVIIe siècle* (Neuchâtel: Griffon, 1954), 287–98.
18. Or, to be fair, 'to deduce rules of medicine more certain than those we have at present'; René Descartes, *Discours de la méthode & Essais* (1637), in *Oeuvres*, ed. Charles Adam and Paul Tannery (reprint, Paris: Vrin, 1982), VI, 78.
19. Graindorge (a physician) to Pierre Daniel Huet, 11 November 1665, in D. S. Lux, *Patronage and Royal Science in Seventeenth-Century France: The Académie de physique in Caen* (Ithaca: Cornell University Press, 1989), 48. Cf. the opinion of Lucantonio Porzio, in Maurizio Torrini, *Dopo Galileo: Una polemica scientifica (1684–1711)* (Florence: Olschki, 1979), 176–9.
20. Thus Guarini speaks in Geminiano Montanari, 'Discorso del vacuo recitato nell'Accademia della Traccia (1675)', in M. L. Altieri Biagi and Bruno Basile, eds., *Scienziati del seicento* (Milan: Ricciardi, 1980), 512–36, on 519–20.

21. Quoted from J. L. Heilbron, *Galileo* (Oxford: Oxford University Press, 2010), 269, 274. Campanella made a similar appeal; *Galileo*, 196.
22. Descartes to Mersenne, 11 October 1638, in Descartes (n. 12), 878, 884.
23. Elizabeth to Descartes, 21 February 1647, in Descartes (n. 12), 2400.
24. Descartes to Mersenne, 23 December 1630, in Descartes (n. 12), 190.
25. Descartes (1982), 46–54.
26. Pierre Mouy, *Le développement de la physique cartésienne* (Paris: Vrin, 1934), 73–85; E. G. Ruestow, *Physics at 17th and 18th Century Leiden* (The Hague: Nijhoff, 1973), 38, 45, 62–4.
27. M. H. Fisch, 'The Academy of Investigators', in E. A. Underwood, ed., *Science, Medicine and History* (2 vols, Oxford: Oxford University Press, 1973), I, 521–63.
28. Mouy (n. 26), 98–9; Descartes (n. 18), 63; Roger Hahn, *The Anatomy of a Scientific Institution: The Paris Academy of Sciences, 1666–1803* (Berkeley: University of California Press, 1971), 6–8.
29. Mouy (n. 26), 108–13, 145–8.
30. Joseph Glanville, as quoted by Mordechai Feingold, 'The Mathematical Sciences and New Philosophies', in Nicholas Tyacke, ed., *Seventeenth-Century Oxford* (Oxford: Oxford University Press, 1997), 359–448, on 408.
31. More, as quoted by Alan Gabbey, 'Philosophia cartesiana triumphata: Henry More (1646–1671)', in T. H. Lennon et al., eds., *Problems of Cartesianism* (Kingston and Montreal: McGill-Queens University Press, 1982), 171–250, 181, 187n.
32. Tore Frängsmyr, *Svensk idéhistoria. Bildning och vetenskap under tusen år* (2 vols., Stockholm: Natur och Kultur, 2000), I, 150–2.
33. Quoted in J. L. Heilbron, *Electricity in the 17th and 18th Centuries: A Study in Early Modern Physics* (Berkeley: University of California Press, 1979; New York: Dover, 1999), 37 (text of 1672), and 41 (texts of 1687 and 1739), resp.
34. Ruestow (n. 26), 36.
35. Mouy (n. 26), 170–1.
36. J. R. Armogathe and Vincent Carraud, 'La première condamnation des *Oeuvres* de Descartes d'après des documents inédits aux Archives du Saint Office', *Nouvelles de la République des Lettres*, 2 (2001), 103–27, on 104–10.
37. J. L. Heilbron, 'Censorship of Astronomy in Italy after Galileo', in Ernan McMullin, ed., *The Church and Galileo* (Notre Dame: Notre Dame University Press, 2005), 279–332, on 294–6.
38. Frängsmyr (n. 32), I, 153–4.
39. Gabriel Maugain, *Etude sur l'évolution intellectuelle d'Italie de 1657 à 1750 environ* (Paris: Hachette, 1909), 77–9; Heilbron (ref. 33), 36, 108–14; J. L. Heilbron, *The Sun in the Church: Cathedrals as Solar Observatories* (Cambridge: Harvard University Press, 1999), 212–16.
40. Jean Alphonse Turrettini, 1704, as quoted by Jean Starobinski, 'L'essor de la science genevoise', in Jacques Trembley, ed., *Les savants genevois dans l'Europe intellectuelle du XVIIe au milieu du XIXe siècle* (Geneva: Journal de Genève, 1988), 7–22, on 7.
41. Quoted in Heilbron (n. 33), 29.
42. W. E. K. Middleton, 'Science in Rome, 1675–1700, and the Accademia fisicomatematica of Giovanni Giustino Ciampini', *British Journal for the History of Science*, 8 (1975), 138–54.
43. Salvatore Rotta, 'Scienza e "pubblica felicità" in G. Montanari', *Miscellenea seicento*, 2 (1971), 64–210.
44. Marta Cavazza, *Settecento inquieto. Alle origini dell'Istituto delle scienze di Bologna* (Bologna: Il Mulino, 1990), chaps. 1 and 3.
45. Francesco Eschinardi, *Raguagli ... sopra alcuni pensieri sperimentali proposti nell'Accademia fisicomatematica di Roma* (Rome: Tinassi, 1680), 3–6, 24–5.

46. Rotta (n. 43), 67–8, 72, 98–9, 131–3; Cavazza (n. 44), 44–6, 135. Montanari said that he began as a 'passionate Cartesian' and ended as an eclectic corpuscularian; Geminiano Montanari, *Le forze d'Eolo. Dialogo fisicomatematico* (Parma: Poletti, 1694), 112.
47. Cavazza (n. 44), 213–6.
48. Silvano Benedetti, 'L'Accademia degli Aletofili di Verona', in *Accademie e cultura. Aspetti storici tra sei- e settecento* (Florence: Olschki, 1979), 223–6; Salvatore Rotta, 'L'Accademia fisicomatematica ciampiniana: Un'iniziativa di Cristina?' in W. Di Palma, ed., *Cristina di Svezia. Scienza ed alchimia nella Roma barocca* (Bari: Dedolo, 1990), 99–186, on 151–4; Francesco Bianchini, 'Dissertazione ... recitata nella radunanza dell'Accademia degli Aletofili [1685]', in A. Colgerà, *Nuova raccolta di opuscoli scientifici e filologici*, 41 (1785), 3–37, on 4–5, 21.
49. Resp., Paolo del Buono to Leopoldo de' Medici, 4 January 1652, concerning one Father Antinori, in Rotta (n. 43), 137n5; and Borelli to del Buono, 10 October 1657, in Rotta (n. 43), 136n3.
50. P. Giovanfrancesco Bonomi, in Geminiano Montanari, *Pensieri fisico-matematici* (Bologna: Manolesi, 1667), 'Imprimatur'.
51. Maugain (n. 39), 55–62, 70–1, 77.
52. L. A. Muratori, *Opere*, ed. Giorgio Falco and Fiorenzo Forti (2 vols., Milan: Ricciardi, 1964), I, 166–210. Cf. Rotta (n. 43), 68–9, 139–40, on a 'league of philosophers' proposed by Montanari. Neither initiative succeeded.
53. J. A. Schmidt, *Physica positiva* (Bratislava: Brachvogelius, 1721), 13–15; Leibniz to Schmidt, 15 April 1696, in G. W. Leibniz, *Epistolae ad D. Ioannem Andream Schmidium* (Nürnberg: Monath, 1788), 18.
54. Heilbron (n. 33), 140–1, 261–3.
55. Notger Hammerstein, 'Accademie, società scientifiche in Leibniz', in Laetitia Boehm and Ezio Raimondi, eds., *Università, accademie e società scientifiche in Italia e Germania dal cinquecento al seicento* (Bologna: Il Mulino, 1981), 395–419, on 406–8.
56. The physician and historian Mirko Grmek introduced the concept of pathocenosis in 'Préliminaires d'une étude historique des maladies', *Annales: Economies, sociétés, civilizations*, 24 (1969), 1437–83.

CHAPTER 2

GALILEO'S MECHANICS OF NATURAL MOTION AND PROJECTILES

N. M. SWERDLOW

Galileo's first interest was mechanics, the application of mathematics to equilibrium states, statics, and to motion, kinematics and dynamics, the object of which is to reduce these subjects to geometry, without hidden forces or qualities. This interest never left him; it is the subject of much of his greatest work, and his method of analysis for practically everything. What is often called 'Platonism' in Galileo, his appeal to mathematics and idealized conditions, is in fact the abstract mathematical analysis of mechanics, and he came to regard anything outside mechanics, anything not subject to mathematical analysis, anything invoking hidden causes, as not 'within the limits of nature'. His mechanical way of thinking was his own, it was the way his mind worked, and he learned it from only a few earlier mathematicians. Those of interest to him were Italians of the sixteenth century in the tradition called *practical geometry*, considering both pure and applied mathematics, mechanics, and the principles of machines, of whom the most important were Niccolò Tartaglia (c.1500–1557), his onetime student Giovanni Battista Benedetti (1530–1590), the great translator of Greek mathematics Federico Commandino (1509–1575), and his student Guidobaldo del Monte (1545–1607), with whom Galileo corresponded.

Among the subjects in mechanics considered by Galileo, the most important are 'natural motion', the descent of falling bodies including on inclined planes, and the motion of projectiles under an impressed force. He also considered, and made contributions to, the resistance of solid bodies to fracture and the hydrostatics of floating bodies, but here we shall consider the more extensive work on falling bodies and projectiles. His research and writing on these subjects extend over a long period, from the qualitative analysis of his early treatise *De motu*, through the experimental and theoretical researches recorded in his manuscript notes on motion, neither of them published in his lifetime, to the first statement in print of some of his discoveries in the *Dialogue on the Two Great Systems of the World*, and the final, complete exposition in the *Discourses and Mathematical Demonstrations concerning Two New Sciences*. We shall consider all of these, with the condition that in an essay of this length they can be treated only in part, although, we intend, the most important part.

2.1 *DE MOTU*

In 1589 Galileo was appointed to lecture on mathematics at the University of Pisa, the university he had attended to study medicine and left without a degree in 1585. There, if true, occurred his most famous experiment. According to Vincenzio Viviani, Galileo's amanuensis in his last years and his first biographer, in accordance with the common philosophical axiom 'ignorance of motion is ignorance of nature' (*ignorato motu ignoratur natura*), he devoted himself entirely to its study, and then, to the great disturbance of all the philosophers, to show that many of Aristotle's conclusions on motion were false, demonstrated that moveables of the same material of unequal weight move through the same medium, not with speeds proportional to their weights, ascribed to them by Aristotle, but all move with equal speed, by repeated experiments made from the height of the Campanile of Pisa, the Leaning Tower, in the presence of the other lecturers and philosophers and all the students. This story has often been doubted, but it is hard to see how Galileo could resist the temptation, and it is the very example described in *De motu*, written or revised while he was in Pisa. For in *De motu* he denied that bodies, of the same material in the same medium, fall with speeds proportional to their size or weight, that if two stones, one twice the size of the other, are thrown from a high tower, the larger will reach the ground when the smaller is only half-way down. He argued instead that they fall with the same speed from the contradiction that if a larger and smaller body are joined together the composite would, by this reasoning, fall slower than the larger body; but the composite is now larger than the larger body, contradicting the principle that the larger body falls faster. The same argument, less a proof than a paradox, appears nearly fifty years later in the *Two New Sciences*.

As set out in *De motu*, natural motion, whether upward or downward takes place through the essential heaviness or lightness of the moveable. By essential heaviness and lightness is meant specific weight, weight per unit volume, so heaviness and lightness are relative, not absolute, and nothing is devoid of weight. Whatever moves naturally moves to its proper place, and we see that heavy things are closer to the centre of the universe and light are more distant, that the heavy has a natural tendency to move downward and the light to move upward. The cause of natural motion is therefore heaviness and lightness, but we must consider not only the heaviness or lightness of the moveable but also the heaviness or lightness of the medium through which it moves. A moveable will move downward in a lighter medium and upward in a heavier medium, and not move at all in a medium of the same specific weight. This can be seen by the analogy of a balance with equal arms, in which a heavier weight on one side moves downward and a lighter weight on the other side moves upward, and equal weights on both sides are balanced and move neither up nor down. What then is the cause of greater or lesser speed of a moveable in a medium? It is the difference between the specific weight of the body and of the medium, that is, the speed of any body in any medium is proportional to the difference between the specific weight of the body and the medium. Hence, bodies of the same material, as two stones, fall with the same speed in the medium of air, but bodies of different material, as wood and stone, do not fall with the same speed in the medium of air. Similar hydrostatic reasoning concerning the specific weights of bodies and of media extends the argument to speeds of falling and rising bodies of different materials in different media. Galileo makes up small numbers for specific weights and resultant speeds, but the analysis is nevertheless qualitative. Now according to Aristotle, the speed of natural motion in a medium is directly proportional to the weight of the body and inversely proportional to the density of the medium. Both are refuted if the speed is proportional to the difference of specific weight of the body and the medium. The proportionality of speed to weight has already been refuted experimentally and theoretically. As for the inverse proportionality to the density of the medium, Aristotle draws the conclusion that a void cannot exist, since motion in a void would be instantaneous, which is impossible. But if speed is proportional to the difference of specific weight of the body and the medium, the speed in a medium of zero specific weight, a void, would still be finite, depending upon the specific weight of the body alone unreduced by the specific weight of a medium, which also refutes the impossibility of a void.

Thus far, speed of natural motion has been treated without regard to acceleration, change of speed, for which it is necessary to consider the motion of projectiles. First, the Aristotelian theory that projectiles are moved by the surrounding air is refuted by several arguments. Rather, the cause of projectile motion is the removal of heaviness when the moveable is impelled upward, but the removal of lightness when impelled downward. Thus, motion upward is still due to lightness and downward to heaviness. Motive force, namely, lightness (*virtus motiva, nempe levitas*), is retained in a stone no longer touching what moved it, as heat is retained in iron when removed from fire, and the impressed force (*virtus impressa*) gradually gives out (*remittitur*) in the projectile when absent from the projector (*a proiciente absente*), as heat gives out in iron when absent from fire. The stone finally returns to rest, the iron similarly to natural coldness. Motion by the same force is impressed more in a more resistant moveable than in one which resists less, as more in a stone than in light pumice, just as heat is impressed more hotly by the same fire in the hardest, coldest iron than in soft, less cold wood. The idea here is that the motive force, inhering in the moveable (*mobili inhaerens*), reduces the body's natural or intrinsic heaviness, so that it is light, moving upwards, preternaturally or by accident or by force (*preternaturaliter aut per accidens aut vi*); and when freed from the impelling force, the projectile shows its true and intrinsic heaviness by descending. The difference between intrinsic lightness and accidental lightness, removing heaviness through an impressed force, is that the effect of the force is temporary. That is projectile motion. Now, as the body moves upward it moves more slowly, and as it moves downward it moves more rapidly. Why? Its speed decreases as it moves upward because the force that reduces the heaviness of the body gradually weakens until it is just equal to the weight of the body and can no longer move it upward so the body is neither heavy nor light. Then, as the force continues to decrease, the heaviness of the moveable begins to predominate and it begins to descend. But since at the beginning of the descent there still remains much of the upward impelling force—although it is no longer greater than the heaviness of the moveable—which is lightness, the proper heaviness of the moveable is reduced by this lightness and consequently its motion is slower in the beginning. And as the extrinsic force, the lightness, weakens further, the heaviness of the moveable increases, and the moveable moves still faster. This, Galileo writes, is what I believe to be the true cause of the acceleration of motion, and two months after I worked it out, I learned from Alexander (of Aphrodesias in Simplicius, *Commentary on De caelo*, probably not Galileo's direct source) that it was also the opinion of the very learned philosopher Hipparchus, praised by the very learned man Ptolemy in the *Almagest*. And the same is true of a descent not preceded by an upward motion, for when a stone leaves the hand it still has an upward force impressed on it equal to its weight, for since the stone presses down with its weight, it is necessary that it be impelled upward with another equal upward force from the hand, neither greater nor smaller. So when the stone leaves the hand, it still has an upward force impressed on it and falls more slowly at the beginning and increases in speed as the upward force decreases.

Galileo's theory of natural motion is consistent. Natural motion is a result of essential heaviness and lightness, specific weight, in a medium, which is considered only hydrostatically for its specific weight, not for resistance, as from friction or viscosity. The speed of a moveable, whether descending or ascending, is proportional to the difference of specific weight of the moveable and the medium. Thus, bodies of different specific weights in the same medium, or of the same specific weight in different media, fall or rise with different speeds. In order to explain projectile motion, an impressed force producing upward motion is considered a preternatural or accidental lightness that gradually diminishes. Such an impressed force causes a projectile to move upward, with decreasing speed as the force diminishes, until the lightness produced by the force is equal to the intrinsic heaviness of the body, at which point the body is neither heavy nor light. As the upward force, the accidental lightness, continues to diminish, the heaviness of the body predominates and the body descends, at first slowly, then with increasing speed. Hence, in projectile motion also, speed of ascent and descent is the result of lightness and heaviness, and change of speed the effect of an impressed force which increases lightness such that as the impressed force, the accidental lightness, diminishes, the speed of ascent decreases and the speed of descent increases, as the speed of the moveable is determined more and more by its intrinsic heaviness.

He also considers why at the beginning of motion of descent, lighter bodies fall faster than heavier bodies, and the reason is that heavier bodies take on more of the contrary quality, the accidental lightness of the impressed force, just as if a lead ball and wood ball are shot from the same cannon, the lead ball will travel farther and longer; if two weights, of lead and wood, suspended by chords, receive an impetus from equal distance from the perpendicular, the lead weight will swing for a longer time; and if lead and wood are equally heated, the lead will retain the heat longer. Again, the resistance of the medium is not considered. Hence, a heavier body, receiving more of the accidental lightness, does not fall as fast at the beginning of motion as a lighter body, which receives less. The reason for this supposed effect is also the explanation of the length and form of the path of projectiles. Iron balls shot from cannon are carried through a longer distance in the same straight line, as the line of motion forms less acute angles with the horizon, meaning angles closer to the perpendicular. The reason is that when the cannon is turned toward the vertical, the ball, pressing downward with its weight, resists the force of the charge longer and so takes on more of the force of the charge than in a cannon turned toward the horizontal, where the ball offers less resistance and so is fired before the full force of the charge is received. Another example is that a ball (not a cannon ball!) coming towards you can be hit farther than a ball at rest or retreating, since the approaching ball, because of its resistance to changing direction, can take on more of the force hitting it than a ball at rest or retreating, which moves away sooner and takes on less of the force. A second reason is that when a ball is fired perpendicular to the horizon, it cannot turn back and descend until the entire upward force is expended—which appears to contradict the earlier explanation that upward force still remains in a projectile as it begins to descend, although the explanation here concerns only the upward part of the trajectory—but when it is fired less inclined to the horizon, it can depart from the straight line and begin to descend before the entire upward force is expended. Note that what is considered here is not the entire trajectory but only the first, upward part of projectile motion, which is longer as the inclination to the horizon is greater. The form of the trajectory itself is treated as ascent along a straight line, followed by turning through a curve, and then descending along a path of unspecified form.

2.2 Notes on Motion in Galileo Ms. 72

From 1592 until 1610 Galileo lectured on mathematics at the University of Padua. His teaching concerned mostly elementary courses on Euclid and spherical astronomy, but his most important research in this period, before his discoveries with the telescope, was in mechanics. The principal record of this research is notes contained in Firenze Biblioteca Nazionale Centrale Galileo Ms. 72, a composite manuscript written in part in Padua from not long after 1600 until about 1609, and in part in Florence from about 1618 until 1636. This dating of the notes is from Drake (1979), an inventory and photographic facsimile of the manuscript. The entire manuscript is now available electronically in high resolution. In addition to Galileo's hand, there are pages in the hands of his students Niccolò Arrighetti and Mario Guiducci that appear to be copies of his own notes he considered worth preserving. There is a large literature on the manuscript with considerable differences of interpretation, and parts of the manuscript have still not been fully investigated. We consider its contents very selectively.

In about 1604 Galileo investigated the rate of change of accelerated motion, probably by rolling a ball down a very slightly inclined plane. He found that in equal units of time the spaces traversed, and thus the speeds—actually the mean speed through each space—are as the odd numbers $1, 3, 5, 7 \ldots$, and the distances, which are the sum of the spaces, as the squares $1, 4, 9, 16 \ldots$; that is, distances are as the sums of the speeds and as the squares of the times. In the *Two New Sciences*, Salviati describes how he tested the experiments of the author, Galileo, in his presence. In the narrowest side of a beam twelve braccia (cubits) long, half a braccio wide, and three dita (digits) thick, he cut a straight channel a little over a dita wide, and glued within it a smooth strip of parchment. Where 1 braccio $= 24$ dita $= 58.4$ cm, the dimensions are close to length 700 cm, width 29 cm, thickness 7.3 cm, and channel 2.4 cm. A hard, polished bronze ball was made to descend in the channel by raising one end of the beam from one to two braccia from the horizontal; hence, the inclination to the horizontal was from $4.8°$ to $9.4°$. The timing of the descent was achieved by allowing water run from a large bucket through a narrow tube to a small glass while the ball descended, and then weighing the water in the glass; hence, the weight of the water is the measure of time. It was not necessary to know how much water flowed in any unit of time, since the measurements need only be proportional, of the weight of water during any fraction of the descent to a complete descent. He measured the time for the ball to roll down the entire channel, from many trials never finding a difference of even a tenth of a pulse beat; then one-fourth, one-half, two-thirds, three-quarters, and many other fractions of the entire length, repeated a good hundred times. The spaces passed were always found to be proportional to the squares of the times, and this in all the inclinations of the channel in which the ball descended. Was this experimental discovery or confirmation of naturally accelerated motion possible? It has often been doubted, and Galileo is said to have reached his conclusions theoretically, without experiment or observation. However, in a famous article, Thomas Settle (1961) described repeating Galileo's experiment with similarly simple apparatus, the only difference being that the water was measured by volume in a graduated cylinder rather than by weight. The result was confirmation of the relation of distance to the square of the time within 1 millilitre of water, or 1/20 of a second of time, and it appears that Galileo could have done better than that.

On 16 October 1604 Galileo wrote a letter to Paolo Sarpi explaining briefly his recent thoughts on natural motion, that is, naturally accelerated motion, and a longer note in Ms. 72 sets out in more detail what he had in mind. Both have been translated and analysed by Koyré (1939), Drake (1969), and others. He had observed, determined by experiment, that the spaces passed by natural motion are in the squared proportion of the times, and consequently that the spaces passed in equal times are as the odd numbers from one, and other things. What he lacked was a completely indubitable principle to put as an axiom in order to demonstrate these occurrences (*accidenti*), and he was reduced to a proposition which has much of the natural and evident: namely, that the natural moveable goes increasing in speed (*velocità*) with that proportion with which it departs from the beginning of its motion. His principle, which is just assumed and not certain, is that in naturally accelerated motion, as of a falling body, the degree of speed, the speed at any instant, is proportional to the distance from the beginning of the motion. He illustrates this by drawing a line with equal divisions *A*, *B*, *C*, *D*, as in Fig. 2.1, and explaining that the degree of speed of a body falling from *A* at *C* is to the degree of speed at *B* as *AC* is to *AB*. And conversely for projectile motion, thrown upward from *D* to *A*, the body passes through the same proportions of speed, that is, the degree of speed decreases with distance such that the impetus (*impeto*), meaning speed, at *D*, *C*, *B* decreases in the proportion of *DA*, *CA*, *BA*. Hence, if the body goes acquiring degrees of speed in the same proportions in natural fall, what I have said and believed is true.

Note that 'the spaces past in equal times are as the odd numbers from one' also implies that *the speeds increase as the times*, because the speeds through each space are as the spaces passed *in equal times*. And this applies to both the cumulative speed within a space and the instantaneous speed at the end of a space. Thus, the increase of speed corresponds to *both* the increasing spaces and the equal times. Galileo assumes that the fundamental principle for the increase of speed is proportionality to the distance from the beginning of motion, the sum of the spaces, not to the time from the beginning of motion, the sum of the equal times. The letter contains no demonstration that natural motion follows from this principle, presumably because he does not

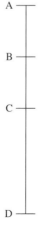

Fig. 2.1.

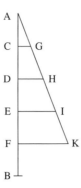

Fig. 2.2.

yet have one, but a longer note, Ms. 72 f. 128, contains an attempt at a demonstration, perhaps as he is writing, as it were, thinking on paper: I suppose (and perhaps I shall be able to demonstrate this) that the naturally falling heavy body goes continually increasing its speed (*velocità*) just as (*secondo che*) the distance from the limit from which it departs increases. Thus, in Fig. 2.2, the body departing from A and falling through line AB, the degree of speed at D is as much greater than the degree of speed at C as distance DA is greater than CA, and thus at every point of the line AB, the body is found with degrees of speed proportional to the distances of the points from the limit A. This principle appears to me very natural, and corresponds to all the experiences we see in instruments and machines that act by striking, in which the striker makes so much greater effect as it falls from a greater height. And this principle assumed, I shall demonstrate the rest.

Now, draw AK at any angle to AF, and draw parallels CG, DH, EI, FK, and the speeds at C, D, E, F will be as these lines so that the degrees of speed at all the points of AF increase continually as the parallels drawn from those points. And since the (cumulative) speed with which the moveable has come from A to D is composed of all the degrees of speed it had in all the points of line AD, and the (cumulative) speed from A to C is composed of all the degrees of speed in line AC, so the (cumulative) speed with which it passed AD has the proportion to the (cumulative) speed with which it passed AC which all the parallel lines drawn from all the points of AD to AH have to all the parallel lines drawn from all the points of AC to AG; and this proportion is that which (the area of) triangle ADH has to triangle ACG, that is, as the square of AD to the square of AC. Therefore the (cumulative) speed with which the moveable has passed line AD to the (cumulative) speed with which it has passed AC has the square of the proportion (*doppia proporzione*) of DA to CA.

What Galileo has done thus far is not rigorous, since a sum of lines does not make an area, but is intuitively reasonable, the sum of the increasing speeds is proportional to the sum of the lengths of the lines (at each instant and) at each distance from A and, if the lines are placed along side each other, their sums, the cumulative speeds, may be represented by the areas of similar triangles, proportional to the square of the distances from *A*. But what follows, the crucial step of the demonstration, either amounts to assuming what is to be proved or has no meaning at all. 'And since (cumulative) speed to (cumulative) speed has the contrary proportion (*contraria proporzione*) of that which time has to time (because to increase the (cumulative) speed is the same as to decrease the time), therefore the time of motion in *AD* to the time of motion in *AC* has the proportion of the square root (*subduplicata proporzione*) of that which the distance *AD* has to the distance *AC*.' (Hence, if we square the ratios of both distances and times), accordingly the distances from the beginning of motion are as the squares of the times, and by division (*dividendo*) the spaces passed in equal times are as the odd numbers from unity, which corresponds to what I have always said and observed with experiences, and all the truths correspond among themselves.

Using the 'contrary proportion' of time and cumulative speed, Galileo now substitutes for the cumulative speed, just shown to be proportional to the square of the distance, the time, which he says is proportional to the square root of the distance. This is what amounts to assuming what is to be proved—or to nothing at all in the 'contrary proportion' of time and speed, which would be applicable to time and different uniform speeds—because he knows from experiment that the distance is proportional to the square of the time or, as it appears here, the time is proportional to the square root of the distance. Hence, by next squaring the ratios of distances and times, he reaches just what he wants, that the distances are as the squares of the times and the spaces passed in equal times are as the odd number from unity, (supposedly) derived from the principle that degree of speed is proportional to distance. Did Galileo know that his demonstration was defective? If not immediately, then eventually, and he also came to understand that his principle that the degree of speed is proportional to distance cannot be correct as it leads to all sorts of contradictions, as that *every* distance is traversed in the *same* time, contrary to observation of bodies falling from unequal heights, as he must have known. Through a later analysis, including this contradiction, by 1608 he determined correctly that the degree of speed is proportional only to time.

Among the notes in Ms. 72 are also experimental and theoretical investigations of the parabolic trajectory of a projectile. It is possible that Galileo suspected this years before he worked out his account of natural acceleration, even in the incorrect form of 1604. In the Second Day of the *Two New Sciences* he explains two methods of describing a parabola. The first is by rolling a small, somewhat warmed and moistened bronze ball along a nearly vertical metal mirror, which leaves a trace of a parabolic line, wider or narrower as the ball is rolled higher or lower. The second is by suspending from two nails along a horizontal line on a wall a fine chain, which forms a curve that he calls both a parabola and close to a parabola. The curve is a catenary, and in Ms. 72 are drawings of what appears to be a catenary within a parabola, which he perhaps believed would be produced by a perfect chain with infinitely small links. Both methods are described in a notebook of Guidobaldo del Monte, in which the curve is said to be similar to a parabola and hyperbola and the trace is made by an inked ball thrown along the plane of a table in a nearly perpendicular position.

A similar comparison of the path of a projectile to the curve formed by a suspended rope is found in Paolo Sarpi's notebooks under the date 1592. It thus appears possible that Galileo was present at the experiment described by Guidobaldo and discussed it with Sarpi as early as 1592.

The recognition, or suspicion, that the curve described by a projectile is a parabola is not in itself certain until it is determined that in natural descent the distance is proportional to the square of the time. Once this is determined, one can assume that the vertical component of projectile motion is that of natural descent, as the square of the time, and test whether the horizontal component is uniform, which would indeed produce a parabola. Now Galileo wrote in the *Dialogue* that motion downward on an inclined plane increases in speed and motion upward, which requires an impressed force, decreases in speed, and thus motion on a horizontal plane without impediments once started would be uniform and perpetual without resistance. In *De motu*, motion on a descending plane is natural motion requiring no force, motion on an ascending plane requires force, and motion on a horizontal plane is neither natural nor forced, requiring the least force of all, less than any other force. It is not said that the motion is uniform and perpetual, but the uniformity would seem to follow from the increasing speed in descending and the decreasing speed in ascending. Hence the horizontal component of projectile motion should be uniform; but how can this be tested? Among the notes on motion is just such a test on f. 116v of Ms. 72, reproduced here in Fig. 2.3, although just what is shown on the page has been subject to different interpretations. The drawing appears to show the vertical heights from which a body descends followed by curves showing how far the body is projected horizontally, as would a ball descending an inclined plane of that vertical height and then continuing off a table to the floor. The drop from the table is the vertical component proportional to the square of the time, and the horizontal distance from the table can be computed and compared with measurement. The computation is done proportionally, and Galileo has by now determined that speed is proportional to time and to the square root of distance, $v \sim t \sim s^{1/2}$. He uses the square of the speed v^2 acquired in the vertical distance s on the plane, $v^2 \sim s$, and the following horizontal distance d in uniform motion, proportional to the speed, is then $d^2 \sim v^2 \sim s$. Hence, for any two distances s and d, $d_2^2/d_1^2 = s_2/s_1$. Galileo's units are *punti* (points), slightly less than 1 mm. From the first measurement, in what must be round numbers, $s_1 = 300$ and $d_1 = 800$. Hence, letting $s_2 = 600$, $d_2 = \left(d_1^2 (s_2/s_1)\right)^{1/2} = \left(800^2 (600/300)\right)^{1/2} = 1131$; the measured d_2 is 1172, and it is noted that the difference is 41. The other trials show similar agreement, for example, for $s_2 = 1000$, the computed $d_2 = 1460$ and the measured d_2 is 1500 with a noted difference of 40. This analysis of f. 116v is by Drake (1973); he later (1985) used a different explanation of the transition from motion on the inclined plane to horizontal projection and different numbers on the page to obtain closer agreement, but the principle is the same, which is our concern here. As noted, the page has also been interpreted quite differently, and it is possible that Galileo took uniform horizontal motion as a theoretically correct assumption without a specific quantitative test.

GALILEO'S MECHANICS OF NATURAL MOTION AND PROJECTILES 35

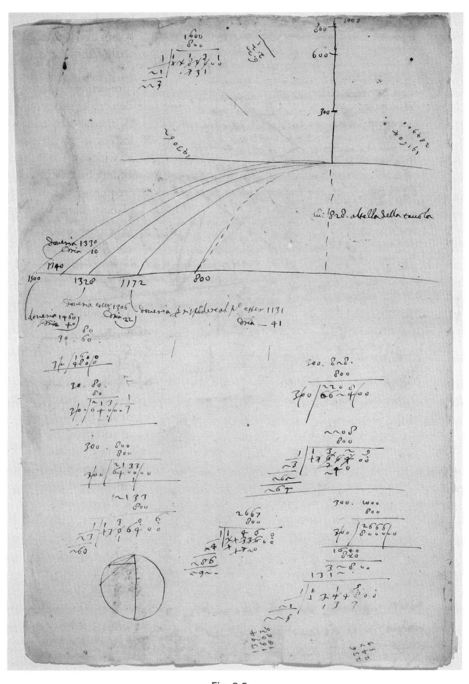

Fig. 2.3.

2.3 Mechanics of Motion in the *Dialogue*

Galileo's first publication of naturally accelerated motion occurs in the Second Day of the *Dialogue on the Two Great Systems of the World* (1632). Salviati explains that it is known to all that the movement of heavy bodies descending is not uniform but, beginning from rest, goes continually accelerating. But the proportion in which the increase in speed (*velocità*) is made has been unknown to all philosophers and was first discovered and demonstrated by our common friend the Academician, who, in some of his writings not yet published, demonstrates that the acceleration of the straight motion of a heavy body is made according to the odd numbers from unity, which in sum is the same as to say that the spaces passed by the mobile beginning from rest have to each other the duplicate proportion (*proporzione duplicata*) of the times in which such spaces are measured, or that the spaces passed are to each other as the squares of the times. It is here that Galileo gives his only published value for the natural motion of fall, which is, not a constant of acceleration, but the distance 100 braccia in 5 seconds, a round number corresponding to a constant $a = 2s/t^2 = 8$ braccia/sec^2, which is never given. Where 1 braccio $= 58.4$ cm, this is 467 cm/sec^2, less than half the correct value. He does not demonstrate that the spaces are as the odd numbers and the distances proportional to the square of the times, which are simply stated as correct, but he does use the principle of speed proportional to time in demonstrating that if accelerated motion is continued for an equal time by uniform motion at the greatest speed, it will pass through twice the distance of the accelerated motion. The demonstration is of interest in showing how Galileo now treats a continuously increasing quantity.

In Fig. 2.4 we take right triangle *ABC*, and dividing *AC* into any number of equal parts *AD*, *DE*, *EF*, *FG*, we draw through *D*, *E*, *F*, *G* straight lines *DH*, *EI*, *FK*, *GL* parallel to the base *BC*. Now, the spaces on *AC* are equal times and the parallels drawn through *D*, *E*, *F*, *G* represent 'the degrees of speed, accelerated and increasing equally in equal times' (*i gradi della velocità accelerate e crescenti egualmente in tempi equali*). Here is the statement that the speed increases as the time. But since the acceleration takes place continuously from moment to moment, and not intermittently from one part of time to another, to represent the infinite degrees of speed beginning from *A* in the interval of time *AD* which precede the degree *DH*, there must be imagined an infinite number of lines, always smaller and smaller, parallel to *DH*, which infinity of lines at last depict to us the area of the triangle *AHD*. And the same is imagined of the infinite degrees of increasing speed corresponding to the infinite lines beginning from *A* and drawn parallel to *HD*, *IE*, *KF*, *LG*, *BC*, meaning that they depict the areas of triangles, as *AKF*, *ABC*. Now, complete the parallelogram *AMBC*, and extend to *BM*, not only the parallels marked in the triangle, but the infinity of those imagined to be extended from all the points of side *AC*. Just as the area of the triangle was the multitude (*massa*) and sum of all the speeds with which such a space was passed in the time *AC*, so the parallelogram comes to be the multitude and aggregate of just as

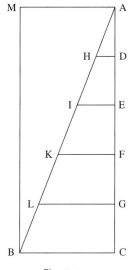

Fig. 2.4.

many degrees of speed, but with each one equal to the maximum BC, which multitude of speeds comes to be double the multitude of increasing speeds of the triangle, since the parallelogram is double the triangle. And thus, if the falling moveable, making use of the degrees of accelerated speed forming the triangle ABC, has passed in so much time over such a space, it is very reasonable and probable (*è ben ragionevole e probabile*) that, making use of the uniform speeds corresponding to the parallelogram, in the same time it would pass with a uniform motion over double the space it passed with the accelerated motion.

The phrase 'it is very reasonable and probable' is (probably) because of the method of proof, taking an infinite number of line segments to form an area. That a moveable with a uniform speed of the greatest speed of a descent passes in the same time over twice the space of the descent is equivalent to a moveable with a uniform speed of half the greatest speed of the descent, or the speed at the midpoint of the time of the descent, moving in the same time over the same space as the descent. Both forms, sometimes referred to as 'double-speed' and 'mean-speed', can be found applied to any kind of change, as of quality, in scholastic writings, although the statements, when geometrical, use only areas of triangles and rectangles without considering them to represent infinite numbers of parallel lines. There is a considerable literature on whether Galileo did or did not know these scholastic writings, for what little evidence there is may be used to support either opinion.

In addition to the treatment of the acceleration of a falling body in the *Dialogue*, there are other applications of the mechanics of motion, two of which in the Second Day are used to refute arguments against the diurnal motion of the Earth. Both are well known, and both are defective. One is to show that the accelerated rectilinear descent of a falling body, a stone falling from the top of a tower, is actually the result of two uniform circular motions, the diurnal rotation of the Earth and motion along

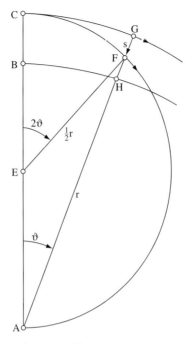

Fig. 2.5.

an arc of a semicircle from the top of the tower terminating at the centre of the Earth. Salviati calls it a *bizzarria*, a curiosity, which it is. In Fig. 2.5, A is the centre of the Earth, $r = AHG = ABC$ is the sum of the radius of the Earth AB and the height of the tower BC, and E is the centre of semicircle CFA of radius $\frac{1}{2}r = EF = EC$. In the time the Earth rotates through ϑ, which is proportional to time, $\vartheta \sim t$, carrying the tower from CB to GH, radius EF of the semicircle rotates through 2ϑ, and F, the intersection of the semicircle with AHG, shows the position of the falling body, so the descent $s = GF$. In triangle AEF,

$$AF = \tfrac{1}{2}r\,\frac{\sin 2\vartheta}{\sin \vartheta} = r\,\frac{\sin \vartheta \, \cos \vartheta}{\sin \vartheta} = r\cos \vartheta,$$

from which

$$s = GF = AG - AF = r - r\cos \vartheta = r(1-\cos \vartheta) = r\,\text{vers}\,\vartheta = 2r\sin^2 \tfrac{1}{2}\vartheta.$$

Thus, the descent s is rectilinear and, where ϑ is small, $\sin^2 \vartheta \sim \vartheta^2 \sim t^2$, that is, $s \sim t^2$, the descent is proportional to the square of the time, although Galileo did not know this. There is, however, a problem. Taking Galileo's value for the radius of the Earth, 3,500 Italian miles of 3,000 braccia, if the height of the tower is an insignificant fraction of the radius of the Earth, we may take the radius of the Earth itself as their sum, $r = 10{,}500{,}000$ braccia (and it would make almost no difference if the tower were an additional 100,000 braccia tall). Where the day is 24 hours, in $t = 5$ seconds $\vartheta = 1'\,15''$, for which $s = r\,\text{vers}\,\vartheta = 0.694$ braccio and for small arcs the acceleration

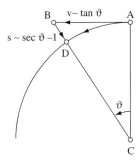

Fig. 2.6.

$a = 2s/t^2 = 0.0555$ braccio/sec$^2 \approx 3.24$ cm/sec^2, which is absurd, things would fall *very* slowly, and as ϑ increases a decreases. This is absolutely fatal to Galileo's theory of semicircular fall; there is nothing that can be done to save it.

The other is to show why the diurnal rotation of the Earth cannot project a body off its surface. In Fig. 2.6, C is the centre of the Earth, the motion of projection is along the tangent AB and of fall along the secant BC through distance BD. Where ϑ is the rotation of the Earth in some time, the motion of projection $v \sim \tan \vartheta$ and the distance to fall back to the surface of the Earth $s \sim (\sec \vartheta - 1) \sim \tan^2 \vartheta$. Thus, s decreases as the square of v, as v is divided by 10, s is divided by 100, and as $\vartheta \to 0$, $v/s \sim \tan \vartheta /(\sec \vartheta - 1) \to \infty$. Before the body can move any *finite* distance along the tangent, it will fall the *infinitely* smaller distance on the secant, and thus the body will not move at all. The problem, which Galileo did not grasp, is that although $v \sim \tan \vartheta$ is a velocity, the distance the body will move in some time, $s \sim \sec \vartheta - 1$ is only a distance, and the speed with which the body would traverse s is unspecified and, unlike $\sec \vartheta - 1$, has no relation to $\tan \vartheta$. He believed that as s is infinitely small, nothing (*niente*), at the point of projection, any speed would be adequate to traverse it, but this is not true, as some sufficient speed must be specified, and that is the error in the demonstration. He also extends the argument to a tendency to descend (*propensione allo scendere*), a speed of descent, that varies with weight, which he later says is not true of falling bodies, the proof of which is the same and has the same problem, but he does say that it is essential for any downward motion that there be *some* weight. This is in fact equivalent to saying that some sufficient speed of descent must be specified, although the reason is only made clear through the Newtonian dependence of weight on acceleration, $w \sim mg$ at the surface of the Earth. And for this reason Galileo's intuition that a body cannot be thrown off the rotating Earth is correct. Centrifugal acceleration at the equator is 3.39 cm/sec^2, 1/289 the acceleration of gravity, 980 cm/sec^2, so obviously nothing will move. What is notable about both demonstrations, semicircular fall and projection from the surface of the Earth, ingenious although defective as they are, is that their purpose is not only to answer arguments against the rotation of the Earth, but to remove hidden forces from nature, to reduce the downward motion of heavy bodies to kinematics, to reduce gravity to geometry.

2.4 DISCOURSES AND MATHEMATICAL DEMONSTRATIONS CONCERNING TWO NEW SCIENCES

Galileo's most complete treatment of accelerated motion is the section *On Naturally Accelerated Motion* in the Latin treatise *On Local Motion* contained in the *Two New Sciences* (1638). The treatise is divided into three parts: *On Uniform Motion* and *On Naturally Accelerated Motion* in the Third Day, and *On the Motion of Projectiles* in the Fourth Day. In setting out his definition of uniformly accelerated motion, he explains that just as we call movement uniform when in equal times equal spaces are passed over, so this motion is uniformly and likewise continuously accelerated when in any equal times equal additions of speed are added in succession (*superaddantur*). Hence, the definition: I call that motion equably or uniformly accelerated which, departing from rest, in equal times adds to itself equal moments (degrees) of swiftness in succession (*temporibus aequalibus aequalia celeritatis momenta sibi superaddit*). This is followed by the question, raised by Sagredo, of whether the acceleration of a falling body is truly continuous, whether it passes through every degree of speed, however small, beginning from rest, since it appears that falling bodies accelerate quite rapidly. Salviati's answer, or better analogy, to confirm that it does is provided by the smaller and smaller impact of bodies falling from smaller heights, showing a decrease of speed, and likewise by the decrease of speed of a stone thrown upward. The next question, again from Sagredo, concerns the cause of acceleration in fall, that when a body is thrown upward, the force impressed (*virtù impressa*) upon it diminishes until it is exactly balanced by the weight of the body; but the force is not yet exhausted, and so the body begins to fall slowly and as the force diminishes still more falls faster and faster. This is none other than the theory of projectile motion in *De motu*, that an impressed force by which the motion of a projectile continues gradually diminishes, which is also the explanation of the acceleration of a falling body, although here the upward force is not said to produce a preternatural or accidental lightness. Salviati remarks that now is not the time to enter into an investigation of the cause of the acceleration of natural motion, concerning which philosophers have had various opinions, and which he dismisses as 'fantasies'. It suffices to investigate and demonstrate some properties (*passiones*) of a motion accelerated, whatever be the cause of its acceleration, such that the moments (degrees) of its speed go increasing after its departure from rest with the simplest proportion, with which the continuation of time increases, which is as much as to say that in equal times makes equal additions of speed. And if it happens that the properties (*accidenti*) which have been demonstrated take place in the motion of naturally descending and accelerated heavy bodies, we may conclude that the assumed definition includes such motion of heavy bodies, and it is true that their acceleration goes increasing as the time and the duration of the motion increases. The point here is that natural acceleration is to be treated purely kinematically, without inquiry into whatever may cause the acceleration.

Sagredo's next question is the most important, whether in a uniformly accelerated motion the increase of speed is proportional to the space passed over, which appears reasonable as the force of impact of a heavy body appears proportional to the height from which it falls. Salviati responds that it is comforting to have such a companion in error, and says that this explanation has so much of the likely and probable that our Author himself remained for some time in the same fallacy. The refutation, somewhat convoluted in expression and frequently misunderstood or doubted, is simply that if speed were proportional to distance, *every* distance would be traversed in the *same* time, which can only happen in instantaneous motion (a superfluous addition since there is no such thing) and is manifestly untrue of falling bodies, for example, bodies dropped simultaneously from different heights do not reach the ground together. And this also refutes the observation that force of impact is proportional to the height of fall, since impact depends upon speed, and the speed, as just shown, is not proportional to the distance of fall. (This is incorrect, since force of impact from natural fall is proportional to the distance of fall. The reason is that the acquired speed is as the square root of the distance and the force of impact, the kinetic energy, is as the square of the speed and thus as the distance.)

Galileo next repeats the definition of uniformly accelerated motion and proposes a single assumption: I accept that the degrees of speed acquired by the same moveable (in descending) over different inclinations of planes are equal when the heights (*elevationes*) of the planes are equal. In the discussion, Salviati says that he wishes to increase the probability of this assumption by an experiment so that it will fall little short of a necessary demonstration. Suspend a lead ball by a thread two or three braccia in length from a nail driven into a wall so that it can swing freely. When the ball is released on one side, it will return on the other side to nearly the same height from which it descended, falling short by a very small interval because of the impediment of the air and the thread. Now drive a nail into the wall below the first nail so that when, after being released on one side, the thread is caught on the nail, and the ball cannot complete its original arc on the other side but is forced into a smaller arc. No matter where the nail is placed, as long as it is above the height from which the ball was initially raised, the ball still reaches the same height, which shows that the speeds acquired in descent on each side are equal. And the longer arc on one side and shorter arc on the other side may be taken as analogous to inclined planes of different inclinations and equal heights, thus providing experimental evidence for the assumption. Let us now take this as a postulate, the complete truth of which will later be established by seeing that other conclusions built upon this hypothesis correspond to and agree precisely with experience.

We return to the treatise with Theorem I Proposition I: The time in which any space is passed over by a moveable with a motion uniformly accelerated from rest is equal to the time in which the same space is passed over by the same moveable carried with a uniform motion, the degree of speed of which is one-half the greatest and final degree of speed of the previous uniformly accelerated motion. In Fig. 2.7, let line AB represent the time in which a moveable with uniformly accelerated motion passes from rest at C over the distance CD, let BE represent the greatest and final speed,

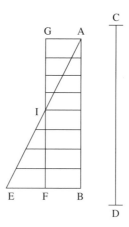

Fig. 2.7.

and the lines parallel to BE meeting AE the speed at each instant. Bisect BE at F and complete parallelogram ABFG, which will be equal (in area) to triangle ABE. If the parallels in triangle AEB are extended to IG, the aggregate (*aggregatum*) of all parallels in the quadrilateral will equal the aggregate of all parallels in the triangle, since those in triangle IEF are equal to those in triangle GIA, and those in trapezium AIFB are common. Since the parallels within triangle AEB represent increasing degrees of the increased speed, while the parallels within the parallelogram ABFG represent just as many degrees of the speed not increased but uniform, so equal distances will be passed over in the same time by two moveables, of which one is uniformly accelerated from rest and the other has a uniform motion equal to half the greatest speed of the accelerated motion. Note that now Galileo does not take an infinite number of lines to form an area, nor does he use the equal areas of the triangle and parallelogram directly, but only the aggregate, the unspecified but equal number, of parallels representing accelerated and uniform speed that may be drawn in each. The proof is again more intuitive than rigorous, but makes its point clearly. Also, the statement of the theorem concerned equal times in accelerated and uniform motion by the same moveable, but the conclusion concerns equal distances by two moveables.

Next is the demonstration of the natural accelerated motion of a falling body, that the distances are proportional to the square of the time. Theorem II Proposition II: If some moveable descends from rest with a uniformly accelerated motion, the spaces passed by it in any equal times are to each other in the duplicate ratio of the same times: namely, as the squares of the same times. In Fig. 2.8, let the flow of time from some first instant A be represented by line AB, in which let there be taken any two times AD and AE; let HI be the line in which the moveable descends uniformly accelerated from point H as the first beginning of motion; and let HL be the space passed in the first time AD, and HM the space through which it will descend in time AE. I say that space MH is to space HL in the duplicate ratio of that which time EA has to time AD, or, let us say, that the spaces MH and HL have the same ratio as have the squares of EA and AD. Let line AC be drawn at any angle with AB, from points D and

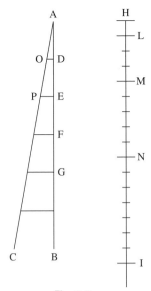

Fig. 2.8.

E let there be drawn parallels DO and EP, of which DO will represent the maximum degree of speed acquired at instant D of time AD, and PE the maximum degree of speed acquired at instant E of time AE. From the preceding Theorem I, it follows that spaces MH and LH are the same as those which would be passed in times EA and DA by uniform motions the velocities of which are as one-half PE and OD. If, therefore, it were shown that these spaces MH and LH are in the duplicate ratio of the times EA and DA, that is, $MH/LH = EA^2/DA^2$, what it intended will be proved. But in Proposition IV of *On Uniform Motion*, it was demonstrated that the spaces passed by moveables carried by uniform motion have to each other the ratio compounded from the ratio of the speeds and the ratio of the times. Here the ratio of the speeds is the same as the ratio of the times, for the ratio one-half PE has to one-half OD, or of the entire PE to OD, such has AE to AD. Hence the ratio of the spaces past is double the ratio of the times, which was to be demonstrated. (That is, since $PE/OD = AE/AD$, so $MH/LH = PE/OD \cdot AE/AD = AE^2/AD^2$.) It is also clear from this that the same ratio of spaces is the duplicate ratio of the maximum degrees of speed, namely, of lines PE and OD, since PE is to OD as EA is to DA. (Again, since $PE/OD = AE/AD$, so $MH/LH = PE/OD \cdot AE/AD = PE^2/OD^2$.)

Corollary I adds that it is clear that in equal times the spaces past successively from rest are as the odd numbers from unity, $1, 3, 5, 7 \ldots$, for this is the relation of the excesses of the squares following each other from unity. Thus, when the degrees of speed increase as the simple numbers from unity, $1, 2, 3, 4$, the spaces passed in the same times increase as the odd numbers from unity, $1, 3, 5, 7$, (which are their successive sums, $1, 1 + 2 = 3, 2 + 3 = 5$, and so on, and the total distances increase as square numbers from unity $1, 4, 9, 16$, which are their cumulative sums, $1, 1 + 3 = 4, 1 + 3 + 5 = 9$, and so on). Corollary II is that if any two spaces are

Fig. 2.9.

passed from rest in any two times, the times will be to each other as either of the two spaces to the mean proportional between the spaces. In Fig. 2.9, letting ST and SV be the two spaces of which the mean proportional is SX, that is, $ST/SX = SX/SV$, the time of fall through ST will be to the time of fall through SV as ST is to SX, or the time through SV is to the time through ST as VS is to SX. For since the spaces are as the squares of the times, $ST/SV = SX^2/SV^2$, from which $(ST \cdot VS^2)/VS = SX^2$, and thus $ST \cdot VS = SX^2$ or $ST/SX = SX/SV$.

Theorem III Proposition III. If the same moveable is carried from rest on an inclined plane and in a perpendicular of the same altitude, the times of the motions will be to each other as the lengths of the plane and the perpendicular. This follows directly from the assumption that the degrees of speed acquired by the same moveable (in descending) over different inclinations of planes are equal when the elevations of the planes are equal. In Fig. 2.10, let the inclined plane AC and the perpendicular AB be of the same altitude above the horizontal BC. Let any lines DG, EI, FL be imagined parallel to the horizontal BC. It follows from the assumption that the degrees of speed of the moveable from the first beginning of motion A acquired in points G and D are equal, since the approaches to the horizontal are equal, and likewise the degrees in points I and E will be the same as well as the degrees in L and F. And if the parallels from all points of AB are drawn to AC, the moments or degrees of speed at the end points of each parallel will always be equal to each other, so that the two spaces AC and AB are passed with the same degrees of speed. But it was demonstrated that if two spaces are passed by a moveable carried with the same degrees of speed, the times of the motions have the same ratio as the spaces. (By Proposition I of *On Uniform Motion*, if a moveable carried by uniform motion with the same speed passes through two spaces, the times of the motions have the same ratio as the spaces. This can be extended from uniform motion to the degrees of speed of uniformly accelerated motion by using as the uniform motion one-half the final speed of Proposition I here.) Therefore, the time of motion through AC to the time of motion through AB is

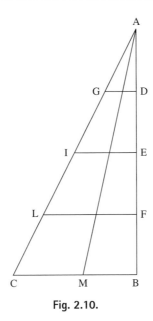

Fig. 2.10.

as the length of plane AC to the length of perpendicular AB. The corollary is that the times of descents on differently inclined planes of the same elevation are to each other as their lengths. For if another plane AM is imagined, from A terminated at the same horizontal CB, the time of descent through AM is to the time of descent through AB as AM is to AB; and as the time through AB is to the time through AC as AB is to AC, thus the time through AM is to the time through AC as AM is to AC.

In the following Theorem IV Proposition IV, it is shown that the times of descent over equal planes unequally inclined are to each other inversely as the square roots of the heights of the planes. Finally, in Theorem V Proposition V, the general case is shown, that the ratio of the times of descent over planes differing in inclination and height is compounded from the ratio of the lengths of the planes directly and the square roots of their heights inversely. Propositions I–V are the fundamental theorems for most of the demonstrations in the rest of the book, including the most famous, Theorem VI Proposition VI, that times of descent along all chords of a vertical circle are equal.

Proposition VI is given no less than three proofs, and there are yet more in Galileo's earlier manuscripts, beginning with (trivial) proofs for speed proportional to distance before he had discovered the law of uniform acceleration. We shall show the third proof, which is the most direct, in Fig. 2.11. On the horizontal AB is a vertical circle with diameter CD. From the highest point D, let there be inclined to the circumference any plane DF. I say that the descent of the same moveable through plane DF and the fall through diameter DC are completed in equal times. Draw FG parallel to AB, which will be perpendicular to diameter DC, and let FC be joined. From Corollary II of Proposition I, the time of fall through DC is to the time of fall through DG as the mean proportional between DC and DG is to DG. And since angle DFC in a

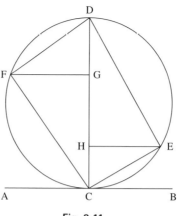

Fig. 2.11.

semicircle is a right angle, by similar triangles *DFC* and *DFG*, $DG/DF = DF/DC$, and *DF* is the mean proportional between *DC* and *DG*. But by Proposition III, the time of descent through *DF* is to the time of fall through *DG* as *DF* is to *DG*. And it was just shown that the time of fall through *DC* is to the time of fall through *DG* as *DF* is to DG. Hence, the time of descent through *DF* and of fall through *DC* have the same ratio to the time of fall through *DG*, of *DF* to *DG*, and therefore they are equal. And the same may be shown for any plane drawn to the lowest point of the circle, as *EC* with *EH* parallel to AB, that the time of descent through *EC* is equal to the time of fall through diameter *DC*. Corollary I is that the time of descent through all chords drawn from end points *C* and *D* are equal. Corollary II is that if from the same point there descend a perpendicular and an inclined plane on which descents are made in equal times, they are both located in a semicircle of which the perpendicular is the diameter.

In Proposition XXXVI Theorem XXII it is shown that if in an arc of a circle not greater than a quadrant a single inclined plane from the lowest point meet any point of the arc, and from the same points two inclined planes meet the arc, the descent along the two planes dividing the arc is completed in less time than in the first plane alone or in the lower of the two planes. In Fig. 2.12, the descent in *AE* and *EC* together is completed in less time than in *AC* or *EC* alone. The proof is, to say the least, lengthy, and the conclusion is correct. In a following scholium it is concluded that if the arc be divided into any number of small segments by chords, as *AD*, *DE*, *EF*, *FG*, *GC*, the descent along adjacent smaller chords is completed in less time than along longer chords, as descent along *EF*, *FG*, *GC* in less time than along *EC*. And the descent is completed in less time as the number of chords is increased and the closer the chords approach the circumference of the circle. Hence the swiftest motion of all is not through the shortest line, namely through a straight line, as *AC*, but through a part of a circle. It is true that the arc of a circle is a swifter descent than through any chord or number of chords, but it is not the curve of quickest descent, which is a cycloid, as later shown by Huygens.

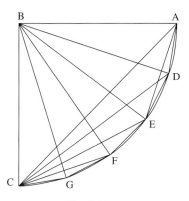

Fig. 2.12.

Following the theorems on accelerated motion on inclined planes, the principal subject concerning motion in the *Two New Sciences* is the parabolic trajectory of a projectile treated in *On the Motion of Projectiles* in the Fourth Day. Galileo begins with an explanation of projectile motion similar to his experiments recorded in Ms. 72. If a moveable is projected on a horizontal plane with all impediments removed, its motion would be uniform and perpetual. But if the plane ends, the moveable adds to its previous uniform motion a downward tendency from its own weight, and there emerges a kind of motion compounded from uniform horizontal and naturally accelerated downward motion which I call projection. Proposition I, the fundamental theorem of projectile motion, of a parabolic trajectory, is more an illustration of this description than a proof, and is applied to a semiparabola, the curve that would be followed by the moveable in Galileo's description. The proof is preceded by two lemmas. Lemma I is that in a parabola the square of the ordinate, the perpendicular from the axis to the curve, is proportional to the distance on the axis from the vertex. Lemma II is that if an ordinate is drawn from the axis to any point on a parabola, and from that point a line is drawn that intersects the axis beyond the vertex of the parabola such that the distance along the axis from the ordinate to the vertex equals the distances from the vertex to the intersection, the line will be tangent to the parabola. Each of these lemmas is actually applied in its converse form: I to show that a curve meeting the conditions of the lemma is a parabola, and II to show that where a tangent to a parabola intersects the axis, the distance from the intersection to the vertex equals the distance from the vertex to the ordinate. These are the only properties of parabolas used in the demonstrations.

Theorem I Proposition I is then: When a projectile is carried by a motion compounded from a uniform horizontal motion and a naturally accelerated downward motion, it will describe a semiparabola in its motion. In Fig. 2.13, imagine a horizontal line or plane *AB* on which a moveable is carried uniformly from *A* to *B*, and at *B* there supervenes a natural downward motion from its own heaviness along perpendicular *BN*. Beyond *B*, let the line, as if it were the measure of time, be divided into equal parts *BC*, *CD*, *DE*, and from points *C*, *D*, *E* draw parallels to *BN*. From *C*,

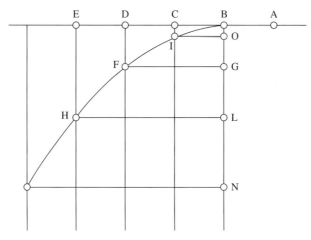

Fig. 2.13.

take some part *CI*, from *D*, take four parts *DF*, from *E*, take nine parts *EH*, thus as the squares of *CB*, *DB*, *EB*. Since the descent of the moveable is as the square of the time, it will successively describe the spaces *CI*, *DF*, *EH*, which are as the squares of *CB*, *DB*, *EB*. Draw lines *IO*, *FG*, *HL* parallel to *EB*, so that *HL*, *FG*, *IO* will be equal to *EB*, *DB*, *CB* respectively, and *BO*, *BG*, *BL* will be equal to *CI*, *DF*, *EH*. Hence, $HL^2/FG^2 = LB/BG$, and $FG^2/IO^2 = GB/BO$, and therefore points *I*, *F*, *H* lie on one and the same parabolic line. The reason is the converse of Lemma I, that if for some line the square of the ordinate, as *IO*, *FG*, *HL*, is as the distance from the vertex, as *BO*, *BG*, *BL*, the line is parabolic, here a semiparabola. The theorem is thus an illustration of the application of the lemma to a body with uniform horizontal motion and naturally accelerated vertical motion.

The theorem is followed by a discussion of three qualifications. (1) The axis of the parabola is directed to the centre of the Earth, but since the parabola widens from the axis, the projectile would never end at the centre of the Earth, or if it did, it would describe some line different from the parabolic. (2) Since the horizontal motion departs farther from the centre of the Earth, it is ascending, and thus the motion cannot be perpetual and uniform but would become slower. (3) The resistance of the medium, which cannot be removed, will disturb the uniformity of the transverse motion and the rule of acceleration for heavy bodies. Each of these objections is taken to be correct, but the effects of the first two are very small. (1) The path of the projectile above the surface of the Earth is very nearly parabolic, since the entire length of its path is minute compared to the distance to the centre of the Earth, and for the same reason, (2) the transverse motion is very nearly at the same distance from the centre of the Earth and thus very nearly uniform. (3) The resistance of the medium, which can arise from multiple causes and have a greater effect, is treated more extensively. Considering the impediment of air, its effect can depend upon the shape, heaviness, and speed of the moveable. For the greater the speed, the greater the obstruction of the air, which affects bodies more as they are less heavy. When a body falls from very great heights, the impediment of the air will take away the power of increasing its speed and reduce it to uniform and equable motion. And motion on a horizontal plane, which should be uniform and perpetual, will be altered and finally stopped by the impediment of the air. Considering the effect of the air on the fall of heavy bodies, two balls of equal size, one ten or twelve times heavier than the other, as one of lead and the other of oak, both descend from a height of 150 or 200 braccia with very little difference in speed, for if there were a great effect of the air, the lead ball would arrive at the ground long before the oak ball, but the anticipation is not one-hundredth part of the entire height. Likewise, the effect of the air on the same moveable is not much greater when it is moved with great speed than when moved slowly. Suspend from two equal threads four or five braccia in length two equal lead balls, remove both from the perpendicular, one by 80° or more, the other by not more than 4° or 5°. When they are set free, the first describes very large arcs of 160°, 150°, 140°, and so on, which gradually diminish, the other passes small arcs of 10°, 8°, and so on, which also diminish. In the time that the first passes its large arcs, the second passes its small arcs, from which it is evident that the speed of the first is 16 or 18 times the speed of the second, and if the greater speed were impeded by the air more than the lesser speed, the oscillations in the large arcs should be less frequent than those in the small arcs. But experience opposes this, for if two friends count the oscillations, one the very large and the other the very small, they will see that they will count not just the tens, but also the hundreds without disagreeing by a single one, or rather by only one moment. This report of near isochronism has often been doubted, but in fact, after a difference of one swing in thirty or so oscillations, the arcs are both small enough to be sensibly isochronous. Galileo takes this as evidence of both the isochronism of pendulums and that the effect of the impediment of the air is proportional to speed, since if the faster pendulum were slowed in a greater or lesser proportion than the slower, the number of oscillations would not continue equal.

The following demonstrations depend upon two principles: (1) that uniform horizontal motion with the speed acquired in vertical fall traverses twice the distance of the vertical fall in the same time, which determines the amplitude of a semiparabola, and (2) that continued accelerated vertical fall in the same time determines the altitude. Both of these also follow from the treatment of naturally accelerated motion in the Third Day. These principles would seem to be confirmed by Galileo's earlier experiments of letting a ball descend an inclined plane and then pass off a table to describe a semiparabola: (1) the height of the inclined plane determining the speed of the ball and thus amplitude of the semiparabola, and (2) the height of the table determining its altitude. In Proposition III Galileo defines technical terms and the relation of impetus, here meaning speed, to distance in the formation of the parabola. In Fig. 2.14, *AB* is the 'altitude' and *BC* the 'amplitude' of semiparabola *AC* described by the naturally accelerated motion of descent through *AB* and the uniform motion through the horizontal *AD*. The speed acquired at *C* is determined by descent through altitude *AB*, and to determine the speed acquired at *A*, extend altitude *CA* upward (*in sublimi*) through the 'sublimity' (*sublimitas*) *AE*, falling through which the moveable

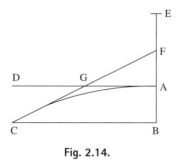

Fig. 2.14.

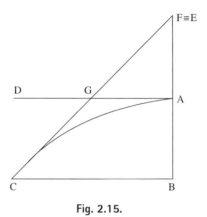

Fig. 2.15.

acquires its speed at A. (The sublimity AE is the distance along the axis of the parabola from the vertex to the directrix, which is equal to the distance from the focus to the vertex. Galileo does not use these properties of a parabola.) In the time of fall through EA, the moveable will traverse double the distance in the horizontal AD. Let CF be a tangent to the semiparabola at C, which, by the converse of Lemma II, will meet the axis at F where $AF = AB$, and it follows that $AG = \frac{1}{2}BC = \frac{1}{2}AD$. In Proposition V, it is shown that AE is a third proportional to AF and AG, that is $AF/AG = AG/AE$. For if EA is the measure of time of fall from E to A and of the speed acquired at A, then by *On Naturally Accelerated Motion* Proposition II Corollary II, the mean proportional AG shows the time and speed of fall through FA or AB. Thus a moveable descending through EA will in the time AG traverse twice the horizontal distance AG, or AD, and the vertical distance AB and describe the semiparabola AC of altitude AB and amplitude BC. Since $AB = AF$ and $\frac{1}{2}BC = AG$, it follows in a corollary that $AB/\frac{1}{2}BC = \frac{1}{2}BC/AE$, and thus in Proposition VI, given the sublimity AE and altitude AB, the amplitude BC may be found from $BC = 2(AE \cdot AB)^{1/2}$.

In Proposition VII, shown in Fig. 2.15, we consider semiparabola AC with amplitude BC twice its altitude AB. Drawing tangent CF, since $AF = AB$, it follows that $FB = BC$ and angle BCF is half a right angle. Since from Proposition V, $FA/AG = AG/EA$, and $AG = FA$, thus $EA = FA$, the high point of fall E and the intersection of the tangent and the axis F coincide. Semiparabola AC is described by a projectile with uniform horizontal speed acquired in fall through FA, the measure of the speed at A from which in time EA it would describe AD, and natural vertical fall through AB in the time and speed shown by $AG = FA$. It is to be shown that the speed required to describe this semiparabola is less than any other with the same amplitude. From Proposition IV, the combined speed at C is as the diagonal FG, equal in square to FA and AG, that is, $FG^2 = FA^2 + AG^2$, and the quite lengthy proof of the proposition is that for any semiparabola of the same amplitude and greater or lesser altitude, EA is greater than FA and the diagonal showing the combined speed greater than FG. The corollary to the proposition is that in the reverse direction, a projectile from C with elevation of half a right angle requires less speed through semiparabola CA than through any with greater or smaller elevation, and with the same speed will have the greatest amplitude.

By now the discussion has turned to the range of artillery shots, leading to the calculation of tables for amplitude and altitude of semiparabolas described with the same speed (*impetus*) and different angles of elevation. It is noted that the maximum range of artillery shots is for an elevation of half a right angle, 6 points or 45°, and that the range of shots at elevations that exceed or fall short of 6 points by equal angles are equal. Thus, where 12 points = 90° and 1 point = 7°, the range of 7 and 5 points, 8 and 4 points, 9 and 3 points, and so on, are equal. Horizontal projection, 0 points, is *tiro di punto in bianco*. Proposition VIII shows that the amplitudes of semiparabolas described by projectiles sent forth (*explosis*) with the same speed with elevations at equal angles above and below half a right angle are equal. In Proposition IX it is shown that the amplitudes of semiparabolas are equal when their altitudes and sublimities are inversely proportional and in Proposition X that the speed (*impetus* or *momentum*) of motion through any semiparabola is equal to the speed of natural fall through the combined sublimity and altitude, that is, in Fig. 2.15 the speed at C is equal to the speed at B of natural fall through FB. Hence, as stated in the corollary, for all semiparabolas for which the sums of the altitudes and sublimities are equal, the speeds (*impetus*) are equal. This goes both ways, that is, the speed or impetus of a projectile from C is *defined* as the speed acquired in fall through the sum of the sublimity and altitude FB. This can be taken as a third principle to the two given earlier: (1) that uniform horizontal motion with the speed acquired in vertical fall traverses twice the distance of the vertical fall in the same time, which determines the amplitude of a semiparabola; (2) that continued accelerated vertical fall in the same time determines the altitude; and now (3) that the initial speed of a projectile is that of the speed acquired in fall from sum of the altitude and sublimity. Proposition XI is given the speed (*impetus*) and amplitude of a semiparabola, to find its altitude and sublimity.

The calculation of amplitude and altitude of semiparabolas described by projectiles sent forth with the same speed and different angles of elevation is shown in Propositions XII and Proposition XIII followed by the tables. In Fig. 2.16 we again consider semiparabola CA with an elevation FCB of half a right angle, 45°,

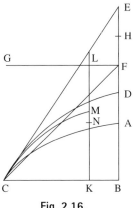

Fig. 2.16.

of amplitude BC, altitude AB, and sublimity FA. By Proposition X, the speed at C is equal to that of a moveable falling through the combined sublimity and altitude FB, which is taken as the speed of projection at C, and taking FG parallel to BC, the combined altitudes and sublimities of all semiparabolas with the same speed at C will fall between parallels FG and BC, that is, equal to FB. Now, take an elevation angle ECB greater than FCB, and suppose a semiparabola tangent at C. If such a parabola has the same amplitude BC, it will require a sum of altitude and sublimity greater than FB and so require a greater speed at C, contrary to the assumption of equal speed at projection. For let $BE = \tan ECB$ and bisect BE at D. Then by the converse of Lemma II, a semiparabola with elevation ECB will be CD, meeting the axis at D. Let FH be a third proportional to BD and one-half BC, that is $FH/\frac{1}{2}BC = \frac{1}{2}BC/BD$, or $FH = (\frac{1}{2}BC)^2/BD$, and FH will be the sublimity from which a moveable falling and deflected horizontally at D will describe semiparabola DC. But $HB = BD + DH$ is greater than $FB = BA + AF$, which thus requires a greater speed at C. We require a parabola with elevation ECB and combined altitude and sublimity equal to FB, for which the amplitude BC must be reduced. Hence, take amplitude CK where $CK/BC = FB/HB$, and CK will be the required amplitude such that semiparabola CM with elevation ECB will have a combined altitude and sublimity $LK = MK + LM = FB$.

The instructions for the computation of amplitude in the lettering of Fig. 2.16 are: Take the tangent of the given angle BCE, to one-half of which add the third proportional of it and of one-half BC, which is DH; then as HB is to FB, let BC be to another (distance) CK, namely, the amplitude sought. What this amounts to is computing the ratio of the amplitude of a semiparabola of arbitrary elevation to the amplitude of a semiparabola of elevation equal to half a right angle with the same sum of altitude and sublimity, that is, described by a projectile with the same initial speed. The steps are: $BD = \frac{1}{2}BE = \frac{1}{2}\tan ECB$. From $DH/\frac{1}{2}BC = \frac{1}{2}BC/BD$, $DH = (\frac{1}{2}BC)^2/BD$. Take $HB = BD + DH$ and for the elevation of half a right angle $FB = BA + AF = BC$. Then the amplitude $CK/BC =$

FB/HB or $CK = BC(FB/HB)$. Galileo's numerical example is for $ECB = 50°$ where $BC = 10,000$. In these units $BE = \tan 50° = 11,918$, $BD = \frac{1}{2}BE = 5959$, $DH = (\frac{1}{2}BC)^2/BD = 4195$, and $HB = BD + DH = 5959 + 4195 = 10,154$. Then amplitude $CK = BC(FB/HB) = 10,000(10,000/10,154) = 9848$ and the amplitude of the entire parabola is 19,696. Given the amplitude, the calculation of the altitude is as follows: In the figure, let LK be bisected at N, so that $KN = \frac{1}{2}KL$, and be met by the semiparabola at M. From the corollary to Proposition V, $(\frac{1}{2}CK)^2 = KM \cdot ML$ and from Euclid 2.5, $(\frac{1}{2}KL)^2 = KM \cdot ML + NM^2$; thus $NM^2 = (\frac{1}{2}KL)^2 - (\frac{1}{2}CK)^2$ and altitude $MK = KN + NM$. Galileo's numerical example is for $ECB = 55°$ for which $CK = 9396$, $(\frac{1}{2}CK) = 4698$, and as in the previous calculation $LK = FB = 10,000$. Thus, $NM^2 = (\frac{1}{2}KL)^2 - (\frac{1}{2}CK)^2 = 25,000,000 - 22,071,204 = 2,928,796$, so that $MN = 1,710$ and the altitude $MK = 6,710$.

These calculations are unnecessarily complicated, and were greatly simplified by Torricelli in *Two books on the Motion of Heavy Bodies Naturally Descending and Projected* (*De motu gavium naturaliter descendentium et proiectorum libri duo*) of 1644, a complete revision and enlargement of the entire treatise *On Local Motion* in the *Two New Sciences* which, while relying upon Galileo's exposition for its fundamental discoveries, referred to again and again with the greatest praise and respect, makes Galileo's demonstrations and procedures obsolete. For example, if we let elevation BCE be ϑ, amplitude BC for $\vartheta = 45°$ be r, and the required amplitude CK be r', then, after some reduction, Galileo's procedure is equivalent to $r' = r/\frac{1}{2}(\tan\vartheta + \cot\vartheta)$, which, through trigonometric identities reduces to $r' = r\sin 2\vartheta$. Likewise, for finding the altitude of the same semiparabola from a given amplitude r', Galileo's computation is $k^2 = (\frac{1}{2}r)^2 - (\frac{1}{2}r')^2$, from which the altitude $a = \frac{1}{2}r + k$. This too can be reduced to a direct computation from ϑ, namely, $a = \frac{1}{2}r \text{ vers } 2\vartheta = \frac{1}{2}r(1 - \cos 2\vartheta) = r\sin^2\vartheta$. Torricelli demonstrates these relations directly, uses $r' = r\sin 2\vartheta$ and $a = \frac{1}{2}r \text{ vers } 2\vartheta$ for his table, and points out that Galileo's lengthy demonstrations, Propositions VII and VIII, that of semiparabolas described with the same impetus, the one with an elevation of half a right angle has the greatest amplitude, and that the amplitudes from elevations equally above or below half a right angle are equal, are obvious from $r' = r\sin 2\vartheta$. Likewise, from $a/r' = \sin^2\vartheta/\sin 2\vartheta$, it follows that $a = \frac{1}{2}r'\tan\vartheta$.

In Proposition XIV, for computation of a third table for finding the altitudes and sublimities of semiparabolas of equal amplitude r, no longer referring to lettering of the figure, Galileo uses $a = \frac{1}{2}r\tan\vartheta$, which follows from the converse of Lemma II. In his example, for $\vartheta = 30°$, where $r = 10,000$, $a = 2887$. For the sublimity, he notes that from the corollary to Proposition V, one-half the amplitude of a semiparabola is a mean proportional between the altitude and sublimity. Letting s be the sublimity, $s/\frac{1}{2}r = \frac{1}{2}r/a$ or $s = (\frac{1}{2}r)^2/a$. The example is $s = 25,000,000/2,887 = 8,659$. But this too can be simplified as Torricelli does for computing the same table, namely, substituting $s = (\frac{1}{2}r)^2/a = (\frac{1}{2}r)^2/\frac{1}{2}r\tan\vartheta$, it follows that $s = \frac{1}{2}r\cot\vartheta$. In Galileo's table, the altitude and sublimity are computed separately, but Torricelli simply inverts the order of $\frac{1}{2}r\tan\vartheta$ for a and s. Salviati remarks that the table shows the minimum speed (*impetus*) is required for the same amplitude at an elevation of 45° where the sum of the altitude and sublimity, each 5,000, which determines the speed, is a minimum of 10,000 and shows that for any other elevation the sum is greater. Further, for equal angles above and below 45°, the altitudes and sublimities are reversed, in his example, for 50° $a = 5{,}959$ and $s = 4{,}196$, for 40° $a = 4{,}196$ and $s = 5{,}959$. These relations are obvious from $a = \frac{1}{2} r \tan \vartheta$ and $s = \frac{1}{2} r \cot \vartheta$, but it is certain that Galileo did not know that, as he says that in the computations we do not take account of some fractions. It is worth noting that Torricelli's entire treatise, Book I on the descent of heavy bodies and Book II on projectiles, is not only a great improvement upon Galileo's treatment, but also provides a clarification of its obscurities, of which there are not a few.

References

The literature on Galileo's mechanics is very large and our references are highly selective. The principal texts are in Favaro's edition *Le Opere di Galileo Galilei* (1890–1909, repr. 1968), *De motu* in vol. 1, the *Dialogue* in vol. 7, the *Two New Sciences* and transcriptions of notes from Ms. 72 in vol. 8. There are photographs of most of Ms. 72 in S. Drake, *Galileo's Notes on Motion*, Supplemento agli Annali dell'Istituto e Museo di Storia della Scienza (1979), Fascicolo 2, in which the pages are arranged in probable order of composition. The entire manuscript can now be found online in high-resolution colour. Search under 'Galileo Ms. 72' or 'Galileo Notes on Motion'. *De motu* is translated by I. E. Drabkin in Galileo Galilei, *On Motion and On Mechanics*, trans. I. E. Drabkin and S. Drake (University of Wisconsin Press, 1979). The *Dialogue* is translated by Drake in Galileo Galilei, *Dialogue concerning the Two Chief World Systems, Ptolemaic and Copernican* (University of California Press, 1967); repr. The Modern Library (2001). The *Two New Sciences* is translated by Drake in Galileo Galilei, *Two New Sciences, including Centers of Gravity and Force of Percussion*, 2nd edition (Wall & Thompson, Toronto, 1989).

Stillman Drake's many papers on these subjects are collected in Stillman Drake, *Essays on Galileo and the History and Philosophy of Science*, 3 vols. (University of Toronto Press, 1999); the papers on mechanics are principally in vol. 2. The papers referred to specifically here, Drake (1969) is at 2.187–207, (1973) at 2.147–59, and (1985) at 2.321–31. There is also much on mechanics in Drake's *Galileo at Work, His Scientific Biography* (University of Chicago Press, 1978, repr. Dover, 1995), and in *Galileo: Pioneer Scientist* (University of Toronto Press, 1990).

Of the more important secondary literature, one may mention in chronological order: A. Koyré, *Etudes Galiléennes* (Hermann & Cie, Paris, 1939), trans. by J. Mepham, *Galileo Studies* (Humanities Press, New Jersey, 1978); A. Koyré, 'A Documentary History of the Problem of Fall from Kepler to Newton', *Transactions of the American Philosophical Society*, 45, 4 (1955), 329–395. T. Settle, 'An Experiment in the History of Science,' *Science* 133, No. 3445 (1961), 19–23. W. Wisan, 'The New Science of Motion: A Study of Galileo's *De motu locali*', *Archive for History of Exact Sciences* 13 (1974), 103–306. There are a large number of papers on these subjects by R. H. Naylor, listed in Damerow *et al.* below. P. Galluzzi, *Momento, Studi Galileiani* (Ateneo e Bizzarri, Rome, 1979). E. D. Sylla, 'Galileo and the Oxford *Calculatores*: Analytical Languages and the Mean-Speed Theorem for Accelerated Motion,' *Reinterpreting Galileo*, ed.

W. A. Wallace, *Studies in Philosophy and the History of Philosophy* 15 (Catholic University of America Press, 1986), 53–108. J. Renn, P. Damerow, S. Rieger, D. Giulini, 'Hunting the White Elephant: When and How Did Galileo Discover the Law of Fall?', *Science in Context* 13 (2000), 299–419; also published in *Galileo in Context*, ed. J. Renn (Cambridge University Press, 2001), 29–149. P. Damerow, G. Freudenthal, P. McLaughlan, and J. Renn, *Exploring the Limits of Classical Mechanics*, 2nd edition, *Sources and Studies in the History of Mathematics and Physical Sciences* (Springer, 2004), contains a very thorough bibliography of literature on Galileo's mechanics. A recent biography of Galileo that treats the mechanics, along with the rest of his work, is J. L. Heilbron, *Galileo* (Oxford University Press, 2010), which also contains an extensive and current bibliography.

CHAPTER 3

CARTESIAN PHYSICS

JOHN A. SCHUSTER

3.1 Introduction: Cartesian Physics or Natural Philosophy

By the term 'physics', René Descartes and other educated men of his generation would have understood 'natural philosophy', the attempt systematically to understand matter and cause in nature, as well as cosmic and terrestrial structure. This was to be accomplished either by means of one or another version of the institutionally dominant neo-Scholastic Aristotelianism of the day, or one of its increasingly popular alternatives, neo-Platonic, Stoic, or atomistic. Descartes certainly would not have identified the word 'physics' with the classical mechanics later thinkers discerned in embryonic form in the work of Galileo and which matured in the next two generations in the work of Huygens, Newton, and others. Nevertheless, Descartes' natural philosophy marks a significant moment in the larger history of physics, as conceived in this volume. His system of natural philosophy was a novel, daring, and intricate construction in that field, with two main sets of historical significances for later physics.

First of all, Cartesian natural philosophy contained several novelties: for example, his treatment of collision, force of motion, and rest, and the attempted mechanization of the theory of light and celestial mechanics, which had unintended technical significances in the formation of the Newtonian and Huygensian dispensations in mechanics. Secondly, the Cartesian system embodied in its own terms a number of values, goals and strategic positionings which it shared with other innovative natural philosophies of the day, and which during the course of the mid and later seventeenth century broadly facilitated and shaped the eventual emergence of classical mechanics. These included his concerted effort to establish the truth of a realist and multi-planetary system version of Copernicanism, and, albeit in his own *sui generis* and ultimately unsuccessful manner, his attempt to mathematicize natural philosophy.

This chapter thus aims to explicate in detail Descartes' physics in a way that also brings into relief these two sets of subtle, largely unintended, but nonetheless significant consequences for Newtonian physics. These goals can be achieved only by a careful genealogical anatomy of Descartes' natural philosophy, taking account not simply of its mature, systematized form, but also the little explored but highly revealing history of how and why it developed as it did. We undertake this in Section 3.2, before Section 3.3 examines the implications for classical mechanics of the Cartesian form of physics.

3.2 The Developmental Anatomy of Cartesian Physics, 1618–44

3.2.1 Successes, Failures, and Fate of Descartes' Early Physico-Mathematics Programme

In November 1618, Descartes, then twenty-two years of age, met and worked for two months with Isaac Beeckman, a Dutch scholar eight years his senior. Beeckman was one of the first supporters of a corpuscular–mechanical approach to natural philosophy. However, it was not simply corpuscular mechanism that Beeckman advocated to Descartes. He also interested Descartes in what they called 'physico-mathematics'. In late 1618 Beeckman wrote that 'There are very few physico-mathematicians', adding that Descartes 'says he has never met anyone other than me who pursues enquiry the way I do, combining Physics and Mathematics in an exact way; and I have never spoken with anyone other than him who does the same' (Beeckman, 1939–53, I 244). They were partly right. While there were not many physico-mathematicians, there were of course others, such as Kepler, Galileo, and certain leading Jesuit mathematicians, trying to merge mathematics and natural philosophy (Dear, 1995).

Physico-mathematics was an actors' term of the day, and taken generically, across the spectrum of different natural philosophers who used it, dealt with the way the traditional mixed mathematical disciplines, such as hydrostatics, statics, geometrical optics, geometrical astronomy, and harmonics, were conceived to relate to the discipline of natural philosophy (Gaukroger and Schuster, 2002). In Aristotelianism the mixed mathematical sciences were interpreted as intermediate between natural philosophy and mathematics and subordinate to them. Natural philosophical explanations were couched in terms of matter and cause—something mathematics could not offer, according to most Aristotelians. The mixed mathematical sciences used mathematics not in an explanatory way, but instrumentally for problem-solving and practical aims. For example, in geometrical optics one represented light as light-rays. This might be useful, but it did not explore the underlying natural philosophical questions concerning 'the physical nature of light' and 'the causes of optical phenomena'. In contrast, physico-mathematics signalled a commitment to revising radically the Aristotelian view of the mixed mathematical sciences, which were to become more intimately related to natural philosophical issues of matter and cause. Paradoxically, the issue was not mathematization. The mixed mathematical sciences, which were already mathematical, were to become more 'physicalized', more closely integrated into whichever brand of natural philosophy an aspiring physico-mathematician endorsed.

In the case of Descartes and his mentor Beeckman, the preferred, if unsystematized natural philosophy of choice was a fragmented and embryonic variety of corpuscular mechanism. Hence what Descartes and Beeckman meant by the programme of physico-mathematics was that reliable geometrical results in the mixed mathematical sciences were to be explained by invoking an embryonic corpuscular-mechanical matter theory and a causal discourse concerning forces and tendencies to motion (Schuster, 2013). Three of Descartes' early exercises in physico-mathematics survive. The most important and symptomatic was his attempt, at Beeckman's urging, to supply a corpuscular-mechanical explanation for the hydrostatic paradox, which had been rigorously derived in mixed mathematical fashion by Simon Stevin (AT X 67–74, 228; Gaukroger and Schuster, 2002). We shall examine this portentous work in some detail before turning briefly to the other two cases.

In 1586 Simon Stevin, the great Dutch maestro of the practical mathematical arts and mixed mathematical sciences, had demonstrated that a fluid filling two vessels of equal base area and height exerts the same total pressure on the base, irrespective of the shape of the vessel, and hence, paradoxically, independently of the amount of fluid it contains. Stevin's mathematically rigorous proof applied a condition of static equilibrium to various volumes and weights of portions of the water (Stevin, 1955–66 I, 415–7).

In Descartes' treatment of the hydrostatic paradox (AT X 67–74) the key problem involves vessels B and D, which have equal areas at their bases and equal height, and are of equal weight when empty (Fig. 3.1). Descartes proposes to show that 'the water in vessel B will weigh equally upon its base as the water in D upon its base'—Stevin's paradoxical hydrostatical result (AT X 68–69).

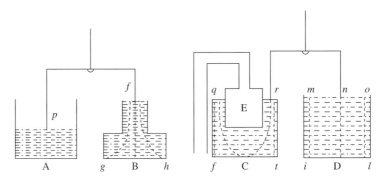

Fig. 3.1. Descartes' 'Hydrostatics Manuscript' (AT X 69).

First, Descartes explicates the weight of the water on the bottom of a vessel as the total force of the water on the bottom, arising from the sum of the pressures exerted by the water on each unit area of the bottom. This 'weighing down' is explained as 'the force of motion by which a body is impelled in the first instant of its motion', which, he insists, is not the same as the force of motion which 'bears the body downward' during the actual course of its fall (AT X 68).

Then, in contrast to Stevin's rigorous proof, Descartes attempts to reduce the phenomenon to corpuscular mechanics by showing that the force on each 'point' of the bottoms of the basins B and D is equal, so that the total force is equal over the two equal areas, which is Stevin's paradoxical result. He claims that each 'point' on the bottom of each vessel is serviced by a unique line of instantaneously exerted 'tendency to motion' propagated by contact pressure from a point (particle) on the surface of the water through the intervening particles. But, while the surface of D is equal to and congruent with its base and posed directly above it, the surface of B is implied to have an area of one-third its base. Hence exemplary points i, D, and l in the base of D are each pressed by a unique, vertical line of tendency emanating respectively from corresponding points, m, n, and o on the surface. In contrast, point f on the surface of B is the source of three simultaneous lines of tendency, two being curved, servicing exemplary points g, B, and h on the bottom of B. Descartes claims that all six exemplary points on the bottoms are pressed by an equal force, because they are each pressed by 'imaginable lines of water of the same length' (AT X 70); that is, lines having the same vertical component of descent. Descartes smuggles the tendentious three-fold mapping from f into the discussion as an 'example', but then argues that *given the mapping*, f can provide a three-fold force to g, B, and h (AT X 70–1).

The 'hydrostatic manuscript' thus shows the young Descartes articulating the programme of physico-mathematics by taking firmly established geometrical results in a mixed mathematical science as a basis for inferring in some way a natural philosophical explanation in terms of matter and cause, thus fusing together sound findings from mixed mathematics and the explanatory realm of his favoured brand of natural philosophizing. Stevin's treatment of the hydrostatic paradox fell within the domain of mixed mathematics rather than natural philosophy. It did not explain the phenomenon by identifying its causes. The account Descartes substitutes for it falls within the domain of natural philosophy, attempting to identify the material bodies and causes in play. Fluids are made up, on Descartes' account, of corpuscles whose instantaneously manifested movements or tendencies to movement are understood in terms of a theory of forces and tendencies. This should explain what causes the pressure exerted by a fluid on the floor of its containing vessel.

These moves implied a radically non-Aristotelian vision of the relation of the mixed mathematical sciences to Descartes' emergent form of corpuscular-mechanical natural philosophy. He aimed to shift hydrostatics from mixed mathematics into the realm of natural philosophy. Indeed, he seems to have believed that from crisp, simple geometrical representations of sound mixed mathematical results one could read out or 'see' the underlying corpuscular-mechanical causes (Gaukroger and Schuster, 2002, 549–550). This proclivity was to be displayed in other parts of his physico-mathematical work, most especially in his geometrical optics. More generally, although Descartes never again directly considered hydrostatical problems, his mature natural philosophy was to develop and rearticulate some of the explanatory resources he first glimpsed in this early work. Throughout his later career Descartes continued to use descendants of the concept of instantaneous tendency to motion analysable into its directional components (later termed 'determinations'). These ideas became central to what may be termed his 'dynamics', the concepts that govern the behaviour of micro-corpuscles, in *Le Monde* (1629–33) and the *Principles of Philosophy* (1644, 1647), his two systematic treatises of natural philosophy.

Descartes' and Beeckman's well-known and frequently commented-upon studies of accelerated free fall also belonged to their physico-mathematical project (AT X 58–61, 74–8, 219–22; Jullien and Charrak, 2002). In contrast to the way this work has traditionally been interpreted, they were not simply striving for a 'Galileo-like' kinematic, or mixed mathematical law of fall, but also, as physico-mathematicians, they were looking for the causal explanation of such a law. This does not mean that the problem of mathematically describing fall was unimportant to Beeckman and Descartes. As physico-mathematicians they certainly wanted to find the descriptive law, if it existed. But simply to find such a law would have been to work superficially and without insight into natural-philosophical issues of matter and cause, as they criticized Stevin for having done in hydrostatics. They did not fail to find candidate laws, nor did they fail to find speculative candidate causes. The problem was that there were too many possible and plausible regimes of natural philosophical causation, in continuously and discontinuously acting modes, and there were too many resulting descriptive laws, which might well be impossible to determine one from another, even if measurement were possible. The problem of fall ended up looking like a poorly defined or unsolveable puzzle in physico-mathematics. This was not hydrostatics, where, as we have seen, Stevin's stunning and paradoxical results led Descartes, at least, to think he had made (quite radical) physico-mathematical capital; nor was it optics, where Descartes would eventually achieve profound physico-mathematical results.

This outcome undoubtedly contributed to his cool and sceptical response to Galileo's kinematics when it appeared eighteen years later. As early as 1619 Descartes could have begun to form the opinion that the highly idealized study of fall, in search of some sort of descriptive, mixed mathematical law, was of no natural-philosophical—that is, physico-mathematical—import. Of course, the search for and discovery of a law of falling bodies would be one of the key exemplars in the crystallization of classical mechanics during the course of the seventeenth century. But the study of falling bodies would play no role in Descartes' formulation of the causal register, the dynamics, that would sit at the heart of his later system of corpuscular mechanics. That dynamics also made a contribution to classical mechanics, but as we shall see, it would be derived physico-mathematically from important work in optics, hence we need to look at the third and final of Descartes' early physico-mathematical case studies, turning to his initial halting steps in rendering optics a physico-mathematical discipline.

In 1620 Descartes attempted in a physico-mathematical manner to find the mathematical law of refraction of light by considering the geometrical representation of its probable causes. He based the endeavour on some passages and diagrams in Kepler's optical masterpiece, *Paralipomena ad vitellionem* (1604), in which Kepler had suggested that light moves with more force in denser optical media and 'hence' is bent toward the normal in moving from a less to a more dense medium (AT X 242–3; Schuster, 2000, 278–285, 287–9). Descartes explored Kepler's speculation, using his newly acquired physico-mathematical style of 'reading' geometrical diagrams representing phenomena for their underlying message about the causal principles at work. On this occasion Descartes found neither a law of refraction nor its natural-philosophical causes. However, seven years later, whilst working with the mathematician Claude Mydorge, he discovered the long-sought law of refraction. This discovery was just one key step in a sustained course of inquiry in optics which brought Descartes' physico-mathematics to a climax and established the principles and exemplars for his emerging corpuscular-mechanical natural philosophy. His foray into optics in the late 1620s and early 1630s was thus one pivot of his natural-philosophical career, and it repays careful study in any exploration of his physics.

3.2.2 The Optical Triumph of the 1620s: From Physico-Mathematical Optics to the Principles of Corpuscular Dynamics

The trajectory of Descartes' work in optics between 1620 and the publication of his *Dioptrique* in 1637 has often been ignored in studies of his optics and physics. Recently, more light has been thrown on his project in optics—mixed and physico-mathematical: Around 1627 he found, by traditional mixed mathematical means, a simple (cosecant) version of the law of refraction. He immediately set to work attempting, in a physico-mathematical manner, to exploit his discovery by reading out of his results causal principles on the natural-philosophical level. This proceeds in two steps. First, from his key geometrical diagram for the new law he tried to read out the principles of a mechanical theory of light as an instantaneous impulse. As early as 1628 he was attempting to demonstrate his law of refraction, using these mechanical conceptions of light which he had obtained by reflecting, physico-mathematically, upon the new law. Then, secondly, as he became involved in writing *Le Monde* from late 1629, he tried to extract further physico-mathematical causal 'insight' from the optical work. Using the principles of his mechanical theory of light elicited from the work on refraction, he reformulated and polished the central concepts of his dynamics of corpuscles—the 'causal register' of his emerging system of corpuscular-mechanism—whose earliest, embryonic manifestations we have seen in his 1619 physico-mathematization of hydrostatics. This more mature elaboration is presented in *Le Monde* (1629–33), where the polished dynamics of corpuscles, itself a physico-mathematical product of the optical work, runs Descartes' vortex celestial mechanics and his corpuscular-mechanical theory of light in its cosmological setting (Schuster, 2005).

It would be ideal if this story could be explicated in detail in strict chronological order. Unfortunately, the materials for reconstructing Descartes' mixed and physico-mathematical optical project are few and scattered. This, combined with the curious and opaque presentation of the law of refraction in the *Dioptrique*, as well as the need to factor in Descartes' physico-mathematical proclivities, dictates that a different strategy must be used to unpack the details of his evolution from physico-mathematical optics to a dynamics of corpuscles. One must start from the end point—the *Dioptrique*, published in 1637 as one of the three *Essais* supporting the *Discours de la Méthode*—working back through scattered earlier hints and clues to uncover the genealogy of the discovery of the law of refraction, and its physico-mathematical exploitation, leading to 'seeing the causes' in a mechanistic theory of light and corpuscular-mechanical natural philosophy.

However, on its surface the *Dioptrique* does not reveal the trajectory of Descartes' struggles in mixed and physico-mathematical optics. Indeed, it has traditionally raised its own problems. For example, Descartes deduces the laws of reflection and refraction from a model involving the motion of some very curious tennis balls. Descartes' contemporaries tended not to see any cogency in this model, nor did they grasp the theory of motion (actually his corpuscular dynamics) upon which it is based. These problems fuelled the question of how Descartes had arrived at the law of refraction. Accusations arose that he had plagiarized the law from Willebrord Snel, though it has long been well established that this is quite unlikely (Korteweg, 1896; Kramer, 1882). Accordingly, in decoding the *Dioptrique* one must first grasp how the tennis-ball model for refraction coherently links to Descartes' theory of light as a mechanical impulse *through* his dynamics of micro-corpuscles. And in order to do that one must understand Descartes' mature dynamics of corpuscles, as first inscribed in *Le Monde* between 1629 and 1633 (Schuster, 2013, Chapter 4).

Descartes' system of natural philosophy in *Le Monde* was concerned with the nature and mechanical properties of microscopic corpuscles and a causal discourse, consisting of a theory of motion and impact, explicated through key concepts of the 'force of motion' and directionally exerted 'tendencies to motion' or 'determinations'. It is this 'causal register' within Descartes' natural philosophical discourse which scholars increasingly term his 'dynamics', as noted earlier (Gaukroger and Schuster, 2002). We have seen that the rudiments of this dynamics of instantaneously exerted forces and determinations dates back to Descartes' work on the physico-mathematics of hydrostatics. In *Le Monde* Descartes teaches that bodies in motion, or tending to motion, are characterized from moment to moment by the possession of two sorts of dynamical quantity: (1) the absolute quantity of the 'force of motion'—conserved in the universe according to *Le Monde's* first rule of nature, and (2) the directional modes of that quantity of force, the directional components along which the force or parts of the force act, introduced in *Le Monde's* third rule of nature (Schuster, 2000, 258–61). It is these that Descartes termed actions, tendencies, or most often, determinations.[1]

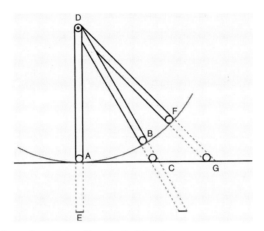

Fig. 3.2. Descartes' dynamics of the sling, in *Le Monde*.

As corpuscles undergo instantaneous collisions with each other, their quantities of force of motion and determinations are adjusted according to certain universal laws of nature, rules of collision. Therefore Descartes' analysis focuses on instantaneous tendencies to motion rather than finite translations in space and time.

Descartes' exemplar for applying these concepts is the dynamics of a stone rotated in a sling (AT XI 45–6, 85; G 30, 54–5) (Fig. 3.2).

Descartes analyses the dynamical condition of the stone at the instant that it passes point A. The instantaneously exerted force of motion of the stone is directed along the tangent AG. If the stone were released with no other hindrances affecting its trajectory, it would move along ACG at a uniform speed reflective of the conservation of its quantity of force of motion. However, the sling continuously constrains what we may call the privileged, 'principal' determination of the stone and, acting over time, deflects its motion along the circle AF.[2] Descartes decomposes the principal determination into two components: one along AE completely opposed by the sling—so no actual centrifugal translation can occur—only a tendency to centrifugal motion; the other, he says, 'that part of the tendency along AC which the sling does not hinder' (AT XI 85), which over time manifests itself as actual translation in a circle. The choice of components of determination is dictated by the configuration of mechanical constraints on the system.

Finally, in approaching Descartes' treatment of the law of refraction, we need to bear in mind his mechanistic theory of light as presented in its natural philosophical context in *Le Monde*. Leaving aside Descartes' theory of elements and his cosmology, his *basic* theory of light within his natural philosophy is that light is a tendency to motion, an impulse, propagated instantaneously through continuous optical media. So, light is or has a principal determination—a directional quantity of force of motion—which can be further analysed into components as and when necessary.[3] Note that in Descartes' theory the propagation of light is instantaneous, but the magnitude of the force conveyed by the tendency to motion constituting a light ray can

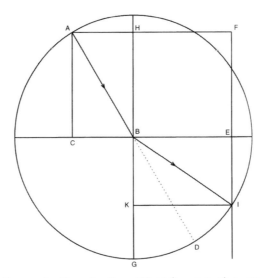

Fig. 3.3. Descartes' diagram for the refraction of light (tennis ball), in *Dioptrique*.

vary—there can be stronger and weaker light rays, all propagated instantly (Schuster, 2000, 261).

Using Descartes' theory of light, and his corpuscular dynamics, one can analyse both his published 'proof' of the law of refraction of light and its underlying rationale in terms of his real theory of light as instantaneous tendency to motion transmitted through the spherical particles of second element (*boules*) which make up his vortices.[4] In the *Dioptrique* of 1637 (Fig. 3.3) Descartes demonstrated his law of refraction using the model of a tennis ball struck by a racket along AB towards refracting surface CBE, which is taken to be a perfectly flat, vanishingly thin cloth (AT VI 97–8; CSM 1, 158–9). The tennis ball's weight and volume are ignored. It moves without air resistance in empty geometrical space on either side of the cloth, and in breaking through it the ball loses, independently of its angle of incidence, a certain fraction—one half—of its total quantity of force of motion. Descartes applies two conditions to the motion of the ball: [a] the new quantity of force of motion is conserved during motion below the cloth; and [b] the parallel component of the force of motion, the parallel determination, is unaffected by the encounter with the cloth. Drawing a circle of radius AB around B, he assumes the ball took time t to traverse AB prior to impact. After impact, losing half of its force of motion, hence half its speed, it must take $2t$ to traverse a distance equal to AB, arriving somewhere on the circle after $2t$. This represents condition [a]. Descartes writes that prior to impact the parallel determination 'caused' the body to move towards the right between lines AC and HBG. For condition [b] he considers that after impact, the ball takes $2t$ to move to the circle's circumference, so its unchanged parallel determination has twice as much time in which to act to 'cause' the ball to move toward the right. He sets FEI parallel to HBG so as to represent that doubled parallel travel. At time $2t$ after impact the ball will be at I, the intersection of FEI and the circle point below the cloth. It follows that (sin i/sin r) = (AH/KI) or 0.5 for all angles of incidence.

Descartes' published proof is superficially kinematic and has the look of a Galilean-type idealization of bodies, motion and space. But if we consider Descartes' corpuscular dynamics and the fact that his tennis ball is virtually a mathematical point in motion, we can translate his proof into the terms of his actual theory of light as instantaneously propagated tendency to motion (Schuster, 2000, 265) (Fig. 3.4). That is, beneath the kinematic and idealized veneer, Descartes is actually not talking about an idealized situation, but rather what he holds as a fully realistic picture of light and its modes of action in his plenist matter theory—a point to which we shall return in Section 3.2.4. Consider a light-ray incident upon refracting surface CBE. Let length AB represent the magnitude of the force of the light impulse. The *orientation* and *length* of AB represent the principal determination of the ray. The force of the ray is diminished by half in crossing the surface. So, to represent condition [a] we draw a semi-circle below the surface about B with a radius equal to one half of AB. As for condition [b], the unchanged parallel determination, we simply set out line FEI parallel to HBG and AC so that AH = HF. The resulting intersection at I gives the new *orientation* and *magnitude* of the force of the ray of light, BI, and the law follows, as a law of cosecants. The case of the light-ray requires manipulation of unequal semi-circles, representing the ratio of the force of light in the two media. In the case of the tennis ball Descartes moved from the ratio of forces to ratio of speeds and hence differential times to cross *equal* circles. But, at the instant of impact, the same force and determination relations are attributed to the tennis ball and the light-ray, and hence Descartes' overt presentation of an idealized, kinematical situation is ultimately deeply misleading as to his actual physics and its principles.[5]

In sum, we see that in both the tennis-ball model and our reconstruction of Descartes' actual theory of light, the two conditions, [a] and [b], allowing derivation of the law of refraction, are actually dynamical premisses—physico-mathematical

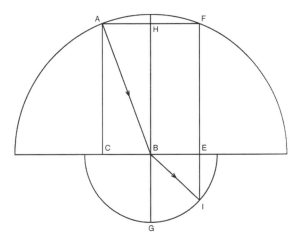

Fig. 3.4. Refraction of light using Descartes' dynamics and real theory of light.

principles—for a mechanical theory of light: [1] the absolute quantity of the force of motion of a light-ray (or its tennis-ball model) is conserved within any given medium, as is the ratio of those quantities of force of motion in refraction phenomena; [2] the component of the force of motion parallel to the refracting surface (parallel determination) of a light-ray (or its tennis-ball model) is unaffected by encountering the interface and is conserved in the second medium. These two dynamical premises of a theory of light may have permitted (from Descartes' perspective) a plausible deduction of the law of refraction, but they generated difficulties. Descartes' account in the *Dioptrique* has problems as soon as he discusses space-filling media, or refraction toward the normal, and more generally with the question of how it happens that the alteration in the normal determination is variable, depending upon the angle of incidence (Schuster, 2000, 270–1). Moreover, the first dynamical assumption—path-independent ratio of the force of light—seems to entail that optical media are isotropic, whilst the second dynamical assumption—conservation of the parallel determination—seems to entail that they are not (Schuster, 2000, 267–70).

All this suggests that Descartes did not obtain his two dynamical premises through a deep inquiry into the conceptual and empirical requirements of a mechanical theory of the propagation and refraction of light in actual media. It seems more plausible to associate the premises closely with the very geometry of the diagrams in which Descartes depicts and constructs the paths of refracted rays, once we understand the underlying dynamical rational of his proof. The question is: Were the dynamical premises *post facto* glosses of geometrical constructions arrived at in some other way? Although Descartes dabbled with the physico-mathematics of refraction of light as early as 1620, recent work shows he discovered the law of refraction independently of any mechanical assumptions and through a process entirely within the bounds of a traditional mixed-mathematics approach to optics. Then, the geometrical diagrams expressing his newly found law suggested to him (in physico-mathematical manner) the precise form and content of his two dynamical premises for a mechanistic theory of light and their mode of relation in explaining refraction.

In Paris in 1626/27 Descartes, collaborating with the mathematician Claude Mydorge, discovered the law of refraction of light. This was accomplished independently of, but in the same manner, as Thomas Harriot, who had first discovered the law around 1598, using only traditional geometrical optics (Fig. 3.5).

Harriot used the traditional image-locating rule to map the image locations of point sources taken on the lower circumference of a half-submerged disk refractometer.[6] (Buchdahl, 1972; Lohne, 1963) This yielded a smaller semi-circle as the locus of image points, and hence a cosecant law of refraction of light. In a letter, the content of which dates from 1626/27, describing an identical cosecant form of the law, Mydorge presented a virtually identical diagram (Fig. 3.6), but flipped the inner semi-circle above the interface as a locus of point sources for the incident light (Mersenne I 404–15; Schuster, 2000, 272–5).

Figure 3.6 closely resembles Fig. 3.4, the derivation of the law of refraction using Descartes' conditions from the *Dioptrique* and his theory of light as instantaneously propagated tendency to motion. It is the key to unpacking the co-evolution of

CARTESIAN PHYSICS 67

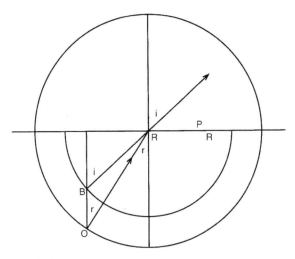

Fig. 3.5. Thomas Harriot's key diagram.

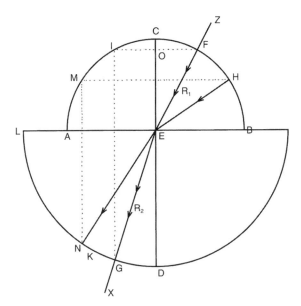

Fig. 3.6. Mydorge's (and Descartes') cosecant form of the law of refraction.

Descartes' theory of light and his dynamics of corpuscles (Schuster, 2000, 275–277). After his discovery of the law of refraction by these purely geometrical optical means, issuing in the cosecant form of the law (Fig. 3.6), Descartes sought to explain the law using a dynamics of corpuscles. Working in the style of his physico-mathematics, he transcribed into dynamical terms some of the geometrical parameters embodied in the cosecant representation. The resulting dynamical principles concerning the mechanical nature of light were, as we have seen: [1] the absolute quantity of the force of the ray is increased or decreased in a fixed proportion, whilst, [2] the parallel component (parallel determination) of the force of a light ray is unaffected by refraction. By October 1628 Descartes used these concepts to explain the law of refraction to his old friend Beeckman. Descartes revealed a remarkable analogy between the causes of the refraction of light and the behaviour of a bent arm balance beam whose arms are immersed in media of differing specific gravity. This analogy expresses precisely the two dynamical principles later used behind the *Dioptrique* proof of the law of refraction (AT X 336; Schuster, 2000, 290–295).

Hence, Descartes' physico-mathematical trajectory explains the puzzle of why he was so focused on keeping the two dynamical premises, despite their dubiousness. It was not just because they allowed 'deduction' of the law of refraction. It was also because, physico-mathematically, they had come from the well-grounded, mixed-mathematical law. For Descartes this physico-mathematical interpretation of the cosecant law-of-refraction diagram arguably became the paradigmatic case of deriving natural-philosophical causes from a well founded and geometrically clearly represented mixed-mathematical result. Furthermore, his two mechanistic conditions for a theory of refraction in turn suggested the two central tenets of his mature dynamics as he composed *Le Monde* (1629–33). The first rule of nature in *Le Monde* asserts the conservation of the quantity of the instantaneously exerted force of motion of a body in the absence of external causes (AT XI 38). This rule subsumed and generalized [1]. The third rule of nature defines what we previously called the principal determination of the instantaneously exerted force of motion of a body, along the tangent to the path of motion at the instant under consideration. In the absence of external constraint, the principal determination would be conserved from instant to instant. Thus, this rule subsumed and generalized [2]. For Descartes, the basic laws of light—itself an instantaneously transmitted mechanical impulse—immediately revealed the principles of the instant-to-instant dynamics of corpuscles (Schuster, 2000, 302–3; 2013, 204–9).

In sum, then, when assessing the physics of Descartes from a developmental perspective, it is crucial to see that the principles of the dynamics of corpuscles governing that system arose from his physico-mathematical project. Epochal results in traditional geometrical optics allowed insight into the realm of causes in a mechanistic theory of light, and were then extrapolated as principles of the emergent corpuscular-mechanical system.

3.2.3 Descartes' Career Inflection, 1628–33

In describing Descartes' elaboration in *Le Monde* of his corpuscular dynamics out of the results of his physico-mathematical optics, we have skipped over a crucial inflection point in his career, in the period 1628–33. Descartes stopped being a practitioner of piecemeal physico-mathematics with corpuscular-mechanical leanings, and became the designer of a systematic corpuscular-mechanical natural philosophy, which enjoyed certain theological and metaphysical groundings and displayed on the technical level residual physico-mathematical genes. Why and how did this happen?

The failure of the dream of universal method

During 1618–28 Descartes not only had physico-mathematical hydrostatics, optics, and piecemeal corpuscular mechanism on his mind, but he also in 1618–20 had hit on a series of grand intellectual projects. In quick succession he envisioned two breathtaking projects reaching beyond physico-mathematics: universal mathematics and universal method. First, he imagined his universal mathematics encapsulating and transcending 'mere' physico-mathematics. Then, in a peak of excitement later in 1619, around the time of his famous dreams on St Martin's Eve, he envisioned his universal method, which was to absorb universal mathematics and move on much further. Traces of the universal mathematics project are embedded in the text of his *Rules for the direction of the mind*, the so-called rule 4B,[7] whilst an early statement of the method was inscribed in a text termed rule 4A which Descartes inserted into the *Rules*, along with most of the texts of rules 1–3 and 5–11, with some small exceptions (Schuster, 1980, 54). The young Descartes, therefore, was not just struggling to work out a physico-mathematics with possible corpuscular-mechanical bearings. He was also a dreamer of gigantic and seductive methodological fancies (Schuster, 1986).[8]

These endeavours came to a critical impasse in the later 1620s. After his optical breakthrough, and working partly in the shadow of Marin Mersenne's cultural battle against both radical scepticism and religiously heterodox natural philosophies, Descartes picked up universal mathematics and method again in detail, and tried to write a unified treatise about his earlier dream of a methodologically sound universal mathematics, the unfinished *Rules for the direction of the mind* (Schuster, 1980, 58–69). Systematic natural philosophy had no part in this vision. Natural philosophy, reduced to the content of solid but piecemeal physico-mathematical results—as in his physico-mathematical hydrostatics or optics—would become just one pursuit within the total range of application of this universal mathematics run by a master method. Unfortunately, the renewed *Regulae* project of the late 1620s did not blossom into the intended magisterial work of method and universal mathematics. It collapsed under its own weight of self-generating mathematical, epistemological, and ontological problems and contradictions—an outcome that shaped Descartes' next, decisive career moves (Schuster, 1980, 73–80; 2013, 334–46).

Between late 1629 and 1633 Descartes was engaged in constructing his first system of natural philosophy, *Le Monde*. At the same time, he was devising two doctrines meant to buttress parts of that natural philosophical system: [1] He worked on the first skeletal lines of his dualist metaphysics, with its strict ontological distinction between mind and matter, the latter construed as a completely space filling matter-extension; and [2] he took up and revised some elements of a voluntarist and creationist theology. He had never previously engaged in projects of this type. Work on the *Regulae* stopped. The method, the universal mathematics, and the physico-mathematics supposedly contained within the *Regulae* were never again further articulated. Nor did

Descartes ever again represent himself simply as a 'physico-mathematician' or practise the style of piecemeal, problem-oriented physico-mathematics that had marked his earlier years. However, continuities persisted. The Descartes of *Le Monde* in 1633 was not the Descartes of the *Meditations* and *Principles of Philosophy* of the mid-1640s; yet, that later Descartes was indeed a systematic philosopher of nature and a metaphysician—just what Descartes became between 1629 and 1633.[9] Similarly, the inflection of Descartes' agenda and identity between 1629 and 1633 had not completely erased his earlier practices or products. They lived on, sifted, revised, and retranslated. His new systematic natural philosophy bore definite marks of his aims and results in physico-mathematics, and his physico-mathematical optics and the principles of the dynamics of corpuscles it suggested and supported became central pillars of his system of mechanical philosophy. We therefore need to look at the work achieved in grounding Descartes' natural philosophical system by his voluntarist/creationist theology and his dualist metaphysical doctrine of matter extension.

Voluntarist/creationist theology and the foundations of Le Monde

The work on voluntarist theology was initiated in 1630, in parallel with the natural philosophical moves leading directly toward *Le Monde*. It had a legitimatory role in the newly emergent natural philosophy, especially in regard to Descartes' new laws of corpuscular dynamics. Starting from the proposition that creation utterly depends upon God's will, Descartes elaborated a doctrine of continuous creation, or at least a doctrine of the necessity of God's continuous sustenance of creation from instant to instant. He stresses the necessity of God's acting at each moment both to conserve the existence of natural bodies and their modes (including force of motion and rest), and to enforce the rules according to which natural change occurs. At the moment of a corpuscular impact, God instantaneously adjusts the quantities of force of motion and the determinations that will characterize the corpuscles concerned in the instant after the impact. God does this by following certain laws and rules of impact that He has framed and 'ordinarily' follows. Consider Descartes' first 'rule of nature' in *Le Monde*:

Each part of matter always continues to exist in the same state as long as other bodies do not constrain it to change that state. If it has a certain size, it will never become smaller, unless other bodies divide it . . . if a body has stopped in a given place, it will never leave that place unless others force it out; and if it has once commenced to move, it will continue along with the same force, until other bodies stop or retard it (AT XI 38).

While this seems to assert the conservation of the motion (or rest) of a body in the absence of external constraints, closer inspection reveals that Descartes is speaking of the instantaneously exerted 'force of motion'. Consistent with our analysis of the development of his optics and his principles of a dynamics of corpuscles, this is the quantity which is conserved. Here in *Le Monde* and later in the *Principles* Descartes is using the term in relation to his voluntarist—or, more properly, his creationist—understanding of ontology:[10] God must continually support (or re-create) bodies and their attributes from moment to moment. This implies that in the final analysis a body in phenomenal translation, in motion, is really being recreated or continually supported at successive spatial points during successive temporal instants. In addition, in each of those instants of re-creation, it is characterized by the divine injection of a certain quantity of 'force of motion'. Similar reasoning, framed for presentation by theological concerns, arrives at the second law in *Le Monde*: the conservation of the total quantity of force of motion in collisions between bodies. At the instant of collision God conserves the total quantity of force of motion and redistributes it amongst the bodies involved.[11]

As regards the third law of nature in *Le Monde*, we have seen that it defines what we have called the principal determination of the instantaneously exerted force of motion of a body, along the tangent to the path of motion at the instant under consideration. Viewed in theological terms, it says that in the absence of external constraint, this particular directional quantity of force of motion would be conserved by God from instant to instant. The instantaneous determination of motion is rectilinear, *because* only a straight line can be grasped entirely in an instant without God having to calculate or observe the path of the body at one or more other instants, past or future. A straight line can be defined in any instant of the body's motion through God's consideration of its present position and the implicit endpoint of the (straight) line along which one would point in saying that 'At this instant the body is in the act of moving in *that* direction'. By contrast, to conserve a curved determination God would have to recalculate the determination at each instant based on memory or prediction of one other point. Thus God would not conserve the body, as Descartes says, 'precisely as it is at the same instant that He conserves it' (AT XI 44; G 30), but rather in a manner also dependent upon consideration of its past or future path.

In sum, since the corpuscular-mechanical universe of *Le Monde* was to be stripped of all Aristotelian forms, qualities, and teleological processes, impact and pressure became the sole causes of natural change. Such a universe of perfectly hard particles foreclosed consideration of the problematical quality of 'elasticity'; but, by the same token, in such a universe impact and pressure were rendered inexplicable by natural causes. At just this point the recourse to theology became necessary in order to provide a rationale for the laws of nature, explicating how bodies interact at the mechanically opaque instant of collision. The answer on this legitimatory level is that material bodies do not in fact interact physically and causally with each other. They appear to do so only on a phenomenal level, this being the expression or effect of the rule-bound ways in which God from moment to moment maintains or alters their forces and determinations of motion.

Descartes did not proceed in *Le Monde* to deal with these issues at a further level of philosophical or theological analysis. He was primarily interested in the reduction of the analysis of phenomenal translation to the consideration of divinely governed instants of time. This procedure was critically important to his enterprise of justifying the principles of his dynamics (which we know had always *in practice* been focused on the dynamical status of corpuscles in motion or tending to motion in terms of their instantaneous quantities and determinations of force). Descartes' discussion of the three laws of nature thus aimed to impart theological backing to his species of dynamics, and to this end he selectively employed resources of voluntarist theology articulated with special attention to the punctiform character of God's conserving concourse.

Of course, for the philosophically scrupulous the conceptual entanglements of Descartes' voluntarist grounding for natural philosophy are obvious, and began to show up in Descartes' thinking later with his most important published works, the *Meditations* and especially the *Principles of Philosophy*. In the latter work he articulated his voluntarist, punctiform dynamics of particles by reframing it within a more consistently deployed neo-Scholastic terminology of substance, mode, and cause. This prompted debate about the correct Cartesian interpretations of these categories amongst his later seventeenth-century followers and detractors, and still promotes discussion amongst leading historians of philosophy (Machamer and McGuire, 2009; Schmaltz, 2008). Nevertheless, for the purposes of the history of physics, and the remainder of our discussion in this chapter, a possible way to comprehend how the voluntarist corpuscular dynamics was meant to work was provided by Gueroult (1980). It suggests a number of distinctions that can be applied to Descartes' voluntarist corpuscular dynamics in both *Le Monde* and in its scholastically articulated form in the *Principles*. One should distinguish between [1] instantaneously exerted force of motion (causal), which is identified with God's own causal action and is discussed in the language of theology and metaphysics; this is equated with [2] an instantaneous force of motion (modal), which is the manifestation of that divine action in the material world, potentially observable in some of its effects, and taken by humans on the level of technical natural philosophical discourse of dynamics as a moment-to-moment possession of a body moving or tending to motion. Finally, [3] our common-sense notions of motion in space and time, unenlightened by either theology or natural philosophy, are seen to be merely appearances caused by God's law-like moment-to-moment causal actions upon matter.

Dualist metaphysics and the grounding role of matter-extension

Descartes' project in dualist metaphysics differed from that in theology, in that he began work on it before the events of mid and late 1629 which crystallized into the project of *Le Monde*.[12] The dualism of mind and body was initially aimed at addressing problems emergent in the *Regulae*; but it soon came to play an important grounding role behind *Le Monde*, as that text next began to take shape. The essential matter theoretical finding of his metaphysics is that the notion of a void space is unclear and that a conceptual analysis of our ideas shows that every extended space is filled with matter—indeed, *is* matter. The impossibility of any void spaces means that if a particle is to move, the 'space' it is about to vacate must be filled simultaneously by another particle of equivalent volume. This further implies that any motion at all entrains an instantaneous circuit of displacement, leading to the filling of the about-to-be-voided space. Vortices became imaginable, and in turn invited detailed dynamical description. So, the vortex mechanics, central to the entire content and structure of *Le Monde* and the *Principles*, was elicited from the matter-extension plenum which itself followed from the initial metaphysical work. Descartes might well have thought of this as an admirably fruitful course of intellectual discovery, from the metaphysics down to the intricacies of the vortex mechanics. The emerging sophisticated vortex celestial mechanics in turn became the core exemplar for working out wide swathes of the natural philosophy—not just the celestial mechanics of planets, but of comets as well, plus the theory of light in cosmic setting, the theories of local (planetary) gravity, and tides, and the behaviour of planetary satellites. It is true that Descartes' embryonic dualist metaphysics eventually resided not so much inside the text of *Le Monde*, as immediately behind it. None of the detailed metaphysical argument presented later in the *Discours*, *Meditations*, and *Principia* appear in *Le Monde*, and presumably the arguments were in only preliminary and very basic form. However, his metaphysically grounded commitment to a theory of matter-extension and a resulting plenist universe is essential to *Le Monde*, as it explicitly was later to the *Principles*.

3.2.4 Vortex Mechanics

The vortex celestial mechanics, as presented in the *Principles of Philosophy* and *Le Monde*, are the 'engine room' of Descartes' system of natural philosophy. As such, the vortex mechanics exemplifies what we may term a 'plenist–realist' style of explanation which permeates the entirety of Descartes' natural-philosophical project (Schuster, 2013, 373–84). This style prohibits mathematical abstraction or idealization of the sort found in the traditional mixed mathematical sciences, and in the mechanics of Galileo which was to appear in 1638. Instead, Descartes' plenist realism favours explanations which, arguably, are inclusive or holistic about the factors taken into account, and in doing so supposedly reflect the 'real complexity' of phenomena in the plenist universe, and the 'real set' of causes in play, not some 'abstract' or 'fictitious' picture. Natural philosophical explanations thus need immediately and completely to grasp the tangle of causes and conditions in play behind any phenomenon in this plenum universe. Explanations must not abstract away from some or most of these causes, issuing in over-simplified (strictly not real) models of phenomena under study. This, however, did not mean that physico-mathematical procedures or results were banned from *Le Monde* and the *Principia*. We have already seen this in Descartes' reliance on the belief that his physico-mathematical optics had cut to the core of that plenist reality, revealing, as far as he was concerned, the underlying dynamics of corpuscles that runs the cosmos.

Descartes starts his vortex theory with an 'indefinitely' large chunk of divinely created matter-extension in which there are no void spaces whatsoever. When God injects motion into this matter-extension, it is shattered into micro-particles and myriads of 'circular' displacements ensue, forming large numbers of gigantic whirlpools or vortices. This process eventually produces three species of corpuscle, or elements, along with the birth of stars and planets. The large, irregularly shaped particles of third element form all solid and liquid bodies on all planets throughout the cosmos, including the Earth. Interspersed in the pores of such planetary bodies are the spherical particles (*boules*) of second element. The second element also makes up the bulk of every vortex, while the spaces between these spherical particles are filled by the first element, which also constitutes the stars, including our Sun. Next, using his theory of the dynamics of corpuscles, Descartes introduces a vortex stability principle. In the early stages of vortex formation, before stars and elements have evolved, the then existing vortical particles become arranged so that their centrifugal tendency increases continuously with distance from the centre (AT XI 50–1; G 33). As each vortex settles out of the original chaos, the larger corpuscles are harder to move, resulting in the smaller ones acquiring higher speeds. Hence, in these early stages, the size of particles decreases and their speed increases from the centre out. But the speed of the particles increases proportionately faster, so that force of motion (size times speed) increases continuously. Fig. 3.7 shows the distribution of size and speed of the particles in any vortex before a central star and the three elements have formed[13] (Schuster, 2005, 46).

It is absolutely crucial to notice that according to Descartes the subsequent advent of a central star—a correlate of the formation of the three final elements—permanently and definitively alters the original size and speed distribution of particles in a vortex. It is the star's disturbing effect on the original size/speed distribution that allows the planets to maintain stable orbits and which also creates the limits of the trajectories of comets. A star is made up of the most agitated particles of first element. Their agitation, and the rotation of the star, communicate extra motion to *boules* of

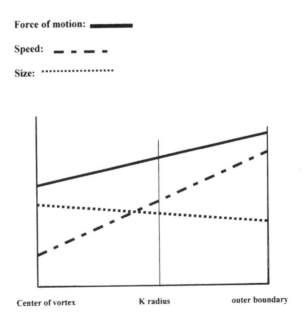

Fig. 3.7. Size, speed, and force of motion distribution of particles of second element, prior to the existence of a central star.

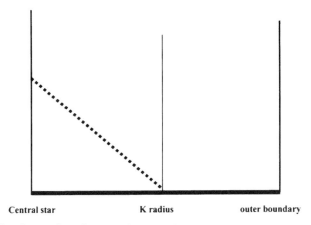

Fig. 3.8. Agitation due to the existence of a central star.

the vortex near the star's surface. This increment of agitation decreases with distance from the star and vanishes at that key radial distance, called K (Fig. 3.8).

This stellar effect alters the original size and speed distribution of the spheres of second element in the vortex, below the K layer. We now have greater corpuscular speeds close to the star than in the pre-star situation. But the vortical stability principle still holds, so the overall size/speed distribution must change, below the K layer. Descartes ends with the situation in Fig. 3.9, with a crucial inflection point at K. Beyond K we have the old (pre-star formation) stable pattern of size/speed distribution, and below K we have a new (post-star-formation) stable pattern of size/speed distribution (Schuster, 2000, 49). This new distribution turns a vortex into a machine that locks planets into appropriate orbits below K and extrudes comets into neighbouring vortices. Celestial vortices behave this way only because a star, made of the first element, happens to inhabit the centre of each vortex, transforming its mechanical parameters and performance. This is Descartes' version of Kepler's emphasis

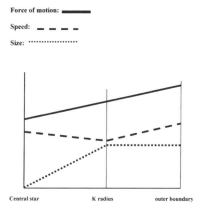

Fig. 3.9. Size, speed, and force of motion distribution of particles of second element in a stellar vortex.

(compared to Copernicus himself) on the physical–causal role of the Sun in orbital mechanics.

The key question in Cartesian celestial mechanics thus becomes this: 'When and why is the centrifugal tendency of orbiting bodies actualized as centrifugal motion, and when and why does that not happen?' The answer has two dimensions. First of all, and unsurprisingly, Descartes' sling exemplar for his dynamics plays a key role. In the vortex, what constrains a planet into a curved path, on analogy to the sling, are the particles of second element, neighbouring it and superjacent to it (Schuster, 2005, 50–52). Secondly, as to why different planets maintain orbits at different distances from the central star, but all within the K radius, the key is Descartes' concept of the 'massiveness' or 'solidity' of a planet. By this he means its aggregate volume-to-surface ratio, which is indicative of its ability to retain acquired motion or to resist the impact of other bodies (Schuster, 2013, 464–67).

The resultant explanation runs as follows. The *boules* of second element making up a vortex also vary in volume to surface ratio, or massiveness, with distance from the central star. This may be gathered from Descartes' stipulations concerning the variation of the size of the *boules* with distance from the central star, illustrated in Fig. 3.9. Fig. 3.10 then represents what we may term the 'resistance curve'—that is, the variation in massiveness of second-element *boules* with distance from a star. A planet is locked into an orbit at a radial distance where a double balance is achieved amongst the relevant vortical mechanical parameters: On the one hand the centrifugal tendency of the planet, a function of its aggregate solidity, is balanced by the resistance to being displaced downward of the second-element *boules* composing the vortex in the vicinity of the planet—that resistance similarly depending on the massiveness or surface-to-volume ratio of the those particular *boules*. On the other hand, a balance is realized simultaneously between the centrifugal force of the subjacent second element *boules* at that radius in the vortex and the resistance to being displaced downward offered by the planet (owing to its degree of massiveness). Hence the condition for a piece of third matter to be in stable orbit in the vortex can expressed as $F^m{}_b = R_{mu}$ and $F^m{}_{ml} = R_b$, where $F^m{}_b$ denotes the force of motion of the orbiting body, R_{mu} denotes the resistance of superjacent layer of *boules* (upper medium) to being

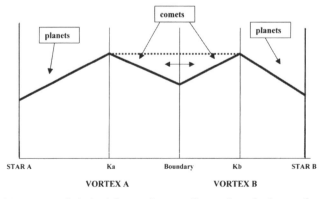

Fig. 3.10. 'Resistance curve', derived from volume:surface ratios of spheres of second element.

extruded downward by the orbiting body, $\mathbf{F^m_{ml}}$ denotes the force of motion of subjacent layer of *boules* (lower medium), and $\mathbf{R_b}$ denotes the resistance of the orbiting planet to being extruded downward by subjacent layer of *boules*, where both $\mathbf{F^m_b}$ and $\mathbf{R_b}$ are functions of the planet's solidity, while $\mathbf{R_{mu}}$ and $\mathbf{F^m_{ml}}$ are functions of the solidity of the superjacent and subjacent boules respectively.

The most 'massive' or 'solid' planet in a star system will be closest to, but not beyond, the K layer—as Saturn is in our planetary system. Comets are planets of such high solidity that they overcome the resistance of *boules* at all distances up to and including K. Such an object will pass beyond the K level, where it will meet *boules* with decreasing volume-to-surface ratios, hence less resistance, and be extruded out of the vortex into a neighbouring one. But, flung into the neighbouring vortex, the comet meets increasing resistance from its *boules* above that vortex's K distance. Picking up increments of orbital speed, the comet starts to generate centrifugal tendency again, eventually being flung back out of the second vortex.

To summarize, then, each vortex is a locking and extrusion device. Given Descartes' theory of the varying solidities of planets and comets, the variation of size and speed of vortical boules of second element with distance from the central star entails that planets are locked into orbits of differing radii. Comets are objects extruded from vortex to vortex, first 'falling' into a vortex and being extruded out.[14] The make-up and dynamical behaviour of the central stars are crucial, not to the bare existence of vortices, but to the creation of planet-locking/comet-extruding vortices. Otherwise, extrusion would be the universal rule. Multiple vortices are conceptually necessary, as each vortex is set in a container made of contiguous vortices, exerting a kind of centripetal backwash at its boundary.

Properly understood, Descartes' vortex mechanics reveals itself as a science of equilibrium, thus underscoring its sources in statical and hydrostatical exemplars, heavily refracted through the *sui generis* style of Descartes' physico-mathematics (Schuster, 2013, 475–77). The forces at work upon a planet can be fully specified only when orbital equilibrium has been attained, though of course, no actual measurements are involved. The rise or fall of a planet (or comet) results from the breakdown of equilibrium and cannot be defined mathematically. Hence celestial equilibrium and disequilibrium are analogous respectively to equilibrium and slippage in systems rigorously treated in the statics of Archimedes or Simon Stevin. However, Descartes' vortex celestial mechanics differs from classical statics. It deals with equilibrium conditions described not in terms of volumes, densities, and specific weights, but rather in terms of Cartesian centrifugal tendency to motion, as well as relative 'solidity'—a property defined by volume-to-surface relations aggregated over the constituent third-matter corpuscles of a planet or comet.[15]

Descartes' vortex celestial mechanics is therefore intimately related to his youthful programme of physico-mathematics. Descartes holds that at the descriptive level, in terms of appearances, orbital establishment and placement are 'statical phenomena'. We saw that in his hydrostatics work of 1619 he insisted that a macroscopic, statical regularity—the hydrostatic paradox—be reduced to corpuscular-mechanical terms. Similarly, in his mature celestial mechanics he insists that beneath the observable radial equilibrium of orbits there resides a corpuscular-mechanical reality, a micro machinery, the behaviour of which is explained by his new dynamical concepts. As the corpuscular-mechanical and dynamical reduction of the hydrostatic paradox was an exercise in physico-mathematics, so too is the vortex mechanics. Taken as a whole, therefore, vortex celestial mechanics is a hybrid entity: part physico-mathematical certainly, but also clearly a piece of generic natural-philosophical discourse, playing the central role in this new corpuscular-mechanical system.

3.2.5 Explanatory Style of Mature Cartesian Physics

It is a mistake to take seriously Descartes' occasional claims to have been able to deduce—as if according to a mathematical ideal of 'demonstration'—his entire system of natural philosophy from absolutely certain metaphysical principles. This folklore arose from the strictly deductivist tone of Descartes' method in both his formal and more off-hand statements about it. In his mature work, after the demise of his detailed project on method in the *Regulae* in 1628, Descartes increasingly came to see that neither the details of particular explanatory models, nor the facts to be explained, could be deduced from metaphysics. *Le Monde*, with its tacit metaphysical grounding, began a process of reflection on the demands of corpuscular-mechanical explanation which is fully expressed at the end of the *Principles* (AT VIII1, 325–9; M 285–8). Descartes tells the reader that when one constructs corpuscular-mechanical models to explain various phenomena, such as gravity, light, magnetism, sunspots, planetary motion, sensory perception, and animal locomotion, nothing should be asserted in any particular model that contradicts any metaphysically derived certainties (concerning the existence of a material plenum, the laws of motion, and presumably the existence of the three elements). Hence, detailed corpuscular-mechanical explanatory models have a necessarily hypothetical character and can achieve at best only 'moral certainty' (Clarke 1977; Buchdahl 1970, 97, 118–26; Sabra 1967, 21–45).

This is not simply a meta-commentary by Descartes on his physics. It reflects his intellectual and discursive processes in forming explanations, since, to reiterate, he had no 'method' for inscribing such models (Schuster, 1986, 62–6). It also highlights the crucial role of factual evidence in Descartes' mature physics. To design any particular middle-level explanatory model, Descartes would have to select, weigh, and appropriately deploy hard empirical evidence, which could include facts needing explanation, or facts lending credibility to the explanatory model offered. In the light of the 'evidence', and the metaphysical 'constraints' on hypothesis-formation, a specific corpuscular-mechanical model for the phenomena would be constructed. Descartes' mature views on the central role of facts and evidence in the framing of explanations in natural philosophy helped inspire the probabilist and hypothetical–deductivist approach to explanation with a broad and loose commitment to a corpuscular-mechanical natural philosophy held widely in the later seventeenth century, at least until Newton began his campaign to install a methodological rhetoric of inductive certainty.

3.2.6 Descartes' Mature Physics: Pursuing Novel Facts and System-Binding Strategies

While *Le Monde* and *Principles of Philosophy* both contain Descartes' vortex mechanics, his mechanistic theory of light in a cosmic setting and his theories of gravity, tides, and satellite motion, *Principles* is much more elaborate on the laws of collision and contains the explicit metaphysical grounding in dualism. However, considered as systems of physics, the two treatises differ in two other ways which are often overlooked. First, the *Principles* deals with a much richer harvest of significant or novel matters of fact than one finds in *Le Monde*. Descartes' pays sustained and detailed attention to the phenomena of magnetism, sunspots, novae, and variable stars. He co-opts and rewrites into mechanistic terms the experimental archive of William Gilbert's *De Magnete* (1600), which he ignored in *Le Monde*. He takes on board the existence and known properties of sunspots, a topic he avoided in *Le Monde*, and he deals extensively with variable stars, only recently discovered, and relates them theoretically to novae which, also, did not appear in *Le Monde*. Secondly, the *Principia* deals with these series of matters of fact in a strategic way entirely absent from *Le Monde*—indeed, the *Principia* contains new system-binding strategies that virtually define the character of Descartes' mature physics.

In the *Principles* Descartes weaves his ranges of newly available, or newly chosen, matters of fact into explanatory and descriptive narratives which have a definite theme and aim: they are *cosmographical* in nature; that is, they have to do with how the relations between heavens and earth are theorized in a natural philosophy (Biro, 2009). Cosmography was an actor's category, used initially in geocentric natural philosophies such as Aristotelianism, where the 'relation' heavens/Earth was not identity of matter and cause, but rather, considerable difference. However, for Descartes and other realist Copernicans, the Earth was a heavenly body and the traditional heavenly bodies were arguably 'like' the Earth. That meant that cosmography became a contested space of great natural philosophical significance. The relation of 'the Earth' to everything else—that is, 'the heavens'—changed. It became the relation of any and all planets, their structures and geneses, to any and all stars, their nature and developmental patterns. Claims about the structure of the Earth could now be exploited cosmographically, for realist Copernican ends. Gilbert had done this in his study of terrestrial magnetism, Galileo with his theory of the tides. Following this strategic lead, Descartes designed the final two Books of the *Principles* as an interrelated set of such radical, realist-Copernican cosmographical threads of narrative-cum-explanation. In these respects it is nothing like *Le Monde*. Descartes' cosmographical strategy has a foundation and a point of leverage: the foundation is his co-optation of Gilbert's claimed facts of magnetism linked to a denial of Gilbert's natural philosophical concepts and aims; the fulcrum is his strategy for handling sunspots (Schuster and Brody, 2013).

William Gilbert's *De Magnete* (1600) was an impressive natural philosophy, grounded in experiments interpreting magnetism as an immaterial power. In his *Principles*, Descartes accepted Gilbert's experiments but explained magnetism mechanistically, based on the movements of two species, right- and left-handed, of 'channelled' or cylindrical screw-shaped particles of his first element. He claimed that magnetic bodies—naturally occurring lodestone, or magnetized iron or steel—have two sets of pores running axially between their magnetic poles—one set accepting only right-handed channelled particles, with the other, directed oppositely, receiving only the left-handed particles. Descartes re-explained Gilbert's experiments, including his use of a sphere of loadstone to demonstrate the properties of magnetized compass needles (AT VIII1 275–311; M 242–72).

However, he did more than appropriate and reinterpret Gilbert's 'laboratory' work. Gilbert called his sphere of lodestone a *terrella*, a 'little Earth', arguing that because compass needles behave identically on the *terrella* as on the Earth itself, the Earth is, essentially, a magnet. Hence, according to his natural philosophy, the Earth possesses a magnetic soul, capable of causing it to spin. Magnetic souls similarly cause the motions of other heavenly bodies. In his *Principles*, Descartes, aiming to displace Gilbert's natural philosophy, focuses on the 'cosmic' genesis and function of his channelled magnetic particles. He argues that the spaces between the spherical corpuscles of second element, making up the vortices, are roughly triangular, so that some particles of the first element, constantly being forced through the interstices of second-element spheres, are forged into relatively stable, longer, cylindrical forms, which are 'channelled' or 'grooved' with triangular cross-sections. All vortical interstitial first-element corpuscles, including these bigger, longer channelled ones, tend to be flung by centrifugal tendency out of the equatorial regions of vortices and into neighbouring vortices along the north and south directions of their axes of rotation, where the large ones receive definitive opposite axial twists. The resultant left- and right-handed screw-shaped first-element particles penetrate into the polar regions of central stars and then bubble up toward their surfaces and drift toward the equator, forming by accretion, Descartes claims, sunspots (AT VIII1 142–8; M 132–6).

Two sets of consequences of the explanatory story so far are crucial to understanding Descartes' strategy in the *Principles*. Firstly, the sunspots formed on a star's surface out of accretions of the curious rimmed and handed first-element particles are asserted to have become third matter. That is, they consist of large corpuscles, irregularly shaped and difficult to set in motion, which constitute the bulk of the matter in planets, comets, and satellites in the cosmos, as we have seen in discussing the vortex celestial mechanics. In the matter theory of *Le Monde*, third-matter corpuscles are said to exist from the moment of creation (AT XI 56–7; G 37), and in that treatise Descartes several times denies the possibility of any transformation of the elements (AT XI 28, 29–30; G 19, 20). But here in the *Principia*, third matter does not exist from the creation, and comes into existence only by means of the cosmic process of sunspot formation. All terrestrial matter is formed on stars out of the stuff of magnetism (Schuster and Brody, 2013).

Secondly, it also follows from Descartes' explanatory narrative that all stars are magnetic, as Gilbert maintained, but in a mechanistic sense, because they are all suited to reception of these oppositely handed, polar-entrant first-element particles. Moreover, for Descartes, planets are also magnetic, as Gilbert claimed, but again the explanation is mechanical. Descartes describes how a star may become totally encrusted by sunspots. This extinguishes the star, its vortex collapses, and it is drawn into a neighbouring vortex to orbit its central star as a planet (AT VIII1 195–96; M 171). This is the only way by which planets are formed, so all planets, including our Earth, bear the magnetic imprint of their stellar origins, possessing axial channels between their magnetic poles accommodated to the right- or left-handed screw particles.

As for sunspots, once Descartes has theoretically reconstituted them on the surfaces of stars in terms of his corpuscular-mechanism and theory of vortices and magnetism, he rederives in terms of his theory their consensually accepted properties.[16] Then, in the pivotal move in this entire explanatory campaign, he uses this theory of sunspots further to explain novae and variable stars and, as just mentioned, the origin and nature of all planets, the Earth included. Variable stars had been discovered only in the late 1630s, after *Le Monde*, and just before Descartes began writing the *Principles*. Alternate creation and destruction of complete sunspot crusts on a star explain its variability. Novae—accepted facts amongst European astronomers since the late sixteenth century—he explains as a sub-class of variables: a nova is a star which has been in an occluded phase and never before observed by humans, which then comes into view for the first time, as far as humans are concerned. Subsequently, it might continue to shine, or quickly or slowly become occluded again (AT VIII1 158–62; M 144–7; Schuster and Brody, 2013).

Descartes' next move, expressing and completing the cosmographical intentions of his system, involves relating the Earth—and, indeed, every single planet in the universe—to a certain pattern of possible stellar development. As indicated above, occasionally, when a star becomes covered with sunspots its vortex collapses and the defunct star is captured by another vortex, becoming a comet or a planet. Here we encounter what is usually termed Descartes' 'Theory of the Earth'. He famously explains, in the first forty-five articles of Book IV ('Of the Earth') of the *Principia*, how from a dead, encrusted star there results the formation of land-masses, with mountains and declivities, the latter filled with water to form oceans subject to tides (another crucial cosmographical phenomenon for Descartes, as for Galileo). But clearly all planets undergo the same processes. The dynamics of spots encrusting and eventually destroying stars is what accounts in matter theoretical and structural terms for each and every planet found in the universe. So, on this breathtaking vision, every planetary object in the cosmos traces its genealogy to a pattern of events that in principle might befall any 'star-in-a-vortex-afflicted-with-sunspots'. This indeed is a grand cosmographical gambit, meant systematically to bind together his system and to establish his brand of realist Copernicanism of innumerable star and planetary systems, all worked by his corpuscular dynamics and vortex mechanics.

Inside the toils of his radical realist Copernican cosmographical explanations *cum* narratives, Descartes did not aim at linear, deductive explanations of each and every particular state of affairs he recognized as a reliably reported matter of fact. Descartes' laws of nature do not function as premises of deductive explanations. Rather, his laws of nature in the *Principia* function as human laws function in the making of legal arguments. The laws are woven, along with carefully selected matters of fact, into flows of argument, narrative lines of description–explanation, of the sort we have just canvassed.[17] Descartes proceeds by asserting a network of basic explanatory concepts involving matter theory, magnetism, vortices, and sunspot formation/dissipation that in principle can explain, via *discursive causal story-telling*, a spectrum of possible empirical outcomes. The causal stories are filled out according to the varieties of observed outcomes by appealing, loosely, to a variety of possible interactions amongst sunspots, vortices, the surfaces of stars, and the 'aether' of old dissipated sunspot material that floats in each stellar vortex near each star.[18] So, although when compared to *Le Monde*, Descartes' mature physics in the *Principia* values novel matters of fact, the system remained relatively closed to novel, deep discoveries at the theoretical level, because unexpected observational outcomes were accounted for at the level of contingent narrative formation, rather than by considering modification to the structure of deep concepts.

3.3 Cartesian Physics and Classical Mechanics: Technical Problems and Shared Goals and Values

This concludes our first task—the exploration of the development of Descartes' physics from a piecemeal physico-mathematics of corpuscular-mechanical leanings to a discursive, cosmographically focused system of corpuscular-mechanical natural philosophy. Our analysis will be completed by investigating the two types of consequence that Descartes' physics had for the emergence of classical mechanics: [1] unintended technical significances, and [2] certain values, goals, and strategic positionings which broadly shaped the eventual emergence of classical mechanics, and which manifested themselves in the work of Descartes and other innovative natural philosophers. We explore [1] and [2] within the three core domains of Descartes' physical inquiry: mechanics, optics, and celestial mechanics.

3.3.1 Descartes' Role in the Attempted Transformation of Traditional Mixed Mathematical Mechanics

In his attempt to transform the traditional mixed mathematical field of mechanics, Descartes was responding in his own way to aims and values characteristic of many natural philosophers of the time. When he spoke about 'mechanics' he was appealing in the first instance to readers' understandings of mixed mathematical statics, hydrostatics, and the study of the simple machines—fields in which he was fully expert and with which he engaged throughout his career. But, as early as 1619 he twice told

Beeckman that he was devising a 'mechanics', which seems to have meant a set of rules about corpuscular behaviour derived from his physico-mathematical projects (AT X 159, 162; Gaukroger and Schuster, 2002, 566–7). That is, somehow the field of mechanics was to become part of the business of natural philosophizing. Such a goal was by no means unprecedented, because from the early sixteenth century there had been attempts—expressed through classificatory arguments, rhetoric about values and aims, or downright technical gambits—to move mechanics into closer contact with natural philosophizing (Hattab, 2005; Laird, 1986).

However, we have seen that Descartes' programme in this regard was quite radical. He worked, over time, in a conceptually precise manner to establish a new set of mechanical principles, which would apply to the corpuscular realm and provide the causal register of his new natural philosophy. The resultant doctrine, which he twice terms a 'mechanics' in the *Principles* (AT VIII1 185, 326; M 162, 285), and which we have dubbed his 'dynamics of corpuscles', was novel, *sui generis*, and had significant evidential basis in his early work in optics and hydrostatics. It also had three unintended relations to the classical mechanics to come:

[1] Descartes' dynamics of corpuscles by no means resembled what post-Newtonian thinkers would mean by classical mechanics. His 'dynamics', as we have seen, was an essential dimension of his system of natural philosophy, not some free standing, emerging master-science of mechanics, intended to displace natural philosophy. Moreover, it did not involve the mathematical treatment of idealized bodies and motions, microscopic or macroscopic. Indeed it rejected that as a relevant concern in natural philosophy. Hence Descartes' dynamics, central as it was to his entire physics, must in the historical long run appear as just one of several daring attempts to transform mechanics and place it at the centre of natural philosophizing.

[2] Nevertheless, Descartes' work on his version of 'mechanics' was shaped by the aim to exploit, enrich, and shift the intellectual role, and evaluation, of traditional mechanics—goals which he shared with other innovators, from the sixteenth-century master engineers such as Benedetti, Tartaglia, and the young Galileo, through the mature work of Galileo and Kepler and up to the origins of Newtonian mechanics. The shared aspirations and values, and the wide variety of technical projects and outcomes, are two sides of the same long-term historical coin.

[3] Descartes' dealings with and about mechanics left technical resources and problems for later thinkers who were developing a more obviously classical mechanics. These concerned inertia (so-called), the laws of collision, and the problem of circular motion. Descartes' claims in these areas, and resultant problems for later physical theorists, are intertwined with one another. This is because they arise from the constraints upon his theorizing exercised by his commitment to a plenum universe, the ubiquity of circular displacements therein, and his adherence to the apparent corpuscular–dynamical revelations of his optics when rearticulated as a theory of light in the plenum cosmos of vortices.

Why no Cartesian law of inertia?

Descartes never stated a law of inertia of Newtonian type, despite occasional later claims that his first and third laws of nature in *Le Monde* (first and second in the *Principles*) conjointly amount to the law of inertia in classical mechanics. In fact, his theory is quite different. In the first place we have seen that his physico-mathematical optical work formed the initial template and motivation for the formation of his first and third laws of nature in *Le Monde*. The laws of behaviour of light, understood by Descartes as an instantaneously transmitted directional quantity of force, provided him an exemplar for framing more general laws of nature, since they involve no complex, curved, or time-dependent physical actions (Section 3.2.2). Upon this basis were superimposed certain conceptual constraints exercised by his emerging system of natural philosophy, including his voluntarist/creationist conception of God's relation to Nature, leading to the final form of the laws (see 'Voluntarist/creationist theology' in Section 3.2.3).

In the dead mechanical world of corpuscles, the continued existence of bodies and their properties is radically dependent upon God's moment-to-moment exercise of his freely willed conserving concourse. It follows that the laws of natural change must deal with divinely regulated instantaneous conservation or alteration of instantaneously exerted forces of motion possessed by bodies in phenomenal translation or merely tending to motion. Moreover, in Descartes' plenum universe no unhindered rectilinear translation is possible; corpuscular collisions occur at every instant; hence there is nothing to be gained by idealizing the situation in the manner of mixed mathematics or Galilean kinematics. A further constraint arises from that fact that although collision involves perfectly inelastic basic particles, it cannot be allowed over time to run down the total quantity of force of motion in the universe. However, this stipulation cannot cover 'determinations', the directional manifestations of force of motion. It probably appeared to Descartes that determination cannot be conserved, since the (scalar) sum of it present in any system depends upon how the grid of components is applied to what we have previously termed the 'principal' determination, denoted by the third rule of nature. If, because of systematic conceptual constraints, rectilinear tendency to motion could be preserved only in instantaneous terms, as the third law of *Le Monde* asserts, so by the same token, cosmic conservation of the total quantity of the force of motion could be formulated only in essentially scalar terms, as the first law of nature in *Le Monde* declares. Moreover, Descartes was in fact fully aware of a principle of inertia very similar to that later asserted in classical mechanics—the one formulated by his original mentor, Beeckman, as early as 1613/14.[19] (Beeckman, 1939–53, I 24–5) But it was precisely this principle which Descartes revised and rearticulated into his first and third laws in *Le Monde*, for the reasons just cited.

Corpuscular collision and its rules

The rules supposedly governing corpuscular collisions were central conceptual elements in any mechanistic philosophy of nature, such as Descartes'. But, how collision was theorized and what rules were asserted depended on the particular mechanistic system in question and its choice of physical exemplars. For example, Descartes' mentor in mechanism, Beeckman, did not believe in any internal moving force in the case of inertial motion, so collisions amongst his inelastic atoms threatened to run down the total quantity of motion in the cosmos over time. His exemplar for formulating laws of collision was the behaviour of the balance beam, interpreted dynamically in the tradition of the medieval science of weights and pseudo-Aristotelian *Mechanica* (Gaukroger and Schuster, 2002, 555–6). Descartes, for his part, paid little attention in *Le Monde* to articulating a collision theory, leaving the matter largely to his second law of nature, which stipulated the conservation of total scalar force of motion in collisions. His corpuscles might be inelastic, but the force of motion, whether looked at in terms of its causal or modal manifestations (see 'Voluntarist/creationist theology' in Section 3.2.3), is conserved, and no run-down of cosmic force could occur.

In the *Principia* the collision theory is both enlarged and conceptually articulated, starting with the third law of nature, which corresponds to law two in *Le Monde*. Now Descartes adds to it a crucial, and seemingly odd, stipulation that a body 'upon coming in contact with a stronger one, loses none of its motion' but 'is turned aside in another direction', thus retaining its quantity of force of motion and changing only its determination (AT VIII1 65; M 61). Collision has become a contest of 'stronger' and 'weaker' forces manifested by bodies, where these terms apply to the force of motion, or of rest (a new conception) possessed by the contending bodies. Seven further rules are elicited, the most telling and paradoxical being the fourth, which asserts the following. If a body C entirely at rest is struck by body B, and body C is even slightly larger than B, the latter cannot move C no matter what its speed of impact; rather B will rebound off C in the opposite direction at the same speed with which it approached. Here we have a contest between forces of motion and rest, where the latter is in effect evaluated in terms of the force of motion B would possess, if, after collision, the total quantity of force of motion were shared out in proportion to the sizes. For in that case C would have more force than B, so in actuality its force of rest (thus evaluated) entirely overcomes the force of motion of B and causes its reversal of direction with its original quantity of force of motion. In other words, B's principal determination is reversed.[20]

What the modern critic should remember about this is that Descartes' rules do not apply in the phenomenal world of macroscopic bodies. Descartes was still elaborating a dynamics of corpuscles. Moreover, in his theory of light he had a perfectly good physical exemplar for this rule. The law-like behaviour of light—an instantaneously transmitted tendency to motion—displayed the underlying dynamics of corpuscles and its rules, as his exploitation of his discovery of the law of refraction had shown. The relevant phenomenon here was reflection of light. The law of reflection would not hold in Cartesian physics unless law three and its following rule four were true: the angle of reflection would not exactly equal the angle of incidence of a ray of light if the (inelastic) reflecting surface of corpuscular objects were to absorb any force of motion from the ray (remembering that in Descartes' optics, transmission is instantaneous, but the strength of a ray can vary, being its instantaneously manifested quantity of force of motion). Hence the contest view is very far from any approach to collision which concentrates on equivalence of action and reaction or attends to the motion of the centre of gravity of the system, though Descartes' manner of dealing with the force of rest may have influenced Newton's path toward his own second law of motion (Gabbey, 1980). The contest view of collision was shaped by Descartes' neo-Scholastic conceptual language in the *Principles*, but also by the more theoretically compelling reason that his exemplar for physical interactions and laws of corpuscular dynamics remained what it had first become in the late 1620s, his own results in a physico-mathematical optics of corpuscular-mechanical tenor.

Circular motion and the failure to focus upon centripetal force

Descartes' physics of ubiquitous vortical displacement circuits places virtually all its conceptual attention on the centrifugal tendencies of bodies actually moving in the plenum. The vortex celestial mechanics focuses on this property, and so does the theory of light in cosmic setting as lines of radial tendency to motion instantaneously propagated from each point on the surface of a star. Both theories are anchored in Descartes' analysis of the instantaneously manifested quantity of force of motion (and its determinations) of a stone in a sling. Of course, Descartes recognized that the sling physically constrains the stone (into its circular path); and that layers of second-element *boules* resist the centrifugal tendency of planets, constraining them at given orbital distances from their stars. But he in no way needed further to conceptualize in his dynamics what came to be called 'centripetal force'. He concentrated on the force instantaneously present 'in' a body (in the 'causal' and 'modal' senses discussed in 'Voluntarist/creationist theology' in Section 3.2.3), and not on the physical process of constraint into curved paths (Gabbey, 1980; Westfall, 1972). Moreover, given the qualitative bent of his mature physics and its propensity to deal in a general but putatively 'law-bound' manner with unavoidably complex situations, he never conceived of quantifying the centrifugal tendency, as Huygens, a mechanist of Cartesian leanings, was to do in the course of his brilliant studies of pendulum motion. When the young Newton first explored the deflection of inertial motion into polygonal approximations of circular paths, he already had to hand a principle of action equalling reaction in collisions and the germ of a generic notion of force as productive of, and measured by, change of momentum (including mere change of direction without alteration of scalar speed)—tools well beyond Descartes' grasp or interest (Herivel, 1965, 6–7; Westfall, 1971, 353–5).

3.3.2 Descartes' Role in the Evolution from Mixed Mathematical Optics toward a Modern Discipline of Physical Optics

Optics was another of the traditional mixed mathematical sciences that evolved into something much more approximating modern form during the seventeenth century. Simple narratives about the field see Newton as the isolated founder of modern physical optics, displacing traditional mixed mathematical geometrical optics. In fact, optics first began to develop away from its ancient and medieval mixed mathematical form within the natural philosophical turbulence and competition of the early and mid-seventeenth century. Like mixed mathematical mechanics, optics began to be reshaped by attempts to render the discipline more physico-mathematical, with closer interaction between optical theorizing and problem-solving on the one hand, and natural philosophical explanation in terms of matter and cause on the other.

Descartes' work in optics was exemplary in this sense, his entire natural philosophical career being characterized by tight mutual articulations between moves in a physico-mathematical optics, and an increasingly systematized corpuscular-mechanistic physics. Yet he did not invent the physico-mathematical gambit in optics nor in any other mixed mathematical field (Dear, 1995), and in optics he very much emulated and attempted to surpass and displace the work of Kepler. The latter had in effect pursued a physico-mathematical agenda, practising geometrical optics under, and in the service of, a neo-Platonic natural philosophy and conception of light, rather than a corpuscular–mechanical one, obtaining brilliant results in the theory of the camera obscura, the theory of vision, and, to some degree, the theory of refraction and the telescope.

As a result of these sorts of early efforts, geometrical optics, the mixed mathematical science of the ancients and Scholastics, began to evolve into a much more obviously physico-mathematical discipline. In the latter half of the seventeenth century the process intensified in the physico-mathematical optical work of Huygens, Robert Hooke, Newton, and a host of others. By 1700 there was a great density of new phenomena, instrumental/practical applications, and new problems. Optics became a physico-mathematical discipline, increasingly independent of any particular natural philosophical system, and indeed relatively independent of the domain of natural philosophizing as a whole. Descartes' physico-mathematical and natural-philosophically relevant optics is therefore both illustrative of the process in its early stages, and also, taking a longer historical perspective, a contribution of the first order to it.

On the technical level, Descartes bequeathed a host of problems, claimed solutions, and puzzles to the evolving field. They include his claim that light is propagated instantaneously; his fundamentally 'ray' rather than 'wave' version of a mechanistic theory of light (Shapiro, 1974); his sine law, which required articulation and correction (Dijksterhuis, 2004); and his views on telescope design and the fabrication of lenses (as planned on paper and as meant to be materially realized) (Burnett, 2005; Ribe, 1997). His greatest achievement, long recognized but only recently explored in its full historical significance, was his explanation of the geometrical and chromatic properties of the rainbow. This was a stunning solution to a mixed mathematical and natural philosophical puzzle of ancient standing. However, it was also something else, portentous, but without further issue in Descartes' career. In the case of his researches on the rainbow, he achieved the only instance in his work where a corpuscular-mechanical model is applied and further articulated with relation to novel experiments which have quantitative implications (Buchwald, 2008). Not only does this prefigure the young Newton's optical work, but both achievements consist in just the kind of continuing interaction between physical models and quantified, often novel, experimental manipulations that characterize the explanatory and practical style, and content, of Newtonian and all subsequent physics.

3.3.3 The Ultimate Context and Driver: Bidding to Establish Realist Copernicanism

There is one other large context into which Descartes' intentions and strategic aims in physics should be set, with regard both to his technical initiatives and to the ways in which his work instantiated larger drivers of the emergence of classical mechanics. This context is the long process of the so-called Copernican revolution, taken as culminating in the emergence of Newtonian mechanics and celestial mechanics. It is obvious that Descartes' physics—like that of Kepler, Galileo, and Newton—taken in its widest acceptation is set almost entirely within the problematic of bold, realist Copernicanism.

In the universities, under the hegemony of neo-Scholastic Aristotelianism, geometrical astronomy had, of course, always been viewed as a mixed mathematical science and routinely held to lack the explanatory power of natural philosophy, because it did not deal with the material and causal principles of planetary motion, but merely with appearance-saving geometrical models. However, when, in the later sixteenth and early seventeenth centuries, two generations after the death of its founder, Copernican astronomy came to be hotly debated, it was not as a set of new calculational fictions, but rather as a system with realistic claims about the physical structure and causal regime of the cosmos, implying the need for a framework of non-Aristotelian natural philosophy adequate to justifying its existence and explaining its physical mechanisms. This radicalized the usually accepted relations between mathematics and natural-philosophical explanation, making the Copernican debate a notable hotspot even within the desperate natural-philosophical struggle of the early seventeenth century in which Descartes was a major player.

Accordingly, in the physics of Descartes, and that of his major contemporaries, we find nothing resembling a smooth, linear, progress toward the crystallization of Newtonian physics. Rather, we see heightened contestation amongst natural philosophers. New doctrines of motion and cause were variously asserted, applied to the heavens or Earth or both, leaving a wide field of problems, apparent technical achievements or failures (depending upon the views of various successors, including Newton), as well as targets of revision or replacement, such as Kepler's treatment of orbital motion under two sets of forces, radial and rotary, or Descartes' vortex mechanics.

That vortex mechanics, as shown above, was a peculiar sort of science of equilibrium, grounded in Descartes' home-grown dynamics of corpuscles, itself emergent from his physico-mathematical optics. For him, this dynamics-as-manifested-in-the-vortex mechanics was the key to a new physics, in the sense of a systematic natural philosophy, which remained purely discursive but was daringly and strategically constructed to win adherents. Cartesian physics was thus one bid amongst several to create a new physics, meant to sit inside, run, and legitimate a Copernican cosmos. It was, like the physics of Galileo or Kepler, a *sui generis*, before the fact (of Newton) and in the end stillborn project. But these various attempts to establish a form of realist Copernicanism within a replacement natural philosophy involving a (hoped-for) more integral use of mathematics display shared attitudes and goals which characterize the intellectual landscape and conditions through which, in these earlier generations of the seventeenth century, classical mechanics was unintentionally taking shape. Newton shared these attitudes and goals, being himself also an innovative natural philosopher, seeking the more integral use of mathematics in natural philosophy, and aiming to refine the (now widely accepted) Copernican view, in order to claim the prize of having established its definitive winning form.

On the plane of technical consequences, it follows from all this that Descartes' vortex celestial mechanics had to become a target of refutation and displacement by Newton in Book II of his *Principia*, in the service of his own mechanics and celestial mechanics. Descartes' vortex fluid of second-element *boules* and interstitial first-element particles offers no resistance whatsoever to the motion of a planet or comet (Gaukroger, 2000). The business of the rotational motion of the Cartesian vortex is to move planets and comets in a rotary manner, leading to the generation of centrifugal tendencies that are the proper objects of analysis on the central question of orbital stability and placement. His focus is on specifying orbital equilibrium or its breakdown, not on solving what seemed to Newton, and later classical mechanics, as exemplary fluid-mechanical problems. It is no wonder, then, that Newton found the Cartesian vortex theory so problematical, and in particular its supposed implications for a properly mathematical and idealized mechanics of vortical fluid motion so unsatisfactory.[21] It is well known, of course, that more highly articulated and quantitative vortex theories were advanced both at the time of Newton's work and on into the eighteenth century on a non-trivial basis. These far outstrip Descartes' earlier qualitative and discursive construction, and are the true competitors to Newtonian celestial mechanics in the eighteenth century (Aiton, 1972). However, it remains true that Newton initially had to confront and, on his own grounds, dispose of Descartes' version of vortex celestial mechanics as part of the making of his own *Principia*.

3.4 Conclusion: Mathematics, Natural Philosophy, and the Path to Classical Physics

The conventional narrative about the emergence of classical mechanics assumes first of all a clear distinction at the time between the world of mathematics and mathematicians and the (merely scholastic) world of natural philosophy and natural philosophers. It then adduces causes for the birth of a mathematicized physics. That reason could be neo-Platonic metaphysics; the rise of a useful practical mathematics and its practitioners; the recovery of ancient mathematics, in particular Archimedes; or the increased cultivation of the mixed mathematical sciences of mechanics, optics, geometrical astronomy, and the like. In contrast, Descartes' trajectory, like that of other physico-mathematically sensitive and ambitious, pro-Copernican, mathematician/natural philosophers, shows that the real but initially unintended growth point for classical mechanics was precisely where such thinkers explored and renegotiated the relations between mathematics and natural philosophy, especially between the now rapidly developing mixed mathematical sciences and the wider field of natural philosophizing. The disciplinary and conceptual boundaries were not fixed, and the range of (shifting) initiatives and outcomes is well illustrated by Descartes' own career.

That is why we have had to look at Descartes' natural philosophy in developmental terms, and with special attention to the aims and outcomes of his early physico-mathematical programme and to his bold concerns with realist Copernicanism. It also explains why we find little in Descartes' work that is in a direct path of development toward classical physics. He was a natural philosopher—a daring, innovative one, with special aptitudes and ambitions for the relevance of mathematics and the existing mixed mathematical fields inside a new system of natural philosophy. His system of physics was one outstanding instance in a spectrum of physico-mathematically coloured, pro-Copernican natural-philosophical initiatives of the time. In the long run, his work, along with that of others, in this genre of natural philosophy had to be displaced in order for classical mechanics to emerge as the new master science and core of a new order of physical knowledge, involving an evolving but essential dialectic of mathematics and experimentation/instrumentation. But his and other related versions of physics marked a stage in the evolution, and eventual dissipation of, natural philosophy, through which elite European knowledge of nature had to pass if there ever was to be a classical mechanics. That is where the texture of unintended consequences comes into play, with positive achievements as well as failures or dead ends being played upon later by Newton and others in the creation of classical mechanics and celestial mechanics.

REFERENCES

AT   *Oeuvres de Descartes* (revised edition, 12 vols.). C. Adam and P. Tannery, eds. Paris: Vrin/CNRS, 1964–76. References are by volume number (in Roman) and page number (in Arabic).

G   *The World and Other Writings*, Stephen Gaukroger, trans. and ed. Cambridge: Cambridge University Press, 1998. References are by page number.

M   *The Principles of Philosophy*, René Descartes. V. R. Miller and R. P. Miller, trans. Dordrecht: Kluwer Academic Publishers, 1991. References are by page number.

CSM   *The Philosophical Writings of Descartes*, vol. 1, John Cottingham, Robert Stoothoff, and Dugald Murdoch. Cambridge: Cambridge University Press, 1985.

AITON, ERIC J. (1972) *The Vortex Theory of Planetary Motion*. London: Macdonald.
BEECKMAN, ISAAC (1939–53) *Journal tenu par Isaac Beeckman de 1604 à 1634*, ed. Cornelius de Waard, 4 vols. The Hague: Nijhoff.
BIRO, JACQUELINE (2009) *On Earth as in Heaven: Cosmography and the Shape of the Earth from Copernicus to Descartes*. Saarbrücken: VDM Verlag.
BUCHDAHL, GERD (1970) *Metaphysics and the Philosophy of Science: The Classical Origins Descartes to Kant*. Cambridge, MA: MIT Press.
——(1972) 'Methodological Aspects of Kepler's Theory of Refraction', *Studies in the History and Philosophy of Science* 3: 265–298.
BUCHWALD, JED Z. (2008) 'Descartes's Experimental Journey Past the Prism and Through the Invisible World to the Rainbow', *Annals of Science* 65: 1–46.
BURNETT, D. G. (2005) *Descartes and the Hyperbolic Quest: Lens Making Machines and their Significance in the Seventeenth Century*. Philadelphia: American Philosophical Society.
CLARKE, DESMOND (1977) 'Descartes' Use of "Demonstration" and "Deduction"', *The Modern Schoolman* 44: 333–344.
——(2006) *Descartes: A Biography*. Cambridge: Cambridge University Press.
DEAR, PETER (1995) *Discipline and Experience*. Chicago: University of Chicago Press.
DIJKSTERHUIS, F. J. (2004) 'Once Snell Breaks Down: From Geometrical to Physical Optics in the Seventeenth Century', *Annals of Science* 61: 165–185.
GABBEY, A. (1980) 'Force and Inertia in the Seventeenth Century: Descartes and Newton', in S. Gaukroger, ed., *Descartes: Philosophy, Mathematics and Physics*. Sussex: Harvester, 230–320.
GAUKROGER, S. (1995) *Descartes: An Intellectual Biography*. Oxford: Oxford University Press.
——(2000) 'The Foundational Role of Hydrostatics and Statics in Descartes' Natural Philosophy', in S. Gaukroger, J. A. Schuster, and J. Sutton, eds., *Descartes' Natural Philosophy*. London: Routledge, 2000, 60–80.
——and SCHUSTER, J. A. (2002) 'The Hydrostatic Paradox and the Origins of Cartesian Dynamics', *Studies in the History and Philosophy of Science* 33: 535–572.
GUEROULT, MARTIAL (1980) 'The Metaphysics and Physics of Force in Descartes', in S. Gaukroger, ed., *Descartes: Philosophy, Mathematics and Physics*. Brighton: Harvester, 196–229.
HARRISON, PETER (2002) 'Voluntarism and Early Modern Science', *History of Science*, 40: 63–89.
HATTAB, HELEN (2005) 'From Mechanics to Mechanism: The *Quaestiones Mechanicae* and Descartes' Physics', in P. Anstey and J. A. Schuster, eds., *The Science of Nature in the Seventeenth Century: Changing Patterns of Early Modern Natural Philosophy*. Dordrecht: Springer, 99–129.
HERIVEL, JOHN (1965) *The Background to Newton's 'Principia'*. Oxford: Oxford University Press.
JULLIEN, VINCENT, and CHARRAK, A. (2002) *Ce que dit Descartes touchant la chute des graves: de 1618 à 1646, étude d'un indicateur de la philosophie naturelle cartésienne*. Villeneuve d'Ascq (Nord): Presses Universitaires du Septentrion.
KNUDSEN, O., and PEDERSEN, K. M. (1968) 'The Link between "Determination" and Conservation of Motion in Descartes' Dynamics', *Centaurus* 13: 183–186.
KORTEWEG, D.-J. (1896) 'Descartes et les manuscrits de Snellius d'après quelques documents nouveau', *Révue de Métaphysique et de Morale* 4: 489–501.
KRAMER, P. (1882) 'Descartes und das Brechungsgesetz des Lichtes', *Abhandlungen zur Geschichte der Mathematischer (Natur) Wissenschaften* 4: 235–278.
LAIRD, W. R. (1986) 'The Scope of Renaissance Mechanics', *Osiris* 2: 43–68.
LOHNE, J. (1963) 'Zur Geschichte des Brechungsgesetzes', *Sudhoffs Archiv* 47: 152–172.

Machamer, Peter, and McGuire, J. E. (2009) *Descartes' Changing Mind*. Princeton, NJ: Princeton University Press.

McLaughlin, P. (2000) 'Force Determination and Impact', in S. Gaukroger, J. A. Schuster, and J. Sutton, eds. *Descartes' Natural Philosophy*. London: Routledge, 81–112.

Mersenne, M. (1932–88) *Correspondance du P. Marin Mersenne*, 17 vols., ed. C. de Waard, R. Pintard, B. Rochot, and A. Baelieu. Paris: Centre National de la Recherche Scientifique.

Prendergast, T. L. (1975) 'Motion, Action and Tendency in Descartes' Physics', *Journal of the History of Philosophy* 13: 453–462.

Ribe, Neal (1997) 'Cartesian Optics and the Mastery of Nature.' *Isis* 88: 42–61.

Sabra, A. I. (1967) *Theories of Light from Descartes to Newton*. London, Oldbourne.

Schmaltz, T. (2008) *Descartes on Causation*. Oxford: OUP.

Schuster, John A. (1980) 'Descartes' *mathesis universalis*: 1618–1628', in S. Gaukroger, ed., *Descartes: Philosophy, Mathematics and Physics*. Brighton: Harvester, 41–96.

——(1986) 'Cartesian Method as Mythic Speech: A Diachronic and Structural Analysis.' in John A. Schuster and Richard Yeo, eds., *The Politics and rhetoric of Scientific Method*. Dordrecht: Reidel, 33–95.

——(2000) 'Descartes *opticien*: The Construction of the Law of Refraction and the Manufacture of its Physical Rationales', in S. Gaukroger, J. A. Schuster, and J. Sutton, eds., *Descartes' Natural Philosophy* London: Routledge, 258–312.

——(2005) 'Waterworld: Descartes' Vortical Celestial Mechanics—A Gambit in the Natural Philosophical Contest of the Early Seventeenth Century', in Peter Anstey and J. A. Schuster, eds., *The Science of Nature in the Seventeenth Century: Patterns of Change in Early Modern Natural Philosophy*. Kluwer/Springer: Dordrecht, 35–79.

——(2013) *Descartes Agonistes: Physico-Mathematics, Method and Corpuscular-Mechanism 1618–33*. Dordrecht: Springer.

—— and Brody, J. (2013) 'Descartes and Sunspots: Matters of Fact and Systematizing Strategies in the *Principia Philosophiae*', *Annals of Science* 70: 1–45.

Shapiro, Alan (1974) 'Light, Pressure and Rectilinear Propagation: Descartes' Celestial Optics and Newton's Hydrostatics', *Studies in History and Philosophy of Science* 5: 239–296.

Stevin, Simon (1955–66), *The Principal Works of Simon Stevin*, ed. Ernst Cronie, *et al.*, 5 vols. Amsterdam: Swets and Zeitlinger.

Vollgraff, J. A. (1913) 'Pierre de la Ramée (1515–1572) et Willebrord Snel van Royen (1580–1626)', *Janus* 18: 595–625.

——(1936) 'Snellius' Notes on the Reflection and Refraction of Rays', *Osiris* 1: 718–725.

de Waard, C. (1935–6) 'Le manuscrit perdu de Snellius sur la refraction', *Janus* 39–40: 51–73.

Weber, Jean-Paul (1964) *La Constitution du texte des Regulae* Paris: Société d'Éditions d'Enseignement Supérieur.

Westfall, Richard (1971) *Force in Newton's Physics: The Science of Dynamics in the Seventeenth Century*. London: MacDonald.

——(1972) 'Circular Motion in Seventeenth Century Mechanics', *Isis* 63: 184–189.

Notes

1. The understanding of determination used here develops work of A. I. Sabra (1967), p. 118–21; Gabbey (1980), pp. 230–320; S. Gaukroger (1995); O. Knudsen and K. M Pedersen (1968), pp.183–186; Prendergast (1975), pp. 453–62; McLaughlin (2000) and Schuster (2000, 2005).

2. The term principal determination was coined by Schuster (2000), 260. It is meant to underscore this important concept, and differentiate this aspect of determination from the other determinations that can be attributed to a body in motion, or tending to motion, at any given moment.
3. In both *Le Monde* and the *Principia Philosophiae*, Descartes explained light mechanically in its full cosmological and matter theoretical context by the centrifugal tendency of the spherical particles of second element constituting each stellar vortex—light consisting in the instantaneous passage through a vortex of lines of tendency to centrifugal motion (AT XI 87–90; G 55–8; AT VIII1 108–16; M 111–18).
4. Everything we are about to say concerning Descartes' dynamics underlying the demonstration of the law of refraction in the *Dioptrique* also applies to his preceding demonstration of the law of reflection (Schuster, 2000, 261–3).
5. Some have claimed that Descartes fell into a contradiction, because his theory of light states that light-rays move instantaneously through any medium, whilst in the tennis-ball model he deals with a ratio of finite speeds. This is mistaken. One simply must distinguish the *speed* of propagation of a light-ray, which is instantaneous, from the *magnitude* of its force of propagation, which can take any finite positive value. The *speed* of Descartes' tennis ball corresponds not to the speed of propagation of light but to the intensity of the force of its propagation.
6. The traditional image locating rule held that in Fig. 3.5 the image of point o, for example, which is immersed in water, will be seen at B, the intersection of the normal from o to the refracting interface, with the line RB, the extension of the path of the ray refracted into the air, back into the water. Willebrord Snel's initial construction of the law of refraction also followed the type of path indicated by the Lohne analysis. (Vollgraff, 1913, 1936; deWaard, 1935–6).
7. So-called by J.-P. Weber (1964), who first showed this. After Weber, others further articulated his findings: for example, Schuster (1980) 51–55, and Gaukroger (1995) 111ff, which builds on a synthesis of Weber and Schuster. See also Schuster (2013), Chapter 5.
8. Many modern scholars now hold that such grand, set-piece doctrines of scientific method, such as Descartes', cannot and do not control and guide living practice in any given field of research, let alone across the entire gamut of disciplines. Descartes' technical achievements in mathematics, the mixed sciences, and natural philosophy cannot and should not therefore be explained as applications of his method. A good example is the contrast between how Descartes actually discovered the law of refraction of light, and its mechanical rationale, as we have examined it, and the fairy tale he tells about this in rule 8 of *Rules for the Direction of the Mind* (Schuster, 2000, 300–02; 2013, 215–220).
9. Although it is not as though Descartes' mature intellectual persona emerged all at once in these years. His career after 1633 is marked by numerous further twists and reorientations, increasingly keyed to public debate and controversy. Works that make clear the continuing patterns of change in Descartes' career after 1633 include most notably Gaukroger (1995), Clarke (2006), and most recently, Machamer and McGuire (2009). Of these, only Gaukroger gives sustained and useful attention as well to the earlier period, down to 1633.
10. Harrison (2002), 69 has importantly pointed out that one should not simply conflate the notion that God's will is primary with the idea that nature is totally dependent upon God: 'the doctrine that places God as the direct cause of what takes place in nature is thus independent of a voluntarism according to which the divine will is above reason.'
11. This law, appearing as the third law of nature in the *Principles*, will be significantly altered and articulated into a number of detailed rules of collision. See 'Corpuscular collision and its rules' in Section 3.3.1.

12. In November 1630 he wrote to Mersenne that he had indeed begun work on what he termed 'a small treatise of metaphysics' early in the year, during his first few months in the United Provinces. The 'principle points' of the work he claimed to be to prove the existence of God and the immortality of the soul, when separated from the body. Descartes to Mersenne, 25 November 1630, AT I 182.
13. In Figs. 3.7, 3.8, 3.9, and 3.10, straight lines are used to represent the functional relations amongst *boules*' sizes, speeds, and distances from the central star, derived from the verbal expressions in Descartes' texts. It is not intended that Descartes necessarily or consistently entertained such linear relations. What is important is the general representation of the force-stability principle and how that relates to Descartes' claims about the size and speed distributions with distance.
14. The term 'falling' is chosen quite deliberately. Descartes makes it clear in discussing the placement of planetary orbits that a planet 'too high up' in the vortex for its particular solidity is extruded sunward, falling (and spiralling) down in the vortex to find its proper orbital distance. Descartes' theory of local fall, and theory of the orbital motion of the Moon, when taken in their simplest and most charitable acceptations, both also make use of this notion of falling in a vortex until a proper orbital level is found (assuming no other circumstances prevent completion of the process, as they do in local fall of heavy terrestrial bodies near the surface of the earth). Ultimately, however, the interpretation of these two theories becomes more fraught, requiring additional interpretative attention, as note 15 makes clear.
15. This point is very important, and helps explain certain difficulties Descartes met, in trying to deal with the local fall of 'heavy' bodies, a phenomenon he tends to want to describe using the vocabulary of classical hydrostatics, which, however, simply will not mesh with his vortex mechanics, a science, at bottom, of 'solidity' or volume to surface relations.
16. AT VIII1 148–50; M 136–9 [1] We see most sunspots in a belt near the equator and not at the poles, because by the time they have managed to stick together into a mass of third-element particles big enough to be visible to our eyes they have covered a considerable distance from the poles. [2] Sunspots are just that: accretions of first-matter magnetic particles into denser clumps of third matter—that is the only way third matter is made in the universe. [3] From the way they come into being it follows they have irregular shapes. [4] Being on the Sun's surface, the spots are carried along by its rotation. [5] Spots sometimes have a dark nucleus surrounded by a lighter area, because at the lighter parts the accumulation of third element is thinner and allows some light to pass through. [6] Bright areas or *faculae* can occur near sunspots. Sunspots restrict the movement of the Sun's first element material, which then tends to surge away at the edges of the spots, thus being more luminous. The central mass of a spot obstructs that tendency to motion; that is, it stops the light. [7] Sunspots can disappear. They become worn away by the rotating matter of the Sun, and disintegrate partially back into first element, and partially into smaller bits of third element, which then become an atmosphere around the sun. This he terms 'aether'. It surrounds each star, and is inherited by a planet resulting from the death of a star, becoming, its atmosphere, as on Earth.
17. In the telling remark that ends Book III (AT VIII1 202; M 177), Descartes asserts that all inequalities of planetary motion can be sufficiently explained using the framework he has provided. Clearly he in no way intends that explanations will proceed by deductions from laws of motion, plus boundary conditions, leading to the exposure and study of various levels and types of perturbations. So, for example, it is not elliptical orbits and their deviations that he wishes to study, leading to refinement of the relevant laws. Rather, he offers

a 'sufficient' (verbal and qualitative) explanation of orbital phenomena and the general facts that no orbit is perfectly circular, and that all orbits display variations over time.
18. Descartes introduces the section of Book III of the *Principles*, dealing with sunspots, novae, and variable stars at Article 101 by stating (AT VIII1, 151; M 139): 'That the production and disintegration of spots depend upon causes which are very uncertain.'
19. Beeckman believed that both rectilinear and circular inertial motion were possible.
20. In correspondence Descartes later further rationalized this approach in terms of a supposed principle of least modal change in a collision (AT IV 185): in rule 4 only the determination of B changes and not the quantity of force of motion of B, or force of rest of C.
21. Descartes' vortical fluid mechanics raises other theoretical problems in relation to his theory of cosmic optics. To explain how the entire disk of the central star is visible to observers on a planet in its vortex, he seems to imply that the radial tendencies of the *boules* lying next to the surface of the star can be considered to give rise to oblique lines of tendency which pass through the vortex in all directions above the tangent planes to each point of the surface. Once constituted, these lines of oblique tendency continue out into the vortex unaltered and without giving rise to any secondary diffusion of oblique tendencies at each intervening *boule*, lest the vortex become entirely filled with light. This theory hardly comports with a classical mechanics understanding of the propagation of pressure in a fluid (Shapiro, 1974, 254–7, 265).

CHAPTER 4

PHYSICS AND THE INSTRUMENT-MAKERS, 1550–1700

ANTHONY TURNER

Instruments were not needed for the book-learnt Scholastic and neo-Scholastic physics prevalent in late Renaissance Europe among conformist, university-trained scholars (Heilbron, 1983), and there was little attempt to change this qualitative investigation of the entire natural world in any fundamental way before the end of the sixteenth century. When some new, experimental and/or mathematical approaches did then begin, in part due to the investigational, operative approach of *magi* such as G. B. Porta, H. C. Agrippa, and John Dee (Clulee, 1988, 68–9), who emerged with the development of neo-Platonic, hermetic studies in the late Renaissance, they immediately ran into the problem of the difficulty of finding craftsmen to produce such apparatus to experiment and explore as was required. This is not to say that no tradition of instrument-making emerged from the later Middle Ages to be developed during the sixteenth century. Rather, it acknowledges that what did emerge was almost entirely concerned with mathematical instruments for time-measurement and cosmology. In the course of the sixteenth century, new instruments were developed for use in land-surveying (*Annals*, 1991), navigation (Waters, 1958), architecture (Gerbino and Johnston, 2009), and warfare (Bennett and Johnston, 1996), and it can be argued that balances have a claim to be considered as precision measuring devices. None of them, however, except perhaps the navigator's compass, which in order to function depended upon a physical force which was not itself understood (was even considered occult), had much to do with physics whether considered as the study of the structure of the natural world in general (the elements and their behaviour), or more specifically as the examination of the action of matter and energy within it.

Such instrument-making workshops as there were in mid-sixteenth century Europe were few, scattered, and small. They had developed during the fifteenth and sixteenth centuries—the period of transition in mathematical instrument-making from an occasional activity of scholar or artisan to a trade carried out in workshops in which it was possible for aspirants to the activity to be trained. At first fragile, mathematical instrument-making as a trade visible in permanent shops and as a recognized, albeit humble, activity at court, became increasingly well established from the mid-sixteenth century onwards. The number of instrument makers and their workplaces increased; they became specialists in their trade, detaching themselves from more general areas of metal-working. The instrument trades originated in a wide variety of other activities such as clock-making, turning, engraving, and founding. During the Renaissance a small number of practitioners of all these crafts concentrated their activity around the making of mathematical instruments, creating a specialized manufacture. The process was a slow one and the association of instrument-making with horology and precision mechanics in general would endure far into the eighteenth century, for with these instrument-making shared not only techniques and skills, but also part of its subject matter. The practical scholar, making his own instruments in default of competent artisans, also continues to be found beyond the time limits of this chapter. Even so, the tendency to specialize is clear.

Viable commerce requires regular demand. One of the characterizing achievements of Renaissance instrument-making throughout Europe was to respond to an increasing demand by the production of semi-standardized instruments while retaining the ability to innovate and develop special instruments to special order. While there is some evidence of quantity or batch production, for example in the making of astrolabes by Georg Hartmann in Nuremberg (Gordon, 1987; Turner, 1987, 39–40), in the sixteenth century, it is likely that in workshops smaller than those of Hartmann, Christoph Schissler or Gualterus Arsenius, stock-in-hand was fairly small—even non-existent—and that even standard instruments were made only once an order had been received for them.

Instrument-making around 1550 existed as a workshop activity in towns and cities (as such it was subject to the regulation of trade exercised by guilds and corporations), and as a privileged activity in princely courts. It might receive and benefit from the exercise of patronage without depending upon it, but increasingly it is identifiable as a distinct activity even while remaining mingled, in the guild and corporation structures of most European cities, with other trades. Although guild regulation seems not to have inhibited innovation (Turner, 2008) and positively maintained standards of quality and education (Epstein, 2008), demarcation disputes as to which guild should control which worker using a particular material could, in some regions, become a severe aggravation and handicap to instrument makers as the development of new areas of experimental investigation produced a demand for new instruments (such as air-pumps and barometers) which required working a wide range of substances.

The typical materials of medieval and early Renaissance instruments were brass and wood. Only minor use was made of glass (mainly for sand-glasses and compass-needle covers, although the latter could also be made of mica), and iron (principally for dividers). Steel was rather little used except for compass-needle pivots, probably obtained from needle-makers, and the points of dividers. Brass could vary considerably in its composition with varying proportions of copper, zinc, and tin, while the absence of large rolling-mills limited the size and smoothness of the plates available. For the stamping and engraving of instruments in wood, hard, close-grained box or fruit-woods were preferable even to oak, although light, close-grained varieties of this could be used. Oak or soft woods could be used for sand-glass stands or compass boxes. Cases for the expensive brass instruments were likely to be made of leather-covered paste-board, and might carry blind-tooled or gold-tooled decoration. On occasion, gold or silver might be used for instruments intended for the very great.

Few such instruments, however, were intended for the purposes of physics, whether for teaching it, demonstrating it, or, still less, for advancing it. Such areas of investigation in physics, such as acoustics and optics, as did escape from a purely theoretical, book-based approach required apparatus that was outside the purview of the traditional mathematical instrument-maker (for the scope of this production in England, see Turner G, 2000; in the Low Countries, Van Cleempoel, 2002). Unfortunately, exactly who made the large mirrors for reflecting sound, light, and heat that were used or described by such as della Porta, Dee, the Digges, Descartes, Cavalieri, and Mersenne remains unknown. Brass or bronze founders we may assume—perhaps those specialists who produced cannon, bells, or mortars and pestles, but clear information is lacking. Similarly absent for the sixteenth century is very much information about the makers of compasses, while nothing at all seems to be known about who armed lode-stones or produced the magnetic equipment used by that pioneer experimental natural philosopher, William Gilbert.

Magnetism and optics were perhaps the first components of natural philosophy to become experimental. Extrapolating from the experience of Galileo leads to the conclusion that for the making of much of the apparatus required the *savant* had to turn to such skilled craftsmen as he could persuade to execute his special, sometimes bizarre and difficult, requirements. None of Galileo's instrument-makers were that by origin, though some of the lens-makers had a connection with mirror-making, spectacle-making, or *pietra dure* work. (Bedini, 1994(1)). That it was a blacksmith who made armatures for Galileo's magnets may give an indication as to such craftsmen as made them elsewhere in Europe, but the shapers of lode-stones and of Gilbert's *terellae* (perhaps the earliest item of physics demonstration apparatus but, given the hardness of magnetite, difficult to make), are unknown.

For the instruments that he might require, the practical investigator of the natural world in the late sixteenth century—who might, *inter alia*, do something akin to what would later be considered as part of experimental physics—had three main possible suppliers: mathematical instrument-makers, compass-makers, and balance-makers. He could also address himself to skilled practitioners in more basic trades, such as founders and blacksmiths. The very practical savant might even make his own apparatus.

Balance-making in the sixteenth century was already an old trade, and one which was highly regulated by both state and local authorities. In Paris, balance-makers had been incorporated since 1324, with their powers of search and the obligation to mark their goods reinforced in 1494. In Rouen the trade was given statutes in 1415 (Machabey, 1949). During the later sixteenth and the seventeenth centuries Lyon would take on importance as a centre for the making of money-scales. Metal-workers first, specialists in a particular instrument second, balance-makers depended upon the metal-working trades. In Nuremberg from the fourteenth century onwards they were masters in the *Rotschmeid*, the association of workers in brass, copper, and bronze (Houben, 1984, 47ff; Lockner, 1981). In Cologne they were associated with the blacksmiths. Whatever their affiliation, if they were not independent as in France, and perhaps also in London, they afforded a source of skilled workmen which *savants* could exploit, in most of the major commercial towns of Europe.

Compass-makers one might expect to be localized in ports. In 1581 Robert Norman located them in Genoa, Venice, Danzig, Seville, Lisbon, La Rochelle, Bordeaux, Rouen, and more generally in Sicily, Flanders, and England. In the Low Countries, Amsterdam was a main centre of production, but compasses were also made in Bruges, Middelburg, Bremen, and elsewhere (ter Kuile and Mörzer Bruyns, 1999, 27–9). In England, compasses were made in London and no doubt in Bristol, and some of the north-east coast ports such as Hull. The product, however, was often defective. William Barlow, in *The Navigator's Supply* (1597), was scathing in his list of faults of construction (Waters, 1958, 28–9). Investigations into magnetic declination, variation, deviation, and dip would have been as much hindered as advanced by the use of apparatus supplied by such makers. It is significant that it was an atypical compass-maker, Robert Norman (*fl.*1560–84; see Bennett, 2004), who discovered magnetic dip. Norman had spent some twenty years at sea, and had noted the inaccuracies of variation compasses. After settling in Ratcliffe, London, as a compass-maker and 'hydrographer', the manufacture of steering- and variation-compasses led him into wider investigations of magnetism as an aid to navigation. These were carried out using his own instruments, some of which—the dip circle—he also designed.

Norman is an anomalous figure: seaman, compass-maker, and experimenter. His researches had significance for natural philosophy but they derived from a concern to advance the mathematical art of navigation. Exceptionally he was one of the few makers who could supply reliable instruments and is an example of that class of men who may be called craftsmen-experimenters, artisan-philosophers, men who, like Cornelis Drebbel, Robert Hooke, and Denis Papin, all of whom originated within this group, would play notable roles in the development of natural philosophy towards a new physics.

The makers of mathematical instruments were, by the end of the sixteenth century, already producing a wider and more diverse range of items than were the specialized balance- and compass-makers. That the trades were not impermeable is illustrated by the career of Emery Molyneux who, from making compasses and running-glasses

(=log-glasses), gained sea experience, possibly on Drake's circumnavigation voyage of 1577–80, before, with the help of Edward Wright, making the first printed English globes (1592). For these he devised a new, impermeable, surface covering and modifications to the mounting. If wood seems to be the link between the three types of instrument Molyneux is known to have made, he would also, if it was indeed he who was apprenticed to William Cooke in the Stationers' Company in the 1550s, have been familiar with printing practices. He is said to have engraved his own plates. In parallel with producing the globes, he developed a new form of cannon (a demi-culverin), and wood, bronze and paper all entered into his production (Crinò and Wallis, 1997; Maxwell, 2004).

Use of a variety of materials would in the course of the seventeenth century become increasingly common in instrument-making workshops. Adaptability on the part of the maker, such as Molyneux displayed, would also become essential to the ambitious philosophical craftsman. If the development of practical research in optics, applying and modifying the far older tradition of theoretical research in the subject, was a major element in the transformation of thought about nature in the seventeenth century, it also constituted a major challenge to traditional mathematical instrument-making. First, it introduced a new material, glass, unfamiliar to these brass- and/or wood-using craftsmen. Secondly, because of this unfamiliarity it brought two new groups of artisans, spectacle-makers and glass-workers, into instrument-making as either collaborators or competitors.

Galileo employed craftsmen with a background in the working of glass and marble in the *pietra dure* workshops of Florence, or who had been spectacle-makers to turn and figure his lenses. In London, Thomas Harriot used the skills of Christopher Tooke, who had earlier been a glass-turner. By the mid-sixteenth century the making of eye-glasses was wide-spread in Europe. A trade already nearly three centuries old, its practitioners had become sufficiently numerous to be incorporated. A guild was established in Nuremberg in 1507. In both Rouen and Paris, the spectacle-makers were associated with the mirror-makers. They were incorporated at Rouen in 1538, in Paris rather later, and were reorganized in 1581. Spectacle-making may have come later to London, perhaps because it could easily be supplied from Rouen and by Flemish peddlers importing products from the Low Countries. But knowledge of the subject was available in London. William Bourne, otherwise a standard mathematical practitioner, drew upon it for his *The Property or Qualityes of Glaces Acordyng vnto ye seuerall mackyng pollychynge and Gryndyng of them*—a tract dedicated to Lord Burghley after September 1572 (Halliwell, 32–47). So did Arthur Hopton in his *Speculum Topographicum or the Topographical Glass* of 1611. It was, however, among the manufacturing Dutch spectacle-makers that the spy-glass and the telescope would originate (van Helden, 1977), and the first examples to be seen in England would come from there.

The glass-workers of Europe provided a source from which skilled workmen could be recruited into instrument-making—a further group to whom investigators of the natural world could turn. By contrast, traditional mathematical instrument-making was slow to absorb the lens- and mirror-making skills needed to meet the demand for telescopes (from c.1615/20) and microscopes (from the mid-century). Although in Paris Daniel Chorez (pre-1585–1659) was making telescopes and microscopes from 1625 onwards, and spyglasses may have been made there from as early as 1611, in London there seems to have been no response from mathematical instrument-makers. None of those active in the first quarter of the seventeenth century are known to have made any optical instruments.

In part, this was probably because of the difficulty of obtaining suitable glass or rock crystal (the development of glass-making in London associated with Sir Robert Mansell would not begin for another decade or so), in part because spectacle-making was less well-developed in London where the makers would not be formed into a guild until 1629, in part because the new instrument would have seemed alien to the traditional London trade. Early optical instrument-making developed outside the existing structures, close to good sources of its raw material. This meant that in the early decades of the seventeenth century it was centred in Italy, where good glass could be had from the area around Florence and from Murano and lens-making took place in Florence with Galileo, followed by his sometime workman Ippolito Francini (1593–1653), Evangelista Torricelli (1608–1647), and Jacopo Mariani (seventeenth century), and in Naples with Felice Fontana. The workshop established by Torricelli came to an abrupt end with his early death, but at almost the same time workshops were established in Rome by Eustachio Divini (1610–1685) and Giuseppe Campani (1635–1715). Both were originally clock-makers who turned to lens-making—Divini in 1646, Campani c.1659 (van Helden, 1999).

Lens-making in Italy, however, was rivalled by activity in the Nuremburg–Augsburg region, which could rely on Bohemian glass. Born in Nuremberg, Johann Morian, 'optical gentleman', had perhaps learnt some of his skill in lens grinding from the local spectacle-makers, but it was in Augsburg that what was probably the first purely optical workshop in Germany was established by Johann Wiesel (c.1583–1662) around 1622. Of Wiesel's training nothing is known, but his activity developed steadily, due in part to the Augsburg art merchant Philip Hainshofer, in part to Wiesel's association with the Capucin, Anton Maria Schyrleus de Rheita, inventor of the erecting compound lens for telescopes (Keil, 2003). Although by the time Wiesel's instruments began to arrive in England in the late 1640s (mainly through the activities of the Hartlib circle; Keil, 1994), research and even some trade manufacture were beginning to develop there, they would have an important influence on it.

A trace of telescope-making can be found in London as early as 1626, when Captain John Smith mentions gunners' brass scales and 'prospective glasses' being made by Bates on Tower Hill, (Taylor, 210). With the improvement of London glass and the establishment of the Spectacle-Makers Company, optical instrument-making could develop—a process aided by the active interest taken in it by Walter Warner (c.1558–1643), a colleague of Harriot, and Sir Charles Cavendish (1591–1654), followed in the 1650s by Sir Paul Neile and Jonathan Goddard. Work in London was stimulated by the news sent through Hartlib of optical researches in Amsterdam, Dantzig, and Paris (Keil, 2003). Although in the latter city activity was still hampered by lack of good glass, stimulated by the interest of Mersenne, Descartes, and Daniel Aubéry, lens polishing was carried on with some vigour by Jean Ferrier in the 1620s and 1630s, and by the Sieur de Bressieux in the 1650s and probably earlier. Ferrier seems to have been a mathematical instrument-maker and mechanician who, like Chorez, was prepared to investigate new fields, and Bressieux a specialist 'opticall workman', who like Ferrier worked for a time for Descartes.

The optician with whom Cavendish, Neile, and Goddard worked in London was Richard Reeve. Like Robert Norman, he is an anomalous figure. He was probably not of artisan origin but issue of a landed family in Berkshire. He seems also not to have belonged to any of the guilds. He was sufficiently well off to live, from 1632 onwards, in fashionable Covent Garden, newly developed by the Duke of Bedford. He is first heard of as a practising optician in 1639, and from then until his death produced a series of notable instruments. More important here, as the master of Christopher Cock and probably of John Dunnett, he stands at the head of a craft affiliation line of optical instrument-makers in the Turners' and Spectacle-Makers' Companies which stretched well into the eighteenth century (Simpson, 1985). This succession of makers remains distinct from the mathematical instrument-makers, as indeed of the other specialized craftsmen mentioned previously.

Pressure for the improvement of optical instruments came primarily from astronomers, but a part of the output of commercial precision opticians was relevant to the concerns of physics. The invention of the microscope raised short-lived hopes that it would allow atoms of matter to be seen. Alongside his telescopes, Wiesel produced eye-glasses, flea-glasses, burning glasses, and perspectives, and so, we may suppose, did other leading opticians. Wiesel had sufficient demand from throughout Europe that in 1650 he produced a priced list of his products (Keil, 2003, 125). Burning glasses were of interest to experimental philosophers not just for enflaming substances, but also for concentrating heat, or when made of metal, for reflecting or concentrating light or concentrating and reflecting sound. Research in this area could of course, as it did for Mersenne, lead *savants* to the makers of musical instruments as a source for apparatus, in particular for monochords, though for his experiments on sound ratios Joseph Sauveur would have recourse to a traditional mathematical instrument-maker, one of the Chapotot family, for the 'sonomètre' that he conceived to pursue the work.

As the natural phenomena investigated by experimental natural philosophers in the seventeenth century multiplied, so the range of devices required to scrutinize them also expanded. Insofar that many of them were new, they could often not be supplied by the existing instrument-making trades. Instrument-making in the seventeenth century was a trade in formation. It should not be assumed that the traditional makers of mathematical instruments, balances, compasses, and the like adapted easily to produce the new apparatus demanded. In some cases they would do so, but in many others quite different craftsmen and skills were required. All, eventually, would coalesce to form that distinct, autonomous manufacture we today call scientific instrument-making, whatever specialist subdivisions there may be within it. But this was a development which largely post-dates the period considered here—one which perhaps begins only as the bespoke shops of the specialist craftsmen of the sixteenth and seventeenth centuries were gradually replaced, from the 1690s onwards, by retail shops holding a wide range of stock available on demand.

Exploration of the nature of air in the seventeenth century provoked a very considerable development of new instruments. Thermoscopes, barometers, thermometers, and air-pumps posed as great problems of manufacture as had the *versorium* and armed lode-stones of around 1600. Initially this meant that they were made by those who devised or experimented on them, or under their direct supervision by commissioned craftsmen. Whether Galileo is considered the inventor or (as seems more likely) not, of the thermoscope, he did engage with it as did his friend and correspondent Gianfresco Sagredo. On learning about an instrument of Santorio's 'with which it was possible to measure cold and heat with a compass [i.e. with a pair of dividers] ... I immediately began to make some which were beautiful and exquisite. The ordinary ones cost me 4 lire each, that is, a flask, a small ampule and a glass hose; and my production was such that in an hour I made as many as ten' (letter to Galileo, 30 June 1612, cited from Bedini, 1994(2), 258). Nearly three years later (15 March 1615), still experimenting, Sagredo wished to compare notes with Galileo '. . . because I believe that the instruments made by you, or by your excellent artisans, are much better than my own'. The situation here, as for precision lens-making, is clear; savants made their own instruments or had them made by craftsmen under their close direction. Thermometers—thermoscopes with an incorporated scale—had already been developed by 1611 (Middleton, 1966, 11–12). A new manufacture could therefore develop in areas where good glass was available and there was demand. Already in the 1620s this was the case in the Low Countries, where a characteristic, half-pear-shaped 'air thermometer' was developed. Shortly afterwards, in London, John Bate illustrated several forms of the instrument in his *The Mysteryes of Nature and Arts* . . . (1634). If, as circumstantial evidence makes probable, this John Bate is the same as the Bates on Tower Hill who made gunners' scales and spy-glasses (above, and cf. Walton, 2004), then one can see here in embryo, as with Daniel Chorez and Jean Ferrier in Paris, the assimilation of the new glass-based instruments into the older tradition of brass- or wood-made mathematical instruments.

As with thermometers, so with barometers; the instrument emanated from researches, spread over several years by savants in Italy, France, and England, into air and the vacuum. In this development, traditional instrument-makers played no part, since it was skilled glass-workers using good-quality glass who were required. Florence was therefore, almost inevitably, the first centre of these researches, with Torricelli the central figure. Exactly how important glass-working skills were to the Torricellian experiment is underlined by the failure of Mersenne and Pierre Chaunut to repeat it in July 1645 in Paris, where appropriate glass tubes could not be obtained, and the success of Pierre Petit and Pascal in October 1646 in Rouen, where not only good glass and glass-workers were available but the experiment, made on a large scale, could be carried out in the glass-works itself.

Skilled glass-working remained essential as continuing researches entailed modifications to the apparatus to make it portable and fit it with a scale. What is now known as the siphon barometer was devised by Pascal, probably in October 1646.

In the following twenty-five years the role of air-pressure in its functioning was established, and the possibility of using the instrument to correlate changes in this pressure with changes in the weather was recognized. But neither in Italy nor in London, where good glass was also available, were traditional mathematical instrument-makers involved in these developments.

Nor would they be until the third quarter of the century. Activity in London may then perhaps be taken as emblematic of what would subsequently occur throughout Europe. In about 1670, the interest of Francis North, Baron Guildford, in weather recording led him to suggest, first to a clock-maker (Henry Jones, c.1642–1695), then, after he lost interest, to a mathematical instrument-maker (Henry Wynne, fl.1662–1709), that barometers could be a commercially viable novelty. So it proved. Wynne 'pursued the manufactury to great perfection, and his own no small advantage; and then others took it up, and few clock-makers, instrument-makers, cabinet-makers and diverse other trades, were without them always in their shops, ready for sale' (North). What needs to be noted here is that although the instrument originated among the learned and was initially developed by them and the glass-makers whom they employed and directed, once it entered the public sphere and was introduced to the traditional mathematical instrument-makers they quickly adopted it and made their own developmental innovations.

The air-pump was closely related to the barometer. Like the barometer it originated as a research instrument for the learned, but unlike the barometer, even during the early stages of development, it had a closer relation with traditional instrument-making. When in the mid-1650s Robert Boyle became interested in the air, he ordered his first pump from the young London instrument-maker Ralph Greatorex (1625–1675; see Bendall, 2008), who had been apprenticed to Elias Allen, the doyen of English instrument-makers, in the Clockmakers' Company. He had thus had a traditional training. But he rapidly extended his range into general mechanics and hydraulics, and had wider interests in natural-philosophical subjects such as botany and chemistry. Greatorex indeed seems to belong among the artisan-philosophers (one of whom, Hooke, would as Boyle's assistant build his second air-pump); he had links with the mathematician William Oughtred, with the Oxford experimental group of the 1650s, and attended early meetings of the Royal Society in 1660. With him one can see, in London at least, the traditional instrument-maker reaching out to the adepts of new philosophy, and himself becoming absorbed into the wider scientific community. It is surely significant that Henry Wynne, the pioneer barometer-maker mentioned earlier, was Greatorex' apprentice, and that the latter had made a 'weather-glass' for Samuel Pepys.

In the later seventeenth century increasing numbers of instrument-makers, such as Francis Hauksbee the elder (1660–1713) in London, Hubin and Hartsoeker in Paris, and the Musschenbroeks in Leiden, took on the profile of the craftsman-experimenter, the artisan-philosopher. All three are examples of new men in instrument-making—men who had not had the traditional training of the mathematical instrument-maker. Hauksbee was the son of a Colchester draper, who began his career making air-pumps and pneumatic engines. Hubin was an English enameller who, established in Paris from at least 1673, made thermometers, hygrometers, hydraulic, and air machines. Closely associated with Papin, he made his air-pump and his 'digestor' to which, after careful trials, he made a series of modifications. The first instrument-making Musschenbroek, Samuel (1639–1681), came from a family of general brass-founders particularly well known for their oil lamps. The making of such everyday items and medical appliances would continue in his, and his successors, production for many decades. In the later seventeenth century two developments thus came together to produce the trade of instrument-making as it would affirm itself in the eighteenth century. Traditional mathematical instruments-makers incorporated into their basic production of sundials, surveying, navigational, and calculating instruments some of the new optical and physics instruments, while craftsmen from other fields, glass-working and hydraulics, general mechanics, entered the domain of instrument-making under the supervision of savants and experimenters in order to make new, exceptional, items for them. As these became generalized, so their makers, and their mentors, might also expand into the more traditional areas of mathematical instrument-making.

The essential element in all this was demand. The development of a new, more wide-ranging instrument trade therefore depended upon the dissemination and popularization of the developing new physics—first in the universities of Europe, then more widely in society at large. The crucial development was the realization that instruments used to discover and validate the tenets of the new philosophy could also be used to demonstrate and so communicate them. Demonstrating to convince was a technique typical of the Royal Society in London virtually from its beginning, and it was his experience of this during a visit in 1674 that led Burcherus de Volder (1643–1709), newly appointed as Professor of Philosophy at Leiden University, to demand official approval to employ experimental demonstrations in his teaching, to supply him with the requisite instruments and a space in which to deploy them.

All this was granted, and Leiden became one of the first universities in Europe to incorporate experimental demonstration into physics teaching (De Pater, 1975). It should be underlined, however, that this was in the context of a modified Cartesian physics, and was mainly related to the subject of air pressure and the vacuum. Although it would underlie the development of the Musschenbroek instrument workshop, it was not without precedent. Already in Paris in the late 1660s, Jacques Rohault (c.1620–1672) had been presenting weekly public lectures on experimental Cartesian physics using an apparatus of appropriate instruments. Such an approach stretches back, in the mathematical disciplines, to the sixteenth century and beyond, but just as the making of the new instruments of physics was grafted onto the older manufacture of mathematical instruments to create a new, more wide-ranging instrument manufacture of instruments related to all the sciences, so teaching by demonstration using adapted versions of the original research instruments established itself in parallel with the instrumentally based teaching of astronomy, cosmology, and practical geometry.

The demand essential to support instrument-making of a wider scope was thus established by the integration of the new physics into the university curriculum and cultivated consciousness. The example of Leiden was gradually followed throughout Europe. That the Cartesian physics within which it had originated was displaced by that of Newton would make no difference for the method of didactic demonstration was independent of the philosophical orientation of the physics taught so long as this was experimentally based.

Although some branches of enquiry in physics had already entered into the repertoire of the mathematical instrument-makers—in particular, in Paris, the skills deployed by Chorez in mounting lodestones continued throughout the seventeenth century in the hands of Jean Poüilly (*fl*.1680s) and Michael Butterfield (1635–1724), who would themselves demonstrate the qualities of the instruments they made in their shops—the Musschenbroek workshop in Leiden was probably the first to specialize in 'philosophical instruments', as physics demonstration apparatus came to be called. Samuel, and his successor Jan van Musschenbroek (1660–1707) were not only manufacturers. Although they worked closely with *savants* in the university to produce instruments tailored to their needs, they quickly mastered the principles involved, explaining them to their clients and exploiting them in their workshop to develop and refine the instruments they produced. They also innovated in trade methods. Because their trade was universal, with a particularly strong market in Germany, they quickly realized the usefulness, as Wiesel had done earlier, of a priced catalogue. The earliest example known dates from 1694, and it was followed by at least nine more. Fixing the prices meant that several instruments could be made together and kept in stock against future demand. This, however, did not preclude the making of bespoke instruments, for which the price would be negotiated between maker and client, as had been general practice throughout the preceding century (De Clercq, 1991, 1997).

Instrument-making in 1700 can be considered as a distinct branch of manufacture among the skilled trades of Europe, with a growing number of makers in London, Paris, and Leiden, which were the centres of international trade. Elsewhere, mainly in regional capitals and particularly in the court cities of Germany and Italy, smaller numbers of makers could be found catering to local markets. Some idea of growth in the 150 years considered here is given by the fact that in Paris at least five times as many instrument-makers are known from the seventeenth century as from the three centuries from 1300 to 1600 together (Beaudouin, Brenni, and Turner, in preparation). For Britain, figures are more precise. In 1551 the names of three instrument-makers are known; in 1601, 14; and in 1701, 151—nearly nine times as many (Clifton, 1995, xv). This growth in numbers reflects the growth in demand partly sketched above, but conceals the greater capacity and range of workshops by the end of the century. If the size of the workshop remained small, perhaps on average only two to three people, with very occasional exceptions such as the large manufactory of Jacob Leupold (1674–1727) in Leipzig, sub-contracting and the putting-out of work mitigated the limiting factor that the smallness of the basic workshop could have been upon production. Unfortunately for the seventeenth century there is virtually no direct information about these matters, and these remarks are based only on circumstantial evidence.

Nonetheless, production did expand, helped by the fact that alongside the guild-controlled workshops of the cities, there were court-based craftsmen such as those employed in the Medici workshops, entrepreneurs mechanicians such as Samuel Morland, and that whole group of technical-professionals or philosopher-artisans such as Hooke, Papin, Hartsoeker, Comiers, and Dalencé, fertile in conception and sometimes in the construction of instruments. Such men intersected with trade craftsmen sometimes as inventors, sometimes as clients, sometimes as advisers. Behind both groups stood the *savants* and the societies into which they increasingly organized themselves in the later seventeenth century. The practice of instrument-making as it impinged upon the nascent science of physics in the later sixteenth and seventeenth centuries was an amalgam of all these elements lacking an easily definable structure, but flexible and adaptable in the service of new ideas about nature, and responsive to the commercial opportunities that these could offer.

References

Annals of Science, xlviii, no. 4 (1991), special issue on early surveying.

DENIS BEADOUIN, PAOLO BRENNI, and ANTHONY TURNER, *A Bio-bibliographical Dictionary of Precision Instrument-Makers and Related Craftsmen in France, 1430–1960* (in preparation).

SILVIO A. BEDINI, 'The Makers of Galileo's Scientific Instruments' in *Atti del Simposio Internazionale di Storia Metodologio, Logica e Filosofia della Scienza Galileo Galilei nella Storia e nella Filosofia della Scienza*, Florence 1967, reprinted in Silvio A. Bedini, *Science and Instruments in Seventeenth-Century Italy*, Aldershot 1994, ch. II.

SILVIO A. BEDINI, 'the Instruments of Galileo Galilei', in Ernan McMullin (ed.), *Galileo, Man of Science*, New York 1967, reprinted in Silvio A. Bedini, *Science and Instruments in Seventeenth-Century Italy*, Aldershot 1994, ch. I.

SARAH BENDALL, 'Greatorex, Ralph', in *Oxford Dictionary of National Biography*, Oxford 2004; online edition January 2008. http://www.oxforddnb.com/view/article/11365. J. A. Bennet, 'Norman, Robert' in *Oxford Dictionary of National Biography*, Oxford 2004. http://www.oxforddnb.com/view/article/11365.

J. A. BENNETT and S. JOHNSTON, *The Geometry of War 1500–1750*, Oxford 1996.

KOENRAAD VAN CLEEMPOEL, *A Catalogue Raisonné of Scientific Instruments from the Louvain School 1530–1600*, (De Diversis Aretibus 65, ns 28), Turnhout 2002.

GLORIA CLIFTON, *Directory of British Scientific Instrument Makers 1550–1851*, London 1995.

NICHOLAS H. CLULEE, *John Dee's Natural Philosophy: Between Science and Religion*, London and New York 1988.

ANNA MARIA CRINÒ and HELEN WALLIS, 'New Researches on the Molyneux Globes' *Der Globusfreund. Journal for the Study of Globes and related Instruments*, xxxv–xxxvii 1987, 11–20.

PETER DE CLERCQ, 'Exporting Sientific Instruments around 1700: the Musschenbroek Dcuments in Marburg', *Tractrix: Yearbook for the History of Science, Medicine, Technology and Mathematics*, iii 1991, 79–120.

PETER DE CLERCQ, *At the Sign of the Oriental Lamp: The Musschenbroek Workshop in Leiden 1660–1750*, Rotterdam 1997.

C. DE PATER, 'Experimental Physics' in Th. H. Lunsingh Scheurleer and G. H. M. Posthumus Meyjes (eds), *Leiden University in the Seventeenth Century: An Exchange of Learning*, Leiden 1975, 309–327.

S. R. EPSTEIN, 'Craft Guilds, Apprenticeship and Technological Change in Pre-industrial Europe', in S. R. Epstein and Maarten Prak (eds.), *Guilds, Innovation, and the European Economy 1400–1800*, Cambridge 2008, 52–80.

ANTHONY GERBINO and STEPHEN JOHNSTON, *Compass and Rule: Architecture as Mathematical Practice in England 1500–1750*, Oxford, New Haven. and London 2009.

ROBERT B. GORDON, 'Sixteenth Century Metalworking Technology used in the Manufacture of Two German Astrolabes', *Annals of Science* xliv 1987, 71–84.

J. H. HEILBRON, 'Aristotelian Physcs' in W. F. Bynum, E. J. Browne, and Roy Porter (eds.), *Dictionary of the History of Science*, reissue, London 1983, 25–27.

GERARD M. M. HOUBEN, *2000 Years of Nested Cup-Weights*, Zwolle 1984.

INGE KEIL, 'Technology Transfer and scientific Specialisation: Johann Wiesel, Optician of Augsburg and the Hartlib Circle', in Mark Greengrass, Michael Leslie, and Timothy Raylor (eds.), *Samuel Hartlib and Universal Reformation: Studies in Intellectual Communication*, Cambridge 1994, 268–278.

INGE KEIL, *Von Ocularien, perspillen und Mikroskopen, von Hungersnöten und Friedensfreuden, Optikern, Kaufleuten un Fürsten. Materialien zur Geschichte der Optischer Werkstatt von Johannn Wiesel (1583–1662) und seiner Nachfolger in Augsburg* (Documenta Augustana Bd 13), Augsburg 2003.

HERMANN P. LOCKNER, *Die Merkzeichen der Nuremberg Rotschmiede*, Munich 1981.

ARMAND MACHABEY JEUNE, *Mémoire sur l'Histoire de la balance et de la balancerie*, Paris 1949.

SUSAN P. MAXWELL, 'Molyneux, Emery', in *Oxford Dictionary of National Biography*, Oxford 2004; online edition January 2008. http://www.oxforddnb.com/view/article/11365.

W. E. KNOWLES MIDDLETON, *The History of the Barometer*, Baltimore 1964 (reprint edition 1968).

W. E. KNOWLES MIDDLETON, *The lHistory of the Thermometer and its Use in Meteorology*, Baltimore 1966.

A. D. C. SIMPSON, 'Richard Reeve: the "English Campani" and the Origins of the London Telescope-Making Tradition', *Vistas in Astronomy*, xxviii 1985, 357–65.

SYBRICH TER KUILE and W. E. J. MÖRZER BRUYNS, *Amsterdamse kompasmakers c.1580–c.1850. Bijdrage tot de kennis van de instrumentmakerij in Nederland*, Amsterdam 1999.

ANTHONY TURNER, *Early Scientific Instruments: Europe 1400–1800*, London 1987.

GERARD L'E TURNER, *Elizabethan Instrument-Makers: The Origins of the London Trade in Precision Instrument Making*, Oxford 2000.

ANTHONY TURNER, ' "Not to Hurt of Trade": Guilds and Innovation in Horology and Precision Instrument Making', in S. R. Epstein and Maarten Prak (eds), *Guilds, Innovation, and the European Economy 1400–1800*, Cambridge 2008, 264–87.

ALBERT VAN HELDEN, 'The Invention of the Telescope', *Transactions of the American Philosophical Society*, lxvii 1977, 1–67.

ALBERT VAN HELDEN, *Istituo e Museo di Storia della Scienza: Catalogue of Early Telescopes*, Florence 1999.

STEVEN A. WALTON, 'Bate, John' in *Oxford Dictionary of National Biography*, Oxford 2004. http://www.oxforddnb.com/view/article/53656.

DAVID W. WATERS, *The Art of Navigation in England in Elizabethan and Early Stuart Times*, London 1958.

CHAPTER 5

NEWTON'S *PRINCIPIA*

CHRIS SMEENK AND ERIC SCHLIESSER[1]

5.1 INTRODUCTION

In the Preface to his *Mathematical Principles of Natural Philosophy* (hereafter, *Principia*) Newton announces a striking new aim for natural philosophy and expresses optimism that the aim can be achieved (Newton, 1726, pp. 382–3):

For the basic problem of philosophy seems to be to discover the forces of nature from the phenomena of motions and then to demonstrate the other phenomena from these forces. It is to these ends that the general propositions in books 1 and 2 are directed, while in book 3 our explanation of the system of the world illustrates these propositions. For in book 3, by means of propositions demonstrated mathematically in books 1 and 2, we derive from the celestial phenomena the gravitational forces by which bodies tend toward the Sun and toward the individual planets. Then the motions of the planets, the comets, the Moon, and the sea are deduced from these forces by propositions that are also mathematical. If only we could derive the other phenomena of nature from mechanical principles by the same kind of reasoning!

Pronouncements such as this often reveal a contrast between the goals of a given scientist, however astute, and the subsequent historical development of the field. But this is not so in Newton's case: Newton's work effectively reoriented natural

---

[1] We thank Bill Harper and George Smith for helpful discussions, and especially Erik Curiel and Niccolo Guicciardini for detailed comments on earlier drafts. The usual caveats apply. Sections 5.3.6 and 5.4.1 overlap with forthcoming publications by one of us: Eric Schliesser, 'Newton and Newtonianism', in the *The Oxford Handbook of British Philosophy in the Eighteenth Century*, edited by James Harris, Oxford: Oxford University Press; and 'Newton and European Philosophy', in *The Routledge Companion to Eighteenth Century Philosophy*, edited by Aaron Garrett, London: Routledge.

philosophy for generations. The *Principia*, which appeared in three editions (1687, 1713, 1726) clarified the concept of force used in physical reasoning regarding motion, and marshalled evidence for one such force: gravity. We read the *Principia* to be guided by Newton's evolving recognition of various challenges to evidential reasoning regarding forces, and his development of the tools needed to respond to these challenges. The mathematical results Newton achieved provided an initial framework in which to pursue the project of discovering forces and finding further evidence in favour of gravity. He called attention to the role gravity played in a wide variety of natural phenomena ('the motions of the planets, the comets, the Moon, and the sea'). He gave a precise, quantitative treatment of phenomena that had previously been the subject of inchoate speculation, such as the perturbing effects of planets on each other. He settled decisively the great unresolved cosmological question of his era: the status of the Copernican hypothesis. He defended a Copernican–Keplerian account of planetary motions, and showed on the basis of universal gravity that the Sun itself moves, albeit not far from the common centre of gravity of the solar system (Prop. 3.12).[2] The impact of universal gravity on the subsequent study of celestial mechanics is hard to overstate. Newton's conception of gravity remains a durable part of celestial mechanics, even though it was augmented in the eighteenth century and corrected by Einstein's theory of general relativity in the twentieth century.

Alongside this achievement in laying the mathematical foundations for mechanics, the *Principia* also exemplifies a 'new way of inquiry'. In its mathematical style and approach to mechanics it most closely resembles Christian Huygens' *Horologium Oscillatorium*, which Newton greatly admired. Just as Huygens had generalized and considerably enriched Galileo's results in mechanics, Newton generalized Huygens' treatment of uniform circular motion to an account of forces applicable to arbitrary curvilinear trajectories. He then used this enriched Galilean–Huygensian approach to mechanics to describe the motions of the planets and their satellites rather than only bodies near the Earth's surface. Newton characterized his methodology as offering more secure conclusions than those reached via the hypothetical methods of his contemporaries. His claim to achieve *greater* certainty, given the centrality of a theoretically defined entity, such as force, to his approach, was controversial at the time, and has remained so.

Newton introduced a striking new goal for natural philosophy—'to discover the forces of nature from the phenomena of motions and then to demonstrate the other phenomena from these forces.' He articulated the challenges to achieving this goal, and developed an innovative and sophisticated methodology for overcoming the challenges. Newton's contemporaries and later readers have had great difficulty comprehending his new way of inquiry, and his own attempts to clarify his new 'experimental philosophy' late in his life, in response to controversy, are often too cryptic to provide much illumination. The growing authority of Newtonian

---

[2] This refers to Proposition 12 in *Principia*, Book 3. We will generally refer to propositions in this way, unless the relevant book is clear from context. All page references and quotations are from (Newton, 1726).

science, and the stunning reach and apparent certainty of the claims it makes, made Newton's methodology one of the most contested areas in eighteenth-century philosophy. Here we will draw on recent scholarship on the *Principia* and emphasize three interwoven aspects of Newton's approach. First, Newton establishes mathematical results that license inferences regarding forces that are robust in the sense that they do not require that claims regarding phenomena hold *exactly*. Second, he identifies the various assumptions needed to define a tractable mathematical model, and then assesses the consequences of relaxing these idealizations. And, finally, the initial idealized model serves as the first step towards a more sophisticated model, with the deviations between the idealized case and observed phenomena providing further empirical input.

This chapter discusses the historical context of the *Principia*, its contents, and its impact, with a primary focus on the issues just described. We should acknowledge at the outset two of the important topics for which we do not have sufficient space for the treatment they deserve. First, and most important from the standpoint of the history of physics, we will leave aside entirely Newton's optical works (Shapiro, Chapter 6). The *Opticks* differs from the *Principia* in much more than subject matter; in it Newton elucidates a sophisticated experimental approach, with an expository style that is accessible and free from the *Principia*'s daunting mathematics. The *Opticks* engendered a research tradition of 'Newtonian philosophy' of a different character than that produced by the *Principia*. Of course, many of Newton's readers interpreted the *Principia* in light of the *Opticks*, especially the more speculative 'Queries', as we will occasionally do here.

Second, we also do not aim to present an account of Newton's overall philosophical views or to place the *Principia* in the context of his other intellectual pursuits. The last half-century has seen a renaissance in Newton scholarship due in part to the assimilation of newly available manuscript sources. These manuscripts reveal that his published work occupied only a fraction of his prodigious intellectual energy. He pursued alchemy, biblical chronology, and theology with the same seriousness of purpose as the work culminating in the *Principia*. He did not regard these pursuits as completely isolated from one another, and neither do we. Earlier generations of historians have often dismissed these other aspects of Newton's thought out of hand, but various scholars have undertaken the ambitious project of understanding the genesis of the *Principia* in relation to these other pursuits.

In the General Scholium added to the second edition of the *Principia*, and in the Queries in the *Opticks*, Newton took the opportunity to publicly announce some of the connections between his work in natural philosophy and broader questions in philosophy and theology, and he also considered making these views more explicit in substantial revisions to parts of the *Principia*. He held his cards close to his chest with regard to his dangerously heterodox theological views and his work in alchemy, and as a result, aside from his closest colleagues, few of his contemporaries or successors had access to his views on these other subjects. There are a number of fascinating and difficult questions that face any attempt to produce a systematic treatment of Newton's philosophical and theological views. But given our aim of elucidating the impact of the *Principia* on the development of physics, we focus here on the *Principia* itself and related published texts that would have been widely available to Newton's contemporaries and successors.

5.2 Historical Context

The seventeenth century saw the emergence of a new systematic approach to natural philosophy, called the 'mechanical philosophy', which offered explanations of natural phenomena in terms of matter in motion without appeal to Aristotelian forms. By mid-century, Hobbes, Gassendi, and Descartes had elaborated on and defended significantly different versions of this view. Descartes' *Principles of Philosophy* (1644) was by far the most influential on Newton, and thus sets the context for his own distinctive natural philosophy. Newton's very title alludes to the earlier *Principia*. And as his title offers an implicit correction to Descartes, so too throughout the *Principia* Newton is often at pains to distinguish his views from those of Descartes and his followers. One crucial part of Newton's intellectual context was his critical engagement with a Cartesian version of the mechanical philosophy.

The ambition and scope of Descartes' *Principles* is breathtaking: it offers a unified physics, cosmology, and geology, including mechanical explanations of everything from magnetism to earthquakes (Schuster, Chapter 3). Planets move as they do because they are embedded in a whirling vortex of subtle matter, according to Descartes, and the interaction among vortices associated with different stars explains various other phenomena, such as comets. This entire system was meant to follow from an austere set of basic principles regarding the nature of bodies and laws governing their motion. Few readers were convinced by Descartes' claim to have deduced these laws of nature from his metaphysical first principles, and even a quick reading of Parts III and IV reveals that the connection between the systematic account of nature offered there and the basic principles in Part II leaves more room for the free play of the imagination than Descartes allows. But it provided a compelling research agenda for the mechanical philosophy. Hypothetical explanations would be judged to be intelligible provided that they invoked only the size, shape, and motion of the fundamental constituents of a system, moving according to fixed laws. In the new mechanical philosophy, the laws of motion, including rules for the collision of bodies, would be the linchpin of theorizing about nature.

Within two decades many leading natural philosophers had grown dissatisfied with Descartes' analysis of collision, which was widely seen as empirically and conceptually inadequate. By adopting Galilean principles, Christiaan Huygens was able to give a satisfactory analysis of collision by the 1660s. (Newton had reached the same conclusions privately.) During that decade a consensus grew around his mathematical treatment when the Royal Society published short pieces by Wallis and Wren leading to similar conclusions (though they differed in their metaphysical presuppositions).

Huygens argued that the quantity preserved in a collision is not mv (as Descartes had supposed) but mv^2, which Leibniz later termed *vis viva*; Mariotte's *Traité de la percussion ou chocq des corps* (1673) represented the decisive rejection of Descartes' laws of collision. Yet, the Cartesian mechanical philosophy, with its appeal to intelligibility and simplicity of hypothetical explanations (including modified vortex theories and the denial of a vacuum), subsisted well into the eighteenth century.

While the mechanical philosophy offered the most unified approach to nature, it was not uncontested. In particular, its commitment to the passivity of matter was repudiated by many important natural philosophers. In his influential *De Magnete*, William Gilbert introduced the idea of an 'orb of virtue' in describing a body's magnetism. Gilbert's work directly influenced Kepler's discussion of the distant action of the Sun on a planet, ruled unintelligible by strict mechanists. Throughout the century there were numerous proposals that posited action at a distance (including theories developed by Roberval and Hooke), some of these inspired by the publication at mid-century of Gilbert's posthumous *De mundo nostro sublunari philosophia nova*. One author closely studied by Newton, Walter Charleton, who helped revive Epicurean theory in England, advocated a view of matter with innate principles of activity akin to attractive powers. At Cambridge University the so-called Cambridge Platonists allowed that matter was passive but under strict control of mind-like spiritual substances. Against the Cartesian identification of indefinite space and extension (and its denial of a vacuum), Henry More developed a view of infinite space (and time) as an immaterial entity that emanated from God. More insisted that all entities, material and immaterial, including God, occupied some place in space. This view was also adopted by Newton.

By the 1660s, when Newton began his study of natural philosophy, the most sophisticated natural philosopher was Huygens. Huygens significantly extended Galileo's study of accelerated motion and developed the idea of relativity principles in his derivation of the laws of elastic collision. In the 1650s he advanced the Galilean programme by creating a theory-mediated measurement of the acceleration of bodies in the first half second of fall at Paris that was accurate to four significant figures, using different kinds of pendulums. Huygens' approach was aided by his ground-breaking mathematical analysis of the pendulum. Huygens' discovery that the cycloid was an isochronous curve opened up precise time-keeping (useful in astronomy, geography, and mechanics) and fuelled the search for a practical solution to finding longitude at sea. It also raised the question of whether gravitation was uniform around the globe. The strength of surface gravity (reflected in the length of a seconds pendulum) varied with latitude, but without apparent systematicity. Before publication of the *Principia*, Huygens' *Horologium Oscillitorium* (1673) represented the state of the art in mechanics, and Newton greatly admired the book.

A second aspect of Newton's intellectual context was the development of predictive astronomy. Kepler's innovations were mostly neglected by Cartesian philosophers, in part due to his problematic mix of neo-Platonism and ideas that were incompatible with the mechanical philosophy. Cartesian philosophers did not develop quantitatively precise versions of the vortex theory to rival Kepler's account; in the context of Descartes' theory it was, after all, not clear whether the planetary orbits exhibit stable regularities or are instead temporary features subject to dramatic change as the vortex evolves. Huygens, despite his work in observational astronomy, including the discovery of Saturn's largest moon, Titan, only published a detailed cosmology in response to Newton's *Principia*. Kepler's work set the agenda for those interested in calculating planetary tables, and his proposals led to a substantial increase in accuracy. His *Rudolphine Tables* received a Twere vindicated when Gassendi, in Paris, observed the transit of Mercury in 1631. The gifted English astronomer, Jeremiah Horrocks observed the transit of Venus in 1639, but his work remained little known during his (brief) life. Despite the success of Kepler's innovations in leading to more accurate planetary tables, his physical account of planetary motion was controversial; Boulliau, for example, dismissed his physical account as 'figments'.[3] When Newton began his study of astronomy with Streete's *Astronomia Carolina* there was an active debate underway regarding the best method of calculating planetary orbits. Kepler had motivated what we call his 'area law' on physical grounds, but Boulliau, Streete, and Wing had each proposed alternative methods for calculating planetary positions with comparable levels of accuracy. It is certainly not the case that astronomers prior to the *Principia* took 'Kepler's three laws' to reflect the essential properties of planetary orbits that should inform any physical account of their motion. The first to single out 'Kepler's laws' was in fact Leibniz (1689), who perhaps intended to elevate Kepler's contributions at Newton's expense.

Newton stands at the convergence of Keplerian astronomy and Galilean–Huygensian mechanics, uniquely able to use the latter to provide a firmer physical footing for the former because of his enormous mathematical talent. The third aspect of Newton's intellectual context is the development of mathematics and the central role of mathematics in his new mode of inquiry. Descartes is again the pivotal figure; Newton's early work was guided by his close study of van Schooten's second edition of the *Géométrie*.[4] Among the central problems in mathematics at the time were the determination of the tangent to a given curve and quadrature (finding the area under the curve), for curves more general than the conic sections. Descartes and others had solved these problems for a number of special cases, but from 1664 to 1671 Newton developed a general algorithm for solving these problems and discovered the inverse relation between finding the tangent and performing quadratures—and in that sense he 'invented the calculus'. His generalization of the binomial theorem and use of infinite series allowed him to handle a much broader class of curves than those treated by Descartes. While this is not the place to review these contributions

---

[3] Here we draw on Wilson (1970) description of the debates in assimilating Kepler's innovations in astronomy and Newton's responses to it; see also (Smith, 2002). (See p. 107 of Wilson's paper regarding Boulliau's criticism of Kepler.)

[4] This edition, published in 1659 (the original appeared in 1637), contained extensive supplementary material, including correspondence between Descartes and other mathematicians and further work by Van Schooten's Dutch students—including Hudde, Heuraet, and de Witt (later the leader of the Dutch republic)—on problems posed by Descartes.

in more detail (see, in particular, Whiteside's *Mathematical Papers* and Guicciardini, 2009), we will briefly describe the distinctive mathematical methods employed in the *Principia*.

Newton's mathematical talents and new techniques enabled him to tackle quantitatively a much wider range of problems than his contemporaries. But equally important was his view that the judicious use of mathematics could be used to reach a level of certainty in natural philosophy much greater than that admitted by the mechanical philosophers. Descartes and others regarded the mechanical models they offered as intelligible and probably accurate, but Newton claimed to be able to achieve more certainty. Newton formulated this view quite stridently in his *Optical Lectures*:[5]

Thus although colours may belong to physics, the science of them must nevertheless be considered mathematical, insofar as they are treated by mathematical reasoning ... I therefore urge geometers to investigate nature more rigorously, and those devoted to natural science to learn geometry first. Hence the former shall not entirely spend their time in speculations of no value to human life, nor shall the latter, while working assiduously with an absurd method, perpetually fail to reach their goal. But truly with the help of philosophical geometers and geometrical philosophers, instead of the conjectures and probabilities that are blazoned about everywhere, we shall finally achieve a science of nature supported by the highest evidence.

Here the level of certainty to be attained contrasts with that of the Cartesian programme and also with that of Newton's immediate contemporaries—Fellows of the Royal Society, such as Hooke and Boyle. The degree of certainty Newton had achieved with his 'New Theory of Light and Colours' (1672) soon became the focus of a contentious debate, drawing in Hooke and Huygens, among others. But Newton continued to advocate the importance of mathematics in his new way of inquiry, and we next turn to a study of the contents of the *Principia* and the essential role of mathematics in enabling his deduction of gravity from the phenomena of celestial motions.

5.3 Overview of the *Principia*

5.3.1 From *De Motu* to *Principia*

Newton took the first steps toward writing the *Principia* in response to a problem posed by Edmond Halley in the summer of 1684.[6] Christopher Wren had offered Halley and Robert Hooke the reward of a 'forty-shilling book' for a proof that

---

[5] The lectures were deposited in October 1674 and perhaps delivered during 1670–72. The quotation is from *Optical Papers*, Volume 1, pp. 87, 89; for further discussion of Newton's position in relation to his optical work and more broadly, see Guicciardini, (2009), Chapter 2; Shapiro, (2002, 2004); and Stein (ms).

[6] See (Cohen, 1971) for discussion of the circumstances leading to the publication of the *Principia*, including Halley's visit. Newton reportedly answered Halley's question during the

elliptical planetary trajectories follow from a force varying as the inverse square of the distance from the Sun. The challenge proved too great for Halley and Hooke, and Halley consulted Newton while on a visit to Cambridge.[7] That November, Newton replied with a nine-page manuscript bearing the title *De Motu Corporum in Gyrum* (hereafter *De Motu*). Newton's results in this brief paper alone would have secured him not only Wren's reward but a place in the history of mechanics, and we will describe its contribution as a prelude to the *Principia*.[8] The most striking contribution is bringing together the Galilean–Huygensian tradition in mechanics with astronomy, unified via the new conception of centripetal force. But committing these initial insights to paper was only the first step in a line of inquiry that Newton would pursue with incredible focus and insight for the next three years.

Halley, Hooke, and Wren had a plausible physical motivation for considering a force whose intensity decreases with the inverse square of distance from its source. Huygens' treatment of uniform circular motion in terms of centrifugal force combined with Kepler's 'third law' implied that the force varies as the inverse square of the distance for an exactly circular trajectory.[9] What they lacked was a conceptualization of force sufficiently clear to allow them to relate this hypothesized variation with distance to a trajectory, and to assess the implications of this idea for an elliptical trajectory. In correspondence in 1679, Hooke had already pushed Newton to take an important step in the right direction, to conceiving of planetary trajectories as resulting from a tendency to move in a straight line combined with a deflection due to an external force.[10] But it was only in *De Motu* that Newton combined this idea with other insights to establish the connection between a given curvilinear trajectory and the force responsible for it.

De Motu's beautiful central result, Theorem 3, starts from a generalization of Galileo's treatment of free fall. Galileo established that under uniform acceleration, the distance travelled by a body starting at rest is proportional to the square of the elapsed time. What Newton required was a precise link between a quantitative

---

visit, but could not find the paper where he had already performed the calculation. We do not know how this earlier calculation compared to the manuscript he produced subsequently.

[7] It is now customary to distinguish between two related problems: the direct problem– *given* the orbit or trajectory, find a force law sufficient to produce it, and the inverse problem— *given* the force law and initial position and velocity, determine the trajectory. It is unclear precisely what problem Halley posed to Newton, but *De Motu* addresses the direct problem.

[8] Here we emulate the effective presentation in (de Gandt, 1995), which also includes a discussion of the *Principia*'s mathematical methods and historical context.

[9] Kepler's third law states that $P^2 \propto a^3$ for the planets, where P is the period and a is the mean distance from the Sun. For a discussion of the understanding of Kepler's 'laws' among Newton's contemporaries, see (Wilson, 1970).

[10] See Hooke's correspondence with Newton in 1679–80 (in *Correspondence of Isaac Newton*, Volume 2), and his earlier work cited there. Hooke did not formulate inertial motion as Newton later would, in that he did not say that bodies move *uniformly* in a straight line. However, Hooke certainly deserves more credit than Newton was willing to acknowledge for pushing him to treat curvilinear motion as resulting solely from inertia and a centripetal force.

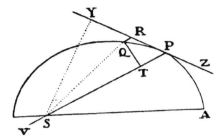

Fig. 5.1 Figure from Proposition 6 in *Principia* (and Theorem 3 in *De Motu*).

measure of the trajectory's deviation from a straight line at each point of the orbit and the magnitude of the force producing this deviation, for forces whose magnitude varies from point to point. The initial draft stated a generalization of Galileo's result as an hypothesis: namely, the deviation produced by *any* centripetal force is proportional to the square of the elapsed time, 'at the very beginning of its motion'. The proof of this result as Lemma 10 in the *Principia* clarifies the importance of the last clause: it is only 'ultimately' (or 'in the limit' as the elapsed time goes to zero) that the proportionality holds. In effect, Galileo's result holds for *finite* elapsed times in the special case of uniform acceleration, but Newton recognized that it is valid *instantaneously* for arbitrary centripetal forces. Newton's next step established that the elapsed time is represented geometrically by the area swept out by a radius vector from the force centre following the trajectory. Theorem 1 of *De Motu* established this result, now known as Kepler's area law, granted Newton's conception of inertial motion and the restriction to centripetal forces that depend solely on the distance to a force centre (that is, central forces). (Newton would later establish the converse as well: namely, that a body sweeping out equal areas in equal times around a given point experiences a net impressed force directed to that point.) Combining these two results leads to an expression relating the magnitude of the force to geometrical properties of the trajectory. In terms of Newton's diagram (Fig. 5.1), the deviation produced by the force acting at point P is represented by the segment QR. This displacement is proportional to the product of the force F acting on the body with the square of the time elapsed, $QR \propto F \times t^2$, as shown by the generalization of Galileo's law.[11] From Kepler's area law, $t \propto SP \times QT$, and it follows that $F \propto \frac{QR}{SP^2 \times QT^2}$.

This result states an entirely general relation between the centripetal force law and the properties of a given trajectory. It holds instantaneously, in the limit as the point Q approaches the point P. But in applying this result, Newton establishes connections between the 'evanescent' figure QRPT and finite quantities characterizing the trajectory, such as the radius of a circle or the *latus rectum* of an ellipse. This leads to an expression characterizing the force law that makes no reference to quantities that vanish in the limit as $Q \to P$. Consideration of the figure QRPT allows Newton to handle the differential properties of the curve geometrically. Along with other special

---

[11] In terms of the *Principia*'s definitions, F is the accelerative measure of the force.

cases, Newton noted that for motion on an ellipse with a force directed at one focus this result implies that the force varies inversely as the square of the distance from the focus.

Newton achieved a great deal in the few pages of *De Motu*. The central achievement is a unification of Kepler's treatment of planetary motion with the Galilean–Huygensian theory of uniformly accelerated motion based on Newton's innovative treatment of force. Theorem 3 allows one to determine a force law sufficient for motion along a given trajectory with a given force centre. Newton generalizes Huygens' earlier treatment of uniform circular motion to arbitrary curvilinear trajectories, and this opens the way for considering a variety of possible central forces sufficient for motion along different plane curves. In *De Motu* Newton took on two different types of problem: projectile motion in a resisting media, and the motion of celestial bodies under an inverse-square force law.

Kepler's area law is strikingly given special standing as a direct consequence of any central force, rather than just being one among many calculational devices in astronomy. Similarly, Newton's work clarifies the status of Kepler's first and third laws. According to Kepler's first law, the planets follow elliptical trajectories. An inverse-square force directed at the focus is sufficient to produce this motion if, in addition, the second law holds with respect to the focus of the ellipse (that is, the radius vector from the focus sweeps out equal areas in equal times).[12] Thus, insofar as Kepler's laws hold exactly for each planet, one can infer an inverse-square force between the Sun and each of the planets. Kepler's third law is a specific instance of a general result linking periodic times to the exponent in the force law (Theorem 2). Furthermore, the ratio of the radii to the periods (a^3/P^2) is the same for all the planets, leading to the conclusion that a *single* inverse-square force directed at the Sun suffices for the motions of the planets. These results led Newton to announce in a scholium that the planets move in their orbits 'exactly as Kepler supposed'. Although he soon recognized that this conclusion is too hasty, he had persuasively answered what may have been Halley's original query. Halley and others would naturally have wondered whether the inverse-square force sufficient for perfectly circular orbits would have to be supplemented by a secondary cause to account for elliptical motion. Newton's results show convincingly that a simple inverse-square force alone is sufficient for Keplerian motion (though these results do not establish that it is *necessary* for Keplerian motion).

The *Principia* grew out of a number of questions provoked by *De Motu*.[13] Two questions would have been particularly pressing to Halley upon reading the manuscript. First, what do Newton's ideas imply regarding a glaring potential

---

[12] Elliptical orbits are also compatible with a force law that varies directly with the distance if the area law holds with respect to the *centre* of the ellipse rather than a *focus*, as Newton showed.

[13] Although we will not trace the details here, the original *De Motu* manuscript was extended step by step in a series of revisions leading up to the *Principia*; see (Herivel, 1965) and Volume 4 of the *Mathematical Papers*.

counter-example to the claim that all the celestial bodies move in accord with Kepler's laws: namely, the Moon? The Moon's motion is far more complicated than that of the planets, and it was unclear whether Kepler's laws hold for the lunar motion as even a rough approximation. The challenge facing Newton was to see what an inverse-square force implies regarding lunar motion. Second, how do the ideas in *De Motu* apply to comets? The nature of comets was the focus of an active debate among Hooke, Halley, Flamsteed, Newton, and others.[14] A central issue in this debate was whether the regularities observed in planetary motion hold for comets as well. In the early 1680s Newton argued against the idea that various sets of observations could be described as those of a *single* body, with a sharp button-hook trajectory around the Sun. Yet in a draft letter to Flamsteed in 1681, he considered the possibility that comets move in this way as a result of an attractive force analogous to magnetism. But the advances of *De Motu* opened up the possibility of a precise quantitative treatment of cometary motion. Newton suggested a procedure for determining cometary orbits that would prove unworkable. A successful account of cometary motion based on an inverse-square force would show that the force effects the motion of bodies other than the planets, over a much wider range of distances from the Sun than those explored by planetary orbits.

5.3.2 Definitions and the Laws of Motion

The opening part of the *Principia* extends and refines the ideas of *De Motu* in several significant ways. First, Newton gives a much more precise characterization of force in the three Laws of Motion.[15] A predecessor of the First Law appears in *De Motu* as Hypothesis 2: 'Every body by its innate force alone proceeds uniformly to infinity in a straight line, unless it is impeded by something extrinsic.' This formulation makes one important advance on earlier formulations of inertial principles such as that due to Descartes. Newton explicitly remarks that the motion will be *uniform*, covering equal distances in equal time, as well as rectilinear. The formulation in the *Principia* marks another important step, in that Newton drops the implication—present in all earlier formulations of the law—that an external impediment must be present to deflect a body from inertial motion. The formulation of the *Principia* requires only that an impressed force acts on the body, but this force is treated abstractly without commitments regarding its mode of operation or source.

The Second Law is implicit in *De Motu*'s quantitative treatment of forces as measured by the deflection from an inertial trajectory.[16] But it is not stated explicitly as it is in the *Principia*:

---

[14] See (Wilson, 1970, pp. 151–160) for a brief overview of the debate regarding comets.

[15] We are reversing the order of exposition in the *Principia*, where the definitions and the scholium on space and time, discussed below, precede the Laws.

[16] Actually, only part of Second Law is used: what is usually called the parallelogram law for the addition of forces, and the generalization of Galileo's treatment of acceleration, discussed above.

Law II: A change in motion is proportional to the motive force impressed and takes place along the straight line in which that force is impressed.

The 'change in motion' is measured by the deflection QR in the diagram above; that is, the distance between the point the body would have reached moving inertially and the point actually reached due to the action of an impressed force. 'Motive force' is introduced in the definitions as one of the three measures of a centripetal force, given by the motion generated in a unit time. Earlier definitions stipulate that quantity of motion is given by the mass times the velocity, and hence the motive force measures the impressed force by the product of the mass of the body and the resulting acceleration. Newton applies the law to cases involving discrete impressed forces, such as impacts, as well as continuously acting forces.

In addition to the motive measure of force, Newton also introduces the accelerative and absolute measures. These three distinct measures quantify different aspects of a force: the absolute measure of force characterizes the overall strength of the force—in the case of gravity this would correspond to the mass of the body producing the force. The accelerative measure characterizes the intensity of the force as a function of radial distance, as revealed in the acceleration it produces on *any* body at that distance.[17] Finally, the motive measure characterizes the force impressed upon a body of a given mass, and is given by the product of the mass of the body with the accelerative measure.

Newton claimed no originality for the first two Laws, and there are antecedents in earlier work, particularly Huygens' *Horologium Oscillatorium*.[18] But there are nonetheless two innovations in Newton's formulation that deserve emphasis. First, Newton treats force as an abstract mathematical quantity independent of any commitments regarding its physical sources. What this allows is the decomposition of a force into arbitrary oblique components, explicitly stated as Corollary 2 of the Laws. The importance of this move is easily overlooked from a modern perspective, but this Corollary is crucial to many of the results in Book 1. For example, Prop. 1.40 establishes that the velocity acquired falling through a given radial distance from a force centre is independent of the path taken as a direct consequence of Corollary 2; Huygens, lacking a similarly abstract conception of force, had to give a more laborious proof of the same result (Proposition VIII of the *Horologium*). The second innovation is the introduction of a concept of 'mass' distinct from the 'weight' or 'bulk' appealed to in earlier work.[19] Mass does not even appear in the initial draft

---

[17] Stein, (1967) argues that for gravity this can be treated as an 'acceleration field', but for other types of interaction this need not be the case.

[18] For more detailed discussion of Newton's formulation of the laws and his conception of force, see, in particular, (McGuire, 1968; McGuire, 1994; Gabbey, 1980), and Chapter 3 of (Janiak, 2008).

[19] Huygens introduced a concept close to Newtonian 'mass' in formulating a law for momentum conservation in elastic impacts in 1669, in response to the Royal Society's competition regarding collision laws. Although Huygens' concept of mass agrees with Newton's in particular cases, Huygens did not formulate the concept at the same level of generality as

of *De Motu*; it emerged only in the course of clarifying the laws.[20] The definitions in conjunction with the first two laws make it clear that mass or quantity of matter is to be measured by a body's response to an impressed force, its resistance to acceleration. Newton then establishes experimentally that mass can be measured by weight.

The Third Law is entirely absent from *De Motu*, although Newton had used it as an illustrative example in lectures on algebra much earlier.[21] There the law is formulated as the principle of conservation of linear momentum, but in work leading to the *Principia* Newton considered several different equivalent statements of the Third Law before deciding on the following:

Law III: That to an action there is always a contrary and equal reaction; or, that the mutual action of two bodies upon each other are always equal and directed to contrary parts.

This formulation emphasizes that forces should be understood as interactions between bodies; to speak of *separate* impressed forces acting on two bodies, which happen to come in an action–reaction pair, is misleading. The 'force' corresponds to a mutual interaction between bodies that is *not* broken down into separate 'actions' and 'reactions', except in our descriptions of it. Given its novelty, Newton's discussion in the scholium following the laws and their corollaries focuses on defending the Third Law. The examples used there illustrate the connection between the Third and First Laws. In order for the First Law to hold for the centre of mass of a closed system of interacting bodies, the Third Law must hold for the interactions among the bodies.[22]

In *De Motu* the centripetal forces are treated just as accelerative tendencies towards a fixed centre, without regard to whether they are produced by a body. The Third Law did not figure in the discussion, and indeed only the accelerative measure of the force is relevant. But in the *Principia* the Third Law supplements the first two laws by allowing one to distinguish apparent from real forces. Given a body in motion, the first two laws allow one to infer the existence of a force producing the motion that may be well defined quantitatively (given a definite magnitude and direction). But the Third Law further requires that the force results from a mutual interaction. The Coriolis force—the apparent force revealed, for example, in the deflection of a ball rolling along the floor of a carousel—illustrates this distinction: the force is well defined quantitatively and could be inferred from observing a body in motion using the first two laws of motion and results from Book 1, but there is no 'interacting body' to be found as the source of the force. The first two laws figure primarily in treating forces from a mathematical point of view whereas the introduction of the Third Law marks an important physical constraint. Although Newton famously abstained from requiring a full account of the 'physical cause or reason' of a force as a precondition

---

Newton did in the *Principia*. See (McMullin, 1978) and Chapter 4 of (Janiak, 2008) for entry points into discussion of Newton's innovative concept of mass.

[20] It first appears in *De Motu Corporum in Mediis Regulariter Cedentibus*; see (Herivel, 1965).

[21] See *Mathematical Papers*, Volume 5, pp. 148–149, f.n. 15.

[22] Newton's illustrative examples involve contiguous bodies, but he does not hesitate to extend the Third Law to cover interactions between non-contiguous bodies as well.

for establishing its existence, any further account of the physical nature of the force would have to satisfy the constraint imposed by the Third Law.

Newton came to consider the implications of the Third Law for planetary motion soon after completing the first version of *De Motu*. His initial results suggested that the motions of the planets resulted from an inverse-square centripetal force produced by the Sun. Similarly, the motion of the satellites of Jupiter suggested the existence of an independent centripetal force produced by Jupiter. How do these two forces relate to one another? How far does their influence extend? How do they effect the motion of a comet? Newton clearly considered such questions before adding what Curtis Wilson has aptly called the 'Copernican Scholium' to a later version of *De Motu*. In this passage (quoted more fully below), Newton argues that the 'common centre of gravity ... ought to be considered the immobile centre of the whole planetary system', and adds that this proves the Copernican system *a priori*. In order to define the common centre of mass of the solar system, Newton applied the Third Law to combine the distinct accelerative tendencies produced by the planets and the Sun. Thus by this point Newton had taken one step towards universal gravity, by apparently treating the forces responsible for celestial motions as a mutual interaction between the Sun and the planets.[23] But, as we will see shortly, with this Newton further recognized that the problem of inferring the force responsible for celestial motions from observations was far more challenging than the problem posed by Halley. The need to face this challenge squarely led to the more elaborate theory of generic centripetal forces developed over the course of Book 1, and dictated the form of the argument for universal gravitation in Book 3.

Before turning to that issue we need to address the part of the *Principia* that has received by far the most attention from philosophers: the scholium to the definitions, regarding space and time. Unfortunately, much of this philosophical commentary has been based on misreading Newton's aim as that of proving the existence of absolute space and time. By contrast, we follow Stein, (1967) and other recent commentators in reading Newton as clarifying the assumptions regarding space and time implicit in the Laws of Motion.[24] The laws draw a fundamental distinction between inertial and non-inertial motion. What conception of space and time is needed for this distinction to be well-founded?

---

[23] Exactly how to characterize this stage in the development of Newton's thought is not entirely clear. As George Smith has emphasized, Newton can infer from Law 4 in *De Motu* that the distances of the Sun and Jupiter to the centre of mass must hold in fixed proportion, with the proportion determined by a^3/P^2 as a measure of the absolute strength of the centripetal force. Thus the step to the two-body solution could be driven by the idea that the centre of gravity should remain stationary, rather than by conceiving of gravity as a mutual interaction subject to Law 3 of the *Principia*.

[24] See (Rynasiewicz, 1995a; Rynasiewicz, 1995b; DiSalle, 2006) for more detailed readings of Newton's arguments in the scholium that are in broad agreement with the line we sketch here. See (McGuire, 1978) for an influential assessment of Newton's debt to Gassendi, More, and others, and the relationship between the views in the Scholium and Newton's theological views.

Newton argued that the laws presuppose a sense of 'absolute motion' that cannot be analysed adequately in terms of the relative positions among bodies. Although he does not identify a target of criticism in the *Principia*, in an unpublished manuscript, 'De Gravitatione', he trenchantly dissects Descartes' relational definition of motion and reveals its inadequacy as a basis for physical reasoning (including Descartes' own).[25] The dispute does not turn on whether there is a distinction between two different senses of motion—motion as defined to be 'merely relative' to an arbitrarily chosen set of reference bodies, versus a body's unique 'true' or 'proper' motion. Rather, the question is whether an adequate definition of the absolute motion can be given solely in terms of relative quantities—spatial and temporal relations with respect to other bodies. Newton annuls Descartes' proposed definition of true motion and further argues that any adequate definition requires reference to the absolute structural properties of space and time.

The required structures are intervals of spatial distance and temporal duration, along with a unique identification of locations over time.[26] Newton characterizes these structures as 'absolute' in several different senses: they are 'immutable' and do not change in different regions or in response to the presence of bodies; they are intrinsic rather than conventional; and they are not defined in terms of the relations between material bodies. Based on these structures, Newton defines absolute motion as 'the translation of a body from one absolute place into another'. By contrast, relative motion is defined with respect to relative spaces, delimited by some particular bodies regarded as immovable. Newton argues that the two are distinguished 'by their properties, causes, and effects'. As an illustration, in the argument that absolute and relative motions differ in their effects Newton invokes the famous example of a bucket of water hanging from a rope, wound up, and then released. When the water in the

---

[25] This manuscript was first published (and translated) in (Hall and Hall, 1962). The date of composition has been the subject of some dispute: it was initially regarded as an early manuscript, composed during Newton's days as an undergraduate, but more recent commentators have argued for a later date given the maturity of the views expressed and the connection between the positions adopted in 'De Gravitatione' and the *Principia*. A great deal of attention has been devoted to this manuscript, because Newton is more forthright in addressing metaphysical issues in it than in his published work.

[26] In more modern mathematical treatments, Newtonian spacetime is described as a differentiable manifold that is topologically $\Sigma \times \Re$. Σ is a three-dimensional space representing 'all of space at a single instant of time', and time is represented by $\Re$ (the real numbers). (To be more precise, these structures are represented by *affine* spaces, in which there is no preferred 'origin'.) Each of these spaces is endowed with separate metrics, such that the following quantities relating events, points in spacetime, are well-defined: (1) the spatial distance between two events at a single instant, and (2) the time elapsed between any two events. In order to distinguish between motion along a 'straight' vs. 'curved' line further structure is needed: the 'kinematical connection' relating locations at different times, which is not fixed by the structures assigned separately to space and time. The kinematic connection can be characterized by its symmetry properties. In what is now called 'full Newtonian spacetime', these symmetries include time-independent translations, rotations of the spatial coordinates, and translation of the time coordinates. See Stein, (1967) and Earman, (1989) for discussions of this approach.

bucket is rotating, it moves away from the axis of rotation, leading to a curved surface. But this curved surface is neither a necessary nor a sufficient condition for *relative* rotation between the water and the bucket: not necessary because initially the surface of the water remains flat despite the relative rotation of the water and the bucket, and not sufficient because after the water has caught up with the rotating bucket, the surface of the water is curved despite the lack of relative rotation of the water and the bucket. As with the arguments from properties and effects, Newton concludes that absolute motion (illustrated here by the dynamical effects of rotation) cannot be analysed in terms of relative motion.[27]

Yet despite this clear distinction between absolute and relative motion, Newton acknowledges that we lack direct access to absolute motion. Our observations are always of relative motions, given some stipulated relative space that we may provisionally take to be absolute space. How then is it possible to determine the true, absolute motions? Newton remarks that the entire *Principia* was composed to elucidate this problem, and 'the situation is not entirely desperate'. Given the connection between forces and absolute acceleration, it is possible to determine whether motions described with respect to a relative space could be taken as absolute motions. Only for the case of absolute motions will the impressed forces exactly match with accelerations. For relative motions there will in principle always be some discrepancy—for example, the Coriolis forces that arise if one choses a relative space that is in fact rotating. This force is not due to a mutual interaction among bodies; instead, it indicates that the proposed relative space cannot be taken as absolute space. Thus Newton's comment that the entire *Principia* is needed to differentiate absolute from relative motions should be taken seriously, for this can be accomplished only by accurately identifying the relevant forces acting on bodies. In the empirical study of motion, absolute motion does not enter directly into the analysis of motions, because it is inaccessible. But the distinction between absolute and relative motions plays a fundamental role in the analysis, given Newton's argument that merely relative ideas are not sufficient to capture the distinction between accelerated and non-accelerated motions.

A final point of clarification concerns the identification of different locations over time. Newton's discussions in the scholium and 'De Gravitatione' suggest a *unique* way of identifying a given position in absolute space over time, which would imply that absolute position and velocity are well-defined quantities. But this is more than is required to fulfil the project of the *Principia* of determining the forces responsible for motions, as indicated by absolute accelerations. Newton states a version of Galilean relativity as Corollary 5 to the Laws: the relative motions of a closed system of bodies are not affected if the entire system moves uniformly without rotation. This makes it clear that absolute positions and velocities are irrelevant to

---

[27] Newton directed this argument against a *Cartesian* account of relative motion, in which motion is defined relative to immediately contiguous bodies, and he apparently did not consider the possibility of a Machian alternative that drops the restriction to contiguous bodies.

determining the absolute accelerations. In modern terms, ensuring that absolute acceleration is well-defined requires a weaker structure than a unique identification of locations over time. This structure, called an *affine connection*, allows one to define the amount of curvature of a spacetime trajectory *without* introducing absolute position and velocity. Inertial trajectories correspond to straight lines through spacetime, and curvature of a path represents acceleration. The next corollary states a further sense in which the structures elucidated in the scholium go beyond what is required:

Corollary 6: If bodies are moving in any way whatsoever with respect to one another and are urged by equal accelerative forces along parallel lines, they will all continue to move with respect to one another in the same way as they would if they were not acted on by those forces.

Thus Newton recognized a sense in which acceleration is also relative: in this case, the accelerations of a system of bodies are judged relative to a common, shared acceleration, which is itself irrelevant to the internal relative motions. This poses a deep challenge to the framework of the *Principia*, as it suggests that in the case of gravity the distinction between inertial and non-inertial motion is not well founded empirically. Einstein clearly recognized the importance of this problem 220 years later, and his efforts to solve it guided his path towards a new theory of gravity.

While Huygens developed a very insightful relational response to Newton in his notebooks, public disputes regarding Newton's views on space and time began within Newton's lifetime in the famous Leibniz–Clarke correspondence. Leibniz initiated a tradition of criticizing Newton from a relationalist perspective. Some of the criticisms in the resultant literature misinterpret Newton as offering something other than a clarification of the assumptions regarding space and time presupposed by the empirical project of the *Principia*. (This is not surprising, as Newton does not address many basic questions about the nature of space and time in the scholium, though he did address some of them in the unpublished 'De Gravitatione'.) Even regarded as such, Newton's account can be criticized as introducing excessive structure, as indicated by Corollaries 5 and 6. These issues were significantly clarified only in the nineteenth century with the emergence of the concept of an inertial frame, and further, in the twentieth century, with the reformulation of Newtonian theory based on Cartan and Weyl's idea of an affine connection following Einstein's discovery of general relativity.

The advent of relativity theory is often taken to vindicate the relationalist critiques of Newton. However, although the transition to general relativity leads to significant conceptual differences, spacetime geometry still underwrites a fundamental empirical distinction between different types of motion. Newton's distinction between inertial and non-inertial motion is replaced by a distinction between freely-falling and non-freely-falling motion in general relativity. In Newtonian terms, an object is 'freely falling' if the net non-gravitational force acting on the object is zero, and the resultant motion is described as a consequence of gravity and inertia. General relativity treats inertia and gravitation as manifestations of a single underlying 'inertio-gravitational field', and freely-falling motion is represented by geodesic curves in a curved spacetime geometry. At least in this sense, that there is a physically distinguished type of motion closely tied to spacetime geometry, Newton's seminal analysis remains valid in contemporary theories.

5.3.3 Book 1

The mathematical theory developed in *De Motu* gave Newton a variety of results that would allow him to infer the centripetal force given an exact trajectory and centre of force. But he soon realized that these results were not sufficient for determining the force responsible for the planetary trajectories, as he noted in the 'Copernican Scholium' (Hall and Hall 1962, p. 280):

> By reason of the deviation of the Sun from the centre of gravity, the centripetal force does not always tend to that immobile centre, and hence the planets neither move exactly in ellipses nor revolve twice in the same orbit. There are as many orbits of a planet as it has revolutions, as in the motion of the Moon, and the orbit of any one planet depends on the combined motion of all the planets, not to mention the action of all these on each other. But to consider simultaneously all these causes of motion and to define these motions by exact laws admitting of easy calculation exceeds, if I am not mistaken, the force of any human mind.

How should one make inferences from observations, given this daunting complexity? Newton recognized the multifaceted challenge to reasoning from the phenomena to claims regarding forces, and Book 1 lays the groundwork for his response. In what follows we will highlight aspects of Book 1 that reveal Newton's novel and sophisticated response to this challenge. But this is only half of the task—discovering the forces from the phenomena of motions—that Newton identified as the basic problem of philosophy in the Preface. We will also note how the results of Book 1 contribute to the second task: that of demonstrating new phenomena from these forces.

Up to the end of section 7, Book 1 buttresses the results stated in *De Motu* without extending them into entirely new domains.[28] Section 1 provides a mathematical prologue—a series of lemmas regarding the use of limits that Newton takes to justify his innovative geometrical techniques. The next nine sections treat the motion of bodies in response to forces, treated as accelerative tendencies directed towards a fixed centre. That is, Newton does not consider the physical origin of the forces, and only the accelerative measure of the force is relevant to these results. Section 2 states a number of results licensing inferences from a trajectory to a force law (*De Motu*'s Theorem 3 appears here as Proposition 6). The next section shows that the trajectories will be conic sections if and only if the force varies inversely with the square of the distance from the focus.[29] Sections 4 and 5 collect a number of geometrical results

---

[28] In the second edition, Newton added important material based on his way of measuring the 'crookedness' of a curve using an osculating circle; see (Brackenridge, 1995) for discussions of these changes and the importance of this different approach.

[29] Newton stated this result cited in the text as Corollary 1 to Prop. 13. Few of his contemporaries were convinced by his terse argument in its favour, though arguably his subsequent

Newton intended to utilize in determining cometary trajectories. He later devised a simpler method (Proposition 3.41), so these results play no role in the *Principia* itself.[30] Section 6 augments an earlier result in Section 3 that gives a solution for the velocity along the trajectory, and addresses the general problem of determining a body's location along a conic section trajectory given an initial position and an elapsed time. Section 7 adds to *De Motu*'s treatment of rectilinear ascent and descent, establishing how to recover results of Galileo and Huygens relating time, distance, and velocity, in this case by treating the rectilinear ascent as the limiting case of motion along a conic section treated earlier. Section 7 starts by considering specific force laws—$f \propto r^{-2}$ and $f \propto r$—but the final result solves the problem of rectilinear motion under *arbitrary* centripetal forces, up to quadrature.[31]

Section 8 builds on this generalization of Galileo's treatment of free fall, and carries Newton beyond what he had achieved in *De Motu*. Propositions 41 and 42 give a general solution for projectile motion under arbitrary central forces, again up to quadrature, derived with some of the most sophisticated mathematics in the entire *Principia*. They are based on a fundamental physical insight: namely, that Galileo's principle that the velocity acquired in falling through a fixed height is independent of the path traversed also holds, suitably generalized, for motion under arbitrary central forces. Although Newton did not use Leibniz's terminology, this is equivalent to conservation of *vis viva*. Results from earlier sections generally draw inferences from a given trajectory to a claim regarding the force law. But the results in Section 8 allow for the study of the effects of a force given only an initial position and velocity (rather than an entire trajectory), and thus Newton has, in a sense, fully solved the 'inverse problem' with these propositions. (We discuss the limitations of these results in Section 5.3.)

The central result of Section 9 is that apsidal precession of an orbit provides a sensitive measure of the exponent of the force law producing that motion. (The apsides are the points of maximum and minimum distance from one focus of an elliptical

---

results in Section 8 do justify this claim (see Guicciardini, 2009, pp. 54–56, 217–223, for discussion). The other propositions describe a variety of properties of motion along conic sections. For example, for the case of a circular trajectory, the force directed at the *centre* varies either as r^{-2} or r. Later, in Section 7, Newton treats direct fall along a rectilinear trajectory as a limiting case of motion along a conic section.

[30] A conclusion to Lemma 19 makes it clear that Newton's mathematical aim was to provide 'a geometrical synthesis, such as the ancients required, of the classical problem of four lines' (Newton 1726, p. 485). The problem in question was the Pappus problem, which requires the construction of a plane curve bearing a specified relationship to n lines. Descartes claimed that his ability to solve the Pappus problem for n lines was one of the main advantages of his algebraic analysis, and Newton clearly intended to counter this argument; see (Bos, 2001) and (Guicciardini, 2009), chapter 5. We thank Niccolò Guicciardini for pointing out to us that Newton's intended rebuke of Descartes misses the mark. Descartes' case for the superiority of his method in fact rested on the claim that in principle he could solve the Pappus problem for $n = 8, 10, 12, \ldots$ lines, but Newton's solution does not generalize to $n > 4$.

[31] 'Up to quadrature' means that there is still an integral to be performed to obtain the full solution, but *in principle* this can be done once the force law is specified.

orbit; in the case of apsidal precession, the body does not form a closed orbit and the apsides shift slightly with each revolution, by an amount given by the apsidal angle.) Proposition 1.45 considers the relationship between apsidal motion and the force law producing it; the first corollary states that for nearly circular orbits if the apsidal angle θ is given by $n = (\frac{\theta}{\pi})^2$, then the force is given by $f \propto r^{n-3}$. This result is robust in the sense that if the apsidal angle is *approximately* θ, the force law is *approximately* $f \propto r^{n-3}$.[32]

Obtaining this striking result required a great deal of mathematical ingenuity, but our main interest is in how it allows Newton to more reliably draw inferences from the phenomena. For the sake of contrast, consider an argument employing Theorem 3 of *De Motu* (Prop. 1.11 of the *Principia*) to infer an inverse-square force from the phenomena of motion—an argument often mistakenly attributed to the *Principia*'s Book 3. What does the theorem imply for an *approximately*, rather than *exactly*, elliptical orbit? How does one establish observationally that the force is directed at the focus rather than the centre?[33] As Smith, (2002) argues convincingly, throughout the *Principia* Newton relies on inferences from the phenomena that are not fragile in the sense of being valid only if the antecedent holds exactly. Instead, propositions like 1.45 establish that an observable quantity (the apsidal angle) can be taken as measuring a theoretical quantity (the exponent of the force law), within some clearly delimited domain, due to the law-like relationship that holds between the two. Inferences based on these results are robust given that the law-like relation holds over some range of phenomena, *not* merely for special cases of exact values for the observable quantities.

Constrained motion, such as motion along an inclined plane or of a pendulum bob, was a central topic in the Galilean–Huygensian tradition. In Section 10 Newton shows how to recover essentially all of the earlier results within his more general framework. A crucial motivation for this section was to assess the validity of Huygens' experimental technique for measuring surface gravity using a pendulum. In *Horologium Oscillatorium*, Huygens had established that a cycloidal pendulum is isochronous—that is, the oscillations have the same period regardless of the starting point of the bob—and placed bounds on how closely a small-arc circular pendulum approximates the isochrone. But Huygens treated gravity as constant (not varying with height) and directed along parallel lines, and it is natural to ask whether his reasoning remains valid for truly centripetal forces that vary with distance and are directed towards a centre of force. The question is pressing because a crucial step in the argument for universal gravitation, the 'Moon test', depends on Huygens' result.

---

[32] The result is also robust in a second sense: namely, that the relationship between precession and the force law holds for orbits of increasing eccentricity; strikingly, though Newton did not show this, the relationship holds *more precisely* as eccentricity increases (Valluri, Wilson and Harper, 1997).

[33] Newton derives two distinct force laws for an object moving along an elliptical orbit: for a force directed at a focus, $f \propto r^{-2}$, and a the centre, $f \propto r$. For an ellipse with eccentricity very close to 1 (such as planetary orbits), the foci nearly overlap with the centre.

In an impressive mathematical display Newton introduces curves more general than the plane cycloid studied by Huygens and shows that the true isochrone for a force law varying directly with distance is given by one of these curves, the hypocycloid. The most striking feature of this section is the attention Newton paid to the assumptions involved in Huygens' measurement.

Newton's results in Sections 2 through 10 establish that motion under centripetal forces is a theoretically important *category* of motion, and that it subsumes the Galilean–Huygensian study of uniformly accelerated motion as a special case. But it is also general in the sense that Newton establishes many results that hold for *arbitrary* central forces.[34] This level of generality is crucial to inferring forces from the phenomena, as the force law itself is the unknown quantity to be inferred from observations. The ability to consider arbitrary central forces leads to a much stronger inference: it opens up the space of competing proposals, rather than restricting consideration to checking the consequences of a single postulated force law. Newton's approach allows him to characterize the physical consequences of a variety of different proposals, providing a rich set of contrasts to compare with observations. Furthermore, it allows for the possibility that specific phenomena can be taken as measuring the parameters appearing in the force law, in the sense that there are law-like connections between the phenomena and parameters within some delimited domain.[35] Establishing such connections requires quantifying over a range of different force laws, which would not be possible without Newton's level of generality.

The generality of the treatment of forces in Book 1 is significant in a different sense: in principle it allows Newton to consider empirical contrasts among the phenomena deduced from distinct force laws.[36] Several results in Book 1 isolate striking features of motion under specific force laws. These results contribute not to the initial inference of a force law but to further assessment of the force law by contrast with other alternatives.

Newton begins Section 11 by acknowledging that the preceding results are all based on an unphysical assumption: they concern 'bodies attracted toward an immovable centre, such as, however, hardly exists in the natural world'. For conceiving of forces physically as arising from mutual interactions, the Third Law implies that interacting bodies will move about a common centre of gravity. Newton shows that a systematic treatment of motion is still possible without the simplifying assumption of the previous sections, and obtains as before solutions for the trajectories of bodies given a force law and initial positions and velocities. In fact, the results of the previous section generalize directly to the case of two interacting bodies: Proposition 58 shows

---

[34] Most of Book 1 considers only $f \propto r^n$ for integer values of n, but several important theorems allow for rational values of n (for example, 1.45).

[35] One of the main themes of Harper, (2011) is that Newton uses the phenomena to measure theoretical parameters in roughly this sense, and that Newton judged a theory to be empirically successful if diverse phenomena give agreeing measurements of the same parameters.

[36] In practice, however, Newton's mathematical methods were sufficient to handle only a handful of force laws.

that for a given force law a body follows the same trajectory around the second interacting body, also in motion, *and* around their common centre of gravity, as it would around an immobile force centre for the same force law. Kepler's area law also holds for either trajectory described with respect to the common centre of gravity *and* also for the trajectory of one body described with respect to the other. Applied to a two-body system with an inverse-square force law, for example, this result shows that each body describes an ellipse with the other body at one focus (Corollary 2). Newton further shows in Proposition 59 that Kepler's harmonic law has an important correction in the two-body case. Proposition 61 solves for the force law from the common centre of gravity that would be needed to treat a two-body problem as an equivalent one-body problem, allowing Newton to apply earlier results (in Propositions 62–63) to find trajectories for arbitrary initial velocities.

Relaxing the simplifying assumption even further to solve for the trajectories of three or more interacting bodies proves much more challenging. In the two-body case the centre of gravity is co-linear with the two bodies, and the distances of each body to the centre of gravity holds in fixed proportion. This consequence of the Third Law is what makes the two-body case fairly straightforward. But it is only in the case of a force law that varies as $f(r) \propto r$ that the general n-body problem is also straightforward due to the Third Law; Proposition 64 shows that in this case the trajectories will be ellipses with equal periodic times with the common centre of gravity at the centre of the ellipsis.

The three-body problem for $f(r) \propto r^{-2}$ is not nearly so simple. Newton strikingly drops the mathematical style of Book 1 and describes in prose the effects of gravitational perturbations to Keplerian motion in a series of twenty-two corollaries to Prop. 1.66, for a situation like that of the Earth–Moon–Sun system. Although Newton characterized his efforts on this problem as 'imperfect' (in the Preface), his results provided an initial step towards lunar theory, treatment of the tides, and an account of precession of the equinoxes (all taken up in Book 3). These results provide some indication of the empirical contrasts between a two-body description of motions and a more realistic account incorporating the effects of multiple bodies.

Sections 12 and 13 relax a further simplifying assumption: namely, that the finite extension of bodies could be entirely neglected, in the sense that there is a single trajectory characterizing the body's motion and that the accelerative tendencies directed toward a body are directed towards a point. These sections reveal a further aspect of Newton's conception of force: he treats the force acting on or produced by a macroscopic body compositionally, as a sum of the forces acting on constituent parts. How does acknowledging the finite extent of bodies, and treating the forces compositionally, affect the results already obtained? The case of highly symmetric bodies is the most tractable mathematically, and Section 12 proves a series of stunning results regarding the net force due to spherical shells on bodies inside or outside the shell. This approach culminates in Propositions 75–76: two spherical bodies interacting via an inverse-square attraction to all parts of both bodies can be described as interacting via an inverse-square attraction directed to their respective centres, with the absolute measure of the force given by their respective masses. Proposition 75 establishes this result for spheres of uniform density, and 76 shows that it holds as well for spheres whose density varies only as a function of the radial distance. Newton further considers the same question for an f(r) ∝ r force law (Props. 77–78) and generalizes to an arbitrary force law (Props. 79–86). But the inverse-square force law is quite distinctive in admitting of the remarkable simplification established by Propositions 75–76, and Newton later emphasized this striking feature of the inverse-square law (in Prop. 3.8). Section 13 goes on to consider how to treat bodies of arbitrary shape. The f(r) ∝ r force law is tractable: bodies of any shape can be treated as if their mass were concentrated at their centre of gravity (Props. 88–89). For the case of an inverse-square force, the general case is mathematically intractable, but Newton obtains several results for special cases, such as the force felt by a body along the axis of rotation of a spheroid (Props. 90–91).

These three sections reveal a quite sophisticated approach to handling idealizations. Rather than acknowledging the unphysical assumptions built into the account of motion developed in Sections 2–10, and then arguing that the mathematical theory nonetheless was a good approximation for some phenomena, as his predecessors had done in similar situations, Newton developed the mathematics to assess the effects of removing the idealizations. Although his results were by no means complete, they could be used to characterize qualitatively the departures from the initial idealized treatment due to many-body interactions and the finite extent of real bodies. And indeed many of the effects identified in these sections were already relevant to assessing the application of the theory to the solar system, and Newton returns to these issues in Book 3.

This sequence of results illustrates Newton's approach to the complexity of observed motions through a series of approximations, which Cohen, (1980) called the 'Newtonian style'.[37] The simplest idealized description of motion obtained in the earlier sections does not exactly describe observed motions, due to its false simplifying assumptions. Even so, the idealized models can be used in making inferences concerning values of physical quantities as long as these inferences are robust in the sense described above. The further results put Newton in a position to assess whether particular systematic deviations from the idealized models can be eliminated by dropping specific assumptions and developing a more complicated model. If the empirical deviations from the ideal case are of this sort, then the research programme can proceed by relaxing the simplifying assumptions. But it is also possible to identify systematic deviations that cannot be traced to a simplifying assumption, which may instead reveal deeper problems with the entire framework of Book 1. Thus Newton's treatment of idealizations allows for empirical results to continue to guide research, even though the identification of the deviations in question presupposes that the simplest idealized models are approximately correct, or, as he says in the fourth rule of reasoning, 'nearly true'.

---

[37] In this brief discussion we draw on more recent studies of the Newtonian style—in particular (Harper and Smith, 1995; Smith, 2001; Smith, 2002; Smith, 2009; Harper, 2011).

5.3.4 Book 2

The final two propositions of *De Motu* concern the motion of projectiles in a resisting medium. Galileo had been able to treat projectile motion only in the idealized case of a projectile that does not encounter resistance, and Newton considered the consequences of including a resistance force that depends on the velocity v of the projectile relative to the medium. This initial discussion of the effects of resistance grew into a treatise on a variety of issues in fluid mechanics, incorporating some of the most challenging mathematics of the *Principia* and its most detailed experimental results.

The derived collection of results has a less coherent structure than Books 1 and 3, though most of the propositions in Book 2 are directed towards three main goals. The first of these is to extend the Galilean–Huygensian study of motion to cover motion in resisting media. A central problem in this area was that of determining the trajectory of a projectile in the air and its dependence on various parameters, and Newton's results constituted substantial progress, albeit not a full solution. A second goal is to provide a framework for understanding resistance forces quantitatively that would allow one to infer properties of resistance forces in actual cases experimentally, in a style similar to the treatment of gravity in Book 3. Finally, Newton gives an empirical argument against the influential vortex theory of planetary motion proposed by Descartes. Unlike Descartes, Newton assesses in quantitative detail the properties of motion exhibited by a body immersed in a fluid vortex, and argued that it was incompatible with the approximately Keplerian motion of the planets.

The first three sections of Book 2 consider the effects of resistance forces that vary with different powers of the relative speed of the object and medium. These sections consider the motion of a body under Galilean gravity (uniform gravity directed along parallel lines) through a medium with resistance proportional to v, v^2, or a linear combination of both terms. The challenge was to recover results like those of Book 1, section 8, where Newton reduced to quadrature the problem of determining the trajectory of a body moving under a specific central force, given the initial position and velocity. Newton made several important steps toward a fully general result of this kind. Proposition 2.10, for example, shows how to find the density of a medium that will produce a given trajectory, assuming Galilean gravity and a resistance force proportional to ρv^2 (where ρ is the density of the medium). This is not a solution of the inverse ballistic problem—that is, a solution for the trajectory given initial position, velocity and the forces. Instead Newton 'adapts the problem to his mathematical competence' (to borrow a phrase from Guicciardini), and the result is fairly limited. Newton showed how to utilize this result to obtain a variety of approximate solutions in the scholium following 2.10.

The historical impact of these sections, and Proposition 2.10, is partially due to the mathematics Newton employed in them. Here Newton explicitly introduces analytical techniques, in the form of a brief introduction to the method of 'moments' he had developed in the 1670s (Lemma 2), and the proof of 2.10 relies on infinite power series expansions. It is not surprising that these sections drew the attention of the best Continental mathematicians. Johann Bernoulli famously found a mistake in the first-edition proof of 2.10 that his nephew Niklaus communicated to Newton after that part of the second edition had already been printed, forcing Newton to insert a correction.[38] Once the priority dispute regarding the invention of the calculus had arisen, this mistake was taken, erroneously, as evidence that Newton had not mastered the calculus by the time the *Principia* was written.

Later in Book 2 Newton reported a series of ingenious pendulum experiments designed to determine features of the resistance forces. In *De Motu* he proposed using projectile motion for this purpose, but this was not possible without a solution for the inverse ballistic problem. But in the case of a pendulum the trajectory is already given and the effect of resistance is merely to damp the oscillations. Propositions 2.30–31 establish systematic relationships between the amount of arc length lost due to resistance with each swing and the parameters appearing in the hypothesized total resistance force. Newton treated the total resistance as a sum of factors depending on powers of the relative velocity: that is, $f_r = c_0 + c_1 v + c_2 v^2$. The beauty of this approach was that Newton could, in principle, determine the relative contribution of these different factors (the coefficients c_0, c_1, c_2) merely by varying the starting point of the pendulum. However, he was only able to reach conclusions about the v^2 contribution to resistance acting on spheres: he argued this term was the dominant contribution to resistance (when v is large), and that it depended on the density of the medium ρ and diameter of the sphere d as ρd^2. The pendulum experiments did not measure the other contributions to resistance, and there were a number of discrepancies in the experimental results that Newton could not account for.[39]

Newton's dissatisfaction led him to substantially revise Book 2 for the second edition and to perform a completely different set of experiments. These experiments involved timing the free fall of globes dropped in water (in a 9-foot trough) and in air (from the top of St Paul's Cathedral). The purpose of these experiments was similar: namely, that of determining the properties of resistance forces. But they were useful in this regard only given a theoretical framework different than that in the first edition. Newton had classified different types of resistance in Section 7 as arising from different properties of a fluid. In the second edition he further gave a theoretical derivation of what he took to be the dominant contribution to resistance: namely, 'inertial resistance' due to the impacts between the particles of the fluid and the body. This result allowed him to predict a time for the freely-falling globes, and differences from the predicted value would then potentially reveal the other contributions to resistance, due to the 'elasticity, tenacity, and friction' of the parts of the fluid.

---

[38] See, in particular, Whiteside's exhaustive discussion in *Mathematical Papers*, Volume 8.
[39] Here we draw on (Smith, 2001), who gives a more detailed analysis of the experiments and their results, an explanation of why the approach is so promising, and an assessment of what may have gone wrong from a modern point of view.

The general approach here is similar to that in Book 3: Newton hoped to use observed phenomena to measure parameters appearing in a general expression for the force of resistance.[40] Doing so would directly respond to longstanding scepticism, expressed forcefully by Galileo, about the possibility of a real science of resistance forces. But despite the similar methodology, the outcome of the two lines of work contrasts sharply. Newton's starting point was fatally flawed, given that there is in fact no separation between an inertial contribution to resistance and other factors such as viscosity. This is vividly illustrated by D'Alembert's 'paradox': in 1752, D'Alembert showed that a body of *any* shape encounters zero resistance to motion through a fluid with zero viscosity. In addition, experiments like the ones initiated in the *Principia* would not have directly revealed the limitations of this starting assumption. Unlike the case of celestial mechanics, it was not possible to develop a full account of the nature of resistance forces from Newton's starting point via a series of successive approximations guided by further experimental results.

The closing section of Book 2 argues that a Cartesian vortex theory cannot satisfy both Kepler's second and third laws, and is therefore incompatible with observed celestial motions. Newton formulates this argument as conditional on an explicit hypothesis that fluid friction leads to a resistance force proportional to the relative velocity. From Newton's point of view the resulting argument has the weakness of depending on this hypothesis, which he presumably had hoped to establish via the earlier experiments on resistance.[41] One notable feature of Newton's argument in this section is that on behalf of the vortex theorist Newton develops a quantitative account of vortex motion (probably the first of its kind) that is not obviously contrived to fail. However, Newton's approach is deeply flawed, as Johann Bernoulli and Stokes would later emphasize, due to its erroneous treatment of torque. Even so, Newton had shifted the burden of proof: vortex theorists would need to show that a vortex is compatible with the observed regularities of celestial motions. In addition to planetary motions, the motion of comets was particularly difficult to account for in a vortex theory; comets cut across the planetary orbits, in some cases (such as the comet of 1682 Newton had observed) moving in the opposite direction as the proposed planetary vortex. In the second and third editions, it was also clear that the vortex theorists would have to further defend the impossibility of a vacuum. Roger Cotes' preface and Newton's General Scholium (both added in the second edition) both appealed to Boyle's air-pump experiments to bolster the claim that space could be freed from (most) matter.

---

[40] Here we emphasize some of the conclusions reached in (Smith, 2001); see also (Truesdell, 1968) for a critical assessment of Book 2 and discussions of the historical development of hydrodynamics in the eighteenth century.

[41] In the scholium to Proposition 2.52, Newton argues that this line of argument is robust in the sense that even if the hypothesis fails to hold exactly the vortex theory still fails to account for Kepler's third law.

5.3.5 Book 3

In the opening sequence of propositions in Book 3 Newton presents an argument for the law of universal gravitation. This line of reasoning is the centrepiece of the *Principia*.[42] It leads to the striking conclusion that every body in the universe attracts every other body with a force that varies as $f \propto \frac{m_1 m_2}{r^2}$, for bodies with masses m_1, m_2 at a distance r. We cannot overstate how shocking this claim was at the time; it far exceeded what many, even those well-versed in the mechanical tradition such as Huygens, expected could be established in natural philosophy. In what sense is the conclusion 'deduced from phenomena', as Newton described it, and what is the character of this line of argument? In particular, how does it differ from a hypothetical approach, as Newton insisted that it did? Finally, how does this opening sequence relate to the remainder of Book 3?

The first three propositions establish that the planets and their satellites, including the Moon, are held in their orbits by inverse-square centripetal forces directed towards their respective central bodies. The premises of these arguments are provided by phenomena stated at the outset of Book 3. 'Phenomena' does not refer to individual observations; rather, Newton uses the term for law-like regularities inferred from the data. The phenomena included the claims that the satellites and the planets satisfy Kepler's harmonic law and area law.[43] Newton then invokes propositions from the opening of Book 1 to determine the explicit form of the force law. But in addition to these phenomena, he notes that the conclusion can be established 'with the greatest exactness' based on Proposition 1.45 and the observation that the aphelia are at rest. The argument does not rely on Kepler's first law as the textbook tradition would have it; as noted above, the argument Newton in fact gives is not fragile in the sense of requiring the phenomena to hold exactly, as an argument based on the first law would. This first step of the argument was already clearly presented in *De Motu*, and it was uncontroversial among Newton's contemporaries. Huygens accepted it immediately. But it was only a first step towards universal gravitation.

The next step was far more striking for Newton's contemporaries. Proposition 4 identifies the centripetal force maintaining the Moon in its orbit with terrestrial gravity. Newton considers what the acceleration of the Moon would be were it brought to near the Earth's surface, as calculated from orbital acceleration in conjunction with the inverse-square variation in the centripetal force. But this turns out to be very nearly the acceleration of terrestrial bodies due to gravity, as measured by Huygens. Newton invokes the first and second 'Rules of Reasoning' to conclude that the two forces should be identified as a single force. A different way of considering the argument is to ask how Newton individuates different fundamental forces. The Moon test establishes that the centripetal force holding the Moon in its orbit and

---

[42] See also (Stein, 1991) for a clear discussion of the structure of Newton's argument in Book 3.

[43] Kepler's first law—that the planets move in elliptical orbits with the Sun at one focus—is *not* stated as a phenomena.

terrestrial gravity agree on all measures (absolute, accelerative, and motive). If we take forces to be individuated by having different measures, or different laws governing their variation, there is then no ground for claiming that there are two distinct forces rather than one. Proposition 5 extends this reasoning to the other planets and the Sun, concluding that the centripetal force responsible for the celestial motions is gravity.

How gravity towards one of the planets varies with the mass of the attracted body is taken up in Proposition 6. Newton first argues that pendulum experiments have established that weight varies with mass for bodies at a fixed distance from the Earth's centre. The corollaries emphasize the contrast here with other forces such as magnetism, in that *all* bodies are affected in the same way at a given distance. He then argues that the satellites of Jupiter have the same acceleration towards the Sun as Jupiter itself, directly from Proposition 1.65 and also based on the absence of observable eccentricity of the orbits that would be the consequence of a difference in acceleration. (That is, the Sun's gravitation does not appreciably displace or distort the motion of the satellites.)[44] Finally, the weights of each part of a planet varies with the mass of that part, or else the response of the planet to the Sun's gravitational field would depend upon its composition (with parts allowed to have different ratios of mass to weight) rather than just its mass. This step of the argument must have also been quite striking for Newton's contemporaries. Many of them would have considered the possibility that gravity is analogous to magnetism, but Newton turned this speculative question into a precise empirical contrast between gravity and other forces.

The final step in Proposition 7 is to the claim that all bodies produce an attractive force directed towards them whose absolute strength is proportional to their masses. The conclusion follows directly from Proposition 1.69. This earlier result establishes that for a central force whose accelerative forces depend only on distance (and *not* on any other properties of the interacting bodies), the absolute measure of force is given by mass.[45] This conclusion follows from the Third Law and the definitions of the accelerative and motive measures of force. For given any pair of interacting bodies A and B, the Third Law implies equality of the motive forces acting on the two bodies. But given that the motive force is the mass of the body times the accelerative force, it follows that the ratio of accelerative forces acting on the two bodies is given by the ratio of their masses. Finally, for a central force independent of any other properties of bodies, at a fixed distance the acceleration towards A or B is the same for all bodies. Hence, for *any* distance, the ratio of the accelerative forces to A and B is given by the

---

[44] There would be *no* 'distortion' of the motion of one of Jupiter's satellites in the idealized case where $d/r \to \infty$, where d is the Jupiter–Sun distance and r is the radius of the orbit of the satellite. In the real case, Newton's theory predicts tidal effects, but these would be quite small for the case of Jupiter.

[45] Newton states this as Corollary 2 to Proposition 1.69; the proposition itself is formulated in terms of inverse-square forces. For a central force whose accelerative measure does not depend upon properties of the interacting bodies, then the absolute measure must be a scalar quantity assigned to each body.

ratio of their masses. The mass thus measures the strength of the accelerative tendency towards each body: that is, it is proportional to the absolute measure of the force.

With this, Newton has completed the initial argument for universal gravitation. However, he clearly endorses a further claim, based on his comments in the Preface but not explicitly argued for in Book 3, that gravity should be taken as one of a few natural powers or general principles of motion. Thus in addition to establishing how gravity depends on masses and distance, Newton treats it as something akin to a 'fundamental force' in roughly the same sense as that term is used in contemporary physics.[46]

Even a reader as sympathetic as Roger Cotes, the insightful editor of the second edition, balked at the application of the Third Law in Proposition 7 (and earlier in corollaries to Proposition 5). Cotes objected that the Third Law itself does not license the conclusion that the force equal and opposite to the one holding a planet in its orbit is a force impressed on the *Sun*. If there were an invisible hand holding a planet in its orbit, to use Cotes' vivid illustration, then the Third Law implies the existence of a force acting on the *hand* rather than on the *Sun*. (This suggestion is not mere fancy on Cotes' part; the then dominant vortex theories similarly involve pressure exerted on the planet by neighbouring bodies.) Newton's argument thus apparently required an hypothesis regarding the nature of gravitation, contrary to the methodological pronouncements added to the second edition's General Scholium. Explicitly, the hypothesis is that the impressed force on the orbiting body is due to a mutual interaction with the Sun. While this is certainly a plausible claim, given that the impressed force is directed towards the Sun and varies with distance from it, Newton is not in a position to offer an argument from the phenomena to this effect as he had the earlier propositions. Furthermore, the evidence offered in defence of the Third Law itself involved only cases of interaction among contiguous bodies, and here Newton has significantly extended its application to cases of non-contiguous interacting bodies.

In what sense, then, can we say that the law of gravity has been 'deduced' from the phenomena, and to what extent is Newton's conclusion warranted? There has been controversy regarding these questions from Newton's time to the present, but one thing is abundantly clear from the text of the *Principia*: Newton did not treat the empirical case for universal gravity as completed after the opening seven propositions. Instead, the opening sequence captures the initial inference to the force law, but the empirical case in favour of universal gravitation *includes* the assessment of its further consequences. From Proposition 8 onwards Newton derives results taking the law of gravity as given in conjunction with a number of further empirical claims. Propositions 8 and 9 immediately note two important aspects of universal gravity (based on the results from Section 12–13 in Book 1): that outside a spherical mass with spherically symmetric density, $f(r) \propto r^{-2}$, and inside a spherical mass of uniform density, $f(r) \propto r$. In the Corollaries to Proposition 8, Newton calculated the masses and densities of other planets and the Sun—and Newton's contemporaries,

---

[46] We follow (Stein, 2002) in emphasizing this point.

such as Huygens (in Huygens, 1690), singled this out as a striking consequence of universal gravity.

The remainder of Book 3 exploits the framework provided by Book 1 to establish a number of distinctive consequences of universal gravitation. Propositions 10–17 elaborate and defend a Copernican–Keplerian account of orbital motions. Newton next considered the shape of the planets and the possibility of determining the Earth's shape by measuring local variations in surface gravity. He added Propositions 18–20 after Halley informed him in 1686 of the discovery that the length of a seconds pendulum varies with latitude. Huygens was already aware of the importance of the pendulum measurements, and immediately recognized the significance of these propositions. Newton's prediction of the shape of the Earth provided a sharp empirical contrast between universal gravity and other gravitational theories, and the contrast was accessible to contemporary pendulum measurements (see Schliesser and Smith, ms).

The most difficult parts of Book 3 concern the question that Halley would have probably posed in response to *De Motu*: what does the inverse-square law imply for the Moon and for comets? The study of the Earth–Moon–Sun system takes up the bulk of Book 3. Newton's main aim was to give an account of the various inequalities in the Moon's motion based on his earlier qualitative treatment of the three-body problem in Prop. 1.66 and its corollaries. But he also argues for two other striking conclusions: first, that the tides result from the gravitational attraction of the Moon, and second, that precession of the equinoxes is due to the Moon's gravitational pull on the Earth's equatorial bulge. The final three propositions (40–42) turn to the problem of comets. Newton had delayed the publication of the *Principia* as he struggled to find a way to use observations of comets to determine their trajectory. He did ultimately find a method and used it to determine the orbit of the comet of 1680/81 as an illustration of the technique.[47]

Many of the results in Book 3 were startling breakthroughs. Yet Newton's success in providing a qualitative dynamical understanding of these phenomena based on gravity was achieved in spite of a variety of errors and illicit assumptions. This is in part due to the difficulty of the problems Newton tackled. His most astute eighteenth century readers would soon discover deep flaws in most of the arguments in Book 3, and correcting these flaws would require significant advances in mathematical physics. But, as we will discuss briefly below, a century of further work on these problems ended up strengthening the case for Newton's claim that these phenomena can all be understood as consequences of gravity.

Taken as a whole, these results transformed the study of celestial mechanics and set the agenda for the eighteenth century.[48] One aspect of this transformation was to

---

[47] We do not have the space to discuss these results; see (Kriloff, 1925) for a detailed reconstruction and defence of Newton's techniques for determining the comet's trajectory, and (Hughes, 1988) for a more schematic overview of Newton's work on comets.

[48] See (Wilson, 1989) for a more detailed account of Newton's contribution to celestial mechanics.

treat celestial motions as problems within gravitational physics. Prior to Newton, predictive astronomy focused on finding calculational schemes that would allow one to compute celestial positions accurately. By contrast, Newton treated all these motions as consequences of the gravitational interactions among celestial bodies. An equally important second part of this transformation was the level of quantitative detail that Newton demanded in a satisfactory treatment of celestial motions. Even in the face of the daunting complexity of observed celestial motions, he emphasized the importance of control over the theoretical description of the system. Rather than taking a rough qualitative agreement between theory and evidence as the appropriate end point of inquiry, he insisted on seeking precise quantitative agreement. Demanding such agreement would make it possible to recognize residual discrepancies between theoretical descriptions and observed motions, discrepancies that might reveal physically important features of the real world that had been initially excluded. This would make it possible to develop ever more accurate models capturing more of the complexity of real phenomena.

The lunar theory provides the clearest illustration of this aspect of Newton's approach.[49] The Moon's motion is enormously complicated compared to that of the planets, and seventeenth-century astronomers struggled to produce a descriptive account of lunar motion with comparable accuracy. Newton thus faced two problems: first, to develop a descriptively accurate account of the Moon's motion, and second, to assess whether all the details of that account could be derived as a consequence of gravity. What was so distinctive and transformative about Newton's approach was that he responded to both problems at once; as he puts it, 'I wished to show ... that the lunar motions can be computed from their causes by the theory of gravity' (Scholium to Proposition 3.35, added in the second edition). In rough outline, Newton's aim was to account for the various known inequalities in the lunar orbit as a consequence of the perturbing effect of the Sun's gravity and other features of the Earth-Moon-Sun system.[50] If the Earth and the Moon were spheres of uniform density and the Sun had no perturbing effects, then the Earth and Moon would each follow elliptical orbits with their common centre of mass at one focus, without apsidal precession. The departures of the Moon's motion from this simple idealized case can reveal what physical features are needed for a more accurate description. The computation of lunar motions from first principles is crucial to this process, as one can seek to identify more subtle effects based on the character of the residual discrepancies remaining once other contributions have been taken into account.

The idea of approaching lunar theory on these terms was a more lasting contribution than the details of Newton's own account. Newton and his advocates often overstated the success of his lunar theory. In fact, he was not able to make substantial progress with regard to accuracy over the ideas of Jeremiah Horrocks. One glaring problem with the lunar theory was Newton's failure to derive the amount of apsidal

---

[49] We thank George Smith for emphasizing to us the importance of Newton's lunar theory as exemplifying the 'Newtonian style'.
[50] Wilson, (1989) gives a concise introduction to Newton's lunar theory.

precession due to the Sun's perturbing effects. He required a number of unmotivated steps to conclude that the rate of apsidal precession was compatible with observations. He also faced a more general obstacle: within his geometric approach it was not possible to enumerate all of the perturbations at a given level of approximation, as one could later enumerate all of the terms at a given order in an analytic expansion. It was only with a more sophisticated mathematics that astronomers could fully realize the advantages of approaching the complexities of the Moon's motion via a series of approximations.

5.3.6 The General Scholium and the Rules of Reasoning

At several points above we have indicated that Newton made many refinements to the *Principia*. In this section we discuss three significant, interconnected changes that influenced fundamentally the content and reception of the *Principia*. First, the relabelling and rewording of nine 'hypotheses' (into 'phenomena' and 'rules of reasoning') at the start of Book 3; second, the addition of the General Scholium, which provided a completely new ending to the *Principia*; third, changes that minimized explicit commitments to atomism. The first two changes are linked because together with Roger Cotes' new preface they explicitly deal with methodological issues and also connect the *Principia* to wider metaphysical and theological concerns. In the General Scholium, Newton provides a list of discoveries as exemplars of the fruits of this method: 'the impenetrability, the mobility, and the impulsive force of bodies, and the laws of motion and of gravitation'. These overlap significantly with his treatment in the third rule of reasoning. There is, therefore, no doubt that the methodological claims of the General Scholium are meant to be read alongside the rules of reasoning.

The three changes are linked in a complex fashion. Atomism is one of the hypotheses that gets dropped in the second edition, as we will show below. This change is connected to a little noticed oddity: the first edition of the *Principia* starts with Halley's laudatory poem (which is clearly modelled on Lucretius' ode to Epicurus) and ends rather abruptly after a technical discussion of how to compute the trajectory of comets and some hints on how these might play a role in the circulation of cosmic materials. We can infer from Newton's reaction to now lost letters from Bentley that Bentley (first and foremost a classicist) imputed Epicurean doctrines to Newton. Bentley heavily edited Halley's poem for the second edition in order to remove any sign of impiety or Epicureanism.[51] The carefully crafted General Scholium tacitly addresses various unnamed critics. In particular, it disassociates Newton from any Epicurean reading. It aligns Newton with—and also created crucial intellectual authority for—the newly influential efforts at creating a Newtonian physical theology (or natural religion) in Great Britain and The Netherlands. So, even though the General Scholium is under 2,500 words it changes dramatically the

---

[51] Much of Halley's original poem was restored in the third edition (see Albury, 1978).

framing of the whole *Principia*. In what follows we discuss it in more detail, but first we discuss the changes to Newton's matter theory and analyse the rules of reasoning.

From a methodological approach the most significant of Newton's changes to the second edition was the relabelling and rewording of nine 'hypotheses' at the start of Book 3 of the *Principia*. Five of these became empirical 'phenomena' that Newton lists just before his argument for the existence of the inverse-square law. (Newton added an original, 'phenomenon 2', to the second edition.) The first two hypotheses were renamed the first two 'rules for the study of natural philosophy'. Hypothesis 3 from the first edition, discussed below, was replaced in later editions with the third rule. The fourth rule was added only in the third edition of the *Principia*. Only one hypothesis (the fourth) survived as an 'hypothesis' in the second and third editions. It is used in a reductio (in proposition 3.11) and thus is not a counter-example against the General Scholium's famous injunction against hypotheses.

One hypothesis, the original hypothesis 3, was dropped entirely; it reads 'Every body can be transformed into a body of any other kind and successively take on all the intermediate degrees of qualities' (*Principia*, p. 198).[52] This transformation thesis is a very broad assertion of the homogeneity of matter. Something like this claim was a staple of the mechanical philosophy and may have also motivated alchemical searches to turn lead into gold. Newton appealed to it only once in the *Principia* (Book 3, proposition 6, corollary 2); he dropped the corollary and reworded the proposition in subsequent editions. Mass as a quantity or measure also presupposes homogeneity of matter in a weaker sense, but does not require the transformation thesis. In *An Account of Sir Isaac Newton's Philosophical Discoveries* (posthumously published in 1748) the leading and most sophisticated Scottish Newtonian, Colin MacLaurin, even goes so far as to suggest that different kinds of matter that have different kinds of resistance to change might well exist (Maclaurin, 1748, p. 100). Something of the spirit behind this dropped hypothesis reappeared in Query 30 of the *Opticks*: 'Are not gross Bodies and Light convertible into one another, and may not Bodies receive much of their Activity from the Particles of Light which enter into their Composition? The changing of Bodies into Light, and Light into Bodies, is very conformable to the Course of Nature, which seems delighted with Transmutations' (Newton, 1730, p. 374).

The dropped hypothesis reflects an important change from the first edition of the *Principia*, where Newton seems committed to atomism. This is not only reflected by the hypothesis, but more clearly in Proposition 3.6, Corollary 3. There Newton relies on the counterfactual assumption that if matter is fully compressed to eliminate all interstitial void spaces, it would be of the same density. In the second edition this was reworded so as to remove the commitment to atomism.[53] Even so, Newton may well have remained committed to atomism throughout his life. For example, in Query

---

[52] See, in particular, (McGuire, 1970) for a discussion of the philosophical significance of this hypothesis and its role in Newton's matter theory.

[53] See (Biener and Smeenk, 2011) for a discussion of a problem with Newton's treatment of matter noted by his editor Roger Cotes. Cotes' incisive criticism of this line of reasoning led

31 of the *Opticks*, Newton freely speculates about 'the small Particles of Bodies' that have 'certain Powers, Virtues, or Forces, by which they act at a distance, not only upon the Rays of Light for reflecting, refracting and inflecting them, but also upon one another for producing a great Part of the Phenomena of Nature'. We should be careful not to conflate Newton's undoubted corpuscularianism with his atomism. Nevertheless, at several places in the *Opticks*, Newton speculates about perfectly 'hard bodies' out of which other bodies are composed as possible and likely (Newton 1730, pp. 364ff; 370; 375–378).

Newton's four rules of reasoning also change the framing of the argument in Book 3 in response to criticisms of the first edition. In particular, all the rules are meant to underwrite steps in the argument for universal gravity. Newton wanted to elucidate and justify the inference and also respond to the common criticism that the entire line of argument was based on the 'hypothesis of attraction'. The four rules can be read as providing norms for causal ascription. We now turn to a careful analysis of the four rules of reasoning. The first two rules are as follows (*Principia*, pp. 794–795):

Rule 1: No more causes of natural things should be admitted than are both true and sufficient to explain their phenomena.

As the philosophers say: Nature does nothing in vain, and more causes are in vain when fewer suffice. For nature is simple and does not indulge in the luxury of superfluous causes.

Rule 2: Therefore, the causes assigned to natural effects of the same kind must be, so far as possible, the same.

Examples are the cause of respiration in man and beast, or of the falling of stones in Europe and America, or of the light of a kitchen fire and the Sun, or of the reflection of light on our Earth and the planets.

We treat these two rules together because their wording ('therefore' in the second rule) encourages understanding the second rule as a consequence of the first. In eighteenth-century discussions they are also often discussed jointly. The first thing to note about these two rules is their focus on how to match causes and effects; Newton's science is causal. Both rules promote causal parsimony and simplicity. Newton advocates a form of reductionism—universal gravity is a single underlying cause for disparate phenomena—without insisting that this cause be reduced to microscopic or physical qualities.

We quote the third rule, but not Newton's lengthy commentary, before discussing them further (*Principia*, p. 795):

Rule 3: Those qualities of bodies that cannot be intended and remitted and that belong to all bodies on which experiments can be made should be taken as qualities of all bodies universally.

While the first two rules promote causal austerity, the third rule promotes a kind of inductive boldness. In particular, it licenses induction for the imputation of properties to very distant and to very small objects. The latter is often called 'transduction' in the literature. As will become fully clear in our analysis of the fourth rule, Newton

Newton to back down in the second edition and acknowledge the hypothetical character of the earlier claim that quantity of matter holds in fixed proportion to quantity of extension.

recognized the limits and dangers of induction (see also (Newton 1730, Query 31, pp. 403–404)). Nevertheless, within the confines of a research programme he advocated bold generalization from the empirically available domain to domains beyond our experimental grasp. David Hume recognized something of Newton's boldness; in *The History of England* he wrote that Newton is 'cautious in admitting no principles but such as were founded on experiment; but resolute to adopt every such principle, however new or unusual' (Hume, 1985, Volume 6, p. 542).

The third rule clearly presupposes rather strong assumptions about the scale invariance of nature. To be clear, Newton had put a lot of experimental and theoretical work into showing that he was allowed to sum 'motions of the individual parts' into 'the motion of a whole' (*Principia*, Definition 2; p. 404) and, in particular, that the mass of a body could be summed from its parts. By analogy in the explication of the third rule Newton now asserts that 'extension, hardness, impenetrability, mobility, and force of inertia of the whole [body], arise from the extension, hardness, impenetrability, mobility, and force of inertia of each of the [body's] parts'. Newton underscores the importance of this compositionality by adding that 'this is the foundation of all philosophy'. The rule therefore licenses quantitative inferences from empirical evidence to parts of nature beyond the reach of our evidence (see, in particular, McGuire, 1970).

We note three things in Newton's discussion of the third rule. First, he now takes an agnostic stance on atomism. This fits well with his removal of the original third hypothesis. Second, the rule contains a not-so-subtle dig at Christiaan Huygens' Cartesian scepticism about the very possibility of universal gravitation of all bodies. For, Newton points out that the empirical argument for the 'principle of mutual gravitation' is far stronger than the argument for the 'impenetrability' of matter (which is presupposed by Cartesians following section 43 of Descartes' *Principia*). The third rule appeals to 'sensation' and 'experiments' and rejects 'reason' as the grounds for asserting impenetrability (and presumably other properties of bodies). Newton clearly intends this as a contrast with the rational insight into the nature of body appealed to by Descartes and his followers.

Third, the rule itself deploys the plural 'bodies'. The plural is used throughout Newton's discussion. This modifies a bit the nature of the inductive leap that Newton advocates. Newton is offering an account in terms of the behaviour of systems of bodies, not an account that has its source in the nature of body. In fact, in the *Principia* Newton never defines the nature of body (McGuire, 1966). It is surely tempting to see Newton as effecting a conceptual reversal, by explicitly formulating the laws of motion and implicitly defining bodies as entities that satisfy the laws.[54] In the gloss to the third rule, Newton is careful to distinguish between essential and universal qualities of bodies presumably to block the implication that if a quality is universal it must also be essential. By 'essential' Newton probably means what we would call 'intrinsic' qualities of bodies, that is, qualities that are presupposed in the very conception or

---

[54] Here we draw on (Brading, 2011), which articulates and defends what she calls the 'law-constitutive' approach to the problem of defining and individuating bodies.

nature of body. In part, the terminology of 'universal qualities' marks Newton's contrast with the Cartesians. But he also hoped to avoid the charge of attributing gravity to bodies as an essential property by calling it universal instead, a claim he made explicit in the third edition.[55]

Finally, Rule 4, which was added to the third edition, and its brief commentary read (*Principia*, p. 796) as follows:

Rule 4: In experimental philosophy, propositions gathered from phenomena by induction should be considered either exactly or very nearly true notwithstanding any contrary hypotheses, until yet other phenomena make such propositions either more exact or liable to exceptions.

This rule should be followed so that arguments based on induction may not be nullified by hypotheses.

This rule has been the subject of considerable recent scholarly attention. Yet, we have been unable to locate a single explicit discussion of it during the eighteenth century! By 'hypotheses' Newton means the kind of proposals offered by mechanical philosophers. The main purpose of this rule is to settle one's attitude toward ongoing research. It encourages one to accept one's going theory as true (or 'very nearly' so). Also, the rule disallows a vantage point outside of ongoing research as providing legitimate sources of principles that could motivate theoretical reinterpretations of one's empirical results (of the sort that mechanical philosophers would promulgate).

But the rule has two further important implications. First, notwithstanding the kind of bold inductive leap that the third rule encourages, this rule is a clear expression of Newton's fallibilism. He knows he could be wrong. This echoes his own 'Author's preface' to the *Principia*: 'I hope that the principles set down here will shed light on either this mode of philosophizing or some truer one' (p. 383). (This was already present in the first edition.) Second, the rule encourages the search for systematic deviations from known regularities. Discrepancies need to be turned into 'phenomena'. As noted above, it is a major methodological innovation and achievement of Newton's *Principia* that systematic discrepancies are both possible sources of much more subtle evidence than previously imagined, as well as guides for refinements to the theory.

We now turn to a more detailed treatment of the General Scholium. While we discuss it roughly in order of Newton's presentation, we will emphasize four aspects: (1) Newton's reiteration, even expansion of his case against the vortex hypothesis; (2) Newton's embrace of a design argument; (3) Newton's treatment of the role of natural philosophy in knowledge of God; (4) Newton's elaboration of his methodology.

The General Scholium opens with three main arguments against the 'hypotheses of vortices'. Newton argues first (echoing the end of Book 2) that the observed Keplerian motion of the planets is incompatible with vortices. Second, he argues that the trajectories of comets, which 'observe the same laws' as planetary motions, are

---

[55] For further discussion of the third rule, see (McGuire, 1970; McMullin, 1978).

incompatible with vortices. Third, he offers an analogical argument that, in accord with all seventeenth-century New Philosophy, relies on the fundamental unity between terrestrial and celestial phenomena; it appeals to evidence from Boyle's vacuum experiments to claim that in the absence of air-resistance all celestial bodies will keep their motion in accord with Newton's laws of motion. We can understand these three arguments as successful burden-shifting.

The General Scholium then turns to arguments for the existence and the nature of our knowledge of God. Newton first argues that while the orbits of celestial bodies are law-governed, neither these laws nor the 'mechanical causes' of his philosophic opponents can be the cause of the orbits themselves. Newton barely gives an argument for his claim that the laws of motion cannot be the cause of the orbits.

We use 'barely' because Newton does claim that it is 'inconceivable' that the laws of nature could account for such 'regular' orbits. But in what follows Newton packs quite a bit into this claim. In particular, it turns out that for Newton the regularity consists not merely in their being law-governed, but also that the trajectories and mutual attractions of the planets and comets hinder each other least. This culminates in his conclusion that 'This most elegant system of the Sun, planets, and comets, could not have arisen without the design and dominion of an intelligent and powerful being' (p. 940). Without further argument he rules out the possibility that these particular three features (that is, (i) law-governed orbits that (ii) hinder each other minimally and that (iii) are jointly beautiful) could be caused by causes other than God. He then offers the 'immense distances' among the planetary systems, which thus avoid the possibility of gravitationally-induced mutual collapse, as another, empirical phenomena that supports his argument from inconceivability. Moreover, given that he could put almost no constraint on the mass of comets, he must have also found it striking that these do not disrupt the motions of the solar system through which they pass. Comets provide a further hint of providential design, in that at aphelia they are sufficiently far apart so as not to disturb each other's motion.

Before we turn to analyzing Newton's argument from (beautiful) design and his views of God, it is worth noting that Newton's position rules out two contrasting, alternative approaches, both discussed later in the General Scholium: i) that God is constantly arranging things in nature. As he writes, 'In him are all things contained and moved; but he does not act on them nor they on him' (p. 941). No further argument is offered against a hyper-active God. ii) That everything is the product of 'blind metaphysical necessity'. This second view is associated with the neo-Epicurean systems of Hobbes and Spinoza. Newton offers an independent argument against that approach: namely, that given that necessity is uniform it seems it cannot account for observed variety. Now, against Hobbes this is a powerful point, but it is only a limited objection against Spinozism, for Spinoza is committed to there being sufficient reason for infinite variety in the modes (E1p16 and E1p28).[56]

---

[56] We cite Spinoza's *Ethics*, as is customary, by part and proposition; the standard English translation is (Spinoza, 1985).

At best Newton has shifted the burden of proof. Because he has the better physics he can claim to have constrained any possible explanation that will account for the observed variety. But it is not insurmountable: all a necessitarian needs to show is how the laws and the 'regular' orbits are possible given some prior situation. Moreover, in the absence of a discussion of initial conditions of the universe and a developed cosmogony, Newton's claim begs the question. One can understand Kant's *Universal Natural History* as taking up the challenge of accounting for the origin of the universe in light of Newton's laws of motion.

Of course, Newton is not merely pressing the existence of variety against the necessitarian; he is also calling attention to the significance of that particular 'diversity of created things, each in its time and place' (p. 942). As we have seen, he argues from (i) law-governed orbits that (ii) hinder each other minimally and that (iii) are jointly beautiful, to the conclusion that an all powerful and intelligent God must have been their cause. Moreover, he uses the distance among planetary systems as a further argument to insist that God must be 'wise'. Regardless of how plausible one finds such an argument to a providential designer, Newton's version is not an anthropocentric argument. In particular, the beauty of our solar system is mimicked by countless other solar systems, too far apart to be of interest to us. Moreover, natural diversity is suited to times and places regardless of human interest.

Such a non-anthropocentric design argument is also available to careful readers of the first edition of the *Principia*, but it had not been highlighted in it. Newton writes, 'Therefore God placed the planets at different distances from the Sun so that each one might, according to the degree of its density, enjoy a greater or smaller amount of heat from the Sun' (Proposition 3.8, corollary 5). It is the only explicit mention of God in the first edition. Newton suggests that the variation in temperature is suitable given the variation in surface gravity and density of the planets. As Cohen notes, Huygens was astounded that Newton was able to calculate the strength of gravity 'the inhabitants of Jupiter and Saturn would feel' (Cohen 1999, quoting Huygens, p. 219).

Much of core of the General Scholium is then given over to articulating the nature of God and the proper way of talking about him. Here we do not do justice to all the complex theological and metaphysical issues that this material poses.[57] We wish to note six points about this material. First, Newton unabashedly argues that natural philosophy includes empirical research into God ('to treat of God from phenomena is certainly a part of natural philosophy', p. 943). (See McGuire, 1978 and Janiak, 2008.) Second, our knowledge of the manner of God's existence is strictly empirical and based exclusively on the argument from beneficial/providential design ('we know him only by his properties and attributes and by the wisest and best construction of things and their final causes', p. 942). Of course, the existence of God may be secured by other arguments and sources, including Scripture. Third, if one denies that God is a beneficial designer, talk of 'God' refers to fate. Fourth, Newton denies that we can have

---

[57] On theological matters see, for example, Snobelen, (2001), Ducheyne, (2007); for useful treatment on Newton's sources see McGuire and Rattansi, (1966) and de Smet and Verelst, (2001).

knowledge of the substance of God. This last point is a consequence of the more general claim that knowledge of substances is not available to human inquirers. (Newton formulates this and related discussions about personhood in very Lockean terms, and this surely encouraged Enlightenment philosophers to link their approaches into a connected programme. More recent scholarship tends to disconnect Newton and Locke.)[58]

The first and the fourth of these points fit with a fifth: in three distinct ways knowledge of God's manner of existing is less secure than the empirical knowledge from which it is derived. (This is emphasized by Stein, 2002.) I. Newton insists that our knowledge of God's substance is even less available to us than knowledge of the substance of bodies. II. Newton insists that we have no knowledge of the manner of God's (immaterial) activity (his manner of understanding, perception, and so on). III. Newton also recognizes that we have a tendency to anthropomorphize God and—strikingly—while he stresses the limitations of this, he seems to think it is the only route available to us when speaking about God. He also makes a number of definitive claims about God's attributes that raise complicated and controversial (theological) questions over the exact relationship between God and space/time; many of these questions were raised and pressed by Leibniz in his famous correspondence with Clarke.

Newton concludes the General Scholium with his discussion of the 'power of gravity' and a speculation on a subtle spirit. On the 'power of gravity' he is explicit about the contours of his knowledge and ignorance of it. His reflections on it provide further insight into his mature methodological views. We offer two main observations: first, he explicitly asserts that celestial phenomena and phenomena of the sea are explained by gravity. Gravity is a single cause for Newton. According to Newton we do not know what causes it, but we are in the position to make some explicit claims about the structure of this unknown cause (p. 943):

> ... this force arises from some cause that penetrates as far as the centres of the Sun and planets without any diminution of its power to act, and that acts not in proportion to the quantity of the *surfaces* of the particles on which it acts (as mechanical causes are wont to do) but in proportion to the quantity of *solid* matter, and whose action is extended everywhere to immense distances, always decreasing as the squares of the distances. Gravity toward the Sun is compounded of the gravities towards the individual particles of the Sun, and at increasing distances from the Sun decreases exactly as the squares of the distances as far out as the orbit of Saturn ...

Despite his inability to infer the underlying cause of gravity from the phenomena, he does have confidence in the constraints that any to-be-identified cause must obey.

Second, at this juncture Newton resists the temptation to offer his own hypothesis, with his famous claim that 'I do not feign hypotheses'. He adds that 'For whatever is not deduced from the phenomena must be called a hypothesis; and hypotheses, whether metaphysical or physical, or based on occult qualities, or mechanical, have

---

[58] For further discussion of Locke and Newton, see (Stein, 1990; Downing, 1997; Domski, 2011).

no place in experimental philosophy' (p. 943). In historical context, Newton is clearly ruling out the demand familiar from the mechanical philosophy for causal explanations in terms of the shape, size, and motion of colliding bodies. But the way he phrases it is far broader. Now, he does not explain in context what he means by 'phenomena'. But as our treatment above suggested, phenomena are law-like regularities inferred from a given body of data. So, 'deducing from phenomena' means something like rigorous inference from well-established, non-trivial empirical regularities. In modern terminology this is something like induction. Confusingly to the modern reader, Newton identifies a second phase of research with induction (that is, 'made general by induction'). But such generalization and systematizing we tend to associate with deduction. So, at the risk of over-simplification: in his methodological statements of the General Scholium, Newton takes for granted the existence of well-established empirical regularities that form the basis of inference to a set of general claims that, in turn, form the basis for deduction. In the *Opticks* he identifies these two stages with the so-called Analytic and Synthetic methods.

Strikingly and confusingly, just after denying any interest in hypotheses, Newton concludes the *Principia* with a bit of speculation about an 'electric and elastic' ether that might account for all kinds of natural forces. He immediately admits that he lacks the experimental evidence to account for its laws, and on that note of ignorance the book closes.

5.3.7 Mathematical Methods

Newton's sheer mathematical genius is evident throughout the *Principia*, as he tackles an astounding range of different problems using a variety of innovative techniques. Yet his mathematical style is immediately jarring for a modern reader, and it would have been nearly as jarring for many readers within a generation of the publication of the first edition. For a variety of reasons, Newton chose to adopt a geometrical style of reasoning in the *Principia* and to suppress many of his most novel mathematical techniques. Although we lack the space to discuss these issues in depth, we will comment briefly on Newton's mathematical prologue, regarding the method of ultimate ratios (Section 1, Book 1), and discuss two examples illustrating the *Principia*'s mathematical style.

In the priority dispute with Leibniz regarding the invention of the calculus, Newton and his proponents claimed that the *Principia* was formulated initially using the methods of fluxional analysis that Newton had discovered in the 1670s, but publicly presented in the superior geometrical style exemplified by Huygens' *Horologium Oscillatorium*. This claim is certainly an exaggeration: the *De Motu* drafts and other manuscripts leading up to the *Principia* all employ the same geometrical limit reasoning as the *Principia*, and there is no evidence of other techniques used initially that were then suppressed.[59]

[59] 'The published state of the *Principia*—one in which the geometrical limit-increment of a variable line segment plays a fundamental role—is exactly that in which it was

The question of whether Newton used the calculus is, more importantly, ill-formed, as Whiteside has emphasized (Whiteside, 1991). The geometrical approach of the *Principia* allowed Newton to handle problems regarding instantaneous variation of quantities now treated with the calculus. Huygens and various others had treated similar problems using techniques like Newton's. At the time the contrast between 'the calculus' and these techniques was partly stylistic and partly substantive. Leibniz argued explicitly in favour of blind manipulation of symbols—'the imagination would be freed from the perpetual attention to figures'—whereas for Newton the attention to figures was a crucial source of certainty for mathematics.[60] Substantively, the two approaches differed in degree of generality and ease of use in particular problems.[61] The contrast between 'the calculus' and Newton's geometric reasoning is thus not as sharp as the question presumes.[62] But, in addition, the question is ill-formed because the *Principia* is not based on a single mathematical method. The synthetic, geometric style Newton chose cannot completely hide the innovative mathematical techniques that informed its composition. Contemporary readers were left frustrated by several lacunae where Newton appealed to unspecified algorithmic procedures, notably quadrature techniques (i.e., ways of finding the area under a curve), and by the overly concise presentation of new ideas, as in Lemma 2 of Book 2.[63] In this sense the *Principia* does establish Newton's mastery of techniques directly related to the priority dispute, even though he did not disclose how he arrived at the results.

The argument in favour of Theorem 3 of *De Motu* described above illustrates Newton's geometric style of reasoning. The most striking contrast with ancient geometry is the treatment of quantities as generated via a continuous motion; this continuity underwrites the evaluation of 'ultimate' ratios that hold as, for example, the point Q approaches P in Figure above. Infinitesimal quantities were handled entirely in terms of limits, whose existence and uniqueness followed from the continuity of the motion generating the quantities. As an illustration, Lemma

---

written' (Whiteside 1970, p. 119). Whiteside further notes that the few attempts at presenting the *Principia* in terms of fluxions are clearly from well after the first edition and seem to have been abandoned quickly.

[60] See Part VI of (Guicciardini, 2009) for a recent discussion; the quotation is from *Mathematische Schriften*, Volume 5, p. 393 (Leibniz, 1962). See also Dunlop 2012.

[61] This is not to say that a calculus-based approach is uniformly easier than Newton's geometric approach; rather, for some problems Newton's methods are particularly illuminating and more powerful, whereas for others a calculus-based approach is more fruitful.

[62] This is in part due to attributing too much to Leibnizian formulations. Leibniz and Newton *both* lacked various key concepts, such as that of a function, introduced by Euler. Projecting Euler's innovations back into Leibniz's work falsely enhances the contrast. See (Fraser, 2003) for an overview of eighteenth-century innovations in mathematics and mechanics, emphasizing the contrast between the work of Euler, Lagrange, and their contemporaries and the earlier work of Leibniz and others.

[63] Newton does remark in a scholium to this Lemma 2 that he had corresponded with Leibniz regarding these methods, which he says 'hardly differed from mine except in the forms of words and notations'. In the third edition, Leibniz's name does not appear, and Newton instead emphasizes that he had discovered these methods in 1672.

10 concerns geometric quantities that characterize an arbitrary curved line and provides a generalization of Galileo's treatment of free fall: namely, that the position varies 'ultimately' as the square of the time. The proof proceeds, roughly speaking, by establishing equalities between ratios of evanescent quantities, which vanish as a point along a curve flows back into an initial point, and ratios of finite quantities. The continuity of the generation of the curve guarantees, Newton argues, that a limiting value for the ratio of evanescent quantities exists, and it is given by the ratio of finite quantities.

Many of the Lemmas in Section 1 had appeared elsewhere, and were used implicitly by Newton's contemporaries such as Huygens. Newton identifies his main innovation as providing proofs of them using limit arguments rather than appealing to the method of indivisibles. Contrasted with Leibniz's symbolic calculus, Newton's geometric approach sacrifices generality but allows him to deal much more directly with the differential properties of curves.[64] Newton also regarded the geometric approach as more intuitive, certain, and direct, as opposed to algebraic techniques which he once characterized as nauseating (*Mathematical Papers*, Volume 4, p. 277). The lemmas have proved controversial from the initial appearance of the *Principia*. Leibniz's marginal notes in his annotated copy of the *Principia* indicate that he doubted the truth of Lemmas 9, 10, and 11, and later Berkeley and others famously attacked the rigor of Newton's approach. Newton did not address foundational questions regarding the calculus that would become the focus of later debates. However, the collection of lemmas is not ad hoc as has been sometimes claimed; Pourciau, (1998) argues that the 11 Lemmas taken together provide a geometric formulation of the main definitions and theorems of the calculus needed in the *Principia*.

Section 1 ostensibly provides the mathematical background needed for the rest of the book, but at several points Newton relies on sophisticated mathematics that was not common knowledge. The inverse problem for an arbitrary force law is solved in Propositions 1.41–42 up to quadrature. But the utility of this result is limited by the analytical techniques available to perform this quadrature, and Newton only states one explicit solution.[65] In Corollary 3 of Proposition 41 Newton applies the general result to an inverse cube force law (attractive or repulsive) to determine the trajectory, and notes that 'All this follows ... by means of the quadrature of a certain curve, the finding of which, as being easy enough, I omit for the sake of brevity' (Newton 1726, p. 532). The quadrature required certainly was not easy for Newton's contemporaries, and the construction of the trajectories Newton specified in the *Principia* was

---

[64] The most serious limitation facing this geometrical approach is that it is difficult to distinguish between infinitesimals of different order. It is relatively straightforward to classify infinitesimals geometrically up to second order, but Newton himself had difficulty with third-order infinitesimals (see, for example, the discussion of Prop. 2.10 in Guicciardini, 1999, pp. 233–247).

[65] The difficulty persists to the present: in modern terms, this line of thought leads to a differential equation which one then integrates to find the trajectory, and the resulting integral is analytically solveable only for a few specific force laws.

unilluminating.[66] This is just one example where Newton invoked analytical techniques that are left completely unspecified. After the first edition, Newton considered augmenting the *Principia* with mathematical appendices that would provide more of the techniques required throughout the text. This case also illustrates a second point: Newton's success in recognizing the crucial physical principles for understanding a problem was not always matched with the mathematical methods needed to exploit it.

Newton's discussion of Kepler's problem in Section 6 includes a mathematical lemma that is absolutely stunning.[67] Newton invokes the lemma to prove that there is no algebraic function that will give the position along an elliptical trajectory as a function of the time; in modern terms, that there is no algebraic solution for x to Kepler's equation: $x - e \sin x = t$.[68] The lemma itself offers an argument for a more general result: that the areas of 'oval' figures cannot be fixed by algebraic equations with a finite number of terms. Consider an arbitrary 'pole' O in the interior of a given oval, with a ray rotating from a given initial point along the oval. Newton introduces the area swept out by the rotating ray geometrically via a point along the ray moving with a velocity proportional to the square of the distance between the pole and the ray's point of intersection with the oval. As the ray rotates about the pole O, the point representing the area swept out moves away from O in an infinite spiral. But any line intersects the spiral infinitely many times, implying that the spiral cannot be specified by an equation of finite degree. Newton concludes that there are hence no ovals such that the area cut off by straight lines can be found via an algebraic equation. He could still handle the problem, using the infinite series expansions that had been the key to his earliest innovations in mathematics, and thus was able to handle a much broader class of curves, including the spiral used in this argument, than Descartes had considered legitimate in the *Géométrie*.

Newton's contemporaries were divided in response to this argument, mainly due to the ambiguity regarding what he meant by 'ovals'.[69] But Newton's approach to the problem is strikingly innovative, in that it utilizes an unprecedented topological argument and proves a result regarding the existence of a particular kind of function. Arnol'd, (1990) argues that Newton's topological, global approach anticipates the pioneering work of Poincaré two centuries later.

---

[66] The difficulty Newton's contemporaries faced in reconstructing these 'easy' results is illustrated by David Gregory's correspondence with Newton regarding this corollary, discussed in chapter 12 of (Guicciardini, 2009). Guicciardini has noted in personal communication that the quadrature required for the inverse-square case is more complicated than the case of the inverse-cube force law, though he thinks that it would have been within Newton's grasp (even though there is no manuscript evidence that he could perform the required integration).

[67] See the discussions of this lemma in (Guicciardini, 2009), chapter 13, and (Pourciau, 2001), on which we draw here.

[68] A curve is defined to be algebraic if it is part of or identical with the solution of some polynomial.

[69] Huygens and Leibniz immediately considered possible counter-examples to the claim; see (Pourciau, 2001).

5.4 Impact

In the *Principia* Newton aimed to establish both that the force of gravity suffices to account for nearly all celestial motions and for many terrestrial phenomena, and to introduce a way of reasoning more securely from the phenomena in natural philosophy. Previously, we distinguished three steps in the argument for univeral gravitation: (1) that the motions of the planets, their satellites, and comets can be accounted for by the force of gravity, (2) that the force of gravity is *universal*, and (3) that gravity is a 'fundamental force'. We also characterized Newton's proposed new way of inquiry and his reasons for taking it to be superior to the hypothetical reasoning favoured by his contemporaries.

Judged by whether Newton persuaded his contemporaries on these issues, the *Principia*'s first edition enjoyed limited success in England and nearly complete failure on the Continent.[70] Halley penned an admiring précis, enclosed with the copy given to James II and published in the *Philosophical Transactions*. Flamsteed and Halley competently utilized Newton's ideas in subsequent research, but many of the members of the Royal Society who apparently accepted the *Principia*'s claims did so without a thorough command of the book. Hooke accused Newton of plagiarism with regard to the inverse-square law, but he lacked the mathematical skill to follow Newton's reasoning in detail. Sir Christopher Wren and John Wallis certainly possessed the technique, but there are no extant records of their assessment or clear signs of influence. It fell to a younger generation of natural philosophers to build on Newton's work—in particular the Scottish mathematician David Gregory, and the talented young Englishman Roger Cotes, who died at the age of 33, shortly after editing the second edition. Cotes contributed to the second edition a clear account of Newton's argument for universal gravitation—a polemical reply to Continental critics. Prior to the appearance of the second edition, few if any natural philosophers in England other than Cotes, Flamsteed, Gregory, Halley, and John Keill could understand Newton's main aims and assess critically whether he had attained them.

On the Continent the *Principia* was, in effect, read alongside Descartes' *Principia Philosophia* and found to suffer by comparison. An influential early French review by the Cartesian, Régis, suggested that the *Principia* was a contribution to mathematics or mechanics, but not to physics:[71]

> The work of M. Newton is a mechanics, the most perfect that one could imagine, as it is not possible to make demonstrations more precise or more exact than those he gives in the first two books ... But one has to confess that one cannot regard these demonstrations otherwise

---

[70] See (Guicciardini, 2003) for a detailed study of the reception of the *Principia* among British and Continental mathematicians, up to about 1736.

[71] The review appeared in the *Journal de Savants* in 1688, and we quote the translation from (Cohen, 1980, p. 96).

than as only mechanical ... In order to make an opus as perfect as possible, M. Newton has only to give us a Physics as exact as his Mechanics.

The view that a more perfect opus must include a mechanical explanation of gravity in terms of action by contact was common in Cartesian circles. The debate between the two *Principia*'s was often framed, as in Fontenelle's *Elogium* (1730), as between two competing hypothetical accounts of celestial motion to be evaluated in terms of their explanatory power. Within this context the reliance on attraction without an underlying mechanism was seen as a crucial flaw.

The denial that the *Principia* was a 'physics' meant that it was understood as not providing a causal account of nature. Even so, it provided an influential framework for addressing a variety of problems in rational mechanics, and had an immediate impact on existing research traditions in mechanics. In Paris, for example, Varignon recognized the significance of Newton's achievement and introduced the *Principia* to the Paris Academy of Sciences. He derived several of Newton's main results using a more Leibnizian mathematical style, starting in 1700 (Blay, 1999). The initial reception of the *Principia*, however, focused mainly on the account of gravity and planetary motions.

Critics of Newtonian attraction attempted to embed aspects of the *Princpia*'s planetary dynamics within a vortex theory.[72] Leibniz proposed a vortex theory in 1689 that leads to motion along elliptical orbits obeying Kepler's area rule (Leibniz, 1689). Despite his claim to independent discovery, manuscripts uncovered by Bertoloni Meli reveal that this theory was developed in response to a close reading of the opening sections of Book 1 (Bertoloni Meli, 1997). As imitation is the sincerest form of flattery, Leibniz's efforts indicate his recognition of Newton's achievement. But Leibniz seems to have appreciated only part of what Newton had accomplished, in effect reading the *Principia* as very close to the first *De Motu* manuscript in that he gleaned the results already contained in *De Motu* from the text, but not much else. The important further results Newton developed in order to treat the complexities of real motions, including departures from Keplerian motion and lunar motion, seem to have escaped Leibniz's notice. Leibniz did not recognize the possibility for further empirical assessment of the theory based on treating deviations from Keplerian motion and the various loose ends in Book 3. But in terms of explanatory power and intelligibility, Leibniz clearly took his own account to be superior to Newtonian attraction. In addition to avoiding attraction, Leibniz argued (in correspondence with Huygens) that his theory, unlike Newton's, could account for the fact that the planets orbit in the same direction and in nearly the same plane.

The strong opposition to attraction spurred the savants of Paris and Basel to develop a number of competing vortex theories, following Leibniz's lead. In particular, Johann Bernoulli criticized most of the mathematical parts of book 2. This was

---

[72] Aiton (1972) gives an authoritative account of the development of vortex theory during this period.

capped by Bernoulli's 1730 apt criticism of Newton's treatment of torque.[73] In the second edition, Newton and Cotes pressed a different, powerful objection to vortex theory, independent of the treatment in Book 2: vortex theories had great difficulty in accounting for the motion of comets, especially retrograde comets. Active work on vortex theories declined in mid-century in light of the work by Clairaut, Euler, and others described below.

Leibniz's later critical comments on the *Principia*, following the acrimonious priority dispute regarding the calculus, focused almost exclusively on Newton's metaphysics and his objectionable reliance on action-at-a-distance.[74] There is still a tendency to regard this issue as the crucial factor in the initial critical response to the *Principia*. Newton almost certainly found this way of framing the debate infuriating, and his responses to Leibniz emphasize the force of empirical considerations.

But there were also important early criticisms on terms that Newton would have accepted: namely, regarding empirical rather than explanatory success. During the first half of the eighteenth century it was by no means a foregone conclusion that Newton's theory would be vindicated empirically. While Huygens, like Leibniz, regarded action-at-a-distance as 'absurd', he acknowledged the force of Newton's argument for the claim that the inverse-square force law governed the behaviour of celestial bodies. However, he rejected the inductive generalization to universal gravity and the conception of natural philosophy as aiming to discover fundamental forces. Like Leibniz, Huygens developed a vortex theory that could account for inverse-square celestial forces without leading to truly universal gravitation. Huygens' argument turned on showing that no central body was necessary to generate planetary vorteces.[75]

But, more importantly, Huygens recognized that there is a crucial empirical contrast between universal gravity and his preferred vortex theory. Research Huygens had performed using pendulum clocks to determine longitude at sea seemed to confirm his theory rather than Newton's. Huygens' theory assumed that the clocks slowed down at the equator due to the centrifugal effects due to the Earth's rotation alone, while Newton also added a correction due to mutual attraction of all the particles within the Earth (see Schliesser and Smith, ms). This developed into a lengthy controversy regarding the shape of the Earth (see Terrall, 2002). Newton's and Huygens' theories both implied that the Earth would have an oblate shape, flattened at the poles to different degrees; by contrast, the Cassinis—a famous Italian family of astronomers at the Paris Observatory—claimed that the Earth has an oblong shape.

---

[73] The criticism appeared in (Bernoulli, 1730), his prize-winning entry in the 1730 Royal Academy of Sciences competition. For a treatment of his detailed criticisms of most of the other results in Book 2 prior to 1730, due primarily to Johann Bernoulli, see Guicciardini (1999), chapter 8.

[74] See, in particular, (Bertoloni Meli, 1997) for an excellent treatment of the interplay between Leibniz and Newton.

[75] Huygens' essay, 'Discourse on the Cause of Gravity', was first published in 1690, and is reprinted in Vol. 21 of the *Ouevres Complétes* (Huygens, 1888–1950).

The controversy could be settled by pendulum measurements far apart, as close as possible to the north pole and the equator. It took a half century to settle the dispute in Newton's favour after the publication of the results of Maupertuis' expedition to Lapland and La Condamine's to what is now Peru. Maupertuis' book on the topic (Maupertuis, 1738) appeared shortly after Voltaire's book-length defence of Newton (Voltaire, 1738), aided by Émilie du Châtelet, who later published an influential translation and brilliant commentary on the *Principia*, and together they helped turn the tide in favour of Newtonianism in France.[76] Adam Smith, who followed French developments closely, believed that the 'Observations of Astronomers at Lapland and Peru have fully confirmed Sir Isaac's system' (Smith, 1982, p. 101).

By 1739 there was no similarly decisive evidence in favour of universal gravity forthcoming from astronomy. The astronomers of Newton's generation did not have the mathematical tools needed to make substantive improvements in accuracy based on his theory. From the second edition of the *Principia* onward, Newton suggested that the explanation of the Great Inequality in the motions of Jupiter and Saturn could be based on their mutual interaction—which provided a stimulus for much technical work for several generations of mathematicians. Within predictive astronomy the *Principia* had the strongest immediate impact on cometary theory. Newton was the first to treat cometary motions as law-governed, enabling predictions of their periodic return. Based on Newton's methods, Halley published a study of the orbital elements of twenty-four sets of cometary observations (*Synopsis astronomiae cometicae*, 1705), and argued that the comets seen in 1531, 1607, and 1682 were periodic returns of one and the same comet.[77] Halley predicted a return in 1758, but the exact time of the return and the expected position of the comet were uncertain. In addition to the inexactness and small number of observations used to determine the orbit, the determination of the orbit was extremely difficult due to the perturbing effects of Jupiter and Saturn.

The time before the comet's return barely sufficed to develop the necessary methods to calculate the orbit based on Newton's theory.[78] Alexis-Claude Clairaut carried out the first numerical integration to find the perihelion of Halley's comet—an incredibly daunting calculation. In November 1758, rushing to beat the comet itself, he predicted that the comet's perihelion would be within a month of mid-April 1759. It was observed to reach perihelion on 13 March. Clairaut argued that this was an important vindication of Newtonian gravitation, but there was vigorous debate within the Paris Academy regarding the accuracy of his calculation.

---

[76] For a detailed discussion of the reception and influence of Newtonianism in France, see (Shank, 2008).

[77] Halley announced this conclusion to the Royal Society of London in 1696, following correspondence with Newton. A revised and expanded version of the *Synopsis* was published posthumously.

[78] This brief summary relies on the clear account given in (Waff, 1995b).

Clairaut's calculation of the comet's orbit was based on the approximate solution to the three-body problem he had found a decade earlier.[79] Clairaut and his contemporaries, most importantly Leonhard Euler and Jean le Rond d'Alembert, advanced beyond Newton's qualitative treatment of the three-body problem (in 1.66 and its corollaries) by using analytical methods to construct a perturbative expansion. These analytical approaches relied on a number of post-*Principia* innovations in mathematics, in particular the understanding of trigonometric series, and it is doubtful whether Newton's geometrical methods could have led to anything like them. One of the most striking limitations of the *Principia*'s mathematical style is the apparent limitation to functions of a single independent variable.[80] Within Newton's lifetime, Varignon, Hermann, and Johann Bernoulli had begun formulating Newtonian problems in terms of the Leibnizian calculus. Euler, Clairaut, and d'Alembert drew on this earlier work, but unlike the earlier generation they were able to make significant advances on a number of problems which Newton had not been able to treat quantitatively.

In 1747 Euler challenged the inverse-square force law due to an anomaly in the motion of the lunar apsides. Newton suggested that this motion could be accounted for by the perturbing effect of the Sun, but close reading of the *Principia* reveals that the calculated perturbative effect was only one-half of the observed motion. Euler preferred a vortex theory, and used the discovery of this anomaly to criticize the supposition of an exact inverse-square attraction. Clairaut and d'Alembert had both developed perturbative techniques to apply to the motion of the lunar apsides earlier, in 1746. Initially they both reached the same conclusion as Euler (Newton's theory was off by one-half), and considered modifying the inverse-square law. But that proved unnecessary; in 1748 Clairaut carried out a more careful calculation and discovered to his great surprise that the terms he had earlier regarded as negligible exactly eliminated the anomaly (see Wilson, 1995). He hailed this result as providing the most decisive confirmation of the inverse law: '... the more I consider this happy discovery, the more important it seems to me ... For it is very certain that it is only since this discovery that one can regard the law of attraction reciprocally proportional to the squares of the distances as solidly established; and on this depends the entire theory of astronomy', (Euler to Clairaut, 29 June 1751; quoted in Waff, 1995, p. 46).

The techniques developed in the 1740s made it possible to assess the implications of universal gravity for a number of open problems in celestial mechanics. Newton suggested (in the second and third editions, 3.13) that the observed inequalities in the motion of Jupiter and Saturn could be accounted for as a consequence of their

---

[79] For overviews of eighteenth-century work on the three-body problem, see (Waff, 1995a; Wilson, 1995)

[80] This independent variable is usually time, but Newton also treats time as a dependent variable in Prop 2.10, in the second and third editions. Generalizing to functions of multiple variables was required for the concepts of partial differentiation and the calculus of variations.

gravitational interaction. But Flamsteed and Newton's efforts to treat the problem quantitatively were not successful, and the Paris Academy sponsored three consecutive prize essays from 1748–52 regarding the inequalities. Clairaut and d'Alembert were actively working on the three-body problem at this time, and served as members of the prize commission (and hence were ineligible to enter). The leading competitors in these contests—Euler, Daniel Bernoulli, and Roger Boscovich—made important contributions to the problem, but a full treatment was only achieved by Laplace in 1785 (Wilson, 1995).

Analytical techniques developed to treat the three-body problem were also applied to the Earth–Moon–Sun system. Newton had proposed (in Prop. 3.39) that precession of the equinoxes is caused by the gravitational attraction of the Sun and the Moon on the Earth's equatorial bulge.[81] In the 1730s the Astronomer Royal James Bradley discovered a further effect called nutation, and published his results in 1748. Nutation refers to a slight variation in the precession of the equinoxes, or wobble in the axis of rotation due to the changing orientation of the lunar orbit with respect to the Earth's equatorial bulge. Bradley's observations provided strong evidence for the gravitational effects of the Moon on the Earth's motion, which was almost immediately bolstered by d'Alembert's analytical solution describing nutation (d'Alembert, 1749). This successful account of precession and nutation provided evidence for the inverse-square law almost as impressive as Clairaut's calculation. But in addition, d'Alembert's innovations in the course of applying gravitational theory to this problem were important contributions in their own right. Paraphrasing Laplace, d'Alembert's work was the seed that would bear fruit in later treatments of the mechanics of rigid bodies.[82]

By mid-century, universal gravitation was deeply embedded in the practice of celestial mechanics. Treating the solar system as a system of point-masses interacting via Newtonian gravitation led to tremendous advances in understanding the physical factors that play a role in observed motions. These advances stemmed in part from developing more powerful mathematical techniques in order to assess the implications of universal gravity for situations that Newton had been unable to treat quantitatively. Just as it is easy to overestimate the empirical case in Newton's favour in 1687, modern readers often mistakenly treat the *Principia* as containing all of modern rational mechanics. But in fact Newton does not even touch on a number of problems in mechanics that had been discussed by his contemporaries, such as the motion of rigid bodies, angular motion, and torque. Several parts of Book 3, including the account of the tides and the shape of the Earth, were flawed as

---

[81] Precession of the equinoxes refers to a measurable, periodic variation in the positions of stars in equatorial coordinates, understood in Newton's time (and now) as a result of the motion of the rotational axis of the Earth.

[82] See (Wilson, 1987) for a detailed discussion of d'Alembert (1749), with a clear account of the contrast between Euler's approach and the impact of d'Alembert's work on Euler, 's (1752) 'New Principles of Mechanics'.

a result. This is not a simple oversight that could be easily corrected. Extending and developing Newton's ideas to cover broader domains has been an ongoing challenge in mechanics ever since.

An important line of thought in the development of rational mechanics was the effort to assimilate and extend the ideas of the *Principia*. But eighteenth-century rational mechanics drew on other, independent lines of thought as well. Pierre Varignon advocated a distinctive approach to mechanics in his *Project of a New Mechanics*, which appeared in the same year as the *Principia*. Newton's work was assimilated to an existing line of research in mechanics, a tradition that had much broader scope. Newton's great contemporaries on the continent—primarily Huygens, Leibniz, and Johann and Jacob Bernoulli—had all made important contributions to a set of longstanding problems in mechanics that Newton did not discuss. These problems involved the behaviour of elastic, rigid, and deformable bodies rather than point masses, and their treatment required concepts such as stress, torque, and contact forces. For example, Huygens, (1673) found the centre of oscillation for a pendulum bob based on what Leibniz would later call conservation of *vis viva*. Jacob Bernoulli treated this problem using the 'law of the lever' rather than Huygens' principle, and then extended these ideas to the study of elastic bodies in the 1690s. This line of work was entirely independent of Newton, and Truesdell, (1968) argues that the impact of Bernoulli's ideas was nearly as significant as the *Principia* itself. The members of the Basel school treated the *Principia* as posing a challenge, to either rederive Newton's results on their own terms or to find his errors. Several of the problematic claims in Book 2 acted as an impetus to particular research areas. Newton's treatment of the efflux problem in Proposition 2.36, for example, partially spurred Daniel Bernoulli's development of hydrodynamics.

The rich interplay of these ideas eventually led to formulations of mechanics such as Euler's *Mechanica* (1736), and his later 1752 paper announcing a 'New principle of mechanics'. This new principle was the statement that $\mathbf{F} = \mathbf{ma}$ applies to mechanical systems of all kinds, discrete or continuous, including point masses and bodies of finite extent. Euler immediately applied this principle to the motion of rigid bodies. There were numerous innovations in Euler's formulations of mechanics, but we emphasize this principle as a warning to those apt to read the work of Euler and others back into Newton. Editions of the *Principia* published during this time, by the Minim friars Le Seur and Jacquier and by Marquise du Châtelet, presented the *Principia* in Eulerian terms and showed how to reformulate some of Newton's results using the symbolic calculus. But by this point the *Principia* itself had largely disappeared from view; it was not required reading for those active in analytic mechanics, and there were better contemporary formulations of the underlying principles of mechanics. The common label 'Newtonian mechanics' for these later treatments, while not entirely unjustified, fails to acknowledge the important conceptual innovations that had occurred in the eighteenth century and the ultimate source of these innovations in the work of Huygens, Leibniz, and the Bernoullis.

5.4.1 Cause of Gravity

One of the central questions of eighteenth-century philosophy was the nature and cause of gravity. In discussing these matters we should distinguish among a) the force of gravity as a real cause (which is calculated as the product of the masses over the distance squared); b) the cause of gravity; c) 'the reason for these [particular–ES] properties of gravity' (*Principia*, p. 943); and d) the medium, if any, through which it is transmitted. Much discussion about Newton conflates these matters. Of course, if the medium can explain all the properties of gravity then it is legitimate to conflate these.

One line of thought made popular by Newton in the General Scholium of the *Principia* is to simply assert that 'it is enough that gravity really exists and acts according to the laws that we have set forth' (*Principia*, p. 943), while famously remaining agnostic about the causes that might explain it. (See Janiak, 2008) On this view one could accept the reality of gravity in the absence of an explanation of it. The significance of this is that future research can be predicated on its existence without worrying about matters external to relatively autonomous ongoing inquiry. While Newton was not the first to defend such an attitude toward inquiry (it echoes his earlier stance in the controversy over his optical research, and during the 1660s members of the Royal Society had investigated experimentally and mathematically the collision rules with a similar stance), his had the most lasting impact.

In his famous correspondence with Leibniz (1715–16), Clarke asserts something similar to Newton's position, though Clarke's argument sometimes suggests a more instrumentalist stance, in which gravity is assumed in order to track and predict effects, namely the relative motion of bodies (Alexander, 1965). In his more revisionary project, Berkeley elaborated this instrumentalist reinterpretation of Newton. For Berkeley (and later Hume), Newton's mathematical science cannot assign causes—which is the job of the metaphysician (Berkeley, 1744, p. 119–120 paragraphs 249–251; for discussion see Schliesser, 2011). Yet most eighteenth-century readers of Newton not only accepted gravity as a causally real force, but were also willing to entertain strikingly divergent positions regarding its causes. This was anticipated by Newton, who already in the first edition of the *Principia* listed at least three different possible mechanisms which could account for attraction (Scholium to 1.69, p. 588):

> I use the word 'attraction' here in a general sense for any endeavour whatever of bodies to approach one another, whether that endeavour occurs as a result of the action of the bodies either drawn toward one another or acting on one another by means of spirits emitted or whether it arises from the action of aether or of air or of any medium whatsoever—whether corporeal or incorporeal—in any way impelling toward one another the bodies floating therein.

The 'action' of bodies 'drawn toward one another' can involve action at a distance. Some of the earliest readers of *Principia* thought that Newton was committed to action at a distance either modelled on Stoic sympathy (as Leibniz dismissively claimed) or on Epicurean innate gravity (as Bentley proposed in now lost letters to Newton). The Stoic sympathy and Epicurean gravity options that interpret attraction as resulting from the nature of bodies go against the previously dominant view of mechanism, which only permitted contact of bodies as acceptable mechanism.

There is eighteenth-century evidence for three accounts of the cause of gravity compatible with Newton's first sense. First, in his editor's preface to the *Principia*, Roger Cotes asserted that gravity was a 'primary' quality of matter and put it on a par with impenetrability and other properties often taken to be essential qualities. However, in the third edition Newton made it clear that he did not accept this position, stating that he is 'by no means affirming that gravity is essential to bodies' (*Principia*, p. 796). Moreover, in famous responses to Bentley's letters, Newton explicitly denied 'innate' gravity 'as essential and inherent to matter' (Newton, 2004, p. 102). Nevertheless, Cotes' interpretation became very influential, and was adopted by Immanuel Kant, among others.

A second one was modelled on Locke's superaddition thesis: that is, God could add mind-like qualities to otherwise passive matter. While gravity is not an essential quality of matter, it is certainly in God's power to endow matter with gravitational qualities at creation. This interpretation was encouraged by Newton in his exchange with Bentley, and it was taken up by many of the Boyle lecturers that developed eighteenth-century physical-theology. It was also made famous in the French-speaking world by a footnote added by the French translator of Locke's *Essay*.

A third way was put forward by Newton himself in his posthumously published 'Treatise of the System of the World'. Curiously, Newton called attention to the existence of this popular, suppressed exposition of his views in the brief 'preface' of the third Book in all three editions of the *Principia*, but it is unclear whether he had a hand in arranging for it to be published the year after his death. In the 'Treatise', Newton offers a relational account of action at a distance. On the view presented there, all bodies have a disposition to gravitate, but it is activated only in virtue of their having this common nature. While there is evidence that the 'Treatise' was read in the eighteenth century, the relational view seems not to have been very popular. But it is compatible with the position adopted by D'Alembert in the widely read *Preliminary Discourse* in describing Newton's achievement: 'matter may have properties which we did not suspect' (D'Alembert, 1751).

Some people attributed to Newton the view that he believed that gravitation is based on the direct will of God. This position was attributed to him by Fatio de Duillier and, perhaps more jokingly, by David Gregory (both of whom were considered as possible editors for a new edition of the *Principia* planned in the 1690s), who were both in his circle especially in the early years after the publication of the first edition of the *Principia*. The position is certainly consistent with Newton's last sense above (assuming God is immaterial), and there are other passages in Newton's writings that seem compatible with it. For example, in a letter to Bentley, Newton writes: 'Gravity must be caused by an agent acting constantly according to certain laws; but whether this agent be material or immaterial I have left to the consideration of my reader' (Newton, 2004, p. 103).

Nevertheless, attributing gravity's cause to God's direct will appears at odds with a very famous passage in the General Scholium, where Newton articulates what he means by God's substantial and virtual omnipresence: 'In him [God] all things are contained and moved, but he does not act on them nor they on him. God experiences nothing from the motion of bodies; the bodies feel no resistance from God's omnipresence.'[83] Whatever Newton means by asserting both God's substantial and virtual omnipresence, he clearly states that God's presence does not interfere with the motions of bodies—either by offering resistance or impelling them. As David Hume aptly noted: 'It was never the meaning of Sir Isaac Newton to rob second causes of all force or energy, though some of his followers have endeavoured to establish that theory upon his authority'.[84]

Finally, ether theories were very popular during the eighteenth century. Sometimes they were put forward in opposition to Newtonian action at a distance (for example, by Euler). But we need to note two facts: first, ether theories had Newtonian precedent: Newton tentatively put forward ether accounts in the closing paragraph of the General Scholium and in a famous letter to Boyle known to eighteenth-century readers. Newton's proposals were not exactly identical: in his letter to Boyle he conceived of an ether as a compressible fluid; in various Queries to the *Opticks*, he emphasizes the different densities of the ether around and between celestial bodies, and he speculates about the need for short-range repulsive forces within the ether. Second, thus, ether theories nearly always include action at a distance over relatively short ranges. One general problem with ether theories is that they require ethers to have neglible mass, which makes them very hard to detect, while being capable of great strength and rigidity in order to transmit light as fast as Rømer had calculated it to be. But Newton clearly did not rule out an immaterial ether composed of spirits of some sort.

References

Aiton, E. J. (1972). *The Vortex Theory of Planetary Motion*. American Elsevier, New York.
Albury, WR (1978). Halley's Ode on the *Principia* of Newton and the Epicurean Revival in England. *Journal of the History of Ideas*, **39**(1), 24–43.
Alexander, H. G. (1965). *The Leibniz–Clarke Correspondence*. Manchester University Press, Manchester.
Arnol'd, Vladmir, I. (1990). *Huygens and Barrow, Newton and Hooke*. Birkhaúser, Boston.
Bernoulli, Johann (1730). Nouvelles Pensées sur le Systéme de M. Descartes. Jombert.
Bertoloni Meli, Domenico (1997). *Equivalence and Priority: Newton versus Leibniz, Including Leibniz's Unpublished Manuscripts on the Principia*. Oxford University Press, Oxford.

---

[83] Newton's footnote to the passage explains he is articulating God's dominion.
[84] This appears in a footnote at the end of 7.1.25, in (Hume, 2000). Regarding Hume's terminology, God is a first cause, whereas laws or forces are secondary causes that act within nature.

Biener, Zvi, and Smeenk, Chris (2011). Cotes' queries: Newton's empiricism and conceptions of matter. In *Interpreting Newton: Critical Essays* (ed. A. Janiak and E. Schliesser). Cambridge University Press, Cambridge.

Blay, M. (1999). *Reasoning with the Infinite: From the Closed World to the Mathematical Universe*. University of Chicago Press, Chicago.

Bos, H. J. M. (2001). *Redefining Geometrical Exactness: Descartes' Transformation of the Early Modern Concept of Construction*. Springer Verlag, New York.

Brackenridge, J. Bruce (1995). *The Key to Newton's Dynamics: The Kepler Problem and the Principia*. University of California Press, Berkeley.

Brading, Katherine (2011). Newton's law-constitutive approach to bodies: a response to descartes. In *Interpreting Newton: Critical Essays* (ed. A. Janiak and E. Schliesser). Cambridge University Press, Cambridge.

Cohen, I. B. (1980). *The Newtonian Revolution: With Illustrations of the Transformation of Scientific Ideas*. Cambridge University Press, Cambridge.

Cohen, I. Bernard (1971). *Introduction to Newton's* Principia. Cambridge University Press, Cambridge.

Cohen, I. Bernard (1999). A guide to Newton's *Principia*. In *The Principia: Mathematical Principles of Natural Philosophy*, pp. 1–370. University of California Press, Berkeley.

D'Alembert, Jean Le Rond (1751). Discourse préliminaire. The Encyclopedia of Diderot and d'Alembert Collaborative Translation Project. Translated by Richard N. Schwab and Walter E. Rex.

de Gandt, François (1995). *Force and Geometry in Newton's Principia*. Princeton University Press, Princeton.

de Smet, R. and Verlest, K. (2001). Newton's Scholium Generale: The Platonic and Stoic legacy: Philo, Justus Lipsius, and the Cambridge Platonists. *History of Science*, **39**, 1–30.

Descartes, René (1983 [1644]). *Principles of Philosophy*. D. Reidel, Dordrecht.

DiSalle, R. (2006). *Understanding Space-Time: The Philosophical Development of Physics from Newton to Einstein*. Cambridge University Press, Cambridge.

Domski, Mary (2011). Locke's Qualified Embrace of Newton's *Principia*. In *Interpreting Newton: Critical Essays* (ed. A. Janiak and E. Schliesser). Cambridge University Press, Cambridge.

Downing, Lisa (1997). Locke's Newtonianism and Lockean Newtonism. *Perspectives on Science*, **5**(3), 285–310.

Ducheyne, S. (2007). The general scholium: Some notes on newton's published and unpublished endeavours.

Dunlop, Katherine (2012). The Mathematical Form of Measurement and the Argument for Proposition I in Newton's *Principia*, *Synthese*, **186** (1), 191–229.

Earman, John S. (1989). *World Enough and Space-Time: Absolute versus Rational Theories of Space and Time*. MIT Press, Cambridge, MA.

Euler, L. (1736). *Mechanica sive motus scientia analytice exposita*. 3 volumes. Royal Academy of Science, St Petersberg, Russia.

Euler, Leonhard (1752). Discovery of a new principle of mechanics. In *Opera Omnia*, Volume II.

Fraser, Craig (2003). History of mathematics in the eighteenth century. In *The Cambridge History of Science* (ed. R. Porter), Volume 4: Eighteenth-Century Science, pp. 305–327.

Gabbey, Alan (1980). Force and inertia in the seventeenth century: Descartes and Newton. In *Descartes: Philosophy, Mathematics and Physics* (ed. S. Gaukroger), pp. 230–320. Harvester Press, Sussex, Totowa, N.J.

GUICCIARDINI, N. (2003). *Reading the Principia: The Debate on Newton's Mathematical Methods for Natural Philosophy from 1687 to 1736*. Cambridge University Press, Cambridge.
GUICCIARDINI, N. (2009). *Isaac Newton on Mathematical Certainty and Method*. MIT Press.
HALL, A. R. and HALL, M. B. (1962). *Unpublished Scientific Papers of Isaac Newton*. Cambridge University Press, Cambridge.
HARPER, WILLIAM and SMITH, GEORGE E. (1995). Newton's new way of inquiry. In *The Creation of Ideas in Physics: Studies for a Methodology of Theory Construction* (ed. J. Leplin), pp. 113–166. Kluwer Academic Publishers, Dordrecht, The Netherlands.
HARPER, WILLIAM L. (2011). *Isaac Newton's Scientific Method: Turning Data into Evidence about Gravity and Cosmology*. Oxford University Press, New York and Oxford.
HERIVEL, J. (1965). *The Background to Newton's Principia*. Oxford University Press, Oxford.
HUGHES, D. W. (1988). The '*Principia*' and comets. *Notes and Records of the Royal Society of London*, **42**(1), 53.
HUME, DAVID (1985). *The History of England*. 6 volumes. Liberty Fund, Indianapolis.
HUME, D. (2000). *An Enquiry Concerning Human Understanding: A Critical Edition*. Volume 3. Oxford University Press, USA.
HUYGENS, CHRISTIAAN (1673 [1986]). *Horologium Oscillatorium*. Iowa State University Press, Ames.
HUYGENS, CHRISTIAAN (1690). *Discours de la cause de la pensanteur*.
HUYGENS, CHRISTIAAN (1888–1950). *Oeuvres complètes*. 22 Volumes. Martinus Nijhoff, The Hague.
JANIAK, A. (2008). *Newton as Philosopher*. Cambridge University Press, Cambridge.
KRILOFF, AN (1925). On Sir Isaac Newton's method of determining the parabolic orbit of a comet. *Monthly Notices of the Royal Astronomical Society*, **85**, 640.
LEIBNIZ, G.W. (1689). Tentamen de motuum coelestium causis. *Acta Eruditorum*, 82–96.
MACLAURIN, COLIN (1748). *An Account of Sir Isaac Newton's Philosophical Discoveries*. Millar & Nourse, London.
MCGUIRE, J. E. (1966). Body and Void and Newton's *De Mundi Systemate*: Some new sources. *Archive for the History of Exact Sciences*, **3**, 206–248.
MCGUIRE, J. E. (1968). Force, active principles, and Newton's invisible realm. *Ambix*, **15**, 154–208.
MCGUIRE, J. E. (1970). Atoms and the 'Analogy of Nature': Newton's third rule of philosophizing. *Studies in History and Philosophy of Science*, **1**, 3–58.
MCGUIRE, J. E. (1978). Existence, actuality and necessity: Newton on space and time. *Annals of Science*, **35**, 463–508.
MCGUIRE, J. E. (1994). Natural Motion and its causes: Newton on the *Vis insita* of bodies. In *Self-motion: from Aristotle to Newton* (ed. M. L. Gill and J. G. Lennox), pp. 305–329. Princeton University Press, Princeton.
MCGUIRE, J. E. and RATTANSI, P. M. (1966). Newton and the 'Pipes of Pan'. *nrrs*, **21**, 108–126.
MCMULLIN, ERNAN (1978). *Newton on Matter and Activity*. University of Notre Dame Press, Notre Dame, Indiana.
NEWTON, ISAAC (1952 (1730)). *Opticks*. Dover Publications, [New York].
NEWTON, ISAAC (1999 [1726]). *The Principia, Mathematical Principles of Natural Philosophy: A New Translation*. University of California Press, Berkeley, California.
NEWTON, ISAAC (2004). *Isaac Newton: Philosophical Writings*. Cambridge: Cambridge University Press.
POURCIAU, B. (1998). The Preliminary Mathematical Lemmas of Newton's *Principia*. *Archive for history of exact sciences*, **52**(3), 279–295.
POURCIAU, B. (2001). The integrability of ovals: Newton's Lemma 28 and its counterexamples. *Archive for History of Exact Sciences*, **55**(5), 479–499.

Rynasiewicz, R. (1995a). By their properties, causes and effects: Newton's scholium on time, space, place and motion, I. The text. *Studies in History and Philosophy of Science Part A*, **26**(1), 133–153.

Rynasiewicz, R. (1995b). By Their Properties, Causes and Effects: Newton's Scholium on Time, Space, Place and Motion, II. The Context. *Studies in History and Philosophy of Science Part A*, **26**(2), 295–322.

Schliesser, E. (2011). Newton's challenge to philosophy: a programmatic essay. *HOPOS: The Journal of the International Society for the History of Philosophy of Science*, **1**(1), 101–128.

Schliesser, Eric, and Smith, George E. (2010). Huygens' 1688 report to the Directors of the Dutch East India Company on the Measurement of Longitude at Sea and the Evidence it Offered against Universal Gravity. unpublished manuscript.

Shank, J. B. (2008). *The Newton Wars and the Beginning of the French Enlightenment*. University of Chicago Press, Chicago.

Shapiro, A. E. (1984). *The Optical Papers of Isaac Newton*. Cambridge University Press, Cambridge.

Shapiro, A. E. (2002). Newton's optics and atomism. *The Cambridge Companion to Newton*, 227–55.

Shapiro, Alan E. (2004). Newton's 'experimental philosophy'. *Early Science and Medicine*, **9**(3), 185–217.

Smith, Adam (1982). *Essays on Philosophical Subjects*. Volume Volume III of the *Glasgow Edition of the Works and Correspondence of Adam Smith*. Liberty Fund, Indianapolis.

Smith, George E. (2001). The Newtonian Style in Book II of the *Principia*. In *Isaac Newton's Natural Philosophy* (ed. J. Z. Buchwald and I. B. Cohen), Dibner Institute Studies in the History of Science and Technology, pp. 249–314. MIT Press, Cambridge, MA.

Smith, George E. (2002). From the phenomenon of the ellipse to an inverse-square force: Why not? In *Reading Natural Philosophy: Essays in the History and Philosophy of Science and Mathematics to Honor Howard Stein on his 70th Birthday* (ed. D. B. Malament), pp. 31–70. Open Court, Chicago.

Smith, George E. (2009). Closing the loop: Testing Newtonian gravity, then and now.

Snobelen, S. D. (2001). 'God of gods, and lord of lords': The theology of Isaac Newton's general scholium to the *Principia*. *Osiris*, **16**, 169–208.

Spinoza, Benedict de (1985). *The Collected Works of Spinoza*. Princeton University Press, Princeton.

Stein, Howard (1967). Newtonian space-time. *Texas Quarterly*, **10**, 174–200.

Stein, H. (1990). On locke, the great Huygenius, and the incomparable Mr Newton. *Philosophical Perspectives on Newtonian Science*, 17–47.

Stein, Howard (1991). From the phenomena of motion to the forces of nature: Hypothesis or deduction? *PSA 1990*, **2**, 209–22.

Stein, Howard (2002). Newton's metaphysics. In *Cambridge Companion to Newton* (ed. I. B. Cohen and G. E. Smith), pp. 256–307. Cambridge University Press, Cambridge.

Stein, Howard (ms.). On metaphysics and method in Newton. Unpublished manuscript.

Terrall, M. (2002). *The Man who Flattened the Earth: Maupertuis and the Sciences in the Enlightenment*. University of Chicago Press.

Truesdell, Clifford (1968). *Essays in the History of Mechanics*. Springer-Verlag, Berlin, New York.

Turnbull, H. W., Scott, J. W., Trilling, L., and Hall, A. R. (ed.) (1959–1977). *The Correspondence of Sir Isaac Newton*. Volume 1–7. Cambridge University Press, Cambridge.

Valluri, S. R., Wilson, C., and Harper, W. (1997). Newton's apsidal precession theorem and eccentric orbits. *Journal for the History of Astronomy*, **28**, 13.

VOLTAIRE, FRANÇOIS MARIE AROUET DE (1738). Eléménts de la philosophie de Newton. Amsterdam.

WAFF, CRAIG (1995a). Clairaut and the motion of the lunar apse: the inverse-square law undergoes a test. In *Planetary Astronomy from the Renaissance to the Rise of Astrophysics: The Eighteenth and Nineteenth Centuries* (ed. R. Taton and C. Wilson), pp. 35–46. Cambridge University Press, Cambridge.

WAFF, C. B. (1995b). Predicting the mid-eighteenth-century return of Halley's comet. In *Planetary Astronomy from the Renaissance to the Rise of Astrophysics. Part B: The Eighteenth and Nineteenth Centuries*, Volume 1, pp. 69–82.

WHITESIDE, D. T. (ed.) (1967–81). *The Mathematical Papers of Isaac Newton*. Volume 1–8. Cambridge University Press, Cambridge, London.

WHITESIDE, D. T. (1970). The mathematical principles underlying Newton's *Principia Mathematica*. *Journal for the History of Astronomy*, **1**, 116–138.

WHITESIDE, D. T. (1991). The Prehistory of the *Principia* from 1664 to 1686. *Notes and Records of the Royal Society of London*, **45**(1), 11.

WILSON, C. A. (1970). From Kepler's laws, so-called, to universal gravitation: empirical factors. *Archive for History of Exact Sciences*, **6**(2), 89–170.

WILSON, C. (1987). D'Alembert versus Euler on the precession of the equinoxes and the mechanics of rigid bodies. *Archive for History of Exact Sciences*, **37**(3), 233–273.

WILSON, C. (1989). The Newtonian achievement in astronomy. In *Planetary Astronomy from the Renaissance to the Rise of Astrophysics. Part A: Tycho Brahe to Newton*, Volume 1, pp. 233–274.

WILSON, C. (1995). The problem of perturbation analytically treated: Euler, Clairaut, d'Alembert. In *Planetary Astronomy from the Renaissance to the Rise of Astrophysics. Part B: The Eighteenth and Nineteenth Centuries*, Volume 1, pp. 89–107.

CHAPTER 6

NEWTON'S OPTICS

ALAN E. SHAPIRO

Newton's *Opticks: Or, a Treatise of the Reflexions, Refractions, Inflexions and Colours of Light* (1704) dominated the science of optics for over a century. His theory of colour and the compound nature of sunlight, and his concept of unequal refrangibility were central to modern optics. More specific contributions, such as the design and construction of reflecting telescopes, the investigation of the colours of thin films, and attempts to mathematize the emission theory of light, added to the central role he played in the development of optics. Moreover, the experimental approach adopted in the *Opticks* was lauded and widely followed in optical science and served as a model for other experimental sciences.

Newton worked on optics throughout his career. In the remarkably fruitful years 1665–66, when he was developing the calculus and the foundations of mechanics, he discovered the compound nature of white light and the unequal refrangibility of rays of different colour—the foundation for all of his optical investigations—and constructed the first successful reflecting telescope. At the end of this period, he also made his first, tentative investigation of the colours of thin films ('Newton's rings'). When he was appointed Lucasian Professor of Mathematics at the University of Cambridge in 1670, he chose to present his optical researches for his inaugural lectures. By 1672 Newton prepared his *Optical Lectures* for publication as the first extended treatment of his new theory, but he subsequently suppressed it. In late 1671 his reflecting telescope was presented to the Royal Society. A month later, encouraged by the enthusiastic reception of his telescope, Newton sent a brief paper to the Royal Society, quickly published in *Philosophical Transactions*, announcing his new theory of light and colour. The theory was not at first well received and over the next few years Newton engaged in an extended series of exchanges, most of which were published in *Philosophical Transactions*. He was so disturbed by the continued intrusion that he withdrew from the public arena of science until late 1675. In December he sent two papers to the Royal Society—one on the colours of thin plates, and a second, speculative paper that contained a physical model of light. Newton did little in optics until he started to compose the *Opticks* in the late 1680s, after he finished the *Principia* (1687). The *Opticks* was largely completed by 1692 but not published until 1704, because of the troublesome section on diffraction.

Since Newton's optical research extended over nearly forty years, we cannot rely solely on the *Opticks* (though it is, to be sure, the most important exposition of his optical research) to properly comprehend his optical investigations; we must have recourse to his earlier writings, published and unpublished. Newton's ideas did not change dramatically in this period, but they did evolve as he encountered new experimental results and theoretical challenges.

6.1 THE REFLECTING TELESCOPE

Newton made his scientific debut in December 1671, when he sent the Royal Society a reflecting telescope that he had designed and constructed. Designs for reflecting telescopes had often been proposed since the beginning of the century, and some reflectors had most likely been constructed but remained unknown to the wider community. In his catadioptrical telescope (Fig. 6.1), as it was called, Newton placed a concave spherical mirror AB at the base of the tube, and near the other end he placed a very small, plane secondary mirror CD inclined to the axis so that it would reflect the rays through a hole in the side of the tube, which contained a plano-convex lens F for viewing the image. This was the first, publically known, successful reflecting telescope. The members of the Society put it to a test in late December and concluded that this diminutive telescope, about 6 inches long, produced a very distinct, colour-free image about two and a half times greater than an ordinary telescope—that is, a refractor—25 inches long.[1]

The telescope was so well received that Newton was nominated to be become a Fellow of the Society. In order to secure the invention for Newton and the English, Henry Oldenburg, the Secretary of the Society arranged for publication in France via Christiaan Huygens. A description was published in the *Journal des sçavans,* together with a letter from Huygens that endorsed the invention (Huygens, 1672; Newton, 1672), and in Jean Baptiste Denis' *Recoeuil des memoires et conferences sur les arts & les sciences* (Denis, 1672). Shortly after the French publications Oldenburg published a fuller description of Newton's telescope in the *Philosophical Transactions* (Newton, 1958, 61–7).

The most troublesome aspect of Newton's telescope was the mirror. He had experimented with the composition of the speculum metal, and cast and polished the mirror himself. It tarnished quickly and had to be polished frequently, and the metal was soft and did not reflect light vividly. Newton was confident that these problems could be solved with the development of new alloys (Newton, 1959, 126–9).

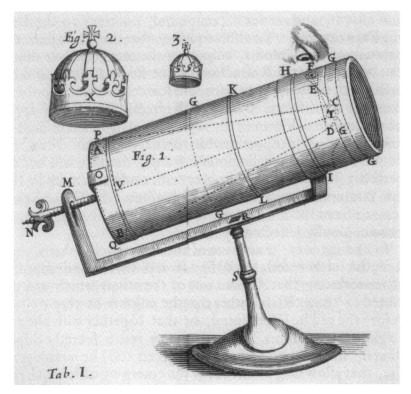

Fig. 6.1. Newton's reflecting telescope, in *Philosophical Transactions of the Royal Society* 7, 81 (25 March 1672), Tab. 1.

Rival designs for a reflecting telescope soon came to the fore. Laurent Cassegrain proposed a telescope with a concave primary mirror but a convex secondary mirror that reflected light back through a hole in the primary where the image is observed with a lens. No claim was made that such a telescope had been constructed. Oldenburg requested Newton's observations on Cassegrain's telescope and a comparison with that proposed by James Gregory in his earlier *Optica promota* (1663). The principal difference between Gregory's and Cassegrain's telescope was that Gregory's had a concave secondary mirror. Newton argued vigorously, if not always justly, for the superiority of his design over both Gregory's—which he had known when designing his own telescope—and Cassegrain's (Newton, 1959, 153–5). Most of his criticism focused on the problems of the curved secondary mirror.

When Denis published an account of Newton's reflector he emphasized the advantages it offered to astronomers because of its small size compared with the large, awkward telescopes, 50 to 100 feet in length, then in use. Astronomers had determined from experience that the longer a telescope the smaller the colour distortion. In his 'New theory' Newton explained that he abandoned work on refracting telescopes and turned to the development of a reflecting telescope as a consequence of his investigation of colour. He discovered that the cause of the colour distortions of lenses was due to the unequal refraction of different colours in lenses and thus was an inherent, unavoidable defect of refractive instruments. Indeed, he argued that chromatic aberration was 'hundreds' of times greater than that caused by the deviation of the shape of spherical lenses from a conic section, which would focus rays perfectly (Newton, 1959, 95–6). Robert Hooke objected to Newton's pessimistic view of the future of refracting telescopes and argued that achromatic lenses could be developed (Simpson, 2009, 423). Since Newton's telescope was announced before his theory of colour was known, the lack of colour aberration was scarcely noted by commentators. His telescope was admired primarily for its small size. Newton was so pleased by the reception of his telescope that he promised Oldenburg he would soon send 'an accompt of a Philosophicall discovery wch induced mee to the making of the said Telescope' (Newton, 1959, 82).

Reflecting telescopes were difficult to construct, and the initial excitement quickly waned. Interest was revived in the early eighteenth century when Newton published a description of a reflector in the *Opticks*. In Bk. I, Pt. I, Prop. 8, he presented a new design for a telescope, which now had a glass mirror silvered on the rear surface rather than a speculum metal mirror and a prism to deflect the light to the eyepiece instead of a secondary mirror. In the long, preceding proposition he invoked experiment and calculation to demonstrate that 'The Perfection of Telescopes is impeded by the different Refrangibility of the Rays of Light' (Newton, 1704, 59).

6.2 Theory of Light and Colour

The fundamental elements of Newton's theory of the nature of white light and colour remained essentially unchanged throughout his career, but exchanges with critics and further investigations compelled him to revise and refine some elements of the theory and alter its formulation. The theory initially met widespread resistance, but by 1704 and the publication of the *Opticks*, it already had gained substantial support. In early February Newton sent Oldenburg the paper that he had promised, which appeared as 'A new theory about light and colour' in the *Philosophical Transactions* in February 1672.

Newton began his paper with an experiment (Fig. 6.2) which, with numerous variations, served as the experimental foundation of much of his optical research. He transmitted a beam of sunlight through a small hole F in his window shutter. He then passed the beam through a prism $ABC\alpha\beta\kappa$ so that it was refracted as much on entering as on leaving the prism—i.e., at minimum deviation—and then he cast its refracted image or spectrum PT perpendicularly onto the wall 22 feet away. According to 'the received laws of Refraction', i.e., Snell's law, he expected this spectrum to be circular, but to his surprise it was oblong. He wrote: 'Comparing the length of this coloured *Spectrum* with its breadth, I found it about five times greater; a disproportion so extravagant, that it excited me to a more than ordinary curiosity of examining,

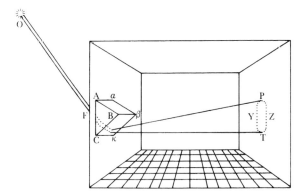

Fig. 6.2. Newton's basic experimental set-up for forming a spectrum *PT*.

from whence it might proceed' (Newton, 1959, 92). By observing the image 22 feet from the prism, he saw a fully coloured, elongated image of the Sun, whereas virtually all his predecessors had observed the image just a few feet away, before the beam fully dispersed, and saw a round white image with colours only on the edges.

Newton rapidly eliminated a series of possible causes of the elongation and finally resolved the problem by introducing an experiment that he called the *experimentum crucis* (crucial experiment). In order to appreciate the response to the *experimentum crucis*, we should note that Newton did not publish a diagram of the experiment (Fig. 6.3) until six months after the publication of the 'New theory'. He took two boards *BC*, *DE* with small holes in them and placed them 12 feet apart. The hole *X* in the first board allowed only a small portion of the beam refracted by the first prism to be transmitted. The transmitted portion of light then passed through the second hole *Y* onto another prism *F* and was cast onto a screen. When the first prism was rotated slightly, different portions of the refracted beam would be transmitted through the hole *X* to *Y*. Newton found that after passing through the second prism, the light

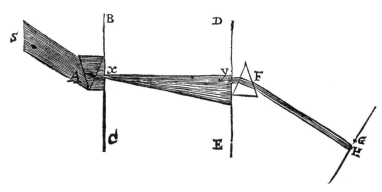

Fig. 6.3. The *experimentum crucis*, in *Philosophical Transactions of the Royal Society* 7, 85 (15 July 1672): 5016. Note that contrary to Newton's instructions, the prisms are not drawn at minimum deviation.

which was refracted the most by the first prism was again refracted more, to *H*, than the light which was refracted the least by the first prism, to *G*. He then concluded:

And so the true cause of the length of that Image was detected to be no other, then that *Light* consists of *Rays differently refrangible*, which, without any respect to a difference in their incidence, were, according to their degrees of refrangibility, transmitted towards diverse parts of the wall. (Newton, 1959, 95)

It should be noted that Newton has thus far not made any assertions about colour, but only about unequal refraction.

Figure 6.4, from Newton's *Optical Lectures*, should help to elucidate his claim that sunlight consists of rays of unequal or different refrangibility, or, as we would say, with different indices of refraction. When a ray of sunlight *OF* is refracted at surface *EG*, it is decomposed into a series of rays, *FP*, *FQ*, *FR*, *FS*, *FT*, and innumerable others, that are each refracted a different amount, or dispersed, though they all have the same angle of incidence. According to the received view there should be only one refracted ray, say, *FR*. Newton would go on to argue that each of these rays obeys Snell's law but with a different index of refraction.

Although Newton carefully avoided talking about colour in his account of the *experimentum crucis*, in the remainder of the paper he 'laid down' his theory of colour in a series of propositions without any proof. He found that rays of different refrangibility exhibit different colours: for example, the most refracted ray *FP* exhibits purple or violet, the least refracted *FT* red, and the mean refracted *FR* green, and so for all the innumerable intermediate colours. 'Colours', he noted, 'are not *Qualifications of Light*, derived from Refractions, or Reflections of natural Bodies (as 'tis generally believed) but *Original* and *connate properties* . . .' (Newton, 1959, 97). Thus, he claimed, whenever colours appear they are only separated from sunlight, in which they are innate; they are never created from some modification of light, such as a mixture of light with shadow, as was then widely held.

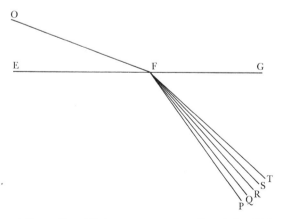

Fig. 6.4. Refraction at the surface *EG* decomposes a ray of sunlight *OF* into rays of different degrees of refrangibility and colour.

He found that no matter how he tried to alter the colour of any particular sort of light-ray, it was immutable. Yet, when rays of different colours are mixed, the colour does appear to change. Such transformations, Newton insisted, are only 'apparent' and not 'real'. When rays are mixed to make a new colour and are then separated from one another, 'they will exhibit the very same colours, which they did before they entered the composition . . .' To resolve this problem he introduced a distinction between simple and compound colours: 'There are therefore two sorts of colours. The one original and simple, the other compounded of these. The Original or primary colours are, *Red, Yellow, Green, Blew* and a *Violet-purple*, together wth Orange, Indico, and an indefinite variety of Intermediate gradations'. Here Newton introduced a fundamental new concept to optics: that of immutable rays of a single colour (and degree of refrangibility). He then announced his most radical conclusion: 'the most surprising and wonderful composition was that of *Whiteness*. . . .'Tis ever compounded . . .' It is 'a confused aggregate of Rays indued with all sorts of Colours' (Newton, 1959, 98).

After setting forth his theory, Newton applied it to explain some obvious phenomena such as the colours produced by prisms and the rainbow. More unexpectedly, he presented a novel explanation of the colours of natural bodies: namely, that their colours arise from the fact that they are disposed to reflect light of one colour in greater quantity than the others. When he illuminated a red mineral, minium, and a blue one, bise, placed next to one another in a dark room with red light, the minium was a luminous and the bise a dim red; and the converse occurred when illuminated with blue light. Since sunlight consists of a mixture of rays of all colours, when it falls on a body the body reflects mostly light of its normal, daylight colour and scarcely any of the rest. Those colours not reflected, Newton explained in his *Optical Lectures*, are absorbed by the body (Newton, 1984, 513–15). Bodies have no proper colour, but appear only of the colour of the light cast on them. Previously the colours of bodies were considered to be intrinsic qualities of the body or, by moderns, to be caused by some modification of the light falling upon them.

Newton concluded his paper by showing how to recombine all the colours of the spectrum into white light just like original sunlight (Fig. 6.5). Behind prism *ABC* he placed a lens *MN*, which causes the light to converge again at the focus *Q*. When a white paper *HI* is inserted and slowly moved toward *Q*, the colours are seen to be gradually converted again into whiteness at *Q* as they are mixed. After the rays

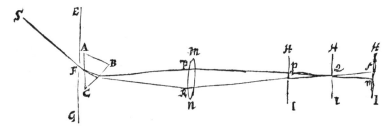

Fig. 6.5. Lens *MN* recombines sunlight, decomposed by prism *ABC*, into white light at focus *Q*. In *Philosophical Transactions of the Royal Society* 7, 80 (19 February 1671/2): 3086.

pass through Q, the colours gradually reappear, but in reverse order, as they separate again. If any of the colours are intercepted, such as at *P* or *R*, the white is changed into other colours. The importance of this experiment is that it showed that when the decomposed colours of sunlight are recombined, they 'reproduced light, intirely and perfectly white, and not at all sensibly differing from the *direct* Light of the Sun' (Newton, 1959, 98).

6.3 Reception

The 'New theory' was not initially well received. A number of factors, many independent of the specifics of his theory, contributed to this. Perhaps the most significant reason was its radical nature. The idea that sunlight is not simple, pure, and homogeneous, but the most compound of all colours, directly opposed a millennia-old tradition of thinking about sunlight. According to Newton, colours are not created by refraction of sunlight in a glass prism but are in the light all along; refraction only decomposes or separates the different colours by bending each a different amount according to their different refrangibility. Previously, colours were thought to be some modification of pure sunlight that were produced when it was refracted or reflected, such as a mixture with shadow or darkness, or a condensation. Newton replaced the concept of modification with a new way of thinking: that of decomposition or analysis by refraction.

A theory that overturned so many received ideas and introduced so many new concepts could barely be outlined, let alone be convincingly demonstrated, in the twelve-page paper that was printed in the *Philosophical Transactions*. In his unpublished *Optical Lectures* and in the *Opticks*, he devoted about ten times as many words to expounding his theory and its experimental support. A theory as revolutionary as Newton's would certainly attract significant opposition even with the clearest of presentations, but the brief sketch that he chose to publish could scarcely be expected to convey the complex concepts and experimental techniques of his theory and convince many.

Thus, two reasons immediately present themselves to explain the difficulties Newton's theory initially encountered: its revolutionary nature and its superficial presentation. Besides its extreme brevity, it contained only three experiments, and of them only one was illustrated. By limiting himself to just three experiments, Newton violated the prevailing methodology of the Royal Society, where it was held that numerous experiments were needed to establish a physical claim. Newton himself was aware of its shortcomings (Newton, 1959, 109, 212). Because of its compact presentation and lack of an illustration, I doubt that anyone initially understood, let alone accepted, the argument of the *experimentum crucis*.

The deliberately provocative stance of Newton's 'New theory' also antagonized the community and stimulated resistance to his theory. He presented his discovery of unequal refrangibility as overthrowing Snell's law of refraction—discovered only a generation earlier and now a new foundation of optics—rather than as an extension of it, and thus needlessly unsettled his audience. Later, in the *Opticks* he adopted a more accommodating approach and presented his discovery as an extension of existing knowledge. If his decision in 1672 to 'lay down' his theory without experimental evidence did not succeed in offending the members of the Royal Society, his challenge to their skeptical philosophy of science surely did. He declared that:

A naturalist would scearce expect to see ye science of [colours] become mathematicall, & yet I dare affirm that there is as much certainty in it as in any other part of Opticks. For what I shall tell concerning them is not an Hypothesis but most rigid consequence, not conjectured . . . but evinced by ye mediation of experiments concluding directly & wth out any suspicion of doubt. (Newton, 1959, 96–7)

This claim to certainty was so contentious that Oldenburg deleted it from the printed version.

Newton's claims as to the certainty of his theory were immediately challenged by Robert Hooke, who was present at the reading of the paper at the Royal Society and was assigned to referee it. Throughout his report, Hooke returned to Newton's claim to certainty and denied that the theory had been demonstrated by 'any undeniable argument' or with 'absolute necessity', or that it was 'soe certain as mathematicall Demonstrations' (Newton, 1959, 110, 113). In his reply to Hooke, Newton rejected the attribution to him of 'a greater certainty in these things then I ever promised, viz: The certainty of *Mathematicall Demonstrations*'. He then clarified his view:

I said indeed that the *Science of Colours was Mathematicall* & *as certain as any other part of Optiques;* but who knows not that Optiques & many other Mathematicall Sciences depend as well on Physicall Principles as on Mathematicall Demonstrations: And the absolute certainty of a Science cannot exceed the certainty of its Principles. Now the evidence by wch I asserted the Propositions of colours is in the next words expressed to be from *Experiments* & so but *Physicall:* Whence the Propositions themselves can be esteemed no more then *Physicall Principles* of a Science. (Newton, 1959, 187–8)

Newton still insisted that as a mathematical science his theory of colour was as certain as geometrical optics and more certain than a qualitative or purely physical account. However, for the first time he admitted the contingency of his experimental principles and abandoned his strong claim for the certainty of experiment. No longer would he assert, as he did in the 'New theory', that his theory was a 'most rigid consequence' deduced from 'experiments *concluding directly* & w^{th} *out any suspicion of doubt*'.

To properly understand Newton's scientific approach, it is necessary to explain his views on scientific method a little more fully. From the beginning of his career Newton was concerned with establishing a new, more certain science to replace the mechanical philosophy—especially as espoused by Descartes and his followers—which he felt was rife with 'conjectures and probabilities' (Newton, 1984, 89). He believed that he could establish a more certain science both by developing mathematical theories and by basing his theories on experimentally discovered properties.

To establish a more certain science, Newton insisted that one must 'not mingle conjectures with certainties' (Newton, 1959, 100). To avoid compromising rigourously demonstrated principles by hypotheses, he developed the techniques of clearly labeling hypotheses as such and setting them apart, as with 'Hypothesis' of 1675 and the queries appended to the *Opticks*.

As part of his campaign to reform science, Newton continually railed against hypotheses, that is, conjectural causal explanations. His condemnations of hypotheses—the most famous being his 'hypotheses non fingo' (I do not feign hypotheses) in the *Principia* (Newton, 1999, 943)—are always aimed at preserving the certainty of scientific principles rather than objecting to the use of hypotheses in themselves. Newton held that hypotheses without any experimental support whatever, such as Cartesian vortices, had no place in science. However, those based on some experimental evidence, though insufficient to establish them as demonstrated principles, could be used to understand properties already discovered and to suggest new experiments, and he used hypotheses in this way. The emission theory of light fell into the second category.

Newton believed that by formulating his theories in terms of experimentally observed properties, or principles deduced from them, without any causal explanations (hypotheses) of those properties he could develop a more certain science. As we shall see, however, in his private work he did use hypotheses to develop theories and predict new properties. When Newton used hypothetical causes such as light corpuscles and the aether in this way, he then purged them (or, at least, he attempted to do so) from his public work and reformulated his theories in terms of experimentally discovered 'properties' such as unequal refrangibility and periodicity. He appears never to have questioned the possibility of constructing a hypothesis-free science. To have denied such a possibility would have been tantamount to denying his conception of science.

Newton's apparent support of an emission theory of light disturbed the three most important early critics of his theory: Hooke, Ignace Gaston Pardies, and Huygens. They were concerned that his colour theory was incompatible with a pulse or wave theory. In replying to Hooke's accusation, Newton did not deny that he believed in the emission theory, but insisted that it played no part in his theory of colour (Newton, 1959, 174). To reassure his critics that his theory does not depend on light corpuscles, he then explained how the wave theory could be accommodated to it. If sunlight was considered to consist of a mixture of waves of various wavelengths ('depths or bignesses') each of which is refracted differently and excites a different colour, then their theories would be compatible with his colour theory without any need to adopt light corpuscles. After offering this suggestion, he then set out his life-long objection to a wave theory: namely, that it would violate the law of rectilinear propagation, as light would bend into shadow.

The existence of unequal refrangibility, and the *experimentum crucis* whereby it was demonstrated, were highly contested in the period after the publication of the 'New theory'. Nonetheless, it came to be accepted rather quickly, particularly by mathematical scientists, since it allowed for the further development of geometrical optics and especially the mathematical treatment of chromatic aberration, which so troubled contemporary telescopes (Shapiro, 1996).

6.4 Refining the Theory of Colour

Both before and after the 'New theory'—in the *Optical Lectures* and the *Opticks*—Newton's argument that sunlight consists of rays unequal refrangibility was built on a large number of experiments and not on a single crucial experiment. Though the steps of the argument are roughly similar in the *Optical Lectures* and the 'New theory', Newton makes no claim in the *Lectures* to have proved 'without any suspicion of doubt' that colours are innate to the Sun's direct light. In his *Optical Lectures* Newton claimed only that the concept that sunlight is composed of innumerable spectral colours is one that no reasonable person could deny (Newton, 1984, 143–5).

Newton utilized two basic arguments to establish that sunlight was a heterogeneous mixture of all the simple spectral colours: The first is founded on similarity: namely, that sunlight that has been decomposed and then recombined into white light is in all respects similar to the Sun's direct light and so should be considered of the same nature. The second is a logical argument that depends on the principal of the immutability of colour. His ideal argument would run something like the following: if the colours of light-rays are absolutely immutable, and if the rays exhibit some colour after refraction, then they must have possessed this same colour before refraction, and the colours are innate to the Sun's direct light, though not yet apparent. The problem confronting Newton is that the colour of the Sun's incident light appears totally different before the first refraction and ever after, once it has been resolved into colours. As Newton himself was ultimately to recognize, it is empirically impossible to prove what may be called the strong principle of colour immutability for the colours of sunlight at the first refraction, since the colours are imperceptible before the first refraction and so may not be compared with the colours after that refraction to see if they have changed. Indeed, if they are compared, it seems as if they have changed. He had no difficulty in establishing a weak principle of colour immutability after the first refraction.

In demonstrating that sunlight consists of rays of different degrees of refrangibility Newton did not face a similar intractable difficulty. Unequal refrangibility can be described by a mathematical law—the sine law of refraction for each particular colour—and that law can be applied in an identical way at every refraction, since there is no intrinsic difference between the first and subsequent refractions. The validity of the sine law of refraction and the immutability of degree of refrangibility can be empirically confirmed or rejected by a series of measurements.

In the 'New theory' Newton did not explain how colours are determined to be simple or compound. Hooke, and later Huygens, challenged Newton on whether there are an indefinite number of primary colours in sunlight or white light, or whether all colours, including white, may be composed from just two primaries as in painters' practice.[2] In his reply to Hooke in June 1672 Newton more clearly defined his concepts of simple and compound colours in terms of refrangibility: 'That colour is *primary or originall wch cannot by any art be changed, & whose rays are all alike refrangible; & that compounded which is changeable into other colours, & whose rays are not alike refrangible*' (Newton, 1959, 180). Experimentally this means that if rays are transmitted through a prism and are all refracted alike, they are simple; but if they are again dispersed, they are compound.

In formulating his theory of colour Newton was confronted by a difficult problem: the same sensible colour can be produced in more than one way. When Hooke and Huygens objected that all colours could be made from only two—yellow and blue—Newton responded: 'But supposing that all colours might according to this experiment be produced out of two by mixture, yet it follows not that those two are the onely *originall* colours . . .' (Newton, 1959, 180). Colours that are sensibly identical, he explained, may be either of a compounded or uncompounded origin. Thus, though two beams of green light 'may to the naked eye appear of the same colour', while one of them is decomposed by a prism into beams of different colour, and the other passes through unchanged: 'I suppose these two Greens will in both cases be granted of a different origine & constitution . . .' (Newton, 1959, 181). Newton is concerned with the physical difference of the two greens, while Hooke and Huygens are concerned with their sensible identity. Underlying the disagreement are two fundamentally different but equally valid meanings of the concept of primary colour. For Newton 'primary' means the irreducible elements of light, and by analysis by refraction there are two different elements in a compound green and an indefinite number of elements, or primary colours, in natural white light. For Hooke and Huygens, however, the primary colours are those elements out of which all other colours can be made.

In January 1673 Huygens wrote to Newton in support of Hooke's contention that all colours could be produced from yellow and blue. He went on to suggest that white too could be composed from just these two colours. He proposed an experimental test that mixed spectral colours. Newton replied in April that 'I believe there cannot be found an experiment of that kind, because as I remember I once tryed by graduall succession the mixture of all paires of uncompounded colours, & though some of them were paler & nearer to white then others yet none could be *truly* called white' (Newton, 1959, 265; italics added). Despite his weak attempt at a rejection, Newton was never again to claim that all colours are necessary to compound white. In all later formulations of his theory the theory was restricted from one about white light in general to one about the Sun's light.

Newton's concession seriously weakened his argument in establishing his theory because it limited the theory's universality—from all white light to the Sun's white light—and forced him to introduce an embarrassing dichotomy between two sorts of white light, one of which was 'unnatural'. One of the two principal arguments Newton invoked to demonstrate his theory was the similarity between direct and decompounded sunlight. So many of his arguments for the compound nature of the

Sun's light have the same form, for instance, that the whiteness formed at the focus of a lens (as we saw above in the 'New theory'), is similar to the whiteness of the Sun's direct light. If it then turned out that there was another sort of white light—say, one composed of only two or three colours—that was sensibly identical to the white light of the Sun, while in other features, such as refrangibility, it was completely different, then an argument appealing to similarity or analogy would no longer be so compelling, particularly if this difference could not be explained.[3]

In a subsequent letter to Huygens in June 1673 Newton reformulated his theory in five definitions and ten propositions with a greater rigour than in his earlier formulations. In the propositions we can see that Newton has been forced by Huygens' criticism to recast his theory into a theory about the properties of sunlight. For example, propositions 4 and 5 state that, 'Whiteness in all respects like that of the *Sun's immediate light* & of all ye usuall objects of our senses cannot be compounded of two simple colours alone', but 'an indefinite variety of them' are required (Newton, 1959, 293; italics added). Thus far Newton has established that white light like the Sun's must be compounded of all the colours, not that the Sun's light is compounded of them. To establish the latter he turns to immutability. Despite the apparent rigour of the argument that Newton outlined for Huygens, he was unable simply to flesh it out with some experimental evidence for his presentation in the *Opticks*. Although the *Opticks* significantly reflects the argument here, the fundamental problem of establishing the strong principle of immutability for the Sun's direct light still remained. Drafts of the *Opticks* show that Newton struggled with this effort but that he could not find an experimental proof, and finally abandoned the attempt to use the principle of immutability to demonstrate that colours are innate to sunlight (Shapiro, 1980). Innateness was not dropped from the *Opticks*, just the attempt to prove it conclusively; likewise immutability remained, but it was demonstrated only in the weak form for homogeneous light. Newton in effect returned to the approach of the *Optical Lectures* and argued that the conclusion was one that no reasonable person could deny. To remove the reader's prejudice against accepting that a heterogeneous mixture could appear simple and homogeneous like the Sun's direct light, he showed many ways to compound white light from all colours that was sensibly like direct sunlight.

6.5 Corpuscular Optics: Explaining Refraction and Dispersion

Newton often appealed to the emission theory of light as heuristic in his optical investigations, but he utilized it most systematically in his quest to explain refraction and chromatic dispersion (the amount that the rays of different colour are separated by refraction, angle *PFT* in Fig. 6.4). His aim was to derive quantitative measures of these effects for different substances by a strict mechanical approach, assuming that light corpuscles are deflected at the interface of different media.

In an essay 'Of Refractions', probably written between 1666 and 1668, Newton calculated a table for the index of refraction of the extreme rays (red and violet) in various media passing into air from water, glass, and 'christall'. From an entry in the table, 'The proportions of y^e *motions* of the Extreamely Heterogeneous Rays', it is clear that he is considering the motion of corpuscles (Newton, 1959, 103, n. 6; italics added; Newton, 1967–81, 1: 559–74.) It is possible to reconstruct his table on this assumption, especially since he utilized the same model in his *Optical Lectures*, though he there suppressed any mention of corpuscles or motions (Newton, 1984, 199–203; Bechler, 1973, 3–6). Newton assumes that when a light-ray in air enters glass at grazing incidence (parallel to the refracting surface), rays of each colour receive the same increase of velocity perpendicular to the refracting surface. This model assumes that the projection parallel to the surface of all spectra are of equal length and that the same colours always occupy equal portions of it: that is, that chromatic dispersion is a property of light and not of the refracting media.

Meanwhile, by early 1672 Newton had deduced another dispersion law on different grounds. He was unable to choose between them on the limited number of measurements that he made.[4] Nonetheless, throughout his career he continued his quest to find a mathematico-mechanical explanation of refraction and dispersion, since the promised pay-off was so high—namely, a mathematical foundation for a theory of colour—and the models so tractable by the new science of mechanics. He would return to it in the *Principia*.

When Newton had developed the concept of force in the *Principia*, he concluded Book I with Section 14 on the analogy between the motion of corpuscles and light. By replacing the action of the aether in an earlier model of refraction with an intense short-range force between the corpuscles of the refracting body and light, he offered a powerful approach to optics and, more generally, to physics. In Fig. 6.6 the force is

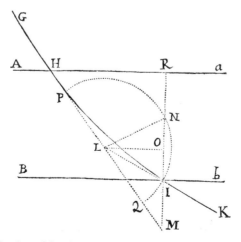

Fig. 6.6. Newton's derivation of Snell's law of refraction, *Principia*, Book I, Proposition 94.

assumed to act in the very small region between the refracting surfaces *Aa*, *Bb* and perpendicular to them. The motion of the particles in this field behaves exactly like that of a projectile 'falling' in the Earth's gravitational field. Newton demonstrated that its path *HI* in the region of the force field is a parabola ('from what Galileo demonstrated'), and that the angles of incidence *QMI* and refraction *MIK* obey Snell's law (Newton, 1999, 622). The derivation yielded an expression for the index of refraction *n* in terms of mechanical parameters. If we let $f(\rho)$ be the force per unit mass, where *i* and *r* are the angles of incidence and refraction, and ρ the distance from the refracting surface, then Newton's result in analytic form is

$$n = \frac{\sin i}{\sin r} = \sqrt{1 - \frac{2\phi}{v^2}}$$

where *v* is the incident velocity and $\varphi = \int_0^R f(\rho)d\rho$.

By at least 1675 Newton had recognized that if the change of motion of the light corpuscles occurs perpendicular to the refracting surface, then Snell's law will always follow (Newton, 1959, 371). Thus, the aim of this demonstration was not to derive Snell's law, but rather to show that corpuscular optics could be brought into the realm of the new mechanics and to explore its physical implications and, in particular, to explain the cause of the different colours and refrangibility of light rays. The most natural explanation of the cause of the different refrangibility of rays of different colour according to this model is that the velocity of the corpuscles varies. Four years after the publication of the *Principia*, Newton realized that this could be tested by observing the colour of the eclipses of Jupiter's moons. When a satellite disappears behind the planet, the slowest colour should be seen last, and when it re-emerges, the fastest colour should be seen first. In August 1691 Newton asked John Flamsteed if he had ever observed any colour changes in eclipses of Jupiter's moons, and the following February he replied that he had not (Shapiro, 1993, 144–7; Bechler, 1973, 22; Eisenstaedt, 1996). This was a serious blow to explaining refraction and dispersion by short-range forces, for it eliminated velocity as a cause of colour and refraction. The model could be applied only with some radical assumptions that conflicted with the principles of terrestrial mechanics. Choosing mass would contradict the motion of projectiles, which is independent of their mass. Allowing the force to vary with the nature of the corpuscle and refracting substance would make the force a selective one like a chemical reaction, which was decidedly unlike any force in the new mechanics.[5] Newton's elegant demonstration based on his concept of short-range forces had to be restricted to monochromatic rays, since colour could not be explained with his new mechanics.

The model was not, however, without a notable success. In 1691 Newton used it to calculate the refraction of light-rays entering the atmosphere and prepared a table of atmospheric refraction that was vastly superior to anything that then existed (Bechler, 1973, 23–6; Newton, 1967–81, 6:422–5, 431–4; Whiteside, 1980). In a Scholium to this section of the *Principia* Newton also suggested that short-range forces acting on light corpuscles could explain diffraction. A few years later, as we shall see, he tried to carry out this program of applying short-range forces to diffraction before he hit a dead end. Newton concluded this Scholium by reminding his readers that he was proposing only an analogy and not arguing that light actually consists of corpuscles:

because of the analogy that exists between the propagation of rays of light and the motion of bodies, I have decided to subjoin the following propositions for optical uses [namely, on geometrical optics], meanwhile not arguing at all about the nature of the rays (that is, whether they are bodies or not), but only determining the trajectories of bodies, which are very similar to the trajectories of rays.[6] (Newton, 1999, 626)

Newton returned to this analogy in the queries added to the *Opticks*.

6.6 THE COLOURS OF THIN PLATES

Newton learned about the colours of thin plates or films from Hooke's *Micrographia* (1665), in which he described the colours seen in sheets of mica. Hooke had conjectured that the appearance of the colours were periodic, though he was unable to measure the thickness of such thin films in order to demonstrate this. Newton's key breakthrough was his insight that if he put a lens (which is really just a segment of a circle) on a flat plane, then by a principle from Euclidean geometry about tangents to circles he could readily determine the distance between them simply by measuring the circle's diameter. If (Fig. 6.7) a convex lens *ABC* is placed on a glass plate *FBG* and illuminated and viewed from above, a set of concentric coloured circles—now known as 'Newton's rings'—produced by the thin film of air *ABCGBF* is seen through the upper surface of the lens. The circles form an alternating sequence of bright and dark coloured rings, and their common centre, the point of contact *B*, is a dark spot. If the diameter of the coloured circles be denoted by *D*, the thickness of the

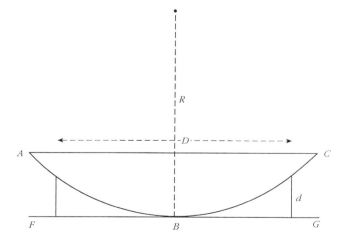

Fig. 6.7. Newton's method for determining the thickness *d* of a thin film of air formed between a spherical lens and a plane.

air film producing that circle by d, and the radius of the lens by R, then $d = D^2/8R$ by Euclid's *Elements*, III, 36.

Newton apparently had this insight while reading the *Micrographia*, and quickly carried out a rough and ready test in 1666 and entered it in his essay 'Of Colours'. To establish that the circles do appear at integral multiples of some definite thickness, he simply had to measure the diameter of successive bright and dark circles. For the first six circles he found that the thickness of the air between the lens and plate increased by integral multiples of the thickness at the first ring: that is, as 1, 2, 3, 4, 5, 6. He then calculated that 'y^e thicknesse of y^e aire for one circle was $1/64000$ inch, or 0,000015625. [wch is y^e space of a pulse of y^e vibrating medium.]' (McGuire & Tamny, 1983, 476–8).[7] His results, though quantitatively wide of the mark, as he later noted, were enough to demonstrate to his satisfaction that the appearance of the colours was a periodic phenomenon, and he succeeded in determining a measure of the periodicity. His method for determining the thickness of the film was in principle valid, and it later allowed him to develop a mathematical theory of periodic colours.

Newton's remark in square brackets shows that from the beginning of his research he was utilizing vibrations in the aether to explain the periodicity of the rings. The essential feature of his aether is its vibrations—stirred up when light corpuscles fall on the film—that reflect the corpuscles at condensations and transmit them at rarefactions. Newton was able to develop this qualitative, mechanical model into a relatively sophisticated mathematical one that agreed with his observations to a high degree of precision. Since this model invoked two hypotheses—light corpuscles and an aether—Newton suppressed the vibrations and their interactions with light corpuscles in all of his formal accounts of his research on the colours of thin films. It was only in his speculative 'Hypothesis' in 1675 that he chose to expound this model fully.

Since one of Newton's immediate aims in 'Of Colours' was to show that the colours of thin films was compatible with his recent discovery of the compound nature of sunlight, he would quite naturally have assumed that those colours in the incident sunlight that were not reflected by the film were transmitted. By examining the transmitted rings, he readily confirmed that the transmitted and reflected rings were complementary. And by examining the rings produced by rays of a single colour, it was possible for him to understand their formation in white light when the colours are not separately visible because of overlapping and mixing. Namely, he was able to see that at the same place some rays are reflected whereas the others are transmitted, and that rays of the same colour are at some places reflected and at others transmitted. At this stage Newton had not fully elaborated these points, especially the second, which requires assigning a particular thickness or vibration length to each colour.

Satisfied with this fundamental result and that his method worked, Newton set it aside until about 1671, when he undertook a serious investigation of the colours of thin films, 'Of ye coloured circles twixt two contiguous glasses' (Westfall, 1964). Newton's primary aim was to examine and describe Newton's rings quantitatively through a series of mathematical propositions and supporting measurements and observations; but he apparently also hoped to confirm his belief in the corpuscular constitution of light and its interactions with the aether. In the following year,

1672, he wrote up his results for submission to the Royal Society, but because of the controversies over his theory of colour he withheld it. When Newton once again felt sufficiently comfortable in revealing his works to the public in 1675, he revised the 'Observations' from 1672 and submitted it with a new companion piece, 'An hypothesis explaining the properties of light', to the Royal Society.[8] Newton did not allow these papers to be published in *Philosophical Transactions*, but they were read and transcribed in the Society's record book. The 1675 version of the 'Observations' was later minimally revised to become the greater part of Book II of the *Opticks*.

Although Newton did not write his account of the conditions for the appearance of the rings and their periodicity as an equation, it is equivalent to

$$d = \frac{D^2}{8R} = \frac{mI}{2},$$

where the first two terms of the equation express the Euclidean theorem cited above for the thickness of the film of air, and I is an interval such that for m odd the ring is bright and for m even dark. The interval I is the length of an aethereal vibration and, later, in the *Opticks* that of a fit.[9] In neither version of the 'Observations' nor in the *Opticks* does Newton introduce this physical interpretation, though it is apparent from 'Of Colours' and 'Of ye coloured circles' that he actually arrived at these results by working with the vibrations. He treats the interval solely as an experimentally determined property of the film—'the interval of the glasses, or thickness of the interjacent air, by which each colour was produced'—and not of light (Newton, 1958, 204). Although Newton did not calculate the value of the interval I in 'Of ye coloured circles', in the 'Observations' he adopted 1/80,000 of an inch—'to use a round number'—for the middle of white light, or yellow (Newton, 1958, 205).[10] In the 'Observations', Obs. 5, he experimentally confirmed that the squares of the diameters of the bright circles increased in arithmetic progression of the odd numbers, and the dark rings of the even numbers.

In 'Of y^e coloured circles' Newton determined the variation of the diameters of the rings when water was placed between the lenses. From his measurements he found that the thickness of the film decreases in proportion to the index of refraction. Thus the earlier equation becomes

$$d = \frac{D^2}{8R} = \frac{mI}{2n},$$

where n is the index of refraction of the film. Newton was probably led to accept this as a general rule valid for any medium, because he was able to deduce it from his model of light particles and aethereal vibrations (Shapiro, 1993, 68). Newton presented the experimental justification for this law in the 'Observations', Obs. 10, though purged of the corpuscular theory.

In 'Of ye coloured circles' Newton attempted to describe the variation of the circles in terms of the motion of the light corpuscles. The paper opens with six propositions to be confirmed in the subsequent observations. The properties of the circles are mathematically described and many of them are interpreted in terms of the 'motion', 'force', and 'percussion' of the corpuscles or rays, though no derivations are presented. Newton's propositions were subsequently contradicted by the observations that follow in the manuscript. At this point Newton undoubtedly recognized that the phenomenon was simply not amenable to a description using corpuscles. In all his later quantitative work on the colours of thin films he worked only with the vibrations set up by the corpuscles. However, in his physical thinking the encounter of the corpuscles with the compressions and rarefactions of the vibrations played a fundamental role.

Of greater theoretical significance, in 'Of ye coloured circles' Newton now explicitly recognized that the vibration length varies for each colour. He made 'many vaine attempts to measure the circles made by coloured rays alone' in order to calculate their vibration lengths, but by a very clever technique that was based on his concept of aethereal pulses he was able to determine, if not absolute values, at least that the ratio of vibration lengths for blue to red was about 9 to 14, and that the intermediate value occurred 'twixt the greene & yellow' (Westfall, 1964, 195–6; Shapiro, 1993, 61). When these results were reported in the 'Observations', Obs. 13, Newton purged any reference to the vibrations. In addition to his observations on Newton's rings and the laws derived from them, he describes related observations of the colours of soap bubbles and plates of uniform thickness. Although he did not at this time develop a mathematical analysis of the colours of plates of uniform thickness, he did observe (Obs. 19) that the colours change when the plate is viewed at different obliquities, which does not occur in Newton's rings. This property would be important in his analysis of the colours of thick plates.

In December 1675 Newton submitted 'An hypothesis explaining the properties of light discoursed of in my severall papers' to the Royal Society, because he hoped that revealing the hypotheses or physical models that underlay his phenomenological theories would make them more intelligible. He insisted, however, 'that no man may confound this with my other discourses, or measure the certainty of one by the other' (Newton, 1959, 364). The 'Hypothesis', which Newton did not allow to be published, is his most openly speculative work and—unlike the thirty-one queries which roam over the scientific landscape—reveals how he used his speculations to explore a single scientific theory. It shows clearly how he was able to control and mathematize speculative mechanical models and arrive at experimentally confirmed laws.

The hypotheses assert the existence of the aether, that it is capable of vibrating, and that light and aether act on one another. The aether, which is diffused through all space, is 'much of the same constitution with air, but far rarer subtiler & more strongly Elastic' and 'a vibrating Medium like air; only the vibrations far more swift & Minute' (Newton, 1959, 364, 366). This aether is almost without resistance, for it resists the motion of light particles only initially, at their emission from a luminous source, and at the boundaries of different bodies, where its density changes. When light particles are emitted they are accelerated 'by a Principle of motion ... till the resistance of the Aethereal Medium equal the force of that principle'. Henceforth the aether offers as little resistance as a vacuum. This is contrary to the principles of Galilean mechanics, and Newton knew it, but he thought 'it better to passe it by' (Newton, 1959, 370). Although the problem of the aether's resistance would vanish when Newton replaced the contact action of the aether with forces, this shows how he was able to elide physical difficulties in order to pursue the mathematical representation of a phenomenon. Newton emphasizes that he considers the particles to be light and not the vibrations, 'I suppose Light is neither this Aether nor its vibrating motion', which is simply an effect of light (Newton, 1959, 370).

The aether has a stiff surface that is responsible for the reflective power of bodies. The constant bombardment of light particles excites vibrations in the surface that are propagated throughout the aether. If a light corpuscle strikes the surface when it is compressed, it will be reflected because the surface is too stiff and dense to let the corpuscle pass; but if a corpuscle happens to strike the surface when it is expanded, it will pass through.[11] This is the physical mechanism that Newton uses to introduce periodicity to a corpuscular theory of light. That he had failed in quantifying the relationship between the corpuscles and the magnitude of the excited vibrations did not hinder him from using it as the basis for describing the periodic colours of thin films. The corpuscles still play a fundamental, if less prominent, role in that one has to keep track of the location of both the corpuscles and vibrations to determine the observed phenomenon.

The periodicity of Newton's rings are now readily explained (Fig. 6.8). At the centre A, where the glasses touch, the corpuscles will be transmitted because the aether in the two glasses is continuous, and a central dark spot will be seen. At a certain thickness BC ($= I/2$) away from the centre the corpuscle will encounter the condensed part of the first overtaking vibration and be reflected, and a bright ring will be seen;

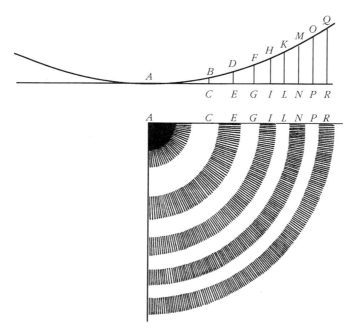

Fig. 6.8. One quadrant of Newton's rings produced with light of a single colour. After his drawing in 'An Hypothesis explaining y^e properties of Light'.

at double that thickness *DE*, it will encounter the rarefied part of that wave and be transmitted, and a dark ring will be seen; at triple the thickness *FG* it will encounter the condensed part of the second wave and be reflected; and so on in arithmetic progression, in agreement with observation. To extend this model to white light, Newton had only to introduce the idea that the rays or particles of different colour vary in 'magnitude, strength or vigour' and so excite vibrations of different size (Newton, 1959, 376). The red vibrations are assumed to be larger than the violet ones and thus to form larger circles, as is observed.

As Newton's diagram of the reflected rings shows, he does not assume that reflection occurs only at the greatest compression but also at thicknesses on either side of it, extending as far as a quarter interval ($I/4$) on either side. The reflection is assumed to be greatest at the thickness corresponding to the greatest compression and to decrease gradually as the wave is less condensed (and so resists the particles' motion less); the possibility of reflection ends altogether at thicknesses corresponding to rarefactions.

In the *Opticks* Newton transformed the aethereal vibrations into the 'fits of easy reflection and transmission'. The fits were now held to be a property of light, and not of the aether or other medium based on his investigation of the colours of thick plates.

6.7 The Colours of Natural Bodies

In his theory of light and colour, Newton, as we have seen, explained that the colours of natural bodies arise from the fact that they are disposed to reflect light of one colour in greater quantity than the others. This explanation says nothing about the cause of the disposition, but in his account of the colours of thin plates he proposed that the colours of bodies are produced in the same way as they are in thin films. He developed this theory in the early 1670s, and it forms the third part of the 'Observations' of 1672 and 1675 and of Book II of the *Opticks*.

Newton opens his theory by arguing that coloured transparent and opaque bodies consist of transparent corpuscles and pores. While he considered the existence of aether to be hypothetical, he does not question the existence of corpuscles composing matter. His reasoning in support of these propositions is straightforward. Since reflection occurs only where there is a difference in optical density, for reflection to occur from the corpuscles composing bodies, the bodies must have pores that are of a different optical density from the corpuscles. Opacity is attributed to multiple reflections caused by the internal parts of the body. Newton's evidence for these claims comes almost entirely from macroscopic bodies, and it is then extended to the imperceptible corpuscles. For example, he argues that the 'least parts' of bodies are transparent from observations that when made sufficiently thin, bodies become transparent; and he argues that opacity arises from a multitude of internal reflections by observing that transparent substances like glass becomes opaque when shattered into tiny pieces. This can be a tricky mode of argument.

Proposition 5 is the core of Newton's theory: 'The transparent parts of bodies, according to their several sizes, must *reflect* rays of one colour, and *transmit* those of another, on the same grounds, that thin plates or bubbles do reflect or transmit those rays: and this I take to be the ground of all their colours' (Newton, 1958, 229). Newton demonstrates this by what would become his second Rule of Reasoning in the *Principia*: namely, that 'the causes assigned to natural effects of the same kind must be, so far as possible, the same' (Newton, 1999, 795). He presents evidence showing that the colours of natural bodies and thin plates are of the same kind, and therefore have the same cause. With this demonstrated, Newton estimated the size of the corpuscles composing various bodies from their colour. He assumed that the corpuscles are of the same optical density as water or glass, 'as by many circumstances is obvious to collect'. In his account of the colours of thin films, Newton had prepared a table of the thicknesses of films of air, water, and glass that produce the various colours of each ring or order. For example, he deduced that the green of vegetation corresponds to the green in the third coloured ring, and from his table he found that the corpuscles of vegetable matter are $17\frac{1}{2} \times 10^{-6}$, or about 1/60,000 inch in diameter, assuming that they are the same density as glass (Newton, 1958, 231). The corpuscles of black and colourless transparent bodies must be less than any of those producing colours, just as the central spot in Newton's rings is colourless and reflects no light.

While it is not possible to see light corpuscles, Newton anticipated actually seeing the corpuscles of bodies. He explained that he deduced the sizes of the corpuscles 'because it is not impossible, but that microscopes may at length be improved to the discovery of the corpuscles of bodies, on which their colours depend' (Newton, 1958, 233). If their magnification could be increased five or six hundred times, we might be able to see the largest corpuscles.

As strange as the explanation of the colours of natural bodies may seem today, it was widely adopted for over a century. It was so surprisingly robust that it was not conclusively disproved until about 1830 (Shapiro, 1993).

6.8 The Composition of the *Opticks*

Newton wrote most of the *Opticks* between 1687 and early 1692. He wrote Book I, Parts I and II, expounding his new theory of light and colour, in 1687. He then appears to have set aside the *Opticks* for about three years, but by the late summer or autumn of 1691 he had considered it—at least for a few months—to be complete. It is most likely that he carried out new research and wrote the remainder of the *Opticks*—that is, Books II and III—in the winter or spring of 1691, or perhaps six months earlier. At some time between late August 1691 and late February 1692, Newton decided to revise his draft significantly. After this effort he brought it close to its published form except for the brief last book on diffraction and the queries which were not prepared for publication until shortly before publication in 1704 (Shapiro, 1993, 138–150).

The composition of Book II in 1690 or 1691 at first went very quickly. Newton made so few changes in the text that he was able to mark up the manuscript of the 'Observations' from 1675 for his amanuensis to copy for the *Opticks*. This formed Parts I and II and much of Part III. At this point it is important to note that until shortly before publication the *Opticks* consisted of four books, since Newton had designated what became the two parts of Book I as separate books; that is, they were Books I and II. The published Book II was then Book III. After revising the 'Observations', Newton was confronted with a decision on how to end his book. At first he planned to follow this material with a new fourth book or part on diffraction, but he was also toying with the idea of a speculative 'Fourth Book'. Newton soon reined in his more speculative tendencies and turned to more empirical optical investigations. He continued experiments on diffraction and also discovered an entirely new phenomenon: coloured rings produced in transparent thick plates. By the autumn of 1691, Newton had completed and written up his investigation of thick plates as Book IV, Part I, which, together with his research on diffraction, Book IV, Part II, was to form the concluding book of the *Opticks*.

Between late August 1691 and late February 1692 Newton removed the two parts of the new Book IV from the manuscript and set about revising them. The part on diffraction was troublesome and remained incomplete until shortly before publication. Within six months, however, he revised the part on the colours of thick plates, incorporated it into Book III because of their affinity to those of thin films, and essentially put it into its published state. During this revision Newton also introduced his theory of fits—an immaterial vibration to explain the physical cause of periodicity in light that replaced his earlier aethereal and corpuscular vibrations.

6.9 The Colours of Thick Plates and the Theory of Fits

Newton's work on the colours of thick plates, undertaken when he was in his late forties, was his last major, successful optical investigation. The heart of Newton's explanation of the colours of thick plates was a complex calculation that could predict the observed dimensions of the rings in all circumstances. He was able to accomplish this by extending to this new phenomenon the mathematical–physical concepts—most notably, periodicity—that he had formulated earlier for the thin films of Newton's rings and by introducing the concept of a path difference between two rays. However, he had to grapple with two fundamental differences between the phenomena of Newton's rings and thick plates: The films in the former were of varying thickness, whereas in the latter they were of constant thickness; and the circles vanished in the former after a thickness of about 1/10,000 inch, whereas in the latter they first appeared at $1/4$ inch.

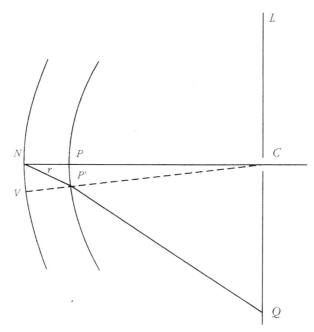

Fig. 6.9. Newton's explanation of the coloured rings produced by spherical mirror PP'VN and projected onto screen LQ, with a hole at C. Incident rays CN are scattered at point N.

Since the colours of thick plates has almost totally vanished from the historical and scientific literature, I will present a brief description of the phenomenon. In Fig. 6.9, in front of a spherical glass mirror that is silvered on the back side NV (a mirror for a reflecting telescope), a white screen LCQ is placed at its centre C and perpendicular to its axis CN. If sunlight is admitted through a small hole in a window shutter (not illustrated) and then passes through a small hole C in the screen and is reflected from the mirror back to the screen, there will be seen four or five coloured circles around centre C like Newton's rings, but larger and fainter. The circles are in fact seen when the screen and window, or light source, are at any distance from the mirror, though they are most distinct when the screen and window are coincident at the centre of the mirror. According to the wave theory of light the rings are produced by the interference of two beams of light that are reflected from the rear surface and scattered by some imperfection on the first surface. Newton attributed the rings to a single ray scattered at the second surface.

For Newton's experimental arrangement, the wave theory expresses the diameters D of the coloured rings by the relation

$$D^2 = \frac{8nR^2}{d} \cdot \frac{mI}{2}$$

where n is the index of refraction of the glass mirror, R its radius, d its thickness, m is an integer, and I is Newton's vibration length, which is equal to half a wavelength. I have presented this analytical solution to show that while Newton conceived of the colours of thick plates by 'the Analogy ... with those of the like Rings of Colours' of thin plates, the analogy is not a close one and his solution required an independent mathematical–physical analysis and careful experiment (Newton, 1704, 295). As we saw previously, the equation describing Newton's rings is

$$D^2 = \frac{8R}{n} \cdot \frac{mI}{2}$$

Both kinds of rings obey the law of arithmetic progression, though, as Newton found, the rings of thick plates are like the transmitted, and not the reflected, rings of thin films; that is, for a given integer m, the light and dark rings are interchanged in the two phenomena. Otherwise, the relations are altogether different. For example, in thin films the square of the diameter of the rings varies linearly with the radius of the lens, whereas in thick plates it varies as the square of the radius of the mirror. It is important to recognize that Newton did not derive his result for the diameters of the rings in analytic form. Rather, by introducing a concept of path difference he was able to transform the problem of thick plates to an equivalent one of thin films and calculate the diameters of the rings.

The thrust of the first seven observations is to establish the analogy of the colours of thick plates with those of thin films, and consequently that they are produced 'much after the manner that those were produced by very thin plates' (Newton, 1704, 295). In particular, his aim is to show that the rings of thick plates are periodic, with the same periods, or vibration lengths, as found for thin films. It should be noted that the diameters of the rings in thick plates are an order of magnitude larger than those in Newton's rings, so that they may be more precisely measured. From some measurements (Obs. 5) of the rings in monochromatic light (which he admitted were 'hard to determine ... accurately'), he determined the ratio of the squares of the diameters of the extreme red and violet rings to be '3 to 2 very nearly'. He judged that this 'differs not much' (less than 4%) from the 14 to 9 determined in Newton's rings (Newton, 1704, 292–3). As tenuous as this conclusion may be, it established that the previously determined vibration lengths could be applied to thick plates.

The aim of Observation 7 is to establish the exact nature of the analogy and to show how the explanation of the colours of thin films may be extended to those of thick plates. To demonstrate that the colours of thick plates, like those of thin films, depend upon both surfaces of the mirror and its thickness, Newton carried out two simple experiments. He rubbed off the quicksilver from the back of the mirror and saw the rings formed as before, except fainter, so that the quicksilver only increases the brightness of the rings and does not cause them. He also found that the rings were not produced in the single surface of a metal mirror. His observation that the colour of the rings changed as the obliquity of the incident light varied was essential for the formulation of his explanation, for it informed him that these rings were like those of thin films of constant thickness rather than Newton's rings with varying thickness. The reason, Newton explained, why the colours of a thin film of constant thickness varied was that at any obliquity only rays of a particular kind are reflected while the others are transmitted.

Newton next addressed the issue of why the rings in thin films became dilute and vanish after the film increased somewhat in thickness, whereas they appear distinctly in a thick plate of $1/4$ inch. His answer was that in thin films the reflection and transmission of the rays occur at the second surface, while in thick plates they occur at the first surface, after the rays pass back and forth through the mirror. The essence of Newton's explanation is straightforward, even though there are subtleties in interpreting the theory of fits and computing the rings' magnitude, which will not be entered into here (see Shapiro, 1993, ch. 4). In Fig 6.9, *LCQ* is a white screen with a small hole *C* placed at the centre of mirror *PP'VN* and perpendicular to the axis *CN*. Ray *CP* falls on the mirror at *P* and proceeds through it to the second surface, where it is reflected at *N*. The perpendicularly reflected rays pass through the first surface again at *P* and form the central white spot *C* on the screen *CQ*. Some rays, however, at *N* are scattered to the first surface at *P'* and after refraction proceed to the screen, where they form coloured rings at *Q*. The scattered rays depict coloured rings, because with a longer, oblique path *NP'* than the perpendicular rays, some sorts will be in fits of easy transmission when they return to the first surface and so will emerge from the mirror, whereas others will be in fits of reflection and be turned back into the mirror. With this consideration, Newton has introduced the concept of path difference. When *NP'* differs from *NP* by half a vibration length, no rays will be transmitted, and there will be a dark ring. When that difference increases to a whole vibration, some rays will again be in fits of easy transmission and depict coloured rings, and so on for a number of successions, until the rings of different colour overlap and blend into white. Thus, the reason why rings can appear at such great thicknesses is that their appearance depends on the difference of the two path lengths *NP*, *NP'*, and not on the absolute length of the path traversed by the rays, as in thin films. Although the paths are very large (approximately 35,000 intervals), the first ring appears when their difference is just one interval, thereby restoring the analogy to thin films.

Perhaps the best way to understand the nature of fits is to follow Newton's example and compare them with the earlier vibrations (Newton, 1704, 280). Qualitatively they are equivalent to the earlier aethereal vibrations. The rarefactions correspond to fits of easy transmission, the compressions to fits of easy reflection, and the intervals of the fits to the vibration lengths.

Despite the difference in the physical cause of the formation of the rings in thin and thick plates, Newton succeeded in drawing a formal analogy between them to calculate the rings' diameters in thick plates. He reduced the two passages of the ray through the mirror to the equivalent case of a single path, as in a thin film of uniform thickness, by considering only the ray *NP'* scattered from the rear of the mirror. In turn, he cleverly transformed the case of a film of uniform thickness to a variable one as in Newton's rings. Because Newton had already established that the same intervals apply to both thin and thick plates, he was able to show that the appearances predicted by his explanation agree with observation.

Newton was justifiably satisfied with the agreement between theory and observation and the consequent confirmation of his extension of the theory of vibrations to the new phenomenon of thick plates. In demonstrating the theory of fits this would serve as his principal evidence that periodicity was a general property of light, or 'propagated to great distances', and not something that was exhibited solely in the tiny realm of thin films (Newton, 1704, Obs. 8, $_2$102). Previously he had established only that periodicity extends a few ten thousandths of an inch in thin films, barely beyond the realm of the microscopic. Now he had demonstrated that the same property continues for at least a quarter of an inch, safely in the macroscopic world. This calculation was the most complex quantitative derivation in all of his work on physical optics. Yet, as his papers show, this agreement was hard won and ultimately necessitated redetermining the vibration length or interval of the fits. One of the reasons that led Newton to revise what was then Book IV of the *Opticks* between late August 1691 and late February 1692 was that he was dissatisfied with the agreement between his calculations and measurements. He redetermined all the parameters that entered into his calculation, and redetermined the value of the vibration length that he had been using for twenty years. The new value, 1/89,000 inch, is a significant 11% smaller than the old value and, as best as can be judged, very close to the modern value of the wavelength of yellow light.

Newton also had a serious methodological reason for revising Book IV, Part I and the end of Book III, Part III. Even before he had completed the first state of Books III and IV, he was struggling with the idea of adding a physical explanation of the cause of the coloured rings of thin films and thick plates. From the time that he had control of the phenomena of thick films, he attributed them to the vibrations in the corpuscles of bodies. He had, however, decided that it was too hypothetical to set forth in the *Opticks*, and suppressed a draft for the conclusion of Part III in the first complete state in which he invoked the corpuscular vibrations. No doubt he excluded these propositions because he came to recognize that corpuscular vibrations were as hypothetical as the aethereal vibrations he had invoked earlier in the 'Hypothesis' and had no place being mingled with the more certain results of the *Opticks*. Evidently he wanted to include an explanation of the physical cause of periodic colours in the *Opticks* and pondered how to do this without violating his own methodology.

Newton resolved the problem by considering the vibrations abstractly without any conjectural vibrating medium in his theory of fits.[12] Newton set forth the theory at the conclusion of Book II, Part III, in nine abstract mathematical propositions with minimal commentary. In the first proposition and definition of the sequence he introduced the fundamental ideas of the theory:

PROP. XII.
Every ray of Light in its passage through any refracting surface is put into a certain transient constitution or state, which in the progress of the ray returns at equal intervals, and disposes the ray at every return to be easily transmitted through the next refracting surface, and between the returns to be easily reflected by it.

DEFINITION
The returns of the disposition of any ray to be reflected I will call its *Fits of easy reflexion,* and those of its disposition to be transmitted its *Fits of easy transmission,* and the space it passes between every return and the next return, the *Interval of its Fits.* (Newton, 1704, $_2$78, $_2$81)

To establish the periodicity of the fits, or that they return at equal intervals, he appeals to his observations of Newton's rings. Moreover, anticipating the results of his investigation of thick plates in 'the next part of this Book', he can state that these vicissitudes return many thousands of times. 'So that this alternation seems to be propagated from every refracting surface to all distances without end or limitation' (Newton, 1704, $_279$). With this conclusion, Newton is preparing the way for the next proposition, which asserts that light already possesses fits even before it encounters a refracting surface and probably possesses them when they are emitted from a luminous source. Newton has phrased both the proposition and the definition to emphasize that the fits are only *dispositions* to be *easily* reflected or transmitted and not states in which they necessarily occur.

Newton attributed fits to light before it falls on bodies because he believed that he had sufficient evidence that periodicity was a general property of light, but his immediate motivation was to explain partial reflection. In Prop. 13 he asserted that 'The reason why the surfaces of all thick transparent Bodies reflect part of the Light incident on them, and reflect the rest, is, that some rays at their incidence are in Fits of easy reflexion, and others in Fits of easy transmission' (Newton, 1704, $_281$). In any modern wave theory, periodicity is an inherent and permanent property of light, but Newton, who thought in terms of light corpuscles, was very cautious in attributing that property to light and considered it to be only probable. In his commentary to Prop. 13 he observed: 'And hence Light is in fits of easy reflection and easy transmission, before its incidence on transparent Bodies. And *probably* it is put into such fits at its first emission from luminous Bodies, and continues in them during all its progress. For these fits are of a lasting Nature . . .' (Newton, 1704, $_282$; italics added). Despite its similarity to a wave theory of light, and despite Newton's genuine belief that the theory of fits was hypothesis-free, it was still wedded to the emission theory in its single-ray theory for explaining the colours of thin and thick plates.

6.10 The Delayed Publication of the *Opticks*: Diffraction and Hooke

Newton's inability to conclude the part on diffraction that he had removed from the first completed state of the *Opticks* was a principal reason that its publication was delayed for twelve years until 1704. The manuscripts related to the *Opticks* allow us to identify the cause of Newton's problem. He had developed a model of diffraction that assumed that the paths of the fringes were identical to, or coincided with, the rectilinear paths of the rays that produced them, but then carried out an experiment that conclusively refuted this assumption. Newton first learned about diffraction from Hooke's account at a meeting of the Royal Society in 1675, but he only dabbled with the phenomenon until he carried out a serious experimental investigation in 1690–91 (Hall, 1990).

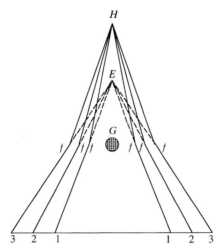

Fig. 6.10. Newton's model for diffraction by a hair G. The model assumes that the paths of the rays and the fringes, f1, f2, and f3, are identical.

Newton's experiments on diffraction focused on two set-ups. In one he stuck the points of two knives with straight edges into a board so that the blades formed a V with an angle of about half a degree. He placed the knives in a narrow beam of sunlight and observed a complex pattern of three fringes, which are nicely illustrated in a well-known diagram in the *Opticks* (Newton, 1704, Bk, III, Plate 1). From his measurements he had already derived some mathematical laws describing the fringes. In the second set-up he illuminated a hair with a narrow beam of light admitted through a small hole about 12 feet away and observed three coloured fringes on each side of a central shadow. Newton made systematic measurements of the fringes at two distances from the hair and deduced a number of laws describing the phenomenon. He reports the distance at which the rays depicting the three fringes passed the hair, but does not explain how he determined those distances. From some of his worksheets for this section of the *Opticks*, it is possible to reconstruct his method of determining these distances, as well as the angle at which the incident rays are deflected (Shapiro, 2001). Underlying Newton's claims is a model of diffraction (Fig. 6.10) that assumes that the incident rays coming from the point H at the centre of the hole are bent away from the hair G at a point a short distance f away from the centre of the hair by a short-range force. The rays then proceed in a straight line and depict a fringe of a given order, which is essentially the assumption that the paths of the rays and the fringes are identical. If the diffracted rays on each side of the hair are projected backwards, they intersect at a point E on the axis of symmetry. In a draft of this part he wrote:

The surfaces of bodies act regularly upon the rays of Light at a distance on both sides so as to inflect them variously according to their several degrees of refrangibility, & distances from the surface. The truth of this Proposition will appear by the following observations. (Shapiro, 2001, 51–2)

Newton did not mention his underlying model in the text of the *Opticks*, because it invoked a hypothesis, which was prohibited by his methodology.

A new experiment that Newton carried out around February 1692, when he had removed the just completed part on diffraction from the manuscript of the *Opticks*, undermined his model for diffraction and forced him to recognize that the paths of the rays and the fringes are distinct. In his experiment with the knife blades that formed a V, Newton measured the distance between the blades when the intersection of the first dark fringes fell on a white paper at various distances. His measurements show that the light forming the same fringe, when it is observed at various distances from the knives, comes from different distances from the edges and is deflected at different angles. Since light propagates in straight lines after it has passed through the blades and been deflected, it cannot be the same light that forms the fringes at different places; that is, the fringes do not propagate rectilinearly. Newton's record of this experiment from 1692 contains no comments on its implication, but when he included it in the *Opticks* he tersely noted: 'And hence I gather that the Light which makes the fringes upon the Paper is not the same Light at all distances of the Paper from the Knives' (Newton, 1704, Bk. III, Obs. 9, $_2$127). This conclusion rendered much of his work on diffraction invalid, and this is no doubt the reason that this part of Book IV was removed from the manuscript. He had intended to return to his diffraction experiments, as he recounts in the conclusion to the published *Opticks*, Book III, Observation 11:

When I made the foregoing Observations, I designed to repeat most of them with more care and exactness, and to make some new ones for determining the manner how the rays of Light are bent in their passage by Bodies for making the fringes of Colours with the dark lines between them. But I was then interrupted, and cannot now think of taking these things into further consideration. (Newton, 1704, $_2$132)

Newton was nearing the end of his active scientific career, and, despite his intentions in the early 1690s, he never resumed his experimental research on diffraction.

In 1696 Newton left Cambridge for London to assume the duties of Warden of the Mint. He no longer had the time and apparently the interest to resume his experiments on diffraction. The *Opticks* in Newton's eyes was still incomplete. Newton could not yet accept publishing the *Opticks* without the part on diffraction. According to the traditional account, Newton delayed the publication of the *Opticks* until after Hooke's death in March 1703, because he feared yet another dispute with him (Hall, 1993, 92). The principal reason for the delay, however, was that there was no *Opticks* to publish. It was incomplete. Until the late 1690s Newton evidently intended to return to his diffraction experiments. He was certainly fearful of another round of controversy as he affirmed in his correspondence and the Advertisement to the *Opticks*: 'To avoid being engaged in Disputes about these Matters, I have hitherto delayed the Printing, and should still have delayed it, had not the importunity of Friends prevailed upon me' (Newton, 1704, [iii]). Yet, neither Newton nor any of his contemporaries specifically mentioned Hooke.

On 15 November 1702, according to a memorandum by David Gregory, Newton finally declared his intent to publish the *Opticks* (Hiscock, 1937, 14). The book appeared by 16 February 1704, when Newton presented a copy to the Royal Society. What prompted him to publish at this time? Westfall suggests that Newton's forthcoming election to the presidency of the Royal Society may have provided the 'crucial stimulus', although he still holds that the 'primary cause' was Hooke's death. Other factors that he mentions are Newton's desire to publish some of his mathematical manuscripts and the recent publication of Huygens's *Dioptrica* in 1703 (Westfall, 1980, 638–9).[13]

When Newton did decide to prepare the book for publication he felt compelled to treat diffraction, since it was an essential part of any account of light. Most of the revisions to the part on diffraction entailed removing passages that were based on the assumption that the paths of the rays and fringes coincided. In Observation 9 he added the measurements on diffraction with the knife blades that had invalidated that model, and elsewhere he also clearly explained that the paths of the rays and fringes were not identical. Compared to the preceding parts of the *Opticks*, this section was rather perfunctory. It recounted a series of experiments with original observations, but there was no attempt to bring some order to the phenomenon. Yet, if Newton had failed to explain diffraction, his predecessors had done little more; their efforts, like Newton's, were largely devoted to describing the phenomenon.

Since Newton published the *Opticks* without a complete investigation into diffraction, which he had hoped would support a corpuscular theory of light in which light corpuscles were acted on by short-range forces of matter, he appended sixteen 'Queries in order to a further search to be made by others' (Newton, 1704, 2132). The opening queries indeed asked whether light was acted on at a distance by forces of bodies. By the time Newton published a second English edition of the *Opticks* in 1717, the queries had grown to 31, and the speculations extended well beyond optics.

References

Bechler, Zev (1973). 'Newton's search for a mechanistic model of colour dispersion: A suggested interpretation'. *Archive for History of Exact Sciences* 11: 1–37.

Denis, Jean Baptiste (1672). 'Description d'une petite lunette par le moyen de laquelle ou voit les objets éloignez aussi distinctement qu'on peut faire avec les plus grands telescopes, inventée par Monsieur Newton'. *Recoeuil des memoires et conferences sur les arts & les sciences* (1 March): 43–49.

Eisenstaedt, Jean (1996). 'L'optique balistique newtonienne à l'épreuve des satellites de Jupiter'. *Archive for History of Exact Sciences* 50: 117–56.

Hall, A. Rupert (1990). 'Beyond the fringe: Diffraction as seen by Grimaldi, Fabri, Hooke and Newton'. *Notes and Records of the Royal Society of London* 44: 13–23.

——(1993). *All Was Light: An Introduction to Newton's Opticks*. Oxford: Clarendon Press.

Hiscock, W. G., ed. (1937). *David Gregory, Isaac Newton and their Circle: Extracts from David Gregory's Memoranda 1677–1708*. Oxford: Printed for the Editor.

HUYGENS, CHRISTIAAN (1672). 'Extrait d'une lettre de M. Hugens . . . touchant la lunette catoptrique de M. Newton'. *Journal des sçavans* (29 Février): 22–4.
MCGUIRE, J. E., and MARTIN TAMNY (1983). *Certain Philosophical Questions: Newton's Trinity Notebook*. Cambridge: Cambridge University Press.
NEWTON, ISAAC (1672). 'Nouvelle lunette catoptrique inventée par M. Newton Professeur des mathematiques dans l'Université de Cambridge'. *Journal des sçavans*, (29 Février): 19–22.
——(1704). *Opticks: Or, a Treatise of the Reflexions, Refractions, Inflexions and Colours of Light*. London: Sam. Smith and Benj. Walford.
——(1958). *Isaac Newton's Papers and Letters on Natural Philosophy and Related Documents*, ed. I. Bernard Cohen. Cambridge, MA: Harvard University Press.
——(1959). *The Correspondence of Isaac Newton*, vol. 1, ed. H. W. Cambridge: Cambridge University Press.
——(1967–81). *The Mathematical Papers of Isaac Newton*, ed. D. T. Whiteside, 6 vols. Cambridge: Cambridge University Press.
——(1984). *The Optical Papers of Isaac Newton. Vol. 1. The Optical Lectures, 1670–1672*, ed. Alan E. Shapiro. Cambridge: Cambridge University Press.
——(1999). *The Principia: Mathematical Principles of Natural Philosophy. Preceded by a Guide to Newton's Principia by I. Bernard Cohen*, trans. I. Bernard Cohen, Anne Whitman and Julia Budenz. Berkeley, CA: University of California Press.
SHAPIRO, ALAN E. (1979). 'Newton's "achromatic" dispersion law: Theoretical background and experimental evidence'. *Archive for History of Exact Sciences* 21: 91–128.
——(1980). 'The evolving structure of Newton's theory of white light and colour'. *Isis* 71: 211–35.
——(1993). *Fits, Passions, and Paroxysms: Physics, Method, and Chemistry and Newton's Theories of Coloured Bodies and Fits of Easy Reflection*. Cambridge: Cambridge University Press.
——(1996). 'The gradual acceptance of Newton's theory of light and colour, 1672–1727'. *Perspectives on Science* 4: 59–140.
——(2001). 'Newton's experiments on diffraction and the delayed publication of the *Opticks*'. In *Isaac Newton's Natural Philosophy*, ed. Jed Z. Buchwald and I. Bernard Cohen, pp. 47–76. Cambridge, MA: MIT Press.
SIMPSON, A. D. C. (2009). 'The beginnings of commercial manufacture of the reflecting telescope in London'. *Journal for the History of Astronomy* 40: 421–66.
WESTFALL, RICHARD S. (1964). 'Isaac Newton's coloured circles twixt two contiguous glasses'. *Archive for History of Exact Sciences* 2: 181–96.
——(1980). *Never at Rest: A Biography of Isaac Newton*. Cambridge: Cambridge University Press.
WHITESIDE, D. T. (1980). 'Kepler, Newton and Flamsteed on refraction through a "regular aire": The mathematical and the practical'. *Centaurus* 24: 288–315.

NOTES

1. Simpson (2009, 421–5) presents a brief account of Newton's telescope with references to the historical literature.
2. The two primaries adopted by Hooke and Huygens are in fact variants of the painters' three primaries that were introduced in the early seventeenth century.

3. The rules of colour mixing, especially the complexities involving the distinction between subtractive and additive colour mixing, and the nature of colour vision, were at the root of this problem, which was not resolved until the nineteenth century.
4. Newton's new law was based on the musical division of the spectrum, and I have called it his 'linear dispersion law' (Shapiro, 1979). He adopted this law in the *Opticks*, Bk. I, Pt. II, Prop. 3, Expt. 7. Had Newton examined a greater range of substances, he would have found that neither law is true. In the eighteenth century it was discovered that there is no law relating dispersion to mean refraction, and that dispersion is a property of matter and not, as he had assumed, light.
5. In fact, Newton did consider the possibility that the optical force behaved like a chemical force (Shapiro, 1993, 142, n. 16).
6. Newton had in fact deduced the two propositions more than fifteen years earlier in his *Optical Lectures* without the corpuscular theory of light (Newton, 1984, 417–19).
7. The angled brackets here are Newton's way of setting off his speculative or interpretive comments.
8. Since Newton left what I call the 'Observations' untitled, it has received various names. In Newton (1958) it was called 'Newton's second paper on colour and light', and in Newton (1959), 'Discourse of Observations'.
9. The thickness of air at which the first bright ring is produced is one half the physical vibration, or pulse length, that I call the interval I. Newton's law for the appearance of rings is the same as that derived according to the modern wave theory except for a factor of 2, because his interval I turns out to be one half of the wavelength λ in the wave theory of light.
10. In the *Opticks*, as a consequence of his investigation of the colours of thick plates, described below, Newton redetermined this value once again.
11. It can be shown that the vibrations travel at twice the speed of the light corpuscles.
12. Newton drew the terms 'fit' from contemporary medical language. A fit is one of recurrent attacks of a periodic ailment—in particular, malaria, which was then a common disease in England, especially in the fens of Cambridgeshire.
13. Newton appended two mathematical papers, 'Enumeration linearum tertii ordinis' and 'Tractatus de quadratura curvarum', to the first edition of the *Opticks*.

CHAPTER 7

EXPERIMENTATION IN THE PHYSICAL SCIENCES OF THE SEVENTEENTH CENTURY

NICO BERTOLONI MELI

7.1 INTRODUCTION: PRELIMINARY REFLECTIONS

To talk about experiments in the physical sciences in the Early Modern period requires a few words of introduction. In this chapter, by experiment I understand a conscious intervention in the course of nature involving some manipulations of appropriate objects aimed at finding out some law-like regularities in nature's behaviour. In this regard I am excluding engineering tests that were aimed at establishing the feasibility of a project, such as the construction of a fountain or a canal, for example, since these projects did not involve primarily nature's law-like behaviour as such but were limited to the project at hand. Moreover, astronomy too is excluded from my work, since celestial objects do not allow for any intervention and are more the objects of observation than experimentation. The situation would be different for microscopy, which involves and in fact often requires highly interventionist techniques such as boiling, delamination, injections, and staining. However, some of the most important microscopic investigations involved disciplines such as anatomy and natural history, which do not fall under the compass of this essay—although physico-mathematicians often collaborated with anatomists and physicians (Heidelberger and Steinle, 1998; Steinle, 2002; Bertoloni Meli, 2008).

The notion of physical sciences too requires clarification. Strictly speaking, the physical sciences related to *physica* or natural philosophy, namely a study of nature that was initially distinct from the mixed mathematical disciplines, such as mechanics and optics, for example. In the course of the seventeenth century, however, we witness an erosion of the boundaries between *physica* and the mixed mathematics, with the emergence of hybrid physico-mathematical disciplines. In this essay I will consider in this respect the physical sciences in a broad sense, including the mixed mathematical disciplines. However, I will also be restrictive in another sense; *physica* involved different aspects of the study of nature, including human and animal anatomy, for example. Such areas were crucial to extensive and sophisticated experimental activities by the standards of any time. In line with the general framework of this volume, however, here I will exclude those researches.

The inclusion of the physical and mixed mathematical disciplines together creates a tension in the treatment and presentation of experiments. Briefly put, the mixed mathematics were often presented in axiomatic fashion, following the Archimedean tradition of *On the Equilibrium of Planes*. In this tradition, as it was perceived in the seventeenth century, experiments were often conceived of as inherently uncertain and therefore they could not be placed at the foundation of a science, lest that science too be tainted with that same degree of uncertainty. To be sure, experiments were still used as heuristic tools, for example, but their role often remained private, concealed from public presentations, or was recast in suitable form for publication, as Galileo did in the *Discorsi* for the science of motion and Huygens in the posthumous *De motu corporum ex percussione* for the collision of bodies.

By contrast, in other areas, such as natural philosophy, the situation was different; there axiomatic formulations were not standard and the problems of presentation could be quite different, such as the reliability of generalizations from single or limited series of contrived experiments. Often, however, there was no objection to making experiments the foundations of knowledge, since no higher authority was recognized: this is very much what William Gilbert and Robert Boyle sought to do with their celebrated work on magnetic and electrical phenomena, the pressure of the air, the air-pump, and Boyle's law. Indeed, Galileo too often relied on this mode of presentation when the matter was not pertinent to an Archimedean presentation, as in many instances in the *Assayer*.

These reflections lead us to a related issue concerning the tension between how experiments were performed and reported in print. Only in some cases do we have manuscript data and private notes on experiments, and even these cannot be read as if they provided unmediated access to the past, since the way those records were taken was itself a historical process following styles and conventions of its own. As to printed accounts, they followed a changing set of styles and conventions dependent on a variety of factors, including disciplinary affiliation, period when they were performed, authorial style, and intended audience. In this essay, apart from a few exceptions, I will focus especially on printed sources, refraining from examining private manuscript material that was not available to the intellectual world of the seventeenth century.

Today, experimental reports in the scientific literature are frequently couched in a codified language. By and large, there was no generally accepted canon for writing experimental reports in the seventeenth century, although certain forms of communication, involving language and style, emerged in specific cases, such as the *Philosophical Transactions* of the Royal Society. Throughout the century we find a variety of styles, involving accounts in the first person emphasizing the direct involvement of the narrator; the impersonal form, emphasizing the neutrality and objectivity of the account; the emphasis on the role of witnesses, often of high social status, certifying the reliability of the experimental report—something one encounters already in 1559 in Realdo Colombo's *De re anatomica*—or the detailed account of the experiment, warts and all, intended to give the reader the impression of being there, a narrative style that has been characterized as 'virtual witnessing' (Licoppe, 1996; Shapin and Schaffer, 1985; Dear, 1995).

7.2 Some Features and Problems of Seventeenth-Century Experimentation

Experiments were certainly not new to the Early Modern period; in fact, I am convinced they can be traced back not only to medieval times but also to antiquity. At the same time, I would argue that experiments became more widespread and prominent in the seventeenth century, reaching more disciplines and an iconic status that they probably lacked in the past. It is no accident that the expression 'experimental philosophy' was coined in the seventeenth century and occurred in the titles of books published at the time. Contemporary practitioners definitely saw their time as one of great changes and of the rise of experiment as a privileged form of knowing, including the production and classification of new phenomena. The seventeenth century witnessed a growing epistemological awareness of the centrality of experiment and the coinage by Francis Bacon and Robert Hooke, for example, of the notion of 'crucial experiment', or an experiment designed to select univocally between competing views. Recent accounts too agree on the centrality of experiments in the Early Modern period. Gone are the days of Alexandre Koyré, who ridiculed the role and imprecision of experiments: a host of more recent historians and philosophers of science have been paying increasing attention to the practice of science in the so-called 'practical turn', marking the shift from the exclusive interest in ideas and theories to the focus on experimentation (Koyré, 1968; Settle, 1961, 1967; Hacking, 1983; Heidelberger and Steinle, 1998; Buchwald, 2006).

However, experiments in the physical sciences—as opposed to chemistry, for example—did not enter the university classroom in the seventeenth century. Rather, experiments were performed by individual scholars, in the context of the newly established scientific academies, and at the margins of university life, such as at private gatherings of professors and selected students. It was mainly in the eighteenth century, with the establishment of university chairs devoted to the experimental philosophy, that experiments began to become officially part of university training.

What made the seventeenth century such a fertile period for the development of experimentation? This is a question that has attracted a great deal of attention by historians and philosophers of science; it has been argued that the rise of experimentation as a form of knowing was due to technological and economic transformations, the Reformation, the coming down of the barriers among different social groups, the heroic intellectual struggles of a few individuals, the erosion of confidence in ancient texts following the discovery of America, or the age of global commerce. Possibly an important cue to unravel the issue comes from coming to grips with humanism and the Renaissance. Whereas in the sixteenth century critical editions of authoritative ancient texts, from Archimedes to Galen, took centre stage on the intellectual scene, their role became far less significant in the following century. Concerns with the emulation of the ancients and their investigations of nature led to a critical reading of ancient sources; the shift from their books to the book of nature led to renewed investigations and the attempt to complete and expand traditional sciences, such as mechanics, or to create new ones, such as the science of motion. Thus arguably a critical engagement with the Renaissance was a key factor in the emergence of experimentation in the seventeenth century (Cohen, 1994).

Let us briefly survey the protagonists of the experimental philosophy: the experimenters themselves. William Gilbert was a prominent London physician. Galileo Galilei and Isaac Newton started as university professors of mathematics in 1589 and 1669, but ended as mathematician and philosopher to the Medici court, and as Warden and then Master of the London Mint, respectively. Evangelista Torricelli was mathematician, though not philosopher, to the Grand Duke of Tuscany; Benedetto Castelli was a Cassinese Monk who taught mathematics at Pisa University and was later professor of mathematics at Rome, and was also employed by the Pope in the Management of Waters in the Papal States. His student Giovanni Alfonso Borelli started teaching mathematics at Messina in 1639 and from 1656 to 1667 he was both professor of mathematics at Pisa and member of the Cimento Academy at the Tuscan Court. Gianbattista Riccioli was a Jesuit professor of mathematics at Bologna. Otto von Guericke was burgomaster in Magdeburg, Christiaan Huygens was an aristocrat of independent means, but he also became a stipendiary member of the French *Académie Royale des Sciences*. Robert Hooke and Christopher Wren were members of the Royal Society and professors of mathematics at Gresham College and Oxford. Moreover, they were also architects who rebuilt London after the Great Fire of 1666. Descartes was a man of means sufficient to devote himself to pursue knowledge, and ended as court philosopher to Queen Christina in Sweden, while his correspondent Marin Mersenne was a member of a religious orders. This brief list highlights that experimenters came from varied professional and social backgrounds, and constituted a heterogeneous body of practitioners from various quarters.

But which were the experiments in the physical sciences that contemporary intellectuals found especially significant, and which defined this area of inquiry in their mind? They ranged across mechanics and the science of motion, optics, and natural philosophy. Moreover, in many instances we find not isolated experiments but entire series of related trials, such as those on the barometer and the void, for example, which continued for decades and grew in originality and complexity. At times, an experimental set-up gave rise to several trials in which individual variables were altered one by one; at other times, experiments were repeated in different material and intellectual settings, taking on a life of their own. Clearly, this rich material can be arranged from a variety of perspectives, though I believe that a partial census of the experiments perceived by contemporaries as the most significant offers considerable advantages and provides a set of concrete cases for reflection.

Experiments ranged from the extensive work on magnetism by Gilbert at the turn of the century in *De magnete*, to Newton's study of the resistance to motion in *Principia mathematica* towards the end of the century. Experiments were becoming more and more prominent throughout the century and enjoyed a high standing. It would be impossible to provide even a brief account of the most notable experiments of the time: Robert Boyle was such a prolific experimenter in so many fields that entire books have been devoted to his work alone. Some experiments sought to provide quantitative data in order to establish relations between magnitudes, such as time, distance, velocity, angle, and so on, while others sought to determine numerical values for their own sake, such as the distance travelled by a falling body in 1 second, or the height of the mercury column in the Torricellian tube. Others still did not look at numerical values primarily or at all, but rather focused on qualitative data, such as the behaviour of a carp bladder or an animal in an evacuated container, such as a Torricellian tube or air pump. Nor was the nature of an individual experiment cast in stone: originally Huygens believed the length of the seconds pendulum to be a universal unit, and only subsequently did he realize that the length of the seconds pendulum is not the same everywhere but depends on the latitude. Thus, measuring the variations in the period of the same pendulum at different latitudes became a tool for investigating the shape and density of the Earth. In this case the discrepancies observed between the expected values and the initial observations became themselves a valuable object of investigation.

Despite the visibility of many experiments in many fields, the main areas in which they were performed are more limited. I have selected three, which in the perceptions of the time stand out for their key role: they involve predominantly experiments on the science of motion, including falling bodies, the inclined plane, and the pendulum; the barometric experiments and the investigations of the vacuum, such as Florin Périer's ascent of the Puy-de Dôme, following Blaise Pascal's suggestion, or Boyle's air-pump experiments; and optical experiments on the nature and behaviour of light, ranging from the investigations of refraction to Newton's experiments on the nature of light.

Despite their visibility and high standing in the seventeenth century, experiments were far from unproblematic and uncontroversial. In fact, it is not easy to identify experiments that were accepted at face value: whether we think of Galileo's inclined plane experiments or Torricelli's barometric trials, Boyle's air-pump investigations or Newton's optical experiments with prisms, we encounter debates and controversies that often the original experimenters found incomprehensible and frustrating.

7.3 Experiments on the Science of Motion

Although it is generally believed that Galileo's works played a pivotal role in the development of the experimental philosophy, exactly what that role his experiments played either in his private investigations or in print is not clear. There are complex reasons for this state of affairs. On the one hand, it seems established now that Galileo relied extensively on experiments as a heuristic tool in his investigation on motion, though the fragmentary nature of his manuscript records makes it hard to reconstruct his procedures and train of thought; on the other, the way Galileo presented his experiments in print differs considerably from his private itinerary. There are also differences between a more informal presentation, as in the *Dialogo sui massimi sistemi* (Florence, 1632, in Galileo, 1967), and the more formal presentation in the *Discorsi e dimostrazioni matematiche* (Leiden, 1638, in Galileo, 1974), in which he sought an axiomatic formulation partly aided by pendulum experiments. The art of presenting experiments in print, especially in mechanics and related disciplines, was not established; thus Galileo had not only to present new result, but also to invent a suitable way of presenting them adapting an Archimedean style. The problematic nature of his enterprise is highlighted by the fact that this specific portion of his work underwent major revisions in the second posthumous edition of the *Discorsi* (Bologna, 1656, discussed in Bertoloni Meli, 2006; Drake, 1978).

The number of experiments performed and narrated by Galileo would be an appropriate subject for a book, and may be too great even for that. Here I shall offer a few reflections on falling bodies, the inclined plane, projectiles, and the pendulum, ending with some celebrated experiments by Newton. It is striking that most celebrated experiments of the seventeenth century on mechanics and the science of motion did not require more than relatively simple objects that were either in everyday's use or could be fabricated with little expense.

7.3.1 Falling Bodies

Experiments on falling bodies were not uncommon in the sixteenth century, especially at Pisa, where Galileo studied. Galileo's engagement with falling bodies dates from his early years in Pisa. Overall, experiments disproved Aristotle's claim that the speed of the falling bodies was proportional to its weight, though exactly how heavy bodies fell remained unclear. For a while, Galileo subscribed to the view that their speed depended on the difference between their specific gravity and the specific gravity of the medium in which they fell—a view he shared with other mathematicians inspired by Archimedes's principle about floating bodies, even though Archimedes dealt with floating rather than motion (Clagett, 1964–1984; Camerota and Helbing, 2000).

It seems that the inclined plane intersected Galileo's work on falling bodies in two ways: by using the inclined plane, Galileo was able to slow down falling bodies and therefore study them more accurately. For example, Galileo realized that bodies fall according to the odd-number rule, whereby in equal intervals of time they cover distances proportional to the odd numbers, 1, 3, 5, 7, and so on. This statement is equivalent to the claim that the distances traversed by a falling body are proportional to the squares of the times, and though the equivalence is straightforward for us it was difficult to prove using the traditional mathematics on which Galileo relied.

Moreover, it is likely through a failed experiment that Galileo came to realize that there was something wrong in his initial views on motion. He believed that the speed of a body would be affected by the inclination of the plane: the closer the plane to the vertical, the greater the speed. Since he believed that the speed depended also on the specific gravity, Galileo thought of combining specific gravity and inclination so as to make two bodies fall at the same time. This could be achieved by having a wooden ball, for example, fall along a steep inclined plane, and an iron ball fall along one closer to the horizon. Alas, the experiment did not work, and it is possible that it was in this way that Galileo came to realize that specific gravity did not work the way he thought, and that all bodies fall with the same acceleration. This instance teaches us that experiment could be significant not only for the positive results they produced, but for the negative ones as well—those going against expectations.

In the *Discorsi* we find an important passage dealing with the speeds of falling bodies. Galileo gave some numbers and introduced an important notion. Simplicio was about to accept that bodies of the same material fall in the same time, regardless of their weight, but admitted his reluctance to accept it. Salviati replied (Galileo, 1974, *Two New Sciences*, 77–80, at 80; Bertoloni Meli, 2006, 103):

But I do not want you, Simplicio, to do what many others do, and divert the argument from its principal purpose, attacking something I said that departs by a hair from the truth, and then trying to hide under this hair another's fault as big as a ship's hawser. Aristotle says: 'A hundred-pound iron ball falling from the height of a hundred *braccia* hits the ground before one of just one pound has descended a single *braccio*.' I say that they arrive at the same time. You find, on making the experiment, that the larger anticipates the smaller by two inches; that is, when the larger one strikes the ground, the other is two inches behind it. And now you want to hide, behind those two inches, the ninety-nine *braccia* of Aristotle, and speaking only of my tiny error, remain silent about his enormous one.

This passage raises an important issue with regard to precision: Galileo here did not pretend perfect agreement between predictions and experimental data, but rather much greater precision than Aristotle, thus introducing the notion of order of magnitude. Moreover, Galileo attempted to explain away small differences in terms of secondary factors, such as a difference in specific gravity or a difference in size affecting air resistance. He tried to argue that buoyancy affects the speed of fall in that a body's gravity is diminished by a factor depending on the density of the medium. For a body with a density 1,000 times greater than air, for example, the diminution is 1 part in 1,000, whereas for a body whose density is only ten times greater than that of air, the diminutista is one tenth. Speeds of fall, other things being equal, would be affected in the same proportion: 'For example, an ivory ball weighs twenty ounces, and an equal amount of water weighs seventeen; therefore the speed of ivory in air is to its speed in water approximately as twenty is to three'. The numbers provided by Galileo were not the results of experiments, but rather seem to be simply plausible values suitable for didactic demonstration. Falling bodies were less problematic than the motion of projectiles, where parabolas seemed to diverge more and more from the point where, if unhindered, they were believed to reach: namely, the Earth's centre (Galileo, 1974; Bertoloni Meli, 2006; Buchwald, 2006).

Mersenne developed a complex attitude to Galileo and his experiments. On the one hand, he came to accept Galileo's odd-number rule on the basis of his own experiments on falling bodies; Mersenne dropped balls from different heights and measured the times with a pendulum. On the other, he challenged Galileo's claims about the distance covered by a falling body in 1 second. We face here an interesting tension between Galileo's and Mersenne's perspectives: Galileo was primarily interested in experiments in order to find regularities in nature in the form of proportions between magnitudes, whereas Mersenne was more interested in finding numerical values. In the *Dialogo* Galileo had stated that a ball would fall 100 braccia in five seconds. Mersenne found this results unacceptably low, and embarked on a series of trials to determine a better value, and researches to figure out the length of the braccio used by Galileo—a useful reminder of how complex these issues could be at a time when unit measures were not standardized (Dear, 1988).

The issue raised by Mersenne had already attracted the attention of other readers. In response to their query, Galileo stated that he had not actually dropped heavy weights from a high tower, but rather extrapolated the results obtained from rolling spheres along inclined planes. However, in a marginal note in his own copy of the *Dialogo*, he also stated that a heavy sphere would fall more than 400 braccia in four pulse-beats, thus substantively changing his previous value. It appears that as if pressed by questions and criticisms, Galileo paid attention to the accuracy of numerical values—something that was not at the top of his agenda previously. The reference to the inclined plane highlights the advantages but also the dangers involved in relying on it, and serves as an introduction to our next section.

The Jesuit mathematician Gianbattista Riccioli performed a series of influential experiments on falling bodies from Bolognese towers; they saw the light in the 1651 *Almagestum novum*. Riccioli measured time with a pendulum and dropped pairs of balls of different sizes and weights, finding that in no instance did they fall in exactly the same time. Discrepancies, however, were small. He also provided an accurate measure of the distance covered by a heavy body in the first second of fall, which differed significantly from the value given by Galileo. Huygens's later measurements based on different methods were in close agreement to Riccioli's. Riccioli, however, accepted Galileo's odd-number rule, and it was his work that marked the broader acceptance of Galileo's results, which had been the objects of criticism from several fronts in the intervening years (Bertoloni Meli, 2006).

7.3.2 The Inclined Plane and Projectiles

Galileo's engagement with the inclined plane dates from his early years and is possibly responsible for his attaining some especially important results, as we have seen in the previous section. Furthermore, about 1590, together with his mentor Guidobaldo dal Monte, Galileo performed an experiment with the inclined plane in order to investigate projectile motion: they rolled an inked ball on an inclined plane so that the ball would mark its trajectory on the plane. At the time, views about projectile trajectories differed, though most authors divided them into three segments—one approximately rectilinear in the direction of firing, one circular or curved downwards, and the third approximately vertical. It is not immediately clear why dal Monte and Galileo performed an experiment and what they were looking for, but what they found almost certainly surprised them: the trajectory appeared symmetrical, like a parabola or a hyperbola. This result required a major reconceptualization of projectile motion—and possibly of vertical fall too, insofar as vertical fall was a component of motion—though in a different sense from what they may have envisaged (Renn *et al.*, 2001).

Galileo reported reflections and experiments on the inclined plane in the *Dialogo*, but it is probably in the *Discorsi* that he reported one of his most celebrated experiments (Galileo, *Two New Sciences*, 1974, 169):

In a wooden beam or rafter about twelve braccia long, half a braccio wide, and three inches thick, a channel was rabbeted in along the narrowest dimension, a little over an inch wide and made very straight; so that this would be clean and smooth, there was glued within it a piece of vellum, as much smoothed and cleaned as possible. In this there was made to descend a very hard bronze ball, well rounded and polished, the beam having been tilted by elevating one of its ends above the horizontal plane from one to two braccia, at will.

Time was measured by weighing the water flowing through a thin tube at the bottom of a large bucket. The result was that the distance had to be reduced to a quarter to halve the time. Galileo accompanied his account with unconvincing claims of remarkable precision—less than a tenth of a pulse-beat, thus displaying a different attitude to discrepancies from the passage quoted above.

It is worth underlining that Galileo did not present his science on motion in the *Discorsi* on experiments. Rather, he sought to give definitions and postulated that would be accepted by our mind as natural; in fact, he felt forced to justify his postulate about bodies falling along inclined planes with some experimental results,

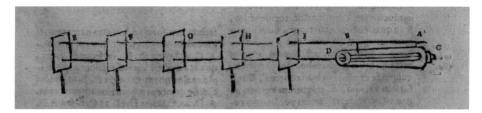

Fig. 7.1. Cabeo's experiment on the trajectory of projectiles.

though he saw this as a defect that undermined the solidity and elegance of his entire construction, and he found a way to replace or supplement those passages with a formal non-experimental proof in the later posthumous edition of his work. What Galileo wished to do was present his new science of motion in axiomatic form in a purely mathematical fashion, arguing that his theory would stand as a mathematical construction regardless of how nature actually behaved. It was only later that he used the inclined plane experiment discussed above; thus the experiment did not appear in a foundational role, but only to show that nature does indeed behave as predicted by his theory.

Galileo's treatment of the inclined plane came under Mersenne's scrutiny. The French Minim reported a series of experimental results challenging Galileo's claims about distances covered by bodies rolling down inclined planes. Mersenne arranged for two balls to fall at exactly the same time vertically and along the inclined plane, observing whether when the ball on the incline was when the other reaches the ground. What Mersenne found was a systematic divergence whereby the bodies on the incline covered a smaller distance than expected. Mersenne's criticism of Galileo's claims about the inclined plane, much like his challenge to Galileo's work on falling bodies, focused on the accuracy of empirical data and was rather thin on theoretical considerations. In this case too, however, he had a point—one that would need about a century to be fully appreciated, since it depended on the much more sophisticated mechanics of the rigid body (Bertoloni Meli, 2006).

We have seen above that Guidobaldo dal Monte and Galileo were seeking to study the trajectory of projectiles through visualization. They were not the only scholars to do so: after the publication of the *Two New Sciences*, both the Jesuit Niccolò Cabeo and a group of scholars and gunners were in correspondence with Torricelli. The experiment (Fig. 7.1) consisted in firing a cannon ball through a set of equally spaced paper screens, which would be perforated by the ball, and the positions of the holes would provide a series of positions at which the ball would have passed at successive instants (Dear, 1995, 127–9).

7.3.3 The Pendulum

The pendulum was the most emblematic mechanical device of the seventeenth century—one that was not only extremely significant in itself, but also for its relevance to a large number of other investigations, from a time-measuring device in many circumstances to a tool for studying the collision of bodies or—as in the second book of Newton's *Principia*—motion in a resisting medium. Although it was due to Galileo that the pendulum became a key device, paradoxically it was Mersenne, in the *Addition* to *Les mechaniques* of Galileo, who formulated the fundamental relation between length and period: namely, that the length is proportional to the squares of the times (Dear, 1988; Smith, 2001; Büttner, 2008).

Galileo's most significant engagements with the pendulum date from the beginning of the century, though it was in the *Dialogo* and then in the *Discorsi* that he formulated the main results. Galileo was convinced that pendulum oscillations are exactly isochronous, meaning that they take the same amount of time regardless of their amplitude. Alas, matters are not quite as Galileo thought, because oscillations are not exactly isochronous, though his claim is pretty accurate for small amplitudes; it seems plausible that Galileo used rather light pendulum bobs—musket balls rather than cannon balls. Since the amplitude of their oscillations decreases rather rapidly, Galileo may have been led to his belief from this specific case (Palmieri, 2009). Galileo also claimed that the bob has the ability to return to the original height from which it had fallen—an important statement on which he relied in his formulation of the science of motion.

In this case too, Mersenne took Galileo's statements and dissected them, and once again he disagreed with them. Mersenne was quite confident that pendulum oscillations are not isochronous, at least in air. This example too shows a pattern similar to the previous ones, with Galileo optimistically charging ahead dismissing small discrepancies as negligible perturbations, while Mersenne was scratching his head in bewilderment.

Of course, it was Huygens in the third quarter of the century who established theoretically and experimentally the conditions for perfect isochronism. In the 1673 *Horologium oscillatorium*—a treatise on the pendulum clock—he showed not only that Mersenne was correct, because pendulum oscillations depend on their amplitude, but also how to address the problem, by constraining oscillations between cycloidal cheeks. The cycloid was a curve that was being investigated by mathematicians at the time for entirely unrelated reasons (Fig. 7.2). Soon after publishing the *Horologium oscillatorium* Huygens asked himself what exactly made oscillations constrained by cycloidal cheeks isochronous: his answer, based on an analysis of the geometric properties of motion, was that under those conditions the force is directly proportional to the displacement. In other words, if the oscillations of a pendulum are constrained by cycloidal cheeks, the pendulum behaves like a spring (Fig. 7.3). In the same years, Robert Hooke had established experimentally precisely that proportion—known as Hooke's law—by attaching different weights to elastic bodies and measuring their displacements (Mahoney, 2000; Bertoloni Meli, 2010).

Huygens relied extensively on the pendulum also in his investigations and demonstrations of the rules of collision, but his formal presentation followed an axiomatic Archimedean style from which experiments were excluded. It was his colleague

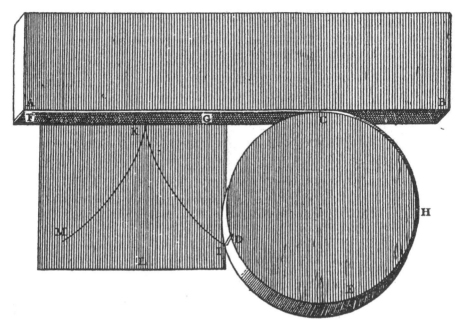

Fig. 7.2. Drawing a cycloid.

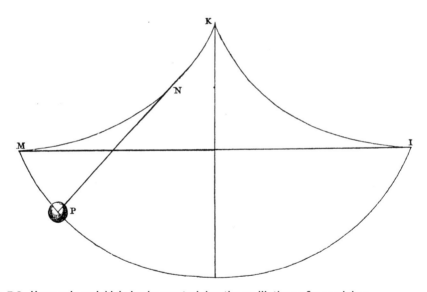

Fig. 7.3. Huygens's cycloidal cheeks constraining the oscillations of a pendulum.

Edme Mariotte who reformulated Huygens's rules of collision as a series of propositions largely equivalent to Huygens's, though based on experiments. This example shows beautifully how experiments could figure in radically different ways, as private heuristic devices or in the formal presentation of a theory.

7.3.4 Further Uses of the Pendulum

The study of collision instantiates an important transformation in the role of the pendulum: from object of study to a powerful tool of investigation. By clarifying a number of issues, such as the exact shape of the cheeks constraining its oscillation, or the determination of the effect of the shape of the bob, the pendulum had become a reliable and robust tool of investigation. No one relied more on it that Isaac Newton.

The pendulum was a key experimental device in the 1687 *Philosophiae naturalis principia mathematica*. Newton's reliance on the pendulum was so extensive that it would be impossible to address here the wide range of topics covered. I shall focus on two aspects: the justification of the notion of mass, and the determination of the speed of sound.

The notion of mass—as opposed to weight—was a novel one. Newton needed to justify its introduction by showing that it was meaningful and measurable—not an easy task in a world still pervaded by Cartesian vortices that entered and exited all bodies through their pores. Newton relied on pendulum experiments to show that the resistance to motion of an oscillating bob was due almost exclusively—probably entirely—to its external surface, suggesting that a subtle matter—if one existed at all—penetrating its pores played no role in motion. In addition, Newton performed a celebrated pendulum experiment to show that all bodies fall with the same acceleration. Rather than experimenting with falling bodies, Newton used pendulums, which can oscillate for a long time and enable much more accurate measurements. Newton used two 11-foot pendulums with wooden round boxes as bobs, and filled one with wood and the other with the same weight of many different substances such as gold, silver, lead, glass, sand, common salt, wood, water, and wheat. The two pendulums oscillated isochronously for a very long time, thus showing that the amount of matter each pendulum bob contained was exactly proportional to its weight. Newton argued that his experiments would have detected a difference in the behaviour of inertial and gravitational mass of less than a thousandth. His conclusion was highly general and significant: mass and weight, though conceptually distinct, were exactly proportional, and the mass in the expression for quantity of motion in the second law (commonly though slightly anachronistically written as $\mathbf{F} = d\mathbf{p}/dt$, where $\mathbf{F}$ is the force, $\mathbf{p}$ the quantity of motion, and t the time) was exactly proportional to the mass responsible for universal gravity.

Newton's determination of the speed of sound was a notable seventeenth-century experiment—one that relates to the topic of the following section on the barometric experiment and the void. Newton performed his experiment in the courtyard of Trinity College, Cambridge, by measuring the time it took the sound generated at one end to produce a fourfold echo. In order to measure the time, Newton used a pendulum with a variable length. After some trials he determined that its length was greater than $5\frac{1}{2}$ inches and less than 8 inches. Since it is the square root of the length that is proportional to the period, the difference in the periods is smaller—more than 920 and less than 1,085 English feet. His conclusion was that sound consists exclusively in the agitation of air. In all probability Newton performed this experiment to refute an hypothesis put forward by Boyle on the existence of a fluid subtler than air responsible for the propagation of sound, since he had heard the sound of a bell in a container evacuated of air. Thus Newton's result provided a quantitative value for the speed of sound and at the same time showed that sound consists exclusively in the motion of air, not a subtler fluid whose existence he questioned. (Newton, 1999; Bertoloni Meli, 2006)

7.4 The Barometric Experiments and their Aftermath

The barometric experiments associated with Torricelli gained an iconic status in the seventeenth century and led to a flurry of investigations. Some of those experiments count among the most spectacular, refined, and philosophically significant of the entire century. However, it is difficult to pinpoint their exact nature and purpose, since reliance on the Torricellian tube could serve different purposes, and the same experiment or measurement could be seen in different lights by different observers. It could be a way to prove the existence of the vacuum against Aristotle's view that nature abhors a vacuum, or to establish the weight of the air, or to measure changes in air pressure.

Experiments with the Torricellian tube travelled from Italy to France, and then, following the construction of air pumps, from Germany to England. Experiments and debates on air pressure and the void were too numerous and involved too many issues—such as the transmission of sound, light, and magnetic effects through a vacuum, for example—to be discussed here, and I shall therefore focus on a few select episodes—notably, Torricelli's original trial, the carp bladder and void-in-the-void experiments, Pascal and Périer's Puy-de Dôme experiment, and the investigations with the air-pump in Germany and England.

Experiments on the void required more elaborate apparatus than those in mechanics and the science of motion; fabricating long tubes of glass, obtaining relatively large quantities of mercury, and especially building an air-pump, were not trivial tasks, especially the last, which has been characterized as seventeenth-century 'big science'.

7.4.1 Torricelli's Barometric Experiments

Despite its iconic status, it is difficult to provide a satisfactory analysis of the background and origin of the Torricellian experiment. News circulated through word of mouth and epistolary exchanges, for many years there were no publications, and many of those publications were in the form of ephemeral pamphlets that did not circulate widely at the time and have largely disappeared today. In the second quarter of the seventeenth century we find several investigations and reflections on the reason why water does not rise higher than about 11 metres—something that seems to have been known from practical and engineering pursuits. The Genoa patrician Giovanni Battista Baliani, Galileo, and the Roman mathematician Gasparo Berti were involved in early investigations and reflections.

The key experiments probably dated from 1644 and became known through correspondence and witnessing. In his letter to the Rome mathematician Michelangelo Ricci of 11 June 1644, Torricelli described the procedure involving filling glass vessels of different shapes, about 2 braccia long, closed at one end. Once filled with mercury, the open end was sealed with a finger, and the tube was carefully turned upside down and placed in a bowl with more mercury. Once the finger was removed, the mercury in the tube descended to a given height regardless of the tube's shape; inclining the tube at an angle, the shape and size of the empty space at the top changed, though the column of mercury remained at the same height. In order to show that the top of the tube was 'perfectly empty', Torricelli added some water at the top of the mercury in the basin. Raising the tube carefully until it reached the water level, the mercury in the tube descended all the way and the water suddenly and abruptly filled to the very top with water. Torricelli believed that we live at the bottom of a sea of air, and interpreted his experiment as evidence that mercury in the tube is sustained by an external weight, due to atmospheric air. Torricelli and many of his contemporaries compared air to wool, and seemed aware that air, like wool, could be compressed and had a tendency to be restored to its original shape (Middleton, 1964).

7.4.2 Pascal, Périer, and the Puy-de Dôme Experiment

One of the most striking and sophisticated experiments of the time was performed by Florin Périer, a magistrate in Clermont and Pascal's brother in law, in 1648. It is unclear whether the original suggestion came from Descartes or Pascal. At any event, it was Pascal who suggested to Périer to climb the Puy-de Dôme, a high mountain near Clermont, in order to test whether the mercury in the barometer was lower at the top. In fact, Pascal asked the impossible of Périer: namely, to climb the mountain several times in a day, in order to ensure that there were no accidental variations in the height of the mercury. Périer, however, went beyond Pascal's suggestion by relying on two barometers—one of them kept in a monastery at the base of the mountain in what was called a 'continuous experiment', being checked by a monk in order to compare its readings with the one carried to the mountain top. Périer's objective was clear: it was conceivable that air pressure would change during the day, thus it was crucial to make sure that this had not occurred. Indeed, the readings of the barometer at the base of the mountain confirmed that the height of the mercury column had not changed. The height of the column of mercury at the top of the mountain, however, was noticeably lower, and in addition, readings of the apparatus at intermediate heights on the way down showed that mercury was, as expected, at progressively lower intermediate heights as they descended. The Puy-de Dôme experiment showed that the height of the column of mercury decreased at higher altitudes, and testifies to the increasing sophistication of control procedures (Pascal, 1937; Dear, 1995; Bertoloni Meli, 2009).

7.4.3 The Carp Bladder and Void-in-the-Void Experiments

Refinements of the Torricellian experiment led to developments aimed at understanding the reason why mercury does not descend. One of the reasons cited by some was that space at the top of the tube was filled with volatile spirits from the liquid, such as mercury or water. In a dramatic demonstration in a glass factory in Rouen, Pascal performed the experiment with wine. Since wine is much more 'spiritous' than other liquids, the space at the top of the tube should have been much larger and filled with such abundant spirits, while water should have stood higher than wine. However, exactly the opposite happened—because the specific gravity of wine is smaller than that of water.

Another experiment, credited to the astronomer and polymath Adrien Auzout, concerned the so-called void-in-the-void (Fig. 7.4). The issue at stake was similar to the one involved in the Puy-de Dôme experiment: namely, altering the outside conditions to ascertain whether the height of the mercury column is affected. Auzout managed to arrange two Torricellian tubes, one inside the other, sealed with a membrane consisting of a bladder. Initially, the second Torricellian tube was inside the evacuated space at the top of the first tube. The mercury in the second tube did not rise at all, while in the first tube it reached a height of 27 inches. Ostensibly, external air balanced the mercury in the first tube, whilst the seal prevented external air from entering the space at the top; therefore there was no air to balance the mercury in the

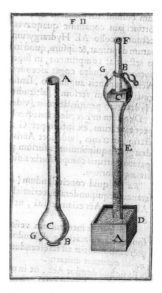

Fig. 7.4. Auzout's void-in-the-void experiment.

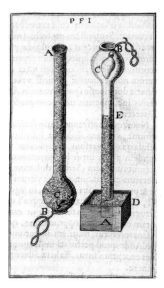

Fig. 7.5. Roberval's carp-bladder experiment.

second tube. Subsequently, the bladder was pierced and air entered the space at the top of the first tube. The column of mercury in the first tube descended completely, while in the second tube it rose to 27 inches. Auzout concluded that air is heavy even in its own sphere, and that it is atmospheric air that is responsible for the behaviour of mercury in the Torricellian tube.

Roberval is credited with another important variant: namely, the introduction into the space at the top of the Torricellian tube of the bladder of a carp, which had been carefully emptied of air. Once inside the evacuated space, however, the carp bladder became inflated. (Fig. 7.5) Clearly, at the very least, there was a lot less air in the space at the top of the evacuated tube than outside, otherwise the carp bladder would not have inflated. Roberval came to believe that it was the pressure of outside air that sustained the column of mercury, and that the top of the Torricellian tube contained greatly evacuated air.

This brief survey of experiments with the Torricellian tube gives a partial but vivid sense of the richness and sophistication reached at mid-century in this research area.

7.4.4 The Air-Pumps of von Guericke and Boyle

The Magdeburg diplomat, burgomaster, and politician Otto von Guericke had a long-standing interest in the void and space. His strategy for dealing experimentally with the issue was different from Torricelli's, however, of whose experiment he apparently became aware only in the 1650s. Guericke's early experiments were inspired by Descartes' claim that matter and extension coincide, and if this were the case, Guericke argued, removing air from a sealed container would lead to the container's implosion. Early attempts to construct a viable air or vacuum pump were plagued by leaking problems, since air was heard whistling through the badly sealed cracks. For Guericke it seemed especially important to perform the evacuation procedures under water, as a way to assure that no leaking occurred. Although in one early attempt the container consisting of tightly wrought copper hemispheres did indeed crash, as Guericke believed one would expect if Descartes was right, he remained unconvinced and performed yet another experiment. This time he succeeded in his intent of constructing a viable air pump that, in his opinion, refuted Descartes' views (Fig. 7.6). The most spectacular public demonstration took place in Magdeburg in 1657, when several pairs of horses failed to separate the copper hemispheres, which could be opened easily when air was reintroduced. In the same year the Jesuit natural philosopher Gaspar Schott published Guericke's important results in *Mechanica hydraulico-pneumatica* (Schott, 1657).

It was a combination of the reports from Italy and Germany that led to a number of experiments in England, chiefly due to Robert Boyle and his assistant Robert Hooke.

Fig. 7.6. von Guericke's celebrated air-pump experiment.

Boyle's extensive experiments involved both the air-pump and more mundane devices, such as differently shaped tubes.

Boyle entered the fray in 1660 with *New Experiments Physico-Mechanical, Touching the Spring of the Air*. It was in the sequel, *A Defence of the Doctrine Touching the Spring and Weight of the Air* (1662), that Boyle presented the law named after him. His formulation was empirical and was based on experiments: in fact, it presented in the form of tables containing the numerical values of experimental results compared with the results one would expect from a simple mathematical proportion. The discrepancies were deemed to be very small.

Briefly put, Boyle used a J-tube open at the tall end and sealed at the short one. The experiment consisted of pouring mercury into the long open leg and measuring both its height and the diminishing volume occupied by the air in the closed short leg. What he found was that the volume of the air compressed in the short leg of the tube was inversely proportional to the pressure of that air, which was the sum of atmospheric pressure and the pressure due to the column of mercury in the long leg. Boyle performed also a symmetrical experiment trying to show the effects of a decreasing spring of the air. He immersed a glass pipe with the diameter of 'a swan's quill' into a container about 70 inches deep filled with mercury, until the pipe emerged by only 1 inch. He then sealed with wax the open end of the pipe above the mercury, and started raising the pipe, measuring on a graduated scale attached to it the space occupied by the air and how much the mercury had risen. The idea is that the air trapped in the tube is compressed by the pressure of the atmosphere minus the pressure of the raised column of mercury or, in other words, the volume of air is inversely proportional to the effective pressure. Since the amount of trapped air is the same, the volume is inversely proportional to the density, and the effective pressure or spring of the air is proportional to the density. The results were again tabulated and compared to the hypothesis that the density is as the spring of the air. It is noteworthy that Boyle, unlike Galileo for the science of motion or Huygens for the rules of collision, established and presented the law named after him on empirical grounds. He was not very sympathetic to mathematical formulations, and did not seek an Archimedean style but rather forged a style of his own.

Inspired by von Guericke and helped by Hooke, Boyle embarked on the construction of an air-pump with a design different from the German one, consisting of a glass receiver attached to a mechanical apparatus for progressively emptying it of air. With it, the English virtuosi managed to perform a number of experiments building and expanding on those we have seen so far, including the void-in-the-void and the carp bladder. In addition, they could investigate a wide range of issues, from the behaviour of marble plates adhering to each other to that of birds and other animals.

Air-pumps—especially Boyle's—showed a versatility that Torricellian tubes lacked: it was easier to insert objects inside the receiver and to experiment with them than in the space at the top of a Torricellian tube. At the same time, they were expensive devices that were difficult to construct and to operate, and it seems that despite detailed illustrations and verbal descriptions, without actually seeing one, they were very difficult to build or operate. Moreover, air-pumps leaked, and at times generated anomalous results that, at least for a period, generated debates as to the implications of the researches obtained with them and of experiments more generally (Shapin and Schaffer, 1985).

7.5 Optical Experiments: The Behaviour and Nature of Light

The seventeenth century witnessed a range of celebrated experiments in optics, especially on refraction and the nature and composition of light. Descartes and Newton were the chief protagonists of those researches, which involved water flasks, prisms, and small and large lenses. In this case too, as for mechanics and the science of motion, experimental apparatus was relatively simple and easy to obtain, though there would have been differences for prisms and lenses in the quality of glass between toys—such as prisms which were sold at fairs—and more reliable tools with glass free of bubbles and other defects.

7.5.1 Refraction and the Rainbow

Whereas the law of reflection of light had been known since antiquity, the law of refraction was discovered only in the seventeenth century. Refraction occurs when a light-ray travels from one medium to another with different physical properties, such as from air to water, for example, or air to glass. The law is usually formulated as stating that, in the transition between two media, the ratio between the sine of the angle of incidence and the sine of the angle of refraction is constant and depends on the nature of the media.

In the second century AD Ptolemy had tried to determine some regularity experimentally, but his work was seen as problematic. Although the Dutch mathematician and astronomer Willebrord Snel discovered the law first, he did not publish it, and the original manuscript with his findings and the path he followed have been lost. It was René Descartes who first published the law of refraction in his *Dioptrique* of 1637—one of the essays accompanying the *Discours de la méthode*. In that work, however, he argued that he had found the law *a priori*, not experimentally, and that experiments had only confirmed the law at a later stage. Descartes' demonstration of the law of refraction was based on a series of doubtful assumptions that are also hard to reconcile with his own views about light. For example, although he believed light to be an instantaneously transmitted pressure, his reasoning was based on the assumption that it travels with different speeds in different media, and specifically that the speed is greater in denser media. His presentation relied on a thought experiment comparing the motion of light to that of a tennis ball hitting at an angle a horizontal surface;

while the horizontal component of the ball's speed is unaffected, the component perpendicular to the surface is affected and can increase or decrease. Descartes showed geometrically that his account led to the sine law (Descartes, 1979; Sabra, 1981, chapter 4).

Regardless of how Descartes attained his law, he did apply it in *Les météores*, also published in 1637, together with a thorough series of experiments to investigate the rainbow, attaining remarkable quantitative results concerning the location of the primary and secondary bow and colours. Specifically, Descartes experimented with water-filled flasks, which he took to represent water droplets generating the rainbow, and determined the angle under which rainbows appear to the viewers (Buchwald, 2008).

7.5.2 Newton's Prism Experiments and Theory of Light

Newton's prismatic experiments count among the iconic ones of their time—possibly of the entire history of science. They are also considered as the first truly significant experiments to be published in a scientific journal, the *Philosophical Transactions of the Royal Society* of 1672. It would be profoundly misleading to believe that Newton's optical experiments were generally accepted as unproblematic; their novelty and subtleties in execution led to prolonged controversies that drove him to abandon for a while the experimental philosophy—or at least publication. Readers may find accounts of those controversies in the extensive literature (Sabra, 1981; Shapiro, 1996).

Newton's experiments form a series of increasingly refined trials. He started by noticing that the spectrum of white sunlight projected by a prism across a long room was not circular but elongated. This seemingly innocuous observation led to renewed investigations with far-reaching conclusions. I will focus in particular on two experiments: the *experimentum crucis* and the recomposition of white light.

In the first, Newton had white light refracted through a prism; a pierced board close to the prism let only a thin beam of rays of one colour through, blocking all the others. Those rays that passed through the aperture reached a second pierced board at a distance of about 12 feet, and behind it was another prism, which refracted the light further, projecting it onto a screen. By rotating the first prism around its axis, Newton allowed different colours to pass through the aperture and be refracted by the second prism, projecting a coloured spot of light onto the screen. He noticed that the position of this spot changed depending on the colour of light that was let through—a phenomenon that led him to conclude that white light consists of rays differently refrangible, where the degree of refrangibility depends on the colour: each coloured ray was refracted by the same amount by the first and second prism. In addition, Newton's *experimentum crucis* showed that whereas the first prism generated a spectrum out of white light, the second left individual colours unchanged. In order to show immutability, however, Newton had to modify the experiment by adding a lens that allowed a sharper separation of the colours (Fig. 7.7).

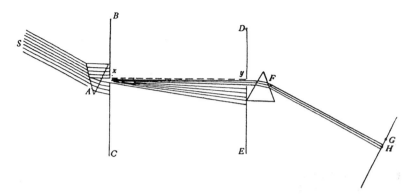

Fig. 7.7. Newton's *experimentum crucis*.

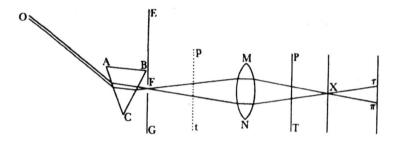

Fig. 7.8. Newton's recombination of white light experiment.

The second experiment consisted of recombining the coloured rays: this time Newton separated white light into a coloured spectrum, and then, by means of a lens, recombined the coloured rays by focusing them. As he expected, he obtained white light again, thus supporting his views that a prism does not generate colour by modifying white light but merely decomposes white light, which can be recomposed by a lens. (Fig. 7.8) White light is a heterogeneous mixture of rays differing in their colour and degree of refrangibility. Some of Newton's optical experiments, with their emphasis on analysis or decomposition and synthesis or recomposition, echo in conception and language chemical experiments (Westfall, 1977; Shapiro, 1984).

7.5.3 Newton's Rings

Robert Hooke's *Micrographia* was one of the most visually striking books of the entire century, with dozens of engravings of seeds, insects, leaves, and mold seen under the microscope in unprecedented and arresting detail. Among the many topics addressed by Hooke was that of the colour of bodies appearing in thin slices of Muscovy glass (mica), soap bubbles, and between two pieces of glass.

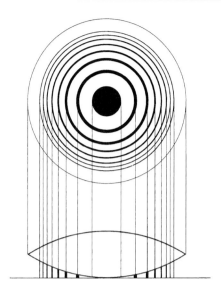

Fig. 7.9. Newton's rings.

Hooke was an able experimentalist who had provided quantitative analyses in a number of areas, such as the study of elasticity, for example, for which he had established the direct proportionality between force and displacement, as we have seen previously. In the case of thin films between pieces of glass, however, it was Newton who provided a quantitative analysis by pressing against a flat slab of glass a lens with a known curvature, up to 51 feet, which enabled him to measure accurately the space between the lens and the slab. Looking down on the apparatus, a series of coloured rings appeared (Fig. 7.9). Newton observed the phenomenon in monochromatic light, and noticed that if the first ring appeared for a given space or thickness of the air film between the two glass surfaces, subsequent rings appeared at odd-numbered multiples, 3, 5, 7, and so on; and at even-numbered multiples, 2, 4, 6, and so on, dark rings appeared. Moreover, the position of the rings depended on the colour of light used (Newton, 1952; Westfall, 1980, 1977).

Of course, in retrospect, Newton's results appear to be powerful evidence in support of a wave theory of light and interference phenomena. Despite the difference with his own rather elaborate interpretation of the phenomenon, his experiments and measurements count as among the most accurate of the period (Shapiro, 1993).

7.6 Concluding Reflections

Examining physical experiments in the seventeenth century is a task of huge complexity, and a partial census of some of the most prominent cases merely skims the surface of this rich field, which includes a wide range of disciplines and a broad set of issues.

Experiments were far more widespread than my brief survey has been able to document, and in many areas they became an integral part of research, like fertile humus for further reflections and investigations. Entire societies, such as the Accademia del Cimento and to a large extent the Royal Society, were devoted to experiments. Nonetheless, the material we have discussed offers food for thought concerning a number of issues (Middleton, 1971; Shapin, 1988).

Motion, barometric experiments, and optics were possibly the major areas of experimentation in the physical sciences of the seventeenth century. Even in each of these three individual domains, however, the role of experiments was far from being straightforward or univocal, as the reasons why experiments were performed could be heavily dependent on theoretical concerns or were largely unrelated to them. Likewise, the form of presentation could vary depending on a variety of circumstances. Mersenne and Boyle, for example, saw experiments as the basis of knowledge, and attributed accordingly a key role to them in their narratives, while others, such as Galileo and Huygens, had a more varied stance and adopted a different style of presentation in different works, relegating experiments to an ancillary role in their works modelled on Archimedes, while relying more explicitly on them in other works.

Experiments developed hand in hand with a growing methodological awareness of their significance and increasing sophistication in conception and execution. However, it is difficult to trace a linear growth of methodological awareness throughout the century. The Puy-de Dôme experiment involved the usage of a 'control' barometer kept at the base of the mountain to ensure that no changes in pressure occurred during the day. Such precautions can be found in other instances, but they did not become routine during the century. Indeed, the main popularizer of the experiment, the anatomist Jean Pecquet, omitted to report the 'control' altogether. (Shapin and Schaffer, 1985; Buchwald, 2006; Bertoloni Meli, 2008).

Experiments rarely provided immediate definitive answers to the questions asked. Whether in the science of motion, or the void and pressure of the air, or the nature and behaviour of light, for every optimistic Galileo there was a skeptical Mersenne challenging and questioning what he had done. Boyle believed that experiments could provide matters of fact on which to build sound knowledge, yet even his alleged matters of fact often proved controversial. At the same time, many experiments eventually did achieve genuinely novel results that were broadly accepted and that opened new horizons for reflection and investigation. It would be impossible to grasp the novelty of seventeenth-century investigations of nature without paying close attention to experiment and the experimental philosophy.

Acknowledgements

I wish to thank Friedrich Steinle for several conversations on experimentation in the early modern period and beyond.

Select Bibliography

Primary Sources

DESCARTES, RENÉ. *Le Monde: Ou Traité de la lumière*, translation and introduction by Michael S. Mahoney (New York: Abaris Books, 1979).
GALILEI GALILEO. *Dialogue Concerning the Two Chief World Systems*, translated by Stillman Drake (Berkeley, CA: University of California Press, 1967^2).
—— *Two New Sciences, Including Centers of Gravity and Force of Percussion*, translated by Stillman Drake (Madison, WI: University of Wisconsin Press, 1974).
HUYGENS, CHRISTIAAN, *The Pendulum Clock*, Translated by Richard J. Blackwell (Ames, IA: Iowa State University Press, 1986).
NEWTON, ISAAC, *Opticks* (New York: Dover, 1952; reprint of the fourth English edition, 1730).
—— *The Principia: Mathematical Principles of Natural Philosophy. A New Translation by I. Bernard Cohen and Anne Whitman, assisted by Julia Budenz* (Berkeley, CA: University of California Press, 1999).
PASCAL, BLAISE, *The Physical Treatises*, translated by Isidore H. B. Spiers and Alexander G. H. Spiers (New York: Columbia University Press, 1937).
SCHOTT, GASPAR, *Mechanica hydraulico-pneumatica* (Frankfurt/M: sumptu heredum J. G. Schönwetteri, 1657).

Secondary Sources

BERTOLONI MELI, DOMENICO. *Thinking with Objects: The Transformation of Mechanics in the Seventeenth Century* (Baltimore, MD: Johns Hopkins University Press, 2006).
—— 'The Collaboration between Anatomists and Mathematicians in the mid-17th Century. With a Study of Images as Experiments and Galileo's Role in Steno's *Myology*', *Early Science and Medicine*, **13** (2008), 665–709.
—— 'A Lofty Mountain, Putrefying Flesh, Styptic Water, and Germinating Seeds. Reflections on Experimental Procedures from Périer to Redi and Beyond', in Marco Beretta, Antonio Clericuzio, and Lawrence M. Principe, eds. *The Accademia del Cimento and its European Context* (Science History Publications, 2009), 121–34.
—— 'Patterns of Transformation in Seventeenth-Century Mechanics', *The Monist*, **93** (2010), 580–97.
BUCHWALD, JED Z. 'Discrepant Measurements and Experimental Knowledge in the Early Modern Era', *Archive for History of Exact Sciences* **60** (2006), 1–85.
—— 'Descartes's Experimental Journey Past the Prism and Through the Invisible World of the Rainbow', *Annals of Science*, **65** (2008), 1–46.
BÜTTNER, JOCHEN. 'The Pendulum as a Challenging Object in Early-Modern Mechanics', in Walter Roy Laird and Sophie Roux, eds. *Mechanics and Natural Philosophy before the Scientific Revolution*, ed. (Dordrecht: Springer, 2008), 223–37.
—— 'Big Wheel Keep on Turning', *Galilaeana*, **5** (2008), 33–62.
CAMEROTA, MICHELE, and MARIO O. HELBING. 'Galileo and Pisan Aristotelianism: Galileo's *De motu antiquiora* and the *Quaestiones de motu elementorum* of the Pisan Professors', *Early Science and Medicine*, **5** (2000), 319–365.
CLAGETT, MARSHALL. *Archimedes in the Middle Ages* (Madison, WI: University of Wisconsin Press; Philadelphia: American Philosophical Society, 1964–84), 5 vols.

COHEN, FLORIS H. *The Scientific Revolution: A Historiographical Inquiry* (Chicago: University of Chicago Press, 1994).
DEAR, PETER. *Mersenne and the Learning of the Schools* (Ithaca, NY: Cornell University Press, 1988).
―― *Discipline and Experience: The Mathematical Way in the Scientific Revolution* (Chicago, IL: University of Chicago Press, 1995).
DRAKE, STILLMAN. *Galileo at Work: His Scientific Biography* (Chicago, IL: University of Chicago Press, 1978).
HACKING, IAN. *Representing and Intervening: Introductory Topics in the Philosophy of Natural Science* (Cambridge: Cambridge University Press, 1983).
HEIDELBERGER, MICHAEL, and FRIEDRICH STEINLE, eds. *Experimental Essays. Versuche zum Experiment* (Baden-Baden: Nomos, 1998).
KOYRÉ, ALEXANDRE, *Metaphysics and Measurement: Essays in the Scientific Revolution* (Cambridge, MA: Harvard University Press, 1968).
LICOPPE, CHRISTIAN, *La formation de la practique scientifique; le discours de l'expérience en France et en Angleterre (1630–1820)* (Paris: Éditions la Découverte, 1996).
MAHONEY, MICHAEL S. 'Huygens and the Pendulum: From Device to Mathematical Relation', in Herbert Breger and Emily Grosholz, eds. *The Growth of Mathematical Knowledge* (Dordrecht: Kluwer, 2000), 17–39.
MIDDLETON, W. E. KNOWLES. *The History of the Barometer* (Baltimore: Johns Hopkins University Press, 1964).
―― *The Experimenters: A Study of the Accademia del Cimento* (Baltimore: Johns Hopkins University Press, 1971).
PALMIERI, PAOLO. 'A Phenomenology of Galileo's Experiments with Pendulums', *British Journal for the History of Science*, **42** (2009), 479–513.
RENN, JÜRGEN, PETER DAMEROW, and SIMONE RIEGER, with an Appendix by Domenico Giulini. 'Hunting the White Elephant: When and How did Galileo Discover the Law of Fall?', in Jürgen Renn, ed. *Galileo in Context* (Cambridge: CUP, 2001, originally published in *SIC*, **13**, 2000), 29–149.
SABRA, ABDULHAMID I. *Theories of Light from Descartes to Newton* (Cambridge: Cambridge University Press, 1981; first published, 1967).
SETTLE, THOMAS B. 'An Experiment in the History of Science', *Science*, **133** (1961), 19–23.
―― 'Galileo's Use of Experiment as a Tool of Investigation', in Ernan McMullin, ed. *Galileo, Man of Science* (New York: Basic Books, 1967), 315–38.
SHAPIN, STEVEN. 'The House of Experiment in Seventeenth-Century England', *Isis*, **79** (1988), 373–404.
SHAPIN, STEVEN, and SIMON SCHAFFER. *Leviathan and the Air Pump: Hobbes, Boyle and the Experimental Life* (Princeton: Princeton University Press, 1985).
SHAPIRO, ALAN E. ed. *The Optical Papers of Isaac Newton* (Cambridge: Cambridge University Press, 1984).
―― *Fits, Passions, and Paroxisms* (Cambridge: Cambridge University press, 1993).
―― 'The Gradual Acceptance of Newton's Theory of Light and Colors, 1672–1727', *Perspectives on Science*, **4** (1996), 59–140.
SMITH, GEORGE. 'The Newtonian style in Book II of the *Principia*', in Jed Z. Buchwald and I. Bernard Cohen, eds. *Isaac Newton's Natural Philosophy* (Cambridge, MA: MIT Press, 2001), 249–298.

STEINLE, FRIEDRICH. 'Experiments in the History and Philosophy of Science', *Perspectives on Science*, **10** (2002), 408–432.

WESTFALL, RICHARD S. *The Construction of Modern Science* (Cambridge: Cambridge University Press, 1977; originally published, 1971).

——*Never at Rest: A Biography of Isaac Newton* (Cambridge: Cambridge University Press, 1980).

CHAPTER 8

MATHEMATICS AND THE NEW SCIENCES

NICCOLÒ GUICCIARDINI

8.1 Revolutions in Mathematics in the Long Seventeenth Century

A few decades ago it was customary to define the years between the publication of Copernicus's *De Revolutionibus* in 1543 and that of Newton's *Principia* in 1687 as the period of the 'Scientific Revolution'. However, the view according to which a radical transformation and momentous leap forward occurred between the mid-sixteenth and late-seventeenth century due to the work of a few giants has been criticized recently. The historian of mathematics, however, is reluctant to abandon this view completely, since it is indeed the case that in the years in question the mathematical sciences underwent changes so deep—in terms of both scope and method—to deserve to be called revolutionary.[1]

The advent of symbolic algebra in the late Renaissance, of analytic geometry with the works of Fermat and Descartes—among others—in the early decades of the seventeenth century, and then the discovery of the calculus in the age of Newton and Leibniz enabled the mathematization of phenomena that had been regarded as well beyond the scope of mathematical treatment by the generation of natural philosophers before Galileo. To give a measure of this change one need only consider the fact that in the late sixteenth century one of Galileo's mentors and one of the most important promoters of the mathematical approach to nature (including work on falling bodies and percussion), Guidobaldo dal Monte (1545–1607), concluded that

---

[1] For further information on the theme treated in this section, see Chapter 1 of this volume, 'Was There a Scientific Revolution?', by John Heilbron.

the motion of bodies was subject to too many irregularities for it to be considered the province of mathematical science (Bertoloni Meli 2006, 33–34), whereas in the early decades of the eighteenth century mathematicians with an expertise in the techniques of the calculus were mathematizing—amongst other things—the motion of projectiles in resisting media, the ebb and flow of tides, the shape of planets, the bending of beams, and the motion of fluids. The present chapter is devoted to this momentous progress—or revolution, if we prefer to call it so—and to the actors responsible for it.

The mathematical revolution in natural philosophy also elicited much criticism and resistance. Indeed, many found the new mathematical methods unrigorous, or were sceptical about the possibility of applying the results achieved by mathematicians to natural phenomena. Mathematicians were subject to criticism from experimentalists who argued that natural phenomena, in all their complexity, could hardly be submitted to mathematical laws. Those I have in mind here are thinkers as removed in time and diverse in their methodologies as Francis Bacon, who depicted nature as a *silva* defying geometrization, Robert Boyle, who favoured experimental philosophy over mathematized natural philosophy, and Alessandro Volta, who had a poor opinion of Charles Augustin Coulomb's success in discovering an inverse-square law for electric and magnetic phenomena.

As we shall see, in overcoming these criticisms the promoters of the use of mathematics had a profound impact on disciplines such as physics, astronomy, cosmology, and optics: they changed the methods used in these disciplines and their very nature, making their results almost inaccessible to laymen. Actors with different intellectual backgrounds and social and academic qualifications contributed to this process: professors of mathematics working in universities, mathematical humanists active at princely courts, engineers, astronomers, and even instrument-makers. Before delving into the complexities of the mutual influences between mathematics and the new science that emerged, whereby new scientific enterprises fostered the development of new mathematical methods and mathematical developments in turn paved the way for new scientific research, it will be worth examining the work of mathematicians in the late sixteenth century.

8.2 The Status of the Mathematical Sciences in the Late Renaissance

Since the publication of Alexandre Koyré's seminal works it has become commonplace to consider the 'mathematization of nature' a key factor in the development of sciences such as mechanics, astronomy, and optics, that took place in the period under consideration in this chapter—a factor as important as the turn toward experimentation praised in the work of Francis Bacon. Indeed, mathematics and mathematicians enjoyed renewed interest and prestige in the second half of the seventeenth century. While Koyré has attributed the rising status of the mathematical sciences in the late Renaissance to a crisis in Aristotelian philosophy and a concomitant shift towards neo-Platonism, Richard Westfall has suggested that the economic importance of technologies such as ballistics, fortifications, water management, navigation, and cartography stimulated mathematical research and offered mathematicians new jobs and opportunities (Koyré 1968) (Westfall 2001).

If we turn our attention first to universities, we find that mathematics was taught in the context of a medieval curriculum based on the Aristotelian subordination of the mathematical sciences to natural philosophy. One of the main tasks of natural philosophy was to determine the causes of change: for instance, those of local motion. We should keep in mind that the notions of cause and change in Aristotelian philosophy possessed a much broader semantic spectrum than those defined by seventeenth-century thinkers such as Descartes, who reduced all change to the local motion of particles, and interpreted change of motion as caused by localized impacts between corpuscles. What is relevant for us is that the language in which Aristotelian natural philosophy was couched was not mathematical but logical: for the natural philosopher's discourse was based on syllogisms. According to Aristotle, it is syllogistic reasoning and not mathematics that reveals the causes of change in nature.

There is no better example to illustrate the subordinate status of mathematics compared to natural philosophy than the relationship existing between cosmology and astronomy according to many Aristotelians. Astronomy was considered part of the so-called 'mixed' mathematical sciences, along with harmonics, mechanics, optics, and so on. The 'pure' mathematical sciences were, of course, arithmetic and geometry—two clearly distinct disciplines, one dealing with discrete multitudes, the other with continuous magnitudes. According to Aristotelians, astronomy as a mathematical science could not—and should not—be confused with cosmology, a part of natural philosophy. In their view, the task of astronomy was to predict the positions of the heavenly bodies via the application of mathematical models. Astronomy did not describe the heavens, nor explain the causes of heavenly motions. Therefore, it was conceived of as a mixed mathematical science, with a status subordinate to that of philosophical sciences like cosmology. It is this subordination of astronomy that was called into doubt by Copernicus, who stated it was mathematics that revealed the true structure of the planetary system. The heliocentric system, Copernicus argued, can be described through a mathematical model that is superior to the geocentric one essentially because the parameters of the planetary orbits are interrelated in the heliocentric system, whereas in the geocentric system each planet is treated separately. The mathematical superiority of the heliocentric model was seen as a sign of its truth. It is undeniable that, as Thomas Kuhn has underlined, Copernicus and the few Copernicans active in the latter half of the sixteenth century saw themselves as restorers of a Platonic position that assigns mathematics a role unthinkable for the Aristotelians (Kuhn 1957). Indeed, *contra* Aristotle, early Copernicans believed that, as Plato had taught, nature was essentially mathematical in structure—hence the power of mathematics to reveal the true nature of the planetary system, even when this conflicts with accepted physics and common sense. The debate on the status of mathematical sciences was thus closely linked to the contested reception of the heliocentric planetary system.

The above debate erupted in a different context in 1547, following the publication of Alessandro Piccolomini's (1508–1579) commentary on pseudo-Aristotle's *Problemata Mechanica*, a text devoted to so-called 'simple machines' (sometimes listed as the lever, the balance, inclined plane, wedge, pulley, and the screw, which attracted the attention of professors of mathematics as well as of engineers) (Piccolomini 1547). Actually, *Problemata Mechanica* played an important role in reviving interest in the applications of mathematics to mechanics, but Piccolomini, like other Aristotelian philosophers, opposed the rising status of mathematics. In his commentary he maintained that mathematics did not possess the deductive purity of syllogistic logic and was not a science because it did not disclose causal relations (Jardine 1997). Piccolomini, Benito Pereira (1535–1610) and others entering this heated science war often stressed the fact that mathematical demonstrations link premises and consequences in ways that are not unique, for usually more than one demonstration can be found for the same theorem and more than one construction for the same problem. Furthermore, geometrical constructions are carried out by deploying auxiliary figures that do not belong to the figures whose properties are under investigation. Arbitrariness in definitions, plurality of methods, and auxiliary figures and lemmas characterize mathematical practice—a clear sign, for Aristotelians, of the lower scientific status of mathematics compared to natural philosophy, a discipline that instead reveals unique causal relationships by focusing on the essential properties of the objects examined. Typically, Jesuits such as Christoph Clavius (1538–1612), who were giving pride of place to mathematics in their innovative *ratio studiorum*, defended the scientific status of mathematics. It is hardly surprising that this academic debate on the status of mathematics took place in a moment in which Copernicanism and an upsurge of neo-Platonic philosophies threatened the Aristotelian *subalternatio* of mathematics to natural philosophy. Yet, well-paid professors of philosophy—in the University of Pisa in the 1580s professors of philosophy were paid many times more than mathematics professors—had to worry about two additional factors that raised mathematics above the status it was assigned in the Aristotelian curriculum: the development of engineering and new trends in mathematical humanism.

Mathematicians were not at work only in universities. We must turn our gaze towards dockyards, arsenals, battlefields, the banks of rivers and canals, and workshops of instrument-makers and cartographers if we wish to find traces of the interesting developments that were taking place in the period of great innovations in the science of war, the regimentation of rivers, and long-distance travels by land and sea. Engineers and polymaths such as the Dutchman Simon Stevin (1548/49–1620), the Italian Benedetto Castelli (1578–1643), the Portuguese Pedro Nuñez (1502–1578), and the English Thomas Digges (1546–1595) found patronage chiefly outside the universities. The mathematics they practiced served a different purpose from that of the discipline that was taught at universities (Biagioli 1989). Their methods were praised for their usefulness, not for their rigour or scientific merits.

The latter half of the sixteenth century was also the period in which humanist philology gave birth to decisive translations and editions of mathematical texts from the Classical tradition. Of particular importance is the work of Federico Commandino (1509–1575), a Humanist, medical adviser, and mathematician who was active in Urbino (Italy) in the service of its Duke and in Rome as the personal physician of a cardinal. Commandino produced commented editions of a great number of works by Apollonius, Archimedes, Euclid, and other Classical authors (Napolitani 1997) (Bertoloni Meli 1992). Relevant here are his commentary of Archimedes' work on floating bodies (1565), which improved upon the previous edition by Tartaglia (1543), his original work on centres of gravity (1565), and his editions of Heron's *Spiritalium liber* (*Pneumatics*) (1575) and Pappus's *Mathematicae Collectiones* (*Mathematical Collections*) (1588).

The result of this philological work was the rediscovery of Greek mixed mathematics in the fields of statics and fluid equilibrium. Against Aristotle, Archimedes' works and the compilation of Greek studies in 'rational mechanics' featured in the eighth book of Pappus's *Collectio* (mainly devoted to the results of Philo of Byzantium and Heron) which dal Monte brought to print after Commandino's death, showed that mathematics could successfully be applied to nature, at least in some well-defined cases of static equilibrium. The problems treated in this Archimedean tradition were the equilibrium in simple machines and that of bodies (shaped, for instance, as paraboloids of revolution) in a fluid. In this context it was important to determine the centre of gravity of solids. Galileo (whose work will be examined in Section 8.3) was introduced to the Archimedean tradition cultivated in Urbino—a tradition that opened new vistas beyond the Aristotelian canon of mixed mathematical sciences—by his mathematics teacher Ostilio Ricci (probably via Tartaglia's edition (1543) of Archimedes' works (Drake 1995, 4)), and by one of Commadino's most gifted pupils, Guidobaldo dal Monte, the author of the most comprehensive and authoritative treatise on the theory of simple machines, *Mechanicorum Liber* (1577). Guidobaldo further developed the Archimedean tradition by publishing two works: *In Duos Archimedis Aequeponderantium Libros Paraphrasis* (1588), which Galileo received as a gift, and *De Cochlea Libri Quatuor* (1615). Around 1590, Galileo and dal Monte carried out a joint experiment on projectile motion by throwing an ink-stained ball over an inclined plane. Guidobaldo concluded that the trajectory described by the ball rolling over the plane was that of a suspended chain, 'similar to a parabola or hyperbola'. Motion perhaps was not as mathematically intractable as he had formerly thought.

8.3 Galileo and Proportion Theory

8.3.1 Early Work on Motion in Pisa

Galileo began developing a mathematical science for the motion of bodies in his early years in Pisa. In 1589, thanks to the support of Cardinal Francesco Maria dal Monte,

Guidobaldo's brother, he was elected 'lettore di matematica' at the local university.[2] Galileo's very first works were devoted to improving Archimedes' treatise on floating bodies (*La bilancetta* (1586)) and to the study of centres of gravity. This initial research located Galileo within a mathematical tradition cultivated by scholars interested in mathematical statics such as Clavius and dal Monte. The project of mathematizing the motion of bodies was considered by many Archimedeans as lying beyond the scope of mathematics. Attempts to mathematize projectile motion had already been made, most notably by the polymath and abacus master Niccolò Tartaglia (1499?–1557) (Fig. 8.1), but the results achieved hardly attained the same level of rigour and exactness as ancient Greek scientists such as Archimedes and Pappus in the field of statics. This Classical tradition is what Galileo's correspondents and mentors were striving to match. Galileo probably benefited from his exchanges in Pisa with professors of philosophy—most notably, Girolamo Borro and Francesco Buonamici—who were experimenting on falling bodies. Guibobaldo himself, while sceptical about the possibility of mathematizing phenomena as variable and irregular as falling of bodies, experimented on the force of falling bodies hitting an obstacle. The so-called 'forza della percossa' was indeed taken to be a measure of instantaneous speed (Drake 1995).

The question young Galileo asked himself was: 'Is it possible to mathematically describe the fall of bodies?' His answer can be found in a set of manuscripts known as *De motu antiquiora*, which he composed before moving to Padua as Professor

Fig. 8.1. Ballistic trajectories according to Tartaglia's theory as developed in the second book of *Nova scientia* (1537), here illustrated in Walther Hermann Ryff, *Der furnembsten, notwendigsten der ganzen Architectur* (1547), p. 313.

---

[2] For further information on Galileo, see Chapter 2 of this volume, 'Galileo's Mechanics of Natural Motion and Projectiles', by Noel M. Swerdlow.

of Mathematics in 1592. Aristotelians assumed that bodies move upwards or downwards depending on their *levitas* ('lightness') or *gravitas* ('heaviness'). Those moving downwards (heavy bodies mostly composed of earth) do so with a speed that is proportional to their heaviness and inversely proportional to the density of the medium through which they move. Perfect void cannot exist, according to this view, since density equal to zero would imply an infinite speed for falling bodies. Galileo's answer was as follows. *Pace* Aristotle, all bodies are heavy. Some move upwards, some downwards, because of Archimedes' law of buoyancy. Tartaglia (1551) and Benedetti (1553) had already conjectured that the speed of falling bodies is proportional to the difference between their specific gravity and that of the medium in which they find themselves. This makes the existence of the void possible. Furthermore, Galileo noted that the rarer the medium, the closer the speeds of vertically falling bodies with different specific weights. He was not far from realizing that in the void all bodies fall with the same velocity (Bertoloni Meli 2006, 45–60).

So how does speed vary? In *De motu*, Galileo assumes that a body falling from rest, after an initial period in which it accelerates, reaches a state in which it falls with constant speed. Acceleration is here conceived of as a transient irregular accident that should not be taken into account in the mathematical treatment of motion. Constant velocity Galileo regarded as the natural state of motion of falling bodies. Of course, uniform rectilinear motion is the easiest to mathematize. As far as 'violent' motion against gravity caused by projection (for instance, the throwing of a stone), Galileo accepted an *impetus* theory–familiar in the Middle Age–according to which the projected body moves against gravity in the direction of projection because it is endowed with a slowly decaying *impetus*. When the *impetus* is completely consumed, the cause of violent motion is no longer present and vertical natural motion occurs. The principle of inertia was yet to come. However, as Galileo realized, the results obtained by applying this theory did not find experimental confirmation. *De motu* was left unfinished, and it is in Padua that Galileo focused his mathematical and experimental efforts on acceleration, no longer seen as a transient, accidental phenomenon, but as a key feature of the rectilinear fall of bodies. Before turning to the results that Galileo achieved in Padua in the period from 1592 to around 1609, it is worth considering the mathematical tools at his disposal, since these determined the theoretical possibilities and constraints within which his well-known mechanics of projectile motion developed.[3]

8.3.2 Proportion Theory

Time, distance and speed—the magnitudes Galileo sought to subject to mathematical treatment—are all continuous magnitudes. Since antiquity, continuous magnitudes had been approached through the theory of proportions, codified in Book 5 of

---

[3] Galileo built his theory of projectile motion on previous results achieved by Medieval natural philosophers as well as by Renaissance mathematicians and ballistic engineers. We will not deal here with this historiographic issue.

Euclid's *Elements*. The Euclidean theory of proportions was the main mathematical tool employed in the mathematization of natural phenomena in the first half of the seventeenth century (Bos 1986).

The theory of proportions applies to magnitudes, where a magnitude is a general concept covering all instances in which operations of adding and comparing have meaning. For instance, a magnitude might be a given length, volume, or weight. Ratios can be formed between two magnitudes of the same 'kind' (for example, between two areas or volumes), while a proportion is a 'similitude' (an *analogy*) between two ratios ('A *ratio* is a sort of relation in respect of size between two magnitudes of the same kind'). So two volumes V_1 and V_2 might be said to have the same ratio as two weights W_1 and W_2. Employing algebraic notation not available to Galileo:

$$V_1/V_2 = W_1/W_2. \tag{8.1}$$

Two magnitudes are 'homogeneous'—that is, of the same kind—when 'they are capable, when multiplied, of exceeding one another'.[4] Note that the theory of proportions does not allow the formation of a ratio between two heterogeneous magnitudes. This is particularly important for kinematics, since it is not possible, for instance, to define speed as a ratio between distance and time.

It should be stressed that ratios are not magnitudes. We tend to read ratios as numbers and proportions as equations (as in eq. (8.1)), but this was not the case in the old Euclidean tradition. The discovery of incommensurable ratios (such as that between the lengths of the side and diagonal of a square) was interpreted as implying that the concept of number was unable to cover the extension of the concept of ratio. Nonetheless, operations on ratios are possible. An operation on ratios that will concern us was called 'compounding'. When A, B and C are homogeneous magnitudes, A/C was said to be the 'compound ratio' of A/B and B/C. Thus, ratios between homogeneous magnitudes in continued proportion could be compounded: that is, ratios such that the last term of the former, A/B, is equal to the first of the latter, B/C. In order to compound two ratios, A/B and C/D, one has to determine the 'fourth proportional' to C, D and B, that is, a magnitude F, homogeneous to B and A, such that $C/D = B/F$. The compound ratio of A/B and C/D is equal to the compound ratio of A/B and B/F, that is, A/F (Bos 2001, 119–134).

The simple example of uniform rectilinear motion ('equable motion') will allow us to understand some characteristics of the mathematization of natural philosophy in Galileo's work. The relationship between time, distance and speed in equable motion might be expressed as

$$s = v \cdot t. \tag{8.2}$$

---

[4] Euclid, *The Elements*, **5**, definition 4. In modern terms, a ratio A/B (when $A < B$) can be formed if and only if there is a positive integer n such that $nA > B$.

This was not possible within the framework of the theory of proportions, since one had to express a magnitude, speed, as being equal to the ratio between two heterogeneous magnitudes, distance and time. One has instead to state a series of proportions allowing only two of the three magnitudes involved (distance, time and speed) to vary. So it is possible to state that 'when the speed is the same and we compare two equable motions the distances are as the times':

$$s_1/s_2 = t_1/t_2, \qquad (8.3)$$

where s_1 (s_2) is the distance covered in time t_1 (t_2). Or, 'when the distance covered is the same, the speeds are as the times inversely':

$$v_1/v_2 = t_2/t_1. \qquad (8.4)$$

A more elaborate mathematical formulation of uniform motion can be achieved through the operation of compounding. The aim here is to state that the distances are in the compound ratio of the speeds and times. This can be achieved as follows. Let us assume that distance s_1 is covered with speed v_1 in time t_1, and that distance s_2 is covered with speed v_2 in time t_2. How can we compare the two motions? The trick consists in considering a third uniform motion which lasts for a time $t_3 = t_1$ and with speed $v_3 = v_2$. We can state:

$$s_1/s_3 = v_1/v_3, \qquad (8.5)$$

and

$$s_3/s_2 = t_3/t_2. \qquad (8.6)$$

The operation of compounding allows one to state that s_1/s_2 is the compound ratio of $v_1/v_2(=v_3)$ and $t_1(=t_3)/t_2$. This fact was expressed by statements such as 'the distance is proportional to the speed and to the time conjointly'. However, in order to be purely Euclidean, to compound the right-hand ratios of eqs. (8.5) and (8.6) we first have to determine three homogeneous magnitudes (let us say segments with lengths K, L and M) such that $v_1/v_2 = K/L$ and $t_1/t_2 = L/M$. So at last we can state that

$$s_1/s_2 = K/M. \qquad (8.7)$$

The theory of proportions was not flexible enough to solve the problems of seventeenth-century natural philosophy. Several proposals aimed at relaxing the rigidity of the Euclidean scheme were advanced (Giusti 1993). As we shall see in Section 8.4, Descartes' geometry replaced the theory of proportions. But this still had to be developed when Galileo was devising his new theory of uniformly accelerated motion—a theory that was disseminated via correspondence and finally printed in the *Discorsi* (1638).

8.3.3 Work on Accelerated Motion in Padua

It was in Padua, where Galileo took up the post of Professor of Mathematics in 1692, that the new science of motion was born. Our aim here is to concentrate on the mathematical tools that Galileo deployed in order to mathematize uniformly accelerated vertical fall, a key element in his study of projectile motion (Clavelin 1974) (Bertoloni Meli 2006, 96–104).

One of the basic intuitions that Galileo had was that in vertical fall bodies do not reach a terminal speed; rather, they accelerate continuously. Acceleration is not—as Galileo believed when he was in Pisa—an accidental and irregular phenomenon that only takes place at the beginning of an object's fall. It is likely that Galileo's study of pendulums (through which he obtained the famous law of isochronism, approximately valid for small oscillations) persuaded him of the need to study motions with a continuously varying velocity.

It thus became necessary for him to mathematize a variable speed. Galileo initially maintained that in vertical fall speed varies directly with the length of the space covered by the falling body from rest. Only after 1604 did he realize that speed varies linearly with time rather than distance. From the extant manuscripts, it is clear that experiments with bodies falling along inclined planes contributed more to this discovery than what was printed in the *Discorsi* (and which we shall shortly turn to examine).

The third day of the *Discorsi* opens with a series of theorems concerning equable (or uniform) motion. Mathematizing constant velocity is indeed an important premiss to Galileo's science of motion, not only because—as he himself realized—equable motion is a state in which a body perseveres until some external cause, such as an impact or gravitation, changes its state, but also because the study of accelerated motion was reduced by Galileo to equable motion.[5] The theorems on equable motion opening the third day of the *Discorsi* are framed in the language of proportion theory. To acquire a flavour of this language one can turn to theorem VI, which reads:

If two moveables are carried in equable motion, the ratio of their distance will be compounded from the ratio of spaces ran through and from the inverse ratio of times. (Galilei 1974, 152)

In symbols we would write:

$$v_1/v_2 = s_1/s_2 \times t_2/t_1. \tag{8.8}$$

Several observations are in order. Firstly, speed is not defined as the ratio of distance over time. As previously noted, this definition goes against the rules of proportion theory according to which ratios can only be formed between homogeneous magnitudes. Galileo, therefore, defines the properties of speed via a series of postulates and theorems. Secondly, the fact that four magnitudes are in the same ratio is verified by showing that the conditions stated in definition 5, book 5, of Euclid's *Elements* are

---

[5] The peculiarities of Galileo's formulations of a principle of 'circular inertia', and their difference from the modern one, is a much debated issue.

met—something which requires rather laborious demonstrations.[6] Thirdly, the compounding of ratios requires the determination of a fourth proportional. The need to recur to these mathematical techniques makes Galileo's simple theorems on equable motion painfully cumbersome and difficult to follow. Finally, it should be noted that the theory of proportions is best suited to express linear relations between continuous magnitudes. Even slightly more complicated functional relationships pose difficulties, as we shall see in the case of the quadratic dependence of distance from time in the case of uniformly accelerated motion.

In order to express the dependence of distance from the square of time in the language of proportion theory, Galileo ultimately resorts to the following corollary, which reduces the quadratic relationship to a linear relationship via the determination of a mean proportional:

if at the beginning of motion there are taken any two spaces whatever, run through in any [two] times, the times will be to each other as either of these two spaces is to the mean proportional space between the two given spaces. (Galilei 1974, 170–1)

In symbols we would first determine the mean proportional s_{mean} between the two given distances s_1 and s_2:

$$s_2/s_{mean} = s_{mean}/s_1, \qquad (8.9)$$

and then translate the corollary as follows:

$$t_2/t_1 = s_2/s_{mean}.[7] \qquad (8.10)$$

In order to appreciate the constraints that proportion theory imposes on Galileo's treatment of naturally accelerated motion, it is worth quoting Theorem 1 at length. This reduces accelerated motion to equable motion via a proposition that had been well known since the Middle Ages:

The time in which a certain space is traversed by a moveable in uniformly accelerated movement from rest is equal to the time in which the same space would be traversed by the same moveable carried in uniform motion whose degree of speed is one-half the maximum and final degree of speed of the previous, uniformly accelerated, motion.

Let line AB [Fig. 8.2] represent the time in which the space CD is traversed by a moveable in uniformly accelerated movement from rest at C. Let EB, drawn in any way upon AB, represent the maximum and final degree of speed increased in the instants of the time AB. All the lines reaching AE from single points of the line AB and drawn parallel to BE will represent the increasing degrees of speed after the instant A. Next, I bisect BE at F, and I draw FG and AG

---

[6] Definition 5 reads: 'Magnitudes are said to be in the same ratio, the first to the second and the third to the fourth, when, if any equimultiples whatever are taken of the first and third, and any equimultiples whatever of the second and fourth, the former equimultiples alike exceed, are alike equal to, or alike fall short of, the latter equimultiples respectively taken in corresponding order.' Translation by D. E. Joyce at http://aleph0.clarku.edu/~djoyce/java/elements/elements.html.

[7] Indeed, allowing ourselves an algebraic language not accessible to Galileo,

$$t_2/t_1 = s_2/s_{mean} = s_2/\sqrt{s_2 s_1} = \sqrt{s_2}/\sqrt{s_1}.$$

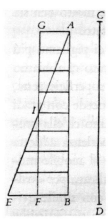

Fig. 8.2. Theorem 1 on naturally accelerated vertical motion, in Galileo's *Discorsi*. Reproduced in (Galilei 1974, 165).

parallel to *BA* and *BF*; the parallelogram *AGFB* will [thus] be constructed, equal to the triangle *AEB*, its side *GF* bisecting *AE* at *I*.

Now if the parallels in triangle *AEB* are extended as far as *IG*, we shall have the aggregate of all parallels contained in the quadrilateral equal to the aggregate of those included in triangle *AEB*, for those in triangle *IEF* are matched by those contained in triangle *GIA*, while those which are in the trapezium *AIFB* are common. Since each instant and all instants of time *AB* correspond to each point and all points of line *AB*, from which points the parallels drawn and included within triangle *AEB* represent increasing degrees of the increased speed, while the parallels contained within the parallelogram represent in the same way just as many degrees of speed not increased but equable, it appears that there are just as many momenta of speed consumed in the accelerated motion according to the increasing parallels of triangle *AEB*, as in the equable motion according to the parallels of the parallelogram *GB*. For the deficit of momenta in the first half of the accelerated motion (the momenta represented by the parallels in triangle *AGI* falling short) is made up by the momenta represented by the parallels of triangle *IEF*.

It is therefore evident that equal spaces will be run through in the same time by two moveables, of which one is moved with a motion uniformly accelerated from rest, and the other with equable motion having a momentum one-half the momentum of the maximum speed of the accelerated motion; which was [the proposition] intended. (Galilei 1974, 165–6)

Some observations can be drawn here. Uniformly accelerated motion is graphically displayed by the abscissae *AB* representing time and ordinates *EB* representing the linearly increasing degrees of speed. Via this theorem Galileo is able to associate an equable motion to an accelerated motion, and therefore to mathematize uniformly accelerated motion in terms of equable motion. Note that Galileo does not identify the distance *CD* traversed by the falling body with the area of the triangle *AEB*. Mathematicians belonging to the generation after Galileo's, such as Evangelista Torricelli, would here be inclined to conceive of the surface of triangle *AEB* as constituted by infinitely many parallelograms whose infinitely small areas (in Leibniz's notation vdt) represent the lengths of the infinitely small spaces (ds) covered by the body in an infinitesimally small interval of time (dt). Galileo never reached this understanding: he was rather comparing the 'aggregate of all parallels' contained in the quadrilateral with those contained in the triangle. It appears, he writes, that 'there are just as many momenta of speed consumed in the accelerated motion ... as in the equable motion'. It is this equality of aggregates that justifies Galileo's 'evident' conclusion that equal spaces will be traversed by the two bodies, one accelerated in free fall, the other moving with constant speed. The infinite and infinitesimal lurk behind Galileo's demonstration, yet its author was anxious about the dangers inherent in reasoning with the infinite, and framed his demonstrations one inch within (or beyond?) the 'limits', so to speak, of what has been termed 'pre-classical' mechanics (Damerow 2004). As we shall see in Section 8.6, these limits were transgressed by those mathematicians who dared to employ infinitesimal magnitudes.

The success of Galileo's mathematical treatment of bodies is illustrated in the fourth day of the *Discorsi*, where, by compounding inertial equable motion in the (horizontal) direction of projection with vertical naturally accelerated motion, he proves that frictionless motion of projectiles takes place along a parabola. A successful answer to those who were sceptical about the possibility of treating the motion of bodies in mathematical terms. Yet, despite Galileo's breakthrough in the mathematization of motion, it was only possible to move beyond the limits of his *Discorsi* by abandoning the strictures of proportion theory. This was precisely the move made by René Descartes.

8.4 Descartes and Analytic Geometry

Few works in the history of mathematics have been more influential than Descartes' *Géométrie* (1637) (Bos 2001). The canon defined in this revolutionary essay was to dominate the scene for many generations, and its influence on seventeenth-century mathematicians cannot be overestimated. In the very first lines of this work, its author claims to be in possession of a method capable of 'reducing any problem in geometry'.[8]

Book 1 of the *Géométrie* explains how to translate a geometrical problem into an equation. Descartes was able to attain this goal by making a revolutionary break with the tradition of the theory of proportions to which Galileo adhered. Indeed, Descartes interpreted algebraic operations as closed operations on segments. For instance, if *a* and *b* are segments, the product *ab* is not conceived by Descartes as representing an area, but rather another segment. Prior to the *Géométrie* the multiplication of two segments of lengths *a* and *b* would have been taken as the representation of the area of a rectangle with sides measuring *a* and *b*. Descartes' interpretation of algebraic operations was a huge innovation (Fig. 8.3). From Descartes' new viewpoint, ratios are considered as quotients, and proportions as equations. Consequently, the direct

---

[8] For further information on Descartes' physics, see Chapter 3 of this volume, 'Cartesian physics' by John A. Schuster.

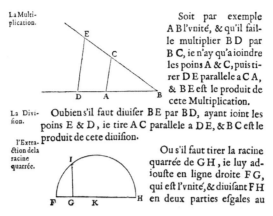

Fig. 8.3. Descartes' geometrical interpretation of algebraic operations. He wrote: 'For example, let AB be taken as unity, and let it be required to multiply BD by BC. I have only to join the points A and C, and draw DE parallel to CA; and then BE is the product of BD and BC'. So, given a unit segment, the product of two segments is represented by another segment, not by a surface. The second diagram is the construction of the square root of GH. Given GH and a unit segment FG, one draws the circle of diameter FG + GH and erects GI, the segment that, as required, represents the square root of GH. Source: Descartes, Géométrie (1637), reproduced in (Descartes 1954, 4).

multiplication of ratios is allowed, contrary to what occurred in the theory of proportions. The homogeneity of geometrical dimensions is no longer a constraint for the formation of ratios and proportions. As far as physical magnitudes are concerned, once a unit is specified for each magnitude, these can be represented through numbers or segments. Since on numbers and segments the four operations are closed, the constraints on the homogeneity of physical dimensions characteristic of the theory of proportions are overcome. So multiplication of time and speed or division of weight by surface's area are all possible. Descartes' approach won great success.

Descartes' method of problem solving was divided into an analytical part and a synthetic (Bos 2001, 287–9). The analytical part was algebraic and consisted in reducing problems into polynomial equations. If the equation was in one unknown, the problem was determinate. The equation's real roots would correspond to the finitely many solutions of the problem. The analysis, or resolution, was not, according to early-modern standards, the synthesis, or solution, of the problem. The solution of the problem must be a geometrical construction of what is sought in terms of legitimate geometrical operations. Descartes accepted the traditional idea that constructions must be performed through the intersection of curves. Therefore, he devised procedures aimed at geometrically constructing the roots of algebraic equations in one unknown via the intersection of plane curves. For instance, Descartes explained how one can construct segments, the length of which represent the roots of third- and fourth-degree equations, via the intersection of circle and parabola.

The synthetic part of the process of problem solving opened up a series of questions. Which curves are admissible in the solution of problems? When can these be considered as known or given? Which curves, among those admissible and constructible, are to be preferred in terms of simplicity? In asking himself these questions Descartes was continuing—albeit on a different plane of abstraction and generality—a long debate on the function and classification of curves in the solution of problems, which from Antiquity had reached early-modern mathematicians such as François Viète, Marino Ghetaldi (Getaldić), and Pierre de Fermat by way of Pappus's *Mathematical Collections* (1588) (Bos 2001, 287–9).

Descartes prescribed that in the construction of problems one had to use what he called 'geometric' (that is algebraic) curves of the lowest possible degree. By contrast, he banished what he called 'mechanical' (that is transcendental) curves such as the spiral and cycloid.

One of the high points of the *Géométrie* is its method for drawing normals to algebraic plane curves. Descartes described this as 'the most useful and most general problem in geometry that I know, but even that I ever desired to know' (Descartes 1954, 95). In order to determine the normal at point C to an algebraic curve, Descartes considered the sheaf of circles passing through C and with centre P located on the axis of the abscissae. Descartes observed that the circles having their centre close to the point where the normal cuts the axis intersect the curve in a second point near C, whereas the circle having its centre at the point where the normal cuts the axis touches the curve at C. Descartes was able to translate this geometric condition into algebraic terms (see Fig. 8.4) (Andersen 1994, 293–4).

The determination of normals to algebraic curves played an important role in Descartes' optics. In *Dioptrique* (1637), published alongside the *Géométrie* as one of the essays accompanying *Discours de la Méthode*, Descartes sets out to discover what shape of lens can eliminate spherical aberrations. His answer is a lens with a hyperbolic surface. The problem Descartes sets himself in Book 2 of the *Géométrie* is to find a lens the surface of which is shaped in such a way that light rays starting from one

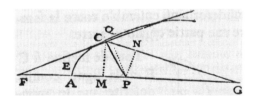

Fig. 8.4. Drawing the normal at point C on the arc AEC of an ellipse in Descartes' *Géométrie* (1637). First Descartes names some segments with algebraic symbols. Let $AM = y$, $MC = x$, $AP = v$, $MP = v - y$, and let $CP = s$ be the radius of a circle with centre P located on the axis AG and passing through a point C on the ellipse. Descartes writes the ellipse equation as: $x^2 = ry - (r/q)y^2$, where r and q are two constants. By applying Pythagoras theorem, he gets that the equation of a circle with radius CP is: $x^2 + (v - y)^2 = s^2$. Elimination of x from the two equations we have just seen leads to a second degree algebraic equation in y: (∗) $y^2 + ((qr - 2qv)/(q - r))y + (qv^2 - qs^2)/(q - r) = 0$. The sought circle touches the curve at C, so that its radius CP is normal to the curve. This translates algebraically in the search for a double root of the equation (∗). Descartes obtains $v = y - (r/q)y + r/2$. This determines $AP = v$ in function of the abscissa $AM = y$, and knowing this one can draw the normal CP, and the tangent, of the ellipse at any point C. *Source*: Descartes, *Géométrie* (1654), p. 94.

point may be refracted when striking it and converge in a second point. By applying the sine law of refraction, Descartes found that surface of this kind are obtained by revolution of the so-called ovals of Descartes around their axis. Cartesian ovals are fourth-degree algebraic curves. Determining the normal to the ovals was, of course, important in order to apply the law of refraction and determine the direction of the refracted rays after hitting the surface.

The success achieved by Descartes in mathematizing plane curves in algebraic terms and applying algebraic tools, such as his method of normals, to optics cannot be overestimated. Still, the *Géométrie* was also marked by some tensions and left certain question open. Descartes' insistence that algebra could be applied to geometry when variables and constants occurring in equations are interpreted as finite segments and the fact that he only deployed polynomial equations seemed to bar the way for the development of techniques of vital importance for seventeenth-century mathematicians—particularly those interested in applying mathematics to natural philosophy. In particular, infinitary techniques and infinitesimals, which proved essential to many innovative works in the seventeenth century, are noticeably absent from the *Géométrie* (or perhaps only obliquely visible in Descartes' method of normals). Consequently, Descartes had little to say about the rectification of curves, the calculation of areas, surfaces, and volumes, and the calculation of centres of gravity. One of the prescriptions of Descartes' that proved to be a serious limitation was his banishment of mechanical (transcendental) curves. In his correspondence with Marin Mersenne in 1638, Descartes was prompted by the Minim friar to consider the cycloid and, in solving a simple mechanical problem, the logarithmic spiral. Descartes discussed these curves, but he could not handle them with the algebraic methods of the *Géométrie*, which were confined to polynomial equations (Jesseph 2007). As the century progressed, the importance of mechanical curves became more and more evident. As we shall see in Section 8.6, these curves naturally resurfaced as solutions for what are nowadays identified as problems of integration, as well as for the solution of differential equations. They proved useful in mechanics, optics, and astronomy. Therefore, a need was felt to overcome the limitations inherent in the mathematical canon of the *Géométrie*. Furthermore, Descartes' physics, unlike his geometrical optics, remained largely qualitative. Most notably, when dealing with planetary motions, Descartes made recourse to explanations in terms of vortices swirling around the Sun that were difficult to translate into any mathematical model.

8.5 ORGANIC GEOMETRY AND MECHANICAL CURVES

It is interesting to observe that Descartes devoted great effort not only to the algebraic study of curves, but also to their mechanical generation via tracing mechanisms (Burnett 2005). This interest in what has often been referred to as 'organic

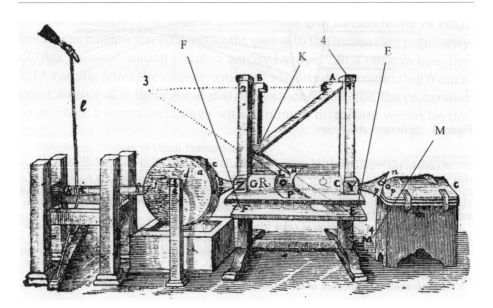

Fig. 8.5. Machine for cutting hyperbolic lenses, in Descartes, *Dioptrique* (1637). Descartes discussed the construction of this and similar machines with artisans such as Jean Ferrier. *Source*: (Descartes 1964–67, vol. 4, 218).

geometry' is rooted in a number of exigencies that shaped the Cartesian conception of geometry and its relation to algebra (Bos 2001). Here I would like to underline the fact that in the *Dioptrique* Descartes provided a detailed description of curve-tracing devices and lens-grinding machinery (Fig. 8.5)—topics he discussed extensively with high-ranking men of letters such as Constantijn Huygens, mathematicians such as Florimond de Beaune, and well-known artisans such as Jean Ferrier.[9]

In his autobiography, the Savilian Professor of Geometry at Oxford, John Wallis, claimed that in early seventeenth-century England mathematics was practiced not as an academic pursuit but as something useful for trade and technology. As Feingold has shown, Wallis's often quoted assertion greatly underestimates the role of Cambridge and Oxford in promoting the mathematical sciences (Feingold 1984, 1, 21). Nonetheless, Wallis's recollections are interesting since they suggest that as historians of mathematics we should search for evidence of the development and application of the mathematical sciences also in the work of traders and artisans; that we should study not only Descartes' peers but also Ferrier's colleagues. Wallis, referring to his studies in Cambridge in the 1630s, wrote:

For Mathematicks, (at that time, with us) were scarce looked upon as Accademical studies, but rather Mechanical; as the business of Traders, Merchants, Seamen, Carpenters, Surveyors of Lands, or the like; and perhaps some Almanak-makers in London. And amongst more

---

[9] For further information on the role of instrument-makers, see Chapter 4 of this volume, 'Physics and the Instrument-Makers, 1550–1700' by Anthony Turner.

than Two hundred Students (at that time) in our College, I do not know of any Two (perhaps not any) who had more of Mathematicks than I, (if so much) which was then but little; And but very few, in that whole University. For the Study of Mathematicks was at that time more cultivated in London than in the Universities. (Scriba 1970, 27)

What kind of mathematics was deployed and improved upon by the practitioners Wallis mentioned, of whom Ferrier was a French representative? And how did these men interact with more theoretically inclined mathematicians, often employed in the universities and later in scientific academies? Historians are just beginning to explore the shores of a continent that has remained hidden in contemporary accounts of the history of the mathematical sciences in the long seventeenth century (Bennett 1986).

Descartes and Ferrier were discussing a topic—the mechanical generation of curves (Braunmühl 1892)—that constituted a trading zone in the commerce of mathematical ideas between humanists, painters, instrument-makers, mechanicians, and pure mathematicians. In this section we shall briefly explore this trading zone and review some of the research that was carried out at the time in the field of organic geometry. Other choices would of course be possible, since other trading zones could easily be identified. Mathematicians were interacting with practitioners in the fields of ballistics, fortification, navigation, and horology, for instance, in many ways. Here it is worth recalling in passing that astronomers' expertise was sought for the determination of longitude; that the study of the motion of falling bodies was not only an academic subject, but one which also aroused the interest of ballistic engineers; and that mechanical theories on corpuscolarism and cohesion addressed issues vital for alchemist, architects and ship-builders. Indeed, as mentioned previously, a complex network of interactions existed between pure mathematicians and practitioners of mathematics in the period under consideration in this chapter, one that represents a huge continent whose exploration is still patchy.

The mechanical generation of curves was clearly already part of the classic geometrical canon: mechanical generations of the conics, conchoids, quadratrix, and spirals occurred in geometrical constructions provided in Greek and Islamic treatises for theoretical purposes. For the mathematical practitioners active in the early modern period a curve-tracing device was often seen not so much as a theoretical construct as an instrument to be applied in one's workshop. The curve traced by an instrument could serve as the conic surface of a lens, the hyperboloid surface of the fusee of a clock, the cycloidal shape of the teeth of a wheel, or the stereographic projection of the lines of equal azimuth of the celestial sphere.

Much research was carried out in the sixteenth and seventeenth centuries on the mechanical construction of conics. Here the probable influence of Islamic mathematicians should be noted, since in the Islamic culture much effort was paid to the construction of trammels and compasses that, amongst other things, were essential in the drawing of sundials and astrolabes (Raynaud 2007). The tracing of conics was of interest to sundial-makers, painters and architects, clockmakers and cartographers. Francesco Barozzi, whose edition of Heron's mechanics (1572) was highly appreciated, described machines for tracing conic sections. Barozzi's work was complemented

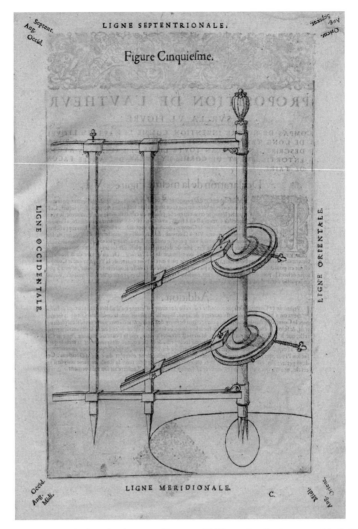

Fig. 8.6. Ellipse-tracing device, in Jacques Besson, *Théâtre des Instrumens Mathématiques et Méchaniques* (1594), Figure 5.

by Jacques Besson's (1540?–1573) *Theatrum Instrumentorum et Machinarum* (1571–72) (Fig. 8.6). Besson was highly esteemed for his description of several ingenious lathes. His discipline, the *ars tornandi*, was further pursued by Agostino Ramelli in *Le Diverse et Artificiose Machine* (1588), by Salomon de Caus in *Les Raisons des Forces Mouvantes avec Diverses Machines* (1615), and by George Andrea Böckler in *Theatrum Machinarum Novum* (1662). Some of this literature on lathes and curve-tracing machines was well-known to Descartes and his acolytes. The young Descartes might have admired many of these machines in the workshop of the famous Rechenmeister, Johann Faulhaber, in Ulm.

Descartes promoted interest in curve-tracing among his Dutch acolytes, who did not fail to stress the usefulness of mathematical machines. One of Descartes' Dutch

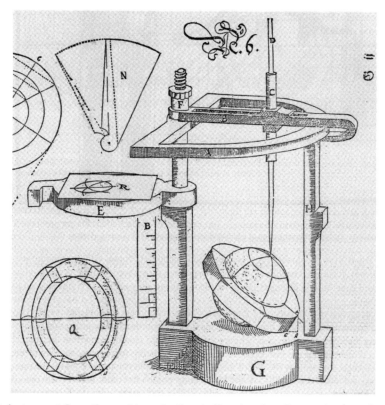

Fig. 8.7. Instrument for orthographic projection, in Hans Lencker, *Perspectiva* (1571).

mathematical disciples, Frans van Schooten (1615–1660), devoted a whole treatise (1646) to the mechanical generation of conics. Right from its title-page he made it clear that he was not simply interested in pure geometry: for his work, he emphasized, was useful not only to geometers, but also to opticians, designers of gnomons, and mechanicians. Van Schooten was Professor of Mathematics at Leiden; he edited Viète's works, translated Descartes' *Géométrie* into Latin, and promoted mathematical research in the Low Countries. One of his disciples was Christiaan Huygens. Van Schooten included Jan de Witt's 'Elementa Curvarum Linearum' in the second Latin edition of Descartes' *Géométrie* (1659–61). De Witt's work was to inspire Isaac Newton, whose mechanism for the tracing of conics—devised in the early 1670s—anticipated an intuition of Steiner's theorem on the homographic generation of conic sections (Guicciardini 2009, 93–101). Alongside professors of mathematics such as Newton and van Schooten stood practitioners developing mechanism for the generations of curves, from Hans Lencker, a Nuremberg goldsmith who in *Perspectiva* (1571) devised an instrument for orthographic projections (Fig. 8.7), to Benjamin Bramer, who authored a treatise in the vernacular entitled *Apollonius Cattus oder Kern der gantzen Geometriae* (1684).

For many centuries the tracing of curves had been the métier of sundial-makers such as Ignazio Danti (1536–1586). The makers of sundials studied the curves traced out by the shadow of the gnomon. Danti is a good example of how broad-ranging the activity of polymaths could be in the late Renaissance. He designed large-scale gnomons for the cathedrals in Florence and Bologna. In addition, he also worked as a cartographer and cosmographer for the Medici and the Pope, and lent his services as an architect, canal- and harbour-engineer, and painter and writer on perspective. Danti taught practical mathematics in Florence and Bologna, and was ultimately rewarded by the Pope with a Bishopric (Biagioli 1989). The design of gnomons—an important topic for the construction of the meridian lines necessary for the reform of the calendar the Catholic Church was promoting—had been dealt with in the ninth book of Vitruvius's *De Architectura*, which was published in two famous editions by Cesare Cesariano (Italian translation 1521) and Daniele Barbaro (Italian translation 1556, Latin edition 1567). The design of sundials occupied many mathematicians in this period, such as Commadino (who edited Ptolemy's *Planisphere* (1558), adding his own treatise on the calibration of sundials), Giovanni Battista Benedetti, and Christoph Clavius. This topic, which required the stereographic projection of circles on the celestial sphere onto the plane of the equator, was related to that of perspective and the generation of curves *per umbras*. The study of the shadows of curves was of course also of interest to painters such as Albrecht Dürer. Pure mathematicians, such as Newton and later Patrick Murdock, studied the shadows generated by the central projection of cubics. Just as conic sections can be generated by centrally projecting a circle, so all cubics—as Newton stated in his *Enumeratio Linearum Tertii Ordinis* (1704) and Murdoch proved in *Neutoni Genesis Curvarum per Umbras* (1746)—can be generated as shadows of one of the five divergent parabolas. An important chapter in the history of projective geometry is therefore related to the practical mathematics promoted by sundial-makers like Danti.

The study of the mechanical generation of curves obliterated the distinction between algebraic and transcendental curves that was central to Descartes' *Géométrie*. Once a curve is defined as the locus traced by a mechanism, the fact that its algebraic representation is not possible via a polynomial equation cannot be taken as a good enough reason to banish it from geometry, as stated by Descartes. The need to study transcendental curves emerged from many sectors of natural philosophy.

One of these curves, the cycloid, deserves our attention. A cycloid is the curve defined by the path of a point on the edge of a circular wheel as the wheel rolls along a straight line without sliding. Its properties attracted the attention of architects and mechanicians. For example, in the 1670s Ole Rømer discussed the use of the cycloid in designing toothed wheels with Leibniz. In the 1630s Mersenne called the attention of mathematicians, including Descartes, to the cycloid, and circulated a method for determining its area and tangent, which he had derived from the work of Gilles Personne de Roberval. Analogous methods are also featured in Torricelli's *Opera Geometrica* (1644). For the finding of the tangent, Roberval had decomposed the motion of the tracing point P, which generates the curve, into two components, and applied the parallelogram law to the instantaneous component velocities of P. The kinematic tracing of tangents played an important role in Newton's early researches on the drawing of tangents to mechanical lines.

In 1658, Blaise Pascal rekindled the interest of mathematicians in the properties of the cycloid with a prize challenge requiring a way of determining the quadrature, cubature, and centres of gravity of plane and solid figures bound by cycloidal arcs. Huygens, François de Sluse, Michelangelo Ricci, and Christopher Wren held back after some initial success. Wallis and Antoine de Lalouvere (1600–1664) vied for the prize but Pascal did not regard their solutions as satisfactory. Many searched for the quadrature and rectification of the cycloid: the results of Pascal and Wren proved particularly decisive.[10] The study of the cycloid was actually the motivation for protracted mutual challenges and accusations between French, Italian, and English mathematicians. The fact that the cycloid attracted so much attention and elicited such heated reactions can be explained partly by the fact that the exigency of dealing with transcendental curves was particularly felt in the mid-seventeenth century.

What is most interesting about the cycloid is that it proved useful not only for mechanical applications, as De La Hire aimed to prove in his *Traité des Epicycloïdes et de Leurs Usages dans les Méchaniques* (1694), but also in the study of natural philosophy. Indeed, in 1659 Huygens discovered that a cycloidal clock is exactly isochronous, while in 1697 the brothers Jacob and Johann Bernoulli, Leibniz, Ehrenfried Walther von Tschirnhaus, Guillaume de l'Hospital, and Newton proved that a cycloidal arc answers the problem of determining the 'brachistochrone'—a curve between two points that is covered in the least possible time by a body that starts at the first point with zero speed and is forced to move along the curve to the second point under the action of constant gravity and assuming no friction.

One of the most important examples of the interaction between mechanical engineering, mechanics, horology, navigation, and natural philosophy is Huygens' *Horologium oscillatorium* (1673), a treatise where the cycloid plays a prominent role. In this work the great Dutch polymath and astronomer used geometrical methods that were comparable in rigour to those employed by Classical Greek mathematicians. Huygens deployed proportion theory and exhaustion techniques reminiscent of the work of Archimedes (Yoder 1988).

Huygens (1629–1695)—an aristocrat whose family was intimate with Descartes, who had studied in Leiden under Frans van Schooten, and who spent many years in Paris as one of the most eminent members of the Académie des Sciences—was much interested in navigation. A clock carried on board a ship and able to keep time in a precise enough way would have enabled the determination of longitude at sea: that is, it would have solved one of the most acutely felt problems that remained open in the art of navigation in the seventeenth century. After devising a balance spring clock,

---

[10] Wren provided a method for calculating the cycloid's arclength. Pascal, in *Histoire de la Roulette* (1658), reviewed the various attempts made to face his challenge and provided a biased history of the study of the cycloid, while publishing his own research on the matter in *Lettre de A. Dettonville à Monsieur de Carcavy* (1658). The *Lettre* proved to be very important for Leibniz's early researches on the calculus. Pascal's publications were badly received in Italy because in them allegations of plagiarism were levelled against Torricelli and in England aroused the criticisms of Wallis.

Huygens found a way of correcting the dependence on amplitude of a simple circular pendulum clock. Galileo's assumption that circular pendulums are isochronous was already known to be wrong, for the period of swing of these clocks slightly increases with amplitude.

Huygens proved that a body sliding without friction along an inverted cycloidal arc has a period of oscillation that does not depend on amplitude. He also studied the evolute and involute of curves and proved a remarkable result: he showed that the evolute of a cycloid is another cycloid. Involutes, it is worth bearing in mind, are mechanically generated as follows: in order to find the involute B of the evolute A, one should wrap a string tightly against A, on its convex side. Then, keeping one end of the string fixed, the other end is pulled away from the curve. If the string is kept taut, then the moving end of the string will trace out the involute B. The evolute of a curve is the locus of all its centres of curvature. Equivalently, it is the envelope of the normals to a curve. The theory of evolutes found many applications in the rectification of curves.

Next, Huygens developed his basic idea for the construction of an isochronous pendulum. He noted that if a pendulum is swung between two cheeks shaped as the arcs of a cycloid, then the bob of the pendulum will trace out the involute of a cycloid— that is, a cycloid (Fig. 8.8). Since the cycloid is an isochrone curve, the pendulum will have a period independent of amplitude. In his late years Huygens devised experiments with clocks carried on ships sailing the Atlantic to the Cape of Good Hope, in order to measure variations in local gravity that might have confirmed his theory of gravitation and disproven the one promoted by Newton in his *Principia* (1687).

Huygens deserves an important place in the history of mechanics in the period between Galileo and Newton. He studied the motion of compound pendulums, found a mathematical expression for acceleration in uniform circular motion, stated the principle according to which the centre of the mass of a system of bodies can rise up to the height from which it was left falling, he applied the principle of Galilean

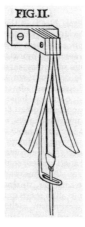

Fig. 8.8. Cycloidal cheeks for the isochronous pendulum clock, in Huygens, *Horologium oscillatorium* (1673), 4.

relativity to the laws governing the impact of bodies,[11] he developed a wave theory of light propagation based on what is still called Huygens' principle (valid for the determination of the advancement of wave-fronts), and developed a theory on the phenomenon of double refraction.

In his last years Huygens corresponded with Leibniz about mathematical topics. Leibniz had first met Huygens in Paris in 1672, and Huygens had been of great help in introducing the young German diplomat to the most advanced mathematical discoveries. But it was Leibniz who now had something new to teach his old mentor: the differential and integral calculus. At first Huygens was sceptical about the new formalism that Leibniz had developed by 1676 and published in 1684–86, but soon he had to admit that the calculus was more powerful compared to the geometrical methods he had privileged all his life. Leibniz was able to show Huygens how to easily find the catenary, the shape that a hanging chain will assume when supported at its ends and acted upon only by its own weight, a transcendental curve as the cycloid.[12] This curve proved useful in architecture, since it is the ideal curve for an arch which supports only its own weight. The correspondence between Huygens and Leibniz in the 1690s is indicative of the change that occurred with the invention of the calculus. The geometrical methods that Huygens promoted in *Horologium oscillatorium* gave way to a more abstract, algorithmic, and general mathematical method that Newton termed the method of series and fluxions and Leibniz the differential and integral calculus. However, even after the creation of calculus, well into the eighteenth century the mechanical tracing of curves continued to play a certain role in the integration of differential equations, as has been shown by Tournès (2009).

8.6 Infinitesimal Analysis

8.6.1 Pre-Calculus

In the mid-seventeenth century, several problems in physics and astronomy called for the development of new mathematical tools capable of handling continuously varying magnitudes in terms of infinitesimals. Acceleration (in uniformly accelerated motion in the case of Galileo, and in uniform circular motion in the case of Huygens) was geometrically represented in terms of infinitesimal deviations from inertial motion acquired in an infinitesimal interval of time. What Galileo had merely hinted at also became clear: that the area subtended by the graph of speed in function of time measures the distance covered. Keplerian astronomy also demanded the deployment of infinitesimals and infinite summations. Johannes Kepler in *Astronomia nova* (1609) proved that planets move in ellipses having the Sun placed at one focus. He

---

[11] Huygens' paper on the impact of bodies was printed in the *Philosophical Transactions* for 1669, while John Wallis and Christopher Wren tackled this issue in papers printed in 1668.
[12] The equation of the catenary in Cartesian coordinates is $y = a \cosh(x/a)$.

also discovered that each planet moves in such a way that the radius vector joining it to the Sun sweeps equal areas in equal times. When the elliptic orbit is known, the position of the planet in function of time can thus be found by calculating the area of the focal sector. This is the so-called Kepler problem, and is equivalent to the solution for x of the Kepler equation $x - e \sin x = z$ (e and z given). Approximation techniques were sought either in terms of geometrical constructions (Christopher Wren proposed one implying the use of the cycloid) or in terms of numerical iterative procedures. In his *Principia*, Newton published a procedure equivalent to the so-called Newton–Raphson method.

Mathematicians interested in methods applicable to continuously varying magnitudes broached problems such as the determination of tangents and of radii of curvature to plane curves; the calculation of curvilinear areas and volumes; the determination of centres of gravity; the rectification of curves, and similar. Other problems that attracted much attention were solutions of what were called 'inverse tangent problems', in which it was required to determine a curve given its local properties—typically the properties of tangents or of osculating circles. We have seen previously how Descartes tackled an inverse tangent problem when he determined the shape of a refracting surface that focuses light rays. Nowadays, inverse tangent problems are solved through differential equations.

Before Newton and Leibniz all these problems were dealt with by a variety of methods that were not ruled by any single theory (Andersen 1994). The contributions made by Kepler, Grégoire de Saint-Vincent, Cavalieri, Torricelli, Fermat, Roberval, Sluse, Hudde, Pascal, Wallis, Neil, J. Gregory, Mercator, and Barrow are generally grouped under the heading of 'pre-calculus'. The diversity of approaches, notations, and methods among these authors, however, renders any such grouping problematic. Still, some shared featured in their research can be identified. Curves were conceived of as polygonals constituted by infinitely many infinitesimal sides, while curvilinear surfaces were seen as composed of infinitely many infinitesimal rectilinear surfaces. These views enabled the determination of tangents and curvatures via limiting procedures (often disguised, as in Descartes' method of tangents, for instance) and that of the areas of curvilinear surfaces and of the volumes of curvilinear solids via infinite summations. Some of these problems were perceived as being easier: thus a variety of well-behaved methods were known for tracing the tangent to some classes of plane curve, while the determination of the area of the surface bounded by a curve or its rectification proved thornier tasks. It was observed that the equation of the tangent to an algebraic curve is algebraic, while the area of the surface bounded by an algebraic curve—or the arclength of an algebraic curve—is often transcendental: consider the area of the surface subtended to the hyperbola: in Leibnizian terms, $\ln a = \int_1^a (1/x) dx$. Indeed, the discovery, due to the Jesuits Grégoire de Saint-Vincent and Alphonse Antonio de Sarasa, of the relationships between the hyperbola and logarithms highlighted the importance of quadrature techniques for the art of navigation, geography, and astronomy. Logarithms, simultaneously introduced by the Scottish nobleman John Napier (1550–1617) and the Swiss craftsman Joost Bürgi (1552–1632) in the second decade of the seventeenth century, greatly improved the composition of numerical tables, as Kepler first appreciated in his *Tabulae Rudolphinae* (1627). Undoubtedly, the need to handle logarithms and trigonometric magnitudes was one of the chief motivations behind the birth of calculus.

A mixture of geometry and algebra characterizes the works of the above-mentioned authors. Their understanding of limit arguments and the convergence of infinite summations was often based on geometrical intuition. These were some of the major results they attained: infinite summations were deployed in order to find curvilinear areas and volumes (Cavalieri, Torricelli, Pascal, J. Gregory); the area subtended to the hyperbola was calculated and it was understood that it measures the natural logarithm (Grégoire de Saint-Vincent, de Sarasa, Mercator); the length of the arc of certain curves was calculated (Wren, Neil); tangents were drawn to algebraic and mechanical lines (Roberval, Fermat, Sluse, Hudde); Gregory and Barrow intuited the inverse relation between tangent and area problems; and finally Wallis developed an arithmetical approach to the quadrature of curves leading to his famous discovery of the infinite product for the calculation of π (Stedall 2002).

It is due to two great thinkers, Newton and Leibniz, that from the multifaceted and unsystematic heritage of the precalculus era two equivalent formalisms surfaced, which enabled—at least in principle—the problems concerning continuously varying magnitudes to be tackled. Newton developed his method of series and fluxions in the mid-1660s, whereas Leibniz discovered differential and integral calculus during his stay in Paris between 1672 and 1676.

The innovations brought by Newton and Leibniz can be illustrated briefly by considering three aspects of their mathematical work: problem-reduction, the calculation of areas subtended to plane curves through an inversion of the process for calculating tangents, and the creation of algorithms. The invention of calculus can thus be seen to lie in these three contributions.

Newton and Leibniz realized that a whole variety of problems concerning the calculation of centres of gravity, areas, volumes, tangents, arclengths, the radii of curvatures, and so on, which had occupied mathematicians in the first half of the seventeenth century, could in fact be traced back to two basic problems (the calculation of the tangent to a plane curve and the calculation of the area of the surface subtended by a plane curve). Furthermore, they fully realized that these two problems were one the inverse of the other (and this is the 'fundamental theorem' of the calculus). They understood that the solution of the former, and easier, problem could be used to answer the latter. Last but not least, Newton and Leibniz developed two efficient algorithms that could be applied in a systematic and general way. It is through these contributions that Newton and Leibniz overcame the limits of precalculus methods.

8.6.2 Newton

Newton conceived of geometrical magnitudes as generated by a continuous flow in time. For instance, the motion of a point generates a line, and the motion of a line generates a surface. The quantities x, y, z, generated by flow are called fluents. Their instantaneous speeds $\dot{x}, \dot{y}, \dot{z}$, are called fluxions. The moments of the fluent quantities are the infinitely small additions by which those quantities increase in each infinitely small interval of time. By conceiving of a plane curve as generated by the motion of a point $P(x, y)$, where x and y are Cartesian coordinates, Newton determined the slope of its tangent as the ratio between $\dot{y}$ and $\dot{x}$; by generalizing the results contained in the appendices to the Latin edition of Descartes' *Géométrie*, he then found the relevant algorithm. This algorithm contains the rules for the differentiation of the sum $(x + y)$, product (xy), power (x^n), and quotient $(1/x)$, as well as the chain rule (Guicciardini 2009).

Newton deployed several techniques of power (even fractional power) expansion of the fluent quantities. The most famous is the series for the binomial elevated to a fractional power that he obtained, generalizing Wallis's results in *Arithmetica infinitorum*, during the winter of 1664–5. Infinite series allowed Newton to express transcendental functions and integrate by termwise integration. A few examples illustrate this fact.

Newton's binomial series is:

$$(1+x)^{m/n} = 1 + \frac{m}{n}x + \frac{1}{1 \cdot 2}\frac{m}{n}\left(\frac{m}{n} - 1\right)x^2$$
$$+ \frac{1}{1 \cdot 2 \cdot 3}\frac{m}{n}\left(\frac{m}{n} - 1\right)\left(\frac{m}{n} - 2\right)x^3 + \cdots . \quad (8.11)$$

Application of the binomial series to negative exponents leads to interesting results that had escaped Wallis. Most notably, Newton wrote,

$$(1+x)^{-1} = 1 - x + x^2 - x^3 + x^4 - \cdots , \quad (8.12)$$

a result that he considered valid when x is small. Newton studied the hyperbola $y = (1+x)^{-1}$ for $x > -1$. He knew that the area under the hyperbola and over the interval $[0, x]$ for $x > 0$ (and the negative of this area when $-1 < x < 0$) is $\ln(1+x)$. By term-wise integration he could express $\ln(1+x)$ as a power series:

$$x - x^2/2 + x^3/3 - x^4/4 + x^5/5 - \cdots . \quad (8.13)$$

Newton somewhat intuitively understood that this series converges for $|x| < 1$ and $x = 1$. In this period, questions regarding the convergence of infinite series were approached without any general theory of convergence. Mathematicians were simply happy to verify by application to numerical examples that the series (8.13) converged when the absolute value of x was smaller than 1. These series allowed Newton to calculate logarithms, he extended his numerical calculations to more than fifty decimal places! Newton obtained also power series representations of the trigonometric functions such as

$$\arcsin x = x + \frac{1}{6}x^3 + \frac{3}{40}x^5 + \frac{5}{112}x^7 \cdots . \quad (8.14)$$

Infinite series were a basic ingredient for squaring curves (that is, integrating functions). Another approach that Newton adopted consisted in applying the fundamental theorem of the calculus. In modern terms, one might say that Newton was aware of the fact that antiderivatives are related to definite integrals through the fundamental theorem of calculus and provide a convenient means of tabulating the integrals of many functions. Further, Newton listed many functions whose area he could determine via methods we now identify as integration by parts and by substitution (Guicciardini 2009).

In the 1670s, for a number of complex reasons Newton somewhat distanced himself from the algorithmic style of his early researches. He became a great admirer of Huygens, whose *Horologium oscillatorium* he read in 1673, and of the ancient geometers. In this context he developed a geometric approach to limiting procedures. He termed this geometric approach the 'method of first and last ratios'. This method is based on postulates or lemmas concerning the limits of ratios and sums of 'vanishing magnitudes', and its purpose is to enable the determination of tangents to curves and the calculation of areas of curvilinear surfaces by geometrical arguments based on limiting procedures. The method of first and last ratios played a prominent role in Newton's *Principia* and is best illustrated through an example.

One of the main problems that Newton broached in the *Principia* was the mathematical treatment of central force motion. In order to deal with central forces by using geometrical methods, a geometrical representation of such forces is required. This result is not easy to attain, since the central force applied to an orbiting body changes continuously, both in strength and direction. Before Newton's work, mathematicians were able to tackle the problem of rectilinear accelerated motion and circular uniform motion. Newton's intuition was that locally one could approximate the trajectory of a body acted upon by a central force either as a small Galilean parabola traversed in a constant force field or as a small circular arc traversed with constant speed v. Therefore, he locally applied either Galileo's law of fall or Huygens' law of circular motion. The second approach required the determination of the radius of curvature ρ of the orbit so that the normal component F_N of the central force F is equal to $F_N = mv^2/\rho$.

Let us see how Newton applied Galileo's law of fall to central force motion (Guicciardini 1999). He does so in Proposition 6, Book 1, of the *Principia*. This proposition applies an hypothesis that had first been suggested to Newton by Robert Hooke. The planet P is accelerated in void space by a central force, and its motion, as Hooke had suggested, is decomposed into an inertial motion along the tangent and an accelerated motion toward the force centre, the Sun S. We thus have a body accelerated by a centripetal force directed toward S (the centre of force) and that describes a trajectory like the one shown schematically in Fig. 8.9. PQ is the arc traversed in a finite interval of time. The point Q is fluid in its position on the orbit, and one has to consider the limiting situation when points Q and P come together. Line ZPR is the tangent to the orbit at P. QR, the deviation from inertial tangential motion, tends to become parallel to SP as Q approaches P. QT is drawn normal to SP. In Lemma 10 Newton states that 'at the very beginning of the motion' the force can be considered constant. In the case represented in Fig. 8.9 this implies that *as Q approaches P the displacement QR is proportional to force times the square of time.*

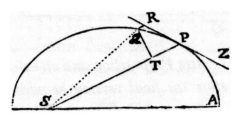

Fig. 8.9. Central force motion in Proposition 6, Book 1, as presented in the first edition of *Principia*. *Source*: Newton, *Philosophiae Naturalis Principia Mathematica* (1687), p. 44.

Indeed, in the limiting situation, QR can be considered as a small Galilean fall caused by a constant force.

Newton could now obtain the required geometrical representation of force. Since Kepler's area law holds for central force motion, as Newton has proven in Proposition 1, the area of SPQ is proportional to time. Moreover, SPQ can be considered a triangle, since the limit of the ratio between the vanishing chord PQ and arc $\widehat{PQ}$ is 1. The area of triangle SPQ is $(SP \cdot QT)/2$. Therefore, the geometrical measure of force is

$$F \propto \frac{QR}{(SP \cdot QT)^2}, \tag{8.15}$$

where the ratio has to be evaluated in the limiting situation where points P and Q come together and $\propto$ is used to mean 'is proportional to'.

Proposition 6 is a good example of the application of the method of first and ultimate ratios. The limit to which the ratio $QR/(SP \cdot QT)^2$ tends is to be evaluated by purely geometrical means. Note that SP remains constant as Q tends to P; therefore one has to consider the limit of the ratio QR/QT^2.

Newton proved that when the trajectory O is an equiangular spiral (in polar coordinates $\ln r = a\theta$) and S is placed at the centre, as Q tends to P,

$$QR/QT^2 \propto 1/SP, \tag{8.16}$$

and thus the force varies inversely with the cube of distance. Whereas, when the trajectory O is an ellipse and the centre of force lies at its centre (Proposition 10), $QR/QT^2 \propto SP^3$, that is, the force varies directly with distance.

Newton considered Keplerian orbits. In Proposition 11, Book 1, of the *Principia* Newton proved that if the body describes a trajectory O, O is an ellipse and the force is directed toward a focus S, then the force varies inversely as the square of the distance. In Propositions 12 and 13, Newton showed that the force is also inverse-square if O is a hyperbola or parabola. To conclude: when the orbit is a conic section and S is placed at one focus,

$$QR/QT^2 \propto 1/L, \tag{8.17}$$

where L is a constant (the *latus rectum*), and the ratio QR/QT^2 is evaluated, as always, as the first or last ratio with the points Q and P coming together. Therefore, the strength of a force that accelerates a body obeying the first two Keplerian laws varies inversely as the square of distance. This is often considered to be the birth of gravitation theory, even though, as experts know, Propositions 11–13, Book 1, played a limited role in Newton's deduction of universal gravitation from astronomical phenomena (Guicciardini 1999).[13]

8.6.3 Leibniz

During his productive stay in Paris (1672–6) Leibniz became acquainted with cutting-edge mathematical research (Hofmann 1974). Following the advice of Huygens, who was one of the most eminent members of the Académie des Sciences, the young German diplomat devoted his attention to many recent mathematical works, including treatises by Pascal, Torricelli, Wallis, and Barrow. By the end of his Parisian period, Leibniz had developed differential and integral calculi, which he published starting in 1684 in a journal he had helped found, *Acta Eruditorum*. In 1684 he published the rules of the differential calculus ($d(x+y) = dx + dy$, $d(xy) = xdy + ydx$, $d(x/y) = (ydx - xdy)/y^2$, and so on), and applied them to some geometrical problems. Most notably, the slope of the tangent of a plane curve whose Cartesian coordinates are x and y was determined by the ratio dy/dx. Two years later, Leibniz published his first paper on the integral calculus, a technique that enabled him to solve problems of quadrature, cubature, rectification, and so on, in a systematic way. Leibniz's calculus aroused criticism from geometric purists such as Vincenzo Viviani and Huygens, and from mathematicians such as Bernard Nieuwentijt, Michel Rolle, and Detlev Clüver, who were worried by Leibniz's puzzling use of symbols for infinitesimal magnitudes, differentials (dx), and even higher-order differentials ($d^n x$). The calculus, however, also had its partisans: initially in Basel, with Jacob and Johann Bernoulli, and later in Paris, with Guillaume F. de L'Hôpital and Pierre Varignon.

Pierre Varignon set himself the task of translating some of the propositions of Newton's *Principia* concerning central force motion and motion in resisting media into calculus terms. In 1700 he published three papers in which he formulated the relation between force F, speed v, and displacement x described in Proposition 39, Book 1, of Newton's *Principia*

$$2 \int F dx = v^2, \qquad (8.18)$$

and the expression of central force in terms of differentials (r, θ, polar coordinates, s trajectory's arc-length),

$$F = \frac{ds d^2 s}{dr dt^2}. \qquad (8.19)$$

---

[13] For Newton's deduction of universal gravitation, see Chapter 5 of this volume, 'Newton's *Principia*', by Smeenk and Schliesser.

He applied formulas (8.18) and (8.19) to the treatment of central force motion (Blay 1998, 108–130).[14] Nowadays we would understand equation 8.18 as the work-energy theorem.

The importance of these translations of Newton's geometrical results into calculus terms cannot be overestimated. There were two main advantages to the application of the calculus rather than geometry to the science of motion. First. As the readers of the *Principia* will know well, the geometric methods that Newton employed can often be applied only to the specific problem for which they were devised. Varignon showed that the results obtained in calculus terms were applicable to whole classes of problems. Second. The calculus allowed a systematic handling of higher-order differentials. Newton was often at pains to classify the infinitesimal order of vanishing geometrical magnitudes (a simple example will suffice to illustrate this point: in Fig. 8.9, QR is a second-order infinitesimal, whereas QT is a first-order infinitesimal, so that the limit of eq. (8.15) for Q tending to P is finite). It should also be noted that despite his preference for geometrical methods, in the *Principia*—and particularly its most advanced parts—Newton himself often employed calculus tools such as integration techniques, the calculation of the radii of curvatures, and infinite series expansions.

Leibniz, Varignon, the Bernoullis, Jacob Hermann, and a group of Italian mathematicians that included Jacopo Riccati and Gabriele Manfredi all proved aware of the advantages of the calculus approach to mechanics, the science of motion, and optics. The calculus problems faced by Leibnizians and their British rivals—including Brook Taylor, David Gregory, Roger Cotes, and Colin Maclaurin—at the beginning of the eighteenth century were often inspired by applications to physical situations. The cases of the brachistochrone and catenaria have already been mentioned. The repertoire of curves that emerged from the solution of differential equations ('inverse tangent problems') soon increased. These mathematicians vied for the definition of the shape of a loaded beam (the elastica) or of a sail inflated by the wind (the velaria), and began studying oscillations of extended bodies, such as those of vibrating strings. The solutions to these problems were hardly of much use to engineers interested in designing ships—for instance—or suspension bridges, yet they showed that the calculus could be conceived of as a promising tool for dealing with rather complex physical facts in mathematical terms. The power of calculus in dealing with transcendental curves and in integrating classes of differential equations was greatly appreciated. Jacob Bernoulli was so fascinated with the logarithmic spiral ($\ln(r/a) = b\theta$), the properties of which he studied in a *tour de force* of mathematical ingenuity in the early 1690s, that he requested this curve be reproduced on his funerary emblem. Bernoulli (and at an earlier date John Collins) also stressed the usefulness of the logarithmic spiral as the stereographic projection to the tangent plane at the pole of the loxodrome: the curve crossing each meridian at a constant angle θ that was sought by mathematicians interested in navigation, such as Pedro Nuñez,

---

[14] As customary in the literature of this period, the constant mass term was not made explicit: equations such as (8.18) and (8.19) were still read as proportionalities.

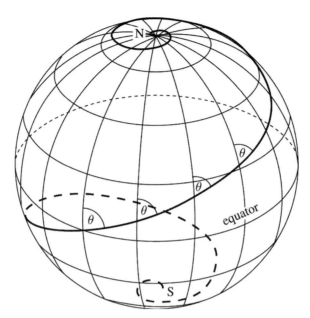

Fig. 8.10. The loxodrome is the curve sought by a mariner who maintains a route at fixed angle with the meridian. *Source*: By Compomat, s.r.l. from Bernoulli, Jacob (1999), p. 340.

Simon Stevin, Willebrord Snel, Gerhard Mercator, Thomas Harriot, and Edmond Halley (Fig. 8.10).

The generality and power of the approach to the science of motion in terms of ordinary differential equations can be appreciated best through an example. In 1711 Johann Bernoulli sent a letter to Varignon in which he discussed a problem that had come to his mind when reading Newton's *Principia* (Bernoulli, Johann (1711)). In Proposition 10, Book 2, Newton had considered a body moving under the action of uniform constant gravity and a resistance proportional to the product of the density of the medium times the square of the speed of the body. Given the trajectory traced by the body, Newton sought to determine the density and velocity at each point of the trajectory (Guicciardini 1999, 233–249). Bernoulli faced a more general problem. He assumed that the trajectory was given, and studied the case in which the body is acted upon by a central force (F) and moves in a medium exerting a resistance (R) proportional to the density (ς) times some power of the speed ($R = \varsigma v^n$). In Bernoulli's case, the density is known and one seeks to determine the central force.

Bernoulli employed formulas to represent central force in calculus terms which were well established by 1711. He stated that the intensity of the tangential component (F_T) of the central force is given by:

$$F_T = F \frac{dr}{ds} = \frac{v^2}{\rho} \frac{dr}{rd\theta}, \qquad (8.20)$$

where r and θ are the polar coordinates, v is the speed (a scalar quantity), ds the infinitesimal arc, and ρ the radius of curvature.[15]

In order to write the differential equation of motion, Bernoulli expressed the tangential component of the total force acting on the body by 'adding or subtracting' to the tangential component of central force F_T (equation (8.20)) the force of resistance $R = \varsigma v^n$ 'according whether the body is moving away from the centre [in this case F_T and R have the same sign] or whether it is approaching it [in this case F_T and R have the opposite signs]' (Bernoulli, J. 1711, p. 503). According to Bernoulli's sign conventions, the tangential component of the total force acting on the body multiplied by $dt = ds/v$ is equal to $-dv$. Therefore, setting $dy = rd\theta$, Bernoulli writes:

$$-dv = F_T dt \pm R dt = \frac{v\, dr\, ds}{\rho\, dy} \pm \varsigma v^{n-1} ds, \qquad (8.21)$$

or

$$\frac{dv}{v} + \frac{dr\, ds}{\rho\, dy} \pm \varsigma v^{n-2} ds = 0. \qquad (8.22)$$

And 'since the relation between dr, dy, ds, ρ and ς is given in terms of r and y', setting $p = ds/(\rho dy)$ and $q = \varsigma ds/dr$,[16] one has the following differential equation:

$$\frac{dv}{v} + p(r)dr \pm v^{n-2} q(r) dr = 0. \qquad (8.23)$$

This is the non-linear (for $n \neq 1, 2$) differential equation that Jacob Bernoulli had already solved in 1695, while Johann had proposed an alternative integration method in 1697. By integrating eq. (8.23), Johann obtained v and, since $F = (v^2/\rho)(ds/dy) = p(r)v^2$,

$$F = p(r)e^{-2\int p(r)dr} \left(\sqrt[2-n]{\mp (2-n)\int e^{-(n-2)\int p(r)dr} q(r) dr} \right)^2. \qquad (8.24)$$

Similar results were achieved by Verzaglia, Varignon, and—in a more creative and independent way—Hermann (Mazzone & Roero 1997).[17]

One cannot but appreciate the generality of Johann Bernoulli's approach to the motion of projectiles in resisting media based on the theory of differential equations.

---

[15] $F = (F_N ds)/(rd\theta) = (v^2/\rho)(ds/rd\theta)$. Note that there is no normal component of the resistive force: it is thus legitimate to set $F_N = v^2/\rho$. As customary, the mass term m is 'absorbed' in these prototype equations/proportions so that, to all effects, $m = 1$.

[16] Both p and q are functions of r and y, but one of the variables can be eliminated by employing the equation of the curve which is given.

[17] It is worth briefly illustrating Johann Bernoulli's method to solve eq. (8.23). Cases $n = 1$ and $n = 2$ considered by Newton (resistance proportional to v or v^2) lead to an easy linear differential equation. For the non-linear cases Bernoulli searched the solution as a product $v = M \times N$. Substitution in the differential equation leads to

$$\frac{dM}{M} + \frac{dN}{N} + p(r)dr \pm M^{n-2} \times N^{n-2} q(r) dr = 0.$$

In their treatment of the matter, Leibnizian mathematicians deployed the wealth of results reached by the Basel school in the field of differential equations and the calculus of transcendental magnitudes. Their approach proved to be more successful compared to the one followed by the British Newtonians. When in 1718 John Keill proposed as a challenge against the Leibnizians the solution of the inverse problem of resisted motion—that is, the problem of determining the trajectory of a body accelerated by constant gravity and by a resistance proportional to density times the square of velocity—Johann Bernoulli replied in papers published in 1719 and 1721 by solving an even more general problem: that of determining the trajectory of a body resisted by a force proportional to density times some power $2n$ of the velocity. This problem had indeed already been solved by Hermann in his *Phoronomia* (1716), due to the use of the differential equation solved by Bernoulli in 1711 (Blay 1992, pp. 322–30).

Since M and N are arbitrary functions, Bernoulli can set

$$\frac{dM}{M} + p(r)dr = 0,$$

which leads to

$$M = e^{-\int p(r)dr}.$$

For this value of M the equation is transformed into

$$\frac{dN}{N} \pm e^{-(n-2)\int p(r)dr} \times N^{n-2} q(r)dr = 0,$$

and therefore

$$\frac{dN}{N^{n-1}} = \mp e^{-(n-2)\int p(r)dr} q dr,$$

which is in integrable form. One obtains the integral as

$$\frac{1}{2-n} N^{2-n} = \mp \int e^{-(n-2)\int p(r)dr} q(r) dr.$$

From this the value for N is obtained:

$$N = \sqrt[2-n]{\mp (2-n) \int e^{-(n-2)\int p(r)dr} q(r) dr}.$$

And multiplying M by N one obtains v:

$$v = e^{-\int p(r)dr} \left(\sqrt[2-n]{\mp (2-n) \int e^{-(n-2)\int p(r)dr} q(r) dr} \right).$$

8.7 Enlightened Mechanics

I began my chapter by stating that the idea of a 'scientific revolution' seems justified from the point of view of the historian of mathematics: one need only compare the qualitative geometric treatment of projectile motion in the works of Tartaglia or dal Monte carried out in the sixteenth century to the calculus approach illustrated in Johann Bernoulli's solution of eq. (8.21) in 1711.

The achievements of the generation immediately following Leibniz and Newton's (that of Varignon, Jacob and Johann Bernoulli, and Hermann on the Continent, and of Cotes, Stirling, and Taylor in Britain) made a lasting impact on the mathematical sciences. Ordinary differential equations became the language employed for the mechanics of point masses, as evidenced by Leonhard Euler's *Mechanica* (1736). The need to extend mathematical methods to the study of continuous, rigid, elastic, and fluid bodies led to the development of partial differential equations. Extremal principles, such as the least action principle, were framed in terms of the calculus of variations, bringing abstraction one step further.

The establishment of calculus as the language of what came to be described as 'analytical mechanics' had consequences that cannot be overestimated. The laws of motion came to be expressed through equations instead of proportions, as had still been the case in the works of Huygens and Newton. Dimensional constants (for example, with a dimension equal to length divided by the square of time) thus became a feature of the equations of eighteenth-century mathematicians, making the expression of explicit solutions much easier. More fundamentally, the calculus enabled the adoption of a systematic approach to the science of motion, making the algorithmic control of approximations possible, as illustrated by the works on planetary perturbations that were carried out in the mid-eighteenth century by d'Alembert, Euler, and Lagrange. Indeed, the series expansions included in the works of these authors on planetary motions enabled systematic control of the order of approximation that could be carried on, say, up to the level of a chosen power of the orbit's eccentricity.

All these advantages should not make us forget that progress came with a price (Gingras 2001). In the first place, the mathematical sciences became the province of professional experts—gifted mathematicians who acquired extraordinary prestige in Continental academies. While Galileo's *Discorsi* could have been read by all cultivated readers interested in natural philosophy, Newton's *Principia*, and even more so the eighteenth-century treatises of Clairaut, Euler, and Lagrange, could be understood by only a handful of well-trained experts in the field.

Secondly, the explanatory power of mathematical models—their ability to describe the causes of natural phenomena—was weakened, according to some critics, by the abstraction of symbolism. For mathematicians practicing the analytic mechanics in vogue at the time—especially in Continental academies—solving a problem meant producing a technique for the integration of a differential equation, rather than any explanatory model.

Thirdly, when considering eighteenth-century analytical mechanics we tend to forget how remote from applications that sort of pure mathematics was: before the expansion of mathematical science to heat, magnetism, light, and so on, as described by Kuhn, mathematical physics was of interest at most for the astronomer (Kuhn 1977). In fact, the mechanical problems faced by, say, Johann Bernoulli or by Euler were pursued for the sake of their mathematical interest, even when they were motivated by mechanical or geometric applications. It has often been said—and with good reason—that analytical mechanics provided the main stimulus for eighteenth-century mathematicians. However, problems which aroused much interest, such as the brachistochrone and the vibrating string, were only faintly connected with applications; they were rather pursued because of their mathematical interest. As Terrall notes:

The most rigorous analytical mathematics that filled the pages of the [French] academy's journal was not necessarily useful in terms of practical applications or public displays. The equations of celestial mechanics or hydrodynamics did not translate easily into engineering design or industrial production. (Terrall 1999, 247)

Again, the example of Johann Bernoulli's approach to projectile resisted motion can illustrate the point. Bernoulli's equation (8.21) remained fundamental for the study of exterior ballistic. However, much experimental and theoretical work was still to be done in order to reach a realistic description of the drag force that could be used with some success in practical applications. Indeed, it was only at the end of the nineteenth century that the equations of fluid dynamics could be used on the battlefield.

In the seventeenth century, mathematical giants such as Huygens and Newton preferred to publish their results in a geometrical language. By contrast, few mathematicians resisted the new drive towards abstraction and symbolism in the eighteenth century: it was left to outsiders such as Louis Bertrand Castel to oppose the mathematical abstractions of the powerful *académiciens*. The turn towards analytical mechanics proved irreversible.

Selected References

Andersen, Kirsti (1994). 'Precalculus, 1635–1665'. In: *Companion Encyclopedia of the History and Philosophy of the Mathematical Sciences*, edited by Ivor Grattan-Guinness, pp. 292–307. London: Routledge.

Bennett, James A. (1986). 'The Mechanics' Philosophy and the Mechanical Philosophy'. *History of Science* 24, 1–27.

Bernoulli, Jacob (1999). *Die Werke von Jacob Bernoulli. Band 5 Differentialgeometrie*. Edited by André Weil and Martin Mattmüller. Basel: Birkhäuser.

Bernoulli, Johann (1711) [with Nicolaus I Bernoulli]. 'Extrait d'une Lettre de M. Bernoulli, écrite de Basle le 10. Janvier 1711, touchant la maniere de trouver les forces centrales dans les milieux resistans en raison composée de leur densités & des puissances quelconques des vitesses du mobile. Addition de M. (Nicolas) Bernoulli, Neveu de l'Auteur de ce Memoire-cy'.

Histoire de l'Académie Royale des Sciences avec les Mémoires de Mathématiques et de Physique, 47–56.

BERTOLONI MELI, DOMENICO (1992). 'Guidobaldo dal Monte and the Archimedean revival'. *Nuncius*, 7, 3–34.

BERTOLONI MELI, DOMENICO (2006). *Thinking with Objects: The Transformation of Mechanics in the Seventeenth Century*. Baltimore: Johns Hopkins University Press.

BESSON, JACQUES (1594). *Théâtre des Instrumens Mathématiques et Méchaniques*. Genève/Lyon: J. Chouët.

BIAGIOLI, MARIO (1989). 'The social status of Italian mathematicians, 1450–1600'. *History of Science* 27, 41–95.

BLAY, MICHEL (1992). *La naissance de la mécanique analytique: la science du mouvement au tournant des XVIIe et XVIIIe siècles*. Paris: Presses Universitaires de France.

BLAY, MICHEL (1998). *Reasoning with the Infinite: From the Closed World to the Mathematical Universe*. Chicago: University of Chicago Press.

BOS, HENK J. M. (1986). Introduction [to] Huygens, Christiaan. *The Pendulum Clock or Geometrical Demonstrations Concerning the Motion of Pendula as Applied to Clocks*. Ames: Iowa State University Press: i–xxxii.

BOS, HENK J. M. (2001). *Redefining Geometrical Exactness: Descartes' Transformation of the Early Modern Concept of Construction*. New York: Springer.

BRAUNMÜHL, ANTON VON (1892). 'Historische Studie über die organische Erzeugung ebener Curven von den ältesten Zeiten bis zum Ende des achtzehnten Jahrhunderts'. In *Katalog mathematischer und mathematisch-physikalischer Modelle, Apparate und Instrumente*, edited by Walther Dyck, pp. 54–88. München: Universitäts-Buchdruckerei.

BURNETT, D. GRAHAM (2005). *Descartes and the Hyperbolic Quest: Lens Making Machines and their Significance in the Seventeenth Ccentury*. Philadelphia, PA: American Philosophical Society.

CLAVELIN, MAURICE (1974). *The Natural Philosophy of Galileo: Essay on the Origins and Formation of Classical Mechanics*. Translated by A. J. Pomerans. Cambridge, MA: MIT Press.

DAMEROW, P., FREUDENTHAL, G., MCLAUGHLIN, P., RENN, J., (2004). *Exploring the Limits of Preclassical Mechanics: A Study of Conceptual Development in Early Modern Science, Free Fall and Compounded Motion in the Work of Descartes, Galileo, and Beeckman*. New York and London: Springer.

DESCARTES, RENÉ; (1954). *The Geometry of René Descartes with a Facsimile of the First Edition*. Edited and translated by D. E. Smith and M. L. Latham. New York: Dover.

DESCARTES, RENÉ (1964–67). *Oeuvres de Descartes*. Edited by C. Adam and P. Tannery. 12 vols. New ed. Paris: Vrin.

DRAKE, STILLMAN (1995). *Galileo at Work: His Scientific Biography*. New York: Dover Publications.

FEINGOLD, MORDECHAI (1984). *The Mathematicians' Apprenticeship: Science, Universities and Society in England, 1560–1640*. Cambridge: Cambridge University Press.

GALILEI, GALILEO (1974). *Two New Sciences, including Centres of Gravity & Force of Percussion*. Translated, with introduction and notes, by Stillman Drake. Madison: University of Wisconsin Press.

GINGRAS, YVES (2001). 'What Did Mathematics Do to Physics?', *History of Science* 39, 383–416.

GIUSTI, ENRICO (1993). *Euclides reformatus: la teoria delle proporzioni nella scuola galileiana*. Torino: Bollati Boringhieri.

GUICCIARDINI, NICCOLÒ (1999). *Reading the Principia: The Debate on Newton's Mathematical Methods for Natural Philosophy from 1687 to 1736*. Cambridge: Cambridge University Press.

GUICCIARDINI, NICCOLÒ (2009). *Isaac Newton on Mathematical Certainty and Method.* Cambridge (Mass.): MIT Press.

HERMANN, JACOB (1716). *Phoronomia, sive de viribus et motibus corporum solidorum et fluidorum libri duo.* Amsterdam: R. & G. Wetstenios.

HUYGENS, CHRISTIAAN (1673). *Horologium Oscillatorium: Sive, De Motu Pendulorum ad Horologia Aptato Demonstrationes Geometricae.* Paris: F. Muguet.

JARDINE, NICHOLAS (1997). 'Keeping Order in the School of Padua: Jacopo Zabarella and Francesco Piccolomini on the Offices of Philosophy'. In: *Method and Order in Renaissance Philosophy of Nature, The Aristotle Commentary Tradition,* edited by Daniel A. Di Liscia, Eckhard Kessler, and Charlotte Methuen, pp. 183–209. Aldershot, U.K.: Ashgate.

JESSEPH, DOUGLAS M. (2007). 'Descartes, Pascal, and the epistemology of mathematics: the case of the cycloid'. *Perspectives on Science* 15, pp. 410–33.

HOFMANN, JOSEPH E. (1974). *Leibniz in Paris 1672–1676: His Growth to Mathematical Maturity.* Cambridge: Cambridge University Press.

KEPLER, JOHANNES (1627). *Tabulae Rudolphinae, quibus astronomicae scientiae, temporum longinquitate collapsae restauratio continetur.* Ulm: J. Saur.

KOYRÉ, ALEXANDRE (1968). *Metaphysics and Measurement: Essays in Scientific Revolution.* London: Chapman & Hall.

KUHN, THOMAS (1957). *The Copernican Revolution: Planetary Astronomy in the Development of Western Thought.* Cambridge [MA]: Harvard University Press.

KUHN, THOMAS (1977). 'Mathematical versus Experimental Traditions in the Development of Physical Science'. In: *The Essential Tension: Selected Readings in Scientific Tradition and Change,* pp. 31–65. Chicago and London: University of Chicago Press.

LENCKER, HANS (1571). *Perspectiva hierinnen auffs kürtzte beschrieben.* Nürnberg: Dietrich Gerlatz.

MAZZONE, SILVIA & ROERO, CLARA S. (1997). *Jacob Hermann and the Diffusion of the Leibnizian Calculus in Italy.* Firenze: Olschki.

NAPOLITANI, PIER DANIELE (1997). 'Le edizioni dei classici: Commandino e Maurolico'. In: *Torquato Tasso e l'Università,* edited by Walter Moretti and Luigi Pepe, pp. 119–42. Florence: Olschki.

NEWTON, ISAAC (1687). *Philosophiae Naturalis Principia Mathematica.* London: J. Streater.

PICCOLOMINI, ALESSANDRO (1547). *In mechanicas quaestiones Aristotelis, paraphrasis paulo quidem plenior . . . Eiusdem commentarium de certitudine mathematicarum disciplinarum.* Roma: A. Blado.

RAYNAUD, DOMINIQUE (2007), 'Le tracé continu des sections coniques à la Renaissance: Applications optico-perspectives, héritage de la tradition mathématique arabe'. *Arabic Sciences and Philosophy* 17(2), 299–345.

RYFF, WALTHER HERMANN (1547). *Der furnembsten, notwendigsten, der ganzen Architectur angehörigen mathematischen und mechanischen Künst eygentlicher Bericht.* Nürnberg: J. Petreius.

SCRIBA, CHRISTOPH J. (1970). 'The autobiograhy of John Wallis, F.R.S.'. *Notes and Records of the Royal Society of London* 25, 17–46.

STEDALL, JACQUELINE A. (2002) *'A Discourse Concerning Algebra': English Algebra to 1685.* Oxford: Oxford University Press.

TERRALL, MARY (1999). 'Metaphysics, Mathematics, and the Gendering of Science in Eighteenth-Century France'. In: *The Sciences in Enlightened Europe,* edited by William Clark, Jan Golinski, and Simon Schaffer, pp. 246–271. Chicago and London: University of Chicago Press.

Tournès, Dominique (2009). *La construction tractionnelle des équations différentielles*. Paris: Blanchard.

Van Schooten, Frans (1646). *De Organica Conicarum Sectionum in Plano Descriptione Tractatus: Geometris, Opticis, Praesertim Vero Gnomonicis & Mechanicis Utilis*. Leiden: cx Officina Elzeviriorum.

Westfall, Richard S. (2001). 'The Background to the Mathematization of Nature'. In *Isaac Newton's Natural Philosophy*, edited by Jed Z. Buchwald and I. Bernard Cohen, pp. 321–39. Cambridge (Mass.): MIT Press.

Yoder, Joella G. (1988). *Unrolling Time: Christiaan Huygens and the Mathematization of Nature*. Cambridge: Cambridge University Press.

PART II

THE LONG EIGHTEENTH CENTURY

CHAPTER 9

THE PHYSICS OF IMPONDERABLE FLUIDS

GIULIANO PANCALDI

9.1 INTRODUCTION

Around 1800, several natural philosophers admitted the existence of 'imponderable' (weightless) fluids: special fluids which, when added to ordinary, 'ponderable' matter were expected to explain a wide range of phenomena. Depending on the author and the theory, up to six such fluids had been introduced to account for the phenomena of cohesion, chemical affinity, heat, light, electricity, magnetism and, according to some, also life, or at least 'galvanism', or animal electricity.

Considering that at about the same time many authors—including some who subscribed to imponderables—were striving to apply experimental, possibly quantitative methods to all branches of natural philosophy, and that chemists were presenting the balance as the key instrument of a new science of the micro-world, the appeal to imponderables already seemed puzzling to a number of contemporaries. However, imponderables prospered and they continued to flourish until well into the nineteenth century—long enough to pose substantial interpretive problems for historians of science up to the present.

The best way to start examining the question of imponderables is probably to consider how they presented themselves to those natural philosophers who, around the middle of the nineteenth century, started questioning their status and legitimacy. Some of the sceptics then tried to understand why the notion had thrived, and why, despite its obvious ambiguities, it had brought with it a conspicuous harvest of empirical results and some very good quantitative science. In the process, sceptics were led to address a number of broader questions, including the role of hypotheses in the physical sciences. Sceptics also came to realize what must have been already apparent to their colleagues of earlier generations. That is, that when discussing imponderables natural philosophers often went beyond trusted, empirical, quantitative, experimental science and entered territories in which philosophy, worldviews, religion, and even politics could claim to have a say.

Joseph Henry, professor of natural philosophy at Princeton, may be chosen to represent the attitude of mid-nineteenth-century sceptics *vis-à-vis* the imponderables. Reflecting on 'so-called' imponderables in 1851, he objected to the appropriateness of the name. However, he found a rationale for continuing to use them, arguing that, by admitting imponderables, scientists could bring within 'the category of the laws of force and motion' several classes of phenomena otherwise not amenable to it. Displaying an instrumentalist penchant quite widespread among supporters of the imponderables, Henry remarked that imponderables did not need to be true to be adopted to explain and predict phenomena. Without assumptions of that kind, Henry insisted, it was often impossible to arrive at the general expressions which constitute science (Henry, 1886: 297).

Known mechanical principles and the imponderables thus coexisted without apparent tensions well into the nineteenth century. As seen by Henry in 1851, however, recent discoveries had revealed the 'intimate connections' of all the phenomena attributed to imponderables—especially heat, light, electricity, and magnetism. The connections led Henry, like several others, to suppose by 'legitimate analogy' that all such phenomena resulted from the action of 'one all-pervading principle' or 'elastic medium' (Henry, 1886: 304). The way out of the uncertainties associated with the imponderables appeared to Henry to be offered by the old hypothesis of a single fluid somehow unifying them all in one special entity: the ether.

It would be unwarranted, however, following Henry, to represent the physics of imponderables from around 1800 as displaying a linear path leading somehow to the theories of the ether that prospered later in the century. In depicting that period, a substantial historical literature has emphasized, rather, a multifaceted story. Interpreters, to begin with, have disagreed conspicuously over the comparative merits and demerits of the main, powerful tradition that put imponderables at the centre of its research programme: Laplacian physics.

To provide an overall picture of the main interpretive issues at stake, I shall begin the present chapter with a discussion of the existing historical literature on the physics of imponderables. I shall then add to the picture by exploring how supporters of imponderables, and sceptics alike, reacted after 1800 to the introduction of new instruments of analysis that followed the introduction of a momentous, new object. The object was the Voltaic battery, announced in 1800, and it was followed by the development of an impressive if disorganized array of new concepts that included sophisticated mathematical tools, unanticipated chemical procedures, and ingenious wood and brass apparatus. The tangled developments accompanying the introduction of these new instruments culminated somehow, after two decades, with the adoption of the notion of electric current.

The goal of the exercise will be to add these new 'instruments of analysis'—Humphry Davy's phrase to describe the battery (Davy, in a manuscript dated 6 August

1800, quoted in Fullmer 2000: 304)—to an assessment of the physics of imponderables and its fortunes during the first two decades of the nineteenth century. From about 1800, I will argue, the new instruments intruded into the domain of imponderables in several unexpected ways, interacting with the attempts made at bringing some order, and possibly a rigorous mathematical treatment, into a highly variegated set of phenomena. I will consider in some detail how the new instruments were addressed by three authors who, according to existing literature, belonged to three quite distinct groups or schools of thought: Jean Baptiste Biot, for the Laplacians, Humphry Davy, for the sceptics, and André-Marie Ampère, for the 'etherians' (Caneva 1980). Artefacts and the connected instruments of analysis, I will argue, have interesting stories to tell about the physics of imponderables, stories that cannot be dissolved entirely into the vicissitudes of different epistemologies, conflicting research schools, and worldviews, or diverse cultural and social contexts.

The new artefacts and the associated instruments of analysis can help us understand why, around 1820, a highly heterogeneous community of natural philosophers, experimenters, and instrument-makers belonging to different countries, mixed walks of life, and diverse religious and social views was brought to recognize some unprecedented electromagnetic phenomena while developing new techniques and interpretive tools that, as in a domino effect, transformed physics in depth. By so doing, that heterogeneous community, unaware, was trying new avenues at the intersection between 'mathematical power, experimental sagacity, and manipulative skill'. An intersection that, a few decades later, James Clerk Maxwell would point at as the place of origin of experimental physics as he knew it, and citizens at large would recognize as the traceable beginnings of a new age: the age of electricity (Maxwell 1871: II, 246; Benjamin 1886).

9.2 Interpreting Laplacian Physics, or the Multiple Uses of Imponderables

As Pierre Simon Laplace put it in 1808, 'the phenomena of nature can be reduced ultimately to actions at a distance between molecules, and these actions must be the basis on which to build a mathematical theory of such phenomena' (Laplace 1809: 338).

According to Laplace, that was demonstrably the case of the attractive and repulsive forces detected in capillarity, electricity, magnetism, and elastic bodies (Laplace 1809: 329). Where no ponderable matter was apparently involved in the phenomena considered, what was needed was to assume that the molecules of as many imponderable fluids were involved, and that they displayed a pattern of action at a distance amenable to mathematical treatment. On the *real* existence of the imponderables Laplace and his followers displayed a range of positions that oscillated from stern declarations of an instrumentalist restraint from the search for true causes, to frequent, more easy-going statements implying the existence of the fluids. The oscillations have puzzled historians of science to the present.

Laplace's programme for terrestrial physics had been conceived as an extension of the successful Newtonian programme that he himself had developed for astronomy in his monumental *Mécanique céleste*. The extension implied Laplace's return, early in the nineteenth century, to a set of interests that he had cultivated in the 1780s in cooperation with Lavoisier. Laplace's new programme for terrestrial physics, however, could not rely on a situation comparable, for consensus, to the one that had characterized positional astronomy since the introduction of Newton's laws of mechanics and the principle of gravitation (Hahn 158–9).

The situation of terrestrial physics early in the nineteenth century must have resembled, rather, the one described in the dismal words of a young follower of Laplace, Jean-Baptiste Biot. In the preface to his *Traité de physique expérimentale et mathématique*, the textbook of Laplacian physics published in 1816, Biot wrote:

All those who have had a chance to carry out researches in several domains of physics must have realized with regret how dispersed the materials of this fine science still are, and how uncertain its general march is. When a result is adopted in one country, a different one is agreed upon in another. If a numerical evaluation is routine in one place, the same is regarded as doubtful and inaccurate in another place. Not even on general principles a universal consensus is attained. If electricity be the subject, the French will argue adopting the two fluids hypothesis, while the British will prefer the idea of one fluid developed by Cavendish, and Volta will make the most brilliant discoveries by admitting the existence of electric atmospheres to which neither will attach a reality whatsoever . . . (Biot 1816: I, ii)

Lack of consensus, a plurality of competing programmes, and substantial national differences: this was the context in which the Laplacians tried to impose their ambitious programme for terrestrial physics, based on the notion of distance forces acting at the molecular level. Despite the daunting task, they could rely—as Robert Fox has shown—on some powerful tools and a few special circumstances (Fox 1973, 1974, 1992).

The tools the Laplacians could avail themselves of were of two kinds. One consisted in appealing to a tradition of assumptions about attractive and repulsive forces at the molecular level that went back to Newton himself, and had been made popular throughout the eighteenth century by British and French Newtonians (Heimann [Harman] 1981; Buchwald 1981; Home 1982; Home 1992; Buchwald 2000; Home 2003). That same tradition had been recently revived within chemistry by the work of Claude Louis Berthollet, one of the two leaders, with Laplace, of the informal but powerful Society of Arcueil (Berthollet 1803; on Arcueil: Crosland 1967). A boost to speculations about molecular forces was also coming, early in the century, from debates over the theory of heat. The camp here was divided into two: those who agreed that the imponderable 'caloric' could exist in a body also in a latent state, that is, without affecting its temperature; and those who denied this (Fox 1974: 106). For both, speculations over what was going on at the molecular level appeared mandatory, thus encouraging the same sort of assumptions that were at the centre of Laplace's programme for terrestrial physics.

Another powerful tool in the hands of Laplace and his followers was mathematics. As John Heilbron has shown, Laplacian physics benefited from, and powerfully contributed to, the early nineteenth-century turn of physicists towards domains traditionally belonging to 'applied mathematics', such as mechanics, optics, and astronomy (Heilbron 1993; see also Heilbron 1981, and Grattan Guinness 1990: II, 440ff). For the many Laplacians who viewed this turn as a fundamental step towards the advent of a new physics, it carried with it a predilection for quantification and measurement that could verge on compulsion.

Of the many examples Heilbron offers to illustrate the 'quantitative spirit' preached by the Laplacians, the long paper published by Biot and Arago in 1806 on the specific gravity and the refractive powers of gases is especially revealing (Biot and Arago, 1806). The paper touched upon the forces acting between common matter and the imponderable fluid of light. It aimed, among other things, at establishing the numerical value of the 'barometric coefficient': a key number for the 'law' describing how the pressure of the atmosphere decreases with height. As Heilbron remarks, in their endeavour Biot and Arago 'worked to six or seven (in)significant figures, corrected in detail for the effects of temperature, pressure, and humidity on their instruments . . .' After several adjustments the coefficient was set at 18,334 metres—a number that 'agreed suspiciously well' (Heilbron 1993: 31) with the coefficient (18,336) established by Louis Ramond de Carbonnières sometime earlier when developing an impressive formula proposed by Laplace himself (Laplace 1805: IV, 290). The formula (Ramond 1811: 265) allowed measurement of heights by using a barometer, and it ran as follows:

$$z = L\left(\frac{h}{H}\right) 18.336^m \left(1 + 0.0028371 \cos 2\Psi\right) \left(1 + \frac{2(t+t')}{1000}\right)$$

$$\left\{1 + \frac{\left(\log\left(\frac{h}{H}\right) + 0.868589\right)\frac{z}{a}}{\log\left(\frac{h}{H}\right)}\right\}$$

In it, z is the vertical distance in metres between two locations, L denotes log, H is $H = h + h'\left(\frac{T-T'}{5412}\right)$, h and T, and h' and T', are the heights and the temperatures of the barometer at the two locations, ψ is latitude, t and t' are the temperatures of the air at the top and bottom of the height to be measured, and a is the Earth's radius, set at 6,366,198 metres. 'People with a taste for heights and calculation were soon measuring all the mountains they could climb, also church towers, flag poles, and the mast of sailing ships'. (Heilbron 1993: 32). The Laplacians' measuring compulsion manifested itself in many other fields, most conspicuously in the French campaign for the adoption of the metric system of weights and measures, as well as in cartography, geodesy, setting the speed of sound, adiabatic processes, and mercantilist mathematics.

The appeal to molecular forces and to the rigour of mathematics were the main but not the only levers on which the Laplacians could count. They relied also on the power—academic and political—of its leaders, and on the many opportunities that

that power made available to them. As Robert Fox has shown, Laplace and Berthollet developed a system of patronage within Napoleonic France that must be taken into account when trying to understand 'the rise and fall' of Laplacian physics (Fox 1974). At its zenith—between 1805 and 1815—the two leaders could count on a rich supply of graduates from the École Polytechnique, whom they supported and encouraged in their careers through the influence they exerted within the major institutions of the country. The school was further sustained through the money that Berthollet and Laplace poured into the experimental researches carried out in the laboratory established in the house of Berthollet in Arcueil, just outside Paris, where Laplace also had his house. Arcueil attracted in succession brilliant young scholars like Jean Baptiste Biot, Étienne Malus, Louis Jacques Thenard, Joseph Louis Gay-Lussac, Siméon Denis Poisson, and François Arago, and was open to every foreign personality prominent in the sciences visiting Paris, including Alexander von Humboldt, Augustin de Candolle, and Alessandro Volta.

The system of patronage developed by Lavoisier and Berthollet helped Laplacian physics to conquer a dominant role in French science. According to Fox, however, it was realized at the cost of establishing an orthodoxy that anchored its adepts to a debatable doctrine of microphysics, and bound them to occasionally artificial displays of mathematical prowess. Worst, the school kept its adepts away from direct confrontation with the experimental results that elsewhere, in the same years, were transforming the physical sciences and opening the road to new fields, such as electrochemistry, or to new concepts, such as energy and energy conservation (Fox 1974: 133–4, 136).

Heilbron conveys a somewhat different assessment of Laplacian physics. Having chosen to focus 'not on the details of the resultant scheme of imponderable fluids, but on its epistemological and pedagogical status, and on the standards of achievement and precision it helped to achieve', Heilbron emphasizes the instrumentalist epistemology and the pedagogy of precision pursued by the Laplacians, and he sees their accomplishments as important steps in the story of the 'quantifying spirit' (Frängsmyr, Heilbron, Rider 1990; Heilbron 1993: 4).

Comparatively less attention has been devoted by historians to the interactions between the basic tenets of Laplacian physics and their potential implications for religion and society. Laplace—who as a young man had been trained for the Church, and had then adopted the secular philosophies of the Enlightenment—was well aware of such implications. He expressed his views in some reflections, not meant for publication, in which he treated the imponderable fluids, 'intellectual qualities', and 'the substances termed spiritual' side by side. Roger Hahn has published these private reflections of Laplace, and they are worth quoting here to introduce the theme of the multiple uses of imponderables:

Matter, which several philosophers view as inert and passive, develops a prodigious activity and astonishing properties in such imponderable fluids as light, heat, magnetism, and electricity. Does not all this seem to indicate common qualities and special properties in all these entities? Are there not between them essential differences based on their nature, and on which the great differences in their behaviour depend? Could it not be that the intellectual qualities perceived by the inner sense belong to them, in such a way that the substances termed spiritual are among those whose effects are disclosed by our outer senses, and whose intimate properties are revealed to us by the inner sense when these properties are placed in circumstances favourable to the development of these properties? (Laplace in Hahn 2005: 175–6, 186–7)

Such highly speculative questions in which Laplace indulged privately could bring the notion of imponderables to bear on issues that were being explored in those years by some of his contemporaries interested in psychology, such as Pierre Cabanis and François Magendie. They had implications for religion as well. Laplace's speculations offered him naturalistic explanations for spiritual substances and for the 'illusions' and 'myths' of traditional religion, which he had discussed in those same private notes. That imponderables could be made to bear on such issues was often mentioned during the first half of the nineteenth century. However, if Laplace was inclined to use imponderables to materialize spiritual substances, others could be interested in trying the other way round.

Most explicit in the latter direction was William Whewell in 1833. In his Bridgewater Treatise devoted to astronomy and general physics he argued in favour of the imponderables, or 'ethers', by placing them at the centre of a world that had been ostensibly designed by the inscrutable 'will and power of one supreme Being, acting by fixed laws' (Whewell 1833: 138–141, 372). Whewell portrayed the luminiferous ether as a complex 'machine', the atmosphere being another such machine. These machines had been set by the supreme being to mediate between mind and matter. The same supreme being was evoked in Joseph Henry's already mentioned assessment of 'so-called' imponderables. Henry supported his instrumentalist epistemology about imponderables by arguing that 'man, with his finite faculties, cannot hope in this life to arrive at a knowledge of absolute truth; and were the true theory of the universe, or in other words, the precise mode in which the Divine Wisdom operates in producing the phenomena of the material world, revealed to him, his mind would be unfitted for its reception' (Henry 1851: 300).

Mediating somehow between mind and matter, as imponderables were supposed to do according to many nineteenth-century observers, we cannot be surprised to find that social and political, as well as religious implications, were attributed to them. Physicists dealing with the imponderables were inclined to explore—and occasionally exploit, privately or publicly—this kind of implication; a circumstance no more surprising than the fact that they shared substantial traits of the cultures to which they were contributing (Wynne 1979, Shapin 1980, Cantor 1981, Noakes 2005, Morus 2009). Worth noticing, rather, is the ease with which imponderables could be bent to serve quite contrasting agendas, like those pursued by Laplace and Whewell.

Writers aiming at more popular audiences than those cultivated by Laplace or Whewell had even fewer problems in accommodating imponderables to their own cherished visions of religion and society. That was remarkably the case of Thomas Dick, a Scot close to evangelicalism, a popular science writer, and a philanthropist. To Dick, Laplace was a 'patron of infidelity' (Astore 2001: 87). Yet in his *The Christian*

Philosopher, first published in 1823 and for some forty years a bestseller in Britain and the USA, Dick used the imponderables and his own amateur curiosity for astronomy (including the nebular hypothesis and the plurality of worlds), electricity, and galvanism, to convey a religious view of science, technology, and society that became popular on both sides of the Atlantic (Dick 1826, Astore 2001: 142). Recent achievements in the science and technology of electricity, chemistry, and magnetism were for Dick 'inseparably connected with the propagation of Christianity through the world'. He clearly associated the special nature and universality of the imponderables with the universalistic messages he wanted to convey, which included, besides Christian religion, the improvement of the middle classes, pacifism, and the abolition of slavery (Dick 1823: 248–250).

Having surveyed the uses of imponderables, and the differences in the historians' judgement about their worth and implications, we focus now on a side of the story that has received comparatively less attention by historians.

9.3 Instruments to Catch the Imponderables

By 1800 an impressive array of instruments and methods that promised to catch, display, or measure the imponderables was already available. A selective survey carried out by John Heilbron (Heilbron 1993: chapt. 2) mentions the following impressive list of instruments and procedures used in the physics of imponderables around 1800.

Biot's and Arago's method for measuring the refraction of a gas; Malus' system for obtaining the index of refraction of Iceland spar; Coulomb's apparatus for measuring the force between magnetic poles; Robison's apparatus for measuring the force between small charged spheres; Coulomb's electrical balance; electrometers of various kinds; the same for electroscopes; Volta's apparatus for showing the influence of geometry on electrical capacity; a long series of thermometers using mercury or gases; Cavendish's and Nollet's procedures for fixing the boiling point of water; Gay-Lussac's apparatus for determining the coefficient of thermal expansion of gases; Dulong's and Petit's apparatus for measuring the expansion of mercury; Lavoisier's and Laplace's ice calorimeter; Delaroche's and Bérard's, and Clément's and Desormes' apparatus for measuring the specific heat of gases; Rumford's differential thermoscope; John Leslie's differential thermometer; William Herschel's set-up for investigating the heating effect of rays beyond the red end of the spectrum; Herschel's apparatus for studying the effects of screens on the heating power of solar rays; Leslie's apparatus for testing the heating effects of the different solar rays.

A list just as long and as impressive could be made out of the instruments and methods meant specifically for electricity. Eighteenth-century electricians and instrument-makers had put endless effort into devising apparatus aimed at producing, detecting, and measuring electrical effects. A book for beginners, compiled by the London instrument-makers George Adams Jr and William Jones, in its fifth edition of 1799 described some forty-eight instruments and apparatus that the serious amateur electrician was expected to know, and possibly to possess (Adams and Birch 1799). After 1800, any such list would include Volta's battery, announced in the spring of that year and immediately replicated in expert and amateur circles across Europe and North America.

The introduction of the battery was the outcome of several research paths, which merged in the work of Alessandro Volta (Pancaldi 2003). In a few months or weeks during the fall of 1799, those different strands catalyzed to produce a radically new apparatus, generating a continuous flow of electricity that was unlike the sparks produced by the typical friction or induction electrical machines of the eighteenth century. The new apparatus was the source of endless surprise and admiration in both expert and amateur circles, but it also conveyed many uncertainties when it came to assess what was going on inside the new instrument, the exact nature of its electrical effects, and their bearing on contemporary views about imponderable fluids. Because the instrument was announced by a physicist who, in the preceding eight years, had been known mainly for his contributions to a field of study known as animal electricity or 'galvanism', for a while many observers were inclined to regard the new apparatus as belonging to the domain of phenomena first described by Luigi Galvani in his work on the 'forces of electricity in muscular motion' (Galvani 1791, Pera 1992, Bresadola 2003). These interpreters often called the battery and its phenomena 'galvanic', and were inclined to view them as confirmation that a special galvanic fluid existed and had to be added to the repertoire of already known imponderables. Volta, and those who opposed the notion of a special 'animal electricity', interpreted the battery instead as proof that the simple contact of two different metals, together with the addition of a third, humid conductor as used in the voltaic battery, could set the electric fluid in motion and produce effects similar to those of common electricity.

It is well-known that scientific instruments tend to have, up to a point, a life of their own, and the voltaic battery was no exception. In the present context, however, it may be useful to consider what the inventor of the battery had to say about imponderables and their possible bearing on his new instrument.

Volta displayed a cautious attitude towards imponderables throughout his career. When teaching, he made allowance for a universal ether involved in electrical phenomena, but he warned his students that the 'laws' by which the ether acted were more likely to be ascertained than its real nature (Volta 1918–1929: IV, 320–353, from lectures notes of 1778–1780). In the late 1790s, during the controversy over animal electricity with Galvani, and some eighteen months before building the battery, he spelled out a thorough instrumentalist attitude concerning the 'arcane force or virtue' that kept the electric fluid in motion in 'galvanic' phenomena (Volta 1918–1929: I, 545, from a publication dated April 1798). Volta's contact theory of electricity, which he claimed could explain what happened inside the battery, as well as the twitching of the legs observed by Galvani in his dead frogs, did not elaborate about the microscopic forces involved. The theory relied, instead, on the observation of macroscopic electrical effects, produced by varying endlessly the conductors employed, the modes of operation, and the instruments used to detect electricity.

After building the battery, Volta presented it as supporting his contact theory, but this did not stop him from elaborating other possible explanations of its effects; explanations that went occasionally beyond the contact theory, which he had elaborated well before the introduction of the battery. Of these other explanations, one was closer to traditional speculations about the imponderables. It claimed that, just as different bodies have different capacities for heat, so different bodies might have different capacities for electricity. If so, Volta thought, when different conductors were placed into contact with each other, the 'imbalance' between the fluids pertaining to each body probably had the effect of setting the electric fluid in motion, generating the typical current produced by the battery. Volta did not publish this hypothesis, which has survived only through his manuscripts (Pancaldi 2003: 243–245).

The only explicit connection between the Voltaic battery and the imponderables remained the fact that, with the battery, the author claimed he had shown that there was no need to postulate one more imponderable, galvanism, which several of his contemporaries were inclined to add to the list. The battery, in any case, had an enormous following among expert and amateur electricians and among chemists, quite independently of the ideas espoused by its author. In what follows I will use the reactions to the battery, and to a number of other electrical instruments and devices introduced in its wake, as a test case to assess the physics of imponderables as it was practiced in different quarters during the period of the 'rise and fall' of Laplacian physics.

9.4 Normalizing Novelty: Laplacians Face the Battery

Voltaic batteries were first built in Paris in the summer of 1800 (Blondel, 1999: 192). An early interest in the new apparatus was shown especially by amateurs and by physicians, the instrument being perceived as linked closely to the galvanic experiments on which amateur groups and physicians' committees had been working in Paris since 1792. The first powerful batteries, accordingly, were set up within medical institutions; but their effects soon attracted the attention of scholars with a broader view of the battery's potential interest. These included Laplace, Fourcroy, and Vauquelin (Blondel 1999: 193). At that time young Jean Baptiste Biot's interests still focused on mathematics, but, as a protégé in Laplace's entourage, his master's long-term interest in electricity, and now in the voltaic apparatus, lured him too.

Biot found himself in a peculiar position with regard to the new field of study. He was among the youngest Parisian scholars to be involved in experiments and interpretations of the battery, and he knew that the apparatus was not a trophy of his powerful masters. The battery came from abroad, it had been introduced by a man belonging to an earlier generation of natural philosophers, and to a research tradition with which Biot was not familiar.

Published in the late spring of 1801, Biot's first electrical paper focused not on the battery, galvanism, or the peculiar fluid set in motion by the Voltaic apparatus: Biot's topic was instead the distribution of (static) electricity on the surface of an ellipsoid of revolution. Laplace himself had recommended the topic to Biot, and had suggested he solve the problem by adopting a mathematical approach he had developed in his *Mécanique céleste* when dealing with the shape of the Earth (Biot 1801a: 23; Laplace 1805: II, 21). The solution provided by Biot introduced a version of what was later known as the potential function for electrostatics (Todhunter 1873: II, 25ff; Heilbron 1999: 498–9). It was based on the assumption formulated by Charles Augustin Coulomb—the acknowledged authority on electricity in Laplace's circles—that the 'molecules of electricity of the same kind repel each other with a force which is proportional to the masses involved, and varies inversely as the square of the distance' (Biot 1801a: 21).

When Biot published his first paper on electrostatics, he must have been already well acquainted with Volta's battery and its 'galvanic' phenomena: only two months later, on 14 August 1801, he read at the Institut a memoir 'on the movement of the galvanic fluid' based on extensive experiments with batteries of various kinds (Biot 1801b). The title of Biot's new presentation evoked the author's intention to try out a mechanical approach to the fluid of the battery which, though called 'galvanic' in the title, was shown to share most of the attributes of common electricity. The molecular premise of Biot's argument was the same as the one adopted in his earlier publication on electrostatics. The molecular level, however, played a marginal role in Biot's second paper. The point was to present a model for the new apparatus based on the notions of the quantity and speed of the electric fluid involved. Reminiscent of the geometrical approach developed in the paper on electrostatic 'potential', Biot insisted on the possibility of representing the fluid's behaviour as depending on the size and shape of the conductors used in the apparatus. Having once adopted this hypothesis—which was in tune with an old tradition of speculations by electricians about the properties of pointed versus round conductors in releasing or retaining the fluid—Biot could claim that 'the diversity of the phenomena observed in different [voltaic] apparatus has its main cause in the different proportions in which the quantity or mass of the fluid is combined with its speed' (Biot 1801b: 273).

No mathematical treatment of the battery was offered by Biot at this stage. However, the imagery of the fluid flowing within and from the battery was so ably combined with his argument about the speed and quantity of electricity that the reader could easily forget the paradox implicit in attributing mass to an imponderable substance. The rich repertoire of experiments discussed in the paper, on the other hand—and the endless variations in the shape and assemblage of the apparatus tried by Biot—pointed to a primacy of the apparatus itself in his search for an interpretive model. As for the chemical effects of the battery, then already well-known and debated in expert circles across Europe, at that stage Biot preferred to regard them as marginal: he would address them in a publication only two years later (Biot 1803).

Before that, Biot's major challenge was to write the official report that an authoritative committee had been asked by general Bonaparte to submit to the Institut National after Volta had presented his battery in Paris before the Institut and the general, who was one of its members (Biot 1804). The committee included Laplace, Coulomb, Hallé, Monge, Fourcroy, Vauquelin, Pelletan, Charles, Brison, Sabatier, Guyton, and Biot himself. The committee and the report were part of a diplomatic exercise thought up by Bonaparte, in the wake of the peace of Lunéville signed between France and her rivals, to celebrate Volta in front of Europe's cultured elites; Volta was by then a very well-known figure, and an Austrian subject from Lombardy who had been working for France's enemies for a quarter of a century, and now was received with all honours in Paris. The political goal pursued by Bonaparte, to be sanctioned with the award of a golden medal to Volta, consisted in conveying a reassuring image of the ambitions of the French towards Europe. The occasion elicited some important exercise in scientific diplomacy as well.

Already in the title, Biot's report emphasized 'citizen Volta's' new experiments and the theory he had used to explain the electricity produced by his new apparatus. As presented by Volta in Paris, the theory asserted that when two different metals were put into contact and then separated, one metal was found to be positively and the other negatively charged. In fact, the extremely small electricity thus excited became discernible only by applying the metal several times to an electrical multiplier or condenser, until the electricity thus accumulated affected an electrometer.

Biot presented this as the 'the main fact, from which all the others follow' (Biot 1804: 197). Volta's contact theory appealed to Biot and the Laplacians for its strictly physical character. Another advantage was that it allowed one to bring under the same domain of common electricity phenomena for which others preferred to introduce a special, 'galvanic' fluid. In pointing out the latter advantage, however, Biot used caution, stressing that the 'cause' assumed by contact theory was still *unknown* (Biot 1804: 208).

Volta, we know, had not linked his theory to speculations about what was going on at the molecular level, but Biot found it comparatively easy to mention such speculations. He remarked that any electricity displayed by a body in excess or defect of its 'natural' quantity must be 'proportional to the repulsive force by which the molecules of the [electric] fluid repel each other, and repel any other molecule one might add to them'. To him, in bodies retaining their natural amount of electricity the repulsive force was 'balanced by the resistance of the air'; a point central also in Biot's earlier publication on the 'potential' function.

Biot stressed that the microscopic, repulsive force was constitutive of the *tension* of the electric fluid. His ambiguities when treating the notions of electric force and electric mass, however, seems to have affected several subsequent authors (Roche 1998: 163–164.) In his report Biot took the opportunity to correct Volta when he assumed that tension was proportional to the divarication of the straws of an electrometer: in Paris the only apparatus appropriate to measuring tension via electric forces was Coulomb's torsion balance (Biot 1804: 199–200n.) Indeed, research on the battery—and the need to reassert the role of Coulomb's electrical researches, in tune with Laplacian physics, *vis-à-vis* the new voltaic apparatus—was the main reason why Biot had a torsion balance built for him by the celebrated artisan Fortin (Blondel 1994: 114–116). Thus it was the introduction of the battery in Paris that offered the opportunity for Coulomb's balance to become, via Biot, the symbol of a sophisticated, Laplacian approach to electricity for many subsequent generations of physicists.

Having evoked the fundamentals of the science of electricity as seen from Paris, Biot in his report introduced a simple algebraic calculation to describe the likely distribution of electricity inside the battery. The calculation assumed that within an insulated battery the electricities of its different elements formed two arithmetic progressions, with the elements placed at equal distances from the extremities of the pile equally and contrarily electrified (Biot 1804: 201).

Biot's treatment was based, by his own admission, on two delicate assumptions. The first was that the electric fluid flowed from one metallic couple of the battery to the next through the soaked conductors even when the two extremities of the instrument were not linked to form a circuit. That this could be the case was proved, according to Biot, by Volta's experiment showing that a single metallic couple was able to electrify the plate of a condenser when touching it through a layer of soaked paper.

The second assumption was more difficult to establish than the first. It assumed that, for any two metals, the difference in the electricity elicited when they were put into contact was constant. Biot claimed that this was the simplest possible assumption (Biot 1804: 205). Much later he wrote that he had heard Coulomb say that he had once verified the assumption, which on that same occasion Biot called a 'law' (Biot 1816: II, 480.) In the report, however, he declared that the committee of the Institut had been unable to set up the 'very delicate' experiments needed to confirm the assumption, and reported instead Volta's ideas about the different 'electromotive force' displayed by different metals.

In conclusion, according to Biot, Volta's contact theory did not satisfy the requirements expected from a genuine physical theory. In order to satisfy them, the *law* regulating 'the distribution and movement of electricity' within Volta's apparatus had to be established. The adoption of 'the most accurate instruments' (i.e., Coulomb's torsion balance) was a prerequisite for such inquiries. So was the adoption of a frame of mind based on the search for a general theory of electrical phenomena. In an historical sketch placed at the end of his report, Biot outlined the lineage that such a general theory must conform to in order to fit into the Laplacian tradition. After the early contributions of Dufay and Franklin, subsequent key contributions were identified by Biot in the works of Aepinus and especially Coulomb, who had established 'the rigorous law of electric attractions and repulsions', demonstrably the same as 'celestial attractions' (Biot 1804: 209). Within this perspective, Volta's instruments—old ones like the electrophorus and the *condensatore*, and the latest one, the battery—appeared as 'fine' and '*lucky*' discoveries, '*enriching*' the theory of electricity. The implication was that the author of those instruments was not acquainted with the methodology expected from a theory of electricity within the Laplacian tradition (Biot 1804: 210).

Played down in the diplomatic exercise of 1801, Biot's dismissive implications were given full voice twenty-six years later, when he wrote an obituary of Volta. This obituary offers materials for a further assessment of how Laplacian physicists defined their methodology, and how they faced the rich harvest of new concepts and new instruments that were being produced in the science of electricity by different research traditions in several countries.

When writing Volta's obituary, Biot still recalled vividly the discussions with Volta that had taken place in 1801 within the committee of the Institut National. During the discussions, Coulomb and Laplace—the 'physicists-geometers', as Biot called them—had tried to convince Volta to adopt their approach to the science of electricity, to no avail (Biot 1827: 460). 'It was in vain', Biot recalled, 'that they showed him the superiority, we would even say the mathematical necessity, of the method that Coulomb had followed in order to obtain his measures, which are the foundation of the entire science [of electricity].' Volta, according to Biot, lacked 'the genuine theory' of his own discoveries, hence his achievements appeared mere 'combinations of experience' (*combinaisons d'expérience*) (Biot 1827: 460–461). The only way to acknowledge Volta's achievement with the battery, while denying him acquaintance with the scientific methodology, consisted in offering a reconstruction of the discovery in which chance and 'happy induction' played major roles. Volta's principle, according to which different metals could set the electric fluid in motion, was described by Biot as 'absolutely new and unexpected'—something that only Volta's 'extreme sagacity' could reveal. Once the principle had been recognized, Volta was able to 'deduct' from it an application that was possibly an even greater 'discovery' than the principle itself. (Biot 1827: 461). In Biot's account, chance figured prominently already in the process that had led Volta to take an interest in galvanic phenomena in the first place. Biot thus extended to Volta a trope already adopted in the histories of Galvanism, which credited Galvani's observations on frogs' legs being excited by electricity to chance. The fact that the current generated by the battery had turned out to be a major agent for chemical decomposition and composition was also, according to Biot, totally 'unexpected'. On the other hand, to explain why French natural philosophers, despite their superior methodology, had not contributed much to recent developments in the science of electricity, Biot had recourse to social context: the wars in which the country had been involved had kept France 'outside the movement of general civilization' (Biot 1827: 463). Such were the efforts made by Biot to defend his view of the true spirit of physics, and French science, while facing up the novelties from research traditions perceived as at odds with the tradition advocated by Laplace and his followers.

Biot's interest in the battery, however, was genuine if we may judge from the eighty pages he devoted to the contact theory, the battery, and their aftermath in his *Treatise* of 1816, 'the textbook of Laplacian physics' (Biot 1816: II, 467–546; the definition is in Heilbron 1993: 142; see also Frankel 1977: 68–72, and Roche 1998: 154–155; on Biot

Laplacian and experimentalist see Garber 1999: 134). The four chapters on the battery brought to new heights Biot's efforts to produce a quantitative treatment of the instrument, while standardizing the measuring techniques. Biot's virtuoso mathematics and his measures allowed him to correct Volta on several points of detail, but he was keen to stick to Volta's theory, and did not discard Volta's measures altogether. After pointing to an important difference in their respective estimates of the charge generated by a battery when isolated and when earthed, Biot stressed that Volta's system for measuring such charges could be regarded as a 'special case' of his own system (Biot 1816: II, 504).

Biot's attitude towards Volta and his battery offers an interesting glimpse into the physics of the Laplacian school. The attitude reveals several different, underlying strata which came to surface depending on need and circumstance. When the identity and pride of the school within the Institut National and, indeed, in Europe was at stake—as when Bonaparte invited the Institut to reward Volta—the Laplacian programme assumed its most general and ambitious form, while at the same time showing accommodation to the novelties coming from afar. In this most general and systematic form, Laplacian physics was best summarized by Biot when he wrote:

... the highest degree of perfection in natural philosophy would be to deduct from observations and experience a small number of general laws, or principles of movement, and then explain how observable things follow from these principles, rigorously and with the same numerical proportion we find in them. (Biot 1858: II, 113–114)

In different circumstances, the school showed it was eager to impart lessons of scientific methodology, and to denounce the limitations of supposedly less sophisticated, though apparently no less effective approaches pursued in other countries. This dogmatic vein was used apparently as an identity device, as a banner to secure or protect contested territory.

When, on the other hand, Laplacians dealt directly with the deep uncertainties affecting research on the imponderables, they could display an instrumentalist vein that bordered on scepticism. Biot expressed this sceptical vein when he wrote: '. . . not only there are many phenomena whose laws are unknown to us, but there are other phenomena whose production is an impenetrable enigma . . .' 'By resorting to such suppositions [= *the imponderables*] we can, up to a point, represent the majority of phenomena.' But there are other phenomena that 'admit such explanations with difficulty, and still others that do not admit them at all.' He concluded that

veritable physicists admit such fluids only as a convenient hypothesis, to which they carefully avoid attaching any idea of reality whatsoever, and keep ready to change or abandon them entirely if facts turn out to be contrary to them. (Biot 1858: II, 113–4)

On similar occasions, not only did Biot show he was aware that the cosmopolitan community of physicists harbored widely different research programmes, as in the statement quoted at the beginning of this chapter, he even showed he was inclined to consider the possibility that the job facing physicists dealing with imponderables might turn out to be endless.

9.5 'A WONDERFUL AND IMPORTANT INSTRUMENT OF ANALYSIS': CHEMISTRY LESSONS FOR PHYSICISTS

'Nothing tends so much to the advancement of knowledge as the application of a new instrument'. Thus Humphry Davy in his *Elements of Chemical Philosophy*; and he added: 'without the Voltaic apparatus there was no possibility of examining the relationships of electrical polarities to chemical attractions' (Davy 1812: 28–29). To Davy, the battery was 'a wonderful and important instrument of analysis' (Davy, from a manuscript dated 6 August 1800, in Fullmer 2000: 304).

When news of the battery first reached Thomas Beddoes' Pneumatic Institution in Bristol in June 1800, young Davy began a long series of experiments with the new apparatus (Golinski 1992: ch. 7; Fullmer 2000; Pancaldi 2009). In the following seven months he published as many letters or reports on the subject, and continued to use the instrument systematically when he moved to the Royal Institution in London in January 1801. The path thus embarked upon culminated in 1807 with Davy's announcement that, with the aid of the battery, he had discovered two new metals, which he called potassium and sodium.

It has been argued that the invention of the battery, and his subsequent work with it, led Davy to abandon the concept of an electric fluid (Harman 1982: 19–20). In this section I will explore how experiments with the new apparatus interacted with Davy's notions about the fluid.

Expectations about the battery ran high in England, where Volta had decided to give his first public announcement of the new instrument. William Nicholson—the first to publish on the subject in London soon after having discovered, with Anthony Carlisle, the chemical effects that accompanied the battery's operations—presented the instrument as an impressive chemical machine, whose power could be measured in the same way as the power of steam engines was being tested in the country of the first industrial revolution (Nicholson 1800: 187). Expectations for natural philosophy and chemistry, and a better understanding of imponderable fluids, ran high also in the circles around Davy. As Thomas Beddoes, Davy's master in Bristol, put it:

> When such experiments shall have been repeated . . . they will serve as the origin of a new and more subtle chemistry in which ethereal fluids [*including electricity*], as they were styled by Bergman, will make the principal figure by their agency . . . We may expect, at least, to see the disputes relating to the existence of these fluids brought to an end; or, rather, philosophers agreeing in their language on these subjects. (Beddoes 1801: 507)

Beddoes' expectations were in tune with his views emphasizing the need to penetrate the structure of metals and the earths, while paying due attention to their interactions with light and electricity (Beddoes 1799: 229).

In those same years, on the other hand, philosophers like the aged Joseph Priestley expressed a desire that the battery might vindicate phlogiston—the old inflammable principle, the supposed substance emitted during combustion and the calcination of metals—while disproving, together with the doctrine of the decomposition of water, the new chemistry of Lavoisier (Priestley 1802: 202). James Watt, for his part, hoped that young Davy's experiments on the gases elicited from water with the help of a battery would demonstrate his theory according to which water was the basis of all gases (Davy to Watt, 29 December [1801], Birmingham City Archives, Soho Collection). Still others expected the battery to become a new, powerful application of galvanism to medical practice, shedding new light on physiology and the science of life (Erasmus Darwin in King-Hele 1981: 328–329, and Darwin 1825: Additional note XII, 79–82).

During Davy's early experiments, still under Beddoes' influence, he showed similar, highly philosophical, and somewhat tangled expectations. Reporting on his researches on 'galvanism', he expressed the view that 'some of the new facts on this subject promise to afford instruments capable of destroying the mysterious veil which Nature has thrown over the operations and properties of ethereal fluids.' (Davy to Giddy, 20 October 1800, in Paris, 1831: I, 73).

However, Davy learnt how to fend the risks of such expectations quickly. Once in London, immersed in the milieu provided by the industrial culture diffused in the circles within and around the Royal Institution, he developed a more expedient approach to the battery. This new approach favoured experimental skills, together with an endless variation of materials and experimental set-up, over the search for the big questions that had engaged an earlier generation of natural philosophers, with their passion for speculation about underlying molecular phenomena. The shift led Davy to dismiss gradually the imagery and the language of imponderable fluids.

In the first paper he submitted to the Royal Society barely three months after his arrival in London, Davy focused on the battery itself rather than on big interpretive issues, and suggested a new combination of its basic components, consisting of one metal and a variety of fluids, instead of the two metals recommended by Volta. In the paper Davy adopted a strictly phenomenological approach to the effects produced by the different combinations, avoiding any suggestion about the causes at stake (Davy 1801). Six years later, in the celebrated Bakerian Lecture of 1807, he accompanied the announcement of the new metals, obtained with the action of a powerful battery on fused potash and soda, with an implicit declaration of indifference for theories:

Potasium [sic] and Sodium are the names by which I have ventured to call the two new substances: and whatever changes of theory, with regard to the composition of bodies, may hereafter take place, these terms can scarcely express an error, for they may be considered as implying simply the metals produced from potash and soda (Davy 1808: 32).

Another five years later, in *Elements*—Davy's only attempt at a systematic treatise, which he never finished—he tried to give definitive form to ideas he had arrived at about electricity through the experiments carried out during twelve years of extremely successful experimental work. In those ideas there was little room left for imponderable fluids. He presented electricity as a power of matter to produce polarities—a power somehow associated with, but not identical to, the power of producing chemical combinations. Davy now avoided the fluid imagery. He kept the appeal to underlying molecular forces to a minimum, and his generalizations remained close to the experimental setting:

Certain bodies which attract each other chemically, and combine when their particles have freedom of motion, when brought into contact still preserving their aggregation, exhibit what may be called electrical polarities; and by certain combinations these polarities may be highly exalted; and in this case they become subservient to chemical decompositions... Bodies combine with a force, which in many cases is correspondent to their power of exhibiting electrical polarity by contact; and heat, or heat and light, are produced in proportion to the energy of their combination. (Davy 1812: 38–39)

To Davy, Volta's battery showed 'the relations of electrical polarities to chemical attractions', and electricity appeared 'to result from the general powers or agencies of matter' (Davy 1812: 105, 100). Electrical and chemical actions 'depended upon the same cause': they were '*distinct* phenomena, but produced by the same power, acting in one case on masses, on the other on particles.' (Davy 1812: 119–120, Davy's emphasis)

As for the controversial theories about the battery itself, Davy retained the nutshell of Volta's physical, contact theory, and combined it with his own chemical view of the apparatus. Within the battery, 'the contact of the metals destroys the electrical equilibrium, and ... chemical changes restore it', so that 'the action exists as long as the decompositions continue'. This, Davy maintained, 'will possibly lead to new views respecting the elements of matter' (Davy 1812: 125–6). He stopped before entering such an uncertain domain, however, convinced as he was that speculations about the electric fluid or fluids were 'hasty generalizations', while 'the application of electricity as an instrument of chemical decomposition, and the study of its effects, may be carried on independent of any hypothetical ideas concerning the origin of the phenomena, and these ideas are dangerous only when they are confounded with facts' (Davy 1812: 128).

To compare Biot's and Davy's approaches to the battery, and the different destinies they foresaw for the physics of imponderable fluids in roughly the same years, is instructive.

Biot's preferred method to analyse the new instrument and the new phenomena attached to it was via quantification. This led him to develop a model that insisted on the speed and quantity of the fluid or fluids involved. The supposed underlying molecular model, with its Newtonian or Coulombian 'laws', he often relegated to the footnotes, but he brandished it effectively when he was defending contested territory, or when he thought fit to impart lessons of scientific methodology. The increasing room he made for instrumentalism, as the amount of new if often undigested evidence on 'galvanic' electricity and its chemical effects accumulated, does not seem to have led him to search for explanatory models alternative to the ones he had adopted when first confronting the battery in 1800–01 under Laplace's guidance. And that despite the notable, first-hand, experimental know-how of the battery he had accumulated over the years, culminating in the work he carried out with Savart on the magnetism of the battery in 1820, after the announcement of Oersted's effect (Biot and Savart 1820; Biot 1824: II, 707–723).

Davy, like Biot, was well aware of the high expectations that the study of the battery was nurturing among natural philosophers dealing with broad physical questions. However, having been trained in a tradition in which natural philosophy mingled more easily with chemistry and medicine than with mathematics and mechanical philosophy, he found it comparatively easy to take a step that proved crucial to his success: to use the battery as an instrument of analysis in his chemical work. The step was favoured by approaches to the battery as a source of energy and as 'a machine' already adopted in London by people like the chemist and instrument-maker William Nicholson, and in metropolitan and industrial circles close to the Royal Institution (on the latter: Unwin and Unwin 2009; Jacob and Stewart 2004: chapters 4 and 5; on Nicholson: Thackray 2008, Golinski 2004, Pancaldi 2003: 228–230).

The step Davy took, as we have seen, was also a step into instrumentalism, in line with the not too dissimilar instrumentalism that Biot himself developed in his long dealings with the phenomena of the battery. But the goals pursued by the two men were conspicuously different, as was their focus. For Biot, the fundamentals remained a mechanical interpretation and mathematical treatment. For Davy, the goal was the kind of experimental expediency that allowed him to use the battery to decompose metals, earths, and other bodies, paving the way for 'the creation of numberless new arts', as his mentor Thomas Beddoes had once put it (Beddoes 1799: 229).

9.6 Enter Ampère: The Mathematician, Electricity, and the Instrument-Makers

On 24 December 1801, on his way back to Lombardy after presenting the battery in Paris, Volta was in Lyon. There he attended a meeting of the local academy of sciences, during which he submitted a paper that asserted the identity of the galvanic fluid and common electricity. After Volta, a young scholar took the floor to present the summary of an ambitious system that promised to connect all of physics while providing a new explanation of the phenomena of heat, electricity, and magnetism, traced back to the 'common laws of mechanics'. The young scholar was André-Marie Ampère (Ampère 1936: 109n).

Though never published, fragments of Ampère's early attempts at a systematization of electrical and magnetic phenomena, including the one presented in Lyon before Volta, have been preserved. They provide interesting hints on how the battery was assimilated amidst the speculations and experimental attempts of a scholar who shared some of the tenets upheld by Biot and the Laplacians, but who dissented from them on several points.

In those early essays Ampère suggested treating together what he called the '*igniforme*' fluids: fluids behaving somehow like fire (*ignis*), and lighter than gases, which—he claimed—could provide a *combined* explanation of the phenomena of electricity, magnetism, heat, and chemical affinity (Ampère 2005, Mss. 203: 1, 82; see also Hofmann 1995: 50–58). Ampère speculated about the 'mutual action' between the fluids, and on their interaction with the bodies they penetrate (Ampère 2005, Mss. 203: 9). To assess this mutual action he introduced the notion of a *ressort* (spring), proper to each fluid, defined as 'the force with which the fluid tends to pass into neighbouring molecules'. The molecules and the fluids, depending on their spring, were supposed to generate 'atmospheres' around each molecule, which interacted side by side with the bodies themselves (Ampère 2005, Mss. 203: 17). Different bodies were supposed to have different capacities for the different fluids, as had been shown for caloric (Ampère 2005, Mss. 203: 12–13). Arguing from caloric and its ability to penetrate bodies, Ampère supposed that the other fluids—especially electricity—were 'carried around' by, or mixed with, the molecules of caloric (Ampère 2005, Mss. 203: 55–56).

Reminiscent of eighteenth-century speculations about caloric, electrical atmospheres, and electrical induction, Ampère's juvenile 'system' promised to offer an explanation for a rich variety of phenomena. The proclaimed roots of the system in the laws of mechanics, on the other hand, rested on a generic appeal to molecules and their ways of interaction, supposedly following mechanical laws. The claim was in tune with Ampère's reputation as a mathematician—a reputation that attracted the attention of astronomers such as Delambre, and mathematicians such as Lagrange, and secured him, after a short period in the provinces, positions in some of the most prestigious scientific institutions of the French capital (Hofmann 1995: Part II).

On the issue of mechanical laws, however, Ampère in his early speculations attacked the core of the Laplacian approach to electricity with vigour: he rejected both Aepinus' one-fluid theory of electricity, and Coulomb's two fluids solution. Above all, he denied that forces could act at a distance in empty space. There was only one possible exception to the rule, and that was gravitation. Knowing how to combine mathematical thinking and religion, Ampère judged the exception reasonable because he found gravitation convenient to 'the means God used to complete the existence of matter by bringing molecules together'. That granted, Ampère argued that 'nothing is less fitting than to arbitrarily establish a stronger attraction between the molecules of a fluid and those of other bodies, and a mutual repulsion between the former', as Laplacians did (Ampère 2005, Mss. 203: 85).

Ampère's early speculations left long-lasting impressions in his mind: in a more lucid formulation, they were still to be found in a manuscript he circulated among his students in the 1820s (published posthumously in Moigno 1852: 302–317).

If Ampère's juvenile system for the imponderables was admittedly very speculative, in the same years he displayed a keen interest in the new experiments and apparatus that were challenging many received views in the field of electricity. Ampère's use of electrical instruments and apparatus throughout his career deserves probably more attention than it has received from historians so far. Evidence of Ampère's interest in instruments, both experimental and demonstrative, is abundant, but it has apparently been overshadowed by his high reputation as a mathematician.

In conjunction with his duties as a professor of physics in Bourg from February 1802, and from 1804 in Paris in various capacities, Ampère was keen to procure the best electrical instruments available, and he often developed a close collaboration with their makers. Early in 1802, still in Bourg, Ampère was procuring instruments produced by the Dumotiez brothers in Paris (Ampère 2005, Corr: <http://www.ampere.cnrs.fr/amp-corr79.html>). In those years the Dumotiez firm was active in the business of 'galvanic' apparatus, and acted as joint publishers of Joseph Izarn's popular *Manuel du galvanisme* (Izarn 1804). Sometime later, as we shall soon see, Ampère embarked on systematic experiments in chemistry and electricity. In September 1820, at a crucial point in his demonstrations of the interactions among currents, Ampère carried out some of his experiments in Dumotiez' workshop, and managed to secure for himself the big battery that Dumotiez had built for the Faculté des Sciences (Ampère 2005, Corr.: <http://www.ampere.cnrs.fr/amp-corr590.html>). From the 1820s he was in touch with another well-known instrument-maker, Louis François Clément Breguet, who would soon specialize in producing electrical instruments (Payen 2008). Sometime later still, Ampère was working with Pixii on improving an electro-dynamic apparatus (Ampère 2005, Corr.: <http://www.ampere.cnrs.fr/amp-corr621.html>; Ampère 1824), and in 1832 he and Pixii circulated in France one of the earliest machines based on Faraday's principle of induction (Ampère 1832; Gee 1993; Beaudouin 2005: 268).

Ampère's fascination with every new experimental result connected with the battery is illustrated by his vivid account of the experiments he himself repeated following Davy's announcement of the discovery of potassium and sodium in 1807. On this occasion he sided with Davy, and was fascinated by his display of experimental dexterity, in contrast to the scepticism towards Davy's work that was common among his colleagues in Paris (Ampère 2005, Corr.: <http://www.ampere.cnrs.fr/amp-corr342.html>). A significant correspondence with Davy ensued, during which Ampère elaborated a series of theoretical contributions to chemistry that, for a while, he rated second only to his contributions to metaphysics and mathematics: it is well-known that Ampère's efforts before 1820 were divided between metaphysics, chemistry, and mathematics (Ampère 2005, Corr., <http://www.ampere.cnrs.fr/amp-corr472.html>).

If we consider Ampère's continued interest in electrical theory *and* electrical apparatus, especially the battery, we gain a better understanding of his groundbreaking work of 1820, when in a few months he laid the foundations of electrodynamics, impinging on the subsequent development of physics.

9.7 From Fluids to Current

By the evening of 25 September 1820, when Ampère described the situation in a letter to his son, Oersted's announcement circulating during the summer that the 'galvanic currents' of a Voltaic apparatus affected the orientation of a magnetic needle had already become 'an important event' in Ampère's life. In the few preceding weeks Ampère had developed a 'grand theory' providing a new explanation of the magnet and tracing all its phenomena back to 'galvanic phenomena'. The theory claimed that the wires carrying electric currents interacted among themselves, just as Oersted had shown that a wire and a magnetic needle interact (on Oersted: Brain et al. 2007). The new theory—as Ampère reported in the middle of the events—had been confirmed through a number of experiments. He had carried out successfully the 'decisive' experiment, 'demonstrating' the theory, thanks to a powerful voltaic apparatus that he had tried first in the workshop of the Dumotiez instrument-makers, then at his home, in the company of Augustin Jean Fresnel, his friend, and finally at the Institut where, apparently, even sceptics had been convinced. Ampère planned to repeat the same experiments in two days time at the Bureau des Longitudes, with Laplace attending. (Ampère 2005, Corr.: <http://www.ampere.cnrs.fr/amp-corr590.html>; on Fresnel and Ampère: Buchwald 1989: 116ff, 261).

In presenting his experiments and his grand theory in September 1820, Ampère used three basic notions and a few sophisticated pieces of apparatus (Ampère 1820).

One notion was Voltaic: the notion of an electromotive force, active whenever two different conductors are put into contact, and responsible for the battery's effects. The second was also an old acquaintance of the electricians: the notion of 'electric tension', conceived as the effect produced by the battery when its two poles are not linked to form a circuit. The third notion was rather new, at least as Ampère presented it: the notion of an 'electric current', conceived as the set of electrical effects that manifested themselves whenever the two poles of a battery were linked by a conductor to form a circuit. According to Ampère, the circuit included the battery itself, and tension and current excluded each other: when there was tension there was no current, and vice versa.

If one considers Ampère's interests prior to 1820, and especially his interest in Davy's chemical work with the battery, his early publications on interacting currents leave little doubt about the likely source of his original approach to the phenomena revealed by Oersted. The discovery of the Oersted effect reinforced the idea, which Ampère regarded as having been already established by Davy's studies, that a battery linked by a wire forming a circuit exerted a special power, that was now revealed by both its chemical and magnetic effects. Ampère's 'electric current' was the likely common cause of the two sets of phenomena. The notion, in turn, revived Ampère's old speculations that promised to connect up all the various strands of physics and provide a new explanation for the phenomena of heat, chemistry, electricity, and magnetism. After Oersted, the effects on the magnetic needle presented themselves as the most direct and easy way to reveal the current.

In the same few pages of his first memoir on the new subject, Ampère introduced together the notion of current and a new instrument, which he suggested calling a 'galvanometer': a device capable of detecting—and hopefully measuring—the current, by means of its effects on a magnetic needle when no traditional electrometer would do, because no tension was present. In his search for the common cause of electrical, chemical, and magnetic effects, Ampère developed an impressive series of new experiments, and conceived a just as impressive sequence of new apparatus to perform them.

To further clarify the differences between what he called electric tension and the electric current, Ampère focused on the different *directions* that had been attributed to the fluid, within the battery and in the conductors connected to it, according to the different chemical effects taking place at the two poles of the battery. In the French tradition, based on the notion of two electrical fluids—the positive one, and the negative—the electromotive power inside the battery had been supposed to set the two electric fluids in motion at the same time in two opposite directions, with the two currents meeting in the part of the circuit most remote from the source of the electromotive effect (Ampère 1820: 64). Adopting this model, Ampère assumed that in a conductor connecting the two poles of a battery the direction of the current went from the pole where oxygen was produced to the one where hydrogen developed.

From Ampère's perspective, the Oersted effect and some hints contained in the Dane's first memoir on the subject invited trials on how currents running in different directions affected a magnetic needle. Such trials required conceptually simple but delicate apparatus, allowing the detection of weak, macroscopic attractions and repulsions (Blondel 1982: 69–87; Williams 1983; Hofmann 2000: 236–242). With this kind of apparatus Ampère decided to explore whether attractions and repulsions acted also between two thin metallic wires in which currents run without interruption in different directions. The 'decisive experiment' alluded to in Ampère's first private report of his feats of 1820 used this kind of apparatus (exemplar the one in Ampère 1820: Planche 1, fig. 1), in conjunction with the powerful battery that he had been able to buy the day before from the Dumotiez workshop, which had been building it for the Faculty of Sciences. The key novelty of the experiments was Ampère's delicate, combined circuitry and mechanics allowing one to detect the weak interactions among currents: a set of apparatus that young Faraday—having seen it only through the plates published in Ampère's early memoirs—judged 'very delicate, ingenious, and effectual', and soon tried to imitate and improve upon (Faraday 1821: 276).

Ampère, we know, had been in touch with the Dumotiez instrument-makers since at least 1802. After 1818, when he moved to a house in Rue des Fossés-Saint-Victor, he lived next door to one of the two workshops that the Dumotiez firm, then run by Hyppolite Pixii, had in Paris (on Ampère's house: Académie des sciences, Institut de France, carton XXV, chemise 377; on Dumotiez' workshops: Beaudouin 2005: 264). From 1820 Ampère's ties to Pixii became closer, so much so that the arrival of a yearly bill from Pixii could jeopardize Ampère's tight family budget (Ampère 2005, Corr., <http://www.ampere.cnrs.fr/amp-corr651.html>); on Ampère's frequent visits to Pixii: Ampère 2005, Mss.: <http://www.ampere.cnrs.fr/ms-ampere-205-7-1.5.html>; and Ampère 2005, Corr., <http://www.ampere.cnrs.fr/amp-corr651.html>). Sometime later Ampère advertised Pixii's 'electrodynamic instruments', built under his supervision, as an essential ingredient for the success of the experiments he recommended, thus perhaps also reducing the burden of Pixii's bills (Ampère 2005, Corr.: <http://www.ampere.cnrs.fr/amp-corr620.html>). In subsequent years, side by side with his energetic campaign in favour of his theory of electric phenomena, the sophistication of the 'electrodynamic' apparatus used by Ampère grew steadily, as is shown by the plates accompanying his publications.

There are no records, on the other hand, to show that Ampère used his instruments to carry out actual measurements (Blondel and Wolff 2009: 3). To him, instruments were, above all, working models for the refinement of physical imagination. The practice reached a climax with the introduction of some crucial 'equilibrium' experiments and apparatus, in which a mobile circuit, interacting with another circuit, was made to reach an equilibrium condition between the several forces or rotation moments involved (Ampère 1826: 10ff). The latter included the magnetic action of the Earth, which Ampère regarded as resulting from numberless electric currents in its interior (Ampère 1826: 199–200).

To avoid confusion with the ordinary attractions and repulsions studied by electricians, Ampère called the interactions between currents which he had discovered 'Voltaic attractions and repulsions' (Ampère 1820: 62.) Mathematician as he was, working side by side and occasionally in direct competition with the Laplacians, soon after his electromagnetic experiments of September 1820 he started developing a mathematical model for the 'attraction between two infinitely small currents' (Ampère MSS, Chemise 158 [carton 8], 10; see also Blondel and Wolff 2009). On 4 December 1820 he announced publicly an early version of his mathematical model for such interactions. To the infinitely small, rectilinear portions of the circuits, or current elements, he attributed the same macroscopic attractions and repulsions that he had detected in the currents explored with his instruments: attraction between current elements of the same direction, repulsion between current elements of opposite directions. The same behaviour he attributed to infinitely small current elements when parallel. He then introduced an angular dependence by further assuming that the force between two current elements was null when an element was placed on a plane perpendicular to the plane of the other element. In analogy with Newtonian, gravitational forces, he further assumed that the force acting between two current elements depended on their distance, that they were central forces, and that there was no elementary torque (Darrigol 2003: 9). With these assumptions, Ampère set the force between two current elements as proportional to

$$\frac{gh \sin \alpha \sin \beta \cos \gamma}{r^2}$$

where g and h are the electrical 'intensities' of the two currents, r is the distance between the mid-points of the two current elements, α and β are the angles between the elements and the line joining them, and γ is the angle formed by the two planes on which the elements lie.

Ampère submitted his 1820 trigonometric force formula to several modifications in subsequent years, to make room for his own new experimental results as well as for the suggestions and criticism of his colleagues. The latter included Biot and Savart, who introduced a 'law' that explained the interactions between a current-carrying circuit and a magnet rejecting Ampère's all-electric explanation. From Biot's perspective, Ampère's all-electric explanation of electromagnetic phenomena—opposed as it was to Coulomb's theory of magnets—amounted to a return to Descartes' discredited vortices.

In the 'definitive' version of Ampère's formula, published in Ampère's highly idealized *Théorie des phénomènes électro-dynamiques, uniquement déduite de l'expérience* (1826), the force between two current elements was made proportional to

$$i\, i'\, ds\, ds'\, (\sin \alpha \sin \beta \cos \gamma - 1/2 \cos \alpha \cos \beta) / r^2$$

where i and i' are the intensities of the currents, and ds and ds' are the lengths of the two current elements.

Meanwhile, some followers of Ampère's made a point of showing that the law of Biot and Savart, as well as Coulomb's law for magnetic poles, could be *deduced* from Ampère's formula for interacting currents (Demonferrand 1823; Blondel and Wolff 2009: 5). Thus, in February 1823, Ampère could claim that 'all the facts not yet fully explained [*in the study of electricity and magnetism*] . . . are necessary consequences of my formula' (Ampère 2005, Corr.: <http://www.ampere.cnrs.fr/amp-corr635.html>).

Ampère's combined use of mathematical demonstrations, appeal to physical models, and recourse to sophisticated apparatus, are worth considering in detail.

When facing the competition and scepticism of the Laplacians in Paris, Ampère naturally saw his mathematics as crucial to the success of his theory of electrical phenomena. When proselytizing abroad, however, he showed he knew very well how important his physical concepts and his delicate instruments were for the diffusion of his views. On learning that someone in Geneva was spreading doubts about his theory, especially about the electrodynamic action of the Earth, he recommended that Genevan scholars should procure the apparatus built for him by Pixii, and promised to start planning a visit there only once his Swiss colleagues had procured such an apparatus (Ampère 2005, Corr.: <http://www.ampere.cnrs.fr/amp-corr620.html>, <http://www.ampere.cnrs.fr/amp-corr621.html>).

Similarly, when writing to Faraday in 1824 and 1825—and regretting what he regarded as the poor reception accorded to his work in England—he inquired about the availability in London of instruments purposely built to show the interactions between currents, as distinct from those used to test traditional electromagnetic effects. On this occasion Ampère presented his instruments to Faraday as crucial to a full appreciation of his work: 'It is difficult without its help [*of a purposely built apparatus*] to repeat all the experiments relating to the laws of the mutual action of two electrical currents which it is so important to study separately and independently from the other kind of action that take place between a magnet and a conducting wire' (Ampère to Faraday, 27 April 1824 and 3 July 1825, in Faraday 1991: 1, 349–350, and 376–379, 378). Faraday, we know, concurred. Spreading the use of that kind of apparatus, while also discreetly promoting Pixii's business, was a frequent theme in Ampère's correspondence. Meanwhile, Pixii's own work in the field of electrodynamics earned him a gold medal from the Academy of Sciences (*Annales de Chimie et de Physique*, 51, 316).

It may be said that, just as Ampère the theoretician managed to combine a penchant for side by side action with an adhesion to Newtonian forces, in the same way he knew how to combine an inclination for mathematical physics with a deep commitment to the kind of model building and instrumental manipulations that he had developed over the years through his cooperation with Pixii. Model building and instrumental manipulation played important roles both in Ampère's most creative efforts, and when he was trying to convince his colleagues.

As Christine Blondel has shown (Blondel 1989), Ampère shared with his Laplacian colleagues the ideal according to which mathematical physics offered the guarantee of a higher status for physics generally, while also offering a sort of remedy for the conflicting views within the expert community, from which he had suffered in Paris. He expressed this ideal in his *Théorie* as follows:

Whatever the physical cause to which we may refer the phenomena produced by this [*electrodynamic*] action, the formula that I have obtained will always continue to express the facts. If one day we will be able to deduce it from one of those arguments by which we have explained so many other phenomena, such as inverse square attractions, or attractions becoming increasingly small at any significant distance from the particles producing them, or the vibrations of a fluid extended everywhere in space, etc., we will have made one more step in this branch of physics. But this research, that I have not yet addressed though I recognize all its importance, will not change anything as far as the results of my work are concerned. Because, in order to adjust to the facts, the hypothesis adopted will always have to agree with the formula that represents them so thoroughly. (Ampère 1826: 8)

Several subsequent generations of physicists subscribed to the ideal of mathematical physics expressed by Ampère in a passage like this. When the international community of physicists and electricians meeting in Paris in 1881 adopted the ampere as the unit of electric tension, they were somehow acknowledging their debt to Ampère for this conception of mathematical physics. Though, towards the end of the nineteenth century 'Ampère's formula' was no longer mentioned in physics' textbooks, while one of his experimental devices—which he regarded as confirming the formula—is still occasionally used in physics laboratory classrooms today. As it happens, however, the explanation attached to the amperian laboratory exercise no longer refers to Ampère's theory (Blondel and Wolff: 2009).

9.8 Conclusion: A New Toolbox for Physicists

The notion of current allowed physicists to avoid unwanted commitment to the speculations accompanying the long tradition of reflections about imponderable fluids with their multiple uses—religious, social, and political—mentioned earlier in this chapter. Uneasy Catholics like Biot and Ampère, Romantics like Davy, Lutherans fascinated by Immanuel Kant's philosophy and the 'scientific sublime' like Oersted, Sandemanians like Faraday, Calvinists like De La Rive: all could converge in adopting an operational, mid-range concept like current that did not imply adhesion to a specific philosophy of nature, nor to a detailed model of the microscopic world, but kept feared materialism at a safe distance. Above all, the notion of current allowed the exploration of the wealth of new phenomena produced by the increasingly sophisticated instruments that physicists were developing with the help of a new generation of instrument-makers specializing in electricity and magnetism. From the 1820s, the new, rich toolbox of instruments available to physicists expanded further thanks to international cooperation in the field of terrestrial magnetism and, from the 1830s, following the take-off of telegraphy as one of the earliest, most fast expanding and cosmopolitan electrical industries (Warner 1994; Hunt 1994).

The convergence on the mid-range notion of current was made easier—if ambiguous—because the notion itself did not entail rejection of the old fluid imagery that, as we have seen, continued to prosper amidst frequent complaints. Ambiguities about the notion of current would not die soon. As late as 1958 Werner Heisenberg, in a discussion following one of his popular lectures on quantum physics, could still remark that 'we [*the physicists*] act as if there really were such a thing as an electric current, because, if we forbade all physicists to speak of electric current, they could no longer express their thoughts, they could no longer speak . . .' (Heisenberg 1961: 33).

The way in which key developments in the physics of imponderables during the first quarter of the nineteenth century were interpreted and translated into the bulk of nineteenth- and twentieth-century physics, for the benefit of discipline building and teaching, did not favour an accurate reconstruction of those earlier developments. Within later rationalizations, the assessment of the comparative merits of the various tools employed by physicists in the period discussed in this chapter favoured mathematical instruments over the instruments stemming from experimental and laboratory practices. In the process, the conspicuous debts that early nineteenth-century physicists and chemists like Davy had contracted with the instrument-makers, and with several branches of industry interested in trying new materials, new processes, and new instruments of precision, were concealed from view. Yet as we have seen especially in the case of Ampère, the latter contributed significantly—just like the virtuoso mathematics of people like Coulomb, Biot, and Ampère himself—to the physicists' imagination.

We now have problems over following what appears as the tortuous path early nineteenth-century authors were pursuing—often moving back and forth as they did between theory and observation, real phenomena and instrumental artefacts, mathematical modelling and actual measures, philosophical or religious concerns and scientific methodologies. However, as historians, the best thing we can do to capture those developments is to try to render as much as we can the winding path physicists were following, rather than adding one more exemplar story, ours, to the stories that later physicists offered while fashioning their own place in history, or trying to mould the *esprit de corps* of future generations of researchers (Morus 2005).

If we want to adhere to what physicists themselves perceived as the signposts they were following in their winding path, however, we may borrow from Maxwell's characterization mentioned at the beginning. In 1871, inaugurating the new chair of experimental physics at the University of Cambridge, after recalling the fruitful collaboration earlier in the century in the field of terrestrial magnetism among mathematicians like Gauss, physicists like Weber, and instrument-makers like Leyser—the equivalent of the Biots, Ampères, and Pixiis we have been considering in this chapter—Maxwell read the signposts guiding physicists as pointing to: 'mathematical power, experimental sagacity, and manipulative skill.' (Maxwell 1871: II, 246). The order on Maxwell's list conveyed a well-established, intellectual, moral, and social hierarchy. On that same occasion, however, Maxwell showed he was fully aware of the novelty that the introduction of *manipulative skills* in a university curriculum represented for many in his academic audience, as well as for what was expected from old, noble natural philosophy.

Having pointed to the possibly disturbing novelty represented by the introduction of manipulative skills in an academic environment, however, Maxwell unhesitatingly stood by it.

REFERENCES

ADAMS J., and BIRCH G. (1799). *An Essay on Electricity*. 5th ed. London: Dillon.
AMPÈRE, A.-M. (1820). Mémoire présenté à l'Académie royale des sciences, le 2 octobre 1820 où se trouve compris le résumé de ce qui avait été lu à la même Académie les 18 et 25 septembre 1820, sur les effets des courants électriques. Suite du mémoire sur l'action mutuelle entre deux courants électriques, un courant électrique et un aimant ou le globe terrestre, et entre deux aimants. *Annales de chimie et de physique*. 15, 59–75, 170–218.
—— (1824). Description d'un appareil electro-dynamique. *Annales de chimie et de physique*. 27, 390–411.
—— (1832). Note sur une expérience de M. Hyppolite Pixii, relative au courant produit par la rotation d'un aimant, à l'aide d'un appareil imaginé par M. Hyppolite Pixii, *Annales de chimie et de physique*, 51, 76–9.
—— (1936). *Correspondance du grand Ampère*. De Launay, L., ed. Paris: Gauthiers-Villars.
—— (2005). Correspondance. Blondel, C., ed., <http://www.ampere.cnrs.fr/correspondance>
—— (2005). Manuscripts. Blondel, C., ed., http://www.ampere.cnrs.fr/manuscrits/
ASTORE, W. J. (2001). *Observing God: Thomas Dick, Evangelicalism and Popular Science in Victorian England and America*. Brookfield: Ashgate.
BEAUDOUIN, D. (2005). *Charles Beaudouin: Une histoire d'instruments scientifiques*. Les Ulis: EDP.
BEDDOES, T. (1799), Specimen of an arrangement of bodies according to their principles. In: Beddoes, T. and Davy, H. (1799). *Contributions to Physical and Medical Knowledge. Principally from the West of England*. Bristol: Biggs.
—— (1801). Experiments with Volta's pile by M. Desormes. *Monthly Review*, 34, 506–507.
BENJAMIN, P. (1886). *The Age of Electricity*. New York: Scribner.
BERTHOLLET, C.-L. (1803). *Essai de statique chimique*. Paris: Demonville. 2 vols.

BIOT, J. B. (1801a). Sur un problème de physique, relatif à l'électricité, *Bulletin de la Société Philomatique*, 21–3.

—— (1801b). Sur le mouvement du fluide galvanique. *Journal de physique*, 264–74.

—— (1803). Recherches physiques sur cette question: quelle est l'influence de l'oxidation sur l'électricité developpée par la colonne de Volta? *Annales de chimie*, 47, 5–42.

—— (1804). Rapport sur les experiences du citoyen Volta, *Mémoires de l'Institut National, Sciences Mathématiques et Physiques*, 5, 195–222.

—— (1816). *Traité de physique expérimentale et mathématique*. Paris: Deterville. 2 vols.

—— (1824). *Précis élémentaire de physique*, 3rd ed. Paris: Deterville. 2 vols.

—— (1827). Volta (Alexandre). In: *Biographie universelle ancienne et modern*. Paris: Michaud (1811–62). Vol. 49, 459–64.

—— (1858), *Mélanges scientifiques et litteraires*. Paris: Lévy.

—— and ARAGO, F. (1806). 'Mémoire sur les affinités des corps pour la lumière et particulièrement sur les forces réfringentes des différents gaz'. *Mémoires de l'Institut*. 7, 301–385.

—— and SAVART, F. (1820). 'Note sur le magnetisme de la pile de Volta'. *Annales de chimie et de physique*, 15, 222–3.

BLONDEL, C. (1982). *A.-M. Ampère et la création de l'électrodynamique (1820–1827)*. Paris: Bibliothèque Nationale.

—— (1989). Vision physique "éthérienne", mathématisation "laplacienne": l'électrodynamique d'Ampère. *Revue d'histoire des sciences*, 42 (1–2), 123–37.

—— (1994), La 'mécanisation' de l'électricité: idéal des mesures exactes et savoir-faire qualitatifs. In: Blondel, C. et al., *Restaging Coulomb*. Florence: Olschki, 99–119.

—— (1999). Animal electricity in Paris: from initial support, to its discredit and eventual rehabilitation. In: Bresadola, M. and Pancaldi eds., G. *Luigi Galvani International Workshop* (Bologna Studies in the History of Science, 7). Bologna: CIS-University of Bologna. 187–209.

—— and WOLFF, B. (2009), A la recherche d'une loi newtonienne pour l'électrodynamique (1820–26). <http://www.ampere.cnrs.fr/parcourspedagogique/zoom/courant/formule/index.php>

BRAIN, R. M. et al. (2007). *Hans Christian Ørsted and the Romantic Legacy in Science: Ideas, Disciplines, Practices*. Dordrecht: Springer.

BRESADOLA, M. (2003), At play with nature: Luigi Galvani's experimental approach to muscular physiology. In: *Reworking the Bench: Research Notebooks in the History of Science*. Dordrecht and Boston: Kluwer, 67–92.

BUCHWALD, J. Z. (1981). The quantitative ether in the first half of the nineteenth century. In: *Conceptions of Ether*, G. Cantor and M. J. S. Hodge, eds. Cambridge: Cambridge University Press. 215–37.

—— (1989). *The Rise of the Wave Theory of Light*. Chicago, IL: The University of Chicago Press.

—— (2000). How the ether spawned the micro-world. In: *Biographies of Scientific Objects*. Daston, L. ed. Chicago, IL: University of Chicago Press. 203–225.

CANEVA, K. L. (1980). 'Ampère, the etherians, and the Oersted connection'. *The British Journal for the History of Science*, 13 (44), 121–138.

CANTOR, G. N. (1981). The theological significance of ethers. In: Cantor, G. N. and Hodge, M. J. S., *Conception of Ether: Studies in the History of Ether Theories*. Cambridge and New York: Cambridge University Press, 135–55.

CROSLAND, M. (1967). *The Society of Arcueil: A View of French Science at the Time of Napoleon I*. Cambridge, MA: Harvard University Press.

DARRIGOL, O. (2003). *Electrodynamics from Ampère to Einstein*. New York: Oxford University Press.

DARWIN, E. (1825). *The Temple of Nature*. London: Jones.
DAVY, H. (1801). 'An account of some Galvanic combinations, formed by the arrangement of single metallic plates and fluids, analogous to the new Galvanic apparatus of Mr Volta'. *Philosophical Transactions of the Royal Society*, 91, 397–402.
—— (1808). 'The Bakerian Lecture: On some new phenomena of chemical changes produced by electricity, particularly the decomposition of the fixed alkalies, and the exhibition of the new substances which constitute their bases; and on the general nature of alkaline bodies'. *Philosophical Transactions of the Royal Society*, 98, 1–44.
—— (1812). *Elements of Chemical Philosophy*. Philadelphia: Bradford.
DEMONFERRAND, J. F. (1823). *Manuel d'électricité dynamique, ou, Traité sur l'action mutuelle des conducteurs électriques*. Paris: Bachelier
DICK, T. (1826). *The Christian Philosopher*. New York: Carvill.
FARADAY, M. (1821–22). Historical sketch of electro-magnetism. *Annals of Philosophy*. NS 2, 195–200, 274–290, and NS 3, 107–117.
—— (1991–). *The Correspondence of Michael Faraday*. James, F. A. J. L., ed. Exeter: IEE.
FOX, R. (1973). Scientific enterprise and the patronage of research in France 1800–1870. *Minerva*. 11, 442–473.
—— (1974). The rise and fall of Laplacian physics. *Historical Studies in the Physical Sciences*, 4, 89–136.
—— (1992). *The Culture of Science in France, 1700–1900*. Aldershot: Variorum.
FRÄNGSMYR, T., HEILBORN, J. L., and RIDER, R. E., eds. (1990). *The Quantifying Spirit in the 18th century*. Berkeley, Los Angeles, Oxford: University of California Press.
FRANKEL, E. (1977). J. B. Biot and the mathematization of experimental physics in Napoleonic France. *Historical Studies in the Physical Sciences*, 8 (1), 33–72.
FULLMER, J. Z. (2000). *Young Humphry Davy*. Philadelphia, PA: American Philosophical Society.
GALVANI, L. (1791), 'De viribus electricitatis in motu musculari', *De Bononiensi Scientiarum et Artium Instituto atque Academia Commentarii*, 7 (Opuscula), 363–418.
GARBER, E. (1999). *The Language of Physics: The Calculus and the Development of Theoretical Physics in Europe, 1750–1914*. Boston, Basel, Berlin: Birkhäuser.
GEE, B. (1993). The early development of the magneto-electric machine. *Annals of Science*, 50, 101–33.
GOLINSKI, J. (1992). *Science as Public Culture: Chemistry and Enlightenment in Britain, 1760–1820*. Cambridge: Cambridge University Press.
—— (2004). 'Nicholson, William (1753–1815)', *Oxford Dictionary of National Biography*. Oxford: Oxford University Press.
GRATTAN GUINNESS, I. (1990). *Convolutions in French Mathematics*. Basel : Birkhäuser. 3 vols.
HAHN, R. (2005). *Pierre Simon Laplace 1749–1827*. Cambridge, MA, and London: Harvard University Press.
HARMAN, P. (1982). *Energy, Force, and Matter. The Conceptual Development of 19th-Century Physics*. Cambridge: Cambridge University Press.
HEILBRON, J. L. (1981). The electrical field before Faraday. In: Cantor, G. N., and Hodge, M. J. S., eds. *Conception of Ether: Studies in the History of Ether Theories*. Cambridge and New York: Cambridge University Press, 187–213.
—— (1993). Weighing imponderables and other quantitative science around 1800. *Historical Studies in the Physical and Biological Sciences*. 24 (1), 1–337.
—— (1999). *Electricity in the 17th and the 18th centuries*. Mineola, NY: Dover.
HEIMAN [Harman], P. M. (1981). Ether and imponderables. In: *Conceptions of Ether*, ed. G. Cantor and M. J. S. Hodge. Cambridge: Cambridge University Press, 61–83.

HEISENBERG, W. (1961). Planck's discovery and the philosophical problems of atomic physics. In: Heisenberg, W., et al., *On Modern Physics*. New York: Potter.
HENRY, J. (1886). 'On the theory of so-called imponderables' (1851). In: *Scientific Writings*, vol. 1. Washington: Smithsonian Institution. 297–305.
HOFMANN, J. R. (1995). *André-Marie Ampère*. Cambridge: Cambridge University Press.
HOME, R. W. (1982). Newton on electricity and the ether. In: Bechler, Z. et al. *Contemporary Newtonian Research*. Dordrecht: Reidel, 191–213.
—— (1992). *Electricity and Experimental physics in Eighteenth-Century Europe*, Aldershot: Variorum.
—— (2003). Mechanics and experimental physics. In: *Eighteenth-Century Science*, Porter, R., ed. Cambridge: Cambridge University Press. 354–74.
HUNT, B. J. (1994). The Ohm is where the art is: British telegraph engineers and the development of electrical standards. In: *Instruments*, Van Helden, A. and Hankins, T. L., eds. *Osiris* 9. Chicago, IL: University of Chicago Press, 48–63.
IZARN, J. (1804). *Manuel du galvanisme*. Paris: Levrault.
JACOB, M. C., and STEWART, L. (2004). *Practical Matter: Newton's Science in the Service of Industry and Empire*. Cambridge and London: Harvard University Press.
KING-HELE, D., ed. (1981). *The Letters of Erasmus Darwin*. Cambridge: Cambridge University Press.
LAPLACE, P. S. DE (1805). *Mécanique céleste*. Paris: Courcier.
—— (1809). 'Mémoire sur les mouvemens de la lumière dans les milieu diaphanes', *Mémoires de la Classe des Sciences Mathématiques et Physiques de l'Institut de France*, 10, 300–342.
MAXWELL, J. C. (1871). Introductory lecture on experimental physics. In: Maxwell, J. C. *Scientific Papers*. Niven, W. D., ed. New York: Dover, 1965, Vol. II, 241–55.
MOIGNO, F. (1852). *Traité de la télégraphie électrique*. Paris: Franck, 2nd edition.
MORUS, I R. (2005). *When Physics became King*. Chicago and London: University of Chicago Press.
—— (2009). Radicals, romantics, and electrical showmen: placing Galvanism at the end of the English Enlightenment. *Notes and Records of the Royal Society*, 63, 263–75.
NICHOLSON, W. (1800). Account of the electrical or galvanic apparatus of Sig. Alex. Volta. *Journal of Natural Philosophy, Chemistry and the Arts*, 4, 179–87.
NOAKES, R. (2005). Ether, religion, and politics in late-Victorian physics: beyond the Wynne thesis. *History of Science*, 43, 415–55.
PANCALDI, G. (2003). *Volta: Science and Culture in the Age of Enlightenment*. Princeton, NJ: Princeton University Press.
—— (2009). On hybrid objects and their trajectories: Beddoes, Davy and the battery. *Notes and Records of the Royal Society*, 63, 247–62.
PARIS, J. A. (1831). *The Life of Sir Humphry Davy*. London: Colburn and Bentley.
PAYEN, J. (2008). "Bréguet, Louis François Clément". *Complete Dictionary of Scientific Biography*, ed. Charles Gillispie, vol. 2. New York: Cengage Learning, 2008. 438–9.
PERA, M. (1992). *The Ambiguous Frog: The Galvani–Volta Controversy on Animal Electricity*. Princeton, NJ: Princeton University Press.
PRIESTLEY, J. (1802). Observations and experiments relating to the pile of Volta. *Journal of Natural Philosophy, Chemistry and the Arts*, I (NS) 198–203.
RAMOND, L. (1811). *Mémoires sur la formule barométrique de la mécanique céleste*, Extrait par Dhombres-Firmas, in *Journal de Physique*, 75 (1812), 253–77.
ROCHE, J. J. (1998). *The Mathematics of Measurement*. London: Athlone.
SHAPIN, S. (1980). Social uses of science. In: Rousseau, G. S., Porter, R., eds. *The Ferment of Knowledge*. Cambridge, London, New York: Cambridge University Press.

THACKRAY, A. (2008). "Nicholson, William". *Complete Dictionary of Scientific Biography*, vol. 10. Detroit: Charles Scribner's Sons, 2008. 107–9.

TODHUNTER, I. (1873). *A History of the Mathematical Theories of Attraction and the Figure of the Earth from the Time of Newton to that of Laplace*. London: Macmillan.

UNWIN, P. R., and UNWIN, R. W. (2009). Humphry Davy and the Royal Institution of Great Britain. *Notes and Records of the Royal Society*, 63, 7–33.

VOLTA, A. (1918–1929). *Opere*. Milan: Hoepli, 7 vols.

WARNER, D. (1994). Terrestrial magnetism: for the glory of God and the benefit of mankind. In: *Instruments*, Van Helden, A., and Hankins, T. L., eds. *Osiris* 9. Chicago, IL: University of Chicago Press, 65–84.

WHEWELL, W. (1833). *Astronomy and General Physics considered with Reference to Natural Theology*. London: Pickering.

WILLIAMS, L. P. (1983). What were Ampère's earliest discoveries in electrodynamics? *Isis*, 74: 492–508.

WYNNE, B. (1979). Physics and psychics: science, symbolic action and social control in late Victorian England. In: Barnes, S. B., and Shapin, S. A., eds. *Natural Order: Historical Studies of Scientific Culture*. London: SAGE. 167–86.

CHAPTER 10

PHYSICS ON SHOW: ENTERTAINMENT, DEMONSTRATION, AND RESEARCH IN THE LONG EIGHTEENTH CENTURY

LARRY STEWART

10.1 A Moment of Power

On travels in the English Midlands in March 1776, while Samuel Johnson sought the acquaintance of a clergyman's widow, his biographer James Boswell recorded a somewhat more momentous meeting. Boswell encountered Birmingham's 'iron chieftain', Matthew Boulton. Boulton then famously described his Soho works, two miles from Birmingham, as a place where, with his partner James Watt, he manufactured 'what all the world desires to have—Power' (McLean, 2009). Pilgrimages to Soho, amid fires and furnaces, were suitably dramatic. Alongside continual consultations about purchasing engine designs, Watt and Boulton, besieged by visitors, became increasingly suspicious of industrial espionage (Jones, 2008a; Harris, 1998). It was little wonder that engines and their principles induced much interest. But what did Boulton actually then mean by 'power'? It could, most obviously, have been a direct reference to the stunning mechanical forces generated from the expansive power of steam, or even the revelation of nature's energy that attracted the early Romantics of the late eighteenth century. Boulton could just as well have meant the economic alchemy that made wealth from forces captured within the cylinders and condenser of James Watt's noisy innovation (Jacob, 1997; Mokyr, 2009). Boswell left, much impressed.

Amid the sounds and soot of Soho, Watt's reputation also attracted attention—not just from the interlopers who tried to circumvent his patent. Some contemporaries thought Watt and Boulton possessed a formula for improvement to be cultivated by a new generation. In the late eighteenth century, fathers sought to fashion their son's future out of an improving spirit. This sometimes meant seeking out the leading agents of mechanical change, especially amidst the teeming work force of up

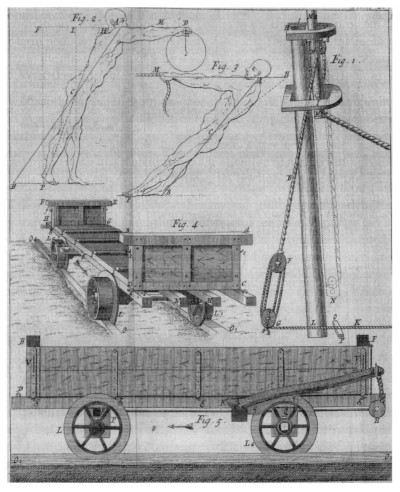

Fig. 10.1. Desaguliers on friction, from *A Course of Experimental Philosophy*. By permission of the Whipple Library, Cambridge.

to one thousand at Birmingham's Soho. Ever hopeful, they looked to Watt on how best to make sons into 'Mechanicks or Civil Engineers'.[1] By the time of Boswell in Birmingham, education in physics and chemistry was at the centre of how Watt defined his personal responsibilities. Despite his earlier experience as an apprentice, Watt had grander ideas for his eldest son. He intended that Gregory should acquire not only French but that he be trained on the Continent in 'Drawing, Mathematicks, Natural Philosophy & chemistry'. Watt had little regard for a university training—about which many expressed scepticism—unless, of course, the aim was to breed sons for the law or the Church. Some may then have been surprised at how the great mechanic responded to their queries. Neither apprenticeship nor scholarship provided much certainty of success.

Watt maintained that basic physical principles were essential to those who would master more than trade.[2] By then, in the late eighteenth century, natural knowledge had become widely accessible (Mokyr, 2009). Watt knew whereof he spoke not just as the product of training by his father and of a brief apprenticeship in London. As a sale of his effects reveals, he was certainly well read (Sotheby's, 2003). He pointed his correspondents in the direction of some of the best-informed publications of the century, most notably to works such as Jean Theophilus Desaguliers' *A Course of Experimental Philosophy* (1734, 1744), in which machines of many kinds were displayed alongside Newton's physics (Fig. 10.1). Watt once wrote to Erasmus Darwin, in 1789, in reference to the accounts of early steam engines by Denis Papin, Thomas Newcomen, Thomas Savery, and Desaguliers. But in doing so Watt also pointedly remarked that he was not interested in 'Calculations or Geometry, my soul abhors them both & all other abstract sciences . . .'[3] It was not simply mechanics, whether in practise or in theory, which thus defined Watt. It was the broad spectrum of his own attachment to the world of the natural philosopher and instrument-maker.

10.2 Emergence of Audience

To any reader in Britain or several places on the Continent, such as The Netherlands, the physical sciences may seemingly have been defined by Desaguliers in the first half of the century. But he was not a solitary actor. Earlier works, like Giovanni Borelli's *De motu animalium* of 1680, had explained muscular motion as a function of bodily statics (Hankins, 1985). In the eighteenth century, Desaguliers adopted an identical motif as he adapted the most recent Newtonian principles to demonstrate the limits of force and the mechanical means of magnifying physical power. Most importantly, Desaguliers, like his early contemporary the Newtonian John Harris of the *Lexicon Technicum* (1704, 1710), did not rely on merely publishing mechanical explanations. Both were Fellows of the Royal Society and scientific lecturers who significantly shaped the expanding realm of public understanding. The first generation of Newton's disciples, among them Harris, Desaguliers, Francis Hauksbee, senior and nephew, and William Whiston, with many others, produced a living by putting science on show. This was a crucial transition. Newton thereby became well known even among those who could hardly be expected to cope with the mathematically abstruse, if magisterial, *Principia Mathematica*.

For sceptics, however, public performance was precisely the problem. To make a theatre of physics was mere pandering to a vulgar understanding. Dramatic experimental theatre obviously could not reach into the depths of the calculus. Indeed, as the anti-Newtonian Roger North once put it early in the century, Newton's reputation would have been secure entirely upon a mathematical foundation. If Newton had 'let dabbling in physics alone, he had done no wrong to his fame . . .'[4] Moreover, Newton's apostles were virtual apostates in physics. North was especially aggravated by what he considered as the emerging sects in philosophy that had been promoted by 'all our second, third, & fourth hand philosophers . . .'[5] For an anti-Newtonian, those disreputable, unknowing disciples, all those 'proletarian scriblasters' who barked at Descartes, those shallow performers and demonstrators who made Newton so accessible, were destructive of a true physics.

In England, public performances proved to be as controversial as they were marketable. By mid-century, there was also a whole lot of sneering going on. In 1747, the engineer and instrument-maker, John Smeaton, sniffed about the 'several Pursuits Designs and Policys of ye common Herd of conjuring Philosophers about Town . . .'[6] And in 1752, John Freke, surgeon to St Bartholomew's Hospital and Fellow of the Royal Society, complained of the lecturer, instrument-maker, and Newtonian promoter, Benjamin Martin, that his works simply were 'as ridiculous as to see a Pygmy attempt to carry a Giant' (Freke, 1752; Schaffer, 1993). To some, popularity was poison.

But public physics was certainly not an English innovation. There were many models, the Dutch and the French among the most obvious. In the 1690s, near Amsterdam, the mathematician and physician Bernard Nieuwentyt was already presenting weekly demonstrations of an air pump. Desaguliers certainly knew of him, translating into English his natural theological *Het Regt Gubruik Der Werelt Beschouwingen* in 1717, just about the start of Desaguliers' career as public lecturer in London. Only shortly before, Newton had nominated Desaguliers as Fellow of the Royal Society, undoubtedly to harness Desaguliers' experimental skill in the contest over Newton's prisms and optical refractions. At that moment Willem 'sGravesande, attached as secretary to the Dutch legation to London, attended Desaguliers's London lectures before returning to an academic life in Leiden. (Vermij, 2003; Van Helden, 1994; Stewart, 1992). 'sGravesande became especially influential in the spread of Newtonian philosophy in The Netherlands, adopting for public experiments the latest Fleet Street apparatus like Francis Hauksbee's electrical generator. Watt had a copy of Desaguliers' translation of 'sGravesande's *Elements of Natural Philosophy* (Uglow, 2002) Desaguliers later retraced these channels of connection, engaging in lecture tours of The Netherlands in 1731 and 1732. Similarly, Petrus van Muschenbroek, 'sGravesande's successor at Leiden, was notably interested in developing demonstrations for amateurs, perhaps derived from his awareness of the growing London scene (Roberts, 1999a; Jacob, 1997).

In the aftermath of Newton's death in 1727, his printed legacy was most directly obvious in the 'Queries' he had added to editions of his *Opticks* (1704, 1706, 1726). It is here that we can see the convergence of inter-particulate attractions, optics, electricity, and chemistry, which established an extensive programme of experimentation throughout much of the century. For the French, and chemists in particular, dramatic demonstrations were generated from light and prisms. Hence, in Paris, Guillaume-Francois Rouelle, as an apothecary, had learned the skills of display at the Jardin du Roi (Rappaport, 1960; Mayer, 1970; Jacques, 1985; Lehman, 2010; Perkins, 2010). He drew large crowds of auditors to his lectures, including notable Enlightenment figures such as Jean-Jacques Rousseau and Denis Diderot, and ultimately he taught chemistry to the young Antoine Lavoisier (Baretta, 1993; Roberts, 1995). Cross-Channel communication was not simply furthered by the publications of Newtonian promoters like Voltaire. Voltaire's polemic about the physics of London's universe in contrast to the vortices of Cartesian Paris was amply, and sarcastically, answered by the abbé Nollet: '... one ought to recognize the truth wherever it shows itself, and not affect to be a Newtonian in Paris and a Cartesian in London' (Nollet, 1775; Home, 1992; Voltaire, 1733). It was, significantly for our purposes, the experimental interests of the French— especially Charles Dufay and the abbé Nollet—which effectively transformed the relationship between philosopher and instrument-maker, as in an electrical theatre, helping make physics part of the public culture of the eighteenth and nineteenth centuries (Golinski, 1992; Morus, 2005, 2006). Both were especially influenced by the demonstrations of electric sparks in London by Stephen Gray. In London in 1734 both Dufay and Nollet witnessed electrical displays, laying the foundation for Nollet's subsequent career as public showman and purveyor of instruments to the notable (Schaffer, 1992, 1997; Bensaude-Vincent and Blondel, 2008). The lecturing trade was a cosmopolitan one (Bertucci, 2007). Nollet's scheme was 'to make physics speak to the eyes & to share it with the most able persons who will know its worth' (Gauvin, 2002). Consequently, as he put it quite explicitly, 'Le goût de la Physique devenu Presque general, fit souhaiter qu'on en mit les principes à la portée de tout le monde'. (Nollet, 1775). Nollet was so successful in translating the public display of physical principles into a promotion for instruments that he attracted numerous enthusiastic clients for his apparatus. Voltaire was clearly seduced by more than his philosophic partner Madame de Châtelet as they explored experimental philosophy together. Voltaire complained that Nollet virtually bankrupted him by offers of apparatus (Lynn, 2006).

One thing was clear among the globes of static electrical machines. The experience of experiment depended on the availability of laboratory instruments. The mission of Nollet proved overwhelmingly successful largely because of the grand range of influential and wealthy clients. His market was international in scope. Jean Jallabert of Geneva bought from Nollet a large experimental collection, especially of pneumatic apparatus and the electrical machines which became all the rage in the middle of the century (Benguigui, 1984). Similarly, the Genevan philosopher Horace-Benedict de Saussure later collected an array of instruments from many sources including, from London, an electrometer of the Neapolitan dealer and writer Tiberius Cavallo and a magnetometer designed by Gowin Knight, both Fellows of the Royal Society (Archinard, 1989). De Saussure also contrived electrometers and magnets, significantly some of which could be dismantled and readily transported to difficult spaces. Experimental apparatus was not confined to drawing-room spaces or gentlemen's societies. De Saussure, for example, measured electrical charges at various altitudes on Chamonix or Mont Blanc (Archinard, 1989). Likewise, in the first half of the century, on the less imposing hills and heaths around London, through electrical wires on the bridges, or tied to the tip of the dome of St Paul's cathedral, and among the cabinets of Huguenot workshops of Spitalfields, experiments ignited enthusiasm and attracted witnesses.

The market in experimental instruments was often explicitly focussed on display and public demonstration. Such an instrumental commerce further drew skilful practitioners from the provinces. One of the most long-lasting was Benjamin Martin, who came from Chichester in the 1750s as a publisher, lecturer, and instrument-maker to London's Fleet Street, just a few doors from the Royal Society. In Fleet Street especially, the competition would have been fierce, but the breadth of his venture provided as much in the way of potential wealth as it did overreach. Martin readily recognized the possibilities for an experimentally inclined schoolteacher. In 1737, in the preface to his *Bibliotheca Technologica*, Martin remarked that 'Men, in General, are not so much wanting in capacity as in time and pocket, for gaining literary accomplishments. Huge tomes, and great prices, are, like mines, the property of the great' (Millburn, 1976). Thus Martin's strategy was to attack the problem of cost not by publishing large works, but by making books, instruments, and lectures accessible across the range of the market. Access to the Royal Society, alas, remained firmly denied to Martin's Grub Street sympathies. If snobbery could overwhelm, so too could markets. Martin ultimately went bankrupt in the 1780s. When Martin moved to the metropolis, James Watt then also quickly recognized that the growing public interest in experimental philosophy provided much opportunity for a young man of skill. Among Watt's papers there is a surviving catalogue from Martin, quite possibly from the spring of 1756, which reveals the extent of the wares he had on offer. Hence, there was to be found a considerable range from an expensive orrery at £150, to much cheaper digesters, magnets, and prisms from 5 shillings (Millburn, 1976).[7] To put physics on show, such an armoury of apparatus would prove necessary.

The spectacle of experiment would likewise have meaning for Watt. At the very moment of Nollet's Parisian forays into public exhibitions of the dramatic shocks of electricity through a line of two hundred Carthusian monks, in London the philosophical market was expanding rapidly. The instrument-maker John Neale, variously of Fleet Street and Leadenhall Street, saw publication as a means of advertising his wares. In 1747 he had printed *Directions for Gentlemen, who have Electrical Machines, How to Proceed in Making their Experiments*. Revealing 'experiments usually exhibited in publick' without 'enquiring into, or defining what electricity is', Neale set out an account of the application of the static electrical generator. The device was sufficient to raise astonishment in uninformed minds. Yet, Neale had a keen sense of the marketplace. He also offered to give a course of lectures on a patented globe, thereby applying himself to those concerned with navigation, and to those confounded by the longitude in particular. But it was his rumination on the uses of the electrical machine, if not the theory, that reveal his understanding of what had, by mid-century, become an international experience. He was well-versed on the Leiden jar developed by van Musschenbroek in The Netherlands. In London, Neale repeated the Parisian electrical demonstrations of Louis-Guillaume le Monnier, as well as an experiment proposed to Neale by John Canton, FRS, master of the Academy in London's industrial Spitalfields (Neale, 1747; Taylor, 1966; Wallis, 1982; Ashworth, 1981). Canton was part of a network corresponding with Joseph Priestley on electrical matters and the production of a battery. Neale may actually have been bankrupt by 1750, not because of his trade as a watch and globe maker (while member of the Skinners' Company) but from an abortive astronomical publishing venture. Significantly, Neale was briefly to take on the young James Watt, who was desperate to circumvent restrictions placed by the London guilds on the unconnected, while Watt in frustration wandered London seeking an apprenticeship in 1755 (Clifton, 1995). While his was a relatively brief period of training, the encounter with Neale would at least have drawn Watt directly into the universe of the lecturer. Neale was by then obviously well aware of the uses of instruments as demonstrations devices. This likewise incidentally engaged Watt himself after his return to Scotland. In 1763 he made repairs on a model Newcomen engine used in the natural philosophy class of John Anderson at the University of Glasgow (Musson and Robinson, 1969). Anderson was a preeminent exponent of experimental method.

Devices of demonstration put physics on show. Desaguliers' influence lingered, but the demand for science shows continued to expand well after his death in 1744 (Schaffer, 1983). As Desaguliers had once put it, in the introduction to the second edition of his *Course of Experimental Philosophy*, 'a great many Persons get a considerable Knowledge of Natural Philosophy by way of Amusement; and some are so well pleas'd with what they learn that way, as to be induc'd to study Mathematicks, by which they at last become eminent Philosophers' (Desaguliers, 1745, I; Stewart, 1992; Schaffer, 1994). Demonstration was an inducement to engagement. Yet, it was possible, and still is, to dismiss entertainment as no serious method of access to the sciences. In our world of puffery, publicity, and consumer culture, there may be something to such a suspicion. But for the eighteenth century, when individuals went out of their way, and paid substantial sums, to subscribe to works of physics and to witness public demonstrations, such scepticism is very wide of the mark. Indeed, Desaguliers was as serious as he was obviously self-serving about the lecturing trade. He reversed the track of learning, in other words from amusement *to* engagement, and refused to reduce entertainment to mere indulgence. For those who sought a living from performance, apparatus thus proved essential. A healthy investment in instruments might set showmen apart from some competitors; for others, apparatus became something of an addiction.

10.3 DEVICES OF DEMONSTRATION

Collectors had long been central to the early-modern sciences. Cabinets of curiosities had once helped make the reputation of a gentlemen or a prince (Freedberg, 2002). By the eighteenth century, physics, chemistry, astronomy, or the new science of electricity, required more devices to explore the boundaries of nature. In drawing rooms and libraries, cabinets full of instruments increasingly replaced collections of curiosities as the mark of philosophic sophistication (Altick, 1978, Anderson, 1984). Moreover, this was an index of another crucial distinction. While curiosities could induce a philosophic contemplation, instruments were not simply to be admired; they were to be used. Practice became paramount in the physical sciences. Of this, whether in America or in Europe, there are obviously many examples (Delbourgo, 2006). But one significant feature of the Enlightenment was the way in which instruments merged with the promotion of science shows—not only did lecturers attract subscribers by advertising their apparatus, but the array was also essential in engaging those auditors to the point where some in the audience themselves privately undertook further experiments and observations. Witnessing in this manner proved more valuable than reading. As Jan Golinski has shown, the thermometer and barometer were among the most commonly available devices (Golinski, 1999) (Fig. 10.2). Interestingly, the exact measures of physical phenomena were in transformation during the eighteenth century, especially in establishing scales to define the rarefaction of atmospheres and the expansion of liquids. To purchase such precise apparatus, in the shops of London or Paris, as in Amsterdam, was at least to hint at application. The relationship between lecturers and audience evolved out of these devices of demonstration.

For those in the provinces who lived at the fringes of the philosophic world, it was instruments that gave them access. To take one relatively obscure example from mid-century, the Scottish mathematical teacher and ultimately peripatetic lecturer James Dinwiddie explored the then contested phenomenon of heat. He remarked in his public lectures that 'The numbers on the scales of Thermometers only inform us of the Variations of heat between two certain points which are known but as scales have been variously marked it has caused Considerable confusion in comparing the experiments made by one Thermometer with those performed by another.'[8] This was especially meaningful, for example, regarding the highly confusing and contested matter of caloric and its precise composition which lurked in the laboratories of experimental chemists throughout the century. Dinwiddie gave up teaching boys mathematics in a school in Dumfries to take advantage of the growing opportunities for experimental philosophers. It was devices that turned him into a peripatetic philosopher. His fascination with instruments turned on their capacity to reveal and to reflect the universal forces of nature wherever such explorations may take place. What was witnessed in Dumfries was, in his experience, no different than in Dublin or Paris—or, for that matter, in Peking or Calcutta. Uniformity and universality were twin hurdles that the early-modern physicist faced. The same, surely, must have been true for any observer or experimentalist who had travelled far beyond the portals

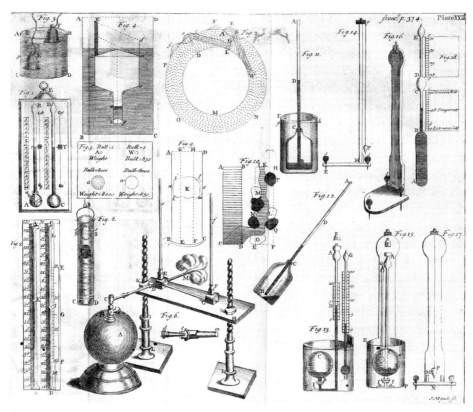

Fig. 10.2. Desaguliers on air thermometers and barometers, from *A Course of Experimental Philosophy*. By permission of the Science Museum, London.

of a gentleman's club in London, Paris, or Amsterdam (Ferreiro, 2011). Instruments made the local universal—and therefore, like many an eighteenth-century travelogue, might tell a profound tale.

For Dinwiddie, what then had begun in the small, if bustling, ship-building town of Dumfries revealed how showmanship followed instruments. Indeed, it is Dinwiddie's singular lack of prominence that now gives us hints of the everyday manufacture of the physics shows. While teaching at the Dumfries Academy, Dinwiddie had become fascinated with practical mathematics and, within a short time, was drawn to the then hot topics of electricity and magnetism. These were the very passions that consumed experimentalists from Paris to Padua, Leiden or Amsterdam, from Bologna to Birmingham. And his reading was from the most recent works of Gray, Dufay, and Canton's friend Joseph Priestley. Priestley's attitude to the instrumental turn was particularly significant in attempting to ensure that readily available, everyday, equipment might be turned to experimental use. He was, however, suspicious of the instrument-makers' market. In 1767 Priestley pronounced that philosophers had far more lofty aims than mere instrument-makers 'who are seldom men of science, and whose sole aim is to make their goods elegant and portable' (Priestley, 1767; Schaffer,

1993). But this seriously underestimated the extent to which even experimentalists like Priestley were themselves dependent upon the design and manufacture of apparatus. There was a tension in Priestley's attitude no doubt, between his own domestic laboratory and its common contrivances, and those spaces increasingly populated by elaborate or expensive devices of drama and display. Yet, even these distinctions can be misleading. Priestley's much wished-for democratization of experiment might be hostage to cost. Indeed, even before Priestley planned his happy move to Birmingham, 'to dedicate his time altogether to physical pursuits', it was rumoured that 'A subscription is to be made by some opulent individuals to defray the expence of his experiments' (Bentham, 1968, I; Rutt, 2003, I). He may have seen in instruments a polemical end; yet he nonetheless depended on the very wealth of his patrons that he did not himself possess.

Dinwiddie did not take Priestley to heart. Dinwiddie's priorities were ones not of novel experiment, but of a profitable performance. By 1774 it appears he was in London briefly, looking at apparatus from a 'Mr. Martin', as well as buying the publications of Priestley on airs and of the lecturer James Ferguson on astronomy.[9] By then, Dinwiddie was already long convinced that demonstration devices might go a long way to helping teach basic principles of astronomy, just as the itinerant James Ferguson was then making his mark. As early as 1765, Dinwiddie had been engaged in what appears to have been his first significant purchase, of an orrery by a subscription raised among the worthies of Dumfries.[10] It was not long before he was incurring much personal debt in securing further apparatus. The seductions of instruments brought him much grief. If they took him to the edge of bankruptcy, they also expanded the limits of nature. He thus discontinued teaching boys to seek his fortune as a public lecturer. This was surely not the best strategy for someone of limited means. Indeed, without the kind of patrons Priestley proved able to attract, such as the wealthy Earl of Shelburne, Dinwiddie had to gain sufficient auditors in country towns by advertising in the newspapers. Throughout Ireland and Britain he was long active, if not especially secure, in the theatre of physics and chemistry. Shortly after his visit to London in 1774, he appears to have obtained telescopes, prisms, portable air pumps, and electrical machines. The details on these are slight, but we know that within a decade his portable apparatus contained glass tubes, electrical devices, and static generators sufficient to make figures dance and sparks to discharge for his audience.[11] It is often difficult to date the traces of Dinwiddie, yet it is probable that by the 1780s he was in manufacturing towns lecturing broadly on mechanics, on pendulums, pulleys and pumps, and machines for reducing friction.[12]

As he wandered from town to town, Dinwiddie scoured the philosophical news often reported in the daily papers. We know that in the mid-1780s he was paying particular notice to the rage for aerostatics, to the daring of the balloonists who were capturing crowds in the tens of thousands on both sides of the Channel. Of course, such witnesses were not always paying subscribers to balloon ascents, but there were surely enough enthusiasts given access to the enclosures of the inflating balloon to make the physical and financial risk worthwhile to the adventurers. For someone versed in the chemistry or specific gravities of airs, the opportunities were obvious.

Dinwiddie was soon among the performers of such trials in Dublin, ultimately dragging his inflatable apparatus through English towns, and putting on a performance at Burlington Gate, London, in the early winter of 1783 (Lunney, 1988; Lynn, 2010; Thébaud-Sorger, 2009; Kim, 2006).[13] Little could have proved more spectacular than these inflatable devices, particularly as many crashed and burned before even getting off the ground. His wandering continued apace. He was in Paris shortly before the Revolution, at a time when the circus of hot-air balloons attracted as much attention as political disputes about the price of bread. He was alarmed by what he then called the 'Buffooneries of the Boulevards', the sidewalk shows of buskers, comics, and musicians, not unlike those in Desaguliers' Covent Garden, which seemed only to distract from serious business. But Dinwiddie also attended the Académie des Sciences, saw the large plate electrical machine at the Lyceum, and looked over the collection of elaborately decorated apparatus of the abbé Nollet.

Demonstration machines made physics real to an amateur audience. Equally fascinating to Dinwiddie in Paris were the surviving devices of the mechanic Jacques Vaucanson of drills, pile drivers, and textile machines on which Dinwiddie made many notes, perhaps in view to explaining them in his lectures.[14] Mechanics was important, not merely because of the work machines might be set to do, but because they provided a means by which it was possible to reveal fundamental principles in physics—just as had once been the case with Desaguliers. As Simon Schaffer has pointed out, in the open world of the lecturers, mechanical apparatus asserted an authority and mathematical certainty that was otherwise unattainable if simply reduced to claims of sensation rather than exact measurement. Precision refined credibility. This was the objective of the Cambridge mathematician George Atwood, before 1780, in the design of a machine that would allow mechanical measurement of friction and reaction (Fig. 10.3). Interestingly, this reflected the contemporary models of the engineer John Smeaton to define the limits of waterwheels, along with Atwood's machine revealed to the Royal Society. Significantly, the London instrument-maker George Adams made Atwood's device part of his offerings, and versions of it were sold worldwide. Some existed in private collections, and demand was consistent. Other versions were offered for resale by the instrument-makers William Jones and Edward Nairne when they purchased Adams' stock in 1793 (Schaffer, 1994).

Philosophical instrument-makers essentially manufactured the market for public performance. Without their apparatus lecturers would surely have had much less with which to seduce subscribers. But their revelations, such as optical refractions against walls or the blue lights glowing from evacuated cylinders in darkened lecture-rooms, however dazzling, were not to be reduced to mere amusement. Demonstration devices had serious aims. In 1793, when William Jones secured many of the lots in the auction of the personal collection of Lord Bute, he immediately offered many for resale. One such was the so-called 'philosophical table'. This had been created, rather like the Atwood machine, to demonstrate mechanical principles, but in this case derived 'according to THE MATHEMATICAL ELEMENTS OF NATURAL PHILOSOPHY of W. J. Gravesande, Desagulier's [sic] edition . . .' This particular version was ultimately bought by Frances Wollaston who was, from 1792, Cambridge Professor of

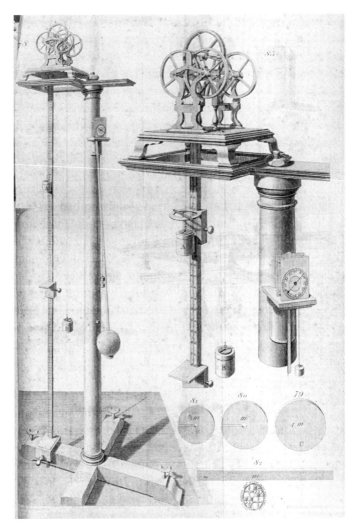

Fig. 10.3. Atwood's machine, from *A Treatise on the Rectilinear Motion and Rotation of Bodies*, Cambridge, 1784. By permission of California Institute of Technology.

Natural and Experimental Philosophy (Turner, 1967). Jones knew his market. Very shortly after the demise of the unlucky Benjamin Martin in 1782 Jones, who had once been Martin's pupil, was cultivating a wide clientele. Both had assumed an aggressive marketing strategy, but Jones was perhaps less over-extended, especially in publishing projects, and held no fear in holding expensive pieces in stock to be proposed to the wealthy (Millburn, 1973, 1976). His first publication, *The Description and Use of a New Portable Orrery*, following Martin's model, was intended to demonstrate astronomical principles to students at limited cost. He consciously intended to attract the custom of public showmen. By 1784 Jones proposed a portable orrery explicitly for itinerant lecturers, but he also offered larger orreries and planetariums modelled on those of Martin and of the provincial lecturer James Ferguson. Orreries were a constant subject of fascination, included in demonstrations of mechanics and astronomy (Ferguson, 1768).

Jones did not let any opportunity escape him. Within a decade, the well-connected Jones was writing to Dinwiddie, then a lecturer in Calcutta, where he had settled since leaving China and Lord Macartney's embassy. Dinwiddie was offered a magnificent orrery that Jones had managed to purchase at the Bute auction. This, he announced, was the best such device at the price 'for a public seminary or a popular lecture'. The so-called 'Grand Orrery', proposed to Dinwiddie in 1795, was designed with a glass pane 'that opens as a door, to admit the hands of the lecturer or pupil' (Millburn, 1973, 1985; Morrison-Low, 2007)[15] (Fig. 10.4). This expensive apparatus was, Jones suggested, to shift from a private collection to that of a cash-rich showman, and offered at half its original price of £400. Despite his longstanding addiction to apparatus Dinwiddie, then contemplating a return to Britain, was divesting himself of devices in India. He did not take up Jones' proposition.

Fig. 10.4. A 'Grand Orrery' by Thomas Wright, c.1733. By permission of the Science Museum, London.

Instrument-makers had to manage their customers carefully. For those who produced experimental apparatus there was clearly a broad spectrum of use, from private laboratories, to the workshops of amateurs, to the burgeoning trade in public lectures. As Simon Schaffer has argued, some makers needed to offer clearly sophisticated philosophical instruments for the investigator along with those that 'are of no further utility than to surprise or amuse the illiterate spectator' (Schaffer, 2006). But science on show did not always allow for clear or distinct objectives. Indeed, there were many makers who took care to promote their wares as worthy of many pursuits, in a spectrum from amusement, to experiment, even to utility. Instruments served several purposes, and public presentations depended as much on skill with devices as well as on their design. Hence, the Swedish engineer Marten Triewald, who had once undertaken electrical research before the Royal Society, had amassed numerous pieces of demonstration apparatus by mid-century. Likewise, the engineer John Smeaton was elected to the Royal Society in 1753 entirely on his reputation as a 'maker of Philosophical Instruments' (Heilbron, 1979).[16] He so impressed Joseph Priestley that Smeaton's air-pump was used in Priestley's experiments on the specific gravities of different types of air. Paola Bertucci has similarly pointed out that the instrument-maker Edward Nairne offered apparatus designed by Tiberius Cavallo, by William Nicholson, who was once an agent to Josiah Wedgwood, and by the innovative Italian Allesandro Volta (Bertucci, 2001).

The eighteenth-century trade in instruments and scientific theatre was international in scope, and it crossed disciplinary and social boundaries as well as geographic ones (Withers, 2007). Indeed, one may argue that it was particularly the role of showmen that helped make the instrument trade so ubiquitous. We know that as early as 1783 Josiah Wedgwood had bought Nairne's Patent Electrical Machine. Nairne later marketed his improved apparatus as a medical device as much as an experimental one. Soon the young Tom Wedgwood, a promising chemist, engaged Humphry Davy to purchase electrical apparatus from Nairne in London. Indeed, visitors to London frequented Nairne's shop (Bentham, 1971, III).[17] Thus, in 1777 the Danish professor of astronomy and mathematics Thomas Bugge had bought from Nairne magnetic and pneumatic apparatus alongside electrical ones (Bertucci, 2001; Bennet, 2002). Cross-Channel commerce expanded rapidly throughout the century, part of which might be attributed to the Enlightenment experience of the Grand Tour. Sir William Hamilton, ambassador to the Court of Naples, had purchased Nairne's powerful machine, as did the Duke of Tuscany. From philosophical Florence, the British collector Lord Cowper ordered Nairne's new device, perhaps following the fashion of Lord Bute, whose own vast store of instruments had also contained a medical electrical apparatus (Bertucci, 2001; Morton and Wess, 1992). And van Marum in Haarlem was continually in touch with Nairne, even for new innovations in thermometers.[18]

The spectrum of public drama in the physical sciences demonstrably engaged many levels of European society. Not only was the cultivation of physics a mark of gentility and learning, it was also a measure of curiosity and possible utility from the English Midlands to The Netherlands. Lissa Roberts has shown how from 1777 the Amsterdam

Society, Felix Meritis, championed the 'Enlightenment values of utility and virtue'. Instruments were at the core of Felix Meritis. Indeed, one of the requirements for initiation into its membership was a donation of a scientific instrument or the financial equivalent (Roberts, 1999b). Experimental engagement, promoted by lecturers, moreover infected the landed as much as the scholarly and the industrial. This was one of the reasons Watt was at the centre of enquiries concerning mechanical training. And, as Watt pointed out, physical principles ought not to be left to artisans or guilds. Watt reflected a wide intellectual movement. Down the social ladder from gentlemen's clubs, from its origins in 1717, the Spitalfields Mathematical Society in east London had collected a large philosophical library and, as in Amsterdam, donations being required of its officers. In Spitalfields they obtained, by the end of the century, a substantial collection of apparatus from some of the best London makers. Spitalfields' own craftsmen could have access to this array of apparatus—including the simple yet effective device by Dollond to show refraction through different liquids. It is difficult now to know when they began to form their cabinet. However, by 1784 they had two double-barrelled air pumps, possibly by Adams, three electrical machines, one of them portable, a hydrostatic balance, along with various microscopes, telescopes, and mathematical instruments (Spitalfields, 1784; Stewart and Weindling, 1995). Within the decade the instrumental realm had conquered the 'mathematicians' of industrial Spitalfields. By 1792 the Society required each of its Fellows take turns 'to give a Lecture on Mathematicks, or some branch of natural, or experimental philosophy, or shew some experiment relative thereto' (Spitalfields, 1793). In 1804 they described their Repository of Instruments as 'another road to the curious investigator of the works of nature'. This was virtually a manifesto of Enlightenment. They declared that 'Many members of the community, without this aid, must have been for ever excluded from much valuable information in this department.' Instrumental access reflected the eristic of democratization that Priestley had promoted. This was made even more explicit by the industrialist James Keir in his translation of Pierre-Joseph Maquer's *Dictionary of Chemistry* (Keir, 1789). For ideals such as these, Fellows of the Spitalfields Mathematical Society stepped forward to present lectures, and thereby 'gave the public at large an opportunity of increasing their knowledge, on terms so easy, as to be within the reach of every individual, who has a taste to cultivate, or a curiosity to gratify' (Spitalfields, 1804).

In London's Spitalfields the convergence between lecturers and purveyors of instruments was complete. This was a Society that could once count among its most accomplished John Dollond, who had created the achromatic lens, the experimentalist John Canton of Spitalfields Academy, and Thomas Simpson, weaver and tutor at the Mathematical School. More importantly, perhaps, its cabinet was extensive.[19] At the end of the century we find exactly the additions we may have expected among the conjurors of physics: numerous Leiden jars of varying sizes, a battery made of twenty-five jars, two tin conductors with stands, a gold-leaf electrometer (possibly by Abraham Bennet), a dish for firing spirits, a universal discharger (probably by William Henley), two galvanic troughs, hydrostatic bellows, hydrostatical column and hydrostatic paradox, cylindrical weights for specific gravities, air pump and condenser, a glass vessel with two bladders to shew the action of the lungs, pneumatic trough, gasometer, deflagrating glasses, air thermometer, pyrometer, and eudiometer (Spitalfields, 1804). All of these were demonstration devices. The tie between the instruments and the shows of physics was precise. In this lay the public power of early-modern physics.

10.4 Audience and Degrees of Engagement

The diffusion of knowledge ought to give us some pause about the meanings and boundaries which now seemingly sets physics apart from other scientific disciplines. Such intellectual barriers simply did not exist in the eighteenth century. No-one then saw themselves exclusively, or even primarily, as a physicist above any other scientific passions such as chemistry, electricity, or astronomy. My proposition here is that entertainment provided by lecturers evolved into engagement of many kinds. To identify members of an audience or their reactions to lectures—at least outside universities—has, as we might expect, proved virtually impossible. Yet some traces do exist. One copy of an early syllabus of a course by Desaguliers had a comment scribbled in the margin about 'the man who combines usefulness with pleasure has covered every point' (Cantor, 1983). The unification of usefulness with pleasure was an undoubted selling point. This is why those captivated by experimental drama in a cosmopolitan Europe were further drawn to the instrument-makers' shops. Many sought out apparatus that transformed amusement into their own exploration of nature. Dissemination in the physical sciences clearly had much to do with this commerce in devices (Levere, 2005; Bennett, 2002). But the purchase of apparatus was almost as anonymous as attendance at lectures. Even so, engagement with public knowledge was not to be confined to those we now acknowledge as having exploded the frontiers of early-modern physics or even of chemistry. Disciplinary barriers were elusive. This occurred just as markedly in North America, where Benjamin Franklin and Ebenezer Kinnersley, along with many others, were ultimately joined by Joseph Priestley, a refugee from the Birmingham mob that burned down his house and laboratory in 1790. America generated enlightened sentiment out of electricity (Delbourgo, 2006). In other words, the cultivation of physics was part of the formulation that captured a deep sense of improvement and drew the ambitious to natural knowledge. Across the breadth of empire, the spread of sensual experience that apparatus could impart was essential both to the drama of shows and to the relatively unknown audiences that witnessed them. Instruments defined the broad competition among the showmen that marked the eighteenth century.

In the 1790s, when James Watt replied to enquiries he was implicitly arguing that the dissemination of physical principles was worthy of consideration. It mattered little if this ultimately derived from lectures or from print. All ranks of society could gain something, whether that be natural knowledge or laboratory practise, or merely polite, drawing-room, chat. By the time Watt had arrived in Birmingham in 1774 there were many lecturers wandering the countryside, advertising in the papers their impending performances and seeking sufficient subscribers. There were many examples, far too numerous to note here. Of particular importance among the purveyors of mechanical principles were James Ferguson, John Arden, and John Warltire. Indeed, their paths not infrequently intersected, and particularly in the English Midlands towns where there were many mechanics and artisans bent on self-improvement (Musson and Robinson, 1969; Jones, 2008b; Elliott, 2009; Bruton, 2009).

The Midlands connection is revealing. Demonstrations of Newtonian natural philosophy and experimental speculation followed the trail of lecturers from town to town. John Arden at Beverley in Yorkshire styled himself (much as had Benjamin Martin) a 'Teacher of Experimental Philosophy'. His tour included Birmingham, Manchester, Whitby, and ultimately brought him to the rather more gentrified spa town of Bath. This was, in essence, a symbiotic transition from Grub Street purveyors to genteel performance. While it is unusual to find first-hand accounts from subscribers to public lectures, one J. Wear published his notes of Arden in February 1760. Wear admitted they were imperfect, and he asked his reader's indulgence. But the notes did appear to be close to a template for Arden's later lectures. Over the rest of the century, Arden's scope covered Newton's colours, electricity following from 'the ingenious Mr. Frankland [sic] of Philadelphia', along with barometers, globes, pneumatics, and even mechanical powers which, Wear admitted, as 'did not Much engage my Attention I pass them by'. We know much of this, essentially, as Arden's son published an *Analysis of Mr. Arden's Course of Lectures on Natural and Experimental Philosophy* in both 1782 and 1789—the last course probably performed at Hull in 1790 (Arden, 1789; Musson and Robinson, 1969).[20] It was the demand from the auditors that had caused the son to print his *Analysis* 'to serve by Way of Memorandum or Pocket-Companion, for those Parts which would most likely escape the Memory . . .' (Arden, 1782).

In the booming industrial towns, subscribers were drawn by lecture notices in the local papers often printed weeks in advance. One of the most enduring of the demonstrators was John Warltire, who had been lecturing since the 1760s. It appears that Matthew Boulton then met him in Chesterfield, and by the late 1770s, at the invitation of Josiah Wedgwood, Warltire was lecturing at Etruria. Perhaps this was the beginning of Wedgwood's determination to educate his sons in experimental philosophy. Warltire ultimately gave private lectures to the sons of both Wedgwood and Erasmus Darwin (Coley, 1969). Warltire's guidance provided much practical advice to those drawn to experimental practise—notably on Wedgwood's longstanding obsessions with chemical combinations. While Warltire had a significant following over a long career, he nevertheless apparently died in poverty in 1810 (Warltire, 1769, 1770?; Coley, 1969; Torrens, 2004). To show the mysteries of physics or chemistry was not always a paying proposition.

In the 1760s John Gilbert, steward and canal-builder to the Duke of Bridgewater, had encouraged his son to attend the lectures of Arden in Birmingham. Gilbert was long connected to Boulton, and his son was then Boulton's guest (Musson and Robinson, 1969). This example was typical of the experimental enthusiasm widespread across the landed, industrial, and even some of the labouring classes. Similarly, when in 1771 James Ferguson passed through Newcastle, near Stoke on Trent, Josiah Wedgwood had unfortunately been unable to attend his lectures. He thus offered Ferguson a trade. He would give Ferguson lessons on pottery manufacture in return for dinner discussions on natural philosophy (Millburn, 1983, 1985).[21] At the same time, Wedgwood was able to secure a copy of James Keir's first translation of Macquer's *Dictionary of Chemistry* which he trumpeted to his fellow potter, Thomas Bentley, as a 'Chemical Library!' The *Dictionary* was a mirror to the public market. This is of some interest, as Keir intended to expand Macquer into several volumes, and to refine it explicitly on the grounds that, by 1789:

The diffusion of a general knowledge, and of a taste for science, over all classes of men, in every nation of Europe, or of European origin, seems to be the characteristic feature of the present age . . . But in no former age, was ever the light of knowledge, so extended, and so generally diffused. (Keir, 1789)[22]

This was the mission of the shows in physics as well.

As Watt revealed, the experience of experimental lectures was significant to promoters of manufacturing ventures in the late eighteenth century. Keir's version of Macquer was an effort to make the chemical dictionary more widely applicable in the English industrial circumstance. In the view of Jeremy Bentham, Keir was associated with London lecturers. But for Keir it was the link to Birmingham's Dr George Fordyce that was of more immediate interest in the invention of a metal promising to be of use in the instrument-making trades (Bentham, 1968–71, I, II). The best method, however, of imparting information in physics and chemistry lay in the lecture courses readily available. Jeremy Bentham, with regard to the education of a young Russian student, advised in 1779 that he 'should take one or two of the flying courses of Natural Philosophy given in London . . .' (Bentham, 1968–71, II). There was no lack of choice.

By the 1760s and the 1770s, just as James Dinwiddie was planning his new experimental venture, public lecturers like Ferguson trumpeted performance as essential to the 'foundation of physics'. This reflected the broadly accepted value of spectacle in the physical sciences. In 1760, in the preface to *Lectures on Select Subjects in Mechanics, Hydrostatics, Pneumatics, and Optics*, Ferguson stated:

The method of teaching and laying the foundation of physics, by public courses of experiments, was first undertaken in this kingdom, I believe, by Dr John Keill, and since improved and enlarged by Mr Hauksbee, Dr Desaguliers, Mr Whiston, Mr Cotes, Mr Whiteside, and Dr Bradley, our late Regius and Savilian Professor of Astronomy. Nor has the same been neglected in other countries: Dr James, and Dr David Gregory, Sir Robert Stewart, and after him Mr Maclaurin, in Scotland; Dr Helsham in Ireland, Messiers 'sGravesande [sic] and Musschenbroek in Holland, and the Abbe Nollet [sic] in France, have acquired just applause thereby. (Ferguson, 1760)

By 1770, in his second edition, Ferguson had added the name of Dr Nathaniel Bliss, James Bradley's successor at Oxford, as yet another model (Musson and Robinson, 1969). This gives us some view of the state of physics lecturing in the universities at a time when the itinerant performers were increasingly active and highly dependent on purchases of instruments. Bliss became a purveyor of experimental lectures while Savilian Professor of Geometry at Oxford, but he did not necessarily impress. According to Jeremy Bentham, who was in attendance in 1763:

We have gone through the Science of Mechanics with Mr Bliss, having finish'd on Saturday; and yesterday we begun upon Optics; there are two more remaining, viz: Hydrostatics, and Pneumatics. Mr Bliss seems to be a very good sort of Man, but I doubt he is not very well qualified for his Office, in the practical Way I mean, for he is obliged to make excuses for almost every Experiment they not succeeding according to expectations; in the Speculative part, I believe he is by no means deficient.

Performance mattered. Bentham looked forward with some anticipation to the lectures of Thomas Hornsby, Bliss's successor, who 'made several Improvements and additions to this course, which will render it completer than any of the former ones, either of his own or his predecessors . . .' (Ferguson, 1760; Bentham, 1968, I; McConnell, 2004; Wallis, 2000; Musson and Robinson, 1969). The survival of such comments is unique, but it is nonetheless evident that public lecturers like Ferguson in 1760, as well as some pupils, were aware of the worth of their many contemporaries and those public competitors who turned physics into theatre. Accessibility was the first step in engaging an audience. In defence of sensuous display in his own lectures, James Dinwiddie later once remarked to his auditors that 'such a course may enable you to read the popular productions of Ferguson or a Nollet—but not to understand all [?] the Sublime Investigations of Dalembert [sic] or a Newton.'[23] It was a telling moment that put display on a par with philosophical investigation—but by then this was obviously a widely promoted view.

10.5 Conclusion: Dinwiddie's Way

Auditors provided patronage—but this was in an age fuelled by connection that could take many forms. It was, for example, Dinwiddie's role in the Macartney expedition to China that ultimately helped establish his reputation. Notably, this probably followed from the influence of Keir's friend, Matthew Boulton, whose manufactured goods were among those selected to impress the Chinese Emperor and open China to British trade. It is equally possible that the choice of Dinwiddie to take charge of a range of electrical, chemical, and philosophical instruments may have had something to do with Dr Thomas Percival, physician in Manchester, with whom Dinwiddie was evidently in touch (Berg, 2006; Schaffer, 2006).[24] Dinwiddie reveals not only the geographic reach of the itinerant lecturers on physics, but his traces include some unique insight into how the lecturing networks functioned. Before he left for China and India, Dinwiddie had been notably active in the London market. In December 1790, among his auditors was the American inventor James Rumsey, who had come to London in 1788 seeking British patents for his steam-boats. Rumsey promised Dinwiddie a model, and Dinwiddie soon proposed to 'have something new to offer, when on the Subject of Hydraulics'. As it turned out, however, the model proved far too large to be of much use. Rumsey then devised a steam-powered barge propelled by water forced through tubes. Rumsey claimed: 'This machine may be something novel to an audience, and will, in some degree, compensate for what they might have expected new on hydraulics; more especially as it is so much united with that Science.' Rumsey, of course, was looking for a venue in which to promote his inventions. He told Dinwiddie that he was thinking of engaging in 'Experiment with or upon water, In a proper room for that purpose, and if agreeable, near you; the whole will afford amusement, for an hour, to the Curious in that line, and Speculative Enquirer; they perhaps might deserve the attention of that much time, in Each Week.' It is not known whether Dinwiddie thought this practicable, but in any case he opted to enlist in Macartney's embassy. Rumsey tried to strike out on his own, but died suddenly during a lecture at the Adelphi in December 1792.[25]

Despite the disappointment at the Court in Peking in 1793, Dinwiddie managed to find many an audience in the years he subsequently spent in Calcutta and during his attachment there to the College of Fort William. Throughout the 1790s he managed to keep abreast of many of the scientific developments in Europe, either by correspondence or by access to some leading European journals. London provided yet further connections of note. In 1796, in Madras, he met Admiral George Elphinstone, whom he remembered as one of his London auditors. Audience as much as lecturers travelled throughout the Empire. One of those who also gave Dinwiddie support was Robert Hobart, fourth Earl of Buckinghamshire, Governor-General in India. Evidently, Hobart, an MP in both the British and Irish House of Commons, had seen Dinwiddie lecture in Dublin in the 1780s. While agreeing to subscribe to lectures in India, Hobart had more practical intentions. He asked Dinwiddie for his advice on the manufacture of Indian saltpetre (Thorne, 2004).[26] Dinwiddie's industrial interests had been cultivated in the Midlands, and he surely would have been a useful adviser to the agents of both the Board of Trade and the East India Company. This may also have been what recommended him to Richard, Marquess Wellesley, who later took up the post of Governor-General in Bengal. Wellesley subscribed to Dinwiddie's course of twelve lectures in Calcutta. These were offered on 'Select Subjects in Natural Philosophy, Chemistry & Galvanism' for a fee of 100 rupees, which would permit the 'ladies in the Family' to also attend. Here Dinwiddie displayed the constant theme of the physics shows: 'All lectures will be illustrated by a great variety of interesting Experiments on a very extensive Apparatus, in which will be included the very entertaining and surprising experiments in the new science of Galvanism.' The surviving list of subscribers included William Hunter, surgeon and Secretary to the Asiatic Society, whose manuscript botanical study of India was dedicated to Wellesley, along with the names of several prominent Calcutta merchants (Schaffer, 2009).

Physics defined broadly, as much as the more narrowly conceived revolutions in chemistry and electricity, was a foundation of the European Enlightenment and its cultivation across Empires. Our tale returns, with Dinwiddie, to London when he once again set foot on English soil in April 1807. It had taken him that long to break the ties to India and to divest himself of most, but not all, of his apparatus, some of which he had designed and built. Within a month of landing in London, we find him among an entirely different audience, this time at the lectures of the handsome Humphry Davy at the Royal Institution. Here Dinwiddie was on the opposite side of the demonstrator's bench. He knew full well that electricity was where the action was, and he set up his own Galvanic battery, temporarily contemplating a return to performance.[27] But in the next year he was attending the Institution regularly and recording his reactions in detail. Here Davy adopted an almost casual, deliberately demystifying style, in lectures on experimental chemistry. And Dinwiddie took particular note of Davy's remark that 'Two of the most eminent Chemists in this kingdom have no other Laboratories than common closets with fire places in them'. So too, James Watt worried away at nature's forces in his simple, and much cluttered, garret workshop in Soho. Davy, and Watt, and Boulton, and Priestley, and Dinwiddie all adhered to 'the necessity of public encouragement'.[28] This alone was justification for startling public shows in physics, as in much else. It generated a tide of enthusiasm, and the crush of young ladies and gentlemen soon forced Albermarle Street, the site of the Royal Institution, to be turned into London's first one-way street (Holmes, 2008). Visits to the Institution in London had taken on the lustre of Enlightenment, just as an excursion to Soho had once been to Boswell a generation earlier.

Bibliography

Primary Sources

ARDEN, JAMES (1782). *Analysis of Mr. Arden's Course of Lectures of Natural and Experimental Philosophy*. Second edition. London?
——(1789). *A Short Account of a Course of Natural and Experimental Philosophy*. Hull.
BENTHAM, JEREMY (1968–1971). *The Correspondence of Jeremy Bentham*. Timothy L. S. Sprigge (ed.), I (1752–1776); II (1777–80); Ian R. Chrisitie (ed.), III (January 1781–October 1788). London: Athlone Press.
DESAGULIERS, J. T. (1745). *A Course of Experimental Philosophy*, second edition, corrected. London.
FERGUSON, JAMES (1760). *Lectures on Select Subjects in Mechanics, Hydrostatics, Pneumatics, and Optics: with The Use of the Globes, and The Art of Dialing*. London: Millar.
——(1768). *Syllabus of a Course of Lectures on the Most Interesting Parts of Mechanics, Hydrostatics, Hydraulics, Pneumatics, Electricity, and Astronomy*. Edinburgh.
FREKE, JOHN (1752). *An Essay to Shew the Cause of Electricity; and Why Some Things are Non-Electricable. In which it is also Consider'd Its Influence in the Blasts on Human Bodies, in the Blights on Trees, in the Damps in Mines; and as it may affect the Sensitive Plant, &c. In a Letter to Mr. William Watson, F.R.S.* Third edition. London.

[KEIR, JAMES] (1789). *The First Part of a Dictionary of Chemistry, &c.* Birmingham: Pearson and Rollason; London: Elliot and Kay; Edinburgh: Charles Elliot.

NEALE, JOHN (1747). *Directions for Gentlemen, who have Electrical Machines, How to Proceed in Making their Experiments. Illustrated with Cuts. With a compleat Series of Experiments, and the Management of them, as shewn in his Course.* London.

NOLLET, ABBÉ JEAN ANTOINE (1775). *Lecons de Physique Experimentale: par M. l'Abbe Nollet, de l'Academie Royale des Sciences, de la Societe de Londres, et l'Institut de Bologne, &c Maitre de Physique & d'Histoire Naturelle des Enfants de France, & Professeur Royal de Physique Experimentale au College de Navarre. Tome premier.* Huitieme Edition. Paris: Chez Durand, Neveu, Libraire, rue Galande, a la Sagesse.

NORTH, ROGER (1890). *The Lives of the Right Hon. Francis North, Baron Guilford; The Hon. Sir Dudley North, and the Hon. and Rev. Dr. John North. Together with the Autobiography of the Author.* Augustus Jessop (ed.). London: George Bell.

PRIESTLEY, JOSEPH (1767). *History and Present State of Electricity.* London.

RUTT, JOHN TOWILL (ed.) (2003). *The Theological and Miscellaneous Works of Joseph Priestley, LL.D. F.R.S. &c. Vol. I. Containing Memoirs and Correspondence, 1733–1837.* Reprint. Bristol: Thoemmes Press.

SPITALFIELDS (1784). *The Articles of the Mathematical Society, Meeting at the Sign of the Black Swan, in Brown's-Lane, Spitalfields.* London.

——(1793). *Articles of the Mathematical Society, Spitalfields, London: Instituted 1717.* London.

——(1804). *A Catalogue of Books, Belonging to the Mathematical Society, Crispin Street, Spitalfields.*

VOLTAIRE, (1733). *Letters on England.* Leonard Tancock (trans.). Harmondsworth: Penguin.

WARLTIRE, J. (1769). *Tables of Various Combinations and Specific Attraction of the Substances employed in Chemistry. Being A Compendium of that Science: Intended Chiefly for the Use of those Gentlemen and Ladies Who attend the Author's Lectures.* London: for the author.

——(1770?). *Concise Essays upon Various Philosophical and Chemical Subjects.* London: for the author.

Secondary Sources

ALTICK, RICHARD D. (1978). *The Shows of London.* Cambridge, MA, and London: Belknap Press.

ANDERSON, WILDA C. (1984). *Between the Library and the Laboratory. The Language of Chemistry in Eighteenth-Century France.* Baltimore and London: Johns Hopkins University Press.

ARCHINARD, MARGARIDA (1989). 'The Scientific Instruments of Horace-Benedict de Saussure', in Christine Blondel, Francoise Parot, Anthony Turner, and Mari Williams (eds.), *Studies in the History of Scientific Instruments.* London: Rogers Turner Books, 83–96.

ASHWORTH, WILLIAM B. (1981). 'John Bevis and His *Uranographia* (c.1750)', *Proceedings of the American Philosophical Society* 125 (Feb. 16): 52–73

BARETTA, MARCO (1993). *The Enlightenment of Matter: The Definition of Chemistry from Agricola to Lavoisier.* Canton, MA: Science History Publications.

BENGUIGUI, ISAAC (1984). *Theories electriques du XVIIIe Siecle. Correspondance entre l'Abbe Nollet (1700–1770) et le physicien Genevois Jean Jallabert (1712–1768).* Geneva: Georg.

BENNETT, JAMES A. (2002). 'Shopping for Instruments in Paris and London'. in Pamela H. Smith and Paula Findlen (eds.), *Merchants and Marvels: Commerce, Science, and Art in Early Modern Europe.* New York and London: Routledge, 370–395.

BENSAUDE-VINCENT, BERNADETTE and CHRISTINE BLONDEL (eds.) (2008). *Science and Spectacle in the European Enlightenment*. Aldershot and Burlington, VT.: Ashgate Publishing.

BERG, MAXINE (2006). 'Britain, industry and perceptions of China: Matthew Boulton, "useful knowledge" and the Macartney Embassy to China 1792–94', *Journal of Global History* 1: 269–288.

BERTUCCI, PAOLA (2001). 'A Philosophical Business: Edward Nairne and the Patent Electrical Machine (1782)', *History of Technology* 23: 41–58.

——(2007). *Viaggio nel paese delle meraviglie. Scienza e curiosità nell'Italia del Settecento*. Torino: Bollati Boringhieri editore.

BREWER, JOHN and Roy PORTER (eds.), (1993). *Consumption and the World of Goods*. London and New York: Routledge.

BRUTON, ROGER (2009). 'An Examination of the Extent to which Scientific Lecturers were Contributing to the Dissemination of Knowledge in the mid-Eighteenth Century West Midlands or were Largely a Source of Entertainment', MA Thesis, University of Birmingham.

CANTOR, GEOFFREY (1983). *Optics after Newton: Theories of light in Britain and Ireland, 1704–1840*. Manchester: Manchester University Press.

CLIFTON, GLORIA (1995). *Dictionary of British Scientific Instrument Makers 1550–1851*. London: National Martime Museum and Zwemmer.

COLEY, N.G. (1969). 'John Warltire, 1738/9–1810: itinerant lecturer and chemist', *West Midlands Studies: A Journal of Industrial Archaeology and Business History* 3: 31–44.

DELBOURGO, JAMES (2006). *A Most Amazing Scene of Wonders: Electricity and Enlightenment in Early America*. Cambridge, MA, and London: Harvard University Press.

ELLIOTT, PAUL A. (2009). *The Derby Philosophers. Science and Culture in British Urban Society, 1700–1850*. Manchester and New York: Manchester University Press.

GAUVIN, JEAN-FRANCOIS (2002). 'Eighteenth-Century Entrepreneur. Excerpts from the Correspondence between Jean Antoine Nollet, Etienne-Francois Dutour and Jan Jallabert, 1739–1768', in Lewis Peyenson and Jean-Francois Gauvin (eds.), *The Art of Teaching Physics: The Eighteenth-Century Demonstration Apparatus of Jean Antoine Nollet*. Sillery: Setpentrion.

GOLINSKI, JAN (1992). *Science as Public Culture: Chemistry and Enlightenment in Britain, 1760–1820*. Cambridge and New York: Cambridge University Press.

——(1999). 'Barometers of Change. Meteorological Instruments as Machines of Enlightenment', in William Clark, Jan Golinski, and Simon Schaffer, (eds.), *The Sciences in Enlightened Europe*. Chicago: University of Chicago Press, 69–93.

——(2000), ' "Fit Instruments": Thermometers in Eighteenth-Century Chemistry,' in Frederick L. Holmes and Trevor H. Levere (eds.), *Instruments and Experimentation in the History of Chemistry*. Cambridge, MA, and London, MIT Press, 185–210.

HANKINS, THOMAS L. (1985). *Science and Enlightenment*. Cambridge and London: Cambridge University Press.

HARRIS, J. R. (1998). *Industrial Espionage and Technology Transfer. Britain and France in the Eighteenth Century*. Aldershot and Brookfield: Ashgate.

HEILBRON, J. L. (1979). *Electricity in the 17th and 18th Centuries. A Study in Early Modern Physics*. Berkeley and Los Angeles: University of California Press.

HOLMES, RICHARD (2008). *The Age of Wonder. How the Romantic Generation Discovered the Beauty and Terror of Science*. London: Harper.

HOME, R.W. (1992). *Electricity and Experimental Physics in 18th-century Europe*. Hampshire and Brookfield: Variorum.

JACOB, MARGARET C. (1997) *Scientific Culture and the Making of the Industrial West*. Oxford and New York: Oxford University Press.

JACQUES, JEAN (1985). 'Le "Cours de chimie de G.-F. Rouelle receuilli par Diderot"', *Revue d'histoire des sciences* 38: 45–53.

JONES, PETER M. (2008a). 'Industrial Enlightenment in Practice. Visitors to the Soho Manufactory, 1765–1820', *Midland History* 33 (Spring): 68–96.

——(2008b). *Industrial Enlightenment. Science, Technology and Culture in Birmingham and the West Midlands 1760–1820*. Manchester and New York: Manchester University Press.

KIM, MI GYUNG (2006). ' "Public" Science: Hydrogen Balloons and Lavoisier's Decomposition of Water', *Annals of Science* 63 (July): 291–318.

LEHMAN, CHRISTINE (2010). 'Innovation in Chemistry Courses in France in the Mid-Eighteenth Century: Experiments and Affinities', *Ambix* 57 (March): 3–26.

LEVERE, TREVOR H. (2005). 'The Role of Instruments in the Dissemination of the Chemical Revolution,' *Endoxa: Series Filsosofica* 19: 227–242.

LUNNEY, LINDE (1988). 'The Celebrated Mr. Dinwiddie: an Eighteenth-Century Scientist in Ireland', *Eighteenth Century Ireland* 3: 69–83.

LYNN, MICHAEL R. (2006). *Popular Science and Public Opinion in Eighteenth-Century France*. Manchester

——(2010). *The Sublime Invention: Ballooning in Europe, 1783–1820*. London: Pickering & Chatto, and New York: Manchester University Press.

MAYER, JEAN (1970). 'Portrait d'un chimiste: Guillaume-Francois Rouelle (1703–1770)', *Revue d'histoire des sceinces et de leurs applications* 23: 305–332.

MCCONNELL, ANITA (2004). 'Bliss, Nathaniel (1700–1764)', *Oxford Dictionary of National Biography*, Oxford University Press, <http://www.oxforddnb.com/view/article/2653>

MCLEAN, RITA (2009). 'Introduction. Matthew Boulton 1728–1809, in Shena Mason (ed.), *Matthew Boulton: Selling What all the World Desires*. New Haven and London: Yale University Press and Birmingham City Council.

MILLBURN, JOHN R. (1973) 'Benjamin Martin and the Development of the Orrery,' *British Journal for the History of Science* 6: 378–399.

——(1976). *Benjamin Martin. Author, Instrument-Maker, and 'Country Showman'*. Leyden: Noordhoff.

——(1983). 'The London Evening Courses of Benjamin Martin and James Ferguson. Eighteenth-Century Lectures on Experimental Philosophy,' *Annals of Science* 40: 437–55.

——(1985). 'James Ferguson's Lecture Tour of the English Midlands in 1771', *Annals of Science* 42: 395–415.

MOKYR, JOEL (2009). *The Enlightened Economy. An Economic History of Britain 1700–1850*. New Haven and London: Yale University Press.

MORRISON-LOW, A. D. (2007). *Making Scientific Instruments in the Industrial Revolution*. Aldershot and Burlington: Ashgate.

MORTON, ALAN Q. and JANE A. WESS, (1992). *Public and Private Science. The King George III Collection*. London and Oxford: Oxford University Press and the Science Museum.

MORUS, IWAN RHYS (2005). *When Physics became King*. Chicago and London: University of Chicago Press.

——(2006). 'Seeing and Believing Science', *Isis* 97 (March): 101–110.

MUSSON, A. E. and ERIC ROBINSON (1969). *Science and Technology in the Industrial Revolution*. Toronto: University of Toronto Press.

PERKINS, JOHN (2010). 'Chemistry Courses, the Parisian Chemical World and the Chemical Revolution, 1770–1790', *Ambix* 57 (March): 27–47.

Rappaport, Rhoda (1960). 'G. F. Rouelle: An Eighteenth-Century Chemist and Teacher', *Chymia* 6: 68–101.
Robinson, Eric and Douglas McKie (eds.), (1970). *Partners in Science. James Watt and Joseph Black*. Cambridge, MA: Harvard University Press.
Roberts, Lissa (1995). 'The Death of the Senuous Chemist: The 'New' Chemistry and the Transformation of Sensuous Technology,' *Studies in the History and Philosophy of Science* 26: 503–529.
——(1999a). 'Going Dutch: Situating Science in the Dutch Enlightenment', in Jan Golinski, William Clark, and Simon Schaffer (eds.), *The Sciences in Enlightened Europe*. Chicago and London: University of Chicago Press.
——(1999b). 'Science Becomes Electric. Dutch Interaction with the Electrical Machine during the Eighteenth Century', *Isis* 90: 680–714.
Schaffer, Simon (1983). 'Natural Philosophy and Public Spectacle in the Eighteenth Century', *History of Science* 21: 1–43.
——(1992). 'Self Evidence'. *Critical Inquiry* 18 (Winter): 327–363.
——(1993). 'The consuming flame: electrical showmen and Tory mystics in the world of goods'. in John Brewer and Roy Porter (eds.), *Consumption and the World of Goods*. London and New York: Routledge, 489–526.
——(1994). 'Machine Philosophy: Demonstration Devices in Georgian Mechanics', *Osiris* 9: 157–182.
——(1997). 'Experimenters Techniques, Dyers' Hands, and the Electric Planetarium', *Isis* 88: 456–483.
——(2006). 'Instruments as Cargo in the China Trade', *History of Science* 44 (June): 217–246.
——(2009). 'The Asiatic Enlightenments of British Astronomy', in Schaffer, Lissa Roberts, Kapil Raj, and James Delbourgo (eds.), *The Brokered World. Go-Betweens and Global Intelligence, 1770–1820*. Sagamore Beach: Science History Publications, 49–104.
Sotheby's. *The James Watt Sale. Art and Science*. London: 20 March 2003.
Stewart, Larry (1992). *The Rise of Public Science. Rhetoric, Technology and Natural Philosophy in Britain, 1660–1750*. Cambridge and New York: Cambridge University Press.
——and Paul Weindling (1995). 'Philosophical Threads: natural philosophy and public experiment among the weavers of Spitalfields', *British Journal for the History of Science* 28 (March): 37–62.
Taylor, E. G. R. (1966). *The Mathematical Practitioners of Hanoverian England 1714–1840*. Cambridge: University Press.
Thébaud-Sorger, Marie (2009). *L'Aerostation au temps des Lumieres*. Rennes: Presses Universitaires des Rennes.
Thorne, Roland (2004). 'Hobart, Robert, fourth earl of Buckinghamshire (1760–1816)', *Oxford Dictionary of National Biography*, online edn, 2009, <http://www.oxforddnb.com/view/article/13396>
Torrens, H.S. (2004). 'Warltire, John (1725/6–1810)', *Oxford Dictionary of National Biography*. Oxford: Oxford University Press. Online edn. (2006) <http://www.oxforddnb.com/view/article/38107>
Turner, G. L'E. (1967). 'The auction sale of the earl of Bute's instruments, 1793', *Annals of Science* 23: 213–242.
Van Helden, (1994). 'Theory and Practice in Air-Pump Construction: The Cooperation between Willem Jacob 'sGravesande and Jan van Musschenbroek', *Annals of Science* 51: 477–495.
Vermij, Reink (2003). 'The formation of Newtonian philosophy: the case of the Amsterdam mathematical amateurs', *British Journal for the History of Science* 36 (June), 183–200.

WALLIS, RUTH (1982). 'John Bevis, MD, FRS (1695–1771)', *Notes and Records of the Royal Society* 36 (February): 211–225.

——(2000) 'Cross-currents in Astronomy and Navigation: Thomas Hornsby, FRS (1733–1810)', *Annals of Science* 57: 219–240.

WITHERS, CHARLES (2007). *Placing the Enlightenment: Thinking Geographically about the Age of Reason*. Chicago and London: University of Chicago Press.

NOTES

1. See, for example, Birmingham Central Library [BCL], James Watt Papers, Copy Books, Watt to A. Brown, 19 August 1791; JWP 4/13/24. Daniel(?) Coke to Watt, 29 March 1792; and Watt to John Robison, 20 May 1793. Professor of Philosophy at the University of Edinburgh, and long-time correspondent of Watt, Robison sought advice from Watt regarding his own son, whom he intended 'to breed up to a Manufacturer'. (Robinson and McKie, 1970), Robison to Watt, 7 March 1793.
2. BCL, James Watt Papers, Copy Books, Watt to A. Brown, 19 August 1791; Private Letters, Letter Box 1 (1782–1789), 68. Watt to Aimé Argand, 31 August 1784.
3. BCL, James Watt Papers, Copy Books, Watt to Erasmus Darwin, 24 November 1789.
4. Roger North to Mr Philip Foley, 22 December 1706, in North (1890), 255.
5. British Library, Add. MSS. 32548. 'Works of Roger North', no. 2, 'Loci phisiologici ordine fortuito', fols. 39, 42; North, 1890, II, 320–321.
6. British Library, Add. MSS. 30094. Letters and Papers of Benjamin Wilson, F.R.S. (1743–1788), f. 41. Smeaton to Wilson, 24 August 1747.
7. BCL, JWP C1/7, n.d. This catalogue, in James Watt's papers, provides Martin's list of instruments for sale, with 'proffits alow'd', as marginalia. It may be that Watt was contemplating dealing in Martin's apparatus, or that this was simply Watt's estimates of what Martin might have expected to make. On the Large Orrery, for example, the profit was anticipated to be £10 10s. See the comparable list of Martin's apparatus shipped by the London agent, Joseph Mico, to Harvard in 1765, in Millburn (1976): 130–135.
8. Dalhousie University Library (hereinafter DUL), James Dinwiddie Collection, MS. 2-726, C 60. Lecture notes: 'The effects of heat and mixture on bodies', pp. 29ff. Dinwiddie is now best known as a member of the Macartney expedition to China. (Schaffer, 2006; and Berg, 2006). I wish to thank Simon Schaffer and Maxine Berg for pointing me in the direction of these sources.
9. DUA, James Dinwiddie Collection, MS 2-726, D 28. 'Magnetism and Electricity', April 1776; F 2. Early Experiments, f. 1v, 1777; F 1. Early Experiments, c.1774 September 2029.
10. DUA, James Dinwiddie Collection, MS 2-726, K 2, Dumfries Philosophical Society, Rules and Minutes, 1776–1777; MS 2-726, A 32, List of the Gentlemen Subscribers for purchasing a Philosophical Apparatus, for the use of the School of Dumfries ... [poss. 1774]. I thank Mr Chris Green, whose collection of correspondence contains the original subscription for the orrery, dated 16 July 1765.
11. DUA, James Dinwiddie Collection, MS 2-726, C 5. Lecture 7, Electry [sic]; F 1, Early Experiments, c.1774; F 2, Early Experiments 1777, fols 1v, 3v, 6r, 7r, 8v, 9 r/v, 10v, 12v.
12. DUA, James Dinwiddie Collection, MS 2-726, C 11. See, for example, Mechanics 1, nos. 15–18, 21–25, 30, 42–43; Mechanics 2, nos. 45–47.
13. British Library, Burney Collection Newspapers. See *Morning Post* and *Daily Advertiser*, no. 3387 (Wednesday, 17 December 1783); *St James's Chronicle* or the British *Evening Post*, no. 3574 (31 January 1784–3 February 1784); no. 3592 (13 March 1784–16 March 1784).

14. DUA, Dinwiddie Collection, MS 2-726, B 12. Journals, Paris 1788, pp. 2, 14, 18, 26.
15. University of Guelph, James Dinwiddie Correspondence, XS1 MS A164, no. 66. William Jones to Dinwiddie, 26 March 1795.
16. Royal Society, MSS, Certificates, 2 (1751–56), nos. 442, and 7.
17. Keele University, Wedgwood MSS. L101-18758. Nairne and Blunt to Wedgwood, 17 June 1783; WM 21. Tom Wedgwood to Joseph Wedgwood, 12 April 1800.
18. Haarlem, Rijksarchief in Noord-Holland, Van Marum correspondence, Box 19; also Box 15, John Cuthbertson to Van Marum, 8 May 1789; Van Marum to Tiberius Cavallo, 2 January 1802.
19. It is perhaps worth noting that Spitalfields' apparatus, which recently surfaced in the collections of the Science Museum, London, appears to have been a refraction demonstration made by Dollond.
20. Whitby Literary and Philosophical Society, J. Wear, jun., 'Some Account of the Philosophical Lectures and Experiments: Read and explained at Whitby in February 1760 by Mr. J. Arden', 25 February 1760, fols. 1, 4, 7. I wish to thank Peg Jacob for providing notes on the lectures.
21. Keele University Library, Wedgwood–Bentley letters, V. 129. Josiah Wedgwood to Thomas Bentley, 13 October 1771.
22. Keele University Library, Wedgwood–Bentley letters, V, 167. Wedgwood to Bentley, 30 or 31 November 1771.
23. DUA, Dinwiddie Collection, MS 2-726, C 14, Miscellaneous Subjects, 1792?, fol. 7v.
24. Glasgow University Library, Special Collections, Cullen MSS 99 (1), Thomas Percival to William Cullen, 11 November 1770; British Library, India Office Records, G/12/20, fols. 596–618. 'Articles bought for Macartney expedition January–August, 1792'.
25. Guelph University Library, Special Collections, James Dinwiddie Correspondence, XS1 MS A164, no. 81. Rumsey to Dinwiddie, 24 December 1790.
26. DUA, Dinwiddie Collection, MS 2-726, Journals B 25, 22 January 2027, 31, February 6, 9, 10 July 1796.
27. DUA, James Dinwiddie Collection, MS 2-726, Journal B-73, esp. London, Saturday, 16 May and Friday, 12 June 1807.
28. DUA, James Dinwiddie Collection, MS 2-726, Royal Institution, J 2, 29 April 1809.

CHAPTER 11

INSTRUMENTS AND INSTRUMENT-MAKERS, 1700–1850

ANITA McCONNELL

11.1 INTRODUCTION

Scientific instruments in the 150 years between 1700 and 1850 enjoyed such rapid advances in design and technology that the period can usefully be divided at about 1800, though the date varied with the progress or otherwise in the lands concerned. Anthony Turner's *Early Scientific Instruments: Europe 1400–1800* (1987) describes and illustrates products of the eighteenth-century European workshops, and Gerard Turner's *Nineteenth-Century Scientific Instruments* (1983) continues the story.

Throughout our period the leading craftsmen were becoming better-educated, many having a good grasp of mathematics. The composition of metals and alloys became better understood, and new materials emerged from the chemistry laboratory. The old tripartite division still held good: mathematical instruments for basic measurement and calculation; optical instruments for enlarging or extending the range of human vision; philosophical instruments for detecting and measuring those aspects of our surroundings which human senses could not perceive quantitatively, namely air pressure, temperature, humidity, terrestrial magnetism, electricity, and gravity. Apparatus for time measurement and for weighing, essential for certain physical experiments and operations, was generally, though not exclusively, constructed by specialist horologists and scale makers and is not dealt with here.

This chapter deals mainly with instruments and their makers in Britain, with a brief survey of the situation in western Europe and in the United States of America. During the period 1700–1850 both 'Germany' and 'Italy' consisted of a multitude of small territories, each with its own prince, duke, bishop, or other magnate, and the terms are used here to denote German-speaking and Italian-speaking regions.

11.2 THE EUROPEAN POLITICAL SITUATION

Britain under the Hanoverian kings[1] enjoyed domestic peace; apart from the Jacobite rebellion all wars were fought on land and sea beyond its shores. British troops took part in the War of the Spanish Succession (1702–13), the War of the Quadruple Alliance (1718–20), and the War of Jenkins' Ear, which merged with the War of the Austrian Succession (1740–8). During the 1750s Britain was fighting in Canada. France and Britain were at war between 1756 and 1763, and in the 1790s there were campaigns in the Low Countries and the Caribbean. The various peace treaties ending these conflicts often led to exchange or surrender of territory. The American War of Independence, 1775–83, brought about the most extensive territorial loss to Britain—that of the eastern American states plus the lands bounded on the western side by the Mississippi river and to the south by Spanish possessions. Thereafter, the United States of America set about creating its own scientific institutions.

Across Catholic Europe, most schools and universities were run by religious orders. The Jesuits especially enjoyed a good reputation despite the limitations imposed by their church's refusal to accept the Copernican heliocentric system. The exceptions were the Venetian Republic's secular University at Padua and the City of Bologna's University, which despite being within the Papal States was controlled by the city itself. Here the science of astronomy was freely pursued. During the decade 1750–60 the Jesuits fell from Papal favour and were supressed in many countries. This brought their universities under secular control, at a time when many of these small states wished to map their territories. Aware that they needed to fix a grid on which to centre such maps, these rulers built and re-equipped their principal observatories. This activity came about at an extremely favourable time for the leading British instrument-makers.

The French Revolution led to the execution or emigration of many of the upper classes. The naval and military officer classes were decimated. The first stage of the Napoleonic Wars ran from 1793 to 1802, then recommenced in 1803 and continued up to the Waterloo campaign. Between 1808 and 1814 troops were active in the Peninsula War. By 1812 the French Empire included the Low Countries, Switzerland, northern and central Italy, and Dalmatia. Throughout the period under review, the East India Company sent naval and infantry troops to India to pursue their territorial expansion at the expense of both French landholdings and those under Indian rule.

Such near-continuous warfare increased demand for naval and military instruments, while from time to time it interrupted the supply of imported materials and the delivery of British instruments overseas. Long-distance trade and voyages of exploration continued, the latter especially providing a good testing-ground for new inventions. It is against this belligerent background that a wide variety of British and foreign customers, funded privately or by public resources, made their way to the shops of British scientific instrument-makers. Entering the shop, some customers might find what they sought already available, new or secondhand. Others carried a sketch of what they had designed, hoping that the craftsmen could, as many did, see how the drawing could be constructed in three-dimensional metal. At the same time, education and learning, as well as technical improvements, both called for and met the needs of customers in what might be termed 'civilian' life. Roy Porter (1985, 2) stresses the difference between the market openings in England, where customers for instruments came from across the social spectrum, with the situation in Italy and Germany, where the nursery for top craftsmen was the patronage of royal courts.

11.2.1 The British Instrument-Makers: Workshop Practice

Having decided on their needs, would-be purchasers made their way to an appropriate instrument-maker's shop, There was still some specialization, and certain makers had a reputation for the excellence of a particular type of instrument. At the beginning of the eighteenth century, most of the first-rate British craftsmen were in London, but if his needs were modest, the amateur might find a provincial maker in towns such as Edinburgh, Glasgow, and York, or in one of the expanding ports like Bristol and Portsmouth. For those without access to London, friends or relations might be prevailed upon to negotiate for the desired item, which on completion would be despatched by road or ship, and later by canal.

The structure of the trade followed the City of London's medieval custom of guilds, regulating apprenticeships, usually of seven years, conferring the freedom to trade under a man's own name, inspecting the quality of work, and prosecuting any shopkeepers who had not come through the formal apprenticeship. Many instrument-makers belonged to the long-established Grocers Company and the later Spectaclemakers Company. The practice of apprentices being bound to a relation in another company, or gradually changing the focus of their trade over the generations, led to some notable instrument-makers being freemen of the Clockmakers, Goldsmiths, or indeed other small guilds whose origin lay far from the world of scientific instruments. The influence of the guilds was declining in the late eighteenth century. By the nineteenth century it had become more of a social organization, ignoring those instrument-makers, including immigrant craftsmen, who lived beyond the City boundary and therefore escaped its control. In the *Directory of Scientific Instrument Makers, 1550–1851* Clifton (1995) lists several thousand known craftsmen within Britain. Women played a minor role in this craft (Morrison-Low, 1991). A few, such as Janet Taylor (*OxDNB*), had their own businesses, while others were widows, daughters, or other relatives who ran the firm until a son was of age, or they found a buyer.

In London, the best-known shops were in the City of London, with a tail-off westwards towards the City of Westminster and eastwards to the busy wharves and quays below London Bridge. Several hundred craftsmen (and a few women) were active in London between 1700 and 1850. Baker (2009) gives a detailed socioeconomic analysis of London makers between 1700 and 1750. The most famous London names in the period considered here were Graham (w.1713–51), Bird (w.1745–76), two generations of Sisson (w.1722–83), Short (w.1764–68), the three generations of the Adams family (spanning 1734–1817), the two brothers Troughton (1756–1826) and the later partnership of Troughton & Simms (1826 to the twentieth century), Nairne, later Nairne & Blunt (spanning 1749–1826), the Dollond family—two generations plus additional members brought in through marriage (spanning 1750–1866), Watkins—expanding to Watkins & Hill and then joining with Elliott (spanning 1747 to the twentieth century), Ramsden, succeeded by Berge (w.1762–1819), in York, the Cooke family (w.1827–78). The numerous Adie families (w.1823 to the twentieth century) started in Edinburgh and expanded to Liverpool and London. It should be said, however, that by 1850 commercial practice led some family firms to convert to limited companies. Although the name was retained as having value, control passed into the hands of shareholders.

For businesses of any size, much of the small repetitive work was put out to 'chamber masters' who laboured in sheds and rooms tucked into the maze of back streets on the margins of the City and in the growing suburb of Southwark, south of the Thames. For large or completely novel instruments, access and floor loading made this impractical; and in addition, the master craftsman had to be in hand to watch progress and possibly intervene to change the design.

If scientific instruments are to provide measurements that can be relied on by all who employ them, there must be accepted standards of weight and measure. In the early eighteenth century much time and effort went into creating a standard of length—the Yard. Dimensions taken from this measure provided those for standards of weight—the Troy and Avoirdupois Pounds, and of capacity—the Bushel. The problem was not that no such standard yard existed, but rather that there were several ancient bars, crudely made, now battered, and corroded. The government commissioned Graham to oversee the production and division of a new standard yard, and he was assisted by Jonathan Sisson and John Bird. They prepared a number of so-called standard yards, but even these bars were found to vary slightly. A committee of the House of Commons attempted to overhaul the standards in 1758 and ordered Bird to prepare a new Yard, but the Bill to legalize it lapsed at its final stage (Simpson 1993, 179–81). Bird made further copies which were distributed round official bodies including the Royal Society, the Mint, and the Ordnance Survey, as well as official bodies in France. Further copies were made for private use, by instrument-makers and others. As time passed, the accuracy of the primary Yard was criticized, and this matter became more important when the Parliamentary Standard Yard was lost in the fire which destroyed the Palace of Westminster in 1834.

How should a new measure be defined? For it was accepted that all the copies of the primary yard differed to a very small degree. By this time there were two possibilities—a measure based on a fraction of the Earth's meridian, or one based on the length of a pendulum swinging seconds at the latitude of London. The arguments over this were not resolved until after the 1850s.

One technique that improved markedly during the eighteenth century was the circular or straight-line graduation of scales (Bennett, 1987). Rulers, the arcs of quadrants, octants, sextants, theodolites, and large astronomical quadrants and circles all needed to be divided with the utmost care and accuracy. This was a time-consuming task, calling for a skilled hand and also good daylight, with steady temperatures. It could take one or more seasons to graduate a large (say 8- or 10-foot radius) astronomical instrument. This skill was so necessary that John Bird, who had been making mural quadrants for the Royal Observatory, was prevailed on to write an illustrated description of his method, which was published by the Board of Longitude and earned Bird the considerable reward of £500.

Smaller instruments for navigation or surveying took a matter of hours. As demand increased, it was not surprising that instrument-makers looked at the tools employed by watch makers to cut gears. Ideas for cutting accurate divisions by mechanical means were in the air by 1760, and the first practical 'dividing engine' was constructed by Jesse Ramsden (Brooks, 1992; McConnell, 2007, 39–51), It was so easy to operate that Ramsden's wife and his apprentices could divide sextants and other small instruments in a short time. Ramsden continued to experiment, and his second engine was in action a few years later. The first was sold to a private buyer in France, but by this time it had been seen, and similar engines soon appeared in the workshops of John Troughton and John Dollond. Ramsden brought the second engine to the attention of the Board of Longitude, and after its products had been examined by Bird and another expert, the Board awarded Ramsden £500 and bought the engine from him, allowing it to remain in his workshop. The limbs of sextants divided on this engine were engraved with Ramsden's initials and a foul anchor (McConnell, 2007, 39–51).

By the 1850s the manufacture and marketing of scientific instruments was changing; exploration and emigration generated customers in India, Australia, Canada, Africa, and elsewhere. A new trade arose: that of the local agent who would supply and repair British-made instruments. To meet the expanded demand, instrument firms began to move outside the crowded cities. New factories were built and power-driven machinery installed; semi-skilled workmen oversaw the production of standard instruments and their replacement parts.

11.3 Materials Available in the Craftsmen's Workshops

11.3.1 Brass and Other Alloys

The principal raw materials of the instrument trade were, as in the previous period, brass, glass, and wood. Fine brass (composed of copper with small proportions of tin and zinc) was produced as rolled bars or sheets in the region round Stolberg, near Aachen, in present-day western Germany, where calamine (zinc ore) was mined.

As it was shipped from ports in the Low Countries it was often referred to as 'Dutch' or 'Hollands' brass. In Britain, brass made with copper ores from the famous Parys mine on Anglesey was manufactured at Cheadle, in Staffordshire. The mills around Cheadle produced fine-quality brass, rolled from plates cast between smooth granite tables. Birmingham brass included shruff (old brass) and was cast in iron moulds, and was of lower quality.

The various types of brass depended on the ratio of copper to zinc and tin, and their manufacturing procedures. Small quantities of additives—arsenic, bismuth, and nitre—modified the hardness, workability, and colour of the brass (Nicholson).[2] Holland (1853) gives the compositions for compounds of metals (707–14) including solders; lacquers (729–31), including a lacquer for modifying the colour of brass of philosophical instruments. Brass of a harder composition was the best material for making large fine and especially tubular screws as, notwithstanding its compact texture, it was easily worked on the lathe. For horological works and mathematical instruments generally, fine brass combined the qualities of tractability in cutting and durability in wear to a surprising degree, while it could be turned easily. Several alloys containing nickel were produced in Germany in the eighteenth century, the most attractive being the so-called German silver, composed of 60% copper with 20% each of nickel and zinc. Nickel alloys were used for electro-plating from the 1830s.

John Edwards (*d*.1783), a mathematician turned amateur metallurgist, reported that in seeking to make an untinted speculum metal which would take a high polish, he tried various additives: silver, platina, iron, copper, bismuth, zinc, and sulphate of barium. Edwards was aware of the unpleasant nature of hot arsenic fumes, but could not avoid them in the course of his trials. Eventually he died from their effect, but his reports of experiments (Edwards, 1783) were studied by Amici (see under Italy).

To make tubes for telescopes and other philosophical instruments, a sheet of brass was cut, the edges tongued-and-grooved, wired to hold the tube in position, then soldered along the join with borax and spelter solder. The tube was pickled in dilute sulphuric acid, then drawn to elongate it, stiffened, and brightened. A machine for drawing tubes was patented by John Lover Martin in 1782 and was installed in Ramsden's workshop where small telescopes and silver-plated tubes for opera-glasses were manufactured (McConnell, 2007, 68–9).

11.3.2 Glass

Two qualities of glass were produced, mainly in glasshouses in London and around Newcastle. The major part was 'crown' glass, used for bottles, lantern glasses, small window panes, and various containers. The instrument-makers used crown glass for sandglasses, tubes for thermometers and barometers, glass covers for compasses and magnetometers, laboratory glassware, for parts of vacuum apparatus, and for the electrical machines, now coming into use by medical practitioners and experimental philosophers. Certain glasshouses also produced the clear 'crystal' or 'flint' glass from which fine drinking vessels and mirrors were made. This 'crystal' served instrument-makers for lenses and prisms. Within any pot of flint glass there was usually only a small layer which was sufficiently free of streaks and bubbles to satisfy the lens-grinders, but experiments to improve the method of production were held back by the government tax imposed on all glass as it was being made, so that even the reject was taxed. (G. L'E. Turner 2000). Several manufacturers worked in secret to resolve the problem. First, before 1800, was the potter Josiah Wedgewood (1730–1795), followed by the French glasshouse owner Aimé-Gabriel D'Artigues (1773–1848), both of whom discovered by experiments that prolonged stirring of the melt removed the defects which otherwise rendered much of the pot unfit for optical purposes (Schofield; D'Artigues).[3] Glassmaker Pierre-Louis Guinand (1748–1824) worked with German optical instrument-maker Joseph von Fraunhofer (1787–1826), who brought the production of optical-quality glass to near-perfection. Following their deaths the secret was purchased by the French glassmaker Georges Bontemps (1801–1882), who in 1837 agreed to share it with the Birmingham firm of Chance Brothers (Morrison-Low, 2007, 185–6; Chance, chap. VII).

11.3.3 Wood

Wood was another component of many instruments. The use of European walnut declined as a fashionable wood for furniture, including clock and barometer cases, after the Naval Stores Act of 1721, by which timber from British colonies in the Americas was imported duty free. The trade in mahogany grew rapidly thereafter, exceeding importations of American walnut by the 1730s. Until the 1760s Jamaica supplied over 90% of the total. Mahogany from Honduras first arrived in 1763, and from Hispaniola a few years later. Cuban mahogany was principally an early nineteenth-century introduction, becoming the dominant source after 1850. By mid-century, mahogany was the usual material for instrument frames and stands; fitted instrument boxes were made from mahogany or oak. Ebony, from the Caribbean, was employed for octant frames. Porous boxwood was employed for barometer cisterns. Pear and other fruit woods, whose close grain allowed fine divisions, was favoured for graduated rules and the arcs of octants and sextants.

11.3.4 Other Materials

Other materials of more specialized use included steel, needed for bearings on large heavy apparatus. Gold and silver could be engraved with the finest divisions. Silver soon tarnished and gold was costly, but from 1806 the alternative was platinum, imported from South America. Platinum was harder than these two metals, did not tarnish, and was less costly than gold; it was available as inlays to graduated arcs on the finest sextants and circles. Mercury and its vapour were unrecognized as slow killers in the workshop, where it had multiple uses.[4] It was employed in the process of gilding, and for backing mirrors, in barometers and in some clock-pendulums.

Ivory, imported in quantity for general ornamental work, served in the construction of magnetometers—instruments which had to be constructed entirely without iron. Bone and ivory were employed for scale plates, especially those of barometers and thermometers. To give a firm grip, telescope tubes were sometimes covered in parchment or vellum, and later in shagreen (the tanned rough skin of rays or sharks). Mica, imported in translucent sheets from the Russian Urals, was preferred for glazing ships' lanterns and as marine compass cards. The axes of dip needles rolled on agate planes, while agate or precious stones such as ruby were fitted as bearings on marine compass pivots. Other materials which made occasional appearances were tortoiseshell, mother-of-pearl, and after the 1850s, ivorine, rubber, and ebonite.

11.4 The British Market

In 1700 there was, strictly speaking, only one professional scientist in Britain: the Astronomer Royal, based at the Royal Observatory, Greenwich, which had been founded in 1675. His salary was paid from the government's Civil List. The first Astronomer Royal, John Flamsteed (1646–1719), was in post when our time-period opens, and he was followed by Edmund Halley (1656–1742), James Bradley (1692–1762), Nathaniel Bliss (1700–64), Nevil Maskelyne (1732–1811), John Pond (1767–1836), and George Airy (1801–92), who was still active at the end of the period.[5] Bliss achieved little, holding office for only two years. The other Astronomers Royal all made contributions to the improvement of instruments in more than one field—most notably Maskelyne and the super-energetic Airy, a prolific contributor to all the physical sciences.

Science played a minor role at the two English universities, where professors were expected to provide their own apparatus. At Oxford, Sir Henry Savile (1549–1622) (*OxDNB*) founded professorships of astronomy and geometry in 1620. The Savilian Professor of Astronomy was to make nightly observations of the skies, while the Savilian Professor of Geometry should demonstrate the art of surveying, in the fields around Oxford. At Cambridge, Sir Thomas Plume (1630–1704) (*OxDNB*) endowed the Plumian Professorship of Astronomy and Experimental Philosophy, to include applied knowledge of mathematics—such as surveying, navigation, and the use of the globes (Atwood 1776). The first Plumian professor was Roger Cotes 1682–1716) (*OxDNB*). Instruments were acquired by 1703, and in 1707 an observatory was constructed over the great gate at Trinity College, though it was removed in 1797 (Price, 1952). Christ's College and St John's College also set up short-lived observatories in the eighteenth century (Howes, 1986, 67–8). In London, the salaries of the professors of Gresham College, inaugurated in 1597 and still active today, were met from the rental incomes of shops round the Royal Exchange.

The first Astronomers Royal were expected to provide their own instruments, and subsequently these were funded by the Board of Ordnance. The purpose of the Royal

Observatory was first and foremost to improve navigation, and a subsequent aim was to devise a means of finding longitude at sea. This it did by compiling measurements of the times of meridian transit of certain prominent stars. Under Maskelyne's direction the Observatory began in 1767 to publish an annual *Nautical Almanac*, in which the Greenwich times of these transits were published for several years ahead. By comparing transit times given in the *Almanac* with those observed on board, the ship's longitude could be calculated. The Astronomers' most conspicuous purchases were the large astronomical mural quadrants (so-called because they were secured to a wall within the observatory) and the portable long-focus telescopes which were installed in the Observatory. But we should not forget the ancillary instruments such as thermometers and barometers, which were needed to correct the observations, the sextants which served for local surveying, and the mathematical instruments employed for drawing and calculating.

The Hydrographic Office of the Admiralty was not set up until 1795, and the first Hydrographer was Alexander Dalrymple (1737–1808) (*OxDNB*), previously Hydrographer to the East India Company. Up to that time, British and French maritime charts were the product of surveys by private individuals, or copies of foreign charts captured from enemy vessels (Fisher, 2001). The compass-maker Gowin Knight (1713–1772) (*OxDNB*) was practising as a physician in London when he began research to improve magnets and compasses. He developed techniques for magnetizing steel bars by a repetitive process of stroking them with other magnets to build up their strength, meanwhile keeping his procedures secret. Cheaper and more reliable than natural lodestone, these artificial magnets played a vital role in subsequent investigations into magnetic phenomena, in popular and educational performances, and in navigational practices. George Adams senior (*OxDNB*) made a variety of angle-measuring instruments for navigation (Millburn, 2000, 69–72) including Knight's instruments. Knight introduced many innovations to marine compasses, negotiating lucrative contracts to supply the Royal Navy with magnets and compasses. Experienced mariners, however, found that the compasses performed badly at sea, particularly in stormy weather. This was a complaint frequently voiced against instruments made by craftsmen who were unfamiliar with conditions at sea.

Other navigational instruments were improving by this time (Mörzer Bruyns, 23–36). The ebony-framed reflecting octant was affordable and widely used up to about 1800. By 1760 the sextant, with its wider arc, was demonstratively better, but its all-brass construction made it a costly instrument, given that each officer, in naval and merchant vessels, had to purchase his own. Armed with a sextant with which to measure the angular distance between the Moon and a particular star, and a copy of the current *Nautical Almanac*, seafarers with some grasp of mathematics could now find their longitude at sea. John Harrison's (1693–1776) (*OxDNB*) development of the marine timepiece, which was proceeding at this time, culminated with the production of the marine chronometer. Chronometers were further developed by watchmakers John Arnold (1736–1799) and Thomas Earnshaw (1749–1829), but until their cost became more affordable to private mariners the so-called Lunar Distances method continued to serve the majority (Betts, 1998, 311–28).

The level of precision required in constructing and graduating the large astronomical instruments was met only by the most skilful instrument-makers. It is not therefore surprising to find dynasties of master and apprentice, sometimes extending over several generations. Within our period the London maker John Bird (1709–1776) (*OxDNB*) had been apprenticed to the clock and instrument-maker George Graham (*c*.1673–1751) (*OxDNB*), trained by Thomas Tompion (1639–1713) (*OxDNB*). Bird trained Jonathan Sisson (1690–1747) (*OxDNB*), who in turn trained his son Jeremiah Sisson (1720–1783) (*OxDNB*). Another dynasty began with John Troughton (1716–88), apprentice of Thomas Heath (*c*.1698–1773), who took his nephew, also John (1739–1807), as his apprentice; he trained his brother Edward Troughton (1756–1835), who later entered into partnership with his former apprentice William Simms (1793–1860) as Troughton & Simms—a business which passed to William's sons (*OxDNB*; McConnell, 1992). The highly regarded Jesse Ramsden (1735–1800) (*OxDNB*; McConnell, 2007) was already an adult when he learned the trade from Mark Burton (*w*.1765–*d*.1786), a highly respected maker of smaller instruments. Ramsden added to this skill in brass-work an understanding of lens-grinding, so that he was able to produce entire instruments, rather than subcontracting the optical parts to the famous family firm of Dollond (*OxDNB*). Ramsden's business passed to his long-term employee Matthew Berge (*c*.1753–1819), after whose death its output and reputation faded. Edward Nairne (1726–1806) (*OxDNB*), whose reputation lay principally as a maker of surveying, electrical, and other philosophical instruments, entered into partnership with Thomas Blunt (*d*.1823), who passed on the business to his sons Thomas (*d*.1824?) and Edward (*d*.1826). George Adams (1709–1772) was also apprenticed to Thomas Heath, and in turn trained his sons George (1750–1810) and Dudley (1762–1801) (*OxDNB*; Millburn, 2000). The Adams family relied on subcontractors to fulfil their contracts for quantities of gunnery and drawing instruments for the Board of Ordnance, as well as gaining a reputation for their microscopes and surveying instruments.

Bird's skill in graduating the large arcs of his astronomical instruments was rewarded by the Board of Longitude with a cash prize for the publication of his method, which relied on the use of a beam compass and an understanding of mathematics. When the government's existing ancient standard yards (the legal measure of length) were found to be corroded and variable in their length, Bird's skill, and that of Jonathan Sisson, were called on for the preparation and graduation of new yards. These standards in turn were destroyed in the fire which swept through the Houses of Parliament in 1834, and William Simms was commissioned to prepare new bars for the Imperial standards, a time-consuming and exacting task. These bars were compared with the metric bars which had been adopted in France and other European countries, to assist with mapping and charting and for international trade.

Graduation of the smaller arcs of theodolites and sextants was mechanized by Ramsden, whose first so-called circular dividing engine was constructed in 1770, with a second improved version working by 1773. (Brooks, 1992; McConnell, 2007, 39–51) With an eye to the prize awarded to Bird for his method of dividing large arcs, Ramsden appealed to the Board of Longitude and received a similar award for describing its construction; and in addition he was allowed to keep the engine, which remained the Board's property. The engine divided arcs up to a radius of 10 inches within minutes, in contrast to the hours required to do it by a method akin to that of Bird. Within a short time, Peter Dollond, John Troughton, and others in Britain and France were able to improve on Ramsden's engine, considerably reducing the cost of production for sextants, protractors and other surveying instruments. Ramsden, followed by his contemporaries, also built straight-line dividing engines for graduating rules, thermometers, and other linear scales. Larger observatory quadrants and circles were still divided by hand. When the crates containing these large instruments were manhandled for transport to distant places the brass structure itself might be distorted. Conscious of this possibility, the Spanish government insisted that Thomas Jones (1775–1852) (OxDNB), maker of a mural circle despatched to the Naval Observatory at Cadiz in 1834, should travel there to set up and divide it on site.

Military men also played a role in the development of instruments for the physical sciences. Prominent among these was William Roy (1726–90) (OxDNB) of the Royal Engineers, whose military surveys in Scotland and along the English south coast led in 1780 to his appointment as head of the British team employed on the Anglo-French triangulation of the distance between the Observatories of Greenwich and Paris. For this project, Ramsden designed and constructed a large theodolite, 3 feet in diameter and capable of great accuracy, though unwieldy in the field. Jean-Dominique Cassini (1748–1845), France's representative, employed a new 12-inch diameter repeating circle designed by Jean Charles de Borda (1733–99) and constructed by Etienne Lenoir (1744–1832), arguably Paris's leading maker (A. J. Turner, 1989). While single observations by Borda's circle were less accurate, its repeating action smoothed out the errors, so that the end result matched that of Ramsden's theodolite (Martin and McConnell, 2008). In 1822–25 a second cross-Channel triangulation was undertaken, directed by Henry Kater (1777–1835) (OxDNB) and the French astronomer Jean-François Arago (1786–1853), to arrive at a more accurate measurement.

11.4.1 Optical Instruments

Spectacles for reading and distance had been imported into England for more than a century before the arrival of the telescope and, together with hand-lenses, they continued to be numerically the major product of opticians. Telescope and microscope lenses had been made since the invention of spyglasses in 1608, and procedures were little changed for the construction of doublet and triplet telescopes (King, 1955; repr. 1976, 1980) Even the arrival of achromatic telescopes in the 1750s, where lenses made from glass of differing refractive powers were combined in a single cell, were ground and polished by similar techniques, though increasing mathematical competence was required to calculate the curvature of the lenses. The Dollond family, whose working lives spanned the years from 1750 to 1871, cornered the market in good-quality flint glass, ensuring that the manufacturers supplied only flawed glass to foreign buyers

(Daumas, 157–9). Their monopoly ended when several continental glasshouse owners found the way to produce sufficiently large pieces of unflawed glass to make lenses for big refractors[6] (see Section 11.2).

Astronomical telescopes with long focal lengths presented an inverted image. Nautical telescopes, fitted with broad objective lenses for maximum visibility at night, did away with the field and erecting lenses, increasing the light-gathering power but inverting the image (Warner, 1998). Optical range-finders, which consisted of telescopes with a split lens, where the distance was found by getting the subject in focus on both halves of the lens, were used at sea and on land.

A variety of microscopes were available for medical, botanical, chemical, and industrial use (G. L'E. Turner, 1989). Reading microscopes were fitted to the graduations of those instruments which needed to be read to a high degree of accuracy, such as mathematical rules, astronomical arcs, quadrants, and circles, theodolites, and barometers. The pocket instrument known as a dynameter was invented by Ramsden around 1775 to measure precisely the image seen through a telescope objective and so calculate the power of its magnification. This value enhances astronomical observations and calculations (McConnell, 2007, 174–7).

The craftsmen most represented in the microscope trade are John Marshall (d.1723), John Cuff (1708–1772), and Benjamin Martin (1704/5–1782). Schickore (2007, 14) disagrees with the view of many historians that eighteenth-century microscopy was not a thriving enterprise, that no momentous discoveries were made, and that there were no significant advances in design and manufacture. In 1759, Martin sought to improve the optics by making reflecting microscopes. The first attempts to remove chromatism and aberration by fitting doublet lenses to microscopes were unsuccessful (Shickore, 2007, 110–1). Much development came from men who were not trained instrument-makers: the Scottish scientist Sir David Brewster (1781–1868) (*OxDNB*) and Joseph Jackson Lister (1786–1869). Gerard Turner's *OxDNB* memoir on Lister gives the best summary of his achievements in microscope lens design. Eventually, improved glass and manufacturing techniques enabled compound lenses to be adopted without loss of clarity (G. L'E. Turner, 1983, 165–8). The specimen could then be presented, lit from above or below, to a range of magnifications. Apart from its use in the medical and natural history laboratory, the microscope was welcomed in polite and family society as opening a window on the marvels of nature.

In 1829, William Nicol (c.1771–1851) discovered the polarizing effect on light transmitted through a calcite prism (Morrison-Low, 1992; G. L'E. Turner, 1983). Microscopes fitted with such 'crossed nicols' assisted mineralogists to identify crystalline minerals in thin rock samples. Seen through a glass prism, spectra of light from various sources—especially the Sun—proved of considerable interest to astronomers, chemists, and other scientists.

One of the most valuable uses of prisms was in the form of lighthouse lenses (A. J. Turner, 1989, 59–60). The French polymath George Louis Leclerc Buffon (1707–1788) had proposed a polyzonal lighthouse lens, but the first practical example, designed by Augustin-Jean Fresnel (1788–1827) and constructed by François Soleil, consisted of rings of glass prisms which magnified the light thrown by the Argand oil lamp to an unprecedented degree (Brenni, 1996, 7). For rotating lights, the prisms were arranged into facets which concentrated and directed the beam as the lantern assembly swung round.

11.4.2 Philosophical Instruments

During the later eighteenth century chemists such as Joseph Black (1728–99) (*OxDNB*), Henry Cavendish (1731–1810), and Joseph Priestley (1733–1804) (*OxDNB*) were working on the chemistry of gases and investigating the constituents of acids and salts. Their laboratories were stocked with glass and metal flasks, and also thermometers, barometers, and balances. Henry Cavendish accumulated a major collection of such apparatus. Another large and valuable collection of manuscripts and scientific apparatus belonging to Joseph Priestley was destroyed when rioters ransacked and burnt his house near Birmingham. (Timmins 1890)

William Roy laboured for many years to perfect the mercury barometer as a means of measuring the heights of mountains. The difficulty did not lie in transporting such a fragile instrument, for in 1695 Daniel Quare (1649–1724) (*OxDNB*) had patented a portable barometer. He replaced the usual boxwood cistern by a soft leather bag which, when compressed by a screw plate, forced mercury to the top of the tube and thus prevented it from oscillating in the tube and shattering the glass. After carriage to the desired location, the screw was released, allowing the mercury to return to the appropriate level. Sturdy barometers on this principle served Roy and others in Britain who were using the barometer to measure altitude. (Continental physicists preferred the Deluc siphon barometer—see below, under Switzerland.) The problem facing all these scientists was to ascertain the precise decrease of atmospheric pressure with height, given that the atmospheric pressure itself might be changing during the time that a barometer was carried up a mountain. Comparison between their findings was made more complicated by the custom in France and Switzerland of measuring altitude in toises and height of mercury in Paris inches,[7] while British scientists measured altitude in feet and height of mercury in English inches.

The two English universities of Oxford and Cambridge did not include demonstrations of practical physics in their lectures on natural philosophy, though some of the professors bought telescopes and other apparatus which they set up for their own researches. A contrary situation obtained in Scotland's four ancient universities, where mathematics and astronomy heralded the Enlightenment, with its welcoming attitude to the teaching of earth sciences and medicine, with the consequent demand for scientific instruments. This was satisfied mostly by importing them from London, pending the slow rise of a Scottish instrument-making tradition (Bryden, 1972; Clarke *et al.*; Morrison-Low, 2002). Alexander Adie (1775–1858) (*OxDNB*) designed and constructed instruments for professors at Edinburgh University, and developed a range of meteorological and surveying instruments. Members of his family subsequently established themselves in London and Liverpool (Clarke *et al.*, 25–84). The arrival of medical practitioners in the new spa towns, principally Bath, created a demand for ancillary medical apparatus such as microscopes, thermometers, barometers, and the fashionable electrical apparatus (Fara, 2005).

The investigation of the earth sciences in general during the later eighteenth and nineteenth centuries attracted physicists from the private and public spheres. How to measure the earth's gravity, how it varied locally (once the flattening at the poles had been demonstrated to general satisfaction), and the consequent effect on a plumbline, became a matter of concern to surveyors and astronomers needing to set up their apparatus with reference to the true vertical. The Honourable Henry Cavendish (1731–1810), probably one of England's wealthiest and best-equipped physicists, devoted much time to the construction of an apparatus by which he hoped 'to weigh the Earth' (Hutton, 1821). He was followed in this pursuit by Nevil Maskelyne, who in 1774 left the comfort of the Royal Observatory to spend the summer on Rannoch Moor in Scotland. Here he measured the deflection of a plumb line on either side of the prominent hill of Schiehallion, with reference to the overhead passage of a particular star, showing that even a small hill could deflect the plumbline of an astronomical instrument, and that such an effect must be taken into account in all astronomical and cartographic work (Howse, 1989, ch. 12, 129–141). Much later, during the Survey of India, came the realization that invisible subsurface geology had a similar effect on the plumb line.

During the early years of the nineteenth century the sea ice in the Greenland Sea retreated, encouraging Royal Navy ships and whaling vessels to venture further than usual into Arctic waters. This brought them close to the north magnetic pole, then located in the Boothia Peninsula. In these regions a magnetic needle was strongly attracted by the near-vertical lines of magnetic force, and hardly at all by the horizontal force, rendering a ship's compass virtually useless. This situation encouraged an investigation of terrestrial magnetism, with the construction of needles suspended so as to measure the angle of dip. Delicate horizontal needles were already in use in Britain and Europe, for measuring the small angle between geographic and magnetic north (Basso Ricci *et al.*, 15–19; English Foreword). Both types of magnetic needle were regularly observed on land (for neither could function onboard a moving ship), and gradually the horizontal and vertical lines of terrestrial force were mapped.

At the same time, maritime expeditions were sent out from Britain and France to distant parts of a world whose oceans were still partly unknown. Marine charts indicated depths only in those coastal waters where ships needed to approach safely and cast anchor, so naval expeditions were sometimes instructed to sound the depths of the open sea or to test the subsurface temperature, for general information and comprehension. By and large, these early attempts failed, partly because they were made in ignorance as to the scale of the ocean's depths, and partly because the materials from which the sounders and thermometers were constructed did not function under pressure, nor could they be lowered vertically or hauled in from any depth, due to the friction between hemp rope and water.

In the nineteenth century, serious attempts were made to overcome these handicaps (McConnell, 1982). When brass mechanical sounders were devised around 1804, increasing pressure with depth caused their mechanisms to jam. Over the years from

1802 the Massey family (*OxDNB*), who were also watch-makers, patented various improvements. The topography of the ocean bed—hitherto supposed to be a great formless abyss—began to take shape only in the last quarter of the nineteenth century, with closely spaced soundings for laying submarine telegraph cables. When the silk merchant James Six (1731–1793) (*OxDNB*) devised a 'deep-sea' maximum–minimum registering thermometer he had no knowledge of the sea nor of the effects of pressure on a fragile liquid-in-glass instrument (Austin and McConnell, 1980). Meteorologists welcomed his design, but its use at sea came only in the later nineteenth century.

The first successful marine barometer accompanied Captain James Cook (1728–1779) on his circumnavigation of 1772–75 (McConnell, 2005). Adapted by Edward Nairne from portable barometers then in use, observations at sea became possible because its constricted tube prevented the mercury from continual oscillation with the motion of the ship. Such marine barometers were supplied to the expeditions sent out from France, Russia, Spain, and Britain during the early 1800s, and collected useful records. The whaling captain William Scoresby the younger (1789–1857) (*OxDNB*) investigated the subsurface pressure and temperature, and made important contributions to marine science during his northern voyages of 1807–22 (McConnell, 1986).

Scientific instruments were brought into manufacturing and industry—on a very small scale at first, but increasing considerably thereafter as industries expanded and legislation was imposed on a number of procedures and products (Wetton, 1994). The modernizing brewer kept a thermometer handy to test his mash, while at the end of the process Excise officers were on hand with their hydrometers, manufactured in large quantities, to gauge the alcohol content of beers, wines, and spirits. Specialist slide-rules and calculating instruments were designed for architects, ship-builders, surveyors, cartographers, navigators, brewers, and distillers. Merchants needed scales and balances, controlled by standards of weight and measure. Bankers and traders needed coin balances (Jenemann, 1997; Crawforth, 1979).

Schoolmasters and travelling lecturers provided themselves with instruments to illustrate their topics—simple orreries, cases of magnetic and electrical toys, mathematical and drawing instruments, and wooden or cardboard models of brass apparatus. Millburn (1976) has focused on the life of Benjamin Martin (1704/5–1782), who was one such 'country showman'. Emphasizing the importance of this rise in the popular interest in science, the *British Journal for the History of Science* devoted a special issue to science lecturing in the eighteenth century.[8]

In the private world, a number of rich amateurs took an interest in the physical sciences, first among them being King George III (*OxDNB*), who commissioned a large variety of instruments, partly for the instruction of his children (Morton and Wess). John Stuart (1713–92), third Earl of Bute, while active in political life, was also a notable collector of telescopes, as was his brother James Stuart Mackenzie (1719–1800). George Spencer (1739–1817), fourth Duke of Marlborough, and Sir George Shuckburgh-Evelyn, baronet (1751–1804), both acquired large costly astronomical instruments for their private observatories. Other aristocrats, notably Sir Joseph Banks (1743–1820) (*OxDNB*; Chambers), had wide scientific interests.[9] The Honorable Henry Cavendish spent most of his time in his laboratory in London, investigating the components of air, thermometry, and that continuing matter of 'the weight of the Earth'. Cavendish discussed the design of apparatus needed for his experiments with Edward Nairne, his preferred instrument-maker.

Such wealthy amateurs naturally turned to the leading makers for their apparatus, often wrote up their findings, and were elected to the Royal Society. Lower down the scale, anyone in the family could learn to use the camera lucida, an optical device which helped the would-be artist to make rapid and accurate sketches—a valuable skill in the days before photography (Hammond and Austin, 1987). Lessons could also be learnt by playing with kaleidoscopes (Morrison-Low, 1984, 82–100), with sets of magnetic toys (Fara, 2005), and with simple orreries, which demonstrated the motions of Earth, Sun, and Moon, and the causes of eclipses (King and Millburn).

The surveys of the Earth's gravity and magnetism undertaken by Henry Kater (1777–1835) (OxDNB), formerly of an Indian regiment, and Edward Sabine (1788–1873) (OxDNB), of the Royal Artillery, gave considerable impulse to the development of magnetic instruments. These form a distinct group, by virtue of the materials of their manufacture (a steel needle, a hard stone or a jewelled bearing, silk suspension, mica), the craft processes employed—and therefore the craftsmen—and the problems specific to their use. Nairne was a pioneer, and was followed by a line of London craftsmen who included Thomas Charles Robinson (1792–1841), Henry Barrow (1790–1870), John Dover (1824–81), and Alfred Walter Dover (1857–1920) (McConnell, 1985). These instruments could yield more accurate measurements of terrestrial magnetism, both as it varied from place to place around the Earth, and with time, and with the variation of gravity. Physicists had to struggle to confront new concepts of the Earth's inner structure. Such instruments were devised in Britain, France, and Germany to measure the direction and strength of the Earth's field on land, revealing that terrestrial magnetism varies from place to place around the Earth, and with time.

The descriptive 'scientist' was not used during the time-span of this chapter, the usual term being 'natural philosopher'. Their interests ranged from the relatively simple study of thermometry, observations of local weather and climate, to the more costly experiments with vacuum apparatus. Amateurs with a taste for deeper, perhaps riskier, investigations chose to explore the new world of electricity. As 'experimental philosophers' they and some medical practitioners opened up a branch of physics which also called for expensive apparatus and space for its operations. Edward Nairne, whose own scientific knowledge led to his election to the Royal Society, established a reputation for his electrical apparatus. Wealthy men with an income derived from family or industry preferred to reside outside the smoky air of towns, and where they had an unobstructed view of the sky, built their own astronomical observatories, furnished with the finest instruments which space and a deep pocket allowed. When their political or business interests took them away from home, such men could afford to employ salaried observers to maintain regular observations on their chosen aspect of astronomy. Pride in their achievements and discoveries led many of these natural and experimental philosophers to publish their findings and thus contribute to the general expansion of knowledge.

11.5 Makers and Markets in Continental Europe and the United States of America

11.5.1 France

The French royal observatory in Paris, founded in 1667, shortly before that at Greenwich, was directed by a dynasty of famous observers: the Cassini family, originally from Perinaldo in what is now Provence. The first to work in Paris, Giandomenico Cassini (1625–1712), had previously been employed at Bologna. A meridian line was projected through the observatory building, its purpose being to provide, by astronomical observations, the origin for a grid of latitudes and longitudes on which to construct a map of France. The Académie royale des sciences differed from the Royal Society in that it consisted of an appointed body of paid scientists, generally working on government programmes. The observatory directed the cartographic surveys; and the astronomers and academicians both needed instruments.

In 1736 the Académie sought to resolve the argument over the shape of the Earth—whether oblate or prolate—and sent two expeditions, one to the Vice-royalty of Perú (present-day Ecuador) (Lafuente and Mazuecos, 1987) and the other to Lapland (present-day Sweden) (Martin, 1987), to measure long arcs of the meridian. Apart from two zenith sectors made by Graham, the other instruments were the finest available from French workshops.

Exhaustive accounts of the colleges, universities, and institutions teaching medical, botanical, technical, engineering and mining, with full bibliographies, are in volume XI of *Histoire de la pensée*, edited by Taton (1964). Several military and other colleges in Paris had observatories (Howse, 1986, 14–20; 1994, 207–18); that of the Marine Academy at the major Channel port of Brest was established after the Revolution. Elsewhere, observatories associated with universities or local societies were established at Dijon (1780), Lyon (1701), Marseille (1702), Montpellier (1741), Strasbourg (1771), and Toulouse 1731) (Howse, 1986, 11–22).[10] Many private observatories were owned by wealthy aristocrats who lost their fortunes or were executed during the Revolution (Hahn, 1964).

There was less royal encouragement for hydrographic surveying and, as in Britain, much knowledge was derived from charts produced elsewhere. French seafarers were the first outside Britain to adopt Hadley's octant (Mörzer Bruyns, 33–4). A description in French appeared in 1739, the first examples were made in Paris by Pierre Lemaire, and makers in port towns followed suit.

Some very fine astronomical instruments emerged from Parisian workshops, despite the impedance of the antiquated guild structures where the instrument-makers were strictly controlled by narrowly specialized guilds. As many instruments incorporated brass, glass, and wood, the three guilds which controlled individual use of each of these materials persecuted the workmen, sometimes even seizing and destroying items under construction. (A. J. Turner, 1989, 1998). Unlike the major London craftsmen with their bank accounts, the French craftsmen had difficulty raising enough money to buy materials and pay workmen while building large apparatus. The guilds did not survive the Revolution, after which time education and encouragement made life easier. Makers of compasses worked at the main ports, and as the navy recovered from the Revolution (which had decimated the officer class), some sextants, octants, and sextants were made there.

In 1775 the Académie des sciences announced a prize for a memoir describing the best method of constructing magnetic compasses, One of the winners was Charles-Augustin de Coulomb (1736–1806). In his design, the magnet was suspended by a silk thread—a virtually friction-free arrangement which enabled the slightest motion to be detected (Basso Ricci 1997, 16–17, 26–7). Geomagnetic instruments, embodying variations on the Coulomb silk suspension and made entirely from non-magnetic materials—ivory, or more usually, wood—gave far more accurate results than previous instruments by which the magnet cap rotated on a sharp point.

Étienne Lenoir (1744–1832) (A. J. Turner, 1989) had developed a successful business making reflecting and repeating circles and other surveying apparatus before the Revolution. He survived that turmoil, and was then called on to deal with the listing and disposal of numerous instruments seized from the aristocratic collections. Ramsden's first dividing engine had been bought by the aristocrat Jean-Baptiste Gaspard Bochart de Saron (1730–1794), and following his execution it was handed over to Lenoir with instructions to redivide its 360-degree circle into the Republican 400 grads, and to instruct young craftsmen in its operation. Lenoir was commissioned to make a number of metric standards of length, some in platinum. He was also involved in the manufacture of parabolic reflectors for lighthouses. His importance brought him several ministerial 'pensions'—effectively salaried posts—raising him from his pre-Revolution status as a humble craftsman to enjoy recognition and respect in the higher circles of academicians and government.

Lenoir's son Paul-Étienne (1776–1827) was running the business by 1819—a time when there was a great demand for instruments to survey France's Napoleonic conquests in Europe. Paul-Étienne developed and published on his range of improved surveying and geodetic apparatus, taking part in trade exhibitions. His early death, his father then being elderly, brought the Lenoir business to a close.

Jacques Canivet (d.1774) was the only Parisian maker active in mid-century to construct large astronomical instruments—in his case, a 6-foot mural quadrant. The optician Robert-Aglaé Cauchoix (1776–1845) produced telescope lenses which were sought after by astronomers outside France. The Alsace-born François-Antoine Jecker (1765–1834) based his workshop organization on that of his former employer, Jesse Ramsden, and he and his brother, and their descendants, established a successful manufactory in Paris (McConnell, 2007, 63). Henri-Prudence Gambey (1787–1847), who had worked as a journeyman under Lenoir, constructed his own dividing engine, producing excellent astronomical, magnetical, surveying, and navigational instruments (Brenni, 1993, 11–13). Many Gambey instruments were bought for colleges and government institutions in the USA (Warner, 1990), and he benefited from the so-called Magnetic Crusade of the 1840s when the mathematician Karl Friedrich Gauss (1777–1855) was organizing the installation of magnetic instruments at observatories round the world, for simultaneous observation at certain times.

One French instrument-maker whose fame spread far beyond his native France was Jean-Antoine Nollet (1700–1770). While studying divinity in Paris, Nollet sought to broaden his general education, but found the instruments he needed too expensive. Self-taught as a maker of physics instruments, his skilful productions funded his own work and composing his popular six-volume *Cours de physique experimentale* (1743–8), which gave full directions and illustrations for his apparatus. Nollet rose to high academic status, but was less successful when he took up the new science of electricity and came into conflict with Benjamin Franklin.

In 1790 Republican France passed a law to replace the multitude of ancient local measures with a standard equivalent to one ten-millionth of the length of a meridian of longitude between the equator and the north pole (a dimension which was itself still rather uncertain). The new metre was created as a platinum bar. Weights and measures, and briefly the Republican calendar, its timekeeping, and its division of the circle, were decimalized. As metric measures were adopted elsewhere in Europe, these countries needed to procure examples of each others' standards for intercomparison.

The Revolution, which was responsible for the execution of France's great chemist, Antoine Laurent Lavoisier (1743–1794), also led to the confiscation of the many cabinets of instruments belonging to aristocrats—a practice which in fact brought many fine French and foreign instruments into the public domain and served to instruct a new generation of craftsmen. Lavoisier's vast collection of several thousands of instruments, tools, vessels and chemicals diminished over the years as it moved around, some of it passing into the Charles cabinet (Beretta, 2003, 318). After the Second World War the remaining items arrived at the Conservatoire des arts et métiers in Paris (Beretta, 2003). Daumas (1953, and 1972, 147) refers to 'two or three outstanding collections'—that of the Collège du Roi, afterwards the Collège de France, whose vast collection of apparatus was intended for demonstration for public education. The Collège de Navarre held a broad-based collection of 235 items. The cabinet of the physicist Jacques-Alexandre Charles (1746–1823) was even more numerous, amounting to 330 items in 1792. In 1807 they were handed over to the State and remain in the Conservatoire des arts et métiers.

As Napoleon marched across Europe during the 1790s, other collections, including those of the Dutch Stadholders, fell victim to what de Clercq (1997, 134) aptly describes as 'the French scientific conquest'. Sent back to Paris, these instruments were dispersed—some into the newly-created Conservatoire des arts et métiers, and others to the École Polytechnique or to colleges in Paris and the provinces. Here they served to instruct and inspire new craftsmen who were responding to the need for instruments for military and naval use.

During the nineteenth century the French instrument trade profited from the expanding number of schools and colleges in Italy, and in North and South America, many of which purchased mathematical instruments, microscopes, and other optical apparatus (Brenni, 1996; Payen, 1985). The guild structure had been swept away by the Revolution; now with better training and in a new political atmosphere, French craftsmen were able to establish profitable and long-lived businesses. The Soleil dynasty, known just before the turn of the century and which came to include the Dubosc family, produced all the apparatus needed to study, measure, and demonstrate optical phenomena (Brenni, 1996). The families of Lerebours, with Secretan (Brenni, 1994), and Froment (Brenni, 1995) all flourished over several generations.

In 1841 the mathematician and historian of science Jean-Baptiste Biot (1774–1862) published on the application of polarimetry, thereby opening the way to medical saccharimetry. Cauchoix, Lerebours, and other French makers were soon marketing these simple instruments (Davis and Merzbach, 1994).

11.5.2 The Low Countries

In 1950 Rooseboom listed 250 instrument-makers in the Low Countries prior to 1840, but apart from a few who are well-known, most names occurred on a single item or once in an auction catalogue. Other 'makers' were not active commercially; the grinding and polishing of lenses for microscopes or telescopes, a task requiring time and patience rather than expensive machinery, was undertaken as a sideline by rich and poor alike.

Nautical instrument-makers clustered in port towns; from 1735 octants began to replace wooden backstaffs, a change which speeded up from 1747 when the VOC (Dutch East India Company) decided to provide each company vessel with an octant and instructions for its use (Mörzer Bruyns, 2009, 34). The main manufacturers were Johannes van Keulen (1733–1801) and his son Gerard in Amsterdam, and Jacobus Kley (1716–1791) in Rotterdam.

Other craftsmen served professors and students in the five old universities: Leiden, founded in 1575; Franeker, 1585; Groningen, 1614; Utrecht, 1636; and Harderwijk, 1648. The first demonstrations of physics depended on the appointment of a professor, provided either with his own apparatus or with university funding for its purchase. Again, Leiden led the way, establishing a physics theatre in the late 1670s, the other universities following during the early 1700s.

The most famous supplier of medical and scientific instruments and apparatus to the universities, to the royal family (Clercq, 1988; 1997), and to private individuals was the Musschenbroek business (Clercq, 1997). Established in Leiden as a brass foundry, the first family member to manufacture scientific apparatus was Samuel Musschenbroek (1640–1681), who around 1675 began production of microscopes and air-pumps. The premises and workforce seem always to have been modest in size. Business continued under Samuel's younger brother Johan, or Jan (1660–1707). While apparatus for demonstrating the principles of mechanics was the main output, the workshop also supplied mathematical, astronomical, and medical apparatus, and balances. After Jan's death his former assistant Antoni Rinssen worked at the same premises until 1765, after which the workshop tools and stock were bought by

Jan Paauw junior, who became the leading Leiden instrument-maker, continuing to produce apparatus similar to that of the Musschenbroeks.

The Musschenbroek business was considerably aided by the publications of Willem Jacob 'sGravesande (1688–1742). Trained as a lawyer, 'sGravesande chose to become a teacher of physics, publishing his two-volume *Physices Elementa Mathematica* in 1720–21. This coherent series of lectures and demonstrations became a standard text elsewhere in Europe, increasing Musschenbroek's sales. 'sGravesande acknowledged the debt the designs of his apparatus owed to demonstrations that had been given by Francis Hauksbee (1688–1763), William Whiston (1667–1752), and John Thomas Desaguliers (1683–1744). He had met Desaguliers during an earlier diplomatic mission to London in 1715, the year in which he was elected Fellow of the Royal Society.

The links to the universities, where foreign professors were regularly employed, ensured that word of the firm's apparatus spread far beyond the Low Countries (Clercq, 1997, 162–72). Musschenbroek's apparatus was sold to Landgrave Carl of Hesse Kassel and to the ancient university of Marburg in Hesse. A few items were sold to Sweden, while others went to the Italian secular universities of Padua and Bologna, both of which began physics demonstrations before 1740. The Russian Academy of St Petersburg probably also bought instruments from Leiden, though the evidence is uncertain, as some apparatus may have arrived in later years (Ryan, 1991).

11.5.3 Portugal

Under the modernizing rule of Sebastião José de Carvalho e Melo (1699–1782), better known by his later title as the Marquês de Pombal, the Jesuits were suppressed, allowing scientific education to catch up with that in France, the Low Countries, and Britain. In 1766 A Colégio dos Nobres in Lisbon began teaching boys aged between seven and thirteen years, and in due course King José V contributed a cabinet of physics, including machines, apparatus, and instruments (Amaral). But the children had to learn mathematics before they could get to grips with the Latin text of 'sGravesande's two-volume textbook *Elementa Physicae*. When the Colégio closed, its Paduan professor, Giovanni Antonio dalla Bella (1730–1823), moved to Coimbra, taking the collection with him. Reform of the university at Coimbra in 1772 had created a faculty of philosophy, where natural history, experimental physics, and practical and theoretical chemistry were taught, following the ideas of 'sGravesande and Nollet. Teachers were recruited from outside Portugal, some apparatus was made locally, others were purchased, mostly from London makers including Adams, Dollond, Ramsden, Nairne, Martin and Champneys (Carvalho, 1978).

When French troops arrived in 1810 this collection suffered from 'the French scientific conquest'. A superb six-lens achromatic telescope by Dollond and other treasures were taken to Paris. Other instruments were lost in the period between their obsolescence and becoming of historic interest. Despite these ravages, the Gabinete da Fisica at Coimbra presents the finest example of an eighteenth-century teaching collection, fully illustrated in *Ingenuity and Art*.

In 1798 a royal maritime society for cartography and hydrography was established in Lisbon. Subsequently it was decided that nautical instruments should be manufactured, and the Swabian-born maker Jacob Bernard Haas (1753–1828), who had been working in London, was invited to move to Lisbon. He produced a wide range of instruments, and was followed in this business by João Frederico Haas (1792–1876). Two young men, Gaspar José Marques (1775–1843) and José Maria Pedroso, were sent to London as apprentices to Ramsden (Estacio dos Reis 1991). Pedroso set up as an instrument-maker in Lisbon, and then accompanied the Portuguese government to Rio de Janeiro in Brazil, ultimately returning to Lisbon (Estacio dos Reis, 2006, 30–1).

11.5.4 Italy

Throughout the period of this chapter, Italy was divided into numerous states, some under foreign rule. There were no colonies, no important navy or merchant fleet, nor any clock-making industry, in the strict sense. In the south, the large estates did not encourage cartography. Technical innovation began in the northern regions of Lombardy, Tuscany, and the Veneto, where mechanics repairing foreign instruments might make simple copies, which being unsigned are difficult to date.

The Bolognese nobleman Luigi Ferdinando Marsigli (1658–1730) presented his city with a new scientific institute in 1711, to which he donated his own instruments. He also encouraged the rebuild of the ancient observatory tower, which was furnished with new instruments, some from Rome, others from London (Baiarda et al., 1995). Other observatories—principally Padua (Pigatto, 2000), Rome, Brera, in Milan, (Tucci, 2000), Palermo (Serio and Chinnici, 1997) and Naples (Stendardo, 2001)—were modernized in the late eighteenth and early nineteenth centuries with the acquisition of English and German instruments, to support their regional cartographic surveys.

The Lusverg, or Lusuerg, family, originally from Bavaria, arrived in Rome around 1668 where the family remained prominent as instrument-makers until the late eighteenth century. Domenico Lusuerg (1669–1744) constructed many instruments now held in museums in Italy and overseas; a later Domenico (1754–1850) was mechanic to the Collegio Nazareno and the Teatro Fisico of the Università La Sapienza, and Angelo (1793–1858), Luigi, and Giacomo, likewise served the major teaching institutions, their names appearing in inventories and sales (Todesco, 1996).

After the death of the famous Roman optician Giuseppe Campani (1635–1715), whose telescopes were sought after in France and England, several dynasties of Italian opticians made simple telescopes and microscope lenses, in response to the demands of naturalists and the medical profession. The Selva family was established in Venice by 1700 (Lualdi, 2001; 2003). Domenico Selva (d.1758) trained his son Lorenzo (1716–1800), who in turn trained his own sons Giuseppe and Domenico. Lorenzo wrote several books on his productions, which included refracting and reflecting telescopes, microscopes, and surveying levels. Lorenzo bought twenty pieces of English flint glass but found only seven were usable (Lualdi, 2001, 539), and this drove him to experiment, with ultimate success. His second book, published in 1771, concerned the manufacture of flint glass—an achievement which gained him the status of 'ottico publico', a title which passed to his sons. Lorenzo's customers included the principal mathematicians and astronomers in France, Spain, and Portugal. Other Venetian opticians included Biagio Burlini (1709–1771), whose catalogue of 1758 depicts his canal-side workshop (Lualdi, 1999; 2003). His telescopes and other optical items enjoyed a high reputation, whereas Leonardo Semitecolo, whose firm lasted for about a century, produced a great many items of mediocre quality (Lualdi, 2003, 32–4). The difficulty of procuring good-quality flint glass meant that from 1750 to 1800 workshops, observatories, and physics laboratories were mostly equipped with London-made instruments. Thereafter, French optical and physics apparatus became more common.

In Modena, Giovanni Battista Amici (1786–1863), whose father was director of the General Land Office, started around 1808 to make small lenses and mirrors in connection with the repair of surveying instruments, while studying engineering and architecture. Amici's father procured copies of texts in Brera Observatory, including copies of letters written in 1783 by John Edwards (d.1783) to Maskelyne, concerning Edwards' experiments to create a new untinted and highly luminous alloy for telescope mirrors (Edwards, 1783). Amici translated this text into Italian, adding information from Robert Smith's *A Complete System of Optics* (1738) and Jean-Etienne Montucla's *Histoire des mathématiques*,[11] and starting his own experiments on casting and polishing metallic mirrors.

Amici's endeavours were so successful that in 1811 the astronomers at Brera judged his reflecting telescope to be equal to those of Herschel, and awarded him a gold medal. Later that year, his second telescope, of 17 Paris feet focal length and the largest so far made in Italy, was sent to Brera (Proverbio, 1994), Thereafter he produced a number of telescopes, and began to develop reflecting microscopes, his publications adding to his overseas reputation. Between travels throughout Italy and later to France and England, Amici produced a wide range of reflecting instruments for mathematics, microscopy, astronomy, surveying, navigation, and other uses. In 1827 he inaugurated a new astronomical observatory at Modena, for which he provided a passage instrument, an equatorial, and a Newtonian telescope (Meschiari, 2006).[12]

Italian physicists, notably Giovanni Battists Beccaria (1716–81), Alessandro Volta (1745–1827) and Luigi Galvani (1737–1798), pioneered research into electricity and had to devise and construct their own instruments and apparatus for this new field.[13] Angelo Bellani (1786–1852) was concerned with thermometry and meteorology, and built his own apparatus. The topographer, optician, and instrument-maker Ignazio Porro (1801–1875), best known for his distance-meter, moved to Paris and, when Gambey died, set up the Institut Technomatique et Optique, which produced hundreds of optical instruments.

Other branches of the Italian instrument-making craft originated from the appointment of mechanics to service the practical scientific and technical courses set up in the schools and universities. Instruments were copied from French and British examples, the establishment of an industry gradually taking shape in the early nineteenth century, despite the political uncertainties of the time. Some Italian-made instruments have been located in the collections of seminaries, but documentation is lacking and much was destroyed by Napoleon, and later in the conflicts accompanying Italian unification (Todesco, 1994).

Italy being a land of earthquakes and volcanos, attempts were made from the early eighteenth century to discover the sources—or as we now say, the foci and epicentres—of earthquakes (Ferrari, 1992), and primitive seismometers were set up in various religious institutes and schools.[14] These could indicate only at best the approximate time and the direction from which the surface seismic waves arrived. Apparatus which could accurately time and record the strength and duration of the tremors was not available until the later nineteenth century.

11.5.5 Scandinavia

Scandinavia was a small market. In Sweden the observatory of the first university, at Uppsala, functioned from the 1730s, the Swedish Academy of Science from the 1750s. Other requirements came from the mining industry and from land measurement. The Swedish instrument-maker Daniel Ekström spent 1739–40 in London, improving his skill, and on return undertook a variety of instruments, all made by himself and his apprentices. But he found it difficult to make a living (Amelin, 1994). The famous 'Expedition to Lapland'—one of two expeditions sent out by the French Académie des sciences to measure an arc of the meridian and thereby to determine whether the Earth was an oblate or prolate spheroid—took a sector by the London maker George Graham. The University of Kiel (a town then part of Denmark) was established in 1773. Teachers of physics and instrument craftsmen were recruited from Switzerland, Germany, and Piedmont, while Denmark's shipping industry called for compasses and sextants. In Copenhagen, Jeppe Smith (1759–1821) ran the first workshop with wide-scale production of instruments for astronomy and physics, probably modelled on English instruments (Anderson, 1993; Mörzer Bruyns, 2009, 34).

11.5.6 Switzerland

The Swiss city of Geneva was home to a remarkable number of scientists in the eighteenth and early nineteenth centuries—notably Jean André Deluc (or De Luc) (1727–1817), who left his native city in 1773 to settle in England, Jacques André Mallet, Horace Bénédict de Saussure (1740–1799), and Marc Auguste Pictet. Pictet bought instruments from Troughton to assist him in making important comparisons between the French and English measures of length. All three men held posts at the University of Geneva and benefited from the Société de physique et d'histoire naturelle. They travelled freely and could buy instruments in London and Paris, but they were additionally served by the Genevan craftsman Jacques Paul (1733–98), who had been apprenticed to Canivet. Paul's business was continued by his son Nicolas (Talas, 1994; A. J. Turner, 1987; Archinard, 1979).

Before he left Geneva, Deluc was already interested in the atmosphere and in the possibility of measuring altitudes with the mercury barometer. His design for a portable barometer called for a siphon tube, with a tap to secure the mercury in the tube during transport. In 1763, H. B. de Saussure carried Deluc's barometer, packed in its light wooden case, to the summit of Mont Blanc (Archinard, 1980). Paul constructed Deluc's barometers in their various stages of development, and he worked to H. B. de Saussure on hygrometers designed by him.

11.5.7 Germany

Prior to the end of the eighteenth century, what is now Germany was a loose confederation of states of various sizes, separated by customs barriers until 1831 when a customs union, the Deutscher Zollverein, was established. This lack of national government prevented any encouragement or subsidies. There were, however, some notable events: at Dresden, in Saxony, a collection of clocks, mathematical and physical instruments was established in 1728. Over the years, other instruments by various makers were added, reflecting the interests of King Friedrich August III (r.1763–1827). In the early nineteenth century, instruments including electrometers and galvanometers were loaned for teaching and research (Schillinger, 1994). The Bavarian instrument-maker Georg Friedrich Brander (1713–1783), later Brander & Höschel, gained a reputation far outside his region. Brander produced a variety of surveying instruments, as well as instruments for physics and mensuration. His workshop lost its significance after 1800 (Brachner, 1983). Johann Christian Breithaupt (1736–99) founded his workshop in 1762 at Kassel, producing mathematical, astronomical, and geodetic instruments, and also a mural quadrant for Kassel. His son Heinrich Carl (1775–1856) followed him, but a younger son, Georg Augustus (1806–1855), left to work on magnetic instruments under Carl Gauss. Johann Christoph Voigtländer (1732–1797), of Vienna, prospered in his day, but left no family successor (Brachner, 1985).

After the Napoleonic wars the need for surveying, especially in Bavaria, helped to revive the situation in the nineteenth century. Brander's tradition was continued in the Munich workshops of Georg Reichenbach (1772–1826), Joseph Liebherr (1767–1840), Joseph Fraunhofer (1772–1826), and Carl August Steinheil (1801–1870), who all gained a world-wide reputation. These men combined a knowledge of mathematics and artisan skill, and their success was aided by indirect subsidy from the Bavarian government through Joseph von Utschneider.

Fraunhofer spent his life working with the twin aspects of optics: the production of fine glass, and the behaviour of optical rays through glass of differing refractive qualities. While investigating the refractive qualities of the spectrum, Fraunhofer opened the way to the science of spectroscopy. With Utscheidner's encouragement and working in the secrecy of the secularized Benedictbeurn monastery, he was able to corner the market, as no-one was permitted to observe his methods (Jackson, 2000, 43–97).

11.5.8 United States of America

Bedini (1964, 3–4) divides instruments in the American colonies into two groups. Philosophical instruments and apparatus used in colleges for teaching and experimental work were generally imported from England and France until well into the nineteenth century. From 1787, donations and bequests of top-quality apparatus served for teaching physics, mechanics, and natural sciences at Harvard College, near Boston (Cohen, 1950, 133–44). A fire destroyed much of that collection in 1764, and only an old catalogue remains (Cohen, 133–44), but money was donated for the purchase of a large assortment of the best London-made apparatus, substantial orders going to Nairne, Dollond, Sisson, Martin, and others, illustrated in Wheatland (1968).

The practical mathematical instruments essential for navigation and land surveying were also imported from England, brought by the first settlers. Grants of land which needed to be defined and measured, and the need to delineate the state boundaries, encouraged the rise of a native-born craft. Relatively few trained instrument-makers crossed the Atlantic before American independence, but thereafter, makers and dealers in England and France responded to this new and growing market. Many emigrated, or sent their sons to the USA, where they established workshops in the commercial centres and port towns, made instruments, and also imported more specialized items from London.

Clockmakers, accustomed to work in fine brass, also turned their hand to mathematical instruments—the Rittenhouse brothers, David (1732–1796) and Benjamin (1740–c.1820), Andrew Ellicott (1754–1820), and four generations of the Chandlee family being among the most celebrated. Craftsmen accustomed to work in the local hardwoods found that this material was both plentiful and a satisfactory substitute for brass in the manufacture of surveying compasses and related instruments. They were able to study the designs illustrated in many technical books published in French and English during the seventeenth and eighteenth centuries. Navigational instruments—backstaffs and octants—were at this time normally made of wood.

One impediment to American instrument-making was the lack of fine glass for thermometer and barometer tubes, but this became available from the 1840s with the arrival of German glass-makers. As the United States expanded westwards, a network of observers with standardized instruments was recruited, their records contributing to an early understanding of surface weather systems (Fleming, 1990, 1–7, 55–73). Cartographic surveying, inland and around the coast, led to the foundation of the US Coast and Geodetic Survey, which also measured tides and inshore waters. Merchant and naval fleets operated from Pacific, Gulf, and Atlantic ports, and encouraged the settlement of nautical and chronometer dealers in those towns.

Several temporary observatories are known, such as that set up by Mason and Dixon in 1763 when they arrived to settle a boundary dispute between Pennsylvania and Maryland. Others were erected for observing the transit of Venus in 1769. In the early 1800s the US Coast and Geodetic Survey acquired British instruments, but made little use of them (Rufus, 127).

From the 1830s, observatories were built in association with an increasing number of universities and colleges, with Yale, Cincinnati, and Harvard leading the way (King, 1955, 246–9). At first, lenses were imported from Europe to be mounted by American craftsmen, and in the 1850s Alvan Clark (1804–1887) began producing lenses of excellent quality, and was so successful that he and his sons set up the first telescope factory at Cambridgeport, Massachusetts.

References

ALDER, K. (1995), 'A revolution to measure: The political economy of the metric system in France', pp. 39–71 in M. N. Wise, *The Values of Precision* (Princeton, NJ: Princeton University Press)

AMARAL, M. (2000–09), 'Real Colégio dos Nobres' *Dicionario Historico, Heraldico, Biografico*. Vol. VII, 108–111. Ed. electronico 2000–2009, <http://www.arqnet.pt/dicionario/rcolnob.html>

AMELIN, O. (1994), 'Daniel Ekström: maker of scientific instruments in 18th century Sweden', pp. 81–83 in Dragoni, G. *et al.* (1994)

ANDERSEN, H. 'JEPPE SMITH (1759–1821): a Danish instrument maker', pp. 403–417 in Anderson, R. G. W. *et al.* (1993)

ANDERSON, R. G. W., J. A. BENNETT, and W. F. RYAN (eds.), *Making Instruments Count* (Aldershot: Variorum 1993)

ARCHINARD, M. (1979), *Collection de Saussure* (Geneva: Musée d'art et d'histoire de Genève)

—— (1980), *De Luc et la recherche barométrique* (Geneva: Musée d'histoire des sciences de Genève)

ATTWOOD, G. (1776?), *A Description of the Experiments, Intended to Illustrate a Course of Lectures, on the Principles of Natural Philosophy, read in the Observatory at Trinity College, Cambridge* (London: 1776?)

AUSTIN, J. and MCCONNELL, A. (1980), *Reproduction of James Six: The Construction and Use of a Thermometer (1794) with Additional Material* (London: Nimbus Books)

BAIDA, E., BÒNOLI, F., and BRACCESI, A. (1995), *Museo della Specola* (Bologna: Bologna University Press)

BAKER, A. S. (2009), 'The business of life: the socioeconomics of the 'scientific' instrument trade in early modern London', pp. 169–191 in F.-E. Eliassen and K. Szende (eds.), *Generations in Towns: Succession and Success in Pre-Industrial Urban Societies* (Newcastle-upon-Tyne)

BASSO RICCI, M., CAFARELLA, L., MELONI, A., and TUCCI, P. (1997), *Due secoli di strumenti geomagnetici in Italia (1740–1971)* (Rome: Istituto Nazionale di Geofisica)

BEDINI, S. A., (1964) *Early American Scientific Instruments and Their Makers* (Washington: Museum of History and Technology)

BENNETT, J. (1987), *The Divided Circle: A History of Instruments for Astronomy, Navigation, and Surveying*. (Oxford: Phaidon)

BETTS, J. (1998), 'Arnold and Earnshaw: The practical solution', pp. 311–328 in Andrewes *et al.*, *The Quest for Longitude*. (Cambridge, MA: 1998)

BERETTA, M. (2003), 'Lavoisier's collection of instruments: a checkered history', pp. 313–334 in Beretta, M. Galluzzi, and Triarco, C. (2003), *Musa Musaei. Studies in honour of Mara Miniati*. (Florence: Olschki)

BRACHNER, A. (1983), *G. F. Brander, 1713–1783. Wissenschaftliche Instrumente aus seiner Werkstatt*. (Munich: Deutsches Museum)

BRACHNER, A. (1985), 'German nineteenth-century scientific instrument makers; pp. 117–157 in Clercq (1985)

BRENNI, P. (1985), 'Italian scientific instrument makers of the nineteenth century and their makers', pp. 183–203 in Clercq (1985)

—— (1993), '19th century French instrument makers, Pt. I: H.-P. Gambey', *Bulletin of the Scientific Instrument Society* No. 38, 11–13

—— (1994), '19th century French instrument makers, Pt. II: Lerebours and Secretan', *Bulletin of the Scientific Instrument Society* No. 40, 3–6

—— (1995), '19th century French instrument makers, Pt. III: Paul-Gustave Froment, 1815-65', *Bulletin of the Scientific Instrument Society* No. 45, 19–24

—— (1996), '19th century French instrument makers, Pt. XIII: Soleil, Dubosc and their successors', *Bulletin of the Scientific Instrument Society* No. 51, 7–16

BROOKS, J. (1992), 'The circular dividing engine: Development in England 1739–1843' *Annals of science* 49/2, 101–135

BRYDEN, D. J. (1972), *Scottish Scientific Instrument Makers 1600–1900* (Edinburgh: Royal Scottish Museum)

CARVALHO, R. DE (1978), *Historia do Gabinete de Física da Universidade de Coimbra* (Coimbra: University of Coimbra)

CHALDECOTT, J. A. (1987), 'Platinum and palladium in astronomy and navigation', *Platinum Metals Review* 31, 91–100

CHAMBERS, N. (2006), *The Scientific Correspondence of Sir Joseph Banks, 1768–1820.* 6 vols. (London: Pickering and Chatto)

CHANCE, J. F. (1919), *A History of the Firm of Chance Brothers and Co, Glass and Alkali Manufacturers* (London)

CLARKE, T. N., MORRISON-LOW, A. D., and SIMPSON, A. D. C. (1989), *Brass and Glass: Scientific Instrument Making Workshops in Scotland*. (Edinburgh: National Museums of Scotland)

CLERCQ, P. R. DE (ed.) (1985), *Nineteenth-Century Scientific Instruments and Their Makers* (Leiden Communication 221 of the National Museum for the History of Science and medicine 'Museum Boerhaave' and Amsterdam: Rodopi)

CLERCQ, P. R DE (1988), 'Science at court: the eighteenth-century cabinet of scientific instruments and models of the Dutch stadholders', *Annals of science* 45, 113–152

CLERCQ, P. R. DE (1997), *At the Sign of the Oriental Lamp* (Rotterdam: Erasmus publishing)

CLIFTON, G. (1995), *Directory of British Scientific Instrument Makers 1550–1851* (London: Zwemmer)

COHEN, B. I. (1950), *Some Early Tools of American Science* (Cambridge, MA: Harvard University Press)

CRAWFORTH, M. (1979), *Weighing Coins: English Folding Gold Balances of the 18th and 19th Centuries* (London)

D'ARTIGUES, A.-G. (1811) *Sur l'art de fabriquer du flint-glass bon pour l'optique* (Paris)

DAUMAS, M. (1953), *Lavoisier théoricien et expérimentateur* (Paris: PUF)

DAUMAS, M., trans. and ed. M. HOLBROOK (1972), *Scientific Instruments of the Seventeenth and Eighteenth Centuries and Their Makers* (London: Batsford)

DAVIS, A. and MERZBACH, U. (1994), 'The role of the polarimeter in the diagnosis of diabetes mellitus, Biot's bequest to 19th century medical saccarometry'. pp. 143–151 in Dragoni, G. et al. (1994)

DE GRAEVE, J. (1986), *250. Anniversaire pour la mesure d'arcs du méridien et la figure de la terre*. (Aalst: Du Caju)

DRAGONI, G., BERGIE, S., and GOTTARDI, G. (1999), *Dizionario biografico degli scienziati e dei tecnici* (Bologna: Zanichelli)

DRAGONI, G., MCCONNELL, A., and TURNER, G. L'E. (1994), *Proceedings of the Eleventh International Scientific Instrument Symposium, Bologna University, Italy, 9–14 September 1991* (Bologna: Grafis Edizioni)

EDWARDS, J. (1783), 'Directions for making the best composition for the metals of reflecting telescopes, and the method of casting, grinding, polishing, and giving the great speculum the true parabolic figure', 60 pp, appended to the *Nautical Almanac for 1787* (published 1783)

ESTÁCIO DOS REIS, A. (1991) *Uma oficina de instrumentos matemáticos e náuticos (1800–65)* (Lisbon: Academia da Marinha)

—— (2006), *Gaspar José Marques e a máquina a vapor* (Lisbon: Edições culturais da Marinha)

FARA, P. (2005), *Fatal Attraction: Magnetic Mysteries of the Enlightment* (Cambridge: Icon Books)

FERRARI, G. (1992), *Two Hundred Years of Seismic Instruments in Italy 1731–1940* (Bologna: SGA Storia Geofisica Ambiente)

FISHER, S. (2001), *The Makers of the Blueback Charts* (St Ives, Cambridgeshire: Imray, Laurie, Norie and Wilson)

FLEMING, J. R. (1990), *Meteorology in America, 1800–1870* (Baltimore, 1990)

FORBES, E. G. (1975), *Greenwich Observatory, vol. I; Origins and Early History* (London: Taylor and Francis)

HAHN, R. (1964), 'Les observatoires en France au XVIII siècle', pp. 653–658 in Taton (1964)

HAMMOND, J. R. and AUSTIN, J. (1987), *The Camera Lucida in Art and Science* (Bristol: Adam Hilger)

HOLLAND, J. (1853), *Treatise on the Progressive Improvement and Present State of the Manufactures in Metal*, vol. 3 of D. Lardner, (ed.), *Cabinet Cyclopedia* (London: Spottiswoode)

HOWSE, D. (1975), *Greenwich Observatory, vol. III; Its Buildings and Instruments* (London: Taylor and Francis)

Howse, D. (1986), 'The Greenwich List of observatories: A world list of astronmical observatories, instruments and clocks, 1670–1850', whole issue of *Journal for the history of astronomy* 17/4; 'Supplement No 1' (1994), *Journal for the history of astronomy* 25/3, 207–218.

—— (1989), *Nevil Maskelyne: The Seaman's Astronomer* (Cambridge: Cambridge University Press)

HUTTON, C. (1821), 'On the mean density of the Earth' *Philosophical Transactions of the Royal Society* 111, 276–292

Ingenuity and Art (1997) (Lisbon and Coimbra: University of Coimbra)

JACKSON, M. W. (2000), *Spectrum of Belief: Joseph von Fraunhofer and the Craft of Precision Optics* (Cambridge, MA, and London: MIT Press)

JENEMANN, H. R. (1997), *Die Waage des Chemikers/The chemist's balance* (Frankfurt am Main: DECHEMA)

KING, H. C. (1955, repr. 1976, 1980), *The History of the Telescope* (London: Griffin; New York: Dover)

—— and MILLBURN, J. R. (1978), *Geared to the Stars: The Evolution of Planetariums, Orreries and Astronomical Clocks* (Toronto and Bristol: Adam Hilger)

LAFUENTE, A. and MAZUECOS, A. (1987), *Los caballeros del punto fijo* (Serbal: CSIC)

LUALDI, A. (1999), 'Biagio Burlini, un ottico del '700 veneziano', *Nuncius* 14/1, 213–220

—— (2001), 'La familiglia Selva, ottici del '700 veneziano' *Nuncius* 16/2, 531–546

—— (2003) 'Venetian makers of optical instruments in the 17th and 18th centuries'. *Bulletin of the Scientific Instrument Society*, 'Biagio Burlini', No. 76, 35–7; 'The Selva family', No. 77, 13; Leonardo Semitecolo and imitators, No. 78, 32–4

MARTIN, J. P. (1987), *La figure de la terre. Récit de l'expédition française en laponie suédoise (1736–1737).* (Cherbourg: Editions Isoète)

—— (2000), *Une histoire de la Méridienne.* (Cherbourg: Editions Isoète)

MCCONNELL, A. (1980), *Geomagnetic Instruments before 1900* (London: Harriet Wynter)

—— (1982), *No Sea too Deep* (Bristol: Adam Hilger)

—— (1985), pp. 29–52 in Clercq (1985)

—— (1986), 'The scientific life of William Scoresby Jnr, with a catalogue of his instruments and apparatus in the Whitby Museum', *Annals of science* 43, 257–258

—— (1992) *Instrument Makers to the World: A History of Cooke, Troughton & Simms* (York: Sessions)

—— (2005), 'The origins of the marine barometer' *Annals of Science* 62/1, 83–101

—— (2007), *Jesse Ramsden 1735–1800: London's Finest Scientific Instrument Maker* (Aldershot: Ashgate)

MARTIN, J. P. and MCCONNELL, A. (2008), 'Joining the observatories of Paris and London', *Notes and records of the Royal Society* No. 62, 355–372

MESCHIARI, A. (ed.) (2006), *Opere edite.* Vols 1* and 1** of *Edizione nazionale delle opere e della corrispondenza di Giovanni Battists Amici* (Naples: Bibliopolis)

MILLBURN, J. (1976), *Benjamin Martin: Author, Instrument Maker and 'Country Showman'* (Leiden: Science in History Series)

—— (2000), *Adams of Fleet Street: Instrument Makers to King George III* (Aldershot: Ashgate)

MORRISON-LOW, A. D. (1984), 'Brewster and scientific instruments', pp. 59–65 in Morrison-Low, A. D. and Christie, J. R. R. (eds.), *Martyr of Science: Sir David Brewster 1781–1868.* (Edinburgh: Royal Scottish Museum)

—— (1991), 'Women in the nineteenth-century scientific instrument trade', chap. 3, pp. 89–117 in Benjamin, M. (ed.), *Science and Sensibility: Gender and Scientific Enquiry, 1780–1945* (Oxford: Blackwell)

—— (1992), 'William Nicol, FRSE, c.1771–1851: Lecturer, scientist and collector'. *Book of the Old Edinburgh Club*, n.s. 2, 123–131

—— (2002), ' "Feasting my eyes with the view of fine instruments": Scientific instruments in Enlightment Scotland, 1680–1820', chap. 2, pp. 17–53 in Withers, C. W. J. (ed.), *Science and Medicine in the Scottish Enlightnment* (East Linton: Tuckwell Press)

—— (2007), *Making Scientific Instruments in the Scientific Revolution* (Edinburgh: National Museums Scotland, and Aldershot: Ashgate)

MORTON, A. Q. (1990), 'Lectures on natural philosophy in London, 1750–1765: S. C. T. Demainbray (1710–1782) and the "inattention of his Countrymen" ', *British Journal for the History of Science*, 411–434

—— and WESS, J. A. (1993), *Public and Private Science: The King George III Collection* (Oxford: Oxford University Press)

MÖRZER BRUYNS, W.F.J (2009), *Sextants at Greenwich* (Oxford: Oxford University Press and the National Maritime Museum)

NICHOLSON, J. (1825) [vere John Farey], *The Operative Mechanic and British Machinist* (London: Knight and Lacey)

Oxford Dictionary of National Biography (2004 and online) (Oxford: Oxford University Press and www.oxforddnb.com). Cited as *OxDNB*

PAYEN, J. (1985), 'La construction des instruments scientifiques en France au XIXe siècle', pp. 159–182 in Clercq (1985)

PIGATTO, L.(2000), *Giuseppe Toaldo e il suo tempo* (Padua: Bertoncello Artigrafiche)

PORTER, R. (1985), 'The economic context', pp. 1–4 in *Science and Profit in 18th-Century London* (Cambridge: Whipple Museum)

PRICE, D. DE SOLLA (1952), 'The early observatory instruments of Trinity College, Cambridge' *Annals of science* 8/1, 1–12

PROVERBIO, E. (1994), 'From reflectors to refractors: the evolution of Giovan Battista Amici, constructor of astronomical instruments', pp. 213–219 in Dragoni, G. *et al.* (1994)

RATCLIFFE, M. (2009), *Quest for the Invisible: Microscopy in the Enlightenment* (Aldershot: Ashgate)

ROOSEBOOM, M. (1950), *Bijdrage tot de geschiedenis der instrumentmakerskunst in de noordlijke Nederlanden tot omstreeks 1840* (Leiden)

RUFUS, W. C. (1924), 'Astronomical observatories in the United States prior to 1848' *Scientific Monthly* 19/2, 120–139

RYAN, W. F. (1991), 'Scientific instruments in Russia from the Middle Ages to Peter the Great' *Annals of science* 48, 367–384

SCHICKORE, J. (2007), *The Microscope and the Eye: A History of Reflections, 1740–1870*. (Chicago, IL: Chicago University Press)

SCHILLINGER, K. (1994), 'The development of the collections of scientific instruments of the Mathematical-Physical Salon between the years 1750 and 1850 and their use for research and teaching' pp. 101–106 in Dragoni, G. *et al.* (1994)

SCHOFIELD, R. E. (1962), 'Josiah Wedgewood and the technology of glass manufacturing', *Technology and Culture* 3/3, 285–297

SERIO, G. F. and CHINNICI, I. (1997), *L'osservatorio astronomico di Palermo* (Palermo: Flaccovio Editore)

SIMPSON, A. D. C. (1993), 'The pendulum as the British length standard: a nineteenth century legal aberration', pp. 174–90 in Anderson, R. G. W. *et al.* (1993)

STENDARDO, E. (2001), *Museo degli strumenti astronomici* (Naples: Luciano Editore)

TALAS, S. (1994), 'The cabinets of physics in Geneva at the end of the 18th century', pp. 85–91 in Dragoni, G. *et al.* (1994)

TATON, R. (ed.) (1986), *Enseignment et diffusion des sciences en France au XVIII siècle*, vol. XI of *Histoire de la pensée*. (Paris: Hermann)

TIMMINS, S. (1890), 'Dr Priestley's laboratory, 1791'. Reprinted from the Birmingham Weekly Post, 15, 22, 29 Mar., 5 Apr. (Coventry: Arley)

TODESCO, P. (1994), 'The physics laboratories of the ecclesiastical colleges', pp. 173–177 in Dragoni, G. *et al.* (1994)

—— (1996), 'La famiglia Lusverg, dall'600 all'800' *Memorie della Società Astronomica Italiana*, 65, 895–901

TUCCI, P. (2000), *I cieli di Brera. Astronomia da Tolomeo a Balla* (Milan: Università degli Studi di Milano)

TURNER, A. J. (1987), *Early Scientific Instruments: Europe 1400–1800* (London: Sotheby's Publications)

—— (1989), *From Pleasure and Profit to Science and Security. Étienne Lenoir and the Transformation of Precision Instrument-Making in France 1760–1830*. (Cambridge: Whipple Museum)

—— (1998) 'Mathematical instrument-making in early modern Paris', pp. 63–96 in Fox, R. and Turner, A. (eds.), *Luxury Trades and Consumerism in Ancien Régime Paris: Studies in the History of the Skilled Workforce* (Aldershot: Ashgate)

TURNER, G. L'E. (1967), 'The auction sales of the Earl of Bute's instruments, 1793' *Annals of Science* 23/3, 213–242

—— (1983), *Nineteenth-Century Scientific Instruments* (London: Sotheby Publications)

—— (1989), *The Great Age of the Microscope: The Collection of the Royal Microscopical Society through 150 years* (Bristol and New York: Hilger)

—— (1996), *The Practice of Science in the Nineteenth Century: Teaching and Research Apparatus in the Teyler Museum* (Haarlem: Teyler's Museum)

—— (2000), 'The government and the English optical glass industry' *Annals of Science* 57, 399–414

WARNER, D. J. (1990), 'Gambey's American customers' *Rittenhouse* 4/3, 65–78

—— (1995), 'Americans encounter aneroids' *Rittenhouse* 9/4, 120–128

—— (1998), 'Telescopes for land and sea' *Rittenhouse* 12/2, 33–54

WETTON, J. (1994), 'Scientific instrument making in Manchester', pp. 71–79 in Dragoni, G. et al. (1994)

WHEATLAND, D. (1968), *The Apparatus of Science at Harvard* (Cambridge, MA: Harvard University Press)

NOTES

1. George II r.1727–69, George III r.1760–1820, George IV r.1820–30, William IV r.1830–37, thereafter Queen Victoria.
2. Nicholson's *Operative Mechanic* went through numerous English editions, being published also in the USA and in French translation.
3. D'Artigues' book contains a 'Dissertation', pp. 1–6; a text 'Sur l'art de fabriquer du flint-glass bon pour l'optique', pp. 7–55; and a report.
4. There were deposits of liquid mercury or its ores in Spain, Illyria, and Italy.
5. Memoirs of the Astronomers Royal are published in the *Oxford Dictionary of National Biography* (2004 and online).
6. Howse's 'Greenwich List of observatories' (1986) and 'Supplement' (1994) gives—as far as is known—the makers of fixed instruments in observatories across the world up to about 1850.
7. There were several standard bars purporting to represent the ancient French measure known as the 'toise', which was divided into six 'Paris feet' of 12 'inches', each of 12 'lines'. In 1798 the metre, fixed at 0.51374 toise, displaced the toise and its subdivisions. See Section 11.5.1, 'France'.
8. 'Science lecturing in the 18th century', *BJHS* 28 (March 1955), 99 pp.
9. Banks' correspondence with instrument makers is scattered throughout these six volumes.
10. Various dates are given for the foundation of these observatories; they may record the start of building, or the date of the first observations. Provincial and private observatories were often short-lived, but sometimes revived in later years.
11. Montucla published two volumes of this compendious work before his death in 1799. The third, expanded and updated by J. J. Lalande, was published in 1802. It is not clear which editions Amici was reading.
12. The two volumes of Amici's published works include copious notes and published translations into English, and so on.
13. For biographies of these and other Italian scientists see the (in progress) *Italian Biographical Dictionary* and Dragoni et al. (1999), *Dizionario biografico*.
14. These generally consisted of a suspended heavy weight with a lower spike just touching a surface of wax or sand. Inertia caused the motion of the heavy weight to be slightly delayed, and the spike's trail across the wax or sand indicated the direction from which the tremor had come, and its strength.

CHAPTER 12

MECHANICS IN THE EIGHTEENTH CENTURY

SANDRO CAPARRINI
AND CRAIG FRASER

12.1 INTRODUCTION

The publication in 1687 of Isaac Newton's *Mathematical Principles of Natural Philosophy* has long been regarded as the event that ushered in the modern period in mathematical physics. Newton developed a set of techniques and methods based on a geometric form of the differential and integral calculus for dealing with the motion of a single mass-point, and further showed how the results obtained could be applied to the motion of the solar system. Other topics studied in the *Principia* included the motion of bodies in resisting fluids and the propagation of disturbances through continuous media. The success and scope of the *Principia* heralded the arrival of mechanics as the model for the mathematical investigation of nature. This subject would be at the cutting edge of science for the next two centuries.

The view that the entire modern edifice of classical mechanics can be traced back to the Principia was promulgated by Ernst Mach in his famous book *The Science of Mechanics A Critical and Historical Account of Its Development*, first published in 1883. In introducing the period in history following Newton, Mach wrote:

The principles of Newton suffice by themselves, without the introduction of any new laws, to explore thoroughly every mechanical phenomenon practically occurring, whether it belongs to statics or to dynamics. If difficulties arise in any such consideration, they are invariably of a mathematical, or formal, character, and in no respect concerned with questions of principle. (Mach 1883, Eng. trans., p. 256)

Mach's opinion has influenced scholars for over a century. However, while *The Science of Mechanics* is deservedly regarded as a classic of the history of science, it was mainly intended as a contribution to the epistemology of physics, and in this respect it was a work of fundamental importance. Nevertheless, most of its historical comments were taken from secondary sources and it displayed a limited sense for what was achieved in the eighteenth century. Mach was also mistaken in his positivistic conviction that it is possible to neatly divide science into the physical investigation of the phenomena on the one hand, and the development of a mathematical theory on the other. In fact, the appearance in the eighteenth century of new physical principles and modes of description was organically linked to the mathematical elaboration of a coherent theory of mechanics. Since Mach composed his book, historians Emile Jouguet, René Dugas, Stephen Timoshenko, Clifford Truesdell and István Szabó, as well as many others, have drawn a different picture of mechanics in the age of the Enlightenment.

There is now a consensus among historians of science that the generation after Newton was left with a problematic legacy. The Newtonian theory had not been developed enough to deal with systems of interacting bodies subjected to constraints, nor to solve most of the problems in continuum mechanics. In addition, the infinitesimal calculus—the mathematical language of the new mechanics—was then of a relatively recent invention and lacked a well-established and generally accepted foundation. To overcome these difficulties required the efforts of a group of exceptionally gifted scientists: Gottfried Leibniz, Pierre Varignon, the brothers Jacob and Johann Bernoulli and the latter's son Daniel, Jacob Hermann, Leonhard Euler, Brook Taylor, Alexis Clairaut, Jean d'Alembert. and Joseph-Louis Lagrange. Their highly creative approach to a wide range of problems shaped the classical theory handed down to us today.

In the light of these developments, the old image of the period from Newton's *Principia* (1687) to Lagrange's *Analytical Mechanics* (1788) as uniformly dominated by a purely formal revision of the principles must be discarded in favour of a more complex pattern of change. There was at first a period of translation of Newton's mechanics into the analytic language of Leibniz's differential calculus, culminating with the publication of Euler's *Mechanics* (1736). During the 1740s, new principles and mathematical methods were created at an almost unbelievable pace by a handful of mathematicians—Johann and Daniel Bernoulli, Euler, d'Alembert and Clairaut—working in more or less friendly rivalry. The collection of special results that they produced was revised, completed, and unified in the 1750s by Euler. From the 1760s onward the resultant theories were refined and formalized by a new generation of scientists, among whom Lagrange and Laplace figure prominently.

The crucial event of this history was the generalization and completion of pre-existing partial theories, a project to which Euler was the leading contributor. The magnitude of this transformation can be compared in terms of its impact on physical theory with the advent of quantum mechanics in the twentieth century. While the concept of revolution in science has been overused and criticized, it is not an exaggeration to call this period of profound change in mathematical physics the analytic revolution. It was characterized by the creation of theories for systems of mass-points, celestial objects, rigid bodies, fluids and linearly elastic bodies; they were based on a highly effective formulation of the principles of mechanics and expressed by means of sets of ordinary or partial differential equations. It is the ensemble of these theories that forms the core of classical mechanics, usually called 'Newtonian'.

12.2 General Principles

The history of the fundamental principles and theorems of mechanics is still partially uncharted territory. Much work on the eighteenth century has been done by historians in the last fifty years, but the general picture is not yet fully understood. The following account is only a summary, clearly not exhaustive. (In order not to confound the reader with outdated notations, formulae will be given in a slightly modernized form.)

Systems of units. Since in the *Principia* mechanics was formulated in geometric fashion, Newton did not need any rational system of units for physical quantities. In the following decades, Varignon and Johann Bernoulli began the process of transforming mechanics into a purely analytical science. Some of their formulas look strange to a modern student, for they are, in effect, a shorthand for Euclidean proportions: physical constants are missing, numerical factors disappear, and results are expressed in the form of proportionalities. Beginning in the late 1730s, Daniel Bernoulli and Euler used a system of units based on length and weight. In Euler's 'Dissertation on the best way to construct a pulley' (1745) we find an early example of the dimensional checking of a formula. Euler was well aware of the importance of the question of physical units, which is mentioned time and again in his memoirs. He systematized these ideas in his *Theory of the Motion of Solid or Rigid Bodies* (1765a), which remained the most advanced presentation of the problem of dimensions until the appearance of Fourier's *Analytical Theory of Heat* (1822). (See Ravetz (1961b) and Roche (1998).)

The principle of inertia. While Newton regarded the principle of inertia as an axiom, some mathematicians of the eighteenth century hoped to demonstrate it. The argument given by Euler in his *Mechanics* (1736) consists essentially in the remark that, in the absence of external forces, there is no reason for a point moving with a given velocity to change its speed or the direction of its motion. Similar ideas were later put into a more elaborate mathematical form by d'Alembert, leading to a non-trivial functional equation (1768a). However, Euler reverted to inertia as an axiom in the *Theory of the Motion of Solid or Rigid Bodies* (1765a), in which he analysed the question at length. Other remarks are scattered throughout his works: in the 'Investigations on the origin of forces' (1752b), for example, he criticized the locution 'vis inertiae' as denoting something which is not a force.

Euler's most substantial contribution to the subject is his 'Reflections on space and time' (1750), in which he discussed Newton's absolute space. Although modern writers beginning with Mach have criticized the concept of absolute space, Newton and Euler found it to be a useful and meaningful notion. Euler wrote that 'anyone who wishes to deny absolute space will fall into the most serious inconvenience.' Indeed modern mathematical presentations of mechanics define the intrinsic geometry of classical spacetime before stating the laws of motion. Newton, with his famous rotating-bucket experiment, had related absolute space to centrifugal forces; in the 'Reflections,' Euler instead emphasized its connection with inertial motion. In short, for Euler, the existence of inertial motion independent of the distribution of matter in the universe demonstrated the existence of a structure of space itself. Going beyond Newton, he also remarked that the uniform motion of free mass-points defines absolute time. This way of looking at the nature of space anticipated some of the ideas of general relativity.

The principle of relativity. The classical principle of relativity was well known to Galileo Galilei, Christiaan Huygens and Newton. In the eighteenth century Euler systematized its use. He first discussed the matter in his *Mechanics* (1736), where he noted that observers at rest in the absolute space, or uniformly moving along a constant direction with respect to it, experience the same physical laws. Therefore, he remarked, 'we will not be very concerned with absolute motion, since relative motion is ruled by the same laws.' Euler gave a more detailed account of the whole question in the introductory chapters of *Theory of the Motion of Solid or Rigid Bodies* (1765a), in which he showed that 'the same differential equations are obtained both for absolute and relative motion; the distinction is discerned in the integration, for each case must be duly adapted to the initial conditions.' Moreover, he calculated the corrections for a non-inertial observer. A discussion of these matters also occurs in Euler's *Guide to Natural Science* (1862, probably written shortly after 1755) in relation to the problem of establishing a unitary theory of physics. The foundations laid by Newton and Euler for the problem of absolute and relative motion were strong enough to sustain physics until the advent of Maxwellian electromagnetism. (See Bertoloni Meli (1993) and Maltese (2000).)

Vectors. While the use of the word 'vector' to describe physical entities of the eighteenth century is slightly anachronistic, at that time there was already a clear distinction between directed quantities, represented by line segments in diagrams, and pure numbers. In the first decades of the century, the parallelogram rule was applied routinely to the composition of forces and velocities. Euler, in his *Mechanics* (1736), resolved acceleration along the tangent and the normal to the trajectory. (See Dugas (1957) and Radelet-de Grave (1996).)

However, with the discovery of the general laws of rigid body motion around 1750, more complicated vector quantities appeared. In 1759 the Italian mathematician Paolo Frisi demonstrated that two infinitesimal rotations about concurrent axes 'can be composed into one exactly in the same way that two forces, represented by the two sides of a parallelogram, are composed into a third force represented by the diagonal.' The corresponding result for moments of forces came much later. In 1780 Euler found the analytical expression for the moment of a force about any axis through the origin, and immediately recognized its similarity to the well-known formula for the projection of a force along a given direction; this meant that the moment of a force was a line-segment, not just a number (1793). In the last lines of his paper, Euler remarked that 'this marvellous harmony deserves to be considered with the greatest attention, for in general mechanics it can deliver no small development.'

Unfortunately, these two important discoveries lay dormant for years. Euler's papers were read only at the beginning of the next century; their content was then reworked by Laplace, Prony, and Poisson. Independently of Euler, in 1803, Louis Poinsot developed a purely geometric vector theory of moments. Frisi's theorem was rediscovered or appropriated by Lagrange in the first volume of the second edition of his *Analytical Mechanics*, published in 1811. At the beginning of the nineteenth century a new structure of mechanics emerged from these discoveries—one in which most of the fundamental entities were vectors. These ideas played an important part in the creation of vector calculus. (See Caparrini (2002, 2003, 2007).)

Composition of forces. The paradigmatic example of the parallelogram rule for the sum of vectors is the composition of forces. Curiously enough, from the time of Newton to the end of the nineteenth century it was essentially considered a theorem, obtainable from first principles. Every mathematician of some repute, and several of the less famous, tested his ability by looking for a proof. It was for decades the most popular theorem of applied mathematics. Significant proofs were given by Daniel Bernoulli (1728), who considered it a theorem in pure geometry and formulated a demonstration in rigorous Euclidean fashion, and by Daviet de Foncenex (1762), who translated the geometrical problem into a functional equation.

While in the end the search proved illusory, the critique of the different proofs resulted in a deepening of the criteria of rigour in physics. Thus, for example, mathematicians began to wonder if statics had priority over kinematics, or were led to explore the connections between the parallelogram of forces, the law of the lever, and the principle of virtual work. After the discovery of non-Euclidean geometry it was shown that some of the proofs retained part of their validity in spaces of constant curvature. (See Bonola (1912), Benvenuto (1991) and Radelet-de Grave (1987, 1998).)

Newton's second law of motion. Our textbooks usually attribute the law $F = ma$ to Newton, but this formula is nowhere to be found in the *Principia*. In fact, rather than an equation, Newton had a principl—roughly equivalent to the assertion that the force acting on a particle is proportional to the variation of momentum—which he applied to a single, unconstrained mass-point. Newton's principle covered both the case of impulsive forces (force proportional to $\Delta(mv)$) and of continuous motion (force proportional to the acceleration).

After the publication of the *Principia*, it took decades for Newton's *lex secunda* to reach the modern formulation. The analytic form $f = dv/dt$ for a particle under the action of gravity was used by Varignon (1703), while the more general expression

$$f = m \frac{dv}{dt} \tag{12.1}$$

first appeared in Hermann's *Phoronomia* (1716). Euler (1738) discussed the impact of hard bodies as a continuous deformation, thus attempting to subsume impulsive forces under the case of a continuous acceleration. By the end of the 1720s the differential form of the second law was well-known to everyone working in dynamics. However, the points of view concerning its significance differed greatly. In his *Mechanics* (1736) Euler tried unconvincingly to prove the formula from the kinematics of accelerated motion. On the contrary, Daniel Bernoulli remarked that this was a clear example of knowledge derived from experience; one could easily imagine, for example, acceleration to be proportional to the square or the cube of the force (1728). D'Alembert held instead that 'force' was simply a formalism for the exchange of momentum, and the second law its mathematical definition (1743).

The final step in this long process was induced by the latest discoveries in continuum mechanics. In the 1740s d'Alembert and Johann Bernoulli had determined respectively the equations of the vibrating string and of perfect fluids by applying the relation between force and acceleration. Their successes led Euler to postulate that the 'true and genuine method' in dynamics consisted in applying the formulas

$$F_x = ma_x, F_y = ma_y, F_z = ma_z, \qquad (12.2)$$

expressed in orthogonal Cartesian coordinates, to every particle of the system, taking into account both the applied forces and the constraints (1752a). This apparently simple idea yielded a harvest of results in a matter of few years. (See Hankins (1967), Truesdell (1968), Cohen (1971), Blay (1992) and Guicciardini (2009).)

Momentum and moment of momentum. While the intuitive ideas that underlie these two concepts—respectively, the impetus of a moving body and the law of the lever—go back to antiquity, the analytic formulation for systems of bodies was obtained only in the eighteenth century. In 1740 Euler, in his *Naval Science* (1749), demonstrated two particular but important cases of the general principles. The first is the formula

$$\mathbf{R} = M\mathbf{a}_G \qquad (12.3)$$

where $\mathbf{R}$ is the resultant of the applied forces, M is the total mass and $\mathbf{a}_G$ is the acceleration of the centre of mass; the second is the law of moment of momentum in the special case of a rigid body turning about a fixed axis (see the next section on rigid bodies). Shortly afterwards, both Euler (1746) and Daniel Bernoulli (1746) gave the proof of the conservation of angular momentum for a mass-point sliding along a smooth tube rotating in a horizontal plane. (This is an early example of a first integral of the equations of motion.) By that time, several theoreticians were working on the same problems; in a letter to Euler of 23 April 1743, Clairaut remarked as follows: 'I was charmed by your theorem on the conservation of rotational moments, but what struck me is that if I had reflected a little on my equations, I also would have found it.' The conservation of angular momentum was also enunciated by the Chevalier d'Arcy (1752) in the form of the law of areas—a generalization of Kepler's second law of planetary motion.

The general principle that the rate of change of the total angular momentum is equal to the sum of moments of the applied forces appeared—somehow hidden under the formalism—in d'Alembert's *Researches on the Precession of the Equinoxes* (1749).

Euler (1752a) immediately gave a clearer formulation of it. Both d'Alembert and Euler apparently thought that the internal forces do not influence the motion of the system, an assumption that was discussed by subsequent authors. The relation between first principles and general theorems in dynamics was somewhat clarified when Lagrange, (1779) derived the main integrals of motion for a system of mutually gravitating mass-points from the Newtonian laws.

Towards the end of his life, Euler (1776b) wrote down explicitly the two relations

$$\mathbf{R} = \dot{\mathbf{M}}, \mathbf{L}_O = \dot{\mathbf{H}}, \tag{12.4}$$

where **R** is the resultant of the external forces, **M** the total momentum, **L** the sum of the moments of the external forces about a fixed point O and **H** the total moment of momentum about O, as expressing the laws of motion of a mechanical system; it is not clear if he considered them axioms or theorems. With the publication of Lagrange's *Analytical Mechanics* (1788), the conservation of momentum and of moment of momentum became consequences of more general principles. (See Truesdell (1968) and Caparrini (1999).)

Energy. The eighteenth-century conservation of *vis viva* (live force) mv^2 may be interpreted as an early form of our conservation of mechanical energy. There were several formulations of this principle; the best known is the equality between 'actual descent' (that is, kinetic energy) and 'potential ascent' (that is, the height of the centre of mass). Apart from its significance in philosophy (Leibnizians vs. Cartesians), the *vis viva* controversy, which opposed partisans of mv and mv^2, was important in discussions of the fundamental axioms of mechanics during the first half of the century.

The most active propagandist of *vis viva* was Johann Bernoulli, who took the idea from Huygens and Leibniz and elevated it to the rank of general principle for the dynamics of systems. With its help, he was able to solve several tricky problems (1729, 1735). This role of the *vis viva* in early eighteenth-century mechanics can be seen, for example, in the correspondence between Johann Bernoulli and Euler. In a letter of 2 April 1737 to Euler, Bernoulli expressed the hope that in the *Mechanics* (1736), then just out of the press, Euler would make use of *vis viva*; Euler replied that he did not need it, because 'these volumes are not the place to present the theory of living forces. It will however appear in the following volumes, where the motion of bodies of limited extent will be thoroughly considered.'

The proponents of *vis viva* knew that there was something that could be dissipated into microscopic motion. This was in effect the main criticism of Johann Bernoulli to his son Daniel's use of the conservation of *vis viva* in the *Hydrodynamics* (1738): at discontinuities of a channel, a finite part of the *vis viva* of flowing water went into the formation of vortices.

During the 1740s both d'Alembert (1743) and Daniel Bernoulli (1750) demonstrated the conservation of *vis viva* from the relations between force and motion. From that point on, the principle became a first integral of motion; the formulation given in Lagrange's *Analytical Mechanics* (1788) is formally the same as the modern one. In Lazare Carnot's *Essay on Machines in General* (1782) there is a description of the conservation of *vis viva* in more physical terms that might be seen to resemble the modern principle of conservation of mechanical energy as it is presented in textbooks today. (For more on Carnot's essay see (Fraser 1983).)

The concept of mechanical work has a different history. It appeared implicitly at the end of the seventeenth century in the solution of problems in technical mechanics (How can we measure the work of several men? How can we compare it with the work of a horse?), and was incorporated into higher dynamics by Euler and Daniel Bernoulli. The formal relation

$$2\int_1^2 f ds = m v_1^2 - m v_2^2 \qquad (12.5)$$

was derived from the second law of motion by Varignon (1703).

During the first decades of the nineteenth century the conservation of *vis viva* was discussed in every textbook of theoretical or applied mechanics. However, while in books on abstract mechanics this was just a corollary of the basic laws, in texts on applied mechanics it was employed as the basic principle. The two traditions appeared back to back in the second edition of Poisson's *Treatise of Mechanics* (1833). English texts of the period even made frequent use of the word 'energy' to denote the intensity of a force, and in books on mathematical mechanics the terms *vis viva* and kinetic energy were used interchangeably well into the twentieth century. Historian Thomas Kuhn (1959) has argued that the French practice in the early nineteenth century of denoting *vis viva* with a 1/2 factor (with each side of eq. (12.5) being divided by 2) was significant as an explicit recognition of the conceptual priority of work in the equations of mechanics. However, in many of the actual writings of the period, the issue of the 1/2 factor did not arise or amounted to a secondary question, more related to typographical problems than to scientific issues.

It could be argued that several of the ingredients of the general principle of the conservation of energy were already present in mathematical physics at the end of the eighteenth century. It is therefore not a surprise, as Kuhn (1959) has observed, that between 1830 and 1850 twelve scientists, independently, announced that heat and work are manifestations of a single 'force' of nature. Kuhn remarks that 'the history of science offers no more striking instance of the phenomenon known as simultaneous discovery.' Part of the explanation is, obviously, that all of these discoverers shared a common scientific background, rooted in a well-established tradition of engineering textbooks. When faced with the same problem, they reacted in similar ways. The connection between the old conservation of *vis viva* in mechanics and its extension to all of physics is especially apparent in Hermann von Helmholtz's *On the Conservation of Force* (1847), which takes as its starting point Lagrange's mechanics. These few examples suffice to show that in considering the history of the conservation of energy during the nineteenth century, special attention should be paid to eighteenth-century sources. (See Kuhn (1959), Hiebert (1962), Hankins (1965), Pacey and Fisher (1967), Truesdell (1968), Mikhailov (2002), Villaggio (2008), and Fonteneau (2008, 2009).)

Potential. The concept of potential emerged in several widely different contexts. The unifying background for these apparently unrelated developments was the application of partial differential equations to mathematical physics from the 1740s onward.

Gravitational potential appeared implicitly in the form of the integral of force in Johann Bernoulli's formula for the motion of a point under the action of a central force (1712). Bernoulli did not connect his result with the conservation of *vis viva*; this step was taken in 1738 by his son Daniel, who also explicitly gave the form of the potential in the case of Newtonian attraction (1747). After an interval of more than twenty years, Lagrange began to make extensive use of the potential, at least as a formal mathematical entity in the equations of mechanics. He first applied it to the study of astronomical perturbations (1776), then to the general motion of mass-points subjected to mutual gravitational interactions (1779), and, finally, he inserted the potential in the equations of mechanics that bear his name (1782) (see eq. (12.17)).

The potential as stored *vis viva* in an elastic system was first mentioned by Daniel Bernoulli in a letter to Euler of 20 October 1743: 'For a naturally elastic band, I express the potential live force of the curved band as $\int ds/r^2$ [...] Since no one has perfected the isoperimetric method as much as you, you will easily solve this problem of making $\int ds/r^2$ a minimum.' Euler proceeded immediately to develop this suggestion in his 'On the elastic curve' (1744). (This essay is discussed in Section 12.5.) A few years later Euler introduced the potential in fluid dynamics; believing (erroneously) that every stationary flux of an incompressible homogeneous fluid is irrotational, he applied the first elements of differential forms and arrived at 'Laplace's equation'

$$\frac{\partial^2 S}{\partial x^2} + \frac{\partial^2 S}{\partial y^2} + \frac{\partial^2 S}{\partial z^2} = 0, \tag{12.6}$$

where S is the velocity potential (1761a). Moreover, following d'Alembert's lead in the use of complex numbers in two-dimensional fluid dynamics, Euler (1757c) also introduced what we would now call a complex potential. (See Wilson (1995b) and Truesdell (1954).)

Laplace's name is associated with the famous partial differential equation for the potential in a region free of sources. He first wrote the equation in two variables (1779), then in three variables in spherical coordinates (1785) and lastly in three variables in Cartesian coordinates (1789). Finally, he made it central in his *Treatise on Celestial Mechanics* (1798–1825).

12.3 RIGID BODIES

The dynamics of rigid bodies—as opposed to their statics—is a creation of the eighteenth century. In fact, before the end of the 1730s there were few results worthy of notice. In 1673 Huygens had determined the motion of a compound pendulum using an indirect principle equivalent to the conservation of mechanical energy, and in 1703 Jacob Bernoulli had addressed the same problem by the equilibrium of the moments of lost forces. For more complicated systems, nothing was known. Newton had left only a few remarks on finite-sized rigid bodies, most of them erroneous (see Dobson (1998)). As late as the early 1730s, Euler was still unable to write a chapter on rigid bodies for his *Mechanics* (1736).

During the first decades of the eighteenth century the question of finding at least *some* rules for rigid body dynamics surfaced in a number of special problems: oscillations of pendula, naval science, the impact of hard bodies, the rolling of round objects down inclined planes, and the precession of the equinoxes. *Ad hoc* hypotheses had to be put forward in each case, and the theory was usually limited to two-dimensional systems.

The first successful attack on the general problem appeared at the beginning of Euler's *Naval Science*, completed in 1741 but published only in 1749. Euler observed that the motion of the centre of mass and the motion relative to it are quite independent of each other. The simple case of a body with a fixed axis could be treated by considering an equivalent mass-point (that is, having the same moment of inertia about the axis as the body) acted upon by an equivalent force (that is, the same moment about the axis as the totality of forces), leading to the relation

$$\frac{d^2\theta}{dt^2} = \frac{L}{I}, \qquad (12.7)$$

where θ is the angular rotation of the body, L is the moment of the applied forces and I is the moment of inertia of the body, both about the axis of rotation. (It is here that the locution 'moment of inertia' originated.) This equation, as Euler remarked, is formally similar to Newton's second law. Euler also calculated the moments of inertia of several common shapes and demonstrated the theorem of parallel axes (today usually called 'Steiner's theorem').

Lacking a treatment of the general case, the best that Euler could do at this stage of his investigation was to conjecture that in every rigid body there are three orthogonal axes through the centre of mass, about which the body can oscillate without wobbling. This hypothesis made it possible to study the small motions of a rigid body as a superposition of independent pendular oscillations. A few years later (1745) he discovered that the rotation about an axis is stable only if the two appropriate products of inertia are both equal to zero. (For a thorough analysis of *Naval Science*, see the essay by Habicht (1978).)

In the same years, the aged Johann Bernoulli, following an idea proposed by his son Daniel, made a fundamental contribution to two-dimensional rigid-body mechanics (1742b). Bernoulli considered a planar rigid body at rest, set in motion by a force acting on an arbitrary point, and tried to determine the position of the centre of the initial rotation. To solve the problem he applied 'the principle of the lever'—that is, the equilibrium of moments—and obtained the correct solution. He termed the point in question *the centre of spontaneous rotation*.

However innovative, Bernoulli's work also contained significant errors. He believed that a three-dimensional rigid system starting from rest under the action of an impulsive force would begin to rotate about an axis orthogonal to the plane passing through the line of action of the force and the centre of mass. Moreover, in considering the oscillations of a compound pendulum, he tried to reduce the problem to the motion of an equivalent particle, and was thus led to use moments of inertia about the wrong axis. These errors clearly demonstrate that the main properties of rotational inertia were still very imperfectly known. (On Johann Bernoulli's mechanics, see the extensive essay by Villaggio (2007).)

The next giant step forward was made by d'Alembert in his *Researches on the Precession of the Equinoxes* (1749), which contained a correct description of the motion of an ellipsoid of rotation about its centre of mass under the action of the gravitational forces exerted by two distant mass-points. In doing so, d'Alembert obtained the general equations of equilibrium of a mechanical system, applied the principle of moment of momentum to every infinitesimal part of the body and, most important of all, demonstrated the existence of the *instantaneous axis of rotation*, the key concept in the kinematics of a rigid body. However, these results were presented in a very difficult and confused way, making d'Alembert's book a flawed masterpiece. With respect to rigid-body dynamics, the near misses are as relevant as the positive achievements: the role of the moments of inertia was buried under the geometrical symmetry of the body, the instantaneous axis of rotation played a secondary role with respect to the axis of figure and, more importantly, there were no general dynamical equations. (For an analysis of d'Alembert's book, see (Wilson 1987, 1995a), (Nakata 2000) and (Chapront-Touzé and Jean Souchay 2006).)

Stimulated by d'Alembert's discovery of the instantaneous axis of rotation, which made the general kinematics of a rigid body analogous to the simple case of the rotation about a fixed axis, Euler wrote his famous memoir 'Discovery of a new principle of mechanics', published in 1752. The 'new principle' referred to in the title is the set of three equations relating forces and motion for a rigid body turning about its centre of mass. These equations, he wrote, are 'the subject of this memoir, which I finally arrived at after several useless attempts undertaken over a long time.' What Euler achieved here, expressed with his usual clarity, surpassed by far every previous effort. (For an appraisal of Euler's theory of rigid bodies, see (Wilson 1987, 1995) and (Langton 2007). Euler's theory has also been described by Blanc (1948, 1968).)

The core of the memoir was about pure kinematics. Referring every point to three fixed orthogonal axes meeting in the centre of mass, and making use only of the hypothesis of rigidity, Euler determined the formulae for the velocity,

$$\frac{dx}{dt} = \lambda x - \mu z, \quad \frac{dy}{dt} = \nu z - \lambda x, \quad \frac{dz}{dt} = \mu x - \nu y, \tag{12.8}$$

where x, y, z are the coordinates of the point and the coefficients λ, μ, ν depend only on time. From these expressions, the axis of instantaneous rotation could be found easily. A modern reader may interpret immediately λ, μ, ν as the components of the angular velocity vector, but the vectorial character of angular velocity was then unknown. (For an account of the genesis and development of the vector idea in mechanics in the eighteenth and nineteenth centuries, see (Caparrini 2002, 2003, 2007).)

The dynamics followed naturally from the fundamental Newtonian principle that force is proportional to acceleration. Differentiating the velocities, Euler obtained the accelerations, and from the accelerations the forces, and from the forces their moments about the coordinate axes. The final result was a set of formulae expressing the moments of the forces in terms of the moments and products of inertia about the three coordinate axes and of the components of the angular velocity.

But the general problem was far from settled, for these equations were extremely difficult to solve. In effect, if the rotating body is referred to fixed axes, the moments of inertia depend on time and must be determined from the equations themselves. To overcome this difficulty, in 1751 Euler launched a fresh attack on the problem. (This pattern of reiterated assaults over the years was typically Eulerian.) The memoir 'On the motion of a rotating solid body about a mobile axis' was published only in 1767, by which time it had been superseded by more sophisticated works, but it is of interest to historians for reconstructing Euler's chain of thought.

At the very beginning, Euler introduced two different frames of reference with a common origin in the centre of mass—one fixed in space, the other rotating with the body. The expressions for the basic kinematic quantities were first determined with respect to the fixed frame, then transformed into the rotating frame. It was a method based on calculations rather than principles; no use was made of any general principle on the relations between mutually rotating frames. After a *tour de force* of algebra and spherical trigonometry, Euler finally reached the formulae for the accelerations in the moving frame. By the same process he had utilized in his previous memoir, he then obtained a simplified form of the fundamental equations, though still encumbered by the products of inertia.

With complicated but manageable equations at his disposal, Euler was able to get some definite results. He studied the conditions which allow an axis of rotation to be fixed, determined the motion when the resultant moments are zero (leading to an elliptic integral) and, in particular, examined the case in which the body is a solid of revolution.

At the end of the memoir, Euler wrote that all that remained to be done was to apply the 'new principle' to specific cases. As a matter of fact, there was still a long way to go. In 1755 his friend Johann Andreas von Segner, professor of mathematics and physics at the University of Halle, published the booklet *Essay on the theory of rotations*, in which he demonstrated that every rigid body has at least three axes of permanent rotation. This result obviously could lead to further simplifications, as Euler proceeded to show. His memoir 'Research on the mechanical knowledge of bodies' (published 1765) was a fairly complete exposition of the properties of the centre of mass and of the moments of inertia, differing from standard nineteenth-century textbooks mainly in the absence of the ellipsoid of inertia (introduced by Augustin Cauchy in 1827) and in its great attention to foundational issues. It established the existence of rotational inertia, as distinguished from linear inertia.

Several classic results made their first appearance in this essay. Euler remarked that the centre of gravity (the common name of the centre of mass at the time) might be more aptly called centre of inertia, and demonstrated that its position does not depend on the choice of the axes. Turning to the problem of calculating the moments of inertia, he argued that the set of all moments of inertia through the centre of mass should have a maximum and a minimum. Euler showed that the two axes enjoying this property are also axes of permanent rotation, that they are perpendicular to each other, and that there is a third such axis perpendicular to the other two. He gave them their current name: principal axes.

Euler now had at his disposal all the ingredients for a successful theory. The final version was given in the memoir 'On the motion of rotation of solid bodies About a variable axis', (1765c). The main problem consisted in finding the motion of a generic point of the body with respect to a fixed frame. The usual formulae for the velocities followed, then the accelerations. Surprisingly, at this point Euler remarked that the expressions for the accelerations found in the fixed frame were also valid in the moving frame at the given moment, if the fixed axes were instantaneously coincident with the moving ones. While this might at first seems unlikely (in a rotating frame, inertial accelerations must be taken into account), it is absolutely correct. Euler's justification of this assertion amounted in effect to a verbal description of the relation

$$\left(\frac{d\varpi}{dt}\right)_{fixed\ frame} = \left(\frac{d\varpi}{dt}\right)_{rotating\ frame}, \tag{12.9}$$

which apparently he could somehow visualize. In addition, he took as moving axes the principal axes of inertia. The fundamental equations were then found in their definitive form,

$$\frac{dp}{dt} + \frac{c^2 - b^2}{a^2}qr = \frac{P}{Ma^2}, \frac{dq}{dt} + \frac{a^2 - c^2}{b^2}rp = \frac{Q}{Mb^2}, \frac{dr}{dt} + \frac{b^2 - a^2}{c^2}pq = \frac{R}{Mc^2}, \tag{12.10}$$

where p, q, r are the components of the angular velocity, M is the mass of the body, Ma^2, Mb^2, Mc^2 and P, Q, R are respectively the moments of inertia and the moments of the external forces about the coordinate axes.

Having finally arrived a correct theory, in 1760 Euler wove together all the different strands of his research on rigid bodies in his *Theory of the Motion of Solid or Rigid Bodies* (1765), a milestone in the history of mechanics. A modern reader is likely to be surprised by the abundance of special cases Euler explored before he arrived at the fundamental equations. He took this approach not only for didactic reasons, but also because of the difficulty of applying the general formulae to particular problems. Every previous work was made obsolete by the appearance of this fascinating treatise; Johann Bernoulli's problem of the centre of spontaneous rotation, for example, became a simple corollary of a theorem on plane motion.

Before the publication of Euler's book the young Lagrange had made his entrance into the field with a section of his long memoir 'Application of the method presented in the preceding memoir to the solution of different problems of dynamics' (1762). (The previous method referred to in the title was his new δ-method in the calculus of variations, a subject we examine in the next section.) Lagrange's aim was to derive the main properties of rigid motion using a variational principle and to subsume the physics to differential algebra. While he was successful in these respects, the tone of the work is very formal, and there is no critical discussion of the concepts. Thus, for example, the proposition that every infinitesimal displacement of a rigid body about a fixed point can be considered as the superposition of three infinitesimal rotations about three orthogonal axes meeting at the point is considered 'easy to see' and is not proved at all. Some years later, Lagrange returned to the subject with the memoir 'New solution of the problem of the motion of a body of any shape that is not subject to any external accelerative force' (1775), in which he tried to find 'a solution completely direct and purely analytical of the question under consideration.' The approach taken involved a minimum of physical assumptions. Lagrange's point of view on the formulation of mechanics, obviously very different from Euler's, marks the beginning of a period of formalization of the discoveries made in the preceding decades.

By the time that Lagrange's 'New solution' was published, Euler was 68 years old and almost completely blind. The task of reading his younger contemporary's memoir proved to be nearly impossible. Not being able to understand what Lagrange had done, Euler felt that his own solution of the problem had been criticized for being too complicated. He therefore went back through his old work, believing that simplification could only be achieved by a deeper analysis of the geometry. He studied the finite motions of a rigid body, thus initiating the analytic representation of displacements. He also demonstrated by synthetic geometry that any rigid motion with a fixed point is equivalent to a rotation. Euler went on to express the position of a generic point of a free rigid body as a function of its initial position and of the time. Putting these formulae into the principles of momentum and of moment of momentum, he obtained a different form of the fundamental equations, in which the instantaneous rotation is expressed by the rate of change of what we would now call the elements of the orthogonal transformation matrix.

Unbeknownst to Euler, a more elegant theorem on the finite motions of a rigid body had been obtained about ten years before by the Italian poet, mathematician and politician Giulio Mozzi in his short book *A Mathematical Discourse on the Instantaneous Rotation of Bodies* (1763). Mozzi's theory is mainly a generalization to three dimensions of Johann Bernoulli's work on rigid bodies set in motion by impulsive forces. His most important result was the demonstration that any finite displacement of a rigid body is equivalent to a screw motion along a fixed line. This theorem remained unnoticed, and was rediscovered only in the early 1830s. (For more information on Mozzi, see (Marcolongo 1905) and (Ceccarelli 2007).)

The theory of rigid bodies was discussed by Lagrange in the *Analytical Mechanics* (1788) and by Laplace in the first volume of the *Treatise on Celestial Mechanics* (1798). Both accounts drew heavily on Euler. A fuller exposition along Eulerian lines was given by Simé on-Denis Poisson in the *Treatise of Mechanics* (1811). It is mainly from these sources, directly or indirectly, that the scientists of the nineteenth century learned the subject.

12.4 THE D'ALEMBERT–LAGRANGE FORMULATION OF MECHANICS

Jean-Baptiste le Rond d'Alembert was a remarkable savant—almost a man born out of his time, who possessed a deep understanding of technical and foundational issues. In subjects as diverse as the fundamental theorem of algebra, the metaphysics of the calculus, the nature of functions and the principles of mechanics he displayed an acute critical sense, grasping issues that would only become the focus of study much later. These gifts were evident early in his career in his seminal *Treatise on Dynamics* (1743). This book was an investigation of the constrained interaction of bodies: the collision of spheres, the motion of pendula, the movement of bodies as they slide past each other, and various other connected systems. Many of the problems would today be studied as part of engineering mechanics. His basic conception was that of a 'hard body.' Such a body is impenetrable and non-deformable. Assume a small hard sphere hits a wall with a velocity that is perpendicular to the wall. When the sphere hits the wall all motion ceases. The closest modern approximation to d'Alembert's conception is that of a perfectly inelastic body, although it must be emphasized that d'Alembert's point of view was different from the modern one. D'Alembert thought in a Cartesian way of hard bodies as geometrical solids in motion, whose laws of interaction could be determined by deductive reasoning from *a priori* postulates or principles. (Hankins (1970) documents the importance of Cartesian philosophy in d'Alembert's scientific thought.) In this conception dynamics is very similar to geometry, where the properties of the objects under study are derivable from a few postulates that are believed to be necessarily true.

Central to d'Alembert's dynamics was a principle that he enunciated at the beginning of the *Treatise* and which in various later forms became known as 'd'Alembert's principle.' (The account which follows is based on (Fraser 1985).) In its original and most basic form the principle may be illustrated by the example of a hard particle that strikes a wall obliquely with velocity u (Fig. 12.1). We must determine the velocity of the particle following impact. Decompose u into two components v and w, v being the post-impact velocity and w being the velocity that is 'lost' in the collision. D'Alembert's principle asserts that if the particle were animated by the lost velocity alone then equilibrium would subsist. From this condition it follows that w must be the component of u that is perpendicular to the wall. Hence v is the component of u that is parallel to the wall, and the collision problem is solved.

Assume now that two hard bodies m and M approach each other with velocities u and U along the line joining their centres. It is necessary to find the velocities after impact. We write $u = v + (u - v)$ and $U = V + (U - V)$, where v and V are the post-impact velocities of m and M. The quantities u, v and $u - v$ are the impressed, actual and 'lost' motions of the body m; a similar decomposition holds for M. Because v and V are followed unchanged v must equal V. In addition, the application of the lost velocities $u - v$ and $U - V$ to m and M must produce equilibrium. D'Alembert

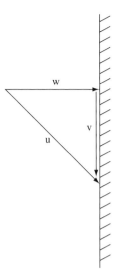

Fig. 12.1. D'Alembert's principle.

reasoned from the very concept of hard body itself that for this to happen we must have $m(u - v) + M(U - V) = 0$. Hence v or V is equal to $(mu + MU)/(M + m)$.

In modern dynamics we would analyse this collision using what are known as impulsive forces. It is assumed that in the collision very large forces act for a very short period of time. Integrating over these forces and using Newton's second law we are able to calculate the changes of velocity that result in the collision. In modern dynamics d'Alembert's principle involves a decomposition of forces or accelerations and is basically a statement combining Newton's second and third laws. By using impulsive forces in the modern principle and integrating, we can produce a decomposition of velocities that looks somewhat (and misleadingly) like the one d'Alembert originally presented. However, d'Alembert's point of view was very different. From the outset in his analysis of the collision of the two bodies he used a decomposition involving finite velocities. There are no forces, and the entire interaction is analysed using the conception of a hard body and the assumption—supported by *a priori* reasoning—that equilibrium would subsist if the bodies were animated by the motions they lose in the collision.

In examples involving forces that act continuously and in which the motion is continuous, d'Alembert analysed the system in a way that has some similarities to the model of instantaneous impulses set out by Newton in the opening proposition of Book One of the *Principia*. The motion is understood to consist of a succession of discrete impulses in which each impulse arises in an interaction of hard bodies or surfaces. The interaction is described in terms of finite velocities and infinitesimal velocity increments where the lost motions are governed by d'Alembert's principle. A clear example of d'Alembert's treatment of continuous forces is the tenth problem of the *Treatise*. Here all of the features of d'Alembert's theory come into play—the concept of hard body, d'Alembert's principle, and the Leibnizian differential calculus.

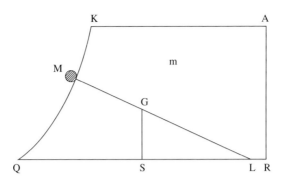

Fig. 12.2. D'Alembert's problem X.

We are given an irregularly shaped object KARQ of mass m which is free to slide along a frictionless plane QR (Fig. 12.2). A body of mass M is situated on the curve KQ which forms the left edge of KARQ. A force acts on M in a direction perpendicular to QR. The two bodies possess given initial velocities (in the particular configuration adopted by d'Alembert it is assumed that the bodies initially possess a common motion to the right). The problem is to determine the motion of the system as M slides down KQ.

In d'Alembert's solution the trajectory traced by M is regarded as a polygon with an infinite number of sides in which the length of each side is infinitesimal. This geometric representation of M's trajectory corresponds to the physical analysis in which M's motion is understood to be the outcome of a succession of discrete dynamical events. The model of a continuous curve as an infinite-sided polygon is also used in the solution to represent the left edge of the body m.

Although d'Alembert's original principle involved a decomposition of velocities, he subsequently extended the principle in certain problems to what was in effect a decomposition of accelerations. He did so in some of the problems of his treatise involving pendulum motion, and adopted a similar formulation in his later researches in hydromechanics and theoretical astronomy. Assume that a body m is part of a constrained system of bodies and is acted upon by an external impressed force. At a given instant let $dv^{(I)}$ be the increment of velocity that would be imparted to the body by the impressed force if the body were free. Let v^+ be the velocity of the body in the next instant, so that $dv = v^+ - v$ is the actual increment of velocity experienced by the body. We have the decomposition of velocities $v + dv^{(I)} = v^+ + w = (v + dv) + w$, or

$$v + dv^{(I)} = (v + dv) + w, \qquad (12.11)$$

where $w = dv^{(I)} - dv$ is the lost velocity of the body. By d'Alembert's principle equilibrium will subsist if each body of the system were subjected to the motion $dv^{(I)} - dv$. In any given problem we now invoke a suitable statical law and obtain a relation among $-mdv$ and $mdv^{(I)}$. In the problems d'Alembert considered this procedure gave rise to a system of differential equations that described the motion of the system.

We may re-express eq. (12.11) in the form

$$dv^{(I)} = dv + w, \qquad (12.12)$$

If eq. (12.12) is divided by dt and multiplied by m this becomes an equation involving forces:

$$\mathbf{F}_a + \mathbf{F}_c = m\mathbf{a}, \qquad (12.13)$$

where $\mathbf{F}_a$ is the applied force acting on m, $\mathbf{F}_c$ is the constraint force on m, and $\mathbf{a}$ is the acceleration of m. (For convenience here and in what follows we use modern vector notation, although this notation was not used in the eighteenth century.) In this formulation d'Alembert's principle states that equilibrium would subsist if each of the bodies of the system were animated by the constraint force; that is, the constraint forces considered as a set of applied forces acting on the bodies result in a system in static equilibrium.

Although Lagrange was influenced by d'Alembert, his own development of mechanics occurred along lines that were significantly different from his older contemporary. Physical hypotheses about the ultimate nature of mechanical interactions were absent, and any adherence to a geometric-differential form of the calculus was rejected altogether. (Fraser (1983) and Panza (2003) explore the foundations of Lagrange's mechanics.) Lagrange's goal was to reduce mechanics to a branch of applied analysis in which the emphasis was primarily on the derivation and integration of differential equations to describe the motion of the system. Following his mathematical philosophy, he eschewed diagrams and geometric modes of representation in favour of a purely analytic approach involving operations, formulas and equations among variables. In his first extended memoir on the principles of mechanics Lagrange (1762) used what he called the principle of least action as the starting point for his analysis of the motion of a dynamical system. This principle was an integral variational law and its application was governed by the calculus of variations, a mathematical subject that Lagrange had pioneered in the very same volume. In subsequent investigations Lagrange abandoned the least action principle. Instead he combined d'Alembert's principle (in the form involving eq. (12.13)) and what is today called the principle of virtual work to arrive at the fundamental axiom of his presentation of mechanics. Thus in his mature theory of mechanics, he used methods and operations derived from the calculus of variations, but he did not develop the subject from an integral variational principle, as he had done in 1762.

Consider a set of external forces acting on a connected system of bodies, and suppose that the system is in static equilibrium under the action of these forces. Consider a small virtual velocity or displacement $\delta \mathbf{r}$ of a given body m of the system. Such a displacement is taken to be compatible with the constraints in the system. As before let $\mathbf{F}$ be the external or applied force acting on m. The principle of virtual velocities asserts that a general condition for equilibrium of the system is given by the equation

$$\sum_m \mathbf{F} \cdot \delta \mathbf{r} = \mathbf{0}. \qquad (12.14)$$

Consider now any constrained mechanical system, not assumed to be in equilibrium. For each body m of the system we have the decomposition $\mathbf{F}_a + \mathbf{F}_c = m\mathbf{a}$ (eq. (12.13)). By d'Alembert's principle the given system would be in equilibrium if each m were animated by $\mathbf{F}_c$, where this force is now understood as an external force acting on m within the connected system. By the principle of virtual velocities we have

$$\sum_m \mathbf{F}_c \cdot \delta \mathbf{r} = \mathbf{0}, \tag{12.15}$$

which we may, using eq. (12.13), express as

$$\sum_m m\mathbf{a} \cdot \delta \mathbf{r} = \sum_m \mathbf{F}_a \cdot \delta \mathbf{r}. \tag{12.16}$$

Eq. (12.16) is a statement of the generalized principle of virtual work and is the starting point for Lagrange's theory of mechanics.

Beginning with eq. (12.16), Lagrange derived a system of differential equations to describe the motion of the system. Using the constraints, one reduces the description of the system to the specification of n 'generalized' variables $q_1, q_2, \ldots, q_m$. (If the system consists of n bodies moving freely then $m = 3n$ and the q_i would be the $3n$ Cartesian coordinates of the bodies.) Each of these variables is independent of the others, and each is a function of time. A system with suitably smooth constraints is then described by the m differential equations

$$\frac{d}{dt}\frac{\partial T}{\partial \dot{q}_i} - \frac{\partial T}{\partial q_i} + \frac{\partial Q}{\partial q_i} = 0, \tag{12.17}$$

Here $\dot{q}_i$ denotes the time derivative of q_i, the function T is what later would become known as the kinetic energy of the system and Q_i is what later would be called the potential, although these terms were not used by Lagrange. Eq. (12.17) became known in later dynamics as the Lagrangian equations of motion.

12.5 Statics of Elastic Bodies

The elastic behaviour of rods and beams emerged as a subject of interest in the early eighteenth century in two related problem of statics. In the problem of fracture one attempted to determine the maximum load that a beam of given material and dimensions can sustain without breaking. Typically it was assumed that the beam was cantilevered to a wall and that the rupture took place close to the wall. In the problem of elastic bending one was concerned with determining the shape assumed by a rod or lamina in equilibrium when subject to external forces. In the example of the elastica, where the elastic rod was treated mathematically as a line, the forces were assumed to act at the ends of the rod and to cause the rod to bend into a curve. The first problem had been considered by Galileo and had attracted the attention of

Varignon and Antoine Parent, among others. Jakob Bernoulli initiated the study of the second problem, and his work became the basis for further researches by Euler. (For histories of elasticity in the eighteenth century see (Truesdell 1960) and (Szabo 1977). The subject of strength of materials is explored by Timoshenko (1953).)

It is important to note that research on elasticity was carried out without the general theoretical perspective that is provided today by the concept of elastic stress. This concept, which underlies such basic modern formulas as the stress–strain relation and the flexure formula, only emerged explicitly in the 1820s in the writings of Claude Navier and Augustin Cauchy. Although one can discern in the earlier work some of the elements that enter into the modern concept of stress, the essential idea—that of cutting a body by an arbitrary plane and considering forces per unit area acting across this plane—was absent.

The divide that separates the modern theory and that of the eighteenth century is illustrated by the problem of elastic bending. Consider the derivation today of the formula for the bending moment of a beam. One begins by assuming that there is a neutral axis running through the beam that neither stretches nor contracts in bending. We apply elementary stress analysis and consider at an arbitrary point of the beam a cross-sectional plane cutting transversely the neutral axis. Elastic stresses distributed over the section are assumed to act across it. Calculation of their moment about the line that lies in the section, is perpendicular to the plane of bending, and passes through the neutral axis leads to the flexure formula, $M = SI/c$, where M is the bending moment, I is the moment of area of the section about the line, c is the distance of the outermost unit of area of the section from the line, and S is the stress at this outermost area.

In the problem of fracture eighteenth-century researchers obtained results that can be readily interpreted in terms of modern formulas and theory. Typically they assumed that the beam was joined transversely to a wall and that the rupture occurred at the joining with the wall. Here the physical situation directly concentrated attention on the plane of fracture—something of concrete significance and no mere analytical abstraction. The conception then current of the loaded beam as comprised of longitudinal fibres in tension is readily understood today in terms of stresses acting across this plane.

By contrast, in the problem of elastic bending researchers were much slower to develop an analysis that connected the phenomenon in question to the internal structure of the beam. Here there was nothing in the physical situation that identified for immediate study any particular cross-sectional plane. In all of Jakob Bernoulli's seminal writings on the elastica the central idea of stress fails to receive clear identification and development.

Although a general theory did not emerge in the eighteenth century, there were many partial results and successful analyses of particular problems. We will consider one such result in some detail: the derivation by Euler of the buckling formula for a loaded column. (An account of Euler's results is given by Fraser (1990).) Euler obtained this result as a corollary to his analysis of the elastica. We are given an elastic lamina oriented vertically, in which the ends A and B are pinned and forces P and $-P$

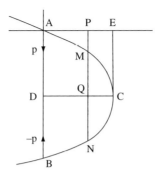

Fig. 12.3. Euler's elastic curve.

act vertically at A and B to produce the stressed configuration depicted in Fig. 12.3. Consider a Cartesian coordinate system in which A is the origin, the positive y-axis is aligned vertically downward, and the positive x-axis extends horizontally to the right. Consider a point M on the lamina. The moment at M exerted by the external force P is equal to Px. This quantity is equated to the Ek^2/R^2, giving rise to the equation

$$Px = \frac{Ek^2}{R^2}. \tag{12.18}$$

Here R is the radius of curvature of the lamina at M, E is a quantity that measures the stiffness of the lamina at M, and k is a constant related to the cross-sectional dimensions of the lamina at M. For x considered as a function of y the quantity R is given in terms of the first and second derivatives of x with respect to y at M. Eq. (12.18) is then a differential equation, and its solution will be an elliptic integral. Euler's primary goal was to undertake a detailed graphical analysis of the geometric forms assumed by solutions of this equation, thus obtaining a classification of the different shapes of the elastic curve.

If the tangent at A to the x-axis is close to $90°$, then the resulting configuration of the elastica provides a model for a loaded column that is beginning to buckle. Euler showed that the size of the force P producing this configuration is

$$P = \pi^2 \cdot \frac{Ek^2}{L^2}, \tag{12.19}$$

where L is equal to AB, the length of the column. A remarkable implication of this result is that a finite force at least equal to eq. (12.19) is required to produce even the smallest bending of the column. In later mechanics eq. (12.19) would become known as Euler's buckling formula, and the value P in eq. (12.19) as the first Euler critical load. Euler believed there was something paradoxical about the result, since a load greater than a specified magnitude was necessary to produce any bending at all of the column.

Euler's project of giving a detailed enumeration of the different transcendental curves that satisfy the differential equation (12.18) failed to become a prominent concern of later research in the theory of elasticity, and does not seem to have been closely connected to the technological aspects of the problem. In engineering statics today one develops a theory for a given object—a beam, strut, or column—by laying down a coordinate system along its unstressed configuration and analyzing small distortions from this position. In Euler's approach the elastic curve provided a unified model for all of these objects and the coordinate system was oriented once and for all with respect to the direction of the external force. Because Euler did not have at his disposal the stress concept, he was not able to connect the quantity Ek^2 to the internal structure of the elastic lamina at M.

12.6 Dynamics of Elastic Bodies: Vibration Theory

In the first part of the eighteenth century researchers investigated a range of problems involving small vibrations: the propagation of pressure waves through continuous media; the vibrations of strings, rings, and rods; the oscillations of linked pendulums and hanging chains; the bobbing and rocking of a floating body. Typically, Newton's second law was applied successfully only to systems involving one degree of freedom, while various special methods were devised to extend the analysis to more general systems. It was not until the 1750s that the analytical equations which express the general principle of linear momentum had become established as the cornerstone of dynamical theory.

Progress in dynamics during the period is illustrated by investigation of the motion of an elastic string. (The following account is based on (Cannon and Dostrovsky 1981). See also (Fraser 1983).) Interest in this problem was motivated by music theory, as well as by the study of engineering structures involving elastic vibration. The detailed physical examination of the corpuscular structure of strings and wires was carried out by investigators such as Willem 'sGravesande (see (Bucciarelli 2003, 39–56) for details). We shall concentrate here on the more purely mathematical advances made in the description of the string's motion. Assume a string is stretched along the z-axis in the $z - y$ plane from $z = 0$ to $z = l$ with tension P and mass density ρ. Consider an element ρdz of the string with coordinates z and y. The string is given a small displacement from its equilibrium position; the problem is to determine the shape it assumes in the ensuing small vibrations and to calculate the period and frequency of these vibrations. In a memoir published in 1732, Daniel Bernoulli first showed using elementary geometry that the force acting on the mass element ρdz equals $P(d^2y/dz^2)dz$ and acts in a direction perpendicular to the z-axis. Intuitively this result is plausible, since the restoring force will be proportional to the curvature (the greater the curvature, the greater the distortion and therefore the greater the restoring force), and for configurations of the string in which dy/dz is small, the curvature is given by d^2y/dz^2. Bernoulli imagined that each mass element of the string vibrates as would a simple pendulum of length L, where L is to be determined. If the displacement of an element ρdz suspended from such a pendulum equals y then the restoring force is given by $-(g/L)(\rho dz)y$, where g is the acceleration due to gravity. This force in turn must equal the actual force calculated above:

$$P(d^2y/dz^2)dz = -(g/L)(\rho dz)y, \quad (12.20)$$

from which we obtain the equation

$$d^2y/dz^2 = (-g/L)(\rho/P)y \quad (12.21)$$

which may now be integrated to give

$$y = \pm c \sin(\sqrt{(g/L)(\rho/P)}z), \quad (12.22)$$

where we have expressed in functional notation what Bernoulli described in geometrical language. (Bernoulli stated that the solution to eq. (12.21) is the curve known as the 'companion to the trochoid' (cycloid)—a characterization of the sine function which survived well into the eighteenth century and is explained by Kline (1972, 351).) Using the end condition $y = 0$ when $z = l$ we obtain a value for L:

$$L = (g\rho/P)(l^2/\pi^2). \quad (12.23)$$

Bernoulli had discovered that the vibrating string assumes the shape of a sine curve. It remained to find the period and frequency of the vibrations. This, however, was now straightforward, since the device used in analyzing the string reduced the problem of finding these values to the analysis of the small vibrations of a simple pendulum of length L. The latter system, involving one degree of freedom, was known from Newton's second law to be governed by the equation

$$\ddot{y} = -(g/L)y. \quad (12.24)$$

The solution of this equation, $y = h\sin(\sqrt{(g/L)}t + \delta)$, immediately yields the desired values for the period T and frequency υ:

$$T = 2\pi\sqrt{(L/g)}, \quad \upsilon = \sqrt{(g/L)}/2\pi. \quad (12.25)$$

Using the value for L given by eq. (12.23) we obtain the final result for the vibrating string:

$$T = 2l\sqrt{(\rho/P)}, \quad \upsilon = \sqrt{(P/\rho)}/2l. \quad (12.26)$$

Bernoulli had therefore derived by mathematical analysis the result known in musical theory of the seventeenth century as 'Mersenne's law', asserting the proportionality of the pitch or frequency to the quantity $\sqrt{(P/\rho)}/2l$.

To solve this problem Bernoulli assumed that the elements of the string undergo small vibrations as simple pendulums all of the same period. Such an assumption was used widely during the period to analyse oscillatory phenomena. Its adoption tended to be combined with certain restrictions on the motion. Thus Bernoulli assumed in his analysis that the elements of the string arrive simultaneously from one side at the equilibrium configuration along the z-axis.

In the 1740s d'Alembert developed an analysis of the vibrating string that proceeded along new and very different lines. His innovative work was made possible by two advances. First, the proportionality of force to the time rate of change of momentum had come to be recognized as a general principle that could be applied to a wide range of physical problems. Second, during the 1730s researchers gained experience with partial differential equations as they investigated such mathematical problems as the construction of orthogonal trajectories to families of curves (see (Engelsman 1984)). The vertical displacement y of the element ρdz of the string depends on both z and t. D'Alembert equated the restoring force $P(d^2y/dz^2)dz$ on ρdz to the time rate of change of its momentum, the latter being given as $\rho dz(d^2y/dt^2)$. For convenience we shall use the notation for partial derivatives that became standard in the nineteenth century. The partial differential equation d'Alembert had obtained may be written in the form:

$$\frac{\partial^2 y}{\partial t^2} = k^2 \frac{\partial^2 y}{\partial z^2}, \qquad (12.27)$$

where $k^2 = P/\rho$. In later mathematical physics eq. (12.27) was called the wave equation. In a brilliant piece of analysis, d'Alembert integrated this equation, showing that its solution is given as

$$y = \Phi(z + kt) + \Phi(z - kt), \qquad (12.28)$$

where Φ is a function that is 'arbitrary', subject to the conditions that it be odd ($\Phi(z) = -\Phi(-z)$), zero at $z = 0$ and $z = l$, and periodic of period $2l$. The configuration of the string at time $t = 0$ is given by the function $y = \Phi(z)$. The class of physical solutions to the problem will therefore depend on the mathematical question of what an arbitrary function is—a question that became a matter of considerable discussion and controversy between d'Alembert and Euler.

In the eighteenth century a function was an object given by a single analytical expression—a formula constructed from variables and constants in finitely many steps using algebraic and transcendental operations and the composition of functions. Such functions were termed 'continuous', in opposition to 'discontinuous' functions—expressions defined piecewise over more than one interval of real numbers. (This definition of continuity is very different from the arithmetic conception of continuity in modern calculus.) D'Alembert insisted that only an equation $y = \Phi(z)$ in which $\Phi(z)$ was an analytical expression was acceptable as an initial solution to the wave equation; hence he would permit only functions that were given by single analytical expressions. Since the initial solution $y = \Phi(x)$ to the vibrating string had been shown to be periodic, one could allow only those functions that were, by virtue of their algebraic form, periodic. (Ravetz (1961a), Youschkevitch (1976), and Fraser (1989) explore the mathematical foundational issues in the debate between d'Alembert and Euler.)

Euler welcomed d'Alembert's derivation and integration of the wave equation as a major advance, but could not accept the restrictions that d'Alembert had imposed on the possible initial solutions. According to d'Alembert, an initial shape of the string that was given by an arc of the parabola $y = cz(l - z)$ would be unacceptable because the expression $cz(l - z)$ was non-periodic. Euler saw no reason why one could not translate the arc of the parabola along the horizontal axis, reflecting it about this axis on every second interval of length l, thus obtaining a new periodic curve. The curve as specified would be given by a periodic function that in analytical terms was defined piecewise over each interval of length l. In the solution $y = \Phi(z)$ the function symbol Φ would now refer to different algebraic expressions depending on the interval of real numbers to which z belonged.

On grounds of physical plausibility and mathematical generality, Euler advocated the acceptance of 'discontinuous' functions as initial solutions to the wave equation. D'Alembert maintained that such functions violated assumptions tacitly invoked in the derivation and integration of this equation. The calculus studied definite, given analytical expressions corresponding to natural modes of generation, and the inclusion of Euler's more general functions violated the basic principles of the subject.

The debate over the nature of arbitrary functions touched the very foundation of analysis and ranks as one of the most interesting episodes of eighteenth-century exact science. Nevertheless, the issues at stake remained somewhat isolated from the mainstream of contemporary mathematical practice. The general functions advocated by Euler raised foundational problems that could not be resolved given the current state of research. D'Alembert's opposition may have seemed obstinate, but it also displayed a clear sense for the spirit of the calculus. The distinguishing property of a 'discontinuous' function—that its algebraic form depended on the interval of real numbers to which the independent variable belonged—undermined the basis of the calculus as a subject explicable by formal analytical principles. To defend the introduction of these objects Euler appealed to the physical model, without any clear specification of the corresponding mathematical conditions. His notion of a more general function was never incorporated into the analytical theory presented in his famous mid-century textbooks, and indeed was at odds with its basic direction.

12.7 Mechanics of Fluids

At the beginning of the eighteenth century, hydrostatics, after the works of Simon Stevin, Evangelista Torricelli, and Blaise Pascal, was reasonably well understood. It was clear that a theory could be based on a few basic principles: the force exerted by a liquid is normal to the surface, it does not depend on the orientation of the surface element, and it is equal to the weight of the column of fluid above the surface. A general and concise formulation of these principles appears at the very beginning of Euler's *Naval Science* (1749). (For general overviews of fluid mechanics in the eighteenth century, see works by Truesdell (1954; 1955), Mikhailov (1983; 1994; 1999), Szabó (1977), Simón Calero (1996), Cross (2002), Darrigol (2005), Blay (2007), and Darrigol and Uriel Frisch (2008).)

On the contrary, the dynamics of fluid motion was just taking its first steps. The only part of the subject that had been thoroughly studied was Torricelli's law for the efflux of water from a hole in the base of the vessel. Pierre Varignon, Johann Bernoulli, Jacopo Riccati, and Jakob Hermann had produced several proofs, which unfortunately mainly showed the need for general principles. A systematic attack on some of the fundamental questions of fluid mechanics, most of them completely new at the time, had been attempted by Newton in Book 2 of the *Principia*: the drag experienced by a body immersed in fluid, the outflow of water from orifices, the viscosity, the propagation of sound in air and of waves on the surface of a liquid. However, his solutions were difficult, not based on a fundamental theory, and sometimes flawed at critical points. The heritage he left to the next century was a list of problems. When, in the Preface to the *Principia*, he wrote 'If only we could derive the other phenomena of nature from mechanical principles by the same kind of [mathematical] reasoning!', he was perhaps thinking in large part of fluid mechanics. As we shall see, this situation was to change completely during the ensuing years, due to a handful of first-rate mathematicians.

Some features of works in fluid dynamics written in the first half of the eighteenth century are likely to confound a modern reader. First of all, principles, expressed in words, were far more important than equations. In addition, since the general principles of mechanics had not yet been discovered, special hypotheses had to be advanced almost for every problem. Results were usually formulated as proportions. Torricelli's law, to cite an important case in point, was expressed by saying that 'the velocities increase in the subduplicate ratio (the ratio of the square roots) of the heights'. Formulas were written in peculiar systems of units; thus, for example, a velocity was usually represented by $\sqrt{v}$, where v denoted the height from which a body had to fall to acquire that velocity. (A general history of systems of units in physics is given by Roche (1998).)

A problem in reading early works on fluid dynamics arises from the absence of the modern concept of internal pressure. In hydrostatics, pressure was the weight of a column of fluid. This exactness became blurred in the passage to dynamics. The word 'pressure' usually denoted the force exerted on the walls of the tube enclosing the liquid, and it was measured in units of length. The first hint of a more general idea occurs in Johann Bernoulli's *Hydraulics* (1742), where essentially it denotes the mutual force between parts of the fluid. A sufficiently clear description of the modern concept finally appears in Euler's memoirs of the 1750s. The importance of this line of development lies in the fact that the pressure in fluids is the first significant case of a contact force, that is, the main concept of modern continuum mechanics.

Despite these differences in methods and formalism, the level of critical thinking throughout the eighteenth century was high. Researchers of the period were fully aware of the importance of such physical phenomena as viscosity and capillarity, of the influence of temperature and of the need for an equation of state connecting temperature, pressure and density. The resistance experienced by a body moving through a fluid was usually assumed to be proportional to the square of the velocity (the 'common rule'), but it was clear that this was only a useful approximation. There were discussions about the composition of a fluid: was it a continuum or a collection of particles? Everybody agreed that matter was made up of small particles, but since it was practically impossible to derive the equations of motion from this hypothesis, researchers turned to the continuum model. While they took into account the results of experiments, they knew that a successful physical theory is simply a representation of reality. Even at this early stage, fluid dynamics had already generated important subfields: theory of tides (Aiton 1955), ship theory (Nowacki 2006 and Ferreiro 2007), and applications to medicine (Zimmermann 1996). The roles of experimenters and theoreticians were clearly divided: physicists did experiments, mathematicians created theories.

In considering these early results, we must always keep in mind that fluid dynamics was a difficult subject. As late as the 1740s, the motion of systems with two or three degrees of freedom was still a matter of discussion. The description of the dynamics of a continuum was a formidable mathematical problem.

The history of modern fluid dynamics properly begins with Daniel Bernoulli's *Hydrodynamics* (1738). (Problems in the period from Newton to Daniel Bernoulli are described by Maffioli (1994) and Mikhailov (1996), and an account of early experiments on fluids is given by Eckert (2006).) In method of research, Bernoulli is the eighteenth century's closest equivalent to a modern physicist. In his work we recognize many of the characteristic features that we now associate with classical theoretical physics: attention to experiments, modelling of phenomena, bold hypotheses, and powerful but unrigorous mathematics. Bernoulli began his work on fluids in the mid-1720s, motivated in part by problems he had encountered during his medical training. (Interestingly, he was probably led to the subject *because* of his medical training.) He composed the work itself at the beginning of the 1730s. In the *Hydrodynamics* we find most of the results on fluids obtained by physicists and mathematicians up to that point, and much more (for details see (Mikhailov 2002; 2005)).

The term 'hydrodynamics' was coined by Bernoulli to denote the union of hydrostatics (equilibrium under the action of forces) and hydraulics (description of motion). The dynamical principle employed was the conservation of mechanical energy in the Huygensian form of the equality between actual descent and potential ascent. This was a bold choice in a period when the principle was a subject of controversy; however, no other general principle for the motion of a system of bodies was then known. Bernoulli also adopted systematically the hypothesis of parallel sections introduced by earlier authors, according to which flow occurs by parallel slices. This was out of necessity, too: a more general flow would have required the use of partial differential equations, not yet known at that time. Bernoulli was fully aware of the limitations of these hypotheses. He believed, for example, that part of the 'living force' was lost because of the microscopic motion of the small particles making up the fluid.

Since today Bernoulli's name is associated with his famous theorem in fluid dynamics, this is probably the result that a modern reader is likely to search for first in the *Hydrodynamics*. However, the original form of the theorem looks quite different from what we are used to seeing. It is found in chapter XII, where Bernoulli set

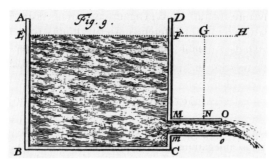

Fig. 12.4. Bernoulli's principle.

himself the problem of the simultaneous determination of pressure and velocity. Not having the full concept of inner pressure at his disposal, he resorted to an ingenious argument. In essence, Bernoulli considered a vessel, constantly kept full of water, discharging through a small horizontal tube (Fig. 12.4). First, he determined the increment of velocity in the horizontal tube by the conservation of energy. Then, to obtain the 'pressure' on a point in the wall of the pipe, he imagined at that point the tube suddenly broke off. The water would therefore flow out of the hole with a known velocity, and the height reached by the jet could be taken as a measure of the pressure.

Another famous result of the *Hydrodynamics* is the derivation of the Boyle–Mariotte law in the chapter on 'elastic fluids' (gases). Bernoulli's gas consists of a virtually infinite number of small spheres 'agitated in very rapid motion'. This model of matter was not unusual at the time, but Bernoulli was among the very few who could obtain definite mathematical results from it. On the basis of simple physical hypotheses, he demonstrated that in gases having the same 'degrees of heat' (temperature) the 'elasticity' is approximately proportional to the density and to the square of the mean velocity of the particles. For almost a century, this was the only result of note in the kinetic theory of gases.

These results do not exhaust the content of the *Hydrodynamics*. Some of the topics look strikingly modern. Bernoulli studied hydraulic machines, defining the work done by a force in a given time and the efficiency of hydraulic engines. In particular, he considered the work done by compressed gases. Turning to applications, he examined the possibility of propelling carts and ships by means of jets: that is, the theory of rockets. These problems of variable-mass dynamics, considered tricky even today, were frighteningly difficult in the 1730s (see (Mikhailov 1976) and (Cerulus 2006)).

The appearance of the *Hydrodynamics* indirectly led to the publication of Johann Bernoulli's *Hydraulics* of 1742. Its complete title is *Hydraulics, Now First Discovered and Proved Directly from Purely Mechanical Foundations, in the Year 1732*. However, despite the date 1732 he actually began writing it after the appearance of his son Daniel's book. Johann had been deeply interested in fluid mechanics at least since 1715, but the finer points of the *Hydraulics* were suggested by Daniel's book. (Indeed, as late as the middle of the 1720s he still believed that the pressure on the pipes was independent of the motion of the fluid.) Disgusted with his father's behaviour, Daniel left the field to which he had contributed so much.

All things considered, it would be wrong to consider Johann a simple plagiarist, for there is enough novelty in his book to make it a masterpiece of fluid mechanics on its own. The *Hydraulics* is essentially a rethinking of Daniel's theory of the relation between velocity and pressure. Though Johann had long advocated the principle of living forces, here he criticizes his son for using this 'indirect' method. In fact, the great achievement of the *Hydraulics* is the use of the 'genuine' method, that is, the proportionality of force to acceleration or what is known as the principle of momentum. Johann considered the resultant of the internal forces acting on an infinitesimal element of fluid, equated it with the acceleration and obtained a differential relation. This apparently simple idea pointed the way for future developments in general mechanics. Perhaps the most important result obtained by this method is Bernoulli's equation for both steady and unsteady flows, in a form easily recognizable by modern readers.

A special place must be assigned to Alexis Clairaut's *Theory of the Figure of the Earth, Derived from the Principles of Hydrostatics* (1743). (For detailed accounts of this work see (Greenberg 1995) and (Passeron 1995).) The first sections are dedicated to the establishment of a fundamental principle for the equilibrium of an incompressible fluid: 'A fluid mass can be in equilibrium only if the forces acting at all places of a canal of arbitrary form mutually destroy each other' (translation from (Truesdell 1956, xx)). This is the final generalization of axioms due to Newton and Huygens. In modern terms, it states that $Pdx + Qdy + Rdz$, where P, Q, R are the components of the external force per unit volume, is an exact differential. From this principle, and from his work on differential forms, Clairaut was led to establish the relations

$$\frac{\partial P}{\partial y} = \frac{\partial Q}{\partial z}, \frac{\partial P}{\partial x} = \frac{\partial R}{\partial z}, \frac{\partial Q}{\partial x} = \frac{\partial R}{\partial y}. \qquad (12.29)$$

Clairaut's treatment is clear, simple, and general. The fluid is referred to a system of orthogonal Cartesian axes, thus foreshadowing the idea of a field of physical quantities, and the result is expressed in purely analytical terms by means of partial derivatives. It was then obvious that a satisfactory theory of fluid motion should include Clairaut's formulation as a special case.

With d'Alembert we reach an intermediate stage in the development of theoretical fluid dynamics, equally distant from the pioneering efforts of the Bernoullis and from the modern theory of Euler. On the one hand, he had a number of brilliant ideas, central for the development of the theory; on the other, he was unable to obtain much from them, and they can now be understood only within the general framework set up by Euler. While d'Alembert was striving for an abstract and general point of view, he was still at some distance from the modern analytical formulation. His philosophy of mathematics led him to obscure the concept of pressure in the derivation of major theorems. In addition, his technical style was intricate, sometimes downright confused, and full of detours. (In recent years, d'Alembert's contributions to fluid dynamics have been re-evaluated by Grimberg (1995; 2002).)

D'Alembert's first separate publication on fluid mechanics was the *Treatise on the Equilibrium and Motion of Fluids* (1744), in which he tried to obtain hydrodynamics from hydrostatics by his general principle of dynamics. In this he was partly successful, but the *Treatise* does not contain anything not to be found in the *Hydrodynamics*. His next publication was a memoir which won a prize at the Berlin Academy in 1746: *Reflections on the General Cause of Winds* (1747). Here he attributed the cause of the winds to the tidal force of the Moon and the Sun—an assertion for which he was ridiculed by Daniel Bernoulli in a letter to Euler ('When one has read it all, one knows as much about the winds as before'). However, the *Reflections* presents some notable results on the purely theoretical side, such as the use of the differential continuity condition and the application of the momentum principle to an element of fluid (the acceleration is equal to the force exerted by the rest of the fluid and by the tidal interaction.)

Up to this point, mathematical fluid dynamics had been a one-dimensional theory. The jump to three dimensions took place in a memoir in Latin that d'Alembert submitted in 1749 for a competition of the Berlin Academy on the subject of the resistance of fluids; but no prize was awarded, and d'Alembert reworked his results as a memoir in French, published in 1752: *Essay on a New Theory of the Resistance of Fluids*. (Euler had been a member of the jury, and this was the beginning of lifelong tension between the two scientists.) In essence, the *Essay* treats of an axially symmetric flow. Its importance consists in the fact that it is the first work where the motion of a continuum is described by means of a field of physical quantities. There are remarkable peaks: the condition of continuity is imposed by the constancy of the volume of a small parallelepiped of fluid in motion, Clairaut's condition is generalized to compressible fluids, and an early version of the dynamical equations for incompressible and axially symmetric flows makes its appearance. Astonishingly, d'Alembert introduced what would now be called a complex potential function for velocity. In pure mathematics, the *Essay* is known for the first appearance of the 'Cauchy–Riemann' equations of complex analysis.

The results obtained by the Bernoullis, Clairaut, and d'Alembert form the basis of the general formulation of fluid dynamics developed by Euler in the 1750s. (We can follow his interest in their results through his correspondence.) However, Euler's theory is much more than the sum of the previous efforts: it represents such a huge step forward that it is difficult today to appreciate the extent of the innovation. It was the classic—and, in a way, definitive—formulation of the elements of fluid dynamics.

The ingredients of Euler's success are easy to determine. First and foremost, he took the principle of momentum, asserting the proportionality of force and acceleration, as the fundamental dynamical principle. He adopted what is today termed the field viewpoint, made constant use of analytical methods, and formulated general results by means of partial differential equations. The theory of fluids lies at the centre of the analytical revolution.

Euler (1757a) was fully aware of what he had achieved, and his opinion deserves to be quoted in full:

Although I envisage here such a great generality, both in respect to the nature of the fluid as well as the forces which act upon each of its particles, I fear not at all the reproaches often justly directed at those who have undertaken to bring to a greater generality the researches of others. I agree that too great a generality often obscures rather than enlightens, leading sometimes to calculations so entangled that it is extremely difficult to deduce their consequences in the simplest cases ... But in the subject I propose to explain, the very reverse occurs: the generality which I undertake, rather than dazzling our lights, will the more reveal to us the true laws of Nature in all their brilliance, and we shall find therein even stronger reasons to admire her beauty and simplicity. It will be an important lesson to learn that principles which had been thought connected to some special cases have a much greater extent. Finally, these researches will require a calculation hardly more troublesome at all, and it will be easy to apply them to any particular case which one might propose. ('General principles of the state of equilibrium of fluids', translation from Truesdell (1956) lxxv)

Euler had been interested in fluid mechanics since the end of the 1720s, but he had decided not to publish anything so as not to compete with his friend Daniel Bernoulli. When he returned to the subject more than twenty years later, at first he limited himself to clarifying problems in hydraulics. (For a general history of hydraulics, see (Rouse and Ince 1957). On Euler's work in hydraulics, see (Eckert 2002, 2008).) However, even these 'minor' papers contain jewels. In the annotations (Euler 1745) which he added to his translation into German of Benjamin Robins' *New Principles of Gunnery* Euler introduced streamlines, stated that a body immersed in a steady stream suffers no resistance ('d'Alembert's paradox'), and presented a clear derivation of Newton's law of resistance. (D'Alembert's paradox derived its name from its appearance in d'Alembert's 'Continuation of researches on the motions of fluids' (1768). See (Grimberg, Pauls and Frisch 2008).) In a book published in 1749 Euler gave a neat proof of Bernoulli's theorem for the steady flow of an incompressible fluid in a circular tube, not much different from those found in modern textbooks. In 1751 he identified the path of each particle of a fluid in steady motion by means of a parameter (Euler 1767a)—an early form of what in modern theory is called a 'material description'.

The real beginning of the modern theory came with the memoir '*Principles of the motion* of fluids', presented in 1752 but published much later (1761), in which Euler systematically represented a fluid as a continuum referred to a fixed set of orthogonal Cartesian axes and governed by partial differential equations. The first part was about pure kinematics: by a reasoning similar to that employed by d'Alembert, Euler obtained the continuity equation for incompressible fluids. In the dynamical part, Euler applied 'Newton's second law of motion' to an element of fluid and obtained the general dynamical equations for ideal incompressible fluids in the special case where the only external force is gravity.

Never content with simply obtaining new results, Euler recast, completed, and organized the theory in a sequence of three papers in French, which became the main source for the study of theoretical fluid mechanics in the second half of the eighteenth century. The first paper (Euler 1757a) presented all that was then known on fluid statics, giving the general equations of hydrostatics,

$$\nabla p = \rho \mathbf{F}, \tag{12.30}$$

where p is the pressure ρ the density and $\mathbf{F}$ the extraneous force per unit mass. In the application of his principles to the atmosphere, Euler implicitly corrected d'Alembert's faulty hypothesis by attributing the cause of the winds to 'different degrees of heat'. The second paper (1757b) was mainly important for giving, in the case of compressible fluids, the equation of continuity

$$\frac{\partial \rho}{\partial t} + \nabla \bullet (\rho \mathbf{v}) = 0, \tag{12.31}$$

and the general dynamical equations

$$\rho \mathbf{F} - \nabla p = \rho \left(\frac{\partial \mathbf{v}}{\partial t} + v \bullet \nabla \mathbf{v} \right), \tag{12.32}$$

where $\mathbf{v}$ is the velocity. From these rock-solid foundations, it was possible to calculate the first exact solutions: rotation about the origin and rectilinear motion in absence of external forces. The third paper (Euler 1757c) contained the Bernoulli equation for the streamlines.

The importance of these works for the general foundations of physics is difficult to overestimate. In the second paper, for example, we find clearly stated the distinction between external and internal forces acting on a mechanical system. On a more advanced level, in the *Guide to Natural Science*—probably written not long after 1755 but not published until 1862—Euler applied his general equations of fluid dynamics to an hypothetical ether filling all space and responsible for every kind of physical interaction. (On this work, see (Radelet-De Grave and Speiser 2004)). This was the first serious attempt at a unitary theory of physics based on ideas that correspond to the modern field concept. Eulerian fluid dynamics would remain the model for such theories for almost two centuries.

The young Lagrange was indirectly responsible for the next advance. In August 1759 he sent to Euler a memoir in which the one-dimensional propagation of sound was represented by a weightless string loaded by mass-points equally spaced. This provided the stimulus for Euler to set to work on wave propagation through a compressible fluid. He sent some of his results to Lagrange in a letter of January 1760. To describe the disturbance propagating through the particles of air, Euler needed a new kind of mathematical description; in fact, the letter contains the first appearance of the material coordinates (that is, the initial coordinates of an element of fluid) for the motion of a continuum. The complete research was published, after a delay of a few years, in a sequence of three papers published in 1766 (Euler 1766a, 1766b, 1766c). They are remarkable for the first appearance of the three-dimensional wave equation and of the material equation of continuity.

Euler had succeeded in describing waves of all kinds by means of the wave equation. He recognized that sound is a succession of pressure pulses, and that its mathematical description—involving the integration of a wave equation—would require 'discontinuous 'functions, or functions defined piecewise using analytic expressions. As we saw in the problem of the vibrating string, such functions were somewhat controversial, and Euler was hesitant to employ them in his theory. However, the new discrete model advanced by Lagrange in 1759 provided a natural motivation to adopt such functions, and therefore to accept the wave equation as a valid mathematical description of a range of phenomena. Lagrange's model gave Euler the confidence to accept discontinuous functions and to embrace the general applicability of the wave equation in mathematical mechanics.

At the beginning of his *Mechanics* (1736), Euler had stated that he planned to write general expositions of the advanced parts of mechanics. With respect to fluid mechanics, he fulfilled this promise with the publication of a comprehensive treatise presented to the St Petersburg Academy in 1766. Since he could not find a printer, he divided the work into four lengthy memoirs which were published in the proceedings of the Academy (Euler 1769–1772). Here in his usual style he presented all the aspects of the subject in a masterfully clear way. The third memoir is a theoretical treatment of hydraulics from the new point of view, and the fourth memoir is the first systematic presentation of mathematical acoustics. The second memoir contains the dynamical equations in material coordinates, which had also appeared in a memoir on the propagation of sound published by Euler in 1766.

Research on the mechanics of fluids was also important for providing impetus to the study of partial differential equations. Thus, for example, in the second French paper on fluids Euler was led to consider 'a very curious analytic question' which is nothing else than the problem of solving the general first-order linear partial differential equation with constant coefficients. Also, in Euler's third memoir on sound there is the earliest example of solution of a partial differential equation by separation of variables. Finally, the now classic linear transformation for reducing the one-dimensional wave equation to the simple form

$$\frac{\partial^2 z}{\partial u \partial v} = 0 \qquad (12.33)$$

first appears in a letter of Euler to Lagrange of November 1762 (Euler 1892).

The next investigator to cultivate the subject at a fundamental level was Lagrange. For almost twenty years he was the only mathematician who could understand Euler's theory to the point of being able to use it as a starting point for further research. His first contributions appeared in a lengthy memoir on dynamics published in 1762. Lagrange derived Euler's dynamical equations for incompressible fluids from his method of using conjointly d'Alembert's equilibrium of the lost forces and the principle of virtual work, with the condition of incompressibility as a constraint. However, his most important contributions to fluid mechanics appeared in a memoir published in 1783. The first part is a concise presentation of Euler's theory. Building on this general background, Lagrange obtained some classic theorems. Preeminent among them are the velocity-potential theorem (under general conditions, a particle once in irrotational motion is always in irrotational motion), the necessary and sufficient condition for a surface to be material (formalizing intuitive ideas of continuity), and the impulse theorem (the flow produced by impulsive hydrostatic pressure is lamellar). (For details, see (Truesdell and Toupin 1960).) These were followed by the introduction of perturbation methods into fluid dynamics.

This memoir of Lagrange closes the century. The new theory became generally known through Jacques Cousin's *Introduction to the Study of Physical Astronomy* (1787), Lagrange's *Analytical Mechanics* (1788), Laplace's *Celestial Mechanics* (vol. 1, 1799) and Siméon-Denis Poisson's *Treatise on Mechanics* (1798). Waves on the surface of liquids and viscosity were left to the next generation.

12.8 CELESTIAL MECHANICS

In the eighteenth and nineteenth centuries the study of the motion of planetary bodies moving under the mutual attraction of gravity was regarded as a branch of physical astronomy. The theoretical study of the gravitational interaction of bodies following definite force laws distinguished this subject from positional or descriptive astronomy. The subject became know as celestial mechanics, after the title of Simon Laplace's major work of 1798. In broad outline, celestial mechanics in the eighteenth century was marked by three major developments. First, the semi-geometrical theory of Newton's *Principia* was recast in analytical terms using the methods and notation of the Leibnizian calculus. Second, several problems involving the motion of systems of more than two bodies were analysed. The greatest effort was devoted to the Earth–Moon–Sun system and the Sun–Jupiter–Saturn system, and the motion of comets also attracted considerable attention. In the case of the Moon, investigation focused on the motion of its apogee, and on its secular or long-term acceleration. Of particular concern for planetary motion was the problem of stability, of showing that the three-body systems of interest were not subject to any irregularities that over the long term would result in a progressive derangement of the system. Third, phenomena such as libration and precession were analysed using the dynamical theory of rigid bodies that had emerged by the middle of the century.

The modern theory of celestial mechanics is based on the formal methods of Hamilton–Jacobi theory—a subject developed by William Rowan Hamilton and Carl Gustav Jacobi in the first part of the nineteenth century, and subsequently applied to celestial mechanics by Félix Tisserand, Charles Delaunay, and Henri Poincaré (see (Nakane and Fraser 2003) and (Fraser and Nakane 2003).) During the eighteenth century, researchers employed differential equations expressing the proportionality of force and acceleration. A major goal was to show that the inverse-square law of gravitation formulated in Book 3 of the *Principia* was sufficient to account for the various planetary phenomena. The general triumph of Newtonianism often identified by historians with eighteenth-century physical science in fact refers rather specifically to the verification in physical astronomy of the inverse-square law. Innovations in research were based on skilful reduction of the differential equations of motion, and on the development of perturbation methods involving the method of variation of arbitrary constants. Some important advances in the formulation of fundamental principles of mechanics occurred in physical astronomy, particularly in the study of the motion of a rigid body.

Commenting on the historical outlook of modern astronomers, Stanley Jaki (2000, 329–330) observes that 'the very tools with which they work (satellite-directed radar techniques) are too different from the methods of their forebears to prompt them to look back to the eighteenth and nineteenth centuries.' Celestial mechanics of this earlier period is a somewhat curious subject to a reader today. In a broad sense everyone acknowledges the subject's success in confirming the inverse-square law of gravitation. However, the basic results in anything resembling their original historical form are not included in modern expositions of celestial mechanics. Furthermore, the focus on special problems (now treated numerically using computers) and the rather formidable technicalities involved in their solution give rise to a subject resistant to easy contemplation and understanding. (For historical accounts of eighteenth-century celestial mechanics, see (Grant 1852), (Waff 1976), (Forbes 1980), (Linton 2004, Chapter 9), and (Taton and Wilson 1995).)

What is perhaps more meaningful to a modern reader than the details of particular problems is the progress made in the formation of fundamental principles and methods of solution. We consider in some detail two problems that were particularly fruitful from this point of view: the motion of the Moon's apogee, and the changes in average motion of Jupiter and Saturn.

12.8.1 Earth–Moon–Sun System

A basic characteristic of the Moon—one that it shares with the Sun—is that its movement along the ecliptic varies in velocity from day to day. However, unlike the case of the Sun, the point of minimum velocity does not remain fixed but advances by an amount of about 3° each tropical or sidereal month. The anomalistic month—the time taken by the Moon to move from apogee to apogee—is longer than the sidereal month—the time taken by the Moon to travel 360° relative to some fixed point on the celestial sphere. (Two additional lunar periods of note are the synodic month, the time between two full Moons, and the Draconitic month, the time between the Moon's circuit from node to node.) The values of the anomalistic and tropical months were know to the Seleucid Babylonians, and the relations between the different lunar periods were tabulated by Ptolemy in the *Almagest* and *Handy Tables*, and built into his geometric models of the Moon. (See (Pedersen 1974, chapter 6) for details.)

In the *Principia* Newton investigated the motion of the Moon about the Earth, and tried to evaluate the gravitational effect on the Moon exerted by the Sun. In Book 1 he had shown that if a particle moves in an ellipse under the action of a central force located at one focus, then the force varies as the inverse square of the distance of the particle from the force centre. He also showed that for nearly circular orbits, a force law of the form $F \propto \frac{1}{r^{2+\epsilon}}$ (where F is the force and r is the distance from the body to the force centre) will result in a rotation of the line of apsides, no matter how small ϵ is. That is, even the slightest departure of the force law from an inverse-square form leads to a steady progression of the line of apsides. The effect of the Sun's gravity on the Moon could be interpreted in terms of a model involving a slight departure of the force acting on the Moon from a strict inverse-square law given as a function of the distance between the Moon and the Earth. A major difficulty with this theory—one that Newton was unable to resolve—was that the apsidal motion predicted by the model was only one half of the observed value of about 3° per month.

During the 1740s the problem of the motion of the Moon occupied the energies of three leading savants: Clairaut and d'Alembert in Paris, and Euler in Berlin. Newton's investigation was recast in terms of systems of differential equations to describe the Moon's motion. The three men each made important progress in the development of a lunar theory. However, like Newton, they arrived at a value for the apsidal motion that was one half of its observed value. (Clairaut, incidentally, was rather critical of Newton's *Principia*, observing (1749, 329) that in the difficult parts Newton 'employs far too few words to explain his principles, whereas he appears to surrender himself complaisantly to details and facts of calculation that readers will have no qualms accepting'.) By the 1740s the problem of the lunar apogee had become a major challenge to the validity of a physical astronomy founded on the inverse-square law of universal gravitation. It should be noted that the concern here was not a subtle variation in the orbital motion, but a phenomenon evident from basic observation and documented as far back as Hipparchus and Ptolemy in antiquity. (Later historical research would show that it was also well known even earlier, by the Seleucid Babylonians.) If the current gravitational theory was unable to account for gross observations, it was unlikely to succeed at predicting the other more subtle variations experienced by the Moon in its motion.

During the 1740s Clairaut speculated that it might be necessary to abandon the inverse-square law. Euler arrived at a similar conclusion, believing that ultimately gravity resulted from some sort of vortical effect of a ponderous ether, and that it was *a priori* unlikely that an exact inverse-square law would be valid. (In other instances where the theory encountered difficulties, Euler was inclined to attribute the discrepancies to the resistance encountered by planetary bodies as they moved through an interplanetary medium.) D'Alembert was more hesitant to abandon the accepted law of gravitation, though he believed that it was possible that an additional force such as magnetism was influencing the motion of the celestial bodies.

In his researches of the late 1740s Clairaut had reduced the analytical description of the Moon's motion to the equation of a rotating ellipse (Linton 2004, 301):

$$\frac{k}{r} = 1 - e \cos q\theta, \qquad (12.34)$$

where r is the distance of the Moon from the Earth, e is the eccentricity of the Moon's orbit, and θ is the mean anomaly of the Moon. q is a constant slightly less than one, which implies that the Moon must travel slightly more than 360° to move from apogee to apogee or perigee to perigee. Clairaut's derivation of eq. (12.34) involved a skilful integration that attracted Euler's admiration (for details see (Linton 2004, 301), as it was an approximate one obtained by neglecting certain terms, and resulted in a value for apsidal motion that was one half its true value. Clairaut believed that the terms neglected in the derivation of this equation did not contribute substantially to this motion. Nevertheless, in 1748 he went back and examined a modified version of the equation which included the missing terms. After a great deal of labour involving successive approximations he made the exciting discovery that the neglected part did in fact contribute the missing part of the apsidal motion

Clairaut's result was rightly heralded as a breakthrough, and led Euler and d'Alembert immediately to revise their lunar theories. All three men produced treatises on the Moon in the early 1750s. Euler's book employed a Cartesian rectangular coordinate system and introduced new methods for integrating the differential equations of motion. More generally, Euler devised a calculus of trigonometric functions that he employed extensively throughout his celestial researches and made a common part of the subject. D'Alembert concentrated on the algebraic development of the theory, producing a literal theory in which the values of the terms were not replaced by approximate numerical values. (The lunar theorist Brown (1896, 239) observes that 'while [Clairaut] worked out his results numerically, d'Alembert considered a literal development and carried out his computations with more completeness.') D'Alembert derived the following equation (see (Linton 2004, 303)) connecting the sidereal month T_s and the anomalistic month T_a:

$$\frac{T_a}{T_s} - 1 = \frac{3}{4}m^2 + \frac{225}{32}m^3 + \ldots \tag{12.35}$$

Here m is the ratio of the mean motion of the Sun to the mean motion of the Moon (a number slightly less than $\frac{1}{12}$). If the cube term in m is neglected, the older value for the motion of the apogee is obtained. Because the coefficient of the cube term is large, the quantity $\frac{225}{32}m^3$ is substantial and must be retained, and by including it, the observed value for the motion of the apogee is derived. (For a detailed history of the lunar apogee problem see (Waff 1976 and 1995).)

12.8.2 Sun–Jupiter–Saturn System

It had become well known by the middle of the century that the values of the average motions in longitude of Jupiter and Saturn were changing over time. The motion of Jupiter was increasing slowly, while the motion of Saturn was decreasing. This observational phenomenon was called the 'great inequality of Jupiter and Saturn'. Explaining it using Newton's theory of gravity was a major focus of research in celestial mechanics in the second half of the century. Euler laid some of the groundwork for the theory, but the most important progress was made by Lagrange and Laplace in the 1770s and 1780s. Lagrange fashioned the method of variation of parameters into a fundamental analytical tool in the theory of perturbations. Laplace succeeded in applying this theory to the motions of Jupiter and Saturn, and deduced that the inequality was periodic with a period of about 850 years. The three-body system consisting of these two planets and the Sun was the dominant gravitational grouping in the solar system and the one subject to the greatest variation. Laplace's result meant in effect that the stability of the solar system had been derived from the inverse-square law of universal gravity—a result that was the crowning achievement of celestial mechanics in the eighteenth century. (For an account of this history see (Morando 1995).)

Linton (2004, Chapter 9) distinguishes two approaches to the treatment of perturbations. In the first, called the method of absolute perturbations, the orbital parameters are taken as constant, and an approximate solution is obtained for the differential equations involving these parameters. This solution is then substituted back into the differential equations, and a second approximation follows. Proceeding in this way, one arrives at a successively more accurate description of the motion of the perturbed body. (Brown (1896, 47) refers to this method as the 'method of successive approximation'.) This was the method that Clairaut followed in his investigation of the Moon's apogee, in which the second approximation was found to be the key to removing the discrepancy between theory and observation.

The disadvantage of the method of absolute perturbations is that the successive approximation procedure can give rise to terms that increase indefinitely with time. In the eighteenth century such terms were referred to as 'arcs of circles', because they appeared as angles that could be regarded as the arcs of a circle of radius 1. It was necessary to avoid these quantities because they were non-periodic and increased indefinitely with time. A familiar refrain in treatises on celestial mechanics of the period was the desirability of obtaining expressions containing no arcs of circles.

A second approach to perturbations is based on the method of variational of arbitrary parameters, introduced by Euler and developed more fully by Lagrange. The basic idea is the familiar one encountered today in a first course in differential equations when one solves a non-homogenous differential equation using variation of parameters. One takes the general solution of the homogenous equation and then regards the constants of integration appearing in this solution as functions of the independent variable. If the expression considered in this way is substituted back into the non-homogenous equation, one is able to solve for the parameters and obtain a solution of the non-homogenous equation. In the astronomical context, one begins by assuming the orbit of the given body is Keplerian, corresponding to unperturbed motion, so that it is specified with respect to six orbital parameters. These parameters are then regarded as functions of time, and the Keplerian solution is substituted back into the differential equations of perturbed motion. The goal is to integrate these equations in order to obtain a solution of the original problem.

Although the method of variation of orbital parameters was introduced by Euler, it was Lagrange who developed the method into a highly effective tool. An exposition of it was presented in his *Analytical Mechanics* (1788) and in the subsequent editions of this work. The method was at the base of the researches carried out by Lagrange and Laplace during the 1780s in their investigation of the great inequality of Jupiter and Saturn. In the case of the Moon the distance from the Earth to the Moon was negligible in comparison with the distance from the Earth to the Sun, and this fact simplified some of the approximations. By contrast, the ratio of the distances of Jupiter and Saturn to the Sun was not negligible, leading to greater complications in the analysis of their mutual motions. Working from some results of Lagrange, Laplace discovered the following general equation connecting the masses m of the planets of the solar system and the lengths a of their semi-major axes (Morando 1995, 139):

$$\frac{m}{a} + \frac{m'}{a'} + \frac{m''}{a''} + \ldots = const. \qquad (12.36)$$

Because the masses m and $m\prime$ of Jupiter and Saturn are very large compared to the other planetary masses, the above relation is approximately $\frac{m}{a} + \frac{m'}{a'} = const$. Using Kepler's third law Laplace was led to the conclusion that the variation in the mean motion n of Jupiter is related to the variation in the mean motion $n\prime$ of Saturn by the equation $\delta n' = -2.33\delta n$. Moreover, this relation accorded with what was known from observation, and Laplace inferred that the changes in the mean motions of Jupiter and Saturn were a result of their gravitational interaction—a conclusion that both he and Lagrange had rejected explicitly at an earlier stage of their investigations. With this insight, Laplace launched an all-out attack on the problem of the great inequality. He had shown previously that the commensurability or near-commensurability of the motions of two bodies in the three-body problem can lead to long-term periodic terms for their mean motions. The ratio of the mean motion of Jupiter to the mean motion of Saturn was very close to 5:2, and Laplace was able to show after very considerable effort that this leads to a long-term periodic expression for the mean motions with a period of about 850 years.

Although Laplace deserved full credit for resolving the mystery of the great inequality of Jupiter and Saturn, it is worth noting how many of the key ideas and methods used by him were drawn from Lagrange's writings. Today it is the results and methods pioneered by Lagrange that tend to attract the attention of mathematicians and physicists, while Laplace's achievement occupies an honoured place in the mausoleum of past science. A very active area of modern research concerns general mathematical solutions of the three-body problem, treated in a qualitative way using methods of analysis and topology. A noteworthy theoretical result in this direction is a problem analysed mathematically by Lagrange (1777) in a lengthy memoir that won the Paris Academy prize for 1772. Lagrange considered a system in which the three distances between the bodies are equal, and found solutions in which one of the bodies occupies what are today called the 'Lagrange points'. Given two massive bodies, the Lagrange points are located in the orbital plane of these bodies, at the vertices of the two equilateral triangles in this plane taken with respect to the line joining the bodies. Although Lagrange's result was primarily of mathematical interest, in the early twentieth century it was discovered that asteroids are present at the Lagrange points of Jupiter's orbit about the Sun (see (Linton 2004, 326) for details).

References

AITON, ERIC J. 1955. 'The contributions of Newton, Bernoulli, and Euler to the theory of the tides', *Annals of Science* 11, 206–223.

ALEMBERT, JEAN LE ROND D'. 1743. *Traité de Dynamique*. Paris: David. Second edition 1758.

———1744. *Traité de l'équilibre et du mouvement des fluides: pour servir de suite au Traité de dynamique*, Paris: David. Augmented second edition 1770.

———1747. *Réflexions sur la cause générale des vents: pièce qui a remporté le Prix proposé par l'Académie Royale des Sciences de Berlin, pour l'année 1746*, Paris: David.

———1749. *Recherches sur la précession des équinoxes, et sur la nutation de l'axe de la terre, dans le systême Newtonien*, Paris: David. Reprinted 1967 by Culture et Civilisation; = Oeuvres, s. I, v. 7, 1–367.

———1752. *Essai d'une nouvelle théorie de la résistance des fluides*, Paris: David. Reprinted 1966 by Culture et Civilisation.

———1768a. Démonstration analytique du principe de la force d'inertie. In *Opuscules mathématiques: ou, Mémoires sur différens sujets de géométrie, de méchanique, d'optique, d'astronomie, &c*, v. 4, 349–357. Paris: Briasson.

———1768b. Suite des recherches sur les mouvements des fluides. In *Opuscules mathématiques: ou, Mémoires sur différens sujets de géométrie, de mé chanique, d'optique, d'astronomie, &c*, v. 5, 132–170. Paris: Briasson.

———2002–. *Œuvres complètes de d'Alembert*, 5 vols. to date. Paris: CNRS Éd.

ARCY, PATRICK D'. 1752. Problème de dynamique. *Histoire de l'Académie royale des sciences avec les mémoires de mathématique et de physique tirés des registres de cette Académie* (Paris) 49 (1747), 344–361.

BECCHI, ANTONIO et al. 2003. *Essays on the History of Mechanics, in Memory of Clifford Ambrose Truesdell and Edoardo Benvenuto*. Basle: Birkhäuser Verlag.

BENVENUTO, EDOARDO. 1991. *An Introduction to the History of Structural Mechanics*. 2 vols. Dordrecht: Springer.

BERNOULLI, DANIEL. 1728. Examen principiorum mechanicae, et demonstrationes geometricae de compositione et resolutione virium. *Commentarii academiae scientiarum Petropolitanae* 1 (1726), 126–142; = Werke, v. 3, 119–135.

———1738. *Hydrodynamica, sive de viribus et motibus fluidorum commentarii: opus academicum ab auctore, dum Petropoli ageret, congestum*. Strasbourg: Dulsecker; = Werke, v. 5, 93–424.

———1728. Examen principiorum mechanicae, et demonstrationes geometricae de compositione et resolutione virium. Commentarii academiae scientiarum Petropolitanae 1 (1726), 126–142; = Werke, v. 3, 119–135.

———1746. Nouveau probleme de mecanique, resolu par Mr. Daniel Bernoulli. Mémoires de l'académie des sciences (Berlin) 1 (1745), 54–70; = Werke, v. 3, 179–196.

———1747. Commentationes de immutatione et extensione principii conservationisvirium vivarum, quae pro motu corporum coelestium requiritur. Commentarii academiae scientiarum Petropolitanae 10 (1738), 116–124; = Werke, v. 3, 160–169.

———1750. Remarques sur le principe de la conservation des forces vives pris dans un sens géné ral. *Mémoires de l'académie des sciences* (Berlin) 4 (1748), 356–364; = Werke, v. 3, 197–206.

———1982–. *Die Werke von Daniel Bernoulli*, 6 vols. to date. Basle: Birkhäuser.

BERNOULLI, JAKOB I. 1705. Démonstration géné rale du centre de balancement ou d'oscillation, tirée de la nature du levier. *Histoire de l'Académie royale des sciences avec les mémoires de mathématique et de physique tirés des registres de cette Académie* (Paris) 5 (1703), 78–84; = Opera, v. 2, 930–936. (Op. XCVIII)

———1744. *Opera*, 2 vols. Geneva: Cramer & Philibert. Reprinted 1967 by Culture et Civilisation.

———1969–. *Die Werke von Jakob I Bernoulli*, 5 vols. to date. Basle: Birkhäuser.

BERNOULLI, JOHANN I. 1729. Theoremata selecta pro conservatione virium vivarum demonstranda et experimentis confirmanda. Excerpta ex Epistolis datis ad filium Danielem. *Commentarii academiae scientiarum Petropolitanae* 2 (1727), 200–207; *Opera*, v. 3, 124–130; = Werke, v. 6, 489–497.

———1735. De vera notione virium vivarum, earumque usu in dynamicis, ostenso per exemplum, propositum in *Comment. Petropolit. Tomi II pag. 200*. *Acta Eruditorum*, May 1735, 210–230; *Opera*, v. 3, 239–260; = *Werke*, v. 6, 521-539.

———1742a. *Opera omnia, tam antea sparsim edita, quam hactenus inedita*, 4 vols. Lausannae and Geneva: Bousquet. Reprinted 1968 by Olms.

———1742b. Propositiones variae Mechanico-dynamicae. In *Opera*, v. 4, 253–313; = *Werke*, v. 6, 367–421. (Op. CLXXVII)

———1742c. Hydraulica. Nunc primum detecta ac demonstrata directe ex fundamentis pure mechanicis. Anno 1732. In *Opera*, v. 4, 387–484. (Op. CLXXXVI)

BERTOLONI MELI, DOMENICO. 1993. The emergence of reference frames and the transformation of mechanics in the Enlightenment. *Historical Studies in the Physical and Biological Sciences* 23, 301–335.

BLANC, CHARLES. 1948. Préface de l'éditeur. In *Leonhardi Euleri Opera omnia*, s. II, v. 3, VII–XXII.

———1968. Préface des volumes II 8 et II 9. In *Leonhardi Euleri Opera omnia*, s. II, v. 9, VII–XXXIX.

———2008–. *Die Werke von Johann I und Nicolaus II Bernoulli*, 1 vol. to date. Basle: Birkháuser.

BLAY, MICHEL. 1992. *La naissance de la mécanique analytique: la science du mouvement au tournant des XVIIe et XVIIIe siècles*. Paris: Presses universitaires de Francc.

———2007. *La science du mouvement des eaux de Torricelli à Lagrange*. Paris: Belin.

BONOLA, ROBERTO. 1912. *Non-Euclidean Geometry: A Critical and Historical Study of its Development*. Chicago: Open Court Publishing Company. English translation by H. S. Carslaw of the Italian edition of 1906.

BROWN, ERNEST W. 1896. *An Introductory Treatise on the Lunar Theory*. Cambridge: Cambridge University Press. Reprinted by Dover 1960.

BUCCIARELLI, LOUIS L. 2003. Coping with error in the history of mechanics. In (Becchi 2003), pp. 39–56.

CANNON, JOHN T. and SIGALIA DOSTROVSKY. 1981. *The Evolution of Dynamics Vibration Theory from 1687 to 1742*. New York: Springer-Verlag.

CAPARRINI, SANDRO. 1999. On the history of the principle of moment of momentum. *Sciences et Techniques en Perspective*, (2)3, 1999, 47–56.

———2002. The discovery of the vector representation of moments and angular velocity, *Archive for History of Exact Sciences* 56, 151–181.

———2003. 'Early theories of vectors', in (Becchi et al. 2003), pp. 179–198.

———2007. 'Euler's influence on the birth of vector mechanics', in *Leonhard Euler: Life, Work and Legacy*, edited by Robert E. Bradley and Edward C. Sandifer, Amsterdam: Elsevier, pp. 459–477.

CARNOT, LAZARE. 1782. *Essai sur les machines en général*. Dijon: Defay.

CERULUS, FRANS A. 2006. Daniel Bernoulli and Leonhard Euler on the jetski. In *Two Cultures: Essays in Honour of David Speiser*, edited by Kim Williams, 73–96. Basle: Birkh áuser.

CHAPRONT-TOUZÉ, MICHELLE and JEAN SOUCHAY. 2006. Introduction générale. In *Oeuvres complètes de d'Alembert*, s. I, v. 7, XIII–CXXXII.

CLAIRAUT, ALEXIS-CLAUDE. 1743. *Théorie de la figure de la terre, tirée des principes de l'hydrostatique*, Paris: David. Second edition 1808.

———1749. 'Du système du Monde dans les principes de la gravitation universelle', *Histoire de l'Académie royale des sciences avec les mémoires de mathématique et de physique tirés des registres de cette Académie (Paris)* (1745), 329–364.

COHEN, BERNARD I. 1971. *Introduction to Newton's 'Principia'*. Cambridge, MA: Harvard University Press.

COUSIN, JACQUES ANTOINE JOSEPH. 1787. *Introduction à l'étude de l'astronomie physique*, Paris: Didot.
CROSS, JAMES. 2002. La meccanica del continuo. In *Storia della scienza*, v. 6, 471–479. Roma: Istituto della Enciclopedia Italiana, 2001–.
DARRIGOL, OLIVIER. 2005. *Worlds of Flow: A History of Hydrodynamics From the Bernoullis to Prandtl*. Oxford: Oxford University Press.
DARRIGOL, OLIVIER and URIEL FRISCH. 2008. From Newton's mechanics to Euler's equations. *Physica D: Nonlinear Phenomena* 237, 1855–1869.
DOBSON, GEOFFREY J. 1998. Newton's problems with rigid body dynamics in the light of his treatment of the precession of the equinoxes. *Archive for History of Exact Sciences* 53, 125–145.
DUGAS, RENÉ. 1957. *A History of Mechanics.* London: Routledge and Kegan Paul Ltd. English translation by J. R. Maddox of the French edition of 1950.
ECKERT, MICHAEL. 2002. Euler and the fountains of Sanssouci. *Archive for History of Exact Sciences* 56, 451–468.
——2006. *The Dawn of Fluid Dynamics: A Discipline Between Science and Technology.* Weinheim: Wiley-VCH.
——2008. Water-art problems at Sanssouci: Euler's involvement in practical hydrodynamics on the eve of ideal flow theory. *Physica D: Nonlinear Phenomena* 237, 1870–1877.
ENGELSMAN, STEVEN. 1984. *Families of Curves and the Origins of Partial Differential Equations.* Amsterdam: Elsevier Science Publishing Co.

Note: in the following entries for Euler the notation (E) refers to the number assigned to the entry by Gustav Eneström in his 1913 index to Euler's writings.

EULER, LEONHARD. 1736. *Mechanica sive motus scientia analytice exposita.* St. Petersburg: Typographia Academiae Scientiarum; = *Opera Omnia*, s. II, v. 1 (E 15)
——1738. De communicatione motus in collisione corporum. *Commentarii academiae scientiarum Petropolitanae* 5 (1730/1), 159–168; = *Opera*, s. II, v. 8, 1–6. (E 22)
——1744. 'De curvis elasticis', First appendix to Euler's *Methodus inveniendi lineas curvas maximi minimive proprietate gaudentes, sive solutio problematis isoperimetrici lattissimo sensu accepti*, pp. 245–310 = *Opera* s. I, v. 24. (E65)
——1745. Dissertation sur la meilleure construction du cabestan. In *Pièces qui ont remporté le prix de l'académie royale des sciences en MDCCXLI*, 29–87; = Opera, s. II, v. 20, 36–82. (E 78)
——1746. De motu corporum in superficiebus mobilibus. In *Opuscula varii argumenti*, Berlin: Haude & Sperner, v. 1, 1–136; = *Opera*, s. II, v. 6, 75–174. (E 86)
——1749. *Scientia navalis, seu tractatus de construendis ac dirigendis navibus exposita*, 2 v., St Petersburg: Typographia Academiae Scientiarum; = *Opera*, s. II, v. 18–19. (E 110–111)
——1750. Reflexions sur l'espace et le tems. *Mémoires de l'académie des sciences* (Berlin) 4 (1748), 324–333; = *Opera*, s. III, v. 2, 376–383. (E 149)
——1752a. Découverte d'un nouveau principe de mécanique. *Mémoires de l'académie des sciences (Berlin)* 6 (1750), 185–217; = *Opera*, s. II, v. 5, 81–108. (E 177)
——1752b. Recherches sur l'origine des forces. *Mémoires de l'académie des sciences* (Berlin) 6 (1750), 419–447; = *Opera*, s. II, v. 5, 109–131. (E 181)
——1754. Sur le mouvement de l'eau par des tuyaux de conduite. *Mémoires de l'académie des sciences (Berlin)* 8 (1752), 111–148; = *Opera*, s. II, v. 15, 219–250. (E 206)
——1757a. Principes généraux de l'état d'équilibre des fluides. *Mémoires de l'acad émie des sciences* (Berlin) 11 (1755), 217–273; = *Opera*, s. II, v. 12, 2–53. (E 225)
——1757b. Principes généraux du mouvement des fluides. *Mémoires de l'académie des sciences (Berlin)* 11 (1755), 274–315; = *Opera*, s. II, v. 12, 54–91. (E 226)

———1757c. Continuation des recherches sur la théorie du mouvement des fluides. *Mémoires de l'académie des sciences (Berlin)* 11 (1755), 316–361; = *Opera*, s. II, v. 12, 92–132. (E 227)

———1761a. Principia motus fluidorum. *Novi commentarii academiae scientiarum Petropolitanae* 6 (1756/7), 271–311; = *Opera*, s. II, v. 12, 271–311. (E 258)

———1761b. Lettre de M. Euler à M. de la Grange ... Recherches sur la propagation des ébranlemens dans une (sic) milieu élastique. *Mélanges de philosophie et de mathématique de la société royale de Turin* 2 (1760/1), 1–10; = *Opera*, s. II, v. 15, 255–263; = *Œuvres de Lagrange*, v. 14, 178–188; = *Opera*, s. IVa, v. 5, 436. (E 268)

———1765a. *Theoria motus corporum solidorum seu rigidorum: ex primis nostrae cognitionis principiis stabilita et ad omnes motus, qui in huiusmodi corpora cadere possunt, accommodata*, 2 v., Rostock: A.F. Ró se. Second edition, 1790; = *Opera*, s. II, v. 3–4. (E 289–290)

———1765b. Recherches sur la connoisance mé canique des corps. *Mémoires de l'académie des sciences* (Berlin) 14 (1758), 131–153; = *Opera*, s. II, v. 8, 178–199. (E 291)

———1765c. Du mouvement de rotation des corps solides autour d'un axe variable. *Mémoires de l'académie des sciences (Berlin)* 14 (1758), 154–183; = *Opera*, s. II, v. 8, 200–235. (E 292)

———1766a. De la propagation du son. *Mémoires de l'académie des sciences* (Berlin) 15 (1759), 185–209; = *Opera*, s. III, v. 1, 428–451. (E 305)

———1766b. Supplément aux recherches sur la propagation du son. *Mémoires de l'académie des sciences* (Berlin) 15 (1759), 210–240; = *Opera*, s. III, v. 1, 452–483. (E 306)

———1766c. Continuation des recherches sur la propagation du son. *Mémoires de l'acadé mie des sciences (Berlin)* 15 (1759), 241–264; = *Opera*, s. III, v. 1, 484–507. (E 307)

———1767a. Recherches sur le mouvement des rivieres. *Mémoires de l'académie des sciences (Berlin)* 16 (1760), 101–118; = *Opera*, s. II, v. 12, 272–288. (E 332)

———1767b. Du mouvement de rotation des corps solides autour d'un axe variable. *Mémoires de l'académie des sciences (Berlin)* 16 (1760), 176–227; = *Opera*, s. II, v. 8, 313–356. (E 336)

———1769. Sectio prima de statu aequilibrii fluidorum. *Novi commentarii academiae scientiarum Petropolitanae* 13 (1768), 305–416; = *Opera*, s. II, v. 13, 1–72. (E 375)

———1770. Sectio secunda de principiis motus fluidorum. *Novi commentarii academiae scientiarum Petropolitanae* 14 (1769), 270–386; = *Opera*, s. II, v. 13, 73–153. (E 396)

———1771. Sectio tertia de motu fluidorum potissimum aquae. *Novi commentarii academiae scientiarum Petropolitanae* 15 (1770), 219–360; = *Opera*, s. II, v. 13, 154–261. (E 409)

———1772. Sectio quarta de motu aeris in tubis. *Novi commentarii academiae scientiarum Petropolitanae* 16 (1771), 281–425; = *Opera*, s. II, v. 13, 262–369. (E 424)

———1776a. Formulae generales pro translatione quacunque corporum rigidorum. *Novi commentarii academiae scientiarum Petropolitanae* 20 (1775), 189–207; = *Opera*, s. II, v. 9, 84–98. (E 478)

———1776b. Nova methodus motum corporum rigidorum determinandi. *Novi commentarii academiae scientiarum Petropolitanae* 20 (1775), 208–238; = *Opera*, s. II, v. 9, 99–125. (E 479)

———1793. De momentis virium respectu axis cuiuscunque inveniendis; ubi plura insignia symptomata circa binas rectas, non in eodem plano sitas, explicantur. *Nova acta academiae scientarum imperialis Petropolitanae* 7 (1780), 191–204; = *Opera*, s. II, v. 9, 387–398. (E 658)

———1862. Anleitung zur Naturlehre. In *Leonhardi Euleri Opera postuma mathematica et physica*, 2 vols., edited by Nikolai Ivanovich Fuss and Paul Heinrich Fuss, v. 1, 279–293. St Petersburg: Eggers; = *Opera*, s. III, v. 1, 16–180. (E 842)

———1892. Letter to Lagrange, 9 November 1762. In *Oeuvres de Lagrange*, v. 14, 217–273; = *Opera*, s. IVa, v. 5, 448–452.

———1911–. *Leonhardi Euleri Opera omnia*, 78 vols. to date.

EULER, LEONHARD, and BENJAMIN ROBINS. 1745. *Neue Grundsätze der Artillerie: enthaltend die Bestimmung der Gewalt des Pulvers: nebst einer Untersuchung über den Unterschied des*

Wiederstands der Luft in schnellen und langsamen Bewegungen aus dem Englischen des Hrn. Benjamin Robins übersetzt und mit den nöthigen Erläuterungen und vielen Anmerkungen versehen von Leonhard Euler, Berlin: Haude; = Opera, s. II, v. 14, 1–409. (E 77)

FERREIRO, LARRIE D. 2007. *Ships and Science, The Birth of Naval Architecture in the Scientific Revolution*. Cambridge, MA: MIT Press.

FONCENEX, DAVIET DE. 1762. Sur les principes fondamentaux de la méchanique. Miscellanea Taurinensia. *Mélanges de philosophie et de mathématiques de la Societé Royale de Turin pour les années 1760–1761*, 299–322.

FONTENEAU, YANNICK. 2008. D'Alembert et Daniel Bernoulli face au concept de travail mécanique. *Bollettino di storia delle scienze matematiche* 28, 201–220.

——2009. Les antécédents du concept de travail mécanique chez Amontons, Parent et D. Bernoulli: de la qualité à la quantité (1699–1738). *Dix-Huitième Siècle* 41, 343–368.

FORBES, ERIC G. 1980. *Tobias Mayer (1723–1762), Pioneer of Enlightened Science in Germany*. Göttingen: Vandenhoeck and Ruprecht.

FOURIER, JEAN BAPTISTE JOSEPH. 1822. *Théorie analytique de la chaleur*. Paris: Didot; = *Oeuvres*, v. 1.

——1888–90. *Oeuvres de Fourier*, 2 vols. Paris: Gauthier-Villars.

FRASER, CRAIG G. 1981. *The Approach of Jean d'Alembert and Lazare Carnot to the Theory of a Constrained Dynamical System*. PhD dissertation, University of Toronto.

——1983a. 'J.L. Lagrange's Early contributions to the principles and methods of mechanics', *Archive for History of Exact Sciences* 28, 197–241.

——1983b. Essay review of (Cannon and Dostrovsky 1981), in *Bulletin of the American Mathematical Society* 9, 107–111.

——1985. 'D'Alembert's principle: the original formulation and application in Jean D'Alembert's *Traité de Dynamique* (1743).' *Centaurus* 28, 31–61, 145–159.

——1989. 'The calculus as algebraic analysis: some observations on mathematical analysis in the 18th century.' *Archive for History of Exact Sciences* 39, 317–335.

——1991. 'Mathematical technique and physical conception in Euler's investigation of the elastica.' *Centaurus* 34, 24–60

FRASER, CRAIG G. and MICHIYO NAKANE. 2003. La meccanica celeste dopo Laplace: la teoria di Hamilton-Jacoib, in *Storia della scienza*, editor-in-chief Sandro Petruccioli, Roma, Istituto della Enciclopedia Italiana, V. 7, pp. 234–243.

FRISI, PAOLO. 1759. Problematum praecessionis aequinoctiorum, nutationis terrestris axis, aliarumque vicissitudinum diurni motus geometrica solutio. In *Paulli Frisii Dissertationum variarum*, 1–81. Lucca: Vincenzo Giuntini.

GRANT, ROBERT. 1852. *History of Physical Astronomy From the Earliest Ages to the Middle of the Nineteenth Century*. London: Henry G. Bohn. Reprinted 1966 by Johnson Reprint Corporation.

GREENBERG, JOHN LEONARD. 1995. *The Problem of the Earth's Shape from Newton to Clairaut. The Rise of Mathematical Science in Eighteenth-Century Paris and the Fall of Normal Science*. Cambridge: Cambridge University Press.

GRIMBERG, GÉRARD. 1995. D'Alembert et les équations d'Euler en hydrodynamique. *Revista da Sociedade Brasileira de Historia da Ciência* 14, 65–80.

——2002. D'Alembert et les équations différentielles aux dérivées partielles en hydrodynamique. In *Analyse et dynamique: études sur l'oeuvre de d'Alembert*, edited by Alain Michel and Michel Paty, 259–315. Sainte-Foy, Québec: Presses de l'Université Laval; Saint-Nicolas, Québec: Distribution de livres, Univers.

GRIMBERG, GÉRARD, WALTER PAULS and URIEL FRISCH. 2008. Genesis of d'Alembert's paradox and analytical elaboration of the drag problem. *Physica D: Nonlinear Phenomena* 237, 1878–1886.

GUICCIARDINI, NICCOLÒ. 2009. *Isaac Newton on Mathematical Certainty and Method.* Cambridge, MA: MIT Press.

HABICHT, WALTER. 1978. Leonhard Eulers Schiffstheorie. In *Leonhardi Euleri Opera omnia*, s. II, v. 21, VII–CXCVI.

HANKINS, THOMAS L. 1965. Eighteenth-century attempts to resolve the vis viva controversy. *Isis* 56, 281–297.

——1967. The reception of Newton's second law of motion in the eighteenth century. *Archives internationales d'histoire des sciences* 78–9, 43–65.

——1970. *Jean d'Alembert Science and the Enlightenment.* Oxford: Clarendon Press.

HELMHOLTZ, HERMANN VON. 1847. *Über die Erhaltung der Kraft.* Berlin: Reimer.

HERMANN, JACOB. 1716. *Phoronomia, sive De viribus et motibus corporum solidorum et fluidorum libri duo.* Amsterdam: Wetstein.

HIEBERT, ERWIN N. 1962. *Historical Roots of the Principle of Conservation of Energy.* Madison State Historical Society of Wisconsin for the Dept. of History, University of Wisconsin.

HUYGENS, CHRISTIAAN. 1888–1950. *Oeuvres complètes*, 22 vols. La Haye: Nijhoff.

——1673. *Horologivm oscillatorium, sive de motu pendulorum ad horologia aptato demonstrationes geometricae*, Paris: Muguet; = *Oeuvres*, v. 18, 68–368

JAKI, STANLEY J. 2000. Review of (Taton and Wilson 1995). *Isis* 91, 329–330.

JOUGUET, EMILE. 1908 and 1909. *Lectures de mécanique. La mécanique enseignée par les auteurs originaux. Première partie La naissance de la mécanique* (1908). *Deuxième partie. L'organisation de la mécanique* (1909). Paris: Gauthier-Villars. Reprinted 1966 by the Johnson Reprint Corporation.

KLINE, MORRIS. 1972. *Mathematical Thought from Ancient to Modern Times.* New York: Oxford University Press.

KUHN, THOMAS S. 1959. Energy conservation as an example of simultaneous discovery. In *Critical Problems in the History of Science*, edited by Marshall Clagett, 73–96. Madison: University of Wisconsin Press.

LAGRANGE, JOSEPH-LOUIS. 1759. Recherches sur la nature, et la propagation du son. *Miscellanea Taurinensia. Mélanges de philosophie et de mathématiques de la Société royale de Turin 1 (1759)*, 18–32; = *Oeuvres*, v. 1, 3–20.

——1762. Application de la méthode précé dent à la solution de différents problèmes de dynamique. *Miscellanea Taurinensia. Mélanges de philosophie et de mathématiques de la Société royale de Turin 2 (1760–61)*, 176–227; = *Oeuvres*, v. 1, 365–468.

——1775. Nouvelle solution du problème du mouvement de rotation d'un corps de figure quelconque qui n'est animé par aucune force accélératrice. *Nouveaux m émoires de l'Académie des sciences et belles-lettres* (Berlin) 4 (1773), 85–120; = *Oeuvres*, v. 3, 579–616.

——1777. 'Essai d'une nouvelle méthode pour résoudre le problème des trois corps', *Recueil des pieces qui ont remporté les prix de l'Académie royale des sciences, depuis leur fondation en M.DCC.XX*, v.9, vi + 126 pages = *Oeuvres* v. 9, 229–331. Lagrange's prize-winning memoir was for the year 1772.

——1782. Théorie de la libration de la lune et des autres phénomènes qui dépendent de la figure non sphérique de cette planète. *Nouveaux mémoires de l'Académie des sciences et belles-lettres* (Berlin) 11 (1780), 203–309; = *Oeuvres*, v. 5, 5–122.

——1783. Mémoire sur la théorie du movement des fluides. *Nouveaux mémoires de l'Acadé mie des sciences et belles-lettres* (Berlin) 14 (1781), 151–198; = *Oeuvres*, v. 4, 695–748.

——1788. *Méchanique analitique.* Paris: Desaint. Second edition, revised and augmented: *Mécanique analytique.* 2 vols. Paris: Courcier, 1811–15; = *Oeuvres*, v. 11–12.

——1867–92. *Oeuvres de Lagrange*, 14 vols. Paris: Gauthier-Villars.

LANGTON, STACY G. 2007. Euler on rigid bodies. In *Leonhard Euler: Life, Work and Legacy*, edited by Robert E. Bradley and C. Edward Sandifer., 195–211. Amsterdam et al.: Elsevier.

LAPLACE, PIERRE SIMON, MARQUIS DE. 1779. Suite des recherches sur plusieurs points du système du monde. *Histoire de l'Académie royale des sciences avec les mémoires de mathématique et de physique pour la même année* (Paris) 78 (1776), 177–310; = *Oeuvres* v. 9, 281–310.

——1785. Théorie des attractions des sphéro ídes et de la figure des planètes. *Histoire de l'Académie royale des sciences avec les mémoires de mathématique et de physique pour la même année* (Paris) 84 (1782), 113–197; = *Oeuvres* v. 10, 339–419.

——1789. Mémoire sur la théorie de l'anneau de Saturne. *Histoire de l'Académie royale des sciences avec les mémoires de mathématique et de physique pour la même année* (Paris) 89 (1787), 249–267; = *Oeuvres* v. 11, 273–292.

——1798. *Traité de mécanique céleste*, vol. 1. Paris: J. B. M. Duprat; = *Oeuvres*, v. 1.

——1878–1912. *Œ uvres complètes de Laplace*, 14 v. Paris: Gauthier-Villars.

LINTON, C. M. 2004. *From Eudoxus to Einstein A History of Mathematical Astronomy*. Cambridge: Cambridge University Press.

MACH, ERNST. 1883. *The Science of Mechanics A Critical and Historical Account of its Development*. First German edition appeared in 1883. Engl.translation 1893 by Thomas J. McCormack, 1893, published by The Open Court Publishing Co., Chicago and London. Many later German and English editions.

MAFFIOLI, CESARE. 1994. *Out of Galileo: The Science of Waters, 1628–1718*. Rotterdam: Erasmus.

MALTESE, GIULIO. 2000. On the relativity of motion in Leonhard Euler's science. *Archive for History of Exact Sciences* 54, 319–348.

MARCOLONGO, ROBERTO. 1905. Notizie sul Discorso matematico e sulla vita di Giulio Mozzi. *Bollettino di bibliografia e storia delle scienze matematiche* 8, 1–8.

MIKHAILOV, GLEB K. 1975. On the history of variable-mass system dynamics. *Mechanics of Solids* 10, 32–40.

——1983. Leonhard Euler und die Entwicklung der theoretischen Hydraulik im zweiten Viertel des 18. Jahrhunderts. In *Leonhard Euler, 1707–1783: Beitr áge zu Leben und Werk*, pp. 279–293. Basle: Birkhäuser.

——1994. Hydrodynamics and Hydraulics. In *Companion Encyclopedia of the History and Philosophy of the Mathematical Sciences*, 2 vols., edited by Ivor Grattan-Guinness, v. 2, pp. 1006–1022. London; New York: Routledge.

——1996. Early Studies on the Outflow of Water from Vessels and Daniel Bernoulli's Exercitationes quaedam mathematicae. In *Die Werke von Daniel Bernoulli*, v. 6, pp. 199–255.

——1999. The origins of hydraulics and hydrodynamics in the work of the Petersburg academicians of the 18th century. *Fluid Dynamics* 34, 787–800.

——2002. Introduction to Daniel Bernoulli's Hydrodynamica. In *Die Werke von Daniel Bernoulli*, v. 5, 17–86.

——2005. Daniel Bernoulli, *Hydrodynamica* (1738). In *Landmark Writings in Western Mathematics 1640-1940*, edited by Ivor Grattan-Guinness, 131–142. Amsterdam: Elsevier.

MORANDO, BRUNO. 1995. 'The golden age of celestial mechanics', in (Taton and Wilson 1995), pp. 211–239.

MOZZI, GIULIO. 1763. *Discorso matematico sopra il rotamento momentaneo dei corpi*. Naples: Stamperia di Donato Campo.

NAKANE, MICHIYO and CRAIG G. FRASER. 2003. The Early History of Hamilton–Jacobi Theory, *Centaurus* 44 (2003), 161–227.

NAKATA, RYOICHI. 2000. D'Alembert's second resolution in *Recherches sur la Précession des Equinoxes*: comparison with Euler. *Historia Scientiarum* 10, 58–76.

NEWTON, ISAAC. 1687. *Philosophiae Naturalis Principia Mathematica*. London. Second and third editions in 1713 and 1726. English translation *Mathematical Principles of Natural Philosophy* by Andrew Motte 1729.

NOWACKI, HORST. 2006. Developments in fluid mechanics theory and ship design before Trafalgar. In *Technology of the Ships of Trafalgar: Proceedings of an International Congress Held at the Escuela Técnica Superior de Ingenieros Navales, Madrid, and the Diputación Provincial, Cádiz, 3–5 November 2005*, edited by Francisco Fernández-González, Larrie D. Ferreiro and Horst Nowacki. Madrid: Escuela Técnica Superior de Ingenieros Navales, Universidad Politécnica de Madrid.

PACEY, A. J. and S. J. FISHER. 1967. Daniel Bernoulli and the vis viva of compressed air. *British Journal for the History of Science* 3, 388–392.

PANZA, MARCO. 2003. 'The origins of analytic mechanics in the 18th century', in H. Niels Jahnke (Ed.), *A History of Analysis* (Providence, RI: American Mathematical Society; London: London Mathematical Society), pp. 137–154.

PASSERON, IRÈNE 1995. *Clairaut et la figure de la Terre au XVIIIe siècle*, Thèse, Université Paris 7.

PEDERSEN, OLAF. 1974. *A Survey of the Almagest*. Odense: Odense University Press.

POINSOT, LOUIS. 1803. *Elemens de statique*. Paris: Voland.

POISSON, SIMÉON-DENIS. 1811. *Traité de Mécanique*. 2 vols. Paris: Courcier. Second edition, revised and augmented: Paris: Bachelier, 1833.

RADELET-DE GRAVE, PATRICIA 1987. Daniel Bernoulli et le parallélogramme des forces. *Sciences et Techniques en Perspective* 11, 1999, 1–20.

——1996. Vers la représentation vectorielle de la force. In *Conferenze e seminari dell'Associazione Subalpina Mathesis*, 1994–1995, edited by E. Gallo, L. Giacardi and O. Robutti, 92–111. Turin: Associazione Subalpina Mathesis.

——1998. Les études de la loi de composition des forces depuis I.H. Westphal (1817) et C. Jacobi (1818) jusqu' à G. Darboux (1875). In *Tecnica e tecnologia nell'architettura dell'Ottocento: atti del quarto seminario di storia delle scienze e delle tecniche*, Venezia, 11 e 12 novembre 1994, edited by Pasquale Ventrice, 109–151. Venezia: Istituto Veneto di Scienze, Lettere ed Arti.

RADELET-DE GRAVE, PATRICIA and DAVID SPEISER. 2004. Comments on the Anleitung zur Naturlehre. In *Leonhardi Euleri Opera omnia*, s. III, v. 3, CIX–CXX.

RAVETZ, J. R. 1961a. 'Vibrating strings and arbitrary functions', in *The Logic of Personal Knowledge Essays Presented to Michael Polanyi on his Seventieth Birthday*, pp. 71–88 (no author, London: Routledge and Paul).

——1961b. The representation of physical quantities in eighteenth-Century mathematical physics. *Isis* 52, 7–20.

ROCHE, JOHN. 1998. *The Mathematics of Measurement: A Critical History*. London: Athlone Press.

ROUSE, HUNTER and SIMON INCE. 1957. *History of Hydraulics*. Iowa City: Iowa Institute of Hydraulic Research, State University of Iowa. Revised second edition 1963.

SEGNER, JOHANN ANDREAS VON. 1755. *Specimen theoriae turbinum*. Halle: Gebauer.

SIMÓN CALERO, JULIÁN. 1996. *La génesis de la mecánica de los fluídos, (1640–1780)*. Madrid: Universidad Nacional de Educación a Distancia. English translation: *The Genesis of Fluid Mechanics, 1640–1780*. Dordrecht: Springer, 2008.

SZABÓ, ISTVÁN. 1977. *Geschichte der mechanischen Prinzipien*. Basle and Stuggart: Birkhäuser Verlag. Second edition 1979. Third edition 1987. Corrected third edition 1996.

TATON, RENÉ and CURTIS WILSON (Editors). 1995. *Planetary Astronomy from the Renaissance to the Rise of Astrophysics. Part B: The Eighteenth and Nineteenth Centuries*. (*General History of Astronomy*, Volume 2.) Cambridge: Cambridge University Press.

TIMOSHENKO, STEFAN. 1953. *History of Strength of Materials: With a Brief Account of the History of Theory of Elasticity and Theory of Structures*. New York: McGraw Hill.

TRUESDELL, CLIFFORD A. 1954. Editor's Introduction: Rational fluid mechanics, 1687–1765. In *Leonhardi Euleri Opera omnia*, s. II, v. 12, VII–CXXV.

———1955. Editor's Introduction: I. The first three sections of Euler's Treatise on fluid mechanics (1766). II. The theory of aerial sound, 1687–1788. III. Rational fluid mechanics 1765–1788. In *Leonhardi Euleri Opera omnia*, s. II, v. 13, VII–CXVII.

———1960 'Editor's Introduction: The rational mechanics of flexible or elastic bodies 1638–1788.' In *Leonhardi Euleri Opera omnia*, s. II, v. 11, pt. 2.

———1968. *Essays in the History of Mechanics*. Dordrecht: Springer.

TRUESDELL, CLIFFORD A. and RICHARD TOUPIN. 1960. The classical field theories. In *Handbuch der Physik*, 226–793, edited by S. Flügge, v. 3, pt. 1. Berlin: Springer.

VARIGNON, PIERRE. 1703. Manière générale de déterminer les forces, les vîtesses, les espaces et les temps, une seule de ces quatre choses étant donnée dans toutes sortes de mouvements rectilignes variés à discrétion. *Histoire de l'Académie royale des sciences avec les mémoires de mathématique et de physique tirés des registres de cette Académie* (Paris) 2 (1700), 22–27.

VILLAGGIO, PIERO. 2008. General introduction. In *Die Werke von Johann I und Nicolaus II Bernoulli*, v. 6, pp. 1–283.

WAFF, CRAIG B. 1976. *Universal Gravitation and the Motion of the Moon's Apogee: The Establishment and Reception of Newton's Inverse-Square Law, 1687–1749*. PhD dissertation, Johns Hopkins University.

———1995. 'Clairaut and the motion of the lunar apse: The inverse-square law undergoes a test', in (Taton and Wilson 1995), pp. 35–46.

WILSON, CURTIS. 1987. D'Alembert versus Euler on the precession of the equinoxes and the mechanics of rigid bodies. *Archive for History of Exact Sciences* 37, 233–273.

———1995a. The precession of the equinoxes from Newton to d'Alembert and Euler. In (Taton and Wilson, 1995, 47–55).

———1995b. The work of Lagrange in celestial mechanics. In (Taton and Wilson, 1995), pp. 108–130.

YOUSCHKEVITCH, A. P. 1976. 'The concept of function up to the middle of the 19th century', *Archive for History of Exact Sciences* 20, 37–85.

ZIMMERMANN, VOLKER. 1996. Introduction to Daniel Bernoulli's medical writings. In *Die Werke von Daniel Bernoulli*, v. 6, pp. 1–46.

CHAPTER 13

LAPLACE AND THE PHYSICS OF SHORT-RANGE FORCES[1]

ROBERT FOX

13.1 THE NEWTONIAN LEGACY

Laplacian physics can be interpreted as an attempt to realize a supposedly Newtonian ideal of a science that would account for all phenomena—on the molecular as well as the terrestrial and celestial scale—in terms of attractive or repulsive central forces acting between the particles of matter. Its instigator was Pierre Simon Laplace, the descendant of a long-established but relatively undistinguished Normandy family who caught the eye of the mathematician Jean le Rond d'Alembert on his arrival in Paris (in 1768, at the age of 19) and rose to prominence as a regular contributor to the Académie royale des sciences in the early 1770s and then as a leading member of the Académie from his election in 1773 (Gillispie, 1997; Hahn, 2005). Over the next twenty years, as his status grew, Laplace came to see himself as completing Newton's unfinished business. And there certainly were passages in Newton's writings that lent substance to his conception both of his own destiny in physics and of the character of his programme for the physical sciences. In the preface to the *Principia* in 1687 Newton referred briefly to the molecular forces of attraction that accounted for the cohesion of solids and liquids. And later in the book, he explored mathematically the consequences of forces that diminished at a rate greater than the normal inverse square of distance and hence were effective over no more than a very small range. However, it was only in the lengthy Query 23 of the Latin edition of the *Opticks* (1706) and in the revised version of the Query, number 31 in the second edition of 1717–18, that Newton returned to the subject. There he speculated about the existence of certain 'attractions' that extended to such small distances as to 'escape observation' (Newton, 1706: 322; Newton, 1717: 351). And he went on to cite the attractions as a possible cause not only of cohesion but also of the capillary rise of liquids, crystallization, and chemical reactions. Newton offered no explanation for these forces beyond his speculation in the General Scholium of 1713 about the effect of a 'most subtle spirit' pervading all matter and the accompanying tentative suggestion that the spirit might account for inter-particulate attraction and other phenomena, including those of electricity and light and the passage of sensation along the nerves (Newton, 1713; Newton, 1729: 393). Nevertheless, the existence of the forces was widely accepted, and they slipped easily into the canon of eighteenth-century Newtonianism.

The programme that Laplace formulated and pursued for over thirty years between the 1790s and his death in 1827 built on these Newtonian foundations. But it also incorporated theories with a less direct Newtonian pedigree. The most important of these were the theories of the imponderable property-bearing fluids of heat (commonly known as caloric), light, electricity, and magnetism. Although beliefs about such fluids were imprecise and certainly not universally shared, the prevailing view during the eighteenth century was that the imponderables consisted of particles which were thought to be mutually repulsive but also to be attracted by ordinary ponderable matter with forces acting over very short, usually described as 'insensibly small', distances. Imponderable fluids in this sense were not to be found in Newton's writings. Yet they had a Newtonian ancestry of sorts in the model of gas structure that Newton had advanced (briefly and as no more than a mathematical hypothesis) in the *Principia* (Newton, 1687: 301–3). According to this model, the particles of gases were stationary and subject to repulsive forces between them that accounted for pressure and the other characteristic properties of the gaseous state. This was precisely the model on which Benjamin Franklin based his theory of the electric fluid, beginning in the 1740s. In a way that was easily carried over into theories of the other imponderables, Franklin believed that almost all the phenomena of electricity could be explained in terms of a repulsive force between the small weightless particles of the electric fluid and a counter-force of attraction between the particles of the fluid and those of ponderable matter (Franklin, 1941, esp. 'Opinions and conjectures concerning the properties and effects of the electrical matter, arising from experiments and observations made at Philadelphia, 1749': 213–36; Cohen, 1956: esp. 285–362). By 1760, Franz Ulrich Theodor Aepinus had postulated a similar structure for his fluid of magnetism (Aepinus, 1759; Aepinus, 1979: 239–41 and, for Roderick Home's comment on Aepinus's debt to Franklin, 107–36), and by 1780, variants on the model were part of standard doctrine with regard to the property-bearing fluids of heat and light as well (Schofield, 1969: esp. 157–90).

Laplace first engaged with questions relating to the imponderable fluids in the context of his early experimental work on heat in the 1770s. He did so under the influence of his slightly older contemporary and friend Antoine Laurent Lavoisier. Although Lavoisier was always cautious in his statements about the nature of heat, he was easily taken for a convinced believer in the caloric theory, especially in his *Traité élémentaire de chimie* (1789), and his work on heats of reaction and animal heat did much to advance the theory's plausibility. The careful agnosticism that Laplace and Lavoisier expressed in their joint 'Mémoire sur la chaleur' of 1783 makes it difficult to determine Laplace's position at this time (Lavoisier and Laplace, 1780: 357–9). But even if, as is possible, Laplace favoured the vibrational theory (according to which heat consisted in the vibration of the ordinary particles of matter), he was converted to the caloric theory, probably soon afterwards and certainly by the mid-1790s.

It was Laplace's achievement to gather together these various strands of eighteenth-century theorizing and to restate them in a mathematical form that promised a way of quantifying theories, especially of the properties and action of the imponderables, that had hitherto been vague and qualitative at best. The result was a style of physics which, though broadly Newtonian in origin or at least inspiration, was distinctively Laplace's own. A clear indication of what later emerged as the true Laplacian programme appeared in the first edition of Laplace's *Exposition du système du monde* in 1796. Here, in good Newtonian fashion, Laplace cited optical refraction, capillary action, the cohesion and crystalline properties of solids, and chemical reactions as all being the result of an attractive force exerted by the ultimate particles of matter, referred to as *molécules intégrantes* or simply *molécules* (not to be confused, then or in this chapter, with 'molecules' in the modern, chemical sense). He looked forward to the day when further experiments would allow the law governing the force to be determined and when, as he put it, 'we shall be able to raise the physics of terrestrial bodies to the state of perfection to which celestial physics has been brought by the discovery of universal gravitation' (Laplace, 1796: vol. 2, 196–8). In Laplace's view, there was good reason to believe that the molecular forces might themselves be gravitational in nature, though with the proviso that, on the molecular scale, they could not be expected to obey the simple inverse-square law. That departure from the norms of universal gravitation, as he argued, was a predictable consequence of the shape of the individual molecules and was entirely compatible with Newtonian principles.

It is important to stress the contrast between the ambitiousness of Laplace's vision of his programme and the familiarity of the established eighteenth-century beliefs from which the programme was constructed. In physics, through the eighteenth century, short-range inter-particulate forces had been applied with more or less success in treatments of optical refraction, capillary action, surface tension, and crystal structure. Likewise in chemistry, they had been invoked in a continuing tradition of work on affinities; for those working in this tradition, chemical phenomena were to be explained in terms of Newtonian-style forces at the molecular level that accounted for the course of chemical reactions and that it was goal of the chemist to quantify (Thackray, 1970: esp. 199–233). Although both British and French Newtonians had contributed to this legacy, it was the French, in particular Alexis Claude Clairaut and Georges-Louis Leclerc, comte de Buffon, who (next to Newton himself) constituted the richest resource for Laplace, and it is by examining their work that we can see most clearly the continuity between Laplace and his Newtonian precursors. Clairaut, for example, confidently ascribed refraction, the roundness of liquid drops, and the elevation and depression of liquids in capillary tubes to gravitational forces that became large at very small (molecular) distances. Buffon, too, had discussed such forces, in his case with special reference to the laws of chemical affinity, which he saw as providing a 'new key' to exploring 'the most profound secrets of Nature' (quoted in Thackray, 1970; 216).

Although Clairaut and Buffon both reinforced their Newtonian credentials by stating explicitly that the forces they postulated were of the same nature as those operating between celestial bodies, they disagreed fundamentally with regard to the relationship between force and distance. The debate in which this disagreement surfaced arose because of the discovery, in 1745, of a substantial discrepancy between the predicted and observed periods of the apogee of the Moon (Heilbron, 1993: 150–7). To remove the discrepancy, Clairaut suggested two years later that the force law should contain a term inversely proportional to the fourth power of the distance, $1/r^4$, in addition to the usual term proportional to $1/r^2$. This additional term, he argued, would not only remove the anomaly with regard to celestial phenomena but also be consistent with the existence of molecular forces that were very large at insensible distances. Buffon's response was one of concern at what he saw as a loss of elegance and simplicity and hence as a threat to the Newtonian system. He duly upheld the familiar $1/r^2$ law, though some years later he admitted that such a law would be modified at short range by the shape of the particles of matter. The details of the debate do not concern us here, and it is sufficient to note that it ended abruptly in 1749, when Clairaut found that, after all, he could reconcile the motion of the Moon's apogee with a simple inverse-square law. With regard to the short-range forces, however, he was careful not to concede victory to Buffon, and the form of the law governing such forces remained an open question at the end of the century, having attracted little further interest. It was left for Laplace to resurrect the problem and to direct attention once again to the issues left unresolved by Clairaut and Buffon.

13.2 THE PROGRAMME DEFINED

From the time he began to articulate his vision of a unified physics embracing the molecular, terrestrial, and astronomical realms in 1796, Laplace never wavered, at least publicly, in his commitment to the programme. Indeed, his commitment became even more assertive with the publication of the fourth volume of the *Traité de mécanique céleste* in 1805. By now, he had gained an influential ally in Claude Louis Berthollet, his intimate friend since the early 1780s, his next-door neighbour at Arcueil on the southern outskirts of Paris from 1806, and a man who saw chemistry in precisely the Newtonian terms that Laplace sought to apply more particularly in physics. The core perception underlying Berthollet's *Recherches sur les lois de l'affinité* of 1801 and his much longer, two-volume *Essai de statique chimique* of 1803 was of a quantified chemistry based on the measurement of the forces between the particles of matter. Like Laplace, Berthollet dug deep in eighteenth-century Newtonianism, both British and French, drawing heavily on the Keills and Hauksbee in Britain and, among French chemists, on Buffon, Macquer, Guyton de Morveau, and Lavoisier. But, also like Laplace, he did far more than reiterate a conventional view. In the end, he did not succeed in answering many (or any) of the outstanding problems of the eighteenth century, and the systematic determination of chemical affinities that good Newtonian chemists saw as their goal remained (in Mimi Gyung Kim's words) an 'elusive dream' (Kim, 2003: esp. 392–438 on Berthollet). Berthollet could even be said to have put an end to the dream by laying bare the true complexity of chemical reactions—a complexity far exceeding that of the interactions between planets and stars. When, as he believed, not only the mass of the reactants but also the quantity of caloric present had to be taken into account, quantification of the processes of chemistry became virtually impossible. Nevertheless, by his rigorous, critical reworking of the Newtonian principles on which his theory rested, Berthollet had succeeded in creating, if only on paper, a potentially coherent system out of a jumble of rather incoherent beliefs.

Despite the daunting challenges of affinity chemistry as articulated in the *Essai de statique chimique*, the forward-looking programmatic tone of the book, reinforced by Berthollet's confident opening remarks about the identity of gravitational and chemical forces (Berthollet, 1803: vol. 1, 1), can only have reinforced Laplace's will to pursue the problems of molecular physics. They seem also to have encouraged René Just Haüy, recently appointed to the chair of mineralogy at the Muséum national d'histoire naturelle (after a precarious passage, as a priest, through the Revolution and the Reign of Terror), to develop a model for acid–alkali reactions that drew on the model of inter-particulate forces. In his discussion of the forces, which appeared in the second edition of his *Traité élémentaire de physique* in 1806 (though not in the first edition, published three years earlier), Haüy even stated that he had treated the reactions 'according to the ideas of the celebrated Laplace' (Haüy, 1806: vol. 1, 43; Mauskopf, 1970: esp. 184, where the phrase appears). The debt, which certainly extended to Berthollet as well, was a substantial one, and it reflected a level of personal contact manifested in the research on capillary action that Haüy undertook at about this time at Laplace's request.

In Laplace's own work, the immediate fruits appeared in the fourth volume of the *Mécanique céleste* (1805) and in two supplements to the volume, published in 1806 and 1807 (Laplace, 1799–1825: vol. 4, 231–81; supplements to the volume paginated independently). They took the form of lengthy mathematical treatments of optical refraction (in Book X, in the main body of volume 4) and capillary action (in the supplements). The assumptions underlying both treatments rested on the existence of short-range attractive forces acting either between the particles of ordinary matter and those of the imponderable light (in the case of refraction) or between the particles of ordinary matter (in the case of capillary action). Laplace's choice of these two subjects for his first sorties into the realm of molecular physics was understandable: both were manifestations of action at very small distances that had attracted the attention of eighteenth-century Newtonians, as well as of Newton himself, and both bore with them problems that remained unsolved in 1805 (Heilbron, 1993: 150–65). Mathematically, refraction had proved particularly difficult, and the study of the molecular forces that caused the phenomenon, although of such obvious interest to Newtonians, had made little progress (Cantor, 1983: 25–90). By contrast, Clairaut's treatment of the theory of capillary action, which Laplace regarded as the only discussion of the phenomenon worthy of serious consideration, had been perceptive. But its brevity and incompleteness left Laplace with an opportunity for a more detailed study. This included, most importantly, an answer to the problem of discovering the law relating inter-particulate force and distance, which remained in much the same state it was in after the inconclusive confrontation between Clairaut and Buffon nearly sixty years earlier. Laplace's demonstration that the form of the force function for insensible (as opposed to very small) distances was unimportant in the case of capillarity allowed him to proceed with his calculations and so made the uncertainty about the law appear less troubling. To that extent it was a minor triumph.

By now, Laplace believed that there was far more to be achieved by a study of short-range forces than the mere tying up of loose ends, and between 1805 and 1807 he appears to have immersed himself totally in the subject. In this period he spoke four times on capillary action before the Institute and engaged others in work designed to confirm and enlarge on his theoretical work. He asked Haüy, the Parisian engineer Jean Louis Trémery, and Joseph-Louis Gay-Lussac, then a young protégé of Berthollet, to undertake experiments on capillary action, and he engaged Jean-Baptiste Biot, his closest disciple at that time, in an experimental investigation of refraction in gases (Laplace, 1799–1825: first supplement to vol. 4, 52–5). The paper of March 1806 that resulted from this last investigation, by Biot and his younger collaborator François Arago, served Laplace's interests well. As Laplace hoped, it fulfilled an initial purpose by confirming the unproved assumption, which was fundamental to his treatment of the refraction of light as a function of altitude and temperature in the *Mécanique celeste*, that the refracting force of air (a Newtonian measure of the attraction of different gases for the particles of light) was proportional to its density (Biot and Arago, 1806: 322–4 and 347–9).[2]

But the paper did much else as well, and this wider theoretical significance was certainly not lost on Biot and Arago themselves. On the nature of light, they unquestioningly used the language of what has been variously described as the emission, corpuscular, or projectile theory. (I shall use the terms emission and emissionist in referring to the theory, though with no strong preference for one or other of these terms.) However, they did so without explicitly committing themselves beyond a statement that their work accorded better with the theory than with Christiaan Huygens' wave theory (Biot and Arago, 1806: 343–4). More originally, and more importantly, they argued for the potential of the Laplacian model for the investigation of matter on a molecular scale hitherto inaccessible to scientific investigation (Biot and Arago, 1806: 338–46). Essential to this extension beyond the field of conventional optics was the analogy they drew between the short-range forces governing chemical reactions and those, between ponderable and imponderable matter, that caused the refraction of light (Fox, 1971: 196–202; Frankel, 1977: 59–64). Although the analogy stopped short of any notion of the identity of the two types of force, it was then an easy step to speculate that the determination of the forces at work in refraction (significantly referred to, in the title of the 1806 paper, as the 'affinities' of matter for light) might be used as a probe into the innermost properties of matter and a way of determining the composition of chemical compounds. The high refractive power of ammonia, for example, might be accounted for by the presence of hydrogen, which had the highest refracting force of any of the gases studied. In suggesting that the physicist measuring the refractive powers of different gases was acting in exactly the same way as a chemist measuring the tendency of a given base to react with a range of acids, Biot and Arago were associating themselves with Berthollet's affinity chemistry; in both cases, the goal was a quantitatively determined series in which substances were ordered in a way that would allow them to be classified and their presence in compounds to be recognized (Biot and Arago, 1806; Fox, 1971: 200, for a key passage from the 1806 paper). In the event, the promise of an optical method of investigating the composition and chemical properties of matter was never fulfilled; it fell victim some years later to a re-examination by Arago and Petit of its underlying principles, including the emissionist theory of light on which it rested (Arago and Petit, 1816). But that future fate should not lead us to underestimate the hopes that were vested in what many contemporaries, Haüy among them, saw as a possible alternative to the difficult path that Berthollet had charted for his own purely chemical approach (Haüy, 1806: vol. 2, 178 and 183–9).

13.3 Arcueil and the Institute

The Laplacian programme could never have exerted the influence that it did that in the years of the Consulate and Napoleon's First Empire had it not been for Laplace's personal authority within the scientific community. In the extent of this authority Laplace was only matched in France by Berthollet, and the two formed a powerful alliance reinforced by the proximity of their houses at Arcueil and, above all, by their shared unwavering zeal for Newtonian science. As someone temperamentally addicted to power, Laplace did not hesitate to use his position in the elite of the Parisian scientific community to promote his beliefs. In this, wealth too played its part. After his appointment as a Senator in 1799, he rose quickly within the Senate, achieving the rank of chancellor in 1803 and an annual salary of 72,000 francs (Hahn, 2005: 131–9). This was enough to effect a dramatic change in his personal life-style and that of his family and to make possible the purchase of the estate at Arcueil, to which he moved in 1806. Material comfort and the income to support it also opened doors to polite society and created possibilities for influence and patronage that no contemporary *savant* in France could match. Laplace, along with Berthollet (now also highly placed in the Senate), duly exploited the opportunities. The young graduates from the École Polytechnique whose careers the two men were able to promote—Biot, Arago, Gay-Lussac, and Siméon-Denis Poisson chief among them—were the beneficiaries both of the facilities at Arcueil (including Berthollet's private laboratory and, from 1807, the regular meetings of the informal but highly select Société d'Arcueil) and of assistance in the securing of the key junior positions on which their careers and freedom to pursue research depended (Crosland, 1967).

A crucial weapon in Laplace's armoury was the system of prize competitions organized by the First Class of the French Institute—the class that had replaced the Académie des sciences when the academies of the Ancien Regime were resurrected in the post-revolutionary settlement of 1795. Laplace served on, and presumably dominated, the five-man committee that in December 1807 proposed a mathematical study of double refraction as the subject for the prize for mathematics to be awarded some two years later. The choice of the subject bore an unmistakable Laplacian thumbprint, and the intention was clear. A study of double refraction would extend Laplace's treatment of refraction in the fourth volume of the *Mécanique celeste* in 1805, while also promising the elucidation of a phenomenon that had defied satisfactory explanation ever since it was noted in crystals of Iceland spar and first described by the Dane Erasmus Bartholin in 1669. Huygens had admitted that the geometric wave construction that he used to analyse double refraction left many observations unexplained; in particular, a theory such as Huygens' that postulated a longitudinal wave motion in an all-pervading ether was incapable of explaining the behaviour of the ordinary and extraordinary rays when passed through a second crystal. Newton was quite properly critical of Huygens' wave-theory account, but his own explanation, in terms of the 'sides' of rays of light, was vague and difficult to reconcile with the emission theory that he favoured (Newton, 1730: 328–36 and 345–9 (Queries 25, 26, and 29)). Even by the end of the eighteenth century little progress had been made, leaving the theory of double refraction with the enticing status of a long-notorious anomaly. Now, though, it was an anomaly that promised to become increasingly troublesome, because of the emergence of double refraction as an important exploratory tool in the investigation of crystal structure. Here once again Haüy, close to Laplace and respected by him, was a key figure. His five-volume *Traité de minéralogie* (1801) had established him as the leader of his generation of French mineralogists, while his access to the mineralogical collections of the École des mines, which he had curated since 1794, provided him with the specimens and technical support that his continuing research required. Hence his statements of what had been achieved in the study of crystals through the study of double refraction and what might be achieved carried weight (Haüy, 1801: vol. 1, 229–35, and vol. 2, 38–51 and 196–229; Haüy, 1806: vol. 2, 334–55; Mauskopf, 1976: 55–68, esp. 56–60). If that was not enough, his cautious endorsement of Newton's explanation of the phenomenon sealed its attractiveness to Laplace (Haüy, 1806: vol. 2, 354–5).

Laplace's role in the setting and judging of the competition on double refraction was such that it came as no surprise that the prize went to Etienne Louis Malus, who had recently caught the eye of Laplace and was soon to become one of the most brilliant and engaged members of the Arcueil circle (Chappert, 1977; Buchwald, 1989:

3–107, 325–7, 346–57, both essential sources on all aspects of Malus's scientific work, though with significant differences of interpretation). It is perfectly conceivable, in fact, that Laplace conceived the subject of the competition with Malus specifically in mind (Frankel, 1974). At all events, from a Laplacian point of view, Malus was an ideal candidate, possessing as he did the mathematical tools of one of the ablest early graduates of the École Polytechnique and a concern for experimental precision, even though he was virtually self-taught in optics. In December 1807, when the prize was set, Malus had only recently emerged from the obscurity of almost a decade of hard military service, first in the Egyptian campaign of 1798–1801 and then in a series of postings in provincial France during which he turned to optics as a refuge from the troubling memories of his time in Egypt and very probably from health problems resulting from the plague that he contracted there. His background meant that he did not come to the competition with the full panoply of Laplacian credentials. This showed in his 'Traité d'optique'—a lengthy paper that he presented before the First Class of the Institute in April 1807 (Malus, 1811a). In the 'Traité', Malus built on the analytic geometry that he had learned as a student from Gaspard Monge in fashioning a purely analytical treatment of reflexion and refraction. While the treatment was perfectly compatible with an emissionist theory in the Laplacian manner, it rested essentially on the notion of rays, without entailing a commitment to a belief in the corpuscles of light or to any other theory.

Laplace, who was one of the Class's referees for the paper, was impressed, and in a paper on double refraction in January 1808 he took the opportunity of praising Malus's work before the Class (Laplace, 1809: esp. 302 and 309). He also set the agenda for potential entrants in the prize competition by establishing certain principles that he considered to be established beyond reasonable doubt. Malus's experiments, as Laplace insisted, had confirmed Huygens' construction for double refraction. But this did not imply the adequacy of Huygens' wave theory for the explanation of the phenomenon, any more than it made the case for the emission theory. For Laplace, now deeply engaged in his programme, the task ahead was clear: it was simply to arrive at an explanation for double refraction in terms of short-range forces of the kind that Newton had invoked in explaining ordinary refraction. Since Laplace himself had already made inroads into the problem, as he showed in his paper of January 1808, and since he left no doubt as to his own belief in the physical reality of the molecular forces, it would have been an act of remarkable folly to approach the subject of the competition in any way that was at odds with Laplace's. And Malus broadly conformed. His use of the language of the emission theory (his reference to 'molécules lumineuses', for example) and of mathematical techniques appropriate to the theory did not amount to an explicit endorsement of Laplace's preferences; the same was true of his cautious statement of the difficulties attending Huygens' wave theory—a position he distinguished from his defence of Huygens' construction and his rejection of Newton's criticisms of it. But few would have read his entry for the competition as anything other than a reinforcement of Laplacian orthodoxy (Malus, 1811b; Chappert, 1977: 157–202; also, for a succinct statement of Jed Buchwald's different view of Malus's corpuscularianism, Buchwald, 1989: 64–6).

By the time Malus received his 3000-franc prize in January 1810 he had made another signal contribution through his discovery of polarization by reflexion (Malus, 1809a). Malus made the discovery by chance in the autumn of 1808. Looking through a crystal of Iceland spar at light reflected from the windows of the Luxembourg Palace, he noticed that as he rotated the crystal the intensity of the two images of the reflected light varied. In Malus's hands, and no doubt with Laplace's encouragement, the observation moved immediately to the forefront of physical debate. As it happened, Malus succumbed to his long history of ill health (in February 1812) before he could complete the publication of the fruits of his supremely skilful application of mathematical analysis and experimental dexterity in the form of a unified theory embracing both polarization and double refraction. How confidently this theory would have endorsed the emissionist doctrines remains an open question. But Buchwald's reconstruction of aspects of Malus's reasoning that are less than clear or straightforward in the sources available to us shows how easily an optical theory founded on the treatment of light as rays, such as Malus's, could be interpreted by contemporaries to imply not only that the rays had a physical reality (as Malus believed) but also that they consisted of a stream of corpuscles, which Malus's analysis did not require (Buchwald, 1989: Part 1 'Selectionism', and the associated appendices). Laplace himself would have been glad to interpret Malus's theory in this way, and he would certainly have perceived the result as another blow against the wave theory.

Malus's work offered so much that could be exploited in favour of Laplacian physics that his premature death must have come as a grievous blow to Laplace. However, there was no question of abandoning the programme. There were still outstanding questions to be tackled. Foremost among these were the problems concerned with the behaviour of elastic surfaces (Todhunter, 1886–93: vol. 1, a classic account of mathematical studies of elasticity in this period). Laplace had already stated that the theory of such surfaces might be established in terms of short-range inter-particulate forces of repulsion, and in 1809 a competition on the subject had been set by the First Class of the Institute. Napoleon appears to have suggested the subject, following a demonstration to him of the vibratory states of stroked glass plates by the German musician and physicist Ernst Florens Friedrich Chladni. But Laplace, whose zeal for his programme was then at its height, would certainly have had a hand in the matter as well. He must therefore have been disappointed that, by the closing date in October 1811, the only serious entry made no mention of the inter-particulate forces on which he believed any sound theory should rest. The entry was by Sophie Germain, an independent self-taught mathematician who had refined her skills through correspondence with Gauss and Legendre (Bucciarelli and Dworsky, 1980: esp. 20–97; Rowlinson, 2002: 107–10). From a starting point rooted in Euler's analysis of the bending of rods and Lagrange's *Méchanique analitique* (1788), Germain's treatment had deficiencies that Lagrange observed and made known to his fellow members of the Académie's panel of judges, who also included the presumably critical Laplace. Predictably, Germain was unsuccessful. When the competition was reset, with a closing date of October 1813, she competed once more but received only an honourable mention. And it was only when the subject was set yet again that she was finally awarded the prize, in January 1816.

Despite Germain's success, the judges (who included Laplace, Poisson, and the sympathetic but never uncritical Adrien Marie Legendre) still had reservations, and it is not hard to imagine that Laplace hoped for a different outcome. His preference would have been for a treatment based on the changes in the inter-particulate forces with distance. This was the approach that Poisson adopted (independently of the competition and, as a member of the committee that proposed the prize, somewhat irregularly) in a paper that he read to the First Class of the Institute in August 1814 (Poisson, 1812). What Bucciarelli and Dworsky have called Poisson's 'molecular mentality' stood a world away from Germain's analysis of the relations between elastic force and the changing shape of a deformed surface (Bucciarelli and Dworsky, 1980: 65–76; also Rowlinson, 2002: 107–10 and, on the contrast between the approaches of Poisson and Germain, Grattan-Guinness, 1990a: vol. 2, 461–70). But, far from backing down, she returned to the fray some years later, using a reworked version of her theory to underline the distance between her own approach and that of Poisson, the 'habile géomètre' and 'savant auteur' whom she criticized but whose name she never mentioned (Germain, 1821: v–x).

The Laplacian cause was better served by another competition, this time on the theory of heat, which the Institute set during the last years of Napoleon's rule (Fox, 1971: 132–48; Heilbron, 1993: 106–17). As the committee that proposed the subject made clear in announcing the prize in January 1811, it was hoped above all that the competition, which asked for a detailed experimental study of the specific heats of gases, with special reference to the law governing their variation with changing volume, would lead to a decision on an important point in the caloric theory. The point was one that had to be resolved before even the simplest mathematical treatment of the theory could be undertaken. Hence it was of great interest to Laplace, and although he did not sit on the committee that set the competition or on the panel of judges, Berthollet and Gay-Lussac (who did) were present on his behalf. The underlying aim, as the announcement of the competition made clear, was to determine whether it was possible for some caloric to exist in a body in a latent, or combined, state (that is, without being detected by a thermometer) or whether all of the caloric was present in its 'sensible' form and therefore as a contribution to the body's temperature. The issue, which had been a live one since Laplace's early work on heat in the 1780s, had divided calorists into two groups; the supporters of the former view looked chiefly to Lavoisier, Laplace, and Joseph Black as their authorities, while those who advocated the latter, such as Adair Crawford and John Dalton, followed the Scottish pupil of Black, William Irvine (Fox, 1971: 20–39, 69–79, 105–21, and 134–55; Chang, 2004: 64–8 and 168–71). Predictably, victory went, in January 1813, to two young men, François Delaroche and Jacques Etienne Bérard, who had performed their experiments at Arcueil and endorsed the position long favoured by Laplace by upholding the distinction between latent and sensible caloric (Delaroche and Bérard, 1813; Heilbron, 1993: 106–17). The only other competitors, Nicolas Clément and Charles Bernard Desormes, who deviated from the Laplacian view, were no lightweights, and their experiments were intricate and carefully performed (Clément and Desormes, 1819). But Clément and Desormes were outsiders to the Laplacian circle. They had little chance, and were fortunate to receive even the consolation of an honourable mention.

13.4 THE FALTERING OF THE PROGRAMME

At the end of the First Empire, Laplacian physics remained, in many respects, in a powerful position. It was being taught as standard doctrine not only in secondary and higher education but also, crucially, at the École Polytechnique, still a source of some of the nation's ablest young physicists. Haüy's *Traité élémentaire de physique*, commissioned at Napoleon's direction as a textbook for use in secondary schools (1803, with several subsequent editions; see Josep Simon's chapter in this volume), Jean Nicolas Pierre Hachette's *Programmes d'un cours de physique* (1809), intended for pupils at Polytechnique, and the school's annually published outline syllabuses all followed orthodox Laplacian lines. Among the most striking evidence of the firmness of the orthodoxy is a set of notes taken at Alexis Thérèse Petit's lectures at Polytechnique in 1814–15 by none other than the future positivist Auguste Comte (Fox, 1974: 107–8). At this stage, Petit, who was to abandon the orthodoxy in the few remaining years of his life (he died prematurely in 1820), showed no sign of questioning the existence of the imponderable fluids of caloric, light, electricity, and magnetism. And he gave other aspects of Laplacian physics, notably the treatment of capillary action, great prominence.

Despite the strong Laplacian presence in textbooks and syllabuses, the uniformity imposed by the programme was by no means complete, nor did the programme determine the boundaries of good physics. In the first place, there were those, outside the Arcueil circle and usually outside the Parisian 'establishment' of science, who opposed Laplace or, more commonly, worked independently of him. Such outsiders included significant figures, among whom were Sophie Germain, whose gender and position outside the academic world were enough to exclude her from Laplace's circle, Clément and Desormes, both of whom were as involved in industrial activity as they were in scientific research, and Joseph Fourier, still enmeshed in administrative duties in the provinces but known for a mathematical treatment of the propagation of heat in solids that challenged Laplace's approach to thermal physics as early as 1807 (Grattan-Guinness and Ravetz, 1972; Herivel, 1975, esp. 149–91; Dhombres and Robert, 2000: 443–620). Moreover, by the time of the Bourbon Restoration, Fresnel was beginning his serious engagement with the wave theory of light after nearly a decade as a civil engineer working far from Paris and with scientific interests focused rather more on chemistry than on physics. And there were many more whose research was taking them in directions irrelevant to Laplace's ideals. For example, the work of Gay-Lussac and Thenard (both members of the Arcueil circle) on the alkali metals and electrochemistry, Gay-Lussac's experiments on the combining volumes of gases (though they had subversive implications), and J. P. Dessaignes's study of phosphorescence (which won the Institute's prize competition for physics in 1809) simply did not bear on the Laplacian programme, nor, in any direct way, on the theories on which the programme was based.

Laplace could probably have lived amicably enough with the growth of these diverse lines of research and even with a measure of opposition to his ideas. But as the Empire drew to a close and finally collapsed with the defeat at Waterloo in June 1815, there were other, more worrying straws in the wind. Guided research activity of the kind that yielded the work of Malus and Delaroche and Bérard slackened after 1812, and regular meetings of the Société d'Arcueil ceased in 1813. The death of Laplace's only daughter in that year brought with it a period of depression, and from 1815 circumstances converged to undermine his influence in the scientific community. His worsening financial position played a part in this. With the fall of the Empire, both Laplace and Berthollet had lost the salaries they had enjoyed as senators, which had contributed to incomes of about 100,000 francs each in 1814; Berthollet's income fell abruptly to 24,000 francs, and Laplace's probably diminished by a comparable amount, leaving both men with incomes still several times that of a professor at the École Polytechnique but inadequate for the sustaining of a private research school (Crosland, 1967: 69–74 and 400; Fox, 1974: 135n). Another weakness lay in the decline of Laplace's personal reputation, which suffered from his pliability with regard to the changing political regimes. In 1814, as a member of the Senate, he voted for the overthrow of Napoleon, who had done so much to advance his interests, and in favour of a restored Bourbon monarchy. After conveniently absenting himself from Paris during Napoleon's temporary return to power in the Hundred Days—an episode that clearly embarrassed him—he remained loyal to the Bourbons until his death. But along the way he became a target for liberal resentment, in particular because of his reputation (possibly exaggerated) for an undue compliance with the reactionary policies of the Bourbon regime on the freedom of the press (Hahn, 2005: 190–204).

However, the most serious assaults on Laplace's standing remained those launched in physics. Germain's victory in the competition on elastic surfaces in 1816 was an early sign of Laplace's reduced capacity to control events. But far more serious were the elaboration and growing acceptance of Fresnel's wave theory of light between 1815 and the early 1820s. In the face of this challenge, which was championed by Arago, a member of the Arcueil circle since about 1810, Laplace's response was muted. He could do little as Fresnel's star rose. A series of papers on diffraction between 1815 and 1818 propelled Fresnel to the forefront of Parisian science, both in the Académie and in the more intimate circles of the dissidents. Early converts to his wave-theory treatment of diffraction included not only Arago but also Petit (by now Arago's brother-in-law) and Ampère, while Gay-Lussac appears at least to have been sympathetic. In an atmosphere of intense interest in what contemporaries immediately perceived as a far broader challenge to Laplacian principles, the Académie des sciences responded in January 1817 (probably at the instigation of Laplace and Biot) by setting a prize competition on precisely Fresnel's favoured area of research: diffraction.

Biot in particular would have looked forward with some confidence to a victory for the emissionist theory, since he and his young collaborator C. S. M. Pouillet had already based their own preliminary explanation of diffraction on emissionist principles and published it as a supplement to Biot's four-volume *Traité élémentaire de physique*, which was in many respects a *summum* of Laplacian physics, especially with regard to optics (Biot, 1816: vol. 4, 743–75). But even with a five-man panel of judges that included Laplace, Biot, and Poisson, Fresnel won the competition in March 1819. His prize-winning essay distilled an approach and conclusions drawing on his studies of a wide variety of optical phenomena, including polarization and stellar aberration as well as diffraction. The core of Fresnel's work, though, remained his concern with vibrations in an all-pervading ether, which he saw as the foundation for an optics that would avoid the constraints and paradoxes of the emissionist theory. It was from this model, with waves conceived as longitudinal vibrations, that he developed his principle of interference and fashioned a mathematical treatment to account for the interaction between waves in the vicinity of a diffracting edge (Fresnel, 1821–2).

Fresnel's model stood as a challenge not only to the emission theory but also and more generally to the whole notion of property-bearing fluids on which Laplacian physics rested. Laplace himself can only have been disappointed. Yet he remained somewhat detached from the resulting debate. As the debate proceeded, Fresnel's theory of diffraction came to be measured less against Laplace's writings than against the theory of Biot and Pouillet. And it was Biot and Poisson, not Laplace, who continued open resistance to the wave theory into the 1820s, fighting vainly against the powerful unifying conception of the ether that Fresnel elaborated mathematically in his own work. It is even conceivable that Biot and Poisson, possibly Laplace as well, were responsible for a seven-year delay in the publication of Fresnel's prize-winning paper and deliberately mislaying other papers of his. But the evidence amounts to no more than an unsubstantiated recollection some years later by William Whewell (Whewell, 1837: vol. 2, 408–11). In any case, such underhand tactics would have been incapable of stemming the tide of scientific opinion that was running against Laplacian principles. The tide was running strongly in favour of physicists working in ways that either opposed or simply ignored these principles. Fresnel went on to modify his model, moving (from 1821) towards one based on transverse waves and applying his insights and mathematical methods to polarization and double refraction. With regard to thermal phenomena, Dulong and Petit had incorporated a critical discussion of contemporary explanations of heats of chemical reaction in terms of a fluid of heat in announcing their law of atomic heats in 1819 (Petit and Dulong, 1819; Fox, 1968–9: esp. 9–16), and in 1822 Fourier's *Théorie analytique de la chaleur* offered a book-length demonstration of how the propagation of heat could be analysed mathematically without reference to the nature of heat.

New discoveries also played their part in undermining the whole notion of physical science founded on the notion of short-range forces. Hans Christian Oersted's observation of the magnetic effect associated with the passage of an electric current along a wire in 1820 raised two immediate difficulties for physics cast in this form. One was that the force on the compass needle in the vicinity of the wire was rotational (and not central), the other that it implied (contrary to Coulomb's, and hence also to Laplacian, belief) that electricity and magnetism could interact with each other. André-Marie Ampère (an early convert to Fresnel's wave optics who had never been associated with Laplace's circle) and Biot both investigated the phenomenon, and the two men soon came into conflict (Blondel, 1982; Caneva, 1980: esp. 54–9 and 106–7; Grattan-Guinness, 1990a: vol. 2, 917–67, esp. 923–7 and 945–7; Hofmann, 1995; Darrigol, 2000: 1–41). Ampère's explanation, which embraced not only electromagnetism but also the forces between magnets, rested on a notion of physical fluids, allied to a belief in the 'unifying conceptual role' of an all-pervading ether, very different from Coulomb's two fluids of magnetism, which Biot incorporated in his own explanation (Caneva, 1980: esp. 125, where the quoted phrase appears). Whereas Ampère treated electromagnetic and magnetic forces in terms of interactions between current-carrying conductors, Biot conceived electromagnetism as a purely magnetic phenomenon, explained by forces between the tiny magnets that he supposed to be arranged in a circular fashion around any current-carrying wire. Biot's theory, expounded in a succession of papers presented to the Académie and, in its definitive form, in the third edition of his *Précis élémentaire de physique expérimentale* in 1824 (Biot 1824, vol. 2: 766–71), was seriously flawed, and Ampère's demonstration of its weaknesses and of the properly Newtonian character of his own 'etherian' theory ensured its subsequent neglect (Ampère, 1826: 180–8 for a flavour of Ampère's criticism of Biot; Hofmann, 1995: 230–5 and 268–82).

Although the new departures in electrodynamics bore on aspects of physics to which Laplace himself had not paid particular attention, Ampère's criticism of Biot's work set him firmly at odds with the Laplacian cause. Biot had made himself vulnerable by trying to 'wedge' Oersted's discovery into one of the 'conceptual pigeonholes' of Laplacian physics (the expressions are James Hofmann's; see Hofmann, 1995: 278), and he had failed. Less confrontationally but in much the same spirit, Ampère's work also presented a challenge to Poisson, whose belief in two-fluid theories of both electricity and magnetism and the independence of electrical and magnetic phenomena seems to have been as firm in 1824, when he treated the theory of magnetism in two lengthy papers, as it had been in 1812, when he published his work on the distribution of electricity on the surface of a conductor (Poisson, 1811; Poisson, 1821–2a; Poisson, 1821–2b; Home 1983; Grattan-Guinness 1990a: vol. 1, 496–513; Hofmann, 1995: 113–18 and 324–50). Since Poisson's work on both magnetism and electricity bore the unmistakeable imprint of Laplace, his contributions could only be interpreted as those of an unrepentant disciple, and to that extent Ampère's respectful but critical handling of the two papers on magnetism assumed the status of yet another attack on the Laplacian programme in general.

Laplace must certainly have interpreted Ampère's exchanges with Biot and Poisson in that light and recognized that they signalled a still mounting scepticism with regard to his style of theorizing across the board of physics. Nevertheless, he had one more major contribution to make. And he duly made it, developing the most elaborate version of one of his flagship theories: the caloric theory of gases. He did so in a series of papers that he presented before the Académie des sciences and published in the Bureau des longitudes's *Connaissance des tems* between 1821 and 1822 and (in a definitive form) in the fifth volume of the *Mécanique céleste* in 1823 (Laplace, 1799–1825: vol. 5, 87–144, (in Book XII, published independently in April 1823, despite the date of 1825 on the title-page); Fox, 1971: 165–77). His theory drew on early work, including the treatment of gravitation that he had published in the first volume of *Mécanique céleste* in 1799. Now, almost a quarter of a century on, he adapted his expressions for the gravitational forces between spherical bodies to a discussion of the standard Newtonian model of gas structure. The modification that the forces between gas-particles were forces of repulsion, rather than of attraction, and that they were operative only over a short range was easily integrated, as was the essential Newtonian assumption that the repulsive force was inversely proportional to the distance between the particles. From that point, however, his argument moved on to shakier ground. In his first paper, read to the Académie des sciences in September 1821, Laplace used a discussion of the equilibrium of a spherical shell taken at random in a gas in showing that the pressure of a gas, P, was proportional to $\rho^2 c^2$ where ρ was its density and c the quantity of heat contained in each of its particles. The argument had a highly speculative cast. One unfounded assumption was that the repulsive force between any two adjacent gas particles was proportional to c^2. Even more suspect was Laplace's model of the dynamic equilibrium that he believed to exist when the temperature was constant and the particles of the gas both radiated and absorbed caloric at an equal rate. Postulating the simplest of mechanisms, Laplace pictured the radiation from any particle as resulting from the mechanical detachment of the particle's caloric by incident radiant caloric, the density of which, $\pi(t)$, he took as a function and hence a true measure of the temperature, t, alone. Setting the fraction of incident caloric absorbed equal to q (a constant depending solely on the nature of the gas) and assuming (quite gratuitously) that the quantity of caloric detached was proportional both to c and to the total 'density' of caloric in the gas, ρc, Laplace could deduce that

$$\rho c^2 = q \pi(t)$$

The result was rich in consequences. With P α $\rho^2 c^2$, it led easily to Boyle's law, as well as to Dalton's law of partial pressures. It could also be shown that since the function $\pi(t)$ was independent of the nature of the gas, all gases must expand to the same extent for a given increment in temperature, as Dalton (1801) and Gay-Lussac (1802) had observed. The next step, of assuming $\pi(t)$ to be proportional to the (absolute) temperature as measured on the air thermometer, appeared soon after the paper of September 1821 and was axiomatic in the definitive version that Laplace presented in the *Mécanique céleste* in 1823.

The fact that Laplace's theory was consistent with the main gas laws lent it a measure of superficial plausibility as an attempt to give the caloric theory the quantitative character that it had always conspicuously lacked. Yet it would have escaped no-one that the assumptions he made concerning the dependence of P and $\pi(t)$ on ρ and c were not independently verifiable; they were clearly determined by the requirement that the deductions made from them should agree with the gas laws. More importantly, while the expressions involving c and the other hypothetical entities that entered into his calculations created the air of mathematical rigour and precision, the reality was that much of Laplace's argument rested on far simpler premises than he intimated. This point emerged in 1823 from two papers by the loyal Poisson (Poisson, 1823a; Poisson, 1823b). In the papers, Poisson reviewed some key aspects of Laplace's theory of heat, including his identification of the error in Newton's expression for the velocity of sound in air, $v = \sqrt{P/\rho}$. More than twenty years earlier, Laplace and Biot had explained the error by the quasi-adiabatic effect of heating and cooling (unknown to Newton) in the areas of successive contraction and rarefaction that constituted a sound wave; then, in 1816, Laplace had arrived at the correct expression $\sqrt{\gamma P/\rho}$, where $\gamma = c_p/c_v$, the ratio between the specific heats of the air at constant pressure and constant volume, and finally in 1822 experiments by Gay-Lussac and the chemist and industrialist Jean-Jacques Welter (almost certainly undertaken at Laplace's request) had provided the independent determinations of γ that served to confirm the theory (Fox, 1971: 161–5 and 170–7; Kuhn, 1958: 81–6 and 168–77; Grattan-Guinness, 1990a: vol. 2, 811–20; Heilbron, 1993: 166–84). Now, in 1823, Poisson derived the familiar expressions for adiabatic changes in volume, $TV^{\gamma-1} =$ constant, and $PV^\gamma =$ constant. It was a remarkable succession of achievements for the Laplacian school. Yet Poisson's papers of 1823 carried a sting in the tail, since he arrived at his equations without any mention of Laplace's highly suspect mechanisms, now shown (by implication) to be unnecessary. Poisson worked simply from the premiss that the heat content of a gas was a function of its pressure and density alone. Even to a reader convinced of the physical reality of caloric, that more limited assumption must have appeared a far surer starting point than the ones that Laplace offered in the *Mécanique céleste*.

Given Laplace's diminished standing and the difficulties besetting his style of physics, it is not surprising that his work on caloric was received with almost total indifference. It aroused neither overt opposition nor support, and it stimulated no further research. The personal doubts that Laplace must have felt may have been reflected in his apparent preference for Fourier over Biot in the election for one of the two posts of permanent secretary of the Académie des sciences in 1822 (Fox, 1974: 126). Yet he was not ready to abandon his programme. Words that appeared in the fifth volume of the *Mécanique céleste* in the following year were not those of someone who had lost heart. Following a brief résumé of the properties of caloric, Laplace wrote:

By means of these assumptions, the phenomena of expansion, heat, and vibrational motion in gases are explained in terms of attractive and repulsive forces which act only over insensible distances (*distances imperceptibles*). In my theory of capillary action I related the effects of capillarity to such forces. All terrestrial phenomena depend on forces of this kind, just as celestial phenomena depend on universal gravitation. It seems to me that the study of these forces should now be the chief goal of mathematical philosophy. I even believe that it would be useful to introduce such a study in proofs in mechanics, laying aside abstract considerations of flexible or inflexible lines without mass and of perfectly hard bodies. A number of trials have shown me that by coming closer to nature in this way one could render these proofs no less simple and far more lucid than by the methods used hitherto. (Laplace, 1799–1825, vol. 5: 99)

The passage presents a Laplace looking forward to further implementation of his programme. And that eye on the future was given further substance in 1824 in a brief preliminary 'Avertissement' to the fifth edition of *Exposition du système du monde*, in which Laplace wrote that he intended to make molecular forces the subject of a special supplementary treatise (Laplace, 1824: vol. 1, v).

The withdrawal from the fifth edition of the chapter on molecular attraction and a related chapter on universal gravitation that had appeared in the fourth edition (Laplace, 1813: 308–57) was evidently part of Laplace's plan for the fuller treatment to come. But the intention was never fulfilled. No supplementary treatise ever appeared. In the octavo version of the sixth edition of the *Système du monde* (which appeared in 1827, the year of Laplace's death) the two chapters were still missing, and they were then simply reinstated in the quarto edition, confusingly also designated the sixth, in 1835 (Laplace, 1835). No one seems to have noticed. With the great days of Arcueil an increasingly distant memory, there were few who chose even to attack, let alone endorse, Laplace's approach to physics. Internal dissension among those who had once been faithful followers only aggravated the problem. The animosity between Arago and Biot was especially bitter, extending as it did not only to disagreement on the nature of light but also to a wide range of other ideological and scientific issues (recently examined by Theresa Levitt; Levitt, 2009). Things on this particular front came to a head after 1822, when Biot virtually withdrew from the scientific community, following his defeat by Fourier in the election in the Académie—a defeat for which he plausibly laid some of the blame on Arago. Thereafter, he remained aloof from the public life of science, though not inactive, for some years, leaving Poisson to carry the banner of Laplace's programme into the late 1820s and 1830s.

Yet even Poisson was far from being an uncritical admirer. His papers of 1823 laid bare the gratuitousness of the detailed mechanisms in Laplace's version of the caloric theory, and Ivor Grattan-Guinness has shown how markedly Poisson's mathematical techniques departed from those of Laplace through the 1820s and 1830s (Grattan-Guinness, 1990b: 59–62). It is typical of Poisson's view of Laplacian principles at this time that when he returned to the problem of capillarity in 1831, he retained the fundamental principle of 'molecular attraction' but did not hesitate to criticize and correct a number of shortcomings in Laplace's theory (Poisson, 1831: 1–8). In a similar vein, in his *Théorie mathématique de la chaleur* four years later he followed a clear opening commitment to caloric rather than wave motion in an ether for the understanding of the transmission of heat (Poisson, 1835: 7), with over 500 pages of dense mathematical argument in which he totally ignored—as he had done in 1823—Laplace's elaborate definitive version of his theory.

13.5 THE LAPLACIAN LEGACY

The evidence of the gathering dismemberment of the once coherent structure of Laplacian physics after 1815 is abundant, with respect to teaching as well as to research. With Petit (1815–20) and then Dulong (1820–30) in the chair of physics at the École

Polytechnique, the statement that light would be treated as an 'emission from luminous bodies' disappeared from the annual syllabuses, while the term 'caloric' came to be systematically replaced by the more neutral 'heat' (Fox, 1974: 125–6). More generally, in reaction against the extravagance of Laplacian hypotheses, physicists displayed an understandable caution with regard to theorizing, encapsulated in Fourier's famous statement of 1822 that 'First causes (*les causes primordiales*) are unknown to us; but they are subject to simple unvarying laws which can be discovered by observation and the study of which is the object of natural philosophy' (Fourier, 1822: 1). Berthollet's chemistry suffered in a similar way and for similar reasons. It fell victim not to a frontal assault but rather to the difficulty of building on Berthollet's precepts for an impossibly complex form of affinity chemistry and to a recognition that Dalton's atomic theory and the electrochemistry of Davy and Berzelius offered more fruitful foundations for research.

In this ebbing away of power and influence, age also took its toll. By the early 1820s, both Laplace and Berthollet were comfortably in their seventies, and Berthollet died in 1822, leaving no committed pupil or collaborator to continue his work. Other, broader trends have to be taken into account. The last decade of Laplace's life coincided with a growing caution among physicists with regard not just to the imponderable fluids but to unobservable entities in general. The caution sometimes had the appearance of full-blown positivism, and the suspicion of models of the kind that characterized Laplace's approach to physics can certainly not be dissociated from the positivist turn in philosophy. It is surely no coincidence, for example, that Auguste Comte began articulating his philosophy in precisely these years, taking Fourier's rational mechanics as a model of how physics could be pursued—and pursued effectively—without reference to unobservable causal entities of which caloric was a prime example. Yet it would be wrong to associate the decline of Laplacian influence uniquely with the rise of positivism. Comte's philosophical prescriptions may (and, in my view, should) be seen as an historically significant expression of contemporary mistrust of excessively bold theorizing. But they exerted little influence on the contemporary course of physics. When Gabriel Lamé and Jean Marie Constant Duhamel pursued their mathematical analyses of heat transfer in the late 1820s and 1830s, for example, they were simply following Fourier's principle of ignoring causes or at least of reducing speculation on causes to a minimum (Lamé, 1836–7: vol. 1, ii–iii for a flavour of the methodology pursued by both Lamé and Duhamel; also, for a classic study, Bachelard, 1927: 89–132). They were not positivists in the Comtean sense, any more than Jean-Baptiste Dumas was when he expressed caution, amounting to scepticism, with regard to the existence of atoms and even advocated erasing the word 'atom' from science in the seventh of his *Leçons sur la philosophie chimique* in 1836 (Dumas, 1837: 258–90, esp. 290). By inference and implication, they were responding to the demise of a style of physics that they and many others regarded as having run its course.

So what, if anything, can we identify as a distinctively Laplacian legacy? In heat, electricity and magnetism, and optics, the tradition of analysis in terms of particulate mechanisms just faded; although both Fresnel and Augustin Louis Cauchy retained and worked with significant vestiges of a molecular ether, work in anything approaching a Laplacian style gave way in the middle decades of the century to field theories and the new science of energy physics. That, at least, was the general pattern. However, the pattern was complex. Early classical thermodynamics as it was formulated about the mid-century essentially broke with or bypassed molecular hypotheses. Yet through the 1850s and 1860s the emerging (or re-emerging) kinetic theory of gases made such hypotheses a tempting line of enquiry in certain explorations of the new science, notably by August Karl König, Rudolf Clausius, and James Clerk Maxwell (Brush, 1976: vol. 1, 165–89). The chemistry of affinities enjoyed no such revival; once it was no longer sustained by Berthollet's personal influence, its failings as a realistic mode of chemical enquiry ensured its definitive demise. Of studies of capillarity, on the other hand, it could be said that Laplace's ideas were part of an enduring legacy; in John Rowlinson's words, they are recognizably a 'legitimate ancestor' of much subsequent work in the field (Rowlinson, 2002: 102) and can properly be said to have 'dominated' the subject for at least the rest of the nineteenth century (Heilbron, 1993: 165).

A similar though necessarily more nuanced case can be made with regard to the elasticity of solids. Here, as Rowlinson argues, tensions between molecular and macroscopic interpretations endured for the rest of the nineteenth century. Non-molecular approaches certainly grew in favour, most notably (and influentially) among practising engineers, for whom the mechanical properties of solids were of special interest (Rowlinson, 2002: 110–25). But Poisson, deeply if not unconditionally loyal to the Laplacian flame, did not yield. Turning once again to elasticity, he devoted much of his energy in his later years (he died in 1835) to the pursuit of what Grattan-Guinness has described as a 'modified molecularism' (Grattan-Guinness, 1990a: vol. 2, 968–1045, esp. 1015–42). In this he was working, as he so often did, against the grain, remaining wedded to the general thrust, though not always the detailed elements, of Laplace's approach. Classic work by him in the late 1820s—notably a long paper that he read to the Académie des sciences in 1828—reaffirmed his commitment to 'molecular actions' not only as the key to the understanding of elasticity but also as essential if 'analytical' mechanics in the Lagrangian manner were to be extended to embrace the 'physical mechanics' to which he was wedded (Poisson, 1829: 361).

It was a characteristically bold declaration. But Poisson's discussion of the exact form of the molecular forces was vague and did not lend itself to sustained development. Crucially, it failed to win support from such contemporaries in the field as the engineer Claude Louis Navier and the mathematician Cauchy, both of them former students of his at the École Polytechnique. Even though Navier's approach, begun in 1819, entailed the assumption of inter-particle forces, its whole tenor was macroscopic, while Cauchy's priority was the analysis of the elasticity of a continuous medium, albeit with occasional rather unconvincing reversions to a molecular model (Grattan-Guinness, 1990a: vol. 2, 969–1015; Rowlinson, 2002: 110–26). So if something resembling the Laplacian model survived in the study of elasticity, it assumed a secondary role and no longer defined the cutting edge of research on the subject.

In 1852, Lamé warned of the dangers inherent in the quest for a unification of physics founded on uncritical acceptance of the principle of Newtonian universal gravitation extending from celestial phenomena to capillarity and elasticity on the terrestrial scale (Lamé, 1852: 332–5; cited in Rowlinson, 2002: 119–20). Though written as a concluding comment to a work in which Lamé had actually used a molecular as well as a continuum model, it was as categorical a rejection of the unifying ambitions of the Laplacian programme as could be imagined.

While Laplace's authority diminished markedly in the last decade of his life, leaving his ideal of a physics of short-range forces to survive only in the very limited, fragmented form it had by the mid-century, we should avoid the temptation to define his legacy purely in terms of an ambitious programme that could be said eventually to have failed. In a longer perspective, Laplace can be seen as a major figure in the so-called second scientific revolution, provided we allow that revolution to have the extended profile that Stephen Brush and Enrico Bellone, among others, have claimed for it. Brush sees the revolution (commonly regarded as an affair of the very end of the nineteenth century and early twentieth century) as lasting from the beginning of the nineteenth century until about 1950 (Brush, 1976: vol. 1, 35–51). For Bellone, in his book *The World on Paper*, the revolution was, if anything, even more protracted (Bellone, 1980). It consisted in the transition from the quest for explanation in terms of qualitative mechanical models of the kind favoured in the seventeenth century to an ideal of understanding expressed in mathematical terms and tempered by empirically controlled theory. The revolution in this sense really got under way in the late eighteenth century, just when Laplace was emerging as a major force in science and beginning to fashion his vision of a quantitative physics founded on a union of mathematics and fine experiment. Both it and Laplace's work encapsulated the 'quantifying spirit' of the late Enlightenment (Frängsmyr, Heilbron, and Rider, 1990), and its legacy of confidence in the combined power of mathematical methods and observational precision lived on through the foundational years of thermodynamics and electrodynamics and into the age of non-deterministic physics in the early twentieth century.

This is not to claim originality or revolutionary intent for everything that Laplace did. Far from it. Laplace was proud to see himself as attending to Newton's unfinished business and (albeit rather patchily) to acknowledge his debt to the work of others, notably of Lagrange in mathematics. In Charles Gillispie's judgement, he saw himself as a 'vindicator' and improver of the science he inherited from the past rather than as an innovator or revolutionary (Gillispie, 1997: 271–8). Yet the mark he left on physics, as on astronomy and probability, was profound and lasting. His mathematical virtuosity and attention to experimental detail, allied to an immense power of concentration and a prodigiously broad command of contemporary and past science, made him a towering figure in his day. (See Fig. 13.1.) Through his own work and that of the *protégés* and colleagues he inspired and encouraged, he did much to bring together the then disparate traditions of mathematics and *physique expérimentale*. Hence in a longer perspective, he stands as one of the leading pioneers of what we know as the discipline of mathematical physics. No one in the early nineteenth

Fig. 13.1. Pierre Simon, Marquis de Laplace, in later life. The words beneath the portrait convey the honours that Laplace received, including the Grand Cross of the Legion of Honour, the highest rank in the order, to which he was appointed in May 1825. Courtesy of Académie des sciences/Institut de France.

century perceived more clearly the power of the union of mathematical and experimental techniques, and no one worked harder or more effectively to bring the union to fruition.

References

Aepinus, F. U. T. (1759). *Tentamen theoriae electricitatis et magnetismi*. St Petersburg: Typis Academiae scientiarum.

——(1979). *Aepinus's Essay on the Theory of Electricity and Magnetism. Introductory Monograph and Notes by R. W. Home. Translation by P. J. Connor*. Princeton, NJ: Princeton University Press.

Ampère, André Marie (1826). *Théorie des phénomènes électro-dynamiques uniquement déduite de l'expérience*. Paris: Mequignon-Marvis and Brussels: Dépôt général de la librairie médicale française.

ARAGO, DOMINIQUE FRANÇOIS and ALEXIS THÉRÈSE PETIT (1816). 'Sur les puissances réfractives et dispersives de certains liquides et des vapeurs qu'ils forment', *Annales de chimie et de physique*, 1: 1–9.

BACHELARD, GASTON (1927). *Etude sur l'évolution d'un problème de physique. La propagation thermique dans les solides*. Paris: Librairie philosophique J. Vrin. Reissued in 1973.

BELLONE, ENRICO (1980). *A World on Paper. Studies on the Second Scientific Revolution*, trans. Mirella and Riccardo Giacconi. Cambridge, MA and London: MIT Press.

BERTHOLLET, CLAUDE LOUIS (1803 [an XI]). *Essai de statique chimique*, 2 vols. Paris.

BIOT, JEAN-BAPTISTE and FRANÇOIS ARAGO (1806). 'Mémoire sur les affinités des corps pour la lumière, et particulièrement sur les forces réfringentes des différens gaz', *Mémoires de la Classe des sciences mathématiques et physiques de l'Institut national de France*, 7, first semester 1806 [published 1806]: 301–387.

——(1816). *Traité de physique expérimentale et mathématique*, 4 vols. Paris.

——(1824). *Précis élémentaire de physique expérimentale*, 3rd edn., 2 vols. Paris: Deterville.

BLONDEL, CHRISTINE (1982). *A.-M. Ampère et la création de l'électrodynamique (1820–1827)*. Paris: Bibliothèque nationale.

BRUSH, STEPHEN G. (1976). *The Kind of Motion We Call Heat. A History of the Kinetic Theory of Gases in the 19th Century*, 2 vols. continuously paginated. Amsterdam, New York, and Oxford: North-Holland.

BUCCIARELLI, LOUIS L. and NANCY DWORSKY (1980). *Sophie Germain: An Essay in the History of the Theory of Elasticity*. Dordrecht, Boston, and London: D. Reidel.

BUCHWALD, JED Z. (1989). *The Rise of the Wave Theory of Light. Optical Theory and Experiment in the Early Nineteenth Century*. Chicago and London: University of Chicago Press.

CANEVA, KENNETH L. (1980). 'Ampère, the etherians, and the Oersted connexion', *The British Journal for the History of Science*, 13: 121–138.

CANTOR, GEOFFREY (1983). *Optics after Newton. Theories of Light in Britain and Ireland, 1704–1840*. Manchester: Manchester University Press.

CHANG, HASOK (2004). *Inventing Temperature. Measurement and Scientific Progress*. New York: Oxford University Press.

CHAPPERT, ANDRÉ (1977). *Etienne Louis Malus (1775–1812) et la théorie corpusculaire de la lumière. Traitement analytique de l'optique géométrique, polarisation de la lumière et tentative d'explication dynamique de la réflexion et de la réfraction*. Paris: Librairie philosophique J. Vrin.

CLÉMENT, NICOLAS and CHARLES BERNARD DESORMES (1819). 'Détermination expérimentale du zéro absolu de la chaleur et du calorique spécifique des gaz', *Journal de physique*, 89: 321–346 and 428–455.

COHEN, I. BERNARD (1956). *Franklin and Newton. An Inquiry into Speculative Newtonian Experimental Science and Franklin's Work in Electricity as an Example Thereof* [*Memoirs of the American Philosophical Society*, vol. 43]. Cambridge, MA: Harvard University Press.

CROSLAND, MAURICE P. (1967). *The Society of Arcueil. A View of French Science at the Time of Napoleon I*. London: Heinemann.

DARRIGOL, OLIVIER (2000). *Electrodynamics from Ampère to Einstein*. Oxford: Oxford University Press.

DELAROCHE, FRANÇOIS and JACQUES ETIENNE BÉRARD (1813). 'Mémoire sur la détermination de la chaleur spécifique des gaz', *Annales de chimie et de physique*, 85: 72–110 and 113–182.

DHOMBRES, JEAN and JEAN-BERNARD ROBERT (2000). *Fourier. Créateur de la physique mathématique*. Paris: Belin.

DUMAS, JEAN-BAPTISTE (1837). *Leçons sur la philosophie chimique, professées au Collège de France. Recueillies par M. Binau* [*sic*, misprint for *Bineau*]. Paris: Bechet jeune.

FOURIER, JEAN-BAPTISTE JOSEPH (1822). *Théorie analytique de la chaleur*. Paris: Firmin Didot, père et fils.
FOX, ROBERT (1968–69). 'The background to the discovery of Dulong and Petit's law', *The British Journal for the History of Science*, 4: 1–22.
——(1971). *The Caloric Theory of Gases from Lavoisier to Regnault*. Oxford: Clarendon Press.
——(1974). 'The rise and fall of Laplacian physics', *Historical Studies in the Physical Sciences*, 4: 89–136.
——(1990). 'Laplacian physics', in R. C. Olby, G. N. Cantor, J. R. R. Christie, and M. J. S. Hodge (eds.), *Companion to the History of Modern Science*. London and New York: Routledge, 278–294.
FRÄNGSMYR, TORE, J. L. HEILBRON, and ROBIN E. RIDER (1990). *The Quantifying Spirit in the Eighteenth Century*. Berkeley, Los Angeles, and Oxford: University of California Press.
FRANKEL, EUGENE (1974). 'The search for a corpuscular theory of double refraction: Malus, Laplace and the prize competition of 1808', *Centaurus*, 18: 223–245.
——(1977). 'J. B. Biot and the mathematization of experimental physics in Napoleonic France', *Historical Studies in the Physical Sciences*, 8: 33–72.
FRANKLIN, BENJAMIN (1941). *Benjamin Franklin's Experiments. A New Edition of Franklin's Experiments and Observations on Electricity*, ed. I. Bernard Cohen. Cambridge, MA: Harvard University Press.
FRESNEL, AUGUSTIN JEAN (1821–2). 'Mémoire sur la diffraction de la lumière', *Mémoires de l'Académie royale des sciences de l'Institut de France*, 5 (1821–1822 [published 1826]), 339–475.
GERMAIN, SOPHIE (1821). *Recherches sur la théorie des surfaces élastiques*. Paris.
GILLISPIE, CHARLES COULSTON, with the collaboration of Robert Fox and Ivor Grattan-Guinness (1997). *Pierre-Simon Laplace. A Life in Exact Science*. Princeton, NJ: Princeton University Press. Originally published as an article in C. C. Gillispie (ed.), *Dictionary of Scientific Biography*, 16 vols. (New York: Charles Scribner's Sons, 1970–80), vol. 15, 273–403.
GRATTAN-GUINNESS, IVOR (1990a). *Convolutions in French Mathematics, 1800–1840. From the Calculus and Mathematics to Mathematical Analysis and Mathematical Physics*, 3 vols. continuously paginated. Basel: Birkhäuser.
——(1990b). 'Small talk in Parisian circles, 1800–1830: mathematical models and continuous matter', in Gert Konig (ed.), *Konzepte des mathematisch Unendlichen im 19. Jahrhundert*. Göttingen: Vandenhoeck and Ruprecht, 47–63.
——in collaboration with J. R. Ravetz (1972). *Joseph Fourier 1768–1830. A Survey of his Life and Work, based on a Critical Edition of his Monograph on the Propagation of Heat, presented to the Institut de France in 1807*. Cambridge, MA, and London: MIT Press.
HACHETTE, JEAN NICOLAS PIERRE (1809). *Programmes d'un cours de physique. Ou précis de leçons sur les principaux phénomènes de la nature, et sur quelques applications des mathématiques à la physique*. Paris: Veuve Bernard.
HAHN, ROGER (1990). 'The Laplacean view of calculation', in Frängsmyr, Heilbron, and Rider *Quantifying Spirit in the Eighteenth Century*, 363–380.
——(2005). *Pierre Simon Laplace. A Determined Scientist*. Cambridge, MA and London: Harvard University Press.
HAÜY, RENÉ-JUST (1801). *Traité de minéralogie*, 5 vols. Paris: Louis.
HAÜY, RENÉ JUST (1806). *Traité élémentaire de physique*, 2nd edn., 2 vols. Paris: Courcier.
HEILBRON, J. L. (1993). *Weighing Imponderables and other Quantitative Science around 1800*. Supplement to *Historical Studies in the Physical and Biological Sciences*, 24, part 1.
HERIVEL, JOHN W. (1975). *Joseph Fourier. The Man and the Physicist*. Oxford: Clarendon Press.
HOFMANN, JAMES R. (1995). *André-Marie Ampère*. Cambridge: Cambridge University Press.
HOME, RODERICK W. (1983). 'Poisson's memoirs on electricity: academic politics and a new style in physics', *The British Journal for the History of Science*, 16: 239–259.

Kɪᴍ, Mɪ Gʏᴜɴɢ (2003). *Affinity, that Elusive Dream. A Genealogy of the Chemical Revolution.* Cambridge, MA, and London: MIT Press.

Kᴜʜɴ, Tʜᴏᴍᴀs S. (1958). 'The caloric theory of adiabatic compression', *Isis*, 49: 132–140.

Lᴀɢʀᴀɴɢᴇ, Jᴏsᴇᴘʜ-Lᴏᴜɪs (1788). *Méchanique analitique.* Paris: Veuve Desaint.

Lᴀᴍᴇ́, Gᴀʙʀɪᴇʟ (1836–7). *Cours de physique de l'École Polytechnique*, 2 vols. (the second in two separately paginated parts). Paris: Bachelier.

——(1852). *Leçons sur la théorie mathématique de l'élasticité des corps solides.* Paris: Bachelier.

Lᴀᴘʟᴀᴄᴇ, Pɪᴇʀʀᴇ Sɪᴍᴏɴ (1796 [an IV]). *Exposition du système du monde*, 1st edn., 2 vols. Paris: Imprimerie du Cercle-social.

——(1799 [an VII]–1825). *Traité de mécanique celeste*, 5 vols. Paris: J. B. M. Duprat.

——(1809). 'Mémoire sur les mouvements de la lumière dans les milieux diaphanes', *Mémoires de la Classe des sciences mathématiques et physiques de l'Institut de France.* Paris: J. B. M. Duprat. Année 1809 [published 1810], 10 300–342

——(1813). *Exposition du système du monde*, 4th edn. Paris: Veuve Courcier.

——(1824). *Exposition du système du monde*, 5th edn., 2 vols. (octavo). Paris: Bachelier.

——(1827). *Exposition du système de monde*, 6th edn. [octavo], Paris and Brussels. On the difficulty of determining what constituted the true sixth edition, see Gillispie 1997: 282.

——(1835). *Exposition du système du monde*, 6th edn. [quarto], Paris: Bachelier.

Lᴀᴠᴏɪsɪᴇʀ, Aɴᴛᴏɪɴᴇ Lᴀᴜʀᴇɴᴛ, and Pɪᴇʀʀᴇ Sɪᴍᴏɴ Lᴀᴘʟᴀᴄᴇ (1780). 'Mémoire sur la chaleur', *Mémoires de mathématique et de physique, tirés des registres de l'Académie royale des sciences.* Année M.DCCLXXX: 355–408. The paper, read before the Académie in June 1783, was not published until 1784, when it appeared in the delayed volume of the *Mémoires* for 1780.

Lᴇᴠɪᴛᴛ, Tʜᴇʀᴇsᴀ (2009). *The Shadow of Enlightenment. Optical and Political Transparency in France, 1789–1848.* Oxford: Oxford University Press.

Mᴀʟᴜs, Eᴛɪᴇɴɴᴇ Lᴏᴜɪs (1809a). 'Sur une propriété de la lumière réfléchie', *Mémoires de physique et de chimie, de la Société d'Arcueil*, 2: 143–158.

——(1809b). 'Sur une propriété des forces répulsives qui agissent sur la lumière', *Mémoires de physique et de chimie, de la Société d'Arcueil*, 2: 254–267.

——(1811a). 'Traité d'optique. Première partie. Des questions d'optique qui dépendent de la géométrie', *Mémoires présentés à l'Institut national des sciences, lettres et arts par divers savans et lus dans ses assemblées. Sciences mathématiques et physiques*, 2: 214–302. This text differs in certain important respects from the earlier version, published in *Journal de l'École Polytechnique*, 14e cahier, vol. 7 (1808): 1–44 and 84–129.

——(1811b). 'Théorie de la double réfraction', *Mémoires présentés à l'Institut national des sciences, lettres et arts par divers savans et lus dans ses assemblées. Sciences mathématiques et physiques*, 2: 303–508.

Mᴀᴜsᴋᴏᴘғ, Sᴇʏᴍᴏᴜʀ H. (1970). 'Haüy's model of chemical equivalence: Daltonian doubts exhumed', *Ambix*, 17: 182–191.

——(1976). *Crystals and Compounds. Molecular Structure and Composition in Nineteenth-Century French Science* [*Transactions of the American Philosophical Society*, new ser., vol. 66, part 3]. Philadelphia, PA.

Nᴇᴡᴛᴏɴ, Isᴀᴀᴄ (1687). *Philosophiae naturalis principia mathematica.* London: Joseph Streater (by order of the Royal Society).

——(1706). *Optice. Sive de reflexionibus, refractionibus, inflexionibus and coloribus lucis libri tres.* London: Samuel Smith and Benjamin Walford.

——(1713). *Philosophiae naturalis principia mathematica*, 2nd edn. Cambridge: Nicholas Crownfield.

——(1718). *Opticks. Or, a Treatise of the Reflections, Refractions, Inflections and Colours of Light*, 2nd edn. London: W. and J. Innys, With a prefatory 'Advertisement II' dated 16 July 1717.

——(1729). *The Mathematical Principles of Natural Philosophy*, trans. Andrew Motte. 2 vols. London: Benjamin Motte.

——(1730). *Opticks. Or, a Treatise of the Reflections, Refractions, Inflections and Colours of Light*, 4th edn. London: William Innys.

PETIT, ALEXIS THÉRÈSE and PIERRE LOUIS DULONG (1819). 'Recherches sur quelques points importans de la théorie de la chaleur', *Annales de chimie et de physique*, 10: 395–413.

POISSON, SIMÉON DENIS (1811). 'Mémoire sur la distribution de l'électricité à la surface des corps conducteurs', *Mémoires de la Classe des sciences mathématiques et physiques de l'Institut impérial de France. Année 1811, Première partie* [published 1812]: 1–92.

——(1812). 'Mémoire sur les surfaces élastiques', *Mémoires de la Classe des sciences mathématiques et physiques de l'Institut imperial de France. Année 1812, Première partie* [published 1814]: 167–225.

——(1821–22a). 'Mémoire sur la théorie de magnétisme', *Mémoires de l'Académie royale des sciences de l'Institut de France*, 5 (1821–2 [published 1826]): 247–338.

——(1821–22b). 'Second mémoire sur la théorie du magnétisme', *Mémoires de l'Académie royale des sciences de l'Institut de France*, 5 (1821–2 [published 1826]): 488–533.

——(1823a). 'Sur la vitesse du son', *Connaissance des tems . . . pour l'an 1826*: 257–277.

——(1823b). 'Sur la chaleur des gaz et des vapeurs', *Annales de chimie et de physique*, 23: 337–352.

——(1829). 'Mémoire sur l'équilibre et le mouvement des corps élastiques', *Mémoires de l'Académie royale des sciences de l'Institut de France*, 8: 357–570.

——(1831). *Nouvelle théorie de l'action capillaire*. Paris: Bachelier père et fils.

——(1835). *Théorie mathématique de la chaleur*. Paris: Bachelier.

ROWLINSON, J. S. (2002). *Cohesion. A Scientific History of Intermolecular Forces*. Cambridge: Cambridge University Press.

SCHOFIELD, ROBERT E. (1969). *Mechanism and Materialism. British Natural Philosophy in the Age of Reason*. Princeton, NJ: Princeton University Press.

THACKRAY, ARNOLD (1970). *Atoms and Powers. An Essay on Newtonian Matter-Theory and the Development of Chemistry*. Cambridge, MA: Harvard University Press; London: Oxford University Press.

TODHUNTER, ISAAC (1886–93). *A History of the Theory of Elasticity and of the Strength of Materials from Galilei to the Present Time*. Edited and completed by Karl Pearson, 2 vols. (2nd in two parts, separately paginated). Cambridge: Cambridge University Press.

WHEWELL, WILLIAM (1837). *History of the Inductive Sciences*, 3 vols. London and Cambridge.

NOTES

1. I am grateful for the opportunity of reviewing and reworking material that I have published elsewhere on Laplace's physics and the school that he and Berthollet led at Arcueil. Mike Schwartz, Contracts Copyright and Permissions Supervisor at Princeton University Press, has kindly allowed me to draw freely on two sources: Fox (1974) and my contributions to Gillispie (1997). Lizzy Yates, on behalf of the Taylor and Francis Royalties Department, has granted similar permission with respect to Fox (1990).

2. The quantity measured in the experiments was refractive index, μ. Laplace followed Newton in defining 'refracting force' as $(\mu^2 - 1)$—a measure of the increase in the square of the velocity of the particles of light on their passage from air to the refracting medium. This force, conceived as gravitational in nature, was thought to vary with the density ρ of the substance of the medium. For Laplace, as for Newton, 'refractive power', equal to $(\mu^2 - 1)/\rho$, was the quantity most closely related to the nature of the substance and characteristic of it alone.

CHAPTER 14

ELECTRICITY AND MAGNETISM TO VOLTA

JED Z. BUCHWALD

14.1 Speculation

By the end of the first decade of the twentieth century, electromagnetism, based on propagated actions between fundamental particles, or electrons, exhibited the three cardinal characteristics of a widely held physical scheme in the modern era: it was intensely mathematical; it had deep roots in the laboratory; and it was based on a physical hypothesis that connected closely to these other two desiderata. Until well into the eighteenth century the subjects of electricity and magnetism, which were usually kept apart from one another, were scarcely mathematical at all, had only the flimsiest of connections to the laboratory, and were above all loci of elaborate speculations.

   Unlike optics, neither electricity nor magnetism had roots, as subjects, in mathematics. The effects associated with them had traditionally been treated among Aristotelians as parts of physics, requiring therefore the elaborate taxonomic dissections of their place in the lexical schemes of the schoolmen that were also applied to, for example, bewitchment or pain. By virtue of their placement here, rather than in the realm of mixed mathematics, these effects were not considered to be the sorts of things that exhibited the incorruptible permanence thought to be essential for the use of geometry. Although that kind of view had radically changed by the second half of the seventeenth century, vestiges of it nevertheless remained that continued to separate electricity and magnetism from mathematics, and that made it difficult as well to unite them fruitfully to the conception that reliable knowledge can be generated by forcing nature to perform uncommon tasks—the philosophy, that is to say, of experiment.

During this century the very idea of what an experiment might be was only slowly and with difficulty developed and propagated. On the other hand the antique speculative tradition, which sought to plumb the world's essence, remained very much a desideratum despite the rapidly growing rejection of Scholasticism. Seventeenth-century scholars did not usually engage in penetrating dissections of the proper natures of things, or at least these kinds of discussions were not so common as they had once been. Neither did they quantify or produce controlled laboratory effects in our two areas.

Until the late 1600s electricity as a subject reduced to what has been aptly termed the 'amber effect', in which light objects move towards rubbed amber. At century's beginning the Englishman William Gilbert broadened the class of objects that could produce the effect and at the same time introduced a fundamental distinction between it and the properties of the lodestone, or magnet, that resulted in a separation of the two subjects that prevailed until nearly the middle of the nineteenth century. In many respects still a Scholastic, Gilbert considered magnetic actions to be the effects of similarities or contrarieties between a specific magnetic nature or soul that some bodies may have. Gilbert's traditionalism in this respect was, however, balanced by two comparatively novel characteristics of his work: first, his great interest in producing working analogues, or actual models, of the magnetic Earth using spherical lodestones—a form of investigative experimentalism that permeates his work and that does not sit altogether well with his concern for essential natures; second, his insistence that the electric effect does not involve such things, that it derives instead from the purely mechanical action of a sticky effluvium emitted by certain kinds of bodies when rubbed. In his words, 'Electrical motions become strong from matter, but magnetick from form chiefly'. Magnetic bodies come together or push apart in mutual sympathy or antipathy by their very natures; rubbed electrical bodies send out tentacles to rein in their passive neighbours.

Gilbert's Scholastic understanding of magnetism contrasts markedly with what seems to be a quasi-mechanical understanding of electricity, the latter being more congenial to the post-Scholastic way of thinking about nature. However, a modern glancing back at Gilbert's work might vice versa be confused by his sophisticated experimental manipulation of terellae as opposed to the comparative poverty of his electrical work. This reflects two things: first, that there was as yet no firm union between physical discussion and experimental manipulation, no consensual understanding of how to generate trustworthy knowledge about physics from experiment, and, second, the continuing belief that the business of the natural philosopher is to provide understanding of causes, however novel the causes may be in a particular case.

This did not change radically during the century, though the causes offered for electricity did mutate quite markedly, and though quite sophisticated observations were made, particularly during the 1620s by the Italian Niccolò Cabeo, who objected to Gilbert's effluvial gripping and offered an explanation based on motions of the air stimulated by the rushing effluvia. Cabeo's conception excited some experimental work that employed the novel air-pump after mid-century, but by that time the immense, encompassing scheme of the French mathematician and philosopher, René Descartes, was rapidly bringing all such things within its purview, which if anything had the effect of entrenching the traditional attempt to provide understanding through speculation, with experiment adding at most a demonstration or illustration of things thought on other, prior grounds to hold true.

Every effect in the Cartesian scheme reflects a motion of the space-filling (indeed, space-defining) continuum. Space can be divided in several ways, producing as it were particles of various shapes and sizes. Screw-shaped magnetic particles may thread their way through appropriately shaped bores in certain bodies, driving out air between the bored objects (and so forcing them together) if they are aligned with their threads twisted in opposite ways, or else forcing bodies with parallel twists together as the screwed particles rush from one into the next. Electric bodies had their own peculiarly shaped channels which tended to confine active particles; these could be freed by rubbing to lodge in similar bodies and then return home.

Descartes' mechanical structure, which referred everything to motions or confinements of shaped particles, obviously differed considerably from Gilbert's gluey effluvia, since for Descartes, gluiness had itself to result from motion, confinement, and shape. Nevertheless, Cartesian knowledge remained similar to Scholastic knowledge in one important sense: both had their seat in the *a priori*, respectively concerning essential nature or primary qualities. Both embraced descriptions of the empirical world, but neither constructed itself out of such things. It is therefore not surprising that Cartesians did not generate a vibrant experimental programme. In many ways this did not change substantially until well into the eighteenth century, despite the increasing spread of English experimental philosophy under the influence of Newton and his followers.

14.2 EXPERIMENT INTRUDES

14.2.1 Kinds of Electric Objects

Neither Newton's mathematical natural philosophy nor his signal development of experiment in optics had much immediate influence on prevailing opinions concerning electricity and magnetism. For one, canonical experimental devices like the Newtonian prism had not as yet been produced for these subjects. Because argument could not orbit about such a device and its behaviour the subjects remained without centres in the laboratory. For another, nothing like the Newtonian mass-point had been developed for either electricity or magnetism, which left the subjects without the kind of clarifying foundation that Newton had so thoroughly exploited in the first and third books of his *Principia*. This at first left the subjects for the most part where they had been during the previous century—in the realm of explanation and demonstrations designed to illustrate speculation.

Demonstration experiments were however developed, in electricity, to a high art. At the London Royal Society Francis Hauksbee, under Newton's chairmanship, produced the first of a long line of eighteenth-century electrical machines—in his case a spinning, evacuated glass globe that was excited by holding a hand to it. Limp threads hung inside the globe pointed stiffly inwards on excitation (when the globe also glowed), and this Hauksbee took, *contra* Descartes, to show literally the presence of taut threads of electric matter penetrating inwards—a view challenged a few years later by Stephen Gray, whose own work Hauksbee used as he saw fit.

Gray spent some time trying to produce electrification in the usual ways in metals, long thought to be impossible. Two decades later (1729) he discovered that he could do so if the metals were brought into the vicinity of an already excited glass object. Pursuing this line of investigation, he found that he could communicate the electric effect to long distances provided that the communicating wires were themselves suspended by something (silk) that did not work as a good communicator. Gray, initially an outsider to the burgeoning Newtonian community, thereby fashioned what some historians regard as the first central experimental development in electricity, one that could only with difficulty be fit into prevailing effluvial conceptions. Perhaps more important than this ill fit between effluvia and travelling virtue, Gray had produced the first device (his communicating wires) that could be used to fabricate new knowledge. Together with the descendants of Hauksbee's spinning globe, Gray's wires might be said to have for the first time constituted the electric laboratory. This made it possible for the subject to be dealt with by the increasingly numerous proponents of experimental knowledge in essentially the same manner that, for example, they dealt with the air-pump or Newton's prisms: as a subject that must not be constructed on the basis of demonstrative, *a priori* knowledge; as something that instead had to be thoroughly based on the behaviour of devices fabricated in the laboratory.

Gray's wires and the burgeoning production of electric machines were for the most part used either to entertain in variants of old demonstration 'experiments' or else to construct processes similar to ones that had been generated for the past century. However, in the early 1730s Charles François Dufay, then Intendant of the Jardin du Roi, became aware of Gray's work. Taking off from it, Dufay produced in his laboratory two classes of objects in respect to electricity: those that can be electrified by friction and those that cannot be. In respect to electrification by contact, he discovered the eponymously named 'Rule of Dufay', according to which nearly anything could show electric effects by touching it to an already excited body, that metals are strongest in this respect, but that the object had to rest on a third body that was an electric *per se* of sufficient thickness in order to be excited.

Dufay's two classes—electrics *per se* and non-electrics, corresponding to bodies that could or could not, respectively, be electrified by rubbing—together with his rule produced that essential characteristic for all laboratory-based science: an instrumentally-founded classification of objects in respect to the subject under investigation. Further experimental work, as well as conceptual developments, could be molded about this framework. Arguments could be developed that relied upon these distinctions and that connected strongly to devices whose behaviour could now, in some respects at least, be treated as comparatively unproblematic.

In the early 1730s Jean Antoine Nollet became Dufay's assistant. In 1746 he produced a grand electric synthesis in his *Essai sur l'électricité des corps* that, while strongly connected to Dufay's classifications and rules, nevertheless exhibited that same spirit for speculative systematization which had been for so long the rule and model in natural philosophy. Nollet envisioned a world filled with electric stuff. Like all things Cartesian, Nollet's electric matter defined a space and acted primarily by displacing other matter, electric or otherwise, from its path. Present in all bodies, this electric fabric can be set into motion by rubbing (of electrics) or by contact (with non-electrics). Once stimulated, this catholic material flows out from the excited object; but since there can never be any voids in nature, at the same time other electric matter must flow back into the object, thereby keeping space filled. Basing his intuition on impressive German productions of brush discharge, Nollet conceived that the outgoing, or effluent, flows emerge from comparatively few points on the body's surface, each fanning out therefrom like a fast-moving jet of liquid. The incoming, or affluent, streams move in much more slowly because they penetrate over the broad reaches of the body's surface between the points of effluence.

Nollet's scheme, much more closely tied to reproducible effects than most others before it, concentrated particularly on the motions of small objects near electrically excited bodies. These motions could be nicely mapped in his system, as could Dufay's influential discovery that small bodies move away from electrified ones, after first moving to them, on contact (stimulated to emit effluvia by the fast effluent streams of the exciting electric, the small bodies are carried away by mechanical interactions between the two streams). However, like every other system before it whose purpose was primarily to produce understanding, Nollet's was only peripherally related to the fabrication of novel facts in the laboratory. It could explain everything then known; it could not, or at least certainly did not, impel further experimental work, though it was capable of purely qualitatively accommodating nearly anything that involved electrically stimulated motions. The spirit of system was still an overwhelmingly powerful presence among natural philosophers; prestige and material rewards continued to accrue to the successful systematizer, and Nollet became famous.

14.2.2 The Leiden Jar Renovates the Electric Laboratory

Gray's novelties, Dufay's rules and classifications, and the proliferation of increasingly large and elaborate electric rubbing machines had certainly produced a regime under which electric experimenters operated in a commonly agreed manner. Bodies electric and non-electric, action followed by contact and repulsion, communication of effect by non-electrics, these were used as the basis for further laboratory claims without usually generating controversy. The rapid and wide acceptance of Nollet's system, which strikingly embodied these rules and devices, testifies to their comparatively unproblematic status by the 1740s. But an explanatory scheme bound so closely to a specific set of rules and devices as to amount nearly to an intellectual embodiment of them not only has difficulty birthing new and unrelated processes, it also has difficulty incorporating them. The fabrication of an electric object (the Leiden jar) whose behaviour had little to do with previous devices, and nothing to do with the kinds of things that captured Nollet, accordingly posed difficulties.

Invented by Ewald von Kleist in 1745 (according to one historian as a result of his search for 'a portable sparking machine'; Heilbron, 310), whose reports did not enable its reproduction, the device was independently fabricated by Andreas Cunaeus, who informed the Leiden professor of natural philosophy, Pieter van Musschenbroek, of it. The latter generated a recipe for constructing the device, which enabled its reproduction throughout the laboratories of Europe. Nollet, for one, generated the effect with little difficulty.

The Leiden jar itself went through many variations, but in its early form consisted of a bottle or globe of glass partly filled with water; a metal wire in contact with an excited electric dipped into the water. To excite the device required holding the globe in one hand only. Subsequently, to activate it required touching the wire with one hand while continuing to hold the device in the other, producing thereby a powerful shock, one vastly larger than any frictional machine had by itself ever produced.

The central difficulty that the jar posed for Nollet's system derived from its method of excitation. Recall the Rule of Dufay, according to which objects are excited by placing them on electrics *per se*—which is precisely what is not done with the Leiden jar. Because Nollet's system was so closely integrated with Dufay's classifications and requirements, the Leiden jar posed immediate and powerful difficulties. Nothing in his scheme could have led Nollet to anticipate the powerful bottle; nothing in it enabled him easily to accommodate that power. He was in the end able to deal with the jar only by making special allowance for it. This again illustrates the major characteristic of his system: namely, that it was intended to provide comprehensive understanding rather than knowledge aimed at productions in the laboratory.

In 1747 an obscure American from Philadelphia named Benjamin Franklin produced a new approach to electricity that, unlike Nollet's (which was just then becoming influential) kept far from system and that was powerfully bound to the laboratory, in particular to the Leiden jar. The foundation of Franklin's scheme was its entirely novel conception of how electrified objects interact with one another. In Nollet's system, as indeed in essentially all effluvialist schemes, the electric matter is ubiquitous, lying in bodies and flowing through the space between them. When a body becomes electrified it shoots out effluvia, and takes them in, setting up a perpetual commotion. Although the effluvial matter (electricity) never vanishes from the universe, nor is it now being created, nevertheless the amount of electricity that a body possesses has no bearing on effluvialist accounts, which either require the amount to be constant or else make no direct use of quantity. These systems might therefore be said to conserve electric matter, but only in a sense that has essentially no experimental consequences.

Franklin took his stand on what might be called the laboratory conservation of charge. According to him, whenever one object loses a quantity of electric matter some other object must gain an equal quantity. That single principle of conservation, creatively applied, soon produced an avalanche of work, primarily because it integrated closely to the Leiden jar. According to Franklin, glass as a body remains perpetually saturated with electricity. If electricity is thrown onto one side of the jar, then a precisely equal amount must flow out the other side to maintain saturation. Glass, however, resists this shift, which reveals itself as the tremendous shock that occurs when the inner and outer surfaces are brought mediately into contact with one another. Franklinists could create and explain a vast range of experiments that depend upon some object giving electricity to another object. An example drawn from the Franklinist presentation in the first edition of the *Encyclopaedia Britannica* conveys the power:

Place a person on a cake of wax, and present the wire of the electrified phial to touch, you standing on the floor and holding it in your hand. As often as the subject touches it, he will be electrified plus; and any one standing on the floor may draw a spark from him. The fire, in this experiment, passes out of the wire into the subject; and, at the same time, out of your hand into the outside of the bottle. Give him the electrical phial to hold, and touch the wire; as often as you touch it, he will be electrified minus, and may draw a spark from any one standing on the floor. The fire in this case passes from the wire to you, and from him into the outside of the bottle.

Systematists like Nollet had nothing at all comparable to deploy in the laboratory, since their explanations were almost always singular and after the fact. Franklin's plus-and-minus, grounded in laboratory charge conservation, made possible the transformation of electricity into a quantitative, experimental science.

14.2.3 Quantity, Intensity, and Newtonian Calculations

In the early 1750s Nollet nevertheless responded forcefully to Franklin's claims; Franklinists had difficulties answering all of his critiques. On the whole, those who continued to think that the Leiden jar should be treated as an addition to the standing body of electric effects remained with Nollet; those who became convinced that electric science should revolve about the jar became Franklinist. The spread of Franklinism in the face of powerful resistance was in no small measure due to its association with a novel and compelling technology, the lightning rod, which, in conjunction with the theory's intense concentration on jar processes, seemed to endow it with the kind of manipulative capacity (and practical utility) that Nollet's system lacked. The power of pointed rods, strongly advocated by Franklinists, became a subject of political argument during the 1770s, but by that time effluvialism had waned markedly among natural philosophers, and the laboratory focus that underpinned Franklinism had become a widespread desideratum.

Franklinism had never been without its problems, most of which stemmed from Franklin's embrace of the traditional notion that electric matter must in many respects behave very much like ordinary matter: it must, among other things, extend markedly through space. Franklin had recourse to these subsidiary aspects when faced with demands coherently to accommodate such effects as electric repulsion, which is to say the sorts of things that the effluvialists had built their own systems upon. Here there was still no question of producing novel experiments but of embracing effects that the opposing system could handle. To do so, Franklin conceived that electricity thrown onto a body formed about it an extended, mechanically capable atmosphere, and that the atmospheres of two electrified bodies do not mingle but rather push one another apart. Franklin was well aware that bodies electrified negatively also repel one another, and he simply refused to provide an explanation here. His scheme was built upon the Leiden jar as the canonical laboratory device; it could not easily deal with bodily motions engendered by electrification, nor could it deal with induction phenomena, since it tended to assimilate the latter to the same cause that accounted for the former: namely, the mechanical behaviour of electric atmospheres.

Franklin was no more of a mathematician than his adversary Nollet. However, Franklinist doctrine, with its grounding in laboratory charge conservation, was much more amenable to quantification than the explanatory effluvialism of Nollet, except for those parts of it (its deployment of mechanical properties) that remained traditional. Mathematics was brought to bear when in the late 1750s a comparative outsider, Franz Aepinus, removed the atmospheres and provided an avenue for further laboratory investigation and mathematization. While thinking about a puzzling experiment brought to his attention by Johan Wilcke, Aepinus realized that it could be understood if an air gap could act like the glass in a Leiden jar. Experiments undertaken with Wilcke confirmed this hypothesis, which rapidly led Aepinus to abandon electric atmospheres, since the primary Franklinist locus for electric matter (namely glass) could now be replaced by something as insubstantial as air.

In 1759 Aepinus's *Tentamen* was published, in which electric science became quantitative. Aepinus treated electricity as a Newtonian fluid—that is, as a fluid whose parts are self-repulsive according to some force that acts directly between them and that depends upon the distance. In addition, he argued that there is an electric repulsion between ordinary material particles, and an attraction between them and the electric fluid. Without specifying the form of the force law, Aepinus was able through judiciously chosen assumptions to obtain quantitative results for a considerable range of experiments. Aepinus produced in addition (indeed, this was his primary focus in the *Tentamen*) a new and influential account of magnetism, in which he insisted on the separate existence of a Newtonian magnetic fluid. This magnetic fluid differs from electricity in that it moves with great difficulty through bodies like iron, which therefore tend to hold magnetic charge in position.

Aepinus did not know the law of force between electric or magnetic particles, but even if he had he would probably not have proceeded much farther. He generated calculations that could be linked to restricted sets of experiments, but they depended upon highly limiting assumptions concerning the disposition of the fluids. Moreover, he did not, indeed could not, progress very far in connecting laboratory measurements to his calculations. To do so required, at the least, knowing the force law, but even that would not in itself be enough. New sets of techniques had to be developed to fit the new electric (and magnetic) science that Aepinus had forged. In particular, the goal of experimental electric science had to be clarified; the landscape of the laboratory had to be redrawn.

For effluvialists the laboratory was a place for making things move about under electric influence; for Franklinists it was a place for revealing the transfer of electricity from one body to another. Aepinus's reduction of electricity to a Newtonian fluid made both of these goals subsidiary ones. Instead, the problem that his work (at first only implicitly) placed at the centre of electric science was this: to calculate and to measure the distribution of electricity over the surfaces of conducting bodies, the old non-electrics now having become objects that simply did not impede the fluid's motion. Aepinus could not solve this problem, however, and he concentrated instead on loose computations of forces given very simple, assumed distributions. However, with the central question of the subject now shifted from the nature and behaviour of electric stuff to the form of the Newtonian force that governed it, instruments were developed over the next quarter-century to probe that question.

Aepinus's views became influential in part because of a successful application of them to a device, the electrophore, that was invented by the Italian Alessandro Volta in the late 1770s. This instrument—a dielectric covered by tin foil and rubbed against a grounded plate—seemed to be able to electrify alternately without requiring re-excitation. Considered to pose a great puzzle to Franklinist science, the electrophore was explained by Volta himself after he had assimilated the hitherto neglected work of Aepinus as an instance of induction: a process in which charged bodies influence one another through electric force without actually exchanging any electric matter.

In the late 1760s the Scottish natural philosopher John Robison produced (according to his later account) a device to measure this force; his device balanced it against gravity. In the mid-1780s the French Academician and engineer, Charles Augustin Coulomb, built an instrument based on balancing electric force against the torsion in a twisted wire. With his torsion-balance electrometer, Coulomb obtained results that convinced him and many (though not all) of his contemporaries that the force law followed the inverse square, precisely like gravity.

Coulomb was also able to argue that the electric force immediately outside a charged conducting surface must be proportional to the charge density there. This enabled him to give meaning to experiments in which he used his electrometer to measure the force over the surfaces of two charged spheres placed near one another: these numbers were now the canonical goal of all electric science, for they represented the electric distribution. But Coulomb could go no further; he did not know how to compute the distribution from the geometry of the experiment and from the force law.

Despite knowledge of the force law and the corresponding relegation of the precise nature of electricity to comparatively unimportant status, the undeveloped mathematical state of the subject revealed itself in the difficulty of untangling two different aspects of electric matter: on the one hand, the quantity of it on a given conductor, and on the other, its power to produce electric effects, its intensity. It had been recognized since the 1740s that conductors of different surface areas had different 'capacities' for electricity, in the sense that conductors electrified by the same power could acquire different amounts of electricity.

This distinction between amount of electricity and electric tension became sharper, and was quantified, by Volta around 1780 after he had assimilated Aepinus's views and had therefore come to think of electricity as working almost entirely through influence or force. Volta concentrated on the puzzling connection between electric power, or tension, and electric quantity. Grounding his work in the laboratory, he hypothesized that quantity and tension were proportional to one another, with the constant of proportionality representing the capacity of the conductor. Volta's relation was soon turned, particularly in England, to the production of new electric devices, ones that charged by influence.

The connection between tension and the electric force proper remained obscure for quite some time, despite the fact that a reclusive English natural philosopher, Henry Cavendish, had gone quite far in clarifying it in the early 1770s. Long before Coulomb, Cavendish, aware to some extent of Aepinus's concepts, had produced an experiment designed to show that the electric force must obey the inverse square. Cavendish's now-famed null experiment used the property of such a force that its value inside a region surrounded by a spherically symmetric distribution must vanish. Beyond that, Cavendish attempted for the first time actually to compute electric distributions under certain circumstances, for which he developed a way of simulating the operation of tension: conceive of an infinitely long, infinitely thin canal, filled with electric matter, and connected between two charged conductors. These last will not have the same amount of electricity, Cavendish argued, but they will be electrified to the same degree, in the sense that each of them would, when alone connected through a canal to a standard test conductor, transfer the same amount of electricity to it. In this way Cavendish was able to calculate relative capacities for pairs of disks or spheres, and also to develop an appropriate instrument for measuring them to high accuracy.

In one way, electric science ceased to develop after the 1790s for it no longer produced what were thought to be intriguing laboratory novelties. In another way, namely, in its technical structure, it changed markedly during the first quarter of the nineteenth century, at first in the hands of the French mathematician and physicist Siméon Denis Poisson. Relying on the mathematics of spherical harmonics developed by his colleague Adrien-Marie Legendre, and upon the concept of a potential function introduced by his mentor the Marquis de Laplace, Poisson was able to calculate the electric distribution over neighbouring charged spheres that Coulomb had measured a quarter-century before. To do so he relied on Coulomb's relation between force and charge density, as well as upon the condition that the potential function, whose gradient yields force, must be constant within and on a conductor. Poisson's analysis was, however, troubled by difficulties that derived from his continuing insistence that electricity distributes itself in a layer with finite, and varying, thickness near the surface of conductors. Only in 1828 did the English mathematician George Green completely remove physical considerations from the subject, reducing it in effect to a formal exercise in finding appropriate solutions to the Laplace equation in given circumstances. With Green's work the old electric science ceased to be an object of direct interest to physicists, at least insofar as interactions between conducting bodies were concerned. By that time, however, an entirely new subdiscipline had developed, one that derived from the work of Volta at the turn of the century and from the discovery made by the Danish natural philosopher Hans-Christian Oersted in 1820: namely, the magnetic effect of the electric current.

Further Reading

Two major monographs remain essential for understanding developments in electricity and magnetism during these centuries. One of them, R. Home's *Aepinus' Essay on the Theory of Electricity and Magnetism* (Princeton, NJ: Princeton University Press, 1979) combines an extensive essay on this influential work, with a translation. The other, John Heilbron's *Electricity in the 17th and 18th Centuries: A Study of Early Modern Physics* (Berkeley, CA: University of California Press, 1979), provides a comprehensive overview of developments throughout these two centuries. Christa Jungnickel and Russell McCormmach analyse the life and work of Henry Cavendish in their *Cavendish. The Experimental Life* (rev. edn., Bucknell, 1999). On Alessandro Volta, see Giuliano Pancaldi, *Volta: Science and Culture in the Age of Enlightenment* (Princeton, NJ: Princeton University Press, 2003). Full references to the works discussed here can be found in these books.

PART III

FASHIONING THE DISCIPLINE: FROM NATURAL PHILOSOPHY TO PHYSICS

CHAPTER 15

OPTICS IN THE NINETEENTH CENTURY

JED Z. BUCHWALD

15.1 THE EIGHTEENTH-CENTURY BACKGROUND

Two fundamental physical images governed speculation in optics, and occasionally even mathematization, from the eighteenth through the nineteenth centuries: namely, the conception of light as a sequence of material particles moving through a void, on the one hand, and the conception of light as a mechanical disturbance in an all-encompassing medium, on the other. The latter image, in a myriad of forms, had by far the greater number of adherents until well into the eighteenth century, and has its roots in René Descartes' comprehensive mechanical system. In 1690 the speculative Cartesian optical medium acquired a novel character when the disturbance it was supposed to carry was bound to geometry by the Dutch polymath Christiaan Huygens.[1] To do so he introduced a physico-mathematical rule, eponymously termed Huygens' Principle, that governed the propagation of the optical disturbances, and according to which each point on the surface of a propagating pulse of light itself constitutes a secondary source, with the overall pulse being the common tangent to all of these secondaries. Huygens thought his disturbances to constitute what we now term longitudinal pulses, which are isolated disturbances that parallel the direction of their propagation. These pulses had no periodic properties, and indeed Huygens' theory was able to deal neither with colours nor with certain curious phenomena that will shortly be critical for us here and that he had himself discovered on passing light through exotic crystals brought from Iceland. Huygens, however, successfully produced a thoroughly geometrical theory, buttressed by careful experiment, for the peculiar double images produced by these Iceland crystals. It is important to note, though, that the computational tools of the day were inadequate to probe the recesses of Huygens' claims, and his construction for double refraction remained controversial until the beginning of the nineteenth century, when, as we shall see, its confirmation in Paris set in motion a significant chain of events.

During the eighteenth century a considerable amount of speculative natural philosophy was produced, but mathematical optics remained for the most part bound to the physical concept that Huygens' system had in fact demoted from physical primacy; namely, the ray of light. The ray itself had long been the foundation of geometrical optics, and in the seventeenth century it had acquired a new physical reality within the first system mentioned above; namely, as marking the track of the particles out of which Isaac Newton built light. This conception of light's structure was extraordinarily influential, inasmuch as it formed an essential part of Newtonian natural philosophy, if not of Newton's mathematical optics. But mathematical optics of any kind was not extensively pursued in new ways during the eighteenth century, and certainly no novel experimental or mathematical results were produced during the period that attracted widespread attention, though significant instrumental developments certainly did occur. Moreover, throughout much of the eighteenth century, in a number of loci the differences between the Newtonian theory and systems based on motion through a medium were not altogether clear-cut, not least because elements of both appear in Newton's own, widely-read *Opticks* (the first edition of which was printed in 1704), in what many readers evidently found to be a confusing amalgam.[2]

Despite (or perhaps because) of the concentration on physical principles during the period, until the last quarter of the eighteenth century very little work in any area of natural philosophy associated with the laboratory was quantitative, and even less attempted to integrate quantitative detail with precise experimental situations whose accuracy could be specified. This was particularly true for investigations of electricity and heat, and it was also true in comparatively obscure areas such as the optics of crystals. This began to change radically in France during the last quarter of the century. Charles Coulomb in electricity and the Lavoisier–Laplace collaboration in heat exemplify this change: here we find a growing concern with quantitative structure coupled to careful, and often elaborate, experiments designed explicitly to concentrate on quantity. Indeed, by the turn of the century in France, work that was not quantitative, and experiments that were not carefully contrived and mathematically analysed, stood little chance of receiving much attention. This new desideratum was best learned by example, and one place to learn it was at the École Polytechnique in Paris. All four of the major French participants in the early years of the optics controversies that reshaped the discipline in fundamental ways—namely, Jean Baptiste Biot, Etienne Louis Malus, François Arago, and Augustin Jean Fresnel—attended the École in the 1790s and early 1800s.

15.2 RAY OPTICS, THE DISCOVERY OF POLARIZATION, AND THE BIOT–ARAGO CONTROVERSY

During the first decade of the nineteenth century, interest in optics, and particularly in novel optical experiments, became quite strong in France. Stimulated in part by the English chemist William Hyde Wollaston's apparent confirmation of Huygens' construction for the double refraction of Iceland crystal, Laplace had Malus, in whom he reposed a considerable amount of confidence for work that Malus had already done in constructing a mathematics for systems of light rays, to undertake a thorough experimental investigation of the subject. After producing at Laplace's instigation a mathematical *tour de force* in which he translated Huygens' construction into algebra, Malus pressed ahead with a careful experimental investigation which showed convincingly that the construction is extremely accurate. Here Malus deployed both the engineering and mathematical training that he had acquired at the École Polytechnique. His work required an acute combination of analysis with cleverly designed and deployed apparatus, yielding in the end what was, at the time, the most accurate optical measurement that had ever been made. Neither he nor Laplace, however, concluded that Huygens' pulse theory of light, which they carefully and thoroughly distinguished from the Newtonian alternative, must therefore be accepted. Instead, both argued, in different but equally peculiar ways, that the resultant formulae are in fact compatible (and perhaps even uniquely compatible, if one takes Laplace at his word) with the mathematics of particles and forces. This debatable claim was quite persuasive among Laplace's associates and students, as well as in certain quarters in England, and for more than a decade and a half optics remained closely bound to the particle theory, as we will see.[3]

Indeed, the persuasive claims of Newtonian optics were furthered in no small measure by Malus' own discovery of an entirely new optical process—the first such discovery since the seventeenth century. Huygens had already noted that light emerging from doubly-refracting crystals seems to have some sort of asymmetry associated with it, since on entry into a second crystal it is not equably divided in two again. In 1809 Malus found, through a series of acute experiments due initially to a serendipitous observation, that this property (which in 1811 he named polarization) did not require a crystal, but that reflection at a particular angle from any transparent body can also produce it. This discovery stimulated a great deal of experimental and theoretical work during the next decade, undertaken especially by Arago and Biot in France, as well as by David Brewster and, somewhat later, John Herschel in Great Britain. Indeed, the most heavily pursued area of quantitative experimental research in optics during the 1810s orbited about the many instrumental novelties and consequent research opportunities opened by Malus' discovery, particularly when the new light form was passed through thin crystal slices, producing beautiful and complicated colours. This work was not based on the Newtonian theory *per se*, nor was it in any easy sense simply instrumental, even though strongly connected to the new device—his polarimeter—that Malus had designed to produce and to measure the new optical property. It was nevertheless hypothetical.

In this scheme the ray of light (not the Newtonian optical particle) provided the fundamental theoretical tool. Practitioners of ray optics, for whom we will shortly introduce a different name appropriated from Thomas Young (the English polymath who, we shall see below, invented a scheme for waves similar to the first one deployed by Fresnel) considered the ray to exist as an individual object that could be counted, and that rays collected together in groups, or bundles, to form beams of light. In this system the ray itself was the central physical object, and the appropriate mathematics involved ray-counting, or what amounted to a species of ray statistics. The character of the system appears strikingly in Malus's own conception of polarization. The intensity of a beam of light is measured numerically by the number of rays that it contains. Unlike the individual ray, which is too weak, a beam can be seen, and its intensity can be manipulated, if not measured directly, using Malus' polarimeter to sort out the rays of different orientations that comprise it, according to the following way of thinking.

Every ray, Malus insisted, has an inherent asymmetry about its length. Think of it rather like a stick to which a crosspiece is nailed at right angles. Given the direction of the ray, the orientation of the crosspiece in a plane at right angles to the ray determines its asymmetry. As Malus understood the concept, 'polarization', properly speaking, does not apply to the individual rays in a beam but only to the beam as a collection of rays. A beam may be polarized in a certain way, but the individual rays that make it up are not themselves said to be polarized, though each has a certain asymmetry. If the asymmetries of the rays in a given beam point randomly in many directions then the beam is, in Malus' understanding, 'unpolarized'. If, on the other hand, one can group the rays in a beam into a number of sets, each of whose elements shares a common asymmetry, or even if this can be done only for a certain portion of the rays in the beam, then the beam is said to be 'partially polarized'. If all of the rays have the same asymmetry then the beam is just 'polarized'. According to Malus' way of thinking, his polarimeter picked out those sets of rays within the beam that had specific asymmetries.[4]

We shall hereafter refer to Malus, and to those who thought like him, as 'selectionists', since they conceived of polarization as a process in which the rays in a beam are selected and have their asymmetries altered in direction. Selectionism was, on the one hand, not at all coincident with the Newtonian theory, since selectionists could and did draw the distinction in controversy with their wave opponents, but, on the other hand, it was nevertheless thoroughly hypothetical, which is easy to see, because on wave principles it is in fact unsustainable—in wave optics light beams cannot be thought of as collections of discrete rays. More to the point, selectionist principles could be, and indeed certainly were, used to develop mathematical laws that had direct application in complex experiments. These laws were neither mere summaries of experimental results, though they were certainly tied directly to particular kinds of instruments, nor were they simply pulled out of the air. On the contrary, they were deduced directly from the fundamental principles of selectionism, and they are incompatible with laws for the same kinds of phenomena that are implied by the principles of wave optics. However, the two devices (the crystal and the polarimeter) that could be used to examine polarization at the time depended critically upon the eye to judge the presence, absence, or even intensity of light, and in these kinds of experiment the unaided eye has limitations. The limitations were sufficient to preclude any experiment until well into the 1840s that could tell the difference between the selectionist and wave formulae for the most widely influential phenomenon; namely, the partial reflection and refraction of light at the surfaces of transparent media. Indeed, the relationships that Fresnel obtained for calculating the quantities of polarized light reflected and refracted remained without experimental support for decades, thereby generating a pointed controversy concerning them, so long as only the eye could be used to compare optical intensitites.[5] Nor was this the only area in which the difference could not be told.

Malus did not live long enough to develop his own system completely, though its outlines were quite apparent to many people at the time. Arago, for one, saw clearly that Malus' work depended on the division and grouping of light-rays into related sets, and this aided him in explaining a phenomenon involving Newton's rings that he himself discovered shortly before Malus' death. Arago was aged just 23 at the time of his election to the astronomy section of the *Institut de France* in 1809, and his work on Newton's rings two years later, at the age of 25, represented the only research for which he could claim sole responsibility. He had reason to be jealous and intensely proud of his results. From the outset, Arago fully adopted Malus' terminology and understanding of polarization, and with these ideas in mind he decided to make his mark by examining the polarization of Newton's rings. These coloured bands occur when light passes through the narrow gap between, for example, two lenses pressed hard together, and had been extensively investigated by Newton. Working with lenses at the observatory, Arago thought to examine the polarizations associated with the rings. When he did so he rapidly discovered an apparent exception to the rules that Malus had offered for the polarization of reflected light—an exception that Malus himself found to be quite troubling when Arago told him about it. However, as Arago pursued his discovery he did not abandon Malus' understanding, but instead supplemented it by drawing a new distinction between the formation of rings and the generation of their polarization. This, as we shall see in a moment, captured him in a very important way. However, on 11 August he described another, eventually highly influential, discovery he had made that was later termed 'chromatic polarization', involving the generation of coloured patterns by the passage of polarized light through crystal sections.[6]

What happened next proved to be critically important. Between 11 August and the following spring Arago continued to pursue his exciting new discoveries, though we do not know precisely what he was doing during this period. Burdened with heavy teaching and administrative duties he did not, he wrote later, have the time to gather his new work together for a public reading before a disaster occurred. All of a sudden, seemingly out of nowhere, Biot intruded on Arago's field of research and read a note on chromatic polarization that at once thoroughly stripped Arago of his leadership in the new field. Arago demanded that notes he had earlier deposited be examined to show that he had already done what Biot claimed. The Institute appointed Burckhardt and Bouvard to look into this contentious issue, and in April the investigators announced that 'the declaration made by M. Arago [is] most exactly true'. But these notes were actually published only years later, and Arago did not read anything else to the Institute until the following December, eight months after the debacle with Biot over priority. In the meantime the active Biot himself read an extraordinarily long memoir on chromatic polarization to the Institute, followed six months later by an even lengthier discussion that had a major impact on most of his French, and eventually his British, contemporaries.

Arago had lost control of the field he himself had founded. Yet the notes that he had deposited, and which he forced the Institute to examine, scarcely mention the subject of chromatic polarization, which is what concerned Biot. Moreover, Arago's two subsequent memoirs in the general area, as well as his contemporary unpublished notes, are very different in character from Biot's memoirs. Unlike Biot's work, this material is entirely qualitative and yet nearly devoid of any concern with the principles of Newtonian optics. It is almost entirely involved with the overall features of what happens to rays, rather than with why it happens to them or with representing mathematically, in the fashion of Malus, precisely how it happens. But what the notes do contain, though only in a highly undeveloped form, is a general theory that tries to unite the polarization effects of double refraction and thin crystals with those of reflection.

The theory is rather vague and completely non-quantitative, but it does try to unify very different phenomena. Unlike Biot's pre-emptive work in the area, Arago's was very broad in scope. He never attempted to generate formulas from it, and his work does not contain numerical, much less tabular, data of any kind. Biot, by contrast, produced formulas very early on in his work on chromatic effects (though he had nothing like Arago's unifying theory), and his lengthy papers are filled with extensive tables. Of the two, Arago was working in the more traditional, qualitative manner; he was seeking broad principles to encompass several classes of phenomena. Tabular data and formulas do not fit well that kind of endeavour. Biot turned instead to very sharply limited assumptions and pointedly sought to generate formulas from them for specific cases, while attempting to marry his mathematics to quantitative experiment. He made no effort in his early work to link these results to wider classes of phenomena in any firm way.

Biot's first work in this area therefore follows the new pattern of the late 1700s, a pattern that was firmly established in optics by Malus. This pattern was rapidly becoming a standard one. It insisted upon the careful tabular presentation of numerical data and the generation of formulas that are capable of encompassing the material at hand, with little immediate concern to reach out to other, even closely related, phenomena. The differences between Biot's and Arago's work therefore hinged upon changing canons of experimental reporting and investigation, canons that had first appeared in optics in sharpest relief in Malus' work.

During the next several years Biot not only gained fuller control over the subject that Arago had created, but also published long and intricate memoirs that linked it to Newtonian optics. Biot's rapid progress in chromatic polarization culminated in an immensely detailed book, a text that symbolized in concrete form the astonishing success of his endeavour.[7] To many people Biot became that theory's primary exponent. Arago could not have been overjoyed. Even before these events Arago had expressed some doubts about the Newtonian system, or at least about several aspects of it. Little wonder that he came to dislike it violently. Arago was accordingly well prepared to react when in September 1815 he received a long letter from Augustin Fresnel, the nephew of Léonor Merimée (once teacher of design at the École Polytechnique, and by then permanent secretary of the École des Beaux Arts). Fresnel approached Arago with just the kind of quantitative work that Biot could produce and that he, Arago, could not.

Fresnel, himself a graduate of the École Polytechnique a decade before, had briefly visited Arago in Paris in the previous July while on his way to internal exile at his mother's home for having greeted Napoleon's return from exile by joining the Duc d'Angoulême's resistance. Fresnel was already pondering optics by then, and asked about diffraction. Arago gave him a list of English authors, including Thomas Young, which Fresnel could not read, though his brother Leonor could. Fresnel's fall letter to Arago advanced an optical theory similar to the one that had already been discussed by Young, and it contained precise experiments, numerical detail, and beautiful formulas.[8] Most importantly, it seemed to Arago to show something that neither Biot (nor Arago himself) could have predicted: namely, that the coloured fringes produced by light that passes the edge of a narrow object the diffraction fringes move away from the diffractor along hyperbolic paths. Though Fresnel himself did not emphasize this discovery in his first letter to Arago, Arago seized on it and at once pressed Fresnel to improve his observations, to make the discovery indubitable. In passing, he remarked that Fresnel's theory was essentially the same as the Englishman Thomas Young's, though Arago did not at the time also realize that Young had already pointed out the hyperbolic law (which had in any case been remarked for a different configuration than Fresnel's as an empirical generality by Newton in his *Opticks*). What gripped Arago was not so much the excitement of a new discovery as the opportunity to make use of it to redress the recent wrongs he had suffered at the hands of Biot, whereas Fresnel was deeply perturbed by Young's priority, and this in the end stimulated him to even greater exertions. Their two worries nicely intersected for a time.

15.3 THOMAS YOUNG AND INTERFERENCE

In 1799 Thomas Young in England had begun a series of publications that substantially extended the quantitative power of medium theories of light. Young, like Leonhard Euler before him, associated colour with wave frequency, but he went far beyond Euler in his use of the assumption (not least because Euler's "frequencies" referred to arithmetically-separated pulses of light).[9] Trained as a medical doctor, Young had during his studies become deeply interested in acoustics, and especially in phenomena of superposition. This eventually led him to the *principle of interference*, according to which continuous waves of the same frequency and from the same source will, when brought together, produce regular spatial patterns of varying intensity. That principle had not been well understood for waves of any kind until Young began his investigations, and it can indeed be said that the study of wave interference in general began with Young himself, though he did not pursue it extensively outside optics. Many difficult problems had to be solved by Young, including the conditions of coherence that make detectable spatial interference possible at all. Furthermore, the principle of superposition, according to which waves combine linearly and which is a necessary presupposition for the principle of interference, was itself quite problematic at the time and also had to be developed and argued for by Young.

Young applied his principle of interference to the diffraction of light by a narrow body, as well as to the case of light passing through two slits, though in the latter case it seems that he did not carry out careful measurements. In all cases he explained the fringe patterns that he observed by calculating the path difference between a pair of rays that originated from a common source. He did not calculate with waves themselves, but rather assigned periodicity to the optical ray, which accordingly retained a signal place in Young's optics. Although Young was certainly quite familiar with Huygens' work, he did not utilize the latter's reduction of rays to purely mathematical artifacts, for that was bound to Huygens' principle, which Young found difficult to accept.

In any event, Young's optics did not generate extensive immediate reaction. Indeed, his principle of interference was sufficiently difficult to assimilate that no other applications to new phenomena were forthcoming. The most famous, or (in retrospect) infamous, reaction, was that of Henry Brougham. Brougham vehemently objected to Young's wave system as an alternative to the Newtonian scheme of optical particles, and he also objected to the principle of interference itself even as a mathematical law applied to rays. Most contemporary optical scientists were more interested in the physics than in the mathematics of light, and in this area Young's ether posed as many qualitative problems as did the alternative system of light-particles. Although Young's work did not stimulate extensive discussion in France, it was nevertheless known there.

15.4 Fresnel, Interference, Diffraction, and Arago

Arago soon brought Fresnel to Paris and participated in new experiments with him, particularly in ones that seemed to hold out the possibility of casting doubt on some aspect of the Newtonian theory, and so ultimately on the worth of Biot's work.[10] This was the origin of a famous mirror experiment, in which interference occurs between rays that do not pass near material edges and so cannot presumably be affected by the forces that might otherwise be used to explain the formation of fringes. Fresnel's own views underwent considerable development during the next three years, and Arago evidently made certain that he had control over when and where the work was reported. Then, on 17 March 1817, the Academy publicly announced that it had decided to offer a prize on diffraction. By this time Fresnel had extensively changed his original theory, having evolved it from one that was based on comparing rays two at a time to one that was based on wave fronts, Huygens' principle and elaborate integral methods. These developments were stimulated by a succession of increasingly exact experiments in which Fresnel modified his early theory in the face of countervailing observations. His final results reached an extraordinarily high degree of accuracy in placing the loci of diffraction fringes. To do that required Fresnel to develop a series of observational techniques that were designed to provide just the right sort of data for him to deploy numerical methods for approximating his theoretical formulas.

Figure 15.1 is adapted from a diagram drawn by Fresnel himself. In it, C represents the source of a spherically symmetric front AMm' that is intercepted by a screen AG. Adapting Huygens' principle, Fresnel conceived that each point on the front itself emits a spherical wave, albeit with an amplitude that decreases with inclination to the line joining that point to the source C. Introducing z as the distance along AM from the edge A of the diffractor, Fresnel could then represent the amplitude ψ of a disturbance with wavelength λ sent to an arbitrary screen point P in the following way:

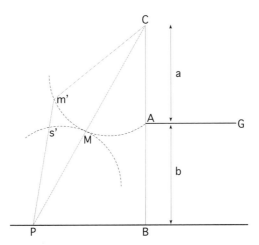

Fig. 15.1. Fresnel's structure for calculating diffraction.

$$\Psi = \sin\left[2\pi\left(t - \frac{CM + m's'}{\lambda}\right)\right]$$

$$m's' \approx \frac{z^2(a+b)}{2ab}$$

whence

$$\Psi = \sin\left[2\pi\left(t - \frac{CM}{\lambda}\right) - \pi z^2 \frac{a+b}{ab\lambda}\right]$$

which decomposes into

$$\Psi = \cos\left[\pi z^2 \frac{a+b}{ab\lambda}\right] \sin\left[2\pi\left(t - \frac{CM}{\lambda}\right)\right]$$

$$+ \sin\left[\pi z^2 \frac{a+b}{ab\lambda}\right] \sin\left[2\pi\left(t - \frac{CM}{\lambda}\right) - \frac{\pi}{2}\right]$$

Fresnel could conclude from this that the square of the resultant from all of the secondaries on the front, pairing up all terms with cosine amplitudes and all terms with sine amplitudes, can be computed from the following sum:

$$\left(\int \cos\left[\pi z^2 \frac{a+b}{ab\lambda}\right] dz\right)^2 + \left(\int \sin\left[\pi z^2 \frac{a+b}{ab\lambda}\right] dz\right)^2$$

or, using a change of variables, to find

$$\int \cos\left(\frac{1}{2}\pi z^2\right) dz \text{ and } \int \sin\left(\frac{1}{2}\pi z^2\right) dz$$

There were two major difficulties with this result with which Fresnel grappled. One was how to establish an appropriately general coordinate system for calculation. This arises in the simplest case when the diffracting object, or aperture, has two edges, for then two limits are involved, and this in effect requires computing values for all points of a surface. The other, which arises in all cases, including that of the semi-infinite plane, where only one boundary occurs (and where, accordingly, the surface just mentioned reduces to a line), is simply how best to calculate useful values for these integrals. Fresnel sought the quickest route to application, and that was by numerical integration (instead, for example, through a series expansion, though Cauchy later developed divergent ones for these integrals). In an astounding computational *tour de force*, Fresnel tabulated the integrals in steps of 0.1 from 0 to 5. His computational errors in doing so amounted to a mean of only 0.0003, and the differences between his values and more accurate ones computed using the series later produced by Augustin Cauchy still amounted to only 0.0006.

The appearance in physical equations of solutions that could only be evaluated by series expansions or by numerical integrations was by this time not altogether unusual. They had emerged quite directly in astronomical problems, and they were soon also to appear in problems involving elasticity and heat flow. Nevertheless, what was unusual was the presence of such things as a *fundamental* expression of the underlying physics. For Fresnel's integrals, unlike, say, Legendre polynomials in astronomy, or, later, Fourier series in thermal processes, were not produced as solutions to given differential equations. On the contrary, they were asserted by Fresnel without his having had in hand the partial-differential equation of which they were meant to be the solutions, much less the methods and techniques necessary to solve such a thing under appropriate sets of boundary conditions. Fresnel, one might say, had discovered the solution to what would later be termed the 'reduced wave equation', or after 1860 as the 'Helmholtz equation', without having any idea at all what that equation was. This character of Fresnel's wave optics contrasted strikingly with that of an optics whose physical imagery at least was grounded on particles and forces, for there the fundamental differential equations were well-known and were, in addition, ordinary.

Of course, there was no way at all to treat what amounted to an n-body problem in Newtonian optics. Consequently, one has the seeming paradox that the only way mathematically to deal with diffraction required a method whose underlying physics and associated fundamental mathematics remained altogether unknown, whereas the alternative physical scheme, in which the physics was very well understood indeed, yielded no effective mathematics at all. Yet this very difference was precisely what aided the assimilation of Fresnel's mathematical methods, and eventually of the physical conceptions that they brought along with them. For no-one on the prize commission, which included Laplace and Siméon Denis Poisson, who were arch-proponents of particle physics, objected that Fresnel's integrals should not be exploited as, in effect, phenomenological expressions of the empirical facts of the matter. During the remainder of the century, Cauchy, George Green, George Stokes, Hermann Helmholtz, Lord Rayleigh, Gustav Kirchhoff, and Arnold Sommerfeld, to mention only a few, grappled with the creation of physics for both sound and light based on the partial-differential wave equation in three-space and its useful solutions. It may with only slight exaggeration be said that all of them were attempting to discover just what these original solutions of Fresnel's were solutions to, and how they could in retrospect be justified or amended.

Laplace, Poisson, and even Biot were not overly disturbed by Fresnel's success in developing formulas for diffraction, not least because they did not think that diffraction was a major topic in optics. Certainly it was an important topic, but in their eyes its application was limited to what happens near edges or when light rays are made to interact with one another in certain kinds of situations. They certainly did not believe that all of optics had to be reconstructed on a new foundation, for they continued to think of the optical ray as a fundamental, irreducible element. They rejected waves. Indeed, acceptance of Fresnel's diffraction formulae without a simultaneous acceptance of Fresnel's wave theory remained quite common until the late 1830s. This does not, however, mean that the Newtonian theory *per se* retained its power. It means rather that a distinction was drawn between that theory and the assumption that rays are individual, countable objects, which remained tacitly unquestioned. Fresnel's work certainly raised many doubts about light-particles, but for quite some time it did not raise corresponding doubts about the physical identity of the light-ray.

15.5 A New Mathematical Optics and the Wave–Particle Debates

Diffraction had not been the exciting topic in optical research throughout most of the 1810s. Polarization, and chromatic polarization in particular, had captured the centre of attention and had made Biot's reputation. This was the subject that Biot had ripped out of Arago's hands in 1811. In mid-July 1816 Arago apparently suggested to Fresnel that he should examine the fringes produced by the interference of the two polarized beams that emerge from a crystal. Fresnel's first discussion of chromatic polarization, which was based on the principle of interference, was not an improvement over Biot's, primarily because at that time Fresnel did not know how to compute the resultant of more than two interfering rays. And though Fresnel submitted this work, which was handed to Arago and Ampère to report on, no report appeared for five years.

Arago's 'report', when it finally did appear, was a polemic directed at Biot. He had delayed five years in writing it probably because it took him that long to understand fully how Fresnel's work might be used to defeat Biot. Biot was given the papers on which Arago was ostensibly reporting, and he saw at once what Arago had done. Remarking that Fresnel himself 'had not proposed as the basic purpose of his work to show that what he calls my theory of mobile polarization is, in many points, insufficient and inexact', Biot concluded by complaining that the report 'deviates from the rules generally established in scientific societies for assuring the equity of their decisions'. The controversy that ensued was nasty, and at times vicious, though Fresnel was himself somewhat on its outskirts. He had written his original memoirs not to attack Biot directly but rather to present his own theory. Arago turned that around, and Biot knew it. However, in the heat of his reply, Biot challenged aspects of Fresnel's work, thereby opening himself to a powerful attack from Fresnel. This confrontation between Biot and Arago, and Biot's subsequent failure to clarify the nature of his theory in exchanges with Fresnel, marks an epoch in the history of optics. Scarcely two years after Fresnel had won the prize for his diffraction memoir, the Institut, now returned in the aftermath of the monarchy's Restoration to its original name as the Royal Academy, ordered printed, over Biot's explicit and public objections, Fresnel's account of chromatic polarization. The members of the Academy did not accept Arago's report, but apparently only because Arago had not insisted that they do so (and he in any case published it almost immediately in his own journal).

Fresnel's final theory of polarization, more than his account of diffraction, broke fundamentally with selectionist optics. As we saw above, contemporary French understanding of polarization considered it to be an essentially static, spatial process in which the rays in a given group have their asymmetries aligned in certain ways. Time does not enter into this scheme, and, according to it, a beam of observably unpolarized light is always just unpolarized, no matter how small a time interval one might consider. For several years Fresnel had found it extremely difficult to discard this notion that common light must not show any signs of asymmetry at all, though he could not use ray-counting procedures to explain why not. He accordingly tried hard to build a scheme in which polarization consists of a temporally fixed combination of directed oscillations: one along the normal to the front, and the other at right angles to it. In common light the transverse component vanishes altogether; in completely polarized light the longitudinal component disappears. This, like its ray-based counterpart, is an inherently static, spatial image. Fresnel was not successful in building a quantitative structure on this basis. Then, in 1821, he set his static image into motion.

The core of Fresnel's new understanding of polarization referred the phenomenon to the change (or lack of change) *over time* of a directed quantity whose square determines optical intensity. This quantity must always lie in the wave front (and is therefore 'transverse' to the ray in optically isotropic media), and in reflection and refraction it can be decomposed, with the components in and perpendicular to the plane of reflection being affected in different ways. Common light consists of a more or less random rotation and amplitude change over time of this directed oscillation, and not, as Fresnel originally understood it, of a spatially fixed (longitudinal) disturbance. Rays of light are mathematical abstractions: they are merely the directions joining the centre of the wave to the front itself. In a purely analytical sense one can link a ray to the asymmetry in the front at the point of the front which the ray contains. Accordingly, in Fresnel's wave theory one can say that a ray is polarized, if one wishes, because polarization refers to the asymmetry at a point in the front, and to each such point there corresponds only one ray. But, and this is a signal characteristic of the scheme, the rays, being mere lines, cannot be counted. A beam of light is not a collection of discrete rays, which means that it cannot be dissected in Malus's fashion. The vast gulf between the two conceptions made it extraordinarily difficult for people who did not think about polarization in the same way to communicate their experimental results to one another without leaving an inevitable residuum of ambiguity that could, and often did, lead to controversies.

Fresnel soon drew startling implications from his new theory of polarization. He produced novel surfaces for doubly refracting crystals that linked polarization to wave speeds and thence to the directions of rays. These surfaces, which Fresnel obtained only after several false starts and much experimental work, opened a new, and unexplored, realm for mathematical and physical analysis, which was soon avidly pursued in England, Ireland, and France by an emerging cadre of wave analysts. Much of this work concentrated on drawing implications from Fresnel's surfaces and associated mathematics. For example, Humphrey Lloyd in Ireland confirmed the existence of a striking phenomenon known as conical refraction that William Rowan Hamilton had deduced; in England, George Airy examined mathematically the special type of double refraction that occurs when light passes through quartz oblique to its optic axis, again deploying with finesse Fresnel's new mathematics for polarization. The late 1830s and the 1840s were also the period during which vigorous debates occurred between partisans of the old and new optics, debates that often raised issues concerning the respective physical foundations of the two theories.

Most wave partisan attacks against the alternative optics looked directly to the latter's putative foundation in particles and forces. Humphrey Lloyd, for example, remarked that the 'emission [Newtonian] theory' of light is essentially useless because 'it is an aggregate of separate principles' concerning the behaviour of optical particles.¹¹ According to Lloyd, wave optics is nothing like the unamalgamated aggregate of emission (particle) optics, and this is why it can generate new physics. Among the new physics that Lloyd had in mind was partial reflection and refraction. Here, he explained, Biot's physics for optical particles, which Lloyd developed in some detail, simply could not yield formulae—nothing useful emerged from it at all. Fresnel was much better. Unlike emission theorists, who could not generate formulae from their forces and particles, Fresnel could, Lloyd asserted, from the properties of the medium, or ether, in which optical waves were supposed to subsist. Ironically, Fresnel's mathematics for partial reflection and refraction was the one aspect of his new wave theory that escaped experiment for many decades, primarily because the photometric measurements required could be done only by using the eye to judge degrees of illumination, which left considerable scope for argument over results.¹²

Lloyd's brief for the power of the wave theory, and for the intuitive sagacity of its partisans, could not convince anyone on the other side who had spent much time engaged in research, and in fact it did not do so. Neither did William Whewell's remarks in a similar vein produce any conversions, for the fact was that in the early years of wave optics it remained possible still to think in old ways, based on rays of light, because the new system had yet to flower on paper and in the laboratory.¹³ During the 1830s, however, wave partisans produced practical analytical tools, and some laboratory tools as well, that enabled them to establish a dynamic research tradition, within which ether physics took its place as one element. Wave partisans, for reasons that did not have much to do with the comparative abstract superiority of their system, also controlled important journals and built a close network of like-minded people through university and (especially) professional associations. Ray partisans did not achieve anything like this; they did not even try to do so. By the 1840s an entire universe of wave devices was being generated—one in which the instruments themselves were increasingly built around the behaviour of wave front and phase, the principal concepts of the new theory. This universe of devices offered no point of entry to the ray physicist, for whom front and phase had no fundamental significance. By then, which was quite some time after the large majority of work in optics had shifted to wave methods, ray physics might be said to be objectively weak in comparison to wave physics.

15.6 The Physical Structure of Wave Optics through Mid-Century

In ray optics the concepts of particle and force served less to generate new mathematics and experiments than to provide a physical foundation. Similarly, in wave optics the ether served for the most part rather to suggest and to justify than to produce new physics. For example, Fresnel's route to a mathematics for diffraction hardly used the ether at all, except for the signal purpose of convincing him in the first place that light did not have to be thought of as a collection of physically distinct rays. In fact, the minor use that Fresnel did there make of the ether prevented him, for nearly five years, from accepting that Huygens' principle had anything useful to say about diffraction. And when he did develop his final theory the ether served him primarily as a strong physical foil with which to counter the physical absurdities (as he saw it) of the optical particle. What it most certainly did not serve him as was a generator of mathematical theory, which in this area at least emerged despite the physical image that Fresnel held of the ether, not because of it.

The same can be said, only even more strongly, of Fresnel's work on polarization. Here he was stymied by his inability to disentangle different types of wave propagation from one another, and to associate only one of these types with polarization. The old way of thinking about polarization—one that Fresnel had much trouble abandoning—linked nicely to the natural image of an optical disturbance as a pressure wave in the ether. The asymmetries that are the core of polarization phenomena required something very different, and the only possibility, Fresnel saw very early on (at Ampère's suggestion), was a transverse wave. This raised problems (such as how the planets could move through such a comparatively rigid thing), but these issues had little effect upon Fresnel in the early years, because even given transverse waves he still did not see how to produce a unified account for polarization. Only after he conceived the entirely novel idea that the asymmetry in the wave front can vary in direction with time as the front propagates did he discover how to construct an entirely new account. And in this development the ether was again more of a stumbling block (as it had been in diffraction) than it was an aid to creation.

But, one might object, there was certainly at least one area where Fresnel did use the ether in much the fashion that the common image of the period suggests—as, that is, a model for producing new mathematical theory: namely, in his creation of the wave surface for biaxial crystals, those which have two optical axes, whereas Iceland spar has only one. His final account, written in 1824, presents an elaborate structure for the ether—one that can, it seems, be used to generate the very difficult series of surfaces that lead to Fresnel's biaxial generalization of Huygens' construction for Iceland crystal. And it was after all this generalization that led, in the hands of Hamilton and Lloyd, to the discovery of conical refraction, which the philosopher William Whewell and others crowed over as concrete evidence of the wave theory's power.

One would nevertheless be wrong to see creative ether dynamics even here, because Fresnel used the image of an elastic ether for precisely one purpose only: to justify physically the assertion that the wave speeds that correspond to an oscillation in a given direction can depend upon that direction. Upon this assertion, and this alone, he built his theory. It led, after several false steps and much hard experimental work, to Fresnel's creation of the surface of elasticity (which is substantially what we now term the optical indicatrix), from which followed the so-called normal surface for front speeds, and thence (through some inspired guessing rather than through rigorous deduction) the wave surface. The ether appears nowhere in this sequence.

However, in his final, published version Fresnel did devote a few pages to suggesting a foundation for ether dynamics—a foundation that he certainly knew to have been flawed, but one that at least seemed to profound a physical basis for the complicated series of steps which derived from the surface of elasticity, a surface that had its physical seat, after all, in ether itself. What, then, was this basis? The ether, for Fresnel (and for many later optical scientists as well) was a system of point masses that exert central, repulsive forces upon one another. This much had almost certainly been in the back of Fresnel's thoughts for many years. But what did he, or even could he, do with this image? Creatively he did scarcely anything at all with it. But he did provide the elements of a foundation for the wave properties of such a system in a way that provides a useful contrast to what the mathematician Cauchy did in the 1830s.

Fresnel developed his ether dynamics on two levels, the highest was closely tied to the immediate demands of optical theory, while the lower was intended to provide a foundation for the upper. The latter asserts essentially two things. First, that the vector reaction to the displacement of any given point is a linear function of the displacement's components. Second, that the coefficients governing the reaction are constants. From this, Fresnel was able rigorously to generate his 'surface of elasticity'. To prove it, however, he made the questionable assumption that the reaction generated by a given point's displacement can be calculated by holding every other point fixed. He was himself entirely aware that the assumption was not a reasonable one, but only by making it could he reach the surface of elasticity from some facsimile of ether dynamics.

This again shows that for Fresnel the behaviour of the ether did not have to be known in order to formulate his new optical theory, including that part of it which does impinge most directly upon the ether's properties: namely, double refraction. But one must not take this too far. Without the ether as a physical underpinning it would have been extremely difficult for contemporary physicists to use the wave theory without feeling that it lacked a firm physical foundation, whereas the alternative to it, which was based on sets of rays, did at least have an apparently firm physical basis in the image of the optical particle governed by forces. Indeed, one of the criticisms that wave theorists often threw at their predecessors was that the ether was a much sounder physical foundation than was the optical particle—in part because, as Fresnel had begun to show, and as Cauchy showed with much greater rigour, one could actually generate from the mechanical structure of the ether theorems that lead to formulas, and not only in double refraction.

It is unlikely that had Fresnel lived he would have continued working much as he had in constructing the wave theory between 1815 and his death in 1827. On the contrary, both published work of his as well as manuscript evidence indicates that the ether was becoming increasingly important to Fresnel as a generator of theory. One phenomenon in particular seemed to him to require an intimate knowledge of the ether's behaviour: the dispersion of light. In 1823 Fresnel advanced a qualitative explanation of dispersion, based on ether dynamics, that became immensely influential in the 1830s, particularly for Cauchy. Dispersion, according to Fresnel, depended on the spacing and forces between the mutually repelling particles of the ether. The clear implication of Fresnel's remarks was that theory had to address these two factors (spacing and force) in order to deal quantitatively with dispersion. This was precisely what Cauchy had begun to analyse, in a different context, in 1827, the very year that Fresnel died. And Cauchy's mathematics for dispersion set a programme of research that was pursued in France, Britain, and Germany during the 1830s and (in Germany and France) into the 1850s. During the 1830s, in fact, optical theory became for a time nearly synonymous with Cauchy's ether dynamics.[14]

Cauchy's early ether dynamics dates to 1830. By that time he was well prepared to see the possibilities in Fresnel's suggestions, and in particular immediately to correct the great lacuna in Fresnel's dynamics: namely, the assumption that the ether lattice remains essentially rigid even when one of its elements is displaced. Assuming only that the displacement is small in comparison with the distances between the points, Cauchy was able to generate a differential equation in finite differences for the motion of an arbitrary lattice element in function of the differences between its displacement and that of every other element in the lattice. This equation became so common in optics articles and texts during the ensuing decade that it should be called canonical. To produce from these intricate expressions a theory of dispersion whose constants reflected ether properties was no easy task, as the almost impenetrable mound of computations and approximations that Cauchy eventually published in 1836 would seem to show. In essence, Cauchy first imposed symmetry conditions on the lattice and then calculated the differences in the displacements by means of a Fourier series. After a very great deal of tedious work these elephantine calculations birthed what some of his contemporaries considered to be a very small mouse: a series for the refractive index that has since been known eponymously as 'Cauchy's series'. In fact, Fresnel had himself generated precisely the same series, though he had never published it, though in his (unlike Cauchy's) the physical meanings of the associated constants remained undeveloped.[15]

Nevertheless the series did seem to work empirically, though precisely what even that meant was open to question, since it had so many freely disposable constants that at least one English physicist (Samuel Earnshaw) felt that it amounted to an identity. Moreover, there was more than one way to extract a dispersion formula from Cauchy's general structure, and even then it was not clear just what the manipulation of constants meant physically. Still, there can be no doubt but that many of Cauchy's contemporaries, including William Rowan Hamilton, were deeply impressed by his ability to obtain a dispersion formula.

Articles on Cauchy's theory poured forth, particularly in Britain, and a very great deal of thought was devoted to it. But, despite the structure's difficult birth of a dispersion law, it was in other respects almost entirely barren. Insofar as the laboratory was concerned, Cauchy's structure was an alien being. He used Fresnel's wave surface as a sort of intellectual laboratory against which to test its results, but nothing empirically novel ever emerged. It remained an exercise in the highest reaches of abstract model-making, or rather it did so until a moody Irish mathematician named James MacCullagh found its Achilles heel.

Many years earlier, Arago had discovered the phenomenon of optical rotation, in which the plane of polarization of light is rotated on passage through quartz. This discovery had in fact been an important clue for Fresnel in his development of a new understanding of polarization. Cauchy's ether lattice had in principle to embrace all optical phenomena, including this one, the procedure being to effect suitable adjustments of the constants that describe the state of the lattice in equilibrium. The equations of the structure are so complicated and difficult to penetrate with any degree of ease that this gargantuan claim seemed entirely plausible, particularly in view of Cauchy's novel success for dispersion, and his early success for double refraction. But in 1841 MacCullagh unequivocally demonstrated that the very structure of the equations forbids optical rotation: the kind of lattice—or better put, the kind of mathematics for the lattice that Cauchy had created—fails without hope of salvation to capture a very important phenomenon indeed.

As far as MacCullagh was concerned, this ended what he in any case thought to be the building of fanciful structures to encompass things that are already known perfectly well (since, like Earnshaw, MacCullagh did not consider the dispersion law to be much more than a sort of interpolation, something with about the same empirical force as a Taylor series expansion). In Britain, Cauchy's structure did indeed disappear rapidly, but not simply because MacCullagh had dealt it the analytical equivalent of a mortal blow. In fact, the structure could be salvaged; Cauchy himself never had any doubt that it could be, and his several French (and later German) supporters agreed with him. Indeed in 1847, six years after MacCullagh's mathematical tirade, a French abbé by the name of Moigno produced a hefty *Repertoire d'Optique Moderne* which was essentially a long argument for the wonders that Cauchy had produced—one that simply treated optical rotation as a troubling little problem that could be made to go away. The problem could be overcome, but only by creating a new mathematical structure for the lattice that made Cauchy's earlier equations seem the soul of simplicity by comparison.

To solve the problem Cauchy in effect created the mathematical theory of periodic structures using concepts that had recently been introduced by Auguste Bravais in crystallography. In terms of his old equations, what this meant was that the 'constants' had now to be treated as periodic functions of position. Cauchy was able to demonstrate, after a truly acrobatic performance with the vast array of relations that periodicity provided, that there are new low-order terms that can capture optical rotation. This took place in 1850, nearly a decade after MacCullagh's critique. By that time the early, powerful influence of Cauchy's conception of the ether lattice had vanished entirely in Britain and was becoming less common in France, though it continued to crop up in Germany, which was only just then beginning to contribute in substantial numbers to research in mathematical optics.

The physical models that gripped theorists of the optical ether, as well as the problems that they addressed, have their origins in Fresnel's work, as we have seen, but here we find for the first time physicists attempting to grapple quantitatively with the models themselves rather than relying exclusively on the more fundamental principles that the models were designed to encompass. In this respect the period of the 1830s during which Cauchy's structure was most avidly pursued resembles somewhat the late 1890s and early 1900s, during which quantitative models for the microstructure of matter were first built that could yield testable optical formulas. Indeed, there was no earlier attempt by physicists to develop on this scale the implications of a detailed model for things that cannot be observed. Until the advent of the electron, these kinds of investigation remained unusual and even highly controversial.

By the early 1840s at the latest there were few physicists or mathematicians who disputed the wave theory's fundamental principles, and of equal importance, many were by then capable of applying it at the high level of mathematical detail that it required. During this period, investigations based on the wave theory began to evolve into two related but distinct areas. Work designed, like Cauchy's, to pursue the implications of ether mechanics continues. However, the fact is that by the early 1850s very few published papers in optics directly concern mechanical deductions; rather, the vast majority of them involve the working out of the wave theory's mathematical principles and their applications.

Ether mechanics proper has received the lion's share of historical attention, though it really has the mouse's share of contemporary optical production. Relativity has so thoroughly coloured modern views that practically the entire Victorian era has often been treated as a sort of prelude to it. This is incorrect, if only because no-one in the nineteenth century knew what was coming in 1905. This is why Cauchy's ether mechanics was something of an aberration rather than a central part of optics: ether mechanics was simply not at the centre of most optical work during the period, though there were times when it did move closer in as it seemed to promise something useful. The preponderance of work concerned such questions as the proper form of the solutions to the wave equation, how to build optically significant equations for various kinds of media, and so on—some of which did touch on the ether, but most of which was substantially independent of it.

There is, however, no doubt but that a considerable amount of powerful effort was devoted to ether mechanics, particularly before 1870 or so, though after the early 1840s a second trend emerged that differed considerably from Cauchy's. It emerged first in the work of James MacCullagh and of George Green, followed in the 1850s and 1860s by that of George Stokes and others. Here lattice equations played no role at all; indeed, they were conspicuously avoided. Instead, the ether's equations were obtained by manipulating a macroscopic potential function, which took on energetic significance after the 1850s and which (in the case at least of Stokes) could be motivated by considerations of the microworld. These two methods—what one might call the molecular and the macroscopic—shared the assumption that matter affects the ether by altering the coefficients that determine its equations of motion (though the theories diverge from one another over whether to alter density or elasticity coefficients).

However, even during the 1840s some physicists thought this assumption was inadequate and believed ether and matter had to be treated as distinct systems that are dynamically connected to one another. This way of thinking disappeared, or at least became a conviction rather than a programme, until the discovery of anomalous dispersion in 1870, on which more below.

15.7 Optical Problems: Huygens' Principle, the Diffraction Integral, and the Nature of Unpolarized Light

Although Fresnel had gone far in developing the principles of wave optics, as well as its mathematical foundation, many thorny issues remained. There were, as we just saw, problems concerning the physics of the ether itself, problems that were dealt with in different ways throughout the century. Other difficulties, which were even more fundamental, stimulated considerable research. Perhaps the central problem left by Fresnel in the foundations of optics concerned Huygens' principle. We saw above how he used the principle to produce the integrals that he then applied with great success to diffraction. However, Fresnel's integrals were asserted by him without his having had in hand the partial-differential equation that they were meant to be solutions of, much less the methods and techniques necessary to solve such a thing under appropriate sets of boundary conditions. Even worse, Fresnel had been forced to conjecture, rather than to deduce, the form of the factor that governs the contribution of a given secondary wavelet to the succeeding front along a specific direction (which was later termed the 'inclination factor'). Moreover, Poisson had challenged the very foundation of Fresnel's integrals—his particular use of Huygens' principle.[16] All these difficulties derive from the fact that Fresnel did not possess a partial differential equation for wave optics, much less a method for obtaining its general solution.

One might say with some justification that the problem Fresnel faced had its origins in the uncertain physical underpinnings of wave optics, but in fact, the major difficulty was essentially mathematical. Although the three-dimensional wave equation had been developed during the eighteenth century, its solutions had not been treated in a form suitable for questions concerning the general behaviour of entire wave fronts, not least because the mathematics for boundary conditions remained undeveloped. Poisson was the first to investigate these sorts of problems. Starting with a closed surface at a given time, and with a given distribution of velocity and condensation, Poisson deduced from the wave equation an expression for the disturbance at any point of space and at any time in terms of an integral over the originally specified surface. In effect, Poisson examined the case of an isolated, spherically symmetric pulse, and he specified particular conditions that it must satisfy. These conditions, it seemed to him, were not easily satisfied by the demands Fresnel placed on Huygens' principle. Poisson admitted the empirical cogency of Fresnel's diffraction integrals, but he rejected Fresnel's justification for them.

Poisson's work on the wave equation accordingly had nothing to offer wave optics—quite the contrary, it seemed to pose problems for it, not the least of which was Poisson's further claim that Fresnel's inclination factor was itself highly problematic. In 1849 Stokes addressed this latter problem, which, ironically, he attacked by using the very solution that Poisson had himself developed. Stokes in fact attempted to develop a theory for the diffraction of a vector wave and from it to deduce the very same inclination factor that Fresnel had long before conjectured, and that Poisson had criticized.[17] However, Poisson's solution was not suitable for addressing the general questions raised by Fresnel's use of Huygens' principle. For his solution began with an aperiodic front and found subsequent fronts by integration over this initial one. Huygens' principle as used in diffraction theory, however, requires considering an infinitely long, periodic disturbance. Indeed, arguments based on periodicity lay at the heart of Fresnel's analysis, and were precisely what Poisson refused to accept.

The route to a useful, general method of front integration, based on the solution properties of the wave equation, was first developed by Hermann Helmholtz in 1860.[18] To analyse the behaviour of organ pipes, he assumed periodicity and reduced the equation to the form $\nabla^2 w + \left(\frac{2\pi}{\lambda}\right)^2 w = 0$ (wherein λ represents the wavelength). The critical step in Helmholtz's analysis involved his use of Green's theorem, which permitted him to express the disturbance at a point as integrals over a bounding surface; these integrals contained the values of the function and its normal derivatives over the surface. Helmholtz's discussion was limited to spherical, harmonic waves, and he did not extend it to diffraction since he was not considering optical disturbances. As a result, difficult questions concerning the boundary conditions to be applied under these circumstances did not arise.

In 1882 Gustav Robert Kirchhoff applied Helmholtz's use of Green's theorem to wave optics. Kirchhoff first extended Helmholtz's result to the general case of an arbitrary, infinitely long disturbance, developed surface integrals suitable for diffraction theory, and then specialized to the case of a purely harmonic form.[19] Kirchhoff's formulae, however, required assuming inconsistent boundary conditions over the diffracting surface, and this inconsistency generated a great deal of subsequent discussion, culminating in a reconsideration of the problem on the basis of electromagnetic theory by Arnold Sommerfeld in 1896, though issues concerning Kirchhoff's analysis remain to this day. Sommerfeld's theory, which made special use of Riemannian spaces, succeeded well for diffraction by perfectly reflecting screens, but still left problems for black, or perfectly absorbing ones. These several analyses, as well as others of the period, reduced Huygens' principle to, at best, a statement concerning the differential terms that appear in general diffraction integrals. It had been stripped nearly altogether of direct physical significance. In the late 1930s, however, B. B. Baker at the University of London and E. T. Copson at the University of St Andrews published an extensive discussion that attempted to retrieve, insofar as possible, the physical content of the principle.[20]

Huygens' principle and the diffraction integrals were not the only problems that troubled wave scientists as they sought to develop and to expand the dominion of the system. Polarization itself posed particularly hard problems. In Fresnel's new optics, polarized light consisted of a directed quantity, located in the wave front proper, whose magnitude and direction change in a calculable manner over time as the front moves through space. To provide an appropriate mathematics for this entity, Fresnel employed a method of orthogonal decomposition: any disturbance could be specified by providing the parameters that characterize its components along a given pair of mutually orthogonal axes. In general, each component has the form $a \cos(\omega t + \varphi)$, where a specifies the component's amplitude and φ its phase. This worked extraordinarily well for handling all the forms of polarized light known in Fresnel's day, as well as for predicting the character of a form that had not yet been investigated (namely, elliptically polarized light).

This system could not, however, easily deal, in the form introduced by Fresnel, with unpolarized or even partially polarized light, because these kinds of polarization do not have stable values of amplitude and phase. Indeed, in Fresnel's optics unpolarized light is defined as light whose phase at least varies randomly over time, while partially polarized light must in some fashion have restricted variations. In neither case can one directly employ for purposes of analysis the decomposition that Fresnel had introduced for polarized light.

In modern terms, Fresnel constructed wave optics on the basis of an amplitude formulation, in which the amplitude and phase of the wave are used directly for purposes of analysis. Yet amplitude and phase cannot be detected directly, they can only be inferred from experiments that work with what can be observed, namely optical intensities and angles. Since unpolarized and partially polarized light do not have stable amplitudes and phases, any system that seeks to incorporate them into a consistent general scheme must work with intensities and angles, and not with amplitudes and phases, with what may be called an intensity formulation. Developed by George Gabriel Stokes in 1849, the new system characterized polarization in terms of four parameters, each of which could be directly observed. The Stokes parameters did not, however, have much contemporary significance (and were not for that matter pursued extensively by Stokes himself) since at the time there were no pressing physical questions to which they could be applied.[21]

15.8 Optics after 1840

Geometric representations of complex numbers had been developed in 1797 by the Norwegian Caspar Wessel, and in 1806 by a Swiss, Jean-Robert Argand. They had not, however, found a natural use in many physical problems. Indeed, wave optics was the first area of physics in which complex numbers provided specific benefits that could not otherwise be obtained. They appeared for the first time in Fresnel's 1823 analysis of internal reflection. When light strikes within, for example, a glass prism placed in air beyond a specific angle of incidence, then it is totally reflected. Fresnel's equations for the ratio of light reflected to light incident yield complex numbers precisely at and above this incidence. In seeking to interpret this apparent failure of his expressions to retain physical meaning, Fresnel developed the first productive (albeit hardly rigorous) use for such things.

He reasoned in the following manner. First of all, Fresnel remarked, the complex expression that he did obtain under these circumstances has the characteristic that the sum of the squares of its real and imaginary parts is equal to the square of the incident intensity. This suggested to him the following interpretation. The light must be totally reflected, and it undergoes some phase shift as well. Using the general decomposition that Fresnel had originally produced for diffraction, such a phase shift can be understood by separating the resultant wave into two parts that differ in phase from one another by 90°. One of the parts has the same phase as the incident wave, and the amplitudes of the parts must be the cosine and sine of the phase of their resultant. In the case of total internal reflection there are two terms (one imaginary, the other real) whose real squares sum to 1. Suppose, Fresnel reasoned, that each of these terms represents the amplitude of one of the parts of the usual quarter-wave decomposition. Then the light will be completely reflected, and it will also be shifted in phase by an angle whose tangent is the ratio of the real to the imaginary part of the complex expression for the reflected amplitude, if we assume that the real part is the one that has the same phase as the incident wave.

In subsequent years Fresnel's interpretation was taken up and further developed by, among others, Cauchy, who himself made fundamental contributions to complex function theory. Cauchy analysed reflection from metals under the assumption that, in them, the index of refraction itself becomes complex, but that the usual Fresnel expressions for the reflected amplitudes remain the same in form. The Irish mathematician James MacCullagh independently hit on the same idea. The Frenchman Jules Jamin was the first to compile tables of the metallic constants for various metals and wavelengths. This approach reached its fullest mathematical form in the German Friedrich Eisenlohr's treatment of the subject, and his equations were widely used during the latter part of the nineteenth century.[22]

The topic of metallic reflection raised several issues that puzzled many scientists during the last quarter of the nineteenth century. Extensive experimental work by, among others, Georg Quincke had shown that for all known metals the real part of the index of refraction must decrease with frequency, implying that metals should (were it observable) exhibit precisely the opposite dispersion of ordinary transparent bodies.[23] Nor was this the only difficulty, for the same measurements also indicated that the wave speed in metals should be more than five times greater than the speed of light *in vacuo*, which raised a number of questions concerning how the specific characteristics of matter affected optical phenomena. Our account of wave optics can close appropriately with a brief discussion of the major changes that began to occur in the

1870s and 1880s, as scientists in Germany shifted their attention from the behaviour of the optical ether itself to issues concerning the interaction between ether and matter.

In 1870 the Danish scientist Christiansen observed a phenomenon that came to be known as 'anomalous' dispersion. Using the aniline dye fuchsin dissolved in alcohol, he determined that the refractive index of the solution increases from the B to D spectral lines, decreases from D to G, and then increases again after G.[24] The existing interest, in Germany, in metallic reflection fuelled concern with Christiansen's discovery, for the peculiar optical properties of metals could be related to his anomalous dispersion. Indeed, scarcely two years after the discovery the German Wolfgang Sellmeier developed a quantitative mechanical theory for it according to which all forms of dispersion are due to the interaction of ether vibrations with the natural oscillatory frequencies of material molecules. The guiding ideas of his account were that the wave equation of the ether is itself unaffected by the presence of matter particles, but that energy must be abstracted from an ether wave in order to displace the massy particle, which is tied elastically to a fixed location. Employing an analysis based solely on energy considerations, Sellmeier deduced a formula for dispersion that yielded the major features of the phenomenon.[25]

The critically important aspects of Sellmeier's theory were its tacit assumptions that neither the elasticity nor the density of the ether itself should be manipulated, its properties remaining effectively fixed, though Sellmeier was not dogmatic on this point. For him these assumptions were primarily conveniences, for he held that ether and matter together actually constitute a dual lattice of point masses. All refractive phenomena then implicate a mechanical resonance action in which ether waves must move massy particles of matter as well as the substance of the ether itself. What Sellmeier's theory lacked was a mathematical representation of what occurs at resonance between the frequency of an ether wave and that of a material particle, which is precisely where absorption takes place.

In 1875 Helmholtz appropriated Sellmeier's basic ideas and reformulated them mathematically in mechanical equations of motion that were capable of dealing with absorption as well as dispersion. Helmholtz's equations were based on the commonly received idea that optical absorption involves the transformation of light energy into the 'inner, irregular motion of the molecules' of matter: that is, into heat. To effect this transformation a force is necessary, and this is what was missing in Sellmeier's theory. Helmholtz's account formed the basis for a great deal of subsequent optical theory, eventually appearing (in changed form) in early electromagnetic optics as well.[26]

Helmholtz was well aware that the fact of absorption (the exponential decrease in light amplitude with distance) requires the presence in differential equations of a term proportional to a velocity (more generally, to an odd-order time derivative), but he also knew that this term cannot appear in the ether's equation of motion since, *in vacuo*, no absorption occurs. Helmholtz's solution to the problem was at once obvious and unprecedented: he constructed a distinct equation of motion for matter that itself contains a velocity-dependent force. The idea was that the material particles, driven by ether waves, are subject to two other forces that emanate from the surrounding matter particles: an harmonic force of restitution and a frictional force of resistance. As the matter particles absorb energy from the ether waves, the optical energy decreases, and the absorbed energy is converted into thermal motion by the frictional force. Just how the latter transformation occurs was foreign to Helmholtz's theory. Neither of the two material forces were supposed to act on an entire molecule. Rather, each molecule consists of a massive central core, which hardly moves when struck by an ether wave, together with a light, moveable particle; the latter is resisted frictionally in its motion, and the harmonic forces tie it elastically to the massive core.

The mutual actions of ether and matter that cause energy transformations between them must, Helmholtz reasoned, satisfy the principle of action and reaction: whatever force represents the action of matter on ether in the latter's equation of motion must appear with the opposite sign in the former's equation of motion. Helmholtz assumed on mechanical grounds that this mutual action is directly proportional to the difference between the displacements from equilibrium of ether and the light, moveable particle of matter. He further treated both media as effectively continuous and interpenetrating, so that the ethereal and material displacements are continuous functions of time and distance. This gave him partial differential equations. The equation of motion of the ether consists of the usual one for an incompressible, isotropic, elastic continuum, to which a term is added to represent the action of matter upon ether. The material equation of motion contains three forces: the ether term (with reversed sign), the frictional force, and the harmonic action. These *twin equations* constitute the mathematical structure of Helmholtz's theory, for they lead at once to a wave equation that is easily applied to all forms of dispersion and to absorption—one that agreed well with laboratory results when proper choices were made for the various constants, as Helmholtz showed in some detail.

Helmholtz's theory was immensely influential, and not only in Germany. During the next fifteen years, numerous German physicists, including Eduard Ketteler, Eugen Lommel, and Woldemar Voigt, used it in one way or another to construct mechanical theories of phenomena in physical optics. These theories, like Helmholtz's, generally gave little detailed consideration to the actual molecular structure of matter, preferring instead to employ simple and usually *a priori* terms in the material equations.[27] Among the British, Lord Kelvin (1884), in his acclaimed *Baltimore Lectures*, based most of the intricate mechanical models for which he soon became famous directly on the Helmholtz–Sellmeier model. Kelvin's work, however, differed substantially from contemporary German accounts in that his goal was to construct a continuum representation for ether and matter.

Helmholtz's twin equations, and the German and British treatments of them before the 1890s, are purely mechanical. By 1878 the idea underlying the twin equations had already been used in electromagnetic theory. H. A. Lorentz in the Netherlands had tentatively supposed that the ether's properties are in themselves invariant and that optical effects are due to the effect of inner electrical motions of material particles. He reaffirmed this idea when he deduced for the first time an electromagnetic formula for dispersion by actually constructing two linked sets of equations: one for the ethereal polarization, and the other for the motion of a moveable charge in a molecule with a fixed central core. The links between the two equations were, on the one hand, the polarization in the invariant ether effected by the moveable charges, and on the other, the driving force exerted by the ether polarization on the charge.[28]

The correspondences between this and Helmholtz's (1875) mechanical theory are manifest. However, Lorentz viewed the ether as itself a polarizable substance, with the result that he drew no clear distinction between material and ethereal polarization. In any case, Lorentz's 1878 theory apparently had comparatively little influence, except perhaps in Holland, for it involved the sort of detailed microphysical computations and presuppositions which, even as late as the mid-1890s, few German physicists were willing to employ. In order for Helmholtz's twin-equation approach to acquire an electromagnetic significance that most German physicists could easily grasp and approve, the equations had to be reinterpreted in a way that preserved both their formal structure as linked systems and their relative independence of detailed microphysical calculations. This was accomplished by Helmholtz himself. His new interpretation worked with Lorentz's to stimulate a thoroughgoing transformation in the foundations of optical science—one that increasingly sought explanations in the structure of molecules and eventually of atoms themselves.[29]

Bibliography

BAKER, B. B., and COPSON, E. T. (1939). *The Mathematical Theory of Huygens' Principle.* Oxford: Clarendon Press.

BEER, AUGUST (1853). *Einleitung in die Höhere Optik.* Braunschweig: Friedrich Vieweg.

BIOT, JEAN BAPTISTE (1814). Recherches expérimentales et mathématiques sur les mouvements des molécules de la lumière autour de leur centre de gravité. Paris.

BUCHWALD, JED Z. (1980a). 'Experimental investigations of double refraction from Huygens to Malus', *Archive for History of Exact Sciences* 21: 245–278.

——(1980b), 'Optics and the theory of the punctiform ether', *Archive for History of Exact Sciences* 21: 245–278.

——(1989). *The Rise of the Wave Theory of Light: Optical Theory and Experiment in the Early Nineteenth Century.* Chicago: University of Chicago Press.

——(2012). 'Cauchy's theory of dispersion anticipated by Fresnel,' in *A Master of Science History*, Archimedes, 30: 399–416.

CANTOR, GEOFFREY (1983). *Optics after Newton: Theories of Light in Britain and Ireland, 1704–1840.* Manchester: Manchester University Press.

CAUCHY, AUGUSTIN LOUIS (1839). 'Sur la quantité de lumière réfléchie, sous diverses incidences, par les surfaces des corps opaques, et spécialement des métaux,' *Comptes Rendus* 8: 553, 658, 961.

CHAPPERT, ANDRÉ (1977). *Etienne Louis Malus (1775–1812) et la théorie corpusculaire de la lumière.* Paris, Vrin.

——(2004). *L'Edification au XIXe Siècle d'une Science du Phénomène Lumineux.* Paris, Vrin.

CHEN, XIANG (2000). *Instrumental Traditions and Theories of Light: The Uses of Instruments in the Optical Revolution.* Dordrecht: Kluwer.

CHRISTIANSEN, C., (1870). 'Ueber die Brechungverhältnisse einer weingeistigen Lösung der Fuchsin; brieflicher Mittheilung.,' *Annalen der Physik und Chemie* 141: 479–480.

Dalmedico, Amy Dahan (1992). *Mathématisations. Augustin-Louis Cauchy et l'École Française*. Paris: A. Blanchard.

Dijksterhuis, Fokko Jan (2004). *Lenses and Waves. Christiaan Huygens and the Mathematical Science of Optics in the Seventeenth Century*. Dordrecht: Kluwer.

Eisenlohr, Friedrich (1858). 'Ableitung der Formeln für die Intensität des an der Oberflächen zweier isotropen Mittel gespiegelten, gebrochenen und gebeugten Licthes,' *Annalen der Physik und Chemie* 104: 346ff.

Helmholtz, Hermann von (1859). 'Theorie der luftschwingungen in Röhren mit offenen Enden,' *Journal für reine und angewandte Mathematik* 57: 1–72.

——(1875). 'Zur Theorie der Anomalen Dispersion,' *Annalen der Physik und Chemie* 154: 582–596.

——(1893). 'Electromagnetische Theorie der Farbenzerstreuung,' *Annalen der Physik und Chemie* 48: 389–405.

Jamin, Jules (1852), 'Memoir on metallic reflection,' in *Scientific Memoirs*, ed. Richard Taylor, Vol. 5. London: Richard and John Taylor, pp. 66–101.

Ketteler, E. (1885). *Theoretische Optik gegründet auf das Bessel-Sellmeiersche Princip*. Braunschweig: Friedrich Vieweg.

Kipnis, Nahum (1991). *History of the Principle of Interference of Light*. Basel: Birkhäuser Verlag.

Kirchhoff, Gustav Robert (1891). 'Zur Theorie der Lichtstrahlen [1882]', in *Gesammelte Abhandlungen (Nachtrag)*, ed. L. Boltzmann. Leipzig: Barth, pp. 22–53.

Kline, Morris (1972). *Mathematical Thought from Ancient to Modern Times*. New York, Oxford University Press.

Lloyd, Humphrey (1834). 'Report on the progress and present state of physical optics,' *British Association for the Advancement of Science Reports*: 295–413.

Lorentz, H. A. (1935–39a). '[1875] Sur la théorie de la réflexion et de la réfraction de la lumière', in *H. A. Lorentz: Collected Papers*, ed. P. Zeeman and A. D. Fokker, vol. 1. The Hague: Nijhoff, pp. 193–383.

——(1935–39b). '[1878] Concerning the relation between the velocity of propagation of light and the density and composition of media', in *H. A. Lorentz: Collected Papers*, ed. P. Zeeman and A. D. Fokker, vol. 2. The Hague: Nijhoff, pp. 1–119.

——(1935–39c). '[1892] La théorie électromagnétique de Maxwell et son application aux corps mouvants', in *H. A. Lorentz: Collected Papers*, ed. P. Zeeman and A. D. Fokker, vol. 2. The Hague: Nijhoff, pp. 165–343.

MacCullagh, James (1880 [1843]). 'On the laws of metallic reflexion, and on the mode of making experiments upon elliptic polarization', in *The Collected Works of James MacCullagh*, edited by J. H. Jellett and S. Haughton. Dublin: Hodges, Figgis, pp. 230–248.

Poisson, Siméon Denis (1866–70). '[1823] Extrait d'un mémoire sur la propagation du mouvement dans les fluides élastiques,' in *Augustin Jean Fresnel: Oeuvres Complètes*, ed. E. Verdet and H. de Senarmont, vol. 2. Paris: Imprimerie Impériale, pp. 192–205.

Quincke, Georg (1866–72). 'Optische Experimental-Untersuchungen', *Annalen der Physik und Chemie*, many issues.

Sellmeier, Wolfgang (1872). 'Ueber die durch Aetherschwingungen erregten Körpertheilchen und deren Ruckwirkung auf die ersten, besonders zur Erklärung der Dispersion und ihrer Anomalien', *Annalen der Physik und Chemie*: 399–421, 520–49 (145); 386–408, 525–44 (147).

Stokes, George Gabriel (1849). 'On the dynamical theory of diffraction,' *Transactions of the Cambridge Philosophical Society* 9: 1–62.

——(1901 [1852]). 'On the composition and resolution of streams of polarized light from different sources', in *Mathematical and Physical Papers*, vol. 3. Cambridge: Cambridge Unviersity Press, pp. 233–258.

Verdet, E., and Senarmont, H. de (eds.) (1866–70). *Augustin Jean Fresnel: Oeuvres Complètes.* 3 vols. Paris: Imprimerie Impériale.
Whewell, William (1837). *History of the Inductive Sciences from the Earliest to the Present Time.* London: John W. Parker.
Young, Thomas (1855). *Miscellaneous Works.* London: John Murray.
Shapiro, Alan E. (1973). 'Kinematic optics: a study of the wave theory of light in the seventeenth century', *Archive for History of Exact Sciences* 11: 134–266.

Notes

1. On seventeenth-century wave optics, see Shapiro (1973) and on Huygens, Dijksterhuis (2004).
2. On which, see Cantor (1983).
3. On Malus' discoveries see Buchwald (1980a) and Chappert (1977).
4. On Malus' theory see Buchwald (1989), chap. 2.
5. On which, see Chen (2000).
6. On Arago and the ensuing controversy with Biot, see Buchwald (1989), chaps. 3–4.
7. Biot (1814).
8. This letter, as well as Fresnel's works discussed below, are all reprinted in Verdet and Senarmont (1866–70).
9. Young's papers are collected in Young (1855). On his optics, see Kipnis (1991).
10. Full references to the original literature referred to in this and the succeeding two sections can be found in Buchwald (1989).
11. Lloyd (1834).
12. On which, see Chen (2000).
13. Whewell (1837).
14. On Cauchy's work in optics, see Buchwald (1980b) and (2012); Dalmedico (1992).
15. For a simplified, canonical deduction of the dispersion equations, see Beer (1853), pp. 187–215. Beer, in fact, generated the 'Cauchy' dispersion formula much as Fresnel had in his unpublished manuscript.
16. Poisson (1866–70).
17. Stokes (1849).
18. Kline (1972), pp. 693–694 on Helmholtz (1859).
19. Kirchhoff (1891).
20. Baker and Copson (1939).
21. Stokes (1901) [1852].
22. Cauchy (1839), Eisenlohr (1858), Jamin (1852), MacCullagh (1880).
23. Quincke (1866–72).
24. Christiansen (1870).
25. Sellmeier (1872).
26. Helmholtz (1875).
27. For an extensive contemporary account, see Ketteler (1885).
28. Lorentz (1935–39a), Lorentz (1935–39b).
29. Helmholtz (1893), Lorentz (1935–39c). For a synoptic overview of nineteenth-century optics, see Chappert (2004).

CHAPTER 16
....

THERMAL PHYSICS AND THERMODYNAMICS

....

HASOK CHANG

16.1 INTRODUCTION
....

Heat is a subject that has commanded people's attention through the ages, for practical as well as scientific reasons. It is still a major subject in introductory physics textbooks and courses, though the science of heat is now presumed reduced to classical or quantum-mechanical principles through statistical reasoning. This chapter covers the development of the physics of heat while it existed as a truly independent subject, which is to say, up to the mid-nineteenth century. Most attention will be paid to the important yet relatively neglected parts of the history, while well-known areas will be covered briefly with references to existing secondary literature.

The study of heat began to flourish in the late eighteenth century, particularly in the chemical communities of Scotland and France. Intense theoretical and experimental activity continued in this field in the first half of the nineteenth century, mostly in the tradition of the material theories based on the basic assumption that heat was (or at least could be conceptualized as) an all-pervasive, weightless and elastic fluid, most commonly called 'caloric'. Great advances were made in caloric-based theoretical treatments of thermal phenomena, which became more quantitative and systematic. Experimental knowledge developed continually both in extent and precision, often quite independently of theory. Also significant among the nineteenth-century developments was the relocation of the study of heat from chemistry to physics, partly prompted by the increasing interest in heat engines. In this chapter we survey some of the significant themes in the development of thermal physics up to the establishment of classical thermodynamics. Much of the early achievement in this field was lost when the assumptions of the existence and conservation of caloric were rejected in favour of energy conservation, but many elements survived in interesting forms, and the new thermodynamic theory brought a great deal of lasting insights whose significance reached far beyond the study of heat itself.

16.2 Thermometry and Calorimetry

The scientific study of heat started with the invention of the thermometer. That is a well-worn cliché, but it contains enough truth to serve as our starting point. The thermometer did not simply tell the temperature. It revealed many deeper things, including the distinction between temperature and heat, which became clear through cases of heat gain/loss without temperature change, or *vice versa*. These phenomena were major subjects of research well into the nineteenth century: changes of state (melting, boiling, and so on); temperature changes produced by chemical reactions and mixtures; the heating and cooling of gases by expansion and compression; and the heating of solids by percussion and friction. Once the distinction between temperature and heat was recognized, many important theoretical questions arose regarding latent heat and specific heat, which will be discussed in Section 16.3. There was a good deal of mutual reliance between theory and experiment in this field; some rough methods of thermometry and calorimetry were prerequisites before any serious theorizing could begin, just as some rough concepts of heat and temperature were required for making and using thermometers and calorimeters.

Galileo and his contemporaries were already using thermometers around 1600. By the late seventeenth century thermometers were very fashionable but still notoriously unstandardized. An important pioneer in practical thermometry was Daniel Gabriel Fahrenheit (1686–1736), who was responsible for the use of mercury in thermometers, as well as the curious scale which still carries his name. Fahrenheit's small and reliable thermometers were very popular, especially among physicians. Yet his most renowned customer, the Dutch physician and chemist Herman Boerhaave (1668–1738), noticed that Fahrenheit's own mercury and alcohol thermometers did not agree with each other—which turned out to be a general problem, also noted by the best thermometer-maker of eighteenth-century France, the naturalist R. A. F. de Réaumur (1683–1757). There are many surprising facts in the history of thermometry. For example, Anders Celsius (1701–1744) made the centigrade scale familiar to us today, but in his original thermometer the freezing point of water was marked as 100°, and the boiling point 0°. Generally, up to the late eighteenth century there was no standard method of making thermometers, even in a given country, and the disorder was such that the Royal Society of London in 1776 appointed an illustrious committee chaired by Henry Cavendish (1731–1810) to make definite recommendations about the fixed points of thermometry. (For a comprehensive general history of thermometry, see Middleton 1966). Before the thermodynamic absolute temperature concept came into experimental use (see Section 16.8), practical thermometry reached a state of near-perfection in the hands of Henri Victor Regnault (1810–1878) at the *Collège de France* in Paris, whose work on thermometry was a foundational part of his government-commissioned study to determine all of the empirical data necessary for the study of the steam engine. Regnault reviewed, criticized, and improved all relevant techniques of thermal physics used by his predecessors, and the description of his procedures and results took up three entire volumes of the *Mémoires* of the Paris Academy, published in 1847, 1862, and 1868–70. Regnault had an immense influence on experimental physics in general, in France and beyond.

There were three major problems in the establishment of thermometry, all of which reflected deep epistemological issues (see Chang 2004 for further details). First of all, which natural phenomena can be used as the 'fixed points' of thermometers? The philosophical difficulty here was knowing which phenomena happened at fixed temperatures before one had a reliable thermometer. In fact all manner of phenomena were proposed and used as fixed points: greatest winter cold and summer heat (Accademia del Cimento, *c*.1640); the melting point of butter (Joachim Dalencé, 1688); deep caves (Halley, 1693) and the Paris Observatory cellars (de la Hire, 1708); 'blood heat' (Newton, 1701) and the healthy body temperature (Fahrenheit, 1720); 'water hottest to be endured by a hand held still' (John Fowler, *c*.1727). By the mid-eighteenth century the freezing and boiling points of water became the most popular fixed points, but questions remained about the fixedness of these points, most of all because of supercooling and superheating phenomena. In the end it was impossible to establish fixed points firmly before making thermometers; rather, thermometers were made on the basis of unjustified fixed points, and such imperfect thermometers helped the development of thermal physics which eventually allowed a better knowledge of fixed points.

Secondly, which thermometric substance expands uniformly with real temperature? Many substances were in use, including air, alcohol (known as the 'spirit of wine'), and mercury (quicksilver). Initially thermometers were made without much thought as to the exact pattern of the thermal expansion of the fluid used in it; the simple division of the interval between fixed points into equal lengths embodied an implicit assumption of uniform thermal expansion. But it was soon apparent that different fluids had different *patterns* of expansion, so that thermometers made with different fluids disagreed in their readings between (and beyond) the fixed points, even if they were graduated to agree with each other at the fixed points. In the eighteenth century the choice of thermometric fluids became a serious question. Again, the difficulty is an epistemological one: without a trusted method of measuring temperature, how can one ascertain the pattern of the thermal expansion (that is, the relationship between volume and temperature)? Jean-André De Luc (1727–1817), Genevan geologist, meteorologist, mountaineer, and theologian, nearly generated a consensus on this matter by employing the method of mixtures, which had been pioneered by Brook Taylor and Joseph Black. Mix equal amounts of freezing water (at 0° centigrade by definition) and boiling water (at 100°, again by definition) in an insulated vessel; if a thermometer inserted in that mixture reads 50°, it indicates real temperature; such mixtures could be made in various proportions, in order to test thermometers at every point on the scale between two fixed points. De Luc (1772) carried out such tests systematically, and concluded that mercury was the best available liquid. However, the method of mixtures took the specific heat of each thermometric fluid to be constant, and this was a groundless assumption (see Section 16.3). Coming into the nineteenth century, theoretical opinion gravitated to the view that the thermal behaviour of gases was simple and uniform (see Section 16.4). Regnault's work gave partial experimental vindication to this view, but his verdict was a very cautious one: air thermometers were the only ones that had not clearly failed his test of 'comparability': namely, the requirement that two thermometers of the same type should agree with each other in every situation, if that thermometer type is to be trusted.

Thirdly, once a thermometric scale is established, how can it be extended beyond its original domain? This became an important practical question in industrial processes involving temperatures beyond the melting point of mercury and glass, and expeditions to cold places where mercury froze (in particular, Siberia and northern Canada). In fact, the freezing point of mercury was a point of great contention in the late eighteenth century, as that is precisely where mercury thermometers broke down. This pattern was exhibited again and again in low-temperature physics: when a previously unknown degree of cold was achieved, the physical regularity underwriting the best available thermometric method broke down; in this respect, what happened to the mercury thermometer at the freezing point of mercury ($-40°$C/$-40°$F) was like what happened to an electric-resistance thermometer at the point where the metal in use went superconducting. At the high-temperature end, the most intriguing phase from the early history was the work of Josiah Wedgwood (1730–1795), who founded the porcelain company that is still flourishing. Unhappy that he had no means of ascertaining exactly how hot his furnaces were, he created a clay pyrometer exploiting the fact that clay pieces shrank when exposed to high temperatures. But Wedgwood assumed that the contraction of clay was linear with temperature, and this turned out to be infelicitous. However, it would be wrong to think that Wedgwood did not do better because he was an uneducated artisan; he was dealing with a fundamental philosophical problem: namely, the impossibility of making an assured study of the thermal behaviour of a substance in a temperature range where there was no previously established thermometric method. Wedgwood investigated this problem in a methodical and sagacious way; his high-temperature thermometer was rejected only because it disagreed with various other thermometers, which agreed roughly among themselves.

Unlike thermometry, calorimetry has not yet received a great deal of attention from historians and philosophers of science. However, there are deep philosophical problems here too. There were no direct methods of determining the amount of heat transferred from one body to another, not to mention the total amount of heat contained in a given body. The only way of estimating the amount of heat absorbed or emitted was to observe the intensity of its effect. Two effects usable for this purpose were state-changes and temperature-changes. But inferring the amount of heat from the state change it produces (for example, how much ice is melted) required a prior knowledge of the relevant latent-heat value, and inferring it from temperature change (how much the temperature of a body of water is raised, for example) required prior knowledge of the relevant specific-heat value. Either way, the measurement of heat relied on the measurement of heat, so calorimetry suffered from a fundamental epistemological circularity.

Calorimetry based on the change of state was developed by Antoine-Laurent Lavoisier (1743–1794) and Pierre-Simon Laplace (1749–1827) in the form of their 'ice calorimeter', and it was employed very effectively in the investigations forming the basis of their famous *Mémoire sur la chaleur* (1783). In this instrument a metallic chamber containing a hot body was surrounded by crushed ice; the body was allowed to cool to the melting point of ice, and the amount of ice melted in that process was measured by collecting and weighing the resultant water. With this instrument Lavoisier and Laplace studied various phenomena ranging from the specific heat of solid bodies to the amount of heat produced by animals. The chief theoretical assumptions in this technique were firstly that heat was conserved, and that the latent heat required in melting a unit weight of ice at its melting point was always the same. Although these assumptions would have been difficult to verify actually, there seemed no particular reason to doubt them and no one challenged them. Unfortunately the Laplace–Lavoisier ice calorimeter suffered from practical defects, especially that not all of the water produced by the melting of the ice could be gathered to be weighed, since it tended to become trapped in the small spaces between the pieces of ice.

Since the Laplace–Lavoisier ice calorimetry was the only plausible attempt during this period to use changes of state for heat measurement, its practical failure meant that scientists had to rely on temperature changes in order to measure heat. If a certain amount of heat was given to a reference body of specific heat c resulting in the raising of its temperature by an amount t, it could be deduced that the amount of heat received by the body was ct. This technique, of course, makes calorimetry entirely dependent on thermometry. In addition, the procedure involved two major assumptions: perhaps it was not so problematic to assume the conservation of heat, but more contentious was the assumption that the specific heat of the reference body was constant throughout the range of temperatures involved in the experiment. This assumption was easily and commonly challenged, as we will see below, but in order for calorimetery to get off the ground at all, it had to be upheld at least as an acceptable approximation.

16.3 SPECIFIC AND LATENT HEATS

The notion of latent heat was at the centre of attempts to understand the phenomena in which heat and temperature were seemingly decoupled. When heat became 'latent'—that is, absorbed into a body without raising the temperature—what exactly happened to it? Why did it cease to register in thermometers (or be perceivable by our bodies)? The Scottish physician and chemist Joseph Black (1728–1799) was the most famous discoverer of latent heat, although the independent contribution of Johan Carl Wilcke (1732–1796) in Sweden has also been widely recognized. (Generally on the discovery of latent and specific heats, see McKie and Heathcote, 1935, and Roller, 1957, section 2.) Black's notion was that the heat apparently spent in melting ice, for instance, was not destroyed but merely converted into a different state, postulated to lack the power of affecting the thermometer. Although Black himself remained agnostic about the metaphysical nature of heat, his concept of latent heat fitted well into the notion popular among chemists that heat was to be treated as a substance that entered into chemical combinations with ordinary matter. This view was advanced, for instance, by Lavoisier, who went as far as including 'caloric' in the table of chemical elements in his authoritative textbook of the new chemistry, *Traité élémentaire de chimie* (1789). On this chemical view of heat, the latent caloric that entered into combination with particles of matter was the cause of increased fluidity when solids melted into liquids and liquids evaporated into gases; this latent caloric would become sensible again in condensation or congelation. The absorption and emission of heat in ordinary chemical reactions were also explained in the same manner. The notion of the chemical combination of caloric with matter was even incorporated into the terminology, of 'combined' versus 'free' caloric, used alongside the more phenomenological terminology of 'latent' and 'sensible' caloric.

In opposition to the chemical view of heat was that of William Irvine (1743–1787), a pupil and collaborator of Black's in Glasgow, who refused to postulate two different states of caloric. (See Fox, 1971 for a thorough discussion of the origin and development of Irvinist caloric theory.) According to Irvine (1805)[1] there was only one kind of caloric, and the amount of caloric contained in a body was the product of its capacity for caloric and its 'absolute temperature', which would be zero degrees at the total absence of caloric. This idea will be referred to as 'Irvine's hypothesis of heat capacity'. In Irvine's view, latent heat phenomena were consequences of changes in bodies' capacities for caloric, not due to any changes in the state of caloric itself. For instance, the change of state from ice to water involved an increase in heat capacity, which meant that more heat was needed just to keep the body at the same temperature. This was explained by an analogy of a bucket which suddenly widens; the level of water contained in it would go down, and more water would have to be put in just to keep the water level where it was before. The heat of chemical reactions was explained similarly, by pointing to presumed differences between the heat capacities of the reactants and the products. Irvine's hypothesis of heat capacity constituted a great advance on previous theories of heat, since it specified a precise relationship between heat capacity, temperature, and total heat, with theoretical elegance and great explanatory power.

Irvine backed up his theory by measurements which showed, for instance, that water indeed had a higher capacity for caloric than ice. But how did he measure the heat capacities? This was done by identifying the phenomenologically measured *specific heat* with the more theoretical notion of *heat capacity*. The standard method of measuring specific heat was the technique of mixtures: typically, a body at a relatively high temperature, say a, would be immersed into a reference body at a lower temperature, b; the specific heat, c, of the first body would be calculated from the resultant temperature of the mixture, m, as follows:

$$(a - m)c = (m - b)c_o$$

where c_o is the specific heat of the reference body, commonly taken as 1 for the unit amount of water.

Irvine's ideas were highly controversial and never commanded a consensus, but they were adopted and elaborated by a number of influential scientists, especially in Britain. Perhaps the most effective advocate of Irvine's ideas was the Irish physician and chemist Adair Crawford (1748–1795), who was active in London in later life but had attended Irvine's lectures while he was a student in Glasgow. Crawford applied Irvine's ideas in order to arrive at a new solution to the longstanding puzzle on how warm-blooded animals generated heat in their bodies (see Mendelsohn, 1964). Crawford's influential treatise on animal heat (1779, 1788) served as an important conduit of Irvine's ideas to other scholars, including John Dalton (1766–1844), the originator of the chemical atomic theory. Other important Irvinists included Sir John Leslie (1766–1832), Professor of Mathematics and then of Natural Philosophy in Edinburgh, and the Scottish chemist John Murray (d.1820), whose textbooks were quite influential in the early decades of the nineteenth century. As it turned out, most of the predictions made on the basis of Irvine's hypothesis were not confirmed, and Irvinism was also theoretically opposed by followers of Lavoisier's chemical caloric theory. Hence Irvine's ideas did not retain much support beyond the first decade of the nineteenth century, but his legacy did have an enormous relevance in the debates that continued. The rejection of Irvinism left a vacuum which needed to be filled in order to restore definiteness to the theoretical treatment of many crucial issues: the relation between latent and specific heat, the total heat content of bodies, the heat content of the vacuum, the adiabatic heating and cooling of gases, the definition of temperature, and so on.

Despite the theoretical controversy surrounding Irvine's hypothesis of heat capacity, the more phenomenological notions of latent heat and specific heat were shared by nearly all investigators. Much effort went into their experimental determination, partly so that the data could be used in testing the rival theories. But long periods had to pass before measurement practices were developed sufficiently to provide precise enough data for most theoretical purposes. The difficulties were not only matters of technical control, but also of theoretical interpretation. Aside from the basic challenges in thermometry and calorimetry discussed in Section 16.2, there were difficulties peculiar to the measurement of the specific heat of gases, so small compared to those of the reference bodies and of the containers used to hold the gases. The first recorded attempt was made by Crawford in the 1780s; as his Irvinist explanation located the origin of animal heat in respiration, it became important to measure the heat capacities of inhaled and exhaled gases. Crawford filled a bladder with various gases at high temperatures and plunged it into a vat of cold water. His results lacked precision, as the changes in the water temperature were very small and inaccurately measured, and improvements made by others during the next two decades did not go very far.

Moving the state of specific-heat measurements beyond these initial attempts was one of the important goals of the 1812 prize competition sponsored by the First Class (Mathematical and Physical Sciences) of the French Institute, which had replaced the old Paris Academy of Sciences in 1795. The competition was won by François Delaroche (1775–1813) and Jacques-Étienne Bérard (1789–1869). They used a constant-flow method, which involved passing a heated gas under a constant pressure through a spiral copper tube immersed in a calorimeter filled with water (Delaroche and Bérard, 1813). From the amount of heat imparted to the water in unit time, and the rate of flow of the gas, Delaroche and Bérard were able to calculate the specific heats (by volume) of various gases under constant pressure. This ingenious method, along with the care they took in eliminating the sources of heat loss, won Delaroche and Bérard wide acclaim, and their results on the specific heat of various gases were considered the best available for several decades. Their work established beyond doubt that the specific heats of various gases were different from each other, the volumetric specific heats ranging from 0.903 for hydrogen to 1.258 for carbon dioxide (taking that of atmospheric air as 1). The work by the other pair of contenders in the competition, Nicolas Clément (1778/9–1841) and Charles-Bernard Desormes (1777–1862), both industrial chemists, was based on less accepted notions and did not receive as much attention. The outcome of the 1812 prize competition had complex and significant theoretical implications, as will be discussed in Section 16.4. Although the basic conceptual difficulties regarding specific and latent heat remained unresolved, improvements continued in laboratory techniques. Some important advances, including the use of the law of cooling, were made by Pierre-Louis Dulong (1785–1838) and Alexis-Thérèse Petit (1791–1820), whose work (1816; 1817) on cooling, thermometry, and specific heats in the late 1810s came to be regarded by many as models of precision experimentation. The experimental work on specific and latent heats, like much else, was nearly perfected by Regnault, who had initially been drawn into thermal physics from organic chemistry through his fascination with the Dulong–Petit law of 'atomic heat', which he showed to be only approximately true.

Two significant points of debate illustrate the interesting questions raised by the concepts of specific and latent heat in their interrelationship. One crucial question affecting thermal measurements was whether the specific heat of a given substance depended on its temperature. This is an issue that illustrates very nicely the various ways in which calorists[2] disagreed sharply with each other on important points of detail. Dalton was one of those who suspected that a change in volume would be cause for a change in specific heat. Dalton thought that there would be a positive correlation between heat capacity and temperature, since the thermal expansion of a body would create more room for caloric to fit in. This seems to have been a common assumption. De Luc (1786–87) had analysed the capacity of bodies for 'fire' into 'geometric' and 'physical' components; the former was an indication of the amount of 'pores' in a body into which fire could fit, and the latter was an indication of the nature of those pores which determined how much fire could fit into a given amount of geometric space. For a given substance in a given state the physical capacity would be constant, but the geometric capacity would increase with the expansion of volume.

This sort of consideration raised a serious problem for the method of mixtures for the measurement of specific heat, which was founded on the assumption that the specific heats involved were not temperature-dependent.

A different kind of consideration regarding the temperature dependence of specific heat existed for those who subscribed to the chemical view of heat, and this is a point that came to play a crucial role in later investigations. Since the chemical theorists postulated that caloric could exist in a latent state that did not register in thermometers, they always had to worry about whether the caloric added to a body all remained sensible. Lavoisier and Laplace had given a clear and widely followed opinion on this issue as early as their 1783 memoir: when a body expanded upon heating one part of the absorbed caloric remained free, and was responsible for the raising of temperature; the rest of it, going latent, was responsible for expansion. Then the next question was what proportion of the absorbed heat went latent, and whether that proportion was a function of temperature. If it was, that would be a reason for specific heat to vary with temperature. The increasingly popular opinion on this issue in the first decade of the nineteenth century was that specific heat would *decrease* with increasing temperature. This view was advanced clearly by the abbot–mineralogist René-Just Haüy (1743–1822), in the second edition (1806) of his physics textbook commissioned by Napoleon for the newly established *lycées*. Haüy argued as follows: at lower temperatures the molecules of matter would be closer together and hence exert stronger attractive forces of affinity on each other; therefore a larger portion of the absorbed caloric would have to be spent on expansion, leaving less caloric available for raising the temperature. In other words, at higher temperatures a lower proportion of absorbed heat became latent so that a higher proportion remained available for heating; hence less total heat would be needed for raising the temperature by a unit amount. This conclusion was precisely the opposite of Dalton's, and there was no clear way of deciding this issue at the time. One of the fundamental problems, again, was that of circularity: in order to reach an experimental conclusion about the temperature dependence of specific heat, a precise thermometer was required; however, testing the correctness of thermometers required the knowledge of the temperature dependence of specific heat.

The other major point of contention involving latent and specific heats in the early nineteenth century was the determination of the absolute zero point of temperature. Although all calorists would have shared a notion of the absolute zero of temperature at which a body would become devoid of all caloric, only Irvinism made it possible to attempt the determination of this quantity. Since the product of absolute temperature and heat capacity gave total heat, the measurement of heat capacity and total heat allowed one to deduce the absolute temperature. Even the Irvinists could not measure the total heat of given bodies, but for the present purpose it was sufficient to measure the *change* of total heat in two successive states of a given body. For instance, with the melting of ice, one could set up an equation for total heat before and after, assuming the conservation of heat:

$$c_i x + L = c_w x$$

where c_i is the heat capacity of ice, c_w is the heat capacity of water, L is the latent heat of fusion, and x is the absolute temperature of ice/water at the melting point. (Since x is 0° on the centigrade scale, the absolute zero of temperature would be x degrees below zero in the centigrade scale.) Similar equations could be set up for any change of state, or for any chemical reaction in which there was an absorption or emission of heat. In each case the absolute temperature could be deduced from measured values of the latent heat (absorbed or emitted) and the heat capacities involved. Measurements were carried out for this purpose by Irvine and Crawford, and other Irvinists, including Dalton and Murray, often cited these measurements and also made their own.

For the chemical calorists, the very basis of these calculations was suspect, since their own notion of specific heat did not support the Irvinist equation for total heat. If total heat were to be given as the simple product of specific heat and absolute temperature, the specific heat would have to be constant throughout the temperature range. This was deemed to be unlikely, or at least unverified, as we have seen above. For the Irvinists, however, the expression for total heat was valid at each given temperature even if the heat capacity varied with temperature, and they could regard the ability to determine absolute temperatures as one of the advantages of their theory. As the theoretical arguments were inconclusive, the debated hinged on an empirical question: were the values of absolute zero obtained from various reactions consistent with each other? Most Irvinists believed that they had obtained consistent results, which served to enhance their belief in the Irvinist caloric theory. Their opponents, starting with Lavoisier and Laplace (1783), thought otherwise. The Irvinists generally responded by blaming the apparent discrepancies on the unreliability of heat capacity measurements. Since the calculated value of the absolute zero was highly sensitive to the heat capacity values ($x = L/(c_w - c_i)$, from the above equation), no clear experimental verdict on this question could be reached until the accuracy in the measurements of specific heat became very high. As it turned out, the better results such as those obtained by Delaroche and Bérard (1813) tended to refute Irvine, which contributed to the eventual demise of Irvinism. The chemical calorists, however, never did find a way of determining the absolute zero, and the topic gradually disappeared from the list of important problems.

16.4 THE PHYSICS OF CALORIC AND GASES

With the firm establishment of the caloric theories of heat in the last decades of the eighteenth century, scientists' ambition grew to develop a general theory of heat by pinning down the properties of caloric. As Robert Fox (1971) has discussed in great detail, much of this latter phase of caloric physics focused on gases. A clear majority of the calorists in the nineteenth century shared the belief that the nature of caloric was exhibited most clearly in the thermal behaviour of gases, rather than liquids or solids. This view had been anticipated by the German polymath Johann Heinrich Lambert

(1728–1777), and earlier by the French physicist Guillaume Amontons (1663–1705). However, the preoccupation with gases came to be widely shared only in the context of caloric theories. One recognized difficulty in that context was the impossibility of isolating caloric in a pure form—a fact usually attributed to its extremely strong affinity for ponderable matter. However, gases (including vapours) were considered to be quite close to a pure form of caloric. Particularly for those who adopted the 'Newtonian' picture in which the macroscopic behaviour of matter were to be explained by the attractions and repulsions between microscopic particles, there was assurance that the behaviour of gases would be due mostly to the actions of caloric, since in gases the particles of ordinary matter were separated so far from each other that the forces of affinity between them must be negligible. This assumption of the special status of gases, combined with the belief in the simplicity of nature, set the dominant tone in much of the investigations into thermal phenomena in the first half of the nineteenth century.

The major experimental impetus in this direction came early in the century, in the nearly simultaneous and independent works of Dalton and Joseph-Louis Gay-Lussac (1778–1850) in 1802, showing that all gases expanded by the same proportion when they were heated from one fixed temperature to another. Although this seems like a commonplace observation in retrospect, the experiments actually required considerable improvement in technique, including the drying of gases. Together with the much older Mariotte's (or Boyle's) law, Gay-Lussac's and Dalton's discovery was largely responsible for the impression that gases in general followed simple laws. It also led many people to assume that the thermal expansion of any type of gas must be equable or uniform, with the consequence that gas thermometers were to be regarded as the true thermometers of nature (see Section 16.1). As both Dalton and Gay-Lussac themselves stated clearly, this additional assumption of uniform expansion was logically groundless, and also empirically unverifiable in the absence of an independently validated thermometric standard, but it was endorsed widely.

The assumption of simplicity was not the only major reason for the early nineteenth-century preoccupation with gases. Part of the attention originated from the greater experimental and theoretical challenges that gases presented in comparison to liquids and solids, due to their great elasticity and compressibility. A whole new arena of investigation was opened up by the recognition that the pattern of thermal expansion could be influenced by varying the external pressure on gases, and the fact that externally enforced compression and decompression of gases resulted in temperature changes. This meant that pressure entered as a crucial variable in the thermal physics of gases, in addition to the others that applied to all bodies (volume, density, temperature and total heat). The considerations arising in this way ended up yielding tremendous insights applying not only to the behaviour of gases but also to general theories of heat.

One subject of intense dispute was adiabatic heating and cooling: namely, changes of temperature in a body that has no communication of heat with the outside. Assuming the conservation of heat, latent heat provided the only obvious way of explaining these phenomena; consequently, adiabatic phenomena provided an important arena in which competing theories of latent heat could be compared and tested. The modern notion that adiabatic phenomena are clear examples of the interconversion of heat and mechanical work was not available in this period, and caloric theories had no obvious difficulty in explaining these phenomena. The first observations of adiabatic temperature changes occurred in the course of experimenting with the air-pump, in fact, going back to Robert Boyle (1627–1691). The heating of a gas by mechanical compression and cooling by expansion were commonplace observations, as was the heating that occurred when air rushed into a vacuum. Eighteenth-century commentators on these phenomena included many illustrious names: Lambert; the physicist Marc-Auguste Pictet (1752–1825) and the naturalist and meteorologist Horace Bénédict de Saussure (1740–1799), both in Geneva; William Cullen (1710–1790), the teacher and predecessor of Black; and Erasmus Darwin (1731–1802), the grandfather of Charles Darwin. A good deal of motivation for the early discussions came from meteorology, but in France it was the Laplacian work on the speed of sound that created the bulk of interest. Laplace thought that the well-known discrepancy between the observed value of the speed of sound in air and the theoretical value predicted by Isaac Newton (1642–1727) in the *Principia* could be removed by considering adiabatic temperature changes that must take place as the propagation of sound waves compressed and decompressed little pockets of air. Laplace's protégés Jean-Baptiste Biot (1774–1862) and Siméon-Denis Poisson (1781–1840) ably filled out the details (Biot 1802; Poisson 1808). They could not actually calculate the correction factor, but Poisson made the best of that situation by deducing that *if* this correction factor were to resolve the discrepancy between the theoretical and the observational values of the speed of sound, the amount of temperature change in adiabatic compression or expansion would have to be 1° for a change of volume by 1/116 of the original volume, which served as the standard figure for quite some time.

By the first decade of the nineteenth century, adiabatic phenomena were widely recognized in all major centres of research on heat. However, there was hardly an agreement on their explanation. The Irvinists initially seemed to have an advantage. On the basis of their common assumption that the heat capacity of a body of gas decreased when its volume decreased, it was easy to explain why compression would lead to higher temperatures on the basis of Irvine's hypothesis of heat capacity. Dalton, for example, advanced such an explanation in a paper of 1802. The case of heating by air rushing into a vacuum was more challenging, but Dalton dealt with it by means of another common assumption, that the heat capacity of a given volume of evacuated space was larger than the heat capacity of the same volume of air. The heat capacity of vacuum may seem like a strange idea, and indeed it makes no sense if we are considering specific heat by weight. However, there is no such problem if the heat capacity is considered by volume, and the notion made eminent sense if one assumed that the vacuum could and did contain some caloric. The latter was a relatively common assumption at the time, at least since Lambert. The filling of the evacuated chamber by air can then be conceptualized as a decrease in the heat capacity of that space, which would raise the temperature unless heat could escape (in fact, if anything, the total quantity of caloric in that space increased, since some additional caloric was brought in by the incoming air). Similar explanations of adiabatic phenomena were given by other Irvinists, including Murray. Leslie (1804) gave a somewhat different treatment, on the basis of his belief that perfect vacua were never attained; therefore Leslie conceptualized the phenomenon as that of the compression of the very thin air that was already in the 'evacuated' chamber, and in effect reduced this problem to the more familiar case of heating by compression.

The Irvinist explanations did not go unchallenged. Gay-Lussac led the critique in his 1807 paper with an experimental argument against the assumption that the vacuum contained any caloric. He put a thermometer in a Torricellian vacuum, varied the volume of that empty space, and observed no temperature changes. This attack was followed by Delaroche and Bérard's (1813) more general assault on Irvinism, in the prize-winning memoir on specific heats mentioned in Section 16.3. Delaroche and Bérard determined the specific heat of steam as 0.8470 (taking that of water as 1), by weight; on the other hand, it is a straightforward consequence of the Irvinist explanation of the latent heat of evaporation that the heat capacity of steam should be greater than that of water. Similarly, their values for specific heats of gases refuted Irvinist predictions concerning the heat released or absorbed in chemical reactions. For instance, the formation of water was accompanied by the release of a good deal of heat, so Irvinist principles implied that its heat capacity should be smaller than the weighted average of the specific heats of the reactants, hydrogen and oxygen; Delaroche and Bérard's data indicated just the opposite. In France at least, these results were widely regarded as a decisive refutation of Irvinism.

How did the chemical caloric theory explain adiabatic phenomena? An intuitive picture that had a broad appeal was that gas molecules were like sponges soaked in caloric; as external pressure was applied and the volume of the gas was diminished, some of the caloric would be 'squeezed out'. The Irvinists could understand this by theorizing that the compression reduced the heat capacity of (the molecules of) the gas. The chemical calorists, on the other hand, needed to assume that the caloric squeezed out of molecules by compression existed in a latent state before the squeezing. This, fortunately, squared nicely with the view that thermal expansion was due to latent or combined caloric, discussed in Section 16.3. If the expansion were to be forcibly undone, the caloric that had gone latent in effecting the expansion would have to be returned to the sensible state. The vacuum experiments, however, continued to present theoretical difficulties. Gay-Lussac's experimental argument that the vacuum contained no heat had undermined the basis of the Irvinist explanation but did nothing to pave the way to an alternative explanation.

Clément and Desormes (1819), the unsuccessful contestants for the 1812 prize competition, chose to adhere to the notion that the vacuum contained caloric. Clément and Desormes made considerable progress in the Irvinist analysis of adiabatic phenomena, by treating the vacuum as an entity with a fixed heat capacity, which can mix with gases. Then the compression of a gas can be understood as the destruction of a measure of vacuum, and the absorption, by the air, of the heat previously contained in the destroyed vacuum. In the case of the evacuated-receiver experiment, all of the vacuum is destroyed leaving its heat to the air that enters from the outside. This way of thinking, combined with the standard Irvinist principle of heat capacity, allowed Clément and Desormes to make experimental determinations of many important quantities—by mixing various amounts of air and vacuum of various initial temperatures, recording the resulting temperatures, and measuring the heat capacity of air at various temperatures. Clément and Desormes's results were most interesting. Several determinations made on their principle all indicated values close to 0.4 for the ratio of heat capacities (by volume) of vacuum and air. The agreement between different determinations gave them confidence about the correctness of this result. They also estimated that the total quantity of caloric contained in a vacuum near room temperature (12.5° C) would raise the temperature of an equal volume of air under atmospheric pressure by 114°; this implied that compressing air adiabatically by 1/114 of its original volume (which would destroy 1/114 of the original vacuum) would raise its temperature by 1 degree. Clément and Desormes noted that this value was very close to that calculated by Poisson in 1807, which was 1/116. In addition, they produced a value for the absolute zero, −267.5° C, which was quite different from most of the earlier values given by Irvinists but uncannily close to −266.66° C, which was the value they obtained following Amontons's idea of extrapolating the thermal expansion curve until the point where the volume of the gas would vanish. As Fox (1971, p. 147) argues, all in all the confidence that Clément and Desormes seem to have felt about the quality of their own work is quite understandable, even if we feel compelled to attribute the pleasing results they obtained to fortunate coincidences or possibly a biased selection of data. However, the Institute's prize committee did not favour their work, and the eventual publication of their memoir in 1819 had little success in reviving the interest in the heat capacity of vacuum and in Irvinist ideas in general.

As the theoretical attention on gases intensified and the measurements of their specific heats improved, the difference between two kinds of specific heats came to be recognized more and more clearly. The distinction made was between (1) specific heat under constant pressure (commonly denoted c_p), that is, the amount of heat required in heating a gas by 1 degree while letting it expand under constant pressure; and (2) specific heat under constant volume (c_v), the amount of heat required in heating a gas by 1 degree while confining it to a fixed volume. The difference between these two specific heats was noted already in Crawford's measurements; however, Crawford did not do much with the distinction, since the numerical difference he detected was very small. Dalton also made a statement that c_p was greater than c_v, but did not elaborate on the distinction extensively. It was probably Haüy (1806) who made the first clear theoretical argument that there had to be a difference between the two specific heats, as a logical consequence of Lavoisier and Laplace's assumption that when a body is heated different portions of caloric were used in expanding it and in raising its temperature. Then it follows that even an expansion at the same temperature requires an input of heat. If one conceptualizes heating under constant pressure as a two-step process consisting of heating under constant volume and then expansion under constant temperature, it follows that c_p has to be larger than c_v, since $c_p = c_v + c_t$, where c_t is the amount of heat required for the expansion at constant temperature; this was, after all, the 'latent heat of expansion'. This kind of reasoning would become standard in later French caloric theory, as discussed below.

The actual measurement of the difference between the two specific heats presented many difficulties, particularly on the side of measuring c_V (Delaroche and Bérard's constant-flow method, for instance, only gave a measure of c_p). The first credible estimate came in the form of the ratio c_p/c_V (commonly denoted γ), not by direct measurement but through the theoretical discussion concerning the speed of sound. Following on the works of Biot and Poisson discussed previously, Laplace himself came into the scene with a paper published in 1816, in which he asserted that the correction factor on Newton's speed of sound was the square root of γ, and that the value of γ was 1.5 (see the reconstruction of Laplace's argument in Fox, 1971, pp. 161–165); the value required to bring theory and experiment into exact agreement was 1.43, not far from Laplace's theoretical estimate. When Gay-Lussac and Jean-Joseph Welter (1763–1852) finally made a reasonable experimental determination of this quantity and obtained the value of 1.3748, this was close enough to Laplace's value that a good deal of confidence was inspired about the reliability of this whole set of results.

With the gradual demise of Irvinism, the theoretical lead in the understanding of gases was taken in the Lavoisier–Laplace tradition. Soon after the execution of Lavoisier in 1794 during the Revolutionary Terror, Laplace became the clear leader in French thermal physics as well as much else in French science. In association with the chemist Claude-Louis Berthollet (1748–1822), also a former collaborator of Lavoisier's, Laplace dominated the physical sciences in France for quite some time. Both of these leaders subscribed to a broadly Newtonian point of view, in which all phenomena would be explained by the action of central forces operating between point-like particles. The overarching ambition of 'Laplacian physics' (as Fox has termed it) was 'to raise the physics of terrestrial bodies to the state of perfection to which celestial physics has been brought by the discovery of universal gravitation' (quoted in Fox, 1974, p. 95). In the first decade of the nineteenth century Laplace began to make concrete attempts in this direction by creating theories of optical refraction and capillary action based on short-range forces (negligible at macroscopic distances). He could not make much headway in specifying the form of the force laws involved, but he managed to show that the exact form of the laws did not matter.

In the physics of heat, Laplace's strategy was similar. According to Lavoisier, caloric repelled itself (while caloric and ordinary matter had a mutual attraction), but in Newton's tradition such action could be analysed only in terms of action-at-a-distance forces operating between particles. This meant that caloric itself had to consist of particles (so it was a 'discrete fluid'), and that there had to be a universal repulsive force between caloric particles that was a function of distance between them. Again, there were no clues about what the shape of this force function was, so in Laplace's analyses it was simply written as an unknown function, except that it was assumed to be vanishingly small at macroscopic distances. This force function could not do much work other than giving rise to a host of other symbols which represented its various integrals, to be rendered into a constant (though of unknown value) at a convenient place by the introduction of a definite integral.

Therefore Laplace was forced to take his analyses of heat onto the level of forces between molecules, rather than forces between caloric particles themselves. Laplace's mature view on caloric theory was developed in various papers published in the early 1820s, and summarized in the fifth and last volume of his *Traité de mécanique céleste* (1825), published just two years before his death. The basis of this picture was a molecular construction of matter, and a dynamic equilibrium of caloric radiation between the molecules. The idea of radiative equilibrium was probably adapted from Prévost's work, discussed in Section 16.6, but Laplace was not satisfied until he had provided a microscopic explanation as to why any caloric bound in a molecule would be radiated away from it. In the end, he attributed that to the repulsive force exerted by the caloric contained in neighbouring molecules. But it was difficult to conceive of latent caloric (or combined caloric) to be so easily disengaged from the molecules, so Laplace took the rather extraordinary step of putting *free* caloric into molecules. Latent caloric was postulated to have lost its repulsive force and did not enter this picture of radiative equilibrium at all; free caloric existed within molecules but retained its repulsive force; the radiated caloric existing in the intermolecular spaces was designated as the 'free caloric of space', to be distinguished from free caloric within molecules. By that point he had abandoned the standard Lavoisierian view of latent caloric.

With the ontology thus settled, Laplace proceeded with derivations. A crucial and very useful assumption which Laplace made was that the repulsive force between two adjacent molecules would be proportional to the product of the amounts of caloric contained in them. Thus the basic force equation obtained a semblance to Newtonian gravitation, though it still contained an unknown function of distance. A further constraint was given by the assumption that all molecules in a body of gas in internal equilibrium contained the same amount, c, of free caloric, and that the repulsive force between two molecules would be proportional to c^2. From such assumptions Laplace managed to derive the familiar gas laws, and also results supporting his longstanding argument that the air thermometer gave the true measure of temperature. However, these achievements seem to have been neglected largely. As Fox (1974) documents, Laplace's influence was waning during the time when he was working out the details of his mature caloric theory. With that context in mind, it is not surprising that his deductions, based on broad speculative assumptions rather than any details about the force function between caloric particles, failed to inspire confidence. Even the faithful Laplacians declined to pursue this line of investigation much further.

16.5 The Movement of Heat

The main focus of attention in the caloric theories was on the interaction of heat with ordinary matter. In parallel, another tradition of thermal physics grew which paid more attention to the motion of heat and its transfer between bodies. The first half of the nineteenth century witnessed great advances in this area. As a preliminary step, the three different modes of heat transmission recognized to this day—conduction, convection, and radiation—were identified by 1800. Benjamin Thompson (1753–1814), better known as Count Rumford to contemporaries and posterity alike, did much to clarify the distinction between conduction and convection, conceptualizing the former as the transfer of heat between the molecules of a body, and the latter as the transfer of heat effected by the transport of the molecules themselves within a fluid. Although the term 'convection' was coined only in 1834 by William Prout (1785–1850) in his *Bridgewater Treatise*, the conceptual distinction between conduction and convection was universally recognized early on as valid, illuminating, and uncontroversial. From this understanding also followed some practical consequences: for instance, the unexceptionable explanation that materials such as eider-down were good insulators of heat because they obstructed the movement of air molecules trapped in them, retarding the convective propagation of heat.

A topic that caused more excitement and controversy was radiation—the transmission of heat across macroscopic distances that was apparently instantaneous and not reliant on a material medium. The radiation of heat was a phenomenon that had been observed for a long time, in ordinary facts such as the intense and direct heat felt from a fire or hot metallic objects. The more scientific interest in radiant heat seems to have been generated usually by the recognition that it could be reflected by shiny surfaces, even when it was not accompanied by visible light. There is evidence of that observation from as early as the seventeenth century in the works of the Accademia del Cimento in Florence, and later by Saussure and Lambert, but it was probably Pictet (1790) who did more than anyone to draw the attention of physicists to radiant heat. First of all, Pictet's results were dramatic; in one experiment he demonstrated the radiation of heat between two concave metallic mirrors separated by more than 12 feet. His experiments were also relatively precise and systematic, and linked up with a highly developed theory of heat. Besides, Pictet was a key leader of a thriving and cohesive community of natural philosophers in Geneva who also communicated actively with scholars in other major centres of research. Pictet's lead was followed by a great deal of useful experimental work by Rumford and Leslie on the rates of heat radiation from various types of surfaces.

For explaining the movements of heat more generally, the early decades of the nineteenth century witnessed the development of a tradition of mathematical and phenomenological analyses eschewing physical and metaphysical accounts of the nature of caloric, spearheaded by Joseph Fourier (1768–1830). When Fourier started his work on heat theory around 1805 he was based in Grenoble, serving as the prefect of the department of Isère after accompanying Napoleon on his Egyptian expedition. In 1807 he presented to the French Institute a memoir containing some important results, including a version of the famous diffusion equation. Not receiving much response to this work, Fourier managed to prompt Laplace to propose an Institute prize-competition on the subject of heat conduction in 1811, which he won handily with a revised version of his 1807 memoir. Publication was delayed (Fourier [1822] 1955), but Fourier's work stimulated a good deal of interest soon enough, and in the end became very influential both in thermal physics and mathematics.

The power of Fourier's work lay not only in a thorough and innovative mathematization of the subject, but even more so in a conscious and explicit narrowing of focus. The narrowing was so extreme that Fourier's analyses dealt with hardly any of the traditional concerns of theories of heat. It is not simply that he refrained from making commitments about the metaphysical nature of heat, which Black, Lavoisier, and numerous other calorists had also done to various degrees. The starting point of his analysis was simply that there be some initial distribution of heat, and some specified temperatures on the boundaries of the body being considered; by what mechanisms these initial and boundary conditions might be produced and maintained were not his concerns. Fourier's notion was that the theory of heat proper should deal only with what is *not* reducible to the laws of mechanics. So the domain of Fourier's theory of heat excluded whatever was treatable by considerations of the forces exerted between particles of matter and caloric; all such mechanical issues Fourier was happy to leave to Laplace and his school for their corpuscularian analysis. Significantly, this meant that the expansive effects of heat, among other things, fell outside Fourier's domain. The only class of thermal phenomena left, then, was the *movement* of heat. It is not clear whether Fourier thought that the flow of heat was not reducible to mechanical actions between caloric particles. What is clear from his published work is that he did not consider such a reduction of heat flow to mechanics to be plausible, at least at that stage.

In sum, Fourier succeeded by defining a very specific class of problems, which he proceeded to solve mathematically. The starting point of a typical problem treated by Fourier was a disturbance in the equilibrium of heat, which causes a flow of heat from places of higher temperatures to places of lower temperatures. In all cases, Fourier noted a dissipation of heat, and the solution of each problem consisted of the temperature distribution as a function of time, and the rate (and direction) at which heat passed through each point in the body. Any effects of heat transfer were not considered, and were perhaps even ignored as inconvenient factors only complicating the analysis. Hence Clifford Truesdell (1980, p. 47) quips that Fourier treated only 'workless dissipation' in his theory.

There were many reasons for the popularity of Fourier's work, one of which was philosophical. Its affinity to positivist philosophy can be seen in Ernst Mach's appraisal ([1896] 1986, p. 113): 'Fourier's theory of the conduction of heat may be characterized as an ideal physical theory ... The entire theory of Fourier really consists only in a consistent, quantitatively exact, abstract conception of the facts of conduction of heat—in an easily surveyed and systematically arranged *inventory of facts*'. The compatibility with positivist thinking was clear in Fourier's own time as well. As documented by Fox (1971, pp. 265–266), Fourier attended the lectures of Auguste Comte (1798–1851) on positivism in 1829; Comte for his part admired Fourier's work, so much as to dedicate his *Cours de philosophie positive* to Fourier (and to Henri Marie Ducrotay de Blainville (1777–1850), the anatomist and zoologist). Fox (1974) also identifies Fourier as an intellectual leader of the younger French physicists who staged a silent revolt against Laplacian physics, an important part of which was a positivistic indifference or hostility toward Laplacian hypothesizing about microphysical forces and structures.

Since temperature and quantity of heat were two of the fundamental variables treated in his theory, Fourier did need to have a definite conception of the relation between the two. It was quite a simplified one, based on the working assumption that the specific heat of a given substance was not a function of temperature. It is not clear how much Fourier was aware of the experimental and theoretical arguments for the temperature dependence of specific heat. There is also no apparent worry in Fourier's work about whether any of the heat flowing around in conductors would go into a latent state. These physical assumptions certainly made Fourier's equations simpler than they would have been otherwise. Mach ([1896] 1986, pp. 113–114) made the extraordinary statement that 'in mechanics and in the theory of conduction of heat it is, really, only *one* great fact in each domain which is ascertained'. For mechanics, what he had in mind was universal gravitation; what was the one great fact in the theory of heat conduction? This goes back to the study of the cooling of hot bodies in air by Isaac Newton. Newton's law of cooling stated that the rate of cooling was proportional to the temperature difference between the hot body and the surrounding air. This law held a wide appeal for its intuitive plausibility, though it would be shown to be only approximately true by the works of Biot, Dalton, Dulong, Petit, and others. Fourier allowed that it may not be strictly true, but assumed that it would be true for small temperature differences, and used it as a basis for developing his analyses of the dissipation of heat. As Truesdell (1980, p. 50) emphasizes, it is not exactly Newton's law of cooling that Fourier was using as a basis of their theories. First the law had to be generalized to all heat transfer rather than just the cooling of macroscopic bodies in air, and that generalization is what Mach called the one great fact: 'the velocity of equalization of small differences of temperature is proportional to these differences themselves.' Fourier also had to make a version of this principle that was adapted for continuous media. Making that adaptation gives Fourier's diffusion equation, as follows.

For any point (or an infinitesimal region) within a continuous body, the rate at which heat flows into it would be proportional to how much its temperature deviates from the mean temperature of its immediate surroundings. For the one-dimensional case the latter deviation is expressed as the second-order partial derivative of the temperature function with respect to the spatial coordinate. Intuitively one can see that the second derivative indicates the curvature of the temperature-position curve; positive curvature would correspond to the surroundings on the average being at a higher temperature than the point itself. The rate of heat transfer would also be proportional to the internal heat conductivity of the material. And the change of temperature would be given by the amount of heat transfer divided by the specific heat of the substance (by weight), and its density. As a result, Fourier ([1822] 1955, p. 112) obtained the following equation:

$$dv/dt = (K/CD)d^2v/dx^2$$

where t and x are the time and space coordinates, v the temperature, K the internal conductibility of the substance, C its specific heat, and D its density; in modern notation we would write the derivatives as partial. For the three-dimensional case we add terms containing the second derivatives in the other spatial coordinates. This basic equation was then modified into various forms suitable for the various shapes of bodies which Fourier considered. There was a predictable reaction from Laplace, who took interest in Fourier's work but considered it incomplete at best, since Fourier's derivation of the diffusion equation did not include any considerations of the microphysical mechanisms of heat transfer. Already in a paper of 1810 Laplace set himself the task of remedying this defect. As discussed in Section 16.4, Laplace postulated a radiative heat exchange between the molecules within a continuous body. From those considerations Laplace managed to derive Fourier's diffusion equation, and regarded his derivation as supplying the true foundations of Fourier's result.

Regardless of the arguments about the physical basis of the diffusion equation, the difficulty of its solution gave rise to active debates and great advances in mathematics (see Grattan-Guinness, 1990, vol. 2, ch. 9, 12). Fourier found an innovative route to the solutions, drawing on the earlier works of Brook Taylor (1685–1731), Daniel Bernoulli (1700–1782), Leonhard Euler (1707–1783), and Jean le Rond d'Alembert (1717–1783) on the analysis of vibrating cords. These analyses had yielded solutions in the form of sinusoidal standing waves, and Fourier realized that he could find solutions to the diffusion equations in terms of infinite series of sine and cosine functions. This method would eventually develop into what we now call Fourier analysis, with invaluable applications in numerous branches of physics and engineering. Therefore, Fourier's work on heat conduction holds great interest not only for its role in the development of the theory of heat, but also for its place in the history of mathematics and general mathematical physics. Fourier made use of earlier mathematical works and certainly gave back as good as he took. Fourier's work marked a significant step in the development of modern mathematical physics, in which techniques of solving certain definite types of equations are developed without particular regard to physical applications. Then the resultant knowledge of how to solve a given type of equation can be used for treating various physical situations which bear structural similarities to each other, though they may have hardly anything in common in material terms. Fourier was fond of the structural similarities which allowed the application of the same mathematical techniques to the analyses of apparently diverse physical phenomena, and also saw his work on heat as contributing generally to mathematical analysis. This was to be borne out very nicely when the diffusion equation was later adapted for the analysis of the propagation of electric waves for the purpose of telegraphy.

16.6 Debates on the Nature of Heat

Despite the positivistic trend in thermal physics fostered by Fourier, debates on the physical and metaphysical nature of heat did not cease. It is well known that Rumford made strenuous arguments in favour of the view that heat consisted in motion, in which he was supported by various others, including Humphry Davy (1778–1829), Thomas Young (1773–1829) and André-Marie Ampère (1775–1836), as well as the little-known John Herapath (1790–1868). Concern about the nature of heat pervaded

Rumford's numerous investigations in thermal physics and technical innovations concerning the use of heat (see Brown, 1967, Rumford 1968). When he married the widow of Lavoisier he boasted that her second husband was going to do away with the caloric concept that her first husband had invented, which was probably not helpful for the short-lived marriage. Most famous among Rumford's experiments directed against the caloric theory was the 'cannon-boring experiment', in which he showed that an apparently indefinite amount of heat could be generated by friction when a solid brass cylinder was hollowed out by a horse-driven drill in the manufacture of cannons; anticipating an Irvinist explanation, Rumford argued that this heat could not be generated by any reduction of heat capacity, as the ground-up metal was shown to have the same heat capacity as the block metal. Rumford also made an experiment demonstrating that heat had no appreciable weight, and considered that an argument against any material theories of heat. (For both of these experiments, and a commentary on them, see Roller, 1957, sections 3 and 4.) The calorists took due notice of these arguments, and Rumford was not an easy man to ignore—either in London, where he founded the Royal Institution, or in Bavaria, where he was made a Count of the Holy Roman Empire, or in his native America, or in Paris, where he spent his last years. But there were sufficient caloric-based explanations of Rumford's experiments, and he failed to persuade the majority of chemists and physicists. It did not come as much of a surprise to most that caloric, a classic imponderable fluid, did not have weight. And the chemical calorists had no trouble arguing that the mechanical agitation in the cannon-boring was liable to shake off some of the combined/latent caloric in the metal to render them free/sensible.

There was, however, more to Rumford's theory of heat than meets the casual retrospective eye. For instance, he made a controversial claim that water, air, and most likely other liquids and gases, were absolute non-conductors of heat, all of the heat transmission in them being due to convection. It may seem that there was no great theoretical issue hinging on this debate, but it was in fact an important part of Rumford's general attack on the caloric theories. He thought that the conduction of heat in solids occurred by the transmission of harmonic vibrations between the molecules. Fluids were unable to sustain such vibrations due to the mobility of their molecules, so heat conduction in them was an impossibility. That reasoning had no force for calorists such as Leslie, since in their view conduction was a flow of caloric between molecules; there was no reason why this intermolecular flow should not happen in liquids. Hence, if Rumford had admitted conduction in fluids, he would have been pressed to assent to the notion that there was such a thing as caloric flowing through ponderable matter. This theoretical necessity led him to argue that the small amount of apparent conduction observed in fluids was actually due to radiation.

Radiation was a subject dear to Rumford's heart, and he joined in on an interesting controversy arising from Pictet's striking experiments on the reflection and concentration of radiant *cold* (see Chang, 2002). Pictet placed cold objects (an ice cube, for instance) at the focus of one concave mirror, and found that the thermometer placed at the focus of the opposite mirror descended immediately. These experiments impressed his Genevan colleague Pierre Prévost (1751–1839) so much that the latter found himself forced to construct an entirely new theoretical framework for understanding the exchange of heat between bodies, which he elaborated over a long period beginning in 1791. Pictet had satisfied himself with the explanation that the apparent radiation of cold was only a consequence of the radiation of heat from the thermometer to the cold object; generally there would always be a radiation of heat from a relatively warm object to a relatively cold object. Prévost generalized this picture further by postulating that *every* object radiated caloric at a certain rate depending on its temperature. A thermal equilibrium would occur when the amount of radiant caloric emitted by a body according to its temperature was balanced out by the amount of caloric received from its surroundings. In other words, there is always caloric radiation coming in and out of every body—when it seems that there is no radiation, that means only that a dynamic equilibrium has been reached.

Pictet's and Prévost's explanations of the radiation of cold satisfied most people, but they failed to convince Rumford, who understood radiant heat as a wave phenomenon, the propagation of the molecular vibrations through an all-pervading ether. Each body would emit rays of a characteristic 'frequency' determined by its temperature, and the rays would have the power to bring the temperature of the receiving body closer toward the temperature of their source. Therefore the exact same rays would act as 'calorific' rays if they are received by a body colder than their source, and as 'frigorific' rays if received by a hotter body. Rumford made an explicit analogy between calorific–frigorific radiation and the propagation and resonance of sound. After successfully repeating Pictet's experiment in 1800, Rumford wrote to Pictet: 'the slow vibrations of ice in the bottle cause the thermometer to sing a lower note' (quoted in Brown, 1967, p. 204). Drawing on his earlier experience in the study of the radiative and reflective power of different surfaces, Rumford performed some experiments to support his view that the radiation of cold was as much a positive phenomenon as the radiation of heat, and that heat and cold were only relative designations. However, his work in this direction does not seem to have been sufficient to overturn the consensus in favour of Prévost.

Among those who subscribed to material theories of heat, the fact that radiant heat was capable of reflection immediately suggested a parallel with light—a parallel that was only to be strengthened when it was discovered that radiant heat was also subject to refraction, and even polarization. These discoveries resonated with the longstanding conjecture that light and heat (and sometimes electricity as well) were different manifestations of the same ultimate entity, which had been advanced, for instance, by James Hutton (1726–1797), Scottish geologist and intimate friend of Black. The study of radiant heat gave a concrete arena in which speculations about the relation between heat and light could be elaborated. The opening of the nineteenth century witnessed a great revival of debates regarding the relationship between light and heat, due to the announcement of the heating effects found beyond the red end of the solar spectrum by the renowned astronomer William Herschel (1738–1822), the discoverer of Uranus. Herschel stumbled upon infrared rays in the course of investigating the differences in the heating powers of different colours in the solar spectrum. Initially he thought that the heating effect observed in the dark space beyond the red indicated the existence of 'invisible light'. This seemed cogent initially, since the infrared rays from the Sun (and also from terrestrial sources) were shown to obey the same laws of reflection and refraction as light. Further experiments, however, converted Herschel to the belief that what he was observing in the infrared region was not invisible light (which he came to see as a contradiction in terms), but caloric rays. What particularly convinced him in this direction were experiments on transmission, in which some substances were seen to transmit light but absorb heat, or *vice versa*. This he saw as the *separation* of light and heat. Herschel's considered view was endorsed by many—for instance, in the influential textbooks by John Murray and by Thomas Thomson (1773–1852)— and sometimes infrared heating was even taken as the most convincing experimental proof of the real existence of caloric. However, vexing ontological questions needed to be resolved before any certain conclusions could be reached (see Chang and Leonelli, 2005). This is illustrated nicely by the case of John Leslie, Herschel's first major critic on infrared rays. In a move that baffled many observers, Leslie denied the existence of infrared rays altogether, blaming the observed heating effects to poor experimental techniques on Herschel's part. Leslie's argument was motivated by his view that heat in general was the effect of the combination of light with ordinary matter.

More damaging and lasting criticism of Herschel's view came from those who accepted his experimental results but disputed his interpretation, favoring instead a Rumford-like new ontology which regarded both light and radiant heat as waves. This points to an entire phase of thermal physics, which Stephen Brush (1976) has elucidated and dubbed 'the wave theory of heat'. The crucial impetus for this development came from the great revival of the wave theory of light in the early nineteenth century, first attempted by Young but achieved more effectively by Augustin Fresnel (1788–1827). For the wave theory of light the phenomena of interference and polarization counted as important evidence, and those who pursued the parallel between light and radiant heat then attempted to demonstrate those phenomena for radiant heat as well. It was Bérard who first claimed to have observed the polarization of radiant heat, in 1813, but a convincing verification of that claim had to wait until the 1834 work of James David Forbes (1809–1868), Scottish geologist and the student and successor of Leslie in Edinburgh. The idea of radiant heat as an ethereal vibration had been advanced by Rumford, and it was easily revived. Now it seemed plausible to treat both light and radiant heat as aspects of one and the same wave phenomenon, which interacted in various manners with substances and sense-organs according to its wavelength. Much experimental and theoretical work in the consolidation of this new view was carried out by the Italian physicist Macedonio Melloni (1798–1854), who had initially shared Herschel's view that light and radiant heat were distinct entities (Chang and Leonelli, 2005).

16.7 Heat as a State Function

The indisputable major event in thermal physics in the mid-nineteenth century was the emergence of thermodynamics. This is not a whiggish statement: on the one hand, it was already a common view at the time that the arrival of thermodynamics was a major development; on the other hand, the original form of thermodynamics, except in the underlying conceptions that Clausius had (but not shared by others), had very little to do with our modern understanding that heat is only a manifestation of the kinetic energy of the molecules of ordinary matter. (The later developments by which thermodynamics came to be understood in terms of statistical mechanics is discussed in Chapter 25, 'The Emergence of Statistical Mechanics'.) The arrival of classical thermodynamics continues to raise non-trivial historiographical and philosophical questions: what was it that brought down the dominance of the caloric theory and ushered in the new science of energy and entropy? If the caloric theory was good enough in 1800, why was it no longer adequate in 1850? How did the concept of energy arise, and why did it not come earlier?

It is important to note that there were several different competing theories and traditions active throughout the period, so the story is not simply that of the demise of the caloric theory and the rise of thermodynamics. We have already noted that there were at least two major eighteenth-century calorist traditions (Irvinist and chemical). Early nineteenth-century developments brought in two important new traditions, which were very different from each other and also did not map neatly onto the Irvinist–chemical distinction: the phenomenological tradition of Fourier focused on the movement of heat, and the microphysical tradition of Laplace focused on the forces between particles of caloric and matter. On the anti-calorist side Rumford engendered two different strands of work, though neither found wide acceptance until after his death: the cannon-boring experiment presaged the interconversion of heat and work, and his work on radiation, except for his advocacy of 'frigorific rays', fed into the wave theory of heat discussed above.

The line of theoretical development which led directly to classical thermodynamics (and the concomitant demise of the caloric theory) started not from the anti-calorist side, but from within the caloric theory. Nor did it come from the tradition of Fourier, which might seem akin to classical thermodynamics in its macroscopic focus and its attention to the movement of heat. Rather, where we must look is in a curiously macroscopic strand of the latter-day Laplacian tradition. One central ingredient in this type of macroscopic analysis was the assumption that heat was a state function. There can be some confusion or at least ambiguity as to what exactly that means. Sometimes it is equated with the assumption of heat conservation, but it is more specific than that. From Poisson (1823b, p. 337) we have perhaps the most careful and informative formulation:

It would not be possible to calculate the total quantity of heat contained in a given weight of a gas ... however, one may consider the excess amount of heat that this gas contains over what it would contain under a certain arbitrarily chosen pressure and temperature. That excess amount, denoted by q, will be a function of p [pressure], r [density], and q [temperature], or simply of p and r because the three variables are related amongst each other by [$p = \alpha r (1 + aq)$; α, a constants]; therefore we have $q = f(p, r)$, where f is a function whose form needs to be determined.

The assumption that q was a state function gave exactly the kind of useful theoretical constraint that Irvine's hypothesis of heat capacity had provided. The rejection of Irvinism deprived the caloric theory of a good deal of empirical content, and the state-function assumption was very helpful in restoring it. Poisson (1823a) put the assumption to effective use in deriving what we now recognize as the adiabatic gas law, PV^r = constant, in the course of furthering Laplace's work on the speed of sound. Poisson imagined a body of gas being heated under constant pressure (hence expanding), absorbing a certain amount of heat, Q. Then he imagined the same gas heating up further by adiabatic compression back to its original volume. Finally, it would be cooled at constant volume back to its original temperature, giving up a certain amount of heat, Q'. If heat is a state function, then the gas must contain exactly the same amount of heat as it started with, when it is restored to its original state. Therefore it follows that Q is equal to Q', since the second stage in the above process is adiabatic and there is no input or output of heat. The rest of the derivation was completed by noting that Q was proportional to c_p and Q' to c_v (where c_p and c_v are specific heats under constant pressure and constant volume, as discussed in Section 16.4).

The power of the state-function assumption was amply illustrated in Poisson's work, but its most remarkable application came in *Réflexions sur la puissance motrice du feu*, the 1824 monograph by the engineer and army officer Sadi Carnot (1796–1832). Today Carnot is celebrated, with good reason, as a major precursor of thermodynamics whose pioneering work was sadly neglected until after his death. His remarkably original work, achieved in isolation from the most of the scientific establishment of his day, was a synthesis of insights arising from several different contexts. (For much valuable analysis, see Fox's introduction and commentary in Carnot [1824] 1986.) There is little doubt that the primary context of Carnot's work was power engineering rather than theoretical physics. Carnot's ultimate aim was to improve the efficiency of steam engines, efficiency being understood as the amount of mechanical effect extracted from a given amount of fuel. Although Carnot's thinking was very abstract, it was clearly directed by this practical question. The basic theory that Carnot needed in order to answer his question was not in place yet, so he had to do some fundamental thinking on his own; and in doing so he produced ideas whose applicability went far beyond his original problem. When it comes to specific insights Carnot drew from the realm of power engineering, the two of most importance are both traceable to that renowned improver of steam engines, James Watt (1736–1819). Watt's installation of the separate condenser went back to 1765, but the theoretical lesson Carnot gathered from it was still relatively fresh: it is not a mere presence of heat that produces mechanical work, but the *flow* of heat from a hot place to a cold place. The other insight originating from Watt concerned the 'expansive principle': namely, that the hot steam should be allowed do further work 'on its own' after its introduction into the cylinder, by virtue of its natural tendency to expand. This phase was formally incorporated into Carnot's thinking as the adiabatic-expansion stroke in his famous cycle (see Cardwell, 1971, p. 52). Not restricting himself to steam engines, Carnot also made an explicit analogy to water engines, which went along nicely with his notion that caloric produced mechanical work in the course of 'falling' from a place of higher temperature to a place of lower temperature.

In the realm of heat theory, perhaps the greatest influence on Carnot was the sidelined Clément (now finally settled with a chair at the *Conservatoire des Arts et Métiers* in 1819), rather than any of the luminaries of the *Académie* or the *École Polytechnique*. Carnot followed Clément and Desormes's 1819 work on the steam engine in analyzing the expansive phase after the 'cut-off' of steam as an adiabatic expansion (with cooling), rather than an isothermal one (that is, with constant temperature) to which Boyle's law could be applied. Carnot also believed 'Clément and Desormes' law' (also advanced by Watt, earlier), according to which a given weight of steam at a given temperature contained the same amount of heat, no matter at which pressure (or temperature) it was formed. Finally, no discussion of Carnot would be complete without a mention of the shadow of his father. Lazare Carnot (1753–1823)—one of the most prominent military and political leaders of France in the Revolutionary and Napoleonic periods—was also a keen mathematical physicist, as Charles Gillispie (1971) has described in great detail. It is probably not far-fetched to speculate that Sadi was trying to extend to heat engines Lazare's work on the efficiency of mechanical engines. For instance, Lazare's point that there should be no percussion in machines (in other words, motion should be transferred between parts that maintain the same velocity) is mirrored in Sadi's insight that every time heat is transferred directly across finite temperature differences there will be a waste of potential to generate mechanical work.

Sadi Carnot's approach can be viewed as a happy medium between Fourier's and Laplace's. Like Fourier, he declined to speculate about the nature of heat and the microscopic mechanisms of thermal phenomena, and instead focused on the phenomenological and macroscopic movement of heat. On the other hand, he followed Laplace in investigating the *effects* of heat in altering the states of material bodies mechanically. Carnot ([1824] 1986, pp. 64–66) set this synthesis in stone by adopting an axiom that the production of mechanical effect in a heat engine was always due to a flow of heat (or a restoration of a disturbed thermal equilibrium). It seems that a large part of Carnot's originality arose in the stimulating context of taking a very practical problem and thinking it through in highly abstract and idealized terms. Since he sought to create a general theory of all heat engines rather than only steam engines, he felt compelled to specify his problem without reference to any particular substances or mechanisms: 'the steam serves simply as a means of transporting the caloric' (Carnot [1824], 1986 p. 64). At this level of abstraction, the working of a heat engine is divided into two parts: one in which heat is absorbed by the working substance from a hot body, and the other in which the absorbed heat is released into a cold body.

Addressing considerations of maximum efficiency, Carnot was quick to note that the flow of heat does not always produce mechanical effect—as in the cases of 'workless dissipation' that characterized most of Fourier's problems. Carnot ([1824] 1986, p. 70) further noted that any temperature change that was not associated with a volume change involved a wasteful flow of heat, and that all such temperature changes were due to a flow of heat between bodies at different temperatures. So, while the working substance in the engine is absorbing heat, its temperature should be equal to that of the heat source (presumed constant), or only infinitesimally lower. When releasing heat, the substance should have the same temperature as the heat sink (or only infinitesimally higher). So an additional stage was needed, to get the substance from the temperature of the heat source to the lower temperature of the heat sink. The only non-chemical way of achieving that without involving a heat transfer was adiabatic expansion. This fitted in nicely with Watt's expansive principle as well, since the practical advice there was that steam should be allowed to expand on its own while cooling. Now just one more innovation was needed. In order to discern purely the mechanical effect of the *passage* of heat through the engine, Carnot needed to consider a situation in which the working substance in the end gave up exactly the amount of heat that it receives. This can be achieved, on the assumption that heat content is a state function, by stipulating that the substance should return exactly to the state in which it began. So Carnot's ideal heat engine operated in cycles, and this is one of the factors that set his analysis apart from most of his contemporary writers on the theory of the steam engine, who merely considered the mechanical effect while the steam was moving the piston and neglected the work one had to supply in order to bring the substance back to its original state.

Carnot's description of his ideal cycle was rather complex, but it did contain the essential elements that we associate with his name: the isothermal communication of heat to and from the working substance; changes in the temperature of the substance by adiabatic compression and expansion; mechanical effect produced in the expansive strokes and spent in the compressive strokes; and a net gain of mechanical effect, due to the expansions taking place at higher temperatures (and pressure) than the compressions. One significant point that Carnot ([1824] 1986, pp. 68–70, 76–77) advanced was that his ideal heat engine would have the same efficiency regardless of the nature of the working substance. This conclusion followed from the mere assumption that the cycle is reversible. According to D. S. L. Cardwell (1971, p. 198), the reverse operation of an engine was not such an unusual concept among engineers at the time; particularly with a water engine it is easy to imagine running it backwards to bring water to a higher place by expending mechanical work. With Carnot's analogy of the fall of caloric, this is quite easy to imagine for heat engines as well. So Carnot says, if one ideal heat engine can produce mechanical effect W by the fall of caloric Q between two given temperatures, and another engine can operate more efficiently (that is, produce $W' > W$ from the same Q), then we could create mechanical effect from nowhere as follows. Take the more efficient of the two engines, and produce mechanical effect W' by letting Q fall from the higher to the lower temperature; take a part of the mechanical effect produced that is equal to W, run the other engine backwards with it, bringing Q back up to the higher temperature. Then we have created an amount of mechanical effect equal to $W'-W$ out of nowhere, which is impossible.

Having thus established that there is a unique ideal efficiency for the fall of caloric between two given temperatures, Carnot ([1824] 1986, pp. 91–93) then set out to estimate that efficiency. He conceived the question as one of comparing the amount of

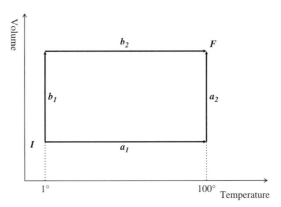

Fig. 16.1. Carnot's argument concerning the temperature dependence of engine efficiency.

heat needed in producing the same amount of motive power in two ideal heat engines operating at different places on the temperature scale. The engine that needs to use less heat would be the more efficient one, so the question was reduced to estimating the amount of heat absorbed in isothermal expansion at different temperatures. Now Carnot imagined heating a sample of air from 1° C to 100° C, allowing it to expand under constant pressure. The end result of that process could also be achieved in two other ways (see Fig. 16.1): (a) by increasing the temperature to 100° C under constant volume first, followed by isothermal expansion; (b) by isothermal expansion at 1° C to the desired volume, and then increasing the temperature under constant volume. Either way, the total caloric required would be the same, because the initial and final states of the gas are exactly the same in both cases. Here enters the assumption of heat as a state function. Then Carnot argued that the temperature-increase phase required more caloric in process b than in a (compare b_2 and a_1 in the figure), because the specific heat was higher at higher volume (or, lower density); this latter assumption about specific heats was inferred from the experimental results by Delaroche and Bérard (1813). Finally, since the total heat added in processes a and b were the same, it followed that the isothermal phase a_2 must absorb more heat than that b_1. That is to say, isothermal expansion required more heat at higher temperatures, and therefore the engine efficiency was lower at higher temperatures.

Carnot's use of the assumption that heat is a state function was crucial in the above derivation, and in many other parts of the *Réflexions*. In one footnote he stated the assumption explicitly, from which we can see that it was a very similar one to Poisson's version of the general idea, defining the state in terms of the body's 'density, temperature, and mode of aggregation' (Carnot, [1824] 1986, p. 76). For Carnot's modern admirers, this assumption, and the associated belief in the conservation and materiality of heat, are the most unfortunate aspects of his work. Consequently, much has been made of the fact that Carnot's belief in caloric and its conservation was always shaky, and finally renounced very clearly in some unpublished manuscripts composed in his last years. However, it is clear that these 'mistaken' assumptions did very useful work for Carnot. Besides, Carnot's ideas as expressed in the *Réflexions* are what inspired by the originators of classical thermodynamics, while the later manuscripts were made public only in 1878 and did not have much scientific impact.

The practical conclusions that Carnot drew regarding engine efficiency did not help the engineers very much, as they either merely affirmed well-known practical wisdom (such as higher efficiency attainable with higher pressures and with the expansive use of steam), or predicted quite small effects (such as higher efficiency at lower temperatures). However, in the course of this work he also derived, almost as a digression, some very interesting and significant results concerning the physics of ideal gases. For instance, Carnot argued that the amount of heat absorbed in isothermal expansion was independent of the nature of the gas, and that the difference (not ratio) between the two specific heats (c_p and c_v) was the same for all gases and for all densities of a given gas. He also derived a relation between the addition of heat and the increase of volume in isothermal expansion, and a relation between specific heat and volume. These results all followed from the assumption that heat was a state function, and the assumption that the efficiency of all ideal engines was independent of the specific mechanisms involved. It is not entirely clear why the academic physicists ignored Carnot's work, though it must have had something to do with Carnot's isolation from the academic world.

The revival of Carnot's work was initially due to Emile Clapeyron (1799–1864), civil engineer and graduate of the *École Polytechnique*, whose speciality was the design and construction of steam locomotives. In 1834 Clapeyron published, in the *Journal of the École Polytechnique*, a restatement of Carnot's theory which was both analytically sharper in its mathematical formulations, and intuitively more appealing thanks to the use of Watt's indicator diagram (plotting the state of the working substance by its pressure and volume), which has become the universal mode of representing the Carnot cycle in modern textbooks. Clapeyron did not amend Carnot's basic ideas much at all, but the benefits of the clarification he made in Carnot's ideas were quite clear. To start with, Clapeyron reformulated Carnot's description of the cycle as follows (see Fig. 16.2). The substance starts in a certain state, characterized by its volume, pressure, and temperature (T_1). First, it expands isothermally at temperature T_1, absorbing a certain amount of caloric (Q) from a heat source; during this stroke,

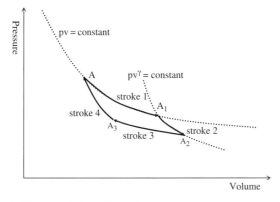

Fig. 16.2. Clapeyron's diagram of Carnot's cycle.

mechanical effect (W_1) is produced due to the expansion of the substance. Second, it expands adiabatically, cooling to the temperature of the heat sink (T_2) in the process; during this stroke further mechanical effect (W_2) is produced. Third, the substance is compressed isothermally at T_2, until it has released to the heat sink the same amount of heat it absorbed in the first stroke; some mechanical effect (W_3) must be expended in this compression. Finally, the substance is compressed adiabatically; this stroke should return the gas to its original volume, and in that case the temperature and pressure of the gas also will be restored to their original values, assuming that the total heat content of the gas will be the same as at the start; further mechanical effect (W_4) is expended in this adiabatic compression. At the end of this closed cycle, the net mechanical effect W, equal to $W_1 + W_2 - W_3 - W_4$, has been produced by the fall of the amount Q of caloric from temperature T_1 to T_2.

We can see from the above description that Clapeyron not only retained a belief in caloric and its conservation, but gave an even more prominent and explicit place to the assumption of heat as a state function than in Carnot's original formulation. Starting with the characterization of the state of a gas by its volume (v), pressure (p), temperature (t) and quantity of heat (Q), Clapeyron further assumed that knowing two of those quantities would allow the other two to be determined. This was because there were two relations holding between these variables: the ideal gas law ($pv = R(267 + t)$, in Clapeyron's notation), and also a relation expressing Q as a function of the other variables, as yet unspecified. Deriving and applying that expression for Q was an essential part of Clapeyron's work. On his state-function assumption, Q would be a function of p and v only, if t were fixed; then one can add the assumption that the product pv is a linear function of t, due to the ideal gas law. From those considerations, Clapeyron ([1834] 1837, p. 358) arrived at the following result:

$$Q = R(B - C \log p)$$

where R is a constant and B and C are functions of temperature. This equation was essentially given by Carnot already, but Clapeyron formulated it in a more straightforward way, and used it more effectively as well. The variable C in the equation, later to be known as Carnot's function, was especially important, as Clapeyron was able to show that ($1/C$) was proportional to the efficiency of the ideal Carnot engine. From this equation for total heat Clapeyron gave new derivations of Carnot's results concerning the amount of heat given off in an adiabatic compression, and the relationship between the two specific heats of gases.

16.8 THE EMERGENCE OF CLASSICAL THERMODYNAMICS

Although the Carnot cycle is a cornerstone of almost any presentation of thermodynamics today, it was not at all an easy transition for physicists to go from Carnot's theory to what we know as classical thermodynamics. *Carnot's* cycle was not the modern *Carnot* cycle, and its transformation into the modern form was an essential part of the emergence of thermodynamics. A striking illustration of the difference between Carnot's theory and thermodynamics is the fact that Carnot's theory lacked the two best-known elements of classical thermodynamics: namely, its first and second laws. It could be argued that Carnot's reasoning already embodied the second law in an implicit form, but the first law was clearly absent. The emergence of thermodynamics and its two laws is a well-known story (for example, Cardwell, 1971; Smith, 1998), so I will not try to tell it in detail. However, a brief overall summary must be given here, and there are also a few lesser-known aspects of the story that will be useful to highlight. The laws of thermodynamics are in fact something of a misnomer, since they have broad roots and consequences reaching far beyond the domain of thermal physics. It is best to acknowledge that these laws are entirely general in their scope and cannot rest on any empirical generalizations concerning heat and its relation to mechanical work. What we can acknowledge, nonetheless, is that the concepts of energy and entropy were first articulated clearly in the context of thermodynamics.

The first law of thermodynamics, otherwise known as the conservation of energy, has very deep roots. The term 'energy' is very old, and originally had very vague meanings. The notion of conservation is perhaps even older, appealing to the intuition that nothing real in nature is created out of nothing or disappears into nothing. Various conservation principles had been formulated in science, including Descartes' conservation of force, Leibniz's conservation of *vis viva*, and Lavoisier's conservation of weight in chemical reactions. Heat itself was seen as a conserved quantity in the caloric theories, and the great innovation at the establishment of thermodynamics was the doctrine of the *inter*-conversion of heat and mechanical work, and the consequent *non*-conservation of heat itself. The credit for this innovation goes equally to James Prescott Joule (1818–1889) and Julius Robert Mayer (1814–1878). Travelling as a ship's physician in the East Indies, Mayer noticed that the venous blood of his patients was redder in the tropics than in cooler climates, which he thought was because in warmer surroundings the body needed to generate less heat, and therefore needed to use less energy, or burn less oxygen (on Mayer, see Caneva, 1993). This chance observation led Mayer to formulate a general doctrine of the interconversion of various forms of energy, though he had some trouble getting his ideas noticed by physicists, and his 1842 paper was published by the chemist Justus Liebig. James Joule's approach to the subject was more experimental and quantitative than Mayer's, and he focused on measuring the 'mechanical value of heat'—namely the ratio between the amount of mechanical work that was destroyed and the amount of heat that was produced as a result—through his famous 'paddle-wheel' experiment and other related experiments. With his origin in a provincial brewing family, Joule also had some initial difficulty in being accepted in the academic physics community. However, he did receive opportunities to present his findings to the meetings of the British Association for the Advancement of Science in 1844 and 1845, which gave him visibility and introduction to some better-established scientists.

Later, there was a quite a fraught dispute regarding the priority between Joule and Mayer, especially surrounding John Tyndall's public championing of Mayer in 1862. Whatever the exact truth of the matter, this priority dispute pales in importance when it is viewed in the context of Thomas Kuhn's (1977) claim that between 1830 and

1850 as many as twelve different people in various European countries independently arrived at the idea of energy conservation. It is difficult to avoid the impression that a combination of the experiences of power engineering and the long-lasting thema of conservation was bound to yield something like the energy concept. (The concept of energy also had deep cultural resonances, discussed in Chapter 17, 'Engineering Energy'.) The first general and systematic statement of energy conservation was probably due to Hermann Helmholtz (1821–1894), who presented his classic paper 'Die Erhaltung der Kraft' to the Physical Society of Berlin in 1847.

It is even trickier to pinpoint the origin of the second law, although the story involves fewer protagonists. A key figure here is William Thomson (Lord Kelvin) (1824–1907), in Glasgow. Thomson was closely guided by Carnot's work, even after he was convinced by Joule to renounce the existence of caloric and the conservation of heat. Thomson first encountered Carnot's work through Clapeyron's paper while working as an apprentice in Regnault's laboratory in 1845, and his work in thermodynamics began with an attempt to understand and extend Carnot's ideas. Rudolf Clausius (1822–1888) followed Thomson's rendering of Carnot, and produced his own synthesis of Carnot and Joule. Both Thomson and Clausius focused on the fact that in a Carnot engine mechanical work was produced in the process of heat being transferred from a warmer to a colder place, but modified Carnot's reasoning by noting that not all of the heat arrived at the colder place, the lost portion being turned into mechanical work. In this process the system approached thermal equilibrium, lessening the potential for humans to extract mechanical work even though no energy was destroyed in an absolute sense. Thomson and Clausius agreed that 'heat cannot, of itself, pass from a colder to a hotter body', which was the formulation of the second law of thermodynamics that Clausius gave in 1850. In the terminology introduced by Clausius, we now say that the entropy of a system either increases or at best remains the same; the latter holds if the transfer of heat only takes place across two parts of a system which are at the same temperature. Isothermal heat-transfer also constituted a condition of reversibility, and Clausius and Thomson agreed that the maximally efficient Carnot engine would be reversible. If there were a heat engine more efficient than a reversible one, then one could use such an engine to run a reversible one backwards, effecting a net transfer of heat from a colder place to a hotter place; but this was deemed impossible, as it violated the second law of thermodynamics.

One notable feature of classical thermodynamic theory is its highly abstract nature. In that vein, a development of great importance was Thomson's establishment of the concept of absolute temperature (see Chang and Yi, 2005). Thomson was certainly appreciative of the achievements of his mentor Regnault in thermometry, but he was unhappy about tying the definition of a fundamental physical quantity to a particular substance (air, in this case). This was Thomson's main concern, rather than counting temperature from an absolute zero. Starting in 1848, he tried out a succession of definitions based on the thermodynamics of ideal heat engines, and by 1854 he had arrived at the notion that the ratio of two absolute temperatures was as the ratio of the quantities of heat taken in and given out at those temperatures in a Carnot cycle. But there were difficulties with using such definitions for experimental work, since it was not possible even to approximate an ideal Carnot engine in reality. More generally, it is not trivial to connect an abstract concept with concrete operations in order to make physical measurements possible. In the end, Thomson argued that an ideal gas thermometer would indicate his absolute temperature, and that the deviation of actual gas thermometers from the ideal could be estimated by means of the Joule–Thomson effect (temperature change in a gas pushed through a small hole or a porous plug). However, the measurement of this effect required measurements of temperature, so there was a problem of circularity. Thomson and Joule forged ahead by measuring the Joule–Thomson effect with an ordinary thermometer. Later investigators, particularly Hugh Longbourne Callendar (1863–1930), refined and justified Thomson's practice as an iterative one: the correction of the gas thermometer based on the Joule–Thomson effect is itself subject to an error; however, after the first-order correction is made to the thermometer, the Joule–Thomson effect measurements can be corrected, resulting in second-order corrections; and so on. The first-order corrections were already small enough, and Callendar trusted that higher-order corrections would be negligible. With the operationalization of the absolute-temperature concept, thermal physics had reached a notable point of maturity and synthesis. Ironically, however, a whole other set of transformations had just got seriously under way, with a new microscopic energy-based understanding of heat and temperature and the rendition of the entropy concept in statistical terms (see Chapter 25).

Bibliography

SANBORN C. BROWN (1967), *Benjamin Thompson, Count Rumford: Count Rumford on the Nature of Heat* (Oxford: Pergamon Press). This also contains many useful excerpts from Rumford's papers.

STEPHEN G. BRUSH (1970), 'The Wave Theory of Heat: A Forgotten Stage in the Transition from the Caloric Theory to Thermodynamics', *British Journal for the History of Science*, 5 (1970), 145–167.

——(1976), *The Kind of Motion We Call Heat: A History of the Kinetic Theory of Gases in the 19th Century*, 2 vols. (Amsterdam: North-Holland Publishing Company).

KENNETH L. CANEVA (1993), *Robert Mayer and the Conservation of Energy* (Princeton, NJ: Princeton University Press).

DONALD S. L. CARDWELL (1971), *From Watt to Clausius: The Rise of Thermodynamics in the Early Industrial Age* (Ithaca, NY: Cornell University Press). Also 1989 reprint (Ames, IA: Iowa State University Press).

NICOLAS-LÉONARD-SADI CARNOT ([1824] 1986), *Reflections on the Motive Power of Fire* (a critical edition with the surviving scientific manuscripts), trans. and ed. by Robert Fox (Manchester: Manchester University Press). Originally published as *Réflexions sur la puissance motrice du feu et sur les machines propres à développer cette puissance* (Paris: Bachelier, 1824).

HASOK CHANG (2002), 'Rumford and the Reflection of Radiant Cold: Historical Reflections and Metaphysical Reflexes', *Physics in Perspective*, 4 (2002), 127–169.

——(2004), *Inventing Temperature: Measurement and Scientific Progress* (New York: Oxford University Press).

——— and SABINA LEONELLI (2005), 'Infrared Metaphysics: The Elusive Ontology of Radiation (Part 1)'; 'Infrared Metaphysics: Radiation and Theory-Choice (Part 2)', *Studies in History and Philosophy of Science*, 36 (2005), 477–508, 686–705.

——— and SANG WOOK YI (2005), 'The Absolute and Its Measurement: William Thomson on Temperature', *Annals of Science*, 62 (2005), 281–308.

BENOIT-PIERRE-ÉMILE CLAPEYRON ([1834] 1837), 'Memoir on the Motive Power of Heat', trans. by Richard Taylor, in *Scientific Memoirs, Selected from the Transactions of Foreign Academies of Science and Learned Societies and from Foreign Journals*, 1 (1837), 347–376. Originally published as 'Mémoire sur la puissance motrice de la chaleur', *Journal de l'École Polytechnique*, 14 (1834), 153–190.

NICOLAS CLÉMENT and CHARLES-BERNARD DESORMES (1819), 'Détermination expérimentale du zéro absolu de la chaleur et du calorique spécifique des gaz', *Journal de Physique*, 89 (1819), 321–346, 428–455.

ADAIR CRAWFORD (1779), *Experiments and Observations on Animal Heat, and the Inflammation of Combustible Bodies* (London: n. p.). Second edition, 1788.

FRANÇOIS DELAROCHE and JACQUES-ÉTIENNE BÉRARD (1813), 'Mémoire sur la détermination de la chaleur spécifique des différens gaz', *Annales de chimie et de physique*, 85 (1813), 72–110, 113–182.

JEAN-ANDRÉ DE LUC (1772), *Recherches sur les modifications de l'atmosphère*, 2 vols. (Geneva: n. p.).

———(1786–87), *Idées sur météorologie* (Paris: la Veuve Duchesne).

PIERRE-LOUIS DULONG and ALEXIS-THÉRÈSE PETIT (1816), 'Recherches sur les lois de dilatation des solids, des liquides et des fluides élastiques, et sur la mesure exacte des températures', *Annales de chimie et de physique*, 2d ser., 2 (1816), 240–263.

———(1817), 'Recherches sur la mesure des températures et sur les lois de la communication de la chaleur', *Annales de chimie et de physique*, 2d ser., 7 (1817), 113–154, 225–264, 337–367.

JAMES DAVID FORBES (1858), A review of the progress of mathematical and physical science in more recent times, and particularly between the years 1775 and 1850 (Edinburgh, 1858); a separate publication of Forbes's Dissertation of a similar title for the 8th edition of the *Encyclopaedia Britannica*, Vol. 1 (1860), 794–996.

JEAN-BAPTISTE-JOSEPH FOURIER ([1822] 1955), *The Analytical Theory of Heat*, trans. by Alexander Freeman (New York: Dover). Originally published as *Théorie analytique de la chaleur* (Paris: Didot, 1822).

ROBERT FOX (1971), *The Caloric Theory of Gases from Lavoisier to Regnault* (Oxford: Clarendon Press).

———(1974), 'The Rise and Fall of Laplacian Physics', *Historical Studies in the Physical Sciences*, 4 (1974), 89–135.

CHARLES COULSTON GILLISPIE (1971), *Lazare Carnot, Savant* (Princeton, NJ: Princeton University Press).

IVOR GRATTAN-GUINNESS (1990), *Convolutions in French Mathematics, 1800–1840* (Basel: Birkhäuser Verlag).

RENÉ-JUST HAÜY (1806), *Traité élémentaire de physique* (Paris: Courcier). First edition, 1803; third edition, 1821.

WILLIAM HERSCHEL (1800), 'Investigation of the Powers of the prismatic Colours to heat and illuminate Objects . . .' etc., *Philosophical Transactions of the Royal Society*, 90 (1800), 255–283, 284–292, 293–326, 437–538. This is a series of four papers, each with a different title.

WILLIAM IRVINE and WILLIAM IRVINE (1805), *Essays Chiefly on Chemical Subjects* (London: J. Mawman).

THOMAS S. KUHN (1977), 'Energy Conservation as an Example of Simultaneous Discovery', in *The Essential Tension* (Chicago, IL: University of Chicago Press), 65–104.

PIERRE-SIMON LAPLACE (1825), *Traité de mécanique céleste*, vol. 5 (Paris: Bachelier).
ANTOINE-LAURENT LAVOISIER (1789), *Traité élémentaire de chimie* (Paris: Cuchet, 1789). English translation by Robert Kerr (1790), available as a reprint, *Elements of Chemistry* (New York: Dover, 1965), with an introduction by Douglas McKie.
—— and PIERRE-SIMON LAPLACE (1783), *Mémoire sur la chaleur* (Paris: Gauthier-Villars). This was also published in 1784 in the *Mémoires* of the Paris Academy (for the year 1780), pp. 355–408. English translation by Henry Guerlac, *Memoir on Heat* (1982).
JOHN LESLIE (1804), *An Experimental Inquiry into the Nature, and Propagation, of Heat* (London: J. Mawman).
ERNST MACH ([1896] 1986), *Principles of the Theory of Heat, Historically and Critically Elucidated*, trans. by T. J. McCormack, P. E. B. Jourdain and A. E. Heath, ed. by Brian McGuinness, with an introduction by Martin J. Klein (Dordrecht: Reidel). Originally published as *Die Prinzipien der Wärmelehre* (Leipzig: Barth, 1896); this translation is from the second German edition of 1900.
DOUGLAS MCKIE and N. H. DE V. HEATHCOTE (1935), *The Discovery of Specific and Latent Heats* (London: Arnold).
EVERETT MENDELSOHN (1964), *Heat and Life: The Development of the Theory of Animal Heat* (Cambridge, MA: Harvard University Press).
W. E. KNOWLES MIDDLETON (1966), *A History of the Thermometer and Its Uses in Meteorology* (Baltimore: Johns Hopkins Press).
MARC-AUGUSTE PICTET (1790), *Essai sur le feu* (Geneva: n. p.). English translation by W. B[elcombe], *An Essay on Fire* (London: E. Jeffery, 1791).
PIERRE PRÉVOST (1791), 'Sur l'équilibre du feu', *Journal de Physique*, 38 (1791), 314–332.
SIMÉON-DENIS POISSON (1823a), 'Sur la vitesse du son', *Connaissance des Tems pour l'an 1826* (published in 1823), 257–277.
——(1823b), 'Sur la chaleur des gaz et des vapeurs', *Annales de chimie et de physique*, 23 (1823), 337–352.
DUANE ROLLER (1957), 'The Early Development of the Concepts of Temperature and Heat: The Rise and Decline of the Caloric Theory', in James Bryant Conant, ed., *Harvard Case Histories in the Experimental Sciences*, 2 vols. (Cambridge, MA: Harvard University Press), vol. 1, 117–214.
RUMFORD [BENJAMIN THOMPSON] (1968), *Collected Works of Count Rumford*, ed. by Sanborn C. Brown, Vol. I. *The Nature of Heat* (Cambridge, MA: Harvard University Press).
CROSBIE SMITH (1998), *The Science of Energy: A Cultural History of Energy Physics in Victorian Britain* (London: Athlone Press).
WILLIAM THOMSON (1848), 'On an Absolute Thermometric Scale Founded on Carnot's Theory of the Motive Power of Heat, and Calculated from Regnault's Observations', *Proceedings of the Cambridge Philosophical Society*, 1 (1848), 66–71. Reprinted in William Thomson, *Mathematical and Physical Papers*, vol. 1, 100–106.
CLIFFORD A. TRUESDELL (1980), *The Tragicomical History of Thermodynamics 1822–1854* (New York: Springer, 1980).

NOTES

1. Irvine did not publish his views, but they spread through his teaching and personal contacts. Some of his essays on heat were published posthumously by his son (also named William), but only in 1805.
2. I follow Robert Fox's usage of this term to designate those who did their thermal physics on the basis of caloric.

CHAPTER 17

ENGINEERING ENERGY: CONSTRUCTING A NEW PHYSICS FOR VICTORIAN BRITAIN

CROSBIE SMITH

In 1959 Thomas Kuhn published a path-breaking paper entitled 'Energy conservation as an example of simultaneous discovery', in which he identified a dozen men of science and engineering, widely scattered across Western Europe, who, 'within a short period of time (c.1830–50), grasped for themselves essential parts of the concept of energy and its conservation'. Kuhn's approach to the history of energy did much to shift historical focus away from nineteenth-century debates over competing priority claims for national heroes and individual geniuses. In doing so he added momentum to the rapid development of history of science into a mature, professionalized historical discipline during the second half of the twentieth century. Seeking common features among his twelve 'simultaneous discoverers', Kuhn found that four of them had specific concerns with engines (waterwheels and heat engines), while others with wider natural philosophical interests (especially electromagnetic phenomena) also exhibited engineering interests.[1]

In the half-century that has elapsed since Kuhn's paper, historians of physics have challenged almost every feature of his analysis, including the appropriateness of the 'simultaneous discovery' model and his assumption that the principle of energy conservation was a matter of 'discovery' like that of a precious gem hidden in nature. Indeed, his labelling of the twelve 'discoverers' as 'pioneers' suggested that each individual was sleepwalking towards the principle. Moreover, just as gems are neither discovered in nature nor endowed with intrinsic value, so 'energy conservation' had to be crafted by human agency and taken to the scientific and engineering market places of Europe and beyond. For the modern historian of physics, therefore, 'discovery' becomes 'construction': energy principles are formulated, articulated, and promoted by human individuals and groups whose own approaches and perspectives are shaped by the local and national cultures within which they practice their sciences.[2]

An enduring feature of Kuhn's study, however, has been his recognition of the 'concern[s] with engines'. A range of later scholarship has explored the significance of engines, especially heat engines, for the history of energy and thermodynamics: Robert Fox's work on Sadi Carnot, Donald Cardwell's studies of thermodynamics from James Watt to Rudolph Clausius, Ivor Grattan-Guinness's and Norton Wise's explorations of the sciences of 'work' in French and British contexts, Smith and Wise's contextual biography of William Thomson (later Lord Kelvin), and Ben Marsden's examination of the history of air-engines, for example.[3] A common thread in much of this scholarship is the highlighting of the measurement of work done by engines of every variety (human, animal, hydraulic, steam, and air) but especially through the increasing use of indicators and indicator diagrams.[4]

In this chapter I argue that the construction of the new sciences of thermodynamics and energy in Britain occurred not simply within the broad contexts of industrialized engineering but that the new industries of marine engineering and the new sciences were, in specific local contexts on the Thames and on the Clyde, integral to one another. In the first section I therefore begin with an account of James Thomson's marine engineering networks centred on the Thames at Millwall. In the second section I move to the context of his brother William's Glasgow College laboratory. And in the third section I shift to the shipbuilding yards and marine engineering works of the Clyde at Glasgow, still in its relative infancy as the producer of the British Empire's ocean steamers.

17.1 James Thomson: Invisible Apprentice

From his teenage years, James Thomson had a passion for practical engineering and invention. While never patented, his earliest inventions sought to minimize waste of useful work. In 1836, for example, after observations of Clyde steamers whose paddle wheels 'wasted' power by lifting water out of the sea, he tested a self-adjusting paddle blade arranged to shed the water as soon as the blade surfaced. Again, in 1840 he built models of vessels designed to utilize the downward flow of a river to power mechanical legs which would replace the human labour used to propel barges upstream with poles.[5] That same year, when he was aged 18, he graduated Master of Arts in mathematics and natural philosophy at Glasgow College, and took on the post of secretary to the 'Committee on Exhibition of Models and Manufactures' for the annual meeting of the British Association for the Advancement of Science (BAAS), due to assemble in Glasgow for the first time in September.

The exhibition gave pride of place to two great symbols of Clyde invention: the Newcomen steam engine model that had inspired James Watt's separate condenser during his association with the College in the previous century, and the marine engine of Henry Bell's *Comet* (1812) that had come to represent, in Britain at least, the dawn of a steamship age.[6] Indeed, James' father, now mathematics professor at the College, had witnessed for himself while a student the spectacle of one of the *Comet*'s early voyages on the river, having just endured a three-day passage from his native Ireland by a smack carrying lime to the iron works of Scotland. For the Thomson family, steam engines and steam ships represented, in practice and in principle, a new Victorian age of progress and improvement.[7]

James' involvement with the models and manufactures exhibition related first and foremost to the British Association's 'Mechanical Science Section' (Section G). Among the distinguished list of contributors to the Section was the Scottish-born engineer William Fairbairn, who presented no fewer than three papers. But heading the contributions were two papers by shipbuilder and naval architect John Scott Russell. His first presentation treated the question of temperature in relation to the most effective condensation in marine steam engines, while the second concerned the 'most economical' proportion of engine power to tonnage in steamers, especially in those designed for ocean voyages. As an example, Russell estimated the coal consumption of two transatlantic steamers. The first, of smaller power, consumed less coal per day but took longer in all weathers than the second. The other ship, of greater power, showed its economic value in foul weather by completing her voyage in 25% less time. Both papers focused on quantitative estimates of marine engine performance with the goal of minimizing 'a loss of fuel and power'.[8]

The timing of Russell's British Association papers was highly significant. A couple of months earlier, the first Cunard steamer, *Britannia*, had inaugurated a regular mail and passenger service between Liverpool, Halifax, and Boston. In the same month as the British Association meeting the Peninsular and Oriental Steam Navigation Company (P&O) launched a new line from Southampton to Alexandria, with plans for connecting steamship lines eastwards from Suez to India and China. Most ambitious of all, the Royal Mail Steam Packet Company (RMSP) won an Admiralty mail contract (worth £240,000 per annum) in March 1840 to carry the Royal Mails to and from the Caribbean territories, including a link across the isthmus of Panama to connect with the Pacific Steam Navigation Company (PSNC), whose first two steamers had left England in July to take up permanent station on the west coast of South America. Many of these services, ironically, depended on sailing vessels which shipped prodigious tonnages of (mainly) high-quality South Wales steam coal to coaling stations at intervals along the routes.[9]

A graduate of Glasgow College, Russell's original ambitions had been to fill a chair of mathematics or natural philosophy in Scotland, but in the late 1830s his enthusiasm for naval architecture (mathematical and experimental), as well as invention, led to an appointment as manager of Caird & Company, marine engine builders at Greenock on the lower Clyde, following the early death of the head of the firm, John Caird. At the same time, Russell's experimental researches into the hull forms of least resistance placed him centre stage in the first of the British Association's research committees on ship science. As well as experiments with iron-hulled vessels on the Forth and Clyde Canal, he constructed an experimental tank in the grounds of Virginia House—a former ship-owner's mansion in Greenock—which provided a meeting place for Clyde shipbuilders interested in hull design. In the early 1840s Russell played a leading role in executing the Greenock firm's share of RMSP's unprecedentedly large order for fourteen ocean steamers. Constructing engines and boilers for four of the steamers, Cairds subcontracted the wooden hulls to neighbouring Clyde shipyards. As the ships took shape, Russell performed a special role to coordinate substantial design changes with rival contractors on the Clyde and Thames as well as in Leith, Cowes, and Bristol.[10]

In the period 1840–42, meanwhile, a knee injury prevented James Thomson from pursuing an apprenticeship in railway engineering. Remaining with his father in Glasgow College, he attended the civil engineering classes of the new Regius (Crown-appointed) professor of civil engineering and mechanics, Lewis Gordon, who had built up a wide repertoire of knowledge from extensive travels on the Continent. Gordon's experience of water turbines and dynamometers for measuring work done ('mechanical effect') combined to inspire much of James' later professional work.[11] Even before his younger brother William went up to Cambridge as an undergraduate in 1841, James began a regular correspondence which relayed important maritime news from the Clyde. 'I suppose you have heard of the dreadful explosion of the telegraph steamer at Helensburg[h]', he wrote to William in April. 'It was a new steamer with a high pressure engine; and it went at about 18 miles an hour'. And in the same letter he recorded that among the damage caused by recent stormy weather 'the Precursor, a large steamer [Robert] Napier has just finished for the India trade, was driven a-shore from her moorings in Gare Loch'. Similar in tonnage to Napier's steamers for Cunard, the *Precursor* marked a bold attempt by Calcutta and London merchants to upstage P&O on Eastern seas.[12]

As his health improved in 1842, James and a former class-mate, J. R. McClean, moved to the West Midlands town of Walsall, where McClean had oversight of practical improvements to canal locks. In this context, James and William debated apparent losses of power when water, which might have been deployed to drive a waterwheel, filled a lock in the form of eddies and spray without doing useful work. Soon after, James began work in the drawing office of the Horseley Iron Works at Tipton in the industrial Black Country to the north of Birmingham. The firm claimed to have constructed in 1821 the first iron steamer to put to sea, the *Aaron Manby*, which had been assembled on the Thames and steamed from there to Paris to serve on the Seine. This venture may also have inspired Manchester-based William Fairbairn to construct an iron-hulled canal steamer, *Lord Dundas,* which he steamed by sea from the Mersey to the Clyde around 1830 prior to undertaking experiments with the vessel on the Forth and Clyde Canal.[13] When the Horseley Iron Works collapsed within a few months of James' arrival, his father agreed to pay £100 for the privilege of his eldest son entering a 'premium apprenticeship' with the prestigious Fairbairn Company.[14]

Fairbairn, whose wealth and reputation derived from his Manchester works for the construction of mill machinery, boilers, and stationary steam-engines, had established a shipbuilding yard at Millwall in 1835. As he later explained:

> The system . . . of building ships of 100 to 250 tons burden in an inland town, taking them to pieces, and having to rebuild them at some convenient seaport, was in itself a process that could not be long maintained . . . [A]fter mature consideration, London being the seat of government, and a railway communication having been determined on [from Manchester to London], I arrived at the conclusion that the metropolis held out more encouraging prospects for the formation and extension of this new business than Liverpool. Having foreseen from what had already been done that iron shipbuilding must of necessity increase, I came to the determination to establish works in London, and for this purpose I bought a plot of land at Millwall, and with one of my own pupils, Mr Andrew Murray, who was given a small share in the business, entered on the premises early in the year 1835. The following year we had orders for twelve vessels for navigating the Ganges, for the East India Company, and four others for different parts of Europe.[15]

Fairbairn had borrowed upwards of £50,000 to establish the Millwall shipbuilding 'factory'. Although successful in attracting orders at first, the increasing competition from neighbouring shipbuilders who had moved into iron construction threatened the profitability of the yard. Around 1837, for example, Robert Napier's cousin David had begun building iron steamers on a site just upstream from Fairbairn's. Moreover, as Fairbairn himself later admitted:

> We made many blunders as to prices &c. in a business which we had yet to learn, and the rapid increase of the demand for iron vessels, and the consequent necessary outlay and extension of the works in buildings, tools, &c. trenched so hard upon our limited capital as to hamper us for a long time.

The distance between Manchester and Millwall too proved to be a problem, especially as Fairbairn's primary commitment lay with the Manchester works. He further admitted that with regard to Murray that 'it could hardly be expected that so young a man could exercise all the judgement and precaution of a person whose training had attained greater maturity'.[16]

Oblivious to the financial burdens threatening the yard, James conveyed a sense of excitement at his involvement with maritime life beside the Thames. In November 1843, despite a prolonged and debilitating cold, he told his undergraduate brother that he had seen in the East India Docks the *Great Northern*, 'the screw steam ship made at Londonderry', which was 'rigged quite like a sailing vessel, and of course has no paddle boxes'.[17] Named with evident resonances to Brunel's *Great Western*, the 1,750-ton wooden-hulled vessel had attracted an estimated 20,000 spectators at her launch in the north of Ireland, while in 1843 the *Illustrated London News* ran a special feature highlighting her dimensions as exceeding those of 'any steam vessel ever built'.[18] For

James, however, she was yet another example of the enormous range of engineering options opening up in the mid-1840s for projectors of ocean steam navigation.

Very unwelcome news, however, soon tempered James' excitement and optimism. 'Fairbairn's Works here are to be sold', he told William. 'He is going to retire and his sons are to carry on the establishment in Manchester. I suppose I shall go there in about a fortnight, though I don't know yet.'[19] The apprentice's transfer to the Manchester Works was nevertheless delayed, and it seemed that the boiler-making part of the site might be retained. Armed with drawings of a modified Morin dynamometer, James tried to persuade one of Fairbairn's sons to use such a dynamometer to measure exactly mechanical effect supplied from the engine to neighbouring firms and thus to avoid the work done by the engine from simply going to waste. Fairbairn then asked him to remain 'to make out drawings for the 580 [hp] marine engines which are to be made'. And to his brother he reported that he had 'also been seeing a good deal of shipbuilding.'[20]

James now commenced working in the drawing office. His personal notebook shows that he collected detailed data on the costs per horse-power of marine engines constructed not only by Fairbairns in Manchester and London, but also by other Thames builders including Maudslay Son & Field, Miller & Ravenhill, and Penn, as well as by Boulton & Watt in Birmingham and Robert Napier in Glasgow.[21] He also engaged in a discourse with William about the merits of oscillating or vibrating engines for large steamers. In this arrangement the whole cylinder oscillated about a fixed horizontal axis in such a way that the piston drove the crank directly rather than by means of connecting rods. In contrast, the 'side-lever' engine, favoured by Robert Napier and given pride of place in the early Cunard steamers, operated as a variation of the beam engine by means of the piston moving one end of a heavy beam which in turn connected at its opposite end, through a system of connecting rods, to the crank.

The Fairbairn site gave the apprentice engineer an unrivalled vantage point from which to witness shipbuilding and marine engineering on the Thames. At the adjacent site, David Napier constructed fast iron river-steamers. Downriver at Blackwall were the different yards of Mare, Pitcher, Green, and Wigram. Many of the Thames yards were engaged in the building of hulls to be fitted with engines from the Admiralty-approved firms of Penn across the river at Greenwich, Maudslay upstream at Lambeth, or Seaward & Capel at nearby Limehouse. William Pitcher's yard at Blackwall also received a regular flow of RMSP repair work, including the mail steamer *Thames* after grounding in the Caribbean.[22] Her presence there would have generated intense discussion across the maritime communities of the Thames as to whether or not the fourteen large RMSP vessels (of which the *Thames* was one) were fulfilling the promise of reliable and regular ocean steam navigation. Although the engines came from a variety of builders, all seem to have been of the side-lever (beam) type and were driven by steam at about 6 pounds per square inch (psi).

RMSP Company managers themselves had become increasingly desperate for solutions to costly, and at times dangerously large, consumption of coal. They pursued a number of options in an attempt to save the fledgling line from collapse, including weighing up the merits of oscillating engines, testing the calorific value of coal from different mines, and encouraging their seagoing engineers to use expansive working. In early August 1843, for example, the RMSP secretary, Captain Edward Chappell, received from his superintendent engineer, Mr Mills, a report on 'the results of the trial of the Black Vein Coal at Woolwich Dock Yard'. The experiment deployed one and a half tons of coal (3,360 lb) to evaporate 30,720 lb of water distributed in sixteen separate tanks. Each pound of coal therefore turned 9.142 lb of water into steam.[23]

Although the Company had no control over the coal shipments during the long and arduous voyage across the Atlantic, the number, small size, and variety of the vessels chartered ensured that RMSP was unlikely to suffer from a failure in delivery. In the spring of 1844, for example, the Secretary accepted the following coal freights:

Lord Canterbury	599 tons at 16/- for Bermuda
Ann	190 tons at 16/- for Bermuda
Trial	158 tons at 15/- for Bermuda
Marie Antoinette	160 tons at 15/- for Havana
William	240 tons at 15/- for Grenada
Cuba	291 tons at 15/- for Grenada
Demerara	288 tons at 14/7 for Jamaica

Within a month, a further five sailing vessels had been accepted for Bermuda and West Indian stations. But the appetite for coal was far from satiated. On a single day, less than a week later, Captain Chappell ordered freights to be obtained for St Thomas (1,500 tons), Jamaica (1,000 tons), Havana (800 tons), and Bermuda (800 tons), another 4,100 tons in total from the South Wales mines.[24]

The mail steamers themselves stood at the centre of all these concerns. Even with the Secretary's unrelenting efforts to ensure the supply of fuel, RMSP vessels did on occasion run short of coal. In June 1843, for example, Captain Boxer was called before the managers to report 'how the Trent became short of coals on the passage home from Bermuda', while in July the following year Captain Strutt passed Fayal in the Azores without taking on coal and reached Southampton with only two tons remaining in the bunkers. In the managers' opinion, this omission was tantamount to 'running unnecessary & unjustifiable risk, which might have produced very unpleasant consequences'.[25]

From the beginning, RMSP, in part at the behest of the Admiralty, conducted rigorous steaming trials as each new ship joined the fleet. Last of the original fourteen steamers, the *Avon* prepared for a formal trial in early August 1843. Mr Mills attended the RMSP Court of Directors in person to report her machinery, 'upon trial, to be in a thorough state of effectiveness; and that the consumption of coal did not exceed one ton [per] hour'. This performance roughly equated to $5\frac{1}{2}$ lb coal per horse-power per hour. Six days later the Secretary recorded a 'trial of the Avon's engines and boilers round the Isle of Wight in which the performance was very satisfactory not only to the Managers, but to Messrs Maudslay, Field & Pitcher who were present' as expert witnesses. Also aboard was Mr Kingston, the Admiralty Surveyor, who 'expressed himself well satisfied with the performance of the machinery and boilers'.[26]

A few months later the managers instructed Mr Mills to 'adopt some means of lessening ... [the] consumption of fuel' by the *Avon*'s sister *Severn*. By altering her furnaces and 'chimney' (funnel), the engineer expected to diminish the draught of the furnaces and thus effect 'a considerable reduction in the consumption of fuel, with acceleration of speed'. As a further measure, Captain Chappell ordered the shipment of 60 tons of patent fuel (in addition to 620 tons of coal). The patent fuel was 'to be wholly consumed between Southampton and Madeira; and ... a minute account be transmitted, from the latter place, as to the consumption per day, generation of steam, &c &c'.[27] The results of the changes were scarcely encouraging. By March 1844 the Secretary recorded the 'expenditure' of coal on the *Avon*'s second voyage to be 'at least $8\frac{1}{2}$ lbs per horse power per hour'. The *Severn* also appeared to suffer a similar tendency. He thus ruefully reflected that if this statement were correct 'it would really appear a serious consideration if it would not be a better plan to withdraw these ships altogether, or only to employ them in the shorter voyages'.[28]

RMSP mail steamers were consistently put through their paces before an audience of expert witnesses and not simply for the benefit of high-ranking dignitaries from Parliament and the City. In October 1843, for instance, the recently purchased Caird-built *Actaeon* underwent trials in the Thames prior to entering service in the West Indies as an inter-island vessel. These trials tended to focus on meeting the Admiralty's stringent standards with respect to the safety and seaworthiness of the boilers, engines, and hull while also ensuring that the machinery delivered of its best results. Accompanied on the trip by the Admiralty surveyor (Kingston), RMSP's superintendent engineer (Mills), and Messrs Maudslay, Field & Company, the Secretary recorded that the 'new boilers gave an abundant supply of steam and a Diagram taken by the Indicator showed a most efficient performance of the [side-lever] engines'.[29]

Attempts to obtain the best performance from machinery also continued while RMSP ships were in service. Captain Chappell reported in September 1843, for example, that the managers had questioned the *Dee*'s master, Captain Hemsley, 'as to sundry delays on his voyage and as to what attention had been paid to working the steam expansively'. Such expansive working, known to steam engineers since the days of Watt, reduced fuel consumption by admitting steam for only part of the stroke and allowing the steam for the rest of the stroke to expand in the cylinder under its own pressure. The managers had also called 'the attention of Mr Mills to the great expenditure of coal in the *Dee*'—a massive $39\frac{1}{2}$ tons per day over her first 18 months in service, although reduced to $34\frac{1}{2}$ tons per day for the last voyage.[30] Some conception of this rate of consumption can be gained by noting that every ten days a mail steamer such as the *Dee* would require one or even two sailing ship loads of the best Welsh steam coal!

Deeply conscious of this 'coal question', the managers suggested that they be authorized to reward captains and engineers 'where striking cases occur, evincing unusual care and economy in the management of stores or expenditure of coals.' Captain McDougall of the *Medway* was a case in point. The managers recommended presenting a reward to him 'for his great attention to the expansive working of the steam and other means of economizing fuel [and] that this be made known to all the captains and engineers.'[31]

While RMSP were saddled with a large fleet constructed to a single design, they did begin to consider adopting oscillating engines for a new steamer in 1844.[32] P&O, on the other hand, had from the start opted for a far more varied range of steamships. Aware of the high costs of contracting sailing ships to transport vast quantities of coal to the Mediterranean and beyond, the Company now planned to adopt the oscillating engine for their next generation of long-distance steamers on order from Thames and Mersey yards. James Thomson thus invited his mathematical brother to help solve a question 'which will require a good deal of applied mathematics and one which is of great importance at present'.[33]

The problem he set William was 'to find the friction of a vibrating engine as compared with that of some others such as those with slides or beam engines.' 'Many people', he claimed, 'are beginning to give up the prejudice which they had against vibrators; and, for my own part, I am much inclined to think that for both large and small steam boats they are far preferable to any other kind of engine'. He admitted, however, that 'The friction which frightens people is that produced by the pressures of the piston rod on the packing box and of the piston on the cylinder [neck] which are necessary for giving motion to the cylinder'—pressures not significant in conventional slide or beam engines. At the same time he cited 'The fact that there is scarcely any wearing in those rubbing parts . . . [as] quite sufficient proof that there is very little friction [in the vibrating engine], but at the same time it would be an interesting and useful application of theory to determine this *a priori*.' On his own quantitative estimates, he reckoned that the resistance to the motion of the cylinder due to its inertia—'which is usually insisted on most'—was 'extremely small'. But he now looked to his brother to offer 'any tolerably easy and tolerably correct way' to 'represent the total loss from friction.'[34]

In response to William's 'long letter about the vibrating cylinder engine', James assured him that concern raised there about the likelihood of 'snapping the piston rod across in a heavy sea' was unwarranted. '[I]f there was any fear of that', he explained, 'the friction would be very great', whereas the evidence pointed to the opposite conclusion. Furthermore, even if 'a wave stopped the paddle wheel in half a revolution, this would cause scarcely any greater change of velocity in the motion of the cylinder than what actually occurs when the engine is going regularly.' Indeed, he felt satisfied that the 'side-pressure necessary to give motion to the cylinder is extremely small in comparison to the pressure transmitted to the crank'—that is, the force needed to move the cylinder backwards and forwards on its gudgeon was small compared to the driving force of the steam engine delivered to the paddle shaft. In the so-called 'direct action marine engines with guides', the side pressure, he claimed, is 'almost as great as the pressure on the crank', implying much greater friction and therefore much greater wear and waste of power.[35]

Although the two brothers were in agreement about geometrical (graphical) ways of representing the friction at any part of the stroke, the apprentice engineer announced that he had more immediate priorities. 'I can scarcely ever find time for writing letters', he apologized. 'Several times of late I have been working in the drawing office from six in the morning till eight in the evening.' His time was wholly devoted to 'copying some drawings for Mr [Robert] Murry [sic] my companion in the office.' Because the 'Peninsular C$\underline{\text{o}}$ had offered a prize of £100 for the best plans for a steam boat of 1200 tons', Murray was 'trying for it, but had not time to copy his drawings before giving them in.'[36]

The submission paid off. Indeed, there is good evidence that the P&O Board planned to make the steamer occupy a pre-eminent role in the building of their eastern steamship empire. In mid-June 1845 the Board had arranged an entertainment at the Albion Hotel, London, for Sir Henry Pottinger, who had returned to Britain from China the previous year. Appointed by Foreign Secretary Palmerston as envoy and plenipotentiary, as well as superintendent of British trade, Sir Henry had taken the leading role in forcing China to agree to the treaty, signed aboard a British warship at Nanking in 1842, to cede Hong Kong to Britain and open five ports (Canton, Amoy, Foochow, Ningpo, and Shanghai) to British trade. From 1843 until 1844 Sir Henry served as Hong Kong's first governor under British rule. P&O staged the celebration to mark the 'occasion of opening the monthly steam communication with China.' Cabinet ministers, East India Company Directors, and the heads of several departments of Government made up the list of very distinguished guests. P&O then honoured Sir Henry by assigning his name to the Fairbairn steamer building for the China mail service.[37]

With the same ink with which he told his brother of the P&O project, James asked William 'who it [is] that has proved that there is a definite quantity of mechanical effect given out during the passage of heat from one body to another.' William, it seemed, had mentioned on a previous occasion Emile Clapeyron's 'Memoir on the motive power of heat', published in translation in Richard Taylor's *Scientific Memoirs* in 1837. Now James, surrounded by some of the world's largest prime movers consuming vast quantities of fossil fuel, confessed to being in the process of writing an article for the *Artisan* magazine about its proposal as to the possibility of 'working steam engines (theoretically) without fuel by using over again the heat which is thrown out in the hot water from the condenser.'[38] Although the tone of James' remarks is sceptical, such a perpetual recycling of heat would, *if realized in practice*, eliminate most of the problems of the ocean steamship.

Even without Clapeyron's printed paper to hand, James plunged into a distinctive reading of the working of heat engines: '... during the passage of heat from a given state of intensity to a given state of diffusion a certain quantity of mec[hanical] eff[ect] is given out whatever gaseous substances are acted on, and that no more can be given out when it acts on solids or liquids'. That was all he could prove, he continued, because, in the cases of solids or liquids, he did not know whether the fall of a certain quantity of heat would actually produce a certain quantity of mechanical effect, and that the same mechanical effect *'will give back as much heat'*. That is, 'I don't know that the heat and mec eff are interchangeable in solids and liquids, though we know they are so in gases.'[39]

Recalling recent discussions with Professor Gordon centred on waterwheels, measurement of mechanical effect, and most probably the motive power of heat, James then drew William's attention to the analogy between waterwheels and heat engines:

'The whole subject you will see bears a remarkable resemblance to the action of the fall of water.' Whether we let water 'fall from one level to another' or whether we 'let heat fall from one degree of intensity to another', in both cases a definite quantity of mechanical effect 'is given out but we may get more or less according to the nature of the machines we use to receive it.' A water mill, for example, 'wastes part by letting the water spill from the buckets before it has arrived at the lowest level', while a steam engine 'wastes part by throwing out the water [condensed from steam] before it has come to be of the same temperature as the sea.'[40]

Echoing an 1842 essay on overshot wheels written for Gordon, James also stressed that in such a machine 'much depends on our not allowing the water to fall through the air before it commences acting on the wheel', while in a steam engine 'the greatest loss of all is that we do allow the heat to fall perhaps from 1,000° to 220°, or so, before it commences doing any work. Practical constraints existed regarding the utilization of mechanical effect from such high levels: 'we have not the materials by means of which we are able to catch the heat at high level', and in any case 'if we did generate the steam at 1,000° a great part of the heat would pass unused up the chimney.' And a waterwheel near the source of a stream would 'waste all the tributary streams which run in at a lower level.'[41]

Within three months of writing this letter, James had returned to Fairbairn's Manchester works, where he reported 'many great engines' under construction. Visiting his family in Glasgow, he found himself diagnosed with 'a quickness of the pulse'. His doctor recommended that he cease to think of this life and 'prepare himself for the other world'. Retaining a correspondence link with Robert Murray at Millwall, he instead focused on devising economical modes of distilling fresh from salt water aboard ship. Resuming his friendship with Gordon, he also began development of a vortex turbine (horizontal waterwheel). Meanwhile, William had completed his Cambridge degree in January 1845, following which he spent several months in Paris. In addition to gaining practical skills working in Victor Regnault's physical laboratory on the properties of gases and vapours (including steam), he read Clapeyron's paper in its original French version (1834), and searched in vain for Sadi Carnot's treatise on the motive power of fire (1824).[42]

17.2 Robert Mansel: Invisible Undergraduate

By the autumn of 1846 William Thomson, aged just 22, had been elected to the chair of natural philosophy in Glasgow College. In the 1840s the University of Glasgow attracted the sons of several Clyde shipbuilding and marine engineering families, including John Scott, William Simons, James Robert Napier, John Elder, and John Caird.[43] But within Professor Thomson's first undergraduate class, possibly even his very first undergraduate, was Robert Mansel who, at the age of 20, was only a little younger than his professor. A native of Glassford, Lanarkshire, Mansel had already studied at the Andersonian Institution, well known for its industrial and practical courses, and at the Glasgow Mechanics Institution. His own contexts were thus far removed from that of elite, gentrified Cambridge. By the end of the 1846–47 session, Thomson selected Mansel to be his experimental assistant at a time when the new professor had begun a radical overhaul of the apparatus and instrumentation for lecture demonstrations, and had started to introduce precision experimental research into the venerable rooms that constituted the College territory of natural philosophy.[44]

When William Thomson joined the Glasgow Philosophical Society in December 1846, he was aligning himself with a group of reform-minded academics (including Glasgow University's professor of chemistry, Thomas Thomson, the professor of astronomy, John Pringle Nichol, and various Andersonian professors) and industrialists (including cotton manufacturer Walter Crum and mill engineer Charles Randolph). Randolph had worked for William Fairbairn's business in Manchester and Robert Napier's works in Glasgow, where he had been trained by Napier's long-serving manager David Elder. Professor James Thomson had joined the Society in 1839 and his son James in 1841. Lewis Gordon had become a member in 1840, and at some point soon after appears to have presented his reflections on the Carnot–Clapeyron theory to the Society.[45]

Towards the end of 1846 the *Mechanics Magazine* published a detailed paper authored by James Stirling and recently presented to the Institution of Civil Engineers. His brother Robert, a Presbyterian minister in Ayrshire, had originally patented an air-engine in 1816 but was granted a new patent in 1840. James Stirling now made a particular case for its future adoption as a marine engine:

It would be needless, in its present state, to speculate on the probable results of the introduction of this engine for marine purposes. There every ton of fuel that is saved is not only a direct gain to the extent of its market price, but in long voyages an indirect gain is effected of at least twice as much more, from the increased accommodation which it allows for the carriage of merchandise. Nay, when the value of time is so well understood as at present, or when the rapid communication of intelligence is a matter of such importance, surely no slight objection ought to prevent the employment of any means by which a ship may be enabled to hold on her course without interruption, for twice or thrice the length of time she now does.[46]

With this kind of promise it is not hard to understand James Thomson's excitement at the possibilities opened up by an engine which deployed air, rather than steam, as its working substance.

Early in 1847 William Thomson wrote to his fellow professor of natural philosophy, James David Forbes, at the University of Edinburgh with the news that, in clearing out his predecessor's stockpile of apparatus, he had come upon a model Stirling engine. In the very same paragraph he raised the question of loss of mechanical effect when heat conducted through a solid. If, he argued, we considered the case of a fire inside a hollow conducting shell using all its heat to melt ice at 32° F on the outside of the shell, then, since the melting of ice involved neither expenditure nor gain of mechanical effect, there was 'really a loss of effect in conduction of heat through a solid'. Such a loss of power, he admitted, 'seems very mysterious' but added that perhaps it was no more so than power 'lost in the friction of fluids . . . by which there does not seem to be any heat generated, nor any physical change effected'.[47] Fluid friction, and the apparent loss of power, had long been a concern in his correspondence with James, dating back to their reflections on canal locks at Walsall. Very soon after, William embodied the question in Stirling's air-engine, read through the eyes of a natural philosopher familiar with Carnot's theory.

For his first paper presented to the Glasgow Philosophical Society, William offered a 'Notice of Stirling's air-engine' in April 1847. Carefully restricting himself to matters of natural philosophy and thus avoiding any trespass upon the sensitive disciplinary territory of engineering, William began by explaining, from Carnot's perspective, that the work due to the transmission of heat depended 'on the difference between the temperature of the air in the cold space above and the heated space below the plunger.' Given that the temperature difference (analogous to a fall of water) was far greater in an air than a steam engine, it followed that the former would deliver 'a much greater amount of mechanical effect . . . by the consumption of a given quantity of fuel.' Thomson then deployed the Stirling engine to illustrate 'general physical principles' concerned with the relations of heat and work. In particular, he focused on questions raised by an apparent production of ice without expenditure of work.[48]

He considered three cases in turn. First, if the external heat source below the plunger were removed, the engine would continue to turn 'forward' on account of the temperature difference until, with the transfer of heat from below to above the plunger, the temperature became equalized. Second, if the engine were now (manually) cranked forward, heat would continue to be transferred from below to above the plunger. Work would therefore be required to effect the transfer. Third, if the upper space were instead held at constant temperature (say 32° F) and the lower space at the same temperature, then the engine could be cranked without expenditure of work, apart from that needed to overcome friction in an actual machine.

In this hypothetical case, he supposed the upper part kept at constant temperature by a stream of water at 32° and the lower constant in a basin of water at the same temperature. Turning the engine forwards would transfer heat from the lower to the upper part, resulting in the gradual freezing of the water in the basin—without expenditure of work. Thus ice-making, consistent with Carnot's theory, seemed to involve no temperature difference and no expenditure or production of mechanical effect. Conversely, a large temperature difference could, as he had told Forbes, melt ice with the apparent loss of mechanical effect otherwise delivered by an engine utilizing the high-temperature source.

Two months after William Thomson presented this paper he met for the first time James Joule at the Oxford meeting of the British Association. He interpreted Joule's experimental findings in terms of the conversion of mechanical effect into heat, thereby offering an answer to his and James' long interest in what happened in cases of fluid friction where the work which might have been done seemed to disappear.[49] At this point, Mansel became the Glasgow professor's experimental assistant. Over three sessions, and especially in the course of the 1847–48 session, Mansel and the Thomson brothers laid much of the experimental groundwork for a new science of the motive power of heat, or 'thermo-dynamics', which they quickly broadened into a science of energy.

James' notebook on the motive power of heat opened, in April 1848, with preliminary reflections on the theoretical and practical possibilities of air or gas engines through the lens of Carnot's theory. By the close of the month he recorded in detail a meeting in Glasgow with the Rev Robert Stirling at which he 'had a good deal of conversation with him regarding the air-engine.' Now he was 'going on with some improvements in his air-engines.' With this knowledge, James noted the importance of witnesses to their discussion. Thus William and Mansel attended from the beginning, while the talk took place 'afterwards in the presence of my father.' As he put the matter, 'I told him [Stirling] particularly not to tell me anything that he did not regard as entirely public, because I had some ideas on the subject myself.'[50] It was therefore evident that James was aiming to develop an air or gas engine himself, or in alliance with his immediate circle.

In the course of the discussions, James found that Stirling 'does not understand his own engine; not knowing at all the way in which the heat is expended in generating the work.' Indeed, Stirling appeared to have believed that a key to the engine's economy lay in its practical arrangements which allowed, at least in theory, the heat to be recycled without loss. In other words, he took the view that the furnace 'is useful merely to give a small supplement to the heat returned by the respirator ['economizer' or 'plunger'] so as to make up for incidental losses due to practical imperfections of the apparatus, such as conduction of heat, incomplete absorption &c [and] not that the removal of some heat from the fire is essentially connected with the development of work.'

James, however, 'told him that some transference of heat from the furnace to the water by means of the changes of temperature of the air is essential to the action of the engine; that otherwise it would be theoretically a perpetual motion.' In response, Stirling declared that 'there are plenty of theoretical perpetual motions if we [leave] friction resistances &c out of consideration' to which James replied that 'there are these, but not perpetual sources of power.' Stirling, it seems, paused at this point to reflect before admitting 'that perhaps what I [James] had said was correct and that he had never thought particularly on the difference between a perpetual motion and a perpetual source of power.'

This discussion, centred on James' reading of the air-engine through Carnot's theory in the presence of the engine's inventor, coincided with William's presentation of his paper 'On an absolute thermometric scale, founded on Carnot's theory of the motive power of heat, and calculated from the results of Regnault's experiments on the pressure and latent heat of steam' at the April meeting of the Glasgow Philosophical Society.[51] The meeting with Stirling also provided the context for James' and William's return to the subject of ice-making first addressed, as we have seen, in William's 1847 paper to the Glasgow Philosophical Society. Since that paper, the brothers had suspected that the original conclusion was incomplete because water expanded on freezing and might thus be made to produce work, seemingly for no expenditure of power. If, however, the freezing point of water were to be lowered by pressure, the reasoning might be rendered complete.

In May 1848, James' notebook recorded a theoretical calculation of the lowering of the freezing point of ice by pressure. By October, he added that 'William and I have examined the investigation on the last page. The principles and the numerical result are extremely nearly true . . .'[52] The principles included Carnot's, but the reasoning became public only with William's presentation of James' paper, 'Theoretical considerations on the effects of pressure in lowering the freezing point of water' to the Royal Society of Edinburgh in early January 1849.

Embodied in the working of a Stirling air-engine, James' reasoning refined his brother's earlier conclusion that the making of ice involved no expenditure or production of work because it took place without temperature difference. He explained that William's inference

at first appeared to me to involve an impossibility, because water expands while freezing; and therefore it seemed to follow, that if a quantity of it were merely enclosed in a vessel with a movable piston and frozen, the motion of the piston, consequent on the expansion, being resisted by pressure, mechanical work would be given out without any corresponding expenditure; or, in other words, a perpetual source of mechanical work, commonly called a perpetual motion, would be possible.

Thus, in order to avoid such an inference, 'it occurred to me that it is necessary farther to conclude, that the freezing point becomes lower as the pressure to which the water is subjected is increased.' This conclusion then, consistent with Carnot's theory, modified William's inference by predicting that such production of external work from the freezing of water required a temperature difference.[53]

For the experimental measurement, Mansel constructed an exceptionally sensitive thermometer which prompted William Thomson to inform his natural philosophy class in January 1850 that it was 'assuredly the most delicate that ever was made, there being 71 divisions in a single degree of Fahr[enhei]^t.' A few days earlier he had given Forbes the sequence of experimental work almost as it happened:

The ether thermometer w^h Mansell [sic] has made for me is so sensitive that more than 2 i[nches] (nearly three I think) correspond to 1° Fahr. We have divided the tube very roughly, & we find that somewhere about 70 divisions correspond to 1° Fahr. I have found it quite impossible as yet however, & with the means I have at my command to get anything but the rudest estimate of the value of my divisions in Fahr. degrees. The estimate I mention was made by comparing our ether therm^r with one of Crichton's, but there are great difficulties, I find, in the way of making any comparison at all, & it will be impossible, without other means of comparison, to attain any satisfactory accuracy.[54]

Later in the same letter Thomson admitted that at this point he was not obtaining the hoped-for agreement with James' theoretical prediction. Everything, however, changed while he continued to compose the letter. 'As soon as we got the thermometer (hermetically sealed in a glass tube) into Oersted's apparatus [for demonstrating the compressibility of water] everything was satisfactory', he wrote. 'The column of ether remained absolutely stationary until pressure was applied'. Under nine atmospheres of pressure the column of ether 'sank very rapidly' to settle about 7.5 divisions lower. Increasing the pressure to 19 atmospheres took the ether down another ten divisions. Assuming 70 divisions to 1 degree Fahrenheit, the temperature of the ice–water mixture was being lowered by 0.250° Fahrenheit or 0.139° Centigrade. Thomson thus concluded that, with James's prediction for the lowering of the freezing point at 0.1425° C for 19 atmospheres of pressure, 'the agreement is wonderfully satisfactory'. Indeed, he confessed that having only just become aware of the agreement, it 'really surprises me by being so close'. Forgetting earlier doubts about attaining 'any satisfactory accuracy', he expressed his enthusiasm 'as soon as possible to communicate a notice to the Royal Society [of Edinburgh] and to have something published (a few lines would do) in the Philosophical Magazine, so that people may repeat the experiment before the frost goes.'

Within a week, Thomson communicated the news to his natural philosophy class. In one of the experiments, he told the students, '0.246 was the lowering according to experiment; 0.2295 according to theory.' Theory and practice, he concluded jubilantly, matched 'within 1/3000 of a Fahr. degree.' Unlike their predecessors more used to receive established scientific truth through well-worn class-room demonstrations, this student body was now seeing physics fresh from the making, four days before the results had been put before the Royal Society of Edinburgh and several months before the paper appeared in print in the scientific periodical literature.[55]

Throughout the process, however, Robert Mansel was a full participant—if a largely invisible experimental assistant—in this critical and formative period for the theory of the motive power of heat. As he later wrote:

... in the succeeding three [sessions] (when I acted as experimental assistant), that is to say, down to the spring of 1850, we have an interval in which the mechanical theory of heat received its first thorough exposition. It was the salient feature of Sir William Thomson's physical teaching during this period, and I have never forgotten the enthusiasm with which the then young professor entered upon and conducted this subject, nor the generous recognition he bestowed upon the labours of his predecessors and contemporaries whether experimental or mathematical...[56]

17.3 CLYDE-BUILT

On 13 December 1848 John Elder, 24-year-old son of Robert Napier's works manager David Elder, became a member of the Glasgow Philosophical Society and thus joined a network of like-minded scientific reformers that already included Lewis Gordon and the Thomson brothers. A year later they were also joined by John Napier, one of Robert Napier's two sons at the core of the firm's iron shipbuilding and marine engineering activities. By the early 1850s, John's brother James Robert was a contributor of papers to the society on scientific shipbuilding (including water-tight compartments, ships' compasses, and the accurate measurement of a vessel's speed). Macquorn Rankine, aspiring academic engineer, joined in 1852.[57]

In the early 1840s Elder had served a five-year apprenticeship with Robert Napier's firm but under the direction of his own father and 'working successively in the pattern-shop, moulding-shop, and drawing-office'. Delicate in constitution, John's only higher education was a brief period with Professor Gordon at the University. He then spent about a year as a pattern-maker with the steam-engine builder Messrs Hick of Bolton, Lancashire, and as a draughtsman at the works of the Grimsby docks in Lincolnshire. On his return to Glasgow, probably in late 1848, he rejoined Napier's firm and soon took charge of the drawing office in the shipbuilding department.[58]

Originally a builder of marine engines for wooden hulls constructed by shipyards on the lower Clyde, Robert Napier had developed an iron shipbuilding yard at Govan in 1841–43. He appointed a relation, William Denny, as naval architect with the specific task of instructing his sons in the art of shipbuilding. On Denny's departure, James Napier took over management of the yard. In 1849 the firm delivered the steamer *Bolivia* to the Pacific Steam Navigation Company (PSNC) for service on the west coast of South America. Evidently well pleased, PSNC placed an order with Napiers' for four large iron mail steamers.[59]

With business booming as never before, the Napiers also appointed Robert Mansel to their shipbuilding yard, where he 'acquired a practical knowledge of iron ship building' to add to his exceptional experimental skills. It is probable that for a time Mansel served both in Thomson's class-room and in the yard before being appointed, in 1850, as naval architect—a position which he retained for some thirteen years. During that period he played a role in designing the hulls of the record-breaking Cunard mail steamers *Persia* (1856) and *Scotia* (1862) as well as the iron-clad HMS *Black Prince* (sister to the Thames-built *Warrior*). On departure from Napier's, Mansel founded (with James Aitken) his own yard which, over a 30-year period, built many vessels, including ocean steamers for the Union Steamship Company running to South Africa, and cross-Channel steamers for the London and South-Western Railway.[60]

In transferring from Professor Thomson's natural philosophy class-room and informal research laboratory to the Clyde shipyards and marine engineering works, Mansel brought with him a uniquely direct familiarity with the experimental practices and mathematical principles of the physics of heat. Glasgow Philosophical Society members and undergraduate audiences had certainly gained early access to various components of the new physics (through Thomson's presentations on Stirling's air-engine, for instance). But now John Elder and James Napier in particular came face-to-face in the shipyards with the very person who had played an active and crucial role in rendering a scientific experiment on the depression of the freezing point of water under pressure as the single most credible foundation to date of Carnot's theory of the motive power of heat. Furthermore, Mansel had also been a central participant in Thomson's experiments supporting Joule's claims for the quantitative conversion of mechanical effect into heat.[61] Taken together, these demonstrations provided the twin empirical foundations, built upon accurate measurement and reconciling the insights of Joule and Carnot, for the new 'dynamical theory of heat' or 'thermodynamics' that Thomson first drafted in March 1851.[62]

In August 1851 the first of the PSNC quartet, *Santiago*, prepared to run acceptance trials from the Mersey with PSNC directors on board to witness the performance. Overseeing the trials, James Napier had arranged for 420 tons of coal and 80 tons of other stores to make up the 500 tons weight required by the contract. 'The Santiago has done miserably today', he reported to his father as he struggled with the engines (overheating and with no vacuum) and with the paddle wheels (too deep in the water). 'I trust things will come right [...] to Dublin & back will be a famous trial. I shall take care they burn or waste plenty of fuel that we may be as light as possible.'[63] The owners, however, refused to accept the *Santiago* until Napiers had completed expensive alterations. Thereafter, they claimed penalties for delays in delivery.[64]

The steamer's performance almost certainly prompted fresh endeavours to engineer much more economical forms of heat engine. James Napier, whose specific alterations to the *Santiago*'s hull form might have contributed to her problems, soon began a joint project with Rankine to design an air-engine for marine use and in accordance with thermodynamic principles. Difficulties with the experimental model, however, precluded the construction of full-scale air engines to their design.[65] John Elder, whose primary interest lay in marine engineering rather than shipbuilding, left Napier's yard in the summer of 1852 to become a partner in the firm of Randolph, Elliott & Company. As well-known millwrights of high standing, they had not previously constructed marine engines, even though, as we have seen, Charles Randolph had previously worked for Napier. By January 1853, however, Randolph and Elder had taken out their first patent for a marine double-cylinder (or compound) steam engine.[66]

The first steamship fitted with the new design, *Brandon*, received a favourable report on the trials in the pages of the *Nautical Magazine*, whose readership included shipmasters and officers both in the merchant and naval services. Based on the eye-witness account of Captain George Peacock, who had served PSNC as master and superintendent on the Pacific coasts of Chile and Peru for five years, the report used the *Brandon's* good performance to correct 'a popular error which very generally exists, that there is a great advantage in using high-pressure steam, as being a saving of fuel.' Arguing that the higher the steam pressure the greater the quantities of heat required, the journal concluded that 'it follows that there will be equal or even greater economy of coals raising steam at a low than at a high pressure.' Ironically, this reading of the new engines was precisely the opposite of that offered by the new thermodynamics, and showed just how wide the gulf was between the Clyde 'thermodynamic' network and broader communities of knowledgeable marine engineers, of whom Peacock was fairly typical.[67]

Doubtless aided by Peacock's first-hand evidence, Randolph, Elder and Company now persuaded PSNC to adopt the double-cylinder engines for two new steamers for delivery in 1856. The Line's problems with fuel economy had become increasingly severe. First, the larger Napier-built steamers all consumed prodigious quantities of coal compared to their smaller predecessors. Second, local coal supplies on the west coast of South America tended to be limited and of low quality. And third, the outbreak of war with Russia in the Crimea meant high freight rates for the carriage of coal in sailing vessels from Britain—and a corresponding dearth of tonnage to carry Welsh steam coal half-way round the world.[68]

As a safeguard to both owners and builders, Manchester engineer William McNaught was appointed consulting engineer to ensure that the engines met the terms of the contract, especially with regard to fuel economy. McNaught had begun his engineering career as an apprentice in Napier's Vulcan Works around 1827 (doubtless under the watchful eye of works manager David Elder), worked alongside his own father John in the manufacture and marketing of steam engine indicators from 1838, and took out a patent for adding a second, high-pressure cylinder to traditional beam engines used in the textile industry, before moving, five years later, to Manchester, where he found a larger market among the cotton mills of Lancashire.[69]

When the two steamers, *Valparaiso* and the smaller *Inca*, began trials in the spring of 1856, McNaught reported that their engines were actually *less*, not more, economical than the ordinary single-cylinder engines that had reliably driven cross-Channel and ocean steamships for around two decades. For weeks on end, arguments smouldered among the three parties involved. Elder's investigations into the problems revealed an accumulation of water within the larger of the two cylinders. McNaught, as a practical engineer more familiar with stationary mill engines, placed the blame firmly on faulty workmanship, especially the boring of the cylinders. Meanwhile, the two new steamers began to look like a foolish speculation about to crash in financial ruins for the contractors and owners alike.[70]

Elder, however, had immense confidence in the truth of his, and his Glasgow 'scientific' friends', diagnosis and solution. Better insulation of the cylinders, including steam jacketing, seemed like a practical means of preventing the liquefaction of the steam in the larger cylinder. But it also gained legitimacy from the new thermodynamics. As his friend Rankine later explained, additional heat had to be provided to the steam through the cylinder walls to replace the heat that had been converted into work. Very quietly, Elder began to question the validity of McNaught's continuing doubts. He also attempted to persuade the sceptical owners that he intended 'to have the performance of the ships certified by several leading practical marine engineers', thus implying a lack of confidence in McNaught to provide reliable certification *and* his own confidence that the engines, with due modification, would be certified by 'leading practical marine engineers' (presumably from other Clyde works) as entirely trustworthy both in theory and in practice.[71]

Receiving a negative response from PSNC on the grounds that such multiple witnesses did not conform to the contract specifying McNaught as sole adjudicator, Elder changed tactics. He thus set about taking a select group of witnesses, including PSNC's managing director, on a long trial trip from the Clyde to the Fastnet Rock off the south coast of Ireland. On this voyage they all saw for themselves the superior economy of the *Valparaiso*'s engines, suitably jacketed. And rather than engage in supposed absolute measurements of the coal consumption, Elder also arranged for

McNaught and—crucially—other engineers to make direct comparisons with conventional steamers of similar dimensions and power. In all of these trials, the new engines showed a marked gain in economy over their older rivals. Only in the early autumn of 1857, however, was there a final financial settlement between the parties. The relationship between thermodynamics and marine engineering was thus neither easy nor self-evident.[72]

By the end of the 1850s, belief had eroded doubt in specific contexts. Randolph, Elder and Company's compound marine steam engines had become standard on PSNC steamers. The Line even replaced ordinary engines on some of their older vessels. And local papers such as the *Liverpool Albion* and national papers such as *The Times* soon carried reports of a dramatic 50% reduction in fuel consumption. At the very time Rankine published his *Manual of the Steam Engine and Other Prime Movers* (1859), with a 150-page chapter entitled 'Principles of thermodynamics' described by Maxwell as 'the first published treatise on the subject', Elder addressed the British Association at its Aberdeen meeting with the first of his presentations on the results and consequences of trials with his new compound engines.[73]

As with the prototype *Brandon*, none of Elder's early PSNC steamers used unusually high steam pressures (20–30 pounds per square inch was probably typical). But in 1857 John Scott, Greenock shipbuilder, constructed the steamer *Thetis* to his own account for experimental trials with new designs of boiler and compound engines built by Rowan & Co of Glasgow to operate at exceptionally high pressures (over 100 psi on occasion). The vessel embodied Scott's conviction, consistent with thermodynamics, that high-pressure steam was the key to economy in marine engines. Towards the end of that year, Rankine, now professor of civil engineering and mechanics at Glasgow University, told James Napier that 'Rowan's engine burns ... 1.018 lb. per I.H.P. [indicated horse-power] per hour.' In contrast, RMSP's *Avon* recorded, as we saw, a figure of more than eight times this consumption in the early 1840s.[74] Rankine's measurements coincided with a surge of British Association interest in a range of topics concerned with steamship performance, including tonnage measurement, strength of wrought iron and steel, and fuel consumption.[75]

Founded on 1 May 1857 at the very moment of Elder's hard-won victory over McNaught's scepticism, the Glasgow-based Institution of Engineers in Scotland brought together, in more specialized form, shipbuilders, owners, and marine (as well as civil) engineers. With Rankine as first President and James Napier as a Vice-President to set the agenda, members of the first Council included John Elder and David Rowan (engine builders). Within two years, other Clyde marine engineers had joined, including Randolph, John M. Rowan, and William Simons. In late 1858 a discussion took place 'respecting certain recent trials of steamers showing great economy of fuel', to be followed a month later by Elder's presentation 'On marine steam engines' at a meeting that Rankine himself chaired. Early in 1859 the Institution approved a succession of honorary members, including James Joule, William Thomson, William Fairbairn, and Rudolph Clausius. In the same period, Rankine presented copies of two of his manuals (*Applied Mechanics* and *The Steam Engine*) together with his paper(s) 'On the conservation of energy'.[76] A shipbuilder in his own right from 1863, Mansel joined in that year the now-named Institution of Engineers and Shipbuilders in Scotland, and was its President in the period 1878–80. In 1882 he contributed a long paper 'On some points in the history and applications of the theory of thermodynamics'.[77] Few institutions could have demonstrated so strongly the integration of the new physics and the new marine engineering.

Acknowledgements

I thank the editors for all their support and especially their patience. I also acknowledge the generous support of the AHRC for funding the Ocean Steamship Project (2001–07), without which the research for this chapter would have been impossible. For their assistance with primary materials, I am enormously grateful to archivists in the Glasgow University Archives, the University of Glasgow Library, UCL Archives, Cambridge University Library, Queens University Library, Merseyside Maritime Museum, the National Maritime Museum's Caird Library, and St Andrews University Library. For their intellectual and methodological support as always, I thank especially Ben Marsden and Norton Wise. Thanks also to Steven Shapin for the inspiration ('invisible technicians').

Notes

1. T. S. Kuhn, 'Energy conservation as an example of simultaneous discovery'. In M. Claggett (ed.), *Critical Problems in the History of Science* (Madison, 1959), pp. 321–356. Kuhn's four engineers were Sadi Carnot, Marc Seguin, Karl Holtzmann, and G.A. Hirn. Of the others, James Joule and L. A. Colding had strong engineering interests.
2. Crosbie Smith, *The Science of Energy. A Cultural History of Energy Physics in Victorian Britain* (Chicago and London, 1998), 11–12 (summarizing subsequent criticisms of Kuhn's account).
3. Robert Fox (ed.), *Sadi Carnot. Reflexions on the Motive Power of Fire. A Critical Edition with the Surviving Scientific Manuscripts* (Manchester, 1986); Donald Cardwell, *From Watt to Clausius. The Rise of Thermodynamics in the Early Industrial Age* (London, 1971); I. Grattan-Guinness, 'Work for the workers: advances in engineering mechanics and instruction in France, 1800–1830', *Annals of Science* 41 (1984): 1–33; M. Norton Wise (with the collaboration of Crosbie Smith), 'Work and waste: political economy and natural philosophy in nineteenth-century Britain (I–III)', *History of Science* 27 (1989): 263–301, 391–449; 28 (1990): 221–261; Crosbie Smith and M. Norton Wise, *Energy and Empire. A Biographical Study of Lord Kelvin* (Cambridge, 1989); Ben Marsden, 'Blowing hot and cold: Reports and retorts on the status of the air-engine as success or failure', *History of Science* 36 (1998): 373–420.
4. See, for example, Smith, *Science of Energy*, 34; David Philip Miller, 'Testing power and trust: the steam engine indicator, the "Reynolds controversy", and the relations of engineering science and practice in late nineteenth-century Britain', *History of Science* 50 (2012): 212–250.

5. Elizabeth King, *Lord Kelvin's Early Home* (London, 1909), 129, 181–182; Smith, *Science of Energy*, 32.
6. Jack Morrell, 'Reflections on the history of Scottish science', *History of Science* 12 (1974): 81–94, esp. 88–92; Jack Morrell and Arnold Thackray, *Gentlemen of Science. Early Years of the British Association for the Advancement of Science* (Oxford, 1981), 202–222; Smith, *Science of Energy*, 32–33.
7. S. P. Thompson, *The Life of William Thomson. Baron Kelvin of Largs* (2 vols., London, 1910), vol. 1, pp. 3–4. The Thomson family's commitment to 'whig' values forms a central thread in Smith and Wise, *Energy and Empire*.
8. John Scott Russell, 'On the temperature of most effective condensation in steam vessels' and 'Additional notice concerning the most economical and effective proportion of engine power to the tonnage of hull in steam vessels, and more especially in those designed for long voyages', *Report of the British Association for the Advancement of Science* 10 (1840): 186–187; 188–190, on 189; Smith, *Science of Energy*, 34–35.
9. F. E. Hyde, *Cunard and the North Atlantic 1840–1973. A History of Shipping and Financial Management* (London and Basingstoke, 1975); Crosbie Smith and Anne Scott, ' "Trust in Providence": Building confidence into the Cunard Line of Steamers', *Technology and Culture* 48 (2007): 471–496; Freda Harcourt, *Flagships of Imperialism. The P&O Company and the Politics of Empire from its Origins to 1867* (Manchester, 2006); Stuart Nicol, *Macqueen's Legacy. A History of the Royal Mail Line* (2 vols, Brimscombe Port Stroud, 2001); A. C. Wardle, *Steam Conquers the Pacific. A Record of Maritime Achievement 1840–1940* (London, 1940) (PSNC).
10. 'The late Robert Duncan, shipbuilder, Port Glasgow', *The Marine Engineer* 11 (1889–90): 192–193 (Virginia House and its test tank); G. S. Emmerson, *John Scott Russell. A Great Victorian Engineer and Naval Architect* (London, 1977), 2–14, 23, 308; Morrell and Thackray, *Gentlemen of Science*, 275, 322–323, 505–508; Ben Marsden, 'The administration of the "engineering science" of naval architecture at the British Association for the Advancement of Science, 1831–1872', *Yearbook of European Administrative History* 20 (2008): 67–94; Nicol, *Macqueen's Legacy*, vol. 2, pp. 13–16 (Russell and RMSP). The relationship between Russell and the RMSP project is explored more fully in Crosbie Smith, ' "This great national undertaking": John Scott Russell, the Master Shipwrights and the Royal Mail Steam Packet Company', in Richard Dunn and Don Leggett (eds.), *Re-inventing the Ship* (Farnham, 2012), pp. 25–52.
11. See esp. Ben Marsden, ' "A most important trespass": Lewis Gordon and the Glasgow chair of civil engineering and mechanics 1840–1855', in Crosbie Smith and Jon Agar (eds), *Making Space for Science. Territorial Themes in the Shaping of Knowledge* (Basingstoke, 1998), 87–117; Smith, *Science of Energy*, 35–39.
12. James to William Thomson, 1 April 1841, T378, Kelvin Collection, Cambridge University Library (ULC); Harcourt, *Flagships of Imperialism*, 52 (*Precursor*).
13. William Pole (ed.), *The Life of Sir William Fairbairn, Bart*. (Newton Abbot reprint of 1877 publication, 1970), 335–338 (*Aaron Manby*), 133–146 (*Lord Dundas*).
14. Smith, *Science of Energy*, 40. Fairbairn's contributions to the meeting of the British Association in 1840 included 'An experimental inquiry into the strength of iron with respect to its application as a substitute for wood in shipbuilding', *Report of the British Association for the Advancement of Science* 10 (1840): 201–202. He acknowledged that the experiments were superintended by Manchester engineer Eaton Hodgkinson.
15. Pole, *Fairbairn*, 154.
16. Ibid., 154–155; Philip Banbury, *Shipbuilders of the Thames and Medway* (Newton Abbot, 1971), 171–174.

17. James to William Thomson, [July–November 1843], T391–93, T402, Kelvin Collection, ULC.
18. Brian Mitchell, *The Making of Derry. An Economic History* (Derry, 1992), 112–114. William Makepeace Thackeray also witnessed the construction of the large screw steamer on the banks of the River Foyle.
19. James to William Thomson, [1 May 1844], T398, Kelvin Collection, ULC.
20. James to William Thomson, [19 and 21 May and 19 June 1844] T399 and T401, Kelvin Collection, ULC.
21. James Thomson, Note Book No. 4, Manuscripts of James Thomson, Queen's University Belfast. Napier's costs (£43 per hp) were the lowest and Miller & Ravenhill's (£55 per hp) the highest.
22. Entries for 6 and 17 July 1844, RMSP Daily Minutes, vol. 1, University College London (UCL). At this time, Pitcher won the contract for major repairs and maintenance of the RMSP fleet, even though the Line's ships used Southampton as their departure port.
23. Entries for 4 August and 9 November 1843, RMSP Daily Minutes, vol. 1, UCL. A subsequent investigation recommended changes to delivery practices between pithead and shipboard loading in order to maintain the quality of the coal.
24. Entries for 18 March and 3 and 8 April 1844, RMSP Daily Minutes, vol. 1, UCL.
25. Entries for 20 June 1843 and 12 July 1844, RMSP Daily Minutes, vol. 2, UCL.
26. Entries for 3 and 9 August 1843, RMSP Daily Minutes, vol. 1, UCL.
27. Entries for 13, 25 and 30 November and 18 December 1843, RMSP Daily Minutes, vol. 1, UCL.
28. Entries for 22 and 30 March 1844, RMSP Daily Minutes, vol. 1, UCL.
29. Entry for 25 October 1843, RMSP Daily Minutes, vol. 1, UCL. The *Actaeon* was lost by grounding one year later.
30. Entry for 25 September 1843, RMSP Daily Minutes, vol. 1, UCL.
31. Entry for 26 September 1843, RMSP Daily Minutes, vol. 1, UCL.
32. Entry for 8 October 1844, RMSP Daily Minutes, vol. 2, UCL.
33. James to William Thomson, 19 June 1844, T399, Kelvin Collection, ULC.
34. James to William Thomson, 19 June 1844, T399, Kelvin Collection, ULC.
35. James to William Thomson, 4 August 1844, T402, Kelvin Collection, ULC. William's reply does not appear to have survived.
36. James to William Thomson, 4 August 1844, T402, Kelvin Collection, ULC. Robert seems to have been the brother of Fairbairn's young manager, Andrew Murray.
37. William Broadfoot, 'Pottinger, Sir Henry, first baronet (1789–1856)', re. James Lunt, *Oxford Dictionary of National Biography*, Oxford University Press, 2004; online edn, Jan 2008 [http://www.oxforddnb.com/view/article/22626, accessed 8 December 2008]; entry for 13 June 1845, P&O Board Minutes, P&O/1/100, P&O Collection, Caird Library.
38. James to William Thomson, 4 August 1844, T402, Kelvin Collection, ULC.
39. James to William Thomson, 4 August 1844, T402, Kelvin Collection, ULC.
40. James to William Thomson, 4 August 1844, T402, Kelvin Collection, ULC.
41. James to William Thomson, 4 August 1844, T402, Kelvin Collection, ULC.
42. Smith, *Science of Energy*, 42–44.
43. Smith, *Science of Energy*, 29 (Caird), 151 (Elder); Marsden and Smith, *Engineering Empires*, 117–119 (Scott); Smith and Wise, *Energy and Empire*, 24 (Napier); William Thomson's Natural Philosophy Class register 1847–48, Glasgow University Archives (Simons).
44. 'Obituary. Robert Mansel, Shipbuilder', *The Marine Engineer* 27(1005–06): 452; Crosbie Smith, '"Nowhere but in a great town:" William Thomson's spiral of classroom credibility', in Crosbie Smith and Jon Agar (eds), *Making Space for Science. Territorial Themes in*

the Shaping of Knowledge (Basingstoke, 1998), pp. 118–146 (Thomson's reform of natural philosophy at Glasgow).

45. William Thomson, 'Notice of Stirling's air-engine', *Proceedings of the Glasgow Philosophical Society* 2 (1844–48): 169–170. Thomson refers to Gordon's explanation of Carnot's theory presented to the Society at a previous meeting. There is, however, no record of this paper in the first two volumes of the *Proceedings*. The first volume covers the period 1841–44, so it is possible that Gordon addressed the Society in the period 1840–41 shortly after his appointment to the new Glasgow chair and his election to the Society. Membership data are taken from these volumes, esp. vol. 1, pp. 1–3. On Randolph, see Michael S. Moss, 'Randolph, Charles (1809–1878)', *Oxford Dictionary of National Biography*, Oxford University Press, 2004 [http://www.oxforddnb.com/view/article/23115, accessed 14 January 2008]
46. James Stirling, 'Stirling's air-engine', *Mechanics Magazine* 45 (1846): 559–566.
47. William Thomson to J. D. Forbes, 1 March 1847, Forbes Papers, St Andrews University Library; Smith and Wise, *Energy and Empire*, 294.
48. Thomson, 'Stirling's air-engine', 169–170; Smith and Wise, *Energy and Empire*, 296–298. I thank Norton Wise for the concise analysis of Thomson's paper.
49. For example, Smith, *Science of Energy*, 78–79
50. James Thomson, 'Motive power of heat: Air engine', Notebook A14(A), James Thomson Papers, Queen's University Library, Belfast.
51. Smith and Wise, *Energy and Empire*, 299–301. William Thomson had first discussed the possibility of such a scale in a letter to Gordon towards the end of 1847 (p. 300).
52. James Thomson, 'Motive power of heat: Air engine', Notebook A14(A), James Thomson Papers, Queen's University Library, Belfast.
53. James Thomson, 'Theoretical considerations on the effects of pressure in lowering the freezing point of water', *Transactions of the Royal Society of Edinburgh* 16 (1849): 575–580; Smith and Wise, *Energy and Empire*, 298–299 (providing a full exposition of James's reasoning).
54. William Thomson to J. D. Forbes, Forbes Papers, St Andrews University Library; Smith, '"Nowhere but in a great town,"' 126–129 (Mansel and Thomson's instruments).
55. William Smith, Lecture for 17 January 1850, 'Notes of the Glasgow College natural philosophy class taken during the 1849–50 session', MS Gen.142, Kelvin Collection, University of Glasgow; Smith, *Science of Energy*, 96–97; '"Nowhere but in a great town",' 129–130.
56. Robert Mansel, 'On some points in the history and applications of the theory of thermodynamics', *Transactions of the Institution of Engineers and Shipbuilders in Scotland* 25 (1882): 85–132, on 91–92.
57. Membership data on John Elder, John Napier, and Macquorn Rankine is taken from *Proceedings of the Glasgow Philosophical Society* 3 (1848–55).
58. W. J. M. Rankine, *A Memoir of John Elder Engineer and Shipbuilder* (London and Edinburgh, 1871), 4–6.
59. James Napier, *Life of Robert Napier of West Shandon* (London and Edinburgh, 1904), 149–151, 182–183.
60. *The Marine Engineer* 27 (1905–06): 452 (obituary of Mansel). Stephen Fox, *The Ocean Railway. Isambard Kingdom Brunel, Samuel Cunard and the Revolutionary World of the Great Atlantic Steamships* (London, 2003), 163, 190, refers to Mansel as Napier's 'resident naval architect' and designer of the *Persia* and *Scotia*. He cites the same obituary of Mansel as his source, though the obituary in fact uses the weaker phrase 'intimately connected ... with the design and construction of the various mercantile and war vessels' at Napier's, including the Cunarders.

61. Smith, *Science of Energy*, esp. 81. On the reworking of Joule's paddle wheel experiment in late 1847, Thomson noted that 'my assistant [Mansel] was preparing to make an experiment yesterday. To commence with water at $80°$ or $90°$, & to go on grinding (along with another, for relief) for about 4 hours, but I have not heard yet whether it went on.' In this instance there was no doubt as to who was doing the real work.
62. Smith, *Science of Energy*, 78–125.
63. James Robert to Robert Napier, 7 and 8 August 1851, DC90/2/4, Napier Papers, Glasgow University Archives.
64. Napier, *Robert Napier*, 184–185.
65. Marsden, 'Blowing hot and cold', esp. 395–400; Marsden and Smith, *Engineering Empires*, 73–75, 113–116.
66. Rankine, *John Elder*, 6, 31; Napier, *Robert Napier*, 183–186.
67. 'The screw steamer "Brandon"', *Nautical Magazine and Naval Chronicle* 23 (1854): 507–509. See also W. J. M. Rankine, *A Memoir of John Elder. Engineer and Shipbuilder* (Edinburgh and London: Blackwood, 1871). Peacock had had long experience with marine steam engines, including the Royal Navy in the 1830s. The far-from-unproblematic introduction of Randolph and Elder's marine compound engine into PSNC service will form the subject of a separate study.
68. William Just (PSNC managing director) to Macquorn Rankine, 21 October 1870, in Rankine, *John Elder*, 65–67 (compound engine economy in relation to fuel shortages); PSNC Minute Book No.3 (1852–59), PSNC Papers, Merseyside Maritime Museum (numerous entries on coal problems).
69. Entry for 3 May 1855, PSNC Minute Book No.3 (1852–59), PSNC Papers, Merseyside Maritime Museum; Richard L. Hills, "McNaught, William (1813–1881)," in *Oxford Dictionary of National Biography* (Oxford, 2004) (http://www.oxforddnb.com/view/article 37724, accessed 15 January 2008. See also Richard L. Hills, *Power from Steam. A History of the Stationary Steam Engine* (Cambridge: Cambridge University Press, 1989), 157–159.
70. Entries for the period 24 April to 7 August 1856, PSNC Minute Book No.3 (1852–59), PSNC Papers, Merseyside Maritime Museum.
71. Entries for the period 24 April to 7 August 1856; Rankine, *John Elder*, 28.
72. Entries for 16 and 18 October 1856 (Fastnet trip), 5 May 1857 (McNaught's final acceptance award) and 22 September 1857 (final financial settlement between PSNC and the contractors), PSNC Minute Book No. 3 (1852–59), PSNC Papers, Merseyside Maritime Museum.
73. 'The Callao steamship', *The Engineer* 6 (1858): 348 (reprinted from the *Liverpool Albion*); *The Times*, 16 May 1859; Smith, *Science of Energy*, 150–151 (Rankine's *Manual*).
74. W. J. M. Rankine to James Robert Napier, 22 November 1858, DC 90/2/4/38, Napier Papers, Glasgow University Archives;, Rowan, 'Compound engine', 52–59 (published versions of Rankine's 1858 reports); Marsden and Smith, *Engineering Empires*, 119 (*Thetis* summary). Rankine was preparing for the engine builders a formal report on performance measurements.
75. Marsden, 'Naval architecture', 12–18.
76. Entries for 24 November and 22 December 1858 (steam engine economy), 2 and 16 March (honorary membership) and 13 April 1859 (Rankine's donations), General Minute Book Institution of Engineers in Scotland (1857–77), UGD168/1/1, Glasgow University Archives. I thank Ben Marsden for alerting me to these records.
77. Mansel, 'Thermodynamics', 85–132.

CHAPTER 18

ELECTROMAGNETISM AND FIELD PHYSICS

FRIEDRICH STEINLE

In the early nineteenth century, electricity had become a fully respected field of research to such a degree that it formed an integral part of the physical sciences and was present in all research in the physical sciences. Electrical experiments were commonplace, and instrumentation had become fairly common, to be bought ready-made on the shelves of instrument-makers, except particularly huge apparatus, such as M. van Marum's 'Very large electrical machine', now in the Teyler Museum in Haarlem. As to the topics treated, the formerly fierce debate about whether one or two electric fluids should be assumed had run dry and given way to two specific branches of electrical inquiry.

18.1 Volta's Pile

One of the branches was galvanism and, in particular, the effects of the new instrument, proposed by the Paduan physics professor Alessandro Volta in 1800,[1] and quickly connected to his name as 'Volta's pile'—a sequential arrangement of plates of copper, zinc, and moist tissue. The apparatus produced electric and galvanic effects in previously unknown strength and duration. When its two ends were not connected, they acted like electric poles and exhibited effects similar to well-known electric effects such as shocks, sparks, and smell. The most striking feature was that the apparatus, unlike Leiden jars or electric machines, produced those effects repeatedly without being 'charged', Volta even claimed the pile to be a *perpetuum mobile*. Another way to operate the pile was to connect the two ends by a wire, thus 'closing' the pile by a 'connecting wire'. In that mode, no electric effects occurred whatsoever, but rather galvanic effects, for example, on animal tissue. The wire itself, moreover, became hot—even red hot.

Volta had developed the apparatus in the context of his longstanding debate with Luigi Galvani about the cause of the 'galvanic' effects, discovered some twenty years earlier. Galvani had emphasized the importance of animal tissue and had assumed a specific animal electricity. Since the pile did not involve any animal tissue, Volta regarded it as a decisive argument against Galvani's view. That the story was much more complicated would be accepted only some decades later. But the pile was not only powerful in bringing about effects of previously unknown strength, and as an argument against Galvani's animal electricity, but also, and most important, as a trigger of a wave of research throughout Europe.

The reactions to Volta's instrument were characteristically different at different places. In London, the chemical effects came into focus, both those within the pile and those produced between the two ends of the wire—effects which Volta had reported but never really analysed. The surgeon Anthony Carlisle (1768–1840) and his friend William Nicholson (1737–1798), editor of a scientific journal, realized that from a drop of water, hydrogen and oxygen were released, and they immediately interpreted their results as a decomposition of water—an interpretation that led to much discussion. It also led Humphry Davy, a young lecturer and researcher at the Royal Institution of London, to explore those chemical effects further. In the course of intense experimental work, he managed not only to give account of most of the additional chemical reactions, but also to obtain exactly the expected quantities of oxygen and hydrogen. For the crucial point of how to explain the separate places of gas development, he offered a theory of electrochemical dissociation which strongly drew from previous accounts by Berzelius and by von Grotthus, without their names being mentioned, however. His study was widely accepted as providing definite proof of electrolytic decomposition. Not only was he highly honoured in London and Paris, but the *Ecole Polytechnique* received a special donation from Napoleon in order to build a even bigger Voltaic battery.

Most important, the role of the battery itself had changed from an object of research to an instrument for chemical analysis. Pursuing that aspect, Davy produced another discovery in 1806. When soda or potash were treated with a strong Voltaic apparatus—quite a difficult task—there emerged oxygen and some new substances. Davy again interpreted the process as decomposition, and named the substances sodium and potassium. That interpretation not only shattered well-accepted chemical knowledge, but the situation was even more peculiar, since the central means to obtain those processes was the Voltaic battery, which itself was far from being well understood. Only after many long debates did Davy's interpretation become gradually accepted and pursued.[2]

In German countries, galvanism had been intensely treated even before 1800. As a result of the news of Volta and Nicholson/Carlisle, activity intensified. Again the most original contributions came from the young Johann Wilhelm Ritter, who had made himself known by original contributions to galvanism. While deeply inspired by the ideas of the German *Naturphilosophie*, Ritter pursued his own approach in his research, with strong emphasis on empirical and experimental work. Volta's discovery was not too much of a surprise for him, since he had already worked out a close relationship between galvanic and chemical activity. Hence he was much opposed to Volta's view of the pile as solely driven by the contact of metals. Ritter developed new forms of the pile: the 'Ladungssäule' ('charge-pile'), which could be 'reactivated' after exhaustion by being connected to another one, and the 'Trockensäule' ('dry pile'), which worked with (nearly) dry intermediate layers and could thus be handled more easily. Ritter would probably have won the annual galvanism prize of the Paris *Institut* for 1803 if he had not, at the same time, claimed another discovery—the supposed 'electric' poles of the Earth—which finally turned out to be a failure. Ritter formulated clearly their different effects on the open or closed pile, and came near to the distinction of two independent quantities: tension and current.

In Paris, the first reactions to the news of the pile came from chemists, and in the debate about how to interpret the chemical processes the idea of electrolysis came much in favour. But when, in October 1801, Volta presented his apparatus at the *Institut* and was honoured by Napoleon (who had invited him), the young physicist and *Polytechnicien* Jean-Baptiste Biot was charged with writing a report on the pile and focused his interest on the instrument. He was determined to take a different approach than the chemists, and it is significant that in his first publication he did not mention any of the chemists' work, but instead praised Laplace, who had published nothing in the field. As a result, in 1804 he presented a mathematical scheme to calculate the activity of piles of different size, based on the assumption of metal contacts being the only source of the pile's activity, and all chemical effects being just marginal side-effects.[3] Biot's scheme did not attempt to account for the cause of the tensions between different metals, but rather offered a way to calculate how those tensions add up in the complicated arrangement of a pile. Being concentrated on tensions, it was, taken strictly, restricted to the effects of the open pile—a point that Biot disregarded, leading to some confusion. Nevertheless, Biot's theory became prevalent in France for many years, and the physicist Haüy even regarded Biot's report (rather than Volta's invention!) as the beginning of a 'modern era' of Voltaic science. Not that there had not been in France broader approaches as well. Gautherot, for example, had called for a broader experimental exploration of the field, before proceeding to theories. But such a call came from those who had a closer eye on the chemical effects involved. That they found no resonance, and were finally muted, has much to do with the institutional setting. Many of the key positions in the *École Polytechnique*, in the *Institut*, and in journals were soon held by those who promoted a mathematization of the field along Laplacian lines—and that was simply not the academic environment in which more qualitative, exploratory work could well develop.

All important inventions and conceptual or theoretical steps concerning the pile had been carried out in the first decade of the nineteenth century. In the second decade, research activities slowed down considerably, leaving many questions unsolved.

The problem of where the origin of the pile's activity should be located remained open until mid-century. Even the basic concepts had not become clear. The term 'tension', though used by many, was still vague, and that was also true for the notion of electric current. It had been investigated by Volta and by Biot, among others, but it had different meanings here and there, and was always connected with speculations concerning what was happening in the wire. In order to avoid such speculations, many writers preferred to bypass that notion and to talk of the 'connecting' or 'closing' wire of the battery. Much of that vagueness had to do with the lack of means to quantify or even measure the effects of the closed pile. In order to compare different piles, sometimes the thermal effects were used ('the pile could melt a platina wire of such and such diameter'), or the chemical ones ('still after one hour's operation the pile was able to dissolve acidulated water'). Those measures were rough, and had the serious disadvantage that they affected the activity of the pile. For weak piles, they were just too coarse.

Despite the uncertainties on a fundamental level, piles were widely distributed. They belonged to the standard equipment of chemical laboratories and physical cabinets, and there was much communication on new forms of batteries, or improvement of certain points. Piles were even more widely distributed for medical applications. Galvanism had already been applied in medical therapy before 1800, and those activities were enormously enhanced by the invention of the pile. There were not only spectacular treatments of deafness, but also wide use by practitioners. There had also been a prototype of an electric telegraph by Samuel Thomas Soemmering at the Munich Academy. After all, small apparatus were affordable and could be bought ready-made from the shelves of instrument-makers. Their handling became standard and could be learned easily. The widespread distribution of small and medium-size apparatus and of skills was one of the main factors why the spectacular and surprising effect of electromagnetic interaction, when discovered by Ørsted in 1820, could be replicated so easily and quickly.

18.2 MATHEMATICAL APPROACHES TO ELECTRICITY

The other major field of electrical research aimed at precision measurement and mathematization. It was mainly localized in Paris, where it took its origins both in the tradition of high-precision astronomical and magnetic measurements (mostly in the Paris Observatory) on the one hand, and in the tradition of rational mechanics and celestial mechanics on the other.

Based on techniques that had been developed in the Paris Observatory for precision measurements of the magnetic declination and its variations, the Paris engineer C. A. Coulomb had, in the 1780s, refined the technique of measuring very small forces by torsion wires, and applied it to electricity. His confirmation of the $1/r^2$ law of electric attraction and repulsion was based on two experimental arrangements: his famous torsion balance, and an arrangement of an oscillating suspended bar. While the torsion balance became a public icon of the triumph of precision measurement (but was not very intensely used later), the oscillation-bar arrangement became the central means for all later electric and magnetic precision measurements.

While Coulomb himself was mainly focused on measurements, the idea of developing a thorough mathematical to physical domain and basing it on precision measurement was promoted in late eighteenth and early nineteenth century by a small but prominent group led by the astronomer Laplace and the chemist Berthollet. The main point of the agenda was mathematization. The tremendous success of celestial mechanics had led to the hope of similar success in other areas such as optics, elasticity, heat, electricity, and magnetism. Quantification in form of exact measurement, and arrangement of the data in form of analytical expressions, were considered the first steps. Much as in celestial mechanics, the further steps of mathematization were based the on reduction of all physical interactions to the action of central forces, acting at a distance and according to well-formulated force-laws. As centres of the forces, particles of microscopic substances or fluids were assumed, such as particles of light, heat, or electric or magnetic fluid. The 'empirical laws', gained by precision measurement, should finally be explained from the interaction of those particles. The famous picture of Laplace's demon serves well as an icon for the whole programme. For him, from his knowledge of place and motion of all particles and from his infinite calculating power, both the past and future of the world were open. It was during this time and at this place that the notion of physics as a discipline characterized by exact measurement and mathematization gained ground for the first time.

That the Laplacian school could thrive in Paris was due to the dramatic reorganization of the French educational system after the Revolution, so that there was a focus on exact sciences—all the more so because Napoleon, who had secured an influential place for himself in the system, strongly supported that approach. Moreover, Laplace had close ties to Napoleon and had a strong influence on the way in which physical sciences were treated in the new institutions of expert education and academic research—the École Polytechnique and the Institut de France, with its Académie des Sciences. Finally, Laplace and Berthollet were, after a short time with Laplace as a minister in Napoleon's cabinet, given a generous grant that enabled them to gather, around 1806, a group of young researchers in the Paris suburb of Arcueil. As their students and fellows were successively put in key positions of the Paris academic community, the movement soon came to dominate research activities, and for a period of ten to fifteen years, French physical science was mainly Laplacian.[4]

Electricity was a promising case for the Laplacian programme, not least because Coulomb's excellent precision measurements were available. Indeed, already in 1801 Laplace made Biot work on giving Coulomb's theory of electricity a more thorough and generalizable form, but in the autumn of that year Biot left the topic and turned to the Voltaic pile. It was only ten years later that the topic was taken up again by Siméon-Denis Poisson (1781–1840), who also had joined Laplace's group in his youth, and who had been active in other fields such as heat theory and elasticity theory.[5] His work in electricity was only a rather short excursion from his major research in other fields, but it was instrumental in his long-desired election as a member of the First Class of the Institut. In January 1812 a prize competition was announced for studies of the distribution of electricity on conducting surfaces. A few months later, Poisson delivered a first paper on the subject, and towards the end of the year a second one,[6] by which time he had already reached his goal of being elected to the Institut.[7]

The challenge was to develop mathematical tools which allowed the inverse-square law to be applied to complicated arrangements. Poisson took (as Biot had done) the most powerful mathematical instrument that Laplace had developed for celestial mechanics—a mathematical function V, which later was named potential function and whose partial derivatives gave the force. The inverse-square law was already deeply built into the very form of V:

$$V = \int dm\,(x',y',z')/r'$$

where dm is the mass of an element, located at the point (x',y',z'), and r' is the distance of that point to a (fixed) point outside. The integration covered the whole body of which dm was an element. In order to transfer that tool to the field of electricity, assumptions on the nature and behaviour of the electric fluid had to be added. Poisson analysed the distribution of the electric fluid on the surface of conducting bodies of various shape, such as ellipsoids or spheroids, or a combination of spheres. In comparing his theoretical approach with Coulomb's data, he found only small differences (there are, of course, no considerations whatsoever as to error intervals or the like) and took that as a confirmation not only of his theory, but of the applicability of the Laplacian approach to electricity in general.

The importance of Poisson's theory lay in elaborating the mathematical methods: solving integral equations by expansion into power-series, and dealing with new types of integrals. Within general electric research, Poisson's studies remained rather isolated. Even among Paris academicians, only few were able to follow his analysis. For the 'electrician' in general he had little to offer. The theory related only in a few points to empirical results, and he did not contribute to lively questions of the day, such as Voltaic electricity. And even in the context of common electricity, there were important points which he did not even touch upon, such as the question of what kept the freely moveable electric fluid within the boundaries of conducting bodies. It is significant for the power and limits of the Laplacian school that Poisson, when he published, in 1824, a treatise on magnetization and again developed forceful mathematical methods for that domain, did not say much concerning magnetic questions in general, and electromagnetism in particular, which for many years had been one of the most prominent topics at the Paris Academy.

In the second decade of the nineteenth century, Laplacian physics began to decline. Not only did special funding cease after the turmoil of the Napoleonic era, but an increasing number of people considered the programme to be too narrow. In 1807,

Jean Baptiste Joseph Fourier (1768–1830) presented a new and un-Laplacian theory of heat—which, characteristically enough, could not be published until much later. In 1815, Augustin Jean Fresnel offered a similarly un-Laplacian theory of light, and in 1819 Pierre Louis Dulong (1785–1838) and Alexis Thérèse Petit (1791–1820) combined their presentation of the law of specific heat with a strong criticism of Laplace's theory of caloric. To be sure, no-one called into question the ideal of mathematization. But as the Laplacian approach was, after all, based on a deeply metaphysical view of the world, alternatives were sought. It is significant that most of the protagonists of the counter-movement, such as Fresnel, Dulong, Petit, and François Arago (1786–1853), had themselves grown up in the Laplacian framework. The prospect of opening new horizons would make Paris a special place when news of an interaction between electricity and magnetism arrived in 1820.

18.3 Ørsted and the Discovery of Electromagnetism

The question of a possible interaction between electricity and magnetism had arisen soon after Volta's report of the pile—after all, the pile showed so many interactions between electricity and other physical powers such as heat, light, and chemical action that the question of whether magnetism would also be affected was inevitable. Besides unsuccessful trials by Ritter in 1801, and some indications achieved by Italian scholars but unacknowledged by the authors themselves (1802 and 1804), two Paris academicians, N. P. Hachette and C. P. Desormes, undertook a high-sensitivity trial in 1805. They made a very large pile float on water so as to minimize friction and to see whether there was any alignment due to terrestrial magnetism or any other reactions to magnets. Fitting to Biot's account of the pile, they considered it sufficient to work with the pile in its 'open' state. In their experiment they did not obtain any indication of an action of magnetism onto the pile. From what became known a little later, this negative result is not surprising, since only currents but not static electricity interact with magnetism—indeed, had the Paris group experimented with a closed pile (which from the experimental perspective would have been only a small variation), they would very probably have discovered electromagnetism. This negative result is a striking case of how fixation of perspective to a specific account can block even small variations and thereby sometimes prevent major discoveries. The Paris physicists took the result as significant. There were no further experiments on electromagnetic interaction, and the non-existence of such an action was taken for granted. This view was corroborated indirectly, moreover, by Poisson's success in mathematizing both electricity and magnetism, with an account based on the assumption of non-interacting imponderable fluids.

Not everyone in Europe, however, gave up the idea. In particular, for those who were close to the ideas of German *Naturphilosophie* (such as Ritter, Schelling, and

Oken, to name but a few), the interaction (for some, even the identity) of all forces of nature was a major heuristic programme. One of those to pursue it was Hans-Christian Ørsted. Son of an apothecary in rural Denmark, he was interested in experimental research from early on, and on a European tour visited all major centres of physical research. Back in Copenhagen, his academic career advanced rapidly, and he became one of the most influential figures in Danish academia from the 1820s on. A close friend of Ritter, he had reported Ritter's discoveries to the Paris academy in 1803, and knew first-hand the reservations against all *Naturphilosophie*-like approaches—reservations that he had himself fuelled in 1803 when he presented Ritter's purported 'electric pole' experiment and completely failed to reproduce the effect.

In 1812 Ørsted published a small book, *Ansichten chemischer Naturgesetze* (*Views of the Chemical Laws of Nature*), in which he expressed, among many other things, his belief that all forces of nature interact with each other, and in particular rehearsed his conclusion that there must be an action between electricity (or galvanism) and magnetism. At that time he did not pursue the topic experimentally, but took it up in 1820, stimulated by his experimental lectures at Copenhagen university. It was in this context that he made his famous discovery, the demonstration of which required only elementary apparatus (Fig. 18.1). When a wire was held above a magnetic bussole and connected to the battery, the needle was turned away from its normal north–south position by an angle that varied with experimental parameters. Ørsted carried out various series of experiments to determine these dependencies. He varied many parameters of the original arrangement—distances, materials, arrangements, battery power, polarity, and geometrical constellation—and was able to formulate quite generalized results, such as: 'The pole, above which negative electricity enters [the wire], is deflected towards the west, the one, below which it enters, towards the east.' Such

Fig. 18.1. Oersted's discovery of electromagnetism, 1820 (a later representation).

claims already comprised many individual experiments, and formed strong generalizations; note, for example, that the claim quoted does not specify the poles of the needle, and indeed is valid for both types of pole. The quote also illustrates one of the most serious difficulties that Ørsted (and all others during that early phase) had to face: the motion of the needle's poles had always to be referred to the compass directions, since there was no concept available that allowed referring them in a more general manner to the wire. One of the most striking features of the electromagnetic action was that, if only the wire was placed beneath instead of above the needle, the deviation of the needle changed its direction—an effect that could not, so Ørsted emphasized, be grasped with the basic notion of attractive or repulsive forces.

Ørsted was well aware of the spectacular nature of his discovery, and did everything to make it known as quickly and broadly as possible. His peculiar strategy was to write his findings immediately in Latin (which was becoming less common by then), to print it at his own expense, and to send it directly to a large number of researchers throughout Europe, some of whom were editors of scientific journals. He thus not only reduced the usual time between submission to a journal and publication, but at the same time made his discovery known throughout Europe. His strategy worked. Within a few weeks the news spread, and the topic became of intense interest for the majority of experimental physicists. The text was quickly translated into several languages—French (in two independent versions), English, German, Italian, and Danish—and without delay it was published in several journals.

Ørsted's text was clearly structured in four elements of quite different length. He gave (1) a brief historical view of his experiments and named the witnesses of the experiment, (2) briefly described his experimental set-up, (3) gave an extensive account of numerous experiments and their results, and (4) ended with some thoughts as to the causes of those results in terms of what he called the 'electric conflict'. By far the longest part was devoted to the experiments and general formulations of their outcomes. In his explanatory account, Ørsted took recourse to the notion of 'electric conflict', by which he meant, as he emphasized, a label for the otherwise unknown process in the wire that caused the effects. He stated, first, that that conflict obviously had its effects also in the space around the wire and, second, that it should be regarded as performing circles around the wire, or rather spirals, since it also had a longitudinal component. Only such circular action, he emphasized, could account for the various effects of which an account of attractive and repulsive forces obviously did not work. However, he did not specify how that circular or spiral action should be considered in detail.

As to the reception of those results, some peculiar points should be highlighted. First, and notwithstanding serious doubts that the report faced at many places, replication of the effect worked nearly everywhere and silenced those doubts immediately. This is all the more remarkable as there was no theory or explanation visible that could account for the results. On the contrary, they were apt to show strikingly how much the existing theories were limited in scope. The central point for easy replicability was that the experiment required only standard apparatus—piles, bussoles, wires—that were present everywhere. Second, many commentators and editors highlighted the challenge that the effect posed with its puzzling features, of which no account could be given within the traditional framework of attraction and repulsion. Third, the importance of the finding was realized all over Europe, and research started immediately at very many places. Many researchers even found a new era of research opened, and the Royal Society immediately awarded its Copley Medal to Ørsted. Fourth, the experimental activity that started at many places aimed at formulating regularities and laws of the new effects, and thus had largely an exploratory character—after all, there were no theories available that otherwise could guide experimentation. Fifth, the reading and reception of Ørsted's text were split in characteristic ways: While his experiments were discussed and pursued most intensely, his account to explain them by means of the 'electric conflict' was more or less ignored. Probably that concept appeared too vague or was too reminiscent of the first version of 1812. Nobody seemed to realize that the concept had changed considerably, and had a sort of instrumental character without any longer involving the former speculative elements. Sixth, while most researchers acknowledged that Ørsted's findings were the fruit of a well-directed programme of research inspired by Naturphilosophie, a few others (such as Gilbert in Leipzig) presented the result as a pure chance discovery. They were not ready to acknowledge that such an approach as *Naturphilosophie* could lead to anything useful. Finally, despite basic similarities of the first reactions to Ørsted's report, the research lines that developed out of them at various places quickly branched away from each other, most strikingly in Paris and London, as will be discussed in the next sections.

18.4 ELECTROMAGNETISM IN PARIS

In Paris, Ørsted's results met the established Laplacian programme in a particularly fragile state. François Arago, as one of those who had become increasingly critical of the programme, had learned of the experiments and participated in a successful replication on a visit to Geneva in August 1820. Completely excited about them, he presented a report to the Academy immediately after his return. Moreover, he arranged for a French translation of Ørsted's letter to be printed in the *Annales de Chimie et de Physique*, of which he had assumed editorial and financial responsibility since 1815, together with Joseph-Louis Gay-Lussac (1778–1850). The academicians' reaction, however, was rather frosty and reluctant, and some even surmised another 'reverie allemande'. Not only did such an effect appear very unlikely from the received theories, but there were still some bad memories connected with Ørsted's name. Only when Arago replicated the effect before the Academy did it become a serious challenge for exactly those theories. Unfortunately, Biot, the leading physicist who could respond from the Laplacian perspective, was not in town. The Laplacian reaction would be delayed.

Some quickly realized this situation. Besides Arago himself, feverish activity came, unexpectedly, from André-Marie Ampère.[8] Unlike most Paris academicians,

Ampère had had no Paris education, but had grown up mostly self-educated in the countryside near Lyon. Some of his interests, such as metaphysics, Kantian philosophy, and psychology were unusual for Paris academicians—and he even was active in writing poems. His personal life in Paris had seen a very unhappy second marriage, and his closest friends still lived in Lyon. He was professor of mathematics at the École Polytechnique, and had become a member of the Academy on the strength of his publications on mathematics and chemistry. Physics, in particular experimental physics, was not his field at all. His speculations on a unified ether theory of electricity and magnetism, which he had never elaborated, lay twenty years in the past. Of course, Ørsted's report revived that old idea and touched his metaphysical interests in the underlying structure of the world. Like Arago, moreover, he was unhappy with the too strict framework of Laplacian physics. But there were other reasons for his unexpected interest. It was immediately clear that the field would attract much interest, and all serious activity could be sure of wide attention. Since Ampère wanted to improve his position in Paris, this was a welcome aspect. And indeed, when four years later he obtained a position at the Collège de France, his work on electromagnetism played an essential role. For the time being, however, Ampère was determined to occupy the field before Biot came back. No wonder his activities were extremely hectic, to say the least.

18.5 Ampère's Entry: Forming a Direction of Research

The most puzzling aspect of Ørsted's effects was that the magnetic needle set itself 'across' the wire, and that its behaviour depended in some complicated way on the spatial arrangement. Ampère started his research by attempting to grasp that behaviour and to formulate regularities.[9] Soon he derived two 'general facts'. First, the needle was always turned into a *rectangular* position to the wire; second, once it had achieved that position, it was attracted (or repelled, when set antiparallel). In order to specify which of the two possible positions the north pole of the needle took, Ampère introduced the notions of 'right' and 'left' of the current, illustrated by a picture of a man placed on the wire, with his face turned to the needle. Later, that formulation would be called Ampère's 'swimmer'-rule. Only by such a complicated idea could he formulate the needle's behaviour independently of the compass directions, and was the first to do so. Although those effects required a central position of the needle, Ampère claimed (though never elaborated) that *all* electromagnetic effects could be 'reduced' to them—in other words, be understood as superpositions. He envisaged an explanation on a phenomenological level and explicitly kept away from considerations about underlying causes involved—about what was going on in the wire, for example. His experimental activity in that context was exploratory and broadly oriented, its main principle being the systematic variation of a broad range of experimental parameters. Such an exploratory experimental activity was typical for most researchers during those early days of research on electromagnetism.

In that first and intense phase, three further achievements came about. Ampère noticed, probably by chance, that the pile itself exerted an effect on the wire. In order to include that effect in his first 'general fact', he introduced the notion of a current *circuit* that comprised both the pile and the closing wire—an idea that was a very unusual and alien to Biot's electrostatic theory of the pile. Although Ampère introduced the concept in a strict instrumentalistic understanding, it would shortly afterwards stimulate him to develop a physical theory of electric current. Second, he proposed to use the magnetic action of the pile to measure its activity. As he learned from Arago, the Genevan professor Charles Gaspard de la Rive (1770–1834) had already considered such an idea. But with his 'general fact', Ampère had a better means to pursue the point. He introduced a scale and proposed the 'galvanometer' as an instrument to measure the current or whatever the process in the wire was named. Given the problems in quantifying the effects of the pile, it is no wonder that he was proud of his proposal. Third, he exerted considerable effort into establishing the mechanic *reversibility* of the new effects. Since the battery wire moved a magnet, a fixed magnet should move a moveable wire! Here he was confronted with the severe technical problem of providing easiest mobility while maintaining excellent electrical contact. The experience gained here became very important for all his future research.

Parallel to this exploratory work, Ampère also pursued theoretical speculations which achieved a fascinating dynamics. First, he considered terrestrial magnetism as possibly being caused by gigantic circular currents within the body of the Earth (Fig. 18.2)—an idea which he quickly generalized: perhaps *all* magnetism was caused by electric currents within magnetic bodies. Since such currents had never been noticed, however, he took an indirect way to establish plausibility by examining whether arrangements of circular currents would show magnetic-like behaviour. And indeed, most of the numerous experiments he designed to this purpose were successful. In contrast to his exploratory work, these experiments were well designed to prove a theory. His expectations were strong: when some experiments failed, he did not blame his hypothesis, but rather sought to 'optimize' the apparatus, to use stronger

Fig. 18.2. Ampère's sketch of his hypothesis on terrestrial magnetism ('bonhomme d'Ampère'), 1820.

batteries, and so on. Finally, he succeeded in demonstrating even the mutual attraction of two spirals of wire without mediation of iron—a totally new effect, going far beyond Ørsted's discovery. He not only took it as a 'definite proof' for his magnetic theory, but opened up another field of research: a theory of interaction of electric currents. Only three weeks after the start of his research he presented the breathtaking scope of his theory. From the interaction of electric currents as core effect, not only electromagnetic effects should be derived, but also terrestrial magnetism, and all the well-known effects of usual magnetism! Although he was well aware that empirical support was still very weak, he ordered that an outline of his theory be printed, and had it widely circulated, even to the Royal Society in London. Having thus committed himself publicly and irrevocably to that particular programme, he immediately dropped his previous exploratory work. All his further research in electromagnetism would be concentrated on elaborating that programme.[10]

In those first weeks, the only other person to present results in Paris was Arago, who carried out research on magnetization by electricity, both galvanic and common. The two exchanged their results, and Arago was crucial in giving Ampère access to laboratory resources at the Obsérvatoire. Ampère started quickly to set up his own laboratory at home at considerable cost. In order to make his 'crucial' experiment with two spirals run, for example, he spent half a month's salary to buy the largest battery Paris had to offer. Ampère also interacted intensely with the Paris instrument-maker Nicolas-Constant Pixii (1776–1861), without whose experience and ideas he hardly would have acquired his new apparatus and technologies.

For the next three months, Ampère gave almost weekly presentations to the Academy, paralleled by numerous publications. In order to establish the core effect of his theory, he (and Pixii) developed the famous 'current balance', by which the interaction of rectilinear currents could be demonstrated (but, it should be noted, not measured!). He developed a physical theory of how the effects of the closed pile related to those of the open pile and common electricity—a theory of currents and tensions.[11] Although still somewhat vague, it led him to switch his own and cautious terminology of 'galvano-magnetic' interactions to Ørsted's term of 'electromagnetism'. Later he invented the term *electrodynamics* (as opposed to *electrostatics*) for his theory. He left the development of the galvanometer to others, as he did with electromagnets and with his proposed electromagnetic telegraph. The next step he envisaged for his theory was a mathematical force-law; but before he could derive a conclusive result, Biot entered onto the scene.

18.6 THE LAPLACIAN RESPONSE: BIOT AND SAVART

Only six weeks after the news of Ørsted's discovery had reached Paris, Biot was back, and immediately began research. The competition with Ampère was immediate and bitter: when Biot, in his first publications on the subject, gave a sketch of the state of the field, he mentioned neither Ampère nor any of his achievements. Biot's approach was straightforwardly shaped by the Laplacian programme: 'The first thing which we must determine is the law according to which the force emanating from the conjunctive wire decreases at different distances from its axis'.[12] Following that precept, he set up, together with the young Félix Savart (1791–1841), a Coulomb-like apparatus in which a suspended magnet was able to oscillate in a horizontal plane about its vertical axis, and in which the time of oscillation was measured as it depended of the distance to a nearby electrical wire, running in vertical direction (Fig. 18.3). Although Biot was familiar with magnetic measurements of that type from his former collaboration with Alexander von Humboldt (1769–1859) on terrestrial magnetism, there arose serious

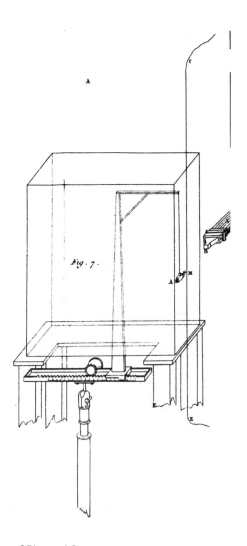

Fig. 18.3. The apparatus of Biot and Savart, 1820.

experimental problems. The power of the battery, for example, decreased quickly during its operation. After all, this was the first example of an electromagnetic *measurement*, and neither Ørsted nor Ampère had presented any (nor would they ever do so). After only two weeks of extremely intense work, the two collaborators presented their results, and another session followed six weeks later. From measurements with a straight wire, they claimed to prove an inverse-square law for the action of the magnetic elements of the wire onto those of the magnet.[13] Further experiments with a bent wire led them to suppose a dependency on an angle. In sum, the force exerted from a current element on a particle of a magnet was supposed proportional to sin ω/r^2, where ω is the angle formed by wire and the line from the current element to the magnet. In its modern and expanded notation, $d\mathbf{B} = (\mu_0/4\pi)\left(I\,\mathbf{dl}\times\mathbf{r}\right)/r^3$, this is still known as the Biot–Savart law. The way in which it was first established was far from straightforward; indeed, in its first formulation there was a mistake, which Biot acknowledged some years later.

Biot and Savart never really faced the problem of the totally un-Laplacian direction of electromagnetic forces. Biot simply stated that those forces were directed perpendicular to the plane in which both the wire and the magnetic particle laid, but said nothing concerning how that would relate to the common Laplacian notion of central forces. He could do so only because he restricted himself to those two particular arrangements. Others who explored more generally the various positions of the needle could not just bypass the fundamental problems involved here. But Biot never carried out such an exploration. In striking contrast to Ampère and many others, he based his account on only two experimental arrangements. As to microscopic theories, he retained the view that all new interactions were strictly magnetic and took place between particles of magnetic fluids. The only role of the electric fluid was to induce in the wire a peculiar sort of magnetism. But the two types of fluids remained, in good Laplacian tradition, strictly separated. With those two points—force-law and microscopic explanation—the essential goals of the Laplacian programme were achieved. Biot's deep commitment to that programme was one of the several reasons why he, after his second paper, more or less stepped out of the field.

18.7 Developing Electrodynamics, 1821–26

Although Ampère's and Biot's searches for a force-law proceeded along quite different pathways, they shared some basic concepts. Both of them dealt with central forces between infinitesimal current elements, following an inverse-square law. The forces were simply stipulated as proportional to the currents. In a significant step beyond tradition, the forces were supposed to depend not only on distance but on some *angles* between the currents. While Biot, dealing with the action of a current element onto a point, introduced one angle, Ampère looked for the interaction of two current elements and thus had to introduce *three* angles. He intended even to *measure* how the force depended on the angle γ, and designed a 'Coulombian' instrument to count the oscillations, not of a magnet, but of a well-suspended battery wire. But that first attempt to measure the forces between wires was a failure: if the instrument was constructed at all, it most probably did not deliver stable data. In the long run, Ampère would bypass the problem: with his famous technique of zero experiments he simply avoided any measurement whatsoever. In general, he would proceed by exploring the mathematics and the basic principles (such as the addition of actions) in all consequences and by designing specific experimental arrangements which would act somewhat like boundary conditions. His first force-law,

$$F = g h (\sin \alpha \cdot \sin \beta \cdot \cos \gamma + n/m \cdot \cos \alpha \cdot \cos \beta)/r^2$$

presented in December 1820, involved still an unknown numerical constant n/m, which he introduced for generality, but surmised to be zero (g and h are the intensities of the currents and r the distance between the elements). Although Ampère was preceded by Biot in presenting his formula, he showed it to be more general by deriving Biot's observed result from it. What is more, Biot's measurements provided indirect empirical support for Ampère's formula—a not untypical move for Ampère. With this success, only four months after the news of Ørsted's discovery had arrived, Ampère had firmly established his general approach, and his feverish activity slowed down somewhat.

His research of the next five years was devoted mainly to theory, and his numerous experiments were always designed to provide some empirical input to the evolving theoretical system. One main question was whether the supposed electric currents within magnets should be considered as macroscopic and concentric to the axis of the magnet, or rather as microscopic and scattered all over the magnet. The question became urgent in October 1821 when Michael Faraday (1791–1867), then a relatively unknown chemical assistant in Davy's laboratory in London, announced not only a new effect—electromagnetic rotation—but also explicitly addressed problems of Ampère's magnetic theory.[14] He had made the closing wire of a battery rotate round the pole of a magnet, and *vice versa*, and had shown that the pole of a bar magnet was not located right at the end (as was the case with coils), but at some point inside. Ampère took the challenge and came up both with a qualitative explanation of the rotations and with a refined theoretical view. He now argued for the view of microscopic currents, with the additional assumption that not all of them were directed perpendicular to the axis of the magnet. In the months to follow, he elaborated that view with many other experiments of his own.

Another main task was the mathematical elaboration of the force-law itself. In an attempt to generalize the inverse-square relation, Ampère rewrote his formula in 1822 as:

$$dF = i\, i'\, ds\, ds'(\sin \alpha \cdot \sin \beta \cos \gamma + k \cos \alpha \cdot \cos \beta)/r^n$$

which, as a result of extensive work, he finally reformulated thus:

$$dF = -\,i\,i'\,ds\,ds' \cdot r^{1-n-k}/(1+k) \cdot \partial^2(r^{1+k})/\partial s\,\partial s'$$

where i, i′ are the currents, ds, ds′ the line elements, k the (renamed) unknown parameter, and ∂/∂s the derivations in the directions of the lines. Such a shift from trigonometrical methods to calculus not only reflected a general tendency in French mathematics, but made the formula more easily accessible to the necessary integrations. The formula had now *two* numerical parameters: n and k. From another equilibrium experiment Ampère derived a relation: $2k + n = 1$. If n was taken as 2, as traditionally assumed, it resulted that $k = -\tfrac{1}{2}$, contrary to what he had surmised for a long time. With considerations of such a type, and experiments for well-selected, specific cases, Ampère successively elaborated and systematized his theory. He applied it to numerous specific arrangements of wires, such as circles, closed circuits of whatever shape, or solenoids. He designed many experiments to demonstrate the principal features of the interaction of currents with other currents, with magnets and with terrestrial magnetism. When, in 1824, he secured the long-desired position as professor of physics at the Collège de France, he regularly used a whole table of those experiments in his lectures. Not that he went all that way alone. Besides the continuing lively exchange with Arago and Fresnel, he collaborated with his former student Félix Savary (1797–1841) for the mathematical part, and with the Versailles professor Jean Baptiste Firmin Demonferrand (1795–1844) for a presentation in a handbook. Numerous letters with international correspondents such as Davy, Erman, de la Rive (father and son), and Faraday concerned mostly experimental questions, but did not essentially touch the course of his evolving theory.

The stabilization of his theory was paralleled by an increasing disregard of broad experimental work and of problematic experimental results. The most striking episode happened in 1822. For the question of macroscopic or molecular currents, the process of magnetization was, of course, an essential point: would the external magnet (or currents) create new currents within the iron, or would they just align existing but randomly oriented ones? Triggered by conversations with Fresnel, Ampère set up an experiment to decide whether in a ring of copper macroscopic currents would be induced by a nearby coil or magnet. A first trial in July 1821 produced a negative result which fitted well into Ampère's increasing preference of the view of microscopic currents. When he repeated the experiment with a stronger apparatus in August 1822, however, he indeed obtained an effect, and realized that this was the induction of currents by magnets. But as a consequence of his struggle with Faraday's rotations, he concentrated on his magnetic theory. Although the positive result of the induction experiment again opened the way for both interpretations of magnetization, it did not provide any positive hint concerning which of them should be preferred. Thus Ampère declared only that the result did not refer to his theory, and decided not to pursue it further.[15] A decade later, when Faraday again discovered electromagnetic induction and gained great publicity, Ampère bitterly complained about his former disregard of the result.

In 1826 he finally presented a systematic and summarizing account: his well-known *Théorie mathématique des phénomènes électro-dynamiques, uniquement déduite de l'expérience*.[16] Given that title, and the large number of experiments during six years, it has been surprising ever since that he based his presentation mainly on just four experiments, one of which he had, as he confessed, never carried out. But the point is significant: He felt his conceptual and mathematical frame was so rigid and strong that the four well-designed experiments could provide a totally firm connection to the empirical world. It is that aspect of stringent argumentation which would later lead James Clerk Maxwell (1831–1879) to his famous characterization of Ampère as the 'Newton of electricity'. Much like Newton, moreover, there are in Ampère's presentation no traces left of how his system had grown. And there is a further point which Maxwell could not know. Like Newton with regard to gravitation, Ampère constantly kept alive the idea that all the electrodynamic forces, acting at a distance, should finally be explained by an ether theory. Even more than Newton, Ampère kept those considerations to himself, and scarcely mentioned them in print.[17]

Ampère's electrodynamics was by far the most elaborate account of electromagnetic effects in the 1820s, and remained so until mid-century. Nowhere else was research done so ardently and with comparable results, as most of the other researchers had been active in the field only for shorter periods. As a theoretical approach, many (such as Wollaston, Berzelius, August Heinrich Jakob Althaus (born 1791), Erman, and Johann Josef von Prechtl (1778–1854)) considered a sort of 'transversal magnetism', but that notion remained notoriously unclear. Of course, one could hardly imagine another place where such a particularly mathematical development would have found enough resources to set it off. It is not by chance that Ampère's theory in the end had some quite 'Laplacian' features, at least regarding its basic concepts. When Ampère more or less left the field after 1826 (in order to concentrate on his other main subject: the classification of sciences), his theory had finally, if with some delay, gained ground in France. In Britain it was announced and was well known, but neither Barlow nor Faraday, as the main figures in electromagnetism in the late 1820s and 1830s, adopted its general approach. Elsewhere it was received with silence or explicit objection (such as by Ørsted and Berzelius). Only when mathematical physics gained more solid ground even outside France towards the middle of the century, Ampère's electrodynamics would be taken up again both in Britain and Germany.

18.8 Michael Faraday: 'Chemical Assistant in the Royal Institution'

Electromagnetism was also treated on the other side of the English Channel, and most spectacularly developed at a specific place: the laboratory of the Royal Institution in

London. Here, in the best-equipped resource for experimental research in Britain, Davy worked on a variety of fields, such as chemical analysis, voltaic piles, the miner's safety lamp, and the copper sheeting of ships. Among his laboratory assistants, he had in 1813 hired the young Michael Faraday.[18] Raised as the son of a blacksmith from an even more modest background than Davy, Faraday had, after his apprenticeship as a bookbinder, worked in that business for a short period. He had a strong interest in natural philosophy, and had taught himself, sometimes together with others, a great deal in various fields, and carried out chemical experiments. As a gift of a customer in his master's bookbinding shop, he had received (otherwise unaffordable) tickets for Davy's chemistry lectures Davy at the Royal Institution. By sending elaborated lecture notes to Davy, he managed to make Davy aware of him, and when shortly afterwards the position of assistant became free, it was offered to Faraday, who accepted it and gave up his profession. The Royal Institution became his 'home' for the next fifty years, with just one interruption when he travelled with Davy, as assistant and servant, through Europe in the war years 1813–15. Davy highly estimated Faraday's assistance, and encouraged him to undertake own research. Faraday's laboratory practice was formed by his work with Davy, whom he deeply admired. From 1816 on, Faraday published several minor papers on chemical analysis and on alloys of steel, even in the *Philosophical Transactions*. The year 1821 saw some most important events for him. He improved his position at the Royal Institution (he became Superintendent of the House), he married his beloved Sarah Barnard (1800–1879), and he made his confession to the Sandemanian church, to which Sarah also belonged. The Sandemanians were a small sect with an intense religious life, based not on sophisticated theology but on a most literal construing of the Bible.[19] Having thus stabilized his personal life in highly significant ways, he undertook a research initiative in a new field, which interested him greatly: electromagnetism.

18.9 ELECTROMAGNETIC ROTATIONS

Faraday had already assisted in some of Davy's electromagnetic experiments, sometimes together with Wollaston. His own research on electromagnetism began with an elaborate review article, for which he read the literature extensively and replicated most of the experiments—after all, he had access to both a rich library and one of the best-equipped laboratories of Europe. In the resultant text he gave a nearly comprehensive (he did not consider the German publications, since he could not read them) and often rather critical account of the previous work, and did not hesitate to criticize authors, even Ampère, for lack of clarity regarding the notion of current, and so on. What was characteristic already was Faraday's intense use of visual sketches and pictorial devices to express general statements that abstracted from individual experiments.

The 'Historical sketch' offered Faraday the opportunity to identify open problems—one of which was the lack of a general regularity for the behaviour of the needle, particularly in those cases in which the needle had no central position. This is where he began his own research. It is an historical irony that Faraday, without knowing it, set out on the very same problem on which Ampère, a year earlier, had stopped his exploratory research in favour of his electrodynamic hypotheses. By careful and broad experimentation in September 1821, Faraday found more complex motions than previously recognized, and again used pictorial means to express general features. This led him in the end to propose, as an appropriate means to represent those motions, an imagined circular path of a magnetic pole around the wire. That very idea might have been, of course, inspired by both Ørsted and Davy, who had considered some circular action. But Faraday now attempted and succeeded in turning that idea into a real motion. The continuous electromagnetic rotation of a wire round a magnetic pole, or *vice versa*, was a spectacular achievement. Faraday went even further on the conceptual level and argued that such a motion should be regarded as the elementary electromagnetic phenomenon to which all other electromagnetic effects, including Ampère's electrodynamic attractions, could be 'reduced'—a quite ambitious claim indeed. Faraday published his discovery immediately,[20] and even sent small rotation apparatus to many places throughout Europe. Although the joy of his first discovery was soon darkened by (unjustified) rumors of his having plagiarized Wollaston's ideas, he won an international reputation. His achievement even posed a serious challenge to the most advanced account of electromagnetism: Ampère's theory had dealt solely with attractions and repulsions, but had not envisioned rotations, and Ampère had to seriously revise some points.

Due to other obligations, Faraday could only sporadically return to electromagnetic research over the ensuing ten years. On those rare occasions he searched unsuccessfully, as others did, for a supposed reversal of Ørsted's effects: for an effect of magnetism onto electricity.[21] He also dealt with a mysterious effect, reported by Arago in 1825 (see the next section). In 1823 Faraday discovered the liquefaction of gases, and in 1827 he published his only monograph—significantly enough, a handbook of chemistry. His election as a Fellow of the Royal Society in 1824 distanced him somewhat from Davy, who was at that time President. Davy had opposed Faraday's membership—partly since he wanted to break with the former President's custom of patronage, and partly since he had a hard time seeing the young Faraday rising while his own scientific productivity was in decline. Although Faraday was appointed director of the laboratory of the Royal Institution in 1825, he was still not free to choose his own research agenda. His many tasks included presenting public lectures, which he liked to do, and for which he became famous. In that respect he was an even greater success for the Royal Institution than Davy had been. In 1829 he accepted a professorship in chemistry at the Royal Military Academy in Woolwich—not the least in order to gain more independence. After Davy's death in the same year, Faraday tried to get rid of many obligations. His entire further research was deeply formed by the fact that he 'swerved incessantly from chemistry to physics', as Tyndall said, for all of his lifetime, and continued to publishing articles in both fields.

18.10 Galvanometers, Electromagnets, and Arago's Effect

It is worthwhile examining three important developments in electromagnetism during the 1820s. First, it is significant and indicative of a long-felt desideratum that immediately after the news of electromagnetism was announced, several researchers independently considered using the effect for detecting and perhaps measuring the still-undefined effect of the Voltaic pile. Such attempts were made by Charles-Gaspard de la Rive in Geneva (who called his device a 'galvanomètre'), by Ampère in Paris (with a 'galvanoscope'), by the Halle professor Johann C. S. Schweigger (with his 'multiplier'), by the Berlin student Johann C. Poggendorff ('condenser'), and slightly later by the Cambridge chemistry professor James Cumming. In the ensuing years, the galvanometer underwent rapid development, and its particular use by the Berlin Gymnasium teacher Georg S. Ohm resulted in a bold proposal to separate, clarify, and quantify the concepts of current, voltage, and resistance (1827). The sensitivity of galvanometers was much increased—in particular through their most successful development by the Florentine professor Leopoldo Nobili (1784–1835) into its 'astatic' version—and they came into broad use.

Another important development took place with respect to electromagnets. Very quickly it had become clear that the electromagnetic effect allowed for magnetization of previously unmagnetic pieces of iron, and that hereby (combined with the amplifying effect of coils) ever stronger electromagnets could be built even to such a degree as to surpass the strength of any known usual magnet. In London, for example, the inventor William Sturgeon worked on electromagnets, published his results, and exhibited the magnets regularly. Not surprisingly, navigation officers became interested, and the Royal Military Academy at Woolwich built one of the strongest magnets available. In the United States it was the Albany professor Joseph Henry who devised ever new forms of electromagnets, and who in this context also worked with ring-like magnets. Electromagnets were present at all research sites, and served both as instrument and as items to be investigated.

A third important line of research originated from another area. In 1825 Arago reported on a strange effect. Starting from the observation that the oscillations of a magnetic bussole were more strongly damped when the bussole was placed in a metal instead of a wooden box, he analysed the effects more closely and found that a metallic disc, rotating above a magnetic needle, would set the needle in motion, even if in rest the disc showed not the least sign of magnetism, and *vice versa*. The effect was observed with a number of non-magnetic materials, and the observation that some non-magnetic materials could act on a magnet if only set in motion came to form a serious puzzle. Many researchers, such as Charles Babbage (1791–1871), John Frederick William Herschel (1792–1871), Peter Barlow (1776–1862), Samuel Hunter Christie (1784–1865), Poisson, and Ampère investigated it, and some of them assumed some sort of temporary induced magnetism as the cause. But that notion was never sharply defined, and Arago's effect remained a significant challenge for several years. All three of these developments came together in important ways when Faraday re-entered the electromagnetic scene.

18.11 Electromagnetic Induction and the Electrotonic State

In August 1831, Faraday started a new volume of his laboratory notebook, and the very first entry gave his discovery of an effect of electromagnetic induction. We do not know how he arrived at the particular experimental arrangement in which two electrically separate sets of coils were wound on a soft iron ring, but it was probably stimulated by considerations on strong electromagnets, as developed by the American Joseph Henry, who came very near to the same discovery.[22] While the effect was spectacular (and Faraday immediately realized that point), it had at the same time some unexpected features. An induced current occurred only in the moment when the inducing current was switched on, or a magnet brought near. Even more mysterious, a reverse current occurred when the inducing current was switched off, or the magnet removed. That transient character (due to which the effect had not been noticed for so long) made the effect as puzzling as Ørsted's effect had been eleven years earlier. But there was an essential difference in the setting. In the autumn of 1820, everybody who started electromagnetic research knew that many others did likewise, and that for all claims of originality no time was to be lost. In contrast, in 1831 Faraday knew himself alone with his discovery. Although he was well aware that the effect would raise wide publicity, he gave himself much time for further research. He announced the effect only twelve weeks later, in November, and his publication came out only in April 1832. In the meantime he had a remarkable period of exploratory research, with a large number of experiments, conceptual attempts, setbacks, variations, and revisions.

Faraday's main goal was to get an overview on which factors were important for the new effect, and to formulate regularities. In simplifying his first, quite complex experimental arrangement, he classified two different types of induction: by currents ('Volta-electric induction') and by magnets ('magneto-electric induction'). He attempted and succeeded to obtain every type in 'pure' form. His regularity for Volta-electric induction was based on the Amperian notion of interacting currents. In order to account for the transient character of the effect, he introduced a new theoretical construct: the 'electrotonic state.' This particular state was supposed to be built up in the induction wire when the inducing current was switched on, and to counteract the induced current, making it stop quickly. When the inducing current was switched off, the electrotonic state ceased and gave rise to the observed reverse current. Despite many efforts, however, Faraday could not obtain a more specific physical characterization of that state.

The case of magneto-electric induction turned out to be even more complicated. The effects depended in a complex way not only on the geometrical constellation, but on the *motion* by which that constellation was achieved. Faraday tried hard to find an appropriate reference frame for the spatial constellations and motions: after all, such a frame was needed to formulate a regularity. His initial approach of using the hypothetical currents which Ampére had supposed to be the cause of magnetism proved to be insufficient: they did not allow formulation of a law that comprised the experimental results. Only then did he try to see whether the 'magnetic curves', formed by iron filings around a magnet, would serve the purpose. Not that those curves were new: they had long been noticed, and only in 1831 Peter M. Roget (1779–1869) had published an elaborate attempt to grasp their properties mathematically.[23] But Faraday used them now in a completely new way, and it turned out that such a concept indeed allowed him to express the regularities of magneto-electric induction. The essential condition of induction was now that there was a motion between wire and magnet in such a way that those curves were *cut* by the wire (Fig. 18.4). By regarding the direction of the Earth's magnetic dip as a magnetic curve, Faraday could even include inductive effects by terrestrial magnetism. But now he had two different and incompatible conceptual frameworks for Volta- and magneto-electric induction. Only the latter type (which involved motion) was dealt with using magnetic curves, whereas only the former needed the assumption of the electrotonic state.

These two frameworks came to clash in his investigations on Arago's effect, to which he devoted much effort. He assumed the effect to be caused by induced currents. When, against his expectations, he obtained the effect also with electric spirals as magnets, he had to interpret that as a purely Volta-electric effect, but one which involved the alien factor of motion. That result was a serious challenge of the whole dichotomy, even more so as he suddenly saw that the effect could be accounted for by magnetic curves. If only such curves were supposed to exist also around wires, they

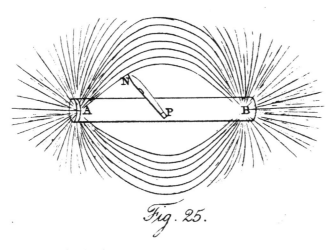

Fig. 18.4. Faraday's law of induction, 1832.

would be cut by the rotating wheel. Such an account undermined the very distinction between Volta- and magneto-electric induction. But it could not explain all those Volta-electric effects which did not involve motion. Only in March 1832, virtually in the last moment before his paper was printed, did Faraday achieve an insight which solved the problem and led to a revision of his concepts.[24] He could now express the regularity of both magneto- and volta-electric induction in *one* single rule which posed the central condition that magnetic curves had to be cut. The only (but crucial) step required was to consider those curves around a wire as moveable, and indeed moving, that is, concentrically expanding from or collapsing towards the wire in those moments, when an electric current was switched on or off. With that step, however, the whole concept of the electrotonic state lost its significance. Since Faraday was not allowed to incorporate changes in his text, the printed paper had the curious feature of introducing the electrotonic state in a whole chapter of the main text, while dismissing it in an attached footnote.[25] Hence the published text showed peculiar signs of a turbulent and hectic research process. His introduction of 'magnetic curves' as an essential tool was the foundation of a development whose far-reaching consequences could be seen only much later, when he would change the term to 'lines of force.'

18.12 Lenz's Law, Electromagnetic Generators, and the Electromagnetic Telegraph

Faraday was not the only one to formulate an induction law. Heinrich Lenz, a physicist at St Petersburg University, was stimulated to begin research when he read Faraday's first paper on induction and realized that the induction law given there was not fully general; but for some reason, he did not see Faraday's second paper in which Faraday spelled out the law in its full generality. Lenz carried out many experiments, and derived a peculiar law for the direction of the induced current in all cases. This law stated that the induced current was always directed in such a way that its effect counteracted the motion that gave rise to the induction. While it did not spell out in detail the conditions under which induction occurred, it gave the correct direction of the induced current in all cases. Lenz's law was well received, and remained a complement to Faraday's induction law.

The discovery of the induction effect pointed to a new source of electricity, as not only Faraday quickly realized. While Faraday deliberately chose not to pursue the question of how that effect could be developed into useful devices, some others did exactly that immediately. After all, the effect opened the perspective to generate electric currents without all the chemical processes and materials involved that made the handling of galvanic apparatus sometimes arduous and, not least, expensive, plus the currents could possibly be generated in much more constant ways than with the always decreasing intensities of galvanic apparatus. The first among those who, probably with such perspectives in mind, devised and built an electromagnetic generator was Hyppolite Pixii, son of Nicolas Constant Pixii, who had for a long time run a large Paris workshop for scientific instruments. His machine, presented in 1832, had a strong horseshoe magnet mounted on a vertical axis and rotating (driven by a hand-crank) below a set of two coils (Fig. 18.5). When the magnet was rotated, the coils gave strong pulses—for example, in the form of sparks. With a slightly later version of the apparatus, Pixii added a commuter to make the electric pulses unidirectional. Despite that success, it would took another fifteen years or so until generators were developed widely enough to really compete with the then established battery technology. At a similar pace as generators, electromagnetic motors were developed and became an important technology in the second half of the eighteenth century.

It is indicative for that development that the earliest major technical application of electromagnetism—telegraphy—worked for a long time with batteries. Electromagnetic telegraphs had been proposed early on—again an indicator of a

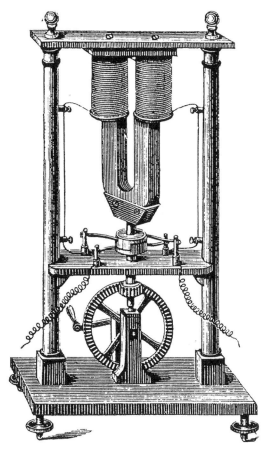

Fig. 18.5. The first electromagnetic generator, Pixii, 1832.

long-felt desideratum that the new effects promised to fill. Optical telegraphy was too dependent on weather and lighting conditions (and perhaps was not sufficiently safe), while the electrochemical telegraph proposed by Soemmering was complicated and impracticable. The first to propose an electromagnetic telegraph was Ampère, in 1820, with a system with many cables and galvanometers attached to them. Following Faraday's discovery of the induction effect, the Göttingen physicists Carl Friedrich Gauss and Wilhelm Eduard Weber developed an electromagnetic telegraph in which both detection and production of the signal was achieved by electromagnetic means, and their machine (1833) worked over a distance of about 1.2 km. Systems that worked with batteries for signal generation and electromagnetism for detection were developed in St Petersburg by Schilling (1832), in Munich by Steinheil (1836), in Pennsylvania by Alten (1836), and in London by Cooke and Wheatstone (1837). The breakthrough of telegraphy was achieved by Morse in the late 1830s and early 1840s with his proposal of a whole system—sources, wiring, sender, receiver, and the code—that was able to bridge rather long distances.

Telegraphy continued to raise fundamental questions—such as how to conceive underwater and transatlantic cables, treated by researchers such as Faraday and W. Thomson and leading to significant conceptual achievements. At the same time, our knowledge of the concepts by which instrument-makers and engineers treated and designed their electrical apparatus is still relatively poor. It is not clear, for example, in what way Faraday's induction law and its central concept of magnetic curves played a role in Pixii's design of the generator, or what considerations and concepts telegraph designers took as their resources. While Sturgeon, in his dealing with electromagnets and motors, seems to have included the concept of magnetic curves, we do not know whether this was an exception or the rule. Given the long and unequivocal repudiation of that concept within the academic community, it would be of considerable interest to study whether those who were in involved in the practical uses of electromagnetism had a different stance.

18.13 Electricities and Chemical Activity

With electromagnetic induction Faraday added a new source of electricity to those of common, voltaic, thermic, and animal electricity (by electric fish, for example). In 1832 he analysed whether all those electricities were identical, or not. Typically enough, he did not approach the question by microscopic theory, but systematically evaluated experimental results, both his own and others'. The result, arranged in a table (that was successively completed in the following decade), was widely acknowledged as a valid demonstration that in all those cases one and the same electricity was involved.

Triggered by that research, Faraday turned to electrochemistry. He built his theory of electrolytic dissociation and his two fundamental laws again on numerous carefully conducted experiments. The first law stated that electrochemical action did not depend on the material or shape of the poles, but just on the quantity of electricity passed through the solution; and as a measure of that quantity he basically took the number of turns of an electrical machine.[26] The law provided, for the first time, a solid base for the old idea of constructing instruments to measure the quantity of electricity by chemical means, and Faraday immediately proposed a 'Volta-electrometer.' The second law stated the proportionality of the amounts of substances deposited to their respective chemical equivalent weights. Later, Faraday would partake in the discussion on the theory of the battery, and not only add powerful empirical arguments for a chemical theory, but also bring that theory into a much more elaborate shape. As another result of his experimentation, he became increasingly sceptical of the view of electricity as a fluid whose particles acted at a distance. The very notion of poles seemed inappropriate, if not misleading, for all electrochemical experiments. Being well aware of the importance of terminology, he sought alternatives. As a result of a long exchange with the Cambridge philosopher William Whewell (1794–1866) and others, he came up with new terms which did not bear that much theory: 'pole' was replaced by 'electrode', being either 'anode' or 'cathode'. 'Electrolyte' designated the substance which was decomposed, 'ions' its particles, being either 'anions' or 'cations'.[27]

Faraday's laws of electrochemistry were based largely on extensive measurement. After all, he was well accustomed to quantitative procedures from chemical analysis. While it is clear, and has often been emphasized, that Faraday was fully incompetent in analytic mathematical methods of his time, it should be highlighted that quantification and precise measurement was sometimes at the core of his research. This is even more clearly illustrated by his research in common (static) electricity, in which he used the most precise instruments available: electrometers and Coulomb's torsion balance.

18.14 Fundamental Concepts: Specific Inductive Capacity and Induction in Curved Lines

Faraday felt that the science of electricity was in a critical state. Different types of electricity (in particular voltaic and common) had to be treated separately; his new concepts in electrochemistry could not be included; and even basic questions such as what held electricity within bodies could not be answered. In order to obtain a more general view, he began a critical revision of the basic concepts—not that he knew in advance where to go. As in his work on electromagnetic induction, he made, in the context of extensive experimental work, various attempts to develop new concepts. Indeed, he had, from 1836 to 1838, another exploratory period, with long phases of uncertainty on a fundamental level. His hesitation to fix and publish his views was increased by the fact that he acted here on a well-studied field, and the concepts to which he was led were at odds with the authoritative mathematical theories.

He began with the simple-looking but deep question of whether matter 'can . . . be charged with one electric force independently of the other?'[28] His finally negative answer was achieved only after many experiments, culminating with his famous 'Faraday cage'. Within that twelve-foot cube with highly conductive walls he could, despite all efforts and with different arrangements, not make any body charged without at the same time making another body outside oppositely charged. He concluded that charge denoted a *relation* between two or more bodies rather than the state of an isolated body. The process of mutual *induction* was assigned a central role, and the insulating media through which it took place between conductors came suddenly into the focus of research. In a most remarkable series of measurements he investigated how induction depended on the properties of the isolating material.

The experimental apparatus consisted of two isolated, concentric metallic spheres, and the space between them was filled with different dielectric materials (gaseous or solid), or even evacuated. The outer sphere was always grounded, whereas the electric state of the inner sphere could be determined via a feedthrough and measured by the most delicate electric instrument available, the Coulomb torsion balance. Faraday had two identical apparatus (one filled with air, the other with other materials), and all measurements were comparative. One apparatus was charged from a Leiden jar and its charge measured. Then the electric power was 'divided' by connecting the spheres of the two apparatus for a moment, and their resulting states were measured. Since in the 'dividing process' the two apparatus were connected in parallel, their resultant states were always equal. When both apparatus were filled with air, the resultant state was exactly half the original state of the first apparatus. When the second had other materials, however, the resultant state was considerably less. The experiments were extremely delicate, and it took Faraday numerous rearrangements and a long training period to obtain stable readings. Every single run had to be done in a very short period of time, since the torsion balance steadily lost its initial charge. Another notorious problem was to keep the spheres isolated when filled with different materials: shellac, for example, is quite hygroscopic. Faraday, moreover, carried out many control experiments to exclude possible sources of error such as imperfections in the electric insulation or included moisture, and to ensure that the two apparatus were really identical apart from the filling material in the space.

Faraday evaluated his results in terms of loss of 'force of induction' in the first apparatus, and gain in the second. When the two apparatus had the same material, the resultant state was exactly the half the original state of the first apparatus, and gain and loss were the same. But when the second apparatus had another material (shellac, for example), it gained less than the first lost. Even more, when the experiment was 'reversed' and the shellac apparatus was first charged and then 'divided' with the air apparatus—that is, when induction through shellac was converted into induction through air—the air apparatus gained much more than the shellac apparatus lost.[29] Faraday interpreted that difference as the greater capacity of the dielectric material to 'allow or cause' the inductive action through its substance, and this is what he called its *specific inductive capacity*. To give it a numerical value was all but straightforward, and required a very detailed and complicated analysis of the apparatus and procedure—all the more since, for technical reasons, the apparatus under analysis could only be filled half when solid material was involved. Faraday obtained a value 2 for shellac, 1.76 for flint glass, and 2.24 for sulphur (always compared to air). For gases of different type, density, pressure, or temperature he found no sensible difference from air.[30]

While Faraday's terminology appears alien to us, two points become clear from the way he used his concepts. He took the 'quantity of electricity' to be constant throughout the experiment, while the 'force of induction' was taken as proportional to that quantity, but dependent on the material through which it acted. We would be tempted to speak of charge Q and tension U, between which there is a constant but material-dependent relation in any capacitor. Only a few years later, Faraday would sharpen and generalize his terminology and speak of 'quantity' and 'intensity' of lines of force—a differentiation that was a core element of his field theory. In 1838, what he had definitely shown was that the induction between two conductors depended on properties (described by the new concept of specific inductive capacity) of the intermediate material, and this he regarded as a strong argument against the action-at-a-distance theories and in favour of his new concepts. He gave a sketch of how to understand that action of the dielectric by a theory of polarization.[31] By another set of experiments, moreover, he came to the conclusion that induction took place in *curved* lines.[32]

The general view at which he eventually arrived was new indeed. The key role was ascribed to 'lines of inductive force' which passed through isolating materials (therefore named 'dielectrics') and caused in them an 'inductive' state of tension. It depended on their *specific inductive capacity*, which gave a measure for the 'conductivity' to the lines. Such a state of tension could be sustained only by isolators and made their particles polarized. Conductors, in contrast, could not sustain that state: the lines of inductive force did not pass them, but ended at their surface. Therefore electric charge appeared only at their external surface. William Thomson would later take those experiments as a first entry into a mathematical formulation of Faraday's ideas.[33]

The concept of lines of inductive force was much more abstract than the concept of magnetic curves. Electric curves could not be presented directly to the eye, but could be established only by measurement and geometrical representation of the data. This changed when Faraday studied discharge in all its types: by conduction, in electrolytes, disruptive (sparks, brush), and even in rarefied gases. Among others, he elaborated the effect of the 'electric brush.' With special techniques developed by Wheatstone, he made here the curved electric lines 'visible'. Thus he had strong support for his unorthodox interpretation of the spark as a sudden release of tension within the dielectric, and not, as traditionally assumed and most visibly suggested, as a stream of some electric substance.

Whereas in his experiments Faraday sometimes collaborated with others and was widely respected for his ingenious discoveries, he stood isolated with his new concepts. Already his 'magnetic curves' of 1832 had found no resonance, and the same was the case in 1839 with his 'lines of inductive action'. No-one took them seriously nor understood what he meant—all the more since his presentation of the concepts was no easy read. For those who in Britain increasingly pushed forward a tendency to mathematization, such as Babbage, Herschel, Whewell, Barlow, or Brewster, his concepts sounded odd. Only in the mid-1840s, when William Thomson would give up his initial 'disgust'[34] and realize how powerful Faraday's concept were, would that situation start to change. The fact that Faraday, despite such negative response and isolation with regard to his concepts, nevertheless continued to work with them and develop them further, is highly remarkable and somewhat unique. It has to do both with increasing 'internal' confirmation and with his personality. Not only did his approach lead to new questions and experiments, but it proved fruitful in embracing an ever-increasing area of phenomena. This is how he understood—not the least on the basis of his religious faith—the obligation of a researcher in 'reading the book of Nature'.[35] His extraordinary acceptance of long periods of conceptual insecurity, moreover, is impossible to imagine without the institutional security at the Royal Institution and at the Woolwich Academy. Also, the embeddedness in his religious faith and in the life of the Sandemanian community in which he and Sarah most actively participated, played a very important role for his personal stability. It is significant that the event which threw him into a deep personal crisis occurred not from the non-reception of his scientific concepts but from his life in his religious congregation when he was expelled from it for a short period in 1844 due to internal conflicts.

18.15 THE FARADAY EFFECT, DIAMAGNETISM, AND THE QUESTION OF POLARITY

Faraday was well aware that his new concepts reached farther than electricity. In 1844 he presented a first account on how they fitted into a view of matter in general. Not surprisingly, he did not favour the Daltonian view of material atoms which were acted upon by external powers, but rather considered the particles of matter as centres of power. He came also back to the question of possible magnetic properties of materials other than iron and nickel. Stimulated by a query of William Thomson, Faraday undertook, moreover, a systematic search for effects of electricity on polarized light, in August 1845. When his experiments gave no result, he switched to magnetism. After many trials with different arrangements and materials, he succeeded with a piece of special heavy glass which he had kept from his 1830 research on optical glass. Light, passing through that piece, had its plane of polarization rotated when the two poles of a strong electromagnet were brought on the same side of the bar. The degree of rotation turned out to be proportional to the strength of the magnet and to the distance passed through the glass. Faraday quickly obtained the effect with a large number of other materials, both solid and liquid. He evaluated the effect within the framework of 'lines of magnetic force', as he now rephrased his former magnetic curves. He found, for example, that the effect only occurred when the path of light was parallel to the lines of force, and he was fascinated by thus having a means to study the path of those lines even within solid bodies. In close analogy to the electric case, he introduced the term 'dimagnetics' (later changed to 'diamagnetics') for those substances through which the lines of force passed without making the material magnetic. The magneto-optical effect (still known as 'Faraday effect') confirmed Faraday's general idea of the convertibility of all forces. In the following years, Faraday would repeatedly come back to that point. Having not found an action of electricity on light, he tried to generate electricity by light or, later, to detect any interaction between gravity and electricity. All those efforts, however, remained unsuccessful.

But the discovery of the magneto-optical effect triggered another and extremely important line of research. Faraday analysed heavy glass for possible further magnetic properties and immediately obtained an effect. A bar of that material was actually directed when put between the poles of a strong electromagnet. But the position it obtained was not along the line between the poles (such as a piece of iron would have), but orthogonal to that line. Faraday spoke of an 'equatorial' (as opposed to the usual 'axial') setting. Again he could quickly present a long list of other materials with such behaviour: it included such different things as Iceland spar and olive oil. Thus, at the end of 1845, he came up with a new classification of all materials according to their magnetic behaviour. Either they showed ordinary magnetism or the new and much weaker 'diamagnetism', which he defined by the equatorial setting between magnetic poles.

The discovery of diamagnetism was spectacular and raised deep questions. Besides Faraday himself, who started a five-year period of research in magnetism,[36] many other scientists were driven to closer investigation, such as Wilhelm Weber (1804–1891, Göttingen), Antoine Becquerel (Paris), Ferdinand Reich (1799–1882, Freiberg, Saxony), Julius Plücker (1801–1868, Bonn), Michele Bancalari (1805–1864, Genoa), Hans Christian Ørsted (Copenhagen), and the young John Tyndall (1820–1893, Marburg, later London). Hence there was a constant flow of new results and theoretical accounts. In contrast to former periods, moreover, there was at least one other scientist who seriously took up Faraday's concepts: William Thomson. For several years there was a strong interaction between those two—a situation that was significantly different from most of Faraday's previous research.

Diamagnetic bodies showed a strange behaviour: being always repelled from the poles but never attracted, they showed no magnetic polarity. Since their motions could not be accounted for by means of the usual magnetic curves, for a short period Faraday introduced a new set of curves, called 'diamagnetic'. When that came out to be inappropriate, he switched to represent the motions not by lines but by location: they always went 'from stronger to weaker points of magnetic action'.[37] It is this context that he first introduced the notion of a 'field' in the sense of 'field of operation'. Thus he not only had two different types of magnetism, but treated them with different concepts. In his attempts to overcome this unhappy state of affairs, Faraday stressed the analogy to electricity and tried to treat the effects in terms of specific differences between the material and its surroundings—diamagnetic behaviour had been quickly shown also for gases. Here the problem arose of how to account for empty space. The situation became even more complicated in 1847, when Plücker, who admired Faraday very much, reported that some crystals were affected by magnets in a much more complicated manner: They were directed into an equatorial *or* axial position, depending on the direction of their optical axis and on the distance of the magnetic poles. Plücker claimed, moreover, to have found polarity in diamagnetic substances such as bismuth. When Faraday pursued those experiments, partly together with Plücker, and later with Tyndall, he considered, following Plücker's proposal, a new type of force: the 'magne-crystallic force'. And to describe its acting, he introduced some lines of minimum resistance for magnetic curves—an idea which had been suggested by Thomson. A coherent conceptualization or classification seemed to go away. What should be taken as a basic effect of diamagnetism—the repulsion, the equatorial orientation, or the orientation along some optical axes? What was the appropriate reference system—the magnetic curve, or the points of weaker intensity? But Faraday was convinced that the now three forces—magnetic, diamagnetic, and magnecrystallic—had to be understood within one and the same framework.

18.16 BEFORE MATHEMATIZATION: QUANTITY AND INTENSITY IN THE FIELD CONCEPT

Only in 1850, after many more experiments, rejections of other accounts, revisions of his own, and fruitful exchanges with Thomson, Faraday obtained a comprehensive view. *All* magnetic behaviour was now classified and understood in terms of lines of magnetic force, of their interaction among themselves and with different bodies. As a crucial point, two quantities had to be differentiated, as in the electric case much earlier (when Faraday had introduced the specific inductive capacity as important factor between them). Faraday emphasized that the intensity of lines of force within a given material was always proportional to the quantity of those lines (measuring the sum of the lines over a cross-section), with a factor that varied with the material, was called 'conductivity', and made all the difference between the types of magnet: *diamagnetics* offered some resistance to the passage of lines of force and made them

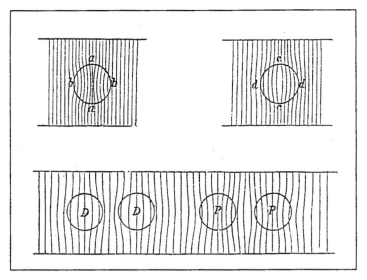

Fig. 18.6. Faraday's general account of magnetism, 1850.

diverge, *paramagnetics* conducted them well and made them concentrate, and *ferromagnetics* were paramagnetics which retained some magnetism (Fig. 18.6). Empty space held an intermediate position between para- and diamagnetics as to conductivity. Magnetic motions were explained by a sort of minimum principle: all bodies, whether dia- or paramagnetic, arranged themselves in such a way as to cause the least possible disturbance of the normal distribution of lines.

It is worthwhile looking in detail at the properties that Faraday now ascribed to the 'magnetic lines of force', as he summarized them in the 26th and 28th series of his *Experimental Researches*, plus some other papers in the *Philosophical Magazine*.

- Lines of force have a definite direction (with an arrow), with opposite features in opposite directions[38]
- The density of lines at any point of space was taken as presenting the intensity of the field, if only one conceived a sort of unity line.
- There is a conservation principle for the number of lines: '... the sum of power contained in any one section of a given portion of the lines is exactly equal to the sum of power in any other section of the same lines, however altered in form, or however convergent or divergent they may be in the second place'.[39] In other words, the number of lines that enter into a given volume always also exit it, whatever shape the volume has. Faraday gave an 'experimental proof of this character of the lines'.[40] What he had formulated here was a conservation principle of utmost importance—lines do not just appear or disappear. He had already formulated a similar principle for the electric 'lines of induction'.[41]
- Bodies were ascribed different 'conductivities' (later, permeability μ) for lines of force, paramagnetic bodies had conductivities higher, diamagnetic bodies lower than empty space. Those conductivities also gave the relation, within a given

material, between the quantity and the intensity of the lines, which meant that in paramagnetic bodies (and a given quantity), the intensity is lower than in air, whereas in diamagnetic bodies it is higher. There was a close analogy here to specific inductive capacities (or later, dielectric constants ϵ). As a further consequence of this and the conservation principle, the field outside a paramagnetic body was necessarily weakened.

- All magnetic motions of bodies could be explained by the principle that bodies always move in the direction of the weaker field, i.e. the region in which lines of force are relatively less dense. Thus motion was also due to the relative difference of conductivities.
- As a most important and fundamental consequence, Faraday stated that the concept of polarity lost its importance, and one could completely do without it. Of course, in handling paramagnetic bodies, it was seen as practically useful, but in principle it could be replaced by lines of force and conductivities.
- Given all sharpness he could now give to the concept of lines of force, and all the generality in which that concept allowed him to formulate the laws of electricity and magnetism, Faraday gave up his former agnosticism on the question of existence and emphasized that he now saw good arguments to ascribe 'physical existence' to those lines (for the whole period since 1831, he had repeatedly and explicitly emphasized that those lines were nothing but a means to describe the distribution of magnetic forces).[42]

In a paper of the same time, moreover, Faraday went further and

- presented the relation of magnetic lines of force and lines of (electric) induction in a most general and highly suggestive scheme.[43]
- Moreover, he ascribed to magnetic curves the tendency to shorten in their length, and to attract each other when parallel.[44]

In sum, he presented a highly developed, highly abstract scheme—a scheme, however, that was formulated with the background of several thousand experiments in mind (or at least in his Diary). The realm and diversity of phenomena which Faraday covered by that framework was by orders of magnitude larger than that of any previous electric or magnetic theory. Significantly, there was no calculation or analytic apparatus involved, but geometrical constellations combined with some principles—in particular, with a powerful conservation principle. At the same time, Faraday's account of magnetic effects stood now in the closest analogy to his treatment of electricity. Paramagnetic materials made the lines of force converge as did electric conductors; diamagnetics and dielectrics offered resistance and made them diverge. This analogy allowed a treatment of electromagnetic interaction solely in terms of lines of force, as demonstrated in this discussion and by Faraday's diagram (Fig. 18.7). Shortly later, the scheme of lines of force became the central point of reference for James Clerk Maxwell.

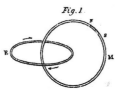

Fig. 18.7. Faraday's diagram of the relation of electric and magnetic lines of force, 1852.

18.17 Faraday and Mathematics

A common picture of the rise of field theory tells us that Faraday was the one to create qualitative concepts, while it was left to Thomson and in particular to Maxwell to give those concepts mathematical form and rigour. While Maxwell himself gave rise to such a picture in some passages of his early work, he also hinted at other aspects that are apt to refine the picture significantly. In particular, in his paper 'On Faraday's lines of force' of 1855 he spoke clearly against the view that Faraday's processes of reasoning were 'of an indefinite and unmathematical character.'[45] Later, and even stronger, in looking back on his own work on field theory in 1873, Maxwell highlighted that 'The way in which Faraday made use of his idea of lines of force . . . shows him to have been in reality a mathematician of a very high order.'[46] While this was written in an obituary and certainly has the character of homage to a great researcher, the way Maxwell specifies that remark is highly instructive. The main item he had in mind was the concept of lines of force, and he pointed to two instances of Faraday's work: first, to Faraday's discovery and conceptualization of electromagnetic induction—that is, to the first occasion on which Faraday introduced the concept of magnetic curves—and second, to the 26th and 28th series of the 'Experimental Researches in Electricity' of 1851, in which Faraday discussed the concept of magnetic lines of force in general terms. The significant point is that Maxwell, rather than contrasting qualitative versus mathematical concepts, contrasted different approaches within mathematics: '. . . we are not to suppose that the calculations and equations which mathematicians find so useful constitute the whole of mathematics. The calculus is but a part of mathematics. The geometry of position is an example of a mathematical science establishes without the aid of a single calculation.' Concerning Faraday in particular, he emphasized that 'Faraday's lines of force occupy the same position in electromagnetic science that pencils of lines do in the geometry of position. They furnish a method of building up an exact mental image of the thing we are reasoning about.' This is what made him not only praise Faraday as 'a mathematician of a very high order', but even as a mathematician 'from whom the mathematicians of the future may derive valuable and fertile methods.' He ascribed to him 'the discovery and development of appropriate and exact ideas, by means of which we may form a mental representation of the facts, sufficiently general, on the hand, to stand for any particular case, and sufficiently exact, on the other, to warrant the deductions we may draw from them by the application of mathematical reasoning.'[47] If one agrees with Maxwell to call the type of such reasoning 'mathematical', the picture of the development of field theory, rather than taking the pathway from qualitative to mathematical, appears rather as a pathway from a specific if not idiosyncratic form of mathematical approach, unusual in his time, to a reformulation of that approach into more known and used mathematical concepts—a reformulation that needed high creativity and led to new concepts, and could scarcely have been achieved by someone less versed than Maxwell both in the analytic approach of the time and in a detailed understanding of the mathematical structure within Faraday's approach.

References

AMPÈRE, ANDRÉ-MARIE (1820), 'Mémoire présenté à l'Académie royale des Sciences, le 2 octobre 1820, ou se trouve compris le résumé de ce qui avait été lu à la même Académie les 18 et 25 septembre 1820, sur les effets des courans électriques', *Annales de Chimie et de Physique* 15 (septembre): 59–76.

——(1826), *Théorie des phénomènes électro-dynamiques, uniquement déduite de léxpérience.* Paris: Mequinon-Marvis.

Arnold, David H. (1983–84), 'The "Mécanique physique" of Siméon Denis Poisson. The evolution and isolation in France of his approach to physical theory (1800–1840),' *Archive for History of Exact Sciences* 28 & 29, S. 243–367, 37–94, 287–307.

BIOT, JEAN-BAPTISTE (1804), 'Rapport sur le expériences du citoyen Volta', *Mémoires de l'Institut National des Sciences et Arts: Sciences mathematiques et physiques* 5: 195–222.

——(1820), 'Sur le magnetisme de la pile de Volta', *Bulletin des Sciences par la Societé Philomatique* (août): 128.

——(1824), *Précis élémentaire de physique expérimentale*. 3. ed. Paris: Deterville.

BLONDEL, CHRISTINE (1982), *A.-M. Ampère et la création de l'électrodynamique (1820–1827)*. Mémoires de la section de sciences. Comités des travaux historiques et scientifiques. Ministère de l'éducation nationale: 10. Paris: Bibliothèque Nationale.

——(1989), 'Vision physique 'etherienne', mathématization 'laplacienne': l'électrodynamique d'Ampère', *Revue d'Histoire des Sciences* 42: 123–137.

BUCHWALD, JED Z. (1977), 'William Thomson and the mathematization of Faraday's Electrostatics', *Historical Studies in the Physical Sciences* 8: 101–136.

CANTOR, GEOFFREY N. (1991), *Michael Faraday: Sandemanian and Scientist. A Study of Science and Religion in the Nineteenth Century*. Basingstoke: Macmillan.

——and GOODING, DAVID C., and JAMES, FRANK A. J. L. (1991), *Faraday*. London: Macmillan.

DARRIGOL, OLIVIER (2000), *Electrodynamics from Ampère to Einstein*. Studies in History and Philosophy of Science. Oxford: Oxford University Press.

DEVONS, SAMUEL (1978), 'The search for electromagnetic induction', *The Physics Teacher* (December): 625–631.

FARADAY, MICHAEL (1821), 'On some new electromagnetical motions, and on the theory of magnetism', *Quarterly Journal of Science* 12: 74–96.

——(1839–55), *Experimental Researches in Electricity. 3 vols. 1839, 1844, and 1855*. London: Taylor.

——(1852), 'On the physical character of the lines of magnetic force', *Philosophical Magazine* (June 1852).

——(1854), 'Observations on mental education', *Lectures on Education Delivered at the Royal Institution of Great Britain*. London, 39–88.

GOLINSKI, JAN (1990), 'Humphry Davy and the "Lever of Experiment"', in H. L. Le Grand (ed.), *Experimental Inquiries: Historical, Philosophical and Social Studies on Experimentation in Science*. Dordrecht: Kluwer, 99–136.

GOODING, DAVID C. (1981), 'Final steps to the field theory: Faraday's study of magnetic phenomena, 1845–1850', *Historical Studies in the Physical Sciences* 11(2): 231–275.

HARMAN, PETER M. (ed.) (1990), *The Scientific Letters and Papers of James Clerk Maxwell, vol. I, 1846–1862*. Cambridge: Cambridge University Press.

HOFMANN, JAMES ROBERT (1995), *André-Marie Ampère: Enlightenment and Electrodynamics*. Oxford: Blackwell.

HOME, RODERICK W. (1983), 'Poisson's memoirs on electricity: Academic politics and a new style in physics', *British Journal for the History of Science* 16: 239–259.

MARTIN, THOMAS (ed.) (1932–36), *Faraday's Diary. Being the Various Philosophical Notes of Experimental Investigation made by Michael Faraday, DCL, FRS, During the Years 1820–1862 and Bequeathed by him to the Royal Institution of Great Britain*. London: G. Bell and Sons.

MAXWELL, JAMES CLERK (1873), 'Scientific Worthies I: Faraday', *Nature* 8 (18 September 1973): 397–399.

MOYER, ALBERT E. (1997), *Joseph Henry: The Rise of an American Scientist*. Washington: Smithsonian Institution Press.

POISSON, SIMÉON DENIS (1812a), 'Mémoire sur la distribution de l'électricité à la surface des corps conducteurs', *Mémoires de la Classe des Sciences Mathématiques et Physiques de l'Institut Imperial de France* 12 (1811): 1–92.

——(1812b), 'Second mémoire sur la distribution de l'électricité à la surface des corps conducteurs', *Mémoires de la Classe des Sciences Mathématiques et Physiques de l'Institut Imperial de France* 12 (1811): 163–274.

ROGET, PETER M. (1831), 'On the geometric properties of the magnetic curve, with an account of an instrument for its mechanical description', *Journal of the Royal Institution of Great Britain* 1 (February): 311–319.

ROSS, SYDNEY (1961), 'Faraday consults the scholars: the origins of the terms of electrochemistry', *Notes and Records of the Royal Society of London* 16: 187–220.

SMITH, CROSBIE and WISE, M. NORTON (1989), *Energy and Empire: A Biographical Study of Lord Kelvin*. Cambridge: Cambridge University Press.

STEINLE, FRIEDRICH (1996), 'Work, Finish, Publish? The formation of the second series of Faraday's "Experimental Researches in Electricity"', *Physis* 33: 141–220.

——(2005), *Explorative Experimente. Ampère, Faraday und die Ursprünge der Elektrodynamik*. Boethius 50. Stuttgart: Franz Steiner Verlag.

VOLTA, ALESSANDRO (1800), 'On the electricity excited by the mere contact of conducting substances of different kinds', *Philosophical Transactions* 90 (part II): 403–431.

WILLIAMS, LESLIE PEARCE (1986), 'Why Ampère did not discover electromagnetic induction', *American Journal of Physics* 54: 306–311.

NOTES

1. Volta (1800).
2. Golinski (1990).
3. Biot (1804).

4. See Fox, this volume, ch. 13.
5. Arnold (1983–84).
6. Poisson (1812a); Poisson (1812b).
7. Home (1983).
8. Hofmann (1995).
9. Steinle (2005) provides a detailed analysis.
10. Blondel (1982).
11. Ampère (1820).
12. Biot (1824), reprint, p. 83.
13. Biot (1820).
14. Faraday (1821).
15. Williams (1986).
16. Ampère (1826).
17. Blondel (1989).
18. Cantor, Gooding, and James (1991).
19. Cantor (1991).
20. Faraday (1821).
21. Devons (1978).
22. Moyer (1997).
23. Roget (1831).
24. Steinle (1996).
25. Faraday (1839–55), (henceforth ERE), 1st. ser. (1832), ch. 3.
26. Faraday (1839–55), 3rd series (1833), §361–367.
27. Ross (1961).
28. ERE 11th ser. (1837), §1169.
29. ERE 11th ser. (1837), §1260–1261.
30. ERE 11th ser. (1837), §1187–1214, 1252–1294.
31. ERE 11th ser. (1837), §1295–1306.
32. ERE 11th ser. (1837), §1215–1231.
33. Buchwald (1977) and (Smith and Wise 1989), ch 7.
34. Darrigol (2000), 116.
35. Faraday (1854).
36. Gooding (1981).
37. Martin (1932–46), entry 8118–8119.
38. ERE, 28th ser. (1851), §3072.
39. ERE, 28th ser. (1851), §3073.
40. ERE, 28th ser. (1851), §§3073, 3109.
41. ERE 12th series (1838), §1369.
42. Faraday (1852), §3264.
43. Faraday (1852), §3265.
44. Faraday (1852), §3267.
45. Harman (1990), 157.
46. Maxwell (1873), 398.
47. Maxwell (1873), 398–9.

CHAPTER 19

ELECTRODYNAMICS FROM THOMSON AND MAXWELL TO HERTZ

JED Z. BUCHWALD

19.1 William Thomson and Faraday

In the 1830s Faraday moved beyond all precedent by seeking in the laboratory for the direct action of a medium in electric effects, and he found it when he discovered what was later termed 'dielectric capacity'. Many of his contemporaries were puzzled by what Faraday had discovered, in major part because he described it in the language of a proto-field theory. As we have seen, Faraday introduced a fundamental distinction that was thereafter to run through all of field physics: the distinction between the *intensity* of a line of force and the total *quantity* or number of lines. These two properties, he asserted, are (in a given medium) proportional to one another. Furthermore, Faraday assumed that quantity is always conserved when the lines of force are redistributed by connecting charged conductors together in various ways, and that lines of electric force must begin and end on a conducting surface. From these propositions he was able to calculate a value for the dielectric's 'capacity (the factor that links quantity to intensity) and to show that it remained the same no matter how the conductors in his apparatus were charged.

This constitutes the true birth of field theory because it provides a new way of dealing both conceptually and quantitatively with electric actions. For Faraday the electric universe (at least) now divided between objects that affect the local state of the field, on the one hand, and the field itself, on the other. The former were of two kinds: either conductors, within which the field's state is somehow completely destroyed, or dielectrics, which sustain the field but alter it. That state, Faraday thought, varies from point to point, depending on the local presence of matter. The electric condition of an object, whether conductor or dielectric, accordingly depends only on the field state at its locus, which in turn depends on the distribution of other objects elsewhere. Faraday's introduction of the dielectric constant accordingly symbolizes the creation of a new field physics. Or, rather, it provides a way of dealing with electric induction. As yet it tells little either about magnetic systems or about how to analyse the moving forces that electrified bodies give rise to. Field theory at this stage was missing something that, in it, could play the role of Newton's laws of motion in dynamics—something that could bind the state of the field to the motions of the objects that affect it. The missing element shortly emerged in the form of energy conservation.

William Thomson (much later, Lord Kelvin) was a young, Cambridge-trained Scotsman when he first considered Faraday's work on electrostatics in the early 1840s. At that time he was less concerned with grasping Faraday's particular concepts than with demonstrating that Faraday's claims for dielectric behaviour could be understood in traditional terms by introducing a model for the dielectric. During the late 1840s and the early 1850s Thomson merged energy conservation with Sadi Carnot's analysis of heat engines, producing (in company with Rudolf Clausius) thermodynamics. At this time Faraday was attempting to extend his field physics from electrostatic to magnetostatics, as we now term it. Here, however, Faraday discovered that, unlike dielectric bodies, there are two classes of magnetic bodies that behave in markedly different ways: one class moves along the direction of increasing field strength; the other class moves in the opposite direction. This, he reasoned, requires that the factors linking intensity to quantity must differ in character between magnetism and electricity. In particular, the magnetic factor must run from zero to one, and from one on up, whereas the dielectric factor is always at least one (with unity being the standard in the absence of matter).

Thomson unified Faraday's understanding of both electro- and magnetostatics with energy conservation in the following way. Taking his stand on integral transformations of the known values for the energy of an electric (or magnetic) system, he showed that the product of quantity by its corresponding intensity represents an energy density for the field—a density that may be considered (the mathematics at least permits it) to subsist in the regions between bodies. Then, he continued, the tendency in all such systems will be to so arrange themselves as to minimize this total field energy. Here, then, one has for the first time a thorough, field-like replacement for the conception of an independent force that acts on objects. Forces certainly do result from Thomson's energy-based analysis, but they are its end-product, not its beginning.

19.2 MAXWELL

Although the distinction between quantity and intensity—or between what we would now term flux and force—underpinned Thomson's merging of field theory with energy conservation, nevertheless he himself never went any further. In particular, he did not attempt to reconstruct all of electrodynamics from the new perspective. That was accomplished by the young James Clerk Maxwell, like Thomson a Cambridge graduate and mathematically adept. In 1856 Maxwell extended Thomson's structure to cover electrodynamics, and this for the first time brought to the fore issues concerning the electric current.

At that time Maxwell in effect introduced what was later termed the 'vector potential', or what he, following Faraday, termed the 'electro-tonic intensity' to represent a function defined at the locus of the current and such that the corresponding electromotive force is given by its rate of decrease with time. This enabled Maxwell to formulate field theory in a manner that, while unfamiliar in retrospect, nevertheless captured the essential relationships between fluxes (quantities per unit area) and forces that are today preserved in macroscopic electromagnetics, with the exception of Maxwell's displacement current. Although, the latter was a critical innovation, his 1856 linking together of all electromagnetic variables through a system of force–flux relations formed the bedrock for his subsequent development of a full-fledged field theory.

The introduction of the displacement current into field physics was not a straightforward affair because it did not (as has often been asserted) follow directly from a perception of a mathematical inconsistency in the field equations that Maxwell had available to him in the early 1860s. Specifically, the partial differential relation that connects conduction current to magnetic force (the so-called 'Ampère law') implies that all such currents must be re-entrant, or closed, which is obviously not the case. One solution to the problem is to alter the field version of the Ampère law by adding to the conduction current a term consisting of the rate of change of electric quantity (or displacement) with time—that is, by adding the 'displacement current' to the conduction current.

Maxwell, however, came to the displacement current rather through a physical model of the ether than through mathematics or more general field-theoretic conceptions (though the concept does undoubtedly have its roots in Faraday's attempts to understand the relationship between electric induction and conduction). Here we find a close binding of field theory to ether models—one that subsisted for about five years or so and that was instrumental in forging the structure of what soon evolved into Maxwellian electrodynamics.

The first form of Maxwell's model represented the electric current (which at this stage still exists only in conductors) by means of ball-bearing-like objects that roll between rotating cells that store angular momentum and kinetic energy. These bearings translate against a frictional resistance in conductors but are not permitted to do so elsewhere. This posed a difficulty for Maxwell, because it meant that the model could not possibly be mechanically consistent: if the particles cannot move in any way at all in a breached portion of a conductor, then the mechanism, it seems, must break. It was to resolve this problem that Maxwell introduced the displacement current. He argued that the rotating cells must be considered elastic, so that they may yield to stress. This permits the bearings between them to shift slightly even within a non-conductor, which allows the mechanism to remain intact without permitting a current (a frictionally-resisted translation of bearings from cell to cell). In other words, the original purpose of the displacement current was precisely to prohibit the occurrence of currents, properly speaking, within dielectrics. During the next decade Maxwell began to think about his theory more in terms of a highly abstract representation of field processes than in terms of the specifics of his model, and in so doing he began to formulate a new structure that reached fruition in 1873, when his influential *Treatise on Electricity and Magnetism* was printed.

19.3 MAXWELLIAN ELECTRODYNAMICS

Maxwell's *Treatise* appeared six years after another extremely influential text was published in Britain: namely William Thomson's and Peter Guthrie Tait's *Treatise on Natural Philosophy*. The two *Treatises* are closely related to one another. In theirs, Thomson and Tait dealt exclusively with mechanics, and they did so in a particularly significant way, for they based it directly on energy functions and either Lagrange's equations or Hamilton's principle. This set a pattern for research among mathematically trained Cambridge 'Maxwellians' during the next quarter-century, because many among them attempted to build theories by discovering appropriate energy functions that, when fed into the Lagrangian or Hamiltonian machinery, led to empirically testable results. Thomson's and Tait's generalized dynamics adapted particularly well to the kind of electromagnetic theory that Maxwell had been developing during the 1860s—one that was not founded *ab initio* on a specific structure for the ether. It is essential to grasp the wide-ranging nature of this uniquely British understanding of dynamical theory in order also to understand how it permitted the creation of a field physics that ultimately had to be abandoned after the introduction of the electron.

The essence of the dynamical method pursued extensively at Cambridge (but elsewhere as well) resided in its assumption that processes can be exhaustively described in terms of energy densities. Different energy expressions, when fed through the dynamical equations, lead to different differential equations and to different boundary conditions. If a substance shows unusual behaviour then the natural procedure to dynamicists is to modify the usual energy expressions and then to follow out the implications of the modification by inserting the new expressions into Hamilton's principle.

Although this may seem to be an unexceptional procedure even today, the modern physicist would raise two principal objections to it. First, one would ask where the energy expression comes from. Second, modern theory allows that one can proceed in this way only when circumstances are such that the microphysical structure of the body does not extract or emit energy that cannot be taken into account in Hamilton's principle. These two objections did not occur to dynamicists for one reason, which marks the divide between their views and modern physics. Dynamicists tacitly assumed that all processes can be represented by continuous energy functions. Those few processes that did not immediately yield to the method (like optical dispersion) required more intricate energy expressions, perhaps ones that depended upon an inherent frequency. The primary goal of research was therefore thought to be the creation of appropriate energy formulae to whose consequences the analyst was then committed. Throughout the 1880s and the early 1890s this was a benefit and not a liability, because the procedure led to important connections between a number of electromagnetic processes that seemed to be otherwise unrelated.

As seen by Maxwellians, electromagnetic theory was accordingly based on the assumption that the seat of electromagnetic processes is a continuous medium, or ether, that is governed by the laws of dynamics. To solve problems in electromagnetism only requires expressions for the energy functions of the ether. We do not need to know its true structure, nor do we need to know how changes in the energy are brought about. The ether has, it was assumed, certain properties that can be altered by the presence of matter. In particular, it possesses two of immediate significance: namely, specific inductive (electric) capacity and magnetic permeability. These two properties were thought to be represented by continuous functions of position. The values of the functions may, and (in the presence of matter) do, change, but the changes are continuous. Maxwellians nevertheless did admit that changes in permeability and in capacity are due to the effects of discrete material particles, and that the values deployed in macroscopic equations are consequently averages over microscopic effects. However, in their view each material particle itself effects a continuous alteration in the ether's properties. As a result, even at the microscopic level continuity is never breached.

At the core of Maxwellian field theory lay its abandonment of the conservation of charge in the previous sense of the phrase. Instead of considering charge to be a special electric substance that can accumulate in bodies, Maxwell treated it as an epiphenomenon of the field. His concept involved the transformation of energy stored in the ether into material form (as heat) through a process rather similar to that of elastic relaxation. This could occur wherever matter was present—though neither Maxwell nor Maxwellians attempted to explain why matter could have this effect, which is represented macroscopically by electric conductivity.

To understand what is involved, imagine a region of the ether that is void of matter but in which an electric field (which stores potential energy in the ether) exists. Place a piece of matter in the region. Since all material substances, according to Maxwellians, have some conductivity, the region of the ether now occupied by matter begins to lose the energy that is stored in the electric field. This energy appears in the matter as heat. The result is the creation of a difference in the values of a certain quantity (the electric displacement, which is the product of inductive capacity by electric field intensity) at the boundary between matter and free ether. This difference represents, at any instant, the electric 'charge' on the boundary. Maxwellian + and − 'charges' are accordingly not individually conserved because there might very well be no charge at all in the universe on these principles. This contrasts with electric fluid theories, which assert only that there may be no net charge in the universe—but the individual positive and negative particles continue to exist. Nevertheless, Maxwellian theory satisfied the very same charge conservation equation that particle theories satisfy, despite this difference. It can do so because of its conception of electric current.

In Maxwellian theory an electric 'current' is the rate at which a portion of the ether is moving. If the ether is quiescent, then no 'current' exists. Electric 'charge' occurs because the ether has moved or is moving through regions in which the ratio of conductivity to inductive capacity varies from point to point. To link the two processes (charge and current) we assume either that the current generates a magnetic field or else that a changing magnetic field generates an ether shift. Either assumption, well formulated, leads with other field equations to the very same expression for charge conservation that particle theories yield. The major point to understand is that, despite this agreement, the Maxwellian current properly speaking is not the rate of change of charge with time: it only may lead to such a change.

These several ideas are present in embryo in Maxwell's *Treatise*, but they are there obscured by the text's novelty and comprehensive character. They and other central concepts are clearly evident in the work of Maxwell's Cambridge followers, as well as others not trained at Cambridge (though the latter differed on various points from Cambridge practitioners). Indeed, the concrete structure of Cantabridgian Maxwellianism was in major part produced as examination students worked out problems under the guidance of their tutors. For our purposes the following beliefs were so developed: 'charge' is nothing but a discontinuity in displacement (which is an ether shift that stores energy); 'current' is nothing but moving ether; to create a new theory, modify the ether's energy function (which amounts to modifying the structure of the ether itself); the effect of matter upon ether is mysterious and must be put off until problems are solved through energy methods; electric conductivity is particularly mysterious and somehow involves the particulate structure of matter; boundary conditions are crucial analytical tools; mechanical models of the ether are important illustrations of energy exchanges, but they are unlikely to reflect the ether's true structure.

Perhaps the best illustration of the power of Maxwellian theory, and of its difference from electromagnetism after the electron, involves the 'Hall effect'. Discovered in 1879 by the American physicist Edwin Hall, this effect is today thought to demonstrate that the electric current consists of negatively charged moving particles. The experiment can be performed easily with modern equipment, and Hall's own technique was not fundamentally different from the modern one. Take a plate, say of copper, and send a current across its length. Attach a sensitive galvanometer across the plate's width, and place the entire device between the poles of an electromagnet, with the field normal to the plane of the plate. The sensitive galvanometer reveals a current while, but only while, the electromagnet is active. This, modern theory argues, directly reveals the deflection of the moving electrons in a magnetic field. Moreover, the direction of the deflection reveals their sign.

Maxwellians, including Hall, thought otherwise. For them there were only two possibilities available. The least radical, which was widely received for some time until Hall refuted it in the laboratory, referred the effect to an action of the magnetic field on the material structure of the metal. This interpretation had the advantage of making the effect less than fundamental, thereby avoiding major alterations in Maxwell's equations proper. The second possibility was more exciting and was the most widespread. If, as Hall insisted, he had discovered a new way to produce an electric current (and not a new way to stress metals), then he had necessarily also discovered a new way to produce an electric field, since in Maxwellian theory an electric current always requires an electric field. The next question was precisely what conditions generated this new field; about this there was much room for discussion, but eventually a widely accepted solution prevailed.

Whenever an electric current exists in the presence of a magnetic field, Maxwellians reasoned, a subsidiary electric field also exists that is at right angles to the current and to the magnetic field. Since a 'current' is simply an ether flow, this meant that Hall's action should also exist in non-conductors: an ether flow in the presence of a magnetic field implies a 'Hall effect'. In fact, since ether flows are much simpler to understand in non-conductors than they are in conductors, theory can better deal with the former than with the latter—though Hall had found the effect only in conductors.

On Maxwellian principles any change in the ether's energy properties implicates a corresponding alteration in the field equations, and *vice versa*. Hall had discovered an effect that required the addition of a new term to one of the field equations, though an admittedly small one. Consequently, the energy properties of the field must also be changed in just the right way to yield Hall's new expression. It was soon discovered that the altered energy densities, applied to dielectrics, yield equations for the Faraday effect (the rotation of the plane of polarization of light passed through a dielectric in a magnetic field). An intricate series of developments within Maxwellian physics ensued as the implications of the new energy terms were followed out during the 1880s, particularly as they applied to the difficult problem of reflection from magnetized, metallic surfaces (the phenomena, that is, of magneto-optics). By the early 1890s it seemed (to J. J. Thomson, for instance) that workable results could be achieved even here.

Nevertheless, the equations that Thomson used, and that seemed to be required by field theory, cannot encompass magneto-optics because they lack a necessary constant, one that is easily provided by a micro-physical theory but that the macroscopic field equations of the Maxwellians could not comfortably embrace. The belief that Maxwellian physics was irreparably flawed did not, however, emerge from improved experimentation (though they might eventually have), but rather from theoretical penetration into the nature of conductivity.

Throughout the 1880s Maxwellians did not consider conductivity to pose a problem. This was not because they were able to incorporate it in their dynamical field equations. They were not able to do so, and in the early 1890s Oliver Heaviside even demonstrated that conductivity cannot be inserted directly into dynamical equations. Rather, Maxwellians avoided the entire question by relegating the subject to an area about which, they were willing at once to admit, they knew little—the unknown mechanism that links matter to the ether—and to do so they deployed John H. Poynting's theorem concerning energy flow through the field. That theorem could be used to trace the energy pattern around a current, and it seemed to indicate that the energy does not flow along the wire but rather radially into the wire, which nicely captured the image of the current as a byproduct of ether processes, since the wire appears rather to act as a sink for electromagnetic energy than as a carrier of it.

Poynting and others deployed this imagery in ways that permitted Maxwellians to bypass questions concerning the nature of conduction, and their techniques worked to the satisfaction of the community for about a decade. Indeed, problems with it did not emerge from within the group of active Maxwellians, but rather from an expert in hydrodynamics—Joseph Larmor—who certainly knew field theory, and who had been trained at Cambridge, but who had done little research in the subject until the mid 1890s. At that time he attempted to answer exactly the kinds of questions that more experienced Maxwellians such as Poynting and J. J. Thomson had long avoided, and this eventually led him into a quagmire. Larmor extricated himself from his difficulties by inventing the electron. At first he did not think of the particle as a substitute for traditional Maxwellianism so much as a way to bypass certain problems in it, but in short order he revamped field physics by separating the electron, as the sole source of fields, from the fields themselves. Larmor was sufficiently persuasive that many Maxwellians eventually came to the conclusion that they could no longer play with field energetics, which was now fixed for all time. Instead, they had henceforth to invent models for material micro-structure, although that form of activity did not flourish at Cambridge in later years.

19.4 HELMHOLTZ

In 1870 Hermann Helmholtz, newly appointed to the chair of physics at Berlin, developed a form of electrodynamics that differed markedly both from the widely accepted (in Germany) theory based on particles that had been developed by Wilhelm Weber, and also from British field physics. Indeed, Helmholtz attempted to produce a completely general set of equations, based on energy considerations, that could accommodate every possible variant of electrodynamics by constructing a 'potential' function. This function had energetic significance, and forces could be deduced from it by variational procedures: variation with fixed spatial coordinates yields an electromotive effect, while variation with the time fixed yields a mechanical effect. The resulting equations contain an undetermined constant k, and Helmholtz claimed that assigning different values to k would yield the corresponding theories for currents in conductors required by existing theories.

Despite Helmholtz's claims, his electrodynamic potential is not conceptually equivalent either to Weber's theory (in which the forces act instantaneously between electric particles) or to field theory (which substitutes fields for particles and in which forces propagate). One reason for this is that Helmholtz's electrodynamics has a different understanding of the relationship between objects as sources from either of the two alternatives. Unlike its competitors, his account never goes beyond objects themselves (to introduce, for example, particles or fields), and it construes all actions as determined immediately and solely by the states of the interacting objects and by their mutual distance. Indeed, in his physics even force, properly speaking, does not exist as an independent entity, because it emerges only via an energy calculation (as it also does, albeit with considerably different meaning, in field theory).

Yet Helmholtz's system was the route through which many, including his student, Heinrich Hertz, eventually adapted aspects of field theory. Indeed, from the outset Helmholtz had kept field theory in mind, but he had early encountered difficulties in assimilating it, as one might expect. He felt that Maxwell's field (which was not completely developed in 1870) could be captured in an acceptable manner by extracting from Maxwell the two concepts that, Helmholtz was convinced, uniquely characterize his scheme: first, the existence of an electrodynamic ether, and, second, the requirement that the disposable constant k must vanish.

To accommodate these two requirements, Helmholtz did not adopt Maxwellian ideas. He instead attempted to adapt traditional understanding of the structure of dielectrics to the ether: to wit, that the ether must (like material dielectrics) be 'polarizable' (meaning that it contains electric and magnetic elements that can be moved physically through restricted ranges by, respectively, electric and magnetic actions), and that changing polarization must be treated (excepting Ohm's law) precisely like an electric current. In this way Helmholtz was able to show mathematically that electrically and magnetically polarizable bodies can be the sites of waves of polarization, their being (in general) both longitudinal and transverse oscillations. If, Helmholtz reasoned in 1870, the disposable constant k in his equations vanishes, then only transverse waves remain. If, in addition, the electric polarizability of the ether is effectively infinite, then one also obtains Maxwell's relationship between electric and optical properties. In later years it was realized (initially by Henri Poincaré) that the requirement on k is redundant if only the polarizability is infinite, but for a decade or more both requirements were usually cited together (in part to distinguish the several ways in which Helmholtz's electrodynamics might relate to Maxwell's).

19.5 HERTZ AND ELECTRIC WAVES

To put Heinrich Hertz, who first produced and detected electric waves, into proper perspective, it is essential to recognize that before his creation in 1887 of the dipole oscillator and resonator—the devices that, respectively, generated and detected electromagnetic radiation—no one knew how to produce freely propagating electric waves. In Britain, optical radiation constituted the only known instance of these sorts of waves, and, therefore, they were generally associated with optical instrumentalities.

Furthermore, at least until the mid-1880s some British Maxwellians (in particular, George Francis FitzGerald) did not even think it possible to generate such waves at all by means of electromagnetic devices. FitzGerald eventually changed his mind about this, but other views militated against any Maxwellian conceiving of a suitable way to generate sufficient power to produce free electric waves that could be detected.

In Germany neither Helmholtz nor anyone else considered how electromagnetic radiation might be artificially produced. Instead, Helmholtz, like his British contemporaries, evidently considered optical radiation to be the paradigm for, and perhaps the only proper instance of, electric waves, except for processes that are confined to or on conducting media. Moreover, the hypotheses that (on Helmholtz's system) might yield electric radiation in non-conducting media raised questions that did not have straightforward answers during Hertz's Berlin years. Indeed, Helmholtz tried to convince his young apprentice to devote himself to their experimental elucidation.

When Hertz began working intensely in 1886 with the extremely rapid oscillations in wires that eventually led him to his experiments with electric waves in air, he initially conceived of wire–wire interactions as involving the direct action of one object on another. When he was able to produce and to detect waves in air, Hertz at first decided simply to assume that the wire–wire action was delayed in time, whatever the cause of the delay might be. However, by the spring of 1888 he had decided that the situation could not be treated in that way. He was by then convinced that his experimental data required a radically different interpretation—one that admitted the active role of a third entity as a mediator. The interaction between conductor A and conductor B, he now believed, was not delayed at all. Indeed, properly speaking, such an interaction simply did not exist. Instead, each of A and B must be thought to interact directly only with a third object, the ether, whose state was entirely specified by the electromagnetic field, and which itself was both ubiquitous and unchangeable. Unlike laboratory objects, the ether in Hertz's conception has no manipulatable properties whatsoever, for its qualities remain invariant (though its condition or state varies).

This conception differed considerably from the one that had been advanced as a possibility by Helmholtz himself, and according to which the ether proper behaves just like a laboratory object. In such a scheme the ether would modify the apparent interaction between A and B by working separately on each of them, while A and B would continue to interact directly and immediately with one another. Hertz was quite familiar with this possibility from Helmholtz's work, and he clearly did not like it, since in 1884 he had produced a version of Maxwell's equations without using the ether at all. As a student of Helmholtz's, Hertz did think of the ether as an object, but he was apparently uncomfortable with its hidden character and wished to avoid introducing it as an entity like all others. He wanted, that is, to remain entirely with laboratory objects proper. In seeking to understand how field theory might be possible, Hertz developed a novel way to multiply interactions between laboratory objects, thereby yielding Maxwell's equations, but not field theory itself, because his scheme was based entirely on leveraging the scope of Helmholtz's electrodynamic potential, which required sources to be directly active entities.

The understanding of electromagnetic radiation that Hertz developed in the spring of 1888 still insisted on the continuing role of the source, but dropped altogether its relation to other sources. Its behaviour is instead specified in respect to a mediating entity: namely the ether, whose state in the immediate neighbourhood of the source is determined by the source's activity. Quite unlike Maxwellian field theory, in Hertz's scheme the source continues to exist as an entity in and of itself, since it is responsible for activating the processes that take place in the field. Where the Maxwellian source in effect merely represents a locus where ether properties change rapidly, the Hertzian source is responsible for activating specific states in an entity (the ether) whose qualities—but not whose states—remain forever the same. On the other hand, the source was not of any more direct interest to Hertz than it was to Maxwellians, except as an emitter or a receiver, because physical activities of note occurred only in the ether itself.

Although the field patterns that Hertz eventually deduced from his version of Maxwell's equations were theoretical constructions, whereas the oscillating electric device that actually produced radiation was a material object, nevertheless for Hertz the intimate structure of the object remained unknown, whereas the inferred field was considered to be known. This inversion encapsulates the originality and power of Hertz's physics. Because Hertz ignored the physical character of the source that produced his radiation—because he boxed it in with a mental quarantine against asking questions about it—he was able to make progress where his British contemporaries had not been able to do so. Although Maxwellians did consider sources to be epiphenomena of the field, they also concentrated closely on the shapes of radiating bodies, for to the British the canonical instance of electric radiation involved what was later termed wave-guidance, in which radiation does not depart from the conducting boundary but, as it were, slips over the surface. For the British the geometry of the surface was accordingly a critical factor in building a theory, and situations that eluded analysis of this sort (such as isolated conductors that yield up their energy to far-distant surroundings) were not thoroughly probed (at least in connection with radiative processes). Furthermore, British analysts already thought that an object like Hertz's oscillator would reach electric equilibrium so rapidly that the radiation it emitted would simply flash away in an essentially undetectable burst. It is therefore hardly surprising that Maxwellian reaction to Hertz's experiments centred principally on his detecting resonator, and not on his (mathematically intractable) oscillator.

Hertz did not think at all about the surface behaviour of his oscillator. Neither did he consider the effects that it produces to be beyond the reach of analysis or experiment. For him the paper analogue of the material oscillator was an inconvenient nuisance, and he immediately reduced it to a pictogram. The very object that enabled Hertz to investigate electric waves accordingly did not exist at all in the mathematical account that he himself developed for its field. The effects of this removal of the experimental object were far-reaching, and can be followed through physics and electrical engineering during the ensuing half-century at least. Hertz's missing oscillators evolved into the antennae of an emerging technological regime; they also evolved into symbols for the unknown entities that were responsible for natural radiation—in particular, Max Planck's resonators.

In 1890 Hertz published two papers on the fundamental equations of electromagnetics that were widely read in Germany and elsewhere during the few years that remained to him. Many contemporary references indicate that these articles had a deep impact on German physicists, which is hardly surprising, since Hertz here introduced many of his German contemporaries to the broad range of electromagnetic processes from the viewpoint of field theory. However, he had already presented the field equations in conjunction with their solutions for the dipole oscillator in 1889. Whereas the 1890 articles contained no diagrams of any kind, the 1889 piece contained several, including one that laid out a temporal sequence of field maps. In the immediate aftermath of Hertz's discovery, this article was frequently used as a basis for understanding hid work, and indeed for developing a pragmatic understanding of a new scientific object: the radiation field.

In the 1890s, before antenna engineers had come into being, Hertz's dipole constituted a new kind of scientific object, one that was at once conspicuously absent from the analytical structure of the effect that it produces, and that was nevertheless physically present as an actual device in the laboratory. Among physicists the dipole never did become an object of great intrinsic interest or significance, because it did not, from their point of view, produce something altogether novel; it just generated, as it were, a kind of artificial light. Nevertheless, for physicists the dipole did serve as a useful tool, as a canonical source for electromagnetic radiation, and it was often inserted without much discussion into radiation calculations during the 1890s and early 1900s. For the evolving coterie of radio engineers during these years, the dipole constituted the sole material method for manipulating the new (and entirely artificial) electromagnetic spectrum. As such it was essential as a technological object, but it remained a tool that was to be used for the effect that it produced, and not itself an object of analysis.

This character of Hertz's physics distinguishes it quite markedly from H. A. Lorentz's electrodynamics, which began to emerge in detailed form in 1892. Lorentz worked from the outset with microphysical entities, eventually with the electron, whereas Hertz never did, and this constitutes an immediate and obvious difference between them. There is, however, another difference, one that runs even deeper. Lorentz always based his electrodynamics on the proposition that the interaction of entity A with entity B is delayed in time—that a change in the state of A at a specific moment occasions an interaction with B at a later time. This kind of physics deploys the fundamental image of interacting objects, albeit microphysical ones. Hertz's physics did not deploy anything like this image, for it embodied a method for building theories that permitted, indeed that impelled, distance both from the object itself and from its connections with other objects. Hertz's premature death in 1894 ended his own development of this sort of physics, but its impact on others was enduring, both for the specific methods he introduced in dealing with electromagnetic radiation, and for the example of how to build a physical theory for certain effects without analysing in detail the object that produces them.

Bibliography

There is considerable literature on the topics discussed here. For William Thomson (Kelvin) see especially Smith, C. and Wise, M. N. *Energy and Empire: A Biographical Study of Lord Kelvin* (Cambridge: Cambridge University Press, 1989). For Maxwell's early work on the basis of Faraday's conceptions see Wise, M. N. 'The mutual embrace of electricity and magnetism' (*Science*, 203 (1979), 1310). Maxwell's route to the concept of displacement is carefully detailed in Siegel, D., *Conceptual Innovation in Maxwell's Electromagnetic Theory. Molecular Vortices, Displacement Current, and Light* (Cambridge: Cambridge University Press, 1991). Maxwell's mature electrodynamics and its development by British and Irish field theorists and experimenters are discussed in Buchwald, J., *From Maxwell to Microphysics*. (Chicago: University of Chicago Press, 1985 and 1988) and in Hunt, B. J., *The Maxwellians* (Ithaca: Cornell University Press, 1991). Warwick, A., *Masters of Theory. Cambridge and the Rise of Mathematical Physics* (Chicago: The University of Chicago Press, 2003) explains the characteristics of Cambridge training that produced the generation of British field theorists after Maxwell. On the unusual Maxwellian Oliver Heaviside see Yavetz, I., *From Obscurity to Enigma. The Work of Oliver Heaviside, 1872–1889* (Basel: Birkhäuser Verlga, 1995). For the development of electrodynamics at the hands of Helmholtz and the production of artificial electric waves by his student Hertz, see Buchwald, J. *The Creation of Scientific Effects* (Chicago: University of Chicago Press, 1994). The best synoptic account of the entire subject is Darrigol, O., *Electrodynamics from Ampère to Einstein* (Oxford: Oxford University Press, 2000). Older literature includes the classic two-volume work by Whittaker, E. T., *A History of the Theories of Aether and Electricity* (London: Longmans, Green and Co., 1910; reprinted in New York by the Humanities Press, 1973). Full references to all the original works mentioned here can be found in these accounts.

CHAPTER 20

FROM WORKSHOP TO FACTORY: THE EVOLUTION OF THE INSTRUMENT-MAKING INDUSTRY, 1850–1930

PAOLO BRENNI

20.1 INTRODUCTION

It is difficult to give a precise and unambiguous definition of the term 'scientific instrument', one that includes a large number of technical artefacts of many different kinds. However, we can say that scientific instruments represent the technology of science and are the material tools used for investigating nature (Van Helden and Hankins, 1994, pp. 1–6; Bud and Warner, 1998; Warner, 1990). In fact, since the seventeenth century, with the advent of the Scientific Revolution, they have become the essential working tools of natural philosophers. The new approach in studying nature considered experiments and the direct observation of natural phenomena to be fundamental practices for philosophers; and experiment and observation required instruments. In the modern age, instruments are directly associated with the research of scientists, but since antiquity they have also been used by practitioners such as surveyors, architects, draughtsmen, cartographers, seamen, and later by gunners in their daily measurement and observation activities (A. Turner, 1987; G. Turner, 1991). Furthermore, at least since the late seventeenth century, instruments have been used as essential aids for teaching science and transmitting knowledge in schools, colleges, and universities. Finally, from the nineteenth century, instruments began to play a more important role in industry and manufacturing, as well as in an ever growing number of technically oriented professions (Sydenham, 1979). So the broad term 'scientific instruments' includes a whole series of artefacts used in research, teaching, professional, and productive activities. In the English language the expression 'scientific instruments' became common only from the late nineteenth century onwards. Previously, together with 'optical and mathematical instruments', it was usual to call the instruments used in physics, 'philosophical' (Warner, 1990).

During the Middles Ages and the Renaissance the production of mathematical instruments such as astrolabes and quadrants was in the hands of highly skilled craftsmen and engravers (A. Turner, 1987; G. Turner, 2000). (From the seventeenth century, with the invention of several new instruments such as telescopes, microscopes, barometers, thermometers, vacuum pumps, electrical machines, and many other types of philosophical apparatus, instrument-makers were required to master an increasing number of techniques and materials.) The skills of metal-workers, founders, cabinet-makers, glass-blowers, mirror-makers, engravers, lacquerers, and so on, became necessary to produce a growing array of different apparatus.

But also in the eighteenth century, with rare exceptions, workshops remained small and employed rather few people. Only a few of British manufacturers, such as Jesse Ramsden (McConnell, 2007), who were involved in the construction of large astronomical instruments, employed workers in large numbers. During the nineteenth century, especially in the second half of the century, instruments were not only needed in laboratories, scientific cabinets, and teaching collections; their systematic use also extended to many professions and into industry (Turner G., 1983). Instrument-making workshops in Europe and in the United States not only multiplied (Brenni, 2002a), but many of them were transformed into large factories. Nevertheless, from a purely economic point of view, the scientific instrument industry[1] was certainly not comparable with those of mining, metallurgy, textiles, or chemicals. Furthermore, the number of workers, the importance of the trade, its social impact, and the extent of infrastructure always remained quantitatively much smaller. But the state of the instrument industry was (and still is) strongly connected with the state of science, technology, and industry in the particular nation concerned. Good instruments called for good collaboration between scientists and technicians, and between laboratories and workshops. Valuable instruments could be successfully produced and commercialized where skilful workers, good tools, machine tools, and raw materials were available, and where trade was free, the economy was healthy and industry was booming. For example, England in the second half of the eighteenth century, France between the early decades of the nineteenth century and the 1890s, Germany between the end of the century and 1914, and the USA for most of the twentieth century, fulfilled most of the criteria, and during these periods, all the countries mentioned had an expanding and successful precision industry. Finally, a strong, highly regarded precision industry has always been an emblem of progress, and although its economic importance may be relatively limited, its weight in terms of prestige in the national and international arenas was certainly very important.

It is often quite difficult to establish reliable comparative studies of the evolution of precision industries in different countries. Industrial statistics for the nineteenth end early twentieth centuries were often very partial and were compiled using very different criteria, and they therefore need to be analysed with extreme care. But by studying statistical data, together with scientific articles, reports of national and international exhibitions, trade catalogues, and industrial and commercial surveys, it is possible to trace a fairly accurate view of the evolution of the instrument industry during the period considered in this essay. It is also interesting to point out that surviving collections of instruments can also be used as powerful historiographical tools (Brenni, 2000). The typologies of instruments and their provenance, as well as the names of the makers, provide very useful data for corroborating the information extracted from the above-mentioned documents.

20.2 INSTRUMENTS

In compiling a profile of the evolution of the precision industry between 1850 and around 1930 it is useful to establish some broad categories of instruments. Because of their extreme variety, precise classification is complex and certainly not very useful in the frame of this article. However, for the period of time considered here, it is possible to divide instruments into a few categories, which also allows a better understanding of some characteristics of the precision industry.

20.2.1 Research and Precision Measurement Instruments

These represent the most sophisticated (and expensive) tools of science (Fig. 20.1). They are used for the investigation of nature, for observing, exploring, and measuring its phenomena, and for establishing the laws which govern them and for testing theories and models (Ambronn, 1919; Cooper, 1946; Sydenham, 1979). Among them we can mention large astronomical instruments (see for example Riekher, 1990), spectrometers, interferometers, delicate analytical balances, high-precision galvanometers and electrometers, calorimetric apparatus, clock regulators, and so on. They were used in metrology, physics, astronomy, geodesy, and chemistry, and by the end of the nineteenth century they were beginning to enter the fields of medical, biological, and physiological research as well as the earth sciences. They are the top of the range in the instrument market, and their production has always remained quantitatively limited when compared with professional and industrial instruments. Therefore, in the early decades of the twentieth century they were made in relatively small number by highly specialized firms, representing the peak of the precision industry. But if the production of these instruments remained limited, essentially supplying the needs of

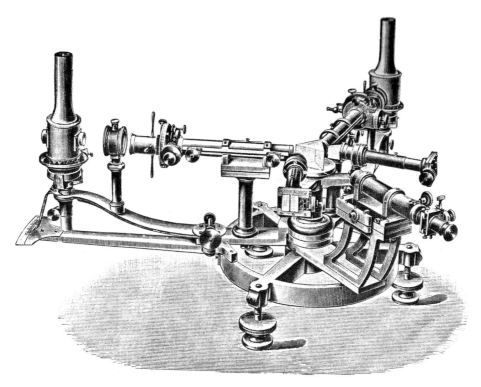

Fig. 20.1. A sophisticated colour-mixing apparatus conceived by H. Helmoltz and A. Koenig and made by Franz Schimdt & Haensch in Berlin. Instruments like this were used in research laboratories. (Franz Schimdt & Haensch, *Katalog II, Spektralapparate*, Berlin, 1930, p. 40.)

laboratories and observatories, the firms that made them enjoyed great prestige, both nationally and internationally.

20.2.2 Didactic and Teaching Instruments

Instruments in this category cannot increase scientific knowledge. They are used rather for the transmission of existing knowledge. Teaching instruments allow us to reproduce and demonstrate physical phenomena in a clear and convincing way, with the possibility of demonstrating laws and principles and reproducing (often in a simplified way) historical experiments (Brenni, 1998; Gauvin and Pyenson, 2002). Many didactic instruments were derived directly from research instruments which, when superseded by better and more modern instrumentation, could, with small modifications, still be used as didactic tools. For example, the 'percussion apparatus' (with several suspended elastic balls) was a research instrument from the end of the seventeeth century to the mid-eighteenth century, but was later simply used as a demonstration device. Others, such as the so-called 'wave machines' (see for example, Holland, 2000) were conceived expressly for teaching (Fig. 20.2). The nineteenth century added an enormous variety of didactic instruments to those proposed earlier

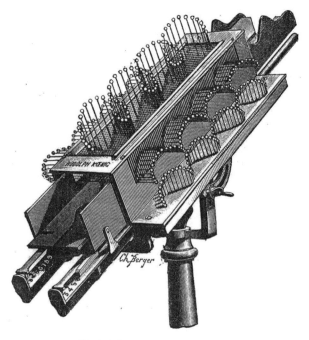

Fig. 130 (h. = 1ᵐ,50).(Nᵒˢ 263-263 a)

Fig. 20.2. The Wheatstone wave machine sold by the acoustical instrument-maker R. Koenig in Paris. This purely didactic instrument, proposed in the 1840s, was used until the beginning of the twentieth century for visualizing sound and optical waves and their phenomena such as interference, polarization, and so on. (R. Koenig, Catalogue des appareils d'acoustique, Paris, 1889, p. 98.)

(Brenni, 2011; G. Turner, 1996). These instruments were needed to equip the growing scientific requirements of schools and universities (Fig. 20.3). However, often the difference between didactic and research instruments depended on the context in which they were to be used. The same vacuum pump could be utilized for a simple pneumatic experiment or for a study of the behaviour of electric discharges in rarefied gases. In the former case, the pump can be considered to be a didactic apparatus, while in the latter it is a research tool. During the interwar years, because of the introduction of new ways of teaching physics and science in general, didactic instruments became more simplified. Many old-fashioned devices, which were more suited to early nineteenth century physics cabinets than to modern teaching laboratories, disappeared from trade catalogues. Modular apparatus, which could perform various demonstrations and experiments using different combinations of a relatively few elements, were introduced. (Heering and Wittje, 2011) Instruments became more user-friendly and their numbers reduced. The characteristics separating them from research instruments became more and more pronounced. However, this transformation, which started in the late 1920s, was slow and was completed only in the second half of the twentieth century.

Fig. 20.3. The Brace Laboratory physics lecture room of the University of Nebraska around 1906. On the table there are several apparatus such as an induction coil, a couple of Leiden jars, and a Tesla transformer. Next to the blackboard is an electric distribution panel, and on the right is a large electrostatic machine. (© University of Nebraska Department of Physics and Astronomy.)

20.2.3 Industrial Instruments

Industrial instruments are used in factories, manufacturing, and electrical and chemical plants, for measuring, testing, recording, and controlling (Cooper, 1946; Cronshaw et al., 1949; Gramberg, 1923; Perazich et al. 1938). Although some instruments, such as steam indicators, manometers, or thermometers and hydrometers had been used in industry since the beginning of the nineteenth century, the role of instrumentation in industrial activities grew extremely fast in the last decades of the century, with the second industrial revolution. Science-based industries such as electrical and chemical manufacture required new instruments, which could also be used by non-specialized personnel and did not require complex manipulations and calibration. For example, electrical measuring instruments for use in laboratories were completely inadequate and far too delicate to be used in electric power plants (see for example: Aspinall Parr, 1903; Brenni, 2009; Edgcumbe, 1918). New direct-reading and pre-calibrated apparatus, which were capable of withstanding the unfavourable environmental conditions of power stations and factories (vibrations, disturbing electromagnetic fields, changes in temperature, and so on) had to be developed and perfected (Fig. 20.4). Optical instruments such as refractometers, polarimeters, and saccharimeters entered into the chemical and food industries, and electric and optical pyrometers became increasingly common for determining the temperature of industrial ovens. The necessity to constantly monitor the time variations of different parameters (currents and voltages, flows, pressures, temperatures, and so on) led to the development of large numbers of recording instruments. In the first decades of the twentieth century the complexity of industrial processes rose constantly, and interconnected instrumental systems were necessary for their control and regulation.

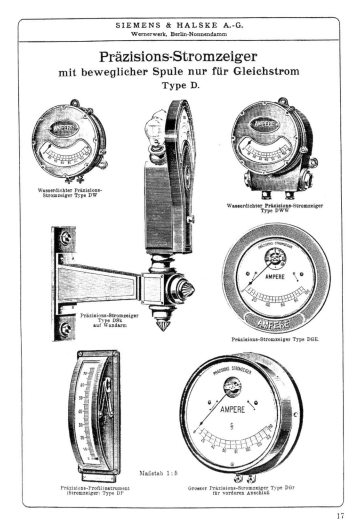

Fig. 20.4. A series of direct-current ammeters proposed by the Siemens & Halske at the beginning of the twentieth century. Solid and direct reading instruments were installed in factories, workshops, and power plant with the boom of electrical industry at the end of nineteenth century. (Siemens & Halske, *Preisliste 51 Messinstrumente und Zubehör für Schalttafeln, Wandarme und Säule*, Berlin, 1908, p. 17.)

20.2.4 Professional Instruments

These instruments are used within the framework of technically or scientifically based activities or professions involving routine measurements and calculations, transmission of signals, quality control and examination, and so on (Anon., 1921; Cooper, 1946). For example, surveying (Fig. 20.5), navigational (Bennett, 1987), meteorological, medical, and surgical instruments all come into this category. The origins

Fig. 20.5. A tacheometer-theodolite manufactured by the Italian firm Salmoiraghi of Milan. A very large number of instruments of this kind were required by engineers, technicians, surveyors, topographers, and road and railroad builders in the nineteenth and twentieth centuries. (A. Salmoiraghi, Istrumenti di astronomia, geometria applicata e ottica, Milano, 1890, p. 38.)

of surveying instruments are rooted in antiquity; others such as mathematical and drawing instruments had commonly been used by practitioners (architects, gunners) since the Renaissance. But in the nineteenth century, the numbers of professional instruments in use boomed due to the growing role of science and technology in old and new professional activities. Medical thermometers and sphygmomanometers, sliding rules and planimeters, electrical test instruments, and so on, are just a few examples.

Military optical instruments, which may be regarded as the professional instruments of warfare, became extremely important in the last decades of the nineteenth century and were of paramount strategic importance in twentieth-century conflicts

(Wright, 1921, pp. 1–8). Until the mid-nineteenth century, gunners used only relatively simple methods for sighting. Cannons were generally fired only at a visible enemy. Distances were simply estimated by observing the fall of projectiles. Gunners pointed their guns by using simple metal sights or mechanical contrivances attached to or placed on the cannon barrel. Improvements in artillery cannons for use in both the navy and the army meant that the firing distance and precision were greatly increased. With these new weapons, whose range could be several miles, it was possible to bombard invisible targets, hidden by obstacles or below the horizon. Therefore, firing became a more complex procedure, and technical problems and instruments for indirect firing control had to be introduced. Distance and precision needed to be increased and, as a result, it was necessary to use theodolites, rangefinders, and optical sighting instruments (Fig. 20.6). These optical weapons, together with telescopes, binoculars, and surveying instruments became essential on the battlefield, and their supply during the First World War required very large quantities of optical glass and the introduction of new mass-production technologies.

FIG. 101.—DIAL SIGHT.

Fig. 20.6. The dial-sight used during the First World War. These types of optical weapons, which were badly needed during the conflict, allowed accurate pointing of a gun in the direction of an invisible target. (*Dictionary of British Scientific Instruments*, London 1921, p. 323.)

During this conflict, other instruments, or technical devices that stemmed directly from scientific instruments, were widely used. These included X-ray and wireless apparatus, telegraphs, telephones, anti-aircraft acoustic locators, sonar detectors, and instruments for aeroplanes and motor cars.

20.2.5 Domestic Instruments

Finally, we should also mention scientific instruments which, since the beginning of the nineteenth century, have become common in household use. Decorative barometers and thermometers, small telescopes, theatre binoculars, stereoscopes, and scientific and optical toys (Fig. 20.7) became very popular, and hundreds of thousands of them were sold in an essentially urban market (see for example, G. Turner, 1987). These were not precision instruments, and often their aesthetic and decorative characteristics were more important than their technical performance.

Of these categories, industrial and professional instruments, such as gas and electricity meters, switchers, large-consumption optical instruments such as photographic cameras and optical weapons, and so on, were the first for which standardized methods of mass production were introduced.

Fig. 20.7. The graphoscope was a typical and very popular domestic optical instrument of the Victorian era. It was used for viewing stereoscopic images and for magnifying photographs, engravings, and so on. (Secretan, *Catalogue, Deuxième fascicule*, Paris, n.d. but c.1875, p. 240.)

20.3 Physics and its Instruments, 1850–1930

It is impossible to separate the history of the production of physics instruments from other kinds of instruments. However, if we consider that most of the instruments used in the different sciences and in different environments (scientific, professional, industrial, and so on) derived from experimental physics, it is useful to retrace some of the fundamental progress made in this field over the period of time that I consider (see for example, Morus, 2005; Nye, 1966).

By the end of the eighteenth century the instrumental arsenal of natural philosophy was very important. Large instrument cabinets could count several hundreds of items of apparatus for studying mechanics, hydrostatics, pneumatics, optics, heat, electrostatics, and for demonstrating their phenomena. Many of these instruments had been conceived and used by scientists and lecturer demonstrators such as W. J. 'sGravesande, J. A. Nollet, W. Whiston, and T. Desaguliers, who translated the highly mathematical and complex physics of Newton into an essentially phenomenological science whose phenomena were illustrated using instruments and experiments (see for example, Gauvin and Pyenson, 2002). In the early decades of the nineteenth century, physics boomed. In 1800 the invention of the electric cell, which allowed a long-lasting electric current to be produced, opened the way to electrochemistry, electrodynamics, and electromagnetism. But if the first practical applications of electricity, such as telegraphy and electroplating, were developed in the first half of the century, the large-scale industrial use of electric lighting, motors, and generators only began around 1880. In the 1860s the Maxwellian theory of electromagnetic fields proposed a general synthesis of optical and electromagnetic phenomena and predicted the existence of waves of oscillating electromagnetic fields. With powerful induction coils (Fig. 20.8) and improved vacuum pumps, the study of electrical discharges in gases proved to be a fruitful field of investigation, and led, for example, to the discovery of X-rays.

On the other hand, systematic research into such optical phenomena as interference, diffraction, and polarization enormously widened the field of optics and its applications, and contributed to a gradual confirmation of the wave theory of light. At the end of the 1850s the spectral analysis of light became one of the most fruitful fields of investigation in physics, chemistry, and astronomy. Later, spectrometry was applied also to infrared and ultraviolet radiations, the otherwise invisible parts of the spectrum.

Acoustics, which for centuries had essentially been studied within the framework of music or mathematics, became a branch of experimental physics, and in the second half of the century employed an array of new instruments for producing, analysing, and synthesising sounds.

The study of heat, and progress in calorimetry and thermometry, also stimulated by the development of steam engines, allowed a better understanding of the physical properties of matter and provided the basis for modern thermodynamics.

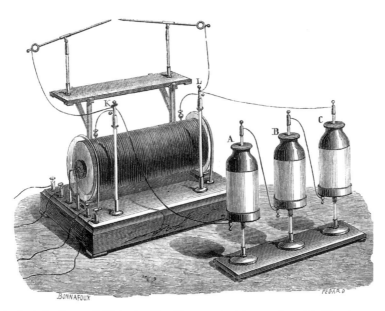

Fig. 20.8. An induction coil (Ruhmkorff coil) connected to a series of Leiden jars and with a spark gap. This kind of transformer was one of the most important instruments in the laboratories of the second half of the nineteenth century. By 1900, powerful induction coils were widely used in wireless and for the production of X-rays. (T. Du Moncel, *Notice sur l'appareil d'induction électrique de Ruhmkorff*, Paris, 1867, p. 48.)

Furthermore, in the same decades, the educational systems in many European countries and in the USA were developing fast. Not only did the number of primary and secondary schools, colleges, and universities increase, with a larger proportion of the population thus having access to a formal education, but physics, and science in general, were taking a more prominent role in the curricula. Because of this, a large number of new scientific instruments were conceived both for the needs of research as well as for general science teaching. By the end of the nineteenth century, specialized instrument-makers had proposed thousands of physical instruments, from the simplest scientific toy to the most sophisticated measuring equipment. In an era when standardization and simplification of production were not the major concerns of the instrument industry, the variety of available apparatus was extraordinary. Manufacturers did not hesitate to illustrate in their catalogues many different instruments for showing exactly the same phenomena or for demonstrating the same effect. Finally, well-equipped laboratories and rich scientific collections (Fig. 20.9) were the pride of scientific and educational institutions (Fox and Guagnini, 1999). On the other hand, progress in electricity, acoustics, and optics generated many devices which, after having been laboratory apparatus, were gradually improved to become industrial instruments or apparatus for domestic use. Phonographs and gramophones originated from early sound-recording devices, instruments for producing and detecting electromagnetic waves gave birth to wireless technology, and optical toys opened the way to moving-image effects and finally to cinematography. However, most of

Fig. 20.9. The collection of physics instruments of the Istituto Tecnico G. Galilei in Florence (around 1900). At the time, large physics cabinets used for didactic purposes could contain several hundred instruments and apparatus. (© Fondazione Scienza e Tecnica Firenze.)

these technologies, which had been introduced in the late nineteenth century, only developed their full technological potential and made their mark on society in the twentieth century.

About 1900, physics, which appeared to be a solid and almost complete science in need of only a few refinements, began to crack. The discoveries of radioactivity, X-rays, the Zeeman effect, superconductivity, and the electron, as well as the birth of quantum theory and of the theory of special relativity, opened wider and often unexpected horizons. Furthermore, several inventions and technological advances provided physicists with new and powerful tools (Brenni, 1997; 2001). Better vacuum pumps, more powerful high-voltage generators, stronger electromagnets, sophisticated cryogenic systems, more penetrating microscopes, and more efficient calculating machines allowed physicists to extend their fields of research. At the beginning of the twentieth century the invention of the first electronic tubes (diode and triode) marked the birth of electronics (Fig. 20.10). With triodes it was possible to amplify direct and alternating current signals, to generate electrical oscillations and to detect electromagnetic waves (see for example, Dummer, 1983). In the first decades of the twentieth century, also due to technological advances during the First World War, electronic instruments began to supersede the old electromechanical types,

Fig. 20.10. Cathode-ray oscillograph proposed by the British firm W. & J. George Ltd in the 1930s. With the progresses of wireless technology and the improvements of thermoionic tubes a new series of electronic instruments entered the laboratories. (W. & J. George, *Physical Laboratory Apparatus and Equipment*, London & Birmingham, 1939, p. 423.)

opening the way for broadcast technology. Over the same period the new physics of atoms, nuclei, and particles was developing fast. Around 1930 so-called 'big science', involving large research teams of physicists, engineers, and technicians, big instrumental systems, and massive financing and political decisions, was just at its beginning.

20.4 Production: Workers, Workshops, Tools, and Machines

In order to better understand the evolution and transformations of the instrument industry in the nineteenth and early twentieth centuries it is necessary to illustrate briefly how workshops were organized, as well as some of the techniques and materials used in the production of instruments.

If we take the example of France about 1860 (Anon., 1864, pp. 755–760; Brenni, 2006), instrument-makers generally started their career by entering into a workshop as an apprentice at an early age (often as young as 12 or 13). The apprenticeship, starting with very humble tasks, could last between three and six years. (Until the end of the nineteenth century the number of schools providing specialized training for precision mechanics was extremely limited and accessible to only a few people.) By 1860 almost all apprentices could read and write, but apart from that their education and knowledge were limited and their potential skills were often unexploited. Frequently, apprentices did not have any kind of contract, but were just employed on the basis of an oral agreement. In Paris at the same time, about half of the approximately 500 apprentices did not receive a salary and were merely housed and fed by their patrons. After the apprenticeship, workers' salaries varied between 2 and 15 francs a day (with an average salary of around 5.5 francs a day). Women, who represented less than 3% of the workforce, earned an average salary of only 3 francs a day. But agreements concerning salaries and treatment varied considerably. Generally, the working day lasted 12 hours with an hour for lunch and one for dinner. Both breaks occurred early in the day, at 11 am and at 4 pm. Clever, intelligent, and experienced workers could become foremen and eventually open their own workshops. Sometimes they remained with their boss, establishing a kind of partnership with them. They might in due course acquire the firm when the latter retired. Marriages between the daughter of the boss and one of his employees were not infrequent, and that often contributed towards maintaining ownership of the firm in the family for two, three, or even more generations. Until the 1880s, with few exceptions, even the leading instrument-makers had rarely followed any kind of academic course. Apart from a basic school education (evening schools), they were often self-taught.

By nineteenth-century standards, instrument making was not a particularly dangerous or uncomfortable job (Delesalle, 1899). Fatal accidents were practically unknown, but among the more common accidents were injuries caused by machine tools (loss of one or more fingers), projection of metal splinters, or burns caused by acids used for cleaning metals, etching glass, or in electric cells. Illnesses could also result from the use of chemicals emitting vapours in small and unsophisticated home workshops.

For most of the nineteenth century, instrument-makers were equipped with rather few machine tools. Today it is surprising to see how many elegant and beautifully made instruments were produced with a selection of relatively simple tools and a few machines. A large part of the work was done by hand, using conventional hand tools such as files and rasps, hand saws, hammers, scrapers, spanners, planes, drills, vices, and so on (see for example, Bouasse, 1917; Hofmann, 1896; Oudinet, 1882). Workshop measurements were made using rulers, callipers, gauges, and compasses of various types. Threads were cut with stocks and dies, master taps, and screw plates. The most important machine tools were lathes of many different kinds (Fig. 20.11) and sizes, accompanied by an impressive series of wood and metal tools and chisels for shaping, cutting, knurling, and making threads and ornaments. Special devices could be applied to lathes for producing conical or spherical elements or decorative patterns, or for dividing circles and cutting gears. An able turner could perform marvels with such contrivances. Probably the most sophisticated machine tools used by instrument-makers were the dividing engines, which were used for tracing the subdivisions on straight lines (linear dividing engine) or on circles and sectors (circular dividing engine). The first efficient and reliable machine of this kind was constructed by Jesse Ramsden in 1775. It was a tool that has since been the object of several technical improvements. (Chapman, 1990, pp. 123–137; McConnell, 2007, pp. 39–51) (Fig. 20.12). Many other important makers too spent a lot of time and energy in

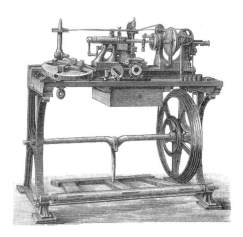

Fig. 20.11. Two typical foot-driven lathes commonly used in workshops at the end of the nineteenth century. (M. Hofmann, *Handbuch der praktischen Werkstatt-Mechanik*, Wien, 1896, pp. 30–31, combined images.)

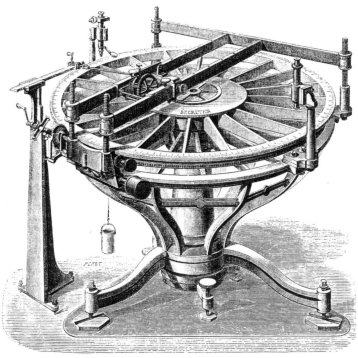

Fig. 20.12. The circular dividing engine used in the workshop of the Parisian firm Secretan in 1874. The quality of the divided scales of many astronomical, navigational, surveying, and optical instruments depended on the performances of such machines. (Secretan, *Catalogue et prix des instruments de sciences*, Paris, 1874, p. 70.)

constructing their own dividing engines, whose performance largely determined the quality of their optical, astronomical, surveying, and nautical instruments. Linear dividing engines were often used for dividing rules or thermometric scales. Such machines allowed non-skilled workers to perform delicate and time-consuming operations which previously could have been successfully carried out only by the most experienced workers. Certainly, instrument-makers often had to design and construct their own tools for manufacturing particular instruments. For example, the Parisian makers Richard invented a series of machines for making and evacuating the elastic metallic Bourdon's tube, which was the core element of manometers and many other instruments (Brenni, 1996).

Prime movers for operating machine tools were introduced in the larger workshops from about the mid-nineteenth century. Small steam engines or, later, gas engines powered lathes, drilling and milling machines, and lens-grinding and polishing machines. In 1860, thirteen steam engines delivering a total of 63 hp were installed in Paris in an instrument-makers' workshops; but their numbers rapidly grew (Anon., 1864, p. 760). Electrification of the most important instrument-making factories with large motors driving the machine tools via a system of axles, pulleys, and belts started only at the end of the nineteenth century in the largest workshops. But there was little incentive to introduce complex machinery, and even at the beginning of the twentieth century, medium–small lathes were still commonly operated by human force (foot lathe). However, for many reasons, mechanization and the introduction of sophisticated machine tools in instrument-making workshops was quite slow. Generally, instrument-making was seen as a form of highly skilled handicraft, and instrument-makers, certainly in France but also in England, found that the use of machines tools was more appropriate for industrial manufacturing than for small precision industry. French observers often noted that foreign instruments, even if good technically, lacked the elegance and artistic taste that characterized Parisian workmanship. A report on the 1867 Paris Universal exhibition stated: 'Indépendamment du point de vue pratique et utile, l'exposition de la classe 12 doit aussi être considérée du point de vue de l'art . . .' (Lissajous, 1868, p. 419). Instruments as works of art were necessarily single hand-made pieces. Secondly, it has to be pointed out that many physics and optical instruments were produced in only limited numbers, which meant that mechanization of the production process was out of the question for both technical and economic reasons. However, the needs of advancing standardization, and the production of instruments to be used in large numbers by the electrical industry (ammeters, voltmeters, electric and gas meters, and so on) or for some professional activity stimulated rationalization and mechanization after the end of the nineteenth century, while the mass production of military instruments and their optical components during the First World War forced makers to systematically adopt machine-tools.

Generally, until the end of the nineteenth century, workshops (Fig. 20.13) were situated together with the shops in the heart of the city, near the centres of learning and research such as schools, universities, and observatories. For Parisian instrument-makers this situation remained unchanged at least until the end of 1900, and only larger factories for the production of high-consumption optics were located outside

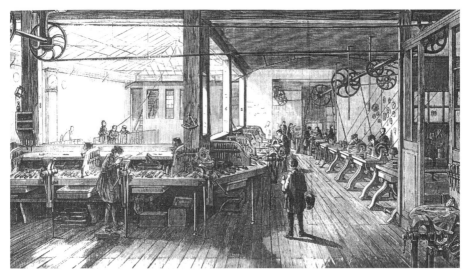

Fig. 20.13. The workshop of the fitters of the instrument-making firm Molteni in Paris. The lathes on the right were activated by a steam engine. (*L'Illustration Journal Universel*, décembre 1854, p. 381.)

the cities (Gourden, 1992). But the need for large premises, the transformation of workshops into factories, the rising price of urban land, and the tendency to move industries to the outskirts, finally pushed most firms to relocate their production to outside cities. In the most important companies, the old workshops were replaced by factories with several departments, dedicated to wood- or metal-working, to lacquering, testing, and packing, and with storerooms for materials, draft rooms for technicians, and offices for the management (Fig. 20.14). However, shops and showrooms generally remained in the urban centre, where potential clients could reach and visit them easily. Many instrument-makers proudly illustrated their premises by inserting engravings and photographs in their trade catalogues (Brenni, 2002c).

20.5 Production: Materials

Instrument-making required a great variety of materials. Wood of various types (mahogany, walnut, boxwood, rosewood, and so on) was employed largely for bases, uprights, stands, boxes, and cases. However, its use declined towards the end of the nineteenth century and almost disappeared in the early decades of the twentieth century, when it was replaced by various metals. Instrument-makers used several different metals, but the commonest ones were copper and especially its alloys brass and bronze. Brass was easy to work, could be well polished, and was ideal for manufacturing a very large number of mechanical organs and elements. Until the

Fig. 20.14. The mechanical workshop of the firm Max Kohl of Chemnitz (c.1920). At the beginning of the twentieth century this firm produced and sold every kind of physics instruments for schools and universities. The factories included numerous workshops as well as testing rooms. (Mak Kohl, *Preisliste N. 100, Band III Physikalische Apparate*, Chemnitz, n.d. but c.1928, p. XIX.)

beginning of the twentieth century brass instruments were generally lacquered. The lacquers (generally yellow-gold solutions of shellac or amber and other resins coloured with saffron, ground turmeric, and dragon's blood) not only protected brass from oxidation, but also conferred an attractive warm goldish colour to the metal.

Iron and steel were ideal for those organs that had to be particularly heavy, strong, or resistant, and, of course, were essential for their magnetic properties in many pieces of electrical apparatus. Precious metals, especially silver, also found various applications in scientific instruments. Silver, which is characterized by its bright shine, was widely employed for divided scales. A fine strip of silver was inserted into brass and then subdivided using a machine. Small and large silvered mirrors for optical and astronomical use were introduced following the discovery of efficient chemical processes for depositing a very thin film of silver onto glass. In the second half of the nineteenth century, silvered glass mirrors, for example, replaced the much heavier and less reflective speculum metal mirrors. In about 1930 silvering was substituted by aluminizing. Aluminium, which tarnishes much more slowly than silver, is vapourized in a vacuum and deposited onto the glass. Gold was sometimes used for gilding precision weights or small parts of electrical instruments in order to avoid oxidation. In the second half of the nineteenth century, platinum and its iridium alloy were adopted for primary and secondary standards of weight and length. The nickel–steel alloy called Invar was invented in 1896. It is characterized by an extremely low coefficient of thermal expansion, and therefore in the twentieth century it found many applications in scientific instruments such as pendulums for precision clocks or for gravity determination or standards of length. For electrical instruments, copper was the ideal metal for wires and conductors, and its alloys with nickel (for example 'Constantan', introduced in the late nineteenth century) were used widely for electrical resistance because their resistivity is constant over a wide range of temperatures. Aluminium was sometimes employed for surveying instruments after the end of the nineteenth century, but its use in scientific and technical apparatus increased significantly only in the last third of the twentieth century, with the introduction of better working and welding techniques for this metal. Mercury, the liquid metal, has played a paramount role in many instruments. Mainly used for thermometers and barometers since the seventeenth century, it found several new applications in the nineteenth century. It was used as a liquid conductor in the contacts of many electrical instruments and pieces of teaching apparatus, and because of its low vapour pressure it has been used widely since the late 1850s in vacuum pumps of various types (Sprengel, Geissler, Gaede, and so on), as well as in manometers. The invention of the low vapour pressure oil 'Apiezon' in the late 1920s led to mercury no longer being used in most vacuum equipment. In the second half of the twentieth century, stricter safety and health regulations tended to ban the use of mercury because of its poisonous vapours and salts.

Glass played a fundamental role in both chemical and physical instruments, as well as in the production of lenses, prisms, and plates used in optical instruments. For a long time, glass was the commonest dielectric and insulating material for Leiden jars, condensers, electrical machines, and apparatus. Around 1850, gutta-percha (extracted from a tropical tree) came into common use for insulating telegraphic cables, and at about the same time, ebonite (or hard rubber) was produced by vulcanizing rubber with a large quantity of sulphur. This became a very common dielectric material in electrical apparatus (plates of electrical machines, insulation of induction coils, and so on), and it was widely used until the first decades of the twentieth century, when it was almost completely abandoned in favour of other early plastics, such as Bakelite.

Up to the twentieth century, glass-blowing had remained an art that called for the skill of highly trained workers (see for example, Threlfall, 1898, pp. 1–89; Ebert, 1897). Physics not only required glass and bars, pipes, serpentines, bells, and vessels of various kinds as the elements of different pieces of equipment, but also barometric and thermometric tubes, hydrometers, balloons for weighing gases, and so on. The introduction of an increasing number of complex items of chemical glassware such as refrigerators and condensers challenged the ability of the best glass-blowers. Among the most intricate glass physics apparatus made in the second half of the nineteenth century were Geissler's tubes (Fig. 20.15), for studying and showing the effects of electrical discharges in low pressure gases. Together with Plücker's, Crookes' and later X-ray tubes, they were manufactured one by one by experienced glassblowers, mostly based in Germany. Automatic or semi-automatic machines for the mass production of electric bulbs and electronic tubes were introduced after the beginning of the

Fig. 20.15. The glass-blowing workshop of A. C. Cossor in London in 1896. The firm produced Crookes and X-ray tubes before becoming a leading manufacturer of thermionic tubes. (*Windsor Magazine*, 1896, p. 375)

twentieth century, but the making of special laboratory vacuum tubes and chemical apparatus remained a handcrafted activity. At the end of the nineteenth century, the German firm Schott produced the highly thermally resistant borosilicate glass which was commercialized under the name Duran. Later known by the brand name Pyrex (introduced by the Corning company around 1915), this glass proved to be ideal in many chemical and physical instruments, and became used universally.

Since the seventeenth century, glass has played a fundamental role in the construction of optical instruments. In the nineteenth century, lenses (Fig. 20.16), prisms, and other optical glass elements were produced one by one using techniques which had changed little since the previous centuries (Threlfall, 1898, pp. 108–181; Henrivaux, 1903, pp. 132–162). Glass blanks were ground and polished by hand, with the aid of metal tools and abrasive powders (such as emery or iron oxide). For example, to produce a convex lens the round glass blank was worked using cast iron and brass concave cups of a determined radius of curvature. Roughing, shaping, and finally polishing a good lens was a tedious and delicate task, the difficulty of which grew exponentially with the radius of the blank. Machine tools operated by small steam engines, which could work on several lenses at the same time, were used routinely after the 1850s. However, these lenses were considered to be of inferior quality compared with the hand-made variety, and were generally used for mass-produced domestic instruments such as spectacles and theatre binoculars, or cheap telescopes. Only from the beginning of the twentieth century were lenses and prisms systematically machine-made (Dévé, 1949; Twyman, 1942). Large lenses and mirrors used for astronomical reflectors and refractors were shaped and polished by specially built machines. Nevertheless, the final retouches and the minute corrections

Fig. 20.16. Lens-polishing in the workshop of the French instrument-maker Chevalier (around 1860). Until the late nineteenth century the production of high-quality lenses was done by hand on an 'optician lathe', using a technique similar to the one introduced in the seventeenth century. (A. Chevalier, *L'étudiant micrographe*, Paris 1865, p. 30.)

of curvature, which were necessary after having performed optical tests, were made by hand.

Until the beginning of the nineteenth century, the production of optical glass remained a very empirical technology based on trial and error. It was often a matter of chance to find in a crucible an homogeneous portion of glass, free from bubbles and striae and suitable for optical instruments. Some major progress was made in the first decades of the century. The work of the Swiss glass-maker P. Guinand was further developed by the German optician Joseph Fraunhofer, in Bavaria. Fraunhofer, who improved the processes of fusion and systematically studied the optical properties of glass, was able to improve its production (Jackson, 2000; Jahn *et al.*, 2008; Rohr, 1929). Nevertheless, for the whole nineteenth century, only a handful of manufacturers in Germany, England, and France could produce good optical glass. By 1900 the firms of Schott in Jena and Parra-Mantois (Fig. 20.17) in Paris were able to

Fig. 20.17. Edouard Mantois examining a very large blank of optical glass. At the time, the company of Mantois (later Parra–Mantois) was the most important French supplier of optical glass, and also produced some of the blanks used for the largest telescopic objectives in the world. (© The Randy and Yulia Liebermann Lunar and Planetary Exploration Collection, George Willis Ritchey Collection.)

systematically produce dozens of different types of glass with optical characteristics precisely determined. In any case, the production of glass blanks for large mirrors and lenses used in astronomical instruments remained a lengthy and very delicate affair—one that was also not free from incidents. From the end of the nineteenth century onwards, the growth in photography for domestic, scientific, professional, and military use (and later in cinematography) greatly increased the need for optical glass. But its strategic importance became especially evident during the First World War, when a huge number of optical weapons were needed. England, France, and the USA made enormous efforts towards augmenting its production and improving its quality.

Physical and chemical instruments also required special optical elements such as polarizing and analysing prisms, wave plates, compensators, and so on, made from Iceland spar, quartz, or other crystalline substances, which could be appropriately cut and polished by only a very few specialized firms. These elements were essential in polarimeters, saccharimeters, and crystallographic and petrographic microscopes, which, in the second half of the nineteenth century found more and more applications in scientific laboratories, as well as in the chemical and food industries.

Finally, fused silica and fused quartz have found special applications (crucibles, very thin, resistant wires, and so on) in a few instruments since the second half of the nineteenth century. But in the twentieth century, the use of these materials was extended, for example, in the construction of telescope mirrors and special lenses.

20.6 Advertising, Trading, and Selling Instruments

The mode of production within the scientific instruments industry was quite complicated, and it is difficult now to reconstruct the intricate network involving makers, subcontractors, and retailers (Brenni, 2002b). If we examine the main trade directories of the time, we can see that in the years 1850–1900 Paris possessed several hundred makers. But these figures do not mean much. In fact, they certainly included the most important makers, as well as the smallest retailers, and the craftsmen who simply repaired instruments. The statistics were not sufficiently precise to distinguish between them. Over this period the list containing the most important makers of international renown, who conceived and produced original instruments of high quality, was certainly limited to about fifty names. Furthermore, a signature on an instrument did not necessarily indicate the name of the real maker. Often it would just indicate the name of the retailer. In fact, the latter could ask the maker to leave a blank label on the instrument, so that the retailer could then engrave his name on it, before selling the instrument.

For example, it can often happen that sophisticated physical or electrical instruments were signed and sold by obscure or minor makers or retailers. However, by careful investigation (with reference to historical documents), it can be shown that these artefacts had been originally manufactured by a famous English, French or German firm. These practices, which were not uncommon at least until the First World War, can today be confusing. In other cases, for example with mass-produced instruments for domestic use, such as stereoscopes, the maker stamped his name on the artefact, which in reality was supplied by one or more subcontractors, whose names we will never know. On the other hand, celebrated and highly skilled makers could be directly involved in supervising and participating in the construction of their most important instruments. In Paris, division of labour had been common practice since the first decades of the nineteenth century. Specialized subcontracted craftsman, sometimes working at home and not in the workshops, made and lined boxes, painted dials, lacquered brass, polished wood, and so on.

Towards the end of the nineteenth century the figure of the famous maker who was at the same time a self-taught scientist, skilled technician, craftsman, salesman, and manager who directed a group of workers and apprentices and was characteristic of the preceding decades, was becoming an old-fashioned concept and was disappearing. Because of the complexity of the market, the multiplication of products, and the increased importance and variety of activities, larger firms had to hire engineers, draftsmen, administrators, and salesman, and thereby acquired a more complex structure and hierarchy. Many firms that for decades had been in the hands of the same family had to reorganize, or were absorbed by larger companies.

The second half of the nineteenth century saw the birth and rapid development of modern advertising and selling (Brenni, 2002c). The expansion of commercial markets, the growing number of industrial competitors, the new needs of a fast-evolving society and the rapid expansion of urban centres required new methods of attracting and conquering potential customers. This evolution was particularly visible in the fields of domestic and everyday commodities such as food, furniture, clothes, pharmaceuticals, and so on. Nevertheless, in the very specialized field of instrument-manufacturing, new marketing strategies were also being introduced. In the second half of the nineteenth century, trade catalogues and advertising, which had been widely used since the eighteenth century and before, became more and more sophisticated. Until around 1850, most instrument trade catalogues were simple lists, generally not illustrated, giving the names of available artefacts and their prices. But from 1850, illustrated catalogues become more common. Detailed wood engravings, which were often copied from those published in scientific treatises and journals, represented in a clear and hyper-realistic way, the artefacts on offer. Instruments were not only listed but also described, and in several cases, a catalogue could serve as the instruction manual for their use (Fig. 20.18). If we analyse trade catalogues from the second half of the nineteenth century, it seems that many physical instrument-makers were producing an enormous variety of instruments. In reality, most of these firms were constructing only some of the instruments that were illustrated. Many other instruments, which were provided by other makers, were just being retailed, and certain instruments were only made to order. The confusion was also augmented by the fact that often exactly the same engravings were used to illustrate the catalogues of different makers, textbooks, scientific journals, and advertisements. Inserting the name of the firm into the engravings was not necessarily sufficient, since the printing blocks were often corrected in order to modify the name or to delete it (Brenni, 1988). These catalogues also contained obsolete and essentially abandoned instruments, which were merely of historical interest rather than of any didactic or scientific value. But these contributed to an increased size of catalogue, and thereby the perceived importance of the firm. Finally, most of the trade literature from the second half of the nineteenth century showed not only what a maker was actually producing in his workshop, but also what he could produce or could supply.

The most important European and American makers published larger catalogues which were regularly updated and distributed free. By the beginning of the twentieth century, German manufacturers such as Max Kohl or Leybold's Nachfolger produced massive volumes containing several hundred pages, including thousands of illustrations, and these were often translated (into French, English, or Spanish) or were multilingual (see for example, Kohl, 1911; Leybold, 1913; and also Zoller, 2009). These catalogues advertised not only individual instruments, but also specially selected sets

Fig. 20.18. A page of the catalogue of the French instrument-maker Salleron which illustrates and describes the use and the technical characteristics of a laboratory pneumatic pump. (J. Salleron, *Notice sur les instruments de précision*, Paris, 1864, p. 97.)

of apparatus and technical furniture for every type of educational institution and laboratory. Furthermore, they reported extracts from dozens of letters written by teachers, professors, and scientists, praising the quality of acquired instruments. They also often included pictures of the company's workshops and buildings (Fig. 20.19). These testimonials contributed to demonstrating the importance of a firm and helped to consolidate its reputation.

On the other hand, from the second half of the nineteenth century, instrument-makers' advertisements increasingly appeared on the first and last pages of scientific journals, scientific books, and treatises, where they would more likely be seen by potential users of instruments. But the illustrations in scientific treatises themselves,

Fig. 20.19. The front page of a catalogue of the German instrument manufacturer Max Kohl. The images of the factory and of the medals and awards of the most important exhibitions showed the importance and the renown of the firm. (Mak Kohl, *Preisliste N. 150, Band II und III Physikalische Apparate*, Chemnitz, n.d., but 1911, p. V.)

especially if they had a large circulation, such as Ganot (1876), Muller–Pouillet (1886–1898) or Weinhold (1881), also acted as subtle vehicles for promoting instruments made by the best French or German makers.

Advertising also exploited other strategies. The universal, international, and national exhibitions as well as specialized industrial fairs, of which there were many in the second half of the nineteenth century and continuing into the twentieth century, were ideal places for instrument-makers to present their products to a wide public. The most important makers rented large stands displaying hundreds of items, often displayed in elegant cases and cupboards. The stands were decorated with carpets and curtains in the typical late nineteenth century eclectic fashion (Fig. 20.20). In these stands, visitors could discuss with the firm's representatives, examine the

Fig. 20.20. The stand of the instrument-maker Carpentier (successor of Ruhmkorff) at the 'Exposition internationale d'électricité' held in Paris in 1881. International and universal exhibitions were ideal events for presenting to a large public the newest and most sophisticated instruments. (*La Lumière électrique*, vol. 7, 1882, p. 155.)

instruments, and sometimes see them in operation. Scientists and engineers often profited from the big exhibitions (in London, Paris, Chicago, and so on) by 'shopping for instruments'. Between the end of the nineteenth and the beginning of the twentieth centuries, German makers displayed the cream of their production by organizing 'collective exhibitions' within the framework of the world fairs. These exhibitions and their illustrated catalogues (Anon., 1893; Anon., 1900; Anon., 1904) clearly showed the vigour of the German precision industry. Finally, prizes, medals, diplomas, and decorations, which were distributed by the juries of the exhibitions, were proudly displayed and were reproduced in catalogues and on trade cards and letterheads as testimonials of the success of the firm. Special periodic exhibitions, such as those organized annually in Paris by the French Société de Physique, allowed makers to present and often demonstrate their newest apparatus to an interested and highly specialized public.

In the second half of the century, in order to better receive their clients and show off their products, important instrument-makers improved and enlarged their premises. With the transformation of urban workshops in factories to outside the city centres, the space for production, the offices, and the salerooms became distinct entities. Many instrumen- makers organized special permanent showrooms, often located in strategic areas, where all (or at least a very consistent portion) of their products were presented for visitors. The characteristics of exhibitions and fairs, the layout of illustrated and descriptive catalogues, and the organization of sales space, changed constantly during the twentieth century, and continues today with the most up-to-date media and other means of presenting and advertising scientific instruments.

20.7 THE SUCCESS OF FRENCH MAKERS IN THE NINETEENTH CENTURY

On the 1 May 1851 the Great Exhibition of the Works of Industry of all Nations, or The Great Exhibition, opened in London and inaugurated the era of universal and international exhibitions, which proliferated during the second half of the nineteenth century and early twentieth century. These impressive events served as spectacular showcases for the western world and its colonies, illustrating with many thousands of artefacts, carefully divided into classes and groups, the technological progress, industrial production, and artistic achievements of a triumphant bourgeoisie. For the first time ever, a huge selection of manufactured products and goods coming from different countries were exhibited side by side and could be judged and compared under the iron and glass roof of Paxton's Crystal Palace (Anon., 1851). Visitors could see locomotives and machine tools, cast-iron railings and guns, fabrics and lace, glass and pottery, eclectic furniture and knick-knacks, natural raw products and chemicals, and also scientific instruments . These were displayed in class X as a shop-window for science and its more recent advances (Brenni, 2004). Not all of the most important European instrument-makers participated in the exhibition. Many of them considered it to be doomed to failure, others were preoccupied with fears of industrial espionage, and certainly several continental and American makers judged that it was too expensive for them to participate. Nevertheless, in spite of some illustrious absentees, the instruments presented at the Great Exhibition gave a good account of the state of the precision industry in the mid-nineteenth century. The report of the jurors, who carefully examined the instruments on display (Glaisher, 1852a, 1852b) in the Crystal Palace, clearly showed that British makers were no longer alone in leading the instrument market as they had done until the end of the eighteenth century. Other countries, especially France, had developed a precision industry capable of seriously challenging British supremacy (Bennett, 1983; Brenni, 2010). In fact, until about 1800, the large astronomical instruments, sophisticated microscopes and optical apparatus, excellent vacuum pumps, and electrical machines made in London were considered to be the finest available and, in spite of their high prices, were in demand for equipping the best European observatories and scientific cabinets. In the second half of the eighteenth century, French instrument-makers could hardly compete with their English counterparts. But at the Great Exhibition it was clear that the situation had changed significantly. In spite of the fact that many excellent makers were absent, Froment's apparatus, Deleuil's precision balances, Nachet's microscopes, Fastré's precision thermometers, Bourdon's barometer and manometers, and especially Duboscq's optical apparatus, greatly impressed the jurors, and were awarded important prizes.

What had happened in about half a century? With the end of the Ancien Régime, many factors, which have been analysed in detail elsewhere (see for example: Daumas, 1953; Turner A., 1989), had contributed within a relatively few years to stimulating the activity of Parisian makers and to triggering a rise in the French instrument industry. The old guilds, whose strict, old-fashioned rules did not favour the multi-faceted activities of instrument-makers were eliminated, new important scientific and educational institutions were founded, the social status of instrument-makers improved, and the decimal metric system was introduced. At the same time, the revolutionary and then the Napoleonic regimes, which were politically and economically isolated from the rest of Europe, required the nation to be technologically supreme. Furthermore, for a few decades from the end of the eighteenth century, French science was extremely successful, making a great contribution to progress in physics, optics, electricity, and chemistry. All of these factors helped and stimulated the production of instruments in France, which in the first half of the century improved consistently, from a qualitative as well as a quantitative point of view (Brenni, 2006; Payen, 1985; 1986).

By 1850, French precision industry, which was essentially concentrated in the capital and therefore a Parisian industry, was extremely successful in the field of physics, optics, surveying, and astronomy, and was supplying the instruments for equipping most of the scientific collections of continental Europe and America (Warner, 1993). The quality, elegance, and ingenuity of French physics instruments were universally appreciated. H. Ruhmkorff or Breguet produced the best electrical laboratory instruments, the physical apparatus and the precision balances of Deleuil and Salleron were highly praised, Soleil and later Duboscq and his successors Pellin were producing the best instruments for physical optics (spectroscopes, polarimeters, saccharimeters, interferometers) available at the time, Lerebours & Secretan made excellent physical instruments, telescopes, microscopes, and surveying instruments, Chevalier's and Nachet's microscopes could challenge the best English models, Golaz specialized in instruments for the study of heat, and Marloye and later Koenig invented and supplied most of the newest apparatus for studying acoustics (Fig. 20.21). These inventive makers—and we have mentioned only a few of them—not only proposed original instruments but were often pioneers of new technologies such as photography, telegraphy, telephony, sound reproduction, and so on. Economic expansion and industrial growth during the Second Empire, the development of a well-ordered educational system, the extensive European diffusion of lavishly illustrated French physics books and treatises, which indirectly advertised French instruments, and the various universal exhibitions which attracted many potential clients to Paris, were some of the factors that certainly contributed to the startling success of the Parisian

Fig. 20.21. The Helmholtz sound synthesiser manufactured by R. Koenig. In the second half of the nineteenth century this German instrument-maker based in Paris conceived and produced most of the instruments used for the study of physical acoustics. (R. Koenig, *Catalogue des Appareils D'Acoustique*, Paris, 1889, p. 25.)

instrument industry. In 1860 (Anon., 1864) there were 487 instrument-makers (compared with only about forty in 1800). Of these, 413 were manufacturing optical and philosophical apparatus, thirty-eight photographic apparatus, and eleven telegraphic apparatus, and twenty-five were cabinet-makers, rule-makers, and so on. But generally businesses remained quite small. Only sixty-four of them had more than ten workers, while 190 were composed of a single person. The total number of workers was 3,108 (2,494 men, 95 women, 508 apprentices, and eleven children). In the late 1870s the most important instrument firms generally employed no more than fifty workers, while only the producers of large-consumption optics such as spectacles, eyeglasses, and theatre binoculars employed several hundred workers, in different workshops and factories.

On average, more than the 30% of their production was exported, but for the most important makers this figure could rise to 60–70%. In the second half of the century, exports, which largely exceeded imports (to a proportion which, depending on the year, could reach 10:1), was directed to England, Germany, Belgium, Spain, the United States, and to a lesser extent to other countries. The number of Parisian makers reached a peak of more than 500 in the early 1880s, and by 1900 their number had declined and was comparable to that of 1860. The golden age of French instrument-makers lasted until about the last decade of the nineteenth century, when the German precision industry began to break into the international market with all its power.

20.8 British Makers around the Mid-Nineteenth Century

In 1850, British instrument-makers, while still retaining a very solid position, were surprised to see at the Crystal Palace the progress being made by foreign makers, and they discovered that in some areas (such as laboratory instruments and optics) they had lost their world leadership (Bennett, 1985; Clifton, 1995; Hackmann, 1985; Morrison-Low, 2007). Certainly, the English microscopes made by Ross, Beck, and Watson, the theodolites and surveying instruments made by Simms (successor of Troughton & Simms: see McConnell, 1992), Cooke's telescopes, the spectroscopes of Browning, the philosophical instruments of Newman and Griffin, the Negretti & Zambra meteorological instruments, and the electrical apparatus of Watkins & Hill, were excellent examples of the best British production, but they were no longer the only ones of their kind available on the market (Bennett, 1983; Brenni, 2010). British makers were no longer showing the ingenuity and the inventiveness which had characterized them in the past. The designs of their instruments were not evolving, and they showed no great originality. Progress was mainly concerned with technical improvements rather than radical transformations. A general conservatism marked their production. There are several explanations for the decline or at least the stagnation of the British precision industry around mid-century. Because of several changes in the organization of science and of the manufacturing industry, collaboration between scientists and makers was certainly not as close as it has been in the late eighteenth century. The social status of the latter had diminished, and their role in the Royal Society had become less important, as had their contributions to *Philosophical Transactions* and to other significant titles of scientific literature. An increasing professionalization and mathematization of exact sciences, which had become more abstract, widened (at least temporarily) the gap between laboratories and workshops. Following the 1851 Great Exhibition, and forgetting the triumphant rhetoric surrounding the event, the most acute observers pointed out that one cause of the British decline lay in deficiencies of the education system, in which science and technology were neglected. On the other hand, the 'master and apprentice' system, which had worked well for the instrument-making industry in the past, seemed to be steadily less adequate as industrialization advanced. But these fundamental problems continued to be discussed up to the beginning of the twentieth century, without finding a definitive solution.

London was certainly the centre of the instrument industry and trade. Yet, despite the prominence of the capital, several makers were established in provincial cities such as Birmingham, Bristol, Liverpool, Manchester, York, and Sheffield, as well as in Glasgow and Dublin. (Clarke *et al.*, 1989; Morrison-Low, 2007). This constituted a remarkable difference compared with the French instrument industry, which was almost exclusively concentrated in Paris until at least the beginning of the twentieth century, certainly as far as the most important makers were concerned. A few figures can help us to draw rough comparisons between France and the United Kingdom. In 1861, England, Scotland, and Ireland together counted about 5,000 people working in the production of philosophical, surgical, and electrical apparatus and instruments. In 1881 this number was 7,600, and 19,500 in 1891. As far as London was concerned, the 1891 statistics recorded that 976 of these were surgical, 2,352 philosophical, and 4,320 electrical instrument-makers (Hackmann, 1985). The sharp increase during this decade was greatly influenced by the expansion of the electrical industry and the consequent need for electrical instrumentation. Production was absorbed by different sectors of the market, the professional and the state ones being very large, and demand was rising in response to the growth of the Empire and the advance of industrialization. The United Kingdom, contrary to most other countries, had a colonial Empire which could absorb a significant number of instruments used for land surveying, civil engineering, navigation, and so on.

20.9 THE RISE OF GERMANY

In a relatively short period in the late nineteenth and early twentieth centuries, Germany developed a powerful scientific instrument industry which seriously challenged, and in several fields overtook, the long-established position of the French and British instrument-makers (Brachner, 1985; 1986; Shinn, 2001; Zaun, 2002). Up to the 1870s, Germany, which was still divided into several states of differing size and importance, played a secondary role in the world of scientific instruments when compared with France and England. Certainly, there were some excellent German makers who had been able to acquire with their instruments an international reputation in the first decade of the century. For example, the Bavarian mechanical and optical workshops that were the setting for the collaboration between the inventor and engineer G. von Reichenbach, the technician and entrepreneur J. von Liebherr, the mechanic J. von Utzschneider, the glass-maker P. Guinand, and the optician and scientist J. von Fraunhofer, were able to manufacture excellent and innovative astronomical, surveying, and optical instruments which were highly appreciated (Brachner and Seeberg, 1976; Brachner, 1986). Fraunhofer learned from Guinand the art of glass-making, and made enormous progress in the production and testing of optical glass (Jackson, 2000; Jahn *et al.*, 2008). The Bavarian workshop inaugurated a tradition of excellence which was successfully continued by the Merz dynasty of instrument-makers. Later, in the 1850s and 1860s, the Berlin mechanic F. Sauerwald produced many good physical instruments for leading scientists such as H. von Helmholtz, Tyndall, and G. Magnus. However, globally the German precision industry on the international arena could certainly not be compared with those of France and Britain. This is also clear from the study of those instruments preserved today in the many surviving European and American scientific collections, where German instruments acquired before the last third of the nineteenth century are quite scarce. Things change radically with the short Franco-Prussian war which, with the debacle at Sedan in 1870, caused the sudden collapse of Napoleon III's Second Empire. The war was immediately followed by the unification of German states under the Prussian crown. In a few decades the second German Reich became one of the leading political, industrial, economic, and military world powers. Not only did traditional industrial activities, such as mining and metallurgy, boom, but Germany also led the second industrial revolution, with the rapid and successful development of new and innovative chemical and electrical industries.

At the beginning of the twentieth century, Germany held an uncontested position as the continental industrial giant. It was the world's third largest producer of coal and it surpassed the United Kingdom in the production of steel. At the same time, German electrical equipment, together with that from America, dominated the world's markets. The newly founded Empire needed science and technology and therefore industrial development was accompanied by a renewal and an expansion of the educational system. New schools, universities, and polytechnic institutions were founded, while the existing ones were reorganized and modernized. New scientific and technological faculties were created, the old-fashioned physical and chemical cabinets became updated laboratories. For example, between 1860 and 1914 the number of students in scientific faculties increased tenfold. Finally, there were several geopolitical, technical, economic, and strategic reasons for the improvement in the German precision industry. If in the 1860s in Prussia (and several other German states) the community of instrument-makers was small and unable to satisfy the demands of laboratories and industry, thirty years later the situation had changed completely. If the empire needed science and technology, then these would require instruments. Didactic instruments were necessary in the schools and universities, professional and technical instruments were essential in the new science-based electrical and chemical industries, and research and measurement instruments were needed to equip laboratories and observatories. New instruments were also becoming essential to the army (geodetic and topographic apparatus, binoculars, telemeters), in the sophisticated communication technologies (electrical apparatus and standards), and also in everyday life (gas meters, cameras, and so on). After much effort, the necessity to have a strong, active, and united community of instrument-makers capable of promoting activities in the field of instrumentation, to improve precision industry, and to expand business opportunities, led in 1879 to the foundation of the Deutsche Gesellschaft für Mechanik und Optik. Major personalities in the society were instrument-makers such as C. Bamberg, R. Fuess, and H. Haensch. Furthermore, in the 1880s two important journals devoted to scientific instruments and their use were founded.

Zeitschrift für Instrumentenkunde was the official organ of the Deutsche Gesellschaft für Mechanik und Optik, and was a journal exclusively devoted to scientific instrumentation. The journal contained descriptions of some didactic equipment, but most of it was devoted to new and highly sophisticated measurement and research apparatus. *Zeitschrift für den physikalischen und chemischen Unterricht* appeared in 1887. In contrast with the previous periodical, this journal dealt only with new teaching and demonstration instruments and with didactic experiments. *Zeitschrift für den physikalischen und chemischen Unterricht*, which appeared every two months, showed clearly how the art of science teaching was advancing and how important it was in Germany (Brenni, 2011a). No similar publications existed anywhere else before 1900. The British *Journal of Scientific Instruments*, for example, was only launched in 1923. Furthermore, as far as physics was concerned, several treatises on physics and electricity, which were extremely successful and circulated widely abroad (to the point of sometimes being translated), also contributed towards advertising and popularizing German instruments. Among them, the books by Weinhold (1881); Müller–Pouillet (1886–1898), and Frick and Lehmann (1904–1909) are especially notable (Fig. 20.22).

In 1887 the Imperial government founded the Physikalisch-Technische Reichsanstalt (PTR), (Cahan, 1989; Moser, 1962). Creation of the PTR was supported and encouraged by W. von Siemens and by H. von Helmoltz, who was its first director. The roles played by one of the most influential of Germany's industrialists and by its leading scientist were emblematic of the growing importance in Germany of the connections between science and industry. The PTR was essentially divided into two sections. The first was dedicated to research and high-precision measurements in the field of physics, for the purpose of solving fundamental problems of theoretical and technical relevance. For this, it was necessary to have highly sophisticated instruments that generally were not available in other institutions. The second section was devoted essentially to technical measurements useful to industry, testing apparatus and instruments, making special instruments, and studying materials. D. Cahan has observed: 'By the turn of the century, the Reichsanstalt stood on the forefront of institutional innovation in science and technology uniting diverse practitioners and representatives of physics, technology, industry, and the state' (Cahan, 1989, p. 1). In contrast, Shinn (2001) stresses the German origins of 'research technology', which was devoted essentially to the invention, construction, and diffusion of sophisticated apparatus for measurement and control. These instruments would find several uses in both research laboratories and industrial practice.

All these factors certainly contributed to the encouragement, support, and development of the German precision industry. Several firms and workshops that had existed since the mid-nineteenth century or earlier developed greatly and, after supplying a quite limited and essentially local market, now profited from the favourable situation; and thanks to the new momentum they expanded considerably and became known internationally. At the same time, a number of new workshops were created, and within a few years these had acquired a solid international reputation. Several firms rapidly conquered portions of the instruments market, which for decades had

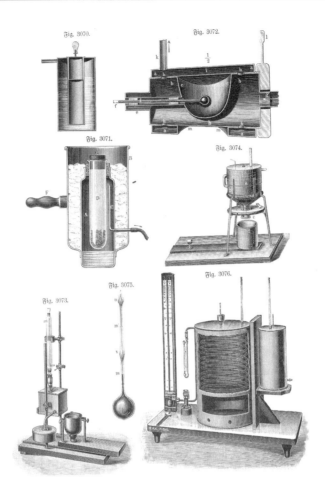

Fig. 20.22. Various calorimetric apparatus illustrated in Frick's *Physikalische Technik*. At the beginning of the twentieth century this very ponderous treatise illustrated and described thousands of physical apparatus used both for teaching and research. (J. Frick, O. Lehmann, *Physikalische Technik*, Braunschweig, 1905, vol. 1, p. 1164.)

been supplied for the most part by French or (to a lesser extent) British makers. Just a few names and examples will give an idea of the spectacular rise of the German precision industry (Zaun, 2002).

An emblematic and oft-repeated success story concerns the collaboration of the instrument-maker Carl Zeiss with the physicist and optician Ernst Abbe (Stolz and Wittig, 1993). Thanks to Abbe's theoretical studies of microscopic imaging, Zeiss microscopes (and other optical instruments) began to be made in accordance with rigorous scientific theory. Furthermore, when the correction of aberrations called for complex lenses made with a new type of optical glass, in 1879 Abbe turned to the chemist and glass technologist O. Schott (Meyer, 1905). With the support of Zeiss and the Prussian government, Schott started a series of glass-melting experiments,

Fig. 20.23. A large crystal refractometer made by Zeiss of Jena at the end of the nineteenth century. In the first decade of the twentieth century the company was one of the leading world manufacturers of instruments for optics, astronomy, surveying, precision mechanics, and so on. (C. Zeiss, *Optical Measuring Instruments*, Jena, 1893, p. 18.)

which were highly successful. After some ten years, Schott was able to systematically produce more than eighty different types of glass, which proved to be essential in achieving the objectives (for example, apochromats) in accordance with Abbe's specifications. The name of Zeiss became synonymous with excellence in microscopes, but the firm's production eventually expanded to embrace a wide range of optical (Fig. 20.23) and astronomical apparatus. Also, the firm of Leitz built its reputation with microscopes before entering the market for projection apparatus and photographic cameras. Fuess made very sophisticated meteorological apparatus, together with optical and mineralogical instruments. Franz Schmidt & Haensch became major producers and exporters of instruments for optical physics, including spectroscopes and spectrometers, polarimeters and saccharimeters, refractometers,

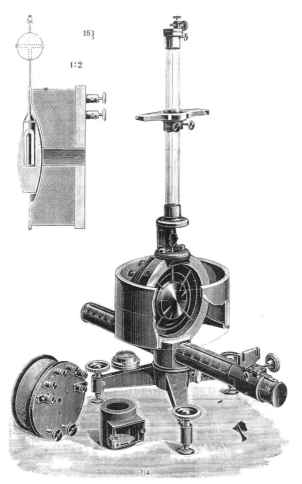

N° 371.
Grand galvanomètre apériodique à miroir.

Fig. 20.24. An aperiodic mirror galvanometer proposed by Hartmann & Braun at the beginning of the twentieth century. The laboratory electrical instruments of this German firm were sold worldwide. (Hartmann & Braun, *Instrument de mesures, Catalogue L*, Frankfurt a/M, n.d. but c.1910, p. 19.)

polarization microscopes, and so on, while Krüss developed photometric and colorimetric apparatus. As far as high-quality didactic and physical instruments were concerned, the firm of Max Kohl, Leybold's Nachfolger and Ernecke became the favoured suppliers for hundreds of schools and universities worldwide. Hartmann & Braun (Fig. 20.24) specialized in electrical measurement instruments for laboratories, while testing and industrial electric apparatus was also manufactured by large industrial companies such as Siemens & Halske and AEG. Older and newer precision firms such as Repsold, Steinheil, Breithaupt, Otto Toepfer, Hans Heele, and Gustav Heyde were leaders in the production of astronomical, surveying, and geodetic instruments (Repsold, 1914). Müller-Unkel, and other firms, following the glass-blowing tradition of H. Geissler, flooded the market with their vacuum and X-ray tubes. Chemical and physico-chemical apparatus were a speciality of the firm Max Woltz, while excellent analytical balances were provided by Sartorius. The list is a long one.

German precision industry was able to profit from several factors, such as an expanding industry and economy, the support of the imperial government, an advanced educational system, good collaboration between scientists and industrialists, and the cohesion of the community of instrument-makers. But a well-organized commercial strategy also made its contribution to the success of the German precision industry. Potential customers also appreciated the production of large, widely distributed, and often multilingual catalogues, massive participation (often as a collective exhibition) in the main exhibitions and fairs, the presence on the stands at such events of well-trained and informed personnel, an important network of European and American representatives, and a stock of immediately available instruments. About the turn of the century, French observers often complained that the marketing of German makers was too aggressive, but they also had to acknowledge that it was more efficient than their own.

By 1891, Germany's precision industry counted 526 instrument-making firms (employing more than ten persons) and a total of over 7614 workers. By 1898, less than a decade later, this had risen to 790 firms with more than 13,600 workers.[2] (Brachner, 1985).

20.10 FROM 1900 TO THE FIRST WORLD WAR

The success of the German instrument industry was very evident at the universal exhibitions in Chicago (1893) and Paris (1900) and at the St Louis World Fair (1904). The *Deutsche Gesellschaft für Mechanik und Optik* organized ambitious collective exhibitions, where the best instrument-makers and the most important institutes (such as the PTR) presented their products (Anon., 1893; Anon., 1900; Anon., 1904). At the 1900 Paris Universal Exhibition, which with over 50 million visitors represented the acme of these events, German industrial products greatly impressed the visitors. In fact, the French writer Paul Morand, a careful observer but certainly a layman as far as science and technology were concerned, wrote in his recollections of the 1900 exhibition:

All Paris rushed to the section of German optics to see the German instruments of precision . . . And the new chemical industries! And the laboratories! . . . Well do I remember the great impression made on me by the huge Helios dynamos of two-thousand horse-power from Cologne, hitched to steam engines and those other generators from Berlin and Magdeburg and the crane that raised twenty-five tons, dominating all the gallery; beside these, the machines of other countries looked like toys. The old gentlemen who were pushed in wheeled chairs through this accumulation of valves, flywheels and regulators, shook their heads, saying, 'This Exhibition is a commercial Sedan'. (Morand, 1931, pp. 78–79.)

The spectre of the battle of Sedan, symbolizing the collapse of France in 1870, constantly haunted France. The fact that France hated Germany, with its growing economic power and political influence and its ever greater lead in the race to modern technology, made the situation especially painful (Williams, 1994, pp. 36–43). It is true that several French makers maintained their reputations and were able to reinforce their position in the international arena. (Anon., 1901–02; Ocagne, 1903). Gautier, for example, remained one of the leading and most respected makers of large astronomical instruments; in 1900 it displayed its most spectacular achievements, including the huge horizontal telescope with a 57-metre focal length and the siderostat with a mirror 2 metres in diameter. Richard was successfully expanding its production of recording apparatus, Ducretet was still a leader in physics apparatus for laboratories and schools, Carpentier and Chauvin & Arnoux were among the best manufacturers of electrical measuring and testing apparatus, and Morin & Genesse and Ponthus & Therrode produced excellent surveying and navigational instruments. But if the French instrument industry was certainly not collapsing, it was certainly losing a significant portion of the world market, and its renown was fading. Its role was not what it had been a few decades earlier, and it was failing to penetrate demanding new areas such as South America, Japan, and Russia, which were flooded with German instruments (Anon., 1899). Furthermore, imports from Germany to France were also growing. From 1890 to 1900 the quantity of imported optical glass increased from about 21 to 60 tons, and that of optical instruments from 22 to almost 90 tons (Schefer, 1917).

It was not until after the closure of the 1900 Paris Universal Exhibition that French instrument-makers followed the example of the Germans by publishing their own collective catalogue (Anon., 1901–02). It is interesting to read the introduction to the catalogue written by the physicist A. Cornu (Anon., 1901–02, pp. V–XII), who did what he could to minimize the successes of the German precision industry and to calm his colleagues. Despite much rhetoric amplifying the past successes of French science and makers, whose skill, intelligence, and 'artistic taste' remained unique, and in spite of an unjustified triumphalism, Cornu pointed out some of the weaknesses of the national precision industry. Among these he mentioned insufficient protection (patent, legislation, marks) of French products in the face of aggressive foreign competition, a lack of cohesion and solidarity between makers, and difficulties in introducing scientific management and methods of production in the workshops. But Cornu's reassuring words could not hide the decline in the French precision industry. This was also apparent from the figures for scientific instrument exports from Germany to other nations (including France), which grew constantly between 1900 and the First World War. But nothing really changed in the first decade of the twentieth century, and the physicist J. Violle, during a conference held in 1915, once again stated the problems previously mentioned by Cornu, though in an even clearer and more direct way. Violle (1915) listed some of the reasons for the German success, while at the same time pointing out the lack of financial resources of French laboratories and focusing his attention on the difficulties with the work-force in the Parisian precision industry: not only very few institutions able to train precision mechanics, who anyway tended to seek employment in the newer industries (automobile, aviation) or in administration, but there was no school at all dedicated to training opticians. Again, in 1916, the chemist and metallurgist H. Le Châtelier (1916) complained about the lack of research and industrial laboratories in France.

Around 1900, on the other side of the English Channel, British commentators, while noting the great progress of Germany, appeared less worried than the French about the situation of their own industry. Great Britain, in spite of the fluctuating continental market, which after having favoured the French then tended towards the German makers, could always rely on a solid domestic client base (both public and private) and on the country's enormous colonial Empire. Furthermore, as far as research and physical instruments were concerned, in the last third of the nineteenth century new firms, together with some older ones which had reorganized themselves, contributed to a revival of the British tradition and produced new and original instruments of modern design that were favourably received by their customers (Williams, 1994, pp. 12–35). Certain names were especially notable. The Cambridge Scientific Instrument Company (CSI) was founded in 1881 by H. Darwin. The story of CSI is an example of the growing interconnections between university research and precision industry (Cattermole and Wolfe, 1987). In fact, because of its sustained involvement with eminent scientists such as Lord Rayleigh, G. Whipple, Vernon Boys, and J. Ewing, to mention just a few, CSI played an important role in the development of instruments for physics and medical science as well as for industrial measuring and control apparatus. In the process, CSI's research on new instruments, for external customers, and the redesign and improvement of old apparatus, became part of the firm's regular activities. Among the instruments concerned were electrometers, oscillographs (Fig. 20.25), pyrometers, extensiometers, thermo-galvanometers, calorimeters, and kymographs.

Another example of fruitful and long-lasting collaboration between a scientist and an instrument-maker was the partnership of W. Thomson (later Lord Kelvin) and the Glasgow optical and philosophical instrument-maker, J. White, who manufactured most of the apparatus that Thomson devised (electrical instruments, magnetic compasses, sounding machines, and so on). Even after White's death in 1884, Thomson's involvement with the firm continued, and in 1900 he became director of the newly founded Kelvin & James White Ltd.

In the 1870s Adam Hilger founded a workshop for the production of optical instruments. The firm was very successful, and in the early decades of the twentieth century Hilger's spectroscopes (Fig. 20.26) and spectrographs were highly appreciated worldwide. In 1881, Baird & Tatlock (originally in Glasgow and then in London) entered the market as a manufacturer of physical, chemical, bacteriological, and other laboratory apparatus. Other firms, such as Casella, and Elliott Brothers, had a longer history and were well established in the 1870s, but in the later nineteenth century they developed

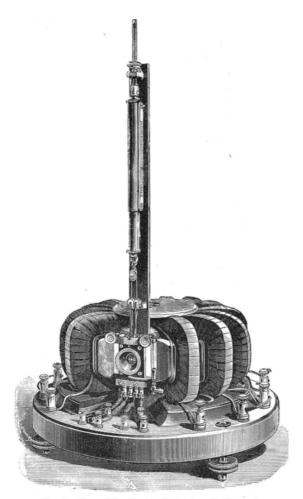

Fig. 5. Type IV. Double Projection Oscillograph.

Fig. 20.25. Duddell's double projection oscillograph made by the Cambridge Scientific Instrument Company. The firm was one of the most innovative in the British instrument industry. (Cambridge Scientific Instrument Company, *Duddell Patent Oscillograph*, Cambridge, 1903, p. 16.)

new instruments and renewed their production. Elliott Brothers, for example, was a leading company in telegraphic equipment and later in electric apparatus for tramways and lighting. About the turn of the century, the firm had some three hundred employees producing high-quality instruments for navigation, surveying, calculating, telegraphy, optics, and mechanical engineering. In addition, an experimental department created in 1895 contributed greatly to improving electrical instruments, and entered into the new market of automobile instruments.

At the beginning of the twentieth century the United Kingdom, while no longer retaining the position of dominance it had held a century before, remained one of

Fig. 20.26. Infrared spectrometer made by Hilger. The spectroscopic apparatus of this British company were used worldwide and equipped a large number of physics laboratories. (A. Hilger, *Catalogue*, London, 1913, p. D14.)

the world's leading producers. From the 1870s, several new companies, and some older ones that had modernized their production methods, contributed to give a new momentum to the British instrument industry. Certainly at the beginning of the twentieth century, the rise of German industry and its successes alarmed British makers. During the Boer War it became apparent that British optical instruments used on the battlefields were inferior to German instruments. From a psychological point of view, Germany's success touched British sensibilities less than it touched the French, which were exacerbated by the still open wound of the debacle of 1870 and the consequent loss of Alsace.

In the United Kingdom the time was ripe for changes and an increasing role for science in governmental policy and decisions (Williams, 1994, pp. 43–60). The existing scientific institutions were judged to be insufficient for the new needs of science and science-based industries. Following the example of the Physikalisch-Technische Reichsanstalt, between 1899 and 1901, the National Physical Laboratory was established, which was intended to be the reference institute for physical standards, for high-precision measurements, and for materials testing. At the same time there began (sporadic) involvement of government with private companies, with inquiries undertaken to improve technical education for precision mechanics and optical worker. But nothing was formalized, partly because of sketchy support from the instrument-makers, whose attitude to governmental initiatives was quite sceptical. Certainly in the years before the First World War, a number of important firms undertook research of various kinds, sometimes in collaboration with government agencies, such as the Board of Trade or the Admiralty, or for other industries or individuals. Finally, in a market of mounting complexity (contacts with government agencies, foreign representatives, joint ventures with other firms, agreements with inventors and other companies, fierce competition, overseas expansion), some of the most important makers began to reorganize their firms by acquiring new expertise. The figure of the accountant became important for several instrument firms which had previously neglected to keep proper accounts as a consequence of ignorance and lack of interest.

20.11 Instrument-Making in the Rest of Europe and in the USA

If during the second half of the nineteenth century and the beginning of the twentieth century, France, the United Kingdom, and Germany were by far the most important producers and exporters of instruments, precision industry was also developing (to a lesser extent) in other European countries and in the USA (see Brenni, 2002a, for a general survey).

20.11.1 Italy, Spain, and Portugal

Until the beginning of the twentieth century the economies of these countries were essentially based on agriculture; their industrialization came late and gradually, and only in a few specific centres. Poor communication, sometimes underdeveloped railway networks, very limited merchant and military navies, and the presence of large landed estates that did not encourage topographical and cartographic surveys, were all factors which seriously hindered the appearance of a strong national instrument industry. However, in the more important cities the presence of prestigious universities, astronomical observatories, and academies, allied to the broadening of the educational system, created a demand for increasing numbers of instruments, which were often supplied from abroad.

In Italy (Brenni, 1985) the scientist and instrument-maker G. B. Amici was the only one capable of achieving international renown over the period from the 1820s to the 1860s. His reflecting and refracting microscopes were highly regarded, and many of them were sold to foreign scientists. But after Italian unification in 1861 a few firms were founded. Notable among these were Filotecnica Salmoiraghi and Tecnomasio Italiano in Milan, and Officine Galileo in Florence. The Salmoiraghi company produced good surveying instruments, and was quite successful in the first half of the twentieth century, when it also exported part of its production to South America. The Officine Galileo manufactured physics and optical instruments (surveying), and after the First World War focused its production on telemeters, periscopes, and other optical weapons, as well as anti-aircraft searchlights. However, despite the efforts of these companies, the large majority of instruments were supplied by foreign companies throughout the remainder of the nineteenth century and much of the twentieth century.

The situation in Spain and Portugal was not very different from that in Italy. In these countries, the presence of a certain number of skilled instrument-makers did not substantially alter their dependence on instruments from abroad, though in general terms their local production did increase and improve over the latter part of the twentieth century. Certain instrument-makers appeared in Spain in the second half of the century as a result of the founding of engineering colleges, such as Recarte in Madrid, Laguna in Zaragoza, and J. Grasselli (see for example Belmar and Sanchez, 2002). After the 1880s one of the most important workshops was that of the Mechanics Department at the Science Faculty in Madrid. This workshop was attached to the faculty's physics laboratory, and it both made and repaired instruments for the university collection. At the beginning of the twentieth century, several companies were established to produce instruments for the electrotechnical sector. A similar situation also existed in Portugal. Various precision instruments were produced by the Industrial Institute, established in Lisbon around 1850. But, as a result of the growing complexity of teaching, experimental, and measuring apparatus in the nineteenth century, the very rich Portuguese collections, such as those in physics and engineering in Lisbon, Coimbra, and Porto, were equipped with instruments made by a succession of British, French, and German makers.

20.11.2 Switzerland

The Swiss precision industry consisted of only a very limited number of firms, but some of these were highly successful on the international market. Excellent surveying and drawing instruments were produced until the last decades of the twentieth century by the firm of Kern in Aarau (founded in 1819) and Wild of Heerbrugg (founded in 1921). In the second half of the nineteenth century the workshop of M. Hipp specialized in telegraphic equipment and electrical apparatus, as well as chronographs and high-precision laboratory clocks. The manufacture of the latter was continued well into the twentieth century by the successors of Hipp. The Société Genevoise d'Instruments de Physique (SIP) of Geneva was founded in about 1860 for the construction of laboratory physical instruments. At the end of the century the SIP became famous for its metrological apparatus (standard of length, comparators) and machines (dividing engine), which were sold successfully all over the world, and after the First World War the company specialized in high-precision industrial machine tools (Brenni, 2011b). Two more firms, one founded around 1860 by J. Amsler in Schaffhausen and the other in 1880 by G. Coradi in Zürich, became famous for

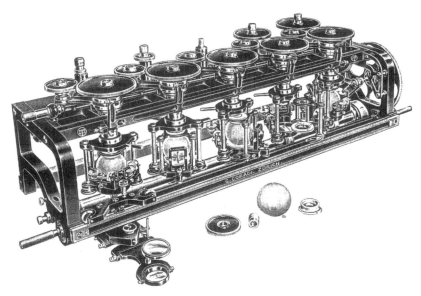

Fig. 20.27. An harmonic analyser made by Coradi of Zürich (c.1920), used for determining the fundamental and harmonic components of a complex waves curve. The firm specialized in the production of very sophisticated mathematical instruments used by physicists and engineers. (J. Marguin, *Histoire des instruments et machines à calculer*, Paris, 1994, p. 162.)

their sophisticated mathematical instruments (planimeters, integraphs, pantographs, harmonic analysers; Fig. 20.27).

20.11.3 Austro-Hungarian Empire

The Austro-Hungarian Empire, while not having a precision industry comparable with any of the great exporting nations and being essentially an importer, possessed a certain number of companies of international renown with large foreign clienteles. In the last decades of the nineteenth century the positive influence of the German precision industry in the Austro-Hungarian Empire was very strong both in terms of style and types of instrument. One of the most important Austrian instrument-makers was an affiliate of the Imperial and Royal Polytechnic of Vienna, which stemmed from Reichenbach's workshops in Bavaria. Over several years, from 1816 onwards, this was organized and managed by G. Reichenbach with the assistance of T. L. Ertel (Brachner, 1985, p. 128). The organization later passed to C. Starke, and in the 1870s Starke's son left the original workshop and founded, with K. Kammerer, Starke & Kammerer. The firm was renowned for its mathematical and surveying instruments. The Voigtländer company, whose origins dated back to the mid-eighteenth century, was famous in the latter half of the nineteenth century for its excellent photographic lenses and optical instruments. Over the last decades of the century the most renowned instrument-makers from the Austro-Hungarian Empire were

Reichert (microscopes), Lenoir & Forster (physics and chemistry apparatus), Nemetz (weights and precision balances), W. J. Rohrbeck's Nachfolger (chemistry and physics equipment)—all of these in Vienna—and Josel and Jan Fric (optical and surveying instruments) of Prague.

20.11.4 Holland

No real industrial revolution took place in Holland, and one cannot speak of industrialization before the 1880s. However, there were a certain number of makers who were capable of producing good instruments, which partially supplied the local and colonial market, though export was very limited (de Clercq, 1985). Generally, the country tended to import from foreign instrument-makers, and Holland's instrument production continued to remain fairly modest in spite of interest from King William I and his government during the 1820s, and later in the 1860s, with the founding of the Society for Industrial Progress. In the second half of the century, however, there were some rather important companies such as Kelman and Son and Van Emden (nautical instruments) in Amsterdam, Logeman in Haarlem (who had gained a reputation for their powerful magnets), Caminada in Rotterdam, and Olland in Utrecht (specializing in meteorological instruments and precision scales). Other companies, such as Kipp & Sons in Delft (teaching and electric apparatus), G. B. Salm in Amsterdam, and J. C. Marius (graduated chemical glass) in Utrecht, while producing a certain number of instruments themselves, essentially devoted their activities to the import of instruments and their distribution to the local and colonial market. By all accounts, it can be estimated that the Dutch companies controlled less than a quarter of their domestic market.

20.11.5 Scandinavia

Sweden and Denmark likewise imported the majority of their instruments during the nineteenth century, though possibly to a lesser degree than the Mediterranean countries. In the second half of the century the most famous instrument-makers included F. G. Jünger, J. Nissen, C. P. Jürgensen, and C. Knudsen, the latter establishing the largest Danish instrument company, from which numerous instruments survive today (Andersen, 1995). C. Weitzmann produced a large variety of didactic instruments, many of them inspired by German models. In addition to these instrument-makers, other companies such as Weilback produced and sold a considerable number of nautical instruments in Copenhagen and other Danish ports.

In the nineteenth century, also in Sweden, there were established a few companies worthy of mention (Pipping, 1977). At the beginning of the 1830s the instrument-maker C. E. Littmann arrived from Munich, where he had worked in the famous Bavarian workshops, and set up his own production in Stockholm. For several years he was the instrument-maker for the Science Academy, and played an important role in fostering Swedish instrument production. Littemann, together with F. G. Berg, specialized in surveying and drawing instruments and, from around 1850, began to export instruments to Russia and the Baltic countries, these being traditional markets for Swedish industry. The second half of the century saw increasing activity due to the work of instrument-makers such as S. Sörensen (who also produced instruments for the Science Academy), G. W. Lyth (mathematical and nautical instruments), and L. M. Ericsson. This latter (like the maker Öller, from whom he learned his trade) specialized in telegraphic instruments, and was later to become one of the most important producers of telephone equipment in the world.

20.11.6 Russia

Unfortunately it is difficult to offer a synthesis of the precision industry in Russia because of the difficulty of obtaining documents, my ignorance of the Russian language, and the paucity of the secondary literature. However, we know of a number of particularly important workshops, which were essentially engaged by state institutions and organizations. In 1811 the Department of General Staff mechanics workshop was founded, producing mathematical, geodesic, optical, and astronomical instruments, and later meteorological instruments and chronometers for both civilian and military use. The optical–mechanical workshop of the naval hydrographic department was created in 1804 to supply instruments to the Russian fleet. By the mid-nineteenth century, these works produced more than 230 types of different physics, surveying, navigational, and hydrographic instruments. A further workshop of special importance was attached to the Pulkovo Observatory, established in 1839. In addition to astronomical and geodesic instruments the Pulkovo workshop developed a close collaboration between scientists and Russian and foreign specialists in producing several of the instruments employed by Mendeleev, and other apparatus used by the commission devoted to geodetic measurement of the 52nd parallel. Other important workshops included those at the St Petersburg Academy of Science and the Toula arsenal.

Other instrument-makers, sometimes originating in Germany and England, also established successful firms in Moscow and St Petersburg. However, in spite of the presence of important companies and institutes (which in some cases exported to eastern European countries), English, French, and German instruments were largely imported into Russia. Only in the 1920s did Soviet Russia fully recognize the crucial role of pure and applied science in the development of the new state, and take steps to support the development of new instruments and to increase their production.

20.11.7 United States of America

The United States, which in the twentieth century played a paramount role in the development of the precision industry, largely met its needs during the nineteenth century by importing instruments. However, rapid and massive industrialization, the conquest of new territories rich in raw materials, and a growing wealth, transformed the nation in the last decades of the century, and by the beginning of the twentieth century the country was in a position to become one of the largest producers and (later) exporters of scientific instruments in the international market.

At the beginning of the nineteenth century, numerous original instruments were produced (nautical compasses, levels, surveying and navigation instruments, solar compasses, and microscopes) together with other apparatus inspired by European instruments but adapted to local requirements. Immigrants (from Germany and elsewhere) played an important role, and while they were initially employed to repair imported instruments, they went on to establish their own companies which shared the traditions of their country of origin with the new knowledge obtained in America. Towards the mid-nineteenth century, many American instrument-makers and dealers such as E. Ritchie and B. Pike and Son presented a wide range of didactic, professional, and experimental instruments in their company catalogues. Some of these instruments were designed and improved by the same makers, while many others were imported from European firms and simply retailed. In fact, the imports from Europe remained very important, and the American market tended to favour French-made physics and optical instruments (Warner, 1993), British navigational instruments, chronometers, and high-quality microscopes, and German-made astronomical and chemical glassware. In the second half of the century, particularly after the Civil War, industrialization in the United States accelerated. Small workshops with a few craftsmen were increasingly replaced by large-scale companies that began to provide mass-produced instead of hand-made instruments. Among these were companies such as Keuffel & Esser (famous for their slide-rules), L. E. Gurley Company (surveying instruments), and James Queen & Co. (physical, optical, chemical, astronomical, and surveying apparatus). Towards the end of the century certain companies began to gain international reputations, such as the Weston Electrical Co. (founded by the English engineer E. Weston at the end of the 1880s), which specialized in electric instruments for industrial measurements, or Bausch & Lomb (established in the 1850s), who produced microscopes and optical instruments (Fig. 20.28).

In the 1890s, highly industrialized Chicago, where new educational and scientific institutions attracted some of America's leading scientists, became one of the most important centres of the precision industry and scientific instrument production in the USA (S. Turner, 2005; 2006). Over just a few years many companies were founded. Some of them, such as the Central Scientific Company (CENCO), the Welch Company, and W. Gaertner (later Gaertner Scientific Corporation), became very important and grew successfully in the twentieth century. So although in around 1900 the United States was still importing a significant number of instruments (especially from Germany), the country's precision industry was growing rapidly. From the beginning of the twentieth century, and especially after the First World War, the United States established itself as one of the leading industrial powers with ever-increasing economical and political influence. In the precision industry, as with other

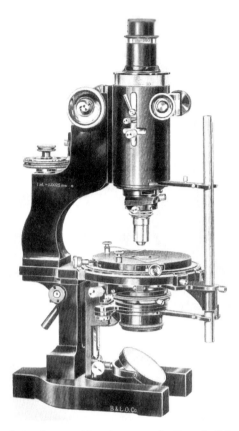

Fig. 20.28. A research photomicrographic microscope by Bausch & Lomb. Founded in 1853, during the first decade of the twentieth century the firm became one of the most important and renowned American producers of optical instruments and optical glass. (Bausch & Lomb, *Microscope and Accessories*, Rocherster, n.d. but c.1930, p. 12.)

industries, American products began to be exported on a massive scale and competed strongly with European products.

This short overview shows how, at the beginning of the twentieth century (outside Germany, France, England, and the USA), very few instrument manufacturers were able to acquire a relevant place in the world market. Some of them partially met local needs, but without reaching a critical mass sufficient for them to have an influence outside their own countries. This does not mean that Canada, Australia, Japan, and some countries in South America did not have important workshops, directed or managed by skilful mechanics and opticians. In Brazil, for example, the company founded by J. M. Dos Reis, in Rio de Janeiro, and which later passed on to J. H. Pazos, built several outstanding astronomical instruments that were displayed successfully at some of the universal expositions. However, their influence was very limited and they did not influence the international market. On the other hand, we have to point out that a few relatively small firms occupying special niches of instrument production (such as the SIP of Geneva with its metrological instruments) managed to win an international reputation and to export an substantial part of their production.

20.12 THE FIRST WORLD WAR

The weaknesses of the British and French instrument industries were illustrated dramatically a few months after the outbreak of the First World War. As M.Williams (Williams, 1994, p. 80) has put it perfectly: 'In some respects it had taken this catastrophe to bring home to individuals the extent to which the industries in Britain and France, firstly depended upon Germany (and her allies) for raw materials and parts and secondly had lost their earlier positions as market leaders to the growing industrial might of Germany.' Certainly the situation was not limited to the precision and optical industries, but the strategic importance of these industries for the army and navy meant that the problems relating to instruments were particularly acute. Britain and British industry were not prepared for a long, global war in Europe, for which a huge number of instruments such as telescopes, binoculars (Fig. 20.29), theodolites, telemeters, and range-finders were required (Williams, 1994, pp. 61–72; MacLoad and MacLoad, 1975). On the other hand, the conflict also required a large number of

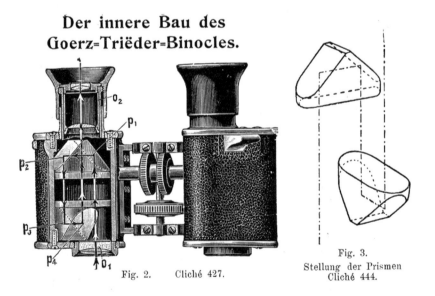

Fig. 20.29. Prismatic binocular illustrated in one of the catalogues of the German firm C. P. Göertz AG. Good binoculars required perfectly-shaped prisms made of high-quality optical glass. Hundreds of thousands of them were manufactured during the First World War. (Goertz C.P., *Triëder Binocles*, Berlin-Fridenau, 1906, p. 9.)

non-optical instruments, such as magnetic compasses, wireless and X-ray equipment, gunnery control calculators, and so on. In the United Kingdom the newly created Ministry of Munitions established the Optical Munitions and Glassware Department (OMGD). In spite of the fact that Chance Brothers, the most important English producer of optical glass, optimistically claimed to be able to supply the military requirements of instrument-makers, its production quickly appeared far too limited and also of low quality. Before the war, in fact, a large proportion of optical glass had been imported from Germany and from France, so that one of the first and main problems was to increase the national production of good optical glass. In 1915 the Ministry of Munitions began to take direct control of an increasing number of firms. The situation was so bad that the civilian population was required to send to the Admiralty and to the War Office any optical instruments that might be useful for the war effort. The instruments were tested, and ultimately more than 33.000 suitable instruments were collected! But the Ministry continued its actions with a view to expanding production and improving manufacture. In 1915, contracts were signed for the production of more than 150,000 instruments. Several firms were subsidized, new factories were built, and existing premises were extended. Nevertheless, shortage of materials, technical difficulties in production, problems of organization, and the need to adapt existing equipment were matters of constant concern. Unsuccessful attempts were also made to acquire instruments from France and the United States, and even from Germany, using an intermediate, neutral Switzerland. Furthermore, the Ministry decided to improve production by introducing new, specially built machine-tools, which were offered on favourable terms to the instrument-makers. This policy contributed, at least temporarily, to a modernization of the technologies used in the precision industry, which was traditionally rooted in a culture of skilled craftsmen and which was quite reluctant to adopt modern machine-tools. For example, due to measures taken by the OMGD, Chance's output of optical glass grew from about 1,000 lbs per month to 14,200 lbs in May 1917, and by the end of the war the firm was able to supply about 4 tons of optical glass per month. Not only did the quantity of glass produced increase enormously, but also the number of different types of glass was greatly augmented. However, if the mechanization and rationalization of factories could reduce manufacturing time (for example, of prisms), there was also some resistance from instrument-makers, who hardly considered themselves industrial manufacturers. However, thanks to the suggestions of the OMGD, it was possible to improve numerous processes such as the production of reticules on lenses, which was a particular speciality of German makers.

The problems generated by the war also raised the question of industrial research (Williams, 1994, pp. 81–88). In 1915 the Council of Scientific and Industrial Research (later transformed into the Department, DSIR) was created, working in collaboration with the National Physical Laboratory. As far as scientific instruments were concerned, the DSIR had to encourage companies and individuals to start specific programmes of research in strategic fields, such as aviation instrumentation, firing and sighting equipment or special photographic cameras. The DSIR also promoted the association of different firms for establishing common research centres. In 1916 the

British Optical Instrument Makers Association was founded to organize and promote the activities of its members, protect their trade interests, maintain collaboration with the Ministry of Munitions and with other industries, and to improve technical education and apprenticeship. All these activities were performed quite efficiently if we consider that, despite an endless series of difficulties, the instrument industry was thoroughly transformed and modernized, and three years after the beginning of the war its production was more than ten times greater than it had been before the war. Finally, during the war the government also refocused its attention on problems related to the education and training of the staff needed for the precision industry. The long debate about the inadequacy of the British educational system, especially in technical and scientific fields, re-emerged. (Williams, 1994, pp. 89–97). Various analyses and proposals (with much rhetoric) were presented and quite a lot of ideas were produced, which were then discussed by the Board of Education and by the London County Council.

Imperial College and the Northampton Polytechnic Institute of Clerkenwell were two of the best candidates for hosting teaching courses at different levels related to optical design, lens computing, and so on. (In the latter, a technical course for instrument-makers had existed since the mid-1890s.) But different opinions and approaches, bureaucracy, and difficulties in the practical organization of the teaching slowed down the realization of such a project. Finally, courses on technical optics started at Imperial College in 1918, and were a success. But unfortunately the plan for a specific department at IC fell victim to the post-war crisis. The teaching of technical optics was absorbed by the physical department and, while indirectly contributing to the optical industry, it was distanced from the trade and from the practical training of specialized workers.

In France, the situation at the outbreak of the war revealed in a dramatic way the differences in performance between the French and German instrument industries. (Williams, 1994, pp. 72–79 and 97–102). The above-mentioned conference of J. Violle in 1915 again pointed out some of the weaknesses of the French precision industries, and showed implicitly that practically nothing was done to reverse the decline, which had been evident since at least 1900. Certainly, French makers with their lack of *esprit de corps* and scientists with their lack of entrepreneurial spirit were partially responsible for the crisis, but the government deserved some of the blame due to its total indifference to the problems of the domestic industry. One of the most common (and recurring) criticisms concerned the absence of a proper system of higher technical education and of a good apprenticeship system for the staff of science-oriented industries. This situation could not be allowed to continue. As far as instruments and optics weapons were concerned, however, the situation in France appeared to be less catastrophic than in Britain. France was the second largest producer of optical glass in the world, but it also imported some instruments and special glasses from Germany. Therefore, certain measures and decisions that had been urgent since the beginning of the war in Britain, appeared, somewhat later, to be now necessary in France. Ultimately, many of the actions taken on both sides of the Channel were similar. The

Ministère de l'armement et des fabrications de guerre was only created at the end of 1916. The government controlled the labour market, encouraged firms to employ women, and offered loans to companies desiring to expand their activities. An unprecedented effort was directed to transforming factories to war production. Before the war, exports of spectacles and eyeglasses still far exceeded imports, though the French had to import (mostly from Germany) binoculars, precision measuring instruments, surveying instruments, and also range-finders from Britain. Certainly, France was still a successful exporter of astronomical instruments, but these represented only a small fraction of the production of the optical industry. Again the role of government had been almost non-existent. For example, the import duties for instruments and raw materials were too low compared with those in other countries. Also, even if, due to the urgent needs of war, it was now far too late to imitate Germany, nevertheless the government had to start supporting the precision industry as a whole, encourage the formation of a collaborating community, and thereby facilitate cooperation between science and industry.

In 1916, after a series of discussions, meetings and consultations between scientists, businessman, entrepreneurs, and politicians, the directors of higher education and of technical education decided to found an Institut d'optique (Fabry, 1922; Mabboux, 1928; Williams, 1994, pp. 97–100). The new school was to have three separate departments, dedicated to theoretical optics, to experimental work on glass and instruments and to professional instruction for specialized workers. As in Britain, the gestation of the institute took a certain period of time, and it could not be opened before the end of the war. But the French had a long-term perspective, and the institute was seen as necessary for expanding certain industries and improving their products in peacetime, and not just for solving the urgent problems caused by the war. The institute should have been one means of regaining the ground lost, especially to Germany, in the two decades before the war. Finally, both in Britain and in France, and to a lesser extent also in other countries such as Italy, the war greatly contributed to triggering a transformation, modernization, and rationalization of the instrument industry and to moving it in the direction already taken by the Germans, well before the conflict. The role of technical education and scientific research was finally fully acknowledged.

Ultimately, the industrial war efforts were important and gave in many cases good results. For example in France the monthly production of optical glass was increased from 4,000 kg to 12,000 kg between 1914 and 1918, while in the same period the number of pairs of prismatic binoculars increased dramatically from 1,500 to 13,000 per month (Partridge, 1919). But there was progress, not only in the production of optical glass and optical weapons, but also other instruments used in chemical and biological laboratories, hospitals, and industry, such as special glassware, porcelains, and X-ray tubes, most of which, before the war, had been imported from Germany (Lindet, 1916).

On the other side of the Atlantic, the United States had to face a shortage of optical glass before the country entered the war (Wright, 1921). In fact, before 1914 most of the optical glass used in the USA, which never really bothered to develop its production to any great degree, was imported from Germany. With the beginning of the war, the German supplies stopped. Instruments for research and teaching were also largely imported from Germany. But after 1914, even France and Britain could not provide enough optical glass for export. It is true that some American firms, such as Bausch & Lomb, Spencer, and Keuffel & Esser, had begun to produce their own glass, but quantities were very limited and quality was not always satisfactory. After the declaration of war in 1917 the Council of National Defence contacted the Geophysical Laboratory and the Carnegie Institution to provide assistance in the production of glass, in cooperation with various other companies. Bausch & Lomb started a series of experiments and research, and by the end of the year the production of optical glass could begin. The glass, made by Keuffel & Esser, proved to be sufficient for requirements. Ultimately, the collaboration of the Army Ordnance Department, the War Industry Board, the Bureau of Standards, which aided in the development of crucibles, and, of course, the cooperation of several firms, worked well, if sometimes chaotically. The United States Geological Survey was helpful in locating sources of good raw materials. When the USA entered the conflict, not only did they lack a supply of optical glass, they also did not have the know-how to produce it. In what we can see as a typical American response, the war production was focused on a relatively small number of types of glass, those essential for optical instruments, and because of the urgency, the development of new and special glasses was neglected. If not all of the glass produced (six different types were manufactured) was of the highest quality, the results were nevertheless quite impressive. In April 1917 the monthly production of 'usable' glass was about 2,850 pounds; it rose to 10,700 pounds in August, and to 42,451 pounds in December. By the end of the war in November 1918 the same production amounted to almost 80,000 pounds of optical glass per month, more than 65% of it produced by Bausch & Lomb. The increase was spectacular. Also due to this effort, the USA emerged from the war much better equipped to compete with the strongest European firms.

The situation in Germany during the war was completely different from that of the allies (Zaun, 2002 pp. 326–332; Krüss, 1920). The German precision industry was in a strong leading position and was able to supply the instruments and optical glass required. Many optical instrument companies were able to orientate their production to optical weapons and military instruments, without major restructuring of their workshops. On the other hand, precision mechanics firms which had had to switch their production from instruments to ammunition, fuses, and the like, had to be fundamentally restructured. The number of workers in the precision industry decreased slightly between 1913 (around 29,500) and 1915 (around 28,200), but with the 1917 'Hindenburg programme', which was supposed to yield an enormous increase in industrial production for the war effort, their numbers rose to more than 46,000! If at first the export of scientific instruments, which could be considered strategic, was forbidden, this prohibition was subsequently eased, and to maintain their sales, companies tended to deal with neutral states. So, at the beginning of the war, exports dwindled to very little, but later they reached almost half their pre-war levels. This strategy allowed Germany to pay for some badly needed materials through the sales of precision instruments. In fact, during the war, the supply of some raw materials became a problem. The shortage of copper and its alloys raised difficulties for instrument-makers, who had to substitute iron or a specially developed zinc alloy. Users often considered instruments made using these replacement metals to be of a lesser quality. Nevertheless, after the war the use of iron and other cheaper metals instead of brass became standard for many components, such as microscope stands.

20.13 The Inter-War Years

Because of the war, allied governments finally came to appreciate the importance of the precision-instrument industry. Also, they now took concrete action to support it. Furthermore, the conflict clearly demonstrated how commercial and technical success was dependent on collaboration between makers, scientists, and technicians. But the return to peace started a period of uncertainty in the industry, and many important changes were necessary.

At the end of the war, Britain's precision industry had large plants, modern machinery, better know-how, a capacity for innovation, and a large potential for production (Williams, 1994, pp. 103–115). But the sudden termination of contracts with the government led to a difficult situation. Large stocks of instruments and optical munitions remained unsold, and the levels of production of instruments in wartime could certainly not be sustained. The spectre of recession, bankruptcy, and unemployment became real, causing protectionist measures to be introduced to avoid collapse. A substantial duty on optical glass and instruments was introduced in 1921, the prices of instruments were reduced, and various reorganizations of labour for skilled and unskilled workers and for women were proposed. The measures were not enough, however. Several firms went into liquidation, and others had to reduce their workforce, also because of the automation being introduced into various branches of production (Williams, 1994. pp. 123–162). Companies adopted different survival strategies. Some of them, with a broad experience of optical engineering, tried to diversify their production by embarking, for example, on the large market for photography and trying to recapture the civilian market. Other firms chose to amalgamate or organized joint ventures with local or foreign companies. But ultimately, in the 1920s, the British instrument industry was struggling to regain its pre-war position. In these difficult years the British Scientific Instrument Research Association was the institution through which the Department of Scientific and Industrial Research continued to support the industry (Fig. 20.30). But the absence of real government patronage clearly showed how dependent the optical and precision industry was on the state. Nevertheless, in the 1920s and 1930s the development of science-based industries, the increasing use of scientifically-based technologies in

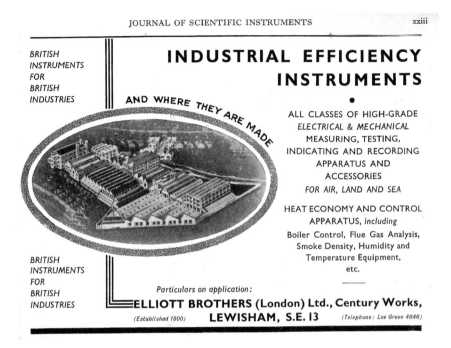

Fig. 20.30. Elliott Brothers advertisement, 1934. The phrase 'British instruments for British industries' encouraged support for national industry. (*Journal of Scientific Instruments*, February 1934, p. XXIII.)

more established industries, and the development of research, all opened up new possibilities and new niche markets for instrument-makers (Twyman, 1924–1925). The chemical, pharmaceutical, and oil industries, electrical and gas companies, car and aeroplane manufacturers, and radio technology, required control instrumentation, apparatus for the continuous recording of processes, and a series of new test instruments (Fig. 20.31). In addition, the growing range of physical and chemical apparatus (spectrometers, interferometers, pyrometers, centrifuges, and so on), which before the war had been used essentially in laboratories, was now finding greater application in other industrial fields too. All these factors represented interesting opportunities for the precision industry. In Britain, rearmament in the mid-1930s gave the industry new momentum and caused it to increase output. But this time the industry was not unprepared, as it had been in 1914. By the beginning of the Second World War, for example, the firm of Chance was producing enough glass to supply also the allied countries.

The situation of the French instrument industry after the war was in many ways similar to that of Britain (Williams, 1994, 139–144 and 162–168). The immediate postwar period, though, was not so difficult, since the industry in France was not as dependent on state intervention as that in Britain. But at the same time, the French government continued after 1918 to support the production of military instruments. On the other hand, various initiatives were taken by wealthy individuals closer to the

Fig. 20.31. A Cambridge Scientific Instrument Company advertisement for a pyrometer. This kind of instrument was greatly needed in many industrial processes, and was mass-produced. (*Journal of Scientific Instruments*, August 1934, p. LXVIII.)

industry. For example, the foundation of the Société d'Optique et de Précision de Levallois (OPL) in 1919 by the Duc A. de Gramont was an important step towards reinforcing the collaboration of the state with the instrument industry (Yvon, 1947). The OPL was created to produce optical instruments, and military range-finders were the first products to be developed. (Before the birth of the OPL, the French army and navy were almost completely dependent on Britain—in particular, the firm Barr & Stroud—for these strategically important instruments.) To these were subsequently added other types of apparatus, such as aerial photographic cameras, gun sights, and also non-military products. In the 1920s the OPL managed to secure profitable contracts both with the French government and with many foreign states. The action of de Gramont was ultimately quite successful in involving the French government in a particular sector of instrument production.

Another important factor in the development of optical instruments in France was the creation of the above-mentioned Institut d'Optique, which inaugurated courses, starting in 1920 (Fabry, 1922; Mabboux, 1928). The courses and lectures were presented by some of the most distinguished French scientists and opticians of the time, such as Fabry, Cotton, Chrétien, and Dunoyer, to mention just a few. In collaboration with the Syndicat patronal des constructeurs d'instruments, the Institut d'Optique inaugurated a new scientific journal, *Revue d'optique théorique et expérimentale*, devoted to articles relevant to the optical and precision industry. The experimental activities of the institute were expanded, and several research projects were carried out. By contributing to the establishment of stronger connections and the transfer of know-how between science and industry, the Institut built up a solid reputation both in France and abroad. In the 1920s and 1930s, partly also because of the catalysing role of the newly founded Institut d'Optique, several new companies were created (REOSC, SOC, Bronzavia, Gallus, SAGEM, and so on). Their production covered a wide range of instruments for military, industrial, and laboratory use. Some successful companies, such as Jobin & Yvon (founded in 1923) resulted from the reorganization of firms whose tradition of excellence went back to the early nineteenth century. Others, such as Ducretet, were absorbed by larger companies, or merged with other firms, while several firms in existence since the mid-nineteenth century simply disappeared or just became retailers.

So, in France the mixture of direct and indirect governmental support for the precision industry, better post-war planning, a more sustained involvement with the military, a determination to acquire French-made instruments, and better collaboration between private and institutional parties, created a situation in the 1920s that allowed the development of a deeply transformed industry, in a less traumatic way than in Britain. Over the same period, the export of French instruments was also favoured by a low rate of exchange for the French franc, which made the competition from foreign makers less dangerous than it had been before the war. The economic crisis hit France later than other countries and generated unemployment and a drop in profits. However, in 1938 the French precision industry employed about 12,000 people. One third of these were working in companies producing instruments for the military sector, one third were producing spectacles, lenses, and drawing instruments, and the remaining third was making laboratory apparatus and instruments for astronomy, geodesy, surveying, and photography.

After the war, the Treaty of Versailles prohibited Germany from production for military use. Many firms that specialized in the manufacture of instruments for the army or for the navy had to diversify their production to survive (Zaun, 2002, pp. 226–232). Companies that were producing physical, electrical, and optical instruments for laboratories, schools, and universities were not affected by this constraint. Others, like Zeiss, which produced high-quality microscopes and astronomical instruments and not just optical weapons, could limit their losses. On the other hand, companies like Goertz, which was one of the most important world manufacturers of optical instruments for military use, and which during the war had several thousand workers, had to develop new products such as photographic optics (Kingslake, 1989), calculating and writing machines, and industrial instruments, and finally had to merge with another optical manufacturer to survive. The years after the war were marked in Germany by mergers and restructuring. For example, Bamberg, the famous maker of astronomical and geodetic instruments, acquired the firms Otto Toepfer und Sohn in 1919, Hermann Wanschaff in 1922, and Hans Heele in 1923. In 1921 it joined the Centralwerkstatte Dessau to form the Askania Werke A.G., which became one of the most important producers of instruments for civilian as well as military use (Fig. 20.32). In spite of the great difficulties experienced in the 1920s, the German precision industry continued to expand. Nevertheless, because of the improved situation in both the French and British industries and because of the expansion in America, it

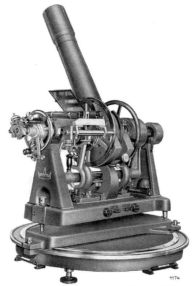

Fig. 14. Meridian Circle, Type Am 100 with Circle Base, Type Aku 90.

No. 26. Meridian Circle Am 90.
No. 27. Meridian Circle Am 100.

Dimensions:
The dimensions of the instruments correspond to those of the Ap 90 and Ap 100 types.

Weight of the instruments:
The weight is the same as Ap 90 and Ap 100, respectively, with the exception of the axis with telescope and microscope mount which weigh about 91 kg. (200 lbs.) and 100 kg. (220 lbs.) respectively.

Fig. 20.32. Meridian circle made by the Askania Werke in the 1930s. The firm was founded in 1921 through the merger of Centralwerkstatt Dessau and Carl Bamberg. In the inter-war period Askania Werke produced a very wide array of precision optical instruments for civilian and military use. (Askania-Werke, *Portable Transit Instruments and Meridian Circles*, Berlin-Friedenau, c.1937, p. 18.)

never regained the overwhelming superiority on the international market that it had before 1914.

The American instrument industry, which had expanded rapidly since the end of the nineteenth century, was poised, after the First World War, to become one of the world leaders in the sector. The demand for scientific instruments in the USA rose exponentially in the first decades of the twentieth century. While the increasing number of laboratories both for teaching and research did much to stimulate the demand for instruments, their introduction on a larger scale in booming industries such as car manufacture, petroleum refining, and electric power generation, also boosted the precision industry (Perazich *et al.* 1938), leading to a roughly ten-fold increase in production between 1909 and 1929. The largest proportion (about 2/3 or more) of these instruments were used within industry. Among these were apparatus for indicating, measuring, recording, and controlling temperature, pressure, fluid flow, electrical current and voltage, liquid level, speed, duration of process, composition of gases, and so on. As far as non-industrial instruments (such as physics, surveying, or astronomical instruments) were concerned, we know that before the First World War, imports exceeded exports, while after the war this situation was reversed. However, the sales of non-industrial instruments to industrial and educational laboratories, while slower than for industrial instruments, also increased noticeably. The peak was reached in 1929, after which the economic crisis caused sales to slow down, though they grew again during 1932–33.

20.14 Conclusion

Between the first decades of the nineteenth century and around 1890, the production of scientific instruments was essentially dominated by the French and the British. But the unification of Germany and the country's subsequent economic, industrial, and military expansion, as well as state support and close cooperation between science and industry in about 1900, raised the German precision industry to a leading position. On the other hand, after the end of the nineteenth century the growing need for industrial and professional instruments encouraged the introduction of new production technologies and machine tools in the largest companies. In France and Britain, where instrument-making was rooted in traditional craft skills, things changed slowly. The increasingly strong performance of the newcomer was certainly in evidence, but in spite of many reports, analyses, and rhetoric, little was done to follow the German example. However, the outbreak of the First World War showed dramatically the technological and productivity gap existing between the countries. France, Britain, and later the USA, were forced to take decisive steps to modernize and expand the production of instruments and optical glass, to improve technical and scientific training and education, and to establish a stronger cooperation between scientists and manufacturers. The involvement of governments in these transformations proved to be essential. At the end of the war the situation was transformed, with American industry occupying a steadily more prominent position in the instrument market. The interwar period was initially characterized by difficulties resulting from the readjustment of production for the civilian market and later by adjustments in response to the economic crisis. In spite of these challenges, the interwar instrument industry reorganized and slowly built its future growth on the new foundations. The Second World War and the second half of the twentieth century were to stimulate even more profound transformations, with the production and trade in instruments being increasingly influenced by finance, marketing, large corporations, politics, and big science.

REFERENCES

AMBRONN, L., 1919. Beiträge zur Geschichte der Feinmechanik, *Beitrage zur Geschichte der Technik und Industrie*, 9, pp. 2–40.

ANDERSEN, H. 1995. *Historic Scientific Instruments in Denmark*, Copenhagen: The Royal Danish Academy of Science and Letters.

ASPINALL PARR, G.D. 1903. *Electrical Engineering Measuring Instruments for Commercial and Laboratory Purposes*, London: Blackie and Son.

BELMAR, A. G., SÁNCHEZ, J. R. B., 2002. *Abriendo las cajas negras Colección de instrumentos científicos de la Universitat de València*, València: Universitat de València.

BENNETT, J., 1983. *Science at the Great Exhibition*. Cambridge: Whipple Museum of the History of Science.

———1985. Instrument makers and the 'Decline of science in England': the effects of institutional changes on the élite makers of early nineteenth century. In P. R. de Clercq, ed., *XIXth Century Scientific Instruments and Their Makers*. Leiden: Museum Boerhaave, Amsterdam: Rodopi, pp. 13–27.

———1987. *The Divided Circle*, Oxford: Phaidon, Christie's.

BOUASSE, H., 1917. *Construction, description et emploi des appareils de mesure et d'observation*, Paris: Librairie Delagrave.

BRACHNER, A., 1985. German nineteenth-century scientific instrument makers. In P. R. de Clercq, ed. *XIXth Century Scientific Instruments and Their Makers*. Leiden: Museum Boerhaave, Amsterdam: Rodopi, pp. 117–157.

———1986, *Die Münchener Optik in der Geschichte Entstehung, Unternehmung, Sternwarten, Lokalitäten, Ausbreitung*. Dissertation Technische Universität München.

———and SEEBERG. M., 1976. *Joseph von Fraunhofer 1787–1826*, München: Deutsches Museum.

BRENNI, P., 1985. Italian Scientific Instrument Makers of the XIXth Century and their Instruments. In P. R. de Clercq, ed. *XIXth Century Scientific Instruments and Their Makers*. Leiden: Museum Boerhaave, Amsterdam: Rodopi, pp. 183–203.

———1988. The Scientific Instrument Makers Illustrated Catalogues. In C. Blondel, F. Parrot, A. Turner, eds. *Studies in the History of Scientific Instruments, Papers presented at the 7th Symposium of the Scientific Instrument Commission, Paris 15–19 September 1987*. London: R. Turner Books, Paris: CSI, pp. 169–178.

———1996. The Richard Family. *Bulletin of the Scientific Instrument Society*, 48, pp. 10–14.

———1997. Physics Instruments in the 20th Century. In J. Krige, D. Pestre eds. *Science in the 20th Century*. Amsterdam: Harwood Academic Publisher, pp. 741–757.

———1998, La funzione degli strumenti scientifici nella didattica fra Settecento e Ottocento. *Studi Settecenteschi*, 18, pp. 421–431.

———2000. Historische Instrumente als materielle Zeugen der Wissenschaftsgeschichte. In: C. Meinel, ed. *Instrument-Experiment Historische Studien*. Berlin: Diepholz Verlag für Geschichte der Medizin, Naturwissenschaft und Technik, pp. 74–81.

———2001. 'Un siècle d'instruments. La transformation des appareils scientifiques et leur utilisation pendant le XXe siècle'. *Sartoniana*, 14, pp. 13–56.

———2002a. La industria de precisión en el siglo XIX Una panorámica de los instrumentos, los constructores y el mercato en diferentes contextos nacionales (with English translation). In J. R. B. Sánchez, A. G. Belmar, eds. *Abriendo las Cajas Negras Coléccion de instrumentos científicos de la Universitat de València*. València: Universitad de València, pp. 53–72.

———2002b, Who made what? The Dilemma of signed instruments. In AAVV, *XVIII International Scientific Instrument Symposium Moscow/St.Petersburg, Russia, 20–25 September 1999 Conference proceedings*. Moscow: Scientific Publishing Centre Engineer, pp. 50–66.

———2002c. 19th Century Scientific Instrument Advertising, *Nuncius, Annali di Storia della Scienza*, 2, pp. 497–514.

———2004. Dal *Crystal Palace* al *Palais de l'Optique*: la scienza alle esposizioni universali. *Memoria e Ricerca*, 17, September–December Issue, pp. 35–63.

———2006. Artist and Engineer: The saga of 19th Century French Precision Industry (The Annual Invitation Lecture), *Bulletin of the Scientific Instrument Society*, 91, pp. 2–11.

———2009. Dal laboratorio all'officina. La nascita della strumentazione elettrica per misure industriali (1870–1900). In V. Cantoni, A. Silvestri, eds. *Storia della tecnica elettrica*. Milano: Cisalpino, Istituto Editoriale Universitario, pp. 101–135.

———2010. La science française au Crystal Palace, *Documents pour l'histoire des techniques*, 19, December issue, pp. 255–265.

———2011, The evolution of teaching instruments and their use between 1800 and 1930. In P. Heering, R. Wittje eds. *Learning by Doing Experiments and Instruments in the History of Science Teaching*. Leipzig: Franz Steiner Verlag, pp. 281–316.

BUD, R., WARNER, D. J. eds., 1998. *Instruments of Science: An Historical Encyclopedia*, New York, London: Garland Publishing Inc.

CAHAN, D. 1989. *An Institute for an Empire, The Physikalisch-Technische Reichsanstalt 1871–1918*, Cambridge: Cambridge University Press.

Catalogue de l'exposition collective allemande d'instruments d'optique et de mécanique de précision Exposition Universelle de 1900 Groupe 3–Classe 15 (Anon., 1900). Berlin: Reichsdrückerei, (reprint Edition Alain Brieux 1984).

CATTERMOLE, M. J. G., WOLFE A. F., 1987. *Horace Darwin's Shop: A History of the Cambridge Scientific Instrument Company 1878 to 1968*, Bristol and Boston: Adam Hilger.

CHAPMAN, A., 1990. *Dividing the Circle: The Development of Critical Angular Measurement in Astronomy 1500–1850*, New York, London: Ellis Horwood.

CLARKE, T., MORRISON-LOW A., SIMPSON, A. 1989. *Brass and Glass Scientific Instrument Making Workshop in Scotland as Illustrated by Instruments from the A. Frank Collection at the Royal Museum of Scotland*, Edinburgh: National Museums of Scotland.

CLIFTON, G., 1995. *Directory of British Scientific Instrument Makers 1550–1851*. London: Zwemmer in association with the National Maritime Museum.

COOPER, H. J. ed., 1946. *Scientitic Instruments*, London, New York: Hutchinson's Publications.

CRONSHAW, H. B. et al., 1949. *The Instrument Manual*. London: United Trade Press.

DAUMAS, M., 1953. *Les instruments scientifiques aux XVIIe et XVIIIe siècles*, Paris: Presses Universitaires de France.

DE CLERCQ, P., 1985. The scientific instrument making industry in the Netherlands in the nineteenth century. In P. R. de Clercq, ed. *XIXth Century Scientific Instruments and Their Makers*. Leiden: Museum Boerhaave, Amsterdam: Rodopi, pp. 204–225.

DELESALLE, P., 1899. *Les conditions su travail des ouvriers and instruments de précision de Paris*, Paris: Syndicat des ouvriers en instruments de précision.

DÉVÉ, C., 1949. *Le travail des verres d'optique de précision*, Paris: Editions de la Revue d'optique théorique et instrumentale (3d edition).

Dictionary of British Scientific Instruments (Anon., 1921), London: Constable and Company.

Die Exportverhältnisse der deutschen Prezisionsmechanik (Anon., 1899), *Deutsche Mechaniker-Zeitung*, 6, pp. 41–43; 8, pp. 61–63, 16, pp. 141–142.

DUMMER, G. W. A., 1983. *Electronic Inventions and Discoveries Electronics from Early Beginnings to the Present Days*. Bristol, Philadelphie: Institute of Physics Publishing (3d edition).

EBERT, H., 1897. *Guide pour le soufflage du verre*, Paris: Gauthier-Villars et fils (2d edition).

EDGCUMBE, K., 1918. *Industrial Electrical Measuring Instruments*. London: Constable and Co. (2d edition).

FABRY, C., 1922. L'Institut d'optique théorique et appliqué, *Bulletin de la société d'encouragement pour l'industrie nationale*, pp. 636–645.

FOX, R., GUAGNINI A., 1999. *Laboratories, Workshop and Sites Concept and Practices of Research in Industrial Europe*. Berkeley: University of California.

FRICK, J., LEHEMANN, O., 1904–1909. *Physikalische Technik oder Anleitung zu Experimentalvorträgen sowie zur Selbstherstellung eifacher Demonstrationsapparate*. Braunschweig: Friedrich Vieweg und Sohn (7th edition).

GANOT, A., 1876. *Traité élémentaire de physique expérimentale et appliqués*. Paris: che l'auteur (17th edition).

GAUVIN, J. F., PYENSON L., 2002. *The Art of Teaching Physics The Eighteenth-Century Demonstration Apparatus of Jean Antoine Nollet*. Sillery (Québec): Septentrion.

German Educational Exhibition World's Fair St Louis 1904 (Anon., 1904). Berlin: Büxenstein.

GLAISHER, J. 1852a. Philosophical instruments and processes, as represented at the great exhibition. In W. Whewell *et al*. ed. *Lectures on the Results of the Exhibition*, London: David Bogue, Philadelphia: A. Hart, pp. 243–301

——1852b. Class X. Philosophical instruments and processes depending upon their use. In *Report by the Juries Exhibition of the Work of Industry of All Nations 1851*. London: W. Clowes and Son, pp. 243–316.

GOURDEN, J. M., 1992. *Le peuple des ateliers Les artisans du XIXe siècle*, Paris: Créaphis.

GRAMBERG, A., 1923. *Technische Messungen bei Maschinenuntersuchungen und zur Betriebskontrolle*. Berlin: Julius Springer.

HACKMANN, W., 1985 The nineteenth-century trade in natural philosophy instruments in Britain. In P. R. de Clercq, ed. *XIXth Century Scientific Instruments and Their Makers*. Leiden: Museum Boerhaave, Amsterdam: Rodopi, pp. 53–91.

HENRIVAUX, J., 1903, *La verrerie au XXe siècle*, Paris: E. Bernard and Cie.

HOFMANN, M., 1896. *Handbuch der praktischen Werkstatt-Mechanik Metall- und Holzdreherei*. Wien, Pest Leipzig: A.Hartleben's Verlag.

JACKSON, M. W., 2000. *Spectrum of Belief Joseph von Fraunhofer and the Craft of Precision Optics*. London, Cambridge: MIT Press.

JAHN, W. *et al*., 2008. *Fraunhofer in Benediktbeuern Glashütte und Werkstatt*. München: Fraunhofer Gesellschaft.

KINGSLAKE, R., 1989. *A History of the Photographic Lens*, Boston: Academic Press Inc.

KOHL, MAX A.G., n.d. (but 1911). *Physikalische Apparate Preisliste Nr.50 Band II und III*. Chemnitz: by the author.

KRÜSS, H., 1920, Die Entwicklung der feinmechanischen und optischen Industrie im Kriege. *Zeitschrift der deutschen Gesellschaft für Mechanik und Optik*, February issue, 1920, pp. 16–17.

LE CHÂTELIER, H., 1916. La science dans ses rapport avec le développement économique du pays, *Bulletin de la société d'encouragement pour l'industrie nationale*, II semestre, pp. 9–12.

LEYBOLD'S NACHFOLGER, n.d. (but 1913). *Einrichtungen und Apparate für den physikalischen Unterricht sowie Übungen in Praktikum nebst Literaturangaben*. Köln.

LINDET, L., 1916. Les effors de l'Industrie française pendant la guerre, *Bulletin de la société d'encouragement pour l'industrie nationale*, II semestre, pp. 13–36.

L'industrie française des instruments de précision Catalogue (Anon., 1901–02). Paris: Syndicat des constructeurs d'instruments d'optique, (reprint Edition Alain Brieux 1980).

LISSAJOUS, J. A., 1868. Instruments de précision et matériel de l'enseignement des sciences. In M. Chevalier, ed. *Exposition Universelle de 1867, Rapport du jury intérnational*, Paris: Imprimerie Paul Dupont, vol. II, pp. 415–420.

MABBOUX, P., 1928. L'Institut d'optique théorique et appliquée, *La Nature*, 1th sem., pp. 107–113.

MacLoad R., MacLoad, K., 1975. War and economic development: government and the optical industry in Britain, 1914–18 In J.M. Winter, ed. *War and economic development*. Cambridge: Cambridge University Press, pp. 165–203.

McConnell, A., 1992. *Instrument Makers to the World: A History of Cooke, Troughton & Simms*, York: William Session Ltd.

——2007. *Jesse Ramsden (1735–1800) London's Leading Scientific Instrument Maker*, Aldershot: Ashgate.

Meyer, W., 1905. Dass Zeisswerk und die Glashütte in Jena. *Zeitschrift für Kunst, Kultur und Unterhaltung*, pp. 557–572.

Morand, P., 1931. *1900 A.D.* Translated from French by R.Fedden. New York: W.Farqhuar Payson.

Morrison-Low, A.D., 2007. *Making Scientific Instruments in the Industrial Revolution*, Aldershot: Ashgate.

Morus, I.R., 2005. *When Physics Became King*. London, Chicago: University of Chicago Press.

Moser, H., ed. 1962. *Forschung und Prüfung 75 Jahre Physikalisch-Technische Bundesanstalt/ Reichsanstalt*. Braunschweig: F.Vieweg & Sohn.

Müller-Pouillet Lehrbuch der Physik un Meteorologie Neunte umgearbeitet und vermehrte Auflage von L. Pfaundler. 1886–1898. Braunschweig: F. Vieweg & Sohn.

Nye, M. J., 1966. *Before Big Science: The Pursuit of Modern Chemistry and Physics 1800–1940*. Cambridge (MA), London: Harvard University Press.

Ocagne d´, M., 1903. *Les instruments de précision Conférence faite au Conservatoire des Arts et Métiers le 15 mars 1903*, Paris, Syndicat des constructeurs d'instruments d'optique et de précision.

Official Descriptive and Illustrated Catalogue: Exhibition of the Work of Industry of All Nations 1851 (Anon., 1851). London:W. Clowes and Son, 1851.

Oudinet, J., 1882. *Principes de la construction des instruments de précision*, Paris: Librairie du dictionnaire des arts et manufactures.

Partridge, J. W., 1919. Les instruments d'optique employés aux armées, *Bulletin de la société d'encouragement pour l'industrie nationale*, 1th semestre, pp. 607–614.

Payen, J., 1985. La construction des instruments scientifiques en France au XIXe siècle. In P. R. de Clercq, ed. *XIXth Century Scientific Instruments and Their Makers*. Leiden: Museum Boerhaave, Amsterdam: Rodopi, pp. 159–182.

——1986. Les constructeurs d'instruments scientifiques en France au XIXe siècle'. *Archives Internationales d'Histoire des Sciences*, 36, pp. 84–161.

Perazich, G., Schimmel, H., Rosenberg, B. 1938. *Industrial Instruments and Changing Technology Report N. M-1*. Philadelphia: Work Progress Administration National Research Project.

Pipping G., 1977. *The Chamber of Physics Instruments in the History of Science Collections of the Royal Swedish Academy of Sciences*, Stockholm: Almqvist and Wiksell.

Repsold, J. A., 1914. *Zur Geschichte der astronomischen Messwerkzeugen von 1830 bis um 1900*. Leipzig: Verlag von Emmanuel Reinicke.

Riekher, R., 1990. *Fernrohre und ihre Meister*, Berlin: Verlag Technik GMbH (2d edition).

Rohr M. von, 1929. *Joseph von Fraunhofer Leben, Leistungen und Wirksamkeit*. Leipzig: Akademische Verlagsgesellschaft.

Schefer, R., 1917. La verrerie d'optique. In *Enquête sur la production française et la concurrence étrangère*, Vol. III, Paris: Association nationale d'expansion économique pp. 225–246.

Shinn, T., 2001, The research-technology matrix: German origins, 1860–1900. In B. Joerges, T. Shinn eds. *Instrumentation between Science, State and Industry*. Dordrecht: Kluwer Academic Publishers, pp. 29–48.

Special Collective Catalogue of Scientific Instruments and Appliances: The World's Colimbian Exposition, German Exhibition. Group 21 (Anon., 1893). Berlin: Deutsche Gesellschaft für Mechanik und Optik.

Statistique de l'Industrie à Paris résultant de l'enquête faite par la Chambre de Commerce pour l'année 1860 (Anon., 1864). Paris: Chambre de Commerce.

STOLZ, R., WITTIG, J., eds., 1993. *Carl Zeiss und Ernst Abbe Leben, Wirkung und Bedeutung*. Jena: Universitätsverlag.

SYDENHAM P. H., 1979, *Measuring Instruments: Tools of Knowledge and Control*. Stevenage UK, New York: Peregrinus in association with the Science Museum, London.

THRELFALL, R., 1898. *On Laboratory Arts*. London: Macmillan and Co.

TURNER, A., 1987. *Early Scientific Instruments Europe 1400–1800*, London: Sotheby's Publications.

——1989. *From Pleasure and Profit to Science and Security: Etienne Lenoir and the Transformation of Precision Instrument Making in France 1760–1830*. Cambridge: Whipple Museum of the History of Science.

TURNER, G. L'E., 1983. *Nineteenth-Century Scientific Instruments*. London: Sotheby, Berleley: University of California Press.

——1987. Presidential Address: Scientific Toys. *British Journal for the History of Science*, 20, pp. 245–258.

——1991. *Gli Strumenti*. Torino: Einaudi.

——1996. *The Practice of Science in the Nineteenth Century Teaching and Research Apparatus in the Teyler Museum*, Haarlem: Teyler Museum.

——2000. *Elizabethan Instrument Makers The Origins of the London Trade in Precision Instrument Making*, Oxford: Oxford University Press.

TURNER, S., 2005. Chicago scientific instrument makers, 1871–1918 Part I: the school market, *Rittenhouse*, 19 (2), pp. 65–128.

——2006. Chicago scientific instrument makers, 1871–1918 Part I: the university and research market, *Rittenhouse*, 20 (1), pp. 1–64.

TWYMAN, F., 1924–25, The vitality of the British optical industry, *Journal of Scientific Instruments*, II (12), pp. 369–380.

——1942. *Prism and Lens Making*, London: Adam Hilger London.

VAN HELDEN, A., HANKINS, T. L. eds., 1994. *Instruments, Osiris, Vol. 9*. Chicago: Chicago University Press.

VIOLLE, J., 1915. *Du rôle de la physique à la guerre. De l'avenir de nos industries physiques après la guerre*, Paris: Berger Levrault.

WARNER, D. J. 1990. 'What is a scientific instrument? When did it become one? And why?', *British Journal of History of Science*, 23, pp. 83–93.

——1993. French Instruments in the United States. *Rittenhouse*, 8, pp. 1–32.

WEINHOLD, A. F., 1881. *Physikalische Demonstrationen Anleitung zum Experimentieren*. Leipzig: Quandt and Händel.

WILLIAMS, M. E. W., 1994, *The Precision Maker: A History of the Instrument Industry in Britain and France 1870–1939*, London, New York: Routledge.

WRIGHT, F. E., 1921. *The Manufacture of Optical Glass and of Optical Systems: A War-time Problem*. Washington: Washington Government printing Office.

YVON, G., 1947. Les industrie françaises de l'Optique, *Revue d'optique*, 25, pp. 177–194.

ZAUN, J., 2002. *Instrumente für die Wissenschaft Innovationen in der Berliner Feinmechanik und Optik 1871–1914*. Berlin: Verlag für Wissenschafts- und Regionalsgeschichte.

Notes

1. In this essay the terms 'scientific instrument industry' and 'precision industry' are used synonymously. However, strictly speaking not all scientific instruments are also precision instruments. For example, teaching and demonstration apparatus and models, or several industrial instruments belong to the first category, but they cannot be considered to be precision instruments like many optical, astronomical, and surveying instruments used for accurate measurement.
2. Following Brachner (1985), with the inclusion of establishments employing less than ten workers, these figures would quadruple.

CHAPTER 21

PHYSICS TEXTBOOKS AND TEXTBOOK PHYSICS IN THE NINETEENTH AND TWENTIETH CENTURIES

JOSEP SIMON

21.1 Textbooks: Past and Future

Textbooks are a classic yet elusive object in the history of science. They received significant attention in the early years of our discipline, and this interest has not diminished. However, it is still not entirely clear what science textbooks are or what status they have as historical objects. Historical studies of science textbooks are still scarce, and much of what we have is historiographically and methodologically limited. The history of physics is no exception in this respect, although a growing interest in the study of physics education has, in recent decades, resulted in a fundamental renovation of the discipline (Kaiser, 2005a and 2005b; Warwick, 2003a; Olesko, 1991 and 2006).

Textbooks might have been expected to become central sources for the writing of a truly international history of science. The study of scientific research in an international perspective has traditionally been described explicitly or implicitly as a matter of leading centres radiating towards passive peripheries (Gavroglu *et al.* 2008). In contrast, science education and textbooks are ubiquitous, for it seems implausible that any country which had a publishing industry would not have structures for the production of science textbooks. Although this is basically true, some national contexts have developed educational and textbook enterprises in science earlier than others, and have been better able to export their production.

In this chapter I focus on the French, German, American, and British cultures of textbook physics. I could have proceeded otherwise, since most countries have had a rich physics textbook tradition. If this possibility is often not brought out sufficiently, it is because textbooks and education have a secondary status in the historiography of science, and because the history of science canon is still nationally biased. Science textbooks have often been exclusively aligned with the training of scientific elites and the enforcement of scientific paradigms. This has done little to advance their study as a key area at the interface of history of science, history of education, and the history of the book, all of which could make a major contribution to more global histories. While research on textbooks has acquired a certain degree of maturity within science education (Choppin, 1992; Johnsen, 2001), such scholarly production has in general not received the attention of historians of science. Furthermore, historians of education have paid little attention to science textbooks, and they have tended to focus on primary education, while historians of science, with few exceptions, have only been interested in textbooks for higher education.[1] Finally, the history of the book is still finding its way into the study of science, and its interaction with the history of education in particular remains rare (Simon, 2011; Topham, 2000; Rudolph 2008; Rose, 2006; Secord, 2000).

Thomas S. Kuhn's *The Structure of Scientific Revolutions* (1962) has done more than any other work to draw attention to textbooks in the history of science, and research using textbooks as sources has hitherto tended to proceed along Kuhnian lines. Little effort has been made to place Kuhn's ideas in context, and to assess critically his agenda as a means of taking this field of research a step forward.[2] Kuhn's work was obviously a product of its time.[3] It reflects particular developments in physics, education, history and philosophy of science, and politics. In fact, textbooks have a low status in contemporary culture—something that illuminates the paradoxes of Kuhn's thought and of subsequent work (Bensaude-Vincent *et al.* 2003; Simon, 2011, 15–18; Brooke, 2000).

In this chapter I argue that the study of textbooks would benefit from greater reflexivity, showing how our views on textbooks have been shaped by events that have established particular hierarchies between scientific research and science education, and between universities and schools. By adopting this approach, as I maintain, we could avoid anachronism and oversimplification in a field that has great potential as a contribution to our understanding of the history of science. Textbooks have had a prominent role in the history of physics, and they can offer new perspectives that go beyond their traditional characterization as mere showcases of scientific paradigms. My chapter is divided into three sections. First, I examine the role that physics textbooks played in the early stages of the professionalization of the history of science.

Second, I offer a general overview of the genesis of textbook physics in the nineteenth century, highlighting major textbooks produced in France and the German states and making some reference to British and American textbooks. Finally, I discuss recent scholarship dealing with textbooks in the history of physics.

21.2 A Very Historico-Scientific History

In 1948, George Sarton issued a programmatic call for a history of physics based on 'The Study of Early Scientific Textbooks'. In referring to 'early textbooks', he meant those 'treatises' published especially before the nineteenth century, which constituted the core of the literature communicating science in particular periods. The aim was to trace 'scientific evolution' by examining the content of these books and tracing changes over time and space in successive editions and translations. Great emphasis was put into the analysis of logical structure, but style and illustration were also considered relevant.

Sarton did not think it necessary to define explicitly what a 'textbook' was, or whether a 'textbook' was any different from a 'treatise'. But he made some reference to educational contexts and, most importantly, drew a clear distinction between early textbooks and those produced from the second half of the nineteenth century onwards. He considered that modern textbooks were too abundant and, surprisingly (in the light of subsequent perspectives on the topic), that the lag between the emergence of new ideas and their integration into modern textbooks was too short to make them relevant to his aims (Sarton, 1948). I. Bernard Cohen, a close collaborator of Sarton, corroborated these remarks, but related his historical perceptions on old textbooks to his contemporary experience as a physics teacher and reader of modern textbooks in the discipline (Cohen, 1948). Although Sarton and Cohen admitted in passing that the authors of early textbooks could be dogmatic or resistant to scientific change, they considered that it was the duty of historians of science to avoid Whig interpretations. Textbook narratives should be considered in their own historical context, the context that shaped their structure and their expository functions and aims.

Between the 1930s and 1950s it was not unusual to find reviews of physics textbooks in *Isis*, by such scholars as Cohen (who succeeded Sarton as editor of the journal), Gerald Holton, and Victor Fritz Lenzen, who combined their history and philosophy of science writing with work in physics and teaching (Dauben *et al.* 2009; Heilbron, 1977; Nowotny, 1990). The books were selected for their particular focus on the history of physics and were characteristic of the literature of the field, which in that period was more scientific than historical (Cohen, 1943; Holton, 1956; Lenzen, 1937). In one of these reviews, Lenzen considered that the abundant historical contents of *Mechanics, Molecular Physics, Heat, and Sound*, a major college physics textbook by Robert Millikan, Duane Roller, and Earnest Charles Watson, made it a valuable contribution to the history of physics. Furthermore, he added, this textbook was a 'notable example of the humanization of science' (Lenzen, 1937).

Similar motivations led Cohen and Holton to participate in the General Education in Science programme developed at Harvard by James B. Conant after the Second World War. They did so along with Fletcher G. Watson, Leonard K. Nash, and Thomas S. Kuhn.[4] All of them had been trained in the physical sciences and were then finding their way as professionals in physics, education, or history and philosophy of science. Their interest in physics textbooks as historical sources ran in parallel with their perception of the need to produce textbooks in all these areas, in particular in the history of science as a means of consolidating the discipline (Bird, 2009; Dauben *et al.* 2009; Dennis, 1997; Jacobs, 2010; Murphy and Valsamis, 1997; Nowotny, 1990).

Cohen and Watson edited the textbook which served as a guide to the General Education in Science course. Holton published two major physics textbooks; one in collaboration with Duane Henry Dubose Roller. Roller, a Harvard PhD in history of science, teamed up with his father—a physicist and textbook author (see above)—in a number of history of science publications, including their contributions to the *Harvard Case Histories in Experimental Science*, edited by Conant and Nash, which soon made their mark on school textbooks by teachers such as Leopold E. Klopfer. Subsequently, he went on to establish a history of science department and a large associated library, which was the foundation for the 'Landmarks of Science' microform collection (Schofield, 1995). This initiative followed the broad lines of Sarton's programme, although it recognized the classic status of nineteenth-century physics textbooks by such authors as Johann Müller, Adolphe Quetelet, Adolphe Ganot, Jules Jamin, John Tyndall, and Adolph Wüllner.

Kuhn's first book, *The Copernican Revolution* (1957) was considered by some contemporaries as a textbook emerging from the General Education in Science programme (Fuller, 2000, n. 90, 219–20), which also had an important influence in his subsequent work, *The Structure of Scientific Revolutions*. However, the making of *Structure* cannot be understood without taking into account two major aspects. One was the climate for school reform in America, which surrounded Kuhn and which strikingly exemplifies the transformation of physics textbook production in the second half of the twentieth century. The other was the cultural pattern that led to the dismissal of textbooks as sources of knowledge—a perception that held sway for a long time and has continued to do so into our own day.

From the 1950s a favourable climate for the development of large-scale school reforms in the sciences started to develop in the USA. This included the reform of curricula, the production and distribution of new textbooks, and the development of teacher-training programmes. Most importantly, it supposed the empowerment of university scientists in the design of the school science curriculum (Rudolph, 2002).

In the nineteenth century, secondary education was the major focus of disciplinary development for the physical sciences. While the sciences had a limited place in university education, their presence was boosted by the development of secondary education within a large number of national contexts. University physicists had always played a significant role in the design of school curricula and the production of textbooks, but the intellectual boundaries between universities and secondary schools were not so well defined as they were later. Secondary-school teachers made a major contribution to the making of physics as a discipline, and the secondary-school curriculum often shaped university physics, not *vice versa* (Stichweh, 1984; Olesko 1991; Simon, 2011).

In the 1950s, American physicists were increasingly concerned about the decline in enrolments in physics courses—a trend that had been apparent since at least the late nineteenth century. This was in contrast with the growing political power of university physicists, based on their close involvement in the war effort (Kelly, 1955; National Research Council/American Institute of Physics, 1955; Rudolph, 2002). The school reforms proposed by physicists were supported by some influential psychologists and educationists. At Harvard, Jerome Bruner expressed his rejection of Conant's programme on the grounds of its non-specialized, amateurish nature. He supported instead the establishment of teaching programmes that would focus on the 'structure' of scientific subjects. Hence he favoured the presentation of subject matter focused less on coverage than on illuminating the relations between the fundamental principles defining a field of knowledge. For Bruner, learning the structure of physics involved learning the subject as physicists understood it. Accordingly, physicists had to take the lead in the development of a new school curriculum that would endorse their perspective on the subject (DeBoer, 1991; Fuller, 2000; Rudolph, 2002).

In the same period the educationist Joseph Schwab characterized nineteenth-century science education as 'embodied in authoritative lecture and textbook, inflexible laboratory instructions, and exercises presenting no problems of choice and application'. He contended that this fundamentally dogmatic approach had survived in American schools up to the twentieth century. According to Schwab, since scientific knowledge could no longer be considered as composed of stable truths, school science and science textbooks should not just present facts; they should also bring out principles of enquiry that constituted the fundamental structure of science (DeBoer, 1991; Schwab, 1958).

The work of Bruner and Schwab, among others, further reinforced the reform movement, which benefited from government funds, stimulated by Cold War competition and events of extraordinary political impact such as the Soviet Union's launch of *Sputnik* in 1957. In this context the Physical Science Study Committee (PSSC) and its ruling team of academic physicists presented themselves as performing a revolution in physics pedagogy aimed at replacing the old-school physics paradigm with a new one based on the principles of modern university physics. Its work included the design of new textbooks and the establishment of an aggressive programme aimed at promoting the adoption of its products in schools and acculturating teachers in their use. In this respect, they clearly aimed at displacing teachers and educators from a field in which they had had a fundamental role in the previous century (Brownson and Schwab, 1963; Rudolph, 2002).

The preparation of the PSSC programme was contemporaneous with Kuhn's writing of the first draft of *Structure*. Both projects arose from the same intellectual milieu, marked by a combined interest in physics and pedagogy and shaped by developments that shook American science and education (Marcum, 2005, 13–5). Kuhn's work matched perfectly the conceptual framework on science, education, and textbooks promoted by the PSSC school reform.[5]

During his time at Harvard, Kuhn had matured and presented his ideas in a series of lectures that constituted the embryo of his book. In 1951, in his first Lowell lecture, entitled 'Textbook Science and Creative Science', he already contended that the 'structure of knowledge in the textbook' masked 'the nature of the creative process' by which knowledge is gained (Marcum, 2005, 30–1). In 1961 Kuhn communicated to historians of science the ideas contained in his forthcoming book. For Kuhn, a special characteristic of science from the early nineteenth century onwards was that education was conducted through textbooks to an extent unknown in other fields of knowledge. Moreover, textbooks presented a surprising uniformity in conceptual structure and only differed in subject matter or pedagogical detail according to their level. Textbook science was the driving agent in the transmission of scientific knowledge through systematic education, and it involved indoctrination. Although this level of systematization was not present before the nineteenth century, there were works which could be considered 'classics', such as the treatises by Aristotle, Ptolemy, Newton, Franklin, Lavoisier, or Lyell, that played a similar role in representing 'universally received paradigms' (Kuhn, 1963).

The success of the PSSC programme was as great as that of Kuhn's classic work. But it was less successful than the programme's propagandists maintained. Although it had a major role in shaping the form that physics textbooks would take during the second half of the twentieth century, the PSSC had competitors among the wide range of approaches to pedagogy and physics that coexisted at the time. Furthermore, the problem of school physics curricula and textbooks was not really new, debates on this topic having taken place periodically during the nineteenth and twentieth centuries.[6]

In the late nineteenth century, the (American) National Education Association (NEA) had already debated the need to promote a closer connection between school and college curricula. Some of its members, such as Edwin H. Hall, professor of physics at Harvard, considered that in spite of the need to use a single standard textbook in physics courses, for their 'connective and comprehensive view of the subject', further emphasis should be given to laboratory training in order to encourage the development of skills of observation (DeBoer, 1991; Hall, 1909).

In the 1920s the physicist Robert Millikan complained that the situation had not changed. He urged that college scientists be made exclusively responsible for teacher training and school curriculum design. The same motivations had led him to publish in 1906 a physics textbook for high schools—a work that went through several editions before the end of the Second World War (Millikan, 1925; Millikan, and Gale, 1906). In his autobiography he recounted his encounter with physics textbooks at a young age, and the standard status of the English translation of Adolphe Ganot's *Traité élémentaire de physique expérimentale et appliquée* (1851) in American high schools and colleges at the time (Millikan, 1951). Ganot's *Traité* embodied precisely the kind of approach that was criticized by most physicists involved in education during the middle decades of the twentieth century. Its rise to standard status and subsequent fall offer an interesting window on the history of the making of textbook physics as a genre.

21.3 TEXTBOOK PHYSICS: A NINETEENTH-CENTURY GENRE?

In 1851, Joseph Lovering, professor of mathematics and natural philosophy at Harvard College, expressed his concern for the lack of appropriate physics textbooks in English ([Lovering], 1851). Lovering saw textbooks as a central tool in the teaching of physics and in its much-needed expansion in education and society in general. In his opinion, reading and recitation from a carefully chosen textbook, allied to lectures and experimental demonstrations, formed the core of a good education in physics. For Lovering, lectures were fundamental, since they allowed the introduction of the most recent developments in the field, which printed textbooks were often not able to incorporate quickly enough. But textbooks were central because they contributed to fix knowledge and to communicate it in a precise and accurate way.

According to Lovering, the major qualities of a good textbook were comprehensiveness, precision in the presentation of factual evidence, a clear and accurate writing style, and, no less important, a distinct type-face. The great challenge of the textbook author was at once to be accurate and to balance the difficult task of judging and selecting from the rapidly growing content of physical knowledge, and to find a pedagogical but engaging expository style. Lovering considered that a physics textbook should not confine itself to presenting the core principles of this field of knowledge, since these were at the time not numerous. More fundamental was the presentation of the 'history of discovery', the scientific instruments used in this process, and the factual evidence and experiments that illustrated the general laws obtained. But a good textbook should still portray the unity of nature and the interconnected character of the physical sciences.

In Lovering's opinion, the shortage of good physics textbooks in English was not due to a lack of major research, but to the type of publications through which British practitioners had communicated it. The monographs and articles composing the series of the *Cabinet Cyclopaedia*, the *Library of Useful Knowledge*, the *Penny Magazine*, and the *Encyclopaedia Metropolitana* had many qualities. But they lacked 'unity of thought', comprehensiveness, and up-to-dateness, and their subject arrangement, narrative, rhetoric, and expository focus made them inadequate vehicles for communicating physics to students, being often too academic or too 'popular'.

According to Lovering, the only proper physics textbooks available in English were also deficient. Examples included those by Golding Bird, lecturer at Guy's

Hospital, London, James William McGauley, professor for the Irish Board of National Education, and Denison Olmsted, professor of natural philosophy and astronomy at Yale College. This was in contrast with the rich French and German traditions of physics textbooks represented by a large number of exemplary authors such as the French Pouillet, Péclet, Pinault, Becquerel, Despretz, Regnault, and Lamé, and the Germans Müller and Peschel.

Indeed, in France and the German states, reforms leading to the establishment or refashioning of structures of secondary and higher education with broad national scope took place in the early nineteenth century, sooner than in other national contexts (Green, 1990). In the same period the making of physics as a discipline was boosted by its inclusion in the curricula of secondary education. These developments went hand in hand with the production of textbooks, which in many cases had a foundational role and were powerful tools for establishing, shaping, and standardizing pedagogy and physics.

In France, a network of secondary schools had come into existence soon after the Revolution (1795–1802). Their curricula were predominantly focused on the sciences, and in this context textbook authors such as Mathurin-Jacques Brisson (1723–1806) and Antoine Libes (1752–1830) linked the *ancien régime* tradition in physics and its public communication with the new Napoleonic framework of formal education. Their physics and pedagogy were closely connected to those of previous authors such as Jean-Antoine Nollet (1700–1770) and Joseph-Aignan Sigaud de Lafond (1730–1810), who in turn had appropriated the Dutch tradition of Newtonian physics, represented by Willem Jacob 'sGravesande (1688–1742) and Pieter van Musschenbroek (1692–1761). The focus on instruments and their description and use in experimental demonstrations characterized the works of French authors in this period (Fournier-Balpe, 1994; Heilbron, 1982).

The production of textbooks was considered a fundamental endeavour during the years of the Revolution and the subsequent Napoleonic regime.[7] Initially, however, it was not obvious to physics teachers that they should adopt texts which they had not written themselves. But, driven by political, commercial, professional, and pedagogical forces, textbooks increasingly moved to a central position. The Napoleonic reforms undertaken between 1806 and 1811 included the preparation of textbooks for every subject in the school curriculum in parallel with the publication of syllabuses.

The first science syllabuses were designed by a committee of university professors, which commissioned a physics textbook by the mineralogist René-Juste Haüy (1742–1822), who had taught this subject at the École Normale. Haüy's *Traité de physique élémentaire* (1803) went through five editions and was translated into English and German. The official textbook recommendations were complemented with a translation of the *Lehrbuch der mechanischen Naturlehre* (1805) by the Berlin secondary-school teacher Ernst Gottfried Fischer. The translation was entrusted to Jean-Baptiste Biot (1774–1862), who also published two physics textbooks which were recommended in successive school curricula, to the preparation of which he contributed.

Biot's two textbooks appeared within a year of each other—a key event that helps to explain how textbook physics developed as a genre during the nineteenth century. His *Traité de physique expérimentale et mathématique* (1816) compressed in four volumes his vision of physics as a discipline. The book opened with a quotation from Newton's *Principia*, which communicated his intention of reducing all physical phenomena to simple laws based on mathematical analysis and experimental precision, following the works of Newton and Laplace, and his own researches. A year later, Biot published his *Précis élémentaire de physique expérimentale* (1817), prefaced this time by a quotation from Francis Bacon's *Novum Organum* and reduced to two volumes. In the preface he explained that he had resolved this time to present the facts in a purely experimental manner, devoid of any complex algebraic calculus, in order to address students interested in physics as a preparation for studies such as medicine or natural history. Medical students, constituted in fact the largest body of physics readers in France for most of the century. From the 1820s they were obliged to take the *baccalauréat-ès-sciences*, which was the same examination attempted by the small number of students wishing to go on to scientific or engineering studies.

Although Biot's shift of emphasis from Newton to Bacon had obvious pedagogical and commercial reasons, it also constituted a key aspect in the shaping of textbook physics in the nineteenth century, since the non-mathematical, experimental focus would characterize most textbooks of the period. Biot's textbooks circulated widely in France and abroad. His *Traité* was translated into German, Italian, and Spanish. And his two textbooks were used by John Farrar, Harvard's professor of mathematics and natural philosophy, to prepare his course on natural philosophy, which had a further edition in the 1840s, edited by Farrar's successor, Joseph Lovering.

Biot's work also inspired a new generation of French authors, who started to publish textbooks from the late 1820s. These authors were typically trained at the École Normale, the institution established to supply French secondary education with teachers.[8] They published the first edition of their textbooks at the beginning of their careers. The publication of a textbook was a useful strategy for developing pedagogical tools for the classroom, but it also had a major role in boosting scientific prestige and in career promotion. Most physics textbooks published in France during the first half of the nineteenth century can be seen as resulting from the intense competition in the fashioning of careers and the circulation of *normaliens* throughout the French educational structure.

In this period the French market for physics textbooks was dominated by five authors: Eugène Péclet (1793–1857), Claude Pouillet (1790–1868), César Despretz (1789–1863), Auguste Pinaud (1812–47), and Nicolas Deguin (1809–60). All were *normaliens* except Despretz. In general, their textbooks were originally designed for use in secondary schools, and their first editions coincided with the expansion of the number of students aspiring to the *baccalauréat-ès-sciences* as a qualification for entry to medical studies. The first editions of Péclet's, Pouillet's, and Despretz's textbooks were published in the 1820s, and those by Pinaud and Déguin shortly after 1837, when the *baccalauréat* requirement was re-established after a brief period of suspension.

From 1829 Péclet lectured in physics at the École centrale des arts et manufactures, newly established (in part by Péclet himself) with the aim of training engineers for senior managerial positions in industry. His textbook, whose first edition was published between 1823 and 1826, owed much to his early experience as a secondary-school teacher. It also presented both his research and his teaching at the École centrale, since he did not draw a clear boundary between his journal publications and textbook writing. Péclet's *Traité élémentaire de physique* went into its fourth edition in 1847, by which time he had risen to the rank of general inspector within the French educational system. However, his refusal to take an oath to the new political regime established after Louis-Napoléon Bonaparte's *coup d'état* of 1851 led him to resign from all his positions and from further textbook writing.

Pouillet too had a career that was cut short by his refusal to swear allegiance to the new political regime. If Péclet represented the connection with French private industry, Pouillet, as the director of the Conservatoire des arts et métiers had a close relation with the milieu of invention and machine and instrument-making. Between 1820 and 1829 he taught physics at one of the leading secondary schools in Paris. He also assisted Jean-Baptiste Biot and Louis-Joseph Gay-Lussac (1778–1850), and obtained a chair of physics at the Paris faculty of sciences. In addition, he held the physics chair at the Conservatoire, becoming the director there in 1832.

In 1827 Pouillet published his *Élémens de physique expérimentale et de météorologie*, in three volumes, which he based on his teaching at the faculty of sciences and dedicated to Biot. From his position at the Conservatoire, Pouillet was a pioneer in physics textbook illustration. The Conservatoire had been established in 1794 as a repository of knowledge about instruments and machines. Pouillet and the teachers of drawing at the Conservatoire were involved in the *Portefeuille industriel*—a compilation of drawings of the major inventions presented to the institution by law. These were used by Pouillet in his textbooks (and copied by many authors), keeping readers abreast of major innovations in instrument design. By 1856 Pouillet's *Élémens* had gone through seven editions, becoming the leading physics textbook in France, and generating translations into Italian, Spanish, and German. Pouillet's writing was directed at the intersection between secondary and higher education, but his three-volume textbook also constituted a comprehensive reference work for the use of researchers, in the fashion of Biot's first treatise. For this reason, in 1850 he published another textbook better suited to secondary-school pupils.

Pouillet's *Notions générales de physique et de météorologie* was cast as a more elementary introduction to the subject in one volume. The book pioneered in France the insertion of illustrations in the text, as well as the use of wood engraving, which allowed for a more accurate and attractive representation of physics instruments than had been possible with copper or zinc plates. So long as these older techniques were used, illustrations had to appear as folded plates at the end of books or in a separate volume. The new page layout had previously been used in British and German textbooks, such as those by Golding Bird and Johann Müller. But French expertise in instrument drawing, engraving, and low-cost printing could not be matched in other countries. This did much to promote the international success of French textbook physics during the second half of the nineteenth century. It also helped to reinforce a disciplinary account of physics focused on the description of scientific instruments.

Pouillet's work also had an important role in the production of a leading German textbook. The *Élémens* was freely translated and edited by Johann Müller (1809–1875) as *Lehrbuch der Experimentalphysik und der Meteorologie* (1839–43). Subsequently, Müller published the *Grundriss der Physik und Meteorologie* (1846)—an abridged version of his *Lehrbuch*, addressed to schools. In the early nineteenth century, French textbooks had a considerable impact on the German textbook tradition. Haüy's *Traité* was translated twice, by the mineralogists Johann Georg Ludolf Blumhof (1771–1825) and Christian Samuel Weiss (1780–1856), respectively. However, translations were in general not literal; instead they adapted the original texts to the different scientific and educational contexts for which they were intended. Weiss, for example, included new content which underlined the contrast between Haüy's Laplacian programme and that of German Romantic philosophy.

Biot's textbooks also occupied a major place in German translations from French originals. His *Traité* was translated by Gustav Theodor Fechner (1801–1887), who used this experience to complement his university training in science while preparing for a medical degree. Although he later turned to experimental psychology, his translations did much to help him secure his first university position in physics at Leipzig. The *Précis* was translated into German by the Berlin secondary-school teacher Friedrich Wolff (1766–1845).

The German states were fertile sources of physics textbooks in the eighteenth century. Important works included those by Johann Andreas Segner (1704–1777), Georg Christoph Lichtenberg (1742–1799), Johann C. P. Erxleben (1744–1777), Johann Peter Eberhard (1727–1779), Christian Gottlieb Krantzenstein (1723–1795), and Karl Wilhelm Gottlob Kastner (1783–1857), among others. Most textbooks were written in the middle of an author's career, as tools for their university or school-teaching. Between the late eighteenth century and the early nineteenth century, around a fifth of all physics professors in universities, technical schools, and academies seem to have written a textbook on their subject. An important though decreasing part of this production was penned by authors with medical training. Many textbooks were written in Latin, though from the 1780s Latin increasingly gave way to German. While the eighteenth-century tradition contributed to shape subsequent physics textbooks, few such works survived beyond this period, owing to the growth of a new educational framework in the early nineteenth century (Clark, 1997 and 1997b; Heilbron, 1982).

In the three decades following the Napoleonic invasion, profound educational reforms took place in the German states, resulting in a wide-ranging structure of secondary schools, technical and vocational schools, and teacher-training.[9] The traditional secondary school (*Gymnasium*) favoured the classical curriculum and monopolized access to the universities through the *Abitur* examination. However, during the middle decades of the century, different course programmes coexisted in these schools, and the sciences found an ever more prominent place. Moreover, new types of school with modern curricula including greater emphasis on science and technology were developed, such as the *Realschulen, Realgymnasien, Oberrealschulen,* and a set of trade schools that subsequently led to the creation of the German Polytechnic (Green, 1990; Müller, 1987).

Between the 1820s and the 1860s, Ernst Gottfried Fischer's *Lehrbuch der mechanischen Naturlehre* was one the books most used in German schools. It placed a relevant emphasis on the role of mechanical approaches in physics and on demonstration and measuring instruments, together with the use of mathematical methods. Other textbooks, such as F. Kries's *Lehrbuch der Naturlehre* and J. Heussi's *Die Experimentalphysik, methodisch Dargestellt,* were less popular or in certain cases did not find a significant market until the late 1840s. Early editions of Fischer's book were used in France and translated into English, Polish, Dutch and Italian, from Biot's French edition. But in the German states, Fischer's textbook had a major role in the mid-nineteenth century as an agent driving educational reforms which led to a more systematic provision of physics cabinets and laboratories in schools (Olesko, 1989, 110–2).

Some of the more successful German textbooks that appeared in the mid-century had a strong experimental focus; they were often produced by professors in technical and secondary schools, and they bridged the gap between school and university teaching.[10] Among them were Hans Anton Brettner (1799–1866), a *Gymnasium* professor of mathematics and physics, Gustav Wilhelm Eisenlohr (1811–1881), professor at the Kalsruhe Polytechnic, and Johann Müller, a secondary-school teacher at Darmstadt and Giessen, and, subsequently, professor of physics and technology at the University of Freiburg. Brettner's *Leitfaden für den Unterricht in der Physik* went through twenty editions between 1836 and 1882. Eisenlohr's *Lehrbuch der Physik* (1836) went into its eleventh edition forty years after its appearance. Müller's abridged textbook for schools, the *Grundriss*, had fourteen editions between 1846 and 1896, and it was translated into English, Dutch, Norwegian, and Danish. Its general course was complemented by a more thorough guide on the practice of lecture demonstrations and laboratory experiments prepared by Joseph Frick, a secondary-school teacher in Freiburg. Frick's *Physikalische Technik* (1850) went into successive editions through to 1909. Decades later, Frick's aims lived on in Adolphe Ferdinand Weinhold's *Vorschule der Experimental Physik* (five editions, 1872–1907). Müller's first textbook had also a very long publication history which fundamentally transformed it, though the book always preserved in its title page the trace of its origins (Müller-Pouillets *Lehrbuch der Physik*). With a growing number of contributors and editors, its five volumes had gone through eleven editions by the 1930s, becoming a standard reference work for students and researchers alike.

During the second half of the nineteenth century, instruction in physics in German universities was increasingly dispensed in seminars. Although the seminars were mainly aimed at training school-teachers, they implemented an educational regime based on the solving of problems and the development of skills in experimental research and mathematical analysis that went beyond the curricula contained in the original textbooks by Brettner, Eisenlohr, or Müller (Olesko 1991). This key turn is well illustrated by the reports on the appointment to the chair of physics at the University of Heidelberg in the 1850s. While the committee, chaired by Robert Bunsen, gave due weight to the success as textbook authors of such candidates as Eisenlohr and Müller, it considered that successful textbook writing and specialized research had become mutually exclusive. Hence the professorship was given instead to the distinguished researcher Gustav Kirchhoff (1824–1887) (Jungnickel, and McCormmach, 1986, 188–9).

An exemplary product of this seminar culture was Friedrich Kohlrausch's *Leitfaden der praktischen Physik* (1870), based on his practical exercises for first-year students in the physico-mathematical seminar at Göttingen. Eleven editions of the book had appeared by the First World War, and a 24th edition was published in 1996. Under successive editorial teams, it was the most successful of a range of German practical physics textbooks that included, in the nineteenth century, Adolph Wüllner's *Lehrbuch der Experimentalphysik* (1862–1907) and Eilhard Wiedemann's *Physikalisches Praktikum* (1890–1924) and, in the twentieth century, Wilhelm H. Westphal's *Physikalisches Praktikum* (1938–1971) and Waldemar Ilberg's *Physikalisches Praktikum* (1967–2001) (Olesko, 2005).

German textbook physics spread internationally in the late nineteenth century and early twentieth century, but its interaction with the French tradition did not cease completely. Wüllner's book, for example, owed much to Jamin's *Cours de physique de l'École Polytechnique* (1858–1866). Jamin's textbook in four volumes went into its fourth edition in 1906 and was a respected work used in teaching and research all over Europe and the Americas. His textbook took over from Gabriel Lamé's coursebook at the École Polytechnique, published in the late 1830s, and was contemporary with one by the *normalien* Émile Verdet based on lectures at the same institution (1868–69). With rather less success, in 1870 Jamin published a one-volume textbook for secondary-school students. His aim was to give a more central position than was the case in existing introductory textbooks to the unification of physics, following on from the acceptance of the mechanical equivalent of heat (Jamin, 1870).[11]

French textbook physics continued to exert an influence internationally in the second half of the nineteenth century. Its strength lay especially in one-volume treatises exploiting the connection between secondary education and university training through the *baccalauréat-ès-sciences*. Among these, Adolphe Ganot's *Traité élémentaire de physique expérimentale et appliquée* (1851) stands out for its global impact. The *Traité* grew from a fruitful interaction between medical and scientific training, instrument-making, and printing practices in mid-nineteenth-century Paris. It offered a comprehensive coverage of the material required by a private teacher preparing candidates for examinations. Ganot's pedagogical experience, combined with the dynamism of the French school context and the international projection of the French book trade, contributed to making the *Traité* an international best-seller. Ganot's *Traité* was in its 25th edition by 1913, was translated into twelve languages, and had a comparable number of editions in English and Spanish. Through its original editions or translations, it became an international standard in school and university teaching.

Ganot's textbook introduced important novelties such as a section of problems related to the main syllabus and to real examinations, and a new visual language in textbook physics. It also displayed the author's conception of physics as a discipline, and its illustrations of instruments and machinery were used for research purposes by leading practitioners such as Zénobe Gramme, William Thomson, and Sebastian de Ferranti. The work's visual dimension benefited from developments in printing and engraving generated by a long interaction between British and French print technicians. Its illustrations were used or copied in many other physics textbooks worldwide. Ganot's *Traité* served as an influential model for textbook physics, strongly focused on the description of instruments and experimental procedures, which survived into the early twentieth century.[12]

Ganot's textbook physics took root in a soil characterized by intense competition between authors. During the second half of the century other major introductory texts were published by former *normaliens*, such as those by Charles Drion and Émile Fernet (six editions, 1861–1877), Augustin Boutan and Joseph Charles d'Almeida (five editions, 1862–1884), Pierre Adolphe Daguin (four editions, 1855–1879), and Augustin Privat Deschanel (1868). Many of these works were translated into several languages. Among them, the English translation of Deschanel's textbook enjoyed great success (seventeen editions, 1870–1913), becoming, like Ganot's textbook, a standard work in Anglophone countries.

Three decades after Lovering's mid-century diagnosis, Silvanus P. Thompson offered a useful update. According to him, for decades British contributions had been limited to the textbooks published during the first half of the century by the medical doctors Neil Arnott and Golding Bird. In contrast, the French had at their disposal those by Verdet, Jamin, Daguin, Fernet, Boutan and d'Almeida, and Ganot, while German originals included those of Müller and Wüllner as well as numerous other authors. For a long time, the best textbooks on the British market were the translations of those by Ganot and Deschanel. The former, though, had become too encyclopaedic and conservative, despite the efforts of Edmund Atkinson (1831–1900), its translator and editor. The latter, on the other hand, had been improved in its English translation thanks to the editorial work of Joseph David Everett (1831–1904) (S. P. T., 1884).

The role of textbook editors was fundamental. The English editions of Ganot and Deschanel owed their success not only to the qualities of the original works but also to their adaptation by editors working in the rapidly expanding world of British schools and colleges. In contrast, textbooks such as Müller's, which was translated into English shortly after its publication, had a short life in Britain because of the lack of such connections.

Thompson's perception was partially inaccurate. In Britain and America, as in many other countries, there was a fertile tradition of textbook-writing in physics, albeit one that started later than in France and the German states and was generally unable to displace the texts of French and German origin. Moreover, the influence of French and German textbook physics in this British and American tradition is unmistakeable. But historians have tended to concentrate their research on late-nineteenth-century British and American physics, and consequently little work has been done on textbook physics in previous decades. The study of this period would certainly offer new ways of understanding the making of physics as a discipline in these national contexts.

Further research is still required, but the list of authors is long. On the British side, authors such as James Renwick, Dionysius Lardner, John Tyndall,[13] Charles Buckmaster, Richard Wormell, Robert Hunt, Jabez Hogg, John Charles Snowball, and Isaac Todhunter were successful in their national market and shaped the teaching of physics in connection with the development of systems of school and university examinations in the second half of the century (Newton, 1983, 1983b and 1983c; Simon, 2011). On the American side, the same can be said of authors such as J. L. Comstock, John Johnston, Denison Olmsted, Benjamin Silliman Jr., Leonard D. Gale, Richard Green Parker, Elias Loomis, George Payn Quackenbos, and Elroy McKendree Avery (Kremer, 2011; Shank, 1952).

Thompson, nonetheless, stressed the distinctiveness of the emerging style of British textbook physics, with its strong focus on the genuinely British principle of energy conservation. He highlighted the value of Balfour Stewart's *Lessons in Elementary Physics* (1870), which through its novel structure did much to spread an understanding of the principles in William Thomson and Peter Guthrie Tait's advanced *Treatise on Natural Philosophy* (1867) (S. P. T., 1884). Its success marks a turning point in British textbook physics, for it was translated into ten languages, including German and Spanish.

In addition, in the context of the expansion of laboratory instruction, British and American publishers did not rely exclusively on translations of German textbooks such as those by Kohlrausch (1873) and Weinhold (1875). They also produced influential textbooks of their own, including Frederick Guthrie's *Practical Physics* (1878) and Edward C. Pickering's *Elements of Physical Manipulation* (1873–1876) (Hentschel, 2002; Kremer, 2011; Simon, 2011).

Textbook physics emerged as a genre in the nineteenth century in the context of educational reforms that enlarged the publics for physics. Physics textbooks were both educational tools and vehicles for communicating knowledge of research physics. Introductory textbooks compressed all knowledge on physics in a single volume, and had a major role in the fashioning of physics as a discipline. Writing an introductory textbook was not a simple matter, nor was it a skill that all physicists possessed. Like research, it required practice and training.

The publication of advanced physics textbooks favoured the development of mathematical physics. But the core of nineteenth-century textbook physics was characterized by its focus on the description of scientific instruments and machines, and experimental practices. This emphasis responded to the need to promote an emerging discipline. But in addition, it was connected to a conception of physics that favoured knowledge about instrument design, experimental practices, and 'applied science' over mathematical and theoretical approaches.

Many nineteenth-century physics textbooks had a long and successful publishing history that continued into the early twentieth century and turned them into classics. However, their structure suffered, in general, from the disciplinary growth and mounting specialization of physics and from the death of their authors. Towards the *fin de siècle* a new style of textbook physics started to emerge. It coexisted for many decades with nineteenth-century patterns, but it became increasingly the work of physicists based in higher education or research institutions, as well as large editorial teams.

21.4 Textbook Physics versus Frontier Physics?

The PSSC project, developed from the 1950s in the US, is an excellent example of a new departure in the production of physics textbooks, one that involved large teams and was driven by academic physicists. This was a relatively new turn, since there were already textbooks produced collectively, such as Müller-Pouillet's *Lehrbuch der Physik*. Furthermore, nineteenth-century physics textbooks that became classics were able to maintain their dominant position because a succession of competent editors assumed responsibility for updating them regularly.

The intervention of academic physicists in school textbooks, as we have seen, was usual in the nineteenth century, but secondary education had its own logic, teachers, and authors. In different periods, academic physicists such as Jamin in France, Stewart in Britain, and Millikan in the US wrote textbooks for the secondary-school market, because they were dissatisfied with the type of physics communicated by most school authors. Like the PSSC members, they thought that textbooks had a major role in the definition of physics as a discipline—not only because they were central to the training of physicists, but also because their contents and approaches contributed to shape those of physics as a field of knowledge. The PSSC members were academic physicists who already had experience as authors of school and college textbooks, and of advanced physics textbooks which developed areas in modern physics such as relativity and quantum mechanics.

But, as John Rudolph has shown, the PSSC was notable for its political support and abundant resources and for its use of managerial techniques developed in large-scale scientific projects connected to the Second World War. This project seems to have raised the production of pedagogical materials for physics, in particular textbooks, to a new level. It was certainly not the first project to promote a radical reform of school curricula and of the structure and approach of textbook physics, but it was certainly one of the most striking examples of this in the twentieth century.[14]

The textbook series produced by the PSSC illustrate the fundamental schism between two traditions and functions—those of the 'treatise' and the 'textbook'—which previously had constituted a single genre. The PSSC textbooks no longer aimed at providing a comprehensive overview of the subject matter of physics. On the contrary, they omitted major areas of physics and concentrated instead on a more limited range of topics. But this was considered the best option in order to be able to focus on more fundamental aspects. It was essential to write a course that would accurately convey how academic physicists worked. This included an emphasis on the role of open inquiry—an essential quality not only of science but also of democratic ideals in American politics (in contrast with what American politicians regarded as Soviet absolutism) (Rudolph, 2002).

The PSSC course also left aside other elements that had been fundamental in textbook physics up to the mid-twentieth century, such as history and technology. The characteristic nineteenth-century focus on scientific instruments and industrial machinery had been preserved and further developed along with the increasingly pervasive presence of technology in twentieth-century society. But the PSSC sought to distinguish clearly between science and technology, with the aim (among others) of circumventing the governmental pressures to continue the technological enterprise boosted by the war projects (Easley Jr., 1959; Rudolph, 2002; White, 1960). Nineteenth-century textbook physics gave an important role to accounts based on the history of discovery and experiment. The PSSC rejected this historical approach, favouring instead a more conceptual presentation of physics that stressed its coherence and unity as a field of knowledge.[15]

In spite of its pedagogical impact, the PSSC was not the only successful textbook project in this period, and other textbooks favoured different approaches. For instance, the (Harvard) Project Physics Course, led from the 1960s by Gerald Holton, F. James Rutherford, and Fletcher G. Watson, prioritized the integration of the history of physics in its textbook series. Holton and Watson had been involved in the PSSC, and they used similar techniques in writing their textbooks. But they also remained true to their previous participation in the General Education in Science programme. Furthermore, they took special care to balance the forces involved in physics textbook design—physicists, teachers, and educators, a triad clearly represented by the three individuals directing the project (Holton, 1967; 1978).[16]

The PSSC reformers presented their programme as a revolution introducing a new paradigm in the teaching of high-school physics. Paradoxically, its course materials, with its textbooks at the core, were not considered easy to use. Teachers had to be trained in how to understand and exploit these textbooks, and the key issue was that without a body of teachers predisposed to empathize with the programme and its aims, its textbooks would be of limited value, as few teachers would accept them, change their pedagogical approach, and make effective use of the new materials (Donahue, 1993; Easley Jr., 1959).

The pattern is Kuhnian, if only because Kuhn and the PSSC reformers shared the same intellectual, cultural, and historical context; and the subsequent success of the PSSC programme endorses some of Kuhn's claims, especially in relation to the power of textbooks to drive the communication of scientific knowledge. It is remarkable that in this context the use of political and economic power by the PSSC in order to displace competing communities of practitioners (teachers and educators) partially undermined the democratic ideology underlying it. Furthermore, the progress of the programme was slow and incomplete. One reason was that the PSSC textbooks competed with many others. Certainly, they greatly influenced subsequent physics textbook production in Anglophone countries and beyond through translation. But in spite of its powerful funding and political support, the PSSC failed to win the battle for pedagogical supremacy.

Moreover, is it not paradoxical that in order to master the proper use of textbooks, school-teachers were subject to the same pattern of training as the one described by Kuhn for scientists and science? Or is it just that the PSSC physicists implemented in their programme measures that were standard in their cultural environment? Indeed, a leading aim of the PSSC was to introduce teachers to modern academic physics and acculturate them into behaving like academic physicists. They thought, like Kuhn, that textbook science was opposed to 'creative or frontier science'. Still, they intended to change textbook physics by making it converge with 'frontier physics', in parallel with the institutional takeover of secondary-school physics by university physics. But if we follow Kuhn, should we consider that academic physicists who were also textbook writers, such as those involved in the PSSC, were not performing creative work and that they were not creative contributors to the making of science? And is it possible to extrapolate from this case to a general discussion on the role of textbook writing in the making of science, including thorough historical evidence and historiographical arguments?

Most current research on physics pedagogy which deals with textbooks follows the lines of Kuhn's approach. Its main focus lies in understanding 'frontier science' and its shaping in academic circles. This is the case, for instance, with the work of Andrew Warwick and David Kaiser, which has been path-breaking in reshaping the study of pedagogy within the history of science. However, their study of training is constrained by its continuities with Kuhn's work, in relation to textbooks and their role (Warwick and Kaiser, 2005a).

In his study of the making of Cambridge mathematical physics, Warwick expresses respect for those educational works that he defines as classic treatises, such as Euclid's *Elements* and Newton's *Principia*. In contrast, he undervalues nineteenth-century textbooks. Warwick shows convincingly that Cambridge training in mathematical physics, based on preparation for examinations, had a capacity to define both pedagogy and physics that textbooks did not have. Textbooks on this analysis were in general partial and unable to absorb developments in the field. Warwick establishes that examinations and private teaching were the driving agents shaping the content and approaches of textbooks in this context, and not vice versa. Subsequently, these textbooks were agents in the diffusion of Cambridge mathematical physics in schools in Britain and the Empire (Warwick, 2003a).

In Warwick's account, textbooks play a secondary role, limited to the passive communication of developments in physics taking place somewhere else. His study offers a fine account of the Cambridge context of higher education. Most importantly, it reminds us of the coexistence of different pedagogical tools, such as examinations, curricula, pedagogical regimes, and textbooks in the shaping of science and education. However, the exemplary character of his study is limited by the specificity of Cambridge, and by his partial focus on a particular context of education. Thus, he dismisses other equally important contexts of knowledge production, such as the school system developing in nineteenth-century England due to a wide range of examination systems with national scope, including the Oxford and Cambridge Locals. In this, Warwick has a conception of the making of scientific knowledge which in the study of science education reproduces the top-to-bottom approach that has been widely criticized in research on science popularization (Simon, 2009 and 2011; Hiltgarner, 1990; Secord, 2004).[17]

In his study of the production and use of Feynman's diagrams in post-war physics, David Kaiser offers a much more sophisticated conceptualization of textbooks—one that in several ways goes beyond Kuhn. This is especially true with regard to the concept of textbooks as rigid repositories of paradigms. Most historians of physics are still interested in textbooks only as sources that show what knowledge was standard in a particular period, and they tend to consider that in times of 'normal science' all textbooks offer the same basic paradigmatic picture. For those who adopt this approach, analysing textbook narratives over time is only a way of locating changes of paradigm.[18]

While Kaiser follows this approach to a certain extent, he is more interested in illuminating the relations between physics and local pedagogical contexts. In his work he has shown that textbooks dealing with Feynman diagrams in post-war America offered a wide variety of perspectives.[19] He considers that Kuhn exaggerated the central role that textbooks play in the training of scientists—observing that other tools such as pedagogical regimes, problem-solving, and lecture notes could often have a more decisive role in this context. Kaiser demonstrates that the writing of advanced textbooks was a creative task. But he also shows that although the circulation of textbooks had an important role in the spread of Feynman's diagrams and their uses, textbooks did not replace other coexisting modes of knowledge communication (Kaiser, 2005a, 253–79).

Subsequently, Kaiser has provided a rich overview of American textbooks of quantum mechanics, including comparisons with other national contexts. He has observed that a process of textbook standardization took place simultaneously with the maturing of this subdiscipline, but also in connection with local pedagogical practices. In the interwar years, quantum-mechanics textbooks displayed a diversity of approaches, especially with regard to philosophical convictions and points of interpretation. However, by the early 1950s, they offered a far more homogeneous perspective, which abandoned philosophical inquiry in favour of a pragmatic approach focused on the solving of more practical and quantitative problems. Although the field had indeed matured in these decades, this homogenization cannot be understood except in the context of changing student enrolments (Kaiser, 2007 and forthcoming).

Josep Simon's study of the production, circulation, and appropriation of Ganot's textbook physics in nineteenth-century France and Britain has points in common with Kaiser's approach, in that it demonstrates the creativity of textbook writing and its important role in the fashioning of scientific disciplines. However, Simon's work puts textbooks at the centre of analysis; it proposes a strong position with regard to education (not only pedagogy) as a powerful driving force in the making of scientific knowledge, attaches greater significance to students and readers, and places the locus of the disciplinary genesis of physics at the frontier between secondary and higher education.

The thrust of Simon's work lies in his proposal for a new way of writing the history of physics through an interdisciplinary analysis of school textbook physics. In bringing textbooks to centre stage, he uses approaches drawn from the history of the book that connect the material production of textbooks with the shaping of the knowledge they contain. Furthermore, he argues that the main reason why secondary-school textbooks, such as those by Adolphe Ganot, can be considered as classic or canonical lies in the number and diversity of their various readerships in different cultural, national, and social spheres, and the active role of readers in providing texts with meaning. He uses historical evidence to transcend the boundaries between the making and communicating of scientific knowledge, and in this way he presents nineteenth-century authors of introductory textbooks as contributing to the making of physics as a discipline (Simon, 2011).

In contrast, the main aim of both Kaiser and Warwick has been to investigate the making of frontier science, through the Kuhnian question of how students in higher education become part of a disciplinary community and come to assimilate the 'paradigms' of their discipline. Their concept of pedagogy as knowledge communication is unidirectional, since they consider that only a selective number of students and readers could appropriate the knowledge acquired in their training, and endow it with original meanings. As a result, their analysis of textbook pedagogy becomes a complementary tool enriching our understanding of scientific research practices. But it ignores the opportunity of going a step further in considering the transformative power of textbooks and their uses on scientific knowledge.

The reworking of Kuhn's approach to pedagogy and textbooks by Kaiser and Warwick has many virtues, however. It matches pretty well how physics was fashioned in post-war American higher education, the very context from which Kuhn's work emerged. But it fails to provide a larger explanatory framework for understanding textbook physics in the long run, and beyond academic circles in leading centres of research. On the other hand, Simon's strong thesis on the making of physics as a discipline through school textbooks still needs to be tested by further case-studies going beyond the nineteenth century and dealing with other local and national contexts.

The complexities and contradictions of this field of study are clearly expounded in a valuable volume on textbooks of quantum physics edited by Jaume Navarro and Massimiliano Badino. The book provides a rich panorama of such textbooks in the first three decades of the twentieth century and sufficient evidence to demonstrate the creative character of textbook writing and the interaction between research and pedagogical practices. This is a much-needed set of case-studies which includes examples of major textbooks by authors such as Arnold Sommerfeld, Max Born, Paul Drude, Wolfgang Pauli, and Paul Dirac, with others by less well represented actors in the history of this fundamental field of modern physics (Navarro and Badino, 2013).[20]

Michael Eckert's analysis of Sommerfeld's *Atombau und Spektrallinien* shows elegantly how the interaction between different spheres of science communication (popular lecturing, university seminars, colloquia and doctoral supervision) can converge in textbook writing, it demonstrates the role of textbooks in building scientific reputation and also their collective dimension as works of research schools (Eckert, 2013). Clayton A. Gearhart's work on Fritz Reiche's *Die Quantentheorie* is particularly interesting for its illustration of the close ties between journal science and textbook science (Gearhart, 2013). Don Howard's study of Pascual Jordan's *Anschauliche Quantentheorie* suggests convincingly the connection of textbook writing with politics—an aspect that is also mentioned in other chapters in the volume (Howard, 2013). Helge Kragh introduces Dirac's *Principles of Quantum Mechanics* as a notable example of textbook writing with the aim of endowing a developing field of knowledge with new conceptual foundations (Kragh, 2013).

All contributions make good use of sources such as reviews, university calendars, biographical memoirs, and private correspondence, in addition to textbooks themselves. But, in general, they fall short in exploiting the potential of tools available to historians of science books, such as the analysis of literary replication of the kind practiced by James Secord, the active role of printers, booksellers, and readers emphasized by Jonathan Topham and Adrian Johns, and the conceptualization of scientific practice proposed by historians interested in the study of science communication (Secord, 2000; Topham, 2000 and 2009; Johns, 1998; Simon, 2009).[21] A key question is obviously how studies focusing on the analysis of textbooks might be able to make an original contribution to leading trends in the history of science. To fully exploit this possibility, it is reasonable to think that a wider range of approaches should be used, combining conventional perspectives in the history of science with others developed in different disciplines, such as history of education, science education and book history.

21.5 CONCLUSIONS

Textbooks have traditionally been standard sources in the history of physics, but in general they have remained marginal in the formulation of large historical questions and the construction of original historical arguments. Kuhn's work helped to draw attention to textbooks and to give them a status as objects of historical study. In their tendency to consider this type of publication as uncreative and dogmatic, Kuhn's ideas were not new. However, he treated textbooks as central agents in the integration of students into the institutionalized and conceptual framework of science. Kuhn's approach has been useful in highlighting the power of textbooks in communicating science and providing a compact characterization of scientific disciplines. Moreover, it has done much to advance the study of physics textbooks produced around the period in which the *Structure of Scientific Revolutions* was conceived, especially those produced close to the cutting edge of academic research. Kuhn's view is partial, however, and it reproduces hierarchies in the organization and public perception of science and education that are tied to particular political, cultural, and social contexts.

It would be naïve to think that the fundamental characteristics of textbooks do not change over time and space, just as physics, education, and print and book trade cultures do. In fact, it is this plurality and temporality that makes them excellent sources on which to base historical questions. Textbooks are different from other types of book because their successive editions are connected to changes in scientific disciplines, educational structures, and pedagogy. They are at the interface between a wide range of forces that converge in education and shape political, social, and economical contexts. They are subject to translations into different languages, which are often active appropriations of their contents and form within different cultures of science and pedagogy. This cross-cultural characteristic and the longevity of certain textbook enterprises have transformed some physics textbooks into scientific, educational, and cultural canons. In this sense, they merit the same attention as works traditionally considered as classics in physics.

If textbooks can contribute to new ways of writing the history of physics, they have to be studied in their own right just like other sources, such as journal papers, laboratory notes, or even popular books, which currently have a higher status in the historiography of physics. Although Kuhn's approach to textbooks squared well with the educational reforms in physics education taking place in mid-twentieth century America and the rise of academic physics as a social and political force, it is inadequate for dealing with other equally fundamental contexts, such as that of school physics. Different questions emerge when one focuses on school textbooks, instead of concentrating exclusively on academic research. The geographical scope and readerships of textbook physics are quantitatively far greater than those of other sources used in the history of physics. This does not mask the fact that there are hierarchies, with contributions that are qualitatively more important than others in the making and development of physics as a discipline. But, in the light of historical evidence, there should be no doubt that, overall, textbook physics is simply physics.

Bibliography

C. Balpe, 'L'enseignement des sciences physiques: naissance d'un corps professoral (fin XVIIie–fin XIXe siècle)', *Histoire de l'éducation*, 73 (1997), 49–85.

B. Bensaude-Vincent, A. García Belmar, and J. R. Bertomeu Sánchez, *L'émergence d'une science des manuels: les livres de chimie en France (1789–1852)*, Éditions des archives contemporaines, Paris, 2003.

A. Bird, 'Thomas Kuhn', in E. N. Zalta, (ed), *The Stanford Encyclopedia of Philosophy*, 2009. <http://plato.stanford.edu/archives/fall2009/entries/thomas-kuhn>

J. H. Brooke, 'Introduction: The Study of Chemical Textbooks', in A. Lundgren and B. Bensaude-Vincent (eds.), *Communicating Chemistry*, Science History Publications, Canton, MA, 2000, 1–18.

W. E. Brownson and J. J. Schwab, 'American Science Textbooks and Their Authors, 1915 and 1955', *The School Review*, 2 (1963), 170–180.

S. G. Brush, 'The Wave Theory of Heat: A Forgotten Stage in the Transition from the Caloric Theory to Thermodynamics', *British Journal for the History of Science*, 2 (1970), 145–167.

———*The Kind of Motion We Call Heat: A History of the Kinetic Theory of Gases in the 19th Century*, North-Holland Publishing Company, Amsterdam, 1976.

——— 'How Theories Became Knowledge: Why Science Textbooks Should be Saved', in Y. Carignan, D. Dumerer, S. K. Koutsky, E. N. Lindquist, K. M. McClurken and D. P. McElrath (eds.), *Who Wants Yesterday's Papers? Essays on the Research Value of Printed Materials in the Digital Age*, Scarecrow Press, Lanham, MD, 2005, 45–57.

H. CHANG, *Inventing Temperature: Measurement and Scientific Progress*, Oxford University Press, Oxford, 2004.

A. CHOPPIN, *Les manuels scolaires: histoire et actualité*, Hachette, Paris, 1992.

W. CLARK, 'German Physics Textbooks in the *Goethezeit*, Part 1', *History of Science*, 35 (1997), 219–239.

——— 'German Physics Textbooks in the *Goethezeit*, Part 2', *History of Science*, 35 (1997b), 295–363.

I. B. COHEN, '[Review of *Physics, the Pioneer Science* by Lloyd William Taylor]', *Isis*, 34 (1943), 378–379.

——— 'Comments', *Isis*, 38 (1948), 149–150.

J. W. DAUBEN, M. L. GLEASON and G. E. SMITH, 'Seven Decades of History of Science: I. Bernard Cohen (1914–2003), Second Editor of Isis', *Isis*, 100 (2009), 89–93.

G. E. DEBOER, *A History of Ideas in Science Education: Implications for Practice*, Teachers College, Columbia University, New York, 1991.

M. A. DENNIS, 'Historiography of Science: An American Perspective', in J. Krige and D. Pestre (eds.), *Science in the Twentieth Century*, Harwood Academic Publishers, Amsterdam, 1997, 1–26.

D. M. DONAHUE, 'Serving Students, Science, or Society? The Secondary School Physics Curriculum in the United States, 1930–65', *History of Education Quarterly*, 33 (1993), 321–352.

J. A. EASLEY Jr., 'The Physical Science Study Committee and Educational Theory', *Harvard Educational Review*, 29 (1959), 4–11.

M. ECKERT, 'Sommerfeld's *Atombau und Spektrallinien*', in J. Navarro and M. Badino (eds.), *Research and Pedagogy*, epubli, Berlin, 2013.

C. FOURNIER-BALPE, *Histoire de l'enseignement de la physique dans l'enseignement secondaire en France au XIXe siècle*, PhD thesis, Université Paris XI, Paris, 1994.

S. FULLER, *Thomas Kuhn: A Philosophical History for our Times*, University of Chicago Press, Chicago, 2000.

A. GARCÍA-BELMAR, J. R. BERTOMEU-SÁNCHEZ and B. BENSAUDE-VINCENT, 'The Power of Didactic Writings: French Chemistry Textbooks of the Nineteenth Century', in D. Kaiser (ed), *Pedagogy and the Practice of Science*, MIT Press, Cambridge, MA, 2005, 219–251.

K. GAVROGLU, M. PATINIOTIS, F. PAPANELOPOULOU, A. SIMOES, A. CARNEIRO, M. P. DIOGO, J. R. BERTOMEU SÁNCHEZ, A. GARCÍA BELMAR and A. NIETO-GALAN, 'Science and Technology in the European Periphery: Some Historiographical Reflections', *History of Science*, 46 (2008), 153–175.

C. A. GEARHART, 'Fritz Reiche's 1921 Quantum Theory Textbook', in J. Navarro and M. Badino (eds.), *Research and Pedagogy*, epubli, Berlin, 2013.

S. GOLDBERG, *Understanding Relativity: Origin and Impact of a Scientific Revolution*, Birkhäuser, Boston, 1984.

A. GREEN, *Education and State Formation: The Rise of Education Systems in England, France and the USA*, Macmillan, Basingstoke and London, 1990.

E. H. HALL, 'The Relations of Colleges to Secondary Schools in Respect to Physics', *Science*, 30 (1909), 577–586.

K. HALL, *Purely Practical Revolutionaries: A History of Stalinist Theoretical Physics, 1928–1941*, PhD thesis, Harvard University, Cambridge, MA, 1999.

—— ' "Think Less about Foundations": A Short Course on Landau and Lifshitz's *Course of Theoretical Physics*', in D. Kaiser (ed), *Pedagogy and the Practice of Science*, MIT Press, Cambridge, MA, 2005, 253–286.

—— 'The Schooling of Lev Landau: The European Context of Postrevolutionary Soviet Theoretical Physics', *Osiris*, 23 (2008), 230–259.

J. L. HEILBRON, 'Éloge: Victor Fritz Lenzen, 1890–1975', *Isis*, 68 (1977), 598–600.

—— *Elements of Early Modern Physics*, University of California Press, Berkeley, 1982.

—— 'Applied History of Science', *Isis*, 78 (1987), 552–563.

K. HENTSCHEL, *Mapping the Spectrum: Techniques of Visual Representation in Research and Teaching*, Oxford University Press, Oxford, 2002.

S. HILTGARNER, 'The Dominant View of Popularization: Conceptual Problems, Political Uses', *Social Studies of Science*, 20 (1990), 519–539.

C. H. HOLBROW, 'Archaeology of a Bookstack: Some Major Introductory Physics Texts of the Last 150 Years', *Physics Today*, 52 (1999), 50–56.

G. HOLTON, '[Review of *Principles and Applications of Physics* by Otto Blüh and Joseph Denison Elder]', *Isis*, 47 (1956), 431–433.

—— 'Project Physics. A Report on Its Aims and Current Status', *The Physics Teacher*, 5 (1967), 198–211.

—— 'On the educational philosophy of the Project Physics Course', in *The Scientific Imagination*, Harvard University Press, Cambridge, MA, 1978, 284–298.

D. HOWARD, 'Quantum Mechanics in Context: Pascual Jordan's 1936 *Anschauliche Quantentheorie*', in J. Navarro and M. Badino (eds.), *Research and Pedagogy*, epubli, Berlin, 2013.

S. JACOBS, 'J. B. Conant's Other Assistant: Science as Depicted by Leonard K. Nash, Including Reference to Thomas Kuhn', *Perspectives on Science*, 18 (2010), 328–351.

J. JAMIN, *Petit traité de physique*, Gauthier-Villars, Paris, 1870.

A. JOHNS, 'Science and the Book in Modern Cultural Historiography', *Studies in the History and Philosophy of Science*, 29 (1998), 167–194.

E. B. JOHNSEN, *Textbooks in the Kaleidoscope: A Critical Survey of Literature and Research on Educational Texts*, Vestfold College, Tønsberg, 2001.

C. JUNGNICKEL and R. MCCORMMACH, *Intellectual Mastery of Nature: Theoretical Physics from Ohm to Einstein. The Torch of Mathematics 1800–1870*, University of Chicago Press, Chicago, 1986.

D. KAISER, *Drawing Theories Apart: The Dispersion of Feynman Diagrams in Postwar Physics*, University of Chicago Press, Chicago, 2005a.

—— *Pedagogy and the Practice of Science: Historical and Contemporary Perspectives*, MIT Press, Cambridge, MA, 2005b.

—— 'Turning physicists into quantum mechanics', *Physics World*, May (2007), 28–33.

—— 'A Tale of Two Textbooks: Experiments in Genre'. *Isis* 103 (2012), 126–38.

—— *American Physics and the Cold War Bubble*, University of Chicago Press, Chicago, [forthcoming], Ch. 4.

W. C. KELLY, 'Physics in the Public High Schools', *Physics Today*, 8 (1955), 12–14.

S. G. KOHLSTEDT, *Teaching Children Science: Hands-On Nature Study in North America, 1890–1930*, University of Chicago Press, Chicago, 2010.

H. KRAGH, 'Paul Dirac and *The Principles of Quantum Mechanics*', in J. Navarro and M. Badino (eds.), *Research and Pedagogy*, epubli, Berlin, 2013.

R. L. KREMER, 'Reforming American Physics Pedagogy in the 1880s: Introducing 'Learning by Doing' via Student Laboratory Exercises', in P. Heering and R. Witje (eds.), *Learning by Doing: Experiments and Instruments in the History of Science Teaching*, Franz Steiner Verlag, Stuttgart, 2011, 243–280.

T. S. KUHN, *The Structure of Scientific Revolutions*, University of Chicago Press, Chicago, 1962.
——'The Function of Dogma in Scientific Research', in A. C. Crombie (ed), *Scientific Change: Historical studies in the intellectual, social and technical conditions for scientific discovery and technical invention, from antiquity to the present (Symposium on the history of science, University of Oxford, 9–15 July 1961)*, Basic Books Inc., New York, 1963, 347–369.
——J. L. HEILBRON, P. L. FORMAN and L. ALLEN, *Sources for History of Quantum Physics: An Inventory and Report*, The American Philosophical Society, Philadelphia, 1967.
V. F. LENZEN, '[Review of *Mechanics, Molecular Physics, Heat, and Sound* by Robert Andrews Millikan, Duane Roller, and Earnest Charles Watson]', *Isis*, 27 (1937), 527–528.
G. LIND, *Physik im Lehrbuch, 1700–1850. Zur Geschichte der Physik und ihrer Didaktik in Deutschland*, Springer-Verlag, Lind, 1992.
[J. LOVERING], 'Elementary Works on Physical Science', *The North American Review*, April (1851), 358–395.
A. LUNDGREN and B. BENSAUDE-VINCENT (eds.), *Communicating Chemistry: Textbooks and Their Audiences, 1789–1939*, Science History Publications, Canton, MA, 2000.
J. A. MARCUM, *Thomas Kuhn's Revolution: An Historical Philosophy of Science*, Continuum, London, 2005.
C. MIDWINTER and M. JANSSEN, 'Kuhn Losses Regained: Van Vleck from Spectra to Susceptibilities', in J. Navarro and M. Badino (eds.), *Research and Pedagogy*, epubli, Berlin, 2013.
R. A. MILLIKAN, 'The Problem of Science Teaching in the Secondary Schools', *School Science and Mathematics*, 25 (1925), 966–975.
——*The Autobiography of Robert Millikan*, London, Macdonald, 1951.
——and H. G. GALE, *A First Course in Physics*, Ginn and Company, Boston, New York, Chicago and London, 1906.
D. K. MÜLLER, 'The process of systematisation: the case of German secondary education', in D. K. Müller, F. Ringer and B. Simon (eds.), *The Rise of the modern educational system: structural change and social reproduction 1870–1920*, Cambridge University Press, Cambridge, 1987, 15–52.
J. T. MURPHY and A. VALSAMIS, 'Obituary: Fletcher G. Watson, 1912–1997', *Bulletin of the American Astronomical Society*, 29 (1997), 1496.
National Research Council/American Institute of Physics, NRC-AIP Conference on the Production of Physicists', *Physics Today*, 8 (1955), 6–18.
J. NAVARRO and M. BADINO (eds.), *Research and Pedagogy: A History of Quantum Physics and its early Textbooks*, epubli, Berlin, 2013. <http://www.edition-open-access.de/sources/index.html>
D. P. NEWTON, 'A French Influence on nineteenth and twentieth-century physics teaching in English secondary schools', *History of Education*, 12 (1983), 191–201.
——'The sixth-form physics textbook, 1870–1980—part 1', *Physics Education*, 18 (1983b), 192–198.
——'The sixth-form physics textbook, 1870–1980—part 2', *Physics Education*, 18 (1983c), 240–246.
H. NOWOTNY, 'Laudatio for Gerald Holton, 1989 Bernal Prize Recipient', *Science, Technology, and Human Values*, 15 (1990), 248–250.
K. T. S. OLDHAM, *The Doctrine of Description: Gustav Kirchhoff, Classical Physics, and the "Purpose of All Science" in 19th-Century Germany*, PhD thesis, University of California, Berkeley, 2008.
K. OLESKO, 'Physics instruction in Prussian secondary schools before 1859', *Osiris*, 5 (1989), 94–120.

——*Physics as a Calling: Discipline and Practice in the Königsberg Seminar for Physics*, Cornell University Press, Ithaca, 1991.
——'The Foundations of a Canon: Kohlrausch's *Practical Physics*', in D. Kaiser (ed), *Pedagogy and the Practice of Science*, 2005, MIT Press, Cambridge, MA, 323–355.
——'Science Pedagogy as a Category of Historical Analysis: Past, Present, and Future', *Science and Education*, 15 (2006), 863–880.
F. PAPANELOPOULOU, A. NIETO-GALAN and E. PERDIGUERO. (eds.), *Popularizing Science and Technology in the European Periphery, 1800–2000*, Ashgate, Aldershot, 2009.
D. PESTRE, *Physique et physiciens en France, 1918–1940*, Editions des archives contemporaines, Paris, 1992.
J. ROSE, 'The History of Education as the History of Reading', *History of Education*, 36 (2006), 595–605.
J. L. RUDOLPH, *Scientists in the Classroom: the Cold War Reconstruction of American Science Education*, Palgrave, New York, 2002.
——'Historical Writing on Science Education: A View of the Landscape', *Studies in Science Education*, 44 (2008), 63–82.
S. P. T., 'Daniell's Physics', *Nature*, 30 (1884), 49–51.
G. SARTON, 'The Study of Early Scientific Textbooks', *Isis*, 38 (1948), 137–148.
R. E. SCHOFIELD, 'Eloge: Duane Henry Dubose Roller, 14 March 1920–22 August 1994', *Isis*, 86 (1995), 80–81.
J. J. SCHWAB, 'The Teaching of Science as Inquiry', *Bulletin of the Atomic Scientists*, 14 (1958), 374–379.
J. A. SECORD, *Victorian Sensation: The Extraordinary Publication, Reception, and Secret Authorship of Vestiges of the Natural History of Creation*, University of Chicago Press, Chicago, 2000.
——'Knowledge in Transit', *Isis*, 95 (2004), 654–672.
P. L. SHANK, *The Evolution of Natural Philosophy (Physics) Textbooks Used in American Secondary Schools before 1880*, PhD thesis, University of Pittsburgh, 1952.
H. SIEGEL, 'Kuhn and Schwab on Science Texts and the Goals of Science Education', *Educational Theory*, 28 (1978), 302–309.
——'On the Distortion of the History of Science in Science Education', *Science Education*, 63 (1979), 111–118.
J. SIMON, 'Circumventing the "elusive quarries" of Popular Science: the Communication and Appropriation of Ganot's Physics in Nineteenth-Century Britain', in F. Papanelopoulou, A. Nieto-Galan and E. Perdiguero (eds.), *Popularising Science and Technology in the European Periphery, 1800–2000*, Ahsgate, Aldershot, 2009, 89–114.
——*Communicating Physics: The Production, Circulation and Appropriation of Ganot's Textbooks in France and England, 1851–1887*, Pickering and Chatto, London, 2011.
R. STICHWEH, *Zur Entstehung des Modernen Systems Wissenschaftlicher Disziplinen: Physik in Deutschland*, Suhrkamp, Frankfurt, 1984.
J. R. TOPHAM, 'Scientific Publishing and the Reading of Science in Nineteenth-Century Britain: A Historiographical Survey and Guide to Sources', *Studies in History and Philosophy of Science*, 31 (2000), 559–612.
——'Rethinking the History of Science Popularization/Popular Science', in F. Papanelopoulou, A. Nieto-Galan and E. Perdiguero (eds.), *Popularizing Science and Technology in the European Periphery, 1800–2000*, Ashgate, Aldershot, 2009, 1–20.
D. TURNER, 'Reform and the Physics Curriculum in Britain and the United States', *Comparative Education Review*, 28 (1984), 444–453.

S. Turner, 'The Reluctant Instrument Maker: A. P. Gage and the Introduction of the Student Laboratory', *Rittenhouse*, 18 (2005), 42–45.

——'Chicago Scientific Instrument Makers, 1871–1918. Part I: The School Science Market', *Rittenhouse*, 19 (2006), 65–128.

M. Vicedo (ed), 'Focus: Textbooks in the Sciences', *Isis*, 103 (2012), 83–138.

A. Warwick, *Masters of Theory: Cambridge and the Rise of Mathematical Physics*, Chicago University Press, Chicago, 2003(a).

——' "A very Hard Nut to Crack" or Making Sense of Maxwell's Treatise on Electricity and Magnetism in Mid-Victorian Cambridge', in M. Biagioli and P. Galison (eds.), *Scientific Authorship: Credit and Intellectual Property in Science*, Routledge, New York, 2003(b), 133–161.

——and D. Kaiser, 'Conclusion: Kuhn, Foucault, and the Power of Pedagogy', in D. Kaiser (ed), *Pedagogy and the Practice of Science*, MIT Press, Cambridge, MA, 2005, 393–409.

S. White, 'The Physical Science Study Committee. (3) The Planning and Structure of the Course', *Contemporary Physics*, 2 (1960), 39–54.

C. S. Zwerling, *The Emergence of the École Normale Supérieure as a Center of Scientific Education in Nineteenth-Century France*, Garland Publishing Inc., New York, 1990.

Notes

1. I am grateful to Dave Kaiser, Rich Kremer, Michael Eckert, John Rudolph, Jaume Navarro, Don Howard, Clayton A. Gearhart, Charles Midwinter, Michel Janssen, Domenico Giulini, and Helge Kragh for allowing me to read work that was still unpublished. The writing of this chapter was possible thanks to a 'Marie Curie' postdoctoral fellowship and to a fellowship of the John W. Kluge Center at the Library of Congress.
2. Notable exceptions to these general trends are, for instance, Bensaude-Vincent *et al.* (2003), García-Belmar *et al.* (2005), and Kohlstedt (2010).
3. This is clearly illustrated by a Focus section on textbooks published in *Isis*. See Vicedo (2012).
4. See Duschl (2005) and Isaac (2011). Also Midwinter and Janssen (2013) and Fuller (2000). Midwinter and Janssen (2013).
5. The 'General education' programmes were then widely regarded as a good niche in which to shelter the emergence of history of science as an academic discipline in the USA.
6. The PSSC and Kuhn play convenient pivotal roles in this narrative for their power to illustrate the arguments deployed in this overview essay. They are no doubt important in textbook history, but there is undoubtedly a vast historical and historiographical universe beyond them.
7. However, see also the differences between Kuhn's and Schwab's thought in Siegel (1978).
8. In the following, the account on French textbook physics is based on Simon (2011).
9. The École Normale soon became a major centre for research as well. Its students constituted an elite who after graduating typically acquired experience as teachers in provincial schools, and the most successful subsequently secured positions in large schools or science faculties in Paris or major capital cities. In this period, the gap between secondary and university education was narrow, and research was also conducted in secondary schools. See Zwerling (1990) and Balpe (1997).
10. The available secondary literature on German education is quite heavily biased towards Prussia. However, it is not within the scope of this essay to deal with comparative distinctions between different German states.

11. Lind (1992) and Jungnickel and McCormmach (1986) are rich sources on German physics textbooks, and I have mainly drawn on them in this part of my essay. But I have also performed my own bibliographical research and analysis of primary sources.
12. Like Jamin, Wüllner also published a second textbook intended for schools (Oldham, 2008, 210–248).
13. This close connection between textbooks, scientific instruments, and illustration seems standard, but there are few thorough studies of the relationship. See Simon (2011); Turner (2005) and (2006).
14. While Tyndall's work as a popularizer has been studied, there are no accounts of his role as teacher and textbook author in the school-driven Science and Art Department.
15. Information about other physics textbooks produced and used in the USA can be found in Holbrow (1999).
16. The substitution of inductivist historical accounts with hypothetico-deductive accounts, in the name of pedagogical efficiency, is also evident in such cases as the English translation of Adolphe Ganot's textbook physics, and in Landau and Lifshitz's *Course of Theoretical Physics* (Hall, 2005 and 2008).
17. See also Heilbron (1987), Siegel (1979), and Turner (1984).
18. Nonetheless, Warwick (2003b) offers an interesting analysis of textbook production and appropriation by readers in his study of Maxwell's *Treatise on Electricity and Magnetism*.
19. A good example of this type of standard approach is found, for instance, in the work of Stephen G. Brush. See Brush (1970), (1976), and (2005). The same approach has characterized the work of historically minded philosophers of science. In spite of his critical rethinking of Kuhnian themes, the work of Hasok Chang illustrates this trend (Chang, 2004). Historians of chemistry have shown a greater interest in rethinking textbooks as sources for research (Lundgren and Bensaude-Vincent 2000; Bensaude-Vincent et al. 2003).
20. Along similar lines, although lacking the historical sophistication of Kaiser and his use of sources such as lecture notes, Stanley Goldberg has provided a useful overview of American physics textbooks dealing with relativity (Goldberg, 1984, 275–305). Analogously, Dominique Pestre has surveyed university physics textbooks in interwar France and compared them with their American, German and British peers (Pestre, 1992, 31–65). On physics textbooks in Soviet Russia, see Hall (1999), 559–764.
21. At the time I submitted this essay, I had been able to read only eight of the twelve papers in the volume. Although some of the papers deal with less well known textbooks, all the authors selected were included in Kuhn *et al.* (1967).

CHAPTER 22

PHYSICS AND MEDICINE

IWAN RHYS MORUS

In 1875 the first Brackenbury Professor of Physiology at Owens College in Manchester, Arthur Gamgee, published the *Elements of Human Physiology*, translated from the fifth edition of Ludimar Hermann's *Grundriss der Physiologie des Menschen*. This was the beginning of the heyday of scientific medicine in the British Isles. Two years later, Michael Foster published his own *Textbook of Physiology*, based on the courses in biology and physiology he had just started teaching at Cambridge following his appointment to a praelectorship in physiology at Trinity College (Geison, 1978; Romano, 2002). Gamgee and Hermann make an interesting pair. Gamgee had studied natural philosophy at Edinburgh under Peter Guthrie Tait and in 1871 spent time in Germany in the laboratories of Wilhelm Kühne in Heidelberg and Carl Ludwig in Leipzig before returning to Britain to take up his Manchester appointment. He was intimately familiar, therefore, with the practices and theories of the latest German laboratory physiology, as well as the latest developments in British physics.[1] Hermann was a student of Emil du Bois Reymond at the University of Berlin during the 1850s before being appointed Professor of Physiology at the University of Zurich in 1868. The first edition of his *Grundriss* had been published in 1863 whilst he was still working with du Bois Reymond in Berlin. Hermann was later to fall out with his master over the question of the existence of 'resting currents' of electricity in healthy muscular and nervous tissue (Finkelstein, 2006).

Hermann's original *Grundriss* and Gamgee's translation are particularly interesting and intriguing in the context of this chapter because of the language they used to describe exchanges of matter and energy in the human body. The terminology was the terminology of political and practical economy, with discussions of income, expenditure, and the value of stock (Hermann, 1875). The point of the physiology of nutrition and digestion as presented was to track the complex quantitative relationship between these values through the different parts of the human system. Just as with a healthy economy, 'the expenditure of the matter of the body is always a little below its income (and this difference is necessary to the existence and persistence of the body), so also the expenditure of energy is always below its income, for the organism always contains a certain store of energy, part of which is potential in its yet unoxidized constituents, part of which is kinetic—its heat.' (Hermann, 1875, p. 215) The language and the model would have been immediately familiar to any reader familiar with contemporary British natural philosophy. This was the way that adherents of the new doctrine of the conservation of energy talked. It was, increasingly, the language of British energy physics (Rabinbach, 1990; Smith, 1998). The choice of terminology was certainly not accidental. Both Hermann and Gamgee were making a point about where physiology as a scientific discipline belonged and about how it should be understood by the medical students at whom their textbooks were aimed.

The relationship between physics and the life sciences—and certainly the relationship between physics and medicine insofar as it has been seriously considered at all—during the nineteenth century has usually been regarded as something of a one way street. In terms of the life sciences, the trajectory has usually been regarded as one leading from vitalism towards new reductionist perspectives as physics and chemistry came to represent ideal practice (Holmes, 2003). So the new science of experimental physiology borrowed its methods and inspiration from physics and scientific medicine in turn helped itself to experimental physiology (Bynum, 1994). Going along with this view of the historical relationship between the physical and life sciences is the assumption that the human body was peripheral to the regime of physics and that it was to become more peripheral during the course of the century. In this chapter I want to take issue with this usual historical trajectory by outlining ways in which the human body remained an important and contested site within physics throughout the nineteenth century for a significant number of protagonists. To this end, I will sketch out a number of historical episodes and controversies during the century in which the body and the processes of life provided common grounds for discussion and debate between medical practitioners, natural philosophers, and physicists. The aim is not to be exhaustive but to explore some of the ways in which medicine and physics intersected around the body during this period.

Debates surrounding galvanism during the opening decades of the nineteenth century provide a useful point of entry into the discussion. The physician Luigi Galvani's experiments during the 1780s, culminating in his claim to have identified an animal electricity specific to living tissue, caused significant controversy and interest across Europe (Kipnis, 1987). As Galvani's fellow Italian and antagonist Alessandro Volta sought to demonstrate that Galvani's animal electricity was a chimera, a range of groups and individuals battled to make the new phenomenon their own. During the 1840s, as the medically trained Hermann von Helmholtz formulated the arguments that would culminate with the publication of *Über die Erhaltung der Kraft* in 1847, the medical and human context remained central to the practical and theoretical development of his ideas. Medicine had similarly provided an important evidential context for Julius Robert Meyer's observations as a ship's doctor with the Dutch East India Company that led to his formulation of a relationship between work and heat. In Britain during the 1840s and 1850s discussions regarding the correlation of physical forces fed into efforts to understand vitalism in relation to other material forces. The context of debate here as well was at least partially medical. Correlation proved a valuable theoretical technology for introducing medical students to chemistry and natural philosophy.

Certainly the conservation of energy provided a significant intellectual resource for medical men—particularly those specializing in the diseases of women—during the second half of the nineteenth century. The language and concepts of conservation provided a set of useful tools for explaining the inherent limitations of the female body and suggesting new therapies for the cure of hysteria and similar nervous disorders. At least some advocates of electrotherapy, for example, turned to the culture of metrology that emerged along with the physics laboratory as a way of bolstering their own therapeutic practices. By showing themselves to share in this metrological culture with its emphasis on precision and standardization they were forging alliances with engineers and physicists who were just as anxious to extend their own realm of authority to the body. The burgeoning instrumentalization of medicine meant that some doctors and physicists increasingly shared practices and technologies. Towards the end of the nineteenth century a broad range of novel therapies emerged—electrotherapy, nerve vibration, and X-ray therapies amongst them—that depended on shared skills and technologies between medicine and physics. By looking at these episodes as instances of contestation, attempted colonization and crossing over of practices and technologies rather than in terms of a one-way street from natural philosophy and physics to medicine I think that a more interesting view of a shared terrain between disciplines emerges.

22.1 BETWEEN VITALISM AND MATERIALISM

Medical men were certainly amongst those who sought to make the phenomena of animal electricity their own in the wake of Galvani's publication of his discoveries in 1791. In England, for example, Thomas Beddoes announced his conviction that these remarkable new discoveries would provide the basis of a wholly 'new system of medicine' (Stock, 1811 p. 72, Jay, 2009). Galvani's animal electricity was hotly debated amongst medical men as a likely candidate for the nervous fluid. One supporter was the radical author and political orator John Thelwall, who announced his conviction that animal electricity was the 'ethereal medium' that completed the 'chain of connection between the divine immortal essence, and the dull inertion of created matter' in an oration to the Guy's Hospital Physical Society in 1793 (Thelwall, 1793). Volta's invention of the pile in 1800 similarly attracted attention from medical men across Europe, who quickly recognized the instrument as a useful tool for investigating the nervous fluid and treating nervous diseases. In Turin, Antonio Maria Vassalli-Eandi experimented with Volta's pile on the bodies of recently decapitated criminals to assess the electric fluid's interactions with the human body. They were interested in establishing whether or not it was possible to use the pile to stimulate various organs—the heart in particular—after death (Vassalli-Eandi, 1803).

The best-known of these kinds of electrical experiments on the dead are probably those conducted by Galvani's nephew, Giovanni Aldini, during his visit to London to defend his uncle's reputation in late 1802 and early 1803 (Fig. 22.1). During the course of his visit Aldini carried out a number of experimental demonstrations in London and Oxford on various animals. Doctors clearly formed an important audience for these demonstrations. He gave public lectures and dissections at the Great Windmill Street Anatomical Theatre, at Guy's Hospital, and at the physician George Pearson's lecture rooms in Hanover Square, for example. In his experiments on the body of Michael Forster, Aldini also collaborated with a number of medical men. Newspaper accounts describe the experiment as having been carried out 'under the inspection of Mr Keate, Mr Carpue, and several other professional gentlemen.' Thomas Keate was the representative of the Royal College of Surgeons, under whose license the experiment was taking place. Joseph Carpue was a London surgeon and known political radical. It seems likely that another London medical man, Charles Wilkinson, was also directly involved. The episode illustrates the extent to which medical men

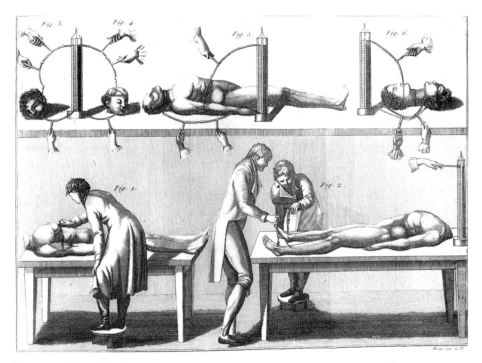

Fig. 22.1. Giovanni Aldini's galvanic experiments on human and animal bodies. (Giovanni Aldini, *Essai Theoretique et Expérimental sur le Galvanisme*, Paris, 1804.)

had their own range of interests in exploring the limits of galvanism. Carpue and Wilkinson were not simply assisting Aldini; they had their own reasons for wanting to experiment (Morus, 1998; 2011; Sleigh, 1998).

Aldini's experiments were widely understood at the time in a broad context that embraced medicine and political radicalism. One satirist, poking fun at the pretensions of ambitious medical men, described how 'in an experiment made on a malefactor who was executed at Newgate, he immediately opened his mouth—doubtless, another application would have made him speak; but the operators, Aldini, Wilkinson, and Co were so much affrighted that they threw down their instruments and took to their heels' (Corry, 1815, p. 60). Other critics, such as the author of the notorious *Terrible Tractoration*, poking fun at the pretensions of experimental natural philosophers, accused Aldini and his fellow experimenters of being no better than medical quacks themselves for their efforts to resurrect the dead. Drawing these kinds of links between experiment, quackery, and demagoguery was a common Tory strategy at the start of the nineteenth century. Some commentators regarded these kinds of galvanic experiments as conveying a straightforwardly materialist message that meshed well with the preoccupations of radical medical men. The materialism was not always self-evident nevertheless. Some, like Charles Wilkinson, argued that electricity itself was not fully material, suggesting instead that it was 'an energizing principle, which forms the line of distinction between matter and spirit, constituting in the great chain of the creation, the intervening link between corporeal substance and the essence of vitality.' (Wilkinson, 1804, p. 298) A similar argument would be put forward a decade later by John Abernethy and fiercely rebutted by the radical materialist William Lawrence—a view supported by the poet Samuel Taylor Coleridge (Jacyna, 1983; Morus, 2011). German *Naturphilosophen* (of whom Coleridge was an avid reader) also argued in similar fashion (Levere, 2002).

Medical men did have a more immanent use for galvanism than its implications for materialist accounts of vitality. Galvanism seemed likely to provide a useful new therapeutic resource. By the beginning of the nineteenth century electricity already had an established history of being used as a therapeutic agent (Rowbottom and Susskind, 1984). The invention of the voltaic pile provided medical practitioners with a new source of the electric fluid. Its promoters advocated electricity as a means of resuscitating the recently drowned, for example. This was one of the reasons that the Royal Humane Society in Britain was interested in Aldini's experiments, and Aldini himself suggested that his researches demonstrated galvanism's potential in that respect. Doctors emphasized that it was medical skill and knowledge that mattered for the safe application of electricity to the human body. They argued that the 'application of so important an agent to the animal economy should not be left to the discretion of one who only comprehends the management of a machine, no more than the treatment of serious disease should be confided to one whose knowledge is limited to the mere compounding of medicines.' (Wilkinson, 1798, p. 81) In other words, interested medical practitioners in this context were certainly not passive recipients of galvanic knowledge. They regarded themselves as active contributors to the making of galvanic knowledge whose medical authority was a key resource.

These kinds of practical medical concerns and experiences were central to Julius Robert Mayer's formulation of his doctrine of the conservation of force in 1842. The story of Mayer's observation of the changed colour of venous blood in the tropics whilst he was acting as ship's surgeon on the Dutch East India Company's ship *Java* is a familiar one. Less often noted is that the observation Mayer made was above all else a medical one. Two points are relevant. Mayer noticed the change in the colour of the crew-members' blood during the course of his duties as ship's surgeon, presumably whilst bleeding them for therapeutic purposes. The observation was also the outcome of his medical experience, since only a doctor used to such procedures would have been likely to recognize the changed colour as either unusual or significant. Mayer realized that the crew-members' blood was unusually red, so that their venous blood looked more like oxygenated arterial blood. His conclusion that the tropical heat meant that the body had less work to do in maintaining its internal heat, and that as a result less oxidation took place in the blood, was based on his medical knowledge of the human body as much as his knowledge of the latest chemical and physiological understanding of the body (Caneva, 1993).

A medical perspective informed the physics of Mayer's German contemporary Hermann von Helmholtz as well. When Helmholtz published *Über die Erhaltung der Kraft* in 1847 he was still acting as an army staff surgeon in Potsdam. He had studied at the University of Berlin with the eminent anatomist and physiologist Johannes Müller. His experimental concerns were those of a medically informed researcher interested in the workings of the human body, and his experimental work had been concerned with trying to trace the relationship between the work done by muscular action, bodily heat, and the oxygenation of food. He was building on the work of Justus von Liebig on the relationship between nutrition and vitality, as well as that of his patron Müller. Where Müller maintained a role for a specific vital force and argued for the distinctiveness of organic as opposed to inorganic matter, Helmholtz, along with others, including Emil du Bois Reymond and Carl Ludwig, aimed at placing the study of organic tissue as part of the physical sciences. In his 1847 essay, Helmholtz argued that postulating a specific vital force violated the principle of the impossibility of perpetual motion, since a vital force was in principle capable of being produced without the expenditure of any other force, whilst itself being capable of doing useful work (Lenoir, 1989).

For Helmholtz, the human body was best regarded as a machine, but he recognized that the language that engineers and physicists used to talk about machinery had its origins in discussions of human and animal bodies. As he acknowledged, the 'idea of work is evidently transferred to machines from comparing their performances with those of men and animals, to replace which they were applied. We still reckon the work of steam engines according to horse-power.' (Helmholtz, 1995) Looking back at the origins of his own interest in the production of work by machines, animals, and humans, Helmholtz pointed to the eighteenth-century fascination with automata—machines that mimicked the actions of living creatures. The fact that it was possible to make machines that did (or appeared to do) what humans and animals did was one of the factors that focused attention on the physical forces animating human beings. This was not, in the end, a particularly reductionist perspective. As much as he was trying to bring to bear the physics of heat and work to bear on the understanding of the human body, he was also attempting to import the understanding of the human body he had acquired through his medical training into physics.

In Britain, physiologists and medical men used the doctrine of the correlation of physical forces as a way of linking organic forces—and their own practices—into a broader network of relationships. The term 'correlation' to describe the relationships between forces was coined by William Robert Grove—first at lectures at the Royal Institution in London, where he was professor of Experimental Philosophy, and then in 1846 in a published essay *On the Correlation of Physical Forces*. Grove suggested that

> the various imponderable agencies, or the affections of matter which constitute the main objects of experimental physics, viz. Heat, Light, Electricity, Magnetism, Chemical Affinity, and Motion are all Correlative, or have a reciprocal dependence. That neither taken abstractedly can be said to be the essential or proximate cause of the others, but that either may, as a force, produce or be convertible into the other, thus heat may mediately or immediately produce electricity, electricity may produce heat; and so of the rest. (Grove, 1846, pp. 7–8)

Grove described his own gas battery as 'a beautiful instance of the correlation of natural forces' because of the way in which it allowed the experimenter to see the gases combining and producing electricity, which was then used to decompose water back into its constituent hydrogen and oxygen in a voltameter (Grove, 1842).

Grove said nothing about the vital forces in the first edition of his essay (it went through several increasingly expanded editions during the course of the century), but others quickly recognized that the language of correlation provided a useful way of discussing the relationship between organic and inorganic forces without overtly describing them in materialist terms. In lectures delivered before the Royal College of Physicians in 1847 the physician and advocate of electrotherapeutics drew on Grove's views of correlation as well as Michael Faraday's electromagnetic researches to provide a safe foundation for electrotherapeutic practice. Bird suggested that 'one of the uses of the electricity so freely developed in the body, especially that existing in the muscles' was 'to excite in the nervous chords the vis nervosa, just as currents, if passing near a bar of iron at right angles to its axis, excite magnetism.' If that was the case, then 'may not such nervous force again induce electric currents in any glandular or other organs, just as magnetism in motion will re-excite electricity? thus accounting for what cannot be questioned, the existence of electric currents in certain organs, exclusively excited by, or depending for their existence upon, the integrity of the nervous influence of the part.' (Bird, 1847, Morus, 1998) This kind of argument made the case for electrotherapy without suggesting the identity of electricity and the vital force.

William Benjamin Carpenter was more explicit in his use of Grove's doctrine in this way. Carpenter was a devout Unitarian—his father had been a Unitarian minister. He started his medical career apprenticed to a local surgeon before studying medicine at Bristol, London, and Edinburgh, graduating MD in 1839. In a paper 'On the Mutual Relations of the Physical and Vital Forces' published in the *Philosophical Transactions* in 1850, Carpenter aimed to show that the general doctrine of the 'Correlation of Physical Forces' propounded by Mr Grove, was equally applicable to those Vital forces which must be assumed as the moving powers in the production of purely physiological phenomena; these forces being generated in living bodies by the transformation of the Heat, Light, and Chemical Action supplied by the world around, and being given back to it again, either during their life, or after its cessation, chiefly in Motion and Heat, but also to a less degree in Light and Electricity. (Carpenter, 1850)

Carpenter's views on correlation fed into his and others' work in mental physiology. They were also incorporated by Grove into later editions of his essay. Correlation provided a language for discussing vitality whilst deferring arriving at conclusions about its origins, which suited Carpenter's Unitarianism (Cantor, 1976).

Physics took its place in medical texts and practices during the first half of the nineteenth century in a variety of ways. Similarly, medicine informed ideas and practices within physics in a number of ways. Running through the various examples discussed in this section is the problem of vitalism. Physics certainly provided resources for radical medical men who wanted to do away with the idea that there was a distinct vital force governing the development and sustenance of living tissue. Electricity, as we have seen for example, could be substituted for the powers of vitality, turning bodies into galvanic machines. But doctors like John Abernethy could turn to the latest electrical theories and experiments to support vitalism as well. Abernethy could cite Humphry Davy's galvanic researches to support his arguments. Davy also, of course, had a medical background, having commenced his career as an apothecary's apprentice. His example, along with those of Mayer and Helmholtz discussed earlier, show how medical training and preoccupations could inform physics too. Medical men carrying out galvanic experiments did not necessarily regard themselves as borrowing the ideas and practices of an alien discipline. They regarded themselves as being firmly situated on their own intellectual territory.

22.2 Energetic Bodies

In the British Isles, early supporters of the doctrine of the conservation of energy made relatively little in the way of claims regarding the place of the human body in their ideas. There was, for example, no systematic treatment of the energies of animal or human bodies in William Thomson and Peter Guthrie Tait's monumental *Treatise on Natural Philosophy* (1867), which provided the conservation of energy with its bible. The situation in Germany was different, of course. There, Helmholtz's researches had solidified the body's place at the centre of articulations of the conservation of energy right from the beginning. British medical men tended to carry on using Grove's language of correlation well into the second half of the century as well. The promoters of energy regarded the correlation of physical forces as 'woefully loose and unscientific' for its lack of mathematical precision, though it was probably its lack of mathematical formulation that made its attractive to those lacking the technical skills needed to fully appreciate the conservation of energy in Thomson and Tait's formulation. Balfour Stewart, in his popular *Conservation of Energy* in 1873, did include the human body within the conservation of energy's ambit, suggesting that 'a human being, or indeed an animal of any kind, is in truth a machine of a delicacy that is practically infinite, the condition or motions of which we are utterly unable to predict' (Stewart, 1873, p. 161).

The doctrine of the conservation of energy did, however, provide an extremely important resource for medical practitioners in one key respect. The conservation of energy provided a new way of talking about the human body's balances and imbalances, particularly in the case of women's bodies (Maines, 1999; Oppenheim, 1991). By the middle of the nineteenth century the language and concepts of electricity had already become common in medical discussions of women's bodies. Thomas William Nunn, for example, suggested in 1853 that

the ovaria, uterus, and mammae form, as it were, a reproductive pile, the circuit being completed by the nervous system. If there be no antecedent ovarian excitement, no impulse is transmitted to the breast. As in the galvanic battery, if no chemical action takes place at one pole, no electric current traverses the wire and no sign is elicited at the other. (Nunn, 1853, p. 3)

In his *Elements of Electro-biology* a few years earlier, Alfred Smee, surgeon to the Bank of England, formulated an elaborate electrical model for the entire human body, proposing various types of electrical apparatus as analogues for the body's organs of sense. Unimpressed, the *British and Foreign Medico-Chirurgical Review*'s reviewer (probably William Benjamin Carpenter) poked fun at Smee's efforts to build a 'living, moving, feeling, thinking, moral, and religious man, by a combination of voltaic circuits' (Anon., 1850, p. 522).

Some doctors specializing in women's diseases and nervous disorders more generally argued that the conservation of energy dictated that only so much energy could be contained in a woman's body, and that diverting that energy from its proper path inevitably led to debilitation. The proper purpose of a woman's body's energy was to be directed towards childbirth and nursing children. Proponents of the conservation of energy's role in their practices, such as the psychologist Henry Maudsley, argued that this was an immutable law of nature rather than a mere social convention. In an article in the *Fortnightly Review* on 'Sex in Mind and Education' he argued that if 'it were not that woman's organization and functions found their fitting home in a position different from, if not subordinate to, that of men, she would not so long have kept that position.' Women's subordinate role was dictated by the conservation of energy:

. . . it is not a mere question of larger or of smaller muscles, but of the energy and power of endurance of the nerve-force which drives the intellectual and muscular machinery; not a question of two bodies and minds that are in equal physical conditions, but of one body and mind capable of sustained and regular hard labour, and of another body and mind which for one quarter of each month during the best years of life is more or less sick and unfit for hard work. (Maudsley, 1874, p. 80)

Men's health was also at risk if they ignored the dictates of the conservation of energy. Masturbation caused spermatorrhaea by squandering energies that were best preserved, resulting in young men who were 'highly nervous, and retiring to such a degree that it becomes noticeable by his friends and companions; he instinctively avoids the society of females, and feels a shame when introduced to them, involuntary flushings add to his discomfort, and he is overjoyed to escape from them.' (Lobb, 1867, p. 36) The American doctor George Beard introduced a new word—neurasthenia—to describe the nervous state produced by the hectic urban life of contemporary industrial culture. Modern industrial man used too much nervous energy, resulting in an epidemic of nervous disorders. Beard dubbed neurasthenia the American disease, and boasted that North America was racing ahead of Europe in the number of its nervous invalids—proof of the nation's superior industrial advancement (Beard, 1880). In Germany, Hermann Griesbach invented the aesthesiometer—an instrument that could measure mental fatigue in terms of reduced sensitivity to stimuli. His follower Theodore Vannod experimented on schoolchildren suffering from 'an excess of cerebral work', showing how the 'appetite disappears, energy diminishes, headache persists, night are agitated, and troubled by insomnia, in short a morbid state occurs which permits us to conclude that there is the presence of a true intellectual fatigue' (Rabinbach, 1990, p. 150).

Measurement and standardization were increasingly important parts of this new culture (Schaffer, 1997). Some medical practitioners were themselves keen to incorporate physicists' habits of precision as well as their language into their own practices. The German electrotherapist Wilhelm Erb argued that 'the human body is nothing more than a large conducting mass of definite resistance; and the laws controlling the distribution of large conducting masses therefore apply to it without any limitation' (Erb, 1883, p. 19). The French electrotherapist Armand de Watteville, running the Electrical Department at St Mary's Hospital in London, agreed that the body was simply 'a vessel bound with poorly conducting material (the skin) unequally packed with non-conducting particles, the interstices being filled up with a saline liquid of fair conducting power' (de Watteville, 1884, p. vi). De Watteville was proud that his proposal to the International Congress of Electricians that the milliweber (later the milliampere) be adopted as the standard unit of measurement in electrotherapy had been accepted. Electrotherapists needed instruments that displayed 'at a simple reading the actual strength of the current passing through the patient's body. This reading, multiplied by the time during which the operation lasts, will give the actual quantity, literally, the *dose* of electricity used.' (de Watteville, 1877, p. 488) Not all medical practitioners agreed. Herbert Tibbits argued that experience rather than measurement was what mattered: 'No man is justified in electrizing a patient without invariably testing electricity upon himself, if only a second at a time, and always when he either increases or decreases the power of the current' (Tibbits, 1877, p. 488).

In his presidential address to the Society of Telegraph Engineers in 1882, Charles Webber called on medical practitioners to join the society. According to him 'the day was approaching when the study of electricity in its physiological functions would be of paramount importance to medical men' (Anon., 1882, p. 173) The electrotherapist William Henry Stone had already spoken before the society in similar terms:

> You can be of very great service to us physicians and physiologists by giving us suggestions, especially in the department with which you are most conversant (more conversant than we are), viz., the department of measurement. Medicine and its kindred departments lend themselves extremely ill to measurement. The tone of mind required in the physician is not that of the mathematician—it is rather judicial than computative. It is often a question of weighing doubtful evidence and of balancing alternatives rather than of solving an equation. (Stone and Kilner, 1882, p. 107)

Stone had been carrying out his own research on the electrical resistance of the human body in a variety of circumstances. Similar research was carried out by Henry Newman Lawrence and Arthur Harries at the Institute of Medical Electricity. The introduction of electrocution in the United States as well as the increasing spate of electrical accidents that accompanied the rise of the electric power industry further stimulated interest amongst doctors and electrical engineers in understanding the electrical characteristics of the human body (Essig, 2003; Morus, 2006).

In France, the physicist Arsène d'Arsonval conducted a number of experiments during the 1890s on the human body's capacity to conduct high-frequency currents at very high potentials that would be considered fatal under most normal circumstances (Fig. 22.2). One of d'Arsonval's main concerns was to ascertain exactly what the physiological effects of these currents were in practice. He demonstrated that electrical currents did not simply pass over the surface of the body but had definite effects on the nerves and muscles. It simply appeared that at such high frequencies the nerves had become insensitive to the electric current's passage, which therefore caused no permanent damage (Morus, 1999). The flamboyant performer Nikola Tesla carried out similar experiments, if in more spectacular fashion. When Tesla visited London in February 1892 he lectured before the Institute of Electrical Engineers on alternating currents of high frequency and high potential, placing his own body in the electric circuit to amazing effect:

> Here is a simple glass tube from which the air has been partially exhausted. I take hold of it; I bring my body into contact with a wire conveying alternating currents of high potential, and the tube in my hand is brilliantly lighted. In whatever position I may put it, wherever I move it in space, as far as I can reach, its soft pleasing light persists with undiminished brightness. (Martin, 1995, p. 200)

The theory of the conservation of energy and the culture of instrumental precision that came to dominate physics laboratories during the second half of the nineteenth century clearly provided important resources for medical practitioners in a variety of ways. The conservation of energy provided some doctors with a useful new way of talking about the human body and its internal balances. In so doing, as will become clear in the next section, it provided medical practitioners with rationales for new kinds of therapies as well. One significant trend in the institutional history of medicine during the second half of the century was one towards increased specialization. The theory of the conservation of energy, and ideas, instruments, and practices more

FIG. 6.—LAMP LIGHTED BETWEEN THE HANDS OF AN OPERATOR PLACED IN A SOLENOID.

Fig. 22.2. The French physicist Arsene D'Arsonval demonstrating the use of the human body as a conduit for the transmission of electric power. (*Electrical Review*, 1894.)

generally appropriated from physics, were useful resources for budding specialists in this respect. Practitioners in the relatively new specialities of women's diseases and nervous disorders in particular turned to the language of conservation to account for their medical interventions. Adopting the language and practices of precision measurement was a way for practitioners to demonstrate their scientific credentials—an increasingly important qualification with the rise of scientific medicine. Whilst doctors turned to physics as an important resource, physicists and electrical engineers turned to the human body as a new field for their discipline to colonize.

22.3 THERAPEUTICS

In what ways did physics contribute to the practice of medicine during the nineteenth century? What new therapies were developed with the instruments and ideas of physics? Medical practice was highly competitive throughout the century and many doctors were keen to investigate alternative ideas that might provide them with an advantage in a highly competitive marketplace for medicine. This was one of the main factors behind the drive towards medical specialization mentioned earlier. In Britain, this trend towards specialization was often highly contested. The traditional view of the elite doctor was of a gentlemanly generalist skilled at treating the whole patient. These gentleman physicians were often strongly opposed to the development of new kinds of institutions and new therapeutic practices, regarding them as inimical to the proper practice of their profession (Peterson, 1978). When the Institute of Medical Electricity was established in London in 1888, for example, the Royal College of Physicians refused to sanction the appointments of William Steavenson and William Henry Stone as consulting physicians, threatening them with loss of membership if they accepted the positions (Ueyama, 1997; 2010). The College's view was that participation in such an institution was incompatible with the social and professional standing of a respectable medical practitioner. As a result, in Britain in any case, new physical therapies were usually developed on the margins of orthodox practice and against considerable opposition from the medical mainstream, though this was less often the case elsewhere in Europe.

The history of medical electricity in Britain is a good example of this trend. From Golding Bird onwards, practitioners attempting to introduce electrotherapy into mainstream hospitals complained about the difficulties of so doing. Bird established an Electrical Room at Guy's Hospital in London in 1836, but it was not until the 1880s that such establishments started to become relatively commonplace. In France and Germany electrotherapy had a far firmer institutional presence throughout the century. Electricity in various forms was usually prescribed for nervous disorders of various kinds. Electricity from various sources was used for therapeutic purposes. The use of electricity from a common electrical machine was termed *franklinization*, the use of electricity from a voltaic battery was called *galvanization*, and the use of electricity from an electromagnetic source was described as *faradization*. A popular therapy was the electric bath, in which the patient was placed in a bath of water through which an electric current was then passed. By the end of the century, small portable generators usually consisting of a battery, an induction coil, and a circuit breaker were quite commonplace. There was a flourishing market throughout the century for electro-medical devices such as electric belts and corsets. Practitioners offered a range of rationales for electrotherapy's efficaciousness, ranging, as we have seen, from the suggestion that electricity was the nervous fluid, though arguments regarding correlation and the conservation of energy to the view that electricity acted as a stimulus for injured or debilitated muscular tissue (Colwell, 1922).

Electrotherapy was often used towards the end of the century in conjunction with massage. The Institute of Medical Electricity offered a therapy in which the patient was massaged whilst lying on a couch in an alternating electric field. The idea was that this would produce induced currents flowing between the patient's body and the masseur (Morus, 1999; Ueyama, 2010). Related to this kind of electrical massage was nerve vibration therapy, developed by Joseph Granville Mortimer. The therapy was based on the view that nerves worked by mechanical vibration. The nerves were like a 'network of mysteriously sensitive threads or filaments stretching throughout the body, capable of vibrating like the strings of a harp or pianoforte, and liable to be thrown into painful action—or "jarred"—by any sudden jerk or strain of the muscles or of the bony organism' (Mortimer, 1884, p. 1). Nervous disorders were caused by 'morbid habits of the nervous system', and Mortimer's percuteur was designed to provide a mechanical counter-vibration that worked in opposition to any unhealthy one and therefore restored the nerves to their usual condition. Earlier versions worked by clockwork but were later replaced by electrical versions. Mortimer attributed his insights to 'Grove's generalization as to the correlation of physical forces, and more recently by Professor Tyndall's beautiful series of experiments with sensitive flames and musical burners' (Mortimer, 1883, p. 21).

When Wilhelm Röntgen announced his discovery of X-rays in 1895 medical practitioners across Europe and North America were quick to recognize their possibilities. Since the photograph that Röntgen included with his original communication was of a human hand made transparent by the rays, it seems likely that he was aware of the possibilities himself as well. In Britain, Alan Archibald Campbell-Swinton was the first to produce a similar photograph, publishing his results in Nature early in 1896. The British Medical Journal commissioned Sydney Danville Rowland, a medical student at St Bartholomew's Hospital, 'to investigate the application of Röntgen's discovery to Medicine and Surgery and to study practically its applications' (Wetherall, 1995, p. 5). At McGill University in Montreal, John Cox, the professor of physics, produced an X-ray photograph that helped doctors locate and remove a bullet from a patient's leg. In Cambridge, the Cavendish Laboratory was soon providing informal X-ray services for Addenbrookes Hospital. X-rays appeared to provide a new therapy as well as a novel and highly effective aid in surgery to remove foreign objects and identify fractures. In November 1896, Leopold Freund in Vienna used X-rays to remove a mole from a young girl's back. X-ray diagnosis and treatment soon became a standard part of the therapeutic armoury of electrotherapy departments in hospitals, as did radiation treatments following Becquerel's and the Curies' discoveries.

René Blondlot's discovery of N-rays in 1903 instigated a similar flurry of medical interest. Investigators rapidly came to the conclusion that there seemed a specific relationship between these new rays and nervous tissue. As a result, medical men took a leading role in further investigating their properties. A number of researchers reported that N-rays seemed to be emitted by the nervous system. Arsène d'Arsonval announced his apparent discovery that N-rays were emitted by Broca's centre in the brain during speech. The flurry did not last, of course. It was not long before N-rays proved a chimera. In context, though, medical practitioners' enthusiasm for the new radiation was not surprising (Nye, 1980). Physics appeared to a number of doctors at the turn of the century as holding the key to an entire panoply of powerful new therapies. Electrotherapy, X-rays, and radiation therapies seemed to be only the tip of the iceberg in this respect. Proponents of physical therapeutics—as this new conglomeration of treatments was often described—were confident that they were at the forefront of research in physics as well as medicine. They were busily institutionalizing themselves as well. Two new journals—*Medical Electrology and Radiology* and the *Journal of Physical Therapeutics*—were established. In 1903 the British Medical Association established a sub-section of medical electricity (Morus, 2001).

Some of these physical therapies proved successful, others less so. Their promoters came from different backgrounds and brought with them a range of skills and assumptions. There were fierce debates as to just what kinds of skills were appropriate for the successful application of these new physical therapies. Many doctors argued that medical skills and knowledge had to remain primary. With regard to electrotherapy, for example, Julius Althaus argued that in 'this remedy more than in any other the mode of application has an all-important bearing upon the result; for it is not electricity that cures diseases, but the Physician, who may cure disease by means of electricity' (Althaus, 1859, p. 344), Others argued that it was the technical skills of the electrician that should be primary. Armand de Watteville insisted that 'measurements of current strength' were 'the essential conditions of a rational application of electricity in medicine' (de Watteville, 1884, p. vi), We can trace the existence of modern divisions of labour such as those between radiology and radiotherapy to these kinds of late Victorian disputes about skill, its possession, and the status of the operator. Hospitals and clinics were not self-evidently the appropriate spaces for the new therapies in all cases. As we have seen, research laboratories like the Cavendish offered their services as diagnostic spaces. X-rays and electrotherapeutic devices could be encountered on fairgrounds and in department stores.

22.4 CONCLUSIONS

The interactions between physics and medicine throughout the nineteenth century were certainly not all in one direction. It was not a simple and straightforward process of medical practitioners adopting ideas, instruments, and practices from the physical sciences. Neither medicine nor physics were unified, monolithic disciplines throughout this period. Medical institutions and medical careers varied significantly across Europe, over the course of the century and within different national contexts. The same can be said about physics. Individual careers and institutional structures, as well as theoretical and practical commitments, varied in important ways between the British Isles, France, and the German lands, for example. There were instances where medical practitioners attempted to make natural philosophical phenomena and practices their own—as in the example of galvanism at the beginning of the nineteenth century. In other episodes, electrical engineers and physicists tried to impose their own authority and institutional practices on the medical treatment of the body. In a competitive medical market-place, physics could provide useful resources for doctors seeking to make reputations for themselves. Similarly, medicine could seem a tempting area for the application of physics. There were periods and places when physicists clearly regarded the medical care of the body as a territory ready for colonization.

Medicine and physics sometimes coincided and sometimes clashed in a variety of ways. The physicist Oliver Lodge and the electrotherapist William Snowden Hedley could share a common language and instrumental concerns in arguing that the nervous system could be modelled usefully on the coherers used to receive radio signals. But at the same time, the gentlemen of the Royal College of Physicians and the professional men of the Society of Telegraph Engineers, who backed the establishment of the Institute of Medical Electricity in 1888, could have radically different notions regarding the proper behaviour of practitioners. We started this chapter with the example of the language the German physiologist Ludimar Hermann used to describe the economy of the human body in his *Grundriss der Physiologie des Menschen* and its translation into English by Arthur Gamgee. Both of their careers illustrate the ways in which medicine and physics could intersect, and the terms they adopted and adapted to describe the body's operations demonstrates their recognition—and their readership's recognition—of the common ground between the disciplines. Making his case to an audience of medical students in 1869, William Robert Grove asked: 'An acquaintance with the motions, sensible and molecular, which the different forces produce in the human body, the means of stimulating them when torpid, of checking them when too active, of apportioning them by diverting forces from one organ to another . . . . . what is all this but physical science?' (Grove, 1869, p. 486).

REFERENCES

ALTHAUS, JULIUS, 1859, *A Treatise on Medical Electricity, Theoretical and Practical; and Its Use in the Treatment of Paralysis, Neuralgia, and Other Diseases*. London: Trubner and Co.

ANON., 1850, 'Instinct and Reason', *British and Foreign Medico-Chirurgical Review*, 6: 522–524.

—— 1882, 'Electricity and the medical profession', *Electrical Review*, 10: 173–174.

BEARD, GEORGE, 1880, *American Nervousness*. New York: G. P. Putnam's Sons.

BIRD, GOLDING, 1847, 'Lectures on electricity and galvanism, in their physiological and therapeutical relations,' *London Medical Gazette*, 39: 705–711, 799–806, 886–894, 975–982, 1064–1072.

BYNUM, WILLIAM F. 1994, *Science and the Practice of Medicine in the Nineteenth Century*. Cambridge: Cambridge University Press.

CANEVA, KEN, 1993, *Robert Mayer and the Conservation of Energy*. Princeton: Princeton University Press.

CANTOR, GEOFFREY, 1976, 'William Robert Grove, the Correlation of Forces and the Conservation of Energy', *Centaurus*, 19: 273–290.

CARPENTER, HUMPHREY, 1977, *J. R. R. Tolkien: A Biography*. London: Allen and Unwin.

CARPENTER, WILLIAM BENJAMIN, 1850, 'On the Mutual Relations of the Physical and Vital Forces', *Philosophical Transactions*, 140: 727–757.

COLWELL, HECTOR, 1922, *An Essay on the History of Electrotherapy and Diagnosis*. London: William Heinemann.

CORRY, JOHN, 1815, *A Satirical View of London at the Commencement of the Nineteenth Century: Comprising Free Strictures on the Manners and Amusements of the Inhabitants of the English Metropolis; Observations on Literature and the Fine Arts; And Amusing Anecdotes of Public Characters*, sixth edn. London: J. Ferguson.

DE WATTEVILLE, ARMAND, 1884, *Practical Introduction to Medical Electricity*, 2nd edn. London: Lewis.

ERB, WILHELM, 1883, *Handbook of Electrotherapeutics*. London: Wood.

ESSIG, MARK, 2003, *Edison and the Electric Chair*. Stroud: History Press.

FINKELSTEIN, GABRIEL, 2006, 'Emil du Bois Reymond vs Ludimar Hermann,' *Comptes Rendus Biologies*, 329: 340–347.

GEISON, GERALD, 1978, *Michael Foster and the Cambridge School of Physiology: The Scientific Enterprise in Late Victorian Society*. Princeton NJ: Princeton University Press.

GROVE, WILLIAM ROBERT, 1842, 'The Gas Battery', *Literary Gazette*, 26: 833.

——1846, *On the Correlation of Physical Forces*. London: S. Highley.

——1869, 'An Address on the Importance of the Study of Physical Science in Medical Education', *British Medical Journal*, 1(439): 485–487.

HELMHOLTZ, HEINRICH VON, 1995, 'The Interaction of Natural Forces,' in David Cahan (ed.), *Science and Culture*. Chicago: University of Chicago Press.

HERMANN, LUDIMAR, 1875, *Elements of Human Physiology* Translated by Arthur Gamgee from the 5th German edition. London: Smith, Elder and Co.

HOLMES, FREDERICK, 2003, 'The Physical Sciences and the Life Sciences', in Mary Jo Nye (ed.), *The Cambridge History of Science, Volume 5: The Modern Physical and Mathematical Sciences*. Cambridge: Cambridge University Press. pp. 219–236.

JACYNA, STEPHEN, 1983, 'Immanence or Transcendence: Theories of Life and Organization in Britain, 1790-1835,' *Isis*, 74: 311–329.

JAY, MIKE, 2009, *The Atmosphere of Heaven: The unnatural experiments of Dr. Beddoes and his sons of genius*. London: Yale University Press.

KIPNIS, NAUM, 1987, 'Luigi Galvani and the Debate on Animal Electricity,' *Annals of Science*, 44: 107–142.

LENOIR, TIMOTHY, 1989, *The Strategy of Life: Teleology and Mechanics in Nineteenth-century German Biology*. Chicago: University of Chicago Press.

LEVERE, TREVOR, 2002 (1981), *Poetry Realized in Nature: Samuel Taylor Coleridge and Early Nineteenth-century Science*. Cambridge: Cambridge University Press.

LOBB, HARRY, 1867, *A Popular Treatise on Curative Electricity, Especially Addressed to Sufferers from Paralysis, Rheumatism, Neuralgia, and Loss of Nervous and Physical Power*. London: Bailliere.

MAINES, RACHEL, 1999, *The Technology of Orgasm*. Baltimore: Johns Hopkins University Press.

MARTIN, THOMAS C., 1995, *The inventions, researches and writings of Nikola Tesla*. New York: Angriff Press.

MAUDSLEY, HENRY, 1874, 'Sex in Mind and Education,' *Fortnightly Review*, 15: 479–480.

MORTIMER, JOSEPH GRANVILLE, 1883, *Nerve-Vibration and Excitation as Agents in the Treatment of Functional Disorder and Organic Disease*. London: Churchill.

——1884, *Nerves and Nerve-Trouble*. London: W. H. Allen and Co.

MORUS, IWAN RHYS, 1998, *Frankenstein's Children: Electricity, Exhibition and Experiment in early Nineteenth-Ccentury London*. Princeton NJ: Princeton University Press.

———1999, 'The Measure of Man: Technologizing the Victorian Body,' *History of Science*, 37: 249–282.

———2001, 'Batteries, Bodies and Belts: Making Careers in Victorian Medical Electricity,' in Paola Bertucci and Giuliano Pancaldi (eds), *Electric Bodies: Episodes in the History of Medical Electricity*. Bologna: Universita di Bologna, pp. 209–238.

———2006, 'Bodily Disciplines and Disciplined Bodies: Instruments, Skills and Victorian Electrotherapeutics', *Social History of Medicine*, 19: 241–259.

———2011, *Shocking Bodies: Life, Death and Electricity in Victorian England*. Stroud: History Press.

NUNN, THOMAS WILLIAM, 1853, *Inflammation of the Breast and Milk Abscess*. London: Henry Renshaw.

NYE, MARY JO, 1980, 'N Rays: An Episode in the History and Psychology of Science', *Historical Studies in the Physical Sciences*, 11: 125–156.

OPPENHEIM, JANET, 1991, *Shattered Nerves: Doctors, Patients and Depression in Victorian England*. Oxford: Oxford University Press.

PETERSON, JEANNE, 1978, *The Medical Profession in Mid-Victorian London*. Berkeley and Los Angeles: University of California Press.

RABINBACH, ANSON, 1990, *The Human Motor: Energy, Fatigue and the Origins of Modernity*. Berkeley and Los Angeles: University of California Press.

ROMANO, TERRIE M. 2002, *Making medicine scientific: John Burdon Sanderson and the culture of Victorian science*. Baltimore MD: Johns Hopkins University Press.

ROWBOTTOM, MARGARET and CHARLES SUSSKIND, 1984, *Electricity and Medicine: History of their Interaction*. San Francisco: San Francisco Press.

SCHAFFER, SIMON, 1997, 'Metrology, Metrication and Victorian Values', in Bernard Lightman (ed.), *Victorian Science in Context*. Chicago: University of Chicago Press, pp. 438–474.

SLEIGH, CHARLOTTE, 1998, 'Life, Death and Galvanism.' *Studies in History and Philosophy of Science*, 29c: 219–248.

SMITH, CROSBIE, 1998, *The Science of Energy: A Cultural History of Energy Physics in Victorian Britain*. London: Athlone Press.

STEWART, BALFOUR, 1873, *The Conservation of Energy*. London: King.

STOCK, JOHN EDMONDS, 1811, *Memoirs of the Life of Thomas Beddoes*. London: J. Murray.

STONE, WILLIAM HENRY and WILLIAM KILNER, 1882, 'On measurement in the medical application of electricity', *Journal of the Society of Telegraph Engineers and of Electricians*, 11: 107–128.

THELWALL, JOHN, 1793, *An Essay Towards a Definition of Animal Vitality*. London: T. Rickaby.

TIBBITS, HERBERT, 1877, 'On Current Measurements in Electro-therapeutics', *Lancet*, i: 488.

UEYAMA, TAKAHIRO, 1997, 'Capital, Profession and Medical Technology: The Electro-therapeutic Institutes and the Royal College of Physicians', *Medical History*, 41: 150–181.

———2010, *Health in the Marketplace: Professionalism, Therapeutic Desires and Medical Commodification in Late Victorian London*. California: Society for the Promotion of Science and Scholarship.

VASSALLI EANDI, A. M., GIULIO, CARLO, and ROSSI, FRANCESCO, 1803, 'Report Presented to the Class of the Exact Sciences of the Academy of Turin, 15 August 1802, in Regard to the Galvanic Experiments made by C. Vassali-Eandi, Giulio, and Rossi, on the 10th and 14th of the Same Month, on the Head and Trunk of Three Men a Short Time after their Decapitation', *Tilloch's Philosophical Magazine*. 15: 38–45.

WATTEVILLE, ARMAND DE, 1877, 'On Current Measurements in Electro-Therapeutics', *Lancet*, i: 488.

WETHERALL, MARK, 1995, *On a New Kind of Rays*. Cambridge: Cambridge University Library.

WILKINSON, CHARLES, 1798, *An Essay on the Leyden Phial, with a View to Explaining this Remarkable Phenomenon in Pure Mechanical Principles*. London: J. Wilford.
——1804, *Elements of Galvanism, in Theory and Practice*. London: John Murray.

NOTE

1. Both of Arthur Gamgee's brothers were also medical men. One of them, Joseph Samson Gamgee, was the inventor of a new form of surgical padding popularly known as Gamgee tissue. It is probable that it was from this source that J. R. R. Tolkien came across the name, which he appropriated for the heroic hobbit Sam Gamgee in *The Lord of the Rings* (Carpenter 1977, p. 21).

CHAPTER 23

PHYSICS AND METROLOGY

KATHRYN M. OLESKO

Physics and metrology, the science and technology of standards of measurement (Kelly, 1816), are so inextricably linked that it is unimaginable that the history of one could be done without the other. Yet until late in the twentieth century their historiographies were separate. Metrology was for a long time the purview of historians of the practice who lovingly and in great detail chronicled the values, conversions, and progression of standards from their customary pre-metric stages to the inauguration of the metre and beyond. The history of physics has for the most part been studied by historians of science who only here and there have taken metrological considerations into account, mostly because the practice of physics became so dependent upon standards after 1800. Although O'Connell (1993) has argued that metrological standards are indispensable for making universals out of particulars, they are not necessarily so; even in the absence of agreed-upon standards a culture of measurement, and even precise measures, can exist. What standards of measurement do is to facilitate the comparability of measurements across different communities.

The exclusion of science and scientists from the historical understanding of metrology is in part justified. Most measures were anthropomorphic or embodied in centuries-old authorities like brass markers on the floor of a church or the wall of the town hall, until the late sixteenth century or so, when Johannes Kepler and others began to tackle customary standards of measure from mathematical and physical perspectives. Only gradually from about the late seventeenth century onward were standards of measurement—primarily of length, weight, and time—reconstructed on the basis of reproducible procedures grounded mostly but not entirely in physical principles. Thus Zupko (1990), who began his study of metrological reform with the eighteenth century, included science and scientists as a part of his story, as did

Hebra (2003; 2010), who delineated in encyclopedic fashion the physical principles underlying conventional instruments, especially the ubiquitous metres of various sorts, that embody more modern standards of measurement.

The transition from customary metrological practices with anthropomorphic foundations to standards verifiable by experimental means was a complex process that involved more than the numerical redetermination and conceptual redefinition of a standard. Where and when they could, nascent nation-states made standards determination their sovereign right, empowered technicians and natural philosophers responsible for standards determination as civil servants, and imposed standards as a means of achieving uniformity and fairness. By the nineteenth century the determination and certification of standards were firmly incorporated into state bureaucracies. Often erroneously associated with the onset of modernity due to its association with objectivity, the physical reformulation of standards was just as permeable to subjective values and meanings as anthropomorphic ones were.

The history of the relationship between physics and metrology, or physical metrology, has not been uniformly developed, but inroads have been made into certain key problems that have shed light on how physics has been practiced. This chapter first examines how physics and metrology came onto the radar screen of historians by considering the impact of some early influential works on the history of measuring practices broadly conceived. It then turns to those metrological issues that to date have received the most extensive historical treatments. One is the construction of standards since the eighteenth century. The metre notwithstanding, historians have tended to focus most of their attention on two other episodes in the history of standards: the attempt to derive a standard of length from the seconds pendulum in the eighteenth and nineteenth centuries, and the determination of electrical standards at the end of the nineteenth century. Both sets of standards engaged physicists or physical principles directly, and both generated controversies that exposed the values and the inner workings of the community of physicists. The second set of issues concerns the historical frameworks within which the relationship between physics and metrology has been embedded. Those frameworks include labour patterns in physics, pedagogical practices, the long-term processes of industrialization and bureaucratization (and so state-building), and the moral dimensions of measuring practices. The history of physics and metrology has thus proven to be a fruitful way to understand large-scale historical processes that link the practice of physics to social, economic, political, and cultural history.

23.1 Physics in Metrology, Metrology in Physics: The Roots

Two early essays on the role of measurement in modern physical science offer insight into how metrology became a part of the concerns of historians of science. When

Koyré (1953) wrote his essay on measurement, theory and its related philosophical concerns dominated the history of science, mostly due to Koyré's own influence on the field. Consequently, he viewed the role of measurement in experimental practice as governed by theory in a teleological sense: 'not only are good experiments based upon theory, but even the means to perform them are nothing else than theory incarnate' (Koyré, 1953, 234). He viewed refined precision measurement as marking the dividing line between medieval and modern, and between the quotidian and the scientific. The metrological standards that made precision measurement possible were in his view evidence of the divine origin of God's work: through measurement, science uncovered the geometry of natural world that God had put their in the first place. He used the example of the isochronism of the pendulum, a property Galileo discovered but never used, to demonstrate how measurement in experiment evolved. Koyré considered Galileo's inclined plane experiments 'worthless' due to errors in time measurement. Marin Mersenne's attempt to perfect the pendulum as a timepiece were better, but limited (he never believed that a seconds pendulum could be constructed). Best, in Koyré's view, were the experiments to determine the value of gravity by the Jesuit Giambattista Riccioli, whose investigations on the isochronism of the pendulum from 1640 and later were motivated by a desire for a precise timekeeper—a goal he failed to achieve. Koyré argued that Christian Huyghens capped these efforts to achieve a precise timekeeper because his 1659 work on the pendulum linked its behaviour to mathematical formulations which in turn could suggest more quantitative experiments. Koyré's narrative and valuation of what was important when reconstructing the history of science were thus guided by notions of progress: greater progress occurred when mathematical theory guided the protocols and goals of experiments. Koyré's sense that metrological refinement was an essential component of experimentation was, however, not widely shared.

For all of the novelty Kuhn introduced into the history of science, his view of the relationship between theory and experiment, and hence of the role of measurement, was of a piece with Koyré's views (Kuhn, 1961). Far more complex was Kuhn's discussion of the *ways* in which measurement functioned in modern physical science. Dismissing the textbook view of reasonable agreement between measurements as a myth, Kuhn distinguished normal measurement (measurements undertaken to achieve agreement between theory and experiment) from extraordinary measurements (anomalous measurements that could not be explained by theory, thereby precipitating a state of crisis in which older theories were held in a state of suspended animation until explanations for the anomaly—perhaps a new theory—could be found). Yet the remarkable features of Kuhn's paper were less the historiographical categories he introduced (such as normal science and crisis) than his strong belief in the directionality of science's history: 'The road from scientific law to scientific measurement can rarely be travelled in the reverse direction' (Kuhn, 1961, 189–90). The metaphors of war he used when discussing the relationship between theory and experimental measurements underscored the conflict between them. Much as he saw measurement as a 'powerful weapon' in 'battle' (Kuhn, 1961, 183–4), though, he failed to address exactly how those weapons were created, and on what basis. That omission is strange, because the dichotomy he set up between traditional science (astronomy, optics, and mechanics)—where measurement was ubiquitous—and Baconian science (heat, electricity, magnetism, chemistry)—where measurements were only gradually introduced over centuries—offered a way to develop the historical significance of metrology. When the Baconian sciences did begin to include measurement, as he argues, objects of scientific investigation were as much the instruments used to study the phenomena as the phenomena themselves. These instruments were complex in the sense that they defined phenomena through measurement *and* they provided a measuring standard, usually idiosyncratic, for doing so. In the case of heat, for instance, before warmth or 'degree of heat' could be measured, the scales of a thermometer had to be determined and calibrated—an exercise in metrology. We could even say that many of the 'meters' of the eighteenth century—thermometers, hygrometers, hydrometers, eudiometers, and so on—were attempts to establish metrological standards using physical principles. Enlightenment instruments of this type, although studied individually, have yet to be treated *en masse* as part of the history of metrological standards.

It was not until nearly three decades after the publication of Kuhn's paper on the function of measurement in modern physical science that historians began to take measurement and metrology more seriously as historical objects of investigation. Three path-breaking volumes, published within a span of five years, created a solid conceptual foundation for future studies in the history of measuring practices (Frängsmyr *et al.* 1990; Heilbron, 1993; Wise, 1995). Articles in *The Quantifying Spirit* (Frängsmyr *et al.* 1990)—the name comes from *l'espirit géometrique* and notions of geometrical order espoused by the likes of Christian Wolff before analysis and algebra became more popular—covered broad areas of mostly applied sciences, including meteorology, geodesy, forestry, and political economy. Authors embedded their stories of order, systematization, measuring, and calculation in the context of centralizing bureaucratic states, thus establishing one of the central contextual frameworks for understanding the significance of the widespread measuring practices in the Enlightenment. Of all the articles in the volume, however, only Heilbron's (1990) on the metric system dealt directly with metrology. Casting his story in the context of revolution, bureaucratic demands, technocratic officials, and the tangled messiness of ancient systems of weights and measures, Heilbron's witty tale outlined the adventurous journeys from Paris to Dunkirk and Paris to Barcelona to measure the length of the meridian through Paris, one ten-millionth part of which would define the metre. Physics was not sidelined in the process: a pendulum still had to be observed, instruments had to be made and calibrated, and for the unit of weight, distilled water at the temperature of melting ice had to be weighed. When considering the entirety of the history of the metre, however, the contributions of physicists were outweighed and overshadowed by the work of astronomers, the travails of their triangulations, and the social and political resistance that had to be overcome to embed the metre in French society and economics (Alder, 2002).

Heilbron made a far more significant contribution to the role of physics in metrology in 'Weighing Imponderables' (1993). The title of this monograph-length article captures its relevance for metrology and its connection to Kuhn's Baconian sciences; for what Heilbron examined were the ways in which imponderable matter—the 'fluids' of air, gases, light, heat, electricity, magnetism—came to be quantified and measured with new instruments introduced at the close of the eighteenth century. The result was an instrumentalism in physics, equally concerned with imponderable fluids and new standards of measurement. Ponderable bodies extend in space and have weight; what about imponderable ones? Coulomb's torsion balance experiments to measure the 'force' of electricity were exemplars of this attempt to capture the world in numbers, as were Cavendish's thermometer, Roy's barometer, Lavoisier's and Laplace's ice calorimeter, and Gay-Lussac's device for measuring the thermal expansion of gases. Mathematical formulations for each grew in complexity as (perceived) precision in measurement increased, sometimes without rhyme or reason except that the British and French wanted to outdo one another on the decimal-place front. Pierre-Simon Laplace and his school made particularly significant contributions to these efforts, especially in the area of optical refraction, capillary rise, and the speed of sound. The metrological thread, however, was largely lost in Heilbron's story with the exception of the last chapter, a reprint of his essay on the metre (Frängsmyr et al. 1990, 207–242; Heilbron, 1993, 242–277). So while metrology was a pressing concern in the Age of Reason, the connections between instrumental 'metres' and the metre as a standard of measurement await examination. We know the outcome. Where all these measurements and instrumentalist impulses took physicists can be found in works such as Buchwald's on optical theory and experiment in the nineteenth century (Buchwald, 1989). By the mid-nineteenth century, experiments and comparisons between them were eased considerably with the adoption of common standards of measurement.

No volume has had a greater impact on the study of precision in measurement, and thus of metrology, than *The Values of Precision* (Wise, 1995). Beginning with the spread of precision in measurement in the eighteenth century, articles in the volume strongly linked the quest for refined measurement to features of modernity customarily associated with Michel Foucault—state power, territorial control, administrative and economic efficiency, and social regulation—an observation that led the editor to conclude that 'precision has thus become the *sine qua non* of modernity' (Wise, 1995, 11). By implication, the same held for standards of measurement as well. The editor's interspersed commentaries argued for the central role of industrialization in the spread of precision in measurement—a conclusion that privileged the British contexts of some of the articles over other national environments where industry was less in evidence. These commentaries also stressed the role that precision in measurement played in achieving agreement within scientific communities—a goal that could not be achieved without common standards of measurement. Four of the ten substantive articles treated the relationship between physics or physicists and metrology (Alder, 1995; Gooday, 1995; Olesko, 1995, and Schaffer, 1995).

Finally, among the early influential works on metrology, Kula's *Measures and Men* (1986) should be singled out. No other book has had a greater impact on the ways in which the historical implications of metrological systems have been expressed. Although exaggerating the separation between the anthropomorphic connotations of older standards of measurement and the rational modernity of the metric system,

Kula's study nonetheless broadened considerably the issues taken into account in examining the history of any standard. He emphasized in particular the persistent and unavoidable moral dimensions of pre-metric standards of measurement: how does the 'just measure', he asked, become that of the 'just man'? (Kula, 1986, 100). Kula viewed past measures as signs filled with signification, good and bad: Cain invented standards of measure (and so measures are associated by implication with a primal crime time), but God's measures were always just (thus by implication conferring grace upon those who did the same). Coming from a Marxist tradition in historical writing, Kula stressed the ways in which conflict over measures were indications of class conflict, judicial rule, and disagreements with authority—thus linking standards of measurement of all types to the exercise of power. Through the adaptation of Kula's ideas—especially to post-metric measures—moral issues have become a persistent dimension of the study of standards of measurement, and by extension, of the study of the behaviour of scientists who participate in their construction.

23.2 APPROACHES TO PHYSICS AND METROLOGY

Of the hundreds of standards that were created before 1900, two in particular have attracted the lion's share of historical attention: the construction of a standard of length based on the seconds pendulum, and the determination of the unit of electrical resistance. Debates over the measurements produced were in each case acrimonious and thus revealing of the assumptions and tacit protocols of scientific practice. These two cases have in particular been principal historical sites for understanding how precision in measurement developed, especially through the analysis of errors. Their stories reveal how physicists (and others) became embedded and even embroiled in community politics, public relations, and national priorities in the eighteenth and nineteenth centuries.

The story of the seconds pendulum began with an anomaly. French expeditionary measurements of the length of the seconds pendulum, mostly in the Atlantic World, revealed in 1672 that the seconds pendulum was shorter at the equator than at higher latitudes by about three millimetres. The finding dampened efforts in Britain and in France to make the seconds pendulum a foundation for a standard of length. Leading natural philosophers—Jean-Dominique Cassini, Jean Picard, Christiaan Huyghens, Philippe de la Hire, and others—challenged the validity of the finding by developing protocols for more refined measurements that would settle the controversy. The problems here were in fact many. In an era grappling with what was the normal and ordinary in nature—assumptions that were every day challenged by objects and reports from global expeditions—the fact that the anomalous length not only came from afar, but also was discovered by an *élève* (assistant) at the Parisian Académie des Sciences, raised questions about the veracity of the report on many fronts.

Dew has viewed the ensuing debate over the length of the seconds pendulum as 'a case study in the construction of scientific authority within the networks of the French Atlantic' (Dew, 2008, 56)—in short, as a problem of establishing trust and maintaining authority in the expanding Atlantic World where data travelled over hundreds of miles (Dew, 2010). This French problem (as it was called) attracted especially the attention of Newton, who was at the time working on the *Principia*. Although Dew mentioned that Newton used the data in Book III of the *Principia*, it was Schaffer who exploited the incident, casting it in broader historical terms as part of the global 'information order' of the *Principia*. While located in Britain, Newton nonetheless participated in the trading of goods (data) taking place in the Atlantic World (and elsewhere) by incorporating French (and other) results into his theory of universal gravitation. The seconds pendulum data said something about the shape of the Earth: the shortening of the seconds pendulum at the equator was the result of weaker gravity, a consequence of the Earth's bulging at the equator and flattening slightly at the poles. In that deft move, Schaffer cast the *Principia*, for the first time, in the ongoing process of globalization that was a part of Newton's world, thus making natural *philosophy* as much a part of the Atlantic World as natural *history* already was (Schaffer, 2009).

At the end of the eighteenth century the French tried once again to make the seconds pendulum a standard of length, only to have it nudged out by the metre. (Jean-Charles de Borda and Charles Augustin de Coulomb still performed seconds pendulum experiments in Paris in 1792, this time with error analysis so accurate that they claimed nothing was wanting from their result.) The British and the Prussians, sceptical of the metre, were more open to the swinging pendulum. Miller (1986) viewed the reform of the British standard of length based on the seconds pendulum between 1815 and the 1830s as an important part of the revival of the physical sciences in the realm. The project engaged the Royal Society of London, the Astronomical Society, the Board of Longitude, and the Admiralty, thus forming a nodal point where disparate groups could cooperate and sustain a community with common interests. Pendulum experiments were designed with greater accuracy, voyages were taken to perform the measurements, and data were acquired that further refined knowledge of the exact figure of the Earth. Data reduction techniques were considered too, with the result that James Ivory, who was responsible for reducing all the data, rejected the method of least squares on the grounds that some of the results the method would incorporate should not be used at all. He developed his own mathematical method for eliminating anomalous data. (The British never developed the love for the method of least squares that the Germans had.) Miller concluded that experiments with the seconds pendulum, though implemented as the basis of a standard of length but eventually rejected after the fire that destroyed the yard, nonetheless were of historical significance because they placed 'large scale, government-supported research in the physical sciences on equal footing with exploration and natural history on naval voyages' (Miller, 1986, 27). As had happened in the seventeenth century, seconds pendulum experiments raised the stature of the physical sciences.

In no state, however, did the seconds pendulum achieve such iconic status as a model of precision measurement and exact experimental physics as it did in Prussia. While in France and in Britain the seconds pendulum heightened controversy over the reliability of precision measurements, raised questions about the dependability of instruments, and alternately united and fractured scientific communities, in Prussia the unit of length based on the seconds pendulum became the standard against which all others were judged, and a source of deep national pride. Olesko cast the story of Prussia's weights and measures reform, which commenced in 1816 and ended in 1839, in the context of post-Napoleonic fiscal reconstruction. Prussia emerged from the crushing blow of the Napoleonic Wars in 1815 with permanent territorial losses that reduced the length of Prussia's main route for trade in the east, the Weichsel River; with knowledge of the introduction of the metric system in and around its western territories while under Napoleon's rule; and with the burden of financial and tariff reforms needed to improve the state's overall fiscal health. Officials linked inaccurate weights and measures directly to problems in trade. Adopting the metric system was not an option. Once the reform was completed, its impact was noticeable. Weights and measures reform helped Prussia achieve financial solvency and administrative consolidation in customs and trade. The model of how Prussia used its weights and measures for tariffs—charging duties on the basis of the weight or size of an item rather than its value—also strengthened the tariff policy of the North German Customs Union (*Zollverein*, established 1834), which adopted the Prussian practice. (Olesko, 1995, 117–122).

A more compelling reason for the esteem held for the seconds pendulum investigations can be attributed to the manner and style in which they were conducted by Friedrich Wilhelm Bessel, the principal investigator of the reform. Two sets of investigations, one conducted at Königsberg in the 1820s and another in Berlin in the 1830s, became exemplars of an exact experimental physics. A major improvement over prior investigations was Bessel's recognition that as the pendulum moved it carried some air with it. So in addition to Newton's hydrostatic correction to a pendulum's motion, Bessel added a hydrodynamic one. Olesko argued that the protocols Bessel brought to bear on the *determination* of the standard were also incorporated into its legal *definition*. Bessel's public reporting of his error analysis spoke to the quality of his data as well as his own honesty and integrity in reporting them. The same held, he stated, for the highly centralized, hierarchical production and circulation of the original standard and its copies in Prussia: no standard was absolutely perfect, and hence the protocols used to produce the standards as well as its intrinsic errors had to be reported *publically*. Bessel's approach circumscribed the exclusive reliance on the material expression of a standard, which was secondary, and placed the *reproducibility* of the standard at the centre of its definition. So in reporting the errors of the standard and its copies, the legal definition of the standard had to specify the limits of the range of those errors—their tolerances (Olesko, 1991).

Olesko pointed out that the importance of publically reporting errors in the system Bessel created compelled a reconsideration of Prussian resistance to the metric system in the 1860s: their stance was not based patriotic sentiment, but on the conviction that the framers of the metric system had not publically reported errors to the degree Bessel had. Hence the metric system lacked the kind of integrity that Bessel had built into the Prussian standard of length. Even in 1868 when the North German Confederation (including Prussia) accepted the metric system for trade, Bessel's protocol ruled. The German metre and kilogramme were not, strictly speaking, the French ones, but standards constructed in relation to the French with ranges of tolerance specified. Error and tolerance thus became key concepts in the North German system of weights and measures, and they had scientific, ethical, and legal dimensions. Thanks to its public importance, Bessel's seconds pendulum investigation became widely emulated in administrative, scientific, and educational circles. (Olesko, 1995, 131–135).

Like the standard of length based on the seconds pendulum, the construction of a unit of electrical resistance was similarly fraught with difficulties stemming from local differences, especially concerning the reporting of errors. Several standards of resistance were developed between the 1840s and 1860s. In Germany between 1841 and 1852, Wilhelm Weber developed an hierarchical system of resistance measures based on absolute units (cgs) of mass, length, and time. Thinking in parallel to the weights and measures used in daily life in northern Germany, Weber founded his resistance measures based on their definition and the protocol used to achieve them. His system was difficult to use, of questionable practical value initially to technicians, and based on complicated calculations. Charles Wheatstone chose a foot of copper wire weighing 100 grammes as a resistance measure in 1843, and Moritz Jacobi constructed a series of resistance étalons in 1848. William Thomson's discovery in the variation of the resistance of copper, however, placed into question its use in a standard. In 1860 Werner Siemens proposed using a column of mercury 1 metre in height and with a cross-section of 1 mm^2 at $0°$ C as a standard easily used in technical areas, especially telegraphy. In 1862 the British Association for the Advancement of Science (BAAS) adopted Weber's absolute unit, while Germans found Siemen's unit more appealing. And so began the controversy over resistance standards—a topic that has been covered from various angles in the literature (Gooday, 2004; Hunt, 1994; Kershaw, 2007; O'Connell, 1993; Olesko, 1996; Schaffer, 1992; 1994; 1995; 1997).

Four main approaches to the controversy over resistance standards have been developed: *social constructionist*, emphasizing community formation; *technical*, centred on the needs of the telegraph industry; *institutional*, focused on scholarship and teaching at Cambridge University and the Cavendish Laboratory, especially under James Clerk Maxwell; and *cultural*, stressing local communities with different practices. Although none of these approaches is mutually exclusive of one other, each one nonetheless tells a different story. O'Connell (1993), who took a social constructionist approach, viewed the controversy over resistance standards as of a piece with other standards controversies: once settlement was attained, each standard became a locus for communities with shared material cultures, shared values, and shared practices. The turning point in the controversy for O'Connell (and others) was the 1858 failure in laying the transatlantic telegraph cable: the failure contributed to the establishment, in 1861, of a BAAS Committee on Resistance Standards which adopted Weber's model for the resistance standard. Their adoption had distinctly British features: the standard was based on the concept of work, then being developed in Britain; and the Committee decided to embody the standard as five pairs of metal coils, each pair of a different metal, thus forming a 'Parliament' of standards that could be compared to one another. Around these five pairs of standards, British scientists manufactured equipment and instruments that embodied the standard, thereby creating a material culture. The alternative to the Parliament of standards was Siemen's mercury column. The 1881 International Electrical Congress elected to define the ohm (the unit of resistance) in terms of the absolute system of units the British had used, but to manufacture the resistance standard as Siemen's metal column.

Hunt (1994; 1996), in reaction to the attention paid to Maxwell on the matter of resistance standards (Schaffer, 1992), asked what role British telegraph engineers, especially submarine telegraph engineers, played in the determination of a resistance standard, as opposed to British physicists. Hunt was determined to gauge how closely science and technology worked together in the construction of the standard, given that one of the primary uses of a resistance standard was the determination of faults in transatlantic cables—an extremely difficult task. The writings of both physicists and telegraph engineers, Hunt argued, showed 'how the demands and opportunities presented by the submarine telegraph industry helped shape the techniques, and literally set the standards, that were to become the ordinary working tools of both physicists and engineers in the last quarter of the nineteenth century' (Hunt, 1994, 49). At stake was the quality of cables—something that electrical measurements could help determine, but only if there was a standard of electrical resistance. Telegraph cables became the main market for electrical measurements. Standardization was thus not only a matter of metrological concern, but also a manufacturing issue: resistance coils as well as telegraph cables had to be made to contract specifications. A coda to his story was added later when he examined how idiosyncrasies in telegraph resistance measures were eliminated, partly through the assessment of character (Hunt, 1996).

Schaffer's many studies of the ohm embed the story in the culture of Cambridge University, specifically in the Cavendish Laboratory, and in the culture of Victorian values (Schaffer, 1992; 1994; 1995; 1997). In addition to the metrological issue of which standard should be chosen and the technical issue of telegraph faults, Schaffer added the importance of the determination of the ratio of electrostatic to electromagnetic units in Maxwell's programme. Not only did the accurate determination of that ratio rest on a resistance standard, it also was an important theoretical consideration: for if that ratio were equal to the speed of light it either constituted evidence for his theory of electromagnetism or, as he later maintained, a comparison between the two values did not discredit his theory. Maxwell, furthermore, wanted to establish the Cavendish as a central authority for standards, and the rational use of instruments as a regular part of academic physics. Yet conducting these measurements at Cambridge was not easy: the university's traditions did not include manual labour in the laboratory, nor a workshop culture. Schaffer's stories of the resistance standard thus illuminated the clash between traditional academic culture and a technical one.

In these studies the German story was largely ignored. Olesko's study was the first to examine the German side of the determination of a resistance standard (1996). Taking her cue from Hunt, she considered academic physics and telegraphy equally in her story. Like Schaffer, she embedded her story in local cultures of precision measurement, which for northern Germany included the protocols of Prussian weights and measures and an obsession with the method of least squares as a necessary technique of data reduction, neither of which were prominent features of the British culture of precision measurement. The very different analytical perceptions of data on either side of the Channel proved to be an obstacle to the German acceptance of British results: they simply were not accurate or precise enough, and hence could not be compared to the German results. But the difference between the two sides was not simply a matter of disagreement over a technique. British labour practices were collaborative in the sense that sources of errors were eliminated and data reduced *in* the laboratory without the public reporting of errors. The release of errors to the public—scientific, technical, and otherwise—was simply not deemed necessary. In Germany, by contrast, the emphasis was on individual labour. In any single laboratory there might be several investigators, but their tasks were hierarchical. All reports of results were accompanied by the range of error determined by the method of least squares. Whereas in Britain, precision in measurement was achieved by synchronizing results so that they clustered around a single value, in Germany, precision represented an agreement between individuals, not among them. Negotiations on the relative value of errors took place *outside* the laboratory, so error perforce had to be reported. The lack of a probabilistic calculation of error in the British case led Germans to conclude that British exactitude was simply specious. Thus the contrast in how errors were determined and reported on each side, and hence how precision was achieved, contributed significantly to the disagreements between the British and the Germans over the proper measure of electrical resistance.

Kershaw (2007) reconsidered the historiography concerning the controversy over the determination of an electrical standard of resistance and embedded these earlier studies, most of which concerned the period before the first International Electrical Congress in 1881, in the context of the congresses that occurred after then. Although he considered the decisions made in 1881 and later to be flawed, he argued that they were adequate for engineering, commerce, and society. By highlighting the strengths of local cultures, especially the ways in which local cultures created strong bases for agreement when finding a solution to collective problems, he was able to bring to the forefront the inadequacies of the system of international congresses where the degree of agreement was less and where the ability to impose standards (he argues) was weaker. His review pinpointed the lingering consequences of having a practical standard, but not a precise one: to wit, the standard would have to be determined again when industrial and scientific demands for precision increased.

Other historical studies of relationship between metrology and physics are relatively new, but also far between. Standards of measurement for the early modern period may not have had as close a connection to physics as later ones did, but they are worth examining in their historical context for the light they shed on how metrological considerations were central to the thinking of antiquarians, philologists, theologians, numismatists, and natural philosophers of the early modern period who linked metrologies to chronologies, conceptions of the mathematical order and structure of society, and Egyptology (through the study of the pyramids). Newton, for instance, was interested in metrology as Warden (1696–1700) and later Master of the Mint (1700–1727) (Shalev, 2002). One interesting question yet to be broached is the relationship between the philological study of ancient weights and measures in the early nineteenth century (see for example, Böckh, 1838) and the widespread reform of weights and measures by scientific means then taking place: did the examination of standards from a deeper past lend legitimacy to contemporary reformulations by augmenting scientific authority with historical ones? Chang (2002; 2004), using an approach that bridged history and philosophy of science, has examined the evolution of thermometric standards, encasing the story in a narrative of scientific progress. His overall concern was not to understand the *history* of the standards but rather to demonstrate the ways in which the history and philosophy of science could be usefully deployed in creating scientific knowledge in instances where science itself failed. Studies such as Chang's could be usefully expanded to include a consideration of thermodynamics, concepts of work, and engineering in the nineteenth century, thereby embedding the story of thermometry in industrialization and the concomitant rise in energy use.

Hentschel's (1993) discussion of Henry Rowland's discovery of the redshift of Fraunhofer lines in the solar spectrum, which culminated in the decision to use the red cadmium line as a standard of wavelength in 1905, artfully demonstrates the strong cooperation of science and industry in turn-of-the-century America. Staley (2002; 2008) has examined in great detail Albert Michelson's attempt to use measurements of the velocity of light as a standard of length—a goal held by many others as well—and here too the connections between scientific work and precision industries in both needing and attaining that goal were crucial. Machine tool industries (with which Michelson closely interacted), the US Coastal Survey, railroads, and any industry using nuts and bolts, gauges and interchangeable parts, and similar manufactured items simply needed greater technical uniformity and material interchangeability than was available at the end of the nineteenth century. In 1893 Michelson determined the length of the metre at Sèvres to be 1,553,163.5 λ of the red line of cadmium. Ironically, whereas Michelson's investigations with his novel interferometer confirmed the invariance of length in industry, it did the opposite in ether drift experiments. Staley argued strongly for the connection between new standards and the new physics taking shape at the turn of the century: discussions of relative and absolute measures were central to electron theory and relativity theory. This connection could with profit also be examined in the history of quantum theory in the first quarter of the twentieth century.

23.3 Broader Historical Issues in Physics and Metrology

With perspicacity Maxwell expressed the connection between large-scale historical processes and metrology in 1872: 'The whole system of civilization may fitly be summarized by a foot rule, a set of weights, and a clock' (Quoted in Gay, 2003, 107). Although we may question whether or not Maxwell's figure of speech accurately identified *what* standards of measure signified, subsequent historical literature has upheld his insight into the symbolic meanings of weights and measures. And if we substitute for 'civilization' the phrase 'civilizing process', then we are well positioned to unravel those meanings by examining several contexts in which standards of measure played a prominent role. Four of them have received extensive historical treatments: labour practices in physics, physics pedagogy, long-term processes of modernity (bureaucratization, industrialization, and the construction of empires), and the moral and ethical connotations of standards of measure.

Traditional standards of measure were rooted in the reality of life and labour: there is hardly a measure of land in any culture that does not make reference to the tilling that could be completed in a unit of time (for example, *Morgen* in north central Europe, including Poland, represented the amount of land that could be tilled with the aid of a work animal in the morning hours). After 1800, standards of measure acquired social meaning not in the labour of the field but in the labour of precise measurement. Thus the construction of standards of measure since about 1800 offers a window on the practice of physics. After the fire in the British Parliament that destroyed the original yard, John Herschel invested the authority of the reconstructed yard not in a natural quality (the seconds pendulum) but rather in the labour invested in producing the new yard (Schaffer, 1997, 442). That window is especially revealing in making visible the so-called 'invisible technicians' (Shapin, 1989). Hentschel's (2005; 2007) studies of Carl Friedrich Gauss's involvement in Hanoverian weights and measures reform led him to investigate, in painstaking detail, the contributions of his mechanic and instrument-maker Moritz Meyerstein, whose only mention in all prior Gaussian literature, including his biographies, was in Olesko (1995). Olesko had pointed out that Gauss, who had only reluctantly participated in the reform, asked Meyerstein to perform the time-consuming and painstaking double-weighing of the Hanoverian pound. But when the state's officials found out, they insisted that Meyerstein's measurements be supervised by Gauss—a signal that they valued the authority of the person over the authority of the technique (Olesko, 1995, 119–121).

Hentschel delved deeper into the archive, even uncovering local documents concerning Meyerstein and his family, and revealed that Gauss and Meyerstein performed extensive comparative measurements to verify Hanoverian weights. The result was a precision of ± 0.001 mg for the so-called 'Meyersteinian' milligramme, a thousand times more precise than the accuracy achieved in the Prussian determination of its standard of weight. Meyerstein was handsomely rewarded for his efforts.

Among their combined innovations was their consideration of specific weights in the determination of a standard of weight, which not even the Prussians had done in their earlier reform. Meyerstein went on to tackle other metrological issues for Hanover. Thus Hentschel's study of Meyerstein's participation in the Hanoverian reform demonstrated how faith in skill replaced trust in persons; how the tasks of the precise measurement of a standard were divided; how labour-intensive and expensive standards determinations were; and how social and economic history and material culture enter into the history of physics and metrology (Hentschel, 2005).

Labour practices in physics begin in instruction, and here too metrology and standards reform during the nineteenth century played key roles in the novel attempts to train the next generation of physicists systematically. Interestingly, among all possible exemplars for laboratory exercises, the seconds-pendulum investigation (used to set the standard of length in Prussia) and the determination of standards of electrical resistance (including the ensuing controversy over them) offered crucial models for the some of the first laboratory exercises in physics in northern Germany and Britain.

Bessel's seconds-pendulum investigation played a crucial role in the mathematico-physical seminar established by Franz Neumann and Carl Gustav Jacobi in 1834. After several bumpy semesters of teaching, Neumann decided in the winter semester of 1838/39 to offer a course on mechanics as an introduction to theoretical physics in response to student needs. Calling the investigation 'the best model to study when combining a theoretical investigation with observation' (Olesko, 1991, 159), Neumann created what Olesko has called the Besselian experiment—an experiment that focused on the limits to the reliability of instrumentation, on the identification of errors, and on the method of least squares to reduce data. The most important topic in the course was gravity, where two-thirds of the discussion—one quarter of the course overall—concerned the seconds pendulum. It could be used, Neumann argued, to measure force, express standards of length and time, and measure gravity—all with the highest accuracy available. Through its example one could furthermore teach series expansions and approximation techniques and solutions to simultaneous equations, as well as the elements of an exact experimental physics, including the analysis of constant errors, instrument construction, experimental protocol, and accidental errors through the method of least squares. Building on the example of the seconds pendulum as a material expression of the force of gravity, these techniques could then be applied to the barometer in aerostatics, the thermometer and capillary tubes in hydrostatics, and bifilar suspension in Gauss's magnetometer (Olesko, 1991, 125–191).

While the emphasis on the model of Bessel's investigation stabilized instruction in the seminar, it also had, Olesko explained, more far-reaching historical consequences. Expressing errors in terms of analytical expressions nearly led to the erasure of the materiality of an instrument. Conveying the techniques of research in this fashion in seminar exercises did not always lead to doctoral theses or original published investigations; often students were forestalled by endless data analysis and data reduction, which became tools of the sceptic. Yet because Neumann's pedagogical innovation coincided with the reform of weights and measures in Prussia—Bessel's investigation became the foundation of a standard of length in 1839—not only did professional values cultivated in the seminar resonate with larger social and economic interests, but also the ensuing national popularity of Bessel's investigation transformed it into an icon of exactitude. The local and national significance of the investigation enhanced the ethos of exactitude in the Königsberg seminar, so much so that by the end of the century that ethos became an *ethic* guiding decisions about scientific right and wrong and professional conceptions of truth and error. It even separated the called from the damned by defining profession identities wherein precision and accuracy became not only technical goals based on skill, but also outward signs of personal and professional identity. The demise of the ethos was the result of the inability of practitioners to live up to the extreme degrees of precision that it demanded (Olesko, 1991, 366–450, 459–60).

In a similar fashion, the determination of standards of electrical resistance shaped a type of laboratory instruction in Britain in the nineteenth century that also cultivated exactitude and defined an exact experimental physics, but was encased in a different set of values. In his richly detailed study of physics teaching laboratories in Victorian Britain, Gooday (1990) emphasized especially William Thomson's teaching laboratory at Glasgow as a site where the supply-and-demand sides of the telegraphic and electrical industries shaped physics instruction decisively. In Thomson's laboratory, student-assisted research on absolute thermal and electrical measurements generated data that were used by the BAAS Committee on Resistance Standards to perfect the Weber's resistance standard as modified by Thomson himself. In order to make this type of practical instruction relevant to the university, it was cast *both* as an instrument of liberal education for accuracy in reasoning, *and* as a means to quantify earlier 'rule of thumb' practices that could now provide training for the global workforce needed for the telegraph industry. Both, Gooday argues, were important for the definition of experimental physics in Britain (1990).

In a similar fashion Friedrich Kohlrausch used the determination of the standard of resistance as an exercise in instruction in exact experimental physics at Göttingen—like Thomson, he too contributed to the debate over standards through his own calculations of resistance—and even used the literature from the debate as part of his own seminar exercises. As in the case of Neumann's use of the seconds pendulum, Kohlrausch's use of the standard of electrical resistance instilled certain cognitive preferences that valued the precise determination of fundamental constants, experimental errors, and experimental data as means to judge the limits of reliability of existing laws (Olesko, 2005).

Neumann, Thomson, and Kohlrausch taught to increasingly diversified students in their physics courses and laboratories, a result of social change and industrial growth in the nineteenth century. The redetermination of metrological standards and the adaption of techniques associated with them to other areas of investigation by all of them were a way to attain standardization in physics instruction that could accommodate broader demographic swaths of students, including those with industrial and other practical interests. The connection between instruction and larger historical processes is but a small facet of the broadly based connections between physics, metrology, and large-scale historical movements. Here the literature is rich, provocative, and suggestive. Historians have cast metrologies dependent upon physical principles in the context of commercialization, bureaucratization, industrialization, statecraft (including good governance, power relations, and social policy)—all of which could be expressed more generally as elements of Weberian modernization. What Scott called the 'hieroglyphics of measurement' (Scott, 1998, 29) aptly expresses the interlocking of metrologies and historical contexts. On the surface we may see seemingly objective and scientific measuring practices, but when we look at the matter more closely we realize that those practices are only the outlines of a complex web of connections, the finer threads of which are woven by the messiness of human interactions in social, political, and economic relations; in commerce, trade, and industry; in the state and the empire; and in international relations. Standards of measurement follow these threads, and keep them taut to a greater or lesser extent.

The state—in the form of its bureaucracies and its civil servants—looms large in this fabric of connections. Once a state sets standards of measurement, monitoring those measurements becomes integrated into the state's portfolio of administrative tasks, and so a part of its power (Scott, 1998, 24–30). Legal metrologies spread precision in measurement, but they also coordinated behaviour and society (Mallard, 1998, 574). Yet power and coordination were achieved only after years of hard technical work. In a series of stellar essays, Ashworth (2001; 2008) detailed how the state revenue initiatives, in particular the imposition of excise taxes, drove increasingly refined measuring procedures and instrumental modifications. The use of the hydrometer to determine the alcohol content of spirits is a case in point. Measurements for alcohol content had to be corrected for temperature, and the proportion of alcohol in a mixture required the determination of specific gravity. Scientific corrections, intended to yield accuracy, did not always reign supreme because the hydrometer was also subject to the forces of trade, manufacturing, and national interest. Ashworth also pointed out that greater revenue did not necessarily follow from greater accuracy. The Hydrometer Act of 1817/1818 incorporated a conception of accuracy dependent upon legislation, community-determined tolerances in measurement, and technical skills in managing the instrument (Ashworth, 2001).

It was difficult enough to manage measures within the boundaries of a state. Empires complicated the task of monitoring physical metrologies. While Master of the Mint, Newton had to deal with the problem of gold-counterfeiting: the Mint began to rely increasingly on gold from Atlantic World trade, especially from Guinea, where its purity could not always be ascertained. Both Newton and Robert Boyle were supporters of the gold system, and plied their craft on gold-assaying: Boyle wanted to use hydrometry to confirm gold content. Bringing the hydrometer into Guinean culture, however, proved problematic, as Schaffer (2002; 2009) has demonstrated. Guineans had their own way of 'assaying' gold—gold was so common there it was used as currency, something that made the British anxious—and the introduction of the hydrometer threatened their customary practices. Schaffer has argued that the hydrometer became a 'moral emblem' and a 'way to get fiscal order on a far away colony' (Schaffer, 2002, 29), with the result that the 'Newtonian mint became an emblematic site of administrative metrology' (Schaffer, 2002, 29, 39).

Beginning in the late eighteenth century, the state created special bureaucracies that institutionalized metrology and provided a means to guarantee its standards. Early on those bureaucracies were scattered, as they were in Prussia at first under Frederick the Great, who eventually created the *Oberbaudepartement* to manage standards as well as provinces newly acquired from Poland (Olesko, 2009, 18–19). The offices that handled British standards after the British Weights and Measures act of 1824 were many, and included the Statistical Department of the Board of Trade (1832), the Factory Inspectorate (1833), the General Registrar Office (1836), the observatory of the British Association for the Association of Science (1842), and other agencies (Schaffer, 1997, 441). By the end of the century, however, disparate boards of oversight and standards development were centralized in a single bureau. The Convention of the Meter included the establishment of the Bureau International de Poids et Mesures (BIPM) at Sèvres in 1875, whose story from then until the twenty-first century has been told from first-hand experience as well as extensive archival research by the physicist and former Director, Terry Quinn (2011). While the French BIPM remained an important leader in international metrological coordination, it was the Berlin-based Physikalisch-Technische Reichsanstalt (PTR), established in 1887 and known later as the Physikalisch-Technische Bundesanstalt (PTB), that became the model for subsequent national metrological institutes. In a foundational book, Cahan established the historical problematic of physical metrology (Cahan, 1989). Written at a time when historians regarded the PTR largely as the setting for experiments on black-body radiation, and therefore as a player in the quantum revolution, Cahan's book recast the PTR in its original historical role: as a standards testing bureau dedicated to 'advancing the art and science of physical measurement' and providing a place of employment for dozens of physicists during the high phase of German industrialization (Cahan, 1989, 5). In particular, Cahan demonstrated that the impulse for the black-body experiments came from the technical needs of the German lighting industry, thereby situating early quantum-theoretical work in practical and metrological as well as theoretical contexts. Quickly rising to become the global centre of precision physical measurement, the PTR inspired imitations abroad: in Britain, the National Physical Laboratory in Teddington (1898); and in America, the National Bureau of Standards in Washington (1901), renamed the National Institute of Standards and Technology in 1988. Both of these institutions await definitive histories at the level of historical scholarship that Cahan brought to bear on the PTR. They are likely to echo Quinn's findings for the BIPM: the search for stable and reproducible metrological standards is linked to the ever finer determination of the physical constants of nature, as demonstrated in twenty-first century international efforts to redefine the kilogramme in terms of Planck's constant—a project in which Quinn played a major role (Crease, 2011, 249–268).

23.4 CONCLUSION: METROLOGY AND MORALS

A common thread running through nearly every study of physics and metrology is the moral dimension of measurement. Especially in the British context, moral values surfaced in discussions concerning metrology. The expression of those values varied. According to Schaffer, for Maxwell the morality invested in precision measurement was a 'natural theology' that allowed him to 'balance the authority of quantitative measurement with the humility of human knowledge before divine will' (Schaffer, 1997, 465). Few physicists were as eloquent in their expression of the moral dimensions of properly performed measurement, but the idea that the intrinsic inability to achieve absolute precision in metrological determinations certainly prompted Germans working on standards and the precision measurements they entailed to acknowledge their human limitations (Olesko, 1991, 366–450).

What Gooday has demonstrated is that such associations are contingent upon context, and that there is an historical evolution of the moral dimensions of measurement. When measurements of or with standards were relatively new—as they were with mid-century resistance standards—the moral dimensions of measurement and of the measurer were all the more important because there was little else on which to base the assessment of those measurements. At that stage, precision became 'mythologized as an uncontestable guarantor of metrological credibility' (Gooday, 1995, 240). In Britain, furthermore, trust in measurements was at the time a matter too of face-to-face contact. Gooday believed that once opposing sides met one another, trust increased even if numerical differences remained. But once instruments, mass produced, embodied the standards that physicists had worked so hard to create, were made accessible to the average citizen, and were widely distributed for use, the issue of morals was bracketed and trust transformed. What was once a laboratory issue became a consumer issue: was the monthly domestic consumption of electricity reasonably calculated? So, Gooday argues, as industrialization permeated society, and measurements become a part of daily life, it is not extreme accuracy that generated trust, but how measurements (say of household use of electricity) were understood within the constraints of time, place, and economics.

I suggest that the historian should seek to locate judgements about (sufficient) accuracy within particular social-historical contexts if they are to understand better how the credence given to claims about accuracy can be linked to the trust accorded to them by specific audiences for their own specific reasons. . . . The business of trusting people, instruments, and materials is in many ways a subject beyond measurement. (Gooday, 2004, 269, 272).

And that is why in any investigation of physics and metrology, historical context and historical contingencies—that which is 'beyond measurement'—need to be brought into consideration. The 'metroscape' (Crease, 2011, 226–248) that has cast its net over the modern world often hides context and contingency, yet both are at work in how metrologies shape human behaviour, values, and economies, and even what we consider to be real.

REFERENCES

ALDER, K., 1995, 'A Revolution to Measure: The Political Economy of the Metric System in France,' in M. N. Wise (ed.), *The Values of Precision*, Princeton University Press, Princeton, pp. 39–71.
—— 2002, *The Measure of All Things: The Seven-Year Odyssey and Hidden Error That Transformed the World*, The Free Press, New York.
ASHWORTH, W. J., 2001, "Between the Trader and the Public:' British Alcohol Standards and the Proof of Good Governance,' *Technology and Culture*, vol. 42, pp. 27–50.
—— 2003, *Customs and Excise: Trade, Production and Consumption in England*, Oxford University Press, Oxford.
—— 2008, 'Metrology and the State: Science, Revenue, and Commerce,' *Science*, vol. 306, pp. 1314–1317.
BÖCKH, A., 1838, *Metrologische Untersuchungen über Gewichte, Münzfüsse, und Maasse des Alterthums in ihrem Zusammenhang*, Veit, Berlin.
BUCHWALD, J. Z., 1989, *The Rise of the Wave Theory of Light: Optical Theory and Experiment in the Early Nineteenth Century*, University of Chicago Press, Chicago.
CAHAN, D., 1989, *An Institute for an Empire: The Physikalisch-Technische Reichsanstalt 1871–1918*, Cambridge University Press, Cambridge.
CHANG, H., 2002, 'Spirit, Air, and Quicksilver: The Search for the 'Real' Scale of Temperature,' *Historical Studies in the Physical and Biological Sciences*, vol. 31, no. 2, pp. 249–284.
—— 2004, *Inventing Temperature: Measurement and Scientific Progress*, Oxford University Press, Oxford.
CREASE, R. P., 2011, *World in the Balance: The Historic Quest for an Absolute System of Measurement*, W. W. Norton & Company, New York and London.
DEW, N., 2008, '*Vers la ligne*: Circulating Measurements around the French Atlantic,' in J. Delbourgo and N. Dew (eds.), *Science and Empire in the Atlantic World*, Routledge, New York and London, pp. 53–72.
—— 2010, 'Scientific Travel in the Atlantic World: the French Expedition to Gorée and the Antilles, 1681–1683,' *British Journal for the History of Science*, vol. 43, no. 1, pp. 1–17.
FRÄNGSMYR, T., HEILBRON, J. L., and RIDER, R. E. (eds.), 1990, *The Quantifying Spirit in the Eighteenth Century*, University of California Press, Berkeley and Los Angeles.
GAY, H., 2003, 'Clock Synchrony, Time Distribution and Electrical Time-Keeping in Britain,' *Past & Present*, vol. 181, pp. 107–140.
GOODAY, G. J. N., 1990, 'Precision Measurement and the Genesis of Physics Teaching Laboratories in Victorian Britain,' *British Journal for the History of Science*, vol. 23, pp. 25–52.
—— 1995, 'The Morals of Energy Metering: Constructing and Deconstructing the Precision of the Victorian Electrical Engineer's Ammeter and Voltmeter,' in M. N. Wise, (ed.), *The Values of Precision*, Princeton University Press, Princeton, pp. 241–282.
—— 2004, *The Morals of Measurement: Accuracy, Irony, and Trust in Late Victorian Electrical Practice*, Cambridge University Press, Cambridge.
HEBRA, A., 2003, *Measure for Measure: The Story of Imperial, Metric, and Other Units*, Johns Hopkins University Press, Baltimore.

———2010, *The Physics of Metrology: All about Instruments from Trundle Wheels to Atomic Clocks*, Springer, Vienna and New York.
HEILBRON, J. L., 1990, 'The Measure of Enlightenment,' in Frängsmyr, T., Heilbron, J. L., and Rider, R.E. (eds.) The Quantifying Spirit in the Eighteenth Century, University of California Press, Berkeley and Los Angeles, pp. 207–242.
———1993, 'Weighing Imponderables and Other Quantitative Science around 1800,' *Historical Studies in the Physical and Biological Sciences*, vol. 24, part 1, Supplement, pp. 1–337.
HENTSCHEL, K., 1993, 'The Discovery of the Redshift Fraunhofer Lines by Rowland and Jewell in Baltimore around 1890,' *Historical Studies in the Physical and Biological Sciences*, vol. 23, no. 2, pp. 219–277.
———2005, *Gaußens unsichtbare Hand: Der Universitäts-Mechanicus und Maschine-Inspector Moritz Meyerstein. Ein Instumentenbauer im 19. Jahrhundert*, Vandenhoeck & Ruprecht, Göttingen.
———2007, 'Gauss, Meyerstein, and Hanoverian Metrology,' *Annals of Science*, vol. 64, pp. 41–75.
HUNT, B. J., 1994, 'The Ohm is Where the Art Is: British Telegraph Engineers and the Development of Electrical Standards,' *Osiris*, new series, vol. 9, pp. 48–63.
———1996, 'Scientists, Engineers, and Wildman Whitehouse: Measurements and Credibility in Early Cable Telegraphy,' *British Journal for the History of Science*, vol. 29, pp. 155–169.
KELLY, P., 1816, *Metrology or an Exposition of Weights and Measures*, Printed for the Author by J. Whiting, London.
KERSHAW, M., 2007, 'The International Electrical Units: A Failure in Standardisation?', *Studies in History and Philosophy of Science*, vol. 38, pp. 108–131.
KOYRÉ, A., 1953, 'An Experiment in Measurement,' *Proceedings of the American Philosophical Society*, vol. 97, pp. 222–237.
KUHN, T., 1961, 'The Function of Measurement in Modern Physical Science,' *Isis*, vol. 52, pp. 161–193.
KULA, W., 1986, *Measures and Men*, trans. R. Szreter, Princeton University Press, Princeton.
MALLARD, A., 1998, 'Compare, Standardize, and Settle Agreement: On Some Usual Metrological Problems,' *Social Studies of Science*, vol. 28, pp. 571–601.
MILLER, D. P., 1986, 'The Revival of the Physical Sciences in Britain, 1815–1840,' *Osiris*, vol. 2 (new series), pp. 107–134.
O'CONNELL, J., 1993, 'Metrology: The Creation of Universality by the Circulation of Particulars,' *Social Studies of Science*, vol. 23, pp. 129–173.
OLESKO, K. M., 1991, *Physics as a Calling: Discipline and Practice in the Königsberg Seminar for Physics*, Cornell University Press, Ithaca and London.
———1995, 'The Meaning of Precision: The Exact Sensibility in Early Nineteenth-Century Germany,' in M. N. Wise (ed.), *The Values of Precision*, Princeton University Press, Princeton, pp. 103–134.
———1996, 'Precision, Tolerance, Consensus: Local Cultures in German and British Resistance Standards,' *Archimedes* vol. 1, pp. 117–156.
———2005, 'The Foundations of a Canon: Kohlrausch's *Practical Physics*,' in D. Kaiser (ed.), *Pedagogy and the Practice of Science: Historical and Contemporary Perspectives*, The MIT Press, Cambridge.
———2009, 'Geopolitics and Prussian Technical Education in the Late-Eighteenth Century,' *Actes d'Història de la Ciència i de la Tècnica*, vol. 2, no. 2, pp. 11–44.
QUINN, T., 2011, *From Artefacts to Atoms: The BIPM and the Search for Ultimate Measurement Standards*, Oxford University Press, Oxford.
SCHAFFER, S., 1992, 'A Manufactory of Ohms: Late Victorian Metrology and Its Instrumentation,' in R. Bud and S. Cozzens (eds.), *Invisible Connections: Instruments, Institutions, and Science*, SPIE Optical Engineering Press, Bellingham, WA, pp. 23–56.

———1994, 'Raleigh and the Establishment of Electrical Standards,' *European Journal of Physics*, vol. 15, pp. 277–285.

———1995, 'Accurate Measurement is an English Science,' in M. N. Wise, (ed.), *The Values of Precision*, Princeton University Press, Princeton, pp. 135–172.

———1997, 'Metrology, Metrication, and Victorian Values,' in B. Lightman (ed.), *Victorian Science in Context*, University of Chicago Press, Chicago, pp. 438–474.

———2002, 'Golden Means: Assay Instruments and the Geography of Precision in the Guinea Trade,' in M. N. Bourget, C. Licoppe, and H. O. Sibum (eds.), *Instruments, Travel, and Science: Itineraries of Precision from the Seventeenth to the Twentieth Century*, Routledge, New York and London, pp. 20–50.

———2009, 'Newton on the Beach: The Information Order of the *Principia Mathematica*,' *History of Science*, vol. 47, pp. 243–276.

SCOTT, J. C., 1998, *Seeing Like a State: How Certain Schemes to Improve the Human Condition Have Failed*, Yale University Press, New Haven & London.

SHALEV, Z., 2002, 'Measurer of All Things: John Greaves (1602–1652), the Great Pyramid, and Early Modern Metrology,' *Journal of the History of Ideas*, vol. 63, no. 4, pp. 555–575.

SHAPIN, S., 1989, 'The Invisible Technician,' *American Scientist*, vol. 77, pp. 554–563.

STALEY, R., 2002, 'Travelling Light,' in M. N. Bourget, C. Licoppe, and H. O. Sibum (eds.), *Instruments, Travel, and Science: Itineraries of Precision from the Seventeenth to the Twentieth Century*, Routledge, New York and London, pp. 243–272.

———2008, *Einstein's Generation: The Origins of the Relativity Revolution*, University of Chicago Press, Chicago.

WISE, M. N. (ed.), 1995, *The Values of Precision*, Princeton University Press, Princeton.

ZUPKO, R. E., 1990, *Revolution in Measurement: Western European Weights and Measures since the Age of Science*, American Philosophical Society, Philadelphia.

PART IV
MODERN PHYSICS

CHAPTER 24

RETHINKING 'CLASSICAL PHYSICS'

GRAEME GOODAY AND DANIEL JON MITCHELL

What is 'classical physics'? Physicists have typically treated it as a useful and unproblematic category to characterize their discipline from Newton until the advent of 'modern physics' in the early twentieth century. But from the historian's point of view, over the last three decades several major interpretive difficulties have become apparent, not least the absence of unequivocal criteria for labelling physicists and their work as 'classical', whether during the nineteenth century or earlier. Some historians have consequently either treated the term as a retrospectively contrived anachronism (such as Olivier Darrigol), or carefully avoided using it in their analyses (such as Jed Buchwald).[1] Nevertheless, current historiographies have not systematically explored the implications of abandoning 'classical physics' as an analytical category. As a result, they arguably overstate the unity of the physics prior to the rise of quantum and relativity theories in the twentieth century. Moreover, many studies into the activities of late nineteenth-century physicists have adopted the perspective of later theoretical developments typically associated with the birth of one type or another of 'modern physics'; for example, the origins of microphysics and, through special relativity, the history of electrodynamics.[2] This focus on theoretical discontinuities has long diverted attention away from important historical continuities in both experimental practice and the applications of physics. We take these reasons as sufficient motivation for rethinking 'classical physics'.

Our analysis builds upon Richard Staley's recent systematic attempt (we believe that it is the first such attempt) to enquire historically into the origins of the distinction between 'classical' and 'modern physics'. Staley fruitfully raises the key questions of how and why theoretical physicists invented, accepted, and used a classical/modern dichotomy to represent their practice.[3] But like Russell McCormmach's nostalgic fictional physicist Jakob, we still remain unsure about the meaning of 'classical physics'.[4] We therefore begin by reviewing Staley's contribution, in particular his thesis that 'classical' and 'modern physics' were invented simultaneously—'co-created', in Staley's terminology—by Max Planck at the Solvay conference in 1911.[5] Through an extension of Staley's methodology, we argue instead that the emergence of these notions took place separately over a period that reached as late as the 1930s, and that this process took different forms in different countries. This leads us to regard the apparent unity of 'classical physics' as the *post hoc* creation of twentieth-century theoretical physicists seeking to consolidate new departures within their discipline. We then explain how this perspective is consistent with present-day physics pedagogy, in which the term 'classical physics' is used to emphasize both continuity and change in theoretical aspects of physics, while making little or no reference to forms of physics with a predominantly experimental or technological basis.

Our review of Staley's approach reveals the interpretive difficulties associated with maintaining a sharp dichotomy between 'classical' and 'modern physics'. The distinction between 'classical' and 'modern physics' neither helps the historian to draw a contrast between different past approaches to physics, nor to demarcate periods of time that possessed an important unity.[6] In the final three sections we develop alternative analytical tools to investigate important non-theoretical continuities in the practice of physics between the late nineteenth and early twentieth centuries through three illustrative examples: the physics of the ether, the industrial connections of physics, and the national context of French experimental physics. We do not claim that these examples necessarily epitomize physics in the period, nor aim to undermine biographically-focused histories of research traditions in theoretical physics (such as Maxwellian or Hertzian). Instead, we wish to show how the history of physics might be enhanced by identifying and revising latent ahistorical presumptions about the unity of physics presupposed by notions of 'classical' and 'modern physics'.[7]

First, we describe how the identification of the ether as a key feature of classical physics has drawn historians' attention towards its changing metaphysical fortunes during the nineteenth century. We argue that characterizing these fortunes in terms of 'belief' and 'non-belief' in the ether distorts the positions taken by practitioners of physics, many of whom took the ether to be irrelevant to their theoretical and/or experimental practice. We draw on historiography employed fruitfully by Jed Buchwald and Andrew Warwick to suggest a more subtle approach in terms of the contingent role played by ontological commitments—whether to a form of ether or any other framework—in guiding and grounding the practice of physics.

Second, we explore the connections between physics and industry that are obscured by the theoretical bias of any dichotomy between 'classical' and 'modern physics'. We characterize these connections in terms of an 'industrial nexus' for physics: a significant range of aims, tools, concepts, and expertise were developed symbiotically between physics and industry. Although this applies notably to the domains of electricity and thermodynamics during the nineteenth century, we extend the thesis to relativity and quantum theory during the early twentieth century. From our survey across this period, we reveal close associations between the development of these areas of physics and their practical–commercial applications that further strengthen our emphasis on the disciplinary continuity of 'industrial physics'.[8]

Finally, we address a particular national approach to physics in the period 1880 to 1930 that likewise resists assimilation into a classical/modern analytical framework. We reveal the *longue durée* continuity in some characteristic aims and approaches of the discipline of French experimental physics through a synthesis of three comparative case studies of French research, due to Mitchell, Atten and Pestre, and Lelong. Without presupposing any trajectory towards a form of 'modern physics', we then explore how these may have changed as a new generation of French physicists began to draw more substantially from the work and approaches of their international peers from the turn of the century onwards. The distinctively anti-metaphysical and heavily empirical character of much French physics, especially in contrast to physics in Britain and Germany, provides further strong grounds to doubt the existence of a monolithic international project for physics in the late nineteenth century, whether labelled 'classical physics' or otherwise.[9] We offer this example as an invitation to historians of physics to explore the ways in which other nationally localized forms of physics also diverged from the model of discontinuous theoretical transition presupposed by a dichotomy between 'classical' and 'modern physics'.

24.1 The Epithet 'Classical' and the Invention of 'Classical Physics'

We begin our review of Staley's work with a nagging problem in the historical study of classical physics: the difficulty in identifying classical physicists. Historians have struggled to identify them in eras when they ought to have been plentiful, according to common characterizations of 'classical physics'.[10] For example, in his *Metaphysics and Natural Philosophy: The Problem of Substance in Classical Physics*, Peter Harman employed the term to distinguish the philosophical assumptions of eighteenth- and nineteenth-century physics from the relativistic and indeterministic doctrines of twentieth century or 'modern physics'.[11] He was forced to reject this distinction as unsuitable because he concluded that the physicists that he studied could not be labelled as 'classical'. Other historians have reached a similar conclusion. In ironic contrast to Russell McCormmach's fictional classical physicist Jakob, Thomas Kuhn considered that Max Planck was not at all radical because he initially sought to integrate quantum analysis into an existing thermodynamical framework.[12] Sympathetic to Kuhn's conclusion, Darrigol has described the diversity of views about Planck's mooted classicism among supporters and critics of Kuhn's account. Following Needell, he judged 'classical' an unilluminating descriptor of Planck's work, at least until a highly retrospective interpretation emerged in the 1910s. Unsurprisingly, given these difficulties, historians have increasingly abandoned direct reference to 'classical physics', and have left it as a term in need of explanation.[13]

This is the point of departure for Staley's work. In *Einstein's Generation*, he charts the unexpected complexity in physicists' divergent uses of the term 'classical' between the 1890s and the 1910s. He describes how, until the turn of the century, it was usually applied to canonical theoretical texts, for example Maxwell's *Treatise on Electricity and Magnetism* of 1873.[14] Physicists only began to label discrete domains of physical theory as classical, such as 'classical mechanics' and 'classical thermodynamics', during the first decade of the twentieth century. This occurred as part of a process to determine methods appropriate for the future development of physics. Unsurprisingly, vehement battles sometimes broke out. Interpretations of 'classical thermodynamics' hinged upon particular stances towards atomism; for example, Ludwig Boltzmann argued strongly that the field should be understood to include a statistical hypothesis about the microscopic motion of molecules, and hence he opposed the macroscopic phenomenological form preferred by anti-atomists (such as the early Planck) and proponents of energetics (such as Georg Helm).[15]

Similarly, Staley argues that the perspective of the entire discipline of pre-quantum physics as 'classical' was an attempt by Max Planck, in his address to the 1911 Solvay conference, to articulate a template for future research in 'modern' theoretical physics.[16] This took place soon after physicists, including Einstein, Lorentz, and Planck himself, interpreted Planck's work as implying the quantization of energy. Debates on the status of the equipartition theorem (the theorem that in any mechanical system at thermal equilibrium, on average the kinetic energy is evenly distributed between each degree of freedom) formed the crucial background to these events. Equipartition was the second of Lord Kelvin's 'clouds on the horizon of the dynamical theory of heat and light', and he recommended its rejection altogether. Rayleigh, Jeans, and Einstein were aware of the difficulties in applying the theorem to the specific heats of gases, and consequently regarded their attempts to construct a law of black-body radiation using the theorem as uncertain theoretical departures, rather than unproblematic extensions of an established technique. Although empirically inadequate, during the 1900s the trio regarded laws based on equipartition as possessing clearer theoretical foundations than Planck's law.

Crucially, none of them described the equipartition theorem as 'classical'. To Staley, equipartition only became 'classical' when Planck extended the concept to include statistical mechanics in his Solvay address. Without speculating unduly on whatever Planck may have wished for, Staley seems to interpret him as inventing a tradition in order to endorse the approach of Rayleigh, Jeans, and Einstein 'as valued and traditional' on the one hand, whilst reducing 'the search for secure foundations to a contrast between the eras before and after quantum theory' on the other. Indeed Staley insists that 'it is hardly to be doubted that interest in the new quantum theory was considerably heightened as a result'.[17] Staley goes on to imply that Planck's address mobilized a new generation of physicists to fashion a clear trajectory and identity for future theoretical physics when faced with the bewildering results of early quantum physics. This gives rise to Staley's thesis that 'modern physics' and 'classical physics' were co-created.[18]

Whilst we find Staley's analysis of the application of 'classical' to individual branches of physics persuasive, we think that the terms 'classical physics' and 'modern physics' were initially coined separately more than a decade later. To begin with, Staley misleadingly identifies Planck's ambiguous invocation of 'classical theory' with 'classical physics'.[19] In his Solvay address, Planck appears to remain concerned with the problems of particular classical branches of physical theory rather than those of 'classical physics' in its entirety: he did not focus explicitly on the demise of a discipline of classical physics. Planck also left open the question of how radically the framework of classical dynamics needed to be modified, and hence did not explicitly proclaim the arrival of 'modern physics':

> The principles of *classical mechanics*, fructified and extended by electrodynamics, and especially electron theory, have been so satisfactorily confirmed in all those regions of physics to which they had been applied, that it had looked as though even those areas which could only be approached indirectly through statistical forms of consideration would yield to the same principles without essential modification. The development of the kinetic theory of gases seemed to confirm this belief. Today we must say that this hope has proved illusory, and that the framework of *classical dynamics* appears too narrow, even extended through the Lorentz–Einstein principle of relativity, to grasp those phenomena not directly accessible to our crude senses. The first incontestable proof of this view has come through the contradictions opened up between *classical theory* and observation in the universal laws of the radiation of black bodies [our italics].[20]

Our interpretation of Planck's Solvay address is consistent with the views Planck expressed in his Nobel lecture 'The genesis and present state of development of the quantum theory', given upon receipt of the Physics prize in 1918. In this lecture he did not frame 'classical physics' as a complete domain that had been superseded by the challenges of quantum theory. He emphasized instead the difficulties of *accommodating* quantum theory into the precepts of classical theories, such as classical electrodynamics and classical mechanics, which he treated as distinct branches of physics. His main concern was that the quantization of energy still proved 'elusive and resistant to all efforts to fit it into the framework of classical theory'.[21]

Neither did the perception that Planck's work on the quanta of radiation carried radical implications lead to the co-creation of 'classical' and 'modern physics' at another time. Niels Bohr considered these implications likely to seem 'monstrous' to a theoretical physicist of the 'older school' and that energy quantization posed a clear and 'pronounced contradiction' to classical theory.[22] When he addressed these issues in his 1922 Nobel prize acceptance speech, he advocated a new approach that broke completely with 'classical physics', but this approach did not yet constitute a 'modern physics':

> It has . . . been possible to avoid the various difficulties of the electrodynamic theory by introducing concepts borrowed from the so-called quantum theory, which marks a complete departure from the ideas that have hitherto been used for the explanation of natural

phenomena. This theory was originated by Planck in the year 1900, in his investigations on the law of heat radiation, which, because of its independence of the individual properties of substances, lent itself peculiarly well to a test of the applicability of the laws of classical physics to atomic processes.

At this time the epithet 'modern' was still used to refer to specific domains of physics, such as X-rays, radioactivity, and electron theory. Even in the late 1920s there are references to 'classical physics' that are not framed as complementary to any new project of 'modern physics'.[23] The specific and systematic application of the term 'modern physics' to quantum and relativistic phenomena did not take place concurrently with the articulation of 'classical physics', but some time later.

Furthermore, Staley does not provide evidence to suggest that Solvay conference participants interpreted Planck's address as setting up a boundary between a 'classical' and a 'forms of modern physics' or that they returned to their countries and spread this terminology amongst their contemporaries.[24] His argument relies instead on the role of the first Solvay conference in shaping an international agenda for physics. We agree with him that the creation of a new form of conference that concentrated on a single subject—quantum theory—was important in focusing international attention on the specific problems from which forms of 'modern physics' emerged, and also with Jungnickel and McCormmach that participants left the conference believing that quantum theory 'entailed something new and important'. But whilst the beginning of a programme to develop a completely new kind of physics can be identified retrospectively with the first Solvay conference, we suggest that it took more time to implement (perhaps as long as another two decades) and was only labelled 'modern physics' much later.[25]

On a conceptual level, Staley's co-creation thesis seeks problematically to explain the origins of an unambiguous present-day understanding of 'classical' and 'modern physics' that maps directly onto the 'non-quantum/quantum' theoretical division of physics that emerged following the Solvay Council.[26] But there is no unique understanding of 'classical' and 'modern physics'. Whilst the 'non-quantum/quantum' division is a necessary means of achieving the cleavage, it can be applied in different ways to emphasize either continuity or change in the metaphysics or practice of theoretical physics. This results in different versions of 'classical' and 'modern physics', each with unique weaknesses, ambiguities, or inconsistencies.

Recent textbooks of 'modern physics' and its branches routinely assume that the classical/modern distinction represents a sharp historical and conceptual break. They assign a watershed moment in the history of physics to the *fin de siècle* when Planck's and Einstein's early work initiated the development of 'modern physics', and use this to construct a clear division between groups of theories. Consider the following representative example due to Robert Eisberg:

Classical physics comprises fields in which the subjects studied are: Newton's theory of mechanics and the varied phenomena which can be explained in terms of that theory, Maxwell's theory of electromagnetic phenomena and its applications, thermodynamics, and the kinetic theory of gases. In modern physics the subjects studied are: the theory of relativity and associated phenomena, the quantum theories and quantum phenomena, and, in particular, the application of the relativity and quantum theories to the atom and to the nucleus.[27]

Yet whilst claiming that 'the theories used to explain the phenomena' with which the fields of modern physics are concerned are 'startlingly different from the theories that were in existence before 1900', the same textbook treats these differences as self-evident. Planck's famous yet modest excursion into quantum analysis the same year is typically assumed to have launched a revolution in the subject. As such, textbooks of modern physics offer no more than an unilluminating and dubious temporal demarcation to support their classification. Neither can classical physics be treated as encompassing Newtonian mechanics, electrodynamics, and thermodynamics by fiat. Textbooks of classical physics subvert Eisberg's neat classification by distinguishing between coexisting classical and modern versions of a branch of physics: for example, classical and quantum electrodynamics.[28]

Textbook authors who make this finer distinction between typically attempt to emphasize the continuity in theoretical *practice* between these classical and modern versions. Firstly, they distinguish between 'classical' and 'modern physics' by subtly identifying the latter with 'quantum-mechanical'. Passing over metaphysical issues related to quantization, they then identify mathematical tools shared between 'modern', quantum-based branches of physics and their 'classical', quantum-free counterparts, then specify mathematically their relevant, mutually consistent domains of applicability (for example, 'low/high' energies or speeds). Hence classical mechanics occupies a central role in the articulation: every other branch of classical physics is established as classical through dependence on mathematical tools that found their initial and canonical applications in classical mechanics (we leave aside the issue of how textbooks of classical physics define 'classical mechanics'). But how is this consistent with the classical form that can be given to the equations of quantum mechanics, or the status of relativity as part of 'modern physics'?[29]

We have identified four conflicting present-day pedagogical purposes for which 'classical physics' is pressed into service. First, to reveal clearly the (familiar) techniques upon which relativity theory and quantum theory are built; second, to focus attention on the rejection of Newtonian absolute spacetime and Laplacian determinism; third, to isolate the physics of everyday objects from the physics of more extreme realms (the 'tiny', 'huge', and 'very fast'), and to prove its adequacy for many experiments and practical applications; and fourth, to provide an antonym for twentieth- and twenty-first-century fields of research in physics that define themselves in terms of their rejection of various common-sense assumptions. Much research remains to be done on how these (and possibly other) purposes may have given rise to different notions of 'classical physics', how these notions became part of physics orthodoxy around the world, when and how they were invested with explicit content, and their links to particular visions of 'modern physics'.[30]

The British context is illustrative of the complexity of this task as well as Staley's overemphasis on the Solvay conference. The first British attempt to articulate the meaning of 'classical physics' seems to have taken place in 1927. This was by the

Quaker pacifist and idealist physicist–philosopher Arthur Eddington, Cambridge University's Plumian Professor of Astronomy, in his Gifford Lectures on the connections between religion and science. These were public lectures at the University of Edinburgh intended by Eddington to explain the results of recent physics to a general audience. His analysis revealed that it remained unclear precisely what constituted 'classical physics' (scare quotes indicated the novelty of the term). But Eddington still left considerable latitude in its content, conceiving of classical physics as vaguely Newtonian whilst permitting great heterogeneity in the nature of energy, light, and electricity:

> I am not sure that the phrase 'classical physics' has ever been closely defined. But the general idea is that the scheme of natural law developed by Newton in the *Principia* provided a pattern which all subsequent developments might be expected to follow. Within the four corners of the scheme great changes of outlook were possible; the wave theory of light supplanted the corpuscular theory; heat was changed from substance (caloric) to energy of motion; electricity from continuous fluid to nuclei of strain in the aether. But this was all allowed for in the elasticity of the original scheme.[31]

Although he had abandoned all previous attempts to hang on to the theories of classical physics (in his sense), Eddington referred to elements of continuity from classical to 'modern physics' and identified key surviving concepts, such as 'waves, kinetic energy, and strain'.

In Britain, Eddington's widely reported lectures probably provided an important stimulus to the shaping of 'classical physics' both inside and outside the physics community. His conception of classical physics, and its relationship to 'modern physics', was not the only one available, however. For example, *The Times*' special science report in 1928 interpreted 'classical physics' as deriving from the work of Maxwell, and opposed it to the 'new physics of quanta and radiation'.[32] Furthermore, some were altogether sceptical of the dichotomy, such as the veteran chemist Henry Armstrong. Writing for *The Times* in 1930, he questioned the neologism of 'classical physics' and demanded an account of the 'modern' in order to make sense of it. He bemoaned the lack of 'good English' in describing the results of scientific workers, and grumbled that

> . . . we are no longer allowed to speak of things as old; they must be 'classical'. The old physics—which was termed good general physics—is now termed 'classical physics'. What is the modern? May it not be dangerous to invite comparison with the classical?[33]

Our re-examination of the co-creation thesis and associated historical enquiries in this section lead us to recommend the extension of Staley's project over a longer period. His identification of a broad range of conceptions of the epithets 'classical' and 'modern' has raised issues of etymology and semantics that ought to shape the direction and methodology of further study into the origins and adoption of the categories 'classical physics' and 'modern physics'. We agree with Staley that the 'rich language in which physicists described the grounds from which the new theory departed should by now have demonstrated the very real possibility that physics might *never* have been described as classical'. Nevertheless, we suggest that this judgement applies not to classical physics, as Staley proposes, 'in the major sense we now recognize', but instead as an actors' category.[34]

We regard this historical contingency in the meaning of 'classical' as a means of establishing broad cultural connections between physics and other knowledge-making enterprises. The range of meanings of 'classical'—ancient, authoritative, perfected, and exemplary—has allowed physicists to use the term 'classical' to bypass thorny epistemological questions surrounding the partially discredited canon from the nineteenth century. A particularly significant claim can be deemed simply 'classical': neither true nor false and yet somehow partially both, which enables credit to be assigned to outmoded yet discipline-making predecessors. Close analogies are apparent in the disciplines of sociology, economics, and even ethology, each of which, like physics, saw mid-twentieth-century revisionists recast their subject's nineteenth-century origins in terms of a 'classical' canon rather than a falsified precursor.[35]

Similarly for the 'modern', a key question that emerges from Staley's analysis is whether 'modern' refers to a particular period in time or to a particular canon. In the period studied in this chapter, physics considered 'modern' by contemporaries included ether theory in the 1880s, electron theory and the study of X-rays and radioactivity in the 1890s, relativity theory in the 1900s, and quantum mechanics by the 1920s. Just as for modern art or modern literature, 'modern physics' came to derive its meaning from a particular set of assumptions and values that 'classical' precursors putatively lacked. 'Modern' could also simply refer to the latest work.

Sensitivity to shifts in meaning between languages, contexts, and over time will be invaluable in understanding cases where what was initially considered 'modern' was later re-designated as 'classical', such as one of the young Einstein's favourite texts: A Föppl's *Einführung in die Maxwellsche Theorie der Elektrizität*, originally published in 1894. This was an introduction to the Maxwellian theory of electricity that drew heavily upon the work of the English Maxwellian and telecommunication theorist Oliver Heaviside. In 1905 a collaborator, Max Abraham, turned Föppl's book into a broader theory of electricity by adding a second volume on the theory of electromagnetic radiation. Abraham described this theory as 'modernische'—an epithet that was dropped by a fourth German edition published in 1920. But the eighth German edition of 1930 was translated into English two years later as Max Abraham, *The Classical Theory of Electricity*.[36] Thus we see one instance of how the boundaries between 'classical' and 'modern' physics could be made, unmade, and remade.

Based on Staley's work, we have established that the emergence of 'classical physics' is a topic for the historian of the early twentieth century (especially the 1920s and 1930s) and not of the eighteenth or nineteenth centuries. This raises the question of how to describe the concepts and practices of physicists during the earlier period. We believe that as a descriptive tool, 'classical' ought perhaps to be abandoned altogether. Even if carefully defined, the term is so ingrained amongst historians and philosophers of physics that it is liable to lead to confusion, either with actors' uses or other current uses. It also clear from the cases that we and Staley examine that classical/modern distinctions were drawn by those concerned largely with problems of theory and metaphysics, and that this is still the case. Their distinctions tend to downplay the practical aspects and applications of physics. Contemporary critique of notions of 'modern physics' based on its narrow focus on physical theory and exclusion of experimentation constituted an implicit denial that quantum (and relativity) theory were sufficiently important to dominate the agenda of physics.[37]

There is one apparent counter-example, however, that merits special attention because *historians* generally consider it to encapsulate a major concern of many turn-of-the-century physicists. Famously identified by Sir William Thomson as a cloud on the prospects of the dynamical theory of heat and light, this is the existence and nature of an all-pervading ether (or aether) of space. We now address whether the alleged abandonment of the notion of the ether might be characterized appropriately in terms of a transition from 'classical' to 'modern physics'.[38]

24.2 Is Ether Theory the Essence of Classical Physics?

Conceived in the late nineteenth century as the medium for communicating many of the forces of nature, including light and other forms of radiation, the ether provided an electromagnetic foundation for the absolute spacetime framework of Newtonian (and arguably of Maxwellian) physics. The few who understood Maxwell's foundational *Treatise on Electricity and Magnetism* dedicated themselves to elaborating the theory of the electromagnetic ether as the basis for a unified physics grounded on the principles of mechanics and hydrodynamics.[39] The supposed failure of Michelson and Morley to detect this ether has been simplistically identified in some naïve accounts as a crucial experiment that enabled the straightforward adoption of the theory of special relativity during the early twentieth century. Hence belief/disbelief in the ether has, on some interpretations, formed the basis of a distinction between 'classical' and 'modern physics'.[40]

As has long been realized, however, the Michelson–Morley experiment of 1887 had no simple epistemic import. Michelson and Morley themselves suspected that their failure to detect ether drift might have been due to errors in their experimental conduct or the inadequacy of their instrumentation, so both of them independently conducted numerous further experiments, Morley over the following five decades. Theoreticians also introduced multiple hypotheses to explain the null result of the Michelson–Morley experiment: Oliver Lodge's notion of 'ether drag', developed during 1889, supposed that motion relative to the ether was difficult to detect because ether clung to matter.[41] Around the turn of the century, the Lorentz–FitzGerald contraction hypothesis posited the contraction of electrons—and hence of all (electromagnetically constituted) matter—in the direction of motion. The idea that the Michelson—Morley experiments spelled the end of the ether is a standard retrospective re-evaluation that emerged only after the development of the theory of special relativity. No experiment ever disproved the existence of the electromagnetic ether, even for Einstein. It simply became marginal to the work of most physicists and engineers.[42]

Nevertheless, the enduring influence of the simplistic interpretation of the 'Michelson–Morley' experiments, bolstered by a particular classical/modern dichotomy, has left an unfortunate tacit imprint on ether historiography. In the so-called 'classical' period—in this case, pre-Michelson/Morley—historians have typically focused on historical figures with a positive ontological commitment to the notion of an ether. This is the approach of contributors to the collaborative volume *Conceptions of Ether*, edited by Cantor and Hodge. The subtitle of this work, 'studies in the history of ether theories', epitomizes the way in which contributors sought to examine the diversity of views and past debates on the nature and properties of the ether, rather than to reconstruct less visible, contemporaneous doubts about its ontological status. One contributor, Dan Siegel, asserted boldly that the ether programme of research 'compelled allegiance through the end of the nineteenth century'.[43] Prominent figures who did not share this conviction (inevitably) received only a passing mention, although as Siegel himself notes, Michael Faraday, a major founder of electromagnetic theory, had serious doubts about the ether's existence.

Similarly, in the 'modern' period that supposedly followed Einstein's key publication in 1905 on *special relativity* (as it *later* became known; note the ambiguous classical/modern status of the interim period between 1887 and 1905) historians have picked on supposedly irrational individuals who persisted in believing in the ether well after Einstein's major publications on special and general relativity. The idea that they were hanging on unreasonably to 'classical' physics lurks behind this judgement of irrationalism. Foremost among the accused is Oliver Lodge, whose commitment to spiritualism was so closely tied to the ether as the repository of the spirit world that he could not abandon one without the other.[44] More subtle is the case of physicist and historian E. H. Whittaker, whose mid-career conversion to Roman Catholicism is interpreted by some as underpinning his emphasis on the continuities between nineteenth-century ether theory and Einstein's theory of the spacetime continuum in general relativity.[45]

The neglect of ether sceptics or their unsympathetic treatment, however, are symptoms of deeper challenges posed by this particular reading of 'classical' and 'modern physics' based on beliefs about the ether (although as the primary inspiration for Maxwell's theory of the electromagnetic field, the neglect of Faraday's position by historians indeed constitutes a significant historiographical gap). These challenges are two-fold: first, a sharp epistemological dichotomy between 'belief' and 'non-belief' is not the best way to differentiate past (metaphysical) stances towards the ether; and secondly, contrary to a common presumption of historical accounts of the development of relativity, there was a wide diversity of views concerning the ontological status of the ether.

We aim to show how the subtle historiography of the interaction between belief and theoretical practice developed by Andrew Warwick (building on Buchwald's work) offers a remedy to the first challenge. To meet the second challenge, we suggest how this historiography might be adapted to illuminate the ether scepticism of non-practitioners, and applied to the historical investigation of alternative theoretical frameworks that found no role for an ether. We offer these techniques as means of re-evaluating the significance of the ether to the practice of physics during the late nineteenth and early twentieth century, over the course of its displacement by an alternative energy-based ontology.[46]

In *From Maxwell to Microphysics,* Buchwald studied the rise, reinterpretation, and (partial) fall of James Clerk Maxwell's theory of electromagnetism in the last third of the nineteenth century. Maxwell's theory, which was first fully elaborated in his 1873 *Treatise on Electricity and Magnetism*, was premised on the ether as the universal medium for communicating electromagnetic interactions. For a network of Maxwellians across Britain, Europe, and the USA in the 1880s and 1890s, Buchwald found that their theoretical practices did not depend on certain knowledge of its structure. To be sure, at least some of them believed in the ether, and hoped to determine its structure eventually, but this was not their immediate goal, even amongst specialists in electromagnetism.[47]

As Warwick reveals in his book *Masters of Theory*, even Maxwellians based at the University of Cambridge—where Maxwell spent his final years until his death in 1879—were not specifically dedicated to securing grounds for belief in the ether. The book is a detailed, localized case study of how a cohort of physicists and mathematicians trained in the Cambridge Mathematics Tripos learned to treat the electromagnetic ether as a universal medium of energy transmission that could be subjected to hydrodynamic calculations. Warwick demonstrates how their metaphysical beliefs in the ether were inseparable from a framework of theoretical practice, and describes the processes of pedagogy and research through which students acquired this framework and maintained these beliefs. These Cambridge Maxwellians used the concept of ether to render the mechanisms of electromagnetic interaction intelligible and to give physical meaning to their theoretical practice. When this proved too difficult they adapted the concept to suit their practice in ways unlicensed by Maxwell's theory.

For example, in his detailed study of the commitment of Cambridge mathematical physicists to Joseph Larmor's post-Maxwellian Electronic Theory of Matter (ETM), Warwick shows how Larmor's development of this theory in the early 1890s lay in his preferred mathematical tools and interests, his most prized tool being the Principle of Least Action.[48] Previously, Larmor had invoked an electromagnetic ether to give physical meaning to the concept of an absolute frame of reference.[49] His reading of a paper by FitzGerald now convinced him that a *rotationally elastic* ether, due to MacCullagh, could provide a dynamical foundation for electromagnetic phenomena; crucially, FitzGerald had manipulated MacCullagh's theory using the Principle of Least Action.[50]

A vigorous correspondence with FitzGerald eventually convinced Larmor to introduce into his mathematical theory discrete charge carriers, or 'electrons', in order to resolve contradictions arising from its physical interpretation.[51] Warwick describes how this constituted a dramatic innovation. Previously, the electromagnetic ether and matter had largely been kept ontologically distinct, which raised unanswered questions about how the two interacted. The Maxwellians worried, for example, about their inability to provide a physical mechanism for conduction, during which electromagnetic energy (in the ether) was converted into heat (in the conductor).[52] Larmor's mature ETM[53] responded to the challenge by reducing matter to mobile discontinuities in the newly reinterpreted ether as 'a sea' populated 'solely by positive and negative electrons'.[54] The development of Larmor's beliefs about the ether thus depended upon his educationally-conditioned selection of mathematical tools and concepts to resolve theoretical challenges posed by electromagnetic phenomena.[55]

Similar conclusions can be drawn from Warwick's analysis of the educationally-induced ontological commitment to ether of Cambridge mathematical physicists trained in the ETM after the turn of the century. Through Larmor's lectures and such textbooks as James Jean's influential *Mathematical Theory of Electricity and Magnetism*, Cambridge students learned to apply the theory to problems set as part of the mathematical *Tripos*, and thus for them the ether became an 'ontological reality'.[56] Warwick describes how their specific commitments to both an ether-based stationary reference frame and ether-derived mass were sustained through the application of Larmor's techniques to original problems well into the second decade of the twentieth century. In other words, Cambridge mathematicians were not clinging on irrationally to a discarded ontology, but modifying that ontology to suit the application of well-established mathematical techniques as part of an active research tradition. Warwick's approach is especially persuasive because it also accounts for the subtle differences between Larmor and his students in the priority that they placed on particular ontological commitments within the ETM.[57]

One case in point was Ebenezer Cunningham, a student of Larmor's and later his colleague at St John's College, Cambridge, who altered the properties of the ether to accommodate developments in the mathematical structure of the ETM.[58] Cunningham had noticed a key property that Larmor had overlooked in his 1900 text *Aether and Matter*: the Lorentz transformations ensured that Maxwell's equations retained the same form in every frame of reference.[59] Warwick explains that, unlike Larmor, Cunningham interpreted this result as implying the impossibility of detecting the Earth's motion through the ether. Further impressed by the mathematical symmetries revealed by Einstein in his famous paper of 1905 'On the electrodynamics of moving bodies', Cunningham dismissed the notion of a unique frame of reference as meaningless. This prompted him not to abandon but to make a radical revision to Larmor's conception of ether, which, according to Warwick, retained the ether's ontological purpose while rendering it compatible with Einstein's new mathematical formalism. Hence minor changes in Cunningham's mathematical practice resulted in major ontological implications.[60]

This account of the changing nature of the ether in the ETM shifts attention to the dynamics of the formation and development of ontological commitments through theoretical practice. In the remainder of this section, we draw on this historiography to reframe scepticism about the ether in terms of independence from, or involvement in, traditions of broader scientific practice. We firstly explain why a 'sceptical agnosticism' was attractive to informed commentators through the example of the electrical researcher and Conservative Prime Minister Lord Salisbury, whose views have been described by Cantor and Hodge as 'symbolic of the attitudes of many'.[61] We then analyse the role of energy conservation in electrical engineering through the example of one of its leading figures, William Preece, in order to illuminate his dismissal of the ether in favour of energy-based ontologies of electromagnetic action. This points the way towards a symmetrical treatment of nineteenth and early twentieth century ontologies through their role in licensing specific practices within distinct traditions.

As a Fellow of the Royal Society and Chancellor of the University of Oxford, Lord Salisbury was invited to become the President of the BAAS when it met at Oxford in 1894. Following a narrow defeat in the 1892 general election, Salisbury had found time to pursue his reflections on the problems of contemporary science. This led him to select the atom and the ether as 'two instances of the obscurity that still hangs over problems which the highest scientific intellects have been investigating for several generations'. He expressed a position that might be called 'sceptical agnosticism' regarding the existence of the ether:

The ether occupies a highly anomalous position in the world of science. It may be described as a half-discovered entity. I dare not use any less pedantic word than entity to designate it, for it would be a great exaggeration of our knowledge if I were to speak of it as a body or even as a substance . . . For more than two generations the main, if not the only, function of the word ether has been to furnish a nominative case to the verb "to undulate".

Salisbury went on to describe the 'most brilliant' theoretical connection established by Maxwell between the speed of light and 'the multiplier required to change the measure of static or passive electricity into that of dynamic or active electricity', which he considered a 'notable extension' to conceptions of the ether. Alluding to Hertz's illustration of 'the electric vibrations of the ether', he interpreted Maxwell's result as establishing with probability that 'light and the electric impulse' were propagated in the same medium. He nevertheless cautioned that 'the mystery of the ether, though it has been made more fascinating by these discoveries, remains even more inscrutable than before' because the interactions between ether and matter remained unknown. Hence Salisbury concluded: 'of this all-pervading entity we know absolutely nothing except this one fact, that it can be made to undulate'.[62]

In their otherwise sensitive survey of 'opposition to ethers', Cantor and Hodge dismiss Salisbury's opinion as 'uninformed'. Yet Salisbury was an Oxford mathematics graduate, possessed all three editions of Maxwell's *Treatise* in his personal library, and had published research from his private laboratory. But whilst his views were grounded in empirical familiarity with the electro-technical practice of domestic lighting, power, and telephony, they were 'uninformed' *by mathematical practice*. This led Salisbury to pose metaphysical questions about the nature of the ether independently of the role it played in mathematical practice, and thereby to reach understandably sceptical conclusions.[63]

In a solely metaphysical context divorced from mathematical practice, Salisbury's complaint that 'even its solitary function of undulating ether performs in an abnormal fashion which has caused infinite perplexity' is highly acute. By the time Maxwell wrote his *Treatise*, he had abandoned the attempt to determine the actual physical structure of the ether. He sought instead to explain electromagnetic processes in terms of the flow of energy through the ether, and only assumed that it was capable of storing kinetic and potential energy, and that the total energy was conserved.[64] The physical structure of the ether had proved difficult to elaborate owing to apparently contradictory mechanical requirements: it had to be both a rigid medium to transmit electromagnetic waves, yet also a non-viscous 'jelly' to allow the unimpeded passage of the planets.[65] Put simply, post-Maxwellian ether-theorists were unable to deliver an internally consistent, ether-based ontology. Furthermore, prior to Larmor's introduction of the electron into the ETM, the mechanism of interaction between ether and matter was rarely discussed.[66]

Although there were exceptions like Ambrose Fleming, electrical practitioners did not entertain questions about the nature of the ether because it played no productive role in their daily practice.[67] Consider the case of William Preece, an informal pupil of Faraday, who rose to prominence in the telegraph industry during the 1880s as the Chief Electrician of the UK Post Office. He came into direct conflict with the Maxwellians, who argued at this time that the 'empty' space (i.e. the ether) around the wire governed the speed and character of electromagnetic signals.[68] Preece and other apprenticeship-trained telegraph engineers and electricians focused on electrical activity within the wires and their insulation without recourse to the surrounding space. A major battle over the operation of lightning rods during 1886–88 with the Maxwellian Oliver Lodge exemplifies these differences in practice. Preece insisted that the key factor in constructing effective lighting rods was to maximize their conductivity by minimizing their resistance to rapid lightning discharges. He thought that the rods behaved basically like a wide drainpipe for direct current, and hence that the discharges travelled *through* them.[69] Lodge argued instead that lightning discharges were very-high-frequency current oscillations, and that the lightning rod dissipated the destructive energy through the ethereal space around it. He accordingly recommended minimizing the rod's coefficient of self-induction. Preece fought back against these kinds of advanced theoretical and metaphysical speculations by emphasizing the tangibility of energy, which engineers measured and manipulated routinely:

The engineer regards electricity, like heat, light, and sound, as a definite form of energy, something that he can generate and destroy, something that he can play with and utilise, something that he can measure and apply . . . something which he can manufacture and sell, and something which the unphilosophic [sic] and ordinary member of society can buy and use.[70]

This energy-based ontology emerged from a framework of physical concepts essential to Preece's electrical practice. For Preece, the engineer's notions of work and power were the keystones to the 'conception of the character of those great sources of power in nature'. His pragmatic, metaphysical reasoning about electricity never strayed far from these notions; he explained that 'the definition of energy is capacity for doing work . . . electricity is something which has a capacity for doing work; [therefore] it is a form of energy'. This parsimonious logic licensed Preece to dismiss laboratory physics as a 'little world' of unintelligible speculation and fictional ethers:

> The physicists—at least some physicists, for it is difficult to find any two physicists that completely agree with one another—regard electricity as a peculiar form of matter pervading all space as well as all substances together with the luminiferous ether which it permeates like a jelly or a sponge . . . The practical man, with his eye and his mind trained by the stern realities of daily experience, on a scale vast with that of the little world of the laboratory, revolts from such wild hypotheses, such unnecessary and inconceivable conceptions, such a travesty of the beautiful simplicity of nature.

Untroubled by the relative motion of the Earth through space or the arcane nature of Maxwell's theory, Preece found that the overarching theory of energy conservation provided a sufficient framework to explain all electrical phenomena: 'no single electrical effect can be adduced that is not the result of work done, and is not the equivalent of energy absorbed'.[71] Preece's disgust at the metaphysical extravagance of the ether was echoed in the response of early twentieth-century experimentalists to the ETM, which they found 'overly speculative and irrelevant' to their practice, according to Warwick.[72]

Generalizing from Preece's example, we suggest that a fruitful way to investigate alternative metaphysical frameworks to the ether is to consider how they were developed and grounded through unique forms of practice, particularly experimental. In nineteenth-century electrical science, for example, studies in this vein might treat the persistence of electrical fluid theories—either the two-fluid version promoted by John Tyndall or the one-fluid version promoted by some telegraph engineers—in the same way that Warwick treated the persistence of the ETM. During this time there was simply no widespread metaphysical consensus on the nature of electricity, and so by taking up our suggestion (and hence casting off any residual traces of a classical/modern dichotomy based on the ether), historians will open up the conceptual space to locate the ether appropriately amongst the discarded ontologies shared by late nineteenth and early twentieth-century science, as well as in relation to the rising quantitative ontology of energy physics. This shared much in common with the industrial physics of transforming and inter-converting the powers of nature, to which we now turn.

24.3 CONTINUITIES IN INDUSTRIAL PHYSICS

The claim that physics and industry shared a deep interdependence is well-worn. Many historians have realized that the growth of industry simultaneously gave rise to, and benefited from, much innovative physics. Bruce Hunt has recently identified and described the connections between thermodynamics and steam power during the eighteenth and early nineteenth centuries, and between electromagnetism, telecommunications, and electrical generation during the late nineteenth and early twentieth centuries. Robert Fox and Anna Guagnini have likewise echoed Hunt's emphasis on the industrial connections of thermal and electrical physics.[73] In this section we develop similarly, under the notion of 'industrial physics', the theoretical, experimental, and instrumental connections between physics and industry during the late nineteenth and early twentieth centuries. Following Hunt, we recommend a synthetic investigation of nineteenth and early twentieth century physics rooted firmly in industrial developments.

As we indicate below, historians still sometimes underestimate the importance and range of the industrial connections of physics (for example, the industrial origin of the problem of black-body radiation). So although we have chosen to focus primarily on the influence of industry on physics, our historiographical proposal is more far-reaching. We suggest that interactions between physics and industry also characterize aspects of relativity and quantum physics, and hence cross the boundary between a retrospective classical/modern dichotomy and its associated periodization. We propose three specific ways in which industry influenced the nature and trajectory of physics.[74] First, key material resources for the practice of physics were either produced by (or shared with) industry; second, industry set some of the key theoretical problems in physics and provided theoretical resources to tackle them; and third, it supplied practical expertise towards the solution of problems in experimental physics.

Victorians publicly endorsed the symbiosis between physics and industry. Even Thomas Huxley, a keen promoter of the efficacy of 'pure science', asserted the common identity of the 'interests' of science and industry in his 1887 jubilee survey of Queen Victoria's reign. He explained that while 'science cannot make a step forward' without, sooner or later, 'opening up new channels for industry', every industrial advance 'facilitates those experimental investigations, upon which the growth of science depends'.[75] The obituary of Lord Kelvin in *The Times* in December 1907 reported approvingly that all Kelvin's scientific enquiries had been pursued with a 'keen eye for practical application'.[76] His career provides a paragon of the symbiosis between physics and industry during the late nineteenth century and early twentieth century. It is representative of the rise of a type of industrial physicist named the 'scientist-engineer' by Sungook Hong. The careers of scientist-engineers offer a means of investigating how the three distinct components of industrial physics that we have identified are entwined. They moved readily between the domains of physics and electrical industry, theorized about the machinery of power and lighting in the lecture theatre and the patent office, and employed electrical apparatus derived from industry in their physical experiments. Prominent scientist-engineers included Ambrose Fleming, Oliver Lodge, Silvanus Thompson, John Hopkinson, and William Ayrton in Britain, Irving Langmuir and Henry Rowland in the United States, Marcel Deprez in France, and Karl Ferdinand Braun in Germany.

While there was considerable heterogeneity in their theoretical practices (the Maxwellians Lodge and Fleming, for example, used ether theory overtly) scientist-engineers all brought crucial expertise to the formulation and solution of the numerous theoretical problems raised by technical or industrial uses of electricity.

The research of Lodge, Thomson, and Rowland combined advanced theorizing on electromagnetism with the concurrent patenting of practical applications of this theory in telecommunications technologies.[77] This iterative process often resulted in the extension and refinement of existing theories of physics. Scientist-engineers also shared a common hardware of electrical measuring instruments and experimental devices. The experiments of Maxwellians on Hertzian waves were based on generically similar electromagnetic measurement devices to those employed by Marie and Pierre Curie to detect alpha and beta radiation, and by German physicists at the PTR in Berlin to investigate radiation phenomena in quantum physics.[78]

The careers of Hong's scientist-engineers testify to the emergence of disciplinary and professional structures in the second half of the nineteenth century that brought natural philosophy into close contact with the electrical engineering of power generation, lighting, and wireless telegraphy. But there are also earlier figures whose work exemplifies the fruitful interactions between practical and theoretical expertise in physics and industry. For example, Hunt has described how Michael Faraday used the phenomenon of retardation in long-distance submarine telegraphy during the 1850s to demonstrate the efficacy of his developing theory of the electrical field. Unlike continental 'action-at-a-distance' theories, Faraday's ideas helped telegraph engineers to reconceptualize telegraph cables as giant capacitors, which enabled them to explain and to some extent to mitigate practical problems of transmission posed by the retardation and distortion of cable signals.[79] In the inverse sense, industrially-acquired practical expertise could also serve as the starting point for physical theorizing. Otto Sibum has shown how James Joule's brewery expertise in delicate temperature measurement enabled him to determine a mechanical equivalent of heat in the 1840s. Joule was among those uniquely placed to perform the kind of thermometric work that would later provide compelling evidence for (what Joule conceived as) the exact inter-convertibility of heat and mechanical work. Although doubted at first by William Thomson among others, Joule's conclusions were eventually accepted as evidence in favour of both the dynamical theory of heat and the principle of the conservation of energy.[80]

Other cases enable the individual components of industrial physics to be examined separately. Industrial processes gave rise to the material apparatus for experimental physics, such as the thermometric and electrical equipment of 1880s physics laboratories. For example, techniques of fault-finding in long-distance submarine telegraphy in the 1860s motivated and facilitated the economic production of material standards of electrical resistance. Similarly, the rise of the electrical lighting and power industry generated the ammeters and voltmeters that came to be widely used for high-speed, low-skill physical measurements.[81] According to Huxley, this was how industry had 'largely repaid' its 'heavy debt' to physics. In his 1887 survey, he reflected on the broader dependency of progress in physics on material resources supplied by industry:

It is a curious speculation to think what would have become of modern physical science if glass and alcohol had not been easily obtainable; and if the gradual perfection of mechanical skill for industrial ends had not enabled investigators to obtain, at comparatively little cost, microscopes, telescopes, and all the exquisitely delicate apparatus for determining weight and measure and for estimating the lapse of time with exactness, which they now command.[82]

Electrical devices found important applications in established branches of physics and investigations into radioactivity and quantum phenomena. For example, in 1898 Marie Curie used electromagnets and electromagnetic measurement devices to determine the charge and constitution of the rays emanating from her samples of polonium and radium.[83] Between 1880 and 1883, Thomas Edison and his team investigated how an asymmetrical deposit of carbon on the glass in his patented light bulbs related to their eventual breakdown.[84] Edison's British associate Ambrose Fleming eventually explained the phenomenon in terms of the tendency of the ionized carbon emanating from a hot electrical filament to flow across the evacuated valve in a unique direction, determined by its negative charge. Fleming then used this 'Edison effect' to produce amplifying 'valves' that became crucial parts of radio receivers and early electronic computers. Soon afterwards, Owen Richardson, the British Nobel prize-winning physicist, and Walter Schottky, a senior German physicist employed by Siemens (and Planck's erstwhile doctoral student), made independent demonstrations of the quantum wave behaviour of electrons based on similar thermionic emission phenomena. This was how the culture of experimentation on incandescent light bulbs (along with Crookes vacuum tubes) contributed to both the development of quantum electron physics and to the inception of semiconductor technology.[85]

There are many examples of industrial problems shaping the theoretical agenda of physics, regardless of whether this agenda might be labelled 'classical' or 'modern'. The exemplary case during the nineteenth century was the problem of the efficiency—or rather inefficiency—of the steam engine. The dissipation or 'waste' of useful power posed a pressing economic concern in industrial Britain and France. Smith and Wise have revealed how William Thomson and his brother James drew upon expertise from Glasgow steam shipyards to frame the problem in the context of the general irreversibility of energy processes. Until this point these kind of asymmetrical processes had not been of much concern to nineteenth-century physicists preoccupied with conservation laws. The macroscopic study of the steam engine's inefficiency enabled them to recognize, however, that whilst energy was always conserved, the generation of mechanical work (to propel machinery) invariably led to the running down of the available 'useful' energy in the universe. Over the following decade James and William arrived at a more precise expression for this approximate statement of the second law of thermodynamics.[86]

Electrical technologists involved in large-scale projects of electrical lighting and telephonic communication, such as Einstein in his later years, shared the challenges of applying electromagnetic theory to quantifying and representing the cyclical and resonant behaviour of alternating currents in large systems. These theoretical applications became crucial both for long-distance power supply and later for high-frequency forms of wireless transmission. Electromagnetic theory itself also benefited. In the United States, the German émigré mathematician-engineer, Charles

Steinmetz, brought vectors and complex numbers into regular everyday usage in American electro-technical laboratories and classrooms, and hence into the mainstream activities of physics. Oliver Heaviside's early training in telegraphy framed his interpretation of Maxwell's *Treatise on Electricity and Magnetism*, which led to two highly productive decades of refinement and rearticulation of Maxwell's theory of electromagnetic waves to deal with the problems of distortion and volume loss in long-distance telephone lines.[87] Although understated by historians of physics, Heaviside tended to treat these projects as integrally interrelated. Indeed, without the stimulus of these practical challenges, why would Heaviside have dedicated so much effort to transforming 'Maxwell's equations' from their original, unwieldy form into one readily usable by his peers?

An unfamiliar instance of the industrial influence on physical theory is the material constitution of magnetism. Familiar for centuries as lodestones and compass needles, the polar attractive and repulsive behaviour of magnets was easy to harness but difficult to explain. Maxwell and his followers offered no theoretical account of the existence of permanent magnets (or magnetization) because this polar behaviour proved uncongenial to a theory that only addressed the properties of fluctuating magnetic fields. Pierre Duhem judged that the very existence of permanent magnets constituted a conspicuous problem for Maxwell's theory, since it was unable (by itself) to account for the existence of an unvarying magnetism fixed in matter.[88] The widespread use of so-called 'permanent' magnets in alternating current machinery from the 1880s soon revealed the variability of their magnetic strength. Following an interaction with neighbouring magnetic fields, magnets did not return (elastically) to their original state but retained persistent effects from the encounter, a phenomenon labelled at that time as magnetic 'memory'.[89] Matthias Dörries has described how the engineer Alfred Ewing and his Japanese team in Tokyo (and also the physicist Emil Warburg in Berlin) developed a theory of molecular 'hysteresis' to explain how this memory constituted the 'permanence' of magnets.[90] A full theoretical explanation of magnetism emerged through the work of the electrical engineer John Hopkinson and instrument-maker Sydney Evershed. While their research ultimately became canonical examples of the physics of molecular magnetism, historical accounts have obscured its industrial origins by focusing instead on the contribution of quantum physics to magnetic theory.[91]

Peter Galison has described how the distinct routes to Einstein and Poincaré's formulations of relativity were embedded in the interconnected material, technological, and epistemological culture of the Patent Office and telegraphic communication.[92] In particular, the problem of simultaneity was initially posed by the growth of transnational railways systems and global telegraphic networks: how could the timing of practical commercial activities at remote locations be synchronized? Galison observes that Einstein was 'not only surrounded by the technology of coordinated clocks', but also located in one of 'the great centres for the invention, production and patenting of this burgeoning technology'. Another source of demand for coordinated clocks was provided by the electrical power industry, which needed a reliable commercial means of measuring the consumption of electricity. Einstein was the son and nephew of electrical engineers who made domestic electricity meters. These meters incorporated two initially-synchronized clocks running at different speeds, which became a motif in Einstein's illustrations of the time-dilating effects of near-light-speed travel. Poincaré likewise pursued research into the laws of electromagnetism and the nature of simultaneity in the midst of a vast trans-European effort to synchronize clocks.

Finally, industrial problems and developments gave rise to the problem of black-body radiation addressed in Planck's early theoretical work on quantum theory. A black body was an idealized object that could absorb all electromagnetic radiation (and hence was black), which made it necessarily the most efficient radiator. David Cahan has described how Planck's research was undertaken between 1893 and 1901 at the Physikalische-Technische Reichsanstalt in Berlin with the financial support of the German state and the Prussian industrialist Werner von Siemens. This support was premised on the study's relevance to the design of efficient electrical incandescent lighting, but ended up producing unanticipated insights into thermodynamics. The large-scale institutional resources made available by the vast PTR enterprise provided Planck with the mass of observational data that posed such a challenge to conventional views of radiation. While the observed variation of radiated intensity with frequency agreed well with the predictions of Wilhelm Wien's law of 1896 for high frequencies, this law completely failed for low frequencies. In order to resolve this 'ultraviolet catastrophe', Planck postulated that light was only emitted in discrete rather than continuously-variable quantities. Hence quantum theory initially took root in the industrial pursuit of efficiency prosecuted with resources at that time available only in German state research.[93]

The examples of James and William Thomson, Steinmetz, Heaviside, Ewing, Hopkinson and Evershed, Einstein, and Planck—figures associated with 'classical' and 'modern physics' alike—serve to establish the ongoing influence of industrial concerns in both posing theoretical problems and resolving them through new techniques. In our final section we use the example of French experimental physics during the same period to exhibit the fertility of a disciplinary approach to the history of physics based on the continuity of *experimental* practice. We show that the temporal continuity of French physicists' distinctive experimental practices undermines both the radical disjuncture in late nineteenth and early twentieth century physics implied by a classical/modern dichotomy, and the trans-national unity presupposed by the notion of 'classical physics'.

24.4 FRENCH PHYSICS AND THE CONTINUITY OF EXPERIMENTAL PRACTICE

Science in France from the late nineteenth to the early twentieth centuries has long posed historiographical challenges that undermine the notion of a unified body of knowledge that might be labelled as 'classical'. Many historians have recognized that the activities of French scientists during the nineteenth century, to quote Elisabeth Garber, do not 'easily fall into line with the work of scientists in the same fields in either Britain or Germany'.[94] Robert Fox, Mary Jo Nye, and Dominique Pestre have all suggested that French scientific research evinced a distinctive style that was responsible for the choices of many French scientists that, as Nye provocatively claims, 'placed them outside what was to become the mainstream of research'.[95] They agree that, whilst there was not a decline in standards in an absolute sense, France's comparative position amongst scientific nations worsened. One of us has summarized (and implicitly endorsed) this consensus as the claim that 'French physics only appeared to decline because the French national style lost out during the transition from "classical" to "modern physics"'. Hence French physics has been evaluated unhelpfully through the prism of an emerging 'modern physics' to which it supposedly failed to contribute.[96]

The historiographical problems posed by nineteenth-century French physics therefore derive at least in part from historians' implicit acceptance of a transition from 'classical' to 'modern physics' that apparently left France 'behind'. As we mentioned in our introduction, this has led historians to focus on the development of key theoretical aspects of 'modern physics' rather than on the experimental interests, aims, and approaches shared by late nineteenth and early twentieth century physicists. Their historiography represents the interests and objectives of physicists during the period as orientated towards theory. Hence the relative lack of recent historical interest in physics in France is explicable because, unlike physics in Britain and Germany, it did not experience the emergence of specialisms loosely defined by theory, but instead remained dominated by experiment until well into the twentieth century.[97]

In this section we offer a critical synthesis of three case studies, due to Mitchell, Atten and Pestre, and Lelong, in terms of the continuity of *one key aspect* of experimental practice in French physics. Each of these studies compares French with British or German research into the same physical phenomena: either electrocapillarity, electromagnetic waves, or X-rays. We reveal striking underacknowledged differences in approaches to experiment, and in interpretations of the relationship between experiment and theory, that would be obscured by any concept of 'classical physics'. Our synthesis reveals continuities in French physicists' distinctive, shared experimental aims and approaches for areas of physics that might fall retrospectively on either side of a classical/modern divide (as in the previous section). We thereby identify a national context for physics for which this periodization is particularly inappropriate.[98]

Although our historiography obviously draws directly from the theme of a French 'national style' articulated clearly by Fox, Nye, Garber, and Pestre, there are three main reasons why we have chosen instead to talk about 'shared approaches'. First, it is unlikely that all of the elements of a national style will be present in a single research project. Faced with this challenge, the pluralist notion of 'shared approaches' grants the historian more flexibility. For example, it might be useful to separate approaches to measurement from approaches to experiment. Secondly, although the community under study may mostly be located in a single country, this need not be the case for all its members. Indeed, in one of our case studies, we treat two Swiss experimentalists as 'French experimental physicists'. And the reverse also applies: we would not expect every French experimental physicist to fit the mould we describe, nor to have adopted all possible shared approaches in a given research context. Finally, a 'national style' is liable to be inferred from scientific products rather than practices. Adopting the notion of a 'shared approach' directs attention towards practitioners' shared practices and goals.[99]

Following the demise of Laplacian physics in the first few decades of the nineteenth century, (the attempt to explain all phenomena in terms of short-range molecular forces), a nationally distinctive institutional and intellectual division between mathematics and experimental physics became widespread in France. Until the 1930s this remained barely affected by the rise of late nineteenth-century specialization in theoretical physics in Britain and Germany (see below).[100] Work in French physics with an explicit theoretical orientation was typically undertaken either by mathematicians holding chairs of mathematical physics, such as Henri Poincaré, or outside a university setting altogether. In the post-Laplacian period, French experimental physicists (*physiciens*) believed that the fundamental nature of phenomena could never be known, and hence that natural knowledge should be based directly on observation and precise measurement. They aimed primarily to discover empirical laws, which they expressed in simple mathematical relationships between macroscopic variables.[101]

Programmatic statements to this effect are easy to find. In a lecture on the nature of electricity given at the Collège de France, the professor of physics of the École Normale, Pierre Bertin-Mourot, described the 'true domain of experimental physics' as 'the attentive observation of phenomena and the experimental investigation of their laws'.[102] The founder of the French Physical Society, Jean-Charles d'Almeida, likewise claimed in his *Cours de Physique* (co-authored with Augustin Boutan) that 'the principal goal followed by the scientist in his research is the discovery of general laws relating phenomena'.[103] Similarly, Josep Simon's detailed study of Adolphe Ganot's textbooks reveals a pervasive focus on laws in the context of 'a compact conjunction of experimentalism, pedagogical demonstration and instrumental analysis'.[104]

Unlike their British and German counterparts, French experimental physicists drew a sharp conceptual separation between facts and laws (supposedly) induced from experiment, and explanations or interpretations. These consisted of attempts to synthesize (descriptive) observations, and they were often framed as provisional because it was possible that these observations might be consistent with many different, mutually incompatible syntheses. Furthermore, *causal* explanations that required commitment to a specific ontology were considered sufficiently uncertain to be labelled 'hypotheses'. D'Almeida and Boutan summarized a widespread French consensus on atoms:

The simplest interpretation of the laws that govern chemical combinations would lead us to suppose that bodies consist of indivisible particles or atoms that withstand all chemical and physical attempts to split them. But this is purely a mental projection that is impossible to actually demonstrate, and against which we should be careful not to attribute the same degree of truth to a fact or a physical law.

Ganot likewise demonstrated sensitivity to the inferior ontological status of 'hypotheses', particularly about the nature of electricity. In fact, he was scornful of causal fluid-based theories of physical phenomena in general, whereas d'Almeida and Boutan merely observed and regretted that the fundamental experimental laws of electrostatics unavoidably expressed some form of fluid hypothesis.[105]

Whilst the role of 'hypotheses' in motivating and guiding research in both French experimental and mathematical physics merits closer examination, it seems that imponderable fluids, micro-molecular mechanisms, and other such inventions were usually invoked for heuristic purposes. Nearly fifty years later a French inspector of public instruction and a researcher at the physical research laboratory at the Sorbonne, Lucien Poincaré, dealt with the ontological status of the ether on exactly this basis. For him, it was not necessary to know whether the ether had an objective existence in order to 'utilize' it because in 'its ideal properties we find the means of determining the form of equations which are valid'. He then went on to claim that 'for scholars free from "all metaphysical prepossession", that was the essential point'.[106]

An influential exemplar of the French search for experimental laws is Gabriel Lippmann's doctoral study of electrocapillarity. This was the name given by Lippmann to the phenomenon of the variation in the surface tension of mercury with the electrical charge at its surface.[107] Lippmann claimed to have established a simple relationship between these variables by measuring the variation in height of a mercury column in a capillary tube (which was directly proportional to the surface tension) with an applied voltage at its surface. His research gained a hugely positive reception amongst French contemporaries. Lippmann's French biographers also praised electrocapillarity, describing it as 'a masterstroke' and 'one of his most celebrated achievements' because it fully realized the French law-based approach to experimental physics. One of them, his friend and later a fellow professor of physics at the Sorbonne, Edmond Bouty, framed the achievement in terms of distinctive national characteristics. Bouty recalled that when he had first read a paper that Lippmann had published on electrocapillarity in the German periodical *Poggendorff's Annalen,* he had initially been 'fooled by the name of the author' and had told himself: 'here is a German who possess all the unique qualities of our race: he deserves to be French'.[108]

Lippmann's findings directly challenged the earlier work of the German physics professor Georg Hermann Quincke. The approaches of the two investigators were diametrically opposed. Whilst attempting to measure the surface tension of various mercury interfaces, Quincke had observed small random perturbations in the height of a mercury column and a gradual decrease in this height over time. He identified the cause as the contamination of the mercury with invisible impurities. Lippmann claimed, however, that the perturbations disappeared whenever the mercury formed part of a *closed* circuit, and suggested that Quincke had failed to observe the disappearance because all his experiments were performed with an *open* circuit. The disappearance of the perturbations allowed Lippmann to assign precise values of surface tension and charge to the mercury interface. In the introduction to his thesis, which was published in the *Annales de Chimie et de Physique* in 1875, Lippmann totally rejected Quincke's causal approach by signalling explicitly how premature 'explanations' impaired the search for laws:

We would have undoubtedly thought of relating these two physical properties of contact surface to each other, [the electrical charge] and surface tension, and of seeking a fixed relationship between them, if we were not so used to considering this latter quantity as a variable, and explaining its variations by the presence of invisible impurities.[109]

Quincke, on the other hand, interpreted Lippmann's work in the context of his own search for the micro-molecular causes of a much wider range of phenomena generated by mercury interfaces. When he read Lippmann's *Annalen* paper, he latched onto the alternative interpretation that Lippmann had given for the open-circuit behaviour of the mercury column, in terms of the gradual loss of charge from the mercury surface. Instead of attempting to directly refute Lippmann's law, Quincke designed experiments to challenge his interpretation. By attributing any discrepancies in their observations to the contamination of the mercury by invisible impurities, Quincke reasserted his own causal hypothesis.

Lippmann responded by retreating to epistemological territory that his French contemporaries found more certain. He simply dropped all such 'explanations' entirely and boasted in his thesis that 'no hypotheses . . . have been invoked . . . [in] the present work; it was in order not to introduce them that I refrained from giving a physical theory, [that is to say] an explanation of the properties that have been observed'.[110] This enabled Lippmann to dismiss Quincke's criticism entirely in a review of the German experimenter's latest research for the *Journal de Physique*. He based his evaluation on the French pursuit of experimental laws, and hence judged the wide variation in conditions investigated by Quincke to be an incoherent way of proceeding. Lippmann's victory in France was total. Quincke's work was rarely, if ever, discussed alongside his own in French textbooks.[111]

Michel Atten and Dominique Pestre have described a similar polarization between the responses of French experimental physicists and British Maxwellians to Hertz's 1887 researches on the transmission of electromagnetic waves. Like elsewhere in Europe, the importance of Hertz's work was recognized swiftly in France. The French Academy of Sciences awarded him its prestigious Lacaze prize following successful public replications by Jules Joubert and Guillebot de Nerville at the 1889 Electrical Congress in Paris. Two members of the Society of Physics and Natural History of Geneva, Édouard Sarasin and Lucien de la Rive, were among those who followed up on Hertz's experiments. First, they sought to create a stationary electromagnetic wave in air through the interference of an incident wave with its reflection (as in optics). They then moved on to investigate 'the regularity and the stability of the [standing wave] phenomenon . . . by varying as many parameters as possible'.[112] These included the sizes of the wave generator (primary oscillator) and the wave detector (circular resonator).

According to Atten and Pestre, the Swiss pair claimed to have employed a 'purely experimental logic by avoiding calculation and any reference to theory'.[113] To their surprise, de la Rive and Sarasin discovered that resonators of different dimensions all responded to the oscillator, albeit with different intensities that fell off with the difference from the primary oscillatory frequency. By expressing this discovery in terms of experimental laws, and by respecting the epistemic primacy of these laws over their tentative interpretations, de la Rive and Sarasin followed two key 'French' approaches to physics. They supplemented the qualitative law that 'the distance between the nodes only depends on the detector and is independent of the generator' with other quantitative laws, such as 'the distance between the nodes for a circular resonator is noticeably proportional to its diameter'. De la Rive and Sarasin then offered an interpretation in terms of the 'selection' of one wavelength by the detector from the multiple wavelengths or spectral band emitted by the generator. They named this phenomenon 'multiple resonance'.[114]

By assuming that the generator emitted a unique frequency, Hertz had shown that his measurements of the distances between the zeroes (nodes) of the stationary wave produced a speed of propagation equivalent to the speed of light. Multiple resonance provoked scepticism in France about this conclusion, which Alfred Cornu voiced cogently in a commentary that followed his presentation of de la Rive and Sarasin's findings before the French Academy of Sciences on 13 January 1890. Whereas Maxwellians such as FitzGerald and Lodge had enthusiastically embraced Hertz's work as a triumphant confirmation of Maxwell's electromagnetic theory of light, Cornu took the experimental laws established by de la Rive and Sarasin to undermine Hertz's calculations of the speed of propagation of the waves. He offered this implication as a cautionary tale to the predominantly British Maxwellians about the proper way of doing science: 'you'll see that it's very prudent to go about things in the manner of MM. Sarasin and de la Rive ... [by] scrutinizing carefully the interesting experimental method devised by M. Hertz before thinking about presenting it as a demonstration of the identity of electricity and light'.[115]

Unsurprisingly, the Maxwellians did not draw the same moral as Cornu. Atten and Pestre describe how the search for a decisive experimental test between Maxwell's theory and its action-at-a-distance counterparts led FitzGerald and his colleague at University College Dublin, Frederick Trouton, to improve (in their view) upon Hertz's experimental set-up, basically by employing a fixed detector and a mobile mirror. Their 'replications' sought to confirm Maxwell's theory rather than to establish and interpret experimental laws. In a short commentary published in *Nature* on 30 January 1890, Trouton claimed to have already discovered 'multiple resonance'. To safeguard Hertz's calculation of the speed of electromagnetic waves, he contended that Hertz's choice of wave generator ensured that the detector registered the 'central' frequency of the spectral band. In other words, Trouton avoided Cornu's sceptical conclusion by committing to an alternative and more favourable theoretical interpretation of de la Rive and Sarasin's experimental findings.[116]

Even when members of the French Physical Society debated the nature of X-rays the epistemic boundary between 'facts' and explanatory 'hypotheses' demarcated by d'Almeida and Boutan remained remarkably stable. The debates focused on the puzzling experimental 'fact' that X-rays could discharge an insulated electrical conductor without contact. This was demonstrated in France in 1896 by Paul Langevin and Jean Perrin at the École Normale and Louis Benoist and Dragomir Hurmuzescu at the Physical Research Laboratory of the Faculty of Sciences (of which Lippmann was Director). The two teams proposed contradictory mechanisms for the discharge (Perrin favoured ionization of the gas, Benoist molecular convection) but, in Benoit Lelong's judgement, 'the protagonists shared a common implicit definition of the boundary between facts and hypotheses. Atoms, ions, and molecules were not experimental facts. "Facts" were in fact mathematical laws between observable parameters. Atomist vocabulary was implicitly forbidden in factual statements, but could be used in interpretations'.[117] Lelong stresses how even Perrin, 'the future champion of molecular reality', had to take pains to reassure readers that statements about 'electric charges' presupposed no hypotheses, and even then Perrin only introduced hypotheses and theories to explain facts and laws.[118]

No such Gallic boundary between facts and hypotheses was recognized by members of Cavendish group at the University of Cambridge working under J. J. Thomson from 1884. They too were investigating the discharge of electricity by X-rays, but with strong ontological commitments to atomism; as Lelong puts it, 'in Cambridge experimental facts were statements about ions'.[119] Unlike Perrin, who resisted the temptation to mobilize experimental facts in support of theories, Thomson had no metaphysical scruples about using Perrin's experimental research for this purpose. He shared the view with Perrin that moving ions created by X-rays were responsible for the conductivity of gases. The Cambridge transformation of Perrin's tentative 'interpretations' into 'factual' knowledge provided an important step towards establishing Thomson's theories. Conversely, when Langevin, who studied the phenomenon of the conductivity of gases at the Cavendish between October 1897 and June 1898, presented Thomson's ideas before the French Physical Society, he was careful to purge them of 'hypotheses' by reformulating Thomson's statements about ions in terms of moving charges.[120]

Our synthesis of these three case studies of French research—into electrocapillarity, electromagnetic waves, and X-rays—reveals that, unlike their British and German counterparts, French experimental physicists typically sought to research experimental laws and approached explanations, especially those based on specific (often micro-molecular) ontologies, with greater epistemological caution. This distinctive difference in the perceived relationship between theory and experiment further undermines the notion of a unified 'classical' approach to physics during the late nineteenth century. This finding leaves open the question of the nature and extent of the transformations undergone by French physics during the first few decades of the twentieth century, and how best to describe them without presupposing some kind of trajectory towards 'modern physics'.

Dominique Pestre's detailed study of physicists and physics in France between 1918 and 1940 suggests that the characteristic aims and approaches exemplified in this section remained recognizable until at least the 1930s. He describes how French textbooks of physics retained a 'historico-inductive' form inherited from the mid-nineteenth century, which consisted of a logical reconstruction of the historical sequence of discoveries and the ways of thinking that resulted in laws of nature. This narrative choice, according to Pestre, was consistent with a view of science that progresses from an ever-increasing stock of immutable facts, 'a solid block, definitively established'. Pestre also explains that the same textbooks integrated new theories and techniques into a Gallic division of physics that was well-established by the mid-nineteenth century: thermodynamics, optics, electricity, and mechanics/acoustics. From the point of view of a British or German commentator, they provided no systematic treatment of the study of matter (for example atomic and nuclear structure) or the emerging quantum physics.

Moreover, Pestre reveals that the handful of French chairs in theoretical physics were often occupied by experimentalists and that only a small percentage of physics students took up the few programmes offered in theoretical physics. In Paris, a Chair of Theoretical and Celestial Physics was created in 1920 following the retirement of Joseph Boussinesq from the Chair of Probability Theory and Mathematical Physics. Of its three occupants until 1937, none of them pursued research in theoretical physics, and only one of them, Eugène Bloch, offered a course on the subject (on quantum theory). In the provinces, only Edmond Bauer occupied a post in theoretical physics (a lectureship at Strasbourg), and Duhem was never replaced as Professor of Theoretical Physics at Bordeaux following his death in 1916. Between 1918 and 1928, Pestre identifies Louis de Broglie and Léon Brillouin as the only theoretical physicists in France. So it appears that the distinctive institutional characteristics of French physics were reformed quite gradually during the first half of the twentieth century.[121]

Nonetheless, French experimental physicists were beginning to soften their hardline epistemological stance against explanatory theories and to diversify their aims and approaches in response to broader international developments. As Lelong has shown, although Perrin and Langevin reluctantly addressed their French colleagues from the viewpoint of a framework that privileged experimental facts and laws over micro-molecular explanations, their work ultimately helped to revise it. Perrin's early twentieth century experimental proofs of the existence of atoms, for example, are not easily explicable with reference to Pestre's account of early twentieth century French physics.[122] And from 1903 onward, when Langevin was appointed professor at the Collège de France, he built a research school in 'a new subfield of experimental microphysics'. Once scientists connected to Langevin (and the Curies) and sympathetic to his approach took over and transformed the editorship of the journal *Le Radium* in 1905, young researchers favourably disposed to ionic physics, such as Paul Villard, found a ready outlet for publication.[123]

To reach a balanced picture of twentieth-century French experimental physics, it will be important to weigh studies of established traditions against the emergence of new ones. Indeed, this will determine whether 'French experimental physics' remains a useful analytical category for historians of physics. It would be misleading to only examine the impact of key turn-of-the-century discoveries and theoretical developments, even if these are likely sources of change, because French experimental physicists often favoured dedicating time and resources to quite different research topics than their foreign counterparts. For this reason, it will be important to identify carefully those areas of physics in which French physicists were actively engaged in research. This would help to guard against inappropriately privileging say, radioactivity over optical metrology, or cathode-ray discharge over the standardization of electrical units.

Jed Buchwald and Sungook Hong have identified six areas with which physicists throughout Europe and America, as well as Japan, were particularly concerned around 1900: the nature of X-rays; the character and behaviour of electrons; the properties of the ether; the statistical description of gases; liquids, and solids; the phenomenon of radioactivity; and the long wavelength regime of electromagnetic waves. These only partially overlap with those areas of physics that Lucien Poincaré identified in 1906 as belonging to a distinctively French version of 'modern physics' (and which a modern-day physicist may be tempted to label as 'classical'). In a book aimed at a wide audience entitled *La physique moderne, son évolution*, he included sections on precision measurement and metrology, experimental research into the statics of fluids, and physical principles, but chose to leave out kinetic theory.[124] This selection reveals a clear bias towards areas of physics for which the discovery of laws and the production of experimental 'facts' were particularly integral, and to which French researchers had recently made key contributions.

For each area of French physics, we recommend tracking continuity or change in the epistemological status of facts and laws in comparison to causal explanations. Established traditions may have encountered new or minority approaches, whereas established approaches may have been applied to new areas of enquiry. Research into piezoelectricity and radioactivity offer illustrative examples. During the nineteenth century, French experimental physicists took great interest in the macroscopic manifestations of the conversion of energy from one form to another—so called 'complex macroscopic effects', for example magneto-optical, electro-mechanical, thermo-electric, and so on. Shaul Katzir has described how, early in their career, Pierre and his brother Jacques discovered and investigated piezoelectricity—'the relations between elastic forces and electric fields in crystals'. In typical French fashion, they pursued 'systematic quantitative experiments to reveal the rules that governed the development of charge by pressure'. Yet the brothers also constructed a micro-molecular mechanical theory based on William Thomson's hypothesis of permanent electrical polarization within bodies, which explained the production of electricity in both pyro- and piezoelectricity in terms of 'mechanical changes of distance between polarized molecules', and defended it from criticism.[125]

In contrast, Pierre Curie and André Debierne pursued investigations of radioactivity in line with the conservative French law-based approach exemplified by the three case studies discussed in this section. Marjorie Malley has argued that their unwillingness to commit to a specific theory hampered their research. Unlike Rutherford, who pursued fully the experimental consequences of his hypothesis of a material 'emanation' from radioactive substances, Curie drew upon thermodynamical analogies based on the transfer of energy to avoid commitment to specific ontological hypotheses. During investigations into the secondary activity excited upon other substances by the radioactive emanation from radium, Curie and Debierne offered only cautious, general speculation on the excited activity and the nature of this emanation ('a radioactive gas'). Instead, they focused on establishing observable properties of the excited activity, such as its quantitative variation under different circumstances (radium as a solid or in solution, in a closed glass vessel or the open air). As far as Rutherford's transmutation theory of radioactivity was concerned, 'they thought he had strayed beyond the established facts by endowing his working hypothesis with the status of reality'.[126]

Our synthesis of studies into late nineteenth century French experimental physics could be extended to include those that tackle other distinctive French approaches. Studies into Lippmann's determination of the ohm by Mitchell and into piezoelectricity by Katzir indicate that we could have drawn similar conclusions about the distinctiveness of late nineteenth century French experimental physics by focusing on an experimental approach adopted by Henri Victor Regnault. Regnault aimed to secure a purely empirical foundation for science through the 'direct' measurement of physical quantities, which entailed eliminating unwanted physical effects through the experimental design, rather than by (subsequently) correcting measurements using theory. An important buttress for this approach fell away as soon as French experimental physicists began to commit themselves to specific ontologies and lose their aversion towards theory.[127]

In this way, local studies of individuals or research schools might be related to broader trends in physics. We would expect further comparative studies in the mould of Lelong to reveal other routes apart from Cambridge ion physics through which new experimental aims, approaches, and results were introduced into France, and how these were transformed in the process. On the other hand, we would not be surprised to discover, at least until an older generation died out, the persistence of the shared approaches described in this section until the 1920s or even 1930s, especially in areas of traditional French strength, for example optics, statics of fluids, and precision measurement. Perrin's biographer Mary Jo Nye follows one of his students in portraying him as 'a figure of compromize between differing viewpoints and traditions', one of which is the mainstream French tradition described in this section. Similar remarks also apply to Langevin, who occupied an intermediate position between Cambridge and Parisian physics, according to Lelong. The coexistence of the old and the new, whether in France or elsewhere, would demonstrate clearly the redundancy of a sharp dichotomy between 'classical' and 'modern physics'.[128]

24.5 Conclusion

So what *is* 'classical physics'? It cannot refer to a discipline as practised and understood during the nineteenth century since no concept of 'classical physics' gained general currency until the early decades of the twentieth century. We suggest instead that the notion was developed by theoreticians during this later period who sought to preserve a restricted role for established theory and techniques whilst setting forth a future research programme based on new forms of theorizing. It is only in this limited sense that classical physics ever existed. Any references to 'classical physics' prior to 1900, therefore, implicitly adopt an anachronistic perspective that was created to legitimize the new foundations for physics proposed within relativity and quantum theory.[129]

As an antidote to this anachronism, we showcased three more fruitful historiographies of late nineteenth and early twentieth century physics. First, following our careful examination of the work of Buchwald and Warwick, we proposed that questions about the ontological status of the ether, whether before or after the Michelson–Morley experiment, should be reframed in terms of its role in sustaining research practices. This focus seems appropriate for studying other discarded theoretical entities during the late nineteenth and early twentieth centuries. Second, we complemented existing historiographies of research traditions in physics by introducing the notion of 'industrial physics' to capture the intimate connection between industry and physical theorizing, experimentation, and instrumentation and apparatus during the same timeframe. We anticipate similarly important connections between physics and chemistry, medicine, geography, and astronomy. Finally, the case of French experimental physics demonstrates that whilst there were clear *longue durée* continuities in physicists' aims and their laboratory practice, these might vary considerably between countries and sites. This undermines any lingering hope of retrieving a transnational 'classical physics'. An especially important task for historians is establishing the importance of such continuities in ensuring either the coherence or fragmentation of physics as a discipline. In order to ascertain the extent of French engagement with British and German approaches to physics during the early nineteenth century, we suggested investigating the changing epistemological status of facts and laws in comparison to causal explanations and (often micro-molecular) hypotheses within French physics.

In summary, this chapter has contributed to two distinct projects: the investigation into how and why physicists invented, accepted, and used a classical/modern dichotomy in physics, and the development of alternative historical tools to reveal important continuities in the practice of physics between the late nineteenth and the early twentieth centuries. Our criticism of Staley's thesis that 'classical' and 'modern physics' were co-created in 1911 should not distract from the methodological importance of his analysis. Nonetheless, we do not subscribe to Staley's (or any other) claim that past notions of classical or modern physics have ever converged on a putative single meaning. We suggest alternatively that the complex history first revealed by

Staley behind different attributions of 'classical' and 'modern' to individual branches of physics is recapitulated in the emergence of the distinction between 'classical physics' and 'modern physics' as entire disciplines. We have suggested when and by whom the terms 'classical physics' and 'modern physics' were first used, and that whilst the inventions of the terms appear to be connected, they were by no means synchronous.

We are therefore now in a position to pose some fundamental questions about 'classical' and 'modern physics'. What factors shaped the interpretation of these terms, and to what extent did different interpretations inspire controversies and broader debates? And why were the terms taken up in the first place? These questions open up the possibility of studying the emergence of 'classical' sciences as part of a broader cultural history of scientific disciplines. The 'transition' from 'classical' to 'modern' was not unique to physics: the need to establish some form of strategically ambiguous Janus-faced connection between the past and the present has parallels with other disciplines in the natural and human sciences.[130] In fact, the persistent widespread use of classical/modern dichotomies serves as an ironic rejoinder to Kuhn's claim that scientific revolutions are rendered invisible by subsequent textbook treatments written from the perspective of the new paradigm.[131] On the contrary, 'modern physics' drew credibility from its continuous emergence from well-established older techniques that continued to prove fruitful in restricted contexts. So far from committing a form of Kuhnian patricide, physicists actively constructed a 'classical' identity for the work of previous generations in order to highlight the origins, nature, and pedigree of their own work. The tensions between continuity and change with which Planck, Bohr, Eddington, and others grappled remain very much alive.

Notes

1. Sources cited in Richard Staley, *Einstein's Generation: The Origins of the Relativity Revolution* (Chicago: University of Chicago Press, 2008) include Allan A. Needell, 'Irreversibility and the Failure of Classical Dynamics: Max Planck's Work on the Quantum Theory, 1900–1915' (Yale University: Unpublished Ph.D. Diss., 1980); Allan A. Needell, 'Introduction', in Max Planck, *The Theory of Heat Radiation* (Los Angeles: Tomash; New York: American Institute of Physics, 1988), pp. xi–xlv; Olivier Darrigol, *From c-Numbers to q-Numbers: The Classical Analogy in the History of Quantum Theory* (Berkeley/Los Angeles: University of California Press, 1992); Olivier Darrigol, 'The Historians' Disagreements over the Meaning of Planck's Quantum', *Centaurus*, 43 (2001), pp. 219–239.
2. Two of the most substantial contributions are Jed Z. Buchwald, *From Maxwell to Microphysics: Aspects of Electromagnetic Theory in the Last Quarter of the Nineteenth Century* (Chicago; London: University of Chicago Press, 1985) and Oliver Darrigol, *Electrodynamics from Ampère to Einstein* (Oxford: Oxford University Press, 2000). Cf. Terry Quinn's recent book on the creation of the Bureau International des Poids et Mesures and its role in the development of standards of measurement, which, despite the

title, does not resuppose a transition from 'classical' to 'modern physics' in its historiography. Terry Quinn, *From Artefacts to Atoms: The BIPM and the Search for Ultimate Measurement Standards* (Oxford: Oxford University Press, 2012).
3. Staley's historiographical stance towards the terms 'classical' and 'modern physics' is closely analogous to Suman Seth's approach to the 'crisis' often associated with early twentieth century theoretical physics. Seth rejects 'crisis' as an analytical tool on the grounds of its circularity, and instead investigates the use of the term as an actors' category. Seth explains that the 'dominant discourse of the new discipline' came from the 'physics of principles' practised by Planck, Einstein, and Bohr, rather than the 'physics of problems' associated with Sommerfeld's school. Suman Seth, 'Crisis and the Construction of Modern Theoretical Physics', *British Journal for the History of Science*, 40:1 (2007), pp. 25–51 pp. (41–2).
4. Russell McCormmach, *Night Thoughts of a Classical Physicist* (Cambridge, MA; London: Harvard University Press, 1982), p. 10. This is a fictional autobiography about how, late in his career during the First World War a German physicist called Jakob experienced with grave concern the threat of military defeat and the revolutionary developments implied by Planck's quantum theory.
5. Richard Staley, *Einstein's Generation* (n. 1).
6. Cf. Heilbron's analogous basis for evaluating the term 'Scientific Revolution' (SR) in chapter 1 of this volume, pp. 7–9, where he writes: 'whether as a period or as a metaphor, SR is an historian's category, an analytical tool, and must be judged by its utility'. The SR survives Heilbron's judgement on both counts.
7. See, for example Jed Z. Buchwald, *The Creation of Scientific Effects. Heinrich Hertz and Electric Waves* (Chicago: Chicago University Press, 1994); Peter Harman, *The Natural Philosophy of James Clerk Maxwell* (Cambridge University Press: Cambridge, 1998).
8. Andrew Warwick, reveals intimate connections between physics and 'applied' mathematics in *Masters of Theory: Cambridge and the Rise of Mathematical Physics* (Chicago: University of Chicago Press, 2003). The authors in Robert Fox and Graeme Gooday (eds.), *Physics in Oxford 1839–1939: Laboratories, Learning and College Life* (Oxford: Oxford University Press, 2005) explore the connections between physics, physical chemistry, and mechanics. For more general discussion see Graeme Gooday ' 'Vague and Artificial': The Historically Elusive Distinction between Pure and Applied Science' *Isis*, 103 (2012), pp. 546–554.
9. For an interpretation of nineteenth century physics as a unified discipline see Jed Buchwald and Sungook Hong, 'Physics', in David Cahan (ed.), *From Natural Philosophy to the Sciences* (Chicago: University of Chicago Press, 2003), pp. 163–195 (p. 165).
10. Staley, *Einstein's Generation* (n. 1), notes that Boltzmann was the only figure in nineteenth century physics known to have described himself as a 'classical physicist' (c. 1899), pp. 364–365; Boltzmann's notion of 'classical physics' derived from the early-nineteenth century French tradition of mathematical physics.
11. Peter Harman, *Metaphysics & Natural Philosophy: The Problem of Substance in Classical Physics* (Brighton: Harvester, 1982), p. 4.
12. Thomas Kuhn, *Black-Body Theory and the Quantum Discontinuity, 1894–1912* (Chicago; London: University of Chicago Press, 1987).
13. Darrigol, 'The Historians' Disagreements' (n. 1). Nevertheless, some historians of popular science continue to use the category literally to describe physics during the nineteenth century without these historiographical scruples. For an orthodox view of classical physics as an identifiable historical entity, see David Knight, *Public Understanding of Science: A History of Communicating Scientific Ideas* (London: Routledge, 2006), Chapter 12. Knight

argues that the origins of classical physics did not derive directly from Newton but more specifically from the work of French mathematical theorists at the turn of the nineteenth century, thereby echoing Boltzmann's interpretation (discussed in Staley, *Einstein's Generation*, see n. 1).

14. Staley, *Einstein's Generation* (n. 1), does not emphasize sufficiently perhaps the longevity of this original narrower author-specific usage of 'classical'. For example, see the reference to the 'classical researches' of Dulong and Petit in Leonard Hill, O. W. Griffith and Martin Flack, 'The Measurement of the Rate of Heat-Loss at Body Temperature by Convection, Radiation, and Evaporation', *Philosophical Transactions of the Royal Society of London. Series B*, 207 (1916), pp. 183–220 (p. 188).

15. Staley, *Einstein's Generation* (n. 1), pp. 355–357 (thermodynamics), pp. 357–360 (mechanics).

16. David Cahan, *An Institute for an Empire: The Physikalisch-Technische Reichsanstalt, 1871–1918* (Cambridge: Cambridge University Press, 1988); and Diana Barkan, *Walther Nernst and the Transition to Modern Physical Science* (Cambridge: Cambridge University Press, 1999) had both previously identified the Solvay conference as a key event in the formulation of 'modern physics', although with less specificity than Staley.

17. Staley, *Einstein's Generation* (n. 1), pp. 375–396, 411–414, quotations at pp. 412, 414. This reconceptualization of the Rayleigh-Jeans law as classical obscures Jean's initial interpretation that it represented a final, unreachable state of nature. See pp. 380–383. On the search for secure foundations for physics, see Christa Jungnickel and Russell McCormmach, *Intellectual Mastery of Nature: Theoretical Physics from Ohm to Einstein. Vol. 2: The Now Mighty Theoretical Physics* (Chicago, London: University of Chicago Press, 1986), Chapter 24.

18. Staley, *Einstein's Generation* (n. 1), pp. 397–422.

19. Likewise, Staley also occasionally alternates between 'classical physics' and 'classical theory'. See *Einstein's Generation* (n. 1), pp. 394, 398, 417. On the possible meanings of 'classical theory' in the *Annalen der Physik* for 1911 and 1912, see Jungnickel and McCormmach, *Intellectual Mastery of Nature* (n. 17), p. 313.

20. Max Planck, [Laws of radiation, and the hypothesis of elementary energy quanta], in A. Eucken (ed.), *Die Theorie der Strahlung und der Quanten: Verhandlungen auf einer von e. Solvay einberufenen Zusammenkunft*, 30. Oktober bis 3 November 1911 (Berlin: Verlag Chemie GMBH, 1913), pp. 77–94 (p. 77), trans. Richard Staley. Staley, *Einstein's Generation* (n. 1), p. 412.

21. See Max Planck's Nobel lecture reproduced in *Nobel Lectures, Physics 1901–1921* (Amsterdam: Elsevier Publishing Company, 1967). 'Max Planck, Nobel Lecture'. ⟨http://nobelprize.org/nobel_prizes/physics/laureates/1918/planck-lecture.html⟩

22. Niels Bohr, 'The Structure of the Atom' (Nobel Prize address 1922), trans. Frank C. Hoyt, *Nature*, 112 (1923), pp. 29–44 (p. 31). This is the British journal *Nature*'s first mention of the term 'classical physics'.

23. For example, see discussion of Levi-Cevita's work on the ballistic theory of light in 'Societies and Academies', *Nature*, 118 (1926), p. 432.

24. Staley's use of Poincaré as an example proves only that developments in quantum theory had revised participants' interpretation of classical mechanics. Poincaré's assent to this interpretation does not imply that he adopted a clear dichotomy between 'classical' and 'modern physics' based on the quantization of energy. See Staley (n. 1), pp. 416–417.

25. Staley, *Einstein's Generation* (n. 1), pp. 399–409; Jungnickel and McCormmach, *Intellectual Mastery of Nature* (n. 17), pp. 318–21 (p. 320).

26. See Staley, *Einstein's Generation* (n. 1), pp. 352, 360, 364, 398, 417, and esp. pp. 375, 392.

27. Robert M. Eisberg, *Fundamentals of Modern Physics* (New York: Chichester: Wiley, 1961), p. 1.
28. The notion of a classical field theory, which cuts across electrodynamics and gravitation, introduces further complexity.
29. As their prefaces generally indicate, the same authors also seem to associate this continuity of practice with an epistemological continuity. Their rhetoric cannot be construed as motivated entirely by pedagogical convenience.
30. Eisberg, *Fundamentals of Modern Physics* (n. 27), p. 1; Kenneth Krane, *Modern Physics*, 2nd edn. (New York: Wiley, 1996), pp. 2–3.
31. Arthur Eddington, *The Nature of the Physical World* (Gifford Lectures, Edinburgh University, 1927) (Cambridge: Cambridge University Press, 1928), Chapter I; Matthew Stanley, *Practical Mystic: Religion, Science, and A. S. Eddington* (Chicago: University of Chicago Press, 2007); Steven French, 'Scribbling on the blank sheet: Eddington's structuralist conception of objects', *Studies in History and Philosophy of Modern Physics*, 34 (2003), pp. 227–259. Commenting on the recurrent themes of Eddington's publications, French notes: 'I have argued that one of the things that Eddington saw as irredeemably 'classical' was the individuality or non-indistinguishability of the particles, and this is only exposed and then replaced in 1927, so in a sense it is not until then that the classical regime finally ends'; personal communication to Graeme Gooday. We thank Steven French for pointing out that the project of structural realism in philosophy of physics complements our project by exploring the mathematical continuities in theoretical physics from the nineteenth to the twentieth century.
32. The journalist noted that 'there is a hint of reconciliation of the present antithesis of waves and particles, the antithesis between the classical physics of Clerk Maxwell and the new physics of quanta and radiation'. ('Scientific Correspondent'), 'The Progress Of Science: New ideas in physics', *Times* (19 November 1928), p. 8E.
33. Henry E. Armstrong, 'Good English', *Times* (22 August 1930), p. 11F.
34. Staley, *Einstein's Generation* (n. 1), pp. 396, 417–418.
35. J. M. Keynes claimed that it was Karl Marx who coined the term 'classical economics' to refer to the works of David Ricardo, James Mill. and their precursors, with whom he was critically engaged. Keynes further extended the scope of 'classical economics' to include John Stuart Mill and several of his contemporaries. See John Maynard Keynes, *The General Theory of Employment, Interest and Money* (London: Macmillan & Co., 1936), Chapter 1, footnote 1. Twentieth-century sociologists have likewise categorized the work of Marx, Weber, and Durkheim as 'classical sociology'. For the origins of the term 'classical ethology', see Gregory Radick, 'Essay Review: The Ethologist's World', *Journal of the History of Biology*, 40 (2007), pp. 565–575.
36. M. Abraham & A. Föppl, *Theorie der Elektrizität: Einführung in die Maxwellsche Theorie*, Vol. 1, Max Abraham, *Elektromagnetische Theorie der Strahlung*, Vol. 2 (Leipzig: Teubner, 1904–5). Max Abraham (trans. Richard Becker; authorized trans. John Dougall), *The Classical Theory of Electricity and Magnetism* (London: Blackie & Son, 1932). Another example of contingent reinterpretations of 'modern' physics through translation is a work by Henri Poincaré's cousin Lucien Poincaré, *La Physique Moderne, son evolution* (Paris: E. Flammarion, 1906), to which we refer below. This was translated into English as *The New Physics and Its Evolution* (London: Kegan Paul, 1907).
37. See F. K. Richtmyer, *Introduction to Modern Physics* (New York; London: McGraw Hill, 1928); and review by L. F. B., 'Modern Physics', *Nature*, 123 (1929), pp. 198–199.
38. William Thomson, 'Appendix B. Nineteenth Century Clouds over the Dynamical Theory of Heat and Light', in his *Baltimore Lectures on Molecular Dynamics and The Wave Theory of Light* (London: C. J. Clay and Sons, 1904), pp. 486–527 (pp. 486–92).

39. Bruce Hunt, *The Maxwellians* (Ithaca, NY; London: Cornell University Press, 2005). Hunt treats Heaviside, Larmor, Lodge, and FitzGerald as the primary Maxwellians. Buchwald identifies a larger international constituency of Maxwellians. Buchwald, *From Maxwell to Microphysics* (n. 2), pp. 73–74.
40. The radical implications of the theory of special relativity are emphasized in Percy W. Bridgman, *The Logic of Modern Physics* (London: Macmillan, 1928), pp. 1–25. See also George David Birkhoff and Rudolph Ernest Langer, *Relativity and Modern Physics* (Cambridge, MA: Harvard University Press, 1923).
41. Harry Collins and Trevor Pinch, 'Two experiments that "proved" relativity', in *The Golem: What Everyone Should Know about Science* (Cambridge: Cambridge University Press, 1993), pp. 27–55.
42. Geoffrey Cantor and Jonathan Hodge, 'Introduction: major themes in the development of ether theories from the ancients to 1900', in Geoffrey Cantor and Jonathan Hodge (eds), *Conceptions of Ether: Studies in the History of Ether Theories 1740–1900* (Cambridge: Cambridge University Press, 1981), pp. 1–60 (pp. 53–4). The empirical falsification of the non-existence of an entity is impossible, as Popper pointed out. Even Einstein granted a place to a rehabilitated form of ether in his theory of general relativity as the form of spacetime manifold. Albert Einstein, *Sidelights on Relativity* (London: Methuen, 1922), pp. 3–24. For a more detailed discussion of Einstein's relationship to the ether, see Ludwik Kostro, *Einstein and the Ether* (Montreal: Apeiron, 2000). See also below for Whittaker's ether-based interpretation of the theory of relativity.
43. Daniel M. Siegel, 'Thomson, Maxwell, and the universal ether in Victorian physics', in Cantor and Hodge (eds.), *Conceptions of Ether* (n. 42), pp. 239–268 (pp. 239–240).
44. Bruce Hunt, 'Experimenting on the Ether: Oliver J. Lodge and the Great Whirling Machine', *Historical Studies in the Physical and Biological Sciences*, 16 (1986), pp. 111–134.
45. H. Dingle, 'Edmund T Whittaker, mathematician and historian', *Science*, 124 (1956), pp. 208–209; G.F.J. Temple, 'Edmund Taylor Whittaker', *Biographical Memoirs of Fellows of the Royal Society of London*, 2 (1956), pp. 299–325; Edmund T Whittaker, *A History of the Theories of Aether and Electricity*, 1st edn. (London: Longman, Green and Co, 1910); 2nd edn. revised and enlarged, Vol. 1: *The classical theories*, Vol. 2: *The modern theories 1900–1926* (London: Nelson, 1951–1953). Whittaker's revised edition aroused controversy by adopting the accretionist approach of crediting the development of special relativity to Henri Poincaré and Hendrik Lorentz, rather than to Albert Einstein.
46. To emphasize often overlooked continuities between this period and later forms of physics, Buchwald emphasizes the role of energy as a unifying theme in Maxwellian theorizing. Buchwald, *From Maxwell to Microphysics* (n. 2).
47. Buchwald, *From Maxwell to Microphysics* (n. 2), pp. 20, 73–74.
48. Larmor considered this 'the clearest, most compact, and most general means of expressing any physical problem'. Warwick, *Masters of Theory* (n. 8), p. 367. Warwick goes on to state that 'Larmor was convinced that this form of expression best revealed the formal mathematical connections that existed between the "different departments" of mathematical physics, and so facilitated the analogical solution of a wide range of problems'.
49. This was required to ensure that the electrostatic potential ascribed to an open circuit possessed a unique, 'true' value. Maxwell had shown that this potential varied according to the chosen frame of reference.
50. More specifically, FitzGerald 'replaced the mechanical symbols in MacCullagh's theory with appropriate electromagnetic symbols and [applied] the Principle of Least Action to the resulting Lagrangian'. Warwick, *Masters of Theory* (n. 8), p. 368.
51. Hunt, *The Maxwellians* (n. 39), pp. 209–222, esp. pp. 217–222.

52. Buchwald, *From Maxwell to Microphysics* (n. 2), pp. 127–186. Hunt, *The Maxwellians* (n. 39), pp. 29, 35–36, 209–210; Warwick, *Masters of Theory* (n. 8), pp. 295–296, 368. Hunt points out on pp. 209–210 that the Maxwellians Heaviside and FitzGerald often referred to Maxwell's theory as a theory of *one* medium (the ether), which drew attention simultaneously to both its strengths and weaknesses.
53. According to Howard Stein, this constituted a version of the 'classical theory of electrons'. H. Stein, ' "Subtler forms of matter" in the period following Maxwell', in Cantor and Hodge (eds.), *Conceptions of Ether* (n. 42), pp. 309–340 (p. 323).
54. The inertial mass of matter was produced by the acceleration of electrons with respect to the ether. And in Larmor's own words, the 'material molecule is entirely formed of ether and of nothing else' Warwick, *Masters of Theory* (n. 8), p. 369.
55. Warwick, *Masters of Theory* (n. 8), pp. 363–376.
56. Warwick argued that this 'leant meaning both to the idea of an ultimate reference system and to the application of dynamical concepts to electromagnetic theory'. Warwick, *Masters of Theory* (n. 8), pp. 396–397.
57. Warwick, *Masters of Theory* (n. 8), pp. 376–398. These differences derived from Larmor's route to the ETM via the physical micro-structure of the ether, which his students had to consider only in a more limited sense. Warwick explains that 'from an ontological perspective . . . they needed to know only that the universe consisted of positive and negative electrons in a sea of ether, the application of the ETM then following as a largely mathematical exercise based on Maxwell's equations and Larmor's electrodynamics of moving bodies' (p. 380).
58. 'The ontological status of Cunningham's ether reflected the mathematical practice inherent in his electrodynamics'. Warwick, *Masters of Theory* (n. 8), p. 425.
59. Larmor initially thought that they transformed Maxwell's equations only to the second order of v/c. He later noticed his error but continued his earlier practice of approximating to the second order because, unlike Cunningham, he believed the transformations to be unjustified physically at higher orders. See Warwick, *Masters of Theory* (n. 8), pp. 374, 411–413.
60. Warwick, *Masters of Theory* (n. 8), pp. 402–403, 409–428 (p. 426). See pp. 404–409 for Warwick's analysis of historiographical problems relating to the 'adoption' of the 'theory of relativity' during the early twentieth century. Warwick's explication of Cunningham's ether is confusing. He appears to suggest that Cunningham introduced a 'plurality of ethers' but quotes Cunningham as saying: 'the aether is in fact, not a medium with an objective reality, but a mental image'. See pp. 424–428; and E. Cunningham, 'The Structure of the Ether', *Nature*, 76 (1907), p. 222.
61. Cantor and Hodge, *Conceptions of Ether* (n. 42), p. 33.
62. 'Inaugural Address by the Most Hon. The Marquis [sic] of Salisbury, K.G., D.C.L., F.R.S., Chancellor of the University of Oxford, President of the British Association', printed in *Report of the Sixty-Fourth Meeting of the British Association for the Advancement of Science Held at Oxford in August 1894* (1894), pp. 3–15; and also appearing in *Nature*, 50 (1894), pp. 339–343. 'Unsolved Problems of Science', in *Popular Science Monthly*, 46 (1894), pp. 33–47. Quotations from pp. 39–41. Pace Cantor and Hodge, Salisbury's scepticism about claims concerning the ether were considered sufficiently important by some Maxwellians that they to referred to them long afterwards. See, for example, Oliver Lodge, 'The Ether of Space', *The North American Review*, 187 (1908), pp. 724–736, esp. p. 727.
63. In his BAAS address Salisbury nonetheless admittedly referred to himself as 'no mathematician'. Salisbury's library is part of the University of Liverpool's Special Collections; the three volumes of Maxwell's *Treatise* are unmarked, however. See

Gooday's discussion of Salisbury's electrical ventures, both practical and otherwise, in Graeme Gooday, *Domesticating Electricity: Technology, Uncertainty and Gender, 1880–1914* (London: Pickering & Chatto, 2008), Chapter 3. Robert Cecil (3rd Marquis of Salisbury), 'On Spectral Lines of Low Temperature', *Philosophical Magazine*, 4th series, 45 (1873), 241–245. Salisbury's assistant regularly encouraged him to publish more of his private researches: Herbert McLeod 'The Marquis of Salisbury, K. G. 1830–1903', *Proceedings of the Royal Society*, 75 (1905), pp. 319–325.

64. Warwick, *Masters of Theory* (n. 8), p. 295.
65. Cantor and Hodge (eds.), *Conceptions of Ether* (n. 42); J. Larmor, *Aether and Matter* (Cambridge: Cambridge University Press, 1900).
66. Salisbury, 'Unsolved Problems of Science' (n. 62), p. 40; Warwick, *Masters of Theory* (n. 8), p. 368. Although Salisbury explicitly aired concerns about the ontological inconsistencies associated with the ether, Larmor's account attained significantly greater levels of technical refinement and clarity.
67. By 1903, however, even Ambrose Fleming's account of wireless telegraphy moved from a strictly Maxwellian focus on the ether as the mediator of electromagnetic waves to Larmor's electron based account; that focussed attention on the loops of strain produced by the motion of the electron as the key factor in generating the radiation of electromagnetic waves. See Sungook Hong, *Wireless: From Marconi's Blackbox to the Audion* (Cambridge, MA: The MIT Press, 2001), pp. 193–197. By the 1920s reference to the ether had become largely superfluous to the electronic theory of wireless telegraphy, and thus was little discussed in texts by radio engineers.
68. Maxwellian physicists—notably John Henry Poynting at the University of Birmingham—were reframing electrodynamics in terms of the directional flow of energy, rather than any particular mechanism of the ether. Buchwald, *From Maxwell to Microphysics* (n. 2), pp. 38–53.
69. See discussion of the debate between Preece and Lodge in Oliver Lodge, *Lighting Conductors and Lightning Guards* (London: Whittaker and Co., 1892), p. 118. Hunt describes Preece's account as the 'drainpipe' theory of lightning conductors. See Bruce Hunt, ' 'Practice vs. Theory': The British Electrical Debate, 1888–1891', *Isis*, 74 (1983), pp. 341–355 (p. 346).
70. William Preece, 'Address to BAAS Section G (Mechanical Science)', *B.A.A.S. Report*, Part II (1888), pp. 790–791.
71. Preece, 'Address to BAAS Section G' (n. 70), pp. 790–791.
72. Warwick, *Masters of Theory* (n. 8), pp. 377, 385.
73. Bruce Hunt, *Pursuing Power and Light: Technology and Physics from James Watt to Albert Einstein* (Baltimore: Johns Hopkins University Press, 2010); Robert Fox and Anna Guagnini, *Laboratories, Workshops, and Sites. Concepts and Practices of Research in Industrial Europe, 1800–1914* (Berkeley: Office for History of Science and Technology, University of California, 1999), pp. 35–40.
74. For a discussion of Boris Hessen's attempt to present the full Marxist case, see Gideon Freudenthal and Peter McLaughlin, *The Social and Economic Roots of the Scientific Revolution: Texts by Boris Hessen and Henryk Grossmann* (Dordrecht: Springer, 2009).
75. T. H. Huxley, 'Science', in Thomas Humphry Ward (ed.), *The Reign of Queen Victoria: A Survey of Fifty Years of Progress* Vol. 2 (London: Smith, Elder & Co., 1887), pp. 322–87 (pp. 330–1). This was republished in the USA as *The Advance of Science in the Last Half-Century* (New York: Appleton & Co., 1888). The essay appears as 'The Progress of Science', in T. H. Huxley, *Collected Essays*, Vol. 1 (London: Macmillan, 1893), pp. 42–129.

76. See discussion of Kelvin's legacy in Christine MacLeod, *Heroes of Invention: Technology, Liberalism and British Identity, 1750–1914* (Cambridge: Cambridge University Press 2007), pp. 66–67.
77. See Stathis Arapostathis and Graeme Gooday, *Patently Contestable* (The MIT Press, 2013); and Stathis Arapostathis and Graeme Gooday, 'Electrical technoscience and physics in transition, 1880–1920', *Studies in History and Philosophy of Science*, 44 (2012), pp. 202–11.
78. On scientist-engineers, see Hong, *Wireless* (n. 67). For use of large industrial machinery in experimentation see Bruce Hunt, 'Experimenting on the ether' (n. 44).
79. Bruce J. Hunt, 'Michael Faraday, cable telegraphy and the rise of field theory', *History of Technology*, 13 (1991), pp. 1–19.
80. Heinz Otto Sibum, 'Reworking the Mechanical Value of Heat: instruments of precision and gestures of accuracy in early Victorian England', *Studies in History and Philosophy of Science*, 26 (1995), pp. 73–106. There were other paths in Europe that led to a similar conclusion, albeit for rather different reasons and with different forms of evidential base, not all of which were industrial. See Thomas Kuhn, 'Energy Conservation as an Example of Simultaneous Discovery', in *The Essential Tension: Selected Studies in Scientific Tradition and Change* (Chicago; London: University of Chicago Press, 1977), pp. 66–104.
81. Graeme Gooday, *The Morals of Measurement: Accuracy, Irony and Trust in Late Victorian Electrical Practice* (New York; Cambridge: Cambridge University Press, 2004); Bruce Hunt, 'The Ohm is Where the Art is: British Telegraph Engineers and the Development of Electrical Standards', *Osiris*, 9 (1994), pp. 48–63; Simon Schaffer, 'Late Victorian Metrology and its instrumentation: a manufactory of ohms', in R. Bud and S. Cozzens (eds.), *Invisible Connections: Instruments, Institutions and Science* (Bellingham, WA: SPIE Optical Engineering Press, 1992), pp. 24–55.
82. Huxley, 'Science' (n. 75), pp. 330–331.
83. Marie Curie, 'Rayons Émis per les Composés de l'uranium et du Thorium' (1898), in Irène Joliot-Curie (ed.), *Oeuvres de Marie Sklodowska Curie* (Warsaw: Państwowe Wydawnictwo Naukowe, 1954), pp. 43–45.
84. For a clear, concise account, see ⟨http://www.ieeeghn.org/wiki/index.php/Edison_Effect⟩
85. Graeme Gooday, 'The Questionable Matter of Electricity: The reception of J.J. Thomson's 'corpuscle' among electrical theorists and technologists', in Jed Z. Buchwald & Andrew Warwick, *Histories of The Electron: The Birth of Microphysics* (Cambridge, MA; London: The MIT Press, 2001), pp. 101–134. See Owen Richardson's Nobel prize lecture, 'Thermionic phenomena and the laws which govern them', reproduced in *Nobel Lectures: Physics 1922–1941* (Amsterdam: Elsevier Publishing Company, 1965). For Schottky's career, see Reinhard W. Serchinger, *Walter Schottky, Atomtheoretiker und Elektrotechniker: Sein Leben und Werk bis ins Jahr 1941.* (Diepholz: GNT Verlag, 2008).
86. Crosbie Smith and Norton Wise, *Energy and Empire: A biography of Lord Kelvin* (Cambridge: Cambridge University Press, 1989).
87. Paul Nahin, *Oliver Heaviside: The Life, Work, and Times of an Electrical Genius of the Victorian Age*, 2nd edn. (Baltimore: Johns Hopkins University Press, 2002); and Arapostathis & Gooday, *Patently Contestable* (n. 77).
88. In his patriotic attack on German science in 1915, Duhem criticized those practitioners who, like German physicists, treated Maxwell's equations as 'orders' whilst using permanent magnets in their experiments. They thereby invoked 'a doctrine whose axioms made the existence of such [magnetic] bodies absurd'. See the reproduction of Duhem's essay 'Some Reflections on German Science' in Roger Ariew & Peter Barker (eds.), *Pierre Duhem: Essays in the History and Philosophy of Science* (Indianapolis: Hackett Pub. Co., 1996), pp. 251–76, quotations from pp. 268, 270.

89. See discussion in James Swinburne, *Practical Electrical Measurement* (London: Alabaster & Gatehouse, 1888), pp. 25–56; and Gooday, *The Morals of Measurement* (n. 81), p. 35.
90. Matthias Dörries, 'Prior History and After-effects: hysteresis and nachwirkung in 19th-century physics', *Historical Studies in the Physical and Biological Sciences*, 22 (1991), pp. 25–55. J.A. Ewing, 'On effects of retentiveness in the magnetization or iron and steel', *Proceedings of the Royal Society*, 34 (1882), pp. 39–45; 'Experimental Researches in Magnetism', *Philosophical Transactions of the Royal Society*, 176 (1885), pp. 523–640. Graeme Gooday, 'Domesticating the Magnet: Secularity, Secrecy and 'Permanency' as Epistemic Boundaries in Marie Curie's Early Work', *Spontaneous Generations: A Journal for the History and Philosophy of Science*, 3 (2009). ⟨http://spontaneousgenerations.library.utoronto.ca/index.php/SpontaneousGenerations/article/view/10615⟩ The term 'hysteresis' is used to characterize systems whose present properties depend in a path-dependent fashion on preceding states. So in this case the distribution of magnetism at any given time depended on the historical sequence of distributions, in other words on previous magnetic influences.
91. John Hopkinson, 'Magnetisation of Iron', *Philosophical Transactions of the Royal Society*, Part II (1885), pp. 455–469; reproduced in Bertram Hopkinson (ed.), *Original Papers by the Late John Hopkinson, Vol. 2: Scientific papers* (Cambridge: Cambridge University Press, 1901), pp. 154–177. On the status of Langevin's theoretical research into magnetism as either 'classical' or 'quantum-mechanical', see L. Navarro and J. Olivella, 'On the nature of the hypotheses in Langevin's magnetism', *Archives Internationales d'Histoire des Sciences*, 47 (1997), pp. 316–45.
92. Peter Galison, *Einstein's Clocks and Poincaré's Maps* (London: Sceptre, 2003), p. 31.
93. Cahan, *An Institute for An Empire* (n. 16), pp. 145–157. Traditional accounts of the Planck black-body story do not mention the PTR or the electrical industry. See for example, Martin Klein, 'Max Planck and the Beginnings of Quantum Theory', *Archive for History of Exact Sciences*, 1 (1962), pp. 459–479; and Kuhn, *Black-Body Theory and the Quantum Discontinuity* (n. 12). In the context of understanding the roots of quantum physics in an industrial setting, one may note Staley's discussion, in *Einstein's Generation* (n. 1), of Solvay's investment in the 1911 conference. See Gooday, *The Morals of Measurement* (n. 81), Chapter 6, for a study of how quantization in energy also arose in the methods used by the electrical supply industry for billing its customers.
94. Elizabeth Garber, *The Language of Physics: The Calculus and Development of Theoretical Physics in Europe 1750–1914* (Boston: Birkhauser, 1999), p. 312.
95. Robert Fox and George Weisz (eds.), 'The institutional basis of French science in the nineteenth century', in *The Organisation of Science and Technology in France 1808–1914* (Cambridge: Cambridge University Press, 1980), pp. 1–28 (p. 26); Mary Jo Nye, 'Scientific Decline: is quantitative evaluation enough?', *Isis*, 75 (1984), pp. 697–708 (pp. 705–708); Dominique Pestre, 'Sur la science en France 1860–1940: à propos de deux ouvrages récents de Mary Jo Nye et Harry W. Paul', *Révue d'histoire des sciences*, 41 (1988), pp. 75–83 (pp. 81–82).
96. Daniel Jon Mitchell, 'Gabriel Lippmann's Approach to French Experimental Physics' (University of Oxford: Unpublished D.Phil thesis, 2010), p. 3, also cf. pp. 234–235. The decline thesis still persists among some historians, however. See Ivan Grattan-Guiness (ed.), 'France', in *Companion Encyclopeadia of the History and Philosophy of the Mathematical Sciences*, Vol. 2 (London: Routledge, 1994), pp. 1430–1442.
97. Matthias Dörries, 'Vicious circles, or the pitfalls of experimental virtuosity', in Michael Heidelberger and Friedrich Steinle (eds.), *Experimental Essays–Versuche zum Experiment* (Baden-Baden: Nomos, 1998); Charlotte Bigg, 'Behind the Lines. Spectroscopic

enterprises in early twentieth century Europe' (University of Cambridge: Unpublished PhD thesis, 2001), pp. 17–18.
98. More specifically to the historiography of French physics, by eliminating the notion of 'modern physics' from our analysis, we invite a re-evaluation of the 'decline' of French physics during the nineteenth century. Whilst French physicists may have marginalized themselves from major theoretical developments (with some notable exceptions), the importance of their practical and experimental contributions to physics remains open to enquiry.
99. We admit that these reasons need further elaboration and defence (and even that the notion of 'approach' remains unsatisfactorily vague), but we believe that 'shared approaches' are no less fit for purpose than 'national style'.
100. Dominique Pestre, *Physique et Physiciens en France 1918–1940* (Paris: Éditions des Archives Contemporaines, 1984), pp. 104–108.
101. Jed Buchwald has raised the question of how optics fits into this picture, especially given that the French maintained strength in both its mathematical and experimental aspects, partly through a shared institutional basis with astronomy. Further studies are required to determine the extent to which the Fresnelian mathematical tradition may have set an agenda for experimental work in optics during this period, and whether this made a difference to its character. For now, we simply note that the transverse wave nature of light was generally regarded as a 'hypothesis' in France, which explains the excitement generated by Otto Wiener's purported photographic proof in 1891.
102. Pierre Bertin, 'Léçon d'ouverture: l'électricité; sa nature', *Revue des Cours Scientifiques de la France et de l'Etranger*, 4 (1866), pp. 41–45.
103. Jean-Charles d'Almeida and Augustin Boutan, *Cours Elementaire de Physique*, 2nd edn. (Paris: Dunod, 1863), p. 2.
104. Josep Simon, *Communicating Physics: The Production, Circulation and Appropriation of Ganot's Textbooks in France and England, 1851–1887* (London: Pickering and Chatto, 2011) pp. 109–33, quotation on p. 132.
105. Mitchell, 'Gabriel Lippmann's Approach' (n. 96), p. 19; Simon, pp. 117–21, 129–30. We thank Jed Buchwald for raising Mascart and Joubert's 1882 *Leçons sur l'électricité et le magnétisme* as an exception. Yet whilst the examples cited in the main text were amongst the most highly regarded textbooks of physics in France, Mascart and Joubert's *Leçons* attracted strong criticism. In a long, savage review for the *La Lumière Electrique*, E. Mercadier pronounced the separation into two volumes of 'pure theory' and 'the examination of phenomena and methods of measurement' as 'dangerous'. In terms consistent with the orthodox French approach to experimental physics, he explained that 'as a result one ends up forgetting that physics, and even mechanics ... are experimental sciences, that is to say that the depend on principles that are in no way a priori axioms, but facts of experience that one generalizes through [subsequent] induction'. E. Mercadier, 'Leçons sur l'électricité et le magnétisme, par MM. Mascart et Joubert', *La Lumière Electrique*, 7:51 (1882), pp. 619–622 (p. 620).
106. Lucien Poincaré, *The New Physics and Its Evolution* (n. 36).
107. Some of the terms in this exposition of Lippmann's work are anachronistic in order to make the explanation more readily intelligible to the general reader.
108. Daniel Berthelot, 'Gabriel Lippmann: la vie d'un savant', *Revue des Deux Mondes*, 10 (1922), pp. 19–46 (p. 24); Edmond Bouty, 'Gabriel Lippmann', *Annales de Chimie et de Physique*, 16 (1921), pp. 156–164 (pp. 156, 158).
109. Gabriel Lippmann, 'Relations entre les phénomènes électriques et capillaries', *Annales de Chimie et de Physique*, 5 (1875), pp. 494–549 (p. 495).

110. Lippmann, 'Relations entre les phénomènes électriques et capillaries' (n. 109), p. 547.
111. Gabriel Lippmann, 'G. Quincke: Ueber elektrische Ströme bei ungleichzeitigern Eintauchen zweier Quecksilberelektroden in verschiedene Flüssigkeiten', *Journal de Physique Théorique et Appliquée*, 4 (1875), pp. 248–251.
112. Michel Atten and Dominique Pestre, *Heinrich Hertz. L'Administration de la Preuve* (Paris: Presses Universitaires de France, 2002), p. 81. Their apparatus was actually one that Hertz took up following its recommendation by Maxwellian theoreticians and its use by Thomson. See pp. 80–81.
113. Atten and Pestre, *Heinrich Hertz* (n. 112), p. 80. We agree with Pestre's subtle analysis, pp. 83–84, of the approach taken by de la Rive and Sarasin and their minimal invocation of theory.
114. Atten and Pestre, *Heinrich Hertz* (n. 112), pp. 79–83; Édouard Sarasin and Lucien de la Rive, 'Résonance multiple des ondulations électriques de M. Hertz', *Comptes Rendus Hebdomadaires de l'Académie des Sciences*, 110 (1890), pp. 72–75. Nonetheless, given the theoretical uncertainty surrounding the production of novel experimental phenomena, we would not be surprised if these 'French' characteristics *appear* more widespread in such cases than they actually are, despite the example of FitzGerald and Trouton (see below). Jed Buchwald has indicated similarities with other German replications of Hertz's experiments, for example. Jed Buchwald, *private communication*.
115. Alfred Cornu, 'Remarques à propos de la communication de MM. Sarasin et de la Rive', *Comptes Rendus Hebdomadaires de l'Académie des Sciences*, 110 (1890), pp. 75–76; quoted in Atten and Pestre, *Heinrich Hertz* (n. 112), p. 87. Atten and Pestre, pp. 30–35, 85–90, give one other example of French doubts, due to the Director of the Bureau of Weights and Mesures, Charles-Édouard Guillaume. He cautioned that the enthusiasm of the moment had resulted in the suppression of practical difficulties associated with the replication of Hertz's experiment, and gave examples from the Parisian replication of Joubert and de Nerville. In his opinion, new experiments were required to determine the validity of particular techniques and results. Nonetheless, the French reception of de la Rive and Sarasin's experiments was not entirely uniform. The professor of physics at the École Normale, Marcel Brillouin, downplayed their significance.
116. Atten and Pestre, *Heinrich Hertz* (n. 112), pp. 43–49, 90–91.
117. Benoit Lelong, 'Ions, electrometers, and physical constants: Paul Langevin's work on gas discharges 1896–1903', *Historical Studies in the Physical Sciences*, 36 (2005), pp. 93–130 (p. 96).
118. Lelong, 'Ions, electrometers, and physical constants' (n. 117), pp. 94–97; Benoit Lelong, 'Paul Villard, J. J. Thomson, and the Composition of Cathode Rays', in Jed Z. Buchwald and Andrew Warwick (eds.), *Histories of the Electron* (n. 85), pp. 135–167, esp. pp. 136–149 (p. 142).
119. Lelong, 'Ions, electrometers, and physical constants' (n. 117), pp. 98–100 (p. 99).
120. Lelong, 'Ions, electrometers, and physical constants' (n. 117), pp. 110–113; Olivier Darrigol, *Electrodynamics from Ampère to Einstein* (n. 2), pp. 300–310, esp. pp. 305–306.
121. See Pestre, *Physique et Physiciens en France 1918–1940* (n. 100), pp. 16–22, 31–65, 104–126, quotation at p. 45.
122. See Mary Jo Nye, *Molecular Reality* (New York: Elsevier, 1972), pp. 145–146, 152–153, 156–157, 161, 165–167 for the French reception of Perrin's experimental proofs of molecular reality, and pp. 52, 58–64 for the roots of Perrin's favourable ontological stance towards molecular 'hypotheses', in particular the influence of his teacher at the École Normale, Marcel Brillouin.

123. Lelong has described the changing approach of the Villard's investigations into the properties of cathode rays. During the late 1890s, Villard's publications tended to enumerate these properties and he 'preferred the production and description of experimental facts' to theorizing about the nature of the rays. After 1906, Villard increasingly articulated theories and made experimental predictions before presenting his experimental results, framed in the language of ions and corpuscles, in support of these theories. Benoit Lelong, 'Paul Villard, J. J. Thomson, and the Composition of Cathode Rays', in Jed Z. Buchwald and Andrew Warwick (eds.), *Histories of the Electron* (n. 85), pp. 135–167, esp. pp. 141–142, 154–156, quotation on p. 142.
124. Buchwald and Hong, 'Physics' (n. 9), p. 164; Poincaré, *The New Physics* (n. 36), pp. ix–xi.
125. Shaul Katzir, 'The discovery of the piezoelectric effect', *Archive for History of Exact Sciences* 57 (2003), pp. 61–91 (pp. 63–71, 75–8), quotations at pp. 61, 66; 'From explanation to description: molecular and phenomenological theories of piezoelectricity', *Historical Studies in the Physical Sciences* 34 (2003), pp. 69–94 (pp. 71–5), quotations at p. 72. Katzir's account leaves open the question of how Curie brothers' perceived the epistemological status of their micro-molecular theory in comparison to the empirical laws. But he offers a clue by explaining that Jacques Curie and Charles Friedel attempted to reinterpret some contradictory empirical findings, due to the German physicist Wilhelm Hankel, in accordance with the Curie brothers' molecular theory.
126. See Marjorie Malley, 'The discovery of atomic transmutation: scientific styles and philosophies in France and Britain', *Isis*, 70 (1979), pp. 213–223, esp. pp. 217–218 (p. 217). We prefer to attribute Curie's metaphysical caution to an experimental approach shared by French experimental physicists (as Malley hints in her conclusion) rather than a 'positivist' philosophy of science. We wonder to what extent Pierre Curie's apparent change in attitude towards hypotheses was influenced by the replacement of molecular theories by phenomenological ones in piezoelectricity. See Shaul Katzir, 'From explanation to description: molecular and phenomenological theories of piezoelectricity', *Historical Studies in the Physical Sciences*, 34 (2003), pp. 69–94 (p. 93). Katzir concludes 'since both molecular and phenomenological realms had potential benefits, advancing simultaneously in both was a reasonable strategy as long as one recognized the hypothetical character of the molecular against the firmer ground of the phenomenological'.
127. Mitchell, 'Gabriel Lippmann's Approach' (n. 96), pp. 33–37, 78; and 'Measurement in French experimental physics from Regnault to Lippmann', *Annals of Science*, 69:4 (2012), pp. 453–82; Shaul Katzir, 'Piezoelectricity and comparative styles', *Studies in History and Philosophy of Modern Physics*, 34 (2003), pp. 579–606; and *The Beginnings of Piezoelectricity. A study in mundane physics* (Dordrecht: Springer, 2006). On Regnault and 'direct' measurement, see Jean-Baptiste Dumas, 'Victor Regnault', in *Discours et Eloges Académiques* (Paris: Gauthier-Villars, 1885), pp. 174–175; Hasok Chang, *Inventing Temperature. Measurement and Scientific Progress* (Oxford: Oxford University Press, 2004), p. 76.
128. Lelong, 'Langevin's work on gas discharges', pp. 123–124, 127–130 (p. 130); Mary Jo Nye, *Molecular Reality* (New York: Elsevier, 1972), p. 170.
129. A useful partial analogy to our case is the common anachronistic use of the word 'science' (vis-à-vis 'natural philosophy') to describe the study of nature during the sixteenth and seventeenth centuries. See Peter Dear, 'What Is the History of Science the History Of? Early Modern Roots of the Ideology of Modern Science', *Isis*, 96 (2005), pp. 390–406.
130. For an example of how other fields were starting to formulate a rejection of a 'classical' canon, see Karl Pearson's 1916 account of the mathematization of evolutionary theory in which he complained of the recurrent 'evil of implicit reliance on a classical theory' in the field of mathematical statistics. Karl Pearson, 'Mathematical Contributions to the Theory

of Evolution. XIX. Second Supplement to a Memoir on Skew Variation', *Philosophical Transactions of the Royal Society of London*, Series A, 216 (1916), pp. 429–457.
131. Thomas Kuhn, *The Structure of Scientific Revolutions*, 3rd ed. (Chicago: University of Chicago Press, 1996), Chapter 11, p. 136 ff. Our thanks to Adrian Wilson for informed discussion on this point. Staley points out that Kuhn's views on this topic were considerably refined in his later volume on black-body radiation, which describes how Planck's views on quantum theory evolved slowly over a long period. Staley, *Einstein's Generation* (n. 1), pp. 349, 375.

CHAPTER 25

THE EMERGENCE OF STATISTICAL MECHANICS

OLIVIER DARRIGOL
AND JÜRGEN RENN[1]

25.1 Mechanical Models of Thermal Phenomena

25.1.1 Heat as a Challenge to Mechanics

Statistical mechanics is the name given by the American physicist Josiah Willard Gibbs to the study of the statistical properties of a large number of copies of the same mechanical system, with varying initial conditions. In this chapter we will outline the history of statistical mechanics in a broad sense, and include any attempt to explain the thermodynamic properties of macroscopic bodies as statistical regularities of systems that encompass a very large number of similar constituents.

Statistical mechanics emerged in the second half of the nineteenth century as a consequence of efforts to account for thermal phenomena on the basis of mechanics, which was then considered to be the most fundamental of the physical sciences. Although phenomena such as light, electricity, magnetism, and heat were apparently not of a mechanical nature, scientists tried to explain them by invisible mechanical entities such as the ether or small particles. Mechanical models of thermal phenomena are part of intuitive physics and have been used since antiquity. For example, the communication of heat from one body to another can be made plausible as the motion of invisible particles, or as the flow of an invisible fluid, or else as the effect of a wave. Such mechanical explanations of heat can also help to understand its links with visible mechanical effects, for instance, the thermal expansion of bodies, the pressure exerted by gases, or the possibility to generate heat by motion. Which of these mechanical models, at a given historical moment, appeared to be most suitable as an explanation of heat depended on various factors including the available knowledge of thermal phenomena, the place of the model in the overall architecture of physics, and its state of elaboration. In the following, we briefly review some of these models and then show how several circumstances, among them the establishment of thermodynamics in the mid-nineteenth century, led to a focus on the model of heat as a motion of particles, which lies at the origin of statistical mechanics.

25.1.2 Heat as a Fluid

The late eighteenth century saw the development of the notion of heat as a fluid—the so-called 'caloric'—in parallel to contemporary ideas on electricity. Heat and electricity share properties that suggest their understanding as 'imponderable fluids'. Both lack a definite shape. As was demonstrated by calorimetric experiments, heat can be stored, just as electricity, in appropriate 'containers' and 'flow' from one body to another, both tend to spread out and fill their containers as much as possible, and both are imponderable; that is, they have no appreciable weight. The fluid model also accounted for the role played by heat in chemical reactions; the caloric there acted in a way similar to other substances, being either bound or set free. At the turn of the century the fluid model lent itself to Laplace's reduction of physics to the play of central forces acting among the particles of ponderable and imponderable substances. Later, in Sadi Carnot's reflections of 1824, it became the basis for discussing the theoretical limits of thermal machines such as steam engines (Brush, 1976; Fox, 1971, chap. 1, 9).

25.1.3 Heat as a Motion

In the first half of the nineteenth century, several developments contributed to the gradual demise of the fluid model and to the widespread acceptance of the notion of heat as a motion. First, in the 1820s the success of the undulatory theory of light affected the understanding of heat, because of the close relationship between light and heat. Second, in the 1830s Italian physicists strengthened the analogies between heat and light, for instance, by revealing the possibility of heat reflection and refraction. Third, in the 1840s James Joule performed careful experiments showing the convertibility of mechanical work into heat. By 1850 the model of heat as a motion was associated with the principle of energy conservation, whose most influential proponents were William Thomson and Hermann Helmholtz (Brush, 1976, chap. 1–3).

Ideas about what kind of motion actually constituted heat varied considerably. Originally, the most common conception was that of a vibratory motion of molecules transmitted by the ether, following André Marie Ampère's suggestion. While this conception accounted both for radiant heat and for heat conduction in matter, it was less relevant to the behaviour of gases. At mid-century major British physicists, among them William Rankine, William Thomson, and James Clerk Maxwell, imagined a vortex motion of gas molecules, in rough analogy with Newton's old explanation of gas pressure in terms of a repulsion between contiguous molecules (at rest). In this picture, heat corresponded to the rotation of the molecules, and elasticity to their centrifugal force.

25.1.4 The Kinetic Theory

Around that time, James Joule, Rudolph Clausius, and others elaborated another conception of heat, as translational molecular motion. The molecules of a gas are assumed to occupy only a small fraction of its volume and to have a rectilinear, uniform motion, occasionally interrupted by mutual collisions or by collisions with the walls of the container. While this model had roots in ancient atomism, its application to gases dates from the eighteenth century. In 1738 Daniel Bernoulli first gave the corresponding explanation of gas pressure, John Herapath rediscovered it in 1820, and John Waterston did so again in the 1840s. The pressure of a gas is roughly assumed to be proportional to the number of collisions of gas particles with the wall and to their momentum. The collision number is itself proportional to the density of the gas and to the velocity of the molecules. Hence, the pressure is proportional to the density of the gas, in conformity with Boyle's law. It is also proportional to the squared velocity of a molecule, in conformity with Gay-Lussac's law if temperature is assumed to depend linearly on squared velocity. Initially, this model had a limited range of applicability. But in contrast to the fluid model, it accounted for the conversion between heat and mechanical work, and in contrast to the vibrational and rotational models, it allowed a quantitative description of the behaviour of gases (Brush, 1976, chap. 1–3; Brush, 1983, sec. 1.5–1.8).

25.2 GAS THEORY AS A BRIDGE BETWEEN MECHANICS AND THERMODYNAMICS

25.2.1 Thermodynamics as a Challenge to Mechanics

The formulation of thermodynamics in the 1840s and 1850s by Rankine, Thomson, and Rudolf Clausius led to a theory of heat which no longer required a mechanical model. The first and second law of this theory are enunciated purely in terms of heat-converting engines, without mention of a caloric substance or of a thermal motion. Nevertheless, the heat-as-motion model acquired a new significance in the context of thermodynamics. Whereas it had been confined to the modelling of the thermal behaviour of gases, it now served a bridge function between two fundamental columns of physics. It made plausible the conversion processes between mechanical and thermal energy in thermodynamic engines and thus offered a mechanical underpinning for the new thermodynamics. The support gained in this way had repercussions for mechanics itself, then still widely conceived as providing a conceptual framework for all of physics. The kinetic theory of gases translated the advances of thermodynamics into challenges for mechanics that eventually led to the creation of statistical mechanics.

In particular, the study of the kinetic theory allowed scientists to confront one of the principal structural differences that distinguish thermodynamics from mechanics: the irreversibility of its laws under time reversal. When two portions of liquid of different temperatures are mixed, for example, the mixture will attain an intermediate temperature. This process is not reversible, that is, the mixture will never spontaneously separate into a cooler and a hotter component. In contrast, the time reversal of any (purely) mechanical process leads to another possible process. This conflict became one of the challenges in the development of the kinetic theory, along with an improved description of the thermal properties of specific physical systems.

25.2.2 Gases as Particles in Motion

The physicist who effectively revived the kinetic theory of gases was Rudolph Clausius, one of the founders of thermodynamics. In 1857, Clausius gave Bernoulli's relation the exact form

$$PV = \frac{1}{3}Nmu^2,$$

where P is the pressure, N the number of molecules, m their mass, and u^2 the mean value of their squared velocity. From this formula he concluded that under Avogadro's hypothesis (according to which the number of molecules in a unit volume is the same for all gases under normal conditions) the average translational energy of a molecule was the same in any gas at a given temperature. Further, he determined the internal energy of the gas as the total translational energy of its molecules. The resultant specific heat at constant volume is $(3/2)R$, where R is the constant of perfect gases. As this value was well below the observed ones, Clausius concluded that the molecules had a rotational and a vibrational motion besides their translational motion. Mutual collisions acted to maintain constant proportions for the average energies of the three kinds of motion (Clausius, 1857).

Although Clausius was aware of the spread of the velocities of the molecules due to collisions, his computations assumed, for simplicity, the same velocity for all molecules. In the approach of his first memoir, he equated the properties of the system as a whole with individual properties of the particles. For instance, he identified the temperature of the gas with the kinetic energy of its molecules. Statistical considerations, such as the formation of average values, only played a minor role in his arguments. Clausius simply assumed that the molecules occupied a negligible fraction of the volume of the gas and that they moved in straight lines with the same velocity until they hit a wall, without treating their mutual encounters in any detail and without taking molecular forces into account. However, the direct identification of properties of the system with properties of its constituents soon brought the model into conflict with empirical properties of gases (Brush, 1976, chap. 4).

25.2.3 The Transport Properties of Gases

The issue that triggered the development of the notion of gases as statistical collections was the understanding of transport properties. One of Clausius's readers, the Dutch meteorologist C. H. D. Buys Ballot, objected that the kinetic theory implied a much faster diffusion of gases than observed, because the molecular velocities computed by Clausius were comparable to the speed of sound in normal conditions. In response, Clausius introduced what amounts to the first non-trivial statistical concept of the kinetic theory: the 'mean free path'. This concept neither describes the property of an individual particle nor the property of a gas considered macroscopically. It makes sense only if the gas is conceived as a statistical collection.

The mean free path is defined as the average length travelled by a molecule before it collides with another. The proportion of the molecules travelling in the direction Ox that hit another molecule in a slice dx of the gas, Clausius reasoned, is given by the fraction $\pi r^2 n dx$ of the surface of this slice occulted by the 'spheres of action' of the molecules, where r is the radius of action, and n the number of molecules in a unit volume. Consequently, the probability that the motion of a molecule remains free over the distance x is

$$W = e^{-\alpha x}, \text{ with } \alpha = \pi r^2 n,$$

and the probability that this molecule experiences its first collision between x and $x + dx$ is

$$W(x) - W(x + dx) = \alpha e^{-\alpha x} dx.$$

The mean free path l is the average $1/\alpha$ of the distance x under the latter probability law. Clausius judged this length to be so small that Buys Ballot's over-rapid diffusion never occurred (Clausius, 1858).

Starting from Clausius's work, James Clerk Maxwell developed the statistical aspects of the kinetic theory of gases much further. In particular, Maxwell took into account the velocity spread generated by the mutual encounters of gas molecules. In 1860 he determined the equilibrium distribution $f(\mathbf{v})$ of the molecular velocities by assuming its isotropy as well as the statistical independence of the three Cartesian components of velocity. These two conditions lead to the functional equation

$$f(\mathbf{v}) = \varphi(v^2) = \psi(v_x)\psi(v_y)\psi(v_z),$$

whose solution has the form

$$f = \alpha e^{-\beta v^2},$$

wherein α and β are two constants. Maxwell later judged this argument 'precarious', because the second assumption remained to be justified. Yet he never doubted the velocity distribution which now bears his name. As we will see, he and Boltzmann later gave better justifications (Brush, 1976; Maxwell, 1860, chap. 4, 5; Everitt, 1975).

The main purpose of Maxwell's memoir of 1860 was to develop new physical consequences of the mean-free-path concept by using a model of the gas as a large number of perfectly elastic spheres. He computed gas interdiffusion, and also internal friction (viscosity) and heat conduction regarded as diffusions of momentum and kinetic energy respectively. The molecules moving from a given layer of the gas, he reasoned, carry their mass, momentum, and kinetic energy over a distance of the order of the mean free path l and thus communicate it through collisions to another layer of the gas. The order of magnitude of the net flux of these quantities in the direction Ox is $lu\partial(nq)/\partial x$, where q stands for m, mu_y, and $(1/2)mu^2$ respectively. The corresponding coefficients of diffusion, viscosity, and heat conduction are lu, $nmlu$, and $(P/T)lu$ up to numerical coefficients which Maxwell obtained by performing the implied free-path averages. George Stokes' values for the viscosity of air and Thomas Graham's for the interdiffusion coefficient of two gases in normal conditions yielded compatible estimates of the mean free path, about 10^{-7} m (the thermal conductivity was still unknown). This convergence naturally pleaded in favour of Maxwell's approach.

Maxwell's understanding of viscosity had a less welcome implication. Since the mean free path is inversely proportional to the density, it follows that internal friction does not depend on the density of the gas, a counter-intuitive result which 'startled' Maxwell. After deriving this and other unwanted consequences, he briefly doubted the overall validity of the kinetic theory. In 1866, however, he careful measured the viscosity of gases with the help of his wife, and thereby confirmed the surprising independence of viscosity from density. This spectacular finding lent much credibility to the kinetic theory in general. At the same time, Maxwell's experiments contradicted the more specific hard-sphere model. The measured viscosity turned out to be proportional to the absolute temperature, whereas the model yielded proportionality to the square root of temperature (Maxwell, 1867).

In reaction to this difficulty, Maxwell switched to a different model, maintaining Clausius's general assumptions but replacing the hard spheres of 1860 with repulsive centres of force. He also gave up the mean-free-path method and turned to a more powerful approach to transport phenomena based on computing the number of collisions of various kinds of molecules and the resulting variations of average molecular properties (see the following subsection). The integration of the relevant equations required the knowledge of the perturbed velocity distribution, except when the collision force varied as the inverse of the fifth power of the distance. Maxwell determined the transport of mass, momentum, and kinetic energy in the gas in this special case. He thus retrieved the Navier–Stokes equation for the motion of viscous gases, and obtained quantitative relations between viscosity, diffusion rate, and thermal conductivity. Viscosity turned out to be proportional to temperature, in agreement with his experiments. To his pleasure, nature seemed to have chosen the mathematically favourable case of the $1/r^5$ force law. Maxwell knew, however, that this law could apply neither to large intermolecular distances for which the force is attractive, nor to small distances for which the structure of molecules comes into play (Maxwell, 1867).

A few years later, Oscar Meyer's and Joseph Stefan's accurate gas-viscosity measurements yielded a temperature-dependence of viscosity at variance both with the $1/r^5$ and with the hard-sphere model. In the 1880s Maxwell's followers elaborated in vain on his transport coefficients. As we may retrospectively judge, this problem eluded their mathematical techniques and physical models.

25.2.4 The Maxwell–Boltzmann Law

A crucial element in the understanding of gases as statistical collections of particles was the equilibrium distribution of velocities. Maxwell and his most outstanding follower Ludwig Boltzmann therefore attempted to establish this distribution as firmly as possible and to generalize it to degrees of freedom other than translation. In 1866, Maxwell provided a new demonstration based on the aforementioned collisions-number approach. In order to determine this number, he considered the trajectory of one of the colliding molecules in the reference system of the other. Call b the distance (impact parameter) between the first asymptote of this trajectory and the second molecule, ϕ the azimuth of the plane of this trajectory, $\mathbf{v}_1$ and $\mathbf{v}_2$ the initial velocities of the two molecules. An encounter for which the impact parameter lies between b and $b + db$ and the azimuth lies between and $\phi + d\phi$ occurs within the time δt if and only if the first molecule belongs to the cylindrical volume $|\mathbf{v}_1 - \mathbf{v}_2| \delta t b db d\phi$. To the number of collisions of this 'kind' per unit time and in a unit volume of the gas, Maxwell gave the natural expression

$$dv = |\mathbf{v}_1 - \mathbf{v}_2| \, b db d\phi f(\mathbf{v}_1) d^3 v_1 f(\mathbf{v}_2) d^3 v_2,$$

where $f(\mathbf{v})d^3 v$ is the number of molecules per unit volume in the velocity range $d^3 v$ around $\mathbf{v}$ (Maxwell, 1867).

A sufficient condition for the velocity distribution to be stationary, Maxwell reasoned, is the equality of the collision number dv with the number

$$dv' = |\mathbf{v}_2' - \mathbf{v}_2'| \, b db d\phi f(\mathbf{v}_1') d^3 v_1' f(\mathbf{v}_2') d^3 v_2'$$

of the same kind (b, φ) for which the final velocities are $\mathbf{v}_1'$ within $d^3 v_1'$ and $\mathbf{v}_2'$ within $d^3 v_2'$. Owing to the conservation of energy and momentum, for a given kind of collision the initial and final velocities are in a one-to-one correspondence that leaves $|\mathbf{v}_1 - \mathbf{v}_2|$ and $d^3 v_1 d^3 v_2$ invariant. Therefore, the equality of dv and dv' requires that

$$f(\mathbf{v}_1)f(\mathbf{v}_2) = f(\mathbf{v}_1')f(\mathbf{v}_2')$$

for any two velocity pairs such that $v_1^2 + v_2^2 = v_1'^2 + v_2'^2$ and $\mathbf{v}_1 + \mathbf{v}_2 = \mathbf{v}_1' + \mathbf{v}_2'$. The only isotropic solution of this functional equation is Maxwell's law. It remains to be shown that the equality $dv = dv'$ is a necessary condition of stationarity. Maxwell's proof of this point was impenetrably concise, even for his most perspicacious readers.

In a Viennese context favourable to atomistics and British methods, Ludwig Boltzmann was naturally fascinated by Maxwell's memoir of 1867 on the dynamical theory of gases. In 1868 and 1871 he generalized Maxwell's collision-number approach to molecules that had internal degrees of freedom and responded to external forces such as gravitation. He thus arrived at the form $\alpha e^{-\beta E}$ of the stationary distribution for the molecular variables on which the energy E of a molecule depends. This law, now called the Maxwell–Boltzmann distribution, has been an essential element of statistical physics to this day. We will return to the means of Boltzmann's generalization (Boltzmann, 1868; 1871a).

25.2.5 The Problem of Specific Heats

The kinetic theory of gases makes it possible to calculate the specific heats of a gas from the energy distribution over the various degrees of freedom. Maxwell and Boltzmann's solution to this problem is the so-called equipartition theorem. In its most general form, this theorem states that to each quadratic term in the expression for the energy of a molecule corresponds an average energy $(1/2)kT$, where k is Boltzmann's constant. In particular, the equipartition implies Clausius's result of 1857 that for a monoatomic gas composed of N point-like particles, the total energy should be $(3/2)NkT$ and the specific heat at constant volume $(3/2)Nk$.

In his hard-sphere model of 1860, assuming the randomness of the impacts between gas molecules, Maxwell showed that collisions between two different sorts of molecules tended to equalize their kinetic energies. This implies the truth of Avogadro's hypothesis (following Clausius's aforementioned reasoning). Maxwell also proved that in the case of non-spherical, hard, elastic molecules, the collisions induce rotations with an average kinetic energy equal to that of translation. To his disappointment, the resultant specific heat was much higher than observed for most gases (Maxwell, 1860).

A few years later, Maxwell became convinced that no consistent mechanical model of the molecules could reproduce the observed specific heats of polyatomic gases. As Boltzmann proved in 1871, the Maxwell–Boltzmann law leads to an average energy $(1/2)kT$, for each quadratic term in the energy of a molecule. As Maxwell argued in 1875, this implies the value $1 + 2/r$ for the ratio γ of the specific heats at constant pressure and at constant volume, wherein r is the number of quadratic terms in the energy function of a molecule. If the degrees of freedom of rotation and vibration of polyatomic molecules are taken into account, the resulting value of γ differs widely from the experimental value. 'Here we are brought face to face with the greatest difficulty which the molecular theory has yet encountered', lamented Maxwell (Maxwell,

1875, p. 433). At the turn of the century, opinions varied on the seriousness of this difficulty. We now know that its solution requires quantum theory.

25.2.6 The Boltzmann Equation and the H-Theorem

By 1871 Boltzmann had proven the stationarity of the Maxwell–Boltzmann law but not its uniqueness. In order to fill this gap, in 1872 he traced the evolution of the velocity distribution of a gas from an arbitrary initial state to equilibrium. If the number of collisions occurring in a spatially homogenous gas is known, he reasoned, the evolution of its velocity distribution can be computed. The variation in a given short time δt of the number of molecules with the velocity $\mathbf{v}_1$ within $d^3 v_1$ is equal to the number of collisions for which the final velocity of one of the colliding molecules belongs to this velocity range, minus the number of collisions for which the initial velocity of one of the colliding molecules belongs to this velocity range. Using Maxwell's formulas for the collision numbers of direct and inverse collisions of a given kind, and taking into account the invariance of the relative velocity $|\mathbf{v}_1 - \mathbf{v}_2|$ and of the product $d^3 v_1 d^3 v_2$, this balance gives the simplest case of 'the Boltzmann equation',

$$\frac{\partial f(\mathbf{v}_1, t)}{\partial t} = \int \left[f(\mathbf{v}_1{}') f(\mathbf{v}_2{}') - f(\mathbf{v}_1) f(\mathbf{v}_2) \right] |\mathbf{v}_1 - \mathbf{v}_2|\, b\, db\, d\phi\, d^3 v_2,$$

where the velocities $\mathbf{v}_1{}'$ and $\mathbf{v}_1{}'$ are the final velocities in a collision with the initial velocities $\mathbf{v}_1$ and $\mathbf{v}_2$, the impact parameter b, and the azimuthal angle ϕ (Boltzmann, 1872).

This equation completely determines the evolution of the distribution f from its initial value. It implies the stationarity of Maxwell's distribution, since the vanishing of the square bracket amounts to Maxwell's stationarity condition. Boltzmann then considered the function (originally denoted E)

$$H = \int f \ln f\, d^3 v.$$

As a consequence of the Boltzmann equation, this function is a strictly decreasing function of time, unless the distribution is Maxwell's. Hence Maxwell's distribution is the only stationary one, and any other distribution tends toward Maxwell's. This is the so-called H-theorem. Boltzmann further noted that the value of $-H$ corresponding to Maxwell's distribution was identical to Clausius's entropy. For other distributions, he proposed to regard this function as an extension of the entropy concept to states out of equilibrium, since it was an ever increasing function of time.

Boltzmann then generalized his equation to more general distributions and systems. The Boltzmann equation has become the central tool for deriving transport phenomena in statistical physics. In Boltzmann's times, however, it could only be solved for Maxwell's $1/r^5$ forces despite Boltzmann's brave efforts in the hard-sphere case. Only at the beginning of the twentieth century did efficient perturbative methods become available for solving the equation in more realistic cases.

25.2.7 Challenges to the Second Law

The bridge established by the kinetic theory between mechanics and thermodynamics not only had repercussions on mechanical thinking by introducing statistical notions into the mechanical description of molecular motion but also affected the understanding of thermodynamics in a similar way. Maxwell used the kinetic molecular theory to 'pick a hole' in the second law of thermodynamics and point to its statistical character. In a letter to Tait of December 1867 and in his *Theory of Heat* of 1871, he argued that a 'finite being' who could 'see the individual molecules' would be able to create a heat flow from a cold to a warm body without expense of work (Maxwell, 1995, pp. 331–332). The being—soon named 'Maxwell's demon' by William Thomson—could indeed control a diaphragm on the wall between warm and cold gas, and let only the swifter molecules of the cold gas pass into the warm gas. In discussions with William Thomson and William Strutt (Lord Rayleigh), Maxwell related this exception to the second law with another obtained by mentally reversing all molecular velocities at a given instant. 'The 2nd law of thermodynamics', he wrote to Strutt in 1870, 'has the same degree of truth as the statement that if you throw a tumblerful of water into the sea you cannot get the same tumblerful of water out again' (Maxwell, 1995, pp. 582–583). Later, Maxwell spoke of 'a statistical certainty' of the second law (Knott, 1911, pp. 214–215). In 1878 he remarked that the dissipation of work during the interdiffusion of two gases depended on our ability to separate them physically or chemically, and concluded: 'The dissipation of energy depends on the extent of our knowledge . . . It is only to a being in the intermediate stage, who can lay hold of some forms of energy while others elude his grasp that energy appears to be passing inevitably from the available to the dissipated state' (Maxwell, 1878, p. 646) (Klein, 1970b).

In 1871, the year before his publication of the H-theorem, Boltzmann emphasized that a mechanical interpretation of the second law of thermodynamics required probability considerations. He also noted that in the mechanical picture the energy of a system in contact with a thermostat fluctuated in time. Even earlier, in 1868, he had noted that special initial states of the gas, for instance one in which all molecules originally lay on the same plane, failed to reach equilibrium. In the introduction of his memoir on the Boltzmann equation, he insisted that 'the problems of the mechanical theory of heat [were] problems of probability calculus', and that the observed regularity of the average properties of a gas depended on the exceedingly large value of the number of molecules (Boltzmann, 1872, p. 317). Yet he stated the H-theorem in absolute terms: the function H 'must necessarily decrease' (Boltzmann, 1872, p. 344). In 1876 Boltzmann's Viennese colleague Joseph Loschmidt remarked that not every initial state of the gas satisfied the theorem. He noted, in particular, that the reversibility of the laws of mechanics implied that to every H-decreasing evolution of the gas system corresponded a reverse evolution for which H increased. Boltzmann's proof of the decrease of H thus seemed to contradict the mechanical foundation of the theory (Boltzmann, 1909, vol. 1, pp. 295, 297, 96; Loschmidt, 1876).

To this 'very astutely conceived' paradox (Boltzmann, 1877a, p. 117), Boltzmann replied (in the more intuitive case of the spatial distribution of hard spheres):

One cannot prove that for every possible initial positions and velocities of the spheres, their distribution must become more uniform after a very long time; one can only prove that the number of initial states leading to a uniform state is infinitely larger than that of initial states leading to a non-uniform state after a given long time; in the latter case the distribution would again become uniform after an even longer time. (Boltzmann, 1877a, p. 120)

Boltzmann's intuition, expressed in the modern terminology of micro- and macro-states, was that the number of microstates compatible with a uniform macrostate was enormously larger than that compatible with a non-uniform macrostate. Consequently, an evolution of the gas leading to increased uniformity was immensely more probable.

25.2.8 The Probabilistic Interpretation of Entropy

To this elucidation of Loschmidt's paradox, Boltzmann appended the remark: 'Out of the relative number of the various state-distributions one could even calculate their probability, which perhaps would lead to an interesting method for the computation of the thermal equilibrium' (Boltzmann, 1877a, p. 121). This is precisely what he managed to do a few months later (Boltzmann, 1877b). The probability he had in mind was proportional to the number of microstates corresponding to a given macrostate. Such a number is ill-defined as long as the configuration of the molecules can vary continuously. That Boltzmann could nonetheless conceive it depended on his peculiar understanding of the continuity introduced in calculus. In his view, integrals and differentials were only condensed expressions for sums with many terms and for small differences. Discrete objects and processes were more basic and more rigorously defined than continuous ones. Whenever Boltzmann faced difficult integrations or integro-differential equations, he studied their discrete counterparts to get a better grasp of the solutions. He did so for instance in his derivation of the H-theorem (Boltzmann, 1909, vol. 1, 84–86, 346–361).

Boltzmann started his state-probability considerations with a 'fiction' wherein the energy of a molecule can only be an integral multiple of the finite element ε. A list of N integers giving the number of energy elements for each molecule defines the microstate of the gas, or 'complexion'. The macrostate is the discrete version of the energy distribution: it is defined by giving the number N_i of molecules carrying the energy $i\varepsilon$ for every value of the integer i. The probability of a given macrostate is proportional to its 'permutability':

$$\Pi = N!/N_1!N_2!\ldots N_i!\ldots.$$

For a given value of the total number $\sum N_i$ of molecules and of the total energy $\sum N_i i\varepsilon$, and in the Stirling approximation for factorials, the permutability is a maximum when N_i is proportional to $e^{-\beta i\varepsilon}$ (wherein β is the Lagrange multiplier associated with the constraint of fixed total energy). Boltzmann next replaced the uniform division of the energy axis with a uniform division of the velocity space, and took the continuous limit of the distribution N_i. This procedure yields Maxwell's velocity distribution. For any distribution N_i the logarithm of the permutability is $-\sum N_i \ln N_i$ in the Stirling approximation (up to a constant), which tends to $-H$ in the continuous limit. Hence the function $-H$, or the entropy of a gas with a given velocity distribution, corresponds to the combinatorial probability of this distribution, as Boltzmann already suspected in his reply to Loschmidt.

In 1878 Boltzmann used the combinatorial approach to explain the existence of a mixing entropy for two chemically indifferent gases. In 1883, after reading Helmholtz's memoirs on the thermodynamics of chemical processes, he showed how his combinatorics, when applied to a reversible chemical reaction, explained the dependence of the equilibrium on the entropy of the reaction. In this context Helmholtz distinguished between 'ordered motion' that could be completely converted into work, and 'disordered motion' that allowed only partial conversion. Accordingly, Boltzmann identified the permutability with a measure of the disorder of a distribution. The mixing entropy thus became the obvious counterpart of increased disorder (Boltzmann, 1878; 1883).

25.2.9 The H-Curve

Boltzmann's probabilistic interpretation of the H-function failed to silence criticism of the H-theorem. In 1894, British kinetic theorists invited Boltzmann to the annual meeting of the British Association, in part to clarify the meaning of this theorem. One of them, Samuel Burbury, offered a terminological innovation: the 'molecular chaos', defined as the validity condition for Maxwell's collision formula. Burbury and Boltzmann also provided an intuitive understanding of this condition: it corresponds to the exclusion of specially arranged configurations, for instance, those in which the velocities of closest neighboring molecules point toward each other (it should not be confused with Helmholtz's molar notion of disorder). As long as the gas remains molecularly disordered, the H-function evolves according to the Boltzmann equation. Boltzmann did not entirely exclude ordered states. He even indicated that an initially disordered state could occasionally pass through ordered states leading to entropy-decreasing fluctuations. However, he judged such occurrences to be extremely improbable (Boltzmann, 1895a; Boltzmann, 1896; 1898; Burbury, 1894, vol. 1, pp. 20–21; Brush, 1976, pp. 616–626).

To this view, Boltzmann's British interlocutors opposed a refined version of the reversibility paradox: H-decreasing and H-increasing states of an isolated gas should be equally frequent, they reasoned, for they correspond to each other by time reversal. In order to elucidate this point, Boltzmann studied the shape of the real H-curve determined by molecular dynamics and discussed its relation with the variations of H given by the Boltzmann equation. The real curve results from the cumulative effect of the rapid succession of collisions in the gas. It therefore has an extremely irregular shape, and does not admit a well-defined derivative dH/dt. The refined paradox of reversibility fails, because it implicitly identifies the decrease of H with the negative sign of its derivative. An accurate statement of Boltzmann's interpretation of the decrease of H reads: for an initial macrostate out of equilibrium and for a finite time of evolution, the number of compatible microstates for which H decreases is much higher than the number of compatible microstates for which H increases. This statement is perfectly time-symmetrical. It means that over a very long time H is for the most time very close to zero, and that the frequency of its fluctuations decreases very quickly with their intensity. Hence any significant value of H is most likely to be very close to a summit of the H-curve. From that point H may increase for some time, but this time is likely to be very short and to be followed by a mostly uniform decrease (Boltzmann, 1895b; Ehrenfest and Ehrenfest, 1911; Klein, 1970a).

The following year Max Planck's assistant Ernst Zermelo formulated another objection to the H-theorem based on Henri Poincaré's recurrence theorem. According to this theorem, any mechanical system (governed by Hamilton's equations) evolving in a finite space with a finite number of degrees of freedom returns, after a sufficiently long time, as close to its initial configuration as one wishes (except for some singular motions). The theorem, Zermelo and Planck argued, excluded any derivation of the entropy law from a mechanical, molecular model. With obvious lassitude, Boltzmann replied that his description of the H-curve was perfectly compatible with recurrences and yet agreed with the statistical validity of the second law because the recurrence times were far beyond human accessibility. Through a simple calculation he estimated these times to have some 10^{10} digits for a gas. He compared Zermelo to a dice player who would declare a dice to be false because he has never obtained a thousand zeros in a row (Poincaré, 1889; Boltzmann, 1896; 1898; Brush, 1976; Zermelo, 1896, pp. 627–639).

25.2.10 Reception of the Kinetic Theory

Zermelo's attack was a symptom of the hostility of many physicists to atomistic considerations. Influential leaders of German physics such as Gustav Kirchhoff, Helmholtz, and Planck favoured a purely macroscopic physics based on differential equations ruling observable quantities. Experimental predictions of the kinetic–molecular theory could often be rederived by purely macroscopic methods. The only important exceptions were the convergent determinations of the mean free path from viscosity, diffusion, and heat conduction, and the compatible estimates of Avogadro's number by Loschmidt (in 1865), Thomson (in 1870), and others from the mean free path and from other phenomena (Brush, 1976, pp. 76–77). To these successes, Boltzmann would have added his explanation of the specific heats of monoatomic and diatomic gases in terms of rigid molecules with spherical and cylindrical symmetry. This was, however, a controversial achievement. Maxwell himself believed that the elasticity of Boltzmann's molecules implied energy equipartition over all their degrees of freedom, including internal vibrations. In sum, empirical success could hardly justify the mathematical and conceptual difficulty of the kinetic theory (Boltzmann, 1876; Brush, 1976; Maxwell, 1877, 356–362).

In Germany, Clausius and Meyer were the only important investigators of the kinetic theory and its experimental consequences. In France, the persistence of Ampère's vibrational theory of heat and the growing distaste for molecular theories prevented the early acceptance of Maxwell's and Boltzmann's ideas. Emile Verdet's lectures long remained the only exception, until at the turn of the century Marcel Brillouin, Poincaré, and Emile Borel entered the scene. The most favourable ground was Britain, owing to the general enthusiasm for dynamical theories. John Tyndall's best-selling *Heat as a Mode of Motion*, published in 1862, popularized kinetic concepts very efficiently. In 1876, Henry Watson's valuable *Treatise on the Kinetic Theory of Gases* appeared. By 1890, British activity was flourishing in this field. In Austria, the traditional interest in molecular theories eased the spread of Boltzmann's theory and prompted Stefan's relevant experimental research. Yet the kinetic theory still lacked a full-fledged account of its conceptual and technical foundations (Brush, 1976; Principe, 2008).

The founders themselves disagreed on the status and achievements of their theory. They did not develop a canonical presentation of its core that could serve as the foundation of a common tradition. Maxwell never wrote a treatise on the kinetic theory. In a letter to Tait he ironically commented on Boltzmann's writings: 'By the study of Boltzmann I have become unable to understand him. He could not understand me on account of my shortness and his length was and is an equal stumbling block to me' (Maxwell, 1995, pp. 915–916). Boltzmann's *Lectures on Gas Theory* came late, in the 1890s, and they covered only some aspects of his and Maxwell's work (Boltzmann, 1896; 1898). As a result, the field remained open for the extensions and reinterpretations that led to modern statistical mechanics.

25.3 FROM KINETIC THEORY TO STATISTICAL MECHANICS

25.3.1 Statistical Mechanics as a New Perspective

Statistical mechanics provides tools for analyzing thermal processes not only in gases but also in general physical systems with microscopic degrees of freedom, whatever their precise constitution may be. At the heart of statistical mechanics is the notion of a virtual ensemble of macroscopic systems, all of which are characterized by the same dynamics but which vary in their initial microscopic configuration. Instead of tracking the statistical behaviour of atomistic constituents of a macroscopic system, statistical mechanics studies the properties of such an ensemble. Different kinds of thermodynamic systems in equilibrium are represented by different statistical ensembles—an isolated thermodynamic system by a 'microcanonical ensemble' in which all members have the same energy; a system which is in contact with a heat reservoir by a 'canonical ensemble', in which the energies of the members are distributed according to an exponential law. Thermodynamic properties are then derived by taking ensemble averages. Due to its generality, statistical mechanics can be employed in classical and, with appropriate modifications, also in quantum physics. For this reason, it played a key role in the transition from classical to modern quantum physics.

Practically all building blocks of statistical mechanics can be found in the numerous publications of Maxwell and Boltzmann. They there appear, however, under perspectives different from that of statistical mechanics as we understand it today. It was only in Josiah Willard Gibbs' *Elementary Principles in Statistical Mechanics*, published in 1902, that a first coherent and autonomous form of statistical mechanics was presented. The now standard terminology 'microcanonical ensemble' and 'canonical ensemble' is due to Gibbs. In the same year, 1902, Albert Einstein published the first of a series of three articles on statistical physics which established, independently of Gibbs, another form of statistical mechanics. This work provided the basis for Einstein's exploration of quantum systems, and also for his analysis of Brownian motion and other fluctuation phenomena as evidence for the existence of atoms. In the same period, the kinetic theory was taken up and further developed, with the result that at the beginning of the twentieth century a variety of approaches were available for dealing with the most diverse problems of statistical physics.

This development of statistical methods was stimulated by the necessity to integrate the growing knowledge of atomistic and statistical processes such as ionic conduction, electronic conduction, and heat radiation. These new contexts of application shifted the emphasis within the kinetic theory of heat from Maxwell's and Boltzmann's questions concerning mechanical foundations to the problem of treating general physical systems in thermal equilibrium. Following this change of perspective, the results of the kinetic theory assumed a new meaning as cornerstones of a more broadly conceived statistical physics. In the following, we sketch the genesis of this new kind of physics by first recapitulating its roots in 'global approaches' to the kinetic theory of Maxwell and Boltzmann, by then discussing the extensions of the kinetic theory to new phenomena, and finally by presenting Gibbs' and Einstein's achievements as the result of a reflection on this development.

25.3.2 Global Approaches

In its earlier and simpler form, the kinetic theory of gases rested on molecular statistics. Boltzmann and Maxwell also developed approaches based on the consideration of the probability of the configurations of the whole system. Whereas the former approach is more intuitive and lends itself to the study of irreversible processes, the latter yields powerful methods for deriving the equilibrium properties of very general systems. Unfortunately, there is no unique, straightforward way to define and derive the global probabilities. This explains why Boltzmann's pioneering considerations on such probabilities were followed by a variety of reinterpretations, including Gibbs' and Einstein's statistical mechanics.

The first example of a global approach is found in Boltzmann's fundamental memoir of 1868. There Boltzmann introduced the distribution $\rho(\mathbf{r}_1, \mathbf{r}_2, \ldots \mathbf{r}_N; \mathbf{v}_1, \mathbf{v}_2, \ldots \mathbf{v}_N)$ such that the product $\rho d\sigma$ gives the fraction of time spent by a system of N point atoms around the phase $(\mathbf{r}_1, \mathbf{r}_2, \ldots \mathbf{r}_N; \mathbf{v}_1, \mathbf{v}_2, \ldots \mathbf{v}_N)$ within

$$d\sigma = d^3 r_1 d^3 r_2 \ldots d^3 r_N d^3 v_1 d^3 v_2 \ldots d^3 v_N$$

after a very long time has elapsed. He then considered the element $d\sigma'$ in phase space made of the phases that the system takes after evolving during a constant time τ from any phase of $d\sigma$. The fraction of time spent by the system in these two elements is obviously equal:

$$\rho d\sigma = \rho' d\sigma'.$$

Since by Liouville's theorem (which Boltzmann rediscovered in this circumstance)

$$d\sigma = d\sigma'$$

and since the time τ is arbitrary, the distribution ρ is constant along the trajectory of the system in phase space. Boltzmann further assumed that the trajectory of sufficiently complex systems filled the energy shell, except for special, highly unstable configurations. This is what the Ehrenfests later called the ergodic hypothesis. Consequently, the distribution ρ must be the uniform distribution over the energy shell, which Gibbs later called microcanonical (Boltzmann, 1868).

From this distribution of the global system, one can derive the distribution of any subsystem that is weakly (thermally) coupled to the complementary system by integrating over the variables of the latter system. When the subsystem is relatively small, its distribution function is approximately proportional to $e^{-\beta H}$, where H is the energy of the subsystem as a function of its phase. The complementary system plays the role of a thermostat, whose absolute temperature is $1/\beta$ in proper units. Boltzmann obtained this general result, corresponding to Gibbs' later canonical law, in 1871. It of course contains the Maxwell–Boltzmann law as the particular case for which the subsystem is reduced to a single molecule (Boltzmann, 1871b, pp. 284–287).

In the same year, 1871, Boltzmann offered an 'analytical proof of the second principle of the mechanical theory of heat' based on the canonical distribution of a system in contact with a thermostat. The idea was to identify the internal energy of the system with the time-average $<H>$ of its energy-function H, and the work provided to the system during an infinitesimal change of the external conditions with the time-average of the corresponding change of its potential energy. Then, the heat exchanged with the thermostat is

$$\delta Q = d<H> - <dH>$$

when measured in energy units. The averages are based on the distribution

$$\rho = e^{-\beta H}/Z,$$

wherein

$$Z = \int e^{-\beta H} d\sigma.$$

Then the product $\beta\delta Q$ is easily seen to be the differential of $\beta <H> + \ln Z$. In other words, there exists an entropy function, which Boltzmann rewrote the following year as

$$S = -\int \rho \ln \rho \, d\sigma.$$

In this circumstance, Boltzmann emphasized the necessity of probability considerations to derive an expression of the entropy on a mechanical basis (Boltzmann, 1871c; Darrigol, 2003).

Despite these interesting results, Boltzmann quickly came to doubt the validity of the underlying hypothesis of ergodicity. He returned to the kinetic–molecular approach, which he believed to be better founded though less general. In 1879, Maxwell revived the global approach and gave it a new twist. Whereas Boltzmann had reasoned in terms of temporal probabilities for a single system, Maxwell adopted the 'statistical specification' of a system, in which the equilibrium properties of a thermodynamic system are to be compared not with those of a single mechanical system but with those of a stationary ensemble of such systems. He proved the stationarity of the microcanonical ensemble for any Hamiltonian dynamics and derived the Maxwell–Boltzmann distribution and energy equipartition from this ensemble. But he did not explain why stationary ensembles represented the thermal properties of macroscopic bodies. He regarded this property as a plausible assumption, to be tested by experiment and perhaps to be justified someday by ergodicity (Maxwell, 1879).

Boltzmann, who welcomed Maxwell's contribution, tried to fill this gap with a heuristic argument. Since, he pointed out, the observed, time-averaged behaviour of a thermodynamic system does not depend on its initial microscopic configuration, almost every system of an ensemble must lead to the same time-averages. Therefore, the time-average of a single system can be replaced with the ensemble average of the time average, which is also the time average of the ensemble average. Hence, for a stationary ensemble, the ensemble average should be equal to the time average. In a non-rigorous manner, this argument justified the use of statistical ensembles as mechanical models of thermodynamics (Boltzmann, 1881).

Yet Boltzmann remained open to other possibilities for constructing mechanical models of thermodynamics. For example, in the 1880s Boltzmann took up Helmholtz's analogy between the so-called 'monocyclic mechanical systems' and thermodynamic systems. Helmholtz had not attempted a realistic description of the mechanical processes responsible for thermodynamic observations. He rather considered simple mechanical systems, with but a few degrees of freedom, for which a function with the properties of the entropy function could be identified. Boltzmann's interest in this approach makes it evident that his principal aim was to identify mechanical analogies apt to clarify the relation between mechanics and thermodynamics. In a memoir of 1885 he introduced and discussed, under the name 'holode', what later came to be called the canonical ensemble. But whereas in modern statistical mechanics the canonical ensemble serves to explore physical properties of quite general systems in thermal equilibrium at constant temperature, Boltzmann merely conceived this ensemble as a mechanical analogue on the same footing as Helmholtz's monocyclic systems. This illustrates a more general fact: although most of the elements of statistical mechanics were anticipated in the work of Boltzmann and Maxwell, they were there embedded in contexts different from that of modern statistical mechanics (Boltzmann, 1885; Klein, 1972a; Renn, 1997).

25.3.3 Beyond Gases

An essential motor of the transition toward modern statistical mechanics was the application of the kinetic theory to systems other than gases. In the limited context of gases, kinetic theory served mostly as a mechanical model for thermodynamics and transport phenomena. Despite a few empirical successes, it retained a precarious character, as is evident from the controversies in which Boltzmann was involved. Yet toward the end of the century it was successfully applied to newer domains of physics involving ions and electrons. In this process the foundations of the kinetic theory were gradually secured and widened.

A first example is the theory of the Dutch physicist Johannes Diderik van der Waals, which represented the first successful attempt at a quantitative understanding of the liquid–gas transition. In 1857, Clausius had already given a qualitative explanation of the three states of matter and sketched a kinetic picture of the transitions between them. Clausius's and Maxwell's kinetic theory, however, was essentially confined to ideal gases, for which the average distance between two molecules is much larger than their diameter, and for which the intermolecular forces are negligible between collisions. In 1873, van der Waals relaxed these two conditions, with the ambition of creating a kinetic theory of liquids. He assumed hard, elastic, spherical molecules, with a rapidly decreasing mutual attraction. The effect of the latter force, he reasoned, boils down to an attraction of the molecules on the fluid's free surface toward its interior. On the one hand, this surface correction implies a correction a/V^2 to the pressure P that balances the dynamical effect of collisions on the surface. On the other hand, this dynamical effect departs from Clausius's ideal value $(1/3)Nmu^2/V$, because the finite extension of the molecules alters their flux near the wall in the proportion $(V-b)/V$, where b is four times the total volume of the spheres. The resulting equation of states is

$$(P + a/V^2)(V - b) = RT.$$

Van der Waals successfully applied this equation to a unified description of the liquid and gas states. This simple, non-rigorous model provides a sound intuition of the most evident properties of real fluids. Maxwell and Boltzmann both contributed to its spread (Brush, 1976; van der Waals, 1873, chap. 7, 11).

Another, later, example for the extension of the methods of the kinetic theory beyond gas theory is Planck's theory of heat radiation of 1896–1900. Planck originally belonged to the denigrators of the kinetic theory and believed that the thermal properties of radiation should be explained without statistical methods. This attitude was rooted in his faith in the absolute validity of the laws of thermodynamics. In his eyes, Boltzmann's interpretation of the entropy law could not be true, for it restricted the validity of this law to a statistical one. Instead, Planck believed that the diffusion of electromagnetic waves by electric resonators was an intrinsically irreversible process, from which the increase of entropy could be deduced. Under the pressure of Boltzmann's criticism, however, Planck soon came to admit the necessity of an additional assumption of 'natural radiation' as the counterpart of molecular chaos, but without accepting Boltzmann's statistical understanding of disorder. According to Planck, the intricacies of the invisible dynamics of resonators, or those of the walls of a gas container, permitted a strictly irreversible evolution of the system, in full harmony with his absolute conception of the second law of thermodynamics. Elementary disorder—a generic name for natural radiation and molecular chaos—thus became the central concept of a non-statistical understanding of the relation between micro- and macro-world (Darrigol, 1988).

Planck's reinterpretation of Boltzmann's kinetic–theoretical reasoning did not stop there. In 1900 his further exploration of heat radiation brought him to apply Boltzmann's relation between entropy and probability, albeit only in a formal way that prevented revolutionary conclusions (Planck, 1900). He justified his new formula for the equilibrium spectrum of thermal radiation ('Planck's black-body law') by means of Boltzmann's counting of complexions, reinterpreted as a quantitative estimate of elementary disorder among resonators of the same frequency v. This procedure allowed finite energy-elements hv to enter the calculation without contradicting the continuous dynamics of the resonators nor the continuous nature of radiation. In this context Planck wrote the formula for the probability interpretation of entropy,

$$S = k \ln W,$$

which can now be read on Boltzmann's grave. Yet he did not admit the statistical validity of the entropy law until 1914. Einstein and Paul Ehrenfest were the first to treat black-body radiation as a thermo-statistical system and thus to arrive at its incompatibility with the laws of classical physics.

Toward the close of the century, the interest in Maxwell's and Boltzmann's kinetic theory rose considerably owing to the ever growing significance of atomistic ideas and statistical methods at the forefront of physics. Prominent examples are Hendrik Antoon Lorentz's electron theory (an atomistic version of Maxwell's electromagnetism), Paul Drude's electron theory of metals, the ionic theory of electrolytic conduction, the kinetic theory of solutions, and the use of atomistic models in inorganic and organic chemistry. Novel opportunities offered themselves for experimental research: there were new kinds of radiation, either waves or elementary particles; new interactions between matter and radiation such as the Zeeman and photoelectric effects, which confirmed atomistic models of matter or suggested new ones; new studies of colloidal suspensions that seemed to reveal molecular motions, and so on. Measures of Avogadro's number could be gained from as many different sources as the study of capillarity, the kinetic theory of gases, experiments with thin layers, and, surprisingly, also from the theory of black-body radiation. The agreement between these measures not only increased confidence in atomistic hypotheses but also made it seem urgent to develop molecular theories bridging the diverse domains of physics (Renn, 1997).

25.3.4 Gibbs' and Einstein's Formulations of Statistical Mechanics

Early in the twentieth century, Gibbs and Einstein proposed two versions of statistical mechanics. They were both convinced that their work did not constitute a fundamentally new contribution but merely a change of perspective on what Maxwell, Boltzmann, and others had achieved. Yet their approaches introduced a new emphasis and also a conceptual organization different from that of the earlier global approaches.

Gibbs' perspective was more mathematical and more abstract than that of his predecessors. As a witness of the late nineteenth-century multiplication of molecular theories, he grew suspicious of special molecular assumptions and instead sought the most general foundation for statistical physics. His strategy was to develop the study of statistical ensembles for their own sake, and then to look for analogies between the properties of such ensembles and the laws of thermodynamics. In most of his elegant *Principles of Statistical Mechanics* of 1902, he described the underlying mechanical system in a formal manner, by generalized coordinates subjected to Hamilton's equations, for he agreed with Maxwell that the violations of energy equipartition made the foundations of molecular dynamics insecure. He introduced and systematically studied the three fundamental ensembles of statistical mechanics: the micro-canonical, the canonical, and the grand-canonical ensemble (in which the number of molecules may vary). He examined the relations between these three ensembles and their analogies with thermodynamic systems, including fluctuation formulas. Unlike Boltzmann, he did not seek an *a priori* justification for the thermodynamic significance of his ensembles. His approach was essentially axiomatic (Gibbs, 1902; Klein, 1972b).

Albert Einstein, who was not aware of Gibbs' *Principles*, approached statistical thermodynamics from still a different point of view, with partially overlapping results but decisively new aims. In his student years, he had been impressed by the rise of molecular physics, despite the skepticism of many German physicists. He especially admired Drude's theory of metals of 1900, which combined Lorentz's electron theory with Boltzmann's kinetic theory. On the basis of his experiences with the new applications of kinetic–theoretical methods, including Planck's work on heat radiation, he perceived Boltzmann's approach as too focused on the mechanical foundation of thermodynamics and too little oriented towards new evidence for the atomistic constitution of matter and radiation. From his perspective, there were gaps in Boltzmann's methods—more so because he only had a fragmentary knowledge of them, mainly through the *Lectures on Gas Theory* (Boltzmann, 1896–1898). Einstein's articles of 1902–04 aimed at providing a new foundation for statistical mechanics, and it was here that he derived the second law of thermodynamics from the canonical ensemble with slowly varying external forces and justified the canonical ensemble as a small sub-ensemble of the micro-canonical ensemble. He interpreted all probabilities physically, as measures of the fraction of time spent by the system in various configurations (Einstein, 1902; 1903; 1904; Renn, 1997).

What distinguished Einstein from his predecessors was a difference of emphasis which turned out to be most consequential. Boltzmann, for instance, had displayed his ability to retrieve the macroscopic laws of thermodynamics and played down the departures from these laws that the molecular structure of matter implied. Einstein did the reverse. He wanted to use statistical mechanics to prove the molecular structure of matter and to probe the non-classical structure of radiation. For this reason, he focused on the fluctuations around equilibrium that were negligible for Boltzmann and inexistent for Planck. He interpreted the probability in the Boltzmann–Planck relation $S = k \ln W$ as the temporal frequency of the fluctuations of a system, and the constant k as the measure of its 'thermal stability'. In 1905 his analysis of Brownian motion showed how fluctuations could become observable on mesoscopic systems such as smoke particles. At the same time, he inverted Boltzmann's relation in order to explore unknown aspects of the dynamics of microsystems. This is how he arrived at the light-quantum hypothesis. Whereas Maxwell and Boltzmann meant to provide a mechanical foundation of thermodynamics, Einstein used statistical mechanics to question this foundation (Büttner, Renn, and Schemmel, 2003).

25.3.5 The Boltzmann Legacy

In their development of statistical mechanics as a synthetic framework, Gibbs and Einstein selected and reinterpreted some aspects of Boltzmann's and Maxwell's work. They neglected other aspects that were essential to Boltzmann or Maxwell, for instance, the discussions of irreversible processes and transport phenomena. More faithful to Boltzmann was his disciple Paul Ehrenfest, who strove with his wife Tatiana to elucidate the interconnections between Boltzmann's various approaches and to clarify the relevant probabilistic assumptions. Their encyclopedia article of 1911 remains an instructive synthesis of Boltzmann's views. They shared his skepticism about the ergodic hypothesis, so much so that Paul imagined a connection between quantum properties and violations of ergodicity. They dwelt on irreversible processes, on the Boltzmann equation, and on the paradoxes of the H-curve. They reproached Gibbs with neglecting this part of Boltzmann's legacy, and hardly noted Einstein's contributions (Ehrenfest and Ehrenfest, 1911; Klein, 1970a).

Statistical mechanics has now become an essential part of the canon of physics. Yet some of its original diversity has survived. Moreover, the nature and solidity of its foundations remain controversial issues. Physicists and mathematicians still discuss the pertinence of the ergodic hypothesis or the split between reversible and irreversible processes. Philosophers still reflect on the relevant notions of probability or on the arrow of time. A century later, statistical mechanics is as open and alive as Maxwell and Boltzmann wanted theories to be.

References

BOLTZMANN, LUDWIG (1868), Studien über das Gleichgewicht der lebendigen Kraft zwischen bewegten materiellen Punkten, *Wien. Ber.*, 58: 517–560, repr. in Boltzmann (1909), vol. 1, 49–96.
—— (1871a) Über das Wärmegleichgewicht zwischen mehratomigen Gasmolekülen, *Wien. Ber.*, 63: 397–418; repr. in Boltzmann (1909), vol. 1, 237–258.
—— (1871b) Einige allgemeine Sätze über Wärmegleichgewicht, *Wien. Ber.*, 63: 679–711; repr. in Boltzmann 1909, vol. 1, 259–287.
—— (1871c) Analytischer Beweis des zweiten Hauptsatzes der mechanischen Wärmetheorie aus den Sätzen über das Gleichgewicht der lebendigen Kraft, *Wien. Ber.*, 63: 712–732; repr. in Boltzmann (1909), vol. 1, 288–308.
—— (1872) Weitere Studien über das Wärmegleichgewicht unter Gasmolekülen, *Wien. Ber.*, 66: 275–370; repr. in Boltzmann (1909), vol. 1, 316–402.
—— (1876) Über die Natur der Gasmoleküle, *Wien. Ber.*, 74: 553–560; repr. in Boltzmann (1909), vol. 2, 103–110.
—— (1877a) Bemerkungen über einige Probleme der mechanischen Wärmetheorie, *Wien. Ber.*, 75: 62–100; repr. in Boltzmann (1909), vol. 2, 112–148.
—— (1877b) Über die Beziehung zwischen dem zweiten Haupsatze der mechanischen Wärmetheorie und der Wahrscheinlichkeitsrechnung respektive den Sätzen über das Wärmegleichgewicht, *Wien. Ber.*, 76: 373–435; repr. in Boltzmann (1909), vol. 2, 164–223.
—— (1878) Über die Beziehung der Diffusionsphänomene zum zweiten Hauptsatze der mechanischen Wärmetheorie, *Wien. Ber.*, 78: 733–763; repr. in Boltzmann (1909), vol. 2, 289–317.
—— (1881) Referat über die Abhandlung von J.C. Maxwell 'Über Boltzmanns Theorem betreffend die mittlere Verteilung der lebendigen Kraft in einem System materieller Punkte (1879) in *Beiblätter zu Annalen der Physik*, 5: 403–417; repr. in Boltzmann (1909), vol. 2, 582–595.
—— (1883) Über das Arbeitsquantum, welches bei chemischen Verbindungen gewonnen werden kann, *Wien. Ber.*, 88: 861–896.
—— (1885), Über die Eigenschaften monozyklischer und anderer damit verwandter Systeme, *Journal für die reine und angewandte Mathematik*, 98: 68–94; repr. in Boltzmann (1909), vol. 3, 122–152.
—— (1895a), Nochmals das Maxwellsche Verteilungsgesetz der Geschwindigkeiten, *Annalen der Physik*, 55: 223–224.
—— (1895b), On certain Questions of the Theory of Gases, *Nature*, 51: 413–415; repr. in Boltzmann (1909), vol. 3, 535–544.
—— (1896), Entgegnung auf die wärmetheoretischen Betrachtungen des Hrn. E. Zermelo, *Annalen der Physik*, 57: 773–784; repr. in Boltzmann (1909), vol. 3, 567–578.
—— (1896–1898), *Vorlesungen über Gastheorie*, 2 vols. (Leipzig, Barth).
—— (1898), Über die sogenannte H-Kurve, *Mathematische Annalen*, 50: 325–332; repr. in Boltzmann (1909), vol. 3, 629–637.
—— (1909), *Wissenschaftliche Abhandlungen* 3 vols. (Leipzig, Barth).
—— (1981–1998), *Gesamtausgabe*, R. Sexl (ed.) (Braunschweig, Vieweg).
BURBURY, SAMUEL (1894), Boltzmann's minimum function, *Nature*, 51: 78.
CLAUSIUS, RUDOLPH (1857), Über die Acht der Bewegung, welche wir die Wärme nennen, *Annalen der Physik*, 100: 353–380.

—— (1858), Über die mittlere Länge der Wege, welche bei der Molecularbewegung gasförmiger Körper von den einzelnen Molecülen zurückgelegt werden; nebst einigen anderen Bemerkungen über die mechanische Wärmetheorie, *Annalen der Physik*, 105: 239–258.
—— (1888), *Die kinetische Theorie der Gase* (Braunschweig, Vieweg).
EHRENFEST, PAUL, and TATIANA EHRENFEST (1911), Begriffliche Grundlagen der statistischen Auffassung in der Mechanik, *Encyklopädie der mathematischen Wissenschaften*, IV/2.II: 1–90.
EINSTEIN, ALBERT (1902), Kinetische Theorie des Wärmegleichgewichtes und des zweiten Hauptsatzes der Thermodynamik, *Annalen der Physik*, 9: 417–433.
—— (1903), Eine Theorie der Grundlagen der Thermodynamik, *Annalen der Physik*, 11: 170–187.
—— (1904), Zur allgemeinen molekularen Theorie der Wärme, *Annalen der Physik*, 14: 354–362.
GIBBS, JOSIAH WILLARD (1902), *Elementary Principles in Statistical Mechanics Developed with Especial Reference to the Rational Foundation of Thermodynamics* (New York, Scribner's sons).
KNOTT, C. G. (1911), *Life and Scientific Work of Peter Guthrie Tait* (Cambridge, Cambridge University Press).
LOSCHMIDT, JOSEPH (1876), Über den Zustand des Wärmegleichgewichtes eines Systemes von Körpern mit Rücksicht auf die Schwerkraft, *Wien. Ber.*, 73: 128–142.
MAXWELL, JAMES CLERK (1860), Illustrations of the dynamical theory of gases, *Philosophical magazine*, 19: 19–32 and 20: 21–37.
—— (1867), On the dynamical theory of gases, *Philosophical Transactions of the Royal Society*, 157: 49–88.
—— (1871), *Theory of Heat* (London, Longmans).
—— (1875), On the dynamical evidence of the molecular constitution of bodies, *Nature*, 11: 357–359, 374–377.
—— (1877), The kinetic theory of gases, *Nature*, 16: 242–246.
—— (1878), Diffusion, *Encyclopedia Britannica*; repr. in Maxwell 1890, vol. 2, 625–646.
—— (1879), On Boltzmann's theorem on the average distribution of energy in a system of material points, *Transactions of the Cambridge Philosophical Society*, 12: 547–570.
—— (1890), *The Scientific Papers of James Clerk Maxwell* 2 vols. (Cambridge, Cambridge University Press).
—— (1986), *Maxwell on Molecules and Gases*, ed. E. Garber, S. G. Brush, and C. W. F. Everitt (Cambridge, Mass., MIT Press).
—— (1995), *The Scientific Letters and Papers of James Clerk Maxwell* (Cambridge, Cambridge University Press), ii.
PLANCK, MAX (1900), Zur Theorie des Gesetzes der Energievertheilung im Normalspektrum, Deutsche Physikalische Gesellschaft, *Verhandlungen*, 2: 237–245.
POINCARÉ, HENRI (1889), Sur le problème des trois corps et les équations de la dynamique, *Acta mathematica*, 13: 1–270.
VAN DER WAALS, JOHANNES DIDERIK (1873), *Over de Continuiteit van de Gas- en Vloeistoftoestand* (Leiden, Stijthoff); English translation in *Physical memoirs* (London, Taylor and Francis, 1890).
ZERMELO, ERNST (1896), Über einen Satz der Dynamik und die mechanische Wärmetheorie, *Annalen der Physik*, 57: 485–494.

STUDIES

BRUSH, STEPHEN (1965–1972), *Kinetic Theory* (New York: Pergamon Press), 3 vols.
—— (1976), *The Kind of Motion we call Heat: A History of the Kinetic Theory of Gases in the 19th century* (Amsterdam, North-Holland), 2 vols.

—— (1983), *Statistical Physics and the Atomic Theory of Matter, from Boyle and Newton to Landau and Onsager* (Princeton, Princeton University Press).
BÜTTNER, JOCHEN, JÜRGEN RENN, and MATTHIAS SCHEMMEL (2003), Exploring the limits of classical physics: Planck, Einstein, and the structure of a scientific revolution, *Studies in History and Philosophy of Modern Physics*, 34: 37–59.
DARRIGOL, OLIVIER (1988), Statistics and combinatorics in early quantum theory, *Historical Studies in the Physical Sciences*, 19: 17–80; continued in *Historical Studies in the Physical Sciences*, 21 (1991): 237–298.
——(2003), The origins of the entropy concept, in J. Dalibard, B. Duplantier, V. Rivasseau (eds.), *Poincaré Seminar 2003: Bose–Einstein Condensation, Entropy* (Basel: Birkhäuser, 2004), 101–118.
EVERITT, C. W. F. (1975), *James Clerk Maxwell. Physicist and Natural Philosopher*. Cambridge. Also as Maxwell, James Clerk in *Dictionary of Scientific Biography* (New York, Scribner's sons, 1970–80), ix: 198–230.
FOX, ROBERT (1971), *The caloric theory of gases from Lavoisier to Regnault* (Oxford, Clarendon Press).
GALLAVOTTI, GIOVANNI (1994), Ergodicity, ensembles, irreversibility in Boltzmann's and beyond, *Journal of Statistical Physics*, 78: 1571–1589.
GARBER, ELIZABETH, STEPHEN BRUSH, and C. W. F. EVERITT (1986), Introduction to (Maxwell, 1986).
KLEIN, MARTIN J. (1970a), *Paul Ehrenfest* (Amsterdam, North-Holland).
—— (1970b), Maxwell, his demon, and the second law of thermodynamics, *American Scientist*, 58: 84–97.
—— (1972a), Mechanical explanation at the end of the nineteenth century, *Centaurus*, 17: 58–82.
—— (1972b), Gibbs, Josiah Willard, *Dictionary of Scientific Biography*, v: 386–393.
—— (1973), The development of Boltzmann's statistical ideas, in E. G. D. Cohen and W. Thirring (eds.), *The Boltzmann Equation: Theory and applications* (Vienna, Springer), 53–106.
PLATO, JAN VON (1991), Boltzmann's ergodic hypothesis, *Archive for the History of Exact Sciences*, 42: 71–89.
—— (1994), *Creating Modern Probability: Its Mathematics, Physics and Philosophy in Historical Perspective* (Cambridge, Cambridge University Press).
PORTER, THEODORE (1986), *The Rise of Statistical Thinking* (Princeton, Princeton University Press).
PRINCIPE, JOÃO (2008), *La réception française de la mécanique statistique*, thèse de doctorat, Université Paris Diderot.
RENN, JÜRGEN (1997), Einstein's controversy with Drude and the origin of statistical mechanics: A new glimpse from the 'love letters', *Archive for the History of Exact Sciences*, 51: 315–354.
SKLAR, LAWRENCE (1993), *Physics and Chance: Philosophical Issues in the Foundations of Statistical Mechanics* (Cambridge, Cambridge University Press).

NOTE

1. An early version of this article was published as 'La nascita della meccanica statistica' in *Enciclopedia Italiana: Storia della scienza*, vol. 7: *L'ottocento* (Rome, Istituto della Enciclopedia Italiana 2003), 496–507.

CHAPTER 26

THREE AND A HALF PRINCIPLES: THE ORIGINS OF MODERN RELATIVITY THEORY

DANIEL KENNEFICK

26.1 INTRODUCTION

In 1900 the field theory of electromagnetism, which owes it origins primarily to the work of James Clerk Maxwell, had been under rapid development for two decades. In the 1880s a number of British physicists, beginning with Oliver Heaviside, had developed Maxwell's work into a successful body of theory which was able to explain a number of important features of electrodynamics. During the 1890s this new theory encountered some difficulties which, as Jed Buchwald, (1985) has shown, were connected with the earlier theory's inattention to the physical nature of the sources of the field, the moving charges themselves. This directed attention towards the problem of microphysics and the nature of the electron and towards a theory of electromagnetism which focused on the reality of charged particles as agents of the field. This was accompanied by a geographical shift away from Britain, whose leading figures came to play a less important role in the development of the theory, to the Continent, in particular to Holland and the German-speaking areas of Europe.

The new Continental theory had important successes, which inspired a hope that was expressed in the term electromagnetic world-view, that all physical phenomena would be expressible in terms of the electromagnetic field. In spite of major achievements by Hertz, Lorentz and others, the new theory still found itself troubled by a number of issues, several of which are now seen to have had a common origin in the subject of relativity theory, and how the new electromagnetic theory was to be reconciled with it. The chief credit for the work on relativity which resolved these problems is usually accorded to Albert Einstein, and the first half of this article will focus primarily on the particular line of thinking which led him to the discovery of what is now called special relativity theory. The contributions of others, particularly Lorentz and Henri Poincaré, will not be neglected, but the adoption of a schema based on the structure of Einstein's theory, while justified by his enormous influence on subsequent research, will inevitably force their contributions into a framework which was not of their making, and therefore tends to distort what they considered to be the meaning of their work.

In the years after 1905 the success of special relativity created a new problem of its own. The old theory of gravity was now seen to be in conflict with the new theory of relativity. It is to Einstein that we owe the resolution of this problem. Although others, such as Max Abraham and Gunnar Nordstrom, also addressed it, their work did not influence Einstein and was made irrelevant by the success of Einstein's theory of general relativity. Therefore it is with less apology that the second half of this paper will focus on Einstein's path to general relativity, with little discussion of alternative efforts to develop a relativistic theory of gravity.

26.2 THE PRINCIPLE OF RELATIVITY

In his paper *On the Electrodynamics of Moving Bodies*, written in 1905, Einstein directly addressed, with his title, one of the largest issues of the time. Electrodynamics aims to describe the motion of charged particles (usually thought of as electrons), whose interaction through the electromagnetic field, as described by Maxwell's equations, affects their respective motions. The problem was so complex because the electromagnetic field theory was not an action-at-a-distance theory. Any calculation of the problem of motion had to take account of the time of propagation of the field effect from the source of the field to the particle which would experience the resulting force. Since the particles themselves were moving and accelerating in response to these forces they would have changed position in the time it took the field effect to propagate, greatly complicating the calculation. The complexity of the mathematics required to handle these issues was encouraging the creation of a new breed of scientist, the theoretical physicist, who specialized in mathematically complex and difficult calculations of this type.

Given that Einstein's paper advertised an attack on this problem of charges in motion and the resulting forces between them, it was unexpected that it began with a long section on the subject of kinematics, the study of the properties of motion itself, divorced from the question of what causes motion. After all, the whole point of electrodynamics was to examine the the causes of motion, rather than the question of how one can measure or correctly describe it. But Einstein showed that the study of the time taken for forces and influences to propagate between particles actually raised a critical issue in kinematics. Kinematics, though fundamental, was regarded as unproblematic.[1] Physicists felt the measurement of motion was well understood. But if the influence of a distant particle can only be felt after the passage of time, so that a local particle responds not to the force as generated from the source's 'current' position but from its past position (how far in the past depending on the propagation time of the influence, the actual position the influenced particle would occupy depending on the speed of the particle itself) might it not be that our measurement of positions, speeds, and accelerations must depend on the speed of this influence, the speed of light? Einstein showed very clearly that if news of distant events takes time to reach us then our measurements of the distances and times to these events must be different for observers moving with different constant velocities. The rules for transforming between the measurements of these observers must be quite different from the transformation relations previously used in kinematics.

Einstein's reformulation of kinematics in the first part of the paper then laid the basis for his resolution of several outstanding problems in electrodynamics, the advertised topic of his paper. A series of known results from electrodynamics theory were shown not to depend in any way on the nature and structure of the electron, or on the properties of the ether, which had been the presumption of theorists such as Lorentz and Abraham, but to arise naturally from the general theory of kinematics itself. Thus Einstein's reformed theory of relativity applied to all matter, of whatever composition and empty space was once more empty. He additionally showed that this new theory of relativity not only had implications for the measurement of the lengths and velocities of moving objects—something which had been at least partially understood by his predecessors—but also for the measurement of time. He showed that a moving clock did not run at the same rate as a clock which was at rest relative to an observer. Finally, he showed that while certain quantities, such as length, mass, and velocity varied with the state of motion of one's frame of reference, there did exist invariant quantities which did not. The use of these invariant quantities permitted a physicist to reconcile the views of the world as seen by observers (or particles) moving at different speeds. In turn, this permitted one to consistently and correctly perform calculations of the problem of motion in cases where the speed of the moving particle was not negligible compared to the speed of the influences between them, the speed of light.

The theory which Einstein presented in his 1905 paper came to be known as Lorentz–Einstein theory, because Einstein's paper had shown how most of the results of Hendrik Lorentz' electrodynamic theory could be derived very simply using Einstein's kinematics, in contrast to the difficult and complex calculations which Lorentz had used to derive them.[2] Lorentz had already appreciated that different

---

[1] For a study of the historical origins of Einstein's work on kinematics, see Martinez, 2009.
[2] For more on Lorentz and the relationship between his theory and Einstein's, see Janssen, 2002.

moving particles might have their own time, which he called local time. He had further shown how to transform between the different local time and space coordinate systems, so that the transformation equations of the new theory of relativity are still known as Lorentz transformations. The most obvious novelty of the new theory of relativity was that these transformations differed from those which one used in the old Galilean theory of relativity, dating back to to the period of the scientific revolution.

One reason for Einstein's success in developing the theory of relativity—indeed, in turning modern field theory in an entirely new direction—was his unique, or at least unusual, style of doing physics. Most physicists of his time were tending to become specialists, who worked within a given subfield of physics. They employed a suite of techniques known to be useful in that field, and pursued problems particular to that field. Accordingly, the problems raised by the failure of electrodynamics to obey the rules of relativity could be viewed as useful if it could provide insight into the particular questions that were important to their subject. For a broad principle like relativity, such field specific information was especially likely in cases where the principle was found to be violated. Thus, if electromagnetism did not obey the usual rules of relativity, perhaps this had something to tell physicists about the specific nature of the elusive medium through which the electromagnetic field was propagated, or even about the specific nature of particles like the electron, which were understood to be the fundamental sources of this field. Thus Lorentz' theory was known as an ether theory and was the rival of Max Abraham's theory, referred to as the electron theory. Abraham's theory was explicitly based upon the electromagnetic world view of physics, in which the endeavour was to reduce all of physics to an epiphenomenon of the electromagnetic field. In particular, Abraham aimed to show that inertia itself was a property of this field.[3] While Lorentz did not subscribe in the same way to this view, it nevertheless influenced him.

Einstein, on the other hand, regarded the principle of relativity as inviolable. It was not to be regarded as a mere pawn, conveniently sacrificed for the gain of a deeper understanding of the nature of matter. If the field theory failed to obey it, and if one believed the field theory was fundamentally correct, then one was required to make a reconciliation. Accordingly, he searched until he found the hidden assumption whose failure to conform to reality was in fact responsible for the apparent contradiction between principle and theory. He was nearly uniquely placed to make this discovery. Men like Lorentz and Abraham were more interested in learning something about the ether or the electron than in saving the principle of relativity.

Another example of sacrificing principle for short-sighted phenomenological success can be found in the earliest days of modern relativity theory. Relativity theory came to the aid of Copernican astronomers in providing an explanation for the failure of humans to notice that the Earth was, in fact, in motion about the Sun. The doctrine of relative motion—what Einstein would call the principle of relativity—says that one cannot tell if a system is moving (unless the motion involves acceleration) without looking outside the system to some object in a different state of motion. But

---

[3] See, for instance, McCormmach, (1970), Goldberg, (1970), and Darrigol, (2000), p. 360.

of course it would have been highly useful to astronomers and physicists who supported the heliocentric model to have been able to prove that the Earth did move. Even relativity theory admits that one can tell the Earth is in motion by observing the stars, but the effects of parallax and stellar aberration were too small for detection with the technology available in the period of the scientific revolution, because of the unexpectedly great distances between ourselves and the stars.

Such was the desire for a proof that the Earth moved, that even Galileo, after whom pre-Einsteinian relativity theory is named, succumbed to the temptation to seek for proof of the Earth's motion in an exception to the relativity principle. Faced with the difficulty of teaching Copernicanism without being able to prove the hypothesis that the Earth moved, he formulated a theory of the tides which argued that they resulted from the failure of the seas to fully keep up with the motion of the Earth, which, being compounded of two circular motions (the Earth's rotation and its orbit of the Sun), was non-inertial in nature. It is in fact true that laboratory experiments, such as Foucault's pendulum, can measure the accelerations involved in the Earth's rotational motion. Indeed Galileo's argument draws attention to the fact that water in a vase can move in response to non-inertial motion of the vase, so that Galileo's vase can take its place alongside Newton's bucket, which we shall discuss later, in the debate over the relative nature of motion in space. It is actually true that our failure to detect the motion of the Earth around the Sun has rather less to do with the principle of relativity than it does with the principle of equivalence, which will also be discussed later. Since everything on Earth falls towards the Sun at the same rate, we experience no relative motion with respect to nearby objects or the Earth as we orbit. Thus Galileo's vase does point out the way which Einstein was to follow towards the theory of general relativity. Nevertheless, the use to which Galileo put it in his own day, though it seemed highly convenient for his argument with powerful anti-Copernicans, was quite erroneous. While there are some patterns in the Earth's ecosphere which appear to respond to the non-inertial nature of the Earth's rotation (for instance, cyclones and anti-cyclones, which rotate oppositely in the northern and southern hemispheres), they were not readily discernible in Galileo's day.[4]

It is difficult to imagine Einstein stooping to such an unprincipled argument, no matter what the stakes. In fairness, by his day relativity had piled up empirical point after point in its favour, compared to what was known to Galileo. Still, having once articulated a principle as powerful as the principle of relativity, he would not easily have let it go for such a qualitative, unquantified, and opportunistic argument as Galileo puts forward in his theory of the tides. Galileo's gambit was to sacrifice relativity as a principle for the convenience of his immediate argument.[5]

---

[4] For a discussion of Galileo's theory of the tides, see Drake, (1970), pp. 200–213.

[5] Of course, one must be careful not to be too sweeping in discussing a physicist's style. Einstein could be a master in looking for the way in which just the right experiment could prove a theory correct. In the case of the kinetic theory of matter, where most of the advocates of this approach had insisted upon the impossibility of finding an *experimentum crucis* which could show whether atoms were real (probably because they feared the results of such

26.3 THE PRINCIPLE OF THE CONSTANCY OF THE SPEED OF LIGHT

The reader new to this subject may by now be puzzled at the audacity of those physicists who were untroubled by the notion that the laws of electromagnetism did not obey the rules of Galilean relativity. Should we not expect all physical phenomena to obey this important principle? Had not Newton and his followers believed that light was made of corpuscles emitted by a source, and would this not suggest that the speed of the corpuscles as measured by those who received them would be their 'natural' speed added to the speed of the source from which they were ejected? This is known as an emission theory of light, and there were still those in the early twentieth century who held this view. Einstein himself tells us that he did so during some of his attempt to reconcile electrodynamics and relativity.

However, during the nineteenth century most physicists converted to a wave picture of light. Their understanding of the analogy with water waves which underpinned the wave theory caused them to read Galilean relativity in a very different way. Galileo, in order to defend the Copernican model of the solar system against the charge that the inhabitants of the Earth would surely be aware if it was in motion, argued that a man on a ship who dropped a ball from the mast would see the ball strike the deck at the foot of the mast whether the ship was in uniform motion or not, because the ball partook of that constant motion and retained it during the period when it was not in physical contact with the ship. But suppose the man were to drop the ball into the sea instead. Now ripples would spread out from the point of impact. At what speed would the wavefront move? It would be at the usual speed for waves of that type on water. Would one need to add the speed of the ball to the speed of the waves to obtain the result as seen either by someone onboard the ship or on the shore? No, the speed of the ball, the source, would be irrelevant to the speed of the waves. All that would matter would be the speed of the medium with respect to the observer. Someone on shore would say the speed of the waves was simply the normal wave speed (in the absence of ocean currents). The man onboard the ship, on the other hand, if the ship was underway, would see the speed of the waves as being their natural speed, minus the speed of the ship. Thus this simple experiment would tell him how fast he was moving, admittedly by interacting with some matter outside the ship. So in the case of light, for which the medium was thought to be the so-called luminiferous ether, it seemed natural to wonder whether experiments with this medium could measure the speed of the Earth as it moved through space.

an experiment), it was Einstein who sought such a breakdown as a means of proving the existence of atoms (or disproving the kinetic theory). Of course, it is worth noting that here also Einstein adopts an especially bold approach whose aim is to turn a model into a principal with universal applicability. According to Pais, Einstein himself used to wonder out loud why his predecessors had not taken this step. In making such arguments he was always at his most careful and precise in estimating the size of the proposed effect.

It was this line of reasoning which lay behind the celebrated Michelson–Morley experiment which set out to measure the ether wind flowing by the Earth as the Earth made its way through space. For those who saw Galilean relativity as a feature of particle motion which did not apply in the same way to wave motion, this experiment seemed an obvious step. To Einstein, and to Henri Poincaré, who like Einstein was always careful to speak of relativity as a principle, the experiment violated the fundamental principle underlying Galilean relativity, that it is impossible to do an experiment inside a closed room, without reference to other material bodies, which will reveal one's state of inertial motion. In the case of Galileo on his ship, the wave experiment involved using an outside material body, the water, to gauge one's velocity. But the ether was intangible and all-pervasive. The bit of ether within one's closed laboratory was somehow part of the same ocean of ether which pervaded all of space. It was this ether, impervious to the effects of normal matter, in which Einstein disbelieved. But for other physicists, like Abraham and Lorentz, relativity was primarily a set of rules for doing velocity addition and for transforming between different frames of reference, the nature of the rules being different for different kinds of interacting entities. For them it was natural enough that the rules of relativity varied for different kinds of systems.

Popular accounts of the development of special relativity often insist upon juxtaposing Michelson and Morley's famous experiment, designed to detect the speed of the Earth's motion through the ether, with Einstein's theory. Yet Einstein himself mentioned the experiment so rarely in his own accounts of relativity that there has been considerable debate about whether he had even heard about it in 1905 when he developed the theory (Pais, 1982, pp. 132–133). He certainly does not mention it in his papers of that year. Why then are people so convinced of the centrality of Michelson–Morley? Because it seemed like a natural route from the position of most late-nineteenth-century physicists, who believed in the ether, to the positions of twentieth-century physicists trained in Einstein's theory. But this does not mean that it actually played a central role in Einstein's route to special relativity. It is true that Einstein introduced the principle that measurements of the speed of light always obtain the same value. This seems to have been the result of the Michelson–Morley experiment. But if Michelson–Morley rules out any theory it is belief in the ether, and if Einstein ever believed in that theory he abandoned it at a relatively young age (Pais, 1982, pp. 130–131). He tells us that he flirted with emission theories, which are not ruled out by Michelson–Morley, before he settled on the belief, the radical departure, that the velocity of light depends neither on the velocity of a putative medium, nor on the velocity of the source, with respect to the observer. He is adamant that he always believed in the principle of relativity and sought only for a way in which to reconcile that with a range of experiments in electromagnetism and optics. When in later years he listed these experiments he essentially never included Michelson–Morley, but we do now have good reason to believe that he did know of the experiment, from one or two references. A recent discovery of a popular presentation given in Chicago shows an account in which he specifically acknowledges Michelson, but casting the experiment as simply being an example of one kind of experiment which solidified his early intuition that the principle of relativity was very fundamental and important (van Dongen, 2009). For Einstein, the experiment was not particularly important because its result was trivial. The principle of the constancy of the speed of light was not urged on him by this experiment, but only became possible to expound when, after years trying to reconcile it with the principle of relativity, he realized that it could be, provided he understood the relativity of simultaneity. It was only to others that the principle of constancy of the speed of light came out of the blue and was made acceptable when understood as a natural consequence of the well-known Michelson–Morley experiment.

For Einstein, a more important consideration was the established experimental fact that Faraday induction worked equally well if one moved the magnet or the coil of wire in which the electric current was induced by the changing magnetic field. This strongly suggested to him that electromagnetism was a phenomenon which, like the rest of physics, did not care which object was 'really' moving.

26.4 Relativity of Simultaneity

In 1905 Einstein had been wrestling with electrodynamics and the puzzle of relativity since he was a student in the late 1890s. Considering his isolation from the academic world, he was quite well read on the latest literature, especially the literature in German (though he had also read some of Poincaré's important works; see Pais, (1982), pp. 128–130). He had considered an emission theory of light as being more consistent with the principle of relativity than the etheric theory, but this concept seemed to be contradicted by the fact that light coming from binary star systems did not seem to be affected, in its time of arrival, by the quite large changes in velocity of the stars. If light was not affected by the speed of the source, and if relativity demanded that it not be affected by the speed of some immaterial medium, then it seemed almost as if all observers would measure a light-beam moving at the same speed, regardless of their own states of motion—an apparently absurd consequence of his logic and his adherence to the relativity principle.

While certain experimental results, such as the Michelson–Morley experiment and the symmetry of Faraday induction, favoured Einstein's position, there were equally well known experimental results that were harder for him to explain. Fizeau's experiments on the speed of light in moving water and the phenomenon of stellar aberration, by which one has to angle one's telescope to compensate for the Earth's motion so that light-waves can make their way down the telescope tube to one's eye, seemed to argue that the motion of the observer was important.

In later recollections Einstein stated that the final breakthrough came after a visit to a friend's house in mid-1905 (Pais, 2008, p. 139). From the solitary acknowledgement in the paper we can assume that the person in question was his close friend Michele Besso. The nature of the breakthrough was the realization that time was dependent on signal velocity because time was in fact defined by our observation of simultaneous events, and that this simultaneity was itself relative for observers in different states of inertial motion. He had been labouring under a presumption which all other physicists, save only Poincaré, also laboured under. The relativity of simultaneity would show Einstein how to resolve the apparent contradiction between his two principles, and how they could together explain all the apparently conflicting experimental results. In his paper, published remarkably rapidly after this breakthrough moment, he could present elegant calculations to explain Fizeau's experiments and stellar aberration. The fact that he does not mention Michelson–Morley anywhere is perhaps explained by the fact that he did not have to do any calculations to explain this result, as it was trivially predicted by his first principle alone.

What Einstein had realized is that not only is space relative, but so is time. Furthermore, when we talk about time we mean nothing more or less than the coincidence of events. While we are tempted to think of time as a thing in itself, in reality it is nothing more than the name we give to the tracking of coincidences between certain events which are perceived to be simultaneous. Einstein realized that if information takes time to travel from place to place, if the soonest we can learn of distant events is to see them when light from the event reaches our eyes, then different people may disagree on whether events at different locations really occur simultaneously. In particular, if we take as an axiom that light travels at a constant speed for all inertial observers, it follows that a person moving with respect to another observer may have news of one of the events before the other. Thus the two observers disagree about whether the events were coincident. They have experienced the relativity of simultaneity. Now, taking his two principles, and armed with the notion that they radically change how motion is measured, Einstein was able to write a paper which began as a study in kinematics and ended with the resolution, by unusually direct and simple means, of a number of outstanding problems in electrodynamics.

Since Poincaré had been the only other physicist to consider the relativity of simultaneity, it is natural to wonder to what extent he anticipated special relativity or could be regarded as a co-discoverer. One could argue that the relation between Einstein and Poincaré is similar to that between Einstein and Planck. Just as Einstein built upon the most radical of Planck's ideas, in an even more daring and radical way, so Einstein took (or independently reformulated in some cases) the more radical opinions of Poincaré on the relativity of space and time, and took them seriously in a more consistently radical way. Poincaré stated clearly that the ether would not prove detectable and that motion was to be understood in a relative rather than an absolute sense—ideas which we associate with Einstein today (Darrigol, 2000). But the truth is that the idea that all motion is relative and that space is not absolute is an ancient one. This issue of principle had, until Einstein, often fallen by the wayside when it came to mathematical expression of the laws of motion. Just as it was Newton's genius to show how the principals of the new physics could be given a precisely mathematical form, so it was Einstein's genius to show how the longstanding aspiration of a truly relativistic physics could be realized in a precise and consistent way, and in a way which solved a number of important outstanding problems in physics.

Another analogy we can draw is between Poincaré and the later Einstein who turned his back on quantum mechanics. Just as Einstein's contemporaries could not comprehend why Einstein would reject quantum mechanics when so much of the theory appeared to vindicate every argument about the old quantum theory he had made throughout his career, so we read Poincaré's statements on relativity theory and find it hard to understand why Poincaré should not have embraced relativity theory even in advance of Einstein. But the truth is that both men drew back from the full implications of these theories whose forms they appeared to have drawn in outline. For instance, Poincaré referred to Lorentz' theory, which we now regard as quite close to the new theory which Einstein would propound as merely 'the least defective' amongst leading theories of the day because it merely 'best accounted for the facts' (translated and quoted in Darrigol, 2000, p. 356)—a formulation which recalls Einstein's attitude to quantum mechanics that it was an incomplete theory which enjoyed the dubious virtue of appearing to coincide with the phenomena.

At any rate, there can be little doubt that Poincaré and Einstein saw eye to eye on the subject of the principles which, in their view, lay at the heart of physics. They would have agreed that these principles have, as Darrigol has put it, an inductive origin, but

the principles are results of experiments boldly generalized; but they seem to derive from their very generality a high degree of certainty. In fact, the more general they are, the more frequent are the opportunities to check them . . . (Poincaré, translated and quoted in Darrigol, 2000, p. 355)

leaving, in the end, no place for doubt as to their correctness. This would certainly have been Einstein's view also. Where Lorentz had painstakingly built up to the correct equations from his microscopic model, and where Poincaré had critically reviewed these efforts and their apparent disregard for the principles, it was Einstein who took the critical step of showing precisely how the principles were to produce the equations clearly and logically.

26.5 MASS–ENERGY EQUIVALENCE

As we have seen, although the gestation of his theory had been long, Einstein published within weeks of the breakthrough insight which permitted him to finish his theory. But his fertile mind then gave birth to an afterthought, published the same year in a famous paper in which he proposed the equivalence of mass and energy. This result is perhaps the best illustration of the power of Einstein's approach to the problem of electrodynamics. Several other physicists had realized that the inertia of energy (the idea that energy, and not just matter as normally conceived, had a resistance to being accelerated) played a role in the study of moving charged particles. Some even came up with the precise and now famous form of Einstein's equation

$$E = mc^2 \qquad (1)$$

but they were all convinced that the phenomenon was peculiar to the electromagnetic field and arose in the process of the complex calculations involving electrodynamics. Some of them speculated that it would turn out that all inertia, and therefore all mass, was electromagnetic in origin, the so-called electromagnetic world view (Darrigol, 2000, p. 360).

The common factor of previous investigations into the inertia of energy was that they concerned only the kinetic energy of charged particles like the electron. It had been noticed that calculations and experiments agreed that a faster electron had more inertia than a slower one. Einstein showed that light also must have inertia in the same proportion to its energy. He made this proposal in order to preserve conservation of momentum within the Lorentzian framework of his theory (a problem which had been noticed earlier by Poincaré). If light could exert pressure and thereby alter the momentum of material bodies, it followed that if a body was to be prevented from moving its own centre of mass without reacting with an outside body, the light must carry mass with it as it flew from one end of the material body to the other (Einstein, 1905b). Although light energy is still electromagnetic in form, Einstein daringly hypothesized that all energy had inertia. He thus completed his escape from the straitjacket of the electromagnetic worldview which had come to dominate the subject in the preceding years. He also pointed the way towards the experimental confirmation of his theory.

26.6 Experimental Tests of Special Relativity

It is well known that relativity theory faced considerable resistance amongst early-twentieth-century physicists who, as many non-physicists still do, had great difficulty accepting it. Einstein's Nobel prize was, if anything, delayed, rather than facilitated, by his groundbreaking work on relativity (Pais, 1982, pp. 509–511). His citation in 1923 does not mention his work on relativity. One reason for the cool reception which his theory received was the scarcity of experimental confirmation. The theory claimed that the predicted effects which departed from Galilean relativity would be noticeable only when the speed of a particle became non-negligible compared to the speed of light. It was extraordinarily difficult to perform direct tests of the theory's main predictions, such as time dilation or the inertia of energy, because it was difficult to find or produce objects moving at the required speeds.

The difficulty of experimentally testing the theory raises the interesting question as to whether the development of the theory itself was influenced by emerging technology. It is certainly suggestive that Einstein worked as a patent clerk and was often called upon to analyse the workings of modern devices concerned with the electromagnetic transmission of signals and electromechanical clocks. Synchronization of clocks, including clocks in different locations, was an important technological challenge of the era. By 1924 the first really popular clock synchronization, from Greenwich, was perfected with the famous 'pip' system broadcast by the BBC. Poincaré also, it has been argued, was very much concerned with the problem of standardization of time made necessary by the advance of rapid travel and communication. Peter Galison, (2003) has argued that this was no coincidence, and that Einstein and Poincaré's ideas were greatly influenced by the increasing technological demand for synchronization of clocks in different locations. But it must be admitted that the level of accuracy in contemporary synchronization did not demand nor require any relativistic corrections, so we cannot go further and say that technology produced a direct demand for a theory of relativity. The paucity of early experimental verification may be one reason why many accounts of the genesis of the theory focus on the Michelson–Morley experiment, which actually took place decades before the theory itself was formulated.

Early attempts to test relativity theory focused on the mass–energy equivalence result and sought to distinguish between the ether theory of Lorentz, with which Einstein's theory was conflated, and the electron theory of Max Abraham. Experiments by Walter Kaufmann appeared at first to favour Abraham's theory, to Lorentz' consternation and Einstein's apparent unconcern. Subsequent experiments by others were said to favour Einstein, though the difficulty of reaching the required precision in the measurements prevented any of these measurements from being considered an *experimentum crucis* (Miller, 1981).

Other efforts centred upon complex calculations which could show how more readily observable phenomena, such as the fine structure of atomic spectral lines, could be shown to depend on relativistic effects, in this case mass–energy equivalence (Pauli, 1958, p. 83). In this case the experiments were more clear cut, but depended for their interpretation on such difficult theoretical work that the claims in favour of Einstein were less likely to be heeded by most of relativity's critics, many (but not all) of whom lacked mathematical sophistication.

Regarding other experimental predictions, the Michelson–Morley experiment was widely taken to be confirmation of length contraction. As for Einstein's main prediction which had not been anticipated by others, time dilation, often known as the transverse Doppler effect, it took many years before this subtle effect could be verified experimentally.

26.7 THE PRINCIPLE OF EQUIVALENCE

The success of relativity theory in the years after 1905 posed at least one major new challenge for physics, mirroring the original problem which had just been solved. Maxwell's field theory of electromagnetism had failed to conform to the old Galilean relativity. Its equations assumed a different form when transformed according to the old relations. Thus different observers would disagree on the results of a measurement. That problem had been solved, since Maxwell's equations were invariant under Lorentz transformations, but Einstein had succeeded in resolving the tensions within electrodynamics by showing kinematically that the Lorentz transforms were, in fact, the correct relativistic transformation equations for all of physics. Now the problem was that Newton's old action-at-a-distance theory of gravity was not invariant under the Lorentz transformations, though it had been under the old Galilean ones. Thus Newton's theory, still in use, failed to conform to the new Einsteinian theory of relativity. Accordingly, in the second decade of the twentieth century several physicists attempted to construct a new relativistic theory of gravity, their goal being a theory whose equations, while agreeing with Newtonian predictions in nearly all instances, would be written in a form which would be invariant under Lorentz transformations.

Perhaps predictably, Einstein took an entirely different route towards a new theory of gravity. He began, instead, by looking for a more general theory of relativity. This quest—so urgent to Einstein that he devoted sometimes enormous effort to it over almost a decade—appeared quixotic to his contemporaries. For Einstein, a general theory of relativity would be one whose field equations (for he expected it to be a field theory, as did his contemporaries) would be invariant under all possible transformations, not merely the Lorentz transformations between two inertial frames of reference. In Einstein's proposed general theory all states of motion, including accelerated motion, would be regarded as valid for the purposes of performing experiments in physics. There were to be no privileged observers in this new theory, and the same set of equations would do for all of them—a feature which Einstein called 'general covariance'.

Other physicists, however, firmly believed that Newton had established that accelerated motion had an absolute character—a judgement based upon a philosophical outlook regarding the reality of the concept of space itself. Thus, someone in a closed laboratory would certainly be able to tell unambiguously if that laboratory was being accelerated through space, say by a rocket engine, by doing quite simple experiments.

Einstein did not dispute this point, but in 1907 he had what he later described as 'the happiest thought of my life'[6] when looking out the window of his workplace at the Swiss patent office in Bern. He realized that a person falling would not feel their own weight. He recalled that Galileo had long ago shown that all objects, no matter what their size, mass, or composition, fall at the same rate of acceleration in the Earth's gravitational field. Accordingly, a painter falling down the side of a house would remain together with all of his gear, unmoving with respect to each other as they fell. Additionally, the surface on which he had been standing would fall with him. It would no longer need to support his weight, nor would it be able to, since it would be falling equally fast. Einstein realized that the man would not feel his own weight, would not see any relative motion in the objects falling with him, and that if he was enclosed (as in an elevator), would no longer have any way to detect if he was on the surface of the Earth, because there would be no apparent gravitational effects. Einstein had discovered the concept of free fall long before the arrival of high-altitude flight.

[6] Translated and quoted in Pais, (1982), p. 178.

So, if accelerating downwards could precisely cancel out any gravitational effects, this insight implied that there was an important equivalence between the effects of gravity and those of acceleration. Einstein realized that if a person such as his painter were in a closed room and did feel the normal effects of gravity, it would be exceptionally difficult for them to perform an experiment which could distinguish between being on a planet or being inside a spacecraft accelerating through space at a particular rate (at least if the room was sufficiently small).

For other physicists, the fact that inertial and gravitational mass were precisely equal was an interesting fact that permitted a pleasing simplification of equations of motion involving purely gravitational forces. For Einstein, it was an example of the kind of extremely well verified feature of nature which deserved elevation to a principle of the Poincaré type and the most serious consideration as a clue to the inner workings of that nature.

GR is so much Einstein's creation that there is little need to talk about the efforts of his contemporaries to develop a relativistic theory of gravitation. Max Abraham's efforts in this regard make a good illustration of why others failed. Abraham was a brilliant physicist who had good insight into the nature of the problems he faced but, as his polemic of 1912 with Einstein shows, he refused to accept the primacy of the equivalence principle (or even the relativity principle), and thus his theory ended up having no influence whatever on the development of gravitational theory, which developed almost exclusively around metric theories of gravity, for reasons which will become apparent. Gunnar Nordstrom, by contrast, profited from responding to Einstein's criticisms of his theory, to the extent that it was to end up by being reformulated, largely by Einstein himself, as a sort of scalar cousin to GR (Norton, 1993). It was not considered a rival to GR, because of its failure to predict correctly the results of the first few tests of GR—in particular, the advance of Mercury's perihelion.

26.8 Spacetime Curvature

It seems probable that when Einstein originally considered a general theory of relativity he thought the process would be much like the development of the special theory. He would identify the correct principles which the theory must uphold, consider carefully those critical experimental results with which it must agree, and identify and analyse the appropriate thought experiments which would, in a clear, logical way permit a mathematically simple theory to emerge. Recall that the mathematics of special relativity had been, in some sense, simpler and more elegant than the calculationally intensive derivations of similar equations by Lorentz. Instead, the experience was one in which Einstein himself was obliged to master more and more arcane topics in mathematics which he had previously ignored or disdained, and ended with his developing a profound respect for mathematics, specifically the search for simplicity therein. As a result he came to feel that the royal road to success in physical theories lay in the quest for mathematical simplicity—a belief which seems not to have left him all the rest of his life (van Dongen, 2010).

The main question which confronted Einstein and everyone else searching for a relativistic theory of gravity at this time was what quantity should be selected as the potential of the new field. Newtonian gravity had long since been recast in a potential form, so the most obvious route was to make use of a suitably modified version of this Newtonian potential. Einstein rejected this route, because it did not permit the retention of the equivalence principle. His earliest attempt was based on his realization that the equivalence principle also demanded the overthrow of one of the two underlying principles of special relativity: the constant light-speed principle. Since gravity must affect light by accelerating it, it followed that the velocity of light was no longer constant in a gravitational field. So Einstein's theory of 1911 actually used c as its potential, its measure of the strength of the field. Therefore, like Newton's theory and like other contemporary efforts such as those of Abraham and Nordstrom, it was a scalar theory. The potential was not a vector, as in the case of the electromagnetic field.

What led Einstein to reject a scalar theory of gravity was again the equivalence principle—in particular, his realization that this principle demanded that spacetime in a gravitational field must be curved. Historians agree that Einstein first came to this insight through his consideration of rotation (Stachel, 1989). If the equivalence principle is correct then a rotational frame of reference, which clearly involves acceleration, will share characteristics of a gravitational field. One of the early debates surrounding special relativity was known as the Ehrenfest paradox after its co-discoverer (along with Max Born), Paul Ehrenfest (both men were to become close friends of Einstein). If one tries to measure both the radius and circumference of a rotating disk, the circumference will be altered by length contraction (it will seem longer because one's co-rotating meter sticks are shortened) while the radius will not (because it lies perpendicular to the direction of motion). Thus the ratio of these two quantities is no longer 2π, and the geometry of the disk is altered. Most physicists of the time focused on the implications that this held for the concept of rigidity in relativity theory (it implied that infinitely rigid bodies could not exist in nature). Einstein latched on to the idea that spacetime could be curved by rotation, and therefore by acceleration and, if the equivalence principle was correct, by gravitation.

It seems likely that he came up with multiple arguments in favour of this conclusion; but at that time he wrote down few of his ideas as he worked on them, except on infrequent occasions when he wrote to colleagues concerning his nascent ideas. From his later correspondence with Nordstrom, we learn of a number of clever thought experiments by which he argues that the equivalence principle, and other fundamental principles, such as conservation of energy, force spacetime curvature upon us when we attempt to construct a relativistic theory of gravity (Norton, 1993). What is clear is that from about 1912 Einstein was convinced that spacetime curvature must play a role in his theory of general relativity.

This line of thinking forced a radical reassessment of the best choice of a potential for the theory. If the mark of a gravitational field was that it altered the geometry of spacetime, this is merely to say that it alters the way in which we measure space and time. Of course, this was a hallmark of the special theory already, in a restricted sense, and this aspect of that theory was brought out nicely in a paper by the mathematician Hermann Minkowksi, in 1907. He showed how Einstein's theory could be rewritten in an inherently four-dimensional form using four-vectors and tensors.[7]

Using Minkowski's concept of spacetime one can show that the rules of length contraction and time dilation can be written concisely in terms of a metric, which can be thought of as a condensed way of writing and generalizing Pythagoras' theorem. If we think of Pythagoras' hypoteneuse as the shortest distance between two points (or two events in spacetime), then the sides of a right-angled triangle can be thought of as the coordinate values associated with that distance (or spacetime interval) in a rectilinear (for instance, Cartesian) coordinate system. By generalizing the coordinate system (and thus dispensing with the Pythagorean right-angled triangle) we can say that the square of the length of the hypoteneuse is equal to the sum of the squares of the coordinate distances (formerly sides of the triangle), plus terms which are cross-products of the coordinate distances (that is to say, terms of the form $dxdy$ as opposed to dx^2), each term having a coefficient necessary to make the derived answer correct for that choice of coordinates. The usual version of the theorem, involving Cartesian coordinates, gives all these coefficients as one where the term is a square, and zero where it would be a cross-product. Use of the Minkowski metric makes it particularly simple to transform between coordinates, because the metric is a tensor, a set of coefficients which obey certain strict transformation rules, rather than a mere matrix of numbers.

Initially, Einstein seems to have been unenthusiastic about Minkowski's approach. In 1907 he preferred a conceptual, phenomenological approach to physics, rather than a mathematical one. But in 1912 he came to appreciate that the metric could play a role very like the gravitational potential in his new theory (Renn and Sauer, 1999). In Minkowski's version of special relativity the metric was physically unchanging. All that changed was the individual coefficients as one transformed between coordinate systems. This can be interpreted as making a change of inertial observer. Each inertial observer sees the same physics and the same laws, but measures different quantities in time and space, depending on their frame of reference, because they are using coordinate systems different from those of other observers, if those other observers are in a different state of inertial motion to them. But in general relativity different observers would actually experience different spacetime geometries. The metric itself, and not just one's coordinate representation of it, would be different in a gravitational field. Thus a generalization of Minkowski's metric representation of special relativity would have just the features of the general theory that Einstein was seeking. But the metric was not a scalar, nor even a vector; it was a four-dimensional symmetric tensor. Thus the new theory would require ten different equations to constrain the metric—four for the square (or diagonal) terms (one each for the four directions of spacetime), and six for the cross-terms, which mixed spacetime coordinates.

---

[7] This represented, at least qualitatively, a return to the quaternions of Hamilton, from which three-dimensional vectors had been derived.

Einstein was familiar with the fact that in the previous century Gauss had developed an algebraic theory of geometry on curved surfaces which differed from traditional Euclidean geometry, which applied only to flat surfaces, in just the kind of way that Einstein's new theory would differ from Minkowski's. Thus special relativity dealt with relativistic spacetime which was flat in the Euclidean sense, while general relativity would deal with curved spacetime. The problem was that Guass' theory was limited to two dimensions. Einstein approached his old school-friend, the mathematician Marcel Grossman, and famously appealed to him: 'Grossman, you must help me or else I'll go crazy.'[8] He wanted Grossman to tell him if there was a theory which described four-dimensional geometries using tensors. Happily, Grossman was able to return with the reply that such a theory already existed—the work of the mathematician Bernhard Riemann, extended by Gregorio Ricci-Curbastro and Tullio Levi-Civita.

We know far more about the subsequent development of GR because Einstein's style of work changed noticeably at this point, from a solitary pursuit to a collaborative one. He worked with Grossman to learn the unfamiliar language of tensor calculus and Reimannian geometry and apply it to gravitational physics, and his research process altered from one involving profound cognition based around thought experiments to one centred on calculations on paper. Pais has described the change in Einstein's style at this time: 'Prior to Einstein's involvement with gravitation, each one of his papers is transparent and self-contained ... From 1907 to 1916, this light touch and this element of closure is missing. His style of writing changes ... [to] reports on work in progress' (Pais, 1982, pp. 194–195).

This change in Einstein's work method is a boon to the historian, because much of his joint effort with Grossman was recorded in the Zurich notebook in 1912 and 1913. What historians have learned about the genesis of GR from this notebook is extensive (Norton, 1989; Renn, 2007a and 2007b). What came publicly from their collaboration was a new theory of gravity, known as the Entwurf (draft) theory from the title of the paper announcing it (Einstein and Grossmann, 1913). As a theory of gravity it appeared far different from anything which had gone before. It also appeared far different from Einstein's hopes and dreams, as one particularly desired feature of the new theory had been dropped entirely.

26.9 General Covariance

The Entwurf theory was not a generally covariant theory. That is to say, its field equations were not valid for all observers. To most physicists this would not have seemed a problem. It had always been understood that some observers experienced physics which was different from that experienced by inertial observers. Inertial observers (those not experiencing acceleration) were privileged in this sense. From whence did their privilege arise? From the fact that they were in uniform motion with respect to

---

[8] Translated and quoted in Pais, (1982), p. 212.

absolute space. For Einstein, this was objectionable. Had he banished the ether but left behind an immaterial void which no-one could ever observe (but for the material bodies which marked its boundaries) and which somehow determined which physical equations we should employ? He had always entertained the hope that his final theory of GR would permit any observer, with a suitable transformation of the equations, to reconcile their measurements with those of any other observer. This was the meaning of general covariance.

In the struggle to develop the Entwurf theory with Grossman, Einstein gave up general covariance. The reasons are complex, but two salient points should be noted. One is that in the Zurich notebook Einstein at one point wrote down something very like the final correct field equations. He most likely was led to reject them because the switch to a metric theory had complicated one of the important touchstones of his programme, which was that the new theory must, in the limit of small velocities and weak gravity, approximate to the old Newtonian theory. But how was a tensor theory to approximate to a scalar theory? At what point (and how) must the complexity of ten equations reduce to just one? Eventually Einstein recognized that if only one of the ten equations contained terms of significant size, than the other nine would make tiny undetectable changes to any measurements one made. But in 1913 he rejected promising approaches because he was uncertain how the Newtonian limit would work.

Faced with his failure to achieve general covariance, Einstein's fertile mind came up with an argument that it was never achievable anyway. It became known as the Hole argument, because Einstein argued that in a hole which consisted of just empty space with no matter, a general relativistic theory would admit of more than one solution, making it an inconsistent theory. Later he realized that there is nothing wrong with this, since no-one can detect these multiple field solutions without introducing some matter into the Hole. But for at least two years the Hole argument convinced Einstein that his search for a generally covariant theory of gravity had reached an imperfect conclusion.

These two years were turbulent for Einstein personally, who moved to Berlin and separated from his wife and children, and for Europe, which witnessed the outbreak of the First World War. Einstein was not ignoring his theory. He made vigorous efforts to test it observationally by obtaining backing for a German expedition to the Crimea in 1914 to observe light-deflection during an eclipse of the Sun. But the outbreak of war doomed the expedition—perhaps fortunately, since Einstein was about to change his theory, and this prediction, radically.[9]

A key figure in the final push to the complete theory was the mathematician David Hilbert of Göttingen. When Einstein moved from Zurich to Berlin in 1914 he left Grossman behind. But in 1915 he paid a visit to Göttingen during which he addressed the mathematicians there on the subject of his Entwurf theory. It seems that Hilbert was very excited by Einstein's approach, which made far more use of quite esoteric branches of mathematics than was typical for the physics of that time. Einstein's interactions with Hilbert seem to have convinced both men that a generally covariant theory of gravity was possible, and Einstein returned to the drawing board in late 1915.

---

[9] Earman and Glymour, (1980a), Crelinsten, (2006).

This time he had the hard-won mastery of tensors and geometry to see the project through, and in November 1915 he discovered the final form of the equations which bear his name. His excitement was intense when, performing calculations now quite familiar from his days working on the Zurich notebook with Grossman, he discovered that his theory precisely explained the well known anomaly in Mercury's orbit. The new theory thus made a triumphant announcement of its birth to the world.

A longstanding controversy concerns whether Hilbert might have actually beaten him by some few days to the final form of the Einstein equations. It is known that Hilbert played a role in Einstein's final successful push, and this step came only after many insights which were Einstein's alone. But firsts are important in physics, and many have argued that Hilbert was first because his paper of early 1916, which discussed Einstein's theory, includes the field equations and was submitted a few days before Einstein's own paper which published them first. Einstein's paper was published so swiftly that some comentators have considered it possible that Hilbert could have learned of the field equations in the interim, and only then included them in his paper. One recent discovery at least suggests this possibility, that Hilbert added the field equations while his paper was in proof, after Einstein's paper had appeared (Corry, Renn, and Stachel, 1997). While this detail does not alter the relative contributions of the two men in any really fundamental way (and the episode did not more than briefly affect their personal relationship), it does cast doubt on those attempts that have been made to lessen the extent to which GR is seen as Einstein's theory. Most historians acknowledge that while special relativity was an idea that was somehow in the air, general relativity would have waited possibly decades to find a discoverer had Einstein not developed it when he did.

Fundamentally, what Einstein achieved was a theory which unified the two principle components of the Newtonian system: the laws of motion, and the law of gravitation. In Newton's *Principia* these are stated quite separately. To make the laws of motion work, you must have a separately stated force law. This force could be gravitational, electromagnetic, or anything else. In GR the gravitational field and the laws of motion—what one might call the inertial field—are inextricably linked. Even in the absence of gravity, the inertial structure of spacetime (the flat spacetime described by Minkowski) still instructs matter how to move and is a perfectly valid solution of Einstein's equations. Indeed, as we shall see, this triumph spelled trouble for Einstein's final principle—the only one of the four guiding principles of his quest for relativity which he did not succeed in fully incorporating into the final form of his theory.

26.10 Mach's Principle

As with special relativity, Einstein promulgated two principles which, in his view, formed the underpinning of his general theory. In the general case the second principle was always referred to by Einstein as Mach's principle. Ernst Mach was a profound believer in the relationalist view of space, which declares that space is not a container into which material objects are put (the substantivalist view), but rather that space only has existence insofar as one is able to measure relations between material bodies. This yields quantities such as position and distance, but these are to be understood as properties of the things themselves and not of some invisible entity known as space. In the relationalist viewpoint, space exists only insofar as objects exist to define it. Without the objects there would be no space.[10]

The argument between these viewpoints is longstanding, and finds its classic modern expression in the debate between Newton and Leibniz. A compelling argument put forward by Newton was that of Newton's bucket, not dissimilar to Galileo's vase, mentioned previously. Newton observed that spinning a bucket of water would produce a curved surface to the water because of the centrifugal force. Someone rotating with the bucket would be able to tell they were in a state of rotational motion simply by looking at the water in the bucket; thus this person could tell his state of motion without looking at an outside body. Mach's response to this was that it could be that the reason for the centrifugal force was the bucket's failure to remain at rest not with the rest frame of space (as Newton and other substantivalists would have it), but with the rest frame of the other matter of the universe. Some force from this matter must be acting on the water in the bucket. The only plausible candidate is the force of gravity, and it seems clear that a notion that gravity might be behind centrifugal forces and the other mysteries of rotation lay at the heart of Einstein's quest for general relativity.[11]

Mach's principle always was, and remains, elusive. Einstein himself, who coined the term, gave it different meanings at different times. It is not entirely clear what Mach would have understood by his own principle. Nevertheless Einstein definitely saw it as expressing some of the key elements of the relationalist programme. It also seems likely that his interest in Mach's principle played an heuristic role in his search for general relativity. It may well have suggested the link between gravitation and rotation which lay at the heart of many of his fruitful insights in developing the theory, and it almost certainly inspired his insight that a general theory of relativity would also be a new theory of gravity.

Ironically, GR failed, in the view of most people, in the attempt to embody either Mach's principle or the relationalist view of spacetime. It is true that where Newtonian gravity had no non-radial component and therefore could not easily find room for the sort of effect that Mach was seeking, GR did introduce gravitomagnetic effects which can allow one body to impart rotation to another body at a distance. The frame-dragging effect itself, however, can only be viewed as Machian in a limited sense. Its modern interpretation is an example of substantivalism at its ripest. As a massive body rotates it drags spacetime around with it, imparting

---

[10] See, for instance, Earman, (1989).
[11] For more, see Babour and Pfister, (1995).

rotation to nearby bodies. To most modern physicists spacetime is the substance which transmits these effects.

Initially, Einstein was sanguine that he had succeeded in his Machian programme. This is most clearly seen in his ground-breaking efforts to address the problem of cosmology through his brand new theory in 1916. He chose a closed universe, at least partly because he felt that an open, unbounded, universe with finite matter posed a problem for relationalists. How could space even exist far from the matter which gave it form? But he astutely realized that a closed universe could not be static in the original form he had given his field equations. Therefore, he added a term, the famous cosmological constant, to permit a static (albeit unstable) solution (Einstein, 1917). Quickly, other theorists, notably Willem de Sitter, showed that the theory did permit solutions which had space but no matter—a direct threat to Einstein's relationalist interpretation of GR. After a vigorous debate Einstein admitted the validity of these solutions (Schulmann *et al.*, 1998, pp. 351–357), and even presented a talk before Lorentz himself, observing that the new spacetime fabric of GR could be viewed as a kind of ether. He comforted himself that this new ether was acted upon by matter and did not just act upon matter, thus removing some of the unphysicality of the old luminiferous ether with its ghost of the absolute frame of reference (Einstein, 1920).

Much later it was shown that the universe is not static, and the cosmological constant was dropped, though it has made a comeback more than once (Earman, 2001). Einstein counselled against it revival. We do not know for certain that he ever called it his 'greatest blunder', but he did write of his 'bad conscience' at having ever introduced it. What is often not appreciated is that with the cosmological constant Einstein also seems to have buried his Machianism. Although this fourth principle, whose status remains in debate, may not enjoy the same foundational role as the other principles he employed in developing his theory, it is hard to deny its heuristic value. The relationalist viewpoint that Mach's principle expressed for Einstein lay at the heart of his unique path to this important theory.

26.11 EXPERIMENTAL PREDICTIONS OF GENERAL RELATIVITY

As with special relativity, a major challenge facing Einstein was to identify scenarios in which his theory would predict novel and measurable experimental results. Almost all the predictions of his theory agreed very closely with those of Newton's theory of gravity—a necessary feature of a new theory because of the enormous empirical success which Newton's theory had enjoyed. Einstein probably considered many possibilities in trying to identify experiments. We know that in letters written shortly after the final development of the theory he emphasized the impossibility of ever observing some features of the theory (examples include frame-dragging and gravitational waves, both proven to exist only in recent times).[12]

In the end, he never did identify tests of his theory beyond three that he had already considered at the earliest stages of his work on the general theory. Two of these were predictions he made on the basis of the equivalence principle alone in 1907. One of these was the famous prediction of the bending of light by the Sun, argued on the basis that light-energy had mass and therefore must fall in a gravitational field, because the equivalence principle demands that all mass responds identically to a gravitational field. The other was the prediction of gravitational redshift of light from the Sun. A third test—which we have reason to believe was in his mind from an early stage, and certainly from 1912 onward—was the possibility of explaining the then largest anomaly in Newtonian celestial mechanics: the advance in Mercury's perihelion. All three cases applied to the vicinity of the Sun, because the Sun has the most intense gravitational field in the solar system. Just as only at high velocities does special relativity differ from Galilean relativity, so does general relativity differ from Newtonian gravity only when a gravitational field is sufficiently strong.

As early as 1913 Einstein was writing to astronomers proposing that they look into making the first two of these tests. They had already done the work in the case of the third, the Mercury perihelion test. This proved fortunate for Einstein, because persuading astronomers to test the theory proved frustratingly difficult. In the case of the light-bending test, this may have been a good thing. Until he finalized his theory in late 1915, he did not appreciate that his original value for the bending of light in the Sun's gravitational field was incorrect. He had simply invoked the equivalence principle to argue that light, having energy and therefore mass, must fall in this field like all other masses. But in addition, his final theory showed that the curvature of spacetime near the Sun, itself an expression of this gravitational field, altered the definition of what straight-line motion would be. In following a local geodesic in this curved spacetime, light would appear to be deflected from the straight-line course as seen by observers in the all-but-flat spacetime of the Earth. Thus pre-1915 efforts, which were all frustrated by circumstance, including the onset of the First World War, would have been testing the wrong theory.

Before any successful tests could be carried out, Einstein was able to use the final version of the theory to explain Mercury's perihelion advance. It turns out that it is the curvature of spacetime predicted by GR which is responsible for the failure of Mercury's orbit about the Sun to describe a closed ellipse. Thus we see why Einstein would not have publicly mentioned the Mercury perihelion advance as a possible test before 1915. He had no way of anticipating that it would prove to be a test which his theory could pass. The unanticipated, though hoped-for, success gave him 'heart palpitations' and caught the attention of many other physicists (Earman and Janssen, 1993). The theory now had the standing for astronomers to begin to take it more seriously. More widespread and better-funded efforts to test the light-bending prediction

---

[12] Einstein to Wilhelm Hort, 29 November 1919, Doc. 181 in Buchwald *et al.*, (2004).

followed, resulting in the famous British expedition of 1919 which confirmed his theory, and an American expedition which allayed most remaining doubts in 1922 (Crelinsten, 2006).

Meanwhile, solar physicists had been wrestling with the problem of the solar redshift since before Einstein had even begun working on his theory, and had initially proved relatively uninterested in Einstein's claims. Moved to consider him more carefully after the success of the Mercury perihelion 'test', they still were inclined to be sceptical. The solar redshift, which sometimes changed to a blueshift, varied from line to line and from place to place on the Sun in a way which seemed difficult to reconcile with Einstein's theoretical prediction. It was apparent that a good part of the line-shifting was due to Doppler-shifting (frequently, as it transpired, blueshift due to rising columns of hot gas in the solar atmosphere), and attention began to focus on the limb of the Sun, where the line of sight is perpendicular to such vertical motions in the gas. But solar spectroscopists were still dubious, it would seem, until after the 1919 eclipse expedition made it two out of three triumphs for GR. Motivated to examine the data through a lens which suggested that they were dealing with an epochal work of genius, rather than another quixotic challenger to Newton, in the early 1920s astronomers quite quickly changed their minds about Einstein's third prediction (Earman and Glymour, 1980b, Hentschel, 1993). Only decades later did a series of technological advances inaugurate an era of precision tests of GR which the theory passed as triumphantly as Newton's theory had its tests in a much earlier epoch (Will, 1985, 1993).

References

BABOUR, JULIAN, and PFISTER, HERBERT (1995). *Mach's Principle: From Newton's Bucket to Quantum Gravity*. Birkhäuser.

BUCHWALD, DIANA KORMOS, SCHULMANN, ROBERT, ILLY, JÓZSEF, KENNEFICK, DANIEL, and SAUER, TILMAN (2004). *The Collected Papers of Albert Einstein Vol 9: The Berlin Years, Correspondence January 1919–April 1920*. Princeton University Press.

BUCHWALD, JED Z. (1985). *From Maxwell to Microphysics: Aspects of Electromagnetic Theory in the Last Quarter of the Nineteenth Century*. Chicago University Press.

CORRY, LEO, RENN, JÜRGEN, and STACHEL, JOHN (1997). Belated Decision in the Hilbert–Einstein priority dispute. *Science* 278, 1270.

CRELINSTEN, JEFFREY (2006). *Einstein's Jury: The Race to Test Relativity*. Princeton University Press.

DARRIGOL, OLIVIER (2000). *Electrodynamics from Ampère to Einstein*. Oxford University Press.

VAN DONGEN, JEROEN (2009). On the role of the Michelson–Morley experiment: Einstein in Chicago. *Archive for the History of the Exact Sciences* 63: 655–663.

VAN DONGEN, JEROEN (2010). *Einstein's Unification*. Cambridge University Press.

DRAKE, STILLMAN (1970). *Galileo Studies: Personality, Tradition and Revolution*. University of Michigan Press.

EARMAN, JOHN (1905). *World Enough and Spacetime: Absolute versus Relational Theory of Space and Time*. MIT Press.

EARMAN, JOHN (2001). Lambda: The constant that refuses to die. *Archive for the History of Exact Sciences* 55, 189–220.
EARMAN, JOHN, and GLYMOUR, CLARK (1980a). Relativity and eclipses: The British eclipse expeditions of 1919 and their predecessors. *Historical Studies in the Physical Sciences* 11, 49–85.
EARMAN, JOHN, and GLYMOUR, CLARK (1980b). The gravitational redshift as a test of general relativity: history and analysis. *Studies in the History and Philosophy of Science* 11, 175–214.
EARMAN, JOHN, and JANSSEN, MICHEL (1993). Einstein's explanation of the motion of Mercury's perihelion. In *The Attraction of Gravitation, Einstein Studies Vol. 5* eds. Howard, Don, and Stachel, John, pp. 129–172. Birkäuser.
EINSTEIN, ALBERT (1905a). Zur Elektrodynamik bewegter Körper. *Annalen der Physik* 17, 891–921.
EINSTEIN, ALBERT (1905b). Ist die Trägheit eines Körper von seinem Energieinhalt abhängig? *Annalen der Physik* 18, 639–641.
EINSTEIN, ALBERT (1915a). Erklärung der Perihelbewegung des Merkur aus der allgemeinen Relativitätstheorie. *Königlich Preußische Akademie der Wissenschaften, Sitzungsberichte* 831–839.
EINSTEIN, ALBERT (1915b). Die Feldgleichungen der Gravitation. *Königlich Preußische Akademie der Wissenschaften, Sitzungsberichte* 844–847.
EINSTEIN, ALBERT (1917). Kosmologische Betrachtungen zur allgemeinen Relativitätstheorie *Königlich Preußische Akademie der Wissenschaften, Sitzungsberichte* 142–152.
EINSTEIN, ALBERT (1920). *Äther und Relativitätstheorie: Rede gehalten am 5 Mai 1920 an der Reichs-Universität zu Leiden*. Springer.
EINSTEIN, ALBERT, and GROSSMANN, MARCEL (1913). *Entwurf einer verallgemeinerten Relativitätstheorie und einer Theorie der Gravitation*. Teubner, Leipzig.
GALISON, PETER (2003). *Einstein's clocks, Poincaré's Maps: Empires of Time*. W. W. Norton.
GOLDBERG, S. (1970), The Abraham theory of the electron: the symbiosis of experiment and theory. *Archive for the History of Exact Sciences* 7, 7–25.
HENTSCHEL, KLAUS (1993). The conversion of St John: a case study on the interplay of theory and experiment. *Science in Context* 6, 137–194.
JANSSEN, MICHEL (2002). Reconsidering a scientific revolution: the case of Lorentz versus Einstein. *Physics in Perspective* 4, 421–446.
JANSSEN, MICHEL, and MECKLENBERG, MATHEW (2007). From classical to relativistic mechanics: electromagnetic models of the electron. In *Interactions: Mathematics, Physics and Philosophy, 1860–1930* eds. Hendricks, V. F., Jorgensen, K. F., Lützen, J., and Pedersen, S. A. Doredrecht: Springer.
MARTINEZ, ALBERTO A. (2009). *Kinematics: The Lost Origins of Einstein's Relativity*. Johns Hopkins University Press.
MCCORMMACH, RUSSELL (1970). H. A. Lorentz and the electromagnetic world view of nature. *Isis* 61, 459–497.
MILLER, ARTHUR I. (1981). *Albert Einstein's Special Theory of Relativity: Emergence (1905) and Early Interpretation (1905–1911)*. Addison-Wesley.
MINKOWSKI, HERMANN (1907). Das Relativitätsprinzip. *Annalen der Physik* 352, 927–938.
NORTON, JOHN (1989). How Einstein found his field equations: 1912–1915. In *Einstein and the History of General Relativity*, eds. Howard, Don and Stachel, John, pp. 101–159. Birkhäuser.
NORTON, JOHN (1993). Einstein and Nordström: some lesser known thought experiments in gravitation. In *The Attraction of Gravitation: New Studies in the History of General Relativity*, eds. Earman, John, Janssen, Michel, and Norton, John, pp. 3–29. Birkhäuser.
PAIS, ABRAHAM (1982). *Subtle is the Lord: The Science and the Life of Albert Einstein*. Oxford University Press.

Pauli, Wolfgang (1958). *Theory of Relativity*. Pergamon Press.
Renn, Jürgen (2007a). *The Zurich Notebook and the Genesis of General Relativity*. Springer.
Renn, Jürgen (2007b). *Einstein's Zurich Notebook: Commentary and Essays*. Springer.
Renn, Jürgen, and Sauer, Tilman (1999). Heuristics and mathematical representation in Einstein's search for a gravitational field equation. In *The Expanding Worlds of General Relativity, Einstein studies vol. 7* eds. Goenner, Hubert, Renn, Jürgen, Ritter, James, and Sauer, Tilman, pp. 87–126. Birkhäuser.
Schulmann, Robert, Kox, Anne J., Janssen, Michel, and Illy, József (1998). *The Collected Papers of Albert Einstein Vol 8: The Berlin Years, Correspondence 1914–18*. Princeton University Press.
Stachel, John (1989). The rigidly rotating disk as the 'missing link' in the history of general relativity. In *Einstein and the History of General Relativity, Einstein studies Vol. 1* eds. Howard, Don, and Stachel, John, pp. 48–62. Birkhäuser.
Will, Clifford (1985). *Theory and Experiment in Gravitational Physics* Cambridge University Press.
Will, Clifford (1993), *Was Einstein Right? Putting General Relativity to the Test*. Basic Books.

CHAPTER 27

QUANTUM PHYSICS

SUMAN SETH

27.1 Introduction

In September 1911, Arnold Sommerfeld, Professor of Theoretical Physics at the Ludwigs-Maximilians-Universität in Munich, gave a speech to the German Society of Scientists and Physicians on 'Planck's Quantum of Action and its General Significance for Molecular Physics'.[1] He opened by noting that this was not the topic that the scientific section of the society had originally suggested. They had proposed a talk on the theory of relativity—the same subject upon which Max Planck had spoken to the group the previous year.[2] Yet, where Planck had compared disagreements and debates over relativity to those that had plagued the Copernican theory several centuries before, Sommerfeld declared Einstein's theory, barely six years old at the time, to have become one of the most secure bases of contemporary physics. The principle of relativity, he claimed, 'scarcely belongs any longer among the real, active questions of physics'. The situation with regard to the theory of energy quanta, by contrast, could not be any more different. Here, Sommerfeld stated, 'the fundamental concepts are still in flux and the problems are innumerable[.]'[3]

If Sommerfeld somewhat overstated the level of agreement over relativity—in 1917 one of his own students would still be working on the question of whether the 'rigid' (Abraham) or 'deformable' (Lorentz–Einstein) theory of the electron was correct—he was not wrong in declaring work on the quantum to be unsettled.[4] Indeed, prior to Sommerfeld's lecture and the subsequent Solvay Conference in November, 1911, only a handful of experimentalists and theoreticians realized that there was, in fact, a problem concerning the quantum of action, leading Planck originally to suggest that talk of a conference on the topic was premature. In June 1910, Planck could only count four people other than he and the recipient of his letter, Walther Nernst, who were 'deeply interested' in the subject.[5] That lack of interest would soon change. Particularly following the end of the First World War a younger generation flooded European universities seeking instruction on the experimental and theoretical methods of the new physics. Locating the point at which quantum theory could be considered as settled and accepted as special relativity, however, is a rather more difficult problem. Certainly it came not after five but rather more than twenty-five years after its putative beginnings. At the earliest, one might suggest—as I do here—that coherence came at the beginning of the 1930s, with the publication of Paul Adrien Maurice Dirac's *Principles of Quantum Mechanics* and Werner Heisenberg's *The Physical Principles of the Quantum Theory*.[6]

The history offered here proceeds in five chronologically ordered sections. The first (Section 27.2) considers quantum physics before it could really be named such, examining the process through which a community of quantum theorists and experimentalists came into being, tracing in detail the roots and fruits of Planck's papers in irreversible processes in nature. The second (Section 27.3) takes up the origin and subsequent development of Niels Bohr's so-called 'planetary model' of the atom, paying particular attention to the extension of Bohr's model by Sommerfeld and members of his school as well to Bohr's use of his principles of correspondence and adiabatic invariance. The third (Section 27.4) considers the post-war years, as the problems of atomic spectroscopy provoked the development of new methodological approaches to quantum theory. Taking the history of Wolfgang Pauli's principle of exclusion as a thread through the maze of issues involved in explanations of atomic structure and spectral lines, the aim is to recover the strikingly disparate approaches favoured in Munich by Sommerfeld and Copenhagen by Bohr. The fourth part (Section 27.5) offers a history of the two distinct new forms of quantum mechanics put forward in the mid-1920s: Heisenberg, Max Born, and Pascual Jordan's matrix mechanics, and Erwin Schrödinger's wave mechanics. The fifth part (Section 27.6), more epilogue than section in its own right, offers an account of the consolidation of quantum mechanics—in spite of considerable and ongoing debate over its make-up and meaning, in Heisenberg and Dirac's widely-read texts. All end-points are arbitrary, but this may be more so than most, for the history of quantum physics is continual. Perhaps all that may be said is that the 1930s marked the end of the opening phase of a story, both the past and present of which is still being debated.

27.2 BUILDING A COMMUNITY OF QUANTUM PHYSICISTS: BLACKBODY RADIATION, THE PROBLEM OF SPECIFIC HEATS, AND THE FIRST SOLVAY CONFERENCE

Comparing the reception of Planck's work with Einstein's proves to be illuminating. Although Einstein's style in writing his famous 1905 paper was non-traditional for an academic work—echoing, it has been suggested, the prose of reports at the patent office where Einstein was a clerk[7]—its very title drew attention to its profound relevance to what was probably the most important theoretical topic of its day: the electrodynamics of the electron, that quintessential charged, moving body.[8] The title of Planck's original papers—'On Irreversible Radiation Processes'—on the other hand, conveyed no information about what would be deemed to be their most shocking result: the idea that energy was not a continuous quantity; that it could only be emitted or absorbed in finite chunks, or quanta. The object that Planck investigated in his theoretical analysis, a 'black body' (by definition, one that absorbs all electromagnetic radiation incident upon it, re-emitting this radiation in a characteristic, temperature-dependent spectrum) was not at the cutting edge of research in physics at the time. Kuhn described black-body theory in the early years of the twentieth century as 'an esoteric specialty of concern to very few',[9] and even those very few—while they would laud Planck's *result* (namely, the black-body equation he derived)—found the *form* of his solution profoundly unsatisfying.[10] Where researchers like Lorentz and Wien hoped to find in Planck's papers a clear statement concerning the mechanism of the interaction between matter and the ether, Planck deliberately eschewed any description of such a mechanism, concentrating instead on general theoretical claims, independent of any statements as to the nature of matter underlying them.

What his contemporaries would have found disappointing in Planck's papers, however, Planck himself saw as a profound achievement. From the beginning of his career, Planck had been impressed by the power of the laws of thermodynamics to describe physical and chemical processes without making any assumptions as to the material bases of phenomena. These laws were to be taken as absolutes, their truth possessing a metaphysical certainty, and the second, in particular, shaped Planck's professional career. The inspiration for the work that would eventually lead to the so-called 'quantum discontinuity' came from studies completed by the Viennese theoretician, Ludwig Boltzmann, on the principle of entropy increase. Boltzmann had shown that if one assumed the atomicity of matter, the second law could possess only a probabilistic certainty. It was enormously unlikely that entropy could spontaneously decrease (Boltzmann noted that it was more likely, in a period of time, that all the inhabitants of a large country would commit suicide purely by accident than that two gases mixed together in a container would spontaneously separate themselves), but that such a decrease was nonetheless *possible*.[11] For Planck, even such an exceedingly improbable violation was anathema. Believing that the problem lay in Boltzmann's assumptions of mechanical atomicity, Planck turned to electromagnetic theory, with the electromagnetic ether to be understood in terms of *continuum* mechanics. In the action of a damped electromagnetic resonator within a mirrored cavity, Planck hoped to find evidence of unidirectional processes—processes, that is, that could never be reversed. Were such processes the fundamental basis for all phenomena, the entropy law could be maintained as an absolute.

There remains considerable historical debate concerning Planck's logic in his five papers, written between 1897 and 1899, 'Über irreversible Strahlungsvorgänge'.[12]

Where did Planck believe the fundamental difference between mechanics and electrodynamics to lie? The problems with a mechanical understanding of an absolute second law were twofold. First was the problem of strict reversibility. The laws of mechanics were completely symmetrical with respect to time; any process possible in one direction (leading to an increase of entropy, for example) was also possible in the reverse direction (meaning that entropy could *decrease*). The second problem involved what was known as the 'recurrence theorem'. Any finite mechanical system could be shown to return arbitrarily closely to its starting position, given enough time. Allow a system to begin in a state of high order and it would (probably) become increasingly disordered, with a corresponding increase in entropy. Yet, allow enough time to go by and it should return arbitrarily closely to its starting point again, with a corresponding decrease in entropy and increase in order. For Boltzmann, one could achieve irreversibility only by building it into one's theoretical system from the beginning. Maxwell's equations would appear, on their face, to be just as time reversible as Newton's or Hamilton's equations. How did Planck imagine that any treatment based in current electromagnetic theory—whether continuous or atomistic—could achieve irreversibility?

Part of the uncertainty on the part of historians has tended to derive from the assumption that Planck's papers were intended to offer a *proof*, in a traditional sense, of irreversibility. If one, instead, understands Planck's aim to have been an exploration of the 'naturalness' of the conditions under which irreversibility could be guaranteed, much of the confusion about the development of his ideas disappears.[13] From the beginning, Planck imposed conditions on the nature of the radiation interacting with the resonator at the centre of his cavity, in essence removing from the start all those kinds of waves that would allow his idealized process to be reversed. After repeated questioning and critique from Boltzmann, however, Planck came to realize that these first hypothesized conditions were inadequate to guarantee irreversibility, even in the simplified case with which he had begun. In his fourth paper, Planck introduced a new and more complex set of conditions that would exclude both simple reversibility and the possibility of recursion. Radiation that did not satisfy these conditions, Planck implied, was somehow 'artificial'—'tuned to the resonator'—and to be opposed to 'natural radiation'.[14]

By the middle of 1898, Planck had achieved most of his original aims, using the concept of natural radiation to determine the conditions under which microscopic processes were irreversible. In a final section of his paper of that year, he proved the existence of a function, analogous to Clausius' entropy function, which increased when the resonator underwent any change. Of course, Planck understood that an infinitude of such functions could exist, and the real question was whether one could determine an expression for the *correct* analogue to the thermodynamic entropy. While he would note the considerable effort he put into this problem, Planck admitted that he could not solve it. Instead, he appears to have worked *backward* from available experimental data on black-body radiation. Such data was extensive, at least in part because the black-body served as a standard against which to compare the radiative emissions of substances being tested for their use in the lighting industry.[15]

Without explanation or further justification, Planck wrote down an expression for the entropy of a resonator and then proceeded to 'derive' the same expression for the energy distribution of radiation of a given wavelength λ at a given temperature put forward earlier by Wilhelm Wien.

At least within certain limits, Wien's law had been confirmed by multiple experiments by the last years of the century. Unfortunately for Planck, however, experimental data appearing to call Wien's law into question appeared in the same month that Planck submitted the summary of his five-paper series to the *Annalen*. Yet the close link he had drawn between Wien's expression and the principle of entropy increase was not one he could give up lightly. Almost seven months passed before mounting experimental criticism and his own identification of a flaw in his earlier reasoning led Planck to offer a purely mathematical 'derivation' for a *new* black-body law. Wien's law, he had shown previously, followed from setting the second derivative of entropy with respect to energy equal to $-\alpha/U$. Since that expression seemed inadequate, the 'next simplest' alternative was to write:

$$\frac{d^2 S}{dU^2} = \frac{-\alpha}{U(\beta + U)}$$

Integrating twice, making use of the expression $dS/dU = 1/T$ and Wien's displacement law, in the form $S = f(U/\nu)$,[16] Planck arrived at the 'Planck distribution':

$$E = \frac{C\lambda^{-5}}{e^{\frac{c}{\lambda T}} - 1}$$

Experimental confirmations followed quickly, while Planck himself began a search for a theoretical derivation of the empirical law he had found. A lecture delivered to the German Physical Society at the end of 1900 offered an outline of a proof based on the combinatorial arguments that Boltzmann had used in his kinetic gas theory.[17]

Highly schematic, Planck introduced as 'the most essential point of the whole calculation' the postulate that the energy of N resonators of frequency ν was made up of 'an entirely determinate number of finite equal parts', the size of which was determined by the 'natural constant' h so that the 'energy element' ε was equal to $h\nu$. A little over three weeks later, Planck submitted a more detailed treatment to the *Annalen* (one similar to that re-produced in physics textbooks today), in which he changed the order of his analysis so that Planck's relation emerges as a result, rather than a postulate of the theory.

The question of what, precisely, Planck meant by the relation $\varepsilon = h\nu$ in the first years of the twentieth century remains an open one. The 'traditional' view has been that Planck knowingly introduced an energy discontinuity into physics. In 1978, however, Kuhn claimed that this view was erroneous: not only was Planck unaware of the import of his papers in 1900 and 1901, it required the work of Paul Ehrenfest and Albert Einstein in 1905 and 1906 to bring the point to his attention and a number of years of further thought before his acceptance of the radical implications of 'his' equation. Olivier Darrigol has tended to support Kuhn's most basic claim (if for different reasons), arguing that the substance and form of Planck's derivations do not indicate a clear awareness of discontinuity. Clayton Gearheart, by contrast, has taken Planck's silence concerning the explicit physical meaning of his energy elements as indicating caution rather than confusion. Since Planck did refer to the assumption of finite energy elements as the 'most essential point' of his calculation in 1900, we are justified in assuming that he was aware of its significance and (presumably) its novelty. Much of the disagreement, as it stands, depends on whether one sees Planck subsequent derivations—particularly that in his 1906 *Lectures on the Theory of Heat Radiation*—as continuations of or deviations from the work of 1900/01. With many of the archival sources relevant to Planck's work lost, the debate seems likely to continue.[18]

What is unarguable is the significance of Einstein's and Ehrenfest's publications in explicating the fundamental and technical implications of the quantum relation. Where Ehrenfest's contributions, however, seem to have derived from attempts to understand Planck's results, Einstein's first paper on the quantum made comparatively scant use of Planck's analyses. What the two men had in common was a profound, possibly unmatched knowledge of Boltzmann's statistical mechanics. Ehrenfest had been Boltzmann's student, while three of Einstein's first five publications dealt with thermodynamics and statistical mechanics.[19]

Today, Einstein's first quantum paper is remembered for its discussion of the photoelectric effect, though the general treatment contained within the article achieved much more than this.[20] Overall, the paper constituted a sustained argument, made through the consideration of a series of problems, for the necessity of what Einstein termed a particular 'heuristic viewpoint concerning the production and transformation of light'. That heuristic viewpoint was the assumption that light can be considered as made up not of waves, but of particles, behaving like the molecules of a gas. As a means of demonstrating the necessity of this iconoclastic hypothesis, Einstein first took up the problem of black-body radiation, imagining a situation where oscillating electrons interacted with the molecules of a gas within a mirrored cavity. Were one to calculate the spectral density of the radiation in the cavity using statistical mechanics and Maxwellian electrodynamic theory, the result (integrated over all frequencies) was decidedly non-physical: the energy density of the radiation was infinite. To avoid this, one needed to employ a different expression for the spectral density, and Einstein proposed Planck's distribution law as the most successful at describing experimental results.

If he cited Planck's expression, however, Einstein displayed no commitment to Planck's arguments. In the next section of his paper he turned his questioning on its head. If he had first asked what kind of distribution resulted from the assumption that light was continuous, he now turned to the assumptions necessary to produce an empirically adequate expression like Planck's or the older and simpler (if more limited) Wien distribution. Beginning with this latter distribution, Einstein was able to show in only a few lines that the expression for the change in the entropy of the radiation for a given change in volume had precisely the same functional form as the same expression for an ideal gas. His conclusion was clear: the same functional relationship held because monochromatic radiation, like a gas, was made up of particles.

The energy of such light particles came in discrete quanta and was dependent on frequency, $E = (R/N_0)\beta\nu$ (R, N_0, β all constants).[21] Applying this viewpoint to the problem of the photoelectric effect, previous puzzles tended to drop away. It had perplexed experimentalists studying the production of electricity through the incidence of light on metals when they realized that the energy of 'photoelectrons' depended only on the frequency of light radiation, rather than its intensity, as one would have expected from standard electrodynamic theory. In Einstein's treatment, this oddity in fact made perfect sense. One had only to imagine that an impinging light particle transferred its energy (dependent on its *frequency*) to electrons within the metal. The maximum kinetic energy of the electrons emitted was thus $E = (R/N_0)\beta\nu - P$, where P was the threshold energy binding the electron to the metal. Higher intensity incident light, by this logic, provided more light particles of a given energy and a larger number of emitted electrons, but not electrons of higher energy.

It was not until the following year that Einstein would engage explicitly with Planck's arguments. When he did so, however, he cut directly to their heart. On the first page of a paper on 'The production and absorption of light', Einstein described his earlier paper and then continued:

At that time Planck's theory of radiation seemed to me in a certain respect the antithesis of my own. New considerations . . . demonstrated to me, however, that the theoretical bases on which Planck's radiation theory rests are different from those of Maxwell's theory and of electron theory. The difference, furthermore, is precisely that Planck's theory implicitly makes use of the light-quantum hypothesis sketched above.[22]

Although few, if any, major figures accepted this last statement—approval of the photon theory, as it came to be known, would have to wait until the mid-1920s—Einstein did put his finger on the central paradox of the older quantum theory. In deriving an expression for the energy of the field within a cavity ($E = (c^3/8\pi\nu^2)\rho(\nu)$), Planck had combined two radically different perspectives. Obtaining the correct answer—one that avoided the 'catastrophe' earlier identified by Einstein—required Planck to use his own distribution as the expression for $\rho(\nu)$, the spectral density. As Einstein showed, however, this could *only* be produced if one assumed that energy was a discontinuous quantity. The factor that preceded $\rho(\nu)$ in the expression above, on the other hand, had been obtained, in Planck's theory, using a continuous electrodynamic theory. Planck's radiation law, Einstein concluded pithily, 'is incompatible with the theoretical foundations which provide his point of departure'.[23]

Ehrenfest's first public discussion of Planck's work was delivered in a lecture to the Vienna Academy toward the end of 1905, although he had had been pondering the peculiarity of Planck's results for more than two years. Like Einstein, Ehrenfest's eventual conclusions tended to break Planck's analysis into two parts: the first concerning the action of resonators and the electromagnetic field; the second concerning the combinatorial derivation of the Planck distribution. In some ways, however, Ehrenfest's may have been the more damaging critique, since he was able to show clearly that Planck's resonators—an analysis of which took up the lion's share of his

1906 *Lectures*—were not, in fact, up to the task for which they were introduced. Since they emitted and absorbed energy at characteristic frequencies, only resonators at the same frequency interacted, producing an equilibrium distribution of intensity and polarization for each colour. For resonators at different frequencies, however, no interaction was possible, so any arbitrary frequency distribution would persist. Non-black radiation would not become blacker over time. Ehrenfest's discussion of Planck's combinatorial arguments could only have made this first conclusion sting more. For the Planck distribution did not require any discussion of resonators at all. Applying a version of Boltzmann's statistical approach to the resonator-free case, Ehrenfest could derive Planck's black-body distribution directly. The only necessary additional hypothesis was the peculiar quantum condition, $\varepsilon = h\nu$.

The question of whether a 'quantum discontinuity' was an unavoidable necessity remained an open one for several years. A number of theorists held out hope that a suitably modified electron theory, for example, could reproduce Planck's results without adopting his analysis entirely.[24] That hope was dashed in 1908. That year Lorentz gave a lecture in Rome in which he demonstrated that no extant electron theory could lead to any result other than the Rayleigh-Jeans law and that law—as experimentalists would quickly explain—was wrong even for bodies that one might observe within the laboratory. Planck's theory, Lorentz would write, 'is the only one that supplies a formula in agreement with the results of experiment, but we can adopt it only by altering profoundly our fundamental conceptions of electromagnetic phenomena'.[25] By 1910, Jeans, Lorentz, Wien, and Sommerfeld could be counted, together with Planck, Einstein, and Ehrenfest as those committed to some form of the quantum hypothesis.

Missing from this list, yet crucial for the further development of quantum physics, is the name of the physical chemist Walther Nernst. In large part, this is due to the fact that Nernst's interest in the quantum arose not because of work on black-body theory and the study of radiative processes, but rather from work on specific heats. If Nernst seems like an outsider from this perspective, however, that impression vanishes when one realizes that both the black-body problems and the problem of specific heats were central to one of the most pressing practical and industrial problems of the late nineteenth and early twentieth century: how to produce the most efficient, effective, and marketable electric lighting systems. Improving the new electric lamp, Barkan has shown, formed Nernst's 'main line of work' for a decade before 1905.[26] She has suggested that the complex issues surrounding the discovery of his heat theorem—a result now known as the third law of thermodynamics—can be understood with reference to 'a very specific and concrete problem: the relationship between radiant heat (or radiant energy) and physical and chemical transformations in the conducting filament, as well as the atmosphere surrounding the filament'.[27] One of the subproblems involved in the question concerned the behaviour of a substance's specific heat at first high and then very low temperatures.

The specific heat of a substance is a measure of how much energy must be added to it to raise its temperature by one degree (that is, $dE/dT = c_v$). A basic rule of thumb for determining any given substance's specific heat had been known since

1819, when Pierre Louis Dulong and Alexis Therese Petit noted that the specific heats of equivalent numbers of molecules of any given substances were essentially equal. Boltzmann would offer a theoretical explanation for this empirical rule in the 1870s, by application of the equipartition theorem.[28] Yet, throughout the nineteenth century, certain deviations from this rule had been noted, most conspicuously for the case of carbon in its various forms, which seemed to approach the Dulong–Petit value asymptotically only at high temperatures. In lectures delivered at Yale in 1906, Nernst put forward early results that suggested that the specific heat for all substances dropped to the same value at absolute zero—a value roughly a quarter that predicted by the Dulong–Petit rule.[29] Early in 1910, following a dedicated programme aimed at determining the behaviour of specific heats for a wide array of substances from room temperature down to around −200° C, he revised his earlier estimate: 'One gets the impression that the specific heats are converging to zero as required by Einstein's theory'.[30]

Much as with early citations to Planck's black-body law, Nernst's reference to Einstein here was a reference to the *expression* Einstein had given for the temperature dependence of specific heats, not for its *derivation*. Nernst was not yet a believer in the necessity of a quantum hypothesis. That point may not be entirely surprising, for Einstein's paper on specific heats had extended the logic of an incipient quantum theory well beyond the bounds of its original application. Planck and most others saw restrictions on energy values as something connected to the structure of radiation. Einstein, however, argued that the peculiarities of a quantum analysis could not be limited to radiative processes, but must include matter as well—indeed, had to apply in all those cases dealing with heat theory at the molecular level. From that assumption, his result followed quickly.

Nernst's papers drew the attention of other researchers to Einstein's work on specific heats—a problem of broader interest than the more esoteric black-body theory.[31] And within a few months the physical chemist was willing not only to mention Einstein's results, but to promote the quantum theory as a whole. Writing to the Belgian industrialist Ernst Solvay in July 1910, he suggested that Solvay fund an invitation-only conference, bringing together leading researchers for a congress in Brussels on Radiation and Quanta.

The Solvay Conference would prove to be a pivotal event in the birth of a *community* of quantum physicists. Nernst and Solvay brought together both those who had commented on quantum issues in the past, as well as leading international researchers for whom many of these issues were foreign. A number of these well-respected figures (most prominently perhaps, the French mathematical physicist Henri Poincaré and the New Zealand-born Ernest Rutherford, who told Niels Bohr of the conference only a few weeks after returning to Manchester) went back to their work as enthusiasts for quantum problems. And the number of publications on such problems increased dramatically after the congress. In 1909, ten authors published on quantum topics. In 1911 that number was close to thirty, and in 1914 was over sixty.[32] In disciplinary terms, however, the conference was as important to the recent past of quantum physics as it was to its immediate future. It was the site, as Staley has argued, for the first coherent articulation of what has come to be the most important historiographical divide of the twentieth-century physical sciences: the schism between 'classical' and 'modern' physics.[33] The structure of the conference aided this historical construction, as tasks were divided up among participants to emphasize the fact that a crisis was underway, one to be solved by the introduction of quantized energies. Some of those invited were asked to present 'reports' on quantum topics, others on the physics that had preceded such work, physics that now required substantial overhaul. The effect was striking. Two months after returning to Paris after the end of the congress, Poincaré published an article in the *Journal de Physique* 'On the Theory of Quanta in which he heralded 'without doubt the greatest and most profound revolution that natural philosophy has undergone since Newton'.[34]

Not all the attendees, however, found the conference useful. Einstein claimed to have learned little that was new at the event that his friend Michel Besso referred to as 'the witches' Sabbath in Brussels'. Few, if any, of his colleagues took seriously his arguments concerning the necessity of treating light as a particulate phenomenon. And Nernst's proselytizing zeal in persuading all the conference participants to understand the problems that contemporary physics faced may have seemed too successful to Einstein, who complained that his peers 'lamented at the failure of the theory without finding a remedy. This Congress had an aspect similar to the wailing at the walls of Jerusalem'.[35] These private comments situate Einstein in the minority, for almost all other participants registered a sense of excitement, rather than jaded *ennui*. In a certain sense, however, Einstein would prove correct. Much that was new was presented at the Solvay Conference—both Planck and Sommerfeld, for example, presented radically novel reformulations of the quantum hypothesis—yet the next major advance for quantum physics did not come from the pen of one of the established scholars invited to attend the august gathering in Brussels. A new era of quantum research began, not with the problems of radiation or specific heats, but with those of chemical structure and what could be understood of the make-up of atoms and molecules from the riddles of spectra.

27.3 The Bohr Model: Principles and Problems in the Older Quantum Theory

Toward the middle of 1911, a 25-year-old Niels Bohr defended his doctoral dissertation at the University of Copenhagen.[36] Its goal was to investigate and extend Lorentz's electron theory as it applied to metallic properties. In the course of his research, Bohr became convinced (as, indeed, did Lorentz somewhat earlier) of the likelihood that 'the electromagnetic theory does not agree with the real conditions in matter' and that recourse to quanta was inevitable.[37] In September, armed with a rough translation of the dissertation, Bohr travelled to Cambridge to work under J. J. Thomson, awarded the Nobel Prize in 1906 for his discovery of the electron. At their initial meeting, Thomson promised to read the young Dane's work, but the promise was never kept. If not unkind, Thomson proved rather indifferent to Bohr's interests, yet the time Bohr spent in the small university town was hardly wasted. He attended lectures by Thomson, Jeans, and the pre-eminent British electron theorist, Joseph Larmor; he worked (admittedly without much luck or guidance) on an experimental problem assigned by Thomson; and he continued his own studies of problems in electron theory. Then, a little over half a year after arriving in England, Bohr took the opportunity to visit that other pre-eminent destination for young physicists, particularly those with any interest in experimental work on radioactivity: Rutherford's laboratory in Manchester.

Bohr later recalled that he had already known of Rutherford's new atomic model while still at Cambridge. Drawing on data produced by Hans Geiger and Ben Marsden in experiments on the scattering of alpha-particles from thin metal foils, Rutherford had argued that cases of wide-angle scattering (some more than 90 degrees) indicated that the point-like alpha particles were being deflected by an atomic kernel of highly concentrated mass and charge. An atom, he proposed, was made up of such a kernel, surrounded by a much larger, diffuse sphere of opposite charge.[38] In 1912 Bohr would take this scattering model—which offered no information on the organization or structure of electrons within the atom—as the basis for his far more complete atomic model. Bohr, having returned to Copenhagen, sent the first iteration of that model to Rutherford in early July 1912, in a document now commonly known as the 'Rutherford Memorandum'.

The memorandum makes clear the fact that Bohr's first interests in atomic modelling had nothing to do with spectral lines. Instead, he was concerned with attempting to explain molecular structure and the periodic structure of the elements. Imagine, he suggested, that an atom is made up of a tiny charged core, as *per* the Rutherford model. Imagine, further, that n equally-spaced electrons orbit this core on a ring. Such orbits will not, in general, be mechanically stable (Bohr left aside the question of *radiative* instability, a crucial point for his eventual publication). To achieve such mechanical stability, Bohr suggested (but did not explicitly prove), one needed an additional condition. For this condition, he proposed a version of Planck's hypothesis: the kinetic energy of an orbiting electron must be proportional to Planck's quantum. Even with this condition, however, Bohr argued (based, it turns out, on a faulty calculation) that one cannot add electrons indefinitely to a ring. For more than seven electrons, Bohr convinced himself, a single ring becomes unstable. The closeness of the number 7 to the period of the periodic table suggested that he had, in his words, 'a very strong indication of a possible explanation of the periodic law of the chemical properties of the elements (the chemical properties is assumed to depend on the stability of the outermost ring, the "valency electrons") by help of the atom-model in question'.[39]

Bohr's papers 'On the Constitution of Atoms and Molecules I, II, and III' were published in the *Philosophical Magazine* in July, September, and November 1913.[40] The latter two parts of the series essentially carried through the programme of the memorandum. The first and most famous paper, however, took up the problem of hydrogen's spectral lines. Bohr had turned his attention to spectra after reading a paper by the Cambridge physicist John William Nicholson, who had used a model somewhat similar to Bohr's to derive remarkably accurate values for lines within the solar corona. Bohr had little respect for Nicholson as a scholar, but he clearly had spectra on the mind when a colleague later suggested that he look up the formula for the Balmer lines for hydrogen. With that formula (particularly in the form $v = C(\frac{1}{4} - \frac{1}{m^2}), m > 2, C = const.$) Bohr had most of the pieces in hand for his publication.

Nonetheless, even Bohr himself would remark on the 'horrible assumptions' used in his first paper. He began with a model of hydrogen comprised of a single electron orbiting a positively charged nucleus. Such a system is mechanically stable, but it is not, as Bohr noted, radiatively stable. The orbiting electron should, if classical physics holds, continuously radiate energy until it spirals into the atomic core. Bohr's first assumption was (simply, if oddly) that certain stable stationary states existed in which no radiative loss occurred. The relationship between the kinetic energy of these stationary states and the rotational frequency of the orbiting electron, ω—harking back to the calculations of the memorandum—was given by the expression $W = \tau h \frac{\omega}{2}$, $\tau = 1, 2, \ldots n$, the factor τ now representing the fact that Bohr was considering not only the ground states ($\tau = 1$) but excited states as well. To recover the Balmer formula, Bohr needed another assumption and introduced Planck's relation for a second time. An electron, moving from a stationary state of higher to lower energy (from W_{τ_2} to W_{τ_1}), will emit radiation of frequency v, given by $W_{\tau_2} - W_{\tau_1} = hv$. In Bohr's notation, Balmer's formula then read:

$$v = \frac{2\pi^2 me^4}{h^3} \left(\frac{1}{\tau_2^2} - \frac{1}{\tau_1^2} \right)$$

The match to empirical data was remarkable. The constant in Balmer's formula came out within 7% of the experimentally determined value. Moreover, Bohr could offer a correction to previous classifications of stellar spectral lines. In 1896, Pickering had discovered a series of lines in the spectrum of the star ζ Puppis. In 1912, Alfred Fowler located the same lines in terrestrial experiments using vacuum tubes. Both men ascribed the lines to hydrogen. Bohr, instead, noted that they fit his formula remarkably well if one set the charge of the nucleus equal to $2e$; that is, if the lines were due to ionized helium. He would soon show, to counter Fowler's objections, that the fit was even better ('in exact agreement', in Bohr's words) if one took into account the rotation of the nucleus around the atom's shared centre of mass.[41]

This empirical accuracy could only have brought the peculiarity of Bohr's premises into sharper relief. Both uses of Planck's relation were puzzling, and the relationship between them was baffling. It had been a basic assumption for all those previously who had sought an explanation of spectral lines in the motion of atomic components that the frequency of a line corresponded to a frequency of rotation or oscillation. Bohr had broken that connection: $v \neq \omega$. Rutherford, who had read Bohr's first paper prior to its publication, put his finger on another oddity, and one (like the riddle of frequencies) that would remain unresolved for some time. An electron, in 'jumping' from one stationary state to another, emits light of a frequency determined by the energy difference between its initial and final orbit. Yet how does it 'know' in advance which orbit it will reach? In Rutherford's pithy phrasing: 'It seems to me that you have to assume that the electron knows beforehand where it is going to stop.'[42] The most common positive reaction seems to have been to credit Bohr's work as a starting point, while remaining critical about its details. Sommerfeld captured a common response in writing to Langevin in mid-1914: 'Clearly a great deal is true in Bohr's model and yet I think that it must be fundamentally reinterpreted in order to satisfy'.[43]

As would be true of almost all those interested in Bohr's theory, however, the re-interpretation would be more on Sommerfeld's side than Bohr's. In a paper published in 1915 on 'The General Dispersion Theory according to Bohr's Model', he concluded that the calculated numerical value for e/m 'is so good that a doubt about the exact correctness of Bohr's hydrogen model is not really possible'. All the more, he continued, since more recent values for the ratio brought it even closer to the theoretical prediction.[44] By the beginning of 1916 Sommerfeld was unequivocal.[45] Experimental evidence showed that the quantized electron-paths he had derived using his generalized model 'correspond exactly to reality' and possess 'real existence'.[46]

The analysis that Sommerfeld offered in papers published at the end of 1915 and the beginning of 1916 depended completely on the exploration and extension of Bohr's model in all its specificity. It had been known for some time that the lines of the Balmer series were not, in fact, singlets. Individual lines appeared to display a fine structure. To Sommerfeld, this suggested that a more complete treatment of the quantum Kepler problem might offer insight into the cause of this line-splitting. He began by extending the quantum conditions defining stationary states to include systems of arbitrarily many degrees of freedom. Where Bohr's condition could be written as a quantization of angular momentum ($p_\theta = ma^2\omega = nh/2\pi$), Sommerfeld's generalized expression was $\int p_n dq = nh$. Where Bohr, for the most part, worked with circular orbits, Sommerfeld began with a calculation of the energy of the elliptical orbit described by an electron moving around a charged centre, proceeding to apply his quantum condition not only to the angular momentum, but also to the radial momentum, p_r. The outcome was that the energy of an orbit seemed to depend solely on the quantum numbers of the two degrees of freedom ($W = -Nh/(n+n')^2$) and the frequencies derived corresponded to the lines of the Balmer series. Successive sections took into account the fact that the charged nucleus was not fixed (and that therefore both it and the electron rotated around a shared centre of mass) and sought to provide an answer for a question raised at the beginning of the paper: how many degenerate lines were collapsed into one another in the Balmer series? Sommerfeld's next paper, published only a month later, added a third component to his method:

quantitative comparison with experiment. Introducing a relativistic correction, he was able to obtain values for the energy difference between new lines that corresponded to the precession of the perihelion of the electron's elliptical orbit, introducing for the first time what is now termed the 'fine-structure constant'. Later sections of the paper extended the analysis to include series other than the Balmer series, including those at X-ray frequencies, and elements other than hydrogen.

For theorists, Sommerfeld's systematic extension of Bohr's model provided a clear programme: begin with a model of the atom and solve for its allowed motions using classical mechanics; then apply the phase-integral conditions to quantize those motions; finally, calculate the frequencies of spectral lines by determining the energy difference between two stationary states and applying Planck's relation.[47] That programme became substantially easier to carry through due to work by Karl Schwarzschild and Paul Epstein, as each developed techniques to explain the Stark effect.[48] Their use and explanation of the methods of Hamilton and Jacobi allowed a rapid extension of the kinds of problem with which quantum theorists could deal. In his famous textbook, *Atombau und Spekrallinien*, Sommerfeld would refer to the Hamilton–Jacobi method as the 'royal road for quantum problems', and the text is filled with the elaborate models analysed using it.[49]

Powerful as Sommerfeld's programme was, however, Bohr's approach to quantum issues throughout the nineteen-teens was strikingly different. Where Sommerfeld and his many students sought to develop and apply methods for the solution of progressively more difficult problems, Bohr turned increasingly to the development of what he saw as the fundamental principles of the new field. At the time that he received word of Sommerfeld's first successes in 1916, Bohr had been working on the page proofs of a paper that was to have been published in the *Philosophical Magazine* in April. The paper, 'On the Application of the Quantum Theory to Periodic Systems', was an attempt, in part, to explain a 'fundamental assumption' of Bohr's theory: the stability of stationary states under the variation of external conditions.[50]

The problem that Bohr came to acknowledge after 1913 was that the application of a field to the atomic system would change the energy of the states. One could thus theoretically manipulate the conditions external to the atom in such a way as to cause the system to emit or absorb energy without making a transition between stationary states and hence directly contradict the principle laid out above. To exclude this possibility, Bohr appealed to a recent result put forward by Ehrenfest. This 'principle of adiabatic invariance', put simply, meant that for periodic systems certain variables exist that remain constant during a suitably slow variation of external parameters.[51] For Bohr's model, it meant that the ratio of the average kinetic energy to the frequency of rotation (that is, T/ω) will remain constant (and hence equal to $h\nu/2$) as long as any applied electromagnetic field is introduced sufficiently slowly. Thus, instead of the system emitting or absorbing energy without a transition, the elliptical electron orbits will smoothly expand or contract in size until they reach a value that corresponds to the new energy level.

For Bohr, Ehrenfest's principle (soon renamed the 'principle of mechanical transformability' to remove the suggestion of a necessary thermodynamic or statistical connection) provided not only stability, but also the requisite connection between the classical (that is, continuous) and quantum (that is, discrete) aspects of his model of atomic structure.

> In fact we have assumed that the direct transition between two such states cannot be described by ordinary mechanics, while on the other hand we possess no means of defining an energy difference between two states if there exists no possibility for a continuous mechanical connection between them. It is clear, however, that such a connection is afforded by Ehrenfest's principle which allows us to transform mechanically the stationary states of a given system into those of another . . .[52]

Bohr would use one final aspect of Ehrenfest's analysis. Since the second round of Planck's papers in 1906, Ehrenfest had continued to work on the extension of statistical mechanics to the quantum realm. In 1911 he had shown that Planck's use of the Boltzmann relation ($S = k \log W$) was valid precisely because Planck had quantized the ratio of an oscillator's energy to its frequency (E/ν); in other words, the system's adiabatic invariant. In addition, Ehrenfest proved more generally, and more crucially, that it was only if the statistical weighting of each division in phase-space depended solely on this ratio, that the validity of the oscillator's statistical thermodynamics could be guaranteed.[53] That result became even more significant in 1916, after Einstein published a paper that derived Planck's law from general probability considerations, analogizing the probability of a transition to the probability of radioactive decay.[54] Bohr seized on both Ehrenfest's and Einstein's ideas, realizing that they implied that once the probabilities for stationary states were known for one system, they were, in effect, known for all systems into which the first could be adiabatically transformed.[55] As Pais has phrased it: 'This clearly brings much improved coherence to the old quantum theory: one still did not know why any system is quantized but now one could at least link the quantization of vastly distinct systems.'[56]

The principle of adiabatic invariance thus resolved two outstanding problems of the quantum theory. First, that of determining the correct method of dividing phase space for a quantum system and second—and perhaps more heuristically useful—providing a justification for the application of specific arguments from mechanics to what were, by definition, non-mechanical systems. The correspondence principle offered, similarly, a means of applying arguments from classical electrodynamics to systems defined by their failure to accord with Maxwell's equations. Although the principle acquired its name in only 1920, Bohr had already used a version of it in his first paper of 1913, as a means of justifying the factor of $\frac{1}{2}$ in the expression $W = \tau h \frac{\omega}{2}$.[57] There, however, he had noted only a correspondence between classical and quantum *frequencies* in the limit of slow vibrations. His model could say nothing about the polarity or intensity of light given off in a transition. While one could determine the frequency, polarity, and intensity of the emitted or absorbed radiation from the equation of motion for the electron in the classical case, for the quantum, Bohr's second postulate meant that one could determine the frequency alone. Bohr therefore generalized the 'correspondence' of his principle in two directions: first to move from frequencies to intensities, and second to move from the limit of high quantum numbers to lower ones. Following Einstein,[58] Bohr further related these intensities to the *probabilities* of transitions between given stationary states.[59]

The 'correspondence' of the 'correspondence principle' thus became that between a quantum transition and a *single* component of the classical spectrum, decomposed into a Fourier series. With this, Bohr could hope to provide information about the intensities, the transition probabilities, and the polarization of the radiation emitted during a transition. And in cases where the relevant Fourier coefficient could be shown to be zero (for example, for the harmonic oscillator for all cases where τ was greater than 1), Bohr's analysis suggested the existence of a selection principle governing the possibility of certain transitions.

The previous description of the meaning of the correspondence principle may overstate its conceptual clarity. In 1923, Hendrik Kramers, Bohr's assistant, wrote that 'it is difficult to explain in what [the correspondence principle] consists, because it cannot be expressed in exact quantitative laws, and it is, on this account, also difficult to apply. [However,] in Bohr's hands it has been extraordinarily fruitful in the most varied fields'.[60] Indeed, use of the principle rapidly became something of a shibboleth within the community of quantum theorists. Proponents vaunted the principle's 'flexibility', often implicitly contrasting this to what was portrayed as the Sommerfeld School's overly rigid mathematical formalisms. Bohr, for example, compared his deployment of a malleable 'rational generalization' to 'the tendency of considering the quantum theory . . . as a set of formal rules'.[61] Ehrenfest, speaking at the Solvay Conference in 1921, spoke of the correspondence principle approvingly as 'variable and groping', and claimed that it was 'not desirable' to cast it in a rigid form.[62] The near-artistry needed to apply the principle, however, did not endear it to all. Sommerfeld saw his own methods as open-ended rather than merely formalistic, and denied that Bohr's was the only means of understanding the relationship between the electromagnetic and the quantum theories.[63] Where Bohr sought an analogy and hence some form of connection between the two, Sommerfeld strove to keep them apart. It would be precisely what he saw as Bohr's gifted intermingling of conceptually disparate worlds to which Sommerfeld would object, but the point of contention was not solely one of mathematical closure or completeness. If he found Bohr's analysis incoherent, this did not mean that Sommerfeld assumed that the only way forward was a fully logical or rational one. On the contrary, he repeatedly stressed not only the incompleteness of his own solution, but the conceptual benefits that could be derived from such an incompleteness. The difference between his and Bohr's solutions was that Sommerfeld's kept classical electromagnetic theory and the world of the quantum largely distinct, while Bohr's principle of correspondence appeared to magically—and unjustifiably—bring them together. Although it remained somewhat muted, Sommerfeld's enthusiasm for Bohr's principle grew during the early 1920s. Nonetheless, as we will see, the basic differences between the schools associated with each figure remained a characteristic of quantum physics until and beyond the development of a fully-fledged quantum mechanics.

27.4 Unsettled Problems in Quantum Spectroscopy: Sommerfeld's Phenomenology, Bohr's Second Theory, and Pauli's Exclusion Principle

Two problems dogged quantum spectroscopy in the first half of the 1920s. How could one explain the complexities involved in the anomalous Zeeman effect, and how should one order X-ray spectra so as to explain the periodic structure of the table of elements? Each would prove so taxing for the methods developed by Sommerfeld in the mid nineteen-teens that he would abandon any hope of understanding their causes in model-based terms. Bohr's response was to develop a second atomic theory, constructing models of electrons penetrating deep into the atomic core in their elliptical orbits in the hope of explaining atomic structure. Neither problem fell to any mode of analysis until Wolfgang Pauli's 'exclusion principle' paper in 1925, which demonstrated an intimate connection between the complex structure of spectra and the make-up of atoms. As Pauli would note in his acceptance speech for his Nobel Prize in 1945, his work would also demonstrate the connection between Sommerfeld and Bohr's quantum theoretical methods.

At that time there were two approaches to the difficult problems connected with the quantum of action. One was an effort to bring abstract order to the new ideas by looking for a key to translate classical mechanics and electrodynamics into quantum language which would form a logical generalization of these. This was the direction which was taken by Bohr's Correspondence Principle. Sommerfeld, however, preferred, in view of the difficulties which blocked the use of the concepts of kinematical models, a direct interpretation, as independent of models as possible, of the laws of spectra in terms of integral numbers, following, as Kepler once did in his investigation of the planetary system, an inner feeling for harmony. Both methods, which did not appear to me irreconcilable, influenced me.[64]

In recovering Pauli's path to his principle, then, one may also recover the structure of quantum spectroscopic studies in the last years of the older quantum theory.[65]

That Bohr's model could shed light on the Zeeman effect seemed clear to many in the nineteen-teens, particularly those with a prior interest in the electron theory. One of the electron theory's early successes had been Lorentz's explanation of the splitting of spectral lines in a magnetic field, discovered by the Dutch physicist Pieter Zeeman in 1896. Yet further experiments showed the existence of what was later termed the 'anomalous' Zeeman effect, which exhibits many more lines and different distances between them, than Lorentz's theory predicted. While empirical rules describing this anomalous line-splitting were developed, theoretical explanations remained largely elusive.

Sommerfeld's first substantial engagement with the problem followed the discovery of the 'Paschen-Back' effect (that for sufficiently strong fields, the multiple lines of the anomalous Zeeman effect reduced to the triplet first predicted by Lorentz.)

His immediate response upon hearing of Bohr's model in 1913 was to ask whether the Dane had applied it to Zeeman phenomena: 'I would like to occupy myself with that'.[66] His early attempts were not crowned with success. His theory could not even reproduce Lorentz's results for ordinary Zeeman splitting and his relativistic treatment could say nothing about the line multiplicities of the anomalous Zeeman effect. By 1919, persistent failure to explain the Zeeman problem in terms of any model led Sommerfeld to posit a 'half-empirical' 'magneto-optic splitting rule' to predict the size of the separations between anomalous lines. Talking of 'A Number Mystery in the Theory of the Zeeman Effect', Sommerfeld confessed to a colleague that he could say nothing about the causes of Zeeman doublets and triplets. In 1920 he went one step further in a programme that sought increasingly to draw empirical rules directly from spectroscopic data, without attempting an explanation in terms of an underlying model. In a paper published in the *Annalen der Physik*, he introduced a new 'selection principle', operating on a new quantum number (termed the 'inner' quantum number) to explain why some transitions—otherwise apparently allowed by known quantum rules—did not occur. In analogy to the azimuthal quantum number, which was associated with the angular momentum of the entire atom (its 'external rotation') Sommerfeld suggested that the new number might correspond to a 'hidden rotation'. Of the number's geometric significance, however, 'we are quite as ignorant as we are of those differences in the orbits which underlie the multiplicity of the series terms'.[67] By 1923, in the third edition of *Atombau*, Sommerfeld was willing to declare the end of model-based accounts and the birth of a new, more phenomenological approach to the problems of the quantum.

Responses to Sommerfeld's new methodology were mixed, ranging from hostility, to bemusement, to enthusiastic approval. Pauli, one of Sommerfeld's former students, was in the latter camp. After receiving a copy of the fourth edition of *Atombau* in 1924, Pauli wrote: 'I found it particularly beautiful in the presentation of the complex structure that you have left all model-based [*modellmässig*] considerations to one side. . . . . One now has the impression with all models, that we speak there a language that is not sufficiently adequate for the simplicity and beauty of the quantum world. For that reason I found it so beautiful that your presentation of the complex structure is completely free of all model-prejudices'.[68] Pauli finished his paper on the exclusion principle—a paper in which 'model-prejudices' are conspicuously absent—within a few months of this letter.

Pauli took up the anomalous Zeeman effect in late 1922, together with Bohr, who had agreed to write a short paper on the topic for the *Annalen der Physik*. Six months later, Pauli was still stumped, writing to Sommerfeld with evident exasperation that it 'would not, but would not come out'.[69] The problem with which he was grappling was his attempt to explain the empirical rules determined previously by Alfred Landé, like Pauli himself a former Sommerfeld student. In the early 1920s, Landé had incorporated and redefined the 'inner' quantum number, j, to obtain a value for the size of line-splittings in a weak magnetic field. Classically, such a field, when applied to a system like Bohr's planetary model of the atom, should cause all electrons to precess together at a given frequency around the field axis. The induced separations between given lines should then be proportional to this precessional (Larmor) frequency (i.e. $\Delta E = m\omega h$, with m integral). In the quantum case, however, Landé determined (empirically, after the fashion of Sommerfeld's a posteriori method) that the energy separation was given by the classical value multiplied by another term, the so-called g-factor (i.e. $\Delta E = g \cdot m\omega h$).[70] In 1921, when Landé first introduced this term, he could offer no physical interpretation of its meaning, other than that it was related to j, now understood as the total angular momentum of the atom.

A model-based explication of the g-factor had to await Werner Heisenberg's publication, in 1922, of his *Rumpf* model of the atom. Infamously introducing half-integral quantum numbers, Heisenberg suggested that the observed splitting could be explained by the existence of a 'coupling' between the angular momentum of a single, outermost electron (governed by k, the azimuthal quantum number) and the sum of the angular momenta of the remaining electrons (given by r), which formed the atomic core (or *Rumpf*). The sum of these two net momenta then gave, according to Landé, the total angular momentum of the atomic system, j. Doublets, in Heisenberg's peculiar vision, arose from a 'sharing' of angular momenta: in effect, the outer electron gave a half quantum's worth of momentum to the core. This *ad hoc* move gave the right number of lines, but did so at the expense of any customary physical understanding.[71] Leaving most of the model's difficulties to one side, Landé deployed a conception similar to Heisenberg's in 1923 to offer an expression for g:[72]

$$g = \frac{3}{2} + \frac{1}{2} \cdot \frac{R^2 - K^2}{J^2 - \frac{1}{4}}$$

Pauli wrestled for months to provide, as he put it in a letter to Landé, 'a satisfactory model-based meaning for such astoundingly simple empirical rules', but did so without substantial success.[73] Frustrated by this atypical failure, he withdrew from work on atomic physics, returning to the field only after accepting a brief to complete an encyclopedia article on the principles of the quantum theory in mid-1924. In the midst of writing that article, Pauli hit upon a novel idea: a means of testing a fundamental assumption of the *Rumpf* model. If an alkali core had a non-zero magnetic moment (as it must when sharing some of the magnetism of the external electron), then g should depend on Z, the atomic number. Landé, tapped for data on the topic, soon returned the results: no dependence on Z, therefore no magnetic moment for a closed shell. The 'sharing' model was dead. The cause of the anomalous Zeeman effect was not to be found in a peculiar coupling of the *Rumpf* with an external electron, but in some peculiar property of the electron alone. 'The doublet structure of the alkali spectrum, as well as the violation of Larmor's theorem comes about through a peculiar, classically non-describable kind of *Zweideutigkeit* [ambiguity, doubled signification] of the quantum-theoretical characteristics of the light-electron'.[74]

In a stroke, Pauli had solved (or, rather, dissolved) the problem of the *modellmässig* understanding of j and the cause of the apparent doubling of the magnetism of the atomic core. The *Rumpf* ceased to play any role in the production of spectral lines and hence the notion of a 'total' angular momentum ceased to have any great significance.

Since the violation of Larmor's theorem—the fact that g was not equal to unity—was the result of a 'classically non-describable' characteristic of the outer electron, no model-based understanding should (or could) be sought. The *Zweideutigkeit* was a quantum property with no classical counterpart.

The 'exclusion principle' proper was predicated on the notion that electronic behaviour (and only that of electrons) was governed by four quantum numbers. The problem that it solved, however, arose not from the complexities of the Zeeman effect, but rather from the attempt to explain the processes of *Atombau* and the reason for the periodic structure of the table of elements. That question had, of course, been in Bohr's mind since his first quantum paper in 1913. It was not until his 'second atomic theory', however, that he could develop what appeared to be a complete solution.[75] Two novel elements—together with the all-powerful correspondence principle—formed the essence of Bohr's second theory: the notion of 'penetrating orbits' and the 'construction principle' (*Aufbauprinzip*). The former made explicit and detailed use of the Bohr model and argued that outer electrons in elliptical orbits about the atomic nucleus must penetrate the orbits of internal electrons, causing deviations in the Coulombic force acting upon them. The latter asserted that quantum numbers were conserved in building up atoms, so that the core of an alkali atom immediately following a rare gas in the periodic table was governed by the same quantum numbers.

Bohr's arguments proceeded by way of both analogy and disanalogy with the Bohr-Sommerfeld model of the hydrogen atom. Characterising each orbit by two quantum numbers, n and k ($n = 1, 2, 3, \ldots, \infty$; $k = 1, 2, 3, \ldots, n$), Bohr took up the question of the shape and size of each n_k orbit for elements with $Z > 1$. As was known from Sommerfeld's work on hydrogenic atoms in 1915, when $n = k$, orbits are circular; elliptical with increasing eccentricity as $n - k$ increases. The area enclosed by each orbit increases with n, so that any 4_k orbit, for example, is larger than any 3_k orbit. It is precisely this aspect that would change in Bohr's second theory.

The notion that electrons 'filled' lower orbitals in a fashion that reproduced the structure of the periodic table (two electrons in the lowest orbital, eight in the next, and so on) was one common to almost all models of atomic structure after the advent of Bohr's first model in 1913. This implied, of course, that outer electrons in elements of higher atomic number than helium were 'shielded' from the full nuclear force experienced by inner electrons. According to Bohr's analysis in the 1920s, however, some orbits for $n \geq 3$ penetrated lower orbitals. In doing so, a 3_2 electron would feel the same force upon it—for this part of its orbit—as an electron in a lower orbital. The effect would be to radically change the size and shape of its trajectory around the nucleus, shifting some elliptical orbits to lower overall energy states and causing the electron to trace out the shape of a rosary, rather than a single, repeated ellipse.

Given that, as Fig. 27.1 indicates, 4_1 and 4_2 orbitals are—contrary to the model of the hydrogen atom—energetically more stable than 3_3 orbitals, one would expect these to be filled first. And so Bohr argued, at least for the first and second elements of the third period of the periodic table. After that, Bohr claimed that a 'simple calculation' demonstrated that the increasing nuclear charge caused the higher 4_1 and 4_2

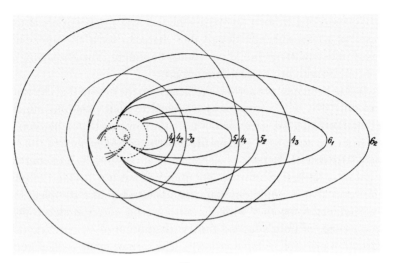

Fig. 27.1. Orbits for multi-electronic atoms.[76]

orbitals to increase in size in relation to a 3_3 orbit.[77] Electrons added to the structure of elements following Calcium ($Z = 20$) thus ceased filling the 4_1 orbits and began filling the 3_3 instead. Since the *Aufbauprinzip* held that the electronic structure of elements with $Z = N + 1$ was identical to that for element $Z = N$ except for the placement of the $(N + 1)^{th}$ electron, one could now explain the puzzling chemical properties of the transition metals—and especially the so-called 'Lanthanides' and 'Actinides'—in the table of elements. The elements from Lanthanum ($Z = 57$) to Lutetium ($Z = 71$) are, in spite of their increasing atomic numbers, chemically near-identical. In Bohr's theory this followed from the fact that each such element possessed an identical number of *valence* electrons in a 6_1 orbital, even as each added more electrons to a previously empty (because previously higher energy) 4_4 orbit. Less convincing were Bohr's explanations of the *periodicity* of the periodic table. Why, for example, were there eight (and only eight) electrons in the second shell? For this, Bohr could only invoke somewhat vague symmetry arguments, which few outside Copenhagen seemed to find believable.[78]

Sommerfeld, originally enthusiastic about Bohr's seemingly miraculous calculations, soon became disillusioned as he and others in Germany realized that the Dane's results did not follow from detailed mathematical analysis at all.[79] Instead, even after being pressed for his method, Bohr could only invoke the power of the correspondence principle and his own intuition. Much more promising to Sommerfeld, at least, seemed to be a new direction proposed by an English researcher, E. C. Stoner, in the *Philosophical Magazine* in October 1924. Further subdividing Bohr's n_k orbitals, Stoner used an extra quantum number, j, and spoke of n_{kj} orbits.[80] Each such n_{kj} orbit was populated by $2j$ electrons. Thus, while in Bohr's conception there were two $n = 2$ orbits—an elliptical 2_1 orbit and a circular 2_2 orbit—each containing four electrons, in Stoner's version there were three orbits. The 2_{11} and 2_{21} ($j = 1$) subgroups each contained two electrons, while the 2_{22} ($j = 2$) contained four. The periodicity of Mendeleev's table could be maintained with no mention of the dynamical character of electronic orbits. The new scheme was, to Sommerfeld's satisfaction, formal, arithmetic, and purely quantum in character: it contained neither argument nor conceptual representation in terms of classical models.

Having passed over Stoner's paper when he first received the relevant copy of the *Philosophical Magazine*, Pauli returned to it after perusing the preface to the fourth edition of *Atombau*. His 'generalization' of Stoner's ideas must have followed almost immediately. In the notation used in his exclusion principle paper, four quantum numbers were necessary. Instead of a single quantum number k, he used two, k_1 and k_2 ($k_1 = 1, 2, 3 \ldots n$; $k_2 = k_1, k_1 - 1$). In place of the once-crucial j, he used m_1, following Sommerfeld in defining this as the (quantized) component of the momentum parallel to an externally imposed field. j was then defined as the maximum value of m_1, given by $j = k_2 - \frac{1}{2}$. Clearly, the maximum number of values m_1 can take is equal to $2k_2$. Stoner's $2j$ identical electrons in each of his n_{kj} orbitals become, in Pauli's formulation, $2k_2$ electrons, each distinguished by the possession of a different value of m_1. The exclusion principle then simply amounts to the fact that there are no such things as equivalent electrons within a given atom. The rule abandons, as Pauli would stress to Bohr, any talk of orbits and provides instead a formal quantum rule connecting the number of terms into which a single spectral line could split with the periodic structure of the table of elements.[81] 'We cannot give a more precise foundation for this rule', he emphasized in public, 'nevertheless, it seems to present itself very naturally'.[82]

Even today, of course, Pauli's principle remains more axiomatic than intuitive. On one point, however—indeed, one of its most confusing assumptions—it soon acquired a more comprehensible gloss. As noted above, Pauli's notation for his four quantum numbers involved splitting the azimuthal quantum number, k, into two different numbers, k_1 and k_2. In this he followed the original notation used by Bohr and Coster and the notation Sommerfeld had accepted in the 1924 edition of *Atombau*. Yet that notation was avowedly peculiar, deriving as it did from two different (and contradictory) explanations of the origins of line splittings. If, for example, one wished to explain the separation between the pair of P orbits and the single S orbit for sodium, one assumed that the two P orbits shared the *same* azimuthal quantum number, which differed from that of the S orbit. If, however, one sought to explain the difference *between* the two P orbits, one assumed that this difference was a relativistic effect due to a *difference* in their azimuthal quantum numbers. Allowing there to be two azimuthal quantum numbers certainly solved this problem in a formal sense, but seemed baffling in *modellmâssig* terms. The azimuthal quantum number traditionally labeled an orbit's eccentricity. As Linus Pauling put it in a later interview, 'How can you have two quantum numbers that describe the eccentricity of the orbit? It only has one eccentricity'. The suggestion by George Uhlenbeck and Samuel Goudsmit was that one could understand all four quantum numbers as electronic degrees of freedom as long as the fourth was understood as a rotation: an electronic 'spin'.

Such a rotating electron was a peculiar entity, since it appeared—if it were to fit the data—to be spinning with a rotational velocity greater than the speed of light.

On the other hand, this seemingly simple assumption (one thoroughly combining classical and quantum analyses) solved an array of the most puzzling results of the older quantum theory, from the 'doublet riddle' described above, to 'the riddle of the statistical weights' (why there are twice as many states observed for a given spectral term than one would expect if the electron had no structure?), to the anomalous Zeeman effect (and more particularly why the g factor should be 2 rather than 1, as predicted classically?)[83] Whether Pauli's result, as he suggested, presented itself 'naturally' or not, we may regard his exclusion principle and the subsequent proposal of the existence of electronic spin—due both to their timing and their methods—as the zenith of the older quantum theory.

27.5 Quantum Mechanics

27.5.1 Crisis?

In reconstructing the paths that led to both matrix and wave mechanics, it is worthwhile beginning by re-emphasising the success of Pauli's exclusion principle in dealing with some of the most pressing difficulties of quantum spectroscopy. In the wake of the new forms of quantum mechanics it became common for both contemporary participants and later historians to stress the insurmountable problems that the older quantum theory had faced—problems that then *required* a radically new approach—rather than the achievements that encouraged more gradual developments. With the benefit of hindsight, historians have described quantum theory before 1925 as 'a thing of rags and patches'[84] which appeared to be 'from the methodological point of view, a lamentable hodgepodge of hypotheses, principles, theorems, and computational recipes, rather than a logical consistent theory.'[85]

Even prior to (but particularly after) the publication of Kuhn's *The Structure of Scientific Revolutions*,[86] periodizations of the history of quantum theory have portrayed the middle years of the 1920s as a time of crisis, induced by the discovery of anomalies that could not be reconciled with extant theories or techniques. This judgement, at least, has more merit than the blanket criticism of work prior to the development of quantum mechanics, since it reflects the views of several of the most prominent figures in quantum theoretical research at the time. Thus Max Born, whose Göttingen school came to rival both Copenhagen and Munich in the 1920s, spoke in 1923 of the disastrous state of work on even the simplest atomic models. Applying powerful mathematical techniques in an attempt to derive the spectral lines produced by excited helium, Born and Heisenberg working in collaboration came up short. Their systematic analysis produced the wrong results, leading Born to describe the outcome as a 'catastrophe'.[87] In the letter to Sommerfeld cited above, Pauli contrasted his approval of Sommerfeld's techniques with the precarious state of other approaches. 'The model-idea', he wrote, 'now finds itself in a difficult, fundamental [*prinzipiellen*] crisis, which I believe will end with a further radical sharpening of the opposition between classical and quantum theory.'

Yet the language of crisis cannot be read transparently. As Forman has shown, 'crisis-talk' was very common within the German physics community in the aftermath of Germany's defeat in World War I and considerably preceded the identification of the particular anomalies and crises that could be said to have provoked the quantum revolution.[88] Both Bohr and Born had called for a fundamental reworking of classical mechanics well before 1924, while Pauli's comment identifies not only a crisis *but also its resolution* (through Sommerfeld's phenomenological methods) within the framework of the older quantum theory. Sommerfeld, meanwhile, neither spoke of a crisis in quantum physics in the mid-1920s, nor acknowledged that a revolution occurred after the development of quantum mechanics. In 1929, he noted explicitly that 'The new development does not signify a revolution, but a joyful advancement of what was already in existence.'[89]

The point here is not to adjudicate the question of who was 'right' in speaking of a crisis. It is, rather, to note that historians should pay more attention to where and why some physicists came to see quantum physics as being at a critical juncture in 1924/25. Doing so brings a somewhat obvious point into sharp relief: it was precisely where researchers were working under the assumption that classical physics should provide insight and a guide for quantum problems that the situation seemed most dire. For the most part, the problems of theoretical spectroscopy in the 1920s had no classical analogues. In any case, Sommerfeld's emphasis on *Gesetzmässigkeiten* rather than *Modellmässigkeiten* intentionally side-stepped the question of the applicability of specific classical models. Born's programme in Göttingen, by contrast, sought to apply the perturbative methods refined by astronomers and mathematical physicists working on problems in celestial mechanics to ever more complex 'planetary' models of the atom. It was this classically rooted programme that stumbled on perhaps the simplest problem after the solution of the hydrogen atom. Bohr's approach, meanwhile—based on the artful application of the correspondence principle—acknowledged both the necessity and the fundamental inapplicability of classical approaches to quantum problems. Fittingly, he seems to have been rather less troubled than Born at 'the fundamental failure of the laws of mechanics in describing the finer details of the motion of systems with several electrons'.[90]

Born's response to the inadequacies of celestial physics as applied to polyelectronic atoms was to formulate a new, explicitly quantum, rule. Classical physics used differential equations to describe continuous processes; in quantum mechanics these would be replaced by difference equations. The discontinuities of quantum jumps between stationary states would be built into the mathematical bedrock of the new mechanics. This mathematical manoeuvre would be central to Heisenberg's *Umdeutung* (reformulation) of quantum theory, yet the set of problems that led the 23-year-old toward what would become known as 'matrix mechanics' were not those of quantized celestial mechanics. Nor, indeed, were they the problems of theoretical spectroscopy. Instead, they involved the attempt to deal with a problem that had found one of its first explanations in the work of Isaac Newton: the dispersion of light.

27.5.2 Matrix Mechanics

Dispersion is an everyday phenomenon: the reason behind the colours of the rainbow. Light, as it passes through a medium, is bent or scattered by an amount dependent, in part, on its frequency. Late-nineteenth-century explanations of the phenomenon depicted it as the interaction between incident light and oscillating charged particles. The frequency of the dispersed light in such a model depends, among other factors, on both the frequency of incident light and the characteristic frequency of oscillation for the charged particles. Therein lay the problem for any quantum treatment, for in Bohr's model, as we have already noted, the frequency of mechanical rotation was not the frequency of emitted light. Early attempts to simply graft a quantum explanation onto the classical model were met with pronounced failure.[91]

In 1921, Rudolf Ladenburg at Breslau offered a promising (if still ad hoc) suggestion.[92] Where earlier attempts began with the classical problem of orbiting electrons, Ladenburg simply ignored the issue of orbits altogether. Instead, he calculated the energy absorption rate for a set of oscillators whose characteristic frequencies were precisely those of an electron's transitions from the ground state in the Bohr model. He then equated this quantity with the energy absorption rate for a set of atoms calculated using Einstein's transition probabilities. The model gave resonance lines at the right location, yet the status of Ladenburg's quasi-classical oscillators was unclear. What was the nature of this infinitude of oscillators, whose frequencies corresponded not to any oscillation within the atom, but rather to the frequencies of jumps between stationary states? Following a correspondence with Bohr, Ladenburg and his collaborator Fritz Reiche coined the term *Ersatz* oscillators, to reflect the fact that these theoretical objects were as much a calculational device as they were a description of intra-atomic processes. In 1924, using perturbational techniques and the correspondence principle, Hendrik Kramers, Bohr's assistant, extended Ladenburg's results to include all transitions, not merely those from the ground state.[93]

In his dispersion papers, Kramers spoke of *virtual* oscillators, drawing the term from a paper completed in the first few weeks of 1924. That paper, which he authored jointly with Bohr and John Slater, an American who had arrived in Copenhagen in December, has achieved a certain notoriety for the fact that it proposed the abandonment of any strict application of the laws of energy and momentum conservation, as well as the principle of causality.[94] Heisenberg later credited it (and one can see from his words the profoundly retrospective character of much crisis-talk) with 'represent[ing] the real high point in the crisis of quantum theory, and, although it could not overcome the difficulties, it contributed, more than any other work of that time, to the clarification of the situation in quantum theory'.[95] At the time the article was published, however, he was less sanguine: 'I do not really see any fundamental progress'.[96] Einstein was far more dismissive of what quickly became known as the BKS paper, famously suggesting that, were the paper correct in abandoning causality: 'In that case, I would rather be a cobbler, or even an employee in a gaming house, than a physicist'.[97]

Einstein had more than one reason for displeasure at the BKS theory, for one may understand the work as perhaps the last sustained theoretical effort by leading quantum theorists to deny the force of the light quantum hypothesis. In late 1922, Arthur Holly Compton had offered decisive arguments against the notion that X-ray scattering could be explained using the wave theory of light.[98] 'The experimental support of the theory', he wrote in 1923, 'indicates very convincingly that a radiation quantum carries with it directed momentum as well as energy'.[99] In the preface to the 1924 edition of *Atombau*, Sommerfeld called the Compton effect 'probably the most important discovery that could be made in the present state of physics'.[100] Slater found the photonic hypothesis appealing as well, suggesting in March 1924 that in the present state of physics 'the discontinuous side is apparently more fundamental' than the continuous theories of classical optics.[101] In a Note to *Nature*, Slater proposed a model whereby distant atoms 'communicated' with one another by means of a virtual radiation field originating from oscillators (like those used by Ladenburg and Reiche) whose frequencies were those of quantum transitions. This field could then function as a guide for the motion of light quanta.[102]

Bohr and Kramers in Copenhagen were not nearly so amenable to the idea of light quanta. In collaborating with Slater, they insisted that he remove all mention of photons from their proposed joint publication.[103] The BKS paper, however, retained the virtual field. An atom in a stationary state continually communicates with other atoms via 'virtual harmonic oscillators corresponding with the various possible transitions to other stationary states'. Transitions between stationary states are assumed to 'induced' by the virtual field, the likelihood of a transition determined by Einstein's transition probabilities. Such transitions, however, have no *direct* effect on distant atoms. After a transition, an atom is in a different stationary state and hence the virtual field associated with it is different. One transition does not *cause* another, and the energy/momentum associated with a transition is thereby only statistically conserved, rather than conserved event by event. The theory offered a means of avoiding the photonic explanation of the Compton effect, but at a price too high for most, abandoning 'any attempt at a causal connexion between the transitions in distant atoms, and especially a direct application of the principles of conservation of energy and momentum ...'[104]

BKS fever ran high for several months among those most closely associated with Copenhagen. Yet by June 1924, largely due to Einstein's criticisms, faith in the denial of causality and conservation had declined. Born (assisted by Heisenberg) would soon use virtual oscillators together with perturbative techniques and his new difference equations to rederive Kramer's dispersion relation without the further heresies of BKS. (The paper bore the title *Über Quantenmechanik*—probably the first published use of the term 'quantum mechanics'.)[105] The final nail in the coffin of the Bohr–Kramers–Slater theory came in 1925, when the experimentalists Walther Bothe and Hans Geiger demonstrated that in scattering experiments, energy and momentum were conserved event by event. Conservation, that is, was direct rather than statistical.

Until early 1925, Kramer's influential studies on the dispersion problem had appeared in publication only in the form of brief, largely detail-free 'Notes' in the journal *Nature*. In his first Note he offered no derivation of his dispersion equation,[106] while his second brief Note offered a mere outline of a derivation. Perhaps more important for later developments, this second Note also included a methodological observation. Contrasting his approach to that put forward by another theorist, G. Breit, Kramers wrote:

[T]he dispersion formula thus obtained possesses the advantage over a formula such as is proposed by Mr. Breit in that it contains only such quantities as allow of direct physical interpretation on the basis of the fundamental postulates of the quantum theory of spectra and atomic constitution, and exhibit no further reminiscence of the mathematical theory of multiple periodic systems.[107]

This phenomenological theme recurred in his complete treatment of dispersion in a paper co-authored by Heisenberg 'On the Dispersion of Radiation by Atoms'. Setting themselves the task of finding the quantum-theoretical 'analogue' of the classical scattering formula, the pair deployed virtual oscillators, the correspondence principle, and difference equations to obtain, as they put it: 'quite naturally, formulae which contain only the frequencies and amplitudes which are characteristic for the transitions, while all those symbols which refer to the mathematical theory of periodic systems will have disappeared'.[108]

It was this emphasis on frequencies and amplitudes—rather than orbits—that provided what we may regard as the philosophical framing of Heisenberg's *Umdeutung*. His paper opened by noting that previous approaches to quantum theory could be 'seriously criticized on the grounds that they contain, as basic element, relationships between quantities that are apparently unobservable in principle, e.g. position and period of revolution of the electron.' The alternative, he declared, was 'to try to establish a theoretical quantum mechanics, analogous to classical mechanics, but in which only relations between observable quantities occur.'[109] Kramers had clearly been important in bringing Heisenberg to this phenomenological position, but by 1925 the critique of atomic modelling could be heard in many locations. Sommerfeld, of course, had turned to phenomenological approaches during precisely the period that both Heisenberg and Pauli were his students and had written his own phenomenological *Umdeutung* of quantum theory—with the help of a nineteen-year-old Heisenberg—in 1921.[110] Pauli, meanwhile, required little persuasion by his supervisor, since he had been pushing the significance of the observability criterion for relativity theory since the late nineteen-teens, and had criticized Heisenberg repeatedly for his unphilosophical thinking on this point. At the end of 1924, while Heisenberg was in Copenhagen, Pauli wrote to Bohr declaring the tenets of his new quantum programme: 'I believe that the values of energy and momentum for stationary states are much more real than "orbits." The (not yet achieved) goal must be to deduce these and all other physically real, observable properties of the stationary states from (whole) quantum numbers and quantum theoretical laws'.[111] Six months later, while Heisenberg was working with them in Göttingen, Born and Pascual Jordan invoked the operationalist definition of simultaneity at the heart of Einstein's 1905 special relativity paper as a model for their analysis of the quantum behaviour of aperiodic systems. 'Only such terms enter into the true natural laws', they wrote, 'that are in principle observable and determinable'.[112]

Heisenberg's embrace of the observability criterion may, under these conditions, almost seem overdetermined. Yet it is important to note that observability did little *work* for Heisenberg in terms of his theoretical practice. As Kragh has argued, 'there was no royal road from the observability principle to quantum mechanics ...'[113] Instead, Heisenberg began with the basic logic of methods, like Kramers', which made use of the correspondence principle and the idea of virtual oscillators. Thus, Heisenberg's paper started by noting that any periodic motion, $x(t)$, can be written as the sum of an infinite number of Fourier components, such that

$$x(n,t) = \sum_{\alpha=-\infty}^{\alpha=+\infty} A_\alpha(n) e^{i\omega(n)\alpha t}$$

In seeking a quantum analogue to the classical term $x(t)$, Heisenberg faced something of a problem. One could not speak of orbits, so how could one speak of a position? The answer was that one, effectively, did not. Instead, Heisenberg largely ignored the summation of terms, associating each individual component of the series with a single virtual oscillator, the amplitude of which was not A_n but was represented as $A(n, n - \alpha)$; and the frequency of which was not ω_n, but $\omega(n, n - \alpha)$, to denote the fact that every frequency corresponded to a transition between two stationary states, labeled by n and $n - \alpha$. This infinitude of terms can then be represented in an array, the numbers n and $n-\alpha$ labeling positions in that array.

Heisenberg then turned to another question: what would the quantum equivalent for $x_n(t)^2$ be? And what of $x_n(t)^3$? The query was not an idle one. In the classical case, one arrived at an expression for the intensity of light at a given frequency by squaring the amplitude of the relevant Fourier component. In addition, having found a more physical problem—the hydrogen atom—too difficult, Heisenberg would soon try to solve the equation for an anharmonic oscillator ($\ddot{x} + \omega_0^2 x + \lambda x^2 = 0$). His suggested solution was not an obvious one. Again, focusing his attention, not on the sum of all terms, but on an individual component of the sum, Heisenberg proposed, in essence, that squaring the expression for position amounted to looking at two successive transitions, from n to $n-\alpha$ and then $n-\alpha$ to $n - \beta$. The amplitude for a given term in the expression for $x_n(t)^2$ was then $\sum_{\alpha=-\infty}^{\alpha=+\infty} A(n, n - \alpha)A(n - \alpha, n - \beta)$. This multiplication rule may have been the most important single contribution from Heisenberg's paper, but it produced an odd result. For squares and cubes of the same quantity, one faced few problems, but if one sought to multiply two different quantities together ($x(t)y(t)$, for example) one faced a puzzling situation. Unlike classical physics, where multiplication was commutative ($xy = yx$), in Heisenberg's new scheme this commutative relationship did not always hold.

On its own terms, Heisenberg could consider the paper only a moderate success. He could solve the problem of the anharmonic oscillator and get reasonable results and he could rederive Kramers' dispersion relation. Yet he could provide no new results, could not even solve the most basic problem of the older quantum theory, the hydrogen atom, and he was puzzled by the strangely non-commutative behaviour of his new expressions. His concluding remarks were remarkably tentative, holding open the possibility that 'this method after all represents far too rough an approach to the physical problem of constructing a theoretical quantum mechanics.'[114]

If the *Umdeutung* paper was somewhat limited, however, it soon inspired important elaborations. Within a few weeks of reading Heisenberg's paper and puzzling over the peculiar multiplication rule within it, Born realized where he had seen a non-commutative relation like it before: Heisenberg's arrays were matrices. Together with Jordan, Born quickly published a paper 'On Quantum Mechanics', developing and explaining the mathematical apparatus for Heisenberg's theory.[115] Emerging from the new formalism was the necessity of the conservation of energy, the energies of stationary states, and Bohr's frequency condition. Born and Jordan also derived what they termed 'the stronger quantum condition' for the position matrix **q**, the momentum matrix **p**, and the unit matrix **1**:

$$\mathbf{pq} - \mathbf{qp} = \frac{h}{2\pi i}\mathbf{1}$$

In November, Heisenberg joined Born and Jordan for a paper titled 'On Quantum Mechanics II', now known as the Drei–Männer–Arbeit (three man paper), extending previous analyses to include systems of arbitrarily many degrees of freedom and developing a matrix-mechanical analogue to their works on perturbation theory.[116] Its legacy has been mixed. Described as 'the first comprehensive exposition of the foundations of modern quantum mechanics in its matrix formulation', Cassidy has nonetheless noted that the method deployed in the paper 'raised the theory to new and (for many readers, then and now) intolerable heights of abstraction'.[117]

Seemingly less abstract was the formalism proposed by a relative newcomer to quantum problems, Paul Dirac, who produced an algebraic formulation of the 'stronger quantum condition' independently of his German colleagues in early November 1925. Dirac saw in Heisenberg's non-commutative relation an analogue to a result from classical dynamics. For variables x and y, the Poisson bracket expression is (where p's and q's are canonical variables of the system):

$$[x, y] = \sum_r \frac{\partial x}{\partial q_r} \frac{\partial y}{\partial p_r} - \frac{\partial y}{\partial q_r} \frac{\partial x}{dp_r}$$

Given this definition, the following relations hold:

$$[q_r, q_s] = [p_r, p_s] = 0; \quad [q_r, p_s] = \delta_{rs} \quad \{\delta_{rs} = 0, r \neq s; 1, r = s\}$$

Dirac's formulation of Heisenberg's multiplication rule then stated that 'the difference between the Heisenberg products of two quantum quantities is equal to $\frac{ih}{2\pi}$ times their Poisson bracket expression'.[118] This meant, in particular, that

$$q_r p_s - p_r q_s = \frac{ih}{2\pi}\delta_{rs}$$

Among the most powerful aspects of Dirac's version of quantum mechanics was that one could take classical results and methods over into the quantum realm with little change. Indeed, for Dirac it was the fact that the mathematical operations in classical and quantum physics obeyed the same laws, rather than the fact that one could recover classical results from the quantum theory as $h \to 0$, that really grounded the 'correspondence' between the two realms.[119] Early in 1926, Dirac (generalizing his approach to form an algebra of 'q-numbers') and Pauli (using matrix mechanics) derived the frequency spectrum for the hydrogen atom, proving that the new techniques could rival previous methods for the paradigmatic problem of the older quantum theory.[120]

27.5.3 Wave Mechanics

Looking back on the mid-1920s Pauli once referred to quantum mechanics as *Knabenphysik*: boys' physics.[121] He had in mind the remarkable youth of many of the contributors to the new field. Pauli, Heisenberg, Jordan, and Dirac were all born in the twentieth century; in 1926, Dirac and Jordan turned 24, Heisenberg 25, and Pauli 26. In this regard, as in so many others, the founder of wave mechanics, Erwin Schrödinger, proved unusual. Born in 1887, Schrödinger did not come to his work on 'Quantization as an Eigenvalue Problem' through the almost-standard route via spectroscopy, but rather through a problem from the era before the advent of Bohr's model: the quantum statistics of gases.[122] He belonged to no school and worked largely independently of the efforts in Munich, Copenhagen, and Göttingen. And where Heisenberg, Pauli, and others were in near-constant personal and epistolary contact, Schrödinger largely worked alone, writing letters comparatively intermittently. This is not to suggest, of course, that Schrödinger was either a crank or a lone genius prior to 1926; rather, his work stands as a useful reminder of the diversity of research in quantum physics in the first quarter of the twentieth century, and of the viability of alternatives to theoretical spectroscopy in spite of the successes of Born, Bohr, and Sommerfeld's programmes.

In 1924, after publishing roughly a dozen papers in as many years on two main topics—the specific heats of both gases and solids, and the magnetic properties of matter—Schrödinger turned to another problem: what Nernst in 1914 had termed the 'degeneracy' of gases. Nernst had argued that, at low temperatures, even ideal gases do not obey the ideal gas law. Instead of gas pressure being proportional to temperature ($PV \propto T$), the pressure of a gas becomes increasingly independent of temperature as T approaches absolute zero, reaching a non-zero value for $T = 0$. Trying to explain why gases should exhibit this behaviour, Schrödinger found reason to criticize previous approaches (most notably Planck's) for their incorrect deployment of Boltzmannian statistics. That Schrödinger's own alternative was hardly immune to criticism is perhaps beside the point, for almost all attempts to deal with the degeneracy problem changed a few months after his paper was published. In July, after translating it himself, Einstein submitted a paper to the *Zeitschrift für Physik* that he had received by mail from an Indian physicist, Satyendranath Bose. In the letter that accompanied the manuscript, Bose noted that he had managed to 'deduce the coefficient $8\pi v^2/c^3$ in Planck's Law [relating the radiation density to the mean energy of an oscillator] independent of the classical electrodynamics.'[123] That is, without requiring the melding of continuous electromagnetic theory with the discontinuities required by a quantum analysis.

Bose's technique involved a new method of counting. Where Planck had asked how many ways one could distribute P energy elements among N oscillators, Bose counted the number of cells of a certain size in phase space that contained 0, 1, 2, ... n light quanta. The novel statistics then gave Planck's black-body law in a matter of lines. In submitting the paper on Bose's behalf, Einstein appended a note, describing the new derivation as an 'important advance'. The method, he continued, 'also yields the quantum theory of the ideal gas, as I will work out in detail elsewhere'.[124] Einstein's 'elsewhere' was in fact two papers presented to the Prussian Academy of Sciences in September 1924 and January 1925.[125] It was not until the second, however, that Einstein put his finger on what it was, in physical terms, that led to Bose's success and that distinguished his statistics from Planck's. In the Bose–Einstein analysis, particles were not assumed to be statistically independent of one another. Precisely what lay behind the 'mutual influence' of gas particles upon one another, however, was, at least for the moment, 'entirely mysterious'.[126]

In the papers that followed Bose's brief letter, Einstein carried to perhaps its most profound conclusion the set of arguments he had been making for two decades on the relationship between waves and particles. His photon theory, of course, which Bose used, suggested that what many had seen in terms of wave phenomena—the nature of light—must also be considered as having particulate properties. One needed to think of radiation, in at least some situations, in the same way that one thought of a gas. With Bose's method in hand, he now reversed that logic, arguing that one could apply the methods just used to deal with a 'radiation gas' to a gas of material particles. Calculating energy fluctuations in a gas obeying Bose's statistics, he recovered an expression functionally identical to one first derived in 1909 in the study of black-body radiation. In both cases, the fluctuation expression was made up of two terms, one corresponding to particulate phenomena, the other to wave phenomena.[127] In the first decade of the twentieth century, Einstein had taken the result as offering support for the notion that light was a particle. Now, in the middle of the third decade, he suggested that matter possessed wave-like properties—that the connection between a material and a radiation gas was 'more than a simple analogy'—and he pointed to the work of a young French theorist, Louis de Broglie, as a further example of the wave–particle connection.[128]

It was thus through Einstein that Schrödinger became aware of de Broglie's idea that an electron revolving around the nucleus in a Bohr orbit is accompanied by a relativistic phase wave. In what has been termed 'the most influential and successful doctoral dissertation in the history of physics',[129] De Broglie combined Planck's relation with Einstein's expression for the rest mass energy of a particle of mass m_0 to define an intrinsic frequency for the electron, $m_0 c^2 / h = \nu_0$. From his insistence that the accompanying wave stay in phase with this internal oscillation, de Broglie could show both that the group velocity of the phase waves was equal to the orbital velocity of the particle and that the orbit itself had to be equal to an integral number of wavelengths. Bohr's condition defining the size of stationary states, in other words, now had a new—undulatory—explanation.

Schrödinger's first thought upon reading de Broglie's work was to follow him in describing Bohr orbits in more detail, attempting to map the behaviour of the waves accompanying an electron in an elliptical orbit.[130] Dismayed by the mathematical complexities that emerged, however, he turned back to gas statistics. In a paper published at the beginning of 1926, Schrödinger sought to offer an alternative to what he portrayed as the artificiality of the Bose-Einstein counting procedure.[131] One could use 'natural' statistics, he argued, if one conceived of the gas as a whole, rather than as the sum of its molecular parts. The argument proceeded through a clear analogy. In deriving Planck's black-body law for radiation, one could either treat light atomistically and then count according to Bose's statistics, or one could (as Debye had in 1910) quantize the degrees of freedom of the electromagnetic ether directly, without reference to light atoms or even resonators, and use Planck's statistical methods. In precise parallel, Einstein's equations for a gas of material molecules—which explained gas degeneracy by noting the tendency of molecules to crowd into the same energy states at lower temperatures—was derived using Bose's statistics. Schrödinger suggested that one could get the same results using more familiar statistics by quantizing the standing matter waves that could be assumed to make up a gas, treated holistically. This meant, as he wrote, 'nothing more than taking seriously the de Broglie–Einstein undulation theory of the moving corpuscle, according to which this is nothing more than a kind of foam [*Schaumkamm*] on the wave radiation that makes up the world's foundation.'[132]

In the first of his papers on wave mechanics, finished only six weeks after 'On Einstein's Gas Theory', Schrödinger drew an explicit connection between the two—seemingly disparate—topics. De Broglie's ideas of 'matter waves' formed the basis of each. Yet, if Schrödinger's earlier work provided an inspiration the path to his eponymous equation was not straightforward. De Broglie's analysis had been a relativistic one and Schrödinger's first thought was to produce a relativistic wave equation. From his correspondence, it seems clear that such an equation was rapidly forthcoming. Solving it, however, was more challenging. Moreover, Schrödinger was soon dismayed to find that his efforts were unsuccessful. His expression—now known as the Klein–Gordon equation—gave the wrong results, failing to describe the fine structure spectrum so well analysed by Sommerfeld a decade previously.[133]

It was thus a non-relativistic equation that Schrödinger published in what would become the first of four papers to bear the title 'Quantization as a Problem of Proper Values'. A reader of this first paper, however, could have been forgiven a certain bafflement at the frankly cursory efforts the author made to motivate his work. One historian has described the derivation as 'not only curiously formal, but straightforwardly cryptical'.[134] Schrödinger himself, in his second paper, described the crucial transformation equation of the first paper as 'in itself unintelligible'.[135] Evidently, he wished for his new discovery to be judged by its fruits, rather than its roots and from this perspective Q1 (as Schrödinger would term it) was eminently successful. From his equation $\left(\nabla^2 \psi + \frac{2m}{K^2}\left(E + \frac{e^2}{r}\right)\psi = 0\right)$, he could derive expressions for Bohr's energy levels where the integralness of quantum numbers emerged, as he put it, 'in the same natural way as it does in the case of the *node-numbers* of a vibrating string'.[136] That is, for the then-paradigmatic case of the hydrogen atom, Schrödinger could offer a simple solution that did not require the introduction of quantization conditions as an additional hypothesis. Such conditions arose as solutions to the wave equation itself.

Schrödinger published the now familiar derivation of his equation in Q2; received at the *Annalen der Physik* toward the end of February 1926. It proceeds by extending, to the quantum realm, an analogy between mechanics and optics first fully developed by William Rowan Hamilton.[137] In the first half of the nineteenth century, Hamilton had shown that the motion of particles and the propagation of light could be described using the same variational formalism. As Schrödinger noted, however (and as Hamilton also knew) the analogy is, strictly speaking, only one between mechanics and *geometrical* optics. No mechanical analogy exists for central elements of *wave* optics, such as amplitude, wavelength or frequency. Geometrical optics, in fact, may be considered merely as a limiting case of a full undulatory theory, valid in the limit where the wavelength is very small relative to other path dimensions. Rather than see this fact as restricting the analogy between mechanics and optics, however, Schrödinger proposed extending the analogy further, regarding classical mechanics as the limiting case of an undulatory quantum mechanics. 'Perhaps', he wrote, 'our classical mechanics' is the *complete* analogy of geometrical optics and as such is wrong and not in agreement with reality . . .'[138]

Having thus motivated a search for an undulatory quantum mechanics on the basis of Hamilton's analogy—a topic that filled Schrödinger's research notebooks since at least 1918—Schrödinger then used the analogy to derive the wave equation itself.[139] Equation in hand, he could then proceed to show the power of his new form of analysis, solving the problems of the Planck oscillator, the rotator with fixed axis, the rigid rotator with a free axis, and the non-rigid rotator in quick succession. The paper ended by promising that a solution to the problem of the Stark effect using perturbative techniques would soon be forthcoming as, indeed, it was in Q3, received in May.[140]

In the months between the second and third instalments of his Quantization papers, Schrödinger turned his attention to the wave equation's rival. In Q2 he had noted that while his programme and that of Heisenberg, Born, and Jordan seemed similar in intent, they differed radically in their starting points and methods. Despite efforts to find one, Schrödinger could not identify a connection. Nonetheless, he generously pointed out that matrix mechanics had certain distinct advantages over his own as then constituted. In particular, it held out the promise of providing information about line *intensities*. Only a few weeks after finishing Q2, however, he had found the link that had evaded him previously. From the 'formal mathematical standpoint', he would write, 'one might well speak of the *identity* of the two theories'.[141]

With that formal equivalence established, Schrödinger became somewhat less generous about his competition (a competition, it should be noted, that had priority in the discovery of the new mechanics). A footnote ostensibly described his first reaction to Heisenberg's work. 'I naturally knew about his theory', he wrote, 'but was discouraged, if not repelled, by what appeared to me as very difficult methods of transcendental algebra, and by the want of *Anschaulichkeit* [visualizability, or graspability]'.[142] Phrased in personal terms, this note, of course, conveyed a clear message to readers: those dismayed by the difficulties of matrix methods and/or frustrated by Heisenberg's phenomenology now had a viable—mathematically equivalent—alternative.

Stressing his own theory's advantages at both the levels of interpretation and calculational simplicity was a smart move on Schrödinger's part. On the question of interpretation alone, Schrödinger stood on strongly contested terrain. Heisenberg dismissed completely the claim that his own theory was less adequate because it possessed a lesser degree of *Anschaulichkeit*. Writing to Pauli in June he was utterly blunt. 'The more I think about the physical portion of the Schrödinger theory, the more repulsive I find it ... What Schrödinger writes about the *Anschaulichkeit* of his theory "is probably not quite right ..." [the quotation marks a reference to Bohr's elliptical mode of speaking], in other words, it's crap.'[143] On the question of ease of calculation, however, there appeared to be little argument. Hans Bethe, in an interview conducted in the 1960s, captured the Sommerfeld School's attitude with some succinctness: 'Sommerfeld said, "Well, of course we really believe that Heisenberg knows better about the physics, but we calculate with Schrödinger".'[144]

Yet practical advantages could not shield Schrödinger from fundamental questions about the meaning of his theory. What, precisely, was ψ?[145] The first three Quantization papers were largely intepretationally neutral on this point. Certainly, conceptions of wave motion pervaded the work, yet as early as Q2 Schrödinger counselled against regarding his equation as a simple depiction of wave-packets acting like particles. For one thing, it was not obvious that one could build such packets that would not spread over time—decidedly un-particle-like behaviour. Perhaps even more profound, ψ was a function in configuration space, not real space. For n particles, ψ was a function in $3n$ space. That is, rather than being a function of three spatial variables (x, y, z), it was a function of $3n$ variables $(x_1, y_1, z_1, x_2, y_2, z_2 \ldots x_n, y_n, z_n)$. In his equivalence paper and then, somewhat differently, in Q4, Schrödinger offered a far more concrete explanation of the meaning of ψ. The function, multiplied by its complex conjugate and by the unit electronic charge ($e\psi\psi^*$) was a measure of the distribution of charge density over space.

What was, for Schrödinger, one of the great strengths of his approach—the fact that he could produce a theory of quantized energy states without invoking any discontinuities—was, for those in Göttingen and Copenhagen in particular, one of its great weaknesses. They had come to see discontinuity as the most fundamental tenet of quantum theory and they pushed Schrödinger to explain those phenomena where wave characteristics seemed to play no part. Even Lorentz had already pointed to the difficulty of explaining the photoelectric effect or the emission of electrons from heated metals using a theory of wave packets.[146] Bohr had similar objections and invited Schrödinger to Copenhagen toward the end of September to discuss them. In their conversations, Bohr gave some ground, acknowledging that one had to accept the undulatory nature of matter as part of a viable interpretation. Yet he refused to accede to Schrödinger's position entirely and to grant that one could reduce all particulate phenomena to the action of waves. In Heisenberg's version of events (written, it should be noted, well after the events described), the arguments between Schrödinger and Bohr became heated; Bohr acting 'like a relentless fanatic', even continuing the conversation after Schrödinger contracted a cold and took to bed. Bohr pointed to Planck's black-body law, insisting that its derivation required the existence of discrete energies and discontinuous changes. 'For we have known for 25 years what the Planck formula means', Bohr says in Heisenberg's reconstruction, 'And also we see the discontinuities, the jumps, quite directly in atomic phenomena, perhaps on a scintillation screen or in a cloud chamber . . . You can't simply wave away these discontinuous phenomena as though they didn't exist'. For Schrödinger's part, Heisenberg portrayed him as equally blunt: 'If we are still going to have to put up with these damn quantum jumps, I am sorry that I ever had anything to do with quantum theory'.[147]

Despite his vehemence, Schrödinger would not win the battle over the meaning of ψ. In July 1926, Born completed a paper on 'The Quantum Mechanics of Collision Processes' which radically reformulated the significance of the term $\psi \psi^*$.[148] What Schrödinger had portrayed as a 'real' density, Born argued was in fact the density of *probability*—the probability that an electron will end up in a given state after a collision. The claim was startling and Schrödinger would have no part of it, rejecting the idea outright in his debates with Bohr in Copenhagen. Born himself made no bones about the oddity of what he portrayed as a 'ghost' or 'pilot' field of 'de Broglie–Schrödinger waves'. One could describe it, he wrote, 'somewhat paradoxically, as follows: The motion of particles follows laws of probability, but the probability itself propagates in accordance with the law of causality'.[149] This would be only one of many 'paradoxes' in what would soon come to be known as the 'Copenhagen interpretation' of the new quantum mechanics.

27.6 THE COPENHAGEN INTERPRETATION AND THE NEW PHYSICS OF THE QUANTUM

By 1930 almost all the conceptual elements that college students today associate with 'quantum mechanics' were in place. That year saw the publication of Heisenberg's text, *The Physical Principles of the Quantum Theory*, based on lectures given in Chicago in 1929. The book's preface declared its aim to be the diffusion of the 'Copenhagen spirit of quantum theory'.[150] This *Kopenhagener Geist* was founded, above all, on two principles: Heisenberg's uncertainty principle, and Bohr's principle of complementarity. On the 23 March 1927 the *Physikalische Zeitschrift* received a copy of a paper by Heisenberg 'On the Perceptual [*anschaulichen*] Content of Quantum Theoretical Kinematics and Mechanics'.[151] The title was a clear broadside at Schrödinger's claims about the greater *Anschaulichkeit* of wave mechanics. The paper itself began by defining what it meant to describe a physical theory as *anschaulich*. Rejecting any talk that might imply the conceptual graspability of classical models, Heisenberg declared *Anschaulichkeit* to be a concept understood in phenomenological and logical terms alone. We believe a theory to be *anschaulich* 'if we can think qualitatively in all simple cases about the experimental consequences of the theory, and if we have simultaneously never observed that the application of the theory contains inner contradictions'.[152] It might seem, Heisenberg suggested, that contemporary quantum mechanics failed the test of *Anschaulichkeit* on both grounds. Following the model of Einstein's theory of special relativity, however, Heisenberg proposed that inner contradictions might be resolved through a more careful examination of fundamental concepts of kinematics and mechanics.

What does it mean, Heisenberg asked, to speak of the position of an electron? To know its position, one must make a measurement: one could, for example, illuminate it and observe the electron through a microscope. How precise such a measurement may be depends on the wavelength of the light used. The more precisely one wishes to know the electron's position, the smaller the wavelength of light used and hence the higher its energy. Yet in allowing light to strike the electron, one had to reckon with the Compton effect. Although he spoke of *wave*lengths, Heisenberg now turned to a fully particulate explanation, involving the momenta and energy of light quanta. Illuminating an electron involves striking it with a quantum of light, which changes the electron's momentum. The higher the frequency of incident light the greater the change in momentum will be. '[T]he more precisely one determines the position', Heisenberg wrote, 'the more imprecisely does one know the momentum, and vice versa'.[153] The simultaneous measurement of any two commuting quantum variables (e.g. position and momentum; energy and time; action variable and angle variable), he would conclude, was not possible to arbitrary accuracy. In mathematical terms, the relation between the uncertainty in an electron's position, Δx and the uncertainty in its momentum, Δp, is $\Delta x \Delta p \sim h$.

Heisenberg submitted his remarkable paper from Copenhagen, but he did so in Bohr's absence. When Bohr returned, he and his young assistant locked horns over the meaning and significance of uncertainty. For Bohr, increasingly convinced of the necessity of both wave and particulate explanations in quantum mechanics, Heisenberg's corpuscular explanation of the action of a microscope was too one-sided. As a result, he noted, Heisenberg had, in fact, made an error in his derivation of the uncertainty relations. Heisenberg had made the locus of uncertainty the act of observation, yet Bohr noted that there was already uncertainty in the design of the microscope itself. Any real-world microscope has an aperture of finite size (say, ε). For light of wavelength λ, the resolving power of the microscope is then $\lambda/2\varepsilon$. There is an inherent uncertainty, in other words, in our determination of the momentum associated with the light used in our measuring instrument. According to Bohr, then, understanding the uncertainties involved in our observation of an electron required using *both* the wave and particulate nature of radiation.[154]

Heisenberg was not so easily convinced. He and Bohr wrangled for weeks, the arguments degenerating until Heisenberg found himself in tears.[155] It was not until May that Heisenberg capitulated. He had steadfastly refused to withdraw the paper from publication. Now, as a compromise, he agreed to add a note in proof, granting Bohr's points.[156] In September, at a conference held at Lake Como, Bohr gave the uncertainty principle a central place in his new, overarching understanding of quantum mechanics. Like Heisenberg he emphasized the havoc that the new physics created for some of the most basic tenets of Western philosophy. One could no longer speak, for example, of a divide between an observer and the observable world: 'an independent reality in the ordinary physical sense can neither be ascribed to the phenomena nor to the agencies of observation'.[157] One could no longer speak, even in principle, of a completely causal universe. Where Laplace could once imagine a Being able to predict the future by knowing the present state of all matter and the laws governing its future development, Bohr and Heisenberg noted that the first piece of knowledge was limited by the uncertainty relations. One could *define* a system mathematically and leave it unperturbed or one could observe and hence alter it, but one could not do both simultaneously. Observation and definition were *complementary*. Related to this complementarity was another: that between wave and particle. In mathematical terms one could speak of wave and particle models as interchangeable. Yet in physical terms one could not speak of phenomena without detailing the experimental apparatus applicable for measurement of such phenomena. And insofar as such apparatus realized any given model of reality, it excluded the realization of any other model. How one designs an experiment, as an observer, conditions that which is observed. And one requires different, mutually exclusive, experimental set-ups in order to capture physical reality in its entirety.

Just as in the case of light we have consequently in the question of the nature of matter, so far as we adhere to classical concepts, to face an inevitable dilemma, *which has to be regarded as the very expression of experimental evidence*. In fact, here again we are not dealing with contradictory but with complementary pictures of the phenomena, which *only together* offer a natural generalization of the classical mode of description.[158]

If Heisenberg's text of 1930 introduced readers to the philosophical peculiarities of quantum theory, another book published the same year eschewed almost all such questions in favour of mathematical clarity and calculational power. According to Kragh, Paul Dirac's *Principle of Quantum Mechanics* rapidly achieved a status for the new mechanics similar to that of Sommerfeld's *Atombau* for the older quantum theory.[159] Deploying the transformation theory that he and Jordan had developed independently of one another,[160] Dirac simply side-stepped the question of the relative physical content of Schrödinger and Heisenberg's physics, introducing each more as a calculational aid than physical fundament and emphasizing his own 'symbolic' approach for its ability to 'go more deeply into the nature of things' and 'to express the physical laws in a neat and concise way'.[161]

The book was based on lectures Dirac had been giving at Cambridge since 1927 and at its heart was the relativistic wave equation that Dirac published at the very beginning of 1928. Unlike the Schrödinger equation, Dirac's was of first order in both space and time, displaying the symmetry required by a relativistic treatment.[162] And unlike the Klein–Gordon equation, Dirac's could be shown to produce the same (empirically confirmed) expression for the fine structure splitting of lines in the hydrogen spectrum. On the other hand, the equation involved 4×4 matrices—and hence four components—only two of which could be associated with the two possible spin values for an electron. Dirac's symmetrical formulation appeared to require the existence of states of negative energy—ignored in any classical solution; not so easily dismissed in quantum theoretical terms. In 1930, Dirac suggested that one should associate these anti-particle solutions with the proton, so named in 1920. The idea met with a storm of protest, since the proton—while it does, indeed, have the same charge, yet opposite sign to the electron—also has a mass roughly 1800 times larger than its purported counterpart. J. Robert Oppenheimer flagged another difficulty when he pointed out that Dirac's theory allowed for the rapid spontaneous decay of the hydrogen atom. Citing both these problems in 1931, Dirac withdrew his earlier suggestion, putting forward the possibility of the existence of an 'anti-electron', a 'new kind of particle, unknown to experimental physics, having the same mass and opposite charge to an electron'. The following year, working with cloud chamber photographs of cosmic radiation (and, by his own account, entirely independently of Dirac's theoretical predictions) Carl Anderson published a paper on 'The Apparent Existence of Easily Deflectable Positives'. In 1936 he would receive the Nobel Prize for his discovery of the positron. The prize for the year before had gone to James Chadwick for his discovery of the neutron, also found in 1932 (leading many to speak of an *annus mirabilis* in particle physics). The decade to follow (indeed, the next three decades or more) saw a veritable cascade of particle discoveries, revealing an array of particles and antiparticles to supplement the electron and proton, once considered the fundamental building blocks of the subatomic world.[163]

The 1930s would end, of course, with the outbreak of the Second World War, which itself culminated in a dramatic and terrible application of quantum mechanics and atomic physics. A field that had roots in technologies that today seem almost prosaic, such as lighting and refrigeration, now found increasingly spectacular and forbidding uses in the construction of weapons for wars both hot and cold.

NOTES

1. Arnold Sommerfeld, 'Das Plancksche Wirkungsquantum Und Seine Allgemeine Bedeutung Für Die Molekularphysik', *Verhandlungen der Gesellschaft Deutscher Naturforscher und Ärzte* 83 (1912).
2. Max Planck, 'Die Stellung Der Neueren Physik Zur Mechanischen Naturanschauung', *Physikalische Zeitschrift* 11 (1910) 922–932.

3. Sommerfeld, 'Das Plancksche Wirkungsquantum', 31.
4. Suman Seth, *Crafting the Quantum: Arnold Sommerfeld and the Practice of Theory, 1890–1926* (Cambridge, MA: MIT Press, 2010) 242.
5. Planck's list was comprised of Einstein, Hendrik Lorentz, Wilhelm Wien, and Joseph Larmor. Planck to Nernst, 11 June 1910, cited in Thomas S. Kuhn, *Black-Body Theory and the Quantum Discontinuity, 1894–1912* (Oxford: Oxford University Press, 1978) 230.
6. P. A. M. Dirac, *The Principles of Quantum Mechanics*, Fourth ed. (Oxford: Clarendon Press, 1981).
7. Peter Galison, *Einstein's Clocks, Poincaré's Maps: Empires of Time* (New York, London: W. W. Norton & Company, 2003) 291–292.
8. Albert Einstein, 'Elektrodynamik Bewegter Körper', *Annalen der Physik* 17 (1905) 891–921.
9. Kuhn, *Black-Body Theory* 228. Kuhn may be displaying a theoretician's bias here. Barkan has argued that the radiation problem was a profoundly important one at the time and the black-body was essential for standardization. For Barkan, the comparative failure to comment on Planck's work 'was due to the fact that it was largely irrelevant to the ongoing experimental and technological application of radiation laws'. Diana Kormos Barkan, *Walther Nernst and the Transition to Modern Physical Science* (Cambridge: Cambridge University Press, 1999) 117. The most important section of Planck's distribution curve lay outside the visible spectrum and was hence much less interesting to the lighting industry.
10. Elizabeth Garber, 'Some Reactions to Planck's Law, 1900–1914', SHPS 7 (1976) 105.
11. Ludwig Boltzmann, *Lectures on Gas Theory*, trans. Stephen G. Brush (New York: Dover, 1995) 444. Orig. *Vorlesungen Über Gastheorie* (1896), 98.
12. Principle contributions to the literature include Martin J. Klein, 'Max Planck and the Beginnings of the Quantum Theory', *AHES* 1 (1962), Martin J. Klein, 'Planck, Entropy, and Quanta', *The Natural Philosopher* 1 (1963), Kuhn, *Black-Body Theory*, Martin J. Klein, Abner Shimony and Trevor J. Pinch, 'Paradigm Lost? A Review Symposium', *Isis* 70 (1979), Allan Needell, 'Irreversibility and the Failure of Classical Dynamics: Max Planck's Work on the Quantum Theory, 1900–1915', Yale, 1980, Thomas S. Kuhn, 'Revisiting Planck', *HSPS* 14.2 (1984), Olivier Darrigol, *From C-Numbers to Q-Numbers: The Classical Analogy in the History of Quantum Theory* (Berkeley, Los Angeles, Oxford: University of California Press, 1992).
13. Seth, *Crafting the Quantum* Ch. 4.
14. Max Planck, *Physikalische Abhandlungen Und Vorträge*, vol. I, 3 vols. (Braunschweig: Friedr. Vieweg & Sohn, 1958) 551–552.
15. Barkan, *Walther Nernst* 116.
16. Wien's law of displacement, not to be confused with his expression for the black-body curve, was a general functional expression relating the density of radiant energy at a given wavelength to temperature. On the early history of such laws, see Hans Kangro, *Early History of Planck's Radiation Law* (London: Taylor and Francis, 1976) 37–47. Wien stated his law in 1893 as follows: ' In the normal emission spectrum from a black body each wavelength is displaced with change of temperature in such a way that the product of temperature and wavelength remains constant'. Quoted in Kangro, *Early History* 45. The form most familiar today, $E_\lambda = \lambda^{-5}\phi(\lambda T)$ was first derived by Joseph Larmor in 1900. Planck derived the distribution law in the form given above in January, 1901.
17. Max Planck, 'Zur Theorie Des Gesetzes Der Energieverteilung Im Normalspektrum', *Verhandlungen der deutschen physikalischen Gesellschaft* 2 (1900). Reproduced in Planck, *PAV I* 698–706.
18. Kuhn, *Black-Body Theory*, Olivier Darrigol, 'The Historians' Disagreement over the Meaning of Planck's Quantum', *Centaurus* 43 (2001), Clayton A. Gearhart, 'Planck, the Quantum, and the Historians', *Phys. Persp.* 4 (2002) 170–215.

19. On Ehrenfest, see Martin J. Klein, *Paul Ehrenfest: The Making of a Theoretical Physicist*, vol. 1 (Amsterdam: North-Holland Publishing Company, 1970). On Einstein and thermodynamics, see Martin J. Klein, 'Thermodynamics in Einstein's Thought', *Science* 157 (1967) 509–516.
20. Albert Einstein, 'Über Einen die Erzeugung und Verwandlung Des Lichtes Betreffenden Heuristischen Gesichtspunkt', *Annalen der Physik* 17 (1905) 132–148, Martin J. Klein, 'Einstein's First Paper on Quanta', *The Natural Philosopher* 2 (1963) 57–86.
21. Note that in this expression, Einstein did not make use of Planck's constant, h. Here T is the gas constant, and N_0 is Avogadro's constant.
22. Albert Einstein, 'Zur Theorie der Lichterzeugung und Lichtabsorption', *Annalen der Physik* 20 (1906) 199–206. Kuhn, *Black-Body Theory* 182.
23. Kuhn, *Black-Body Theory* 185.
24. Suman Seth, 'Quantum Theory and the Electromagnetic World-View', *HSPBS* 35.1 (2004).
25. Kuhn, *Black-Body Theory* 192.
26. Barkan, *Walther Nernst* 3.
27. Barkan, *Walther Nernst* 141.
28. If one assumes that each particle within a three-dimensional lattice has three degrees of freedom, and that the average kinetic energy equals the average potential energy for such particles, then the average energy per mole of substance is $2 \cdot 3 \cdot \frac{kT}{2} \cdot N_0$, giving a value close to that determined by the Dulong–Petit rule of 6cal/mol. See Abraham Pais, *Subtle Is the Lord . . : The Science and the Life of Albert Einstein* (Oxford: Oxford University Press, 1982) 389–394.
29. Walther Nernst, *Thermodynamics and Chemistry* (New Haven: Yale University Press, 1906).
30. Walther Nernst, 'Untersuchungen Über Die Spezifische Wärme Bei Tiefen Temperaturen II', *Sitzungsberichte der Preussischen Akademie der Wissenschaften* (1910). Quoted in Martin J. Klein, 'Einstein, Specific Heats, and the Early Quantum Theory', *Science* 148 (1965) 177.
31. Kuhn, *Black-Body Theory* 214.
32. Kuhn, *Black-Body Theory* 217.
33. Richard Staley, 'On the Co-Creation of Classical and Modern Physics', *Isis* 96 (2005) 530–558.
34. Quoted in Russell McCormmach, 'Henri Poincaré and the Quantum Theory', *Isis* 58 (1967) 43.
35. Quoted in Jagdish Mehra, *The Solvay Conferences on Physics: Aspects of the Development of Physics since 1911* (Dordrecht, Boston: D. Reidel Publishing Company, 1975) xiv.
36. On Bohr's life, see Abraham Pais, *Niels Bohr's Times: In Physics, Philosophy, and Polity* (Oxford: Clarendon Press, 1991). On the pre-history of Bohr's atomic model, see John L. Heilbron and Thomas S. Kuhn, 'The Genesis of the Bohr Atom', *HSPS* 1 (1969), 211–290.
37. Quoted in Heilbron and Kuhn, 'The Genesis of the Bohr Atom', 218.
38. John L. Heilbron, 'The Scattering of Alpha and Beta Particles and Rutherford's Atom', *AHES* 4 (1968).
39. Quoted in Heilbron, 'Rutherford's Atom', 245.
40. Niels Bohr, 'On the Constitution of Atoms and Molecules I, II, and III', *Philosophical Magazine* 26 (1913) 476–502.
41. Cited in Pais, *Niels Bohr's Times* 149.
42. Quoted in Pais, *Niels Bohr's Times* 153.
43. Sommerfeld to Langevin, 1 June 1914. Michael Eckert and Karl Märker, eds., *Arnold Sommerfeld: Wissenschaftlicher Briefwechsel*, vol. 1 (Berlin, Diepholz, München: Deutsches

Museum, Verlag für Geschichte der Naturwissenschaften und der Technik, 2000) 484–485, on 84.

44. Arnold Sommerfeld, 'Die Allgemeine Dispersionsformel Nach Dem Bohrschen Modell', *Arbeiten aus den Gebieten der Physik, Mathematik, Chemie* (1915). Reproduced in F. Sauter, ed., *Arnold Sommerfeld: Gesammelte Schriften*, vol. III, IV vols. (Braunschweig: Friedr. Vieweg & Sohn, 1968) 136–171, on 67.

45. Sauter, ed., *GS III* 136.

46. Arnold Sommerfeld, 'Die Feinstruktur Der Wasserstoff- Und Wasserstoffähnlichen Linien', *Sitzungsberichte der Bayerischen Akademie* (1915) 459.

47. Max Jammer, *The Conceptual Development of Quantum Mechanics* (New York, St. Louis, San Francisco, Toronto, London, Sydney: McGraw-Hill Book Company, 1966) 108. That this was indeed seen as a programme may be discerned from Bohr's comments in Niels Bohr, 'On the Programme of the Newer Atomic Physics (2 December 1919)', *Niels Bohr Collected Works*, ed. J. Rud Nielsen, vol. 3 (Amsterdam, New York, Oxford: North-Holland Publishing Company, 1976) 222. On experiment's crucial role in the development and acceptance of quantum spectroscopy, see Helge Kragh, 'The Fine Structure of Hydrogen and the Gross Structure of the Physics Community, 1916–26', *HSPS* 15.2 (1985).

48. K. Schwarzschild, 'Bemerkungen Zur Aufspaltung Der Spektrallinien Im Elektrischen Feld', *Verhandlungen der deutschen physikalischen Gesellschaft* 16 (1914) 20–24.K. Schwarzschild, 'Zur Quantenhypothese', *Berliner Berichte* (1916) 548. Paul Sophus Epstein, 'Zur Theorie Des Starkeffektes', *Annalen der Physik* 50 (1916) 489–520, Paul Sophus Epstein, 'Zur Quantentheorie der Spektrallinien', *Annalen der Physik* 51 (1916) 1–94, 125–167. Good discussions of the Hamilton–Jacobi method in the context of the quantum theory are given in Jammer, *Conceptual Development* 89–109. Darrigol, *From C-Numbers to Q-Numbers* 102–116.

49. Arnold Sommerfeld, *Atomic Structure and Spectral Lines*, trans. Henry L. Brose (London: Methuen and Co. Ltd., 1923) 610.

50. Niels Bohr, 'On the Application of the Quantum Theory to Periodic Systems (Unpublished Paper, Intended for Publication in the Phil. Mag., April 1916)', *Niels Bohr Collected Works*, ed. Ulrich Hoyer, vol. 2 (Amsterdam, New York, Oxford: North-Holland Publishing Company, 1981) 258.

51. Paul Ehrenfest, 'A Mechanical Theorem of Boltzmann and Its Relation to the Theory of Energy Quanta', *Proc. Amsterdam Acad.* 16 (1913). In Martin J. Klein, ed., *Paul Ehrenfest: Collected Scientific Papers* (Amsterdam: North-Holland Publishing Company, 1959) 340–346.

52. Niels Bohr, 'On the Quantum Theory of Line Spectra, Parts I–III [1918–1922]', *Niels Bohr Collected Works*, ed. J. Rud Nielsen, vol. 3 (Amsterdam, New York, Oxford: North-Holland Publishing Company, 1976) 75.

53. On Ehrenfest and the adiabatic principle see Klein, *Paul Ehrenfest: The Making of a Theoretical Physicist* esp. 264–292.

54. Albert Einstein, 'Zur Quantentheorie Der Strahlung', *Physikalische Zeitschrift* 18 (1917) 121–128. First published in *Mitteilungen der physikalischen Gesellschaft, Zurich* (1916). Translated and reproduced in B. L. van der Waerden, ed., *Sources of Quantum Mechanics* (New York: Dover, 1968) 63–78.

55. For systems of more than one degree of freedom, the ratio of the kinetic energy to the frequency is no longer an adiabatic invariant. In 1917, a student of Ehrenfest's, J. M. Burgers, extended the analysis to arbitrarily many degrees of freedom by proving the adiabatic invariance of Karl Schwarzschild's 'action-angle variables'. Burgers' result ultimately opened the way for Bohr to increase the scope of his model to include the very broad class of conditionally periodic motions.

56. Pais, *Niels Bohr's Times* 190.
57. Bohr, 'On the Constitution . . . ' 13.
58. On the importance to Bohr of Einstein's 1916 papers, which derived Planck's radiation law by assuming that the probabilities of transition could be determined by analogy with the theory of radioactivity, see Jammer, *Conceptual Development* 112–114. Darrigol, *From C-Numbers to Q-Numbers* 118–128.
59. Bohr, 'On the Quantum Theory of Line Spectra, Parts I–III [1918–1922]', 82.
60. Quoted in Pais, *Niels Bohr's Times* 193.
61. Darrigol, *From C-Numbers to Q-Numbers* 138.
62. Ehrenfest quoted in Darrigol, *From C-Numbers to Q-Numbers* 142.
63. See, for example, Arnold Sommerfeld, *Atombau Und Spektrallinien*, 1st ed. (Braunschweig: Friedr. Vieweg & Sohn, 1919) 402.
64. Wolfgang Pauli, *Exclusion Principle and Quantum Mechanics: Lecture Given in Stockholm after the Award of the Nobel Prize of Physics 1945* (Neuchatel: Editions du Griffon, 1945) 8.
65. On Pauli's life, see Charles P. Enz, *No Time to Be Brief: A Scientific Biography of Wolfgang Pauli* (Oxford: Oxford University Press, 2001).
66. Sommerfeld to Bohr, 4 September 1913. Eckert and Märker, eds., *ASWB I* page?
67. (Sommerfeld, 1920b). Reprinted in Sauter (1968a, pp. 523–565, quote on p. 532). Also quoted and discussed in some detail in Forman (1970, p. 191).
68. Pauli to Sommerfeld, 6 December 1924, document 83 in Michael Eckert and Karl Märker, eds., *Arnold Sommerfeld: Wissenschaftlicher Briefwechsel*, vol. 2 (Berlin, Diepholz, München: Deutsches Museum, Verlag für Geschichte der Naturwissenschaften und der Technik, 2004) 176–179.
69. Pauli to Sommerfeld, 6 June 1923. A. Hermann, Karl von Meyenn and V. F. Weisskopf, eds., *Wolfgang Pauli: Wissenschaftlicher Briefwechsel Mit Bohr, Einstein, Heisenberg U. A.* (Heidelberg, New York: Springer, 2005) 97. Quotation from John L. Heilbron, 'The Origins of the Exclusion Principle', *HSPS* 13 (1983) 292.
70. Paul Forman, 'Alfred Landé and the Anomalous Zeeman Effect, 1919–1921', *HSPS* 2 (1970).
71. Not only was the idea incomprehensible in classical terms, it contradicted a basic principle of the quantum theory, Bohr's so-called *Aufbauprinzip*. This 'construction principle' held that quantum numbers were conserved in building up atoms. Assuming, as was standard, that a rare gas atom (where all shells were filled with the maximum possible number of electrons) possessed zero net angular momentum, then the core of an alkali atom which immediately followed it in the periodic table of elements should also have zero net momentum. In Heisenberg's model, by contrast, it possessed a non-zero value: namely, one half.
72. A. Landé, 'Termstruktur Und Zeemaneffekt Der Multipletts', *Zeitschrift der Physik* 15 (1923) 189–205. The capitalized variables bear simple relations to their uncapitalized counterparts. Landé's version was one of many renormalizations, with $K = k - 1/2$; $J = j$ for even multiplets, $J = j - 1/2$ for odd; and $R = r/2$.
73. Pauli to Lande, 23 May 1923. Hermann, von Meyenn and Weisskopf, eds., *PWB* 87–90, on 87. Cf. Heilbron, 'Exclusion Principle', 296.
74. Wolfgang Pauli, 'Über Den Zusammenhang Des Abschlusses Der Elektronengruppen Im Atom Mit Der Komplexstruktur Der Spektren', *Zeitschrift der Physik* 31 (1925) 765–783.
75. Helge Kragh, 'Niels Bohr's Second Atomic Theory', *HSPS* 10 (1979) 123–186.
76. Niels Bohr, 'The Structure of the Atom (Nobel Lecture, Delivered 11 December 1922)', *Nature* (1923) 12.
77. Bohr does not include this calculation. Niels Bohr and J. Rud Nielsen (ed.), *Niels Bohr Collected Works*, vol. 4 (Amsterdam, New York, Oxford: North-Holland Publishing Company, 1977) 303.

78. See, for example, Bohr and Nielsen (ed.), *BCW 4* 303. See also Kragh, 'Niels Bohr's Second Atomic Theory', Heilbron, 'Exclusion Principle'.
79. Kragh, 'Niels Bohr's Second Atomic Theory'.
80. As we shall see below, Bohr also used a third quantum number as a means of classifying and ordering the multitude of X-ray lines. However, he did not use the third number to further subdivide orbits.
81. Pauli to Bohr, 12 December 1924. In Hermann, von Meyenn and Weisskopf, eds., *PWB* 189.
82. R. Kronig and V. F. Weisskopf, eds., *Wolfgang Pauli: Collected Scientific Papers*, vol. II, 2 vols. (New York, London, Sydney: Interscience Publsihers, 1964) 225.
83. For an excellent, near-contemporary summary, see R. H. Fowler, 'Spinning Electrons', *Nature* 119 (1927) 90–92.
84. Edward Mackinnon, 'Heisenberg, Models, and the Rise of Matrix Mechanics', *HSPS* 8 (1977) 144.
85. Jammer, *Conceptual Development* 208.
86. Thomas S. Kuhn, 'Interview with G. Uhlenbeck, 3 March 1962', (AHQP, 1962).
87. Quoted in Darrigol, *From C-Numbers to Q-Numbers* 177.
88. Paul Forman, 'Weimar Culture, Causality, and Quantum Theory, 1918–1927: Adaptation by German Physicists and Mathematicians to a Hostile Intellectual Environment', *HSPS* 3 (1971) 1–115.
89. Arnold Sommerfeld, *Atombau Und Spektrallinien 4th ed* (Braunschweig: Freidr. Vieweg & Sohn, 1924) 4. Suman Seth, 'Crisis and the Construction of Modern Theoretical Physics', *BJHS* 40 (2007) 47.
90. Quoted in Darrigol, *From C-Numbers to Q-Numbers* 178.
91. Jammer, *Conceptual Development* 193. Anthony Duncan and Michel Janssen, 'On the Verge of Umdeutung: John Van Vleck and the Correspondence Principle, Parts I & II', *AHES* 61 (2007) 32.
92. Rudolf Ladenburg, 'Die Quantentheoretische Zahl Der Dispersionselektronen', *Zeitschrift der Physik* 4 (1921). Translated in van der Waerden, ed., *Sources* 139–158.
93. On Kramers, see M. Dresden, *H. A. Kramers: Between Tradition and Revolution* (New York: Springer, 1987). For a detailed reconstruction of Kramer's work on virtual oscillators, see Hiroyuki Konno, 'Kramers' Negative Dispersion, the Virtual Oscillator Model, and the Correspondence Principle', *Centaurus* 36 (1993).
94. Niels Bohr, H. A. Kramers and J. C. Slater, 'The Quantum Theory of Radiation', *Philosophical Magazine* 47 (1924). Reproduced in van der Waerden, ed., *Sources* 159–176.
95. Quoted in Duncan and Janssen, 'On the Verge', 72. Cf. Roger H. Stuewer, *The Compton Effect: Turning Point in Physics* (New York: Science History Publications, 1975) 291.
96. Heisenberg to Pauli, 4 March 1924. Hermann, von Meyenn and Weisskopf, eds., *PWB* 149–150. Cf. Dresden, *Kramers* 202, Duncan and Janssen, 'On the Verge', 71.
97. Quoted in Pais, *Niels Bohr's Times* 237.
98. Stuewer, *The Compton Effect*.
99. Quoted in Jammer, *Conceptual Development* 170.
100. Quoted in Seth, *Crafting the Quantum* 236.
101. J. C. Slater, 'Radiation and Atoms', *Nature* 113 (1924) 307.
102. S. S. Schweber, 'The Young John Clarke Slater and the Development of Quantum Chemistry', *HSPBS* 20 (1990) 339–406.
103. John Hendry, 'Bohr–Kramers–Slater: A Virtual Theory of Virtual Oscillators and Its Role in the History of Quantum Mechanics', *Centaurus* 25 (1981) 200–201.
104. van der Waerden, ed., *Sources* 165.

105. Max Born, 'Über Quantenmechanik', *ZdP* 26 (1924).
106. H. A. Kramers, 'The Law of Dispersion and Bohr's Theory of Spectra', *Nature* 133 (1924). Reproduced in van der Waerden, ed., *Sources* 177–180.
107. H. A. Kramers, 'The Quantum Theory of Dispersion', *Nature* 114 (1924). Reproduced in van der Waerden, ed., *Sources* 199–202.
108. H. A. Kramers and Werner Heisenberg, 'Über Die Streuung Von Strahlen Durch Atome', *Zeitschrift der Physik* 31 (1925). Translated in van der Waerden, ed., *Sources* 223–252, quote on 34.
109. Werner Heisenberg, 'Über Quantentheoretische Umdeutung Kinematischer Und Mechanischer Beziehungen', *ZfP* 33 (1925). Translated in van der Waerden, ed., *Sources* 261–276, quotes on 61 and 62.
110. Arnold Sommerfeld, 'Quantentheoretische Umdeutung Der Voigtschen Theorie Des Anomalen Zeemaneffektes Vom D-Linientypus', *Zeitschrift für Physik* 8 (1922) 257–272.
111. Pauli to Bohr, 12 December 1924. In Hermann, von Meyenn and Weisskopf, eds., *PWB* 189.
112. Quoted in David C. Cassidy, *Uncertainty: The Life and Science of Werner Heisenberg* (New York: W. H. Freeman and Co., 1992) 198.
113. Helge Kragh, *Quantum Generations: A History of Physics in the Twentieth Century* (Princeton: Princeton University Press, 1999) 162.
114. van der Waerden, ed., *Sources* 276.
115. Max Born and Pascual Jordan, 'Zur Quantenmechanik', *ZdP* 34 (1925). Abridged and translated in van der Waerden, ed., *Sources* 277–306.
116. Max Born, Werner Heisenberg and Pascual Jordan, 'Zur Quantenmechanik Ii', *ZdP* 35 (1925). Translated in van der Waerden, ed., *Sources* 321–386.
117. Cassidy, *Uncertainty* 207.
118. P. A. M. Dirac, 'The Fundamental Equations of Quantum Mechanics', *Proceedings of the Royal Society of London A* 109 (1925). Reproduced in van der Waerden, ed., *Sources* 307–320, quote on 14–15.
119. van der Waerden, ed., *Sources* 315.
120. Wolfgang Pauli, 'Über Das Wasserstoffspektrum Vom Standpunkt Der Neuen Quantenmechanik', *Zeitschrift der Physik* 36 (1926), P. A. M. Dirac, 'Quantum Mechanics and a Preliminary Investigation of the Hydrogen Atom', *Proceedings of the Royal Society of London* 110 (1926) 561–580. Early calculations using matrix methods could be seen as a mixed bag, however. Both Pauli and Dirac failed to calculate the intensities of spectral lines—necessary information if one wished to move beyond the hydrogen atom using perturbative techniques. Mara Beller, *Quantum Dialogue: The Making of a Revolution* (Chicago: University of Chicago Press, 1999) 28–29. On the other hand, in 1926 Pauli did manage to use matrix mechanics to solve a problem for which the methods of the older quantum theory had seemed inadequate: perturbations in the energy of a hydrogen atom in crossed electric and magnetic fields. Jammer, *Conceptual Development* 241.
121. Kragh, *Quantum Generations* 169.
122. On Schrödinger's life, see Walter Moore, *Schrödinger: Life and Thought* (Cambridge: Cambridge University Press, 1992). On his work prior to his studies of matter waves, see in particular, Martin J. Klein, 'Einstein and the Wave–Particle Duality', *The Natural Philosopher* 3 (1963) 1–49, Paul A. Hanle, 'The Coming of Age of Erwin Schrödinger: His Quantum Statistics of Ideal Gases', *AHES* 17 (1977) 165–192, Linda Wessels, 'Schrödinger's Route to Wave Mechanics', *SHPS* 10 (1979) 311–340, Christian Joas and Christoph Lehner, 'The Classical Roots of Wave Mechanics: Schrödinger's Transformations of the Optical–Mechanical Analogy', *SHPMP* 40 (2009).

123. Satyendra Nath Bose, 'Plancks Gesetz Und Lichtquantenhypothese', *Zeitschrift der Physik* 26 (1924). Translated (including Bose's note to Einstein) as 'Planck's Law and the Light Quantum Hypothesis' by O. Theimer and Budh Ram, 'The Beginning of Quantum Statistics', *AJP* 44 (1976) 1056.
124. Theimer and Ram, 'The Beginning of Quantum Statistics', 1056.
125. Albert Einstein, 'Quantentheorie Des Einatomigen Idealen Gases', *Sitzungsberichte der Preussischen Akademie der Wissenschaften* (1924) 261–267, Albert Einstein, 'Quantentheorie Des Einatomigen Idealen Gases. Zweite Abhandlung', *Sitzungsberichte der Preussischen Akademie der Wissenschaften* (1925) 3–14.
126. Einstein, quoted in Daniela Monaldi, 'A Note on the Prehistory of Indistinguishable Particles', *SHPMP* 40 (2009) 392.
127. Klein, 'Einstein and the Wave–Particle Duality.
128. Hanle, 'Coming of Age', 176. See also Olivier Darrigol, 'The Origin of Quantized Matter Waves', *Historical Studies in the Physical and Biological Sciences* 16 (1986) 19–24.
129. Edward Mackinnon, 'De Broglie's Thesis: A Critical Retrospective', *AJP* 44 (1976) 1047. L. de Broglie, 'Recherches Sur La Théorie Des Quanta', 1924.
130. Wessels, 'Schrödinger's Route', 322. See also V. V. Raman and Paul Forman, 'Why Was It Schrödinger Who Developed De Broglie's Ideas?', *HSPS* 1 (1969) 291–314.
131. Erwin Schrödinger, 'Zur Einsteinschen Gastheorie', *PZ* 27 (1926).
132. Schrödinger, 'Zur Einsteinschen Gastheorie', 95, Moore, *Schrödinger* 188.
133. Helge Kragh, 'Erwin Schrödinger and the Wave Equation: The Crucial Phase', *Centaurus* 26 (1982) 154–197.
134. Kragh, 'Erwin Schrödinger', 158.
135. Erwin Schrödinger, 'Quantisierung Als Eigenwertproblem, Zweite Mitteilung', *AdP* 79 (1926), Erwin Schrödinger, *Collected Papers on Wave Mechanics*, trans. J. F. Shearer and W. M. Deans, Third (augmented) ed. (Providence, Rhode Island: AMS Chelsea, 1982) 13–40, on 13.
136. Schrödinger, *CPWM* 1.
137. Joas and Lehner, 'Classical Roots'.
138. Schrödinger, *CPWM* 18.
139. On the notebooks in particular, see Joas and Lehner, 'Classical Roots'.
140. Erwin Schrödinger, 'Quantisierung Als Eigenwertproblem, Dritte Mitteilung', *AdP* 80 (1926) 397–400, Schrödinger, *CPWM* 62–101.
141. Erwin Schrödinger, 'Über Das Verhaltnis Der Heisenberg–Born–Jordanschen Quantenmechanik Zu Der Meinen', *AdP* 79 (1926), Schrödinger, *CPWM* 45–61, on 46.
142. Schrödinger, *CPWM* 46.
143. Heisenberg to Pauli, 8 June 1926 Hermann, von Meyenn and Weisskopf, eds., *Pwb* 328–329. Translation from Cassidy, *Uncertainty* 214.
144. Thomas S. Kuhn, 'Interview with Hans Bethe, 01/17/1964', (AHQP, 1964), vol., 9.
145. Edward Mackinnon, 'The Rise and Fall of the Schrödinger Interpretation', *Studies in the Foundations of Quantum Mechanics*, ed. Patrick Suppes (East Lansing, MI: Philosophy of Science Association, 1980), Linda Wessels, 'The Intellectual Sources of Schrödinger's Interpretation', *Studies in the Foundations of Quantum Mechanics*, ed. Patrick Suppes (East Lansing, MI: Philosophy of Science Association, 1980).
146. Lorentz to Schrödinger, 27 May 1926. K. Przibram, ed., *Letters on Wave Mechanics* (New York: Philosophical Library, 1967) 43–54. Lorentz's letter contains a remarkably detailed evaluation of the strengths and weaknesses of Schrödinger's approach.
147. Quoted in Moore, *Schrödinger* 227–228.

148. Max Born, 'Quantenmechanik Der Stossvorgänge', *ZfP* 38 (1926), Max Born, 'Physical Aspects of Quantum Mechanics', *Nature* 119 (1927). On Born's interpretation, see Linda Wessels, 'What Was Born's Statistical Interpretation?', *PSA: Proceedings of the Biennial Meeting of the Philosophy of Science Association* 1980, Volume Two: Symposia and Invited Papers (1980), Pais, *Subtle Is the Lord* . . ., Beller, *Quantum Dialogue* 39–50.
149. Born, 'Quantenmechanik Der Stossvorgänge', 804.
150. Werner Heisenberg, *The Physical Principles of the Quantum Theory* (Chicago: University of Chicago Press, 1930) Preface.
151. Werner Heisenberg, 'Über Den Anschaulichen Inhalt Der Quantentheoretischen Kinematik Und Mechanik', *PZ* 34 (1927), 172–198.
152. Heisenberg, 'Anschaulichen Inhalt', 172.
153. Heisenberg, 'Anschaulichen Inhalt', 175. On Heisenberg's uncertainty principle, see (among many others) Jammer, *Conceptual Development* 343–360, Cassidy, *Uncertainty*, Beller, *Quantum Dialogue* 103–116.
154. Niels Bohr, 'The Quantum Postulate and the Recent Development of Atomic Theory', *Congresso Internazionale dei Fisici, Como* (Nicola Zanichelli, 1927), vol. 2. Reproduced in Jørgen Kalckar, ed., *Niels Bohr: Collected Works*, vol. Six: Foundations of Quantum Physics I (1926–1932) (Amsterdam: North Holland, 1985) 113–136. A great deal has been written on Bohr's principle of complementarity. For a selection, see Jammer, *Conceptual Development* 360–383, James T. Cushing, *Quantum Mechanics: Historical Contingency and the Copenhagen Hegemony* (Chicago: University Press, 1994), Beller, *Quantum Dialogue* 117–144.
155. Cassidy, *Uncertainty* 242–243.
156. That said, as Camilleri has recently reminded us, significant philosophical differences between Bohr and Heisenberg remained. See Kristian Camilleri, *Heisenberg and the Interpretation of Quantum Mechanics;: The Physicist as Philosopher* (Cambridge: Cambridge University Press, 2009).
157. Bohr, 'The Quantum Postulate and the Recent Development of Atomic Theory', *Nature* 121 (1928) 580–590.
158. Bohr, 'The Quantum Postulate and the Recent Development of Atomic Theory', *Nature* 121 (1928) 580–590.
159. Helge Kragh, *Paul Dirac: A Scientific Biography* (Cambridge: Cambridge University Press, 1990) 77.
160. Jammer, *Conceptual Development* 309–323, Kragh, *Paul Dirac* 38–43, Anthony Duncan and Michel Janssen, 'From Canonical Transformations to Transformation Theory, 1926–1927', *SHPMP* 40 (2009) 352–362.
161. Dirac, *The Principles of Quantum Mechanics* viii–ix.
162. P. A. M. Dirac, 'The Quantum Theory of the Electron', *Proceedings of the Royal Society of London, A* 117 (1928) 610–624.
163. See Kragh, *Quantum Generations* 174–205.

CHAPTER 28

THE SILICON TIDE: RELATIONS BETWEEN THINGS EPISTEMIC AND THINGS OF FUNCTION IN THE SEMICONDUCTOR WORLD

TERRY SHINN

This chapter deals with the changing interactions between fundamental physics and the learning and skills situated near engineering and enterprise as related to microelectronics and in particular to semiconductors that occurred over the span of the twentieth century. The term 'changing relations' here signifies: 1) the components (conceptual, material instrumental, organizational, and so on) that are involved in a domain, 2) their shifting relative position in a formal or informal hierarchy, and 3) the characteristics of the system constitutive of fundamental physics and industrial physics at different historical moments. Description and analysis involve two systems: 1) the epistemic system that revolves around properties of matter and explanation, 2) product systems that revolve around 'function' as definition of that which is investigated. My reflections on the evolving relations between things epistemic and matters of function associated with physics will draw on selected episodes in the tide of silicon with reference to an understanding of semiconducting to the invention of transistors and their development. Concretely, this entails study of theories, experiments, models, invention, materials, products, manufacturing markets, and management from Marconi's introduction of Hertzian communication to the 1947 invention by Bardeen, Brattain, and Shockley of the transistor, to the development of the microprocessor in 1970, and finally to the introduction in 2011 of the nanoscale Finfet transistor family by the Intel company.

In discussion of the history of the silicon tide, the unit of analysis to be employed here revolves around what I term 'combinatorials'.[1] Combinatorials include elements such as concepts, materials, instruments, skills, and organization. More specifically linked to the history of the transistor, the combinatorials that figured importantly are intuition, tinkering and individual invention, theory, experimentation, materials, engineering, mathematics logic, programming, management, and the entrepreneurial spirit. The centrality of combinatorials to my analysis does not impinge on the significance of the university or industry as referents. Some combinatorials are historically more closely associated with academia than with enterprise. Theory has been connected with the university; organization is assigned to industry. Nevertheless, such assignment is sometimes misleading, and it conceals the more crucial question of precisely what combinatorial, in connection with what other combinatorial, fuels an innovation, and what is the hierarchic relation between the two. In what measure and why do relevant combinatorials change over time, and what are the circumstances for the introduction of historically new combinatorials? Does the assignment of a combinatorial to the university versus enterprise referent shift?

Combinatorials may be mobilized for two purposes. Combinatorial resources may serve to illuminate a question in order to acquire understanding. Issues of explanation in terms of causes and effects are foremost here. This is the essence of the realm of epistemology. Alternatively, combinatorials can be mobilized to achieve a concrete objective in the absence of issues of understanding. In this scenario, what is wanted is an outcome. In the history of the transistor, for some actors the device was interesting as a window into understanding of the properties and dynamics of the physical world. For other groups involved with the transistor, 'the device is the function'. The device was uniquely an input/output system that produces a wanted effect. In my treatment of the history of the transistor, attention will be drawn to the key distinction between matters epistemic and matters of function. As in the case of combinatorials, the mapping of this unit of analysis onto a university-versus-industry matrix proves complex.

28.1 THE ARTS OF TECHNICAL CONTROL

For the related domains of electronics and microelectronics and early electromagnetic discoveries, for purposes of fundamental understanding of the physical world, for control of physical phenomena, and for techniques of practical communication, what counts as fundamental physics research on the one hand, and what counts as invention, engineering, and technology on the other, can for most practical and analytic aims be distinguished from one another. The basic physics of these strongly related domains involved mathematics, theory, methodology-based laboratory experimentation, and the purification and manipulation of appropriate materials. Physics that concerns the silicon tide thus detects physical effects, describes them, and explains them—in sum, principally cognitive expectations. The associated technology and engineering of electronics revolved around observations of physical effects, their control, the harnessing of effects to reach some specified practical objective, and the introduction of a stable, reliable and cost effective manufacturing system. More recently, technology has also come to entail considerations of management. These latter considerations focus on questions of 'function'—they engage matters of performance and efficiency. The goal is not understanding, but is instead the accomplishment of an objective *per se*.

The decades that stretch between the announcement by Heinrich Rudolf Hertz (1857–1894) of electromagnetic waves and the study of their transmission in space on the one hand, and on the other hand the later development of early efficiently performing high-frequency (HF), very-high-frequency (VHF), and later the ultra-high-frequency (UHF) amplifier tubes was an era when there emerged relatively little cognition about the physics of electron conductivity in the environments of the vacuum tube. For over two decades, many advances in the area of electromagnetic questions was the purview of observations (sometimes almost random and accidental) and of technical control. The subsequent contributions of the physics community to what would first become vacuum tube-based electronics and later solid-state electronics and relevant circuits first consisted of a mix between intuition and tinkering and high physics knowledge, and after that almost exclusively entailed an amalgam of theory and experimental programs.

In 1886 the young German physicist Hertz, whose research included extending James Clerk Maxwell's recent investigations, explored the physical features of electromagnetic waves, and in particular their radiative behaviour. He determined that waves emitted at one site could be detected at a nearby site. Working in Paris at about the same time, the French physicist professor Edouard Branly (1844–1940) independently noted the same phenomenon. However, unlike Hertz, Branly advanced the idea that this newly identified physical phenomenon might be employed for purposes of communication, and in France Branly is sometimes celebrated as the father of radio. While some academics conducted research on the behaviour of electromagnetic waves and their diffusion[2], it was mainly independent inventors who took front stage for the three ensuing decades. They introduced devices that emitted waves, detection apparatus, and increasingly effective antenna systems. However, this group was generally distinct from the laboratory activities of university physics; and robust, critical explanation of effects in the guise of cognition was neither an objective nor a consequence of endeavour. The invention of machines capable of generating electro waves with communication in mind quickly emerged as a prime challenge for a growing number of aspiring inventors, the best known being Guglielmo Marconi (1874–1937). Communication over distances was soon achieved through ever more powerful spark-gap devices, where, using extremely high voltage, a spark leapt across an empty space, therein yielding the necessary wave. A typical device consisted of a circular platform whose outer edge was punctuated with 6- to 12-foot diameter comutators having a series of copper knobs to enhance conductivity. The platform was rotated rapidly, and was fed with around 2,000 volts DC. This yielded a constant transmitting wave. The frequency of the wave could not be controlled, but all the waves were of very low frequency—not always the most efficient frequency for long-distance communication. The receiving devices likewise functioned largely in the absence of cognition. The typical receiver of 1900 centred on a 'coherer', consisting of a cylindrical container filled with very fine nickel and silver filings.[3] For reasons still unexplained today, when subjected to an electromagnetic pulse, this powder becomes conducting. In the receivers of the day, when triggered, an electric circuit was established and registered in the form of a droplet of ink on a moving paper scroll. However, after pulse stimulation the metallic powder had to be restored to its initial state in order to make it respond to a second electromagnetic wave. This was achieved by solidly striking the coherer; and the entire laborious process then recommenced. To repeat, this set-up was entirely empirical. Neither Marconi nor his colleagues grasped why the system functioned as it did, but cognition was not the goal. The set-up worked after a fashion, and it was not long before superior techniques were invented.

Marconi's system also required an antenna for transmission and reception of the electromagnetic waves. This became the principal focus of attention. He tested a myriad different configurations, lengths, and set-ups in terms of antenna–ground potential. He observed that the antenna constituted the key to his ongoing quest for ever-greater distances of communication.[4] The range of his device was 4 miles in 1897. Two years later he could span the English Channel, and in 1902 he established transoceanic communication between Europe and North America. In his 1909 Nobel Prize lecture, Marconi spoke at length about antenna design and theories of wave propagation. It is clear that observations and reflection were not associated with physics. He observed correlations between land and water environments and transmission, conjectured on the possible existence of what we today term 'ground waves', and spoke of emission wavelength and of antenna geometries. These considerations had no internal logic, raised no well-defined questions, and did not seek to couple observation to the small corpus of information on related matters in science. Marconi cited but one academic scientist, John Ambrose Fleming, whose analyses on wave diffusion in alternative environments were grounded in contemporary models of the specific electrical conductivity of different materials. Marconi shared his prize with the German physicist Fernand Braun (1850–1918), based at the University of Strasbourg. Despite Braun's knowledge in the discipline of physics, his publications and 1909 Nobel Prize lecture show that in the domain of radio the voice of academic disciplinary physics was thus alive, albeit only faintly so.[5] Braun's work dealt mainly with antenna configuration, installation, and performance. He recorded highly precise data for the three elements of his experiments, but failed to move beyond description, unable to formulate hypotheses or a model, or to propose a working vocabulary. The status and centrality of physics cognition was soon to change, however.

Voice communication technology, which required controlled emission and reception, were achieved in 1900. Nevertheless, evolution remained mostly disconnected with physics; it was often instead based on inventor trial-and-error and even happenstance observations. In this community of inventors the few explanations that emerged often even proved highly erroneous; but cognition was neither required for the complexification and growth of the electronic-related objective.

Vacuum tubes were introduced at the turn of the century, yet neither was this seemingly highly abstract and complex innovation associated with learning in physics or the physics community. In 1904 the thermionic valve, better known as the vacuum tube, was invented by the technician John Ambrose Fleming (1849–1945).[6] This was a diode that served as a rectifier and yielded modest amplification. The decisive tube breakthrough occurred in 1906 when Lee De Forest (1873–1961) developed what he termed the 'grid electrode' tube. The tube is now known as the triode, which long constituted the backbone of electronics.[7] De Forest's innovation derived from an unnoticed artefact earlier developed by the inventor Thomas Edison (1847–1931), who tried out countless permutations of geometries and metals inside a vacuum envelope, but not critically nor with a specific objective in mind. De Forest's tube at first contained a twisting wire situated between the filament and plate.[8] He, like Edison, continued to tinker with devices, and in 1908 he came across the configuration of a separate grid, cathode, and anode. This device offered considerable gain, and was highly stable and amenable to many different circuits. De Forest's rather entrepreneurial interests with things technical are well viewed in the light of his more than 180 US patents and his numerous involvements in business ventures—mostly unsuccessful.

Permanent innovations in electronic circuitry were also achieved during these decades, and they lay behind the transformation of radio communication from a domain of individual listeners to a societal domain where the broad public benefited from radio and where radio became a large-scale commercial affair. Three novelties proved particularly crucial to technical change and to the industrialization of electronics: the 'regenerative circuit' in receivers (feed-back) allowed for much higher gain. Gain radically modified the world of radio communications. The earlier famous crystal receiver had limited audience capacity to listen to a transmission because the audio output was so weak that headphones were necessary. Regenerative circuits allowed the introduction of loud speakers that enabled group reception. The super heterodyne circuit introduced two benefits—greater sensitivity and frequency selectivity. The many controls that had been required to operate a receiver now disappeared. One simple knob sufficed. Finally, new circuitry permitted broadband radio which was acoustically more comfortable and clearer. Taken together, these modifications transformed, as never before, radio into a mass public activity. The demand for radio receivers expanded hugely, and electronic products became an increasing focus of industrial manufacture. The companies Westinghouse, General Electric, and RCA became large and profitable firms. Indeed, their grip on the sector became so tight through the exercise of patents that for a period, further innovation was stifled. Again, neither the intellectual resources of physics nor the expectation of supplementary cognition underpinned or even contributed to the bulk of this work.

In his carefully documented history of the origins and evolution of technology and enterprise in the Silicon Valley, Christophe Lécuyer argues in the opening chapters of his book *Making Silicon Valley: Innovation and the Growth of High Tech, 1930–1970*[9] that key innovations in vacuum-tube devices were the product of a variety of skills and a particular inventive mentality among the region's amateur radio community. The interests, questions, unorthodox technical experiments, experiences, and intuitions common to the 'ham shack' of amateur radio culture nourished the standing pattern of electronics, and indeed participated in no mean way to the emergence of a new ensemble of combinatorials. In the skill and intellectual environment presented by Lécuyer, one sees that some middle-level physics learning converged with machine and ceramic skills, practice-gained insights in electronic phenomena, and in the tricks useful in the concrete construction of a complex component, and subsequently in its manufacture.

San Francisco was one of the regions in the United States where radio-related activity ran high. This was in part due to the importance of shipping: the port of San Fransisco was a major location for west coast import and export and a centre of shipping managerial operations and ship communications. Technical expertise in ship-to-shore transmission developed with employment for young enthusiasts.[10] Small enterprises also took root to manufacture electronic components, and amateur and commercial transmitters and receivers. In short, the bay area comprised a space where amateur radio capacities could penetrate the world of professional electronics. It is this trajectory that Lécuyer describes for the cases of Charles Litton and Jack McCullough,[11] along with the lasting technical and industrial contributions of the two men in the history of mid-twentieth-century microelectronics. Litton was trained in mechanical engineering at Stanford University. For his part, McCullough had received little formal education; he attended a technical junior college, and he cannot be considered a scientist. He established the radio-component company EIMAC, Eitel-McCullough, Inc.[12] Here a slightly new set of combinatorials emerge different from those that underpinned innovation and product during the first decades of the electronic revolution.

Two key technical problems were addressed and resolved by amateur radio enthusiasts during the 1930: the construction of long-life high-power amplifier tubes, and the development of tubes for ever-higher-frequency operation. At that time the vacuum tubes sold by RCA and other manufacturers were unsatisfactorily limited in power output and were criticized for their short life expectancy. The two aforementioned amateurs from the San Francisco area participated importantly in overcoming these obstacles. Tube construction was an outstanding problem. High-amplification performance depended on achieving a very high vacuum in the glass envelope. It similarly demanded a solid connection between envelope and metal base. Impurities within the vacuum had to be avoided. New metals used for tube elements were called for. Finally, a new tube design was judged crucial by many wishing to upgrade tube performance.

In the framework of Litton Industry, the application of basic mechanical skills and hands-on skills in the use of ovens employed in low-technology metallurgy allowed Litton to make considerable progress in attacking some of the imperfections of tube construction. He designed and built a new vacuum pump adapted to expulsion of air from the glass envelope and also capable of extracting some oily substances that often hampered tube operation. Litton likewise introduced novel heating protocols that helped expel unwanted gases and that solidly connected the base to the envelope. In connection with these measures he replaced the metal tantalum in tube plates with tyrovac, which absorbed residual gas build-up. These processes yielded tubes that proved capable of considerably greater power amplification and whose life expectancy was significantly superior to the devices on offer from the big tube manufacturers. The aforementioned gains were not grounded in the learning of physics or engineering: they were a product of the extension of largely established skills and information to a fresh field, and thus the whole cloth of solutions was generally available.

These ameliorations in power vacuum tube construction helped somewhat in extending the upper-frequency limits of operation. Frequency came to the fore during the 1930s for technical and commercial reasons. During the 1920s, 200-metre wavelength constituted the maximum frequency. This constraint fell to 100 metres, and based on tinkering by amateur radio enthusiasts, the upper limit had fallen to 10 metres by the early 1930s. The decimetric bands of transmission extended the geographic range of reliable radio communication. Shorter wavelengths demanded acute tube changes often beyond those linked simply to construction. The experience, tinkering, and experiments of McCullough contributed to new tube and circuit design that led to the manufacture of innovative tubes—namely, the pentode, which by far surpassed the frequency range and efficiency of the extant triode. Higher-frequency tubes were employed in FM radios, which enjoyed ever-increasing success.

With the entry of the United States into the war in 1941 the military became a key market for electronics, and not least of all extremely high-frequency tubes of the sort required by radar. This demanded tubes capable of operation at 13 cm wavelength or below. The British had designed and built such a tube, and it was now the turn of the Americans. This was swiftly achieved through a combinatorial of physicists and electrical engineers—a new speciality. William Henson, who held a doctorate in physics and electrical engineering from Stanford University and whose professional endeavours straddled academia and enterprise, invented the klystron radar tube. This tube was based on a generic device designed and built by him called the rumbatron.[13] The rumbatron incorporated fundamental electronic principles that permitted production and amplification of UHF signals that were employed in a range of devices. It contributed necessary components for the Stanford linear accelerator, as well as for radar, and clearly established advanced mathematics and physics in the conceptualization, design, and construction of an entirely new generation of apparatus whose domains of application were altogether unprecedented. This tube was manufactured by Varian in response to a huge military market. On the morrow of the war, other UHF tubes also found an expanding market as FM radio in the US shifted frequency from 48 megacycles to around 100 megacycles, thus reducing public demand for an earlier category of vacuum tube. The shift to VHF technologies called for the birth of a new configuration of cognitive, technical, materials, manufacturing, and organizational combinatorials.

28.2 Cognition, Materials, and Coordination

A transistor may be defined as a semiconductor device used to amplify and switch electronic signals. It is made of a solid piece of semiconductor material, with at least three terminals for connection to an external circuit. A voltage or current applied to one pair of the transistor's terminals changes the current flowing through another pair of terminals. Because the controlled (output) power can be much more than the controlling (input) power, the transistor provides amplification of a signal. Today, some transistors are packaged individually, but many more are found embedded in integrated circuits. The transistor is the fundamental building block of modern electronic devices, and is ubiquitous in modern electronic systems.[14]

The transistor was invented in December 1947 by scientists at the Bell laboratory. It is a semiconducting device, neither an electrical conductor nor insulator, that can perform the main functions of vacuum tubes such as rectification, detection, and power amplification. Transistors are extremely small, and are mostly far more robust than vacuum tubes. Operating as integrated circuits, and later as microprocessors, transistors have today become essential components of technology, and for that matter, for many central areas of global culture. In this section, the prehistory of the transistor will be set forth, followed by a description of the transistor's invention and its early introduction into the world of enterprise and fabrication, and to the market.

Advanced theory in quantum physics and learning in crystallography played a key role in the birth of the silicon tide. Insights into phenomena of electron transport associated with progress in solid-state physics during the 1930s introduced the possibility of creating single-crystal diodes and amplifiers, as opposed to vacuum-tube devices. The intellectual resources for such an achievement lay inside the physics community, where mathematical physics and the chemistry, metallurgy, and crystallography learning for control of materials were concentrated. Be it in the environment of the university or industry laboratory, it was principally work in theory and experimentation that established the preconditions for the silicon tide. The tinkering and empiricism of the individual inventor and radio enthusiast became relatively marginalized as combinatorials. The same was largely also true of hands-on skills that prevailed before.

The birth of the transistor constitutes one culminating event in extended and intense research on quantum theory, electron transport, crystal structure and the nature and role of impurities, and of research into the defining properties of insulators and conductors. It was only on the eve of the invention of the semiconductor transistor that agreement over the existence of semiconducting materials arose, with an accompanying preliminary understanding of how they operate. It is clear that the extreme complexity of the transistor, and thus the many cognitive fields that it mobilized, proved very different from the kind of combinatorials present in the pioneering decades of the spark gap, coherer, triode, regenerative circuit, and even the later pentode and klystron. Progress in research associated with the emergence of solid-state physics comprise the preconditions for the transistor—namely, quantum theory and investigation of the optical and electrical behaviour of crystals in terms of their internal structures. Much pre-twentieth-century study of crystals focused on their geometry, and mathematical description became increasingly central. There was then little reflection on the mechanical, optical, and electrical properties of crystals. This changed early in the twentieth century, when scientists observed that crystals often possess regions of internal discolouring, and that this condition is connected with varying optical and electrical behaviour. The investigations of the German physicist Robert W. Pohl (1884–1976) is emblematic of the research carried out that proved directly relevant to the identification of semiconductor materials and to the transistor. Pohl and his team observed the presence of areas of discolouration in rock-salt crystals, and the density of these regions correlated with variations in optical and electron activity. Pohl discovered that discolouration constitutes sites of impurities, where foreign materials are locked into the crystals lattice. These sites were termed 'colour centres'. When crystals were subjected to an electric charge, the position of colour centres migrated from one electrical pole to the other. There appeared to exist a significant link between the areas of impurity and electron transport inside alkali halide materials.

Pohl was a professor of physics at Göttingen University. Due to an earlier observed presence of acute and stable optical and electronic variations in alkali halides, he selected this material for experimentation. Initially, his work was hampered by a need for pure quantities of rock salt, but in 1924 this problem was overcome thanks to the creation of synthetic crystals. The growth of pure crystals would soon emerge as a central component in the evolution of the transistor. Synthetic rock salt was obtained using a crystal that was suspended in a crystal melt, which yielded a single crystal that grew around the initial seed. This process of growing ensured Pohl the supply of relatively pure, standardized, and large crystals needed for his endeavours.[15]

By the end of the 1930s it was generally understood that the presence of colour centres in the form of impurities is necessary to the circulation of electrons, and that the centres act as vacancies that permit circulation—the famous 'F bands'. Pohl had long been a convert to quantum theory, and though he was not involved in theoretical work he nevertheless considered that the mechanism of electron transport could be viewed as an atom-based event and that the Schrödinger equation was imperative for clarification of how electrons flow. He believed that colour centres are areas of atomic defects. Using his knowledge of crystals, in 1938 he built a semiconductor amplifier incorporating a three-electrode alkali halide crystal for AEG company (Allgemeine Elektricitäts-Gesellschaft).

Along with the technique of crystal-growing and experimental study of colour centres and their importance for electron transport, theoretical research on field effects and on electron bands and gaps provided the second major intellectual resource basic to the creation of the transistor. Walter Schottky (1886–1976) was a German physicist who formulated what became known as the 'Schottky effect' and a chain of associated effects that became crucial both to the point-junction and field-effect transistors. Schottky was professor at the universities of Würzburg and later Rostock, and he directed a research laboratory at Siemens. He was the father of electron and ion emission, and was among those who invented the pentode tube. He formulated the Schottky effect in 1914. The effect treats the electronic interactions between a charged point at a distance from a flat metal surface. Schottky identified a force-field barrier as the point approached the surface, moving from the interior toward the exterior. These considerations later became known in the domain of semiconductor physics and the transistor as the 'Schottky barrier', so important to transistor technology. A lowering of the barrier enhances current, and thus power. During research in 1946 and 1947, calculation of the precise position of this barrier and its values constituted one of the most difficult problems. Jumping through this barrier, the electron tunnelling mechanism stands at the very heart of the physical dynamics of microelectronics. In the 1930s Schottky's work began to take into account the recent hypothesis of Fermi layers. The Fermi layer is critical, as it is there that the most relevant processes of electron flow occur. These layers are located through calculation. The equations of the day could not offer a precise solution, and this persisted as a major stumbling block through most of the 1940s and constantly plagued transistor progress. Despite its early imperfections, the Schottky barrier represents a key to semiconductor analysis, design, fabrication, and operations, and the elevated level of theory and mathematics required is staggering.

As late as the mid-1930s the fundamental question of whether 'semiconductors' even existed was still very undecided. In a famous letter, Wolfgang Pauli (1900–1958), the top theoretician of the day, wrote to Rudolf Peierls (1907–1995): 'One shouldn't work on semiconductors. That is a filthy mess. Who knows if a semiconductor exists.'[16] However, in the course of the decade, learning about this elusive material constantly grew, and by 1939 it was not only absolutely sure that semiconductors existed, it had also become clear how they operated. This understanding was inextricably connected to the band theory of solids which explains how insulators differ from conducting metals, and in what ways semiconductors are different from both. This understanding was almost entirely driven by quantum theory and high mathematics. The application of the Schrödinger equation, in a periodic potential comprises the base for band theory and discovery of 'holes'. Calculations covered electron/electron interaction in solids, and wave propagation. Such analysis led to the concepts of conduction gaps and gaps and 'holes'. The physicist Hans Bethe (1906–2005) set up the mathematics for much of the ensuing research, and the tools that he introduced were soon to be applied to investigations of electron transport by Felix Bloch (1905–1983), Werner Heisenberg (1901–1976), Alan Wilson (1934–1991), and Philip Morse (1903–1985) among several others.[17] These scientists calculated that solids possess, or do not possess, layers in which energy distributions of specific values can be transported. Filled band can carry no current: insulators have completely filled conduction bands versus metals, that have partly filled bands. Electron conduction depends on the presence of 'holes'—vacancies—that allow the flow of electrons. In their turn, holes are linked to vacancies in the crystal lattice associated with valence. It is possible to observe conduction with bound electrons, and also without bound electrons. The crucial factors are holes and impurities.

This matter of impurity long proved highly central to the development of the transistors, as they require highly pure materials. For a few years the material was germanium, but this element was soon replaced by silicon. However, as just indicated, semiconduction is also contingent on the presence of impurities. The twin need for exceptionally pure germanium or silicon doped by precise quantities of selected impurities was for several years a key problem—and to some extent it remains a problem, on two levels. First, scientists and engineers had to establish techniques for purification, for growing standardized crystals, and for effective introduction of just the proper amount of selected impurities to obtain the effects required. Second, as transistors became objects of mass production the environment in which fabrication occurred had to be carefully controlled in order to prevent contamination by destructive impurities. Impurity possessed a double, contradictory status. Conducting material is conducting and not an insulator due to the presence of impurities which provide the necessary hole in the crystal lattice.

In the mid-1940s enough was known about semiconducting and semiconductors that Bell Laboratories opened a research group whose mission was the invention of a solid-state amplifier device. This objective was reached in late 1947 with the development of the transistor. Its introduction has heralded a new age, with the rise of microelectronics, and the expression of the latter in computers and a myriad of other digital devices that have significantly transformed countless spheres of contemporary culture.

As indicated above, most of the experimental findings, mathematics, and theory that contributed to an understanding of the electronic behaviour of crystals and to band theory took place inside German universities. How is it, then, that the invention of the transistor occurred inside an industrial laboratory? The reply to this important question is in part linked to historical contingencies; but it is also intrinsically connected to deep structures of research activities that are frequently overlooked. Fundamental research is often viewed as synonymous with academia, and applied research, engineering, and technology are usually automatically seen as synonymous with enterprise. There are sound reasons for this line of thinking. Pure research is indeed congenial with the science norms of 'communism', 'universality', 'disinterestedness', and 'organized scepticism' that Merton proposed as the base of science's values, with scientific disciplines that frame the legitimate fields of specific research activities, and with autonomy—all of which are often institutionalized inside the university.[18] Conversely, the measure of enterprise that is rooted in the profit motive, invention, production, and commercialization of goods is often linked correctly to short-term research focusing on very precise objectives and whose bottom line is 'function'.

The frequent inclination to identify fundamental science with a specific organizational and institutional site—namely, the university—and to automatically identify application-driven research with a different category of organization and institution—namely, enterprise—is to ignore a key feature of contemporary learning-related dynamics. What stands central to fundamental science is 'cognition'. It is safe to say that the bond between science and narrowly epistemological considerations is profound and historically constant. It is admittedly the case that certain environments prove more amenable to epistemological expectation and commitments than others. However, what counts is epistemological drives and not the particular organizational and institutional site in which work occurs.

The pursuit of 'function' in research comprises the contrast with the epistemic turn. Here 'function' refers to achievement of control of matter in order to attain a purpose. It is associated with the idea of control, where control allows realization of an activity, often in terms of technology and a useful product. The realm of 'function' embraces a broad horizon that includes scientific learning, engineering, technology, fabrication, commercialization, the creation and extension of the enterprise, its internal organization, law, and politics. In the world of 'function', epistemic considerations are far from paramount. The expectation is not attainment of original learning, but rather attainment of control of objects as products, and in this environment multiple and often simultaneous technical and non-technical engagements must be sustained.

Correlation between the epistemic and university on the one hand, and between 'function' and enterprise on the other hand, is not inevitable.[19] As evidence of this, the twentieth-century economic successes of ATT/Bell in terms of services and technology components provided economic wealth and introduced a company tradition strongly open to epistemological expectations. The war-driven demand for novel technologies also stimulated Bell to support all forms of research. On the morrow of the Second World War, Bell boasted a laboratory of epistemic expectations with more than 1,000 researchers.[20] There are countless stories where young graduates in physics and mathematics were recruited after 1945, and on arriving at the laboratory were simply told to carry on with projects of interest to them, whatever the focus. This is the epistemic regime in action inside enterprise, and exhibiting the same logic routinely associated with academia, albeit not necessarily correctly so. Bell is far from the sole enterprise to incorporate epistemic laboratories. Another example is Texas Instruments, where in 1988 epistemic research led to the development of quantum dots.[21]

The often-overlooked contrary disjuncture between institutional site and research orientation also typically pertains to enterprise and 'function' investigation.

Beginning in the 1930s, Stanford University increasingly evolved toward 'function' expectations, and MIT too continued its shift in that direction, initiated in the decades following the First World War, in chemistry.[22] Rice University has taken the same turn, as exemplified in its involvement with industry-intended carbon nanotubes.[23]

In the light of this understanding of disjuncture between organizational environment and characteristics of learning (epistemic versus function), it should not be viewed with surprise that Bell Laboratories, following the development of transistors, established a solid-state semiconductor group. The company employed more than 1,000 scientists during the immediate post-war years, and these scientists were encouraged to pursue research projects as they wished,[24] with few instructions and no constraints. This was epistemic research in the strongest sense of the word, and it was enthusiastically supported inside an entrepreneurial environment. In this context, a Bell group set out to explore the operations of semiconductor crystals and possibly to find a solid-state amplifying device. It was directed by William Shockley (1910–1989), a brilliant graduate in quantum physics from MIT,[25] and working with him were John Bardeen (1908–1991), also a brilliant quantum theorist trained at Princeton,[26] and Walter Brattain (1902–1987), who, however, possessed few academic credentials, having only an undergraduate degree from Whitman College in Washington State.[27] Brattain was a radio engineer, and during most of the 1930s he worked at the US National Bureau of Standards on introducing norms for amplifiers. During this period he also became interested in solid-state rectification, and in the use of a sandwich of copper and copper oxide for said purpose. The junction between the two materials permitted the flow of electrons in only one direction. It was because of his experience with solid-state materials and in metals and conduction that Brattain was included on the semiconductor team. Bardeen, however, owed his position to his accomplishments in quantum theory and mathematics.[28] He considered that semiconducting was associated with electron flow as related to Fermi layers, and his intention was to identify and to localize optimal conduction bands, where electron flow is maximal. For his part, Shockley too was a consummate theoretician and mathematician. His direct involvement in the development of the first transistor appears to many historians as less decisive than that of Bardeen and Brattain. Shockley was an efficient group manager for the most part, and he succeeded in maintaining enthusiastic support from Bell headquarters for the project. He had, since the war, also become prominent in Washington in areas related to military technology evaluation, and continued in this capacity throughout the 1950s. These activities frequently diverted his attention from the daily work of research.

For Bardeen and Brattain, the criteria and proof of success was power amplification on the basis of a germanium crystal. As indicated, Bardeen judged that the best path was identification of three-dimensional specifically located conduction bands. Brattain worked to obtain suitable samples of materials and to make empirical tests. In December 1947 the two men undertook one set of experiments where they attached two strips of gold foil to a crystal at calculated junction points. To their surprise and delight, the power meter showed a significant increase. They had succeeded, in part by chance and in part through reflection and calculations, in demonstrating that a semiconductor can increase power in the same way as a vacuum-tube triode. Because of the importance of metal/semiconductor contact and the positioning of contact, this early category of device was called the 'point-contact transistor'. The team of Bardeen, Brattain, and Shockley together received a Nobel Prize in 1956.[29]

Shockley, who already had a reputation for arrogance and authoritarian ways, considered that as team leader he alone should be acknowledged as father of the transistor. Stung by seeing that he did not receive the entire credit for inventing the transistor, he immediately set out to devise a superior solid-state amplifier. He was a truly exceptional theorist and calculator, and within only a few months he solved a corpus of calculations on the Schottky field effect, and therewith proposed the 'junction transistor', which operated on principles very different from the point-contact device.

Bardeen had been hampered by his strategy of trying to identify conduction bands, which entailed analysis for a three-dimensional space. Shockley instead thought of semiconduction in terms of broad regions of electron fields that interact. He conceived a transistor consisting of three layers—two electrically semiconducting negative layers separated by positive layers. Electrons flow between the two negative layers, where a charge applied to the positive layer serves as a control circuit, as with the grid in a vacuum-tube triode.

The three physical parameters crucial to semiconductors are amplification, frequency, power consumption, and stability/reliability. The effectiveness of the junction transistor proved superior to the point-contact in all domains. Nevertheless, even for the junction device, the range of applications still long remained restricted to switching and low-power contexts.

28.3 THE WORLD OF 'FUNCTIONS'

The years 1950–51 saw Bell Laboratories licence manufacturing rights of its transistors to other firms, such as Western Electric and RCA. Bell sought minimal integration in terms of products and components, and thus was happy to turn a profit through the sales of fabrication rights. For a moment, cognitive and highly innovative technology for new electronic components slowed. Only through the introduction of still two other novel combinatorials in the late 1950s and 1970s with the birth of integrated circuits and microprocessors did new species of fundamental innovation recur. Concerning early transistors, a different combinatorial requirement emerged: namely, procedures and techniques entailed in mass fabrication. As transistor fabrication rocketed, a parallel challenge arose: supply of a sizeable volume of high-purity germanium, and later, silicon.

The shift from the theory and mathematics of transistor invention of the 1940s to issues of materials supply and fabrication of the 1950s introduced a new set of combinatorials. Manufacturing mobilized the efforts of engineers, technicians, and also the labour force. On a different register, the epistemic regime was often replaced by the function regime, and this frequently called for a new form of relations between the laboratory and the shop floor, which introduced the combinatorial of management. For almost three decades the epistemic regime became relatively quiescent.

Two superlative articles respectively by Stuart Leslie and by Hyungsub Choi, examining the history of the transistor during the 1950s, set forth the move to the combinatorial of fabrication, function, and many corollary transformations. The Allentown transistor plant located in the Lehigh Valley was a model of high-efficiency fabrication.[30] Transistor manufacture required an exceptionally motivated work-force that operated with precision and conscientious diligence. Transistors were fabricated in a particularly sterile environment, and the facility at Allentown was equipped with one of the first clean rooms. The disciplined labour force hand-soldered transistors and components, and a new kind of testing was adopted: statistical quality control. Because of evolution in product and manufacture, a new management structure was established between research laboratory and factory. It consisted of a kind of interlocking of local research efforts and the transfer of modified devices and procedures to the shop floor. The Allentown point-contact transistors were high-quality and high-performance; but because of the labour-intensive nature of the work they were also very expensive—too expensive. Although of excellent quality, they were often not what the market demanded.

The virtues of Allentown was its introduction of a new work environment, in terms of the physical plant, structuring of work, and not least of all, in terms of laboratory/plant organization and dynamics. In Allentown one sees that it is not simply technical innovation that counts; management, organization, the layout of plant and plant environment, and finally the work force are equally essential.

The trajectory of the transistor at RCA between 1948 and 1960 is both emblematic of the 'function' versus the Bell laboratory 'epistemic' approach and the theme of changing relations between fabrication and laboratory.[31] The RCA story similarly draws attention to the need to grasp that the category 'research laboratory' is not an homogeneous unit, but instead consists of multiple forms of research activity extending to fundamental research (epistemic) to application, development and fabrication-driven integrated research.

In 1948 RCA opted to move into solid-state electronics. The company acquired fabrication rights to the point-contact and the alloy-junction transistor from Bell. The group designated to press RCA's ambitions forward was named the 'tubeless amplifier group', and the reflection and activities of the company and the group were based solely on the demand for amplification. The term 'transistor' was largely absent. The intent was to build a tubeless device, and interest lay exclusively in the performance required and not at all in the physics of the component. The specific physical characteristics and dynamics of the transistor were of little concern, and the transistor existed in the thoughts of RCA uniquely as a means to an end. The transistor was an input/output apparatus that rendered an amplification ratio, and in the company it was considered that 'the device is the function'.

Throughout the early 1950s the notion of 'research' was convergent with fabrication, with mostly empirical attempts to control the behaviour of crystals. Here control for purposes of manufacture was emphasized, as opposed to understanding. In this situation of laboratory/manufacturing relations, processes for preparing crystals and for the introduction of selected impurities were improved. Crystal production was advanced through modifying oven temperature and time. The outcome was positive for RCA. Nevertheless, by the late 1950s the substance of research changed and connections between laboratory and fabrication shifted. Research grew more basic, more physics-oriented, and less closely involved with the daily work and problems of manufacture. The former combinatorial of lab/fab gave way to a more differentiated organizational format. The experience of RCA highlights two important combinatorials. Innovation is not exclusively an issue of science and technology. It is as much a question of manufacture and the shop floor as it is a matter of the laboratory. Second, even in the age of complex technology, endeavours associated with 'function' can be as telling as 'episteme'. This reality was verified in the later emergence of the integrated circuit, and to a somewhat lesser extent the introduction of the microprocessor.

28.4 BIRTH OF SMALL HIGH-TECHNOLOGY SCIENCE AND ENTREPRENEURSHIP

The decade of the 1950s witnessed huge improvements in the transistor. In its early life the power-amplification capability, effective frequency-range, and power-consumption needs of the transistor were very limited. In addition, the device was initially extremely expensive. Its advantages resided in size and increasingly in durability. Early applications were military, such as proximity detonation or miniaturized components in missiles. Nevertheless, transistors also rapidly penetrated communication, where they figured in telephone switching and as repeaters both in transoceanic cables and transcontinental hook-ups. As the frequency range, amplification potential resistance to thermal and vibration factors, and general reliability grew, the variety of uses expanded. In 1955 transistors penetrated the civilian market when the Regency company designed, built, and marketed the world's first transistor radio.[32] Boasting four transistors, the apparatus was a complete technical success. However, the cost of the radio and a measure of hesitancy in the public to try the novel system quickly led to closure of the initiative. The combination of a military and expanding civilian market paved the way for the rise of new small high-technology companies specializing in microelectronics, and in particular transistor innovation, design, fabrication, and marketing.

The 1950s was a period when small high-technology firms took the lead in the science, engineering, manufacturing and marketing associated with the silicon tide. Start-up companies emerged as sites of research and innovation. The firm established by William Shockley, Shockley Semiconductors, is in many ways emblematic of this new species of endeavour.[33] Shockley considered that he had not received the recognition for inventing the new technology that he merited at Bell, and in part because of this injustice he quit Bell, travelled to what would later become known as Silicon Valley, and set up his own firm. He already possessed a reputation for huge achievement. He had led the group that invented a primitive transistor, and he alone invented the far more advanced junction transistor. During his time at Bell, he began to move in the direction of silicon transistors as opposed to germanium transistors. This change was to prove decisive and irreversible. Silicon is easier to control, it is more stable, it allows far faster switching, higher frequencies, and greater power amplification; and the introduction of impurities into the crystal is less uncertain. Shockley, furthermore, accelerated efforts toward diffusion processes of impurity delivery. This permitted greater control of deposition, and the impurities were more evenly spread throughout the silicon wafer.

One central strength of Shockley Semiconductors resided in the strategy to conjoin the combinatorials of fundamental research and fabrication. Shockley's leading idea conjoined the understanding of the transistor in terms of calculation and the processes of fabrication. From the very outset he gathered an immensely powerful team of scientists and engineers who were intended to probe the deep working of the silicon transistor and to engineer a device amenable to efficient manufacture. Shockley brought together eight men who in subsequent years would emerge as the central figures in the silicon tide. They included Sheldon Roberts (1926–) (metallurgy and chemistry), Eugene Kleiner (1923–2003) (industrial engineering), Jean Hoerni (1924–1997) (theoretical physics), Gordon Moore (1929–) (spectroscopy of flames and Moore's law), Jay Last (1929–) (ferroelectric materials and optics), other members of the 'Traitorous Eight', and Bob Noyce (1927–1990). Noyce soon became the most outstanding single personality in the new domain of microelectronics, both for his achievements in research and manufacturing and for his successes in assembling strong engineering and entrepreneurial teams.[34] Shockley Semiconductors soon exploded and disappeared. William Shockley was by most accounts a suspicious, tyrannical, and unstable personality. He came to believe that his staff were plotting against him, and he made the life of his workers quite impossible. Only very reluctantly, the top eight men decided to leave and to venture out to establish their own semiconductor company, Fairchild Semiconductor, directed by Bob Noyce. The talents of the big eight quickly made Fairchild Semiconductor into one of America's most innovative and successful transistor firms.

During the late 1950s Fairchild Semiconductor was recognized as the most competitive firm in the country.[35] Why was this the case? First, the company had the fastest and most stable devices then available; and second, the rate of yield was exceptionally high. At this still early stage of transistor history, manufacturing was still a big problem, and incidents of failed batches were sometimes recurrent. Third, Noyce and his enterprise continually introduced fresh innovations into their product. For example, the planar transistor was manufactured at Fairchild Semiconductor shortly after its invention by Hoerni. Finally, Fairchild had largely cornered the military market, particularly the Air Force. The US military wanted ever better transistors for avionics. They had to be resistant to acceleration, vibration, and heat, and reliability was of foremost importance. Debate continues over the centrality of the military market versus the civilian market during this early period. It is argued by some that the transistor was indeed developed with the military market in mind. This may be true, yet given the extent of fundamental physics required and the long time-lapse needed before mass production, specific market expectations had to be very difficult to calculate. Moreover, military demand fluctuates considerably. On the other hand, other than telephone switching and repeater amplification, what civilian demands could have been anticipated during the 1940s and 1950s? As indicated above, the first civilian application was introduced in 1955 in the form of a transistor radio, and it at first failed. Transistors were then still relatively expensive. So, despite the huge size of the civilian market for silicon devices in the 1960s, during the high tide of Fairchild Semiconductor in the 1950s, that company's privilege with the military is what fuelled profit and what motivated the specifications of new transistor design.[36]

28.5 BACK TO CRAFT

The invention of the integrated circuit in 1958–59 once again points to the centrality of the constant introduction, reintroduction, circulation, and synergy of incapabilities, methods, learning, and so on, in many forms in the processes of the silicon tide, and it demonstrates the alignment of said resources into fresh combinatorials. The transistors manufactured by Fairchild and similar firms during the 1950s were increasingly complex and effective. Fabrication entailed new processes of doping, and thanks to lithographic technology the bundling of several transistors onto a single wafer was made possible. Manufacturing was ever more automated.

It must be understood that transistors constitute but one electronic component whose capability in switching, computation, and amplification are prodigious. By the late 1950s, transistors had begun to replace vacuum tubes in computers, where they served as memory. Despite the innumerable advantages offered by transistors, they remain just one component in a complex electronic circuit, as numerous additional components are also required, such as resistors, capacitors, coils, and so on. For many years these components were carefully attached to a board also containing the precious transistor. Each element was a separate unit, and components were painstakingly soldered to one another. The unit was slow to manufacture, required immense precision, and was not always durable. In important respects, some circuit fabrication may be likened to late eighteenth or early nineteenth-century piece-work. This configuration changed, however, with the invention of the integrated circuit.

Credit for the invention of the integrated circuit is generally attributed to Jack Kilby (1923–2005),[37] and some historians also propose the name of Bob Noyce. Kilby worked as a technician at Texas Instruments,[38] one of the United States' principal transistor firms. Kilby came from a rural background and possessed very little technical training, merely possessing an undergraduate diploma in science from the University of Illinois. More relevant to his way of working, rural-raised Kilby had been an amateur radio enthusiast since his youth, and he enjoyed a kind of intuition for electronics. In this respect he is reminiscent of the earlier generation of electronics pioneers based in California. At Texas Instruments he was nevertheless attached to a research and development group, where he earned a reputation as a clever, hard-working employee. In a self-established and entirely unannounced personal project, he set out to look for a simpler and more robust way to associate the often myriad components involved in transistor circuits. He developed the idea of integrating all components onto a single board, where each element is in electrical contact with appropriate neighbours. This entailed a kind of intuitive appreciation of geometric relations and a logic of economical spacing. Knowledge of electronics was secondary here. It is instead intuition and hand-skill that counts—what can reasonably be done with components in the confines of a given space? This comprises a very specific category of combinatorial quite distinct from the high physics of semiconductor physics! He suggested that many different components, such as resistors, capacitors, and coils, could also be built from silicon or germanium, and could be situated on the same chip as the transistor. The recent technologies of diffusion, vapour deposition, and above all lithography, made this approach possible and even relatively simple. Kilby's integrated circuits constituted standardized, low-cost, solid units. Given some electronic design, transistor manufacturers could henceforth offer clients a range of circuits from a catalogue where the elements could simply be plugged into a complex technical system.

For the purposes of examining relations between academic science and enterprise, the birth and trajectory of the integrated circuit proves highly informative. As indicated at the outset of this essay, the concept of combinatorials of skills, materials, instruments, and ideas whose permutations and centrality are contingent on cognitive, economic, and materials circumstances allows one to better and more fully grasp the nuances and complexities of connections between physics and industry. At a given moment, groups from academic physics stand foremost in the microelectronics sector. The state of knowledge may be extremely elevated. Yet immediately afterwards, unlearned, intuitive-driven, hands-on skills and appreciations for materials can become a dominant factor. Of course, boundaries between academic physics and enterprise persist, yet it is through seeing how new combinatorials emerge between different skills, learning, and so on, that we perceive the deeper dynamics of cognitive and technical evolution. The lessons drawn from the trajectory of Kilby and of the transistor as integrated circuit, underscores just how strangeness of combinatorials provide useful historical insights—here an instance of advanced fundamental physics intersecting with fundamental manual thinking.

28.6 MANAGING INNOVATION

Until recently, among many sociologists of enterprise and also among many entrepreneurs, the concept of managing innovation referred rather strictly to the effective organization of research teams, the distribution of research resources, and the focus of investigation. Today, however, the notion of innovation is directed more broadly to total company organization. Up to the 1960s, transformations in the silicon tide were tightly linked to theory and experimentation, to the development of new materials, to improvement in transistor architecture and construction, and finally to changes in fabrication organization and practices. This situation suddenly changed in the 1960s for a variety of important reasons. The combinatorial of management became paramount.

By that time it had become increasingly clear that most big consumers of transistors did not after all intend to develop internal production capacity. They rejected a policy of internal integration of the component in favour of acquisition from external suppliers. This industrial configuration stimulated the creation of firms specializing in silicon that would service transistor consumer companies. In the 1950s the larger transistor manufacturers included Fairchild, Motorola, Texas Instruments, Westinghouse, and RCA.[39] In the course of the next four decades the number of companies associated with the fabrication, design, and research of silicon-related components grew into the dozens and even the hundreds. Many of the firms remained relatively small. Part of the silicon world became highly specialized, serving a narrow spectrum of clients or developing a product to highly-defined specifications.

Growth in the size and the complexity of the transistor market shifted attention away from the materiality of the transistor *per se* and redirected it to issues of sales and profit. Questions of production cost were not trivial here: product pricing had necessarily to be constantly reduced in order to compete effectively. Transistor firms also strove to expand fields of application by attracting new customers, and this often entailed making transistors a viable substitute for older technologies through savings. Yet perhaps greatest managerial innovation occurred with reference to the selection of profitable clients and the search for truly promising new markets. For many years the military comprised the biggest and richest client, and to a considerable extent, Fairchild Semiconductor grounded its huge success in military purchases. This market was fickle, however, sometimes offering huge profits and at other moments shrinking. Moreover, with the emergence of competition between transistor manufacturers, the military began to impose stringent restrictions which cut earlier bumper profits.

Innovative management practices arose at Fairchild under the direction of Bob Noyce. At this stage in the evolution of the silicon tide, the relevant parameters of management were the broad (not necessarily strictly research-related) internal organization of the firm on the one hand, and on the other, management strategy in the quest for expanding markets. The deeper issue of the entrepreneurial spirit and consequent massive creation of start-ups was not yet prominent. This second expression of management mentality would convulse microelectronics some fifteen years later! Bob Noyce interlaced two combinatorials. His company continued to develop modifications in transistor specifications mainly to satisfy demands of particular clients, and that possibly offered a concept and design base for a generic product. Noyce did not wish to venture into far-reaching research-intensive horizons. Technical change of an established transistor lines yes, but no radical departures! Here Noyce's managerial innovation consisted of conservative material innovation efforts linked to moderately exploratory, moderate-risk entrepreneurship. Materials and money operated hand in hand, mobilizing only limited cognitive and economic investments. Despite this, these combinatorials promoted profit and cognitive/functional advance.

In a second instance of strategic market management characteristic of the day, for Fairchild the horizon of a military market was far too broad, covering a huge range of technical requirements and of interlocutors. There was simply too much material diversity in demand, and too much variation in the possibility for profit. The company again decided to narrow its sights. It now specifically addressed the Air Force market. There were two reasons for this selection. First, aviation imposed extremely high standards on its transistor components, mainly in terms of thermal resistance, resistance to vibration, and overall reliability. Improvement of these particular parameters by Fairchild entailed research and engineering, yet in specific domains and of a limited scope. It could be undertaken without conceiving, designing, building, and fabricating an entirely new species of device. So, continuity with a pinch of innovation. In parallel, since the Air Force demanded somewhat special transistors of the highest quality, the military was prepared to pay a premium. One discerns once again in this policy the growing centrality of organization and strategy-related managerial imperatives in the silicon tide that intersect with moderate technical and cognitive developments.

A final element warrants mention here. Until the early 1960s transistor technology was reserved almost exclusively for military purposes, telephone switching, long-distance amplifiers in telecommunication, and to some extent computation, serving either in calculators or computers. The introduction of transistor technology in the consumer market was at best marginal. Nevertheless, even the possibility of penetrating a broad consumer market began to affect enterprise strategy reflections. While still remote, it began to emerge as a short-term prospect. Discussion of realistic domains of transistor application multiplied along with just how consumers could be attracted to the new products. In the end, it would be structural constraints and opportunities associated with the rise of the computer and the microprocessor that entrenched the logics and the imperatives of management as an unavoidable combinatorial.

28.7 Logic, Control, and Entrepreneurship

The dominant combinatorials in the silicon tide again changed, starting in the 1960s, and the change has been a radical one. While the place of management in the hierarchy of determinants has remained stable, its parameters and expression is today different. Moreover, the combinatorial of skill and intuition that underpinned the integrated circuit, and the combination of theoretical and experimental physics on which the invention of the point-contact transistor and later the junction and MOS devices were grounded, has in many respects now been relatively sidelined. The microprocessor revolution that arose in the late 1960s, and that today continues to dominate silicon microelectronics, is still populated by physics—many practitioners who fathered this innovation hold doctorates in physics. Nevertheless, the substance of the knowledge now deployed has changed greatly.

In terms of technology, the microprocessor is synonymous with the 'information age', whose concept and universalizing potentialities were anticipated by Claude Shannon in his epoch-making article 'A Mathematical Theory of Communication', published in 1948.[40] In that article the twin ideas that communication and intelligibility could more adequately be understood as 'information' was introduced. Since then, information has taken on ideological-like aspects, and much of science and technology incorporates representations of the world and engineering control operations in a perspective of information. Contemporary culture is permeated with digitalized information bits, and the silicon-based components constitute the material substance for the concrete application of said information. The microprocessor constitutes the heart and brain of the computer, and the computer in its enumerable expressions is emblematic of the information age.

Physicists developed the first microprocessors during the 1960s, and thereafter practitioners of the discipline continued to be prominent, but not in the habitual capacity of study of the physical world through theory and experimentation. This silicon technology is instead rooted in the application of logic. Achievement depends less on materials innovation than on the introduction of novel functions. The two key functions here are the processing of information (computation) and the storage of information (memory). Three factors determine effectiveness—speed, capacity, and integration, with other components of the technical system, including human interfacing. These considerations are universes away from the prehistory combinatorials of Marconi and early vacuum-tube technology, from the laboratory investigations of semiconductors mainly of German universities during the 1930s, from the inaugural endeavours of the scientists Bardeen, Brattain, and Shockley, and from the skills of Kilby in inventing the integrated circuit. In this trajectory the presence of particular combinatorials and their hierarchic position crystallized, only to be replaced by others, and then, in different configurations, to reappear.

In the case of the microprocessor the possibility of novel functions is what most counts, where functions are expressed as counting, classifying, identifying, relating, analysing, and so on. All these functions/operations revolve around processes of logic, and the substance of a microprocessor is thus based on the materialization of logic. As will be demonstrated next with a description of the invention of the first widely used microprocessor, the famous Intel 4004, 4008, and 8080, three expressions of logic are mobilized: 1) the formatting of an architecture, where the over all vision of the interactions between the many transistors and circuits comprising the system is conceived in order to obtain a wanted capacity and output; 2) the physical layout of this logic architecture in order to miniaturize the device, to make it work in practical terms, and to enable its compatibility with other components; and 3) the programming of the device so that it performs the designated operations in an efficient and ergonomic fashion. These three tasks require the contributions of scientists possessing different competences.

The early history of the microprocessor is synonymous with the history of the Intel Corporation.[41] The name 'Intel' comprises a kind of abbreviation of 'integrated electronics'. Founded in 1968, the company was a breakaway from Fairchild semiconductors. Just as Fairchild was a breakaway from Shockley, spearheaded by Noyce and a close circle of colleagues, the creation of Intel was also prompted by several members of this group. What prompted Noyce to abandon Fairchild and to set out on his own? Noyce found Fairchild too sluggish. The company was not sufficiently investing in new technologies. In addition, he considered that in-house development of innovation was not enough. Technology also entailed adaptation to client demand. He believed that it is incumbent on enterprise to closely monitor client needs and to aid them in integrating the firm's technology into their products—what came to be known as 'applications engineering'.

Finally, Noyce wanted a company devoted to the expansion of electronic memory. He judged that memory represented the future. It was this combination of considerations in mind that in the autumn of 1968 Noyce recruited Ted Hoff, who was then completing his PhD at Stanford University. Hoff is recognized today as the father of the microprocessor. His dissertation was on electronics, where he specialized in neuro-networks and programming for computer-aided design, and because of his keen interest in memory, Noyce hired him and appointed him applications manager. In this capacity he had two responsibilities. The first was traditional—to assist customers in the best use of Intel integrated circuit components. His second responsibility was, for the time, novel, consisting of deciphering the middle and long-term needs of clients, and then trying to determine the technical devices that would satisfy the anticipated needs. This linkage between applications, the specificities of demand that have a generic components, and designs now represents a key operating tenet and a strong tradition among microprocessor firms.

It was in the course of frustrated discussion with a Japanese client, the Busicom company, that Hoff conceived the microprocessor—a technology that he described even at that time as a 'computer on a chip'.[42] The client called for a device with 40 leads. This was expensive, and possibly not very robust. Moreover, Hoff saw that the device was complicated because it did not efficiently employ its write/memory capacity. He proposed to the Japanese a far simpler component, but there was little interest. Noyce authorized Hoff to pursue his design inclinations, and this led to the birth of the Intel 4004 microprocessor that was first marketed in 1971–72. The inaugural processor bundled 2,000 transistors onto a single chip having a read/write and read-only capacity, and for the time a considerable active and storage memory capacity. The processor was composed mainly of two chips devoted to computation. It was a multi-purpose device capable of numerous functions, such as switch protection and, above all, control. The latter ability allowed a huge range of applications. It became a key component of concept aided design. Most important, the Intel 4004 became the central computational component of large as well as smaller computers. Hoff soon thereafter contributed to developing the Intel 8008 processor, which possesses interrupt capability—the permitted interruption of an application during which a second function can be called up. The 8008 made possible the emergence of early personal computers.

Hoff was responsible for the architecture of Intel microprocessors. He conceived the possible functions given specific circuit constraints, intertwined the logics of circuits, and devised the instruction set and registration for intra-processor communication.[43] One of the foremost challenges was to keep the number of transistors to a minimum. It had been earlier calculated that a processor would require many thousand gates, and as each gate entails three transistors the device would have between 20,000 and 30,000 transistors. It was Hoff's creative architecture and use of logic that succeeded in reducing the number some tenfold—something judged impossible. After the 4004, Hoff moved on, contributing to the Intel 8008. This was an 8-bit device, as opposed to the 4004 processor. It was considerably faster, possessed more memory, and was capable of many additional functions, such as imaging applications. One fascinating aspect of the silicon tide is the intensity of technical curiosity and commitment to innovation among individuals involved with devices. Despite Hoff's universal recognition for fathering the microprocessor, in 1983 he chose to move away from processors. He quit Intel and transferred to Atari. He simply wanted to shift to a new and stimulating domain. Atari was at that time focused on video games and video phones. It was experimenting with a sports-meter that converted mechanical signals into digital information and with simulation. These domains were perceived by Hoff as rich openings for new discoveries. He desired something new where originality was needed.

Italian-born Fernando Faggin developed the layout of the Intel 4004 and later the 8008 and 8080 microprocessors.[44] Faggin received advanced training in mathematics and physics at the University of Padua, and in 1968 he took a position at Fairchild. While there, he developed the polysilicon gate technology that replaced metal gates, and he introduced buried contacts and novel etching protocols. All these innovations later contributed to the manufacture of microprocessors. Like so many other prominent or promising men engaged in the silicon tide, Faggin soon changed companies, looking for opportunities for new experiences and a faster pace. In 1970 he left Fairchild Semiconductor for Intel. He found Fairchild too stayed and slow. Noyce had founded Intel a year before, and that was where things were happening. In addition, Faggin wanted to move away from his work in processor technology and toward design. On arriving at Intel his first task was the creation of the four chips that would form the 4004. Architecture often consists of ideas; according to Faggin, the heart of silicon is transforming ideas into concrete devices, and this is what he did with the 4004 and 8008 line of Intel processors. He designed the location of the many transistor clusters and circuits and the connections between them. This task entails writing the logic that integrates the various elements. In effect, Faggin materialized the vision and introductory endeavours proposed by Hoff.

As indicated throughout this study, use of the concept of combinatorials often offers a useful interpretive device for grasping the history of the transistor and of exploring academic/industry relations in physics-related domains. With the advent of the microprocessor having its specific constellation of knowledge, skills, and profession/relations, another combinatorial arose. People moved often and rapidly from employer to employer. This inclination was already indicated in the case of Hoff, who sought constant cognitive and project stimulation, moving from Fairchild to Intel and then to Atari. The migratory combinatorial is once again present in the trajectory of Faggin. In addition to the move from Fairchild Semiconductor to Intel, Faggin changed firms three more times, and in 1975 founded Zilon, specializing in microprocessor design. The firm was funded by Exxon. While there, Faggin developed the Z80 and then the Z8 microprocessors. The Z80 measured only 180 x 180 mm, and was four times faster and more powerful than the current Intel processors. In 2000 more 50 million Z80 microprocessors were in operation. It boasted twice the number of transistors, and Faggin's original design allowed complete compatibility. The two devices competed effectively with Intel products, and indeed often surpassed them.[45] Zilon rapidly became one of the top firms in the processor world.

Despite his company's indisputable technical and commercial success, in 1983 Faggin left it in order to create a new company: Signet Technology. Faggin here turned to development of a microprocessor product versus a technology. His firm designed 'communication-data managerial systems'. This was a novel concept which connected the telephone to a computer. It was designed for managers who could fluidify and speed their work process through the swift and complete range of data retrieval and transmission capacity of the device while simultaneously speaking on the telephone. Data could be transmitted from work station to work station. This enterprise met with little success, and in 1986 Faggin created still another company: Synattics. He was a co-founder along with Carver Mede, a physicist specializing in neuro-networks and inventor of the silicon retina based at Caltech. The company's ambition was the marketing of a high-performance pattern-recognition system. Although it required more than six years, the team invented the 'touchpad', which allowed computer-operator control of the computer independent of the keyboard, and the capacity to draw through manual contact with the monitor. The originality of the system is that it translated human mechanical input in the form of analogue information into digital information required for the computer.

Why did Faggin leave Intel and then create three firms? One can point to a constellation of motives. He found that by the mid-1970s Intel was too bureaucratic in its operation, and that it was running out of steam. He needed a fresh challenge. Moreover, microprocessors were never Intel's primary concerns. It remained a memory company. He wanted closer contact with the specificities of applications and with clients. The adaptation of technology for specialized operations was important. In this framework, Faggin evolved toward managerial responsibilities. However, the managerial combinatorial of the microprocessor silicon phase differs from the previously discussed managerial orientation. Many of those involved in microprocessor technology created new firms. Here the managerial inclination took the form of pioneering and entrepreneurship as opposed to the earlier intra-firm management concern for efficiency and maintenance of a strong collective, team mentality, and organization. Finally, the microprocessor episode of the silicon tide strongly emphasized component design as opposed to matters of materials and fabrication. The stress lay on research. Zilon, Signet Technology, and Synattics were all strictly research companies that generated visions of the future and ideas, technical design, and prototype devices. The project would subsequently be sold to a different category of firm engaged in product, interfacing, and fabrication and marketing. This new orientation associated with microprocessors comprises yet another historically evolved combinatorial.

The last person involved in fathering the microprocessor at Intel in 1969–70 was Stan Mazor, who trained in physics at the University of San Francisco. He then worked at the University of California at Berkeley and at Fairchild Semiconductor in writing circuit logics and application programs before taking a position as a program engineer at Intel in 1969. His boss was Ted Hoff, who at the time was developing the architecture for the 4004 microprocessor. Mazor is the self-confessed least important figure in the trio.[46] He helped all around, and his most significant contribution consisted in writing the operation codes for the assembled chip and in writing a series of application programs. Decimal-point arithmetic served as a key logic and programming element. It proved no easy matter to make the 4004 central processing unit compatible with peripherals and with control demands. On completing work on the 4004, Mazor subsequently participated in the programming of the 8004 and the 8080.

Although Mazor's contribution to the silicon tide resided principally in writing ever more efficient programs, he was quick to acknowledge that programming alone would not allow future breakthroughs of great significance. For him, it was materials that had to be improved for future big strides forward. Mazor anticipated possibly insurmountable obstacles. In the mid-1990s transistors could be etched that had a size of 0.25 microns. This approached wavelengths where the frequency of light itself became decisive. For Mazor, the question of scale might well limit transistor and microprocessor progress. It was urgent, he declared in an interview, to move forward in materials development and in the domains of control. Architecture, lay-out engineering, and programming are not enough!

Like Faggin and to a lesser extent Hoff, Mazor's professional trajectory also involved considerable mobility, where motivation was always coupled to the wish to become engaged in a new and stimulating microprocessor venture.[47] In this spirit he helped create at least two firms, thus expressing his pioneering and entrepreneurial turn. In 1983 he moved to Silicon Compilers, where he wrote logic for computer-aided design systems and devices. In a variety of guises, computer-aided design and compilers would remain central to Mazor. He soon transferred to Synotsis, and again specialized in computer-aided design applications. Two years later, Computer Aided Training Systems employed him. There, he there became involved in developing programmes for writing models for financial activities. Finally, he was recruited by

T-Systems, where programmes were prepared for purposes of training in a variety of technical tasks. These multiple moves each contained elements of an entrepreneurial inclination.[48]

28.8 WHERE NEXT?

Since the formulation, in 1965, of 'More's law', which anticipated a doubling of transistors' calculating capacity every 18 months, the prediction has been substantiated. But just how long can the trend continue? At what point will obstacles slow or block this growth? What will be the source of stagnation? Will the possible disqualifying of More's remarkable expansion curve be coupled to programming, layout, architecture, or materials?

During the 1990s some science policy-makers inside the US government took the possibility of stagnation in microelectronics very seriously, and they noted that Japanese industry, universities, and the government were actively pursuing programmes intended to circumvent any slowdown in the progress of the transistor's future enhancement. Mihail Roco,[49]—a powerful Washington science-policy militant and spokesman—is emblematic of such reflections and worries. His intention was to inform and to galvanize politicians, scientists, industrialists, and the broad public in order to attack the problem of eventual blockages in microelectronics and to the unthinkable economic consequences if a foreign nation were to outperform America. The group of science-policy spokesmen concluded that the category of learning and of novel devices required for pushing microelectronics, and in particular the transistor and microprocessor, forward lay in development of a national research programme in nanoscience and nanotechnology. They judged that the danger to continued transistor progress was coupled to the scale of components. Present technology would soon be interrupted because of the existence of a lower threshold in scale that could not presently be traversed. The promise of nanoscale research to construct and to control materials on the molecular and even the atomic scale (bottom-up construction) could open the way to future expression of More's law. In 2000 the US National Science Initiative was announced—a research programme intended to invest many hundreds of millions of dollars into nano-research, and not least of all in the search for ultra-high-performance nanoscale electronic components. At the moment of writing this chapter, it is clear that the new materials path advocated by Stan Mazor has been pursued during the last decade. The alternative strategies of architecture, design engineering of layout, and logic/programming have received somewhat less attention as a key breakthrough solution, though they nevertheless certainly remain critically important.[50]

Nanoscale research for the purpose of generating a new species of transistors has been acute over the last decade, and in autumn 2011 Intel introduced a revolutionary transistor: the 'finfet'[51]—fin field-effect transistor. This device is claimed to be 37%

faster than previous components, and can offer a 50% reduction in electrical power consumption. It owes these performance characteristics to nanoscale materials and associated design. Each transistor measures a mere 22 nm, and a 10-nm device is soon to be manufactured. This is a tenfold reduction in scale from the nearest competitors. Each transistor is elevated a few nanometres from the substrate, sitting on top of a sort of fin (hence the name). This geometry is partly responsible for the components' elevated speed. The ebb and flow of different combinatorials alternatively foreground competences associated with university or industrial environment and traditions. With the introduction of the finfet and similar materials science radical innovations, one has gone full circle from the research that led to the point-junction transistor in 1947 by Bardeen, Brattain, and Shockley, who received the Nobel Prize for their successes. As in the case of Intel research on the finfet nano-device, the point-junction transistor was based on theoretical physics coupled to growing expertise in materials. The same combinatorials underpin current progress. As micro-electronics increasingly approaches the molecular and atomic scale, this tandem will probably continue to flourish. Yet as the history of the transistor has clearly indicated, it is very likely that at points in the future, combinatorials of skill, concept, and capacities in architecture, layout, and programming will once again become the principal motors of development.

28.9 CONCLUSION

The underlying intention of this study has been to explore one embodiment of twentieth-century knowledge of physics and of material implementations inside academia, inside industry, between university and enterprise, and as they circulate dynamically between the different poles. The historical trajectory of the transistor, pre-transistor technologies, and late transistor implementations in the form of the microprocessor has been presented here through narration and analysis in terms of two conceptual frameworks: 'combinatorials', and matters epistemic versus matters of function. In the preceding pages, episodes in the genesis and the evolution of transistor-related theory, experimentation, and technological innovation, products, and enterprise have been emblematically referred to as 'the silicon tide' because of the quantitatively massive manufacture and diffusion of transistors and their huge range of applications. It is safe to say that in their innumerable embodiments the transistor constitutes one cornerstone of contemporary material culture.

The combinatorials of skills, materials, instruments, concepts, organization, management, and enterprise that comprise the thread which traverses the entire past of the silicon tide constantly changes configuration, and the hierarchic relations between the elements shift as one moves from the early days of 1890 to the 1930s, from a period of intuition and tinkering to the subsequent era of academic high theory, mathematics, and complex experimentation of the German university over the inter-war decades. During the former epoch, skills were paramount; while during the latter time the combinatorials of concept, materials, and instruments prevailed in a hierarchy of descending importance. Prior to the Second World War, the questions raised by practitioners in the field of radio, and the methods and practices pursued in communications electronics, focused on obtaining wanted technical material effects—otherwise stated, were aimed at achieving some pre-set technical function. For the subsequent period both before and after the war, the principal objective was cognition. The expectation was mainly epistemological in intent.

Combinatorial configurations again shifted radically and on multiple occasions between the 1960s and the end of the century. During the 1950s and more particularly the 1960s, market forces were paramount. This was clearly true in RCA, in Fairchild, and very soon after its creation, even in Intel. Epistemic questions waned. The innovation of the integrated circuit made at Texas Instruments was narrowly function-focused and grounded on largely intuitive appreciations. There was nothing epistemological here! Issues of company organization and management coloured the work of scientist and engineer alike, often to the exclusion of other horizons.

The closing decades of the twentieth century saw the emergence of still another combinatorial plus the redefined renewal of an older one. The earlier emphasis placed on management as the expression of industrial organization was supplanted. Associated with new possibilities of the microprocessor revolution, many individuals began to establish small firms—in short, to become entrepreneurs. This was connected with the rise of a new technical focus in silicon-related work. The microprocessor entailed the introduction of logic as expressed in component architecture, layout, and user-interfacing to micro-electronics systems. Stated differently, there occurred a displacement away from an epistemology of material objects in favour of epistemological expectations in terms of a logic of numerical and symbolic relations and interactions. The sky was the limit here, both in cognitive/functional permutations and in the plausible range of product innovation. It is this combination that in numerous cases fuelled a form of high-risk entrepreneurship, and that stimulated breakthrough in cognition and matters of function. The fathers of the microprocessor, Hoff, Faggin, and Mazor, all fit this picture to varying degrees.

Finally, the decades of the 1940s and 1950s, and one event in the late 1960s, illuminate a frequently overlooked important dynamic of twentieth-century academic physics/industrial physics landscape. One often posits invariant structural connections between forms of intellectual and material practises and goals on the one hand, and institutional/organizational environment on the other. According to dogma, the entrepreneurial environment generates function-related efforts, and the university environment generates epistemological endeavours. Episodes in the silicon tide suggest a need for substantial nuance in comprehension of relations between organization and forms of work. The point-contact and junction transistors were born inside the laboratories of ATT/Bell, and Shockley pursued epistemological objectives in his breakaway firm, Shockley Semiconductor. These were very much industrial settings. Nevertheless, for a period, such enterprise admitted epistemic expectations in a relative absence of function-related plans. To a much lesser degree, the development of the microprocessor by Hoff inside Intel in the late 1960s too serves as an instance of epistemological achievements within an enterprise, though in this case there was certainly also a parallel function-oriented dimension. The lesson to be extracted from these episodes in silicon evolution is that deterministic strong couplings do not constitute an iron cage regarding categories of research and the institutional environment in which the research occurs. This observation invites further exploration of these parameters and relations in additional fields where epistemological expectations and matters of function coexist historically.

Notes

1. Marcovich A. and Shinn T., 'Where is disciplinarity going? Meeting on the borderland', *Social Science Information*, 2011 50 (3–4), pp. 1–25.
2. Cf. the investigations of Ferdinand Braun, following; Nobel Prize 1909.
3. <http://en.wikipedia.org/wiki/History_of_radio>
4. Marconi, J., Nobel Prize Lecture 1909.
5. Braun, F., Nobel Prize Lecture 1909.
6. <http://en.wikipedia.org/wiki/History_of_radio>
7. De Forest, L., *Father of Radio: The Autobiography of Lee de Forest*, River Grove, IL, Wilcox and Follett, 1950.
8. <http://en.wikipedia.org/wiki/History_of_radio>
9. Lécuyer C., *Making Silicon Valley: Innovation and the Growth of High Tech, 1930–1970*, Cambridge, MIT Press, 2007.
10. *Making Silicon Valley*, p. 16.
11. *Making Silicon Valley*, pp. 14–30.
12. *Making Silicon Valley*, p. 36.
13. *Making Silicon Valley*, pp. 56–60. Shinn T., *Research-Technology and Cultural Change: Instrumentation, Genericity, Transversality*, Oxford, The Baldwell Press, 2008.
14. Wikipedia 26 June 2011, <http://en.wikipedia.org/wiki/Transistor>
15. Hoddeson, L., Braun, E., Teichmann, J., and Weart, S. (eds), *Out of the Crystal Maze: A History of Solid State Physics 1900–1960*, New York, Oxford University Press, 1992, Chapter 3.
16. *Out of the Crystal Maze*, Chapter 3, note 217.
17. *Out of the Crystal Maze*, Chapter 3.
18. Merton, R. K., 'The normative structure of science', in Merton, R. K., *The Sociology of Science: Theoretical and Empirical Investigations*, University Press of Chicago, 1973.
19. Marcovich, A. and Shinn, T., 'Respiration and Cognitive Synergy. Circulation in and between Scientific Research Spheres', Minerva, Vol 51, n°1, p. 1–23.
20. Interview with Gil Amelio, 24 March 2000, San Francisco, California. <http://silicongenesis.stanford.edu/transcripts/amelio.htm> Interview with Lester C. Hogan, 24 January 1995, Atherton, California.
21. Reed M. 'Quantum Dots: Nanotechnologists can now confine electrons to point like structures. Such designer atoms may lead to new electronic and optical devices', *Scientific American*, 1992.
22. Etzkowitz, H., *MIT and the Rise of Entrepreneurial Science*, London, Routledge Press, 2002.
23. <http://www.pbs.org/newshour/science/hydrogen/smalley.htm>

24. Interview with Lester C. Hogan, 24 January 1995, Atherton, California. <http://www-sul.stanford.edu/depts/hasrg/histsci/silicongenesis/hogan-ntb.html>
25. Riordan, M. and Hoddeson, L., *Crystal Fire. The Invention of the Transistor and the Birth of the Information Age*, New York, London, W.W.Norton and Company, 1997, p. 79.
26. *Crystal Fire*, p. 80.
27. *Crystal Fire*, pp. 28.
28. Bardeen would receive a Nobel Prize in physics in 1972 for his theory in superconductivity.
29. Nobel Prize lectures by Bardeen, Brattain, and Shockley, 1956.
30. Leslie, S., 'Blue collar science: bringing the transistor to life in the Lehigh Valley', *Historical Studies in the Physical and Biological Sciences*, 32:1, 2001, pp. 71–115.
31. Hyungsub, C., 'The boundaries of industrial research: making transistors at RCA, 1948–1960', *Technology and Culture*, Volume 48, Number 4, 2007, pp. 758–782.
32. *Crystal Fire*, pp. 211–213.
33. *Crystal Fire*, p. 238. *Making Silicon Valley*, pp. 128–134. Interview of Gordon E. Moore, 3 March 1995, Los Altos Hills, California. <http://www-sul.stanford.edu/depts/hasrg/histsci/silicongenesis/moore-ntb.htm>
34. *Crystal Fire*, pp. 247–248, 263–265, 283–284. *Making Silicon Valley*, pp. 129–134, 140–144, 261–266.
35. *Making Silicon Valley*, pp. 135–165. *Crystal Fire*, pp. 270–274. Interview with Gordon E. Moore, 3 March 1995, Los Altos Hills, California. <http://www-sul.stanford.edu/depts/hasrg/histsci/silicongenesis/moore-ntb.html> Interview with Charlie Sporck, 21 February 2000, Los Alto Hills, California. Interview with Harry Sello, 8 April 1995, Woodside, California.
36. DeGrasse, R., 'The military: Short Changing the economy' *Bulletin of the Atomic Scientists*, May, 1984. Holbrook, D., 'Government support of the semiconductor industry: diverse approaches and information flows', *Business and Economic History*, 1995, vol 24 (2), pp. 133–168. Holbrook D. *et al.*, 'The nature, sources and consequences of firm differences in the early history of the semiconductor industry', in Constance E. Helfat, *The SMS Blackwell Handbook of Organizational Capacities: Emergence, Development and Change*, Oxford, Blackwell Publishing, 2003.
37. *Making Silicon Valley*, pp. 156–218. *Crystal Fire*, pp. 256–260.
38. *Making Silicon Valley*, pp. 156–159, 218–223, 268–272. *Crystal Fire*, pp. 209–210, 264–265.
39. 'Blue collar science'. Interview with Wilf Corrigan, 17 October 1998, Los Altos Hills, California. <http://silicongenesis.stanford.edu/transcripts>
40. Claude E. Shannon, 'A mathematical theory of communication', *Bell System Technical Journal*, Vol. 27, pp. 379–423, 623–656, 1948.
41. *Crystal Fire*, pp. 283–285. *Making Silicon Valley*, pp. 8–12, 279–291. Interview with Gordon E. Moore 3 March 1995, Los Altos Hills, California. <http://www-sul.stanford.edu/depts/hasrg/histsci/silicongenesis/moore-ntb.html> Interview with Marcian (Ted) Hoff, 3 March 1995, Los Altos Hills, California. Interview with Dennis Carter, 20 April 2004, Mountain View, California. Interview with Gerry Parker, 6 October 2003, Los Gatos, California.
42. Interview with Marcian (Ted) Hoff, 3 March 1995, Los Altos Hills, California.
43. Interview with Marcian (Ted) Hoff.
44. Interview with Frederico Faggin, 3 March 1995, Los Altos Hills, California.
45. Interview with Frederico Faggin.
46. Interview with Stan Mazor, 9 June 2000, Los Altos, California.
47. Interview with Stan Mazor.
48. Lamy, E. and Shinn, T., 'L'autonomie scientifique face à la mercantilization', *Actes de la Recherche en Sciences Sociales*, no. 164, 2006, pp. 23–49.

49. McCray, W.P, 'From Lab to iPod. A Story of Discovery and commercialization in the Post–Cold War Era Technology and Culture', *Technology and Culture* 2009, Volume 50, Number 1, pp. 57–81.
50. Marcovich, A. and Shinn, T., 'The cognitive, instrumental and institutional origins of nanoscale research: the place of biology', in Carrier, M. and Nordmann, A. (eds.), *Science in the Context of Application*, Dordrecht, Springer, 2011, pp. 221–245.
51. Markoff, J., 'Intel increases transistor speed by building upward', *New York Times*, 4 May 2011.

CHAPTER 29

PHYSICS AND COSMOLOGY

HELGE KRAGH

Although cosmology is very old, dating even to preliterate societies, in a scientific sense it is a relatively recent branch of knowledge. Physical cosmology—here taken to be the study of the universe that does not rely only on astronomical observations but also on physical laws and methods—is even younger, and belongs largely to the twentieth century. While physicists today may consider cosmology to belong to physics, in an historical perspective this is not the case. As the study of the universe at large has developed over the last century, it has been a subject mainly cultivated by astronomers, physicists, and mathematicians—and one should not forget the philosophers. Although the philosophical and conceptual problems related to the universe cannot be ignored or cleanly separated from the more scientific aspects, the following summary account focuses on the work achieved by physicists in transforming cosmology into a research field which relies intimately on fundamental physical knowledge.

Twentieth-century cosmology has far from developed smoothly or linearly, but in spite of many mistakes and blind alleys it has progressed remarkably. It is understandable that physicists take pride in having unravelled at least some of the major secrets of the universe. 'Cosmology has become a true science in the sense that ideas not only are developed but also are being tested in the laboratory', asserted two American astrophysicists in 1988. 'This is a far cry from earlier eras in which cosmological theories proliferated and there was little way to confirm or refute any of them other than on their aesthetic appeal.'[1] Perhaps so, but how did it come to that?

29.1 THE RISE OF ASTROPHYSICS

The physical cosmology of the twentieth century relied on knowledge and methods of astrophysics—a new interdisciplinary research field that emerged in the previous century and did much to change the very definition of astronomy.[2] Still in the early part of the nineteenth century astronomy was conceived as a purely observational science using mathematical methods, and not as a part of the physical sciences. According to the traditional understanding of astronomy, 'astrophysics' was a misnomer and 'astrochemistry' even more so. This changed quite drastically after the introduction of spectroscopy, which from its very beginning in the 1860s was used as a method to obtain information about the physical state and chemical composition of the Sun and other celestial bodies.

Even before the invention of the spectroscope, the Austrian physicist Christian Doppler had argued that if a light source is moving relative to the observer with radial velocity v, one would observe a change in its wavelength given by $\Delta \lambda = \lambda' - \lambda$, where λ' is the measured and λ the emitted wavelength. According to Doppler, the change related to the speed of light c as

$$z \equiv \Delta \lambda / \lambda = v/c$$

The Doppler effect was soon verified for sound waves, whereas its validity for light remained controversial for several decades. Only in 1868 did the British gentleman astronomer and pioneer of astrospectroscopy William Huggins announce that he had found a small shift in wavelength for the light from Sirius, which he took to imply that the star moved away from the Earth. It took another twenty years until the optical Doppler effect was firmly demonstrated for an astronomical body, when the German astronomer Hermann Vogel examined the rotation of the Sun.

The introduction of spectroscopy, based on the seminal invention of the spectroscope by the Heidelberg physicist Gustav Robert Kirchhoff and his chemist colleague Wilhelm Robert Bunsen, effectively founded the physical and chemical study of the stars.[3] With the new method it became possible to identify the chemical elements in stellar atmospheres and to classify stars according to their surface temperature. Astrospectroscopy also led to several suggestions of new elements not known by the chemists, such as helium, nebulium, and archonium. Of these unknown elements, hypothesized on the basis of unidentified spectral lines in the heavens, only helium turned out to be real. The existence of helium in the Sun's atmosphere was suggested by the British astronomer Norman Lockyer in about 1870, and 25 years later the element was detected in terrestrial sources by his compatriot, the chemist William Ramsay. Helium would turn out to be uniquely important to cosmology, but at the time it was just a curiosity, an inert gas supposed to be very rare and of neither scientific nor industrial interest.

The astrochemistry that emerged in the Victorian era opened up new and exciting questions, such as the possibility that atoms might be complex and perhaps exist in more elemental forms in the stars. Lockyer and a few other scientists ventured to extend the perspective of astrospectroscopy to what may be called 'cosmospectroscopy'. For example, in an address to the 1886 meeting of the British Association for the Advancement of Science, William Crookes speculated that the elements had not always existed but were formed in the cosmic past under conditions very different from those known today. He imagined that all matter was formed through processes of 'inorganic Darwinism' and that it was originally in 'an ultragaseous state, at a temperature inconceivably hotter than anything now existing in the visible universe'. According to Crookes, matter had not existed eternally but had come into being in the distant past:

> Let us start at the moment when the first element came into existence. Before this time, matter, as we know it, was not. It is equally impossible to conceive of matter without energy, as of energy without matter . . . Coincident with the creation of atoms all those attributes and properties which form the means of discriminating one chemical element from another start into existence fully endowed with energy.[4]

Cosmic speculations of the type entertained by Lockyer and Crookes were one aspect of the new astrophysics; another was the study of heat radiation and its application to astronomy. Spectrum analysis had its background in Kirchhoff's fundamental investigations of what he called blackbody radiation, the study of which would eventually lead to the hypothesis of energy quantization. Even before Max Planck's law of blackbody radiation, the physics of heat radiation was applied to estimate the surface temperature of the Sun. In 1895 the German physicist Friedrich Paschen used measurements of the solar constant and the Wien displacement law to determine the temperature of the Sun to about 5,000° C, fairly close to its modern value. Just as scientists about 1900 could not imagine the future importance of helium in cosmology, so they could not imagine how important the law of blackbody radiation would come to be in later cosmological research.

29.2 Physical Cosmology Before Relativity

While astronomers took little interest in cosmological questions in the second half of the nineteenth century, physicists, philosophers, and amateur cosmologists discussed such questions on the basis of the fundamental laws of physics, which at the time meant Newton's law of gravitation and the two laws of thermodynamics. It was recognized early on that the second law of thermodynamics, expressing a universal tendency towards equilibrium in any closed system, might have profound cosmological consequences. Indeed, Rudolf Clausius, the inventor of the concept of entropy, formulated the second law as 'the entropy of *the world* tends towards a maximum' (my emphasis). If the entropy of the universe continues to increase towards some maximum value it would seem to imply that in the far future the universe would not only become lifeless but also devoid of structure and organization. When this state had occurred, the universe would stay in it for ever. This is the prediction of the 'heat death', first stated explicitly by Hermann von Helmholtz in a lecture of 1854 and subsequently discussed by numerous scientists and philosophers. Clausius' version of 1868 is as follows:

The more the universe approaches this limiting condition in which the entropy is a maximum, the more do the occasions of further change diminish; and supposing this condition to be at last completely attained, no further change could evermore take place, and the universe would be in a state of unchanging death.[5]

Clausius further pointed out that the second law contradicted the notion of a cyclic universe, the popular view that 'the same conditions constantly recur, and in the long run the state of the world remains unchanged'.

The heat-death scenario was not only controversial because it predicted an end to all life and activity in the universe, but also because it was sometimes used as an argument for a universe of finite age; that is, a cosmic beginning (which was often thought of as a creation). Because, so the arguments goes, if the universe had existed in an infinity of time, and entropy had always increased, the heat death would have occurred already; since the universe is manifestly not in a state of maximum entropy, it can have existed for only a finite period of time. The 'entropic creation argument', discussed and advocated by Christian scientists in particular, was controversial because of its association to religious ideas of a divinely created world.[6] It was severely criticized by philosophers and scientists (including Ernst Mach, Pierre Duhem, and Svante Arrhenius), who denied that the law of entropy increase could be legitimately applied to the universe as a whole. The heated discussion concerning cosmic entropy faded out around 1910, without resulting in either consensus or an improved scientific understanding of the physical state of the universe. Yet the heat-death scenario continued to be part of cosmology also in the later phase of that science.

Newton's law of gravitation had great success in celestial mechanics and possessed an unrivalled scientific authority, and it was generally assumed that the law would be equally successful in accounting for the distribution of the countless numbers of stars that filled the universe, which most physicists and astronomers (following Newton) considered to be infinitely large. However, in 1895 the German astronomer Hugo von Seeliger proved that an infinite Euclidean universe with a uniform mass distribution could not be brought into agreement with Newton's law of gravitation. It would, as he phrased it, lead to 'insuperable difficulties and irresolvable contradictions'.[7] The following year, in an elaboration of his so-called gravitation paradox, Seeliger suggested a way of avoiding it by modifying Newton's law at very great distance r. Instead of keeping to Newton's law of gravitation, he proposed that a body of mass m moving in the gravitational field of a central mass M would experience a force given by

$$F(r) = \frac{GmM}{r^2} e^{-\Lambda r}$$

That is, he introduced an attenuation factor of the form exp(−Λr), where Λ is a very small constant. With this rescue manoeuvre he could escape the gravitational collapse of the infinite Newtonian universe. Other scientists reached the same goal without changing Newton's law. For example, the Swedish astronomer Carl Charlier showed in 1908 that if the assumption of uniformity was abandoned and replaced by a suitable fractal structure of star systems there would be no gravitational paradox. Whether one way or the other, the majority of astronomers found it difficult to conceive a universe that was not spatially infinite.

Only a handful of physicists and astronomers realized the possibility of a closed and finite space of the type that the mathematician Berhard Riemann had introduced in the mid-nineteenth century. Although 'curved' non-Euclidean space was familiar to the mathematicians, it was rarely considered by physicists and astronomers. Perhaps the first to do so was the German astrophysicist Karl Friedrich Zöllner in a work of 1872, and in 1900 Karl Schwarzschild discussed the possibility that the geometry of space might be determined by astronomical measurements. He expressed a preference for space being closed and finite, but did not develop his ideas into a cosmological theory. By and large, curved space as a resource for cosmology had to wait until Einstein's general theory of relativity.

One of the great problems in astronomy and cosmology around 1900 was whether the nebulae, and especially the spiral nebulae, were structures similar to the Milky Way or were perhaps much smaller structures located within it. The first view, known as the 'island universe' theory, dated from the eighteenth century and received some support from some spectroscopic measurements, though without being accepted by the majority of astronomers. According to the alternative view, the Milky Way system was essentially the entire material universe, placed in a possibly infinite space or ethereal medium. 'No competent thinker', said the British astronomer Agnes Clerk in 1890, 'can now, it is safe to say, maintain any single nebula to be a star system of coordinate rank with the Milky Way . . . With the infinite possibilities beyond, science has no concern'.[8] The whole question of island universes versus the Milky Way universe remained unsolved until the mid-1920s, when Edwin Hubble succeeded in determining the distance to the Andromeda nebula, thereby providing firm evidence in favour of the island universe theory.

To the extent that there was a consensus view of the stellar universe in the early twentieth century, it slightly favoured the idea that there was nothing outside the limits of the Milky Way. Based on observations and statistical analysis, eminent astronomers such as Seeliger, Schwarzschild, and the Dutchman Jacobus Kapteyn suggested models of the Milky Way universe conceived as a huge non-uniform conglomeration of stars. Their models of the universe had in common that they pictured the Milky Way as an ellipsoidal disk of dimensions only a few tens of thousands of light years in diameter. Assuming that starlight was not absorbed by interstellar matter, they estimated the mass density of the Milky Way universe to be about 10^{-23} g/cm^3.

29.3 EINSTEINIAN COSMOLOGY

In a pioneering paper of 1917, entitled 'Kosmologische Betrachtungen zur allgemeinen Relativitätstheorie' ('Cosmological considerations on the general theory of relativity'), Einstein applied his new theory of gravitation to the universe at large. In doing so, he was faced with conceptual problems of a kind similar to those which had troubled earlier scientists from Newton to Seeliger. In particular, how can the advantages of a finite universe be reconciled with the necessity for a universe without boundaries? Einstein's solution was to circumvent the problem by conceiving the universe as spatially closed, in accordance with his general theory of relativity based on non-Euclidean geometry. Guided by what little he knew of the available observational evidence, he suggested that the universe was spatially finite but of a positively curved geometry—'spherical' in four dimensions. He also assumed the universe to be static, meaning that the radius of curvature did not vary systematically in time, and that it was filled uniformly with matter of low density.

Einstein's closed universe was formally described by the field equations of general relativity, but in a modified form which included a 'for the time being unknown universal constant'. This constant (Λ), which soon became known as the cosmological constant, expressed the mean density of matter ρ in the universe and was thereby also related to the radius of curvature R. The relations announced by Einstein were

$$\Lambda = \frac{\kappa \rho}{2} = \frac{1}{R^2}$$

where κ denotes Einstein's constant of gravitation, related to Newton's by $\kappa = 8\pi G/c^2$. From a physical point of view the new constant could be conceived as causing a cosmic repulsion balancing the attractive force of gravitation. In this respect, Einstein's constant corresponded to the one that Seeliger had introduced in 1896, but when Einstein published his cosmological theory he was unaware of Seeliger's work. In a letter of August 1918 he explained his reasons for introducing the cosmological constant:

Either the world has a centre point, has on the whole an infinitesimal density, and is empty at infinity, whither all thermal energy eventually dissipates as radiation. Or: All the points are on average equivalent, the mean density is the same throughout. Then a hypothetical constant λ [Λ] is needed, which indicates at which mean density this matter can be at equilibrium. One definitely gets the feeling that the second possibility is the more satisfactory one, especially since it implies a finite magnitude for the world.[9]

Contrary to Einstein's original belief that his closed model was the only solution to the cosmological field equations, later in 1917 the Dutch astronomer Willem de Sitter produced another solution corresponding to an alternative world model.[10] Remarkably, de Sitter's model contained no matter ($\rho = 0$), but was nonetheless spatially closed. Moreover, it followed from the spacetime metric that light emitted by a test body would be redshifted, almost as if the body were moving away from the receiver. In fact, although the 'de Sitter effect' was not a Doppler effect, de Sitter suggested that it might be related to the measurements of nebular radial velocities that Melvin Slipher had been reporting since 1913. Working at the Lowell Observatory in Arizona, Slipher found that the light from most spiral nebulae was shifted towards the red end of the spectrum, indicating a recessional velocity of up to 2,000 km/s. In the 1920s the galactic redshifts attracted increasing attention among astronomers and cosmologists, who suspected some simple relation between the redshifts of the spirals and their distances. The observed redshifts were generally seen as evidence for de Sitter's 'model B', while the undeniable existence of matter in the universe counted against it and for Einstein's 'model A'.

Whatever the credibility of solutions A and B as candidates for the real structure of the universe, from about 1920 there developed a minor industry based on the two models. It was predominantly a mathematical industry, with mathematically-minded physicists and astronomers analysing the properties of the two solutions and proposing various modifications of them. Although the industry was dominated by mathematicians and theoretical physicists, at least some astronomers were aware of the cosmological models and endeavoured to relate them to observations. During the course of their work to understand and elaborate the two relativistic world models, a few astronomers and physicists proposed solutions that combined features of model A and model B. Towards the end of the 1920s there was a tendency to conclude that neither of the two models could represent the real universe, yet finding a compromise turned out to be frustratingly difficult.[11]

If both Einstein's model A and de Sitter's model B were inadequate, and if these were the only solutions, how could cosmology still be based on the theory of general relativity? The alternative of abandoning general relativity and returning to some classical framework was not seriously considered. The 'obvious' solution—to search for evolutionary models different from both A and B—had already been published at the time, but was as unknown to most cosmologists as it was unwelcome. The conceptual climate that governed mathematical cosmology in the 1920s was that of a physically static universe, and the scientists engaged in the field tried hard (but of course unconsciously) to avoid breaking with the paradigm.

Although the mathematical cosmology of the 1920s did not include the physics of matter and radiation, there were a few attempts to adopt a more physical approach to the study of the universe. For example, in 1928 a Japanese physicist, Seitaro Suzuki, developing earlier ideas of Richard Tolman in the United States, investigated the cosmological significance of the relative abundances of hydrogen and helium, the two most common elements in the universe. He suggested, almost prophetically, that the mass ratio could be explained only on the assumption of an early state of the universe at a temperature higher than 10^9 degrees. It is not clear whether Suzuki believed that such a state had actually existed in the early universe.

The first application of thermodynamics to the Einstein universe model was made in a study by the German physicist Wilhelm Lenz, who in 1926 found that the temperature T of an enclosed blackbody radiation would depend on the radius of curvature R, as

$$T^2 = \frac{1}{R}\sqrt{\frac{2c^2}{\kappa a}} \cong \frac{10^{31}}{R}$$

where a denotes the constant in the Stefan–Boltzmann radiation law. Since a can be expressed by Planck's constant h, Lenz implicitly introduced quantum theory in a cosmological context. A more explicit connection was suggested the previous year by Cornelius Lanczos, a Hungarian theorist who, in a study of a periodic Einstein-like world model, was led to introduce a certain 'world period' P, depending on the radius of the universe:

$$P = \frac{4\pi^2 mR^2}{h^2} = \frac{mR^2}{\hbar^2}$$

where m denotes the mass of the electron. Although neither Lanczos nor others developed the idea, Lanczos' suggestion that the quantum nature of microphysics reflected the state of the cosmos is noteworthy. 'The solution of the quantum secrets', he wrote, 'are hidden in the spatial and temporal closedness of the world'.[12] The same kind of micro–macro theorizing would later form the basis of Arthur Eddington's unorthodox 'fundamental' theory, such as he developed in great detail in the 1930s.

29.4 THE EXPANDING COSMOS

The collapse of the static-universe paradigm took place by the complex interaction of two widely separate approaches—one observational and the other theoretical. By the late 1920s, Hubble turned to the problem of the redshifts of extragalactic nebulae and the supposed relationship between redshift and distance. Mostly relying on the redshifts found earlier by Slipher, in an important paper of 1929 he concluded that they varied nearly linearly with the distances of the galaxies—a relationship he expressed as

$$v = cz = Hr$$

Hubble did not really measure recessional velocities v, but understood v as 'apparent velocities'—that is, redshifts that could be conveniently transformed to velocities by means of the Doppler formula. In his paper of 1929 he suggested vaguely that spectral shifts might be interpreted in terms of de Sitter's cosmological model, but his general attitude was to stay away from interpretations and keep to observational data. For the empirical constant H, eventually known as the Hubble constant or parameter, he derived a value of about 500 km/s/Mpc (1 Mpc = one megaparsec $\cong$ 3.26 million light years). Whereas the linear relation of 1929 was not particularly convincing, new and much improved measurements two years later, now published jointly with his assistant Milton Humason, left no doubt that the redshift–distance relation was linear and real.

It is important to realize that Hubble did not conclude, neither in 1929 nor in later publications, that the universe was in a state of expansion.[13] A cautious empiricist,

Hubble adhered to his data and refrained from clearly interpreting them as evidence for an expanding universe. Nor did other astronomers and physicists initially consider the Hubble relation as observational proof that the universe is expanding. It took more than observations to effect a change in the world view from a static to an expanding universe. This change occurred only in the early part of 1930, when it was recognized that both theory and observation strongly indicated that the static universe was no longer tenable.[14]

Unknown to Hubble and most others in the scientific community, the possibility of an expanding universe had already been formulated by theorists—first by the Russian physicist Alexander Friedmann in a paper appearing in *Zeitschrift für Physik* in 1922. In his systematic analysis of Einstein's cosmological field equations, Friedmann showed that the solutions of Einstein and de Sitter exhausted all the possibilities for stationary world models. More importantly, for closed models he found a variety of dynamic solutions where the curvature of space depends on time, $R = R(t)$. These solutions included a class in which the world began at $R = 0$ for $t = 0$ and subsequently expanded monotoneously. Another of the solutions to which Friedmann called attention corresponded to a cyclic universe, starting at $R = 0$ and ending, after having passed $R = R_{max}$, at $R = 0$. He examined world models both with and without a cosmological constant, and in a companion paper of 1924 he extended his analysis by examining also open models with negative space curvature.

The Friedmann equations described all possible dynamic world models satisfying the cosmological principle—that is, assuming the universe to be homogeneous and isotropic at a large scale. However, although the Russian theorist clearly realized the significance of evolutionary or dynamic world models, the emphasis of his work was on the mathematical rather than the physical and astronomical aspects. He did not highlight expanding solutions nor argue that the universe we observe is in fact in a state of expansion. Nor did he refer to astronomical data such as Slipher's galactic redshifts. The mathematical style and character of Friedmann's work may have been one reason why it failed to attract attention, to be rediscovered several years later. Even the few physicists who were aware of his work, such as Einstein, did not see it as a strong argument against the static world. Friedmann's singularly important work is a prime example of what has been called prematurity in scientific discovery.[15]

When five years later the Belgian theorist Georges Lemaître arrived at the conclusion that the universe expands in conformity with the laws of general relativity, he was unaware of Friedmann's earlier work. Motivated by a desire to find a solution that combined the advantages of Einstein's model A and de Sitter's model B, in 1927 he found the same differential equations for $R(t)$ that Friedmann had published earlier. Although from a mathematical point of view Lemaître's work was very similar to Friedmann's, from the point of view of physics and astronomy it differed strikingly from it. Trained in both astronomy and physics, Lemaître wanted to find the solution that corresponded to the one and only real universe such as described by astronomical data and by galactic redshifts in particular. He concluded that the best model was a closed universe expanding monotonously from a static Einstein state of a world radius of about 270 Mpc. As the expansion continued, the mass density would gradually diminish and eventually approach that of the empty de Sitter state.

In striking contrast to Friedmann's paper, galactic redshifts were all-important to Lemaître, who explained that they had to be understood as a cosmic effect of the expansion of the universe. The redshifts were caused by the galaxies being carried with the expanding space in such a way that the 'apparent Doppler effect' expressed the increase in R between emission and reception of light. For the approximate relationship between recession velocity and distance, he found

$$v = \left(\frac{dR}{Rdt}\right) cr = kr$$

where k is a quantity that depends on the time since the expansion began—an early version of the Hubble constant. He estimated from astronomical data that $k \cong 625$ km/s/Mpc. As he further noted, at some time in the future, when z became equal to one, the universe would expand so rapidly that no light from the galaxies would reach us.

Lemaître's theory of 1927 is today recognized as a cornerstone in cosmology and the true foundation of the expanding universe, but at the time it made no more impact than Friedmann's earlier work. It is telling that Einstein (who knew about the works of both Friedmann and Lemaître) in an article of 1929 for the *Encyclopaedia Britannica* maintained his faith in the static universe. 'Through the general theory of relativity', he wrote, 'the view that the continuum is infinite in its time-like extent but finite in its space-like extent has gained in probability'.[16] The expanding universe only became a reality in the spring of 1930, after Eddington, de Sitter, and other leading astronomers became aware of Lemaître's theory and realized how well it fitted with Hubble's empirical redshift–distance relation. Eddington quickly abandoned the static universe, replacing it with Lemaître's expanding model (which for this reason is known as the 'Lemaître–Eddington model'). Einstein too accepted the dynamic solutions of Friedmann and Lemaître, realizing that in an expanding universe the cosmological constant was not necessary. He had for long been dissatisfied with the Λ constant he had introduced in 1917, and now abandoned it for good.

By 1935 the theory of the expanding universe was accepted by the majority of astronomers and physicists, and was the subject of detailed investigations in the scientific literature. It was also disseminated to the public through a number of popular works, such as James Jeans' *The Mysterious Universe* (1930), James Crowther's *An Outline of the Universe* (1931), de Sitter's *Kosmos* (1932), and Eddington's *The Expanding Universe* (1933). Far more demanding were the few textbooks oriented towards the small community of cosmologists, of which the most important were Tolman's *Relativity, Thermodynamics and Cosmology* (1934) and the Otto Heckmann's *Theorien der Kosmologie* (1942).

General acceptance of the expanding universe among leading physicists and astronomers did not extend to the entire scientific community. A considerable minority denied that the universe was in a state of expansion, and consequently had to produce alternative explanations of the observed redshifts. There was a variety of ways in which this could be done on the basis of a static and eternal conception of the universe. Some scientists felt attracted by explanations according to which the redshifts were caused by gravitational mechanisms or by an hypothesis of 'tired light', such as advocated in different ways by Fritz Zwicky, William MacMillan, and Walther Nernst. Although these kinds of non-expansion explanation were fairly popular in the 1930s, they did not succeed in reinstating the static universe at the expense of the new expanding paradigm. Mainstream astrophysicists and cosmologists agreed that the expansion was real, though a few thought that it was not a consequence of the general theory of relativity. From about 1933 to 1948 the alternative cosmological system of the Oxford physicist Edward A. Milne attracted great interest. Milne's cosmos was uniformly expanding, but was not governed by Einstein's equations and did not operate with the notion of curved space. After Milne's death in 1950 his cosmology based on 'kinematic relativity' ceased to attract attention.

As illustrated by the Lemaître–Eddington model, a continually expanding universe does not need to have a finite age. Although Friedmann had formally introduced the idea of a finite-age or 'big bang' universe in his work of 1922, in a physical–realistic sense it dates from 1931. In a brief letter to *Nature* of 9 May that year, Lemaître made the audacious proposal that the universe had once come into being in what he picturesquely described as a huge radioactive explosion of a primeval 'atom' of nuclear density ($\rho \cong 10^{15}$ g/cm^3). Following the original explosion 'by a kind of super-radioactive process', the universe expanded at a furious rate and eventually evolved to the present state of a low-density expanding universe ($\rho \cong 10^{-30}$ g/cm^3). While Lemaître's first communication was brief and purely qualitative, he soon developed it into a proper scientific theory based on the cosmological field equations including a positive cosmological constant. In the first detailed account ever of the 'big bang theory' (a name not yet coined), he described it as follows:

The first stages of the expansion consisted of a rapid expansion determined by the mass of the initial atom, almost equal to the present mass of the universe . . . The initial expansion was able to permit the radius [of space] to exceed the value of the equilibrium radius [of the Einstein world]. The expansion thus took place in three phases: a first period of rapid expansion in which the atom-universe was broken down into atomic stars, a period of slowing-down, followed by a third period of accelerated expansion. It is doubtless in this third period that we find ourselves today.[17]

Lemaître's model of 1931 was a solution to the Friedmann equations, which allowed for several more world models sharing the feature of $R = 0$ at $t = 0$. One such model was proposed jointly by Einstein and de Sitter, who in 1932 presented a theory of a flat, continually expanding universe without a cosmological constant. It followed from the theory that the expansion was just balanced by the gravitational attraction, leading to a 'critical' value of the mass-density given by

$$\rho_{crit} = \frac{3H^2}{8\pi G}$$

The scale factor—a measure of the distance between two galaxies—increased slowly in cosmic time according to $R \sim t^{2/3}$. The Einstein–de Sitter model belonged to the 'big bang' class, since $R = 0$ at $t = 0$, but this was a feature with which neither Einstein nor de Sitter were comfortable, and which they consequently avoided. Like most other physicists and astronomers, they did not consider Lemaître's proposal to be a realistic candidate for the evolution of the cosmos.[18]

The general response to the idea of a 'big bang' in the distant past was to ignore it or reject it as an unfounded speculation. After all, why believe in it? If the universe had really once been in a highly compact, hot, and radioactive state, would it not have left some traces that could still be subjected to analysis? Lemaître believed that there were indeed such fossils from the far past, and that these were to be found in cosmic rays; but his suggestion failed to receive support. In addition to the lack of observational evidence, the proposal of a world born in a radioactive explosion was thought to be contrived and conceptually weird. Few cosmologists in the 1930s were ready to admit 'the beginning of the universe' as a question that could be addressed meaningfully by science. Although the majority of cosmologists wanted to avoid the question, finite-age models of roughly the same kind as Lemaître's were not dismissed entirely. Among the few physicists who took an interest in such models were Paul Dirac in England, Pascual Jordan in Germany, and George Gamow in the United States.

29.5 Nuclear Archaeology and the Early Universe

Lemaître's 'fireworks' universe never received serious attention, but after the Second World War it was independently revived by the Russian-born American physicist George Gamow—a pioneer of nuclear physics—and a small group of collaborators. Gamow's approach to early-universe cosmology differed markedly from that of earlier researchers, in the sense that it focused on nuclear and particle physics with the aim of explaining the build-up of elements shortly after the 'big bang' at $t = 0$. Ignoring the conceptual problems of the 'creation' of the universe, Gamow considered the very early universe to be an extremely hot and compact crucible, an exotic laboratory for nuclear physical calculations. If he could calculate, on the basis of nuclear physics, how the elements were formed in the original inferno, and the calculated element abundances corresponded to those found in nature, he would have provided strong evidence in favour of the 'big bang' origin of the universe. This general research programme had been considered by a few earlier physicists—notably by Carl Friedrich von Weizsäcker in a work of 1939—but without being developed to any extent. The goal of what has aptly been called 'nuclear archaeology' was to reconstruct the history of the universe by means of hypothetical cosmological or stellar nuclear processes, and to test these by examining the resultant pattern of element abundances.[19]

In 1946 Gamow published his first preliminary paper, in which he argued that the problem of the origin of the elements could be solved only by combining the relativistic expansion formulae with the experimentally known rates of nuclear reactions. Together with his research assistant Ralph Alpher, two years later he presented a much-improved version of the 'big bang scenario' which assumed as a starting point a high-density 'soup' of primordial neutrons. Being radioactive, some of these would decay into protons and electrons, and the protons would combine with neutrons to form deuterons, and eventually, by the mechanism of neutron capture, heavier nuclei. Calculations based on this picture were promising insofar as they could be fitted to produce a nuclear-abundance curve not very different from the one known observationally. More or less independently, Alpher and Gamow realized that at the high temperature required for the nuclear reactions (about 10^9 K), radiation would dominate over matter and continue to do so until the universe had cooled as a result of the expansion. To produce a reliable picture of the early universe, they had to take into account both matter and radiation.

In a work of 1948, with Robert Herman, another of Gamow's associates, Ralph Alpher found that the initially very hot radiation would have cooled with the expansion and today appear as low-intensity radiation with a temperature of about 5 K. They argued that this background radiation still filled the entire universe, and thus in principle should be detectable as a feeble fossil from the early radiation-dominated universe. However, Alpher and Herman's brilliant prediction of cosmic background radiation failed to attract the interest of physicists and astronomers.[20] It would be another seventeen years before the background radiation was discovered, and then with dramatic consequences for the development of cosmology.

The research programme carried out mainly by Gamow, Alpher, and Herman focused on the formation of heavier atomic nuclei in the very early universe. Detailed calculations made in the early 1950s resulted in a cosmic helium content between 29% and 36% (by weight), which was in satisfactory agreement with the observed amount of helium in the universe, at the time known only very roughly. On the other hand, Gamow and his group failed to account for the heavier elements of atomic number $Z > 2$, which was seen as a serious problem and even a reason to dismiss the theory. Another problem—which Gamow's theory shared with most other finite-age models—was that it led to a too short time-scale. The age of the universe τ in terms of the Hubble time (the inverse of the Hubble constant, $T = 1/H$) follows from a particular cosmological model—in the case of the Einstein-de Sitter model, being $\tau = 2T/3$. Since apparently reliable astronomical measurements indicated $T \cong 2$ billion years, this led to a universe younger than the Earth! (The problem disappeared in the mid-1950s when it was shown that H is much smaller than believed previously.)

The low appreciation of Gamow's 'hot big bang' model in the 1950s is illustrated by the view of three distinguished British physicists, who in 1956 concluded about the theory: 'At present, this theory cannot be regarded as more than a bold hypothesis'.[21] In fact, by that time the theory had effectively come to a halt. A dozen or so physicists (but no astronomers!) had been engaged in developing Gamow's theory after 1948, but a few years later interest decreased drastically. Between 1956 and 1964 only a single research paper was devoted to what around 1950 had appeared to be a flourishing research programme. The reasons for this remarkable lack of interest are complex, and must be ascribed to both social and scientific factors. Among the reasons were not that the theory was refuted by observations in any direct sense, which it was not. The scientific problems that faced Gamow's theory of the 'big bang'—a name that dates from about 1950—were not the only reason for its decline. Another reason was that it was widely seen as a theory of the *creation* of the universe—a concept which most physicists and astronomers considered to be outside the domain of science, or even pseudoscientific.

It is also worth recalling that although Gamow's theory built on the authoritative general theory of relativity (in the form of the Friedmann equations), it did not follow from it. Some relativistic models, such as the Einstein–de Sitter model, start with a singularity—a 'state' in which all matter and space is concentrated in a single point. But the idea of a nuclear explosion—as first suggested by Lemaître and later in much greater detail by Gamow and Alpher—is a foreign element that added to general relativity theory rather than being part of it. This helps explain why many astronomers, who were in favour of a finite-age universe starting in a singularity, were nonetheless opposed to Gamow's scenario of the early universe. For example, the English–American mainstream cosmologist George McVittie was a supporter of a relativistic evolution universe of, say, the Einstein–de Sitter type, but criticized those 'imaginative writers' who had woven fanciful notions such as the 'big bang' round the predictions of general relativistic cosmology.

29.6 A COSMOLOGICAL CONTROVERSY

At a time when 'big bang' cosmology was still a somewhat immature research programme, it and similar relativistic models were challenged by an entirely different theory of the universe, initially referred to as the 'continuous-creation' theory, but soon to become known as the 'steady-state' model. The basic message of the steady-state theory was that the large-scale features of the universe had always been, and would always be, the same, which implied that there was neither a cosmic beginning nor a cosmic end. However, contrary to earlier cosmological views of this kind, the new theory accepted the expansion of the universe as an observational fact.

The steady-state theory had its beginning in discussions between three young Cambridge physicists, Fred Hoyle, Hermann Bondi, and Thomas Gold, who agreed that the standard evolution cosmology based on Einstein's field equations was deeply unsatisfactory. Their alternative was developed in two rather different versions—one by Hoyle, and the other jointly by Bondi and Gold. Both of the founding papers appeared in summer 1948 in *Monthly Notices of the Royal Astronomical Society*. Although Hoyle's approach differed considerably from that of Bondi and Gold, the two theories had so much in common that they were seen generally as merely two versions of the same theory of the universe. Both papers were characterized by philosophically-based objections to the standard cosmology based on the Friedmann equations. For example, Hoyle considered what he called 'creation-in-the-past' theories to go 'against the spirit of scientific enquiry', because the creation could not be causally explained. Bondi and Gold similarly objected to the lack of uniqueness of the standard relativistic theory:

> In general relativity a very wide range of models is available, and the comparisons [between theory and observation] merely attempt to find out which of these models fits the facts best. The number of free parameters is so much larger than the number of observational points that a fit certainly exists, and not even all the parameters can be fixed.[22]

Bondi and Gold proposed to extend the ordinary cosmological principle into what they called the *perfect cosmological principle*. This principle or assumption remained the defining foundation of the steady-state theory, especially as conceived by Bondi and Gold and their relatively few followers. It states that the large-scale features of the universe do not vary with either space or time. According to Bondi and Gold it was a postulate or fundamental hypothesis: 'We regard the principle as of such fundamental importance that we shall be willing if necessary to reject theoretical extrapolations from experimental results if they conflict with the perfect cosmological principle even if the theories concerned are generally accepted'.[23] One could argue for the perfect cosmological principle philosophically or in terms of the consequences it implied, but not derive it from other physical laws. Although it could be supported by observations, it could not be proved observationally. On the other hand, it could be disproved by observations—namely, if consequences of the principle were unequivocally contradicted by observations. What might look to be an *a priori* principle, and was often accused of being one, was not really of such a nature.

The expansion of the universe apparently contradicts the perfect cosmological principle because the expansion implies that the average density of matter decreases with time. Rather than admitting a contradiction, the steady-state theorists drew the conclusion that matter is continually and spontaneously created throughout the universe in such a rate that it compensates precisely the thinning-out caused by the expansion. Bondi and Gold showed easily that the creation rate must be given by $3\rho H \approx 10^{-43}$ g/s/m^3, where ρ is the average density of matter and H is the Hubble constant. The small creation-rate made it impossible to detect the creation of new matter by direct experiment, yet the hypothesis did have detectable consequences. The new matter was supposed to be created in the form of hydrogen atoms, or perhaps neutrons or electrons and protons separately, but this was nothing but a reasonable assumption.

The simple steady-state theory of the universe led to several definite and testable predictions, and in addition to these, to several consequences of a less definite nature—qualitative expectations of what new observations would be like. Thus, it followed from the theory not only that the matter density must remain constant, but also that it had the definite value of $\rho = 3H^2/8\pi G$, which is precisely the critical density characteristic of the 1932 Einstein–de Sitter model. While in the relativistic model this density implies a slowing down of the expansion, in the steady-state model it corresponds to an exponentially growing expansion. According to this theory, the so-called deceleration parameter—a measurable quantity that expresses the rate of slowing down of the expansion—must have the value $q_0 = -1$. There were a few other predictions, including that the ages of galaxies must be distributed according to a certain statistical law. In other words, the steady-state theory led to unambiguous predictions that could be compared with measurements. Contrary to the class of 'big bang' theories, it was eminently falsifiable.

The theory proposed by Hoyle, Bondi, and Gold was controversial from its very beginning, and widely considered provocative—not least because of its assumption of spontaneous creation of matter, in apparent contradiction to the law of energy conservation. The response to the theory was in part based on observations, but to no less an extent also on objections of a more philosophical nature. As Bondi and Gold had applied methodological and epistemic arguments in favour of the new theory, so its opponents applied similar arguments against it. Foremost among the critics was Herbert Dingle, an astrophysicist and philosopher of science, who since the 1930s had fought a crusade against cosmologies of the more rationalistic kind, such as Milne's. He now felt it necessary to warn against steady-state cosmology, which he accused of being dogmatic and plainly non-scientific. In an unusuallly polemical presidential address of 1953 delivered before the Royal Astronomical Society he charged that the new cosmological model was a *cosmythology*—a mathematical dream that had no credible connection with physical reality:

> It is hard for those unacquainted with the mathematics of the subject, and trained in the scientific tradition, to credit that the elementary principles of science are being so openly outraged as they are here. One naturally inclines to think that the idea of the continual creation of matter has somehow emerged from mathematical discussion based on scientific observation, and that right or wrong, it is a legitimate inference from what we know. It is nothing of the kind, and it is necessary that that should be clearly understood. It has no other basis than the fancy of a few mathematicians who think how nice it would be if the world were made that way.[24]

Dingle judged the perfect cosmological principle totally unacceptable—to be *ad hoc* as well as *a priori*. According to him, the principle had precisely the same dubious nature as the perfectly circular orbits and immutable heavens of Aristotelian cosmology. These were more than mere hypotheses; they were central elements in the paradigm that ruled ancient and medieval cosmology, and as such they were inviolable within the framework of the paradigm. Dingle claimed that the perfect cosmological principle had a similar status.

Despite fundamental disagreements, both parties in the cosmological debate agreed that ultimately the question had to be settled by observation and not by philosophical argument. Bondi, who was greatly inspired by Karl Popper's falsificationist philosophy of science, declared that the steady-state theory would have to be abandoned if observations contradicted just one of its predictions. But he and other steady-state theorists also emphasized that in any conflict between observation and theory, observation was as likely to be found at fault as was theory. A variety of tests, some direct and others indirect, were used in the controversy between the steady-state model and relativistic evolution models. Among the more important were (i) galaxy formation, (ii) nucleosynthesis, (iii) redshift–magnitude relationship, (iv) radio-astronomical source counts, (v) distribution of quasars, and (vi) the cosmic microwave background.

It was agreed that any acceptable cosmological theory should be able to explain the formation and distribution of galaxies—a difficult problem that was approached in different ways by the two competing theories. After much theoretical work the situation was undecided in the sense that the problem was realized to be too complex to warrant any definite conclusion with regard to the two rival conceptions of the world. In other words, galaxy formation failed to work as the test it was hoped to be. Much the same was the case with regard to the problem of nucleosynthesis, where the 'big bang' theories could explain the formation of helium, but not the formation of heavier elements. According to the steady-state theory all elements had to be the products of nuclear reactions in the interior of stars. The first satisfactory explanation of this kind appeared in an ambitious and comprehensive theory published in 1957 by Fred Hoyle in collaboration with William Fowler, Margaret Burbidge, and Geoffrey Burbidge. The so-called B^2HF theory was a landmark work in stellar nucleosynthesis, but weak with respect to the predicted amount of helium and deuterium. By the early 1960s the general view was that nucleosynthesis could probably not yet be used to distinguish unambiguously between the two cosmological theories.

A more promising and relatively straightforward test seemed to derive from the variation of the speed of recession of the galaxies with their distances, from which the deceleration constant and the curvature of space could be inferred. Whereas evolution cosmologies predicted that the rate of recessional velocity was disproportionally greater for distant (older) galaxies, according to the steady-state model the velocity would increase directly proportional to the distance. As mentioned, the deceleration parameter of the steady-state theory was as small as $q_0 = -1$, which distinguished it from most evolutionary models based on the Friedmann equations. Data due to Humason and Allan Sandage at Mount Wilson Observatory indicated a slowed-down expansion, corresponding to a q_0 value considerably greater than -1. This was in agreement with the evolutionary view, but the data were not certain enough to constitute a crucial test, except for scientists (such as Sandage) already convinced about the truth of 'big bang' cosmology. Although Sandage and most other astronomers believed that the accumulated redshift–magnitude observations spoke against the steady-state alternative, the alleged refutation was not clear enough to convince those in favour of the theory.

The most serious challenge to steady-state cosmology came from the new science of radio astronomy. Martin Ryle of Cambridge University, a leader of the new science, soon became an active opponent of the steady-state theory, which, he considered, was ruled out by data from radio sources showing their distribution with regard to intensity. Ryle's group found a distribution which disagreed flatly with the prediction of steady-state cosmology but could be accommodated by the class of 'big bang' theories. Consequently, he concluded that 'there seems no way in which the observations can be explained in terms of a steady-state theory'.[25] Although this conclusion from Ryle's Halley Lecture of 1955 turned out to be premature—the data were not nearly as good as Ryle thought—improved data of 1961 did speak out clearly against the cosmological theory of Hoyle and his allies. The radio-astronomical test was accepted by the the large majority of astronomers as the final overthrow of steady-state cosmology. However, although the radio-astronomical consensus weakened the theory considerably, it was just possible to keep it alive by introducing suitable modifications, which was what Hoyle and a few other steady-state protagonists did. In spite of the solid foundation of Ryle's new data, none of the steady-state cosmologists accepted that they amounted to a refutation of their theory, and none of them converted to the view of evolutionary cosmology because of the verdict from radio astronomy.

29.7 MICROWAVES FROM THE HEAVEN

The steady-state theory of the universe received its death-blow in 1965 with the discovery of a cosmic microwave background of the kind that Alpher and Herman had predicted in 1948. Apparently unaware of the earlier prediction, in 1964 the Princeton physicist Robert Dicke came to suspect the existence of cold blackbody radiation as a relic from a cosmic bounce in which an earlier universe, after having suffered a 'big crunch', was reborn in a 'big bang'. In early 1965, James Peebles, a former student of Dicke's, calculated the temperature of the hypothetical radiation to be about 10 K, and preparations were made in Princeton to measure the radiation. But before they got that far, they learned about experiments made by two physicists at Bell Laboratories. Experimenting with a radiometer redesigned for use in radio astronomy, Arno Penzias and Robert Wilson found an antenna temperature of 7.5 K where it should have been only 3.3 K. They were unable to explain the reason for the discrepancy and only realized that the excess temperature—what they had thought of as 'noise'—was of cosmological origin when they saw a preprint of Peebles' work.

The discovery of the cosmic background radiation, rewarded by the Nobel prize, was serendipitous, insofar that the Penzias-Wilson experiments were not aimed at finding a radiation of cosmological significance and not initially interpreted as such. It is also noteworthy that the discovery and interpretation were made by physicists. None of the discoverers of the microwave background—arguably the most important discovery in modern cosmology—were astronomers, nor specialists in astrophysics. The Bell and Princeton physicists published their work as companion papers in the July 1965 issue of the *Astrophysical Journal*, reporting the discovery and interpretation of fossil radiation from the 'big bang'. While Penzias and Wilson simply announced their finding of an exess temperature at wavelength 7.3 cm, the Princeton group (Dicke, Peebles, Peter Roll, and David Wilkinson) covered the cosmological implications. Neither of the papers mentioned the earlier works of Alpher and Herman.[26]

The 'big bang' interpretation of the observed 7.3-cm microwave background was accepted immediately by the majority of astronomers and physicists, who welcomed it as final proof that the universe had come into being some 10 billion years ago in an explosive event. If the observation of Penzias and Wilson was to be interpreted in this way, the radiation had to be blackbody-distributed, the demonstration of which obviously required more than a single wavelength. The extension to other wavelengths took time, but by the early 1970s there remained no doubt that the spectrum was in fact that of blackbody radiation at a temperature of about 2.7 K. The importance of the 1965 discovery was, first of all, that it provided strong support to the 'big bang' picture and effectively ruled out alternative views of the universe. Whereas the microwave background followed naturally from 'big bang' assumptions, it could be reproduced only from steady-state cosmology by introducing additional hypotheses of an *ad hoc* nature. This strategy was followed by Hoyle, Jayant Narlikar, and a few others, but it made no impression on the large majority of astronomers and physicists, who found it to be artificial and unnecessary.

The microwave background also provided indirect support for the 'big bang' theory by leading to improved calculations of the primordial formation of helium-4 and other very light nuclear species such as helium-3 and deuterium. For example, in 1966 Peebles calculated the helium abundance on the assumption of a radiation temperature of 3 K and arrived at 26–28% helium, depending on the value of the present density of matter. The result agreed excellently with observations, which Peebles took to be further confirmation of the hot 'big bang' model. The work done on primordial production of helium-4 not only amounted to strong support for the 'big bang', but also helped determine the matter density of the universe. As to the cosmic abundance of primordial deuterium, it was only with the launching in 1972 of the Copernicus satellite that it became possible to determine the quantity to $D/H \cong 1.4 \times 10^{-5}$. From this value the American astrophysicists John Rogerson and Donald York derived a value of the density of ordinary (baryonic) matter that indicated an open, ever-expanding universe.

The low density of baryonic matter suggested the existence of large amounts of 'dark matter' in the universe—an idea which had been around for some time, and in the form of 'dark stars' can be found as early as in the second half of the eighteenth century.[27] While studying galactic clusters the Swiss–American astronomer Fritz Zwicky concluded in 1933 that the gravitation from visible matter was not nearly enough to keep the clusters together. He suggested that there must be present large amounts of non-luminating or dark matter, and that this might increase the average density of matter in the universe to about the critical value $\Omega = \rho/\rho_{crit} = 1$, corresponding to space being flat. However, Zwicky's prophetic arguments received little notice, and only in the 1970s did Vera Rubin and others produce convincing evidence that the greater part of matter in the universe must exist in some unknown, dark form.[28] Of course, the discovery of dark matter, quite different from the ordinary kind composed of electrons and nucleons, raised a new question: what is it?

The confidence in the standard hot 'big bang' model that emerged in the late 1960s, primarily as a result of the cosmic microwave background, did not imply that all the main problems of cosmology had been solved: far from it. It meant only that most experts now worked within the same paradigm and agreed upon which were the main problems and how they might be solved. General relativity theory rose to become a strong research area of physics and astrophysics in the same period, and it was part of the programme that the universe at large could be described, and could only be described, by means of Einstein's cosmological field equations. Cosmological models not based on general relativity continued to persist, and new ones were proposed, but being outside the established paradigm they were of marginal significance. What mattered was to decide the cosmological parameters from observations with such a degree of precision that they allowed selection of the best solution to the field equations. This best solution would then describe a cosmological model that most probably would correspond to the real universe.

As Sandage and other observational cosmologists saw it, the relevant parameters were first and foremost the Hubble constant and the deceleration parameter, both of which were quantities that could in principle be derived from observations. The aim of this research programme was epitomized in the title of a paper which Sandage published in 1970: 'Cosmology: a search for two numbers'.[29] However, it proved more difficult than expected to find unambiguous values for the two numbers. Measurements disagreed to such an extent that it was impossible to say with any confidence whether the geometry of the universe was open, flat, or closed. Nor could the age of the universe be pinned down to a value any more precise than very roughly 10 billion years. Most cosmologists assumed the cosmological constant to be zero (that is, non-existing), but their belief was as much grounded in philosophical preferences as in observations. Throughout the period, a non-zero and probably positive cosmological constant remained a possibility. In the absence of firm observational guidance, many cosmologists preferred the flat Einstein–de Sitter model—not because it was confirmed by observation, but because it was simple and not clearly ruled out by observations. It was a compromise model rather than a concordance model.

The change that cosmology experienced in the wake of the mid-1960s manifested itself not only cognitively but also socially. Before that time, cosmology as a scientific discipline scarcely existed, though it did exist as a scientific activity pursued part-time by a small number of physicists and astronomers who did not consider themselves as 'cosmologists'. The number of textbooks was very few, and varied considerably in content and approach, such as Tolman's *Relativity, Thermodynamics and Cosmology* (1934), Bondi's *Cosmology* (1952), and McVittie's *General Relativity and Cosmology* (1956). The decades following the discovery of the microwave background witnessed a growing integration of cosmology into university departments, and courses, conferences, and textbooks became much more common than previously. For the first time, students were taught standard cosmology and brought up in a research tradition with a shared heritage and shared goals.

The number of students increased, connections between physicists and astronomers strenghtened, and new textbooks appeared that defined the content and context of the new science of the universe. While in earlier decades cosmology had to some extent been characterized by national differences, these largely disappeared. Originally, 'big bang' cosmology had been an American theory, steady-state theory had belonged to the British, and the Russians had hesitated venturing into cosmology at all. Now the field became truly international—or nearly so. It was no longer possible to determine an author's nationality from the cosmological theory he or she advocated. For example, the kind of cosmological research carried out by the Russian physicist Igor Zel'dovich and his school was solidly founded on the new standard 'big bang' theory and did not differ from the research by his American and British colleagues. Only in China under the Cultural Revolution was 'big bang' cosmology still, in the 1970s, considered ideologically suspect, and suppressed for political reasons—which was the case for spatially closed models in particular.[30]

The change—some would say revolution—also manifested itself quantitatively, in a rapid and continual increase in publications dealing with cosmological issues.[31] Whereas the annual number of scientific articles on cosmology had on average been about thirty during 1950–62, between 1962 and 1972 the number increased from fifty to 250. As another way of expressing the growth, the annual number of papers on cosmology increased at an average rate of 6.4 papers between 1955 and 1967, while in the period 1968–80 the rate was twenty-one additional papers per year. By 1980 the total number of papers was about 330. Still, compared with other fields of physics and astronomy, cosmology remained a small and loosely organized science, split between the two large sciences of physics and astronomy. For example, there were no scientific societies of cosmology, nor were there any journals specifically devoted to cosmological research or carrying the name 'cosmology' in their title. The increasing number of papers on cosmology were published in the traditional journals of physics, astronomy, and astrophysics. Most of the important papers from the 1960s to the 1980s appeared in *Physical Review*, *Nature*, *Science*, *The Astrophysical Journal*, *Monthly Notices of the Royal Astronomical Journal*, and *Astronomy and Astrophysics*.

29.8 THE FIRST THREE SECONDS

Much of the work within the new framework of 'big bang' cosmology was concerned with the early universe, which physicists wanted to explain in terms of fundamental nuclear and particle physics. In a sense, this was a continuation of the approach adopted much earlier by Gamow and his associates, but since the early 1950s, when this work took place, particle physics had changed dramatically. Progress in high-energy physics offered new possibilities for establishing intimate connections between particle physics and cosmology—two areas of science that entered an increasingly symbiotic relationship. The union was described on a popular level by the eminent particle theorist Steven Weinberg in his best-selling *The First Three Minutes* (1977), in which he describes how the world came into being shortly after the 'big bang'.

The new field of 'particle cosmology' became the playground of specialists in high-energy theory who, more often than not, had neither training in nor knowledge of astronomy. In 1984 more than two hundred scientists—most of them young particle physicists and astrophysicists—convened at a conference on 'Inner Space, Outer Space' at the Fermi National Accelerator Laboratory (Fermilab) outside Chicago. The organizers of the conference spoke of the new revolution in which particle physics promised to reveal some of the deepest mysteries of the universe, and, they said, 'holds forth the possibility of sorting out the history of the universe back to times as early as seconds or even earlier!' Moreover:

Although the earliest history of the universe is only now starting to come into focus, the potential revolutionary implications are very apparent. We may very well be close to understanding many, if not all of the cosmological facts left unexplained by the standard cosmology. At the very least it is by now clear the answers to some of our most pressing questions lie in the earliest moments of the universe.[32]

Eight years after the Fermilab conference the symbiosis between particle physics and cosmology manifested itself in the foundation of a new journal, *Astroparticle Physics*, aimed specifically at research on the borderline between astrophysics, cosmology, and elementary-particle physics.

An early and impressive result of the new research programme in particle cosmology related to the number of neutrino species. In the mid-1970s, two types of neutrino had been detected: the electron neutrino, and the muon neutrino. There might be more neutrino species, but experiments did not reveal how many. In 1977, three particle physicists—Schramm, Steigman, and James Gunn—used cosmological data and theory to argue that the number of neutrino species could not be greater than six, and subsequent refined calculations sharpened the bound to three. This prediction, based solely on cosmological arguments, was confirmed in 1993 when results from CERN, the European centre of high-energy physics, indicated that there were indeed three species of neutrino, and no more.

Particle cosmology has resulted in several other successes of this kind. In addition, advances in high-energy physics has led to a partial understanding of one of the old enigmas of cosmology: namely, why the universe consists of matter with only slight traces of antimatter. Ever since Paul Dirac predicted the existence of antimatter (positrons and antinucleons) in the early 1930s, this exotic kind of matter had occupied a niche in cosmological thinking. Antinucleons were assumed to be as abundant as nucleons in the very early universe, and almost all these would annihilate into photons until the universe had expanded to such a size that annihilation would become rare. The result would be a nearly complete annihilation of matter, in obvious contradiction to observation. The problem could be explained by assuming a slight excess of matter over antimatter in the early universe, but this was merely pushing back the asymmetry to the initial condition of the universe, and hence was not a real explanation. Other discussions of antimatter in a cosmological context assumed the existence of an 'anticosmos' in addition to our cosmos. As the nuclear physicist Maurice Goldhaber speculated in 1956:

Should we simply assume that nucleons and antinucleons were originally created in pairs, that most nucleons and antinucleons later annihilated each other, and that 'our' cosmos is a part of the universe where nucleons prevailed over antinucleons, the result of a very large statistical fluctuation, compensated by an opposite situation elsewhere?[33]

With the emergence in the 1970s of a new class of theories that unified the electromagnetic, weak, and strong forces of nature ('grand unified theory', or GUT) it turned out that such airy speculations were unnecessary. The new unified theories did not conserve the number of nucleons (or, more generally, baryons), and on this basis it proved possible to explain the slight initial excess of baryons over antibaryons.

The most important impact of high-energy physics on cosmology was undoubtedly the introduction, around 1980, of the so-called 'inflationary' scenario—a radically new conception of the very early universe.[34] The principal originator of this highly successful theory or scenario was Alan Guth, a young American particle theorist, who in a landmark paper of 1981 proposed what had happened shortly after the Planck era, given by the time t_{Pl} after $t = 0$ and defined as

$$t_{Pl} = \sqrt{\frac{\hbar G}{c^5}} \cong 10^{-43} \sec$$

Before the Planck time the universe is supposed to have been governed by some as yet unknown laws of quantum gravity, which means that $t = t_{Pl}$ marks the effective beginning of cosmological theory. According to Guth, the universe started in a state of 'false vacuum' that expanded at a phenomenal rate during a very short period of time (about 10^{-30} s), and then decayed to a normal vacuum filled with hot radiation energy. In spite of the briefness of the inflation phase, the early universe inflated by the amazing factor of about 10^{40}. What made Guth's inflation theory appealing was primarily that it was able to explain two problems that the conventional 'big bang' theory did not address. One was the horizon problem—the problem that in the very early universe, distant regions could not have been in causal contact: that is, they could not communicate by means of light signals. So why is the universe so uniform? The other problem, known as the flatness problem, concerns the initial density of the universe, which must have been extremely close to the critical value in order to result in the present universe. Within a few years the inflationary scenario became very popular and broadly accepted as an integral part of the consensus model of the universe. Since its publication in *Physical Review* in 1981, Guth's pioneering paper on inflation has received more than 3,500 citations in the scientific literature.

The original inflation scenario was quickly developed into a number of new versions, some of which were 'eternal' or 'chaotic', meaning that they operated with inflation giving rise to a multitude of ever-reproducing subuniverses. In spite of their great explanatory power, inflationary models were considered controversial by some cosmologists, who complained that they lacked foundation in tested physics and had no connections to theories of quantum gravity. In addition, other critics pointed out that there really was no inflation theory but only a wide range of inflationary models which included so many forms that, taken together, they could hardly be falsified observationally. In spite of these and other objections of a scientific and methodological nature, the inflation paradigm continued to prosper and to remain the preferred theory of the very early universe. As yet, there is no proof that inflation really happened.[35]

While particle physicists and astrophysicists eagerly studied the earliest universe, hoping to understand it down to the Planck time, the future state of the universe was rarely a subject of scientific interest. After all, how could one know what the universe would look like trillions of years from now? Was the heat death discussed in the late nineteenth century still the best answer, or did the new cosmology offer a brighter prospect? As the British astrophysicist Malcolm Longair remarked in an address in 1985: 'The future of our Universe is a splendid topic for after-dinner speculation'.[36] While this was undoubtedly a common view, by that time a few physicists had engaged in the study of the far future of the universe, considering the field more than just after-dinner speculation. What has been called 'physical eschatology' began in the 1970s with the work of Martin Rees, Jamal Islam, Freeman Dyson, and a few others. What these physicists did was to extrapolate the current state of the universe into the far future, conservatively assuming that the presently known laws of physics would remain valid. The favoured scenario in this kind of research was the open, continually expanding case, where the picture would typically start with the extinction of stars and their later transformation into neutron stars or black holes. Even later chapters in the future history of the universe—say 10^{35} years from now—might involve hypothetical processes such as proton decay and evaporation of black holes.

Some of the studies of the far-future universe included speculations on the survival of intelligent life—either humans, or their supposedly much more intelligent descendants (which might be self-reproducing robots rather than beings of flesh and blood). In a lecture entitled 'Time Without End' in 1978, published the following year in *Reviews of Modern Physics*, Dyson argued that in an open universe life might survive indefinitely. Other physicists took up this kind of scientifically informed speculation, which of course appealed greatly to the general public, and which has continued to be cultivated by a minority of astrophysicists and cosmologists.

29.9 THE ΛCDM PARADIGM

Progress in late-twentieth-century cosmology was not limited to the application of nuclear physics and particle physics to the early universe. On the contrary, much of the progress was observational, due to advanced technology and new generations of high-precision instruments. Observations of the cosmic microwave background were greatly improved in quantity and quality with the launching in 1989 of the COBE satellite, carrying instruments specially designed to measure the background radiation over a wide range of wavelengths. Data from the satellite's spectrophotometer produced a very detailed picture of the radiation, confirming that it was precisely distributed as blackbody radiation with a temperature of 2.376 K.

Even more importantly, another of the instruments on the COBE satellite measured tiny variations in the intensity of the microwave background from different directions in space. Since the days of Penzias and Wilson it had been known that the radiation had a high degree of uniformity, but it was agreed that it could not be completely uniform, for in that case the formation of structures, eventually leading to galaxies and stars, would not have occurred. COBE's detector found what was hoped for: small temperature or density variations in the early universe that could provide the seeds from which galaxies were formed. The characteristic density variation turned out to be $\Delta \rho / \rho \cong 10^{-5}$. This was a result of great importance, for other reasons, because it was in excellent agreement with predictions based on the inflation model. As a consequence, the inflationary scenario gained in credibility and was, in some version or other, accepted by a majority of cosmologists. In 2006 the principal coordinators of the extensive COBE project—John Mather and George Smoot—shared the Nobel prize for physics 'for their discovery of the blackbody form and anisotropy of the cosmic microwave background radiation'. This was the first time—105 years after the Nobel award was instituted—that the prize was awarded for cosmological research.

While the detection of density variations in the microwave background agreed with theoretical expectations, the discovery, some years later, of the acceleration of the universe came as a surprise to most astronomers and cosmologists. As early as 1938, Fritz Zwicky and Walter Baade had proposed using the relatively rare supernovae (instead of the standard Cepheid variables) to measure the cosmic expansion as given by the Hubble constant and the deceleration parameter. However, another half century passed before the idea was incorporated in a large-scale research programme—first with the Supernova Cosmology Project (SCP) and then with the rival High-z Supernova Research Team (HZT). Both of the collaborations were international, the first being based in the United States, and the second in Australia. The two groups studied the redshifts of a particular form of supernova known as Type Ia, which could be observed at very large distances.

Although the SCP team originally acquired data which indicated a relatively high-density universe with little or no role for the cosmological constant, in 1998 results obtained by the two groups began to converge towards a very different and largely unexpected picture of the universe.[37] An important part of the new and observation-based consensus view was that the universe is in a state of acceleration, with a deceleration parameter $q_0 \cong -0.75$. Suggestions of an accelerating universe had been made earlier—the first time in 1927, with Lemaître's expanding model—but this was the first time that the idea received solid observational support. The new series of observations showed convincingly that the total mass density Ω_{total} was very close to 1, which implied a spatially flat universe. Since both the ordinary mass density and the density of dark matter were considerably less, the universe was assumed to be dominated by 'dark' vacuum energy. The best data from the early years of the twenty-first century demonstrated a composition of

$$\Omega_{total} = \Omega_{matter} + \Omega_{vacuum} = 0.28 + 0.72$$

where Ω_{vacuum} refers to the vacuum-energy density relative to the critical density. Ω_{matter} includes both baryonic and exotic dark matter.

Most physicists agreed that the dark energy was given by the cosmological constant, though other interpretations were also suggested. It had been known for a long time that Einstein's cosmological constant, interpreted in terms of quantum mechanics, corresponded to empty space having a negative pressure or, expressed differently, that it represents a repulsive force that blows up space. The dark energy associated with the cosmological constant has the remarkable property that the energy density (rather than energy itself) is a conserved quantity. This implies that as the universe expands, the amount of dark energy will increase and dominate ever more. The accelerating universe is a runaway universe. While this had been known theoretically since the 1960s, it was only around 2000 that the hypothetical dark Λ–energy became a reality. As many physicists and astronomers saw it, the cosmological constant had been 'discovered'.

With the emergence of the new picture of the universe, the nature of dark energy became a top priority in fundamental physics. Another and related priority, of a somewhat older age, was to understand the nature of the dark matter that was known to dominate over the ordinary matter in a ratio of about 5:1 (of $\Omega_{matter} = 0.28$; a part of 0.233 is due to dark matter, the remaining 0.047 to ordinary matter). Already, around 1990 it was agreed that the major part of the mysterious dark matter was 'cold', meaning that it is made up of relatively slowly-moving particles unknown to experimenters but predicted by physical theory. The particles of cold dark matter (CDM) were collectively known as WIMPs— 'weakly interacting massive particles'. Several such hypothetical particles have been suggested as candidates for the exotic dark matter, and some are more popular than others, but the nature of the dark-matter component remains unknown. The new picture of the universe, mainly consisting of dark Λ-energy and cold dark matter, is often referred to as the ΛCDM model or even the ΛCDM paradigm.

Cosmologists and astrophysicists were understandably excited over the new developments that promised a new chapter in the history of cosmology. In an article published in 2003 the leader of the SCP project, Saul Perlmutter, gave voice to the excitement:

We live in an unusual time, perhaps the first golden age of empirical cosmology. With advancing technology, we have begun to make philosophically significant measurements. These measurements have already brought surprises. Not only is the universe accelerating, but it apparently consists primarily of mysterious substances . . . With the next decade's new experiments, exploiting not only distant supernovae, but also the cosmic microwave background, gravitational lensing of galaxies, and other cosmological observations, we have the prospect of taking the next step toward the 'Aha!' moment when a new theory makes sense of the current puzzles.[38]

29.10 POSTSCRIPT: MULTIVERSE SPECULATIONS

The early twenty-first century may be the first golden age of empirical cosmology, but it has also seen the return of the kind of higher speculations that traditionally have been part of cosmology. Speculations are still very much alive in cosmological research, but in mathematical forms that make them different from the cosmic speculations of the past. Physical eschatology is one example, and there are many more.[39] For example, there are currently many theories of the universe before the 'big bang'—a notion which according to classical 'big bang' theory makes no sense but can nonetheless be defended on the basis of many-dimensional string theory or other modern theories of quantum gravity. These cosmological theories of a cyclic or eternal universe are speculative to the extent that they build on physics which has not been tested experimentally. On the other hand, they do lead to testable predictions concerning, for instance, primordial gravitational waves and the fine structure of the microwave background radiation.

Perhaps the most controversial of the modern cosmological hypotheses is the idea of numerous separate universes, or what is known as the 'multiverse'—a term first used in a scientific context as late as 1998.[40] Although speculations of other universes extend far back in time, the modern multiverse version is held to be quite different, and scientific in nature. The basic claim of the multiverse hypothesis is that there exists a huge number of other universes, causally separate and distinguished by different laws and parameters of physics. We happen to inhabit a very special universe, with laws and parameters of just such a kind that they allow the evolution of intelligent life-forms. This general idea became popular among some physicists in the 1990s, primarily motivated by developments in inflation theory but also inspired by the anthropic principle and the many-worlds interpretation of quantum mechanics. The main reason why the multiverse is taken seriously by a growing number of physicists, however, is that it has received unexpected support from the fundamental theory of superstrings. Based on arguments from string theory, in 2003 the American theorist Leonard Susskind suggested that there exists an enormous 'landscape' of universes, each of them corresponding to a vacuum state described by the equations of string theory.[41]

The Russian–American cosmologist Andrei Linde is another prominent advocate of the landscape multiverse, the general idea of which he describes as follows:

> If this scenario [the landscape] is correct, then physics alone cannot provide a complete explanation for all properties of our part of the Universe . . . According to this scenario, we find ourselves inside a four-dimensional domain with our kind of physical laws, not because domains with different dimensionality and with alternative properties are impossible or improbable, but simply because our kind of life cannot exist in other domains.[42]

Remarkably, many cosmologists and theoretical physicists have become convinced that our universe is just one out of perhaps 10^{500} universes. According to this most radical hypothesis, things are as they are because they happen to be so, and had they been different, we would not be here to see them. In some of the other universes there are no electrons, the gravitational force is stronger than the electromagnetic force, and magnetic monopoles are abundant. There may even be tartan elephants.

Understandably, the increasing popularity of multiverse cosmology has caused a great deal of controversy in the physics community.[43] Not only is the multiverse a strange creature; it does not help that it is intimately linked to other controversial issues such as string theory and the anthropic principle. The overarching question is whether or not multiverse cosmology is a science. Have physicists in this case unknowingly crossed the border between science and philosophy, or perhaps between science and theology? Almost all physicists agree that a scientific theory has to speak out about nature in the sense that it must be empirically testable, but they do not always agree what 'testability' means or how important this criterion is relative to other criteria. In spite of the undeniable progress of observational and theoretical cosmology during the last few decades, the field cannot, and probably never will, escape its philosophical past.

REFERENCES

ALPHER, RALPH A., and ROBERT C. HERMAN (2001). *Genesis of the Big Bang*. New York: Oxford University Press.
BERNSTEIN, JEREMY, and GERALD FEINBERG, eds. (1986). *Cosmological Constants: Papers in Modern Cosmology*. New York: Columbia University Press.
BRANDENBERGER, ROBERT (2008). 'Alternatives to cosmological inflation', *Physics Today* 61 (March), 44–49.
CARR, BERNARD J., ed. (2007). *Universe or Multiverse?* Cambridge: Cambridge University Press.
——and GEORGE F. R. ELLIS (2008). 'Universe or multiverse?' *Astronomy & Geophysics* 49, 2.29–2.37.
CLAUSIUS, RUDOLF (1868). 'On the second fundamental theorem of the mechanical theory of heat', *Philosophical Magazine* 35, 405–419.
CLERKE, AGNES M. (1890). *The System of the Stars*. London: Longmans, Green and Co.
CROOKES, WILLIAM (1886). 'On the nature and origin of the so-called elements', *Report of the British Association for the Advancement of Science*, 558–576.
DINGLE, HERBERT (1953). 'Science and modern cosmology', *Monthly Notices of the Royal Astronomical Society* 113, 393–407.
EARMAN, JOHN, and JESUS MOSTERIN (1999). 'A critical look at inflationary cosmology', *Philosophy of Science* 66, 1–49.
EINSTEIN, ALBERT (1929). 'Space-time', pp. 105–108, in *Encyclopedia Britannica*, 14th edn, Vol. 21. London: Encyclopedia Britannica.
——(1998). *The Collected Works of Albert Einstein*. Volume 8. Edited by Robert Schulman, A. J. Kox, Michael Janssen, and Jószef Illy. Princeton: Princeton University Press.
GOLDHABER, MAURICE (1956). 'Speculations on cosmogony', *Science* 124, 218–219.
GUTH, ALAN H. (1997). *The Inflationary Universe*. Reading, MA: Addison-Wesley.
HARPER, EAMON, W. C. PARKE, and G. D. ANDERSON, eds. (1997). *The George Gamow Symposium*. San Francisco: Astronomical Society of the Pacific.

HEARNSHAW, J. B. (1986). *The Analysis of Starlight: One Hundred and Fifty Years of Astronomical Spectroscopy*. Cambridge: Cambridge University Press.
HETHERINGTON, NORRISS (2002). 'Theories of an expanding universe: Implications of their reception for the concept of scientific prematurity', pp. 109–123, in Ernest B. Hook, ed., *Prematurity in Scientific Discovery: On Resistance and Neglect*. Berkeley: University of California Press.
ISRAEL, WERNER (1987). 'Dark stars: The evolution of an idea', pp. 199–276, in Stephen Hawking and Werner Israel, eds., *Three Hundred Years of Gravitation*. Cambridge: Cambridge University Press.
JONES, G. O., J. ROTBLAT, and G. J. WHITROW (1956). *Atoms and the Universe: An Account of Modern Views on the Structure of Matter and the Universe*. New York: Charles Scribner's Sons.
KAISER, DAVID (2006). 'Whose mass is it anyway? Particle cosmology and the objects of theory', *Social Studies of Science* 36, 533–564.
KERZBERG, PIERRE (1989). *The Invented Universe: The Einstein–de Sitter Controversy (1916– 17) and the Rise of Relativistic Cosmology*. Oxford: Clarendon Press.
KIRSHNER, ROBERT P. (2004). *The Extravagant Universe: Exploding Stars, Dark Energy and the Accelerating Cosmos*. Princeton: Princeton University Press.
KOLB, EDWARD W. et al., eds. (1986). *Inner Space, Outer Space: The Interface Between Cosmology and Particle Physics*. Chicago: University of Chicago Press.
KRAGH, HELGE (1996). *Cosmology and Controversy: The Historical Development of Two Theories of the Universe*. Princeton: Princeton University Press.
——(2008). *Entropic Creation: Religious Contexts of Thermodynamics and Cosmology*. Aldershot: Ashgate.
——(2009). 'Contemporary history of cosmology and the controversy over the multiverse', *Annals of Science* 66, 529–551.
——(2010). *Higher Speculations: Grand Theories and Failed Revolutions in Physics and Cosmology*. Oxford: Oxford University Press.
——and ROBERT SMITH (2003). 'Who discovered the expanding universe?' *History of Science* 41, 141–162.
——and DOMINIQUE LAMBERT (2007). 'The context of discovery: Lemaître and the origin of the primeval-atom universe', *Annals of Science* 64, 445–470.
LANCZOS, CORNELIUS (1925). 'Über eine zeitlich periodische Welt und eine neue Behandlung des Problems der Ätherstrahlung', *Zeitschrift für Physik* 32, 56–80.
LEMAÎTRE, GEORGES (1931). 'L'expansion de l'espace,' *Revue des Questions Scientifiques* 17, 391–410.
LINDE, ANDREI (2007). 'The inflationary universe', pp. 127–150, in Bernard Carr, ed., *Universe or Multiverse?* Cambridge: Cambridge University Press.
LONGAIR, MALCOLM S. (1985). 'The universe: present, past and future', *The Observatory* 105, 171–188.
——(2006). *The Cosmic Century: A History of Astrophysics and Cosmology*. Cambridge: Cambridge University Press.
NORTH, JOHN (1990). *The Measure of the Universe: A History of Modern Cosmology*. New York: Dover Publications.
NORTON, JOHN (1999). 'The cosmological woes of Newtonian gravitation theory', pp. 271–324, in Hubert Goenner et al., eds., *The Expanding World of General Relativity*. Boston: Birkhäuser.
NUSSBAUM, HARRY, and LYDIA BIERI (2009). *Discovering the Expanding Universe*. Cambridge: Cambridge University Press.

PAUL, ERICH P. (1993). *The Milky Way Galaxy and Statistical Cosmology 1890–1924*. Cambridge: Cambridge University Press.
PEEBLES, P. JAMES E., LYMAN A. PAGE, and R. BRUCE PARTRIDGE (2009). *Finding the Big Bang*. Cambridge: Cambridge University Press.
PERLMUTTER, SAUL (2003). 'Supernovae, dark energy, and the accelerating universe', *Physics Today* 56 (April), 53–60.
RUBIN, VERA (1983). 'Dark matter in spiral galaxies,' *Scientific American* 248 (June), 88–101.
RYAN, MICHAEL P., and L. C. SHAPLEY (1976). 'Resource letter RC-1: Cosmology', *American Journal of Physics* 44, 223–230.
RYLE, MARTIN (1955). 'Radio stars and their cosmological significance', *The Observatory* 75, 127–147.
SANDAGE, ALLAN (1970). 'Cosmology: A search for two numbers', *Physics Today* 23 (February), 34–41.
SCHRAMM, DAVID N., and GARY STEIGMAN (1988). 'Particle accelerators test cosmological theory,' *Scientific American* 262 (June), 66–72.
SEELIGER, HUGO VON (1895). 'Über das Newtonsche Gravitationsgesetz', *Astronomische Nachrichten* 137, 129–136.
SMEENK, CHRIS (2005). 'False vacuum: Early universe cosmology and the development of inflation,' pp. 223–257, in A. J. Kox and Jean Eisenstaedt, eds., *The Universe of General Relativity*. Boston: Birkhäuser.
SULLIVAN, WOODRUFF T. (1990). 'The entry of radio astronomy into cosmology: Radio stars and Martin Ryle's 2C survey,' pp. 309–330, in Bruno Bertotti, R. Balbinot, Silvio Bergia, and A. Messina, eds., *Modern Cosmology in Retrospect*. Cambridge: Cambridge University Press.
SUSSKIND, LEONARD (2006). *The Cosmic Landscape: String Theory and the Illusion of Intelligent Design*. New York: Little, Brown and Company.
TASSOUL, JEAN-LOUIS, and MONIQUE TASSOUL (2004). *A Concise History of Solar and Stellar Physics*. Princeton: Princeton University Press.
WILLIAMS, JAMES W. (1999). 'Fang Lizshi's big bang: A physicist and the state in China,' *Historical Studies in the Physical and Biological Sciences* 30, 49–114.

NOTES

1. Schramm and Steigman (1988), p. 66.
2. For modern histories of astrophysics, see Tassoul and Tassoul (2004), and Longair (2006).
3. Hearnshaw (1986).
4. Crookes (1886), p. 568.
5. Clausius (1868), p. 405.
6. For details about the heat death and the entropic creation argument, and the relation of these ideas to religious thoughts, see Kragh (2008).
7. Seeliger (1895). For history and analysis, see Norton (1999).
8. Clerke (1890), p. 368. On the Milky Way universe, see Paul (1993).
9. Einstein (1998), document 604.
10. English translations of the 1917 papers by Einstein and de Sitter can be found in Bernstein and Feinberg (1986).
11. North (1990); Kerzberg (1989).
12. Lanczos (1925), p. 80.
13. Kragh and Smith (2003).

14. Nussbaumer and Bieri (2009).
15. Hetherington (2002).
16. Einstein (1929), p. 108.
17. Lemaître (1931), p. 399. On Lemaître's road from the expanding to the exploding universe, see Kragh and Lambert (2007).
18. The Einstein–de Sitter model and other early expanding models are treated in Nussbaum and Bieri (2009).
19. On Gamow and his nuclear–archaeological approach to the study of the early universe, see Kragh (1996), pp. 80–141, and Harper *et al.* (1997).
20. Alpher and Herman (2001).
21. Jones, Rotblat, and Whitrow (1956), p. 228.
22. Bondi and Gold (1948), p. 262. For a detailed historical analysis of steady-state cosmology and its development, see Kragh (1996).
23. Bondi and Gold (1948), p. 255.
24. Dingle (1953), p. 403.
25. Ryle (1955), p. 146. See also Sullivan (1990).
26. Peebles, Page and Partridge (2009) is a detailed account of the discovery and exploration of the cosmic microwave radiation as recalled by the scientists who pioneered the work.
27. Israel (1987).
28. For a non-technical account for the evidence for dark matter, as mainly derived from the observation speeds of spiral galaxies, see Rubin (1983).
29. Sandage (1970).
30. On the case of cosmology in China during the Cultural Revolution, see Williams (1999).
31. On the growth of cosmological papers, see Ryan and Shepley (1976) and Kaiser (2006).
32. Kolb *et al.* (1986), pp. ix–xi.
33. Goldhaber (1956), p. 218.
34. On the introduction of the inflation model and its early history, see Guth (1997) and Smeenk (2005).
35. See the extensive criticism in Earman and Mosterin (1999). There are several alternatives to inflation, as discussed in Brandenberger (2008).
36. Longair (1985), p. 188.
37. For a popular account of the events that led to the new world picture, see Kirshner (2004). A more technical account is given in Longair (2006).
38. Perlmutter (2003), pp. 59–60.
39. A critical survey of this class of speculative theories is given in Kragh (2010).
40. For a historically informed introduction, see Kragh (2009). See also the collection of articles in Carr (2007).
41. Susskind (2006).
42. Linde (2007), p. 134.
43. The controversy is analysed in Kragh (2009). For a clear discussion pro and con the multiverse, see Carr and Ellis (2008).

Name Index

A
Abbe, Ernst 619–20
Abernethy, John 683, 686
Abraham, Max 729, 792, 800, 802
Académie Royale des Sciences, Paris 12, 17, 342, 406, 456
Accademia degli Aletofili, Verona 18
Accademia degli Inquieti, Bologna 17
Accademia del Cimento, Florence 12, 489
Accademia Fisico-Matematica, Rome 17–18
Adams, George 275, 309, 334, 335
Adie, Alexander 338
AEG, Germany 621, 869
Aepinus, Franz Ulrich Theodor 286, 407, 439–40
Agrippa, H. C. 96
Airy, George Biddell 333, 457
Aitken, James 524
Aldini, Giovanni 682, 683
Allen, Elias 104
Allentown transistor plant 874
Alpher, Ralph 904, 909
Althaus, Julius 693
Amici, Giovanni Battista 348, 627
Amontons, Guillaume 483
Ampère, André-Marie 285–92, 418, 420, 492, 542–5, 547–50, 551–2, 553, 767
Amsler, Jacob 628–9
Anderson, Carl 851
Anderson, John 305
Aquinas, Thomas 10
Arago, Dominique François Jean 271, 272, 336, 411–13, 418, 446, 447, 449–56, 462, 539, 542–3, 545
Archimedes 230, 231
Arden, Jonn 315–16
Argand, Jean-Robert 466
Aristotle 228
Arnold, John 334
Arnol'd, Vladimir 151
Arnott, Neil 664
Arrhenius, Svante 895
Arrighetti, Niccolò 30
Arsenius, Gualterus 97
Ashworth, William J. 713
Askania Werke A.G., instrument-makers, Germany 642
Atari company 883, 884
Atkinson, Edmund 664
Atten, Michel 745–6, 762n
Atwood, George 309
Augustus, Georg 350
Auzout, Adrien 214
Avery, Elroy McKendree 665
Ayrton, William 737

B
Baade, Walter 916
Babbage, Charles 553
Bacon, Francis 14, 201, 227
Badino, Massimiliano 670
Baird & Tatlock, instrument-makers, Glasgow/London 624
Baker, A. S. 329
Baker, B. B. 465–6
Baliani, Giovanni Battista 213
Bamberg, C. 617
Bamberg, instrument-makers, Germany 642
Bancalari, Michele 563
Banks, Joseph 340
Barbaro, Daniele 246
Bardeen, John 872–3, 881
Barkan, Diana 754n, 821
Barlow, Peter 553
Barlow, William 99
Barnard, Sarah 551
Barozzi, Francesco 243
Barrow, Henry 341
Bartholin, Erasmus 413
Bate, John 103
Bauer, Edmond 748
Bausch & Lomb, instrument-makers, USA 632, 633, 638
Beard, George 688
Beaune, Florimond de 242
Beccaria, Giovanni Battista 348
Becquerel, Antoine 563
Beddoes, Thomas 282, 285, 681
Bedini, Silvio A. 351
Beeckman, Isaac 57, 68, 84, 85
Bella, Giovanni Antonio dalla 346
Bellani, Angelo 348
Bell Laboratories 867, 870, 871–4, 888
Bellone, Enrico 426
Benedetti, Giovanni Battista 14, 232, 246

NAME INDEX

Benedetti, Silvano 24n
Benoist, Louis 747
Bentham, Jeremy 316, 317
Bentley, Richard 140, 160
Bentley, Thomas 316
Bérard, Jacques Étienne 416, 480, 482, 485, 487
Berge, Matthew 335
Berg, F. G. 631
Berkeley, George 150, 159
Bernoulli, Daniel 158, 359, 360, 362–6, 379–80, 385–6, 492, 767
Bernoulli, Jacob 158, 247, 256, 359, 377
Bernoulli, Johann 133, 134, 153–8, 247, 256–9, 261, 359–60, 363–4, 366–8, 383–6
Berthollet, Claude Louis 270, 272, 409–10, 412–13, 416, 418, 424, 487, 537
Berti, Gasparo 213
Bertin-Mourot, Pierre 743
Bertoloni Meli, Domenico 153
Bertucci, Paola 312
Bessel, Friedrich Wilhelm 705–6, 711–12
Besso, Michel 823
Besson, Jacques 244
Bethe, Hans 847, 870
Bianchini, Francesco 18
Biot, Jean-Baptiste 270–2, 276–81, 284–5, 290–1, 345, 411–13, 418–20, 422, 446–51, 456, 484, 535–7, 545–8, 658–9
Bird, Golding 657–8, 660, 664, 685, 691
Bird, John 329, 330, 335
Black, Joseph 338, 416, 475, 478
Blanc, Charles 368
Bliss, Nathaniel 317, 333
Bloch, Eugène 748
Bloch, Felix 870
Blondel, Christine 292
Blondlot, René 692
Blumhof, Johann Georg Ludolf 661
Blunt, Thomas 335
Bochart de Saron, Jean-Baptiste Gaspard 343
Böckler, George Andrea 244
Boerhaave, Herman 474
Bohr, Neils 725–6, 815, 822, 823–34, 837, 838–40, 847–50
Boltzmann, Ludwig 724, 770, 771–86, 816, 817, 819
Bondi, Hermann 905–8, 911
Bontemps, Georges 332
Borda, Jean Charles de 336
Borel, Emile 778
Borelli, Giovanni Alfonso 202, 301
Born, Max 670, 803, 836–7, 839, 840–2, 848
Borro, Girolamo 231
Boscovich, Roger 157
Bose, Satyendranath 843–4
Boswell, James 299–301
Bothe, Walther 839
Boulliau, Ismaël 114
Boulton, Matthew 299–300, 315–16, 317

Bourne, William 100
Boussinesq, Joseph 748
Boutan, Augustin 664, 743–4, 747
Bouty, Edmond 744
Boyle, Robert 11, 16, 104, 200, 203, 216–18, 227, 484, 713
Brading, Katherine 143
Bradley, James 157, 333
Bramer, Benjamin 245
Brander, Georg Friedrich 350
Branly, Edouard 862
Brattain, Walter 872–3, 881
Braun, Fernand 863
Braun, Karl Ferdinand 737
Bravais, Auguste 462
Breguet, Louis François Clément 287
Breit, Gregory 840
Breithaupt, Georg Augustus 350
Breithaupt, Heinrich Carl 350
Breithaupt, Johann Christian 350
Bressieux, Sieur de 102
Brettner, Hans Anton 662
Brewster, David 337, 447
Brillouin, Léon 748
Brillouin, Marcel 778
Brisson, Mathurin-Jacques 658
British Association for the Advancement of Science (BAAS) 510–11, 706, 707
British Optical Instrument Makers Association 636
British Scientific Instrument Research Association 639
Brougham, Henry 452
Bruner, Jerome 655
Brush, Stephen 426, 495, 678n
Bucciarelli, Louis L. 416
Buchwald, Jed Z. 702, 722, 732, 749, 751, 761n, 789
Buckmaster, Charles 665
Buffon, George Louis Leclerc 337, 408–9
Bugge, Thomas 312
Bunson, Wilhelm Robert 663, 893
Buonamici, Francesco 231
Burbidge, Geoffrey 908
Burbidge, Margaret 908
Burbury, Samuel 776
Bureau International des Poids et Mesures (BIPM) 714
Burgers, J. M. 854n
Bürgi, Joost 250
Burlini, Biagio 348
Burton, Mark 335
Busicom company 882
Butterfield, Michael 106
Buys Ballot, C. H. D. 769

C

Cabanis, Pierre 273
Cabeo, Niccolò 208, 433–4

NAME INDEX

Cahan, David 618, 714, 741, 754n, 760n
Caird & Company, Greenock 511
Caird, John 511, 518
Callendar, Hugh Longbourne 505
Cambridge Scientific Instrument Company (CSI) 624, 625, 641
Caminada, instrument-makers, Rotterdam 630
Campani, Giuseppe 101, 347
Campbell-Swinton, Alan Archibald 692
Canivet, Jacques 343
Cannon, John T. 379
Canton, John 305, 313
Cantor, Geoffrey 731, 734, 756n
Cardwell, Donald S. L. 499, 509
Carlisle, Anthony 282, 534
Carnot, Lazare 364–5, 498, 503–4
Carnot, Sadi 497–502, 509, 572, 766
Carpenter, William Benjamin 685–6
Carpentier, instrument-makers, Paris 623
Carpue, Joseph 682
Casella, instrument-makers, UK 624
Cassegrain, Laurent 168
Cassidy, David C. 842
Cassini, Giandomenico (aka Jean-Dominique) 336, 342, 703
Castelli, Benedetto 202, 229
Cauchoix, Robert-Aglaé 343
Cauchy, Augustin Louis 377, 424–5, 455, 460–3, 467
Caus, Salomon de 244
Cavallo, Tiberius 304, 312
Cavendish, Charles 101, 102, 441
Cavendish, Henry 338, 339, 340–1, 474
Celsius, Anders 474
Central Scientific Company (CESCO), instrument-makers, Chicago 632
Cesariano, Cesare 246
Chadwick, James 851
Chance Brothers, glass-producers, UK 635, 640
Chang, Hasok 709
Chappell, Captain Edward 514, 515
Charles, Jacques-Alexandre 344
Charleton, Walter 113
Charlier, Carl 896
Châtelet, Emilie du 155
Chaunut, Pierre 103
Chauvin & Arnoux, instrument-makers, Paris 623
Choi, Hyungsub 874
Chorez, Daniel 101, 103, 106
Christiansen, C. 468
Christie, Samuel Hunter 553
Christina, Queen of Sweden 15
Ciampini, Giovanni Giusti 17
Clairaut, Alexis-Claude 155–6, 157, 359, 363, 386, 393–5, 408–9
Clapeyron, Emile 501–2, 517
Clark, Alvan 352
Clarke, Samuel 125, 147, 159

Clausius, Rudolf 425, 504, 509, 527, 767, 768–70, 778, 782, 894–5
Clavius, Christoph 229, 231, 246
Clément, Nicolas 416–17, 480, 485–6, 498
Clerk, Agnes 896
Cock, Christopher 102
Cohen, H. Floris 7–9, 21n, 22n
Cohen, I. Bernard 131, 146, 653–4
Colombo, Realdo 201
Commandino, Federico 25, 230, 246
Compton, Arthur Holly 839
Computer Aided Training Systems 885–6
Comstock, J. L. 665
Comte, Auguste 417, 424, 490
Conant, James B. 654
Condolle, Augustin de 272
Cook, James 340
Copernicus 228–9
Copson, E. T. 465–6
Coradi, G. 628–9
Cornu, Alfred 623, 746, 762n
Cossor, A. C. 604
Cotes, Roger 134, 137, 140, 152, 154, 160, 256, 333
Coulomb, Charles Augustin 227, 277, 279, 286, 343, 440, 441, 536–7
Cousin, Jacques 391
Cox, John 692
Crawford, Adair 416, 479–80, 482, 486
Crookes, William 894
Crowther, James G. 901
Crum, Walter 519
Cuff, John 337
Cullen, William 484
Cumming, James 553
Cunaeus, Andreas 437
Cunningham, Ebenezer 733
Curie, Marie 739
Curie, Pierre 750, 763n

D

Daguin, Pierre Adolphe 664
d'Alembert, Jean le Rond 134, 156–7, 160, 359–60, 363–5, 368, 372–6, 381–2, 386–7, 393–4, 492
d'Almeida, Jean Charles 664, 743–4, 747
dal Monte, Francesco Maria 230–1
dal Monte, Guidobaldo 25, 33–4, 207, 208, 226–7, 230–1
Dalrymple, Alexander 334
Dalton, John 416, 421, 424, 479, 482, 483, 486
Danti, Ignazio 245
d'Arcy, Patrick 363
Darrigol, Olivier 722, 723–4, 752n, 798, 818–19
d'Arsonval, Arsène 689, 690, 692
D'Artigues, Aimié-Gabriel 332
Darwin, Erasmus 283, 301, 315, 484
Darwin, Horace 624
Daumas, Maurice 344
Davy, Humphry 268–9, 282–5, 287, 312, 319, 492, 534, 551, 552

Debierne, André 750
de Broglie, Louis 748, 844–5
Dee, John 96
De Forest, Lee 864
Deguin, Nicolas 659
Delaroche, François 416, 480, 482, 485, 487
Delaunay, Charles 391
De Luc, Jean André 349–50, 475–6, 480–1
Demonferrand, Jean Baptiste Firmin 549
Denis, Jean- Baptiste 167
Denny, William 524
Department of Scientific and Industrial Research (DSIR), UK 635
Deprez, Marcel 737
Desaguliers, Jean Theophilus 300–2, 305, 307, 314, 346
Descartes, René 10, 13–17, 56–90, 112, 114, 123, 202, 434
 dualist metaphysics 72–3
 Le Monde 60, 61–3, 68–73, 78–85
 mathematics 69–70, 238–45
 mechanics transformation 82–6
 optics 61–8, 86–8, 218–19
 Principles of Philosophy 60, 72, 73, 112
 system-binding strategies 79–82
 Theory of the Earth 81
 voluntarist theology 70–2
 vortex mechanics 73–8, 132–4
Deschanel, Augustin Privat 664
Desormes, Charles Bernard 416–17, 480, 485–6, 498, 539
Despretz, César 659
Dessaignes, J. P. 417–18
Deutsche Gesellschaft für Mechanik und Optik 617–18, 622
Dew, N. 704
Dicke, Robert 909
Dick, Thomas 273–4
Diderot, Denis 303
Digges, Thomas 229
Dingle, Herbert 907
Dinwiddie, James 306, 308–9, 311, 316, 317–19, 324n
Dirac, Paul Adrien Maurice 670, 815, 842–3, 850–1, 903, 913
Divini, Eustachio 101
Dollond, John 313, 330, 336
Doppler, Christian 893
Dörries, Matthias 740
Dos Reis, J. M. 633
Dostrovsky, Sigalia 379
Dover, Alfred Walter 341
Dover, John 341
Drake, Stillman 31
Drion, Charles 664
Drude, Paul 670, 783, 784
Ducretet, instrument-makers, Paris 623, 652
Dufay, Charles François 303, 435–6
Dugas, René 359

Duhamel, Jean Marie Constant 424
Duhem, Pierre 740, 748, 759n, 895
Duillier, Fatio de 160
Dulong, Pierre Louis 419, 423–4, 480, 539, 822
Dumas, Jean-Baptiste 424
Dumotiez, instrument-makers, Paris 287, 288, 289–90
Dunnett, John 102
Dürer, Albrecht 246
Dworsky, Nancy 416
Dyson, Freeman 915

E

Earnshaw, Samuel 461
Earnshaw, Thomas 334
Eberhard, Johann Peter 661
Eckert, Michael 670–1
Eddington, Arthur 728, 755n, 899, 901
Edinburgh, University of 338
Edison, Thomas 739, 864
Edwards, John 331, 348
Ehrenfest, Paul 783, 785, 803, 818–21, 827–9
Einstein, Albert 125, 731, 740–1, 779, 780, 783, 784–5, 790–811, 815–16, 818–19, 821, 823, 838–9, 843–4, 897–9, 901–3
Eisberg, Robert 726–7
Eisenlohr, Friedrich 467
Eisenlohr, Gustav Willhelm 662, 663
Ekström, Daniel 349
Elder, David 519
Elder, John 518, 523–7
Elisabeth of Bohemia 15
Ellicot, Andrew 351
Elliott Brothers, instrument-makers, UK 624–5, 640
Elphinstone, George 318
Epstein, Paul 827
Erb, Wilhelm 688
Ericsson, L. M. 631
Ertel, Traugott Leberecht 629
Erxleben, Johann C. P. 661
Eschinardi, Francesco 17–18
Euclid 233
Euler, Leonhard 156, 157, 158, 260, 359–71, 377–9, 381–2, 387–90, 393–4, 492
Everett, Joseph David 664
Evershed, Sydney 740
Ewing, Alfred 740
Exxon 884

F

Faggin, Fernando 883–5, 888
Fahrenheit, Daniel Gabriel 474
Fairbairn Company 512–18
Fairbairn, William 510–11, 527
Fairchild Semiconductor company 876–7, 879–80, 882, 888
Faraday, Michael 289, 291, 548, 550–2, 554–6, 558–62, 571–2, 685, 731, 738

NAME INDEX

Faulhaber, Johann 244
Fechner, Gustav Theodor 661
Feingold, Mordechai 242
Felix Meritis Society, Amsterdam 313
Ferguson, James 308, 310–11, 315–17
Fermat, Pierre de 240
Fernet, Émile 664
Ferrier, Jean 102, 103, 242–3
Filotecnica Salmoiraghi, instrument-makers, Milan 627
Fischer, Ernst Gottfried 658, 662
FitzGerald, George Francis 580, 732, 746
Flamsteed, John 119, 152, 157, 180, 333
Fleming, John Ambrose 735, 737, 739, 758n, 863, 864
Foncenex, Daviet de 362
Fontana, Felice 101
Fontanelle, Bernard le Bovier de 16, 153
Föppl, August 729
Forbes, James David 495, 519
Fordyce, George 316
Forman, Paul 837
Foster, Michael 679
Foucault, Michel 702
Fourier, Joseph 417, 419, 422, 424, 489–92, 496, 539
Fowler, Alfred 825
Fowler, William 908
Fox, Robert 270, 272, 482, 486–8, 509, 528n, 737, 742
Francini, Ippolito 101
Frängsmyr, Tore 701
Franklin, Benjamin 314, 344, 407, 437–9
Franz Schmidt & Haensch, instrument-makers, Germany 620
Fraunhofer, Joseph von 332, 350, 605, 616
Freke, John 302
Fresnel, Augustin Jean 288, 337, 417–19, 424–5, 446, 451–7, 459–67, 495, 539
Freund, Leopold 692
Fric, Jan 630
Fric, Josel 630
Frick, Joseph 618, 619, 662
Friedmann, Alexander 900, 902
Frisi, Paolo 361
Fuess, instrument-makers, Germany 620
Fuess, R. 617
Fuller, Steve 677n

G

Gaertner Scientific Corporation, instrument-makers, Chicago 632
Gaertner, W. 632
Gale, Leonard D. 665
Galileo 9–10, 11, 14–15, 25–55, 202, 793
  De Motu 26–9, 232
  Dialogue 36–9
  experimentation 200, 204–8, 213
  instrument-makers 98, 100, 101, 103

MS-72 29–35
  proportion theory 230–8
Galison, Peter 740, 800
Galvani, Luigi 275, 348, 534, 680
Gambey, Henri-Prudence 343–4
Gamgee, Arthur 679, 680
Gamow, George 903–5
Ganot, Adolphe 610, 654, 656–7, 663–4, 743–4
Garber, Elizabeth 742
Gassendi, Pierre 114
Gauss, Karl Friedrich 344, 558, 710, 805
Gautier, instrument-makers, Paris 623
Gay-Lussac, Joseph Louis 272, 411, 413, 416, 417, 421–2, 483, 485, 487
G. B. Salm, instrument-makers, Amsterdam 630
Gearhart, Clayton A. 671, 819
Geiger, Hans 824, 839
George III, King 340
Germain, Sophie 415–16, 417, 418
Ghetaldi, Marino 240
Gibbs, Josiah Willard 765, 779, 780, 784–5
Gilbert, John 316
Gilbert, William 79–81, 98, 113, 200, 202, 203, 433
Gillispie, Charles Coulston 426, 498
Giuducci, M. 30
Glasgow Philosophical Society 519, 523, 524
Goddard, Jonathan 101, 102
Goertz, instrument-makers, Germany 642
Goldberg, Stanley 678n
Goldhaber, Maurice 913–14
Gold, Thomas 905–7
Golinski, Jan 306
Gooday, Graeme J. N. 712, 715, 759n
Gordon, Lewis 511, 517–18, 519, 523
Goudsmit, Samuel 835
Graham, George 329, 335
Graham, Thomas 770
Graindorge, André 14, 22n
Grasselli, J. 628
Grattan-Guinness, Ivor 423, 425, 509
'sGravesande, Willem Jacob 302, 309, 346, 379, 658
Gray, Stephen 303, 435, 436
Greatorex, Ralph 104
Green, George 441–2, 455, 463, 465
Gregory, David 151n, 152, 160, 168, 196, 256
Griesbach, Hermann 688
Grmek, Mirko 24n
Grossman, Marcel 805–6
Grove, William Robert 685, 694
Guagnini, Anna 737
Guarini, Guarino 14, 22n
Guericke, Otto von 11, 202, 215–16
Guinand, Pierre-Louis 332, 605, 616
Gunn, James 913
Guth, Alan 914
Guthrie, Frederick 665

H

Haas, Jacob Bernard 347
Haas, João Frederico 347
Hachette, Jean Nicolas Pierre 417, 539
Haensch, H. 617
Hahn, Roger 272–3
Hainshofer, Philip 101
Hall, A. Rupert 7, 123n, 126
Hall, Edwin 576, 577, 656
Halley, Edmond 115–16, 138, 140, 152, 257, 333
Hall, Marie Boas 123n, 126
Hamilton, William Rowan 312, 391, 457, 459, 461, 827, 846
Harman, Peter 723
Harper, William 129n
Harries, Arthur 689
Harriot, Thomas 66, 100, 257
Harris, John 301
Harrison, John 334
Harrison, Peter 93n
Hartmann & Braun, instrument-makers, Germany 621
Hartmann, Georg 97
Hauksbee, Francis 104, 301–2, 346, 435
Haüy, René Just 410, 411, 413, 417, 481, 486–7, 658
Heath, Thomas 335
Heaviside, Oliver 577, 729, 740, 789
Hebra, A. 699
Heckmann, Otto 901
Hedley, William Snowden 694
Heilbron, John 271, 272, 274, 442n, 701, 702
Heisenberg, Werner 293, 815, 832, 836, 838, 840–3, 846–50, 870
Helmholtz, Hermann von 365, 455, 465, 468–70, 504, 578–80, 587, 617, 680, 684, 777, 895
Henry, Joseph 268, 273, 553, 554
Henson, William 866
Hentschel, Klaus 709, 710–11
Herapath, John 492, 767
Hermann, Jacob 256, 259, 359, 383
Hermann, Ludimar 679–80
Herman, Robert 904, 909
Herschel, John Frederick William 447, 553, 710
Herschel, William 274, 494–5
Hertz, Heinrich 579–83, 745–6, 862
Heussi, J. 662
Hilbert David 806–7
Hilger, Adam 624, 626
Hipp, M. 628
Hire, Philippe de la 247, 703
Hobart, Robert 318
Hodge, Jonathan 731, 734, 756n
Hoerni, Jean 876
Hoffmann, James 420
Hoff, Ted 882–3, 884, 885, 888, 889
Hoffwenius, Petrus 16
Hogg, Jabez 665
Holland, J. 331

Holton, Gerald 653–4, 666
Homes, R. 442n
Hong, Sungook 737, 749
Hooke, Robert 115–16, 152, 169, 174, 175, 176–7, 193–6, 201, 202, 209, 216–17, 220–1
Hopkinson, John 737, 740
Hopton, Arthur 100
Hornsby, Thomas 317
Horrocks, Jeremiah 114, 139
Howard, Don 671
Hoyle, Fred 905–9, 910
Hubble, Edwin 896, 899–900
Hubin, instrument maker 104–5
Huggins, William 893
Humason, Milton 899, 908
Humboldt, Alexander von 272
Hume, David 143
Hunt, Bruce J. 707, 708, 736–7, 738
Hunter, William 318
Hunt, Robert 665
Hurmuzescu, Dragomir 747
Hutton, James 494
Huxley, Thomas 737, 738–9
Huygens, Christiaan 112–14, 125, 135, 138, 143, 146, 154, 158, 167, 176–8, 202, 203, 207, 209–10, 247–9, 413, 414, 445–6, 464–6, 700, 703
  Horologium oscillatorium 110, 120, 128–9, 148, 209, 247–9

I

Ilberg, Waldemar 663
Institut de France 413–16, 418–19, 486, 489, 537, 538
Institut d'Optique, France 637, 641–2
Institute of Medical Electricity, London 691
Institution of Engineers in Scotland 527–8
Instituto Tecnico G. Galilei, Florence 596
Intel Corporation 882–5, 886–7, 888, 889
Irvine, William 416, 478–9, 482–4, 507n
Islam, Jamal 915
Ivory, James 704
Izarn, Joseph 287

J

Jacobi, Carl Gustav 391, 711, 827
Jacobi, Moritz 706
Jaki, Stanley 392
Jallabert, Jean 303–4
James Queen & Co, instrument-makers, USA 632
Jamin, Jules 467, 654, 663, 666
Jammer, Max 854n
J. C. Marius, instrument-makers, Utrecht 618
Jeans, James 724, 733, 821, 824, 901
Jecker, François-Antoine 343
Jobin & Yvon, instrument-makers, France 642
Johns, Adrian 671
Johnston, John 665
Jones, Henry 104

NAME INDEX

Jones, Thomas 336
Jones, William 275, 309–11
Jordan, Pascual 671, 840–1, 842, 843, 850, 903
Joubert, Jules 745
Jouguet, Emile 359
Joule, James Prescott 503, 504–5, 520, 527, 738, 766, 767
Jünger, F. G. 630
Jungnickel, Christa 726
Jürgensen, C. P. 630

K

Kaiser, David 668, 669, 670, 678n
Kammerer, K. 629
Kapteyn, Jacobus 896
Kastner, Karl Wilhelm Gottlob 661
Kater, Henry 336, 341
Katzir, Shaul 749, 763n
Kaufmann, Walter 800
Keate, Thomas 682
Keill, John 152, 259
Keir, James 313, 316
Kelman & Son, instrument-makers, Amsterdam 630
Kelvin & James White Ltd, instrument-makers, Glasgow 624
Kelvin, Lord see Thomson, William
Kepler, Johannes 61, 87, 113–14, 116–19, 130, 135, 151, 249–50, 698
Kern, instrument-makers, Aarau 628
Kershaw, Michael 708
Ketteler, Eduard 469
Keuffel & Esser, instrument-makers, USA 632, 638
Kilby, Jack 878, 881
Kinnersley, Ebenezer 314
Kipp & Sons, instrument-makers, Delft 630
Kirchhoff, Gustav Robert 455, 663, 777, 893
Kleiner, Eugene 876
Kleist, Ewald von 437
Kley, Jacobus 345
Klopfer, Leopold E. 654
Knight, Gowin 304, 334
Knudsen, C. 630
Koenig, R. 588, 614
Koening, A. 587
Kohl, Max A. G. 608, 621
Kohlrausch, Friedrich 663, 665, 712
König, August Karl 425
Korteweg, Diederik Johannes 62
Koyré, Alexandre 7, 21, 31n, 201, 227–8, 700
Kragh, Helge 671, 841, 850
Kramer, P. 62
Kramers, Hendrik 829, 838–41, 856n
Krantzenstein, Christian Gottlieb 661
Krie, F. 662
Krüss, instrument-makers, Germany 621

Kuhn, Thomas 228, 261, 365, 503–4, 508, 528n, 652, 654–6, 666–7, 677n, 700–1, 723, 752, 764n, 818, 852n
Kula, Witold 702–3

L

Ladenburg, Rudolf 838
Lafond, Joseph-Aigan Sigaud de 658
Lagrange, Joseph-Louis 359, 362, 364–6, 370–1, 375–6, 389–91, 394–6
Laguna, instrument-makers, Zaragoza 628
Lambert, Johann Heinrich 482–3, 484
Lamé, Gabriel 424, 426, 663
Lancisi, Giovanni Maria 19
Lanczos, Cornelius 899
Landé, Alfred 831–2
Langevin, Paul 747, 748, 750
Langmuir, Irving 737
Laplace, Pierre Simon 157, 269–73, 276, 366, 371, 391, 394–6, 406–27, 441, 477, 481, 484, 486–8, 492, 496, 537, 702
  Arcueil society 270, 272–81, 412–17
Lardner, Dionysius 665
Larmor, Joseph 578, 732–3, 756n, 757n, 824
Last, Jay 876
Lavoisier, Antoine Laurent 270, 272, 303, 344, 408, 477, 481, 487
Lawrence, Henry Newman 689
Lawrence, William 683
Lécuyer, Christophe 865
Legendre, Adrien Marie 416, 441
L. E. Gruley Company, instrument-makers, USA 632
Lehmann, O. 618, 619
Leibniz, Gottfried Wilhelm 19–20, 113, 114, 125, 147, 148–50, 153, 158, 159, 247, 249, 255–9, 359
Leitz, instrument-makers, Germany 620
Lelong, Benoit 747, 748, 750, 762–3n
Lemaître, Georges 900–1
Lencker, Hans 245
Lenoir & Forster, instrument-makers, Vienna 630
Lenoir, Étienne 336, 343
Lenoir, Paul-Étienne 343
Lenzen, Victor Fritz 653
Lenz, Heinrich 556
Lenz, Wilhelm 898–9
Leslie, John 479, 485, 495
Leslie, Stuart 874
Leupold, Jacob 106
Levi-Civita, Tullio 805
Levitt, Theresa 423
Leybold's Nachfolger, instrument-makers, Germany 608, 621
l'Hospital, Guillaume de 247
Libes, Antoine 658
Lichtenberg, Georg Christoph 661
Liebherr, Joseph von 350, 616
Liebig, Justus von 503, 684

NAME INDEX

Linde, Andrei 918
Linton, C. M. 395
Lippmann, Gabriel 744–5, 750, 761– 2n
Lister, Joseph Jackson 337
Littmann, C. E. 630–1
Litton, Charles 865
Litton Industry 866
Lloyd, Humphrey 457–8, 459
Locke, John 147, 160
Lockyer, Norman 893, 894
Lodge, Oliver 694, 730, 731, 735, 737–8, 746
Logeman, instrument-makers, Haarlem 630
Lommel, Eugen 469
Longair, Malcolm 915
Loomis, Elias 665
Lorentz, Hendrik Antoon 469–70, 582, 783, 791–2, 800, 816, 821, 823, 847
Loschmidt, Joseph 774, 777
Lovering, Joseph 657
Ludwig, Carl 684
Lusuerg, Angelo 347
Lusuerg, Domenico 347
Lyth, G. W. 631

M
McClean, J. R. 511
McCormmach, Russell 722, 726, 753n
MacCullagh, James 462, 463, 467
McCullough, Jack 865, 866
McDougall, Captain 515–16
McGauley, James William 658
McGuire, J. E. 72, 143
Machamer, Peter 72
Mach, Ernst 358–9, 490–1, 808, 895
Mackenzie, James Stuart 340
MacLaurin, Colin 141, 256
MacMillan, William 902
McNaught, William 526–7
McVittie, George 905, 911
Magendie, François 273
Magnus, Gustav 617
Malebranche, Nicolas 17
Mallet, Jacques André 349
Malley, Marjorie 750
Malpighi, Marcello 19
Malus, Étienne Louis 272, 274, 413, 446, 447–50
Manfredi, Eustachio 17, 18
Manfredi, Gabriele 256
Mansell, Robert 101
Mansel, Robert 518–23, 524, 528
Marconi, Guglielmo 863, 881
Mariani, Jacopo 101
Mariotte, Edmé 113, 210
Marques, Gaspar José 347
Marsden, Ben 509, 824
Marshall, John 337
Marsigli, Luigi Ferdinando 347
Martin, Benjamin 302, 304, 310, 337, 340
Martin, John Lover 331

Maskelyne, Nevil 333, 334, 339
Mather, John 916
Maudsley, Henry 687
Maupertuis, Pierre Louis 155
Max Kohl, instrument-makers, Chemnitz 602, 608, 610
Maxwell, James Clark 269, 294, 425, 550, 566–8, 572–8, 706–7, 710, 715, 730, 732–5, 740, 767, 769–74, 778–81
Max Woltz, instrument-makers, Germany 622
Mayer, Julius Robert 503, 681, 684
Mazor, Stan 885–6, 888
Mede, Carver 884
Melloni, Macedonio 495
Mercadier, E. 761n
Mercator, Gerhard 257
Mersenne, Marin 66, 69, 103, 202, 206, 208, 209, 241, 246, 700
Merton, R. K. 870–1
Meyer, Oscar 771, 778
Meyerstein, Moritz 710–11
Michelson, Albert 709, 730–1
Millburn, John R. 340
Miller, David P. 704
Millikan, Robert 653–4, 656, 666
Milne, Edward A. 902
Ministry of Munitions, Optical Munitions and Glassware Department (OMGD), UK 635
Minkowski, Hermann 804
MIT 872
Moigno, François 286, 462
Molyneux, Emery 99–100
Monnier, Louis-Guillaume le 305
Montanari, Geminiano 17, 18
Montmor, Henri de 15
Moore, Gordon 876
Morand, Paul 622–3
More, Henry 16, 113
Morian, Johann 101
Morin & Genesse, instrument-makers, Paris 623
Morland, Samuel 107
Morley, Edward 730–1
Morse, Philip 870
Mortimer, Joseph Granville 691–2
Motorola 879
Mozzi, Giulio 371
Müller, Johann 610, 618, 654, 660, 661, 662, 663, 666
Müller, Johannes 684
Müller-Unkel, instrument-makers, Germany 622
Muratori, Ludovico Antonio 19, 24n
Murdock, Patrick 246
Murray, Andrew 512
Murray, John 479, 482, 485, 495
Musschenbroek, Jan van 106, 305, 345–6
Musschenbroek, Pieter (Petrus) van 302–3, 437, 658
Musschenbroek, Samuel 105, 106, 345–6
Mydorge, Claude 61, 66

N

Nairne, Edward 209, 312, 335, 340, 341
Napier, David 513
Napier, James Robert 518, 523–5, 527
Napier, John 250, 523
Napier, Robert 513, 524
Narlikar, Jayant 910
Nash, Leonard K. 654
National Education Association (USA) 656
Navarro, Jaume 670
Navier, Claude Louis 377, 425
Neale, John 304–5
Neile, Paul 101, 102
Nemetz, instrument-makers, Vienna 630
Nernst, Walther 814, 821, 822, 843, 902
Nerville, Guillebot de 745
Neumann, Franz 711–12
Newton, Isaac 20–1, 202, 434–5, 491, 709, 713, 895–6
   Da Gravitatione 123, 134
   De Motu 115–19
   experimentation 211–12, 219–21
   General Scholium 140–8, 407
   Laws of Motion 119–26, 362–3
   mathematics 114–15, 148–51, 245–7
   Optiks 166–96
   Principia 109–15, 119–61, 358, 392–3
   reflecting telescope 167–9
Nichol, John Pringle 519
Nicholson, John William 825
Nicholson, William 282, 285, 312, 534
Nicol, William 337
Nieuwentyt, Bernard 302
Nissen, J. 630
Nobili, Leopoldo 553
Nollet, Jean-Antoine 303, 344, 436–8, 658
Nordstrom, Gunnar 790, 802, 803
Norman, Robert 99
North, Francis 104
North, Roger 302
Noyce, Bob 876–7, 878, 879–80, 882
Nuñez, Pedro 229, 256
Nunn, Thomas William 687
Nye, Mary Jo 742, 750

O

O'Connell, J. 698, 706
Oersted, Hans Christian 287–8, 419, 442, 539–42, 563
Officine Galileo, instrument-makers, Florence 627–8
Ohm, Georg S. 553
Oldenburg, Henry 12, 167–9, 174
Olesko, Kathryn M. 705, 708, 711–12
Olland, instrument-makers, Utrecht 630
Olmsted, Denison 658, 665
Oppenheimer, J. Robert 851
Oughtred, William 104

P

Paauw, Jan 346
Pacific Steam Navigation Company (PSNC) 510, 524–7
Palmer, R. R. 9, 22n
Pappus 230, 231, 240
Pardies, Ignace Gaston 16, 175
Parker, Richard Green 665
Parra-Mantois, instrument-makers, Paris 605–6
Pascal, Blaise 103, 203, 247
   experimentation 213–14
Paschen, Friedrich 894
Pauli, Wolfgang 670, 815, 830–2, 835, 836–7, 840, 843, 869
Paul, Jacques 349–50
Paul, Nicolas 349
Pazos, J. H. 633
Peacock, Captain George 525
Pearson, Karl 763n
Péclet, Eugène 659–60
Pedroso, José Maria 347
Peebles, James 909, 910
Peierls, Rudolf 869
Peninsular and Oriental Steam Navigation Company (P&O) 510, 516–17
Penzias, Arno 909–10
Percival, Thomas 317
Pereira, Benito 229
Périer, Florin 203, 213–14
Perlmutter, Saul 917
Perrin, Jean 747, 748, 750
Pestre, Dominique 742, 745–6, 748, 762n
Petit, Alexis Thérèse 417, 418, 419, 423–4, 480, 539, 822
Petit, Pierre 103
Physical Science Study Committee (PSSC) 655–6
Physikalisch-Technische Reichsanstalt (PTR) 618, 714
Picard, Jean 703
Piccolomini, Alessandro 229
Pickering, Edward C. 665
Pictet, Marc-Auguste 349, 484, 489, 494
Pinaud, Auguste 659
Pitcher, William 513
Pixii, Hippolyte 287, 289–90, 291–2, 557
Pixii, Nicolas-Constant 545
Planck, Max 722, 723–7, 741, 754n, 777, 783, 814, 816–21, 823, 828, 844
Plücker, Julius 563, 564
Plume, Thomas 333
Poggendorff, Johann C. 553
Pohl, Robert W. 868–9
Poincaré, Henri 391, 579, 740–1, 743, 777, 778, 797–8, 822, 823
Poincaré, Lucien 744, 749
Poinsot, Louis 362
Poisson, Siméon Denis 272, 371, 391, 413, 416, 420, 422, 423, 425, 441, 464–5, 484, 497, 537–8

NAME INDEX

Pond, John 333
Ponthus & Therrode, instrument-makers, Paris 623
Popper, Karl 907
Porro, Ignazio 348
Porta, G. B. 96
Porter, Roy 328
Pottinger, Sir Henry 517
Pouillet, Claude S. M. 418–19, 610, 618, 659, 660–1, 666
Poüilly, Jean 106
Pourciau, Bruce 150
Poynting, John H. 577–8
Preece, William 734, 735–6
Prévost, Pierre 488, 493–4
Priestley, Joseph 282–3, 307–8, 312, 314, 338
Prout, William 489
Ptolemy 218, 392

Q

Quackenbos, George Payn 665
Quare, Daniel 338
Quetelet, Adolphe 654
Quincke, Georg 467, 744–5
Quinn, Terry 714

R

Ramelli, Agostino 244
Ramond, Louis 271
Ramsay, William 893
Ramsden, Jesse 330, 335–6, 343, 585, 598
Randolph, Charles 519, 525, 527
Randolph, Elliott & Company 525, 527
Rankine, William John Macquorn 524, 525, 527, 767
Rayleigh, Lord 455
RCA 874–5, 879, 888
Réaumur, R. A. F. de 474
Recarte, instrument-makers, Madrid 628
Rees, Martin 915
Reeve, Richard 102
Regency company 875
Régis, Pierre Sylvain 152
Regnault, Henri Victor 474–5, 476, 480, 504, 750
Reiche, Fritz 671, 838
Reichenbach, Georg 350, 629
Reichenbach, Georg Friedrich von 616
Reichert, instrument-makers, Vienna 630
Reich, Ferdinand 563
Renwick, James 665
Repsold, instrument-makers, Germany 621
Reymond, Emil du Bois 684
Rheita, Anton Maria Schyrleus de 101
Riccati, Jacopo 256, 383
Ricci-Curbasto, Gregorio 805
Ricci, Michelangelo 213, 247
Riccioli, Giambattista 202, 206–7, 700
Ricci, Ostilio 230
Richard, instrument-makers, Paris 600, 623

Richardson, Owen 739
Riemann, Bernhard 805, 896
Rinssen, Antoni 345–6
Rittenhouse, Benjamin 351
Rittenhouse, David 351
Ritter, Johann Wilhelm 534–5, 539
Rive, Charles Gaspard de la 533, 553
Rive, Lucien de la 745–6
Roberts, Lissa 312–13
Roberts, Sheldon 876
Roberval, Gilles Personne de 215, 246
Robinson, John 440
Robinson, Thomas Charles 341
Roco, Mihail 886
Rogerson, John 910
Roget, Peter N. 555
Rohault, Jacques 105
Roller, Duane 653–4
Roll, Peter 909
Rømer, Ole 246
Röntgen, Wilhelm 692
Rouelle, Guillaume-Francois 303
Rousseau, Jean-Jacques 303
Rowan & Co, Glasgow 527
Rowan, David 527
Rowan, John M. 527
Rowland, Henry 709, 737, 738
Rowland, Sydney Danville 692
Rowlinson, John 425
Royal Humane Society, UK 683
Royal Institution, London 550–1
Royal Mail Steam Packet Company (RMSP) 510, 511, 513–16
Royal Society, London 12, 17
Roy, William 336, 338
Rudolph, John 666
Rumford, Benjamin, Count 489, 492–4, 496
Rumsey, James 318
Russell, John Scott 510, 529n
Rutherford, Ernest 822, 824, 826
Rutherford, F. James 667, 750
Ryle, Martin 908–9

S

Sabine, Edward 341
Sagredo, Gianfresco 40, 41, 103
Saint-Vincent, Grégoire de 250
Salisbury, Lord 734–5, 757–8n
Salleron, instrument-makers, Paris 609
Salviati 30, 36, 38, 40, 41
Sandage, Allan 908, 911
Sarasa, Alphonse Antonio de 250
Sarasin, Édouard 745–6
Sarpi, Paolo 30, 34
Sarton, George 653
Sauerwald, F. 617
Saussure, Horace-Bénédict de 304, 349, 350, 484
Sauveur, Joseph 102
Savart, Félix 290–1, 546–7, 549

Savile, Henry 333
Schaffer, Simon 11, 309, 312, 704, 707–8, 713–14, 715
Schickore, Jutte 337
Schissler, Christoph 97
Schmaltz, T. 72
Schmidt, Johann Andreas 19
Schott, Gaspar 11, 216
Schott, instrument-makers, Jena 604, 605–6, 619–20
Schottky, Walter 739, 869
Schott, O. 619–20
Schramm, David N. 913
Schrödinger, Erwin 843, 844–9, 857n
Schwab, Joseph 655
Schwarzschild, Karl 827, 896
Scoresby, William 340
Scott, John 518, 527
Secord, James 671
Seeliger, Hugo von 895, 897
Segner, Johann Andreas von 369, 661
Sellmeier, Wolfgang 468
Selva, Domenico 347
Selva, Giuseppe 347
Selva, Lorenzo 347–8
Semitecolo, Leonardo 348
Seth, Suman 753n
Settle, Thomas B. 30
Shannon, Claude 881
Shapin, Steven 11, 22n
Shinn, Terry 618
Shockley Semiconductors 876, 888
Shockley, William 872–3, 876, 881, 888
Shuckburgh-Evelyn, George 340
Sibum, Heinz Otto 738, 759n
Siegel, Dan 731
Siemens & Halske, Germany 621
Siemens, Werner 706, 707, 741
Signet Technology 884, 885
Silicon Compilers 885
Silliman, Benjamin 665
Simms, William 335
Simon, Josep 669–70, 743
Simons, William 518, 527
Simpson, Thomas 313
Sisson, Jeremiah 335
Sisson, Jonathan 3, 329, 335
Sitter, Willem de 809, 897–903
Six, James 340
Slater, John 838, 839
Slipher, Melvin 898, 899
Sluse, François de 247
Smeaton, John 302, 309, 312
Smee, Alfred 687
Smith, Adam 155
Smith, George 122n, 128, 133n
Smith, Jeppe 349
Smith, John (Captain) 101
Smoot, George 916

Snel, Willebrord 62, 93n, 218, 257
Snowball, John Charles 665
Société d'Arcueil 270, 272–81, 412–18, 537
Société d'Optique et de Précision de Levallois (OPL) 641
Société Genevoise d'Instruments de Physique (SIP) 628
Soemmering, Samuel Thomas 536, 558
Soleil, François 337
Solvay, Ernest 822
Sommerfeld, Arnold 455, 465, 670, 814–15, 821, 823, 826–7, 829, 831, 834–7, 839, 840, 847
Sörensen, S. 631
Spencer, George 340
Spencer, instrument-makers, USA 638
Spinoza, Benedict de 145
Spitalfields Mathematical Society, London 313
Sprat, Thomas 10, 11
Staley, Richard 709, 721–9, 751–2, 753–4n, 823
Starke & Kammerer, instrument-makers, Vienna 629
Starke, C. 629
Steavenson, William 691
Stefan, Joseph 771, 778
Steigman, Gary 913
Steinheil, Carl August 350
Stein, Howard 120n, 122
Steinmetz, Charles 739–40
Stevin, Simon 58, 59, 229, 257
Stewart, Balfour 665, 666, 687
Stirling, James 519
Stirling, Robert 521
Stokes, George G. 134, 455, 463, 465, 466, 770
Stoner, Edmund Clifton 834–5
Stone, William Henry 689, 691
Streete, Thomas 114
Strutt, William 774
Stuart, John 340
Sturgeon, William 553, 558
Sturm, Johann Christoph 19
Suzuki, Seitaro 898
Synattics 884, 885
Synotsis 885
Szabó, István 359

T
Tait, Peter Guthrie 574, 665, 686–7
Tartaglia, Niccolò 25, 231, 232
Taton, René 342
Taylor, Brook 256, 359, 475, 492
Taylor, Janet 328
Tecnomasio Italiano, instrument-makers, Milan 627
Terrall, Mary 261
Tesla, Nikola 689
Texas Instruments 871, 878, 879, 888
Thelwall, John 681

Thenard, Louis Jacques 272, 417
Thompson, Benjamin see Rumford, Count
Thompson, Silvanus P. 664–5, 737
Thomson, James 509–18, 519–23, 739
Thomson, J. J. 577, 578, 747, 824
Thomson, Thomas 495, 519
Thomson, William (Lord Kelvin) 469, 504–5, 509, 511, 516–23, 527, 529–31n, 562, 563, 567, 571–2, 574, 624, 665, 686–7, 712, 730, 737, 738, 739, 766–7, 774
Tibbits, Herbert 688
Timoshenko, Stephen 359, 377
Tisserand, Félix 391
Todhunter, Isaac 415, 665
Tolman, Richard 898, 901, 911
Tompion, Thomas 335
Tooke, Christopher 100
Topham, Jonathan 671
Torricelli, Evangelista 53–4, 101, 202, 212–13, 246
Tournès, Dominique 259
Trémery, Jean Louis 411
Triewald, Marten 312
Troughton, Edward 335
Troughton, John 330, 335, 336
Trouton, Frederick 746
Truesdell, Clifford 158, 359, 490, 491
Tschirnhaus, Ehrenfried Walther von 247
Turner, Anthony J. 326
Turner, Gerard L'E. 337
Tyndall, John 503, 563, 654, 665, 692, 736, 778

U
Uhlenbeck, George 835
Utzschneider, J. von 616

V
van der Waals, Johannes Diderik 782
Van Emden, instrument-makers, Amsterdam 630
van Keulen, Gerard 345
van Keulen, Johannes 345
Vannod, Theodore 688
van Schooten, Frans 114, 245
Varian 867
Varignon, Pierre 153, 156, 158, 255–6, 359, 360, 383
Vassalli-Eandi, Antonio Maria 682
Vaucanson, Jacques 309
Verdet, Émile 663, 778
Viète, François 240
Villard, Paul 748, 762–3n
Violle, J. 623–4, 636
Viviani, Vincenzio 26
Vogel, Hermann 893
Voigtländer, Johann Christoph 350
Voigt, Woldemar 469
Volder, Burcherus de 105
Volta, Alessandro 227, 272, 275–6, 278–82, 285, 312, 348, 440–1, 533, 535–6, 680
Voltaire, François Marie Arouet de 155, 303

W
Wallis, John 152, 242
Warburg, Emil 740
Warltire, John 315
Warner, Walter 101
Warwick, Andrew 668–9, 670, 722, 731–3, 751, 753n, 757n
Waterston, John 767
Watson, Earnest Charles 653–4
Watson, Fletcher G. 654, 667
Watson, Henry 778
Watteville, Armand de 688, 693
Watt, James 283, 299–301, 304–5, 313–16, 324n, 498, 509
Wear, J. 315
Webber, Charles 688–9
Weber, Jean-Paul 93n
Weber, Wilhelm Eduard 558, 563, 578, 706
Wedgwood, Josiah 312, 315, 316, 332, 476
Wedgwood, Tom 312
Weilback, instrument-makers, Denmark 630
Weinberg, Steven 912–13
Weinhold, Adolphe Ferdinand 610, 618, 662, 665
Weiss, Christian Samuel 661
Weitzmann, C. 630
Weizsäcker, Carl Friedrich von 903
Welch Company, instrument-makers, Chicago 632
Welter, J.-J. 422, 487
Wessel, Caspar 466
Westfall, Richard S. 196, 228
Westinghouse 879
Weston, E. 632
Weston Electrical Co, instrument-makers, USA 632
Westphal, Wilhelm H. 663
Wheatstone, Charles 706
Whewell, William 273, 419, 458, 459, 559
Whiston, William 302, 346
White, James 624
Whiteside, Derek T. 115, 149
Whittaker, E. H. 731
Wiedemann, Eilhard 663
Wien, Wilhelm 741, 816, 818, 821
Wiesel, Johann 101, 102
Wilcke, Johan 439, 478
Wild, instrument-makers, Heerbrugg 628
Wilkinson, Charles 682, 683
Wilkinson, David 909
Williams, Mari 634
Wilson, Alan 870
Wilson, Curtis A. 114, 122
Wilson, Robert 909–10
Wise, M. Norton 509, 702
Witt, Jan de 245
W. J. Rohrbeck's Nachfolger, instrument-makers, Vienna 630
Wolff, Christian 701
Wolff, Friedrich 661

Wollaston, Frances 309–10
Wollaston, William Hyde 447
Wormell, Robert 665
Wren, Christopher 115–16, 152, 202, 247
Wright, Edward 100
Wüllner, Adolph 654, 663
Wynne, Henry 104

Y
York, Donald 910
Young, Thomas 448, 451–2, 492

Z
Zeeman, Pieter 830
Zeiss, Carl 619–20, 642
Zel'dovich, Igor 912
Zermelo, Ernst 777
Zilon 884, 885
Zöllner, Karl Friedrich 896
Zupko, R. E. 698–9
Zwicky, Fritz 902, 910, 916

Subject Index

A

Aaron Manby steamer 511
absolute space concept 124, 359–60
accelerated motion 27–54, 113, 120, 235–8, 801–2
    see also motion
acoustics 594
adiabatic heating and cooling 483–6
adiabatic invariance principle 827, 828
affine connection 125
air engine 519–22
air pumps 11–12, 103, 104, 105, 217–18
    experiments 215–18
alcohol content measurement 713
aluminium use in instrument-making 603
amber effect 433
ammeters 590
Ampère's law 573
Andromeda nebula 896
animal electricity see galvanism
anomalous dispersion 468
anomalous Zeeman effect 830–3, 836
Anschauliche Quantentheorie (Jordan) 671
antinucleons 913–14
Arcueil society 270, 272–81, 412–17, 418, 537
Astronomers Royal 333–4
astronomy 228
astrophysics 893–4
astrospectroscopy 893
atomic structure 824–36
    Bohr model 815, 824–9
    construction principle 833
    exclusion principle 815, 830, 833, 835, 836
    penetrating orbits 833
    Rumf model 832
atomism 140–3, 724, 783–4
Austrian instrument-makers 629–30
Avogadro's number 768, 772, 777, 784

B

balance-making 99
Balmer formula 825
barometers 103–4, 307, 338, 350
    marine 340
barometric experiments 212–18
battery see voltaic pile
big bang theory 902–3, 904–5, 908–12
big science 597

Biot-Savart law 547
BKS paper 838–9
black-body radiation 741, 760n, 783, 816, 818, 844–5, 894
background microwave radiation 909–12, 915–16
blood oxygenation 684
Boltzmann equation 773, 774
Boyle's law 385, 421, 767
brass use in instrument-making 330–1, 601–2
Britannia steamer 510
British instrument-makers see United Kingdom
British market 333–41

C

calculus 114, 148–9, 154, 249, 251–9, 260
    Leibniz 255–9
    Newton 251–5
    see also mathematics
caloric theory 416, 420–3, 473, 478, 482–8, 493–503, 766
calorimeters 474
    ice calorimeter 477
calorimetry 476–7
Cambridge, University of 333, 338
    Cambridge Platonists 113
    Cavendish Laboratory 707–8
    chair of experimental physics 294
cannon-boring experiment 493, 496
canonical ensemble 779, 782
capillary action 411, 425
Carnot cycle 498–502
carp bladder experiment 215
Cauchy series 461
Cauchy–Riemann equations 387
celestial motion see planetary motion
centripetal force 86, 116, 126, 135–6
chromatic polarization 449–51, 456
chronometers 334
classical physics 712–14, 751–2
    ether theory significance 730–6
    French experimental practice 741–50, 760–1n
    identification of classical physicists 723–30
    industry interrelationship 736–41
clock synchronization 740–1, 799–800
COBE satellite 915–16
Coimbra university, Portugal 346

cold dark matter (CDM) 917
collision theory 84–6, 112–13, 211
colour 169–78
  natural bodies 186–7
  primary colours 172, 176–7, 197n
  theory of fits 191–3, 198n
  thick plates 188–93
  thin plates 181–6, 220–1
  see also light; optics
combinatorials 861
comets, motion of 119, 138, 144–5, 155–6
Comet steamer 510
compass-makers 99, 334, 343
complementarity principle 848–50, 859n
complex numbers 466–7
Compton effect 839–40
conduction of heat 489, 490–3
conductivity 564, 575, 577–8
conical refraction 459
conic construction 243–4
construction principle 833, 855n
convection 489
Copernican revolution 88–9
Copernican Revolution, The (Kuhn, 1957) 654
copper use in instrument-making 603
Coriolis force 121, 124
correlation 685, 686
correspondence principle 829
cosmology 228, 892
  CDM paradigm 915–17
  astrophysics 893–4
  big bang theory 902–3, 904–5, 908–12
  cosmology before relativity 894–6
  early universe studies 903–5, 912–14
  Einsteinian cosmology 897–9
  expanding universe 899–903, 906–7, 908
  inflation theory 914–15
  microwave background radiation 904, 909–12, 915–16
  multiverse hypotheses 918–19
  particle cosmology 913
  perfect cosmological principle 906–7
  publications 912
  steady-state model 905–9
  textbooks 911
  see also planetary motion
Cours de physique de l'École polytechnique (Jamin) 663
current balance 545
curves, mechanical generation of 243–7
cycloid 113, 209–10, 246–8

D
d'Alembert's paradox 134, 388
d'Alembert's principle 372–6
Dalton's law 421
dark energy 917
demonstrations see public demonstrations
De Motu (Galileo) 26–9, 232

De Motu (Newton) 115–19
Dialog (Galileo) 36–9
diamagnetics 563–5
didactic instruments 587–8
Die Experimentalphysik, methodisch dargestellt (Heussi) 662
dielectric capacity 571
dielectrics 561, 571–2, 579
Die Quantentheorie (Reiche) 671
diffraction 193–6, 453–5, 456
dipole 579, 582
dispersion of light 178–9, 198n, 461–2, 472n, 838–40
  anomalous dispersion 468
dividing engines 330, 335, 359–600
domestic instruments 593
Doppler effect 893, 899
dualist metaphysics 72–3
Dutch instrument-makers 345–6, 630

E
Earth
  Earth–Moon–Sun system 130, 135, 138–40, 157, 392–4
  shape of 154–5, 342
École Normale, France 659, 677n
École Polytechnique, Paris 272, 413, 414, 417, 423–4, 446, 537
Edinburgh, University of 338
Edison effect 739
education 595, 636, 652, 654–5, 712–13
  school curriculum 655
  teaching instruments 587–8
  see also universities
elasticity 366
  dynamics of elastic bodies 379–82
  elastic surfaces 415
  short-range forces approach 415, 425–6
  statics of elastic bodies 376–9
electrical resistance standard 706–8, 712
electric field 575, 577
  see also field theory
electricity 285–7, 432–42, 558–9
  Ampère's studies 285–92
  capacity 441
  conductivity 564, 575, 577–8
  electrical instruments 286–7, 288, 289–90, 291, 341, 559, 590
  electric current 287–93, 573
  electric objects 434–6
  experiments 434–42
  galvanism 275, 533–6, 680
  industrial use 737–8, 738–9
  Leiden jar 436–9
  mathematical approaches 439, 536–9
  physiological effects 686–90
  power supply 739
  public demonstrations 303–5
  sparks 561–2

electricity (cont.)
    static 559
    uniform nature of 558
    see also voltaic pile
electric lighting systems 594, 821
electrocapillarity 744–5
electrochemistry 559
electrocution 689
electrodynamics 287, 291, 545
    development of 547–50, 573–8
electromagnetic generators 556–7
electromagnetic induction 549–50, 554–6, 559–62
    induction coils 596
    lines of inductive force 561
    specific inductive capacity 561
electromagnetic radiation 579–83, 745–6, 862–3
    French response 745–6
    multiple resonance 746
    radio communication 863–7
electromagnetic rotation 548, 551–2
electromagnetic telegraph 557–8
electromagnetic world view 799
electromagnetism 419–20, 547–50, 575, 732, 739–40, 789
    Ampère's studies 543–5
    Arago's effect 553–4
    discovery of 287–9, 539–42
    Laplacian response 545–7
    see also electrodynamics
electromagnets 553
electron
    position of 849
    spin 835–6
    see also atomic structure
electronic instruments 596–7
electronics industry
    future directions 886–7
    innovation management 879–80
    integrated circuit introduction 877–8
    microprocessor revolution 881–6
    radio 863–7
    small high-technology companies 875–7
    transistor manufacture 873–8
electronic theory of matter (ETM) 732–3, 735, 736
electrophore 440
electrotherapy 681, 685, 688–9, 691–2
Élémens de physique expérimentale et de météorologie (Pouillet, 1827) 660, 661
Elements of Physical Manipulation (Pickering) 665
ellipse-tracing device 244
energy 364–5
    conservation of 364, 477, 503, 508, 572, 681, 686–90, 735–6, 739
    interconversion of 503, 749
    mass–energy equivalence 798–9
    medical perspectives 681, 686–90
    quantum theory 816
    see also thermodynamics
engines 509

air engine 519–22
    heat engine 498–500, 509
entropy 496, 504
    heat death concept 895
    increase 816, 894–5
    probabilistic interpretation 775–6, 783
Entwurf theory of gravity 805–7
equipartition theorem 724, 772, 822
equivalence 800–2
    mass–energy equivalence 798–9
ergodic hypothesis 780–1
ether 2, 148, 161, 268, 273, 275, 389, 393, 420, 452
    dynamics 459–60
    in electrodynamic theory 573–81
    as a key feature of classical physics 722, 730–6
    in optical theories 413, 419, 452, 458, 459–64, 469
    structure 459
Euclidean theory of proportions 233
evaporation 485
exclusion principle 815, 830, 833, 835, 836
expansive principle 497–8
experimentation 199–222, 433
    barometric experiments 212–18
    electrical experiments 434–42
    motion experiments 204–12
    optical experiments 218–21

F
falling bodies
    experiments 204–7
    mathematics 232, 234–8
    see also motion
Faraday cage 560
Faraday effect 563, 577
faradization 691
Fermi layers 869, 872
ferromagnetic materials 565
Feynman diagrams 669
field theory 571–6, 580–1
finfet nano-device 886–7
First World War 634–9
fits, theory of 191–3, 198n
fluid mechanics 382–91
    hydrostatics 382, 383
    Torricelli's law 383
forces, composition of 362
France
    education 658
    experimental approach 741–50, 751, 760–1n
    First World War 636–7
    French Revolution 327, 344
    instrument-makers 342–5, 597–8, 600–1, 612–15, 623–4, 636–7, 640–2
    inter-war years 640–2
    metrology 344, 703–4, 714
    textbooks 658–61, 663–4
franklinization 691

friction 300
  in engine design 516, 520
  heat generation 493
  measurement 309, 310

G

galaxy formation 908
galvanism 275, 533–6, 680
  medical perspectives 680, 681–3
galvanization 691
galvanometer 544, 553
gas theory 778
  Boltzmann equation 773, 774
  Boyle's law 385, 421, 483, 767
  caloric theory 416, 420–3, 483–8
  Dalton's law 421
  degeneracy of gases 843
  gases as particles in motion 768–9
  global approaches 779–82
  heat properties 483, 772–3
  kinetic theory 385, 767–78
  Maxwell–Boltzmann law 771–2
  transport properties of gases 769–71
Geissler's tubes 603
general covariance 805–7
general relativity 125–6
  see also relativity theory
Geneva, University of 349
geometry 238–49
  analytic 238–41
  organic 241–9
  see also mathematics
germanium use in semiconductors 870, 872–3
Germany
  education 658, 661–2
  First World War 638–9
  instrument-makers 350, 616–24, 638–9, 642–3
  inter-war years 642–3
  metrology 705–6, 708, 714
  textbooks 661–3
Glasgow, University of 518–19
glass use in instrument-making 331–2, 603
  supply issues, First World War 634–8
glass-workers 100–1, 103–4, 603–6
God, views about
  Descartes 70–2, 83, 93–4
  Newton 113, 144–7, 160–1
gold
  counterfeiting 713–14
  use in instrument-making 332, 602
grand unified theory (GUT) 914
graphoscope 593
gravity 110, 122, 128–31, 135–40, 147, 152–3, 155–7, 895–6
  cause of 159–61
  Einstein's constant 897
  Entwurf theory 805–7
  gravitational potential 366

inverse square law 116, 118–19, 130–1, 156–7, 391–3
  scalar theory 803
  short-range forces 406–27
Great Exhibition, London (1851) 612–13, 615
Great Northern steamer 512
Gresham College, London 333
ground waves 863
Grundriss der Physik und Meteorologie (Müller, 1846) 661, 662
guild system
  Britain 329
  France 342–3

H

Hall effect 576, 577
Halley's comet 155
Hamilton's principle 574
Hamilton–Jacobi theory 391, 827
Harvard Case Histories in Experimental Science 654
Harvard College, USA 351
  General Education in Science programme 654
  Project Physics Course 667
H-curve 776–7
heat 473–4, 477–88, 594
  adiabatic heating and cooling 484–6
  caloric theory 416, 420–3, 473, 478, 482–8, 493–503
  conduction 489, 490–3
  convection 489
  as a fluid 766
  latent heat 477–88
  light relationship 494–5
  mechanical explanation of 765–6
  mechanical work interconversion 503
  as a motion 766–7, 768
  movement of 488–92
  nature of 492–6
  Newton's law of cooling 491
  radiation 489, 493–4, 783
  specific heat 478–87, 772–3, 821–2
  as a state function 495–502
  'wave theory' 495
  see also thermodynamics
heat capacity 478–9, 484–5
heat death concept 895
heat engine 498–500, 509
Heidelberg, University of 662–3
Helmholtz equation 455
Helmholtz sound synthesiser 614
High-z Supernova Research Team (HZT) 916
Hindenburg programme, Germany 638
Hole argument 806
Hooke's law 209
Horsely Iron Works, Tipton 511–12
H-theorem 773, 774–5
Hubble constant 899, 906, 911, 916
Huygens' principle 248, 445, 464–6

hydrodynamics 382–91
hydrogen spectral lines 825
Hydrographic Office, UK 334
hydrometer 713–14
hydrostatic paradox 58–60
hydrostatics 382, 383
hysteresis 740, 760n

I

imponderable fluids 267–9, 407–8, 766
  instrumental approaches 274–6
  multiple interpretations 269–74
  voltaic battery introduction impact 268, 275–85
  see also electricity; heat
impulse theorem 390–1
inclined plane experiments 207
industrial instruments 340, 589
industrial physics 736–41, 751
inertia 84, 360–1, 367, 370, 799
infrared rays 494–5
instantaneous axis of rotation 368
instruments and instrument-makers 96–107, 326–52, 584–644
  Austro-Hungarian Empire 629–30
  balances 99
  barometers 103–4, 307, 338, 350
  Britain 328–30, 615–16, 624–7
  British market 333–41
  catalogues 106, 595, 608–9
  compasses 99, 334, 343
  didactic and teaching instruments 587–8
  domestic instruments 593
  electrical instruments 286–7, 288, 289–90, 291, 341, 559, 590
  electronic instruments 596–7
  First World War impact 634–9
  France 342–5, 597–8, 600–1, 612–15, 623–4
  Germany 350, 616–24
  glass-workers 100–1, 103–4
  graduation of scales 330, 335–6
  industrial instruments 340, 589
  instrument roles 584–5
  inter-war years 639–44
  Italy 347–9, 627–8
  Low Countries 345–6, 630
  marketing 608–12
  materials 98, 100, 330–3, 601–7
  mathematical instruments 99–100
  optical instruments 100–2, 336–8, 347–8, 604–7
  philosophical instruments 338–41
  Portugal 346–7, 628
  professional instruments 590–3
  public demonstrations and trade 303–4, 306–14
  research instruments 586–7
  Russia 631
  Scandinavia 349, 630–1
  Spain 628
  Switzerland 349–50, 628–9

thermoscopes/thermometers 103, 307
tools 598–600
United States of America 351–2, 631–4
workshop practice and organization 328–30, 597–8, 600–1
integrated electronic circuits 877–8
interference 451–5
Invar 603
iron use in instrument-making 602
Italian instrument-makers 347–9, 627–8

J

Journal of Physical Therapeutics 693
Journal of Scientific Instruments 618
junction transistor 873
Jupiter
  eclipses of the moons 180
  great inequality 155, 394–6
  motion of 122, 136, 156–7, 394–6

K

Kiel, University of 349
kinematics 791
kinetic theory 385, 767–78
  gases as particles in motion 768–9
  global approaches 779–82
  Maxwell–Boltzmann law 771–2
  reception of 777–8
  systems other than gases 782–4
  transport properties of gases 769–71
  see also gas theory
Klein–Gordon equation 845
klystron radar tube 866–7

L

labour practices 710–11
Lagrange points 396
Laplace's equation 366
Laplacian school, Arcueil 270, 272–81, 412–17, 418, 537
latent heat 477–88
Lehrbuch der Experimentalphysik und der Meteorologie (Müller) 661
Lehrbuch der Experimentalphysik (Wüllner) 663
Lehrbuch der mechanischen Naturlehre (Fischer, 1805) 658
Lehrbuch der mechanischen Naturlehre (Fischer) 662
Lehrbuch der Naturlehre (Krie) 662
Lehrbuch der Physik (Eisenlohr, 1836) 662
Lehrbuch der Physik (Müller-Pouillet) 666
Leiden, University of 105–6
Leitfaden der praktischen Physik (Kohlrausch, 1870) 663
Leitfaden für den Unterricht in der Physik (Brettner) 662
lens-making 101–2, 240, 242, 336–7, 604–5
Lenz's law 556
Lessons in Elementary Physics (Stewart, 1870) 665

Leiden jar 436–9, 595
light 169–73, 445–6
  absorption 468–9
  constant speed of 794–6
  diffraction 193–6, 453–5, 456
  dispersion 178–9, 198n, 460–2, 468, 472n, 838–40
  experiments 218–21
  heat relationship 494
  interference 451–5
  optical rotation 462
  photon theory 820, 839, 844
  polarization 415, 447–50, 456–7, 466
  public demonstrations 303
  ray optics 447–51
  reflection 458, 467
  refraction 61–8, 93n, 169–70, 174, 178–81, 411–14, 458–60
  see also colour; optics; wave theory of light
light bulb experiments 739
lighthouse lenses 337–8
lightning rod construction 735
Liouville's theorem 780
liquid–gas transition 782
logarithms 250–1
Lord Dundas steamer 511–12
Lorentz–Einstein theory 791
loxodrome 256–7
lunar motion 119, 128–9, 135–40

M

machine tools 330, 335, 359–600
Mach's principle 807–9
Madrid, Science Faculty Mechanics Department 628
Magnetic Crusade 344
magnetic curves/lines of force 561–2, 565–7
  intensity 561, 564–6, 571
magnetic field 576–7
  see also field theory
magnetic memory 740
magnetism 79–82, 432–3, 439–40, 564–5, 740
  experiments 434–42
  hysteresis 740, 760n
  permanent magnets 740
  polarized light response 563
  terrestrial 339, 341, 544–5
  see also electromagnetism; magnetic curves/lines of force
magneto-electric induction 555
magneto-optical effect 563, 577
magnetostatics 572
mahogany use in instrument-making 332
mass 120–1, 211
  centre of 369–70
  mass–energy equivalence 798–9
mathematics 13–15, 226–7, 271
  calculus 114, 148–9, 154, 249, 251–9, 260
  complex numbers 466–7

Descartes 69–70, 238–41, 243–4
electrical research approaches 439, 536–9
geometry 238–49
instrument-makers 99–100
Islamic 243
Leibniz 255–9
Newton 114–15, 148–51, 251–5
optics 446, 447, 456–8
pre-calculus 249–51
proportion theory 230–8
status of in the late Renaissance 227–30
matrix mechanics 838–43, 857n
Maxwell's demon 774
Maxwell–Boltzmann law 771–2
mean free path concept 769, 780
measurement *see* metrology; standard weights and measures
mechanical transformability principle 827–8
Medical Electrology and Radiology journal 693
medicine 679–94
  conservation of energy application 686–90
  therapeutics 690–3
  vitalism and materialism 681–6
Mercury perihelion test 810
mercury use in instrument-making 332, 603
Mersenne's law 380
metre 344, 698, 701–4, 706, 709
metric measures 344, 704, 706
metrology 698–716
  background 699–703
  broader historical issues 710–14
  electrical resistance standard 706–8, 712
  institutionalization 714
  labour practices and 710–11
  moral aspects 714, 715
  seconds pendulum 703–6, 711
  see also standard weights and measures
Michelson–Morley experiment 730–1, 751, 795–6, 800
microcanonical ensemble 779
microelectronics *see* electronics industry; transistors
microprocessor revolution 881–6
microscopes 101, 102, 336, 337, 619–20
microwave background radiation 904, 909–12, 915–16
Milky Way 896
Millwall shipbuilding yard, Fairbairn Company 512
mirror-making 98, 100, 348, 602, 604, 606–7
Models and Manufactures exhibition, Glasgow (1840) 510
modern physics 729, 752
  origins of 725–7
molecular chaos 776
molecular interactions, Laplacian short-range forces approach 406–27
moments 361–2, 363–4, 367
  equilibrium of 367

momentum 51, 363–4
  conservation of 120n, 121, 363–4
Moon, motion of 119, 128–9, 130, 135–40, 157, 392–4
  nutation 157
More's law 886
motion
  accelerated 27–54, 113, 120, 235–8
  experiments 204–12
  Galileo's studies 26–54, 230–2, 235–8
  Newton's Laws 119–26, 362–3
  projectile 27–54, 132–3, 207–8, 231–2
  rigid bodies 366–71
  see also planetary motion
multiple resonance 746
multiplication rule 841–2
multiverse hypotheses 918–19

N
nanotechnology 886–7
National Physical Laboratory 626
Navier–Stokes equation 770–1
navigational instruments 99, 334, 339, 345
Nebraska, University of, Brace Laboratory lecture room 589
nerve vibration therapy 691–2
neurasthenia 688
neutrino species 913
Newcomen steam engine model 510
Newton's rings 181–6, 189–93, 220–1, 449
Notions générales de physique et de météorologie (Pouillet) 660
N-rays 692

O
observatories 327, 333, 341
  France 342
  Greenwich 333–4
  Italy 347, 348
  Paris 342, 536
  United States 351–2
ocean depth recording 339–41
ohm 707, 750
opacity 186
optical indicatrix 459
optical instruments 100–2, 336–8, 347–8, 604–7
  First World War 634–8
  inter-war years 639–44
  military 591–2, 606
optics 445–70
  Cartesian 85, 86–8, 93n
  experiments 218–21
  geometrical optics 846
  mathematical 446, 447, 456–8
  Newtonian 166–96
  ray optics 447–51
  see also light; optical instruments
Optiks (Newton) 166–96
orreries 310–11

orthographic projection instrument 245
Oxford, University of 333, 338

P
Padua, University of 29–30, 327
  accelerated motion studies 235–8
parallelogram rule 361–2
paramagnetic materials 565
Paris Universal Exhibition (1900) 622–3
partial differential equations 390
Paschen-Back effect 830–1
Peace of Augsburg (1555) 10
pendulum experiments 208–12
percussion apparatus 587
periodic table 834–5
photoelectric effect 819, 820
photon theory 820, 839, 844
physical sciences 200
Physical Science Study Committee (PSSC) project, USA 655–6, 666–8
Physikalisches Praktikum (Ilberg) 663
Physikalisches Praktikum (Westphal) 663
Physikalisches Praktikum (Wiedemann) 663
Physikalische Technik (Frick, 1850) 662
physiology 679–80
  see also medicine
piezoelectricity 749
Pisa, University of 229
  falling body studies 26, 204
Planck distribution 818, 821
planetary motion 13, 73–8, 86, 88–9, 94–5n, 391–6
  Descartes' vortex mechanics 73–8, 86, 88–9, 94–5n, 132
  Earth–Moon–Sun system 130, 135, 138–40, 157, 392–4
  great inequality 155, 394–6
  Kepler problem 151, 250
  Newton's *Principia* 113–14, 115–19, 122, 130–1, 135–40, 144–5, 155–7, 392–3
  perception of Earth's motion 792–3
  Sun–Jupiter–Saturn system 394–6
  see also cosmology
platinum use in instrument-making 332, 602
point-contact transistor 873
polarization of light 415, 447–9, 456–7, 466
  chromatic polarization 449–51, 456
  magnetism effect 562–3
Portuguese instrument-makers 346–7, 628
potassium discovery 282, 287, 534
potential 365–6
power 299–300
Practical Physics (Guthrie, 1878) 665
pre-calculus 249–51
Précis élémentaire de physique expérimentale (Biot, 1817) 659, 661
precision instruments 650n
  see also instruments and instrument-makers
primary colours 172, 176–7, 197n

principal determination 63, 65, 68, 71, 93n
Principia (Newton) 109–15, 119–61, 358, 392–3
prize competitions, Institut de France 413–16, 418–19, 485–6, 489, 538
professional instruments 590–3
projectile motion 27–54, 132–3, 231–2
  experiments 207–8
  see also motion
proportion theory 230–8
public demonstrations 302–5, 594
  degrees of engagement 314–17
  demonstration devices 306–14
  knowledge dissemination 314–15
Pulkovo workshop, Russia 631
Puy-de Dôme experiment 213–14
Pyrex 604
Pythagoras' theorem 804

Q

quantum physics 725–6, 836–51
  crisis 836–7
  early development of 815–23
  matrix mechanics 838–43, 857n
  Solvay Conference significance 822–3
  uncertainty principle 848–50

R

radiation
  black-body 741, 760n, 783, 816, 818, 844–5, 894
  electromagnetic 579–83, 745–6
  of heat 489, 493–4, 783
radioactivity 750
radio astronomy 908–9
radio communications 863–7
  gain 864
  transistor radio introduction 875
  vacuum tubes 864, 865–7
rainbow 87, 219, 838
Rayleigh–Jeans law 754n, 821
recurrence theorem 817
redshifts 811, 898, 899, 901
reflecting telescope 167–9, 348
reflection
  metallic 467
  partial reflection and refraction 458
refraction 178–81, 411–14, 459–60
  conical refraction 459
  experiments 218–19
  law of 61–8, 93n, 169–70, 174, 179–80
  partial reflection and refraction 458
  refractive index 431n
relativity theory 125–6, 361, 740–1, 790, 814
  equivalence principle 800–2
  experimental predictions 809–11
  experimental tests of special relativity 799–800
  general covariance 805–7
  general theory 801–2
  Mach's principle 807–9

mass–energy equivalence 798–9
  relativity principle 790–3
  relativity of simultaneity 796–8
  spacetime curvature 802–5
  speed of light constancy 794–6
Revue d'optique théorique et expérimentale 641–2
Rice University 872
Riemannian spaces 465
rigid body dynamics 366–71
Royal Observatory, Greenwich 333–4
  Nautical Almanac 334
Royal Observatory, Paris 342
rumbatron 866–7
Russian instrument-makers 631

S

Scandinavian instrument-makers 349, 630–1
Schottky field effect 869, 873
scientific instruments *see* instruments and instrument-makers
Scientific Revolution 7–21, 227
  absence of in Europe 10
  cadres 15–17
  institutions 17–20
  revolutionary programme 13–15
  second revolution 426
scientist-engineers 737–41
seconds pendulum standard 703–6, 711
seismometers 349
selectionism 448–9
semiconductor development 867–73
  germanium use 870, 872–3
  impurity issues 870
  silicon advantages 876
  see also transistors
sextants 334, 335–6
ship building 510–18, 523–8
  Clyde shipyards 523–8
  coal consumption issues 513–16
silicon 876
  see also semiconductor development; transistors
Silicon Valley 865, 876
silver use in instrument-making 332, 602
Snell's law of refraction 169, 174, 179–80
sodium discovery 282, 287, 534
solar redshift 811
solid state materials 872
  see also semiconductor development; transistors
Solvay conference, 1911 722, 724–7, 814, 822, 823, 829
sound, speed determination 211–12, 484
spacetime curvature 802–5
Spanish instrument-makers 628
spark-gap devices 863
specific heat 478–87, 821–2
  gases 772–3, 777
  see also heat
spectacle-makers 100, 336–7
spectroscopy 893

speed of sound determination 211–12, 484
standard weights and measures 699–716
  background 699–703
  Britain 329–30, 335, 706–8, 714
  electrical resistance 706–8, 712
  France 344, 703–4, 714
  Germany 705–6, 708, 714
  labour practices and 710–11
  moral aspects 714, 715
  seconds pendulum 703–6, 711
Stanford University 872
Stark effect 827, 846
state function, heat as 495–502
statistical mechanics 765–6, 778–9, 784–5
steam engines
  efficiency studies 497, 510, 739
  use by instrument-makers 600
steam ship design 510–18
  coal consumption issues 513–16
steel use in instrument-making 602
Steiner's theorem 367
string theory 918
Structure of Scientific Revolutions, The (Kuhn, 1962) 654, 671, 836
submarine telegraph cables 558, 706–7, 738
Sun
  atmospheric helium 893
  surface temperature 894
sundial-making 243–6
sunspots 80–2, 94n, 95n
Supernova Cosmology Project (SCP) 916
surveying instruments 590–1
Swiss instrument-makers 349–50, 628–9

T

tacheometer-theodolite 591
teaching instruments 587–8
Tecnomasio Italiano instrument-makers, Milan 627
telegraphy 557–8, 738, 740, 757n
  submarine cables 558, 706–7, 738
telescopes 101–2, 336–7, 347–8
  reflecting telescope 167–9, 348
temperature 474
  absolute temperature concept 474–5
textbooks 651–3
  Britain 664–5
  cosmology 911
  editors 664
  France 658–61, 663–4
  genre 657–66
  Germany 661–3
  history 653–7
theodolites 335, 336, 591
theorem of parallel axes 367
thermodynamics 495
  bridge to mechanics 767–8, 774–5
  classical thermodynamics emergence 502–5
  entropy 496, 504, 775–6, 783

  first law 503
  second law 503–4, 774–5
  see also energy; heat
thermometer-makers 103
thermometers 307, 474
thermometry 464–6
  fixed points 474–6
torsion balance electrometer 279, 440, 559, 560, 702
Traité de physique élémentaire (Haüy, 1803) 658
Traité de physique expérimentale et mathématique (Biot, 1816) 659, 661
Traité élémentaire de physique expérimentale et appliquée (Ganot, 1851) 656–7, 663–4
Traité élémentaire de physique (Péclet) 660
transatlantic telegraph cables 558, 706–7
transistors 861, 867–73
  definition 867
  development of 870
  entrepreneurship 875–7
  finfet nano-device 886–7
  future directions 886–7
  innovation management 879–80
  junction transistor 873
  manufacturing issues 873–8
  More's law 886
  planar transistor 877
  point-contact transistor 873
  silicon use 876
  see also electronics industry; semiconductor development
Treatise on Natural Philosophy (Thomson & Tait, 1867) 665, 686
triodes 596

U

uncertainty principle 848–50
United Kingdom
  First World War 634–6
  instrument makers 328–30, 333–41, 615–16, 624–7, 634–6, 639–40
  inter-war years 639–40
  metrology 329–30, 335, 706–7, 714
  textbooks 664–5
United States of America
  First World War 637–8
  instrument-makers 351–2, 631–4, 637–8, 643–4
  inter-war years 643–4
  textbooks 664–5
units, systems of 360
universal mathematical method (Descartes) 69–70
universe
  age of 904
  background microwave radiation 904, 909–12, 915–16
  early universe studies 903–5, 912–15
  entropy increase 894–5
  expanding universe 899–903, 906–7, 908

far-future studies 915
heat death concept 895
inflation theory 914–15
multiverse hypotheses 918–19
steady-state model 905–9
universities 301
  mathematics teaching 228–9
  observatories 342
  see also specific cities
Uppsala, University of 349

V
vacuum heat properties 484–6
vacuum pumps *see* air pumps
vacuum tube construction 865–7
vectors 361–2
velocity-potential theorem 390
vibration theory 379–82
Vienna, Imperial and Royal Polytechnic 629
viscosity, gases 770–1
vitalism 681–6
void-in-the-void experiment 214–15
Volta-electric induction 554
voltaic pile 533–6
  introduction of 268, 275–85, 535
  medical interest 681–2
Vorschule der Experimental Physik (Weinhold) 662
vortex celestial mechanics 13, 73–8, 86, 88–9, 94–5n
  Newton's arguments against 132, 134, 144–5, 154

W
wave equation 381–2, 389, 846
wave mechanics 843–8
wave theory of light 198n, 249, 413, 414, 418–19, 458–64
  Fresnel's theory 417–19, 455, 457–8, 464–6
  Huygen's principle 249, 464–6
  inclination factor 464, 465
  Newton's views 175, 189, 193, 413
  Young's theory 451–2
  see also light; optics
weakly interacting massive particles (WIMPs) 917
Wheatstone wave machine 588
Wien's law of displacement 818, 852n, 894
wood use in instrument-making 332, 601
work, mechanical 365
  heat interconversion 503

X
X-rays 692, 747
  electrical discharge by 747

Y
yard, standard 329–30, 335, 710

Z
Zeeman effect 830–3
  anomalous 830–3, 836
Zeitschrift für den physikalischen und chemischen Unterricht 618
Zeitschrift für Instrumentenkunde 618